H · O · L · T

EARTH SCIENCE

TEACHER'S EDITION

WILLIAM L. RAMSEY

LUCRETIA A. GABRIEL

JAMES F. McGUIRK

CLIFFORD R. PHILLIPS

FRANK M. WATENPAUGH

HOLT, RINEHART AND WINSTON, PUBLISHERS

New York • Toronto • Mexico City • London • Sydney • Tokyo

THE AUTHORS

WILLIAM L. RAMSEY
Former Head of the Science Department
Helix High School
La Mesa, California

CLIFFORD R. PHILLIPS
Former Head of the Science Department
Monte Vista High School
Spring Valley, California

LUCRETIA A. GABRIEL
Science Consultant
Guilderland Central Schools
Guilderland, New York

FRANK M. WATENPAUGH
Former Science Department Chairman
Helix High School
La Mesa, California

JAMES F. McGUIRK
Head of the Science Department
South High Community School
Worcester, Massachusetts

EDITORIAL DEVELOPMENT: Karen Kennedy Gotimer, Stephen Lee
PRODUCT MANAGER: Tom Rooney
FIELD ADVISORY BOARD: Max Callahan, Robert Davis, James Dellisanti, Debbie Deruy, Jim Donatelli, John Eskew, Joseph Hammond, Dave Miller, Robert Moore, Stuart Natof, Barbara Orloff, George Rybak, Marion Stewart, L. Jean Slankard
MARKETING RESEARCH: Erica S. Felman
EDITORIAL PROCESSING: Margaret M. Byrne, Pamela K. Caugherty
ART AND DESIGN: Carol Steinberg, Amy Newberg
PRODUCTION: Judy Baller, Nina R. West, Sandra Moore
PHOTO RESOURCES: Linda Sykes, Rita Longabucco

Photo Credits appear on page 563.
Cover photograph by Shelly Grossman/Woodfin Camp & Assoc.

About the Cover:
The photograph on the cover shows mesas and buttes in Monument Valley. The mesas and buttes are traces of what was once a broad plateau. They rise more than 300 meters above the valley floor. Monument Valley is located on the Utah–Arizona border on the Navajo Indian Reservation.

CONTENTS

INTRODUCTION

Rationale

HOLT EARTH SCIENCE consists of an introductory chapter (Chapter 1: What Is Science?) followed by 18 chapters organized into five units. The units are divided into the study of four major areas of the earth sciences: geology, meteorology, oceanography, and astronomy. Each unit consists of three to six chapters. Chapter Goals at the beginning of each chapter define the learning objectives for that chapter. The chapters are made up of sections, which also have their own objectives. These objectives are tested at the end of the section. The consistent section format is beneficial in the following ways:

(1) It encourages student independence and frees the teacher to aid slower students.

(2) It can easily be adapted for various methods of teaching.

(3) It increases success and enjoyment for students of all ability levels.

(4) It provides optional Investigations and Skill-Building Activities that can be used in a variety of ways. If the facilities are available, a laboratory-oriented course can be built around these optional features.

The authors and editors have devoted particular attention to the readability of HOLT EARTH SCIENCE. The manuscript was tested by the Fry and Dale-Chall reading formulas. In addition, and perhaps more importantly, the text is written in an interesting and lucid style that students will find easy to read.

Components

The optional components that accompany HOLT EARTH SCIENCE include the following:

- Annotated Teacher's Edition of the student text.
- Teacher's Resource Book containing Review Exercises, Skills Exercises, Enrichment Exercises, Laboratories, Text Blackline Masters, Extensions Masters, Overhead Transparency Masters, and Tests.
- Pupil's Edition of Exercises and Investigations containing the Review, Skills, and Enrichment Exercises and Laboratories in consumable form.
- Teacher's Edition of Exercises and Investigations.
- Test Masters containing Chapter, Midterm, and Final Tests in the form of duplicating masters.

Field Testing

The HOLT EARTH SCIENCE program first appeared in 1978. It has been tested in schools around the country. These schools represent a range of locations from urban to rural. For the present edition, comments and suggestions from classroom teachers presently using the program as well as from teachers who reviewed the manuscript have been incorporated. Content-area specialists also reviewed the manuscript.

How To Use the Annotated Teacher's Edition

The annotated Teacher's Edition for HOLT EARTH SCIENCE contains the following information:

- Introduction: The rationale and components of the program are described. Reduced pages of the student text highlighting features are reproduced.
- Teaching Strategies: This section contains a chart listing suggested components to use in a one-module (30-minute class period), two-module (55- to 60-minute class period), or three-module (90-minute class period) course.
- Safety in the Science Classroom: This section provides useful and practical information on the very important topic of safety in science. Guidelines are given for safety in the

classroom, in the laboratory, and on field trips. References, safety equipment suppliers, and audiovisual aids are also listed.

- The Exceptional Student in the Science Class: Suggestions are offered for the effective instruction of exceptional students in the regular classroom.
- Writing a Science Research Report: This section includes specific suggestions and guidelines that students should follow when writing (1) a library research report or (2) a laboratory report.
- Basic Science Skills: This section is designed to help teachers in the development of students' reading and writing skills. These skills are essential to the study of science as well as other subject areas. A description of the skills covered is provided.
- Materials List: The materials necessary to do the Investigations and Skill-Building Activities are listed along with quantities needed and the sections in which they are used. A list of equipment and materials suppliers is also provided.
- Teacher Commentary: The chapter-by-chapter commentary includes an easy reference chart that lists section numbers and titles, approxi-

mate time needed, basic science skills covered, a pacing guide, and a list of audiovisual, print, and computer materials. The commentary itself consists of a Chapter Overview, section background and motivation, hints for performing Investigations and Skill-Building Activities and answers to questions, suggestions for special features, and answers to Chapter Review Questions.

- Teacher Annotations: In addition to the commentary, marginal annotations for the teacher are found throughout this Teacher's Edition. The annotations are divided into the following categories:

    Answers to text questions: Provide answers to questions asked within the body of the text.
    Word Study: Emphasizes the importance of prefixes, suffixes, and roots in science words.
    Reinforcement: Suggests different ways to present content for the benefit of slow learners.
    Enrichment: Provides additional facts and information for advanced students.
    Skills: Identifies particular skills covered in the text.
    Suggestion: Provides tips to help the teacher clarify the concepts for the students.

TEXT FEATURES

CHAPTER OPENER PHOTOGRAPH Each chapter begins with an exciting full-page photograph and descriptive caption. These high-interest photos immediately focus students' attention on the subject of the chapter.

CHAPTER GOALS A list of Chapter Goals clearly tells the students what they will learn in the chapter. These goals are tested in the Chapter Review at the end of the chapter.

SECTION FORMAT Each chapter is divided into several self-contained sections. The first section begins immediately following the Chapter Goals. The easy to follow format helps promote individualized study.

OBJECTIVES Each section starts with a list of section objectives which tell the students what they will learn in that section. These performance objectives are an aid to individualized study and reduce the need for teacher direction.

MOTIVATION A short, intriguing motivational paragraph introduces the concept developed in the section.

MARGINAL DEFINITIONS New science words are defined in the text margin to aid student comprehension and help build a working scientific vocabulary. These words are also printed in boldface type in the text, where they are introduced and used in context.

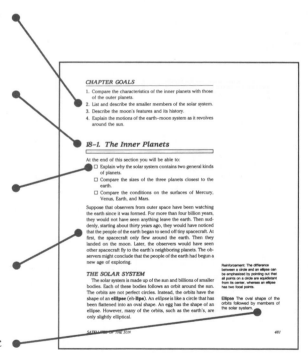

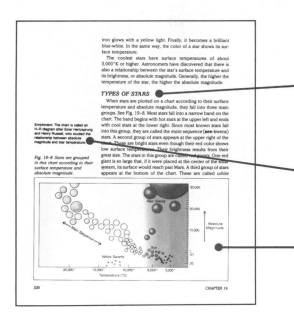

iron glows with a yellow light. Finally, it becomes a brilliant blue-white. In the same way, the color of a star shows its surface temperature.

The coolest stars have surface temperatures of about 3,000°K or higher. Astronomers have discovered that there is also a relationship between the star's surface temperature and its brightness, or absolute magnitude. Generally, the higher the temperature of the star, the higher the absolute magnitude.

TYPES OF STARS

When stars are plotted on a chart according to their surface temperature and absolute magnitude, they fall into three main groups. See Fig. 19–8. Most stars fall into a narrow band on the chart. The band begins with hot stars at the upper left and ends with cool stars at the lower right. Since most known stars fall into this group, they are called the *main sequence* (see-kwens) stars. A second group of stars appears at the upper right of the chart. These are bright stars even though their red color shows low surface temperatures. Their brightness results from their great size. The stars in this group are called *red giants*. One red giant is so large that, if it were placed at the center of the solar system, its surface would reach past Mars. A third group of stars appears at the bottom of the chart. These are called *white*

Enrichment: The chart is called an H–R diagram after Einar Hertzsprung and Henry Russell, who studied the relationship between absolute magnitude and star temperature.

Fig. 19–8 Stars are grouped in this chart according to their surface temperature and absolute magnitude.

530   CHAPTER 19

SUBHEADS Concise subheads appear throughout the section. They can serve as a quick reference or study guide to help students review the section.

TEACHER ANNOTATIONS Throughout the text, helpful annotations provide answers to questions, highlight skills development, and give teachers useful and interesting suggestions for enhancing concept development.

DIAGRAMS AND PHOTOS Illustrations that are attractive as well as functional aid in concept development and reinforce the main ideas of the section.

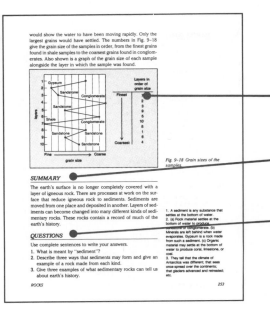

would show the water to have been moving rapidly. Only the largest grains would have settled. The numbers in Fig. 9–18 give the grain size of the samples in order, from the finest grains found in shale samples to the coarsest grains found in conglomerates. Also shown is a graph of the grain size of each sample alongside the layer in which the sample was found.

Fig. 9–18 Grain sizes of the samples.

SUMMARY

The earth's surface is no longer completely covered with a layer of igneous rock. There are processes at work on the surface that reduce igneous rock to sediments. Sediments are moved from one place and deposited in another. Layers of sediments can become changed into many different kinds of sedimentary rocks. These rocks contain a record of much of the earth's history.

QUESTIONS

Use complete sentences to write your answers.
1. What is meant by "sediment"?
2. Describe three ways that sediments may form and give an example of a rock made from each kind.
3. Give three examples of what sedimentary rocks can tell us about earth's history.

1. A sediment is any substance that settles at the bottom of water.
2. (a) Rock material settles at the bottom of water to produce sandstone or conglomerate. (b) Minerals are left behind when water evaporates. Gypsum is a rock made from such a sediment. (c) Organic material may settle at the bottom of water to produce coral, limestone, or coal.
3. They tell that the climate of Antarctica was different; that seas once spread over the continents; that glaciers advanced and retreated; etc.

ROCKS   253

TABLES AND GRAPHS Information is often presented in the form of visually appealing charts and graphs.

SUMMARY A brief paragraph at the end of the section reviews the main ideas of the section and reinforces concept development.

QUESTIONS Several end-of-section questions provide a self-test of the section objectives and encourage students to practice their writing skills.

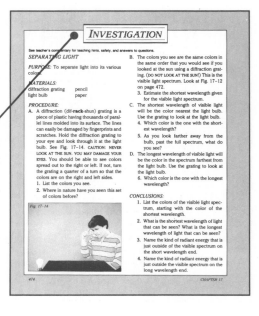

INVESTIGATIONS Most sections are followed by an optional laboratory Investigation that emphasizes the development of good lab techniques and safety procedures. The easy to follow format facilitates the writing of lab reports.

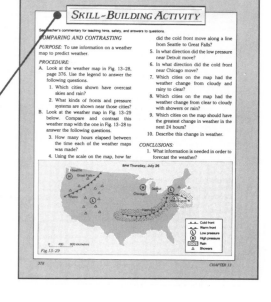

SKILL-BUILDING ACTIVITY These Activities follow the same format as the Investigations while focusing on basic science skills such as measuring, observing, classifying, hypothesizing, comparing and contrasting, and finding cause and effect.

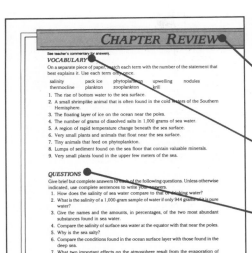

See teacher's commentary for answers.

CHAPTER REVIEW

VOCABULARY
On a separate piece of paper, match each term with the number of the statement that best explains it. Use each term only once.

salinity	pack ice	phytoplankton	upwelling	nodules
thermocline	plankton	zooplankton	krill	

1. The rise of bottom water to the sea surface.
2. A small shrimplike animal that is often found in the cold waters of the Southern Hemisphere.
3. The floating layer of ice on the ocean near the poles.
4. The number of grams of dissolved salts in 1,000 grams of sea water.
5. A region of rapid temperature change beneath the sea surface.
6. Very small plants and animals that float near the sea surface.
7. Tiny animals that feed on phytoplankton.
8. Lumps of sediment found on the sea floor that contain valuable minerals.
9. Very small plants found in the upper few meters of the sea.

QUESTIONS
Give brief but complete answers to each of the following questions. Unless otherwise indicated, use complete sentences to write your answers.

1. How does the salinity of sea water compare to that of drinking water?
2. What is the salinity of a 1,000-gram sample of water if only 944 grams of it is pure water?
3. Give the names and the amounts, in percentages, of the two most abundant substances found in sea water.
4. Compare the salinity of surface sea water at the equator with that near the poles.
5. Why is the sea salty?
6. Compare the conditions found in the ocean surface layer with those found in the deep sea.
7. What two important effects on the atmosphere result from the evaporation of water from the tropical seas?
8. Why can the oceans produce twice the food that land provides?

426   CHAPTER 15

---

9. What is the beginning of the food chain for life in the ocean?
10. Where are the best places for fishing located?
11. Why is upwelling necessary for abundant life in the ocean?
12. The ocean contains a large amount of gold and other minerals, but gold is not taken from sea water. Why is this?
13. What minerals do nodules contain that make them so valuable?
14. Describe the ways energy may be taken from the sea.
15. What is the biggest problem with taking resources from the ocean?

APPLYING SCIENCE
1. Determine the salinity of a sample of drinking water by doing the following: (a) Weigh a clean, dry 500-mL beaker. (b) Measure into the beaker 250 grams of water. (c) Boil until nearly dry. CAUTION: NEVER BOIL UNTIL COMPLETELY DRY! (d) Allow beaker to sit, away from heat, until completely dry. (e) Weigh beaker. The difference in original weight and final weight of beaker is the amount of dissolved material in the water for .25 kilogram of water. (f) Calculate the salinity.
2. Make up a sample of artificial sea water by dissolving 35 grams of rock salt in 965 grams of water. Try to get drinkable water from the salt water by freezing it.
3. In 1984 much of the krill usually present in the Antarctic waters disappeared. Find out why this happened and its effect on the life in the ocean.
4. Use a biology book to find out what plankton look like. Write a list of the ocean food chain in which plankton are involved.
5. Find out the details of recent developments or uses of sea resources such as: sea farming; sea mining; energy from the sea; seaweed products; etc. Report to the class on your findings.

BIBLIOGRAPHY
Britton, Peter. "Cobalt Crusts: Deep Sea Deposits." *Science News*, October 30, 1982.
Britton, Peter. "Deep Sea Mining." *Popular Science*, July 1981.
Matthews, Samuel W. "New World of the Oceans." *National Geographic*, December 1981.
Time-Life Editors. *Restless Oceans* (Planet Earth Series). Chicago: Time-Life Books, 1984.

WATER IN THE OCEAN   427

CHAPTER REVIEW   The two-page Chapter Review tests student recall of the chapter concepts in several ways.

VOCABULARY   Students are asked to match the science words introduced in the chapter with their definitions, or to use the words in context.

QUESTIONS   The Chapter Goals are tested by a series of problems or essay questions that also give students a chance to practice their writing skills.

APPLYING SCIENCE   Students are given an opportunity to explore beyond the scope of the chapter development. Applying Science may involve experiments that can be done at home, consumer-related problems, library research, or written reports.

BIBLIOGRAPHY   A list of relevant books and magazine articles is provided to encourage further student reading on topics of interest.

TEXT FEATURES

SPECIAL FEATURES   Three categories of special features are found throughout the text. They are Careers in Science, Technology, and ¡Compute!

CAREERS IN SCIENCE   These one-page features highlight an interesting assortment of possible careers in physical, earth, and life science.

TECHNOLOGY   The Technology features focus on new and exciting research topics and on technological applications of basic science.

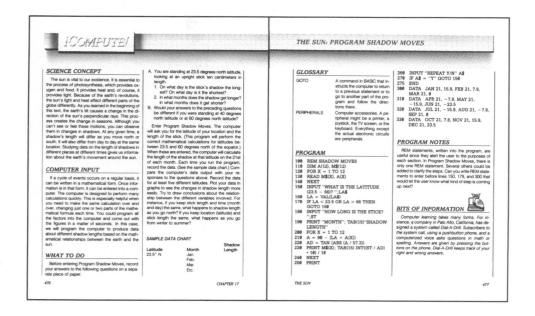

SCIENCE CONCEPT

The sun is vital to our existence. It is essential to the process of photosynthesis, which provides oxygen and food. It provides heat and, of course, it provides light. Because of the earth's revolutions, the sun's light and heat affect different parts of the globe differently. As you learned in the beginning of this text, the earth's tilt causes a change in the direction of the sun's perpendicular rays. This process creates the change in seasons. Although you can't see or feel these motions, you can observe them in changes in shadows. At any given time, a shadow's length will differ as you move north or south. It will also differ from day to day at the same location. Studying data on the length of shadows in different places at different times gives us information about the earth's movement around the sun.

COMPUTER INPUT

If a cycle of events occurs on a regular basis, it can be written in a mathematical form. Once information is in that form, it can be entered into a computer. The computer is designed to perform many calculations quickly. This is especially helpful when you need to make the same calculation over and over, changing just one or two parts of the mathematical formula each time. You could program all the factors into the computer and come out with the figures in a matter of seconds. In this case, we will program the computer to produce data about different shadow lengths based on the mathematical relationships between the earth and the sun.

WHAT TO DO

Before entering Program Shadow Moves, record your answers to the following questions on a separate piece of paper.

A. You are standing at 23.5 degrees north latitude, looking at an upright stick ten centimeters in length.
1. On what day is the stick's shadow the longest? On what day is it the shortest?
2. In what months does the shadow get longer? In what months does it get shorter?

B. Would your answers to the preceding questions be different if you were standing at 40 degrees north latitude or at 60 degrees north latitude?

Enter Program Shadow Moves. The computer will ask you for the latitude of your location and the length of the stick. (This program will perform the correct mathematical calculations for latitudes between 23.5 and 60 degrees north of the equator.) When these are entered, the computer will calculate the length of the shadow at that latitude on the 21st of each month. Each time you run the program, record the data. (See the sample data chart.) Compare the computer's data output with your responses to the questions above. Record your data in graphs to see the changes in shadow length more easily. Try to draw conclusions about the relationship between the different variables involved. For instance, if you keep stick length and time (month and day) the same, what happens to shadow length as you go north? If you keep location (latitude) and stick length the same, what happens as you go from winter to summer?

SAMPLE DATA CHART

Latitude	Month	Shadow Length
23.5° N	Jan.	
	Feb.	
	Mar.	
	Etc.	

GLOSSARY

GOTO	A command in BASIC that instructs the computer to return to a previous statement or to go to another part of the program and follow the directions there.
PERIPHERALS	Computer accessories. A peripheral might be a printer, a joystick, the TV screen, or the keyboard. Everything except the actual electronic circuits are peripherals.

PROGRAM

```
100  REM SHADOW MOVES
110  DIM A(12), M$(12)
120  FOR X = 1 TO 12
130  READ M$(X), A(X)
140  NEXT
150  INPUT "WHAT IS THE LATITUDE
     (23.5 − 66)? ";LA$
160  LA = VAL(LA$)
170  IF LA < 23.5 OR LA > 66 THEN
     GOTO 150
180  INPUT "HOW LONG IS THE STICK?
     ";ST
190  PRINT "MONTH"; TAB(15)"SHADOW
     LENGTH"
200  FOR X = 1 TO 12
210  A = 90 − (LA + A(X))
220  AD = TAN (ABS (A / 57.3))
230  PRINT M$(X); TAB(15) INT((ST / AD)
     ∗ 10) / 10
240  NEXT
250  PRINT
260  INPUT "REPEAT Y/N" A$
270  IF A$ = "Y" GOTO 150
275  END
300  DATA  JAN 21, 15.9, FEB 21, 7.8,
     MAR 21, 0
310  DATA  APR 21, −7.8, MAY 21,
     −15.9, JUN 21, −23.5
320  DATA  JUL 21, −15.9, AUG 21, −7.8,
     SEP 21, 0
330  DATA  OCT 21, 7.8, NOV 21, 15.9,
     DEC 21, 23.5
```

PROGRAM NOTES

REM statements, written into the program, are useful since they alert the user to the purposes of each section. In Program Shadow Moves, there is only one REM statement. Several others could be added to clarify the steps. Can you write REM statements to enter before lines 150, 179, and 300 that would let the user know what kind of step is coming up next?

BITS OF INFORMATION

Computer learning takes many forms. For instance, a company in Palo Alto, California, has designed a system called Dial-A-Drill. Subscribers to the system call, using a pushbutton phone, and a computerized voice asks questions in math or spelling. Answers are given by pressing the buttons on the phone. Dial-A-Drill keeps track of your right and wrong answers.

476   CHAPTER 17

THE SUN   477

¡COMPUTE! These highly motivating two-page features encourage students to explore the use of computers in life, earth, and physical science.

TEACHING STRATEGIES

HOLT EARTH SCIENCE is intended to be used as a one-year course in earth science. However, the flexibility of its design allows it to be used as part of a survey course in conjunction with HOLT LIFE SCIENCE and HOLT PHYSICAL SCIENCE.

When used alone, the HOLT EARTH SCIENCE program can be adapted to class periods of approximately 30, 60, or 90 minutes. The table below recommends which components of the program to use for a 30-minute-class period (1 module), a 55- to 60-minute class period (2 modules), or a 90-minute class period (3 modules).

It should also be noted that, following Chapter 1, the units of HOLT EARTH SCIENCE can be taught in any order without discontinuity. However, it is recommended that the sequence of units and chapters appearing in the text be followed.

PART OF COURSE	1 MODULE	2 MODULES	3 MODULES
Text	All sections except Investigations	All sections	All sections
Skill-Building Activities	Most	All	All
Investigations	As time allows	As many as equipment permits	All
Chapter Review Questions	All	All	All
Applying Science	As time allows	All	All with discussion
Demonstrations	Occasional	As suggested	Maximum effort with discussions
Exercises and Investigations	Occasional use	Regular use (not every day)	Daily use

SAFETY IN THE SCIENCE CLASSROOM

Students have a right to learn how to think scientifically and solve problems through the active use of scientific processes. Students should have ample opportunities to engage in scientific work through laboratory and other "hands-on" activities. However, unless safety precautions are taken to protect the students and others during the performance of certain science activities, science programs could eventually be considered too dangerous to be included in our school curricula.

The responsibility for safety and the enforcement of safety regulations and laws in the science classroom and laboratory are those of the principal, teacher, and student—each assuming his or her share. This means that proper attitudes toward safety should be developed from the outset. The information contained in this section is designed to aid in accomplishing this end. This information is not intended to be all inclusive, but it is designed to alert teachers to potential hazards, to provide suggestions for averting potential dangers, and to list resources for obtaining additional information.

It is imperative that teachers pass along the proper attitude toward safety to their students. To aid in student understanding of the seriousness of safety in the science classroom, students may be required to read, understand, and sign a student safety contract, such as the one that follows, after they have been instructed in safety procedures.

Student Safety Contract

I will:

- Follow all instructions given by the teacher
- Read directions thoroughly before starting an activity
- Protect eyes, face, hands, and body while conducting an activity
- Know the location of first-aid and fire-fighting equipment

- Conduct myself in a responsible manner at all times in a laboratory situation

I, _____, have read and agree to abide by the safety regulations as set forth above and also any additional printed instructions provided by the teacher and/or district. I further agree to follow all other written and verbal instructions given in class.

Date _____

_____
Signature

In HOLT EARTH SCIENCE, throughout the text, students are reminded of safety procedures and the safe use of equipment by specific safety instructions.

School Safety Audit

It is recommended that at the beginning of the school year and throughout the year as the need arises a safety audit of classroom and laboratory facilities be conducted by the teacher. A form could be used to indicate areas needing attention by maintenance staff, etc. As conditions are corrected, a notation could be made on the form, which should be kept on file for future reference.

Storing Chemicals

It is not advisable to store large amounts of potentially hazardous chemicals over long periods of time. In order to help avoid this problem, every effort has been made in the materials list of HOLT EARTH SCIENCE to indicate the minimum amount of each substance needed to perform the activities. After having ordered the necessary chemicals, it is a good idea to compile an alphabetical list of all chemicals on hand in your school laboratory. Include the amount of each and the date it was obtained. Additions to and deletions from this list should be made whenever necessary. Contact your regional EPA solid waste office for information on proper disposal of surplus or dangerous chemicals. To aid in identification of chemicals, a standard labeling system should be used.

Incident/Accident Report

In the science classroom, as well as in any other class, any accident, hazardous incident, or potentially hazardous incident should be officially recorded and reported as soon as possible. A standard report form can be used to facilitate the recording of such incidents and to insure that all necessary information is recorded.

Field Trips

When taking students on field trips to visit sewage treatment plants, power generating stations, and so on, it is advisable for the teacher to visit the area beforehand to note any possible hazards. These potential dangers should then be discussed with the students, and rules for safe conduct should be established. Students should be advised of any special clothing to be worn, such as low-heeled shoes with rubber soles that will prevent tripping on stairs or slipping on wet floors, etc.

Adequate supervision by teachers and/or parents should be insured, and school policies for obtaining proper consent prior to the field trip should be followed.

While actually at the field trip location, it is imperative that students adhere to all necessary safety regulations required in any particular area, such as the wearing of hard hats, protective goggles, and so on.

Laboratory Safety Guidelines

The following is a list of some of the general guidelines that should be followed in the science laboratory at all times. Teachers should review these guidelines with their students and be sure that they are understood.

- All science laboratories should have and use the following safety items: fire extinguishers, first-aid kits, fire blankets, sand buckets, eyewash facilities, emergency shower facilities, approved safety goggles, laboratory aprons, gloves, tongs, respirators, wire gauze.
- Good housekeeping is essential for maintaining safe laboratory conditions.
- Confine long hair and loose clothing. Laboratory aprons should be worn.
- Never have students conduct experiments alone in the laboratory.
- Proper eye protection devices should be worn by all persons who are supervising or observing science activities that involve potential hazards to the eye.
- Make certain that all hot plates and open burners are turned off when leaving the laboratory.
- Every laboratory should have two unobstructed exits.
- There should be an annual, verified safety check of each laboratory.
- One teacher should not have to supervise more than twenty-four students engaged in laboratory activities at one time.
- Master cutoff valves or switches for gas, water, and electricity are advisable for laboratories.
- Emergency instructions of procedures for fires, explosions, chemical reactions, spillage, and first aid should be conspicuously posted near all storage areas.
- Always perform an experiment or demonstration prior to allowing students to do the activity. Look for possible hazards. Alert students to potential dangers. Safety instructions should be given each time an experiment is begun.
- Horseplay or practical jokes of any kind are not to be tolerated.
- Never eat or drink in the laboratory or from laboratory equipment, such as beakers or flasks.
- Exercise great care in noting odors of fumes. Use a wafting motion of the hand.
- Never use mouth suction in filling pipettes with chemical reagents. (Use a suction bulb.)
- Never "force" glass tubing into rubber stop-

pers. Constant surveillance and supervision of student activities are essential.
- A positive attitude toward safety is imperative.

Students should not be afraid to do experiments or to use reagents and equipment, but they should respect them as potential hazards.

Hazardous Waste Disposal

Contact your regional Environmental Protection Agency (EPA) office for information on proper disposal of surplus or dangerous chemicals or solid wastes.

Region 1
Solid Waste Program
John F. Kennedy Building
Boston, MA 02203
(617) 223-5775

Region 2
Solid Waste Section
26 Federal Plaza
New York, NY 10007
(212) 264-0503/4/5

Region 3
Solid Waste Program
Sixth and Walnut Streets
Philadelphia, PA 19106
(215) 597-0980

Region 4
Solid Waste Section
345 Courtland Street NE
Atlanta, GA 30308
(404) 881-3016

Region 5
Solid Waste Program
230 South Dearborn Street
Chicago, IL 60604
(312) 353-2197

Region 6
Solid Waste Section
First International Building
Dallas, TX 75270
(214) 767-2645

Region 7
Waste Management Section
324 East 11th Street
Kansas City, MO 64108
(816) 374-3307

Region 8
Solid Waste Section
1860 Lincoln Street
Denver, CO 80203
(303) 837-2221

Region 9
Solid Waste Program
215 Freemont Street
San Francisco, CA 94105
(415) 556-4606

Region 10
Solid Waste Program
1200 Sixth Avenue
Seattle, WA 98101
(206) 442-1260

THE EXCEPTIONAL STUDENT IN THE SCIENCE CLASS

PL 94-142

Mainstreaming has come to mean the integration of handicapped children into the mainstream of our society and into the educational mainstream wherever feasible. Some students with handicaps have always been mainstreamed. Their exceptionalities are not severe, or they are given the opportunities to achieve normal educational goals under conditions in which their exceptionalities are not thought to be handicapping. Others attend special education classes or receive related services that meet their educational needs.

Federal Law PL 94-142 requires all students with handicapping conditions to be identified, tested, and evaluated by the appropriate agencies or school services. (If you suspect that a student has a handicap never before identified, contact the school officials immediately.) The school service, in agreement with the student and his or her parents, selects the most appropriate and least environmentally restrictive educational plan. The plan can include total integration with related special services, such as speech therapy, or partial integration, with enrollment in selected courses for part of the day and enrollment in special education courses for the remainder.

Handicapping Conditions

Physical impairments include those of an orthopedic nature, visual impairments, blindness, other health impairments or chronic conditions maintained by medication and/or restrictive activity as prescribed by a physician.

Communicative handicaps are any of the conditions that adversely affect a student's ability to effectively use language; hearing impairments, deafness, and speech impairments.

The term "specific learning disability" refers to a dysfunction in any one, or more, of the basic psychological processes involved in understanding or using language. Specific learning disabilities include disorders of auditory language, reading, writing, arithmetic, and nonverbal learning—understanding gestures (receptive) and using them (expressive) and difficulties in motor learning, spatial orientation, and social perception. Such disabilities may be further defined, depending on whether the problem is primarily visual or auditory in nature.

Mild mental retardation refers to a slowed intellectual functioning.

Emotional disturbance refers to a behavioral disorder that hinders a student's overall functioning in social/academic performance.

A student may possess one or more handicapping conditions; such needs are incorporated in an educational plan.

Instructional Management Systems

Be aware of the educational plan selected for each handicapped student in the science class. If a student is partially integrated and returns to the special education teacher or resource teacher for supportive and additional instruction, the resource teacher writes an Individualized Educational Plan (IEP) with input from the course instructors.

Strategies

Management strategies are necessary for optimal student achievement and are individualized depending on the degree of the exceptionality. This approach identifies specific strategies, techniques, and rules that a student can use in coping with the demands of content material.

Regardless of the system selected, the student is maintained as a full-fledged member of the class and is expected to participate in all activities and assignments.

Careers For Persons With Handicapping Conditions

As educational opportunities for handicapped persons have increased, so have opportunities for employment. The Rehabilitation Act of 1973, Public Law 93-112, Section 504, provides for the nondiscrimination of qualified handicapped individuals for employment opportunities in any program receiving federal financial assistance.

Gifted Students

In considering the position of exceptional students in the regular classroom, teachers should not neglect gifted or talented students. These students can often experience frustration if they are not presented with a sufficiently challenging program. In the Pacing Guide for each chapter of HOLT EARTH SCIENCE, suggestions are offered for additional projects, activities, etc., which will help to enrich the course work for your gifted students. These students are identified in the Guide as "Above Average." For example, see page T28. Finally, the Applying Science questions at the end of each chapter provide many opportunities for above average students to do independent research, write reports, or perform optional activities. In the Teacher's Resource Book, many of the Laboratories have optional parts that would be suitable for above average students.

Bibliography

The following books may be helpful in planning a strategy to deal with gifted students. All are available from The Council for Exceptional Children, 1920 Association Drive, Reston, Virginia 22091.

- E. Coston Frederick, *Teaching Content Through Reading: A Human Experience,* Charles C. Thomas Pub., 2600 South First St., Springfield, Illinois, 1984.
- Thinking and Learning Skills: *Volume I: Relating Instruction to Research, Volume II: Research and Open Questions* edited by: Judith H. Segal, Susan F. Chipman and Robert Glaser; Lawrence Erlbaum Associates, Inc., 365 Broadway, Hillsdale, New Jersey, 1985.
- Thaiss, Christopher, "Language Across the Curriculum," ERIC/RCS Digest, Urbana, Illinois, ERIC/NCTE, 1984.
- Alfred F. Collete and Eugene L. Chiappetta, *Science Instruction in the Middle and Secondary School,* C. V. Mosby Co., St. Louis, Missouri, 1983.
- Bill Gearheart and Mel H. Weishahn, *The Exceptional Child in the Regular Classroom,* C. V. Mosby Co., St. Louis, Missouri, 1983.
- A. Harry Passaw, ed. *The Gifted and the Talented: Their Education and Development,* Chicago; University of Chicago Press, 1979.
- Mary M. Dupuis and Eunice N. Askov, *Content Area Reading: An Individualized Approach,* Prentice-Hall Inc., Englewood Cliffs, New Jersey, 1982.

BASIC SCIENCE SKILLS

This section of the Teacher's Edition is designed to help teachers develop in their students the reading, writing, and math skills essential to the study of science.

Science teachers often feel that they do not have time to teach reading and writing skills if they are to cover the science material in the curriculum. Actually, they already spend a great deal of time helping their students develop these skills. The materials in this section will help you to organize, develop, and reinforce these skills as you teach earth science.

Reading and Writing in Science

PREPARATION: Science teachers, as well as teachers in other content areas, have become concerned about "reading level." There are two major methods of determining the reading level of a text. One method makes use of graded word lists, while the other uses a readability scale, which is based on the average sentence length and the total number of syllables. In order to be used in a wide variety of circumstances, these methods must of necessity be mechanical. As a result, nonsense or gibberish can have a low "reading level" when tested by these methods. Coherent expository writing can have a high reading level. Yet this material may actually be more readable (that is, understandable) to students than short, choppy sentences that test at a lower reading level. The final test of "readability" is whether or not students can read and comprehend the text material.

DIAGNOSIS: One method of determining whether a student understands the text material is the Cloze test. The results of this test can help you determine whether the student can work independently, whether he or she requires instruction, or whether, even with prior instruction, the student will have reading difficulties.

To make your own Cloze test, select a 250-word passage from the material you would like to assign. This material must be new to the students. Leave the first sentence complete. Starting with the second sentence, leave a uniform blank for every fifth word. Continue to delete every fifth word until there are 50 blanks. Give this passage to your students. Tell them to fill in each blank with the word that makes the most sense. Score 2 points for each exact original word. Do not count synonyms at this time since it takes more time and requires subjective judgment. If a student scores below 43%, score 2 points for each synonym or appropriate new word. Some students may even use better words than in the original. The level of the material for students can be graded as follows:

58 to 100% correct: The student can study the material independently.

44 to 57% correct: The student needs instruction.

0 to 45% correct: The student will experience frustration.

SAMPLE CLOZE TEST: The following short sample from Chapter 1 page 2 will give you some idea of what a Cloze test looks like.

Stories have been made up to explain volcanoes. For example, the Romans _____ that volcanoes were caused _____ a god. They thought _____ Roman god Vulcan had _____ ovens under the ground. _____ (named after Vulcan) were _____ chimneys of those huge _____ ovens. Scientific study of _____ began when people started _____ make careful **observations** (ob-sur-**vay**-shuns) of _____ earth. *Observations* are any _____ that we gather by _____ our senses. For hundreds _____ years, people had lived in _____ places where old volcanoes _____ rested. They did not _____ that the mountains and _____ were once the scene

_____ fire and explosion. Toward _____ end of the 18th _____ scientists studying the earth _____ to find that volcanoes _____ not unusual geographic features. _____ found rocks believed to _____ formed from lava. Some _____ were found to be _____ volcanoes. Scientists began to _____ that much of the _____ surface was shaped by _____. For the first time, _____ became aware that the _____ surface could change. If _____ could build mountains, then _____ shape of the earth's _____ must not always have _____ the same. By observing _____ changes in the earth's _____, scientists believed they could _____ out how the earth _____. This was the beginning _____ the scientific study of _____ earth.

Skill Development

Many of the following basic science skills are essential to the successful reading and understanding of science materials. Students will have the opportunity to practice these skills throughout HOLT EARTH SCIENCE. Teacher annotations identify places where these skills appear in the text.

USING THE TEXTBOOK: Very early in the course, it is a good idea to review the parts of the text and their uses. Students must be familiar with the table of contents, Glossary, and Index, and know how to use them to find information.

FOLLOWING DIRECTIONS: The importance of following directions cannot be stressed too much in a science class. At times, even the safety of the experimenter is at stake if proper procedures and techniques are not followed. Develop several exercises for the students to show them the importance of following directions.

READING AND CONSTRUCTING GRAPHS AND TABLES: Very often in science, information can be most easily depicted or summarized in the form of a graph or table. HOLT EARTH SCIENCE makes use of these tools in its presentation of material to the students. It is important that students know how to interpret graphs and tables in order to obtain as much information from them as possi-

ble. Students should also know how to arrange data in the form of graphs and tables. Sample data tables are provided in the Investigations and Skill-Building Activities for this purpose.

READING ILLUSTRATIONS: The many photographs and diagrams in the text are designed to help the students understand the concepts better or to help them follow a sequence of steps or directions more easily.

CAUSE AND EFFECT: Scientists are constantly relating cause and effect in their research and experiments. Students should be able to do the same in their reading and investigations. They will learn facts faster if they look for cause and effect relationships between them.

COMPARE AND CONTRAST: Scientists use the reading skills of comparing and contrasting in their work. They analyze the results of experiments, looking for similarities and differences. These may provide the key to a discovery. By looking for similarities and differences in their reading and laboratory work, students will find it easier to remember facts, to relate them to one another, and to apply them to new situations.

SEQUENCING: Sequencing is an important part of science. The continuity of many scientific theories, concepts, and experiments relies on sequencing. Being able to follow a logical order is a valuable skill that will help students perform investigations and recognize concepts. Sequencing will help students predict the outcome of investigations and apply concepts logically to new situations.

CLASSIFICATION: We all use classification in our everyday lives. For example, we use a classification system to find books in a library. Science systematizes knowledge so that it can be studied. Earth science is the study of our planet. The enormous size of this objective makes it necessary to have classification systems. Huge topics such as earth's composition must be broken down into manageable subtopics like minerals, rocks, and resources. By grouping or classifying topics, students can grasp a great deal of information easily.

SCIENCE VOCABULARY: Many new words are

introduced to students for the first time in a science class. In HOLT EARTH SCIENCE, important new terms are printed in boldface type and are defined in the margin. At the end of each chapter, a Vocabulary review tests students' recall of these terms. The Glossary contains an alphabetical listing of all the terms defined in the margins. Students may be confused by new meanings for familiar words, for example, fault, focus, current, and front. It is a good idea to preteach these familiar words in the context of new content. Suggest that students search for the meaning of a new word in the material surrounding it. In the case of long words, show them how to break them down into smaller ones. The Word Study teacher annotations will help you to break words down into roots, prefixes, and suffixes.

PROBLEM SOLVING: Problem solving can be defined as thinking of ways to combine observations with facts to find a solution. Scientists use the scientific method to solve problems. Students should also be encouraged to use the scientific method to solve problems in their everyday lives as well as in science. Students should realize that not every step of the scientific method is applicable to every problem. Knowing how to approach mathematical solutions is also a part of problem solving.

FINDING THE MAIN IDEA: Students may sometimes finish reading a paragraph without knowing what the main idea of the paragraph is. This inability to find the main idea can also prevent students from taking effective notes in class or writing a useful outline. Therefore, it is wise to spend some time teaching students how to find the main idea of a paragraph. The format of each section in HOLT EARTH SCIENCE should help students to write a brief outline of the section for use as a study guide.

WRITING: Students must be able to read the text, understand the concepts, and use problem-solving skills to arrive at conclusions. In addition, they must be able to express those conclusions in writing. The problems students may encounter in writing may fall into one or more of the following categories: (1) difficulty in converting mental ideas into written expression, (2) inability to understand a specified task, (3) lack of organization, (4) lack of familiarity with writing conventions, sentence structure, correct usage, and correct spelling. In HOLT EARTH SCIENCE, students are directed to respond to end of section and chapter questions in complete sentences. In addition, the format of the Investigations and Skill-Building Activities facilitates the writing of student laboratory reports.

WRITING A SCIENCE RESEARCH REPORT

It is important that students are afforded an opportunity to express themselves in writing. Suggestions for library research are presented in the Applying Science section of the Chapter Review. Students are motivated when they select or volunteer to investigate a particular topic. Earth Science offers many such opportunities. No experiment or research is really complete until the conditions and results of the experiment have been stated clearly and conclusions have been drawn from the results. The following guide will help students to research and organize a science report. The two types discussed are *The Library Research Report* and *The Laboratory Experiment*.

The Library Research Report

SELECTING A TOPIC
- Students may select their own topics, subject to teacher approval; teachers may assign topics; or teachers may compile a list of possible topics and allow individual students to select from the list.
- Be certain that titles are specific.

INTRODUCTION
The topic should be stated clearly or defined.

SOURCES
- Information can be compiled as homework and/or class time can be devoted to library research.

- Students should use a variety of sources, including various reference texts, magazines, and newspapers. The use of low-readability texts should be permitted where necessary for student comprehension.
- Librarians and English teachers can be of help regarding the use of the card catalog, *Reader's Guide to Periodical Literature*, microfilm, and so on.
- Encourage students to write letters to specific sources for pertinent information.
- Require a minimum number of sources; for example, no less than five.

COMPILING INFORMATION
- Select one source at a time.
- At the top of a sheet of paper, or on an index card, record all the information necessary for the bibliography.*
- Note page numbers of material used for the report.
- Whenever possible, students should take notes without copying directly from books. Direct quotes should be surrounded by quotation marks and be followed by exact page numbers. This procedure is crucial for accurate footnoting.
- Information from each source should be recorded on a separate sheet of paper or index card. If more than one is needed for a particular source, staple the pages or cards together.

---

*All students should learn how to record sources in a research report. There are several acceptable styles, but the students should be consistent in the style used. Information usually needed for books in a bibliography includes: Author's last and first names, title of book, city where it was published, name of publishing company, year of publication, and pages cited in report. See your school librarian for information regarding other bibliography entries, or use the style of the reading lists at the ends of chapters. Footnotes may be listed either at the end of the report or at the bottom of the page on which the footnoted material appears. Consult an advanced text or a scientific journal for styles. Your school librarian may also be able to help.

WRITING THE ROUGH DRAFT

- Students are now ready to construct a working outline of their reports. Teacher approval is required at this stage. The outline should list the order of topics to be covered in a logical sequence. The teacher can help by giving a review of what an outline should look like.
- Students should follow the approved outline in writing a first draft.
- Information in quotes should now be marked by number, and their sources credited in footnotes. In addition to direct quotes, footnotes should be used when stating someone's opinion or reporting important statistics.

COMPONENTS OF THE FINISHED REPORT

- Title Page
- Introduction
- Outline
- Body of Report
- Student's Conclusions
- Footnotes
- Bibliography

The Laboratory Experiment

SELECTING A TOPIC

- Students can select scientific topics; teachers can assign topics to students; teachers can compile a list of possible topics and allow individual students to select topics from the list.
- Be certain that titles are specific.

INTRODUCTION

The student will clearly state a specific question and a hypothesis.

COMPILING INFORMATION

- The student should research available scientific information regarding the selected area of experimentation.
- The teacher should indicate what portion of

the report should be devoted to gathering pertinent information through library research.
- An outline of an experimental design should next be made. The teacher can review this to be certain that the proposed experiment will answer the question proposed in the introduction, and that the experiment is both practical and safe.

MATERIALS AND PROCEDURES

- List all materials used.
- Explain methods used in laboratory experiments. List equipment used; describe apparatus; how the experiment was set up; and under what conditions it was carried out. Stress the importance of using control experiments and testing only one variable at a time.
- Stress that the procedures section should contain enough information so that another person would be able to repeat the procedure.

RECORD OBSERVATIONS MADE AND DATA COLLECTED

- Include all observations made during the experiment.
- For clarity, data may be presented in the form of charts, graphs, tables, or diagrams.

DRAW CONCLUSIONS

Based on the data and observations recorded above, the student should be able to draw conclusions. Do they agree with the original hypothesis? If not, why?

COMPONENTS OF THE FINISHED LABORATORY REPORT

- Title Page
- Introduction
- Library Research
- Hypothesis
- Materials and Procedures
- Observations and Data Collected
- Conclusions
- Footnotes
- Bibliography

MATERIALS LIST

In order to cut down on costs and avoid having unused chemicals stored in the laboratory, an attempt has been made to specify the minimal amount of each material required for a class of 30 students. If equipment is a problem, students can work in pairs or share the use of certain pieces of apparatus, such as Bunsen burners.

MATERIALS	NUMBER NEEDED FOR 30 STUDENTS	INVESTIGATION CHAPTER–SECTION
Acid, stearic	150 grams	9–1
Alcohol, rubbing	.5 L	12–2
Balance, centigram	15	1–3, 4–3, 15–1, 16–1
Ball, ping pong (half-painted)	15	18–5
Ball bearing, large	15	16–2
Batteries, D cells	15	19–2
Beaker, 150 mL	45	5–3
Beaker, 250 mL	30	2–2, 5–1, 11–1, 11–2, 12–1, 15–1, 15–2, 16–1
Beaker, 400 mL	10	2–1, 10–2, 10–3
Bowl, large	10	11–2
Boxes, cardboard	10	10–5
Bread	2 slices	9–3
Bulb, flashlight (1.5 V)	15	19–2
Bulb, (with socket and cord)	15	10–5, 12–1, 17–2
Burner	15	9–1, 15–1, 15–2
Can, shiny	15	13–1
overflow	10	6–1
Candle	10	11–1
Carbon paper	15 sheets	16–2
Cardboard square	10	11–1
Card deck, (special 8 cards)	30 sets	8–2
Celery seeds	1.5 oz.	2–1
Chalk, (small pieces)	30	8–3
Clamp, tubing	15	15–2
Clay, modeling	3 lbs.	7–2, 8–3, 15–1, 19–1
Clock, wall	1	7–3
Cloth, cotton	45 pieces	1–2, 12–2
Compass, pencil	30	18–2
Container, 10 cm tall	30	10–3
Container, liter	15	16–1
Container, tall (milk carton)	15	15–1

MATERIALS	NUMBER NEEDED FOR 30 STUDENTS	INVESTIGATION CHAPTER–SECTION
Cups, paper (large)	15	7–2
Cups, Styrofoam	30	15–1
Cylinder, graduated (100 mL)	15	1–3, 4–3, 5–3, 11–1, 15–1, 15–2
Diffraction grating	15	17–2, 19–2
Dropper, medicine	15	1–2, 9–3
Flask	10	11–1, 15–2
Food coloring	5 bottles	16–1, 16–3
Glue, paper (in dispenser)	15	3–2
Glycerine	.5 L	12–2
Goggles, safety	30	6–2, 9–1
Gravel, pea	3 lbs.	5–3
Gravel	5 gal. can	6–1, 10–2
Ice cube	15	13–1
Index card	300	1–1, 1–3, 9–1, 12–2
Limewater	200 mL	11–1
Litmus paper, blue	60 strips	11–1
Map from Investigation 3–2	30	4–1
Map, world	30	4–2
Matches	15 boxes	9–1, 11–1
Meter stick	15	7–4, 10–3
Minerals, (samples)	30	8–3
Pan, round (cake or pie)	15	2–1, 10–2, 16–3
Pan, small	15	10–3
Paper	60 sheets	1–1, 2–3, 2–4, 2–5, 4–1, 7–1
	30 sheets	9–2, 10–1, 12–1, 13–4, 16–2, 17–1, 17–2, 18–1, 18–3, 18–4, 18–5
Paper, black	15 pieces	10–5, 19–2
graph	30 sheets	5–2, 10–4, 14–1, 16–4
tracing	30 sheets	3–2, 13–2, 13–3
unlined	30 sheets	18–2
Paper clips, (sm., m., lg.)	2 boxes each	8–1
Pen, felt tip	15	7–3, 14–2
Pencil, colored	15 sets	3–2, 4–2, 5–2
Pennies	30	8–3
Pins	15	16–2, 18–1, 18–3
Plaster of Paris	2 lbs.	7–2
Plastic wrap	1 roll	10–5
Pyrite (fool's gold)	small amount	10–2
Rack, test tube	15	5–1
Razor blade	15	14–2

MATERIALS	NUMBER NEEDED FOR 30 STUDENTS	INVESTIGATION CHAPTER–SECTION
Ring stand (with ring)	15	15–1, 15–2
Rock, basalt	10	4–3
limestone (small pieces)	15	5–1
shale	10	4–3
Rocks, (small)	15	15–1
Rocks, small (heavy, light)	10 each	2–2
Ruler, metric	30	1–3, 9–1, 15–1, 16–4, 18–1, 18–2, 18–3, 19–1, 19–2
Salt	1 lb.	15–1, 16–1
Sand	5 gal can	5–3, 6–1, 6–2, 10–2
Scissors	15	3–2
Shoe box	10	7–3
Shoe box lid	15	6–2
Soil	5 gal can	2–2, 5–3, 6–1, 12–1
Spoon, plastic (small)	15	2–1, 6–2, 9–1
Staplers	several	1–1
Stirring rod	15	11–1, 13–1
Stopper, rubber	30	1–3
Straw	15	11–1, 11–2, 14–2, 15–1, 16–3
Stream table	10	6–1
String	1 roll	12–2, 15–1, 18–1, 18–3
Sugar, cubes	500	7–3
Tape, adding machine	50 m	7–4
masking	3 rolls	5–3, 10–5, 15–1, 19–2
Test tube 13 × 100 mm	15	5–1, 9–1, 15–1, 15–2
Test tube holder	15	9–1, 15–1
Thermometer (Celsius)	30	1–2, 10–4, 10–5, 12–2, 13–1, 15–1
clinical	15	12–1
Tongue depressor	15	7–2
Tubing, glass	15	15–2
Vinegar	.5 L	11–1
Water, carbonated	1 L	5–1
Waxed paper	1 roll	9–3
Wire, #30, iron, 3 m long	15	19–2
Wire screen	30	15–1, 15–2

SOURCES OF EQUIPMENT AND SUPPLIES

Scientific Suppliers

Carolina Biological Supply Co.
2700 York Road
Burlington, NC 27215

Central Scientific Co.
11222 Melrose Avenue
Franklin Park, IL 60131

Delta Education, Inc.
P.O. Box M
Nashua, NH 03061-6012

Edmund Scientific Company
Edscorp Building
Barrington, NJ 08007

Educational Materials and
   Equipment Co.
P.O. Box 17
Pelham, NY 10803

Fisher Scientific Co.
Educational Materials Division
4901 West Lemoyne Ave.
Chicago, IL 60651

Frey Scientific Company
905 Hickory Lane
Mansfield, OH 44905

Hubbard Scientific Company
P.O. Box 104
Northbrook, IL 60062

LaPine Scientific Co.
6001 S. Knox Ave.
Chicago, IL 60629-5496

Macalester Scientific
   Corporation
Route 111 and Everett Turnpike
Nashua, NH 03060

Nasco
901 Janesville Ave.
Fort Atkinson, WI 53538
               (or)
P.O. 3837
1524 Princeton Ave.
Modesto, CA 95352

Sargent-Welch Scientific Co.
7300 North Linden Ave.
Skokie, IL 60076

Scientific Glass Apparatus
   Company
735 Broad Street
Bloomfield, NJ 07003

Stansi Scientific Company
1231 N. Honore Street
Chicago, IL 60622

Wards Natural Science
   Establishments, Inc.
5100 W. Henrietta Rd.
P.O. Box 92912
Rochester, NY 14692

Wall Maps, Charts, Globes

Hammond Incorporated
515 Valley Street
Maplewood, NJ 07040

Hubbard
1946 Raymond Drive
Northbrook, IL 60062

Map Distribution Division
Army Map Service, Corps of
   Engineers
6500 Brooks Lane
Washington, DC 20014

East of Mississippi River
(including Minnesota):
Branch of Distribution
U.S. Geological Survey
1200 South Eads Street
Arlington, VA 22202

West of Mississippi
(including Alaska,
Hawaii, and Louisiana):
Branch of Distribution
U.S. Geological Survey
Federal Center
Denver, CO 80225

Rand McNally Company
Educational Publishing
   Division
P.O. Box 7600
Chicago, IL 60680

Trippensee Planetarium
   Company
301 Cass Street
Saginaw, MI 48602

Audiovisual Suppliers

American Cancer Society
(obtain locally)

American Lung Association
(obtain locally)

Association-Sterling Films
866 Third Avenue
New York, NY 10022

Bausch & Lomb Inc.
Film Distribution Service
P.O. Box 450
1400 N. Goodman St.
Rochester, NY 14692

Bell Telephone Laboratories
(obtain locally)

Churchill Films
662 N. Robertson
Los Angeles, CA 90069

Coronet Films & Video
A Simon and Schuster
   Communications Company
108 Wilmot Road
Deerfield, IL 60015

Encyclopedia Britannica
   Educational Corp.
425 N. Michigan Ave.
Chicago, IL 60611

Film Comm.
108 W. Grand Ave.
Chicago, IL 60610

Houghton Mifflin Company
989 Lenox Drive
Lawrenceville, NJ 08648

International Film Bureau Inc.
332 S. Michigan Ave.
Chicago, IL 60604

McGraw-Hill
1221 Ave. of the Americas
New York, NY 10020

National Bureau of Standards
Dept. of Commerce
Washington, DC 20234

National Geographic Society
17th and M St. NW
Washington, DC 20036

National Oceanic & Atmospheric
   Administration
RAS/DC 472
Rockville, MD 20852

Prentice-Hall Media
150 White Plains Rd.
Tarrytown, NY 10591

Sterling Educational Films
241 E. 34th St.
New York, NY 10016

Time Life Films
1100 Avenue of the Americas
New York, NY 10036

Wards' Natural Science
   Establishment, Inc.
P.O. Box 1712
Rochester, NY 14603

Computer Software

CBS School Publishing
Product Manager
383 Madison Ave.
New York, NY 10017

Cross Educational Software
1802 N. Trenton Street
P.O. Box 1536
Ruston, LA 71270

Dynacomp, Inc.
P.O. Box 18129
Rochester, NY 14618

Earthware Computer Services
P.O. Box 30039
Eugene, OR 97403

Educational Activities, Inc.
P.O. Box 392
Freeport, NY 11520

Educational Materials &
   Equipment Co.
P.O. Box 17
Pelham, NY 10803

Focus Media, Inc.
P.O. Box 865
Garden City, NY 11530

Krell Software Corp.
1320 Stony Brook Rd.
Stony Brook, NY 11790

Minnesota Educational
   Computing Consortium
3490 N. Lexington Ave.
St. Paul, MN 55112

CHAPTER 1 WHAT IS SCIENCE?

SECTION/ TIME NEEDED	MATERIALS (30 STUDENTS)	BASIC SCIENCE SKILLS	PACING GUIDE
1–1. What is Earth Science? (1–2 periods)	210 index cards 60 sheets paper several staplers	Observing (p. 2) Classifying (pp. 3, 4) Observing (p. 6) Classifying (Teacher's Resource Book, p. 1–3)	Slow Learners: Have students describe what earth scientists do from the pictures on the chapter opener page and pp. 4, 5. Above Average: Discuss examples of incorrect observations, such as "The earth is flat" or "The sun goes around the earth." Mainstreaming: Have students make posters of the earth.
1–2. Scientific Thinking (1–2 periods)	30 thermometers (Celsius) 30 pieces dry cloth 15 medicine droppers	Sequencing and Problem Solving (pp. 7–9, 12) Observing, Measuring, Recording Data, Predicting, Hypothesizing (p. 12)	Slow Learners: Write the steps of the scientific method on the board. Use the steps to illustrate how a local problem was solved. Above Average: Place several rocks that sink into water. Then put in a piece of pumice. Explain how this is an example of the way scientific thinking begins. Mainstreaming: Show students advertisements. Discuss whether they are scientifically valid.
1–3. Measuring (1–2 periods)	30 rulers (metric/ inches) 15 graduated cylinders (100 mL) 300 index cards 30 rubber stoppers 15 balances (centigram)	Measuring (pp. 16, 21)	Slow Learners: Have students use the SI system for common measurements, e.g., the length of a pencil, mass of a book, the volume of a soda can. Above Average: Have students estimate and then measure the length and height of various objects, using the metric system. Mainstreaming: Have students measure the length and width of a desk, using parts of the arm and hand: cubit (fingertip to elbow), span (thumb to pinky), palm (width of palm), and digit (width of pointer).

SUPPLEMENTAL MATERIALS

Audiovisual

Metric America, A, AIMS, 16 mm, color, 16 min.
Modern Measure, The SI System, Schloat, 3 color filmstrips.
Myth, Superstition, and Science, NFB, 16 mm, color, 13 min.
Observing and Describing, McGraw-Hill, 186002-9, 16 mm, color, 12.5 min.
Patterns of Scientific Investigation, EBE, 16 mm, color, 22 min.
Sense Perception, Moody, 16 mm, color, 28 min.
Think Metric, Learning Arts, 4 color filmstrips.
What is Science?, Schloat, sound color filmstrip.

Print

Beveridge, W. I. B. *Seeds of Discovery*. New York: Norton, 1981.
Dickson, David. *The New Politics of Science*. New York: Pantheon, 1984.
Goldstein, Thomas. *Dawn of Modern Science*. Boston: Houghton Mifflin, 1980.

Computer

Connections. Krell SoftwareCorp. APL II Series, COM 64; TRS-80 Models III, 4.
Discovering the Scientific Method: Snigs . . . Flirks . . . Blurgs. Focus Media, Inc. APL II Series; TRS-80 Models III, 4.
Discovery Lab: Minnesota Educational Computing Corp. APL II Series.

CHAPTER OVERVIEW

The first section deals with the nature of earth science and describes the four main branches of study that it encompasses. Students will find that understanding earth processes can be important to them individually at this time as well as preparing them to make proper decisions in the future that will help improve our lives on this planet. The remaining sections deal with the scientific method and the SI system of measurement. Even if the students have had some background in these topics, it would be well to include them. You may wish to cover them more quickly if this be the case.

1–1. WHAT IS EARTH SCIENCE?

BACKGROUND: The processes of the earth have long intrigued humans. The earliest records indicate that what could not be understood was attributed to the actions of gods, goddesses, or other spirits. Myths and superstitions arose and had to be put aside prior to any real understanding of the reasons for such things as earthquakes, volcanic eruptions, and violent storms. As more was learned of the composition of the earth and its immediate surroundings, it became obvious that these events were natural and not at all involved with the supernatural. Only through such an understanding can we appreciate the beauty of the earth and marvel at the events that are a natural part of its history.

MOTIVATION: It is proper to start the study of any science with a question. The question "What is earth science?" is asked and the answer explored in the manner of a scientific problem. You may wish to expand on this by looking at Fig. 1–2 (p. 3) and discussing with the class what is required to find an answer to the question asked about it. It may be obvi-

ous to some that a great deal more needs to be known about the mountain. Discuss what kind of observations might be helpful.

Throughout the teacher's edition of the text, there are annotations in the page margins, some of which are designed to help you motivate the students.

INVESTIGATION

Teaching Hints: Observations about the free fall of an object lead to a question that must be answered by obtaining more information by experiment in order to arrive at a conclusion.

You may wish to staple sufficient quantities of the 2- and 4-card sets for the entire class ahead of time. The sheets of paper need only to be of the same size and weight.

It may be well to note that often scientific progress results from someone being able to ask the right question.

Demonstrate the following: Drop a book and a piece of the same size paper separately and simultaneously. Note the effect of air on similarly shaped objects that differ in weight. Then place the paper on the back of the book so that no paper extends over the edges of the book. Drop the book and paper. Note that both reach the floor together. Ask why this occurs. Discuss the answer. (Lack of air friction on the paper.)

Answers to Questions:
1. The 2-card pile is heavier.
2. The 2-card pile falls faster.
3. The 4-card pile is heavier.
4. The 4-card pile will fall faster.
5. The 4-card pile fell faster.
6. Answers will vary.
7. They both weigh the same.
8. They should fall at the same speed.

9. No.
10. The shape of the crumpled paper allowed it to go through the air faster.
11. The folded sheet fell faster.

Conclusions:

1. The two factors are weight and shape.
2. If air were absent, all objects would fall at the same speed.
3. Do an experiment in which air is absent to see if all objects fall at the same speed.

1-2. SCIENTIFIC THINKING

BACKGROUND: Much has been written about the scientific method for solving problems. There are probably as many methods involved that scientists use as there are scientists. Most will use the general steps that are given here and in the sequence that is discussed. Although many problems have been solved by pure chance, the use of the method of scientific thinking outlined in this section greatly improves the probability of answering a question correctly.

MOTIVATION: Because of the high popular interest in the Mount St. Helens eruption on May 18, 1980, it is used as an example to set up the ideas involved in scientific thinking. You may wish to use a more local and recent event to spark interest.

INVESTIGATION

Teaching Hints: The emphasis is on proving or disproving a typical hypothesis by experimentation. Numbering the thermometers by using a masking tape flag at the top will help to keep the thermometers from rolling. Match the thermometers in pairs that give close readings. To have water at room temperature, put aside water in a large container ahead of time. Dry gauze, 3 cm by 3 cm, works well for the dry cloth. If cotton shoelaces are available, 3-cm lengths form cylinders that just fit over the bulb. Fanning the thermometers will cause the wet bulb thermometer to change temperature faster on drier days. This is the principle of the wet and dry bulb psychrometer used to determine relative humidity.

Safety: Keep thermometers away from table edges.

Answers to Questions:

1. Answers may vary. Usually there is no change.
2. Predictions will vary.
3. Wet one cloth with air-temperature water, wait 2 minutes, then take the temperature of both thermometers.

Conclusions:

1. Answers will vary, depending on the hypothesis made.

2. Answers will vary.
3. No. The scientific method is only a guide to scientific problem solving.

1-3. MEASURING

BACKGROUND: The United States has had the metric system as its official system of measurement for more than 100 years. Yet miles, feet, inches, pounds, etc. are the units used by most. *Le Système International*, the SI system, was the result of the 1960 Treaty of the Meter, and the Federal Metric Board was established in 1975 to stimulate the adoption of the use of SI in the United States. The SI units are used in most scientific laboratories throughout the world.

MOTIVATION: The need for a common measuring system throughout the world could be discussed. Ask the students if there is a metric set of tools at home. Those having foreign autos or bicycles at home will no doubt have both the metric and customary tools. Point out the tremendous cost of this nationwide. Note that many items bought by foreign countries are not bought in customary units. For example, the Russians buy wheat by the metric ton (1,000 kg).

INVESTIGATION

Teaching Hints: Students may have been exposed to the SI units previously. It would be well for them to do this investigation anyway, since it also introduces the use of several tools they will need in future work. If students have not had experience in using the balance, have them do the Skill-Building Activity, Using a Balance (p. 22), prior to doing this Investigation.

Index cards measuring 3 × 5 inches work best. You may want to use several sizes of index cards in order to vary the results from group to group. Centigram balances are recommended; less sensitive balances will work. Be sure to choose rubber stoppers that will fit inside the graduated cylinders. You could make up several packs of 10 index cards each in order to cut down on the number of cards needed.

Answers to Questions:

1. Answer will be either 0 or 0.1 cm.
2. Answers will vary.
3. Answers will vary.
4. The difference of #2 and #3.

Conclusions:

1. length: centimeter, cm; mass: gram, g; volume: cubic centimeter, cm³.
2. Weighing 10 cards decreases the error of measuring just one card.
3. The rubber stopper has an irregular shape.
4. The volume of a cube would be measured best by multiplying its length, width, and height. The

volume of a small ball would be measured best by obtaining a graduated cylinder large enough to fit the ball in. Partly fill the cylinder with water, noting the water's level. Place the ball in the cylinder so that it is completely submerged. Note the new level of water. The difference between the levels of water is the ball's volume.

SKILL-BUILDING ACTIVITY

Teaching Hint: This activity is designed to give the students a degree of skill in the use of a balance. Instructions are given for the double pan equal arm balance and the triple beam balance. If you have both kinds, it would be wise to have the students practice using both of them. This Activity is a must for students who have not used a balance before. Since a balance is used in the Investigation, the Skill-Building Activity should be done before the Investigation.

Conclusions:
1. Students should describe in their own words steps 1 through 4 of the Activity.
2. Students should describe steps 5 through 7 of the Activity.

SPECIAL FEATURES

The photo in the Introduction to Special Features shows a computer graphic of the framework of a space shuttle.

Careers: Regarding career choices, point out the continuous need to redefine skills in light of new technology. In an ever-changing society, adaptability may be more important than consistency of choice.

Technology: These segments discuss recent research and technology. Some segments discuss theoretical speculation rather than firm explanations. Students should be encouraged to explore various possibilities to increase their understanding. Magazines such as *Science Digest, Scientific American, Science '86,* and *New Scientist* are probably the most accessible sources.

¡COMPUTE! This section is not meant to teach computer programming but to show how the computer might be included in science education, to provide a degree of computer literacy, and to provide material for discussion even if computers are not available. You may introduce some of the computer magazines directed toward young adults. These magazines explore the many uses of computers and the ways that young adults are involved in computer electronics. They help the reader identify himself or herself as a computer user. *ENTER* and *k POWER* are two such magazines. If these are unavailable locally, write to:

ENTER, One Lincoln Center, New York, NY 10023; (212) 595-3456; or *k-POWER*, 730 Broadway, New York, NY 10003; (212) 505-3580.

CHAPTER REVIEW

Vocabulary:
1. astronomy 2. geology 3. measurement
4. experiment 5. observation 6. meter (m)
7. hypothesis 8. liter (L) 9. scientific theory
10. kilogram (kg) 11. Celsius 12. oceanography
13. scientific method 14. meteorology 15. density

Answers to Questions:
1. Earth science is the study of the planet Earth.
2. The scientific study of the earth began with the observations of rocks that led to the understanding that volcanoes were causing the earth's surface to be changing always.
3. Volcanoes are a part of geology because they change the surface of the earth.
4. Astronomy is a branch of earth science because the earth is a part of the universe.
5. If we can know about when any of the destructive forces of nature are going to happen, people can be warned and lives saved.
6. Scientific thinking starts with a question.
7. Scientific problem solving starts when (1) a question based on observations is stated as a problem. (2) Additional observations are made, then (3) a hypothesis (conclusion) is formed, which explains the observations. (4) An experiment is planned and carried out to check the hypothesis. Based on the results of experiment, (5) the hypothesis is proven correct or modified and retested by experiment.
8. A scientific hypothesis becomes a part of a scientific theory when it has been tested and found to be correct.
9. After a scientific theory has been tested many times and found always to be correct, it is then called a scientific law. An example of a scientific law is the law of gravity.
10. Experiments in earth science usually involve the gathering of more observations about events as they occur in nature.
11. Controlled experiments are considered ideal experiments because scientists can control anything that may affect the experiment.
12. Scientists use SI units for measuring because they are easily communicated worldwide and have the same meaning for everyone.
13. The average value of many measurements reduces the error of measurement.
14. (a) .80 cm; (b) 300 mm; (c) 1.2 cm.
15. (a) .512 kg; (b) 6,120 mL; (c) .450 L.

CHAPTER 2 A UNIQUE PLANET

SECTION/ TIME NEEDED	MATERIALS (30 STUDENTS)	BASIC SCIENCE SKILLS	PACING GUIDE
2–1. Origin of the Earth (1–2 periods)	1.5 oz. celery seeds 10 small spoons 10 round pans 10 containers for water	Hypothesizing (pp. 29–31) Sequencing (Fig. 2–1, p. 30) Reading for Main Ideas (Teacher's Resource Book, p. 2-2)	Slow Learners: Write the steps in the formation of the solar system on the board. Above Average: Use the scientific method to show why the dust cloud hypothesis is accepted. Mainstreaming: Have students make a diagram of Fig. 2–1 (p. 30).
2–2. Birth of a Planet (1–2 periods)	10 250-mL beakers 10 60-mL samples of soil 10 heavy small rocks 10 light small rocks water source	Hypothesizing (pp. 35–36) Inferring (p. 37)	Slow Learners: Place an object such as a marble in a can or box and seal it. Demonstrate indirect evidence by determining the content of the can. Above Average: Discuss the difficulty of obtaining direct evidence of the earth's interior. Mainstreaming: Have students make a diagram of Fig. 2–5 (p. 35).
2–3. Models of the Earth (1–2 periods)	pencil and paper	Problem Solving (p. 39) Comparing and Contrasting (pp. 41–43)	Slow Learners: Have students give examples of models they may own, e.g., airplanes, cars. Above Average: Discuss the need for having different earth models. Mainstreaming: Show students a globe and a map. Discuss the kinds of information each has.
2–4. Position on the Earth's Surface (1–2 periods)	pencil and paper	Measuring (pp. 48–51)	Slow Learners: Have students find the latitude and longitude of their town or city. Above Average: Have students measure latitude and longitude using minutes and seconds. Mainstreaming: Give students different kinds of maps to read.
2–5. Motions of the Earth. (1–2 periods)	pencil and paper	Inferring (pp. 52–53) Comparing and Contrasting, Sequencing (pp. 57–59)	Slow Learners: Use a globe and a light source to show students how the seasons are caused. Above Average: Using a globe and a light source, show how the length of day and the seasons would change if the earth's axis were changed.

SUPPLEMENTAL MATERIALS

Audiovisual

The Cosmos, The Study of the Universe, EDG, videotape, 4 programs, 20 min. ea.

Day and Night, BFA, Super 8 color filmloop.

Earth, Rotation and Orbit, Prentice-Hall, set of 2 color filmstrips.

Focus from Outer Space, Carl Zeiss, Inc., color filmstrip.

Planet Earth, Thorne, GC423, ATSII satellite views, filmloop.

Probing The Universe, MIS, set of 4, color, sound filmstrips.

The Solar System, NFB, 16mm, color, 27 min.

Print

Barnes, Charles W. *Earth, Time, and Life: An Introduction to Geology.* New York: Wiley, 1980.

Bartusiak, Marcia. "The Planet Hunters." *Science Digest,* January 1984.

Maslowski, Andrew. "Geology From Space." *Astronomy,* November 1983.

Sandage, Alan. "Inventing the Beginning." *Science 84,* November 1984.

Tauber, G. E. *Man's View of the Universe.* New York: Crown, 1979.

Computer

Maps and Globes. Micro-Ed, Inc., COM 64, Pet; APL II Series; TRS-80.

CHAPTER OVERVIEW

The chapter begins by introducing the earth as a member of the solar system. Several hypotheses are given to explain how the solar system may have come about. A differentiated model of the structure of the earth is presented, based on studies of earthquake waves. How the planet may have developed is explained in this manner.

The earth globe, projection maps, and topographic maps are used as models of the earth. Each model is discussed in some detail. The student is shown how to read the various maps. A separate lesson is devoted to the use of latitude and longitude as a means of locating positions on the earth. The chapter ends with a discussion of the rotational and orbital motion of the earth and how this motion produces seasons on the earth.

2–1. ORIGIN OF THE EARTH

BACKGROUND: The origin of the solar system and the earth is usually a controversial topic. In this lesson it is presented hypothetically, an explanation based on the facts as presently known and in need of more facts before it can be proven correct or incorrect.

MOTIVATION: Controversial topics usually need little motivation for student attention. The main thing to push is that, if one were to use the scientific method to arrive at an explanation, the hypothesis that seems best must be constantly tested and revised to fit the facts as known. Up to now, the dust cloud hypothesis seems to be the best explanation of the origin of the solar system.

INVESTIGATION

Teaching Hints: Celery seeds in water are used to simulate the way the dust cloud hypothesis may have operated in some respects. The model works the way it does because of the electrostatic charge on the seeds and the surface tension of the water. The forces that operate here are both attractive and repulsive—unlike gravity, which is only attractive.

A 10-inch glass pie dish works best. Whole celery seeds are sold as a seasoning at grocery stores in boxes of 1.5 oz. Whole sesame seeds or pencil shavings from a pencil sharpener work less well but may be used if celery seeds are not available. Provide a suitable container for the waste water and seeds. Seeds should not be poured down a drain.

Answers to Questions:

1. The seeds began to move.
2. The seeds moved toward one another, forming clusters. Some clusters of seeds began to move.
3. The swirling original dust cloud.
4. The seeds are clumping together in larger groups.
5. It would show the forming of planets as the dust cloud condensed.
6. The description depends on the number of seeds but should resemble a circular pattern of groups of seeds clumped together.
7. The pattern will be similar but with fewer, larger clumps of seeds.
8. The seeds clump together as they move in a circle.
9. The clumps of seeds give a circular pattern with very few near the center.

Conclusions:

1. The moving together of the seeds into groups represents the condensing of the dust cloud into planets. These groups are seen to move about in the dust cloud as they form, somewhat as the planets did when they formed.
2. The seeds did not have a circular motion around the center of the pan as did the planets around the sun. There was no particularly large clump of seeds in the center of the pan to represent the sun.

3. The circular motion made the seeds form in rings around the center much as the planets did.

2-2. BIRTH OF A PLANET

BACKGROUND: The exploration of the moon enabled scientists to study moon rocks first hand. This study has given us the information to interpret earth rocks better. Together, these studies have led to a much clearer understanding of the changes occurring constantly on the earth's surface and the role of these changes in the earth's history. The differentiation of the earth into layers is not unique among planets. Space probes to Jupiter have shown satellites in all stages of differentiation. This information has done much to confirm the view scientists hold of the early beginnings of the planet Earth.

MOTIVATION: The Investigation on p. 37 would be an excellent exercise to precede this section rather than follow it. Thus it could serve as a motivation for this section.

INVESTIGATION

Teaching Hints: The student makes observations by doing two experiments. From these observations, a hypothesis is made and checked concerning the way layering of the earth might have occurred.

The soil samples used should have a variety of particle sizes in order for layers to form. All the small rocks must be of similar size. The heavy rocks could be samples of galena, pyrite, hematite, or other metal ores. The light rocks could be samples of limestone, quartz, or other similarly light materials. The idea is to show that the heavy rocks fall through a liquid faster and mire farther into the bottom than the light rocks. Have the students return the rock samples to the container for storage. Provide a container for waste soil and water. This can then be discarded outside at a later time. Emphasize NO SOLIDS DOWN DRAINS.

Answers to Questions:
1. Differences in color and particle size show layering from top to bottom.
2. Particle size is larger at the bottom than at the top.
3. Answers will vary.
4. Some are larger, some are smaller.
5. Most will predict the heavier rock to fall faster.
6. Answers will vary.
1. 2.8 g/cm^3.
2. The crust has a lower density than the earth.
3. The materials in the crust are lighter than the materials under the crust.
4. The density of the earth is an average of the densities of the core, mantle, and crust. Since the earth's density was known to be 5.5 g/cm^3, and the earth's crust was measured to be only 2.8 g/cm^3, the density of the core and mantle must have been much greater than the earth's density.

Conclusions:
1. Yes. As in the experiment, heavier materials would settle inward to give layering.
2. Heavy material would end up in the core.
3. No, the heavier materials would be separated from the lighter materials.
4. Volcanoes are sources of magma, which is formed inside the earth, where heavy metals are found.

2-3. MODELS OF THE EARTH

BACKGROUND: The idea of the earth as a sphere is rather a recent one. Even to this day there are those who still believe in a flat earth. As one moves around in a small area, everything could be explained on the basis of a flat earth. City maps and others are all flat. The most significant evidence for knowing that the earth is like a sphere are the pictures made from deep space by astronauts traveling to and from the moon. Of course, world navigators have always used flat maps and known the earth was round. In this chapter the student learns the ways these maps are useful models of the earth.

MOTIVATION: Discuss Fig. 2-6 (p. 38) with the class. Ask the students if they know of other proofs that the earth has a curved surface. The slow disappearance of a ship over the horizon, astronaut pictures of earth from space, the need for microwave antennas on mountain tops at given intervals to pass straight-line radio waves any distance, Fig. 2-7 (p. 39), etc. could be discussed. Then show an earth globe and ask if they see any problem with using it as a model for the earth. This will then lead into the reading of the text.

INVESTIGATION

Teaching Hints: Topographic maps are very important to earth scientists. This Investigation acquaints the student with topographic maps. If you could obtain a topographic map or an aeronautical sectional chart of your area, have it on your bulletin board for discussion after doing this Investigation.

Answers to Questions:
1. Four.
2. 0 cm, 5 cm, 10 cm, 15 cm.
3. 10 m.
4. It will probably be greater than 30 meters.
5. A black square with a flag symbolizes a school.
6. With hatch marks indicating a depression.

7. With blue symbols similar to those shown on page 46.

Conclusions:

1. By placing the bowl and cup over to one side rather than symmetrically.
2. The contour lines on the steeper side would be closer together.
3. The contour lines would form a V-shape pointing upstream.

2–4. POSITION ON THE EARTH'S SURFACE

BACKGROUND: It would be impossible to convey positions on a globe without having some kind of an imaginary grid to refer to. Over several centuries, the latitude and longitude grid has been most widely used for locating and communicating to others the positions of ships, islands, storms, etc. No such grid would be required if we did not travel beyond the area where we live. Worldwide communications and travel experienced by modern people make it necessary that we become more familiar with the manner of locating positions on earth. In this section the student is given the opportunity to develop a measure of skill in locating positions by latitude and longitude.

MOTIVATION: Use an earth globe to point out the necessity of having some method by which positions can be identified. Note that most of the earth is covered by water and has no landmarks to aid in locating positions. Describe the reasoning behind locating the prime meridian so that it runs through Greenwich, England. (Greenwich is the site of the pioneer observatory to use star positions in exactly locating positions on earth.) You may wish to have the students argue, pro and con, the use of east and west longitudes versus 0- to 360-degree longitudes.

INVESTIGATION

Teaching Hints: The Investigation gives the students first-hand experience at locating positions using latitude and longitude. Fig. 2–24 (p. 51) could be photocopied and enlarged for use by the students so they may mark on it. Caution the students not to mark on the figure in the text. If you have a larger world map containing latitude and longitude lines, more accurate results could be obtained by its use.

Answers to Questions:

All answers are based on Fig. 2–24 and will be less accurate than those taken from a larger map or from the text discussion.
1. Answers will vary.

2. 45°N, 75°W.
3. 7°N, 15°E (answers will vary).
4. 55°N, 5°W (answers will vary).
5. Answers will vary (about 167°W).
6. Australia.
7. Hawaii.
8. Japan.
9. Mexico.

Conclusions:

1. No. You need to know both latitude and longitude to locate any position on earth.
2. By giving both the longitude and the latitude, the position of any point on earth can be described.

2–5. MOTIONS OF THE EARTH

BACKGROUND: Students are almost always aware of the rotation of the earth and its revolution around the sun. However, they seldom understand the relationship of these motions to the sun's apparent daily and yearly motions or of the changing appearance of the night sky. Yet it is just these motions that early astronomers used to derive our modern concept of the solar system. Use the information given in this lesson to help students become better aware of the apparent motion of the sun and other sky objects.

MOTIVATION: Have students look at Fig. 2–25 (p. 52). Ask them to explain how the photo was made. For example, how can they tell that it was the sun moving, not the camera? Ask them to look at the other objects in the picture. Ask, "If the camera wasn't moving, then was it the sun or the earth?" What observations could they make from the earth to find out?

INVESTIGATION

Teaching Hints: Students may want to identify some prominent constellations visible at the time they do the Investigation. Star maps showing the night sky can be found in monthly magazines, such as Sky and Telescope and Astronomy.

Answers to Questions:

1. Q, R, S, T, U.
2. W, V.
3. L, M, N, O, P, A, B.
4. B, C, D, E, F, G, H.
5. G, H, I, J, K, L, M.

Conclusions:

1. The stars in the sky entirely surround the earth. You can see only the stars that are on your side of the earth at night.
2. As the earth revolves about the sun, the stars

remain fixed. Therefore you see some different stars as the earth moves in its orbit.
3. The North Star and others nearby are above the axis of rotation. Thus they appear to move in circles and are visible throughout the year.

CHAPTER REVIEW

Vocabulary:

1. solar system 2. crust 3. mantle 4. core
5. topographic map 6. contour line 7. latitude
8. longitude 9. scale 10. axis 11. standard time
12. summer solstice 13. winter solstice 14. orbit
15. equinox

Answers to Questions:

1. The sun's energy allowed only the heavy materials of the closest planets to condense so close to it, whereas the lighter gases could condense in the very cold regions farther from the sun.
2. The planets contained too small an amount of material to cause the nuclear reactions of a star.
3. Moon rocks have undergone little change since the beginning of their existence. By comparing them to earth rocks, the scientist can imagine the changes that have occurred on earth since its beginning, and thus have an idea of how the earth might have begun.
4. The Moho is the region of great change that occurs between the crust and the mantle.
5. The fact that the earth has layers in which the lightest materials are outermost and the heavier materials are arranged inward is evidence of a once-melted earth.
6. The seasons in the Southern Hemisphere are the reverse of the seasons in the Northern Hemisphere. That is, when it is summer in the Southern Hemisphere, it is winter in the Northern Hemisphere.
7. It is best to keep an entire city on the same time rather than have the eastern part on one time and the western part on a different time.
8. If the axis tilted more, the seasons would have greater differences, whereas, if it tilted less, there would be smaller differences between seasons. If the axis did not tilt at all or if it were tilted at 90°, there would be no seasons.
9. The sun shines for a longer period of time and more directly toward the surface in the summer than it does in the winter.
10. In order that there would be very few people in the area where one day changes to the next.
11. In 13,000 years the earth's axis will be pointing away from the sun on June 22.

CHAPTER 3 PLATE TECTONICS

SECTION/ TIME NEEDED	MATERIALS (30 STUDENTS)	BASIC SCIENCE SKILLS	PACING GUIDE
3-1. Moving Continents (1 period)	none	Comparing and Contrasting (pp. 65–66) Inferring (pp. 66–67) Hypothesizing and Predicting (Teacher's Resource Book, p. 3–2)	Slow Learners: Use a washable marker to show the boundaries of the crustal plates on a globe. Discuss your location on the plate. Above Average: Using Fig. 3–7 (p. 70), have students make a map of what the world will look like millions of years from now. Mainstreaming: Do Investigation (pp. 80–81) as a demonstration by placing the pieces together on the blackboard.
3-2. Action at the Plate Edges (1–2 periods)	15 scissors 15 each of 3 different color pencils 15 paper glue dispensers 30 sheets tracing paper	Compare and Contrast (pp. 72–73) Cause and Effect (p. 76)	Slow Learners: Use two books to demonstrate the three kinds of plate boundaries shown in Fig. 3–10 (p. 73). Above Average: Discuss how the faults, mountain ranges, volcanic islands, trenches, and continental growth relate to the moving plates. Mainstreaming: Use a filmstrip with taped captions or overheads to show how plates move.

SUPPLEMENTAL MATERIALS

Audiovisual

Plate Tectonics, EME, 20-slide set, 35mm color.
The Restless Earth: Understanding the Theory of Plate Tectonics, Science and Mankind Inc., sound-slides, filmstrip, or videocassette (42 min.).
Tectonic Mountains, EBE, S-80108, color filmloop.

Print

Foder, R.V. *Earth in Motion: The Concept of Plate Tectonics.* New York: Morrow, 1978.
McPhee, John. *Basin and Range.* New York: Farrar, Straus, Giroux, 1981.

Miller, Russel. *Continents in Collision* (Planet Earth Series). Chicago: Time-Life Books, 1983.
Raymo, Chet. *The Crust of Our Earth.* Englewood Cliffs, NJ: Prentice-Hall, 1983.
Simmons, Henry. "Before Pangaea." *Mosaic*, March-April, 1981.
Sullivan, Walter. "Tectonics on Venus: Like that of Ancient Earth?" *Science*, January 15, 1982.
West, Susan. "A Patchwork Earth." *Science 82,* June 1982.

Computer

Teach Yourself By Computer. Earth Science Series, TRS-80, Models I, III.

CHAPTER OVERVIEW

This chapter traces the beginning of the theory of plate tectonics as a natural outcome of the discovery of sea floor spreading and continental drift. Plate tectonics is used to explain how continents move and how some mountains, island arcs, and sea floor trenches form at the plate boundaries.

3–1. MOVING CONTINENTS

BACKGROUND: With the publication in the mid-17th century of his treatise summarizing all the events that had occurred on earth from the time of its creation, James Ussher, an Anglican Archbishop in England, was able to satisfy most people of the Western world with an explanation of how the earth came to be as it was at the time. Exact dates had been laid down for the creation of the earth, the building of the ark, the Flood, and other important historical events. The earth was at most a few thousand years old. It was a stable, ordered planet with little change since its creation. Not until the questioning by scholars of the 19th century and the great scientific advances of the 20th century was this early concept seriously challenged. Today we see the earth as an ever-changing planet with a history that spans billions of years. This section gives the modern view and explanation of the earth's history.

MOTIVATION: Use the idea of the jigsaw puzzle (pp. 65–66) to introduce students to the idea of seeing the earth as a puzzle. Cut a piece from a road map of your state that represents much of the area where your students live. Show only this piece to the students. The idea is to ask questions about your immediate area that can be answered by looking at this piece of the map. Then ask a question that can be answered only by using the rest of the map. Produce the map and answer the question. Continue this process until you can point out that the more we learn, the more we realize that our first thoughts are just little pieces of a bigger picture. The explanation of what goes on in our little part of the world must be a part of the explanation for the larger world.

3–2. ACTION AT THE PLATE EDGES

BACKGROUND: Scientists are beginning to see the whole picture of how the continents formed by using the theory of plate tectonics. In this section, plate tectonics is used to explain the manner in which not only the continents but also the mountains, island arcs, and sea floor trenches were formed and are forming.

MOTIVATION: Discuss the introduction paragraph on p. 72. It may be inconceivable to some students that parts of our North American continent could have been a part of other continents at one time in the past. Point out that earth scientists have been puzzled for some time by the fact that certain parts of our seacoasts seem not to be made up of the same rock structure as the rest of the continent. For example, the rock structure of part of southern Florida is remarkably like the structure of a part of the African continent and unlike any of that of the rest of the North American continent. The students will find out in this section how plate tectonics can explain this puzzle.

INVESTIGATION

Teaching Hints: In this Investigation the students will assemble a world map from a jigsaw puzzle of the major crustal plates. Not only will they become more familiar with the names and positions of these plates but by plotting the mid-ocean ridge, some mountains, and ocean trenches, they will realize that these features are related to the plate boundaries.

Copies of Fig. 3–18 (p. 81) can be made and handed out to students. A copy master is also provided in the Teacher Resource Book, Chapter 3, pp. 3–5. Students are to answer questions on a separate sheet of paper, not on the map being assembled. Collect the maps. Hold them for use in several Investigations in the next chapter.

Answers to Questions:

1. Pacific. 2. African, Antarctic, North American, South American. 3. Australian-Indian, Eurasian. 4. Andes, Himalayas. 5. Raised by collision of crustal plates.

Conclusions:

1. Pacific: named for the ocean it holds.
   African: named for the continent it contains.
   Antarctic: named for the continent it contains.

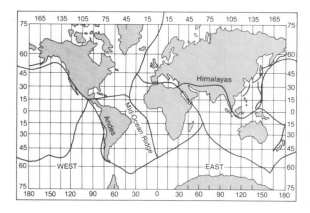

North American: named for the continent it contains.
South American: named for the continent it contains.
Australian-Indian: named for the two continents it contains.
Eurasian: named for Europe and Asia.
2. Mid-ocean ridge, mountains, trenches.
3. Crustal plates move away from each other at the mid-ocean ridge.

TECHNOLOGY

Teaching Hint: The process described in this feature on seismic reflection profiling is difficult to illustrate since the trucks prevent a clear view of the vibrators. However, the photo does depict the caterpillarlike procession of these vehicles and gives some concreteness to this technique.

CHAPTER REVIEW

Vocabulary:

1. sea floor spreading 2. rift valley 3. island arc
4. convection 5. continental drift 6. mid-ocean ridge
7. plate tectonics 8. subduction 9. pillow lava
10. trenches 11. micro-continents 12. asthenosphere

Answers to Questions:

1. Scientists found it hard to believe that giant continents could push their way through the solid crust.
2. The theory of continental drift says that all the earth's continents were once a single land mass that later separated.
3. The discoveries were that (1) the mud and ooze that covered most of the sea floor was missing from the mid-ocean ridge, and (2) the center of the mid-ocean ridge is much warmer than any other part of the ocean floor.
4. Sea-floor spreading says that the mid-ocean ridge is a crack in the earth's crust where molten rock from the earth's interior constantly rises. Cooled by the water, the molten rock becomes solid and is pushed up and outward forming the mountainous ridges.
5. The theory of plate tectonics explains that the continents ride along on the moving plates.
6. Convection currents are caused when heat in the mantle moves out toward the earth's crust.
7. Scientists found that the youngest rocks were closest to the mid-ocean ridge and the oldest rocks were farthest away.
8. Crustal plates are moving away from each other along the mid-ocean ridge.
9. Two crustal plates are sliding past each other along the San Andreas fault.
10. Crustal plates are coming together where the Indian plate is meeting the Asian plate.
11. The Andes, western coastal ranges of North America, and the Atlas range in North Africa were formed by plate collisions.
12. Ocean trenches are long, narrow depressions in the ocean floor. Most of the world's trenches are located near the island arcs in the Pacific Ocean.
13. The Aleutian Islands and the Japanese Islands are examples of island arcs.
14. Ocean plates sink under continental plates because the ocean plate is heavier than the continental plate.
15. Continents crumple when pushed together because one does not sink into the mantle. Both continents remain floating instead.

CHAPTER 4 BUILDING THE CRUST

SECTION/ TIME NEEDED	MATERIALS (30 STUDENTS)	BASIC SCIENCE SKILLS	PACING GUIDE
4–1. Earthquakes (1–2 periods)	crustal plate map from previous Investigation paper and pencil	Cause and Effect (pp. 88–89) Classifying (pp. 90–91) Comparing and Contrasting (Teacher's Resource Book, p. 43)	Slow Learners: Bring in newspaper or magazine articles, local if possible, on earthquakes to discuss. Above Average: Have students do lab activity in the Teacher Resource Book or lab manual, "Determining the Location of an Earthquake." Mainstreaming: Use filmstrips with taped captions.
4–2. Volcanoes (1–2 periods)	30 world maps 30 color pencils	Cause and Effect (p. 95) Classifying (p. 100) Sequencing (p. 99)	Slow Learners: Discuss newspaper or magazine articles on volcanoes. Above Average: Compare the different ways the volcanoes in Fig. 4–16 (p. 101) formed. Mainstreaming: Show students a filmstrip with taped captions on volcanoes.
4–3. Building the Land (1–2 periods)	10 pcs. shale rock 10 pcs. granite rock 10 pcs. basalt rock 10 balances 10 graduated cylinders	Cause and Effect (pp. 105–106) Classifying (pp. 105–106)	Slow Learners: Discuss the shapes and types of local or famous mountains. Above Average: Using vocabulary, have students make crossword puzzles. Mainstreaming: Have students learn vocabulary in margins and read the summary before teaching this section.

SUPPLEMENTAL MATERIALS

Audiovisual

Alaska Earthquake, U.S. Geological Survey, 16 mm, 22 min.

Anatomy of a Volcano, EBE,S-809750, 8 mm filmloop.

The Changing Earth Today, Hawkhill, sound filmstrip or videocassette.

Earthquake Below, NASA,HQ248, 16 mm color, 14 min.

Earthquakes, Lesson of a Disaster, EBE, 16 mm, color, 13 min.

Heartbeat of a Volcano, EBE, 16 mm, color, 21 min.

Lassen's Landscape, Lassen Volcanic National Park, filmstrip.

Volcanic Features, Hubbard, #789, set of 20 color slides.

Print

Gere, James M. and Haresh C. Shah. *Terra Non Firma: Understanding and Preparing for Earthquakes*. San Francisco: Freeman, 1984.

Golden, Frederick. *The Trembling Earth: Probing and Predicting Quakes*. New York: Scribner, 1983.

Harris, Stephen L. *Fire and Ice: The Cascade Volcanoes*. Seattle: Mountaineers Books, 1980.
Raymo, Chet. *The Crust of Our Earth*. Tarrytown, NY: Prentice-Hall, 1983.
Scientific American. *Earthquakes and Volcanoes*, 11-article collection. San Francisco: Freeman, 1980.
Simkin, Tom, and Richard S. Fiske. *Krakatau 1883*. Washington, DC: Smithsonian Press, 1984.
Snead, Rodman E. *World Atlas of Geomorphic Features*. New York: Van Nostrand Reinhold, 1980

Tirbutsch, Helmut. *When the Snakes Awake: Animals and Earthquake Prediction*, Boston: MIT Press, 1984.

Computer

Earth Science. Minnesota Educational Computing Corp., TRS-80, Models I, II.
Teach Yourself By Computer. Earth Science Series, TRS-80, Models I, III.
Volcanoes. Earthware Computer Services, Apl II.

CHAPTER OVERVIEW

The theory of plate tectonics explains the relationship of earthquakes, volcanoes, mountains and plateaus. The movement of crustal plates provides the great forces which strain the earth's crust causing one or another of these events to occur and the resulting land features. Volcanoes are classified by the shape of the features formed while mountains are classified mainly by the way in which they were formed.

4–1. EARTHQUAKES

BACKGROUND: Probably no other natural event has been the cause of more deaths and destruction than earthquakes. Modern technology has given the scientists tools to study these events in great detail. Plate tectonics has provided the ability to explain why they occur. Although the Chinese and Japanese have been successful in accurately forecasting some earthquakes, most still elude our ability to forecast with much degree of accuracy.

MOTIVATION: There is hardly a region on earth where an earthquake will not occur, given sufficient time. There are a number of places in the world where earthquakes occur frequently. Students would like to know where these places are located. By doing the Investigation on p. 94 before reading this lesson, the students will find out where earthquakes most often occur and find that they are related to crustal plate boundaries.

INVESTIGATION

Teaching Hints: Students are asked to plot some representative earthquake locations on a world map. It is readily seen that the earthquake positions describe the boundaries of crustal plates and so must be related in some way to plate tectonics.

The crustal plate map from the Investigation in Chapter 3 is used to plot the earthquake locations. If some students did not do the previous Investigation,

you will need to supply them with a map. A copy master is provided in the Teacher Resource Book, Chapter 4, pp. 4–5.

Answers to Questions:

1. Earthquakes are located along the western coast of North and South America, around the Mediterranean, and along southern and eastern Asia.
2. Earthquakes have no clear relationship to the shape of continents.
3. Earthquakes are found along the edges of crustal plates.

Conclusions:

1. Answers should be similar to answer #1 or #3 above.
2. The pattern found by plotting a number of representative earthquake locations on the world map suggests that many future earthquakes may be expected to occur <u>along the edges of crustal plates</u>.
3. Answers will vary.
4. It is easier to predict where earthquakes will happen. Earthquake predictions can be improved by making as many observations as possible in areas where earthquakes are likely to take place.

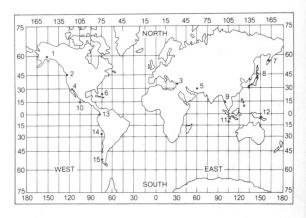

4-2. VOLCANOES

BACKGROUND: Although less destructive than earthquakes at the time they occur, volcanoes last longer and may be more destructive in the long run by the changes in climate for which they may be responsible. Volcanoes can also be helpful. The Hawaiian chain of islands in the Pacific were formed entirely by volcanoes. They provide a place for many people to live and for many others to vacation. How and where volcanoes form is discussed in this lesson.

MOTIVATION: Discuss the nature of a recent volcano such as Mt. St. Helens. Use Fig. 4-14 (p. 99) as a guide. Point out that these events are most often related to crustal plate activity.

INVESTIGATION

Teaching Hints: It is recommended that students do the previous Investigation on plotting earthquakes. In this Investigation the students plot a number of volcano locations on the world map used in the previous Investigation. When compared to the map on which the earthquakes were plotted, they find the locations are very similar. Volcanoes are found to occur most often around crustal plate boundaries.

A world map similar to Fig. 3-18 (p. 81) is recommended, or the crustal plate map previously used for plotting earthquakes may be used. A copy master is provided in the Teacher Resource Book, Chapter 4, pp. 4-5.

Answers to Questions:
1. Atlantic.
2. Pacific.
3. Yes.
4. They are located in the same regions.
5. Volcanoes are located near crustal plate boundaries.

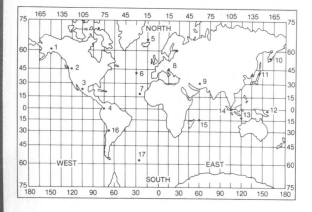

Conclusions:
1. Volcanoes are most likely to occur near crustal plate boundaries.
2. Both earthquakes and volcanoes are generally found in the same places.
3. If enough earthquakes are observed to take place just before volcanoes erupt, a pattern in their frequency may be found. This pattern could be used to predict volcanic eruptions.

4-3. BUILDING THE LAND

BACKGROUND: The movements of crustal plates produce strains in the earth's crust in several ways. The strains result in the formation of mountains and valleys as well as plateaus. How these land features are formed as a result of plate tectonics is explained in this lesson.

MOTIVATION: Mountains are the most noticeable result of strains put on the earth's crust by crustal plate movement. Show how soft plastic can be deformed slightly without breaking. First, apply a slight strain to a piece of plastic large enough for all to see. Ask the class to predict the result of applying a large strain. Point out that rock behaves in a similar manner when responding to strain, plastic deformation or fracturing. See Fig. 4-20, p. 105. Discuss how other common substances, such as glass, metals, modeling clay, etc., would respond to strains.

INVESTIGATION

Teaching Hints: The role of density in isostasy is explored in this Investigation. If your students had a difficult time with density earlier, you may wish to review it at this time. The same technique as was used in Chapter 1, Section 3 Investigation (p. 21) is used to determine the volume of the irregular shaped rocks. Recall that 1 mL and 1 cm³ are equivalent volume measurements.

The three kinds of rock chosen are typical of the earth's crust at the surface and give an average density very near that of granite. You may substitute other rocks available locally. Sandstone makes an excellent substitute for shale. Break the rocks into small enough pieces to fit the graduated cylinder being used. 100 mL cylinders work best. If several classes are to use the same rock samples, provide a separate box for the wet ones of each kind so they may be dried again before being used.

Answers to Questions:
1. Basalt (approx. 3.0 g/cm³), depending on rocks used.
2. Shale (approx. 2.5 g/cm³), depending on rocks used.
3. 2.75 g/cm³, depending on rock samples used.
4. Granite ($2.7-2.8$ g/cm³).

5. The average density of the rocks is much lower than the earth's density.
6. Shale, depending on rocks used.
7. Basalt, depending on rocks used.

Conclusions:

1. The entire earth has the higher density.
2. If the density of the crust is an average of the rocks used in this Investigation, the crust is lighter than the average earth density. Therefore, the material below the crust must be heavier in density, allowing the crust to float on it.
3. Different rocks of the crust have different densities and would tend to float at different levels on the mantle, causing the earth to have an uneven surface.
4. The largest mountains weigh more.
5. As sediments are worn away, the mountains become lighter.

CHAPTER REVIEW

Vocabulary:

1. fault 2. focus 3. epicenter 4. seismograph 5. seismic wave 6. magma 7. volcano 8. lava 9. diastrophism 10. plateau

Answers to Questions:

1. Earthquakes are the result of rock movement on each side of a fault, which is usually caused by the movement of crustal plates.
2. The three ways rock may move along a fault are (1) motion resulting from the separation of plates, (2) motion of rocks being pushed together as in the head-on collision of plates, (3) motion when blocks of rock on each side of a fault move in opposite directions.
3. The focus is the point at which rocks started the movement along a fault and is usually deep underground. The epicenter is the position on the earth's surface directly above the focus.
4. The three kinds of earthquake waves are (1) primary waves (P waves): energy waves sent out due to the back and forth vibrations of the rock; (2) secondary waves (S waves): energy waves sent out due to the up-and-down vibrations of the rock; and (3) surface waves (L waves): the up and down vibrations of the earth's surface, like ripples on water.

5. The Richter Scale uses numbers to express the amount of energy released by an earthquake. Each number represents ten times the energy released by an earthquake measured one number lower. For example, an earthquake registering 4 on the scale releases 10 times the energy of an earthquake registering 3.
6. (a) You should move away from the building out into an open area. (b) Move away from windows, brick chimneys or walls, and get into a doorway or under a desk or table.
7. Magma is red-hot rock found deep underground; when it rises to the surface forming a volcano, it is then called lava.
8. Earthquakes and volcanoes are most likely to occur near active areas around crustal plate boundaries.
9. Two sources of strain are (1) tectonic movements of the crustal plates and (2) differences in density of rocks of the crust and mantle, causing heavier portions of the crust to sink and lighter portions to rise as they float on the heavier mantle.
10. Explosive volcanoes produce tall, narrow cinder cones from the ash and cinders that build up around a vent. Composite volcanoes usually build up a large cone-shaped mountain due to the alternating explosive and quiet flows of ash, cinders, and lava. Nonexplosive volcanoes build up layer on layer of lava forming a mountain that is broad, flat-shaped, with gently sloping sides.
11. The Aleutian Islands formed as a result of colliding crustal plates causing volcanoes, whereas Iceland formed as a result of two crustal plates separating.
12. If there is movement along either side of a crack in the crustal rock, it is called a fault. If there is no movement along such a crack, it is called a joint.
13. A quake registering 7 releases 1,000 times the energy of one registering 4.
14. The Appalachian Mountains are folded mountains. The ridges of the mountains are the part of the fold which is raised and is called an anticline. The valleys of the mountains are the part of the fold which is lowered in the folding and are called synclines.
15. Fault-block mountains are tilted, large blocks of crust separated by faults, and tend not to show any curve to the layers of rock as in the folded mountains.

CHAPTER 5 WEATHERING AND EROSION

SECTION/ TIME NEEDED	MATERIALS (30 STUDENTS)	BASIC SCIENCE SKILLS	PACING GUIDE
5–1. Weathering and Erosion (1–2 periods)	15 test tubes 15 test tube racks 1 L carbonated water 15 small glass containers 15 small pieces limestone rock	Compare and Contrast (p. 115) Cause and Effect (pp. 115–116, 117, 119–121) Classifying, Finding Cause and Effect, (Teacher's Resource Book. p. 5–3)	Slow Learners: Freeze a bottle of water. Discuss what happens. Above Average: Use some local examples to discuss the difference between weathering and erosion. Mainstreaming: Show some objects that have rust, peeling paint, smooth or broken pebbles or rocks.
5–2. Water on the Land (1–2 periods)	30 sheets graph paper 30 black pencils 30 red pencils	Compare and Contrast (pp. 129, 130)	Slow Learners: Use Fig. 5–13 (p. 125) to help discuss the water cycle. Above Average: Discuss the origin of the water in the school or home and where the water will go after being used. Mainstreaming: Demonstrate the water cycle by heating water in a beaker. Place above the beaker a flask containing cold water.
5–3. Water Beneath the Land (1–2 periods)	45 150-mL beakers 15 100-mL graduated cylinders 3 lb. gravel 3 lb. sand 3 lb. soil 1 roll masking tape	Compare and Contrast (p. 140) Cause and Effect (p. 138)	Slow Learners: Show how water enters the ground by adding water to a beaker filled with dry sand. Above Average: Discuss factors that affect the height of the water table and how this affects the depth of many wells. Mainstreaming: Have students learn the vocabulary and read the summary in this section before teaching material.

SUPPLEMENTAL MATERIALS

Audiovisual

Alluvial Fan and Delta, Holt/Ealing, 85-0040/1, color filmloop.

Cavern Formations, BFA, Super 8 color filmloop.

Chemical Changes in Rocks and Minerals, DG, No. 673251, color filmstrip.

Chemical Weathering, Karst Terrains, EME, sound filmstrip.

Conservation of Water, Doubleday Media, 11265, Super 8, color filmloops.

Erosion Leveling the Land, EBE, 2194, 16 mm color film.

Erosion by Water, EME, set of 21, 35 mm color slides.

Groundwater, The Hidden Reservoir, Wiley, 16 mm, color, 19 min.

Investigations in Science, BFA, Earth Science Series, set of 7 on erosion, weathering, soil, Super 8 color filmloop.

Monuments to Erosion, EBE, 16 mm, color, 11 min.

Movement of Groundwater, BFA, Super 8 color filmloop.

Rivers, Meanders, Gorden Flesch, 3515582, Super 8, color filmloop.

Streams and Rivers, Learning Arts, Apple II diskette.

Stream Erosion Cycle, Holt/Ealing, 85-0040/1, color filmloop.

Stream Erosion Cycle, Gorden Flesch, P85-0024/1S2, Super 8, color filmloop.

The Water Series, MIS, 6 sound, color filmstrips.

Water-Soil, Hubbard, HDF 9015, set of 8 Super 8 color filmloops.
Water, The Common Necessity, MIS, 16 mm, color film, 11 min.

Gibbons, Boyd. *Do We Treat the Soil Like Dirt?* Washington D.C.: National Geographic Society, September 1984.
Gunston, Bill. *Water.* Morristown, N.J.: Silver-Burdett, 1982.

Print
Gardner, Robert. *Water, the Life Sustaining Resource.* New York: Messner, 1982.

Computer
Teach Yourself By Computer. Earth Science Series, TRS-80, Models I, III.

CHAPTER OVERVIEW

Without the processes of weathering and erosion, the earth's surface would look much the same today as that of the moon. The presence of a large amount of water on earth's surface apparently makes the difference. Weathering and erosion are constantly tearing down the mountains and would have long ago leveled the earth's surface if it were not for the building processes associated with crustal plate motion. Water and gravity team up to produce the major amount of weathering and erosion of the earth's surface.

5–1. WEATHERING AND EROSION

BACKGROUND: Having studied the forces associated with the building of land features, a study of the forces that tear down these features is in order. It is the cycle of these two opposing processes that is responsible for the ever-changing face of the earth's surface. This section deals with weathering and erosion, the processes which are at work tearing down the land features. It is important to note that the abundance of water on the earth's surface accounts for most of the weathering and erosion that occurs.

MOTIVATION: Discuss the reasons that the earth's surface is constantly undergoing change by comparing the conditions that exist on the earth's surface with those conditions that exist on the moon's surface, where very little change has occurred in several billion years. Locate a rock that has been exposed to the earth's weathering processes for some time. Break it to expose the fresh interior that shows little effect of weathering. Talk about the interface that exists where the process of weathering occurs.

INVESTIGATION

Teaching Hints: The student mimics chemical and physical weathering, similar to that which occurs in nature, with a piece of limestone rock. The students should work in groups of two. Chemical weathering is the result of the reaction of carbonated water on limestone. The use of any clear, colorless, carbonated water will do.

Small baby food bottles make excellent glass containers for the physical weathering part. You may wish to have a few polished rocks available to show what can be done by tumbling rocks for a long period of physical weathering. Provide a waste container for the limestone rocks. Since they have been rinsed in the water, they can be dried and used again.

You may wish to indicate that the bubbles arising from the limestone is the usual chemical test for a carbonate rock or mineral. A good demonstration is to put a piece of marble or limestone in some 3 molar hydrochloric acid, testing for carbon dioxide gas by holding a drop or two of limewater (calcium hydroxide in water) on a stirring rod in the escaping gas and noting the color change to white.

Answers to Questions:
1. Bubbles rise.
2. Bubbles form.
3. They rise through the water.
4. Yes.
5. Yes.
6. Bubbles are forming, so the rock must be dissolving.
7. Chemical weathering.
8. Yes.
9. Physical weathering.

Conclusions:
1. (a) The forming of bubbles and dissolving of the limestone rock in the soda water was an example of chemical weathering.
   (b) The breaking of small pieces from the limestone rock in water is an example of physical weathering. In chemical weathering, the makeup of the rock changes. In physical weathering, the rock is made into smaller pieces without changing the makeup of the rock.
2. The rock in a stream may undergo both chemical and physical weathering. The materials dissolved in the water may be attacking the rock chemically at the same time it is being tumbled down the stream bed.

3. The rock should weather faster in soda water.
4. Yes. The action of water may wear away rock by breaking it into smaller pieces. At the same time, the water may dissolve the rock.

5–2. WATER ON THE LAND

BACKGROUND: This section concerns the role of water in erosion. There is very little if any water escaping from the earth. The water cycle accounts for this fact. At certain stages in the water cycle, water causes erosion. The idea of an earth water budget and a local water budget is introduced. The features of a young river, a mature river, and an old river are compared.

MOTIVATION: Discuss the idea of recycling water by comparing it to the way it is done for astronauts. Notice when you start talking about recovering waste body fluids that the idea sounds revolting to most. However, point out that those living along a water supply such as the Mississippi River must do the same thing. It is becoming more of a problem to recover water from an earth source than what is necessary to do for the astronauts. For example, the rivers are a constant dumping site for waste chemicals. The rivers also receive runoff from agriculture, which contains pesticides and chemical fertilizers. None of this is in the water that the astronauts recycle!

INVESTIGATION

Teaching Hints: The monthly precipitation and evaporation is given for Minneapolis, Minnesota, during a one-year period. This data is then plotted, using the vertical axis for precipitation and the horizontal axis for evaporation. The monthly surplus or deficit is calculated from the table and compared to the graph. The Investigation is best done individually.

You may wish to obtain similar data about your city from the local Weather Service Office, National Oceanic and Atmospheric Administration, or from a local newspaper office, and have the students plot a similar graph to compare with the one for Minneapolis.

Answers to Questions:
The line graph is shown.

1. June.
2. July.
3. Jan., Feb., Mar., Apr., Oct., Nov., Dec.
4. May, June, July, Aug., Sept.
5. The surplus is 30 mm.

Conclusions:

1. A surplus occurs when the precipitation is greater than the evaporation.

2. A deficit occurs when the evaporation is greater than the precipitation.
3. Add up the total precipitation and the total evaporation. If the total precipitation is larger than the total evaporation, there is a surplus. If the total evaporation is larger than the precipitation, there is a deficit.
4. Evaporation is mainly affected by the temperature.

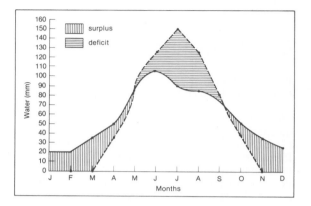

5–3. WATER BENEATH THE LAND

BACKGROUND: We feel the water in the atmosphere as humidity and see it as rain. It is also easy to see how water moves on the surface to form lakes, streams, and rivers. However the water that soaks into the ground to form ground water is not readily observed until it has caused an effect such as a spring, geyser, sinkhole, or an underground cavern that may readily be explored. This section deals with the ways water becomes ground water and some of the effects it has.

MOTIVATION: Compare a coffee percolator to a geyser. Display a coffee percolator and explain that it works by boiling a small amount of water in the base of the stem. Steam thus generated forces a surge of water up the stem at regular intervals as the cycle is repeated. Ask if anyone has seen a "natural percolator." If so, the resulting discussion of geysers can be used to introduce the general topic of ground water. If not, leave the question open and return to it during or after the section.

SKILL-BUILDING ACTIVITY

Teaching Hints: The Skill-Building Activity uses the technique of comparing and contrasting to study permeability and porosity. Sand, fine gravel, and regu-

lar soil should be easily obtained locally. If you set up separate waste containers for the three types of soil being used, the water can be drained from them and, after drying, the soil can be used another time.
CAUTION STUDENTS NOT TO THROW SOLID MATERIALS IN THE SINK

Answers to Questions:

1. Beaker A.
2. Beaker C.
3. Predictions may vary.
4. Predictions may vary.
5. The amount depends on gravel size.
6. The amount depends on the sand used.
7. The amount depends on the soil used.
8. Beaker A.
9. Beaker A.

Conclusions:

1. The gravel had the highest porosity, then the sand, and then the soil.
2. The gravel had the highest permeability, then the sand, and then the soil.
3. A soil sample needs a lot of air spaces between particles to have high porosity.
4. High porosity gives high permeability if the pore spaces are interconnected.

¡COMPUTE!

Teaching Hints: This is a fairly simple program that can be used to help students read programs. Picking out certain kinds of statements and relating conversational English to the program helps demystify programming and computers.

CHAPTER REVIEW

Vocabulary:

1. weathering. 2. erosion. 3. interface. 4. physical.
5. chemical. 6. humus. 7. talus. 8. landslide.
9. creep. 10. slump. 11. water cycle.
12. evaporation. 13. delta. 14. meander.
15. precipitation. 16. organic.

Answers to Questions:

1. Frost action, exfoliation, expansion due to heat, and plant root pressure.
2. Water, oxygen, carbonic acid, and plants are some agents of chemical weathering.
3. Plants cause chemical weathering by the roots giving off an acid that dissolves the rock, allowing the root to grow into the cracks and crevices. As the root becomes larger, it pushes on the rock and breaks it, a form of physical weathering.

4. At the beginning, weathering causes the rock to be broken into small pieces such as sand. At this time there are two layers, the partly weathered rock underneath and a layer of small particles on top. As plants grow in the topsoil, humus is added to the top layer. Gradually a third layer is produced by water dissolving minerals and moving small particles lower. Upon development of the third layer, the soil is said to be mature.
5. Gravity is an agent of erosion because it pulls pieces of rock down to the bottom of a slope.
6. Precipitation may evaporate, run off into streams, or sink into the soil or rock to become ground water.
7. A surplus occurs when the precipitation is greater than the evaporation.
8. Runoff begins as a shallow layer, or sheet, of water flowing downhill. Pebbles and rocks on the surface quickly break the sheet into tiny streams. These tiny streams form shallow channels. Channels merge and are widened and deepened by the water flowing through them. Small streams come together forming larger ones. A group of streams collect runoff from a large area. All the streams from an area flow into one stream or river.
9. A stream which has rapidly running water and a large load would best cut through bedrock.
10. The parts of a stream are the bed, banks, and mouth.
11. Soil or rock that has high porosity usually has high permeability; soil or rock with low porosity also has low permeability.
12. If the level of the surface of the land is below the surrounding water table, the water will flow downhill out onto the surface, producing either a swamp or lake if the water is trapped. If the water is not trapped, it will continue to flow downhill as a stream or river. If the level of the surface is above the water table, the water will remain under the surface since it cannot flow uphill.
13. A spring which disappears is located in a manner such that during a dry period the water table will become lower than the level of the spring. A spring that flows all the time is supplied by ground water sitting on top of a layer of low-porosity soil or rock.
14. Ground water can produce a cavern when it is slightly acidic and flows into cracks and crevices of certain limestone layers found under the surface.
15. (a) Stalactites are formed on the roof of a cavern when the ground water containing dissolved limestone drops from the roof, leaving behind small amounts of the limestone. (b) When the drips strike the floor of the cavern, a small amount of limestone is deposited as a mound called a stalagmite. (c) When the iciclelike formation of the stalactite builds downward and connects with the mound of stone of the stalagmite, a column is formed.

CHAPTER 6 ICE, WIND AND WAVES

SECTION/ TIME NEEDED	MATERIALS (30 STUDENTS)	BASIC SCIENCE SKILLS	PACING GUIDE
6-1. Glacial Ice (1–2 periods)	10 stream tables 10 overflow cans 5 gal. can soil 5 gal. can sand 5 gal. can gravel	Cause and Effect (pp. 149, 155) Compare and Contrast (pp. 151, 152) Finding Cause and Effect (Teacher's Resource Book, p. 6-2)	Slow Learners: Prepare a table comparing and contrasting the characteristics of valley and continental glaciers. Above Average: Find out what glaciers covered the North American Continent during the last ice age (Pleistocene Epoch). Mainstreaming: Have students classify glacial features into two groups: valley and continental glaciers.
6-2. Wind and Waves (1–2 periods)	15 bottles, dry sand 15 shoe box lids 15 spoons 30 goggles	Cause and Effect (pp. 158, 160, 162)	Slow Learners: Discuss local examples of erosion and deposition. Above Average: Have students research the causes of "dust bowl" of the 1930's, and what can be done to prevent another one from occurring. Mainstreaming: Have students classify the erosional and depositional features into two groups: wind, waves.

SUPPLEMENTAL MATERIALS

Audiovisual

Antarctica: Exploring the Frozen Continent, EBE, 16 mm color film, 22 min.

Continental Glaciation, Science and Mankind, set of 52 color, 35 mm slides.

Earth Science, Learning Arts, 4 Apple II diskettes, 48 K.

Erosion by Ice, EME, set of 22 color, 35 mm slides.

Erosion by Water, EME, set of 21 color, 35 mm slides.

Geology, Set I, Shapiro, set of 6 captioned filmstrips.

Mountain Glaciation, Science and Mankind, set of 51 color, 35 mm slides.

The Water Series, MIS, 6 color filmstrips, cassettes.

Print

Fodor, R. V. *Frozen Earth: Explaining the Ice Ages.* Hillsdale, N.J.: Enslow, 1981.

Hoyle, Fred. *Ice: The Ultimate Human Catastrophe.* Continuum, 1981.

McKean, Kevin. "Hothouse Earth." *Discover,* Dec. 1983.

Nixon, Hershell H. *Glaciers: Nature's Frozen Rivers.* New York: Dodd, Mead, 1980.

Ramsey, W. L., C. R. Phillips, and F. M. Watenpaugh. *Modern Earth Science.* New York: Holt, Rinehart and Winston.

Computer

Shore Features. Cambridge Development Laboratory. APL II Series.

CHAPTER OVERVIEW

The role of ice, wind, and waves as forces of erosion changing the surface of the land is explored. Valley and Continental glaciers are discussed and the work they do in changing the existing land features and creating new land features is described. Theories that might account for the ice ages are given. The manner in which wind and waves cause erosion and the kinds of deposits made by each are discussed. A number of shoreline features that result from wave erosion are described. The availability of stream tables will allow the students to do the Skill-Building Activity on p. 155, which is designed to show the cause-and-effect relationship of running water on soil, sand, and gravel. Common materials readily available are used to study "Wind Blown Deposits," in the Investigation on p. 164.

6-1. GLACIAL ICE

BACKGROUND: The fact that glaciers were at work sculpturing the earth's features not long ago is rather obvious to many people who live in the northern half of the United States and in Canada. The reasons for the change in climate to produce large continental glaciers are still not clear. The most recent glacial advance, which covered about two-thirds of the North American continent, occurred about a million years ago.

The glacier left evidence that shows there were a series of climate changes resulting in a number of times in which the glacier advanced and retreated. One of these advances went as far south as southern Illinois and occurred about 200 thousand years ago. In the last 100 thousand years the glacier advanced and retreated four times, going as far south as Wisconsin. It was these advances and retreats of the continental glacier that created the many glacial features that can be seen today.

MOTIVATION: If you live in an area where glacial features are evident, discuss these as time permits. If no glacial features are in your area, you may wish to discuss an area in the continental United States that still has remnants of valley glaciers, such as Glacier National Park, Montana. Another topic that will spark interest is the formation of the North American Great Lakes by glaciation. Information concerning this topic can be found in more advanced earth science texts. A good source would be *Modern Earth Science,* Holt, Rinehart and Winston.

INVESTIGATION

Teaching Hints: The Skill-Building Activity is designed as an open-ended activity to allow the students to set up conditions that will allow them to determine the cause-and-effect relationship of simulated water flow from a rather large source, such as a glacier, as it flows over various terrains.

Stream tables are available from science supply houses. A suitable substitute could be made from flat-bottomed roof guttering. Cut sections approximately 1 m in length. Close both ends to about halfway up. Provide a V-cut in one end for water to spill. Another possibility would be to use a paint roller pan with a hole drilled in the deep end for water to spill. A gallon jug slightly elevated and equipped with a siphon tube could be used as a water source.

Answers to Questions:
1. Answers will vary.
2. The smallest particles, soil, moved.
3. Increasing the slope increased the speed of the water.
4. The larger particles, sand and gravel, moved.
5. The soil moved first, then the sand, and the gravel moved last.

Conclusions:
1. Increasing the slope of a stream would increase the amount of erosion.
2. Increasing the volume of water flowing would increase the amount of erosion.
3. Erosion would decrease the slope of the land. If part of the stream bed were uplifted, the slope would increase. Increased rainfall would increase the volume of the stream.
4. (a) One plan for a fast-moving stream would be to know the rate of erosion. This would tell you how far from the stream your house should be built. (b) For a slow-moving stream, plans for flooding would be a major concern.

6-2. WIND AND WAVES

BACKGROUND: Wind and water are both fluids and abrade land features with the material they carry. Thus they are capable of both physical weathering and of erosion. Wind carries less material than water and causes less abrasion. As a result, a number of rock formations thought to have been caused by wind erosion were actually caused by water when the area was underwater.

MOTIVATION: The Investigation, "Windblown Deposits," on page 164 would make an excellent introductory activity for motivational purposes. If you are located near a shoreline of a large body of water, a lively discussion could be made concerning the kinds of activities that are changing the shoreline, sometimes for the good and sometimes for the bad.

INVESTIGATION

Teaching Hints: The Investigation gives the students an opportunity to study the effect of wind on small particles. The student supplies the wind and is usually surprised to find that it takes quite a lot of wind to move particles of any size.

Safety: Have the students wear eye protection while they are blowing the sand about. Dry, natural sand is to be used. It must have some fine material in it. Try the sand and add some dust if it needs some fine material. Baby food bottles with lids are ideal for dispensing the sand. Provide a container for disposing of the sand. Later a student can sprinkle the sand over the soil in an appropriate place on campus.

Answers to Questions:
1. On the paper.
2. The smaller particles traveled farther than the larger particles.
3. Loess.
4. None of the sand (or very little) is blown out.

Conclusions:
1. Dry, fine dust is most likely to be transported by wind.
2. Sand and dust is blown away by the winds on the dry desert floor.

TECHNOLOGY

Teaching Hint: For further information, see Jon Naer's *The New Wind Power*, Penguin, 1982.

CHAPTER REVIEW

Vocabulary:
1. FALSE, valley glacier. 2. TRUE 3. TRUE 4. TRUE 5. FALSE, windbreak. 6. FALSE, shoreline. 7. FALSE, sea cliff. 8. TRUE 9. TRUE 10. FALSE, sea stack. 11. FALSE, beach. 12. FALSE, spit.

Answers to Questions:
1. Conditions necessary to form a glacier are (a) temperatures such that not all the snow melts each year, (b) thick layers of snow that build up, until (c) the weight of the layers causes movement of the ice downhill.
2. There are two continental glaciers today, one in Antarctica and one in Greenland.
3. The great weight of the layers of ice causes the bottom of a glacier to melt, allowing it to slide more easily over the rock. The grains of ice are under great pressure, allowing each one to slip a tiny amount. Under these conditions, the ice can flow like toothpaste from a tube.
4. One type of theory uses events that occur on earth, such as volcanic eruptions, to account for the change in climate necessary for an ice age. The other type of theory uses events that occur away from the earth, such as energy changes on the sun, to account for the necessary climate changes.
5. An ice age is a cool period in which most of the earth's water is frozen as ice. The last great ice sheet disappeared about 10,000 years ago.
6. It is necessary to be able to predict.when the next cool or warm period is to occur, so that the people of the earth can prepare for it. For example, if a warm period were to occur and melted the icecaps, the ocean level would rise, flooding the cities of New York, London, Tokyo, and many others.
7. Most wind erosion occurs in the deserts. Wind erosion depends on having barren, dry ground for the wind to pick up material. Those conditions are typical of deserts.
8. Wind erosion does not have any effect on things higher than about one meter above the ground.
9. Plants slow down the surface wind, which helps prevent erosion by wind.
10. Longshore currents are caused by waves striking a shoreline at an angle.
11. Sea caves, terraces, stacks, and sea cliffs are all formed as a result of the erosion of rock shorelines by waves.
12. A drop in sea level results in shorelines with sea cliffs and terraces due to new shoreline being exposed to erosion. A rise in sea level results in an irregular shoreline with bays and inlets caused by the flooding of the previous shoreline features.
13. (a) An area eroded by ice shows evidence of scraping and gouging. (b) An area eroded by wind is free of dust and sand. (c) An area eroded by waves shows cliffs or terraces as evidence of previous sea levels.
14. (a) Ice leaves piles of rock and sand. (b) Wind leaves layers of fine dust. (c) Waves build an underwater terrace made of sand and rock particles.
15. (a) High altitude mountain valleys are affected most by ice. (b) Dry, barren desert areas are affected most by wind. (c) Shorelines that arise abruptly out of the ocean are affected most by waves.

SECTION/ TIME NEEDED	MATERIALS (30 STUDENTS)	BASIC SCIENCE SKILLS	PACING GUIDE
7–1. The Rock Record (1–2 periods)	pencil and paper	Sequencing (p. 173) Inferring (p. 175) Inferring (Teacher's Resource Book, p. 7–2)	Slow Learners: Have students focus on reading the past through photos such as the chapter opener, p. 168, and Fig. 7–3, p. 172. Above Average: Have students make a drawing similar to Fig. 7–6 (p. 175). Mainstreaming: Have students read the summary and study the vocabulary words in the margin before teaching this section.
7–2. The Record of Past Life (1–2 periods)	6 sticks, modeling clay 30 common objects 1 kg plaster of paris 15 mixing containers water source	Inferring (p. 177) Classifying (pp. 179–181)	Slow Learners: If available, show students examples of fossils, or make some fossils to show. Above Average: Discuss the limitations of fossils as a complete record of life. Mainstreaming: Tape certain sections. Have students listen to the tape as they read along.
7–3. Age of Rocks (1–2 periods)	500 sugar cubes 10 shoe boxes 10 felt tip pens wall clock	Inferring (pp. 186–187) Compare and Contrast (pp. 189–190)	Slow Learners: Discuss the difference between relative age and exact age, using the ages of students. Above Average: Have students calculate the exact age of some rocks or fossils using the given half-life of various radioactive substances. Mainstreaming: Pile books according to their different copyright dates, with the oldest on the bottom. Discuss the relative and absolute ages of the books.
7–4. An Earth Calendar (1–2 periods)	50 m adding machine tape 10 meter sticks	Classifying (pp. 194, 195) Sequencing (p. 196) Compare and Contrast (p. 193)	Slow Learners: Use Fig. 7–33 (p. 202) to discuss the eras. Above Average: Have students reduce the geologic time scale to one year.

SUPPLEMENTAL MATERIALS

Audiovisual

Earth's Biography, Learning Arts, set of 4, color/sound filmstrips.

Fossils, EBE, 5 color/sound filmstrips.

Geology From Space, NASA, 24 frames, audio tape.

Grand Canyon, Record of the Past, Gordon Flesch, PB5-0263/1S2, Super 8, color filmloop.

History In the Rocks, Hubbard, set of 10 super 8, color filmloops.

Story In the Rocks, Shell, 16 mm, color, sound, 17 min.

Print

Alexander, George. "Going, Going, Gone (Dinosaurs)." *Science 81*, May, 1981.

Brownlee, Shannon. "Cycles of Extinction." *Discover*, May 1984.

Binford, Lewis R. *In Pursuit of the Past: Decoding The Archaeological Record*. New York: Thames and Hudson, 1983.

Colbert, Edwin H. *A Fossil Hunter's Notebook: My Life With Dinosaurs and Other Friends*. New York: Dutton, 1980.

Croucher, R. and A. R. Woolley. *Fossils, Minerals, and Rocks: Collection and Preservation*. New York: Cambridge University Press, 1982.

Macdonald, J. R. *The Fossil Collector's Handbook*. Tarrytown, N.Y.: Prentice-Hall, 1983.

MacFall, Russel P., and Jay Wollin. *Fossils For Amateurs*. New York: Van Nostrand Reinhold, 1983.

Weitzman, David. *Traces of the Past: A Field Guide to Industrial Archaeology*. New York: Scribner's 1980.

Computer

Introduction to Science Series. Focus Media. APL II Series.

Dinosaurs. Cross Educational Software. Apple, IBM.

CHAPTER OVERVIEW

The chapter begins by explaining "uniformitarianism," the idea that the same processes that are acting on the earth now also acted on the earth in the past. The value of fossil study to our understanding of the earth's history is developed. The age of a fossil can be determined by radioactive dating. Relative age, index fossils, and key beds are also discussed. A calendar of earth's history giving the four main eras, the conditions thought to have existed, and the life present is given. Through student Investigations, rock layers are interpreted, mold and cast fossils are made, an analogy to radioactive decay is done, and an adding machine tape model is made of the earth's geologic calendar.

7–1. THE ROCK RECORD

BACKGROUND: Toward the end of the 18th century, James Hutton formulated a theory called uniformitarianism. This created the modern science of geology.

MOTIVATION: Ask students to study Fig. 7–1 (p. 170). Discuss the following questions:

1. Do you think the layers of the canyon have changed much since they were formed? Why?
2. Do you think that each layer of rock was produced at a different time? Why?
3. Which layers are the youngest? Why?
4. Which layers are the oldest? Why?

Answers to these questions may show that the ideas of uniformitarianism are a more natural way of thinking about rock layers.

INVESTIGATION

Teaching Hints: A simulated cross-section of the Grand Canyon is used to study uniformitarianism, the law of superposition, relative age, and unconformity as they relate to the rock layers shown. A portion of the history of the area is revealed as a result of the study of the rock layers.

Answers to Questions:

1. 9.   5. Before.
2. 15.   6. 9.
3. Group B.   7. 14, 15.
4. Yes.   8. 9, 15; 15.
9. The area was covered by water, the weather was good and there was sufficient food in the water to support life. There was no life in the water at this time, or due to the pressure at the depth of layer 7, any remains could have been destroyed.

Conclusion:

1. The clues are (1) the law of supersposition; the bottom layers are generally oldest; (2) uniformitarianism; rock layers are laid down in a level position; (3) relative age, knowing which events happened before others; and (4) unconformity, two layers whose boundary shows erosion.

7–2. THE RECORD OF PAST LIFE

BACKGROUND: What we know of previous life on earth has been derived from the study of fossils. Any remnant of a previous living organism can be termed a fossil. This section will broaden the students' concept of fossils and the information we obtain from their study.

MOTIVATION: Display and discuss a variety of fossils. Ask the students, prior to assigning this section, to bring in any fossils they may have at home. Specimens illustrating the various types of fossils can be purchased from earth science supply houses.

INVESTIGATION

Teaching Hints: Molds of some common objects are produced in clay and then a cast is made using plaster of Paris. The casts are then exchanged among the students to see if others can identify the objects.

Baby food bottles make excellent mixing containers. Students should work in pairs to limit the mess of the experiment. Although plaster of Paris will set up sufficiently to handle in about 20 minutes, you may wish to allow the casts to remain overnight before completing the Investigation. By pressing just an edge of the object into the clay, or by pressing it in at an angle, more difficult casts, similar to what an archaeologist might dig up, can be made of the same objects used by the class.

Answers to Questions:

1. The mold is the same size and shape as the object.
2. The mold is a hollow region, whereas the object is solid. The mold has only one surface; the object has several surfaces. The depressions of the mold are filled with air, whereas the object was solid.
3. The cast is the same size and shape as the original object and is a solid.
4. The cast is made of a different material, may not have the same color, and probably lacks some of the detail of the original object.
5. Answers may vary. It depends on the object as to how much of a cast may be needed to identify it.
6. Answers will vary. Depends on objects used.
7. This is a cast fossil.
8. The shape of the foot, an idea of the animal's size and shape, and comparing the shape of the footprint to that of other animals would identify it.

Conclusions:

1. A mold is an impression of an object made on a surface, whereas a cast is a three-dimensional solid similar to the object.
2. Color, kind of material, completeness, and some fine detail cannot be identified by a cast.
3. The fossil is a mold. It formed when a plant made an impression in mud. The mud hardened into a rock. The original plant has since decomposed.

7–3. AGE OF ROCKS

BACKGROUND: The concept of relative ages of rocks and fossils established in previous sections is reinforced by the introduction of key beds and index fossils at the beginning of this section. The use of key beds, index fossils, and radioactive dating gives considerably more significance to the earth's geologic column and will allow an earth calendar to be devised in the next section.

MOTIVATION: The emphasis of this section is on determining age of rock layers. Have the students suggest ways other things age. Have students note that determining exact age requires more information about the object.

If you have a cloud chamber or a Geiger counter and a radioactive source, demonstrate the nature of radioactivity. This will require some advance planning.

INVESTIGATION

Teaching Hints: The Skill-Building Activity is designed to use sugar cubes for building a model of radioactive decay, to record data concerning the operation of the model, and to make a conclusion as to the half-life of the sugar cube model.

Students should work in groups of three. Wooden or plastic cubes are more durable. Food coloring dispensed in the small dropper type containers could be used in place of the felt tip pen to mark the sugar cubes. You could use poker chips, coins, washers, or buttons; however, because the probability is thus increased, at least 100 would be needed and more shakes required to give suitable results.

Answers to Questions:

1. Answers will vary, about 4 shakes.
2. Answers will vary, about 8 minutes.
3. Answers will vary, same as question 2.

Conclusions:

1. 87 or 88 will have changed.
2. The line should be a decreasing curve from the upper left to the lower right.

7–4. AN EARTH CALENDAR

BACKGROUND: An earth calendar has been established for the earth's history. It is based on the most recent interpretations of the rock record. It is important to point out to the students that the dates for the divisions of the earth's history are approximate and subject to change based on the latest scientific findings and more precise dating techniques.

MOTIVATION: Most people will recognize that 4.5 billion years is a very long time. Point out that, although time is absolute, it is also relative with respect to lifetimes of any living or nonliving thing. Some single-cell animals exist for less than an hour before they divide into two new animals. To such an animal, the human life span is very long. Although we think of the earth's history as very long, this should not interfere with our understanding of some of the major events that occurred.

INVESTIGATION

Teaching Hints: A model of the geologic calendar is made using adding machine tape to make a time line. The relative lengths of time of the four main eras become quite apparent. Some of the conditions that existed on earth at the close of each era are added to the model in order that the student may understand that the occurrence of a natural event can be used as a time boundary.

The students should work in groups of three. Although the first four-fifths of the tape is blank, it is necessary for it to be there in order for the student to appreciate the great lack of knowledge concerning most of the history of the earth. You may wish to extend the table in the Investigation to include Periods and Epochs listed in the table on p. 196. This will add additional time required to complete the work.

Answers to Questions:
1. One mm is equal to 1 million years.
2. One cm is equal to 10 million years.
3. Precambrian time.
4. Cenozoic Era.
5. (a) Paleozoic Era, (b) Mesozoic Era, (c) Cenozoic Era.
6. (a) Invertebrates. (b) Mammals.

Conclusions:
1. The Precambrian time is the longest. It is longer than the three other eras combined.
2. The geologic time scale can be divided by the appearance of certain life forms.

¡COMPUTE!

Teaching Hints: To add a line, bring the program to the screen and enter the new line. If the line is numbered in sequence, the computer will rearrange the statements properly. Remember to save your program again after you've made your addition.

CHAPTER REVIEW

Vocabulary:

1. uniformitarianism. 2. superposition. 3. relative age. 4. unconformity. 5. fossils 6. petrified. 7. index fossils. 8. key beds. 9. geologic column.

10. radioactive decay. 11. half-life. 12. eras. 13. invertebrates. 14. vertebrates.

Answers to Questions:
1. Older rock layers may be found on top of younger layers when great forces bend rock layers enough to push older layers over younger ones.
2. Layers and other marks, such as ripples, found in rock layers in the present and past are caused by the same processes.
3. Fossils of higher forms of life are found in more recent rock layers. Simpler forms of life are found in old rock layers.
4. The age of the rock layer into which the magma flowed is older than the magma flow, and the age of the rock layers involved in the fault are older than the fault.
5. It is necessary that the living thing be frozen or covered soon after death and that it have some hard parts that resist decay to form a fossil.
6. Animals or plants may be preserved by being frozen in ice, trapped in amber or tar, buried under sand, soil, or water, and petrified.
7. Most plants or animals do not become fossils because they are not protected from decaying.
8. Marks left by an animal walking, crawling, or burrowing are the ways trace fossils are formed.
9. (a) A key bed gives only the relative age of rock layers. (b) The presence of an index fossil pinpoints the age to a given short period of time that it lived. (c) Uranium can give the approximate age in years of very old rock. (d) Radioactive potassium can give the approximate age in years of younger rocks. (e) If the rock contains remains of once-living matter, carbon-14 can give the age of young rocks.
10. The geologic column is an arrangement of the known layers of rock in the order of their relative age.
11. Unless the remains of once-living plants or animals are found in the rock, carbon-14 cannot be used to tell age. Many rocks are too old to have any such remains in them.
12. Each era has a beginning and an ending that is marked by changes in life forms and special events that occurred at the time.
13. Precambrian: formation of earth. Paleozoic: rise of insects and other invertebrates. Mesozoic: the breakup of Pangaea, the super continent. Cenozoic: mountain building, volcanic activity.
14. Warm shallow seas were perfect for reptiles.
15. Mesozoic climate was warm and mild. The climate gradually changed to cold at the poles and hot at the tropics during the Cenozoic Era.

CHAPTER 8 MINERALS

SECTION/ TIME NEEDED	MATERIALS (30 STUDENTS)	BASIC SCIENCE SKILLS	PACING GUIDE
8-1. Atoms, Elements, and Compounds (1–2 periods)	90 paper clips, large 90 paper clips, medium 90 paper clips, small	Compare and Contrast (p. 214) Sequencing and Classifying (p. 215) Predicting (p. 214) Classifying (Teacher's Resource Book, p. 8–2)	Slow Learners: Rub a plastic comb or glass rod in wool and use it to pick up pieces of paper. Discuss static electricity, e.g., shocks. Above Average: Have students describe the structure of different atoms using the Periodic Table. Mainstreaming: Use overheads or models to show atomic structure.
8-2. Minerals and Rocks (1–2 periods)	30 sets, 8 marked cards (see instructions below)	Compare and Contrast (pp. 221, 222, 223)	Slow Learners: Have students bring in rocks and try to identify the minerals as in Fig. 8–6, p. 221. Above Average: Contrast rocks and minerals in their makeup. Mainstreaming: Have students study the summary and vocabulary.
8-3. Mineral Properties (1–2 periods)	30 pieces chalk 30 pennies 30 wooden pencils 30 mineral samples	Sequencing (p. 232)	Slow Learners: Have students test for color, luster, streak, fracture, and cleavage. Above Average: Have students identify mineral samples in class. Mainstreaming: Show students samples of minerals that demonstrate each property.

SUPPLEMENTAL MATERIALS

Audiovisual

A is for Atom, GE, 16 mm, color, sound film, 15 min.
Atoms, Learning Arts, set of 2 color/sound filmstrips.
The Atom, Hawkhill, one video cassette, guide.
Building Blocks of Matter, Doubleday Media, Super 8, color filmloop.
Cinnabar, A Game of Rocks and Minerals, Nova Science Corp., 510–2715, 5 games, 3–5 players ea.
Earth Materials, Hubbard, set of 10, Super 8 filmloop.
Geology: Rocks and Minerals, Hubbard, set of 4, color filmstrip.
Matter and Energy, Filmstrip House, set of 4, color/sound filmstrip.
Minerals, Hubbard, set of 20, 35 mm color slides.
Minerals, How They Are Identified, DG 673091, color, filmstrips.
Minerals and Rocks, Creative Visuals, MH-642327-5, color filmstrips.
Rocks, Hubbard, set of 20, 35 mm color slides.

Print

Harari, Haim. "The Structure of Quarks." *Scientific American,* April 1983.
O'Donoghue, Michael. *Color Dictionary of Minerals and Gemstones.* New York: Van Nostrand Reinhold, 1982.
Robbins, Manuel. *The Collector's Book of Fluorescent Minerals.* New York: Van Nostrand Reinhold, 1983.
Schecter, Bruce. "The Moment of Creation." *Discover,* April 1983.
Weinberg, Steven. *The Discovery of Subatomic Particles.* San Francisco: W. H. Freeman, 1983.

Computer

Earth Science. Minnesota Educational Computing Corp. TRS 80, Models I, II.
Earth Science Keyword. Focus Media. APL II Series; COM 64, Pet; TRS-80, Models III, 4.
Identifying Minerals. Focus Media. APL II Series; COM 64, Pet; TRS-80, Models III, 4.

8–1. ATOMS, ELEMENTS, AND COMPOUNDS

BACKGROUND: The nature of the three fundamental particles is discussed and the combinations that go to make up the first 20 elements are given. The joining of atoms to make compounds involves either a sharing or exchange of electrons.

MOTIVATION: Positive and negative electrical fields hold atoms together. A demonstration is outlined in the TE margin on p. 214.

INVESTIGATION

Teaching Hints: Paper clips are used to show that, with just a few atoms, many compounds are possible. The clips should be in chains of 9, 3 clips each of 3 different sizes.

Answers to Questions:
1. Three.
2. An element.
3. There are two different sizes of clips attached.
4. Answers will vary. Only 3 models are possible.
5. Answers will vary. Ten different models are possible. Using s for small, m for medium, and l for large, the arrangements are: l-s-l, l-s-s, l-l-s, l-m-s, m-l-s, l-s-m, m-s-s, s-m-s, m-m-s, and m-s-m.
6. There are 27 possible models: l-l-l-s, l-l-l-m, m-m-m-s, l-l-s-s, l-l-m-m, m-m-s-s, l-s-s-s, l-m-m-m, m-s-s-s, l-s-l-s, l-m-l-m, m-s-m-s, l-s-s-l, l-m-m-l, m-s-s-m, s-l-l-s, m-l-l-s, s-m-m-s, m-l-m-s, m-m-l-s, m-l-s-m, m-s-l-s, m-s-s-m, m-s-l-l, m-l-s-s, m-l-l-s, m-s-s-l.

Conclusions:
1. It increased the number of compounds.
2. It would increase the number of compounds possible.
3. Going from 2 to 4 clips caused many new combinations representing compounds. 90 different clips makes an unlimited number of possibilities. Thus millions of compounds are possible from 90 elements.

8–2. EARTH MINERALS

BACKGROUND: This section focuses on the lithosphere and its chemical makeup. Minerals are seen as the building blocks of rocks. Minerals and rocks make up the lithosphere.

MOTIVATION: Use a handful of soil to show the complex composition of the lithosphere, with its sand, humus, and larger rock particles.

INVESTIGATION

Teaching Hints: Combining of atoms in compounds is related to atomic structure in this Investigation, using prepared cards to represent hydrogen, oxygen, carbon, and nitrogen atoms. A set of 12 cards can be made from three 3×5 index cards. Divide each card into four segments. Label and mark each as shown.

Answers to Questions:
1. One.
2. One.
3. One, six.
4. Two, two.
5. Two hydrogen molecules.
6. Answers will vary. Models should look like b or d.
7. Three.
8. Four.
9. Three.

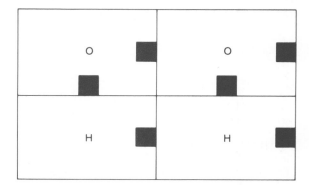

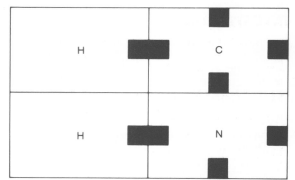

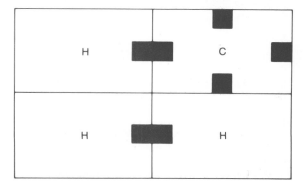

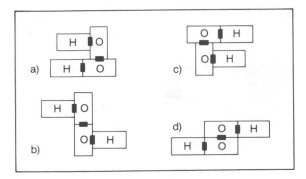

Conclusions:

1. They represent the 4 electrons a carbon atom needs to complete its second shell.
2. NH_3
3. water, H_2O; hydrogen peroxide, H_2O_2.

8-3. MINERAL PROPERTIES

BACKGROUND: This section describes the properties used to identify minerals.

MOTIVATION: Have one or two students examine a pinch of table salt under a hand lens. Discuss the idea of crystals as an orderly arrangement of particles.

INVESTIGATION

Teaching Hints: To use the hardness scale devised by the students, the mineral samples should have a hardness between 1 and 3. Suggestions are: halite, 2; calcite, 3; gypsum, 2; talc, 1; muscovite mica, 2; chlorite, 2; graphite, 1; galena, 2.5; sphalerite, 3.5.

Answers to Questions:

1. Fingernail scratched the wood.
2. Fingernail.
3. Yes.
4. Fingernail.
5. Talc, hardness of 1; gypsum, hardness of 2.
6. Fingernail.
7. Pencil wood.
8. Pencil lead usually. Answers may vary.
9. Pencil lead.
10. Copper penny.
11. Chalk, pencil wood, pencil lead, fingernail, penny.
12. Answers will vary, depending on the mineral sample.

Conclusions:

1. An object is tested by using an object of known hardness to see which scratches the other.
2. Answers will vary, depending on the sample used.
3. The real diamond would scratch the imitation.

TECHNOLOGY

Teaching Hints: Besides Jamestown, another California mining venture is in Napa County. For information, write: California Division of Mines and Geology, 610 Bercut Drive, Sacramento, CA 95814.

¡COMPUTE!

Teaching Hints: If possible, invite a computer systems analyst or engineer to explain digital systems.

CHAPTER REVIEW

Vocabulary:

1. nucleus 2. lithosphere 3. atmosphere
4. mineral 5. electron shell 6. ion 7. gem
8. atomic particles 9. crystal 10. element
11. compound 12. hydrosphere 13. molecule

Answers to Questions:

1. (a) The smallest particle of an element is the atom. (b) A mineral is a nearly pure compound found in nature.
2. A forest is made up of separate trees; a rock is made up of separate mineral grains. Just as trees come in different kinds, the mineral grains in a rock are also different kinds.
3. Quartz is a mineral and not a rock because it contains only one kind of compound.
4. Protons and neutrons are in the atom's nucleus. Electrons are in shells surrounding the nucleus.
5. The electron has a negative charge that is attracted by the positive charge of the proton. Like charges repel, so one electron repels another.
6. An oxygen atom has two more electrons and protons than a carbon atom.
7. (a) The 45 protons are located in the atom's nucleus. (b) It would have 45 electrons. (c) Electrons are found in shells surrounding the nucleus.
8. Two hydrogen atoms provide the two electrons needed in the outer shell of the oxygen atom.
9. Oxygen and hydrogen share the electrons in water, whereas sodium gives up one electron to chlorine, making a sodium ion and a chlorine ion.
10. Hydrogen and oxygen may join to form water, with two hydrogen atoms joined to one oxygen. In hydrogen peroxide, two oxygen atoms are joined together with two hydrogen atoms.
11. Some minerals are poisonous. Also, many minerals cannot be identified by taste.
12. Streak is tested by rubbing the mineral on a hard white surface for its mark. Hardness is tested by seeing if the mineral will scratch other minerals of known hardness.
13. Fracture is an uneven break of a mineral, whereas cleavage is a smooth break of a mineral.
14. See Table 8-3 (p. 230). Apatite, fluorite, calcite, gypsum, and talc. Feldspar is harder.

CHAPTER 9 ROCKS

SECTION/ TIME NEEDED	MATERIALS (30 STUDENTS)	BASIC SCIENCE SKILLS	PACING GUIDE
9-1. Magma and Igneous Rocks (2-3 periods)	Skill-Building Activity: 15 Bunsen burners 30 safety goggles 15 boxes, matches Investigation: 15 test tubes, 13 × 100 mm 150 g stearic acid 30 safety goggles 15 boxes, matches 15 test tube holders 15 index cards 15 plastic spoons 15 metric rulers	Classifying (pp. 239, 240, 240-244) Using Laboratory Equipment (p. 245)	Slow Learners: Display a large piece of an igneous, a sedimentary, and a metamorphic rock, such as granite, conglomerate, and marble. Discuss the formation of each type of rock. Above Average: Discuss the relationship between crystal size and rate of cooling in igneous rocks. Mainstreaming: Have students handle samples of igneous rocks if available.
9-2. Sediments (1-2 periods)	paper and pencil	Sequencing and Inferring (pp. 252, 253, and 254)	Slow Learners: Show samples of sedimentary rocks discussed on pp. 250-251. Above Average: Discuss the formation and geological importance of sedimentary rocks. Mainstreaming: Ask students to bring in samples of rocks.
9-3. Changed Rocks (1-2 periods)	3 sticks modeling clay (different colors) 2 slices, bread 1 roll, waxed paper 15 eye droppers water source	Sequencing (Fig. 9-26, p. 259)	Slow Learners: Show students samples of metamorphic rocks or pictures of marble objects. Above Average: Discuss the formation of metamorphic rock. Mainstreaming: Have students make a rock cycle poster (p. 259).

SUPPLEMENTAL MATERIALS

Audiovisual

Cinnabar, A Game of Rocks and Minerals, Nova, 510-2715, 5 games, 3-5 players each.
Formation of Sedimentary Strata, EBE, 80752, color, filmloop.
History in the Rocks, Hubbard, set of 10, Super 8, filmloops.
Igneous Processes, Listening Library, Super 8, color.
Investigating Rocks, EBE, 9 color/sound filmstrips.
Metamorphism and Coal Formation, Listening Library, Super 8, color filmloop.

Rocks That Form on the Earth's Surface, EBE, 16 mm, sound, color film, 16 min.

Print

Croucher, R. and A. R. Woolley. *Fossils, Minerals, and Rocks: Collection and Preservation.* New York: Cambridge University Press, 1982.
Lye, Keith. *Minerals and Rocks.* New York: Arco, 1979.

Computer

Earth Sciences. Atari, Inc., Atari 800 XL.
Rocks and Minerals Keyword. Focus Media, Inc., APL II Series; TRS-80, Models III, 4.

CHAPTER OVERVIEW

Rocks are classified as igneous, sedimentary, or metamorphic. The way each is formed, its general characteristics, and the story it tells are given, together with examples of each. The rock cycle is discussed. A Skill-Building Activity on the use of the Bunsen burner accompanies Section 1. The Investigations include crystal formation, the profile of sedimentary rock layers, and simulation of metamorphism by using clay models.

9–1. MAGMA AND IGNEOUS ROCKS

BACKGROUND: The reason rocks are classified as igneous, sedimentary, and metamorphic extends beyond knowing the way the rock came into existence. This classification gives the geologist an idea of the history of the rock and of the layer or structure in which it is found.

MOTIVATION: Display samples of igneous rocks or color pictures of them. Ask the students if they know what all of them have in common (the manner in which they formed). Pass around a few of the samples that show obvious crystals. Comment on the time it must have taken for crystals to form. Ask if the students have seen other kinds of rock.

SKILL-BUILDING ACTIVITY

Teaching Hints: This Activity is designed to give the student a good introduction to the use and care of the most typical laboratory burner. Matches are used to light the burner in the Activity. If you have striker lighters you will need to demonstrate their use.

INVESTIGATION

Teaching Hints: The purpose is to show how the rate of cooling of a molten substance affects the crystal size. The solid stearic acid is not corrosive and can be handled similarly to wax. Unlike wax, it gives a crystalline solid when slowly cooled. USP grade can be purchased from any chemical supply house.
   The small test tubes (13×100 mm) were chosen to limit the amount of stearic acid used. It melts at $80°C$. If magnifying lenses are available, these would be excellent to use for looking at the crystals.

Safety: Caution your students that stearic acid can cause a skin burn and not to heat it past its melting point, which will alter the results. It would be wise to show them how to heat a solid in the test tube, and to point it away from all people.

Answers to Questions:
1. The sample on the index card.
2. Answers will vary.

3. Answers will vary. Usually about 5 mm.
4. They are interlocking, long, thin crystals.
5. They are longer.
6. Answers will vary. Usually about 10 mm.
7. The faster the acid cools, the smaller the crystals.

Conclusions:
1. The faster the lava cools, the smaller the crystal size.
2. Small crystals.
3. Large crystals.

9–2. SEDIMENTS

BACKGROUND: Weathering and erosion of rock produce most of the sediment that forms sedimentary rock. As layer on layer of sediment piles up, it is compressed and cemented into solid rock. Chemical and organic sediments are the source of many useful minerals. The Investigation relates layers of sedimentary rock to the history of a given area.

MOTIVATION: Using dilute HCl acid, place drops on sea shells, coquina, limestone and chalk to indicate similar sediments. Have students identify the sediments in sandstone, coquina, and conglomerate.

INVESTIGATION

Teaching Hints: The Investigation uses the profile of sedimentary rocks (p. 252) to point out the variety of information that can be derived from the type of sediment.

Answers to Questions:
1. Shale.
2. Sandstone.
3. Pressure from overlying layers presses the particles closer together.
4. Layer 8.
5. Gypsum.
6. Shale, both layers 2 and 7.
7. Conglomerate.
8. Two times.
9. The speed must have decreased.
10. Shale.
11. Two times.
12. The water supply to the sea was shut off. The water evaporated and left a deposit of gypsum.
13. A chemical deposit (gypsum).
14. One time.
15. The lack of any organic sediment.

Conclusions:
1. 10, 9, and 8, sand; 7, clay and mud; 6, gravel and large round particles; 5, sand; 4, gravel and large round particles; 3, sand; 2, clay and mud; 1, chemicals dissolved in the water.

2. Layer 1: the inland sea was drying up.
Layers 2 and 7: little water was running into the inland sea; few or no currents in the water.
Layers 4 and 6: very rainy weather, much water running vigorously into the inland sea.
Layers 3, 5, 8, 9, and 10: rainy weather, fair amount of runoff carrying the sand grains.

9–3. CHANGED ROCKS

BACKGROUND: Great pressure can squeeze the grains of rock so they no longer resemble their original form. This is the process of metamorphism; the kind of rock produced is called metamorphic rock. The great heat from the presence of magma can also cause metamorphism. The original rock and its resulting metamorphic rock are known as metamorphic pairs. Sandstone and quartzite are metamorphic pairs. The Investigation uses clay models to mimic the process of metamorphism.

MOTIVATION: To focus attention on the effects of pressure and heat on a substance, do the following:
1. Display a piece of bread. Discuss its properties.
2. Crush the bread in your hand. Discuss the changes.
3. Ask the students to predict the results of heating the altered bread in an oven. Display a piece of previously heated bread. Note chemical changes.
4. Mention that greater pressures and temperatures exist deep in the earth. Would these conditions have an effect on rock?

INVESTIGATION

Teaching Hints: Metamorphism is simulated with clay and bread.
At least three colors of clay are needed to show some of the characteristics of metamorphism. Play-Doh will work equally as well and tends to harden if left uncovered. Have the 1 cm cubes prepared in advance. Provide a container for the used clay. Assign someone with a putty knife to clean the floor.

Answers to Questions:
1. Sedimentary rock.
2. The spheres have been flattened.
3. There is now much less space between spheres.
4. The spheres are now gone, the clay is smooth with colors streaked through it. Originally the spheres were easily seen and colors were separate.
5. Particles in the rock appear flattened.
6. Answers will vary.
7. Answers will vary, but fewer drops than in 6.

Conclusions:
1. Pressure reduces the grain size. However, if a chemical change occurs and different crystals form, an increase in size might occur.

2. Compare the denseness, size, and shape of grains.
3. The sedimentary rock would hold more water.

COMPUTER

Teaching Hints: A flowchart symbol has been included in the glossary that does not occur in this chapter. The input/output symbol (a parallelogram) is often used, however. The flowchart appears as a copy master.

CHAPTER REVIEW

Vocabulary:
1. dikes 2. pluton 3. rock 4. batholiths 5. rock cycle 6. sedimentary rock 7. metamorphic rock 8. sediment 9. igneous rock 10. sills

Answers to Questions:
1. Rock that formed when molten material cooled.
2. Extrusive igneous rock: formed from magma that spills out on the earth's crust. Intrusive igneous rock: formed from magma cooling underground.
3. Obsidian, tuff, pumice.
4. Igneous rock that has cooled rapidly is made up of small crystals, whereas igneous rock that has cooled slowly is made up of large crystals.
5. Sills, rock formed from magma intruding in cracks that are parallel to existing rock layers.
Dikes, rock formed from magma intruded in cracks that cut across existing rock layers.
Batholiths, bodies of hardened magma that stretch thousands of kilometers below the surface.
6. The most abundant sediment is mud. Chemical sediments are formed from substances dissolved in the water. Organic sediments are the remains of plants or animals.
7. Pressure may squeeze the sediment, and/or minerals may cement the sediment, making it hard.
8. Sandstone and conglomerate are the most common sedimentary rocks made of rock fragments. Shale, another sedimentary rock, is made of clay.
9. Sediment is formed from dissolved minerals. Gypsum is a rock formed from chemical sediment.
10. Coal formed from organic sediment. The sediment is the remains of plants that lived in swamps.
11. Organic sediment is the remains of plants and animals.
12. Heat may change rock by changing its crystalline structure or by changing its composition.
13. Heat and pressure cause metamorphic rock.
14. Igneous rock comes from molten magma, which might have been metamorphic or sedimentary rock. Metamorphic rock formation excludes melting.
15. No. At present there is no noticeable tectonic movement of crust and little weathering and erosion on the moon.

SECTION/ TIME NEEDED	MATERIALS (30 STUDENTS)	BASIC SCIENCE SKILLS	PACING GUIDE
10-1. People and Resources (1-2 periods)	pencil and paper	Classifying (p. 268) Cause and Effect (pp. 271-273)	Slow Learners: Have students list things that are made of renewable and/or nonrenewable resources. Above Average: Relate population, pollution, and natural resources. Mainstreaming: Tape key passages for students to use while they read.
10-2. Mineral Resources (1-2 periods)	10 shallow dishes or pans 10 large containers 10 samples, mixture of sand, gravel, fool's gold water source	Compare and Contrast (p. 274)	Slow Learners: Have students list things they use that contain metal. Above Average: Discuss how long it has taken for mineral deposits to form and how this relates to conservation and recycling. Mainstreaming: Show students ores and their metals (p. 275).
10-3. Water Resources (1-2 periods)	15 meter sticks 30 10-cm tall containers 15 small pans 15 water containers water source	Cause and Effect (pp. 280-283) Compare and Contrast (p. 284)	Slow Learners: Have students list how they use water. Discuss where this water comes from and what would happen if it ran out. Above Average: Discuss the causes of water pollution and ways of protecting the water supply. Mainstreaming: Discuss the sources and uses of water in your locality.
10-4. Nonrenew- able Energy Resources (1-2 periods)	30 sheets, graph paper	Compare and Contrast (p. 293) Predicting (p. 296)	Slow Learners: Have students calculate their energy consumption per year, using the table on p. 293. Above Average: Discuss the problems of using nonrenewable resources such as fossil fuels. Mainstreaming: Outline the nonrenewable sources of energy.
10-5. Renewable Energy Resources (1-2 periods)	10 cardboard boxes 10 pieces, black paper 10 pieces, transparent wrap 10 100-watt lights 20 thermometers 2 rolls masking tape	Compare and Contrast (p. 302)	Slow Learners: Show pictures of renewable energy sources. Discuss why each is renewable. Above Average: Discuss the value of using renewable energy instead of nonrenewable sources. Mainstreaming: Outline the renewable sources of energy.

SUPPLEMENTAL MATERIALS

Audiovisual

Energy, New Sources, CF, 16 mm color/sound film.
Energy, The American Experience, U.S. Dept. of Energy, 16 mm color sound film.
Energy, The Fuels and Man, NGS, 16 mm color, sound film, 16 min.
Geothermal, Nature's Broiler, U.S. Dept. of Energy, 0519, 16 mm, color, sound film.
Natural Resources, Hawkhill, videocassette, guide.
Nuclear Energy, Perils or Promise?, Science and Mankind, color/sound filmstrip.
Pollution Below: NASA, HQ247, 16 mm color sound film, 14 min.
Pollution in Perspective: Solid Wastes, GE, 16 mm color sound film, 17 min.
Problems of Conservation: Our Natural Resources, EBEC, 16 mm color sound film.

Renewable Energy Resources, Science and Mankind, videocassette, 47 min.
Saving Energy at Home, Ramsgate, 16 mm color sound film, 13 min.
Toxic Wastes, Hawkhill, videocassette, guide.
Up to Our Necks, NBC, 16 mm color sound film, 26 min.

Print

Deudney, Daniel and Christopher Flavin. *Renewable Energy, The Power To Choose.* NY: Norton, 1983.
Herron, Matt. "Falling Water, A Mother Lode of Energy." *Smithsonian,* December 1982.
Nader, Ralph, Ronald Brownstein, and John Richards. *Who's Poisoning America?* San Francisco: Sierra Club Books, 1981.

Computer

Energy Search. McGraw-Hill, APL II Series, TRS-80, Model III.

CHAPTER OVERVIEW

This chapter is concerned with the earth and the way people use its resources. Types of pollution resulting from human activities are discussed in some detail.

Nonrenewable and renewable mineral and energy resources are described. The reasons for conserving and recycling natural resources are developed. The location of usable water, the uses of water, and the ways water becomes polluted are studied.

10–1. PEOPLE AND RESOURCES

BACKGROUND: The earth is likened to a space station which must provide all the needs of the passengers. Population, food supply, and pollution are discussed. The Investigation is designed to alert the student to one form of pollution, trash.

MOTIVATION: Look at Fig. 10–2, p. 269. Use the graph to estimate the population of the earth for the year most of your students were born. At 30, most people are busy earning a living and raising a family. Determine what the population will be when the students are 30 and 70 years old. Ask what they think must be done to supply that many people with food and energy.

INVESTIGATION

Teaching Hints: This Investigation lets the student approximate how much trash his or her town or city produces in a day and relate this to the problems produced.

Place on the board the population of your town. If not known, you could use an average size community of 35,000. Then have the students place individual answers to question 2 on the board. Next, calculate the average amount of the estimates and do questions 4 through 6 as a summary. A metric ton (1,000 kg) is 2,200 pounds and is commonly called a long ton. Remind your students that their calculations do not include sewage and industrial waste disposal. They should understand that waste disposal can be a costly community problem.

Answers to Questions:
1. Answers will vary, usually 0.5–2.5 kg.
2. Answers will vary.
3. More trash by grocers, hospitals, industries, etc. Less by very old and very young people.
4. Answers will vary, based on the answer to question 2 divided by 1,000.
5. Answers will vary. The answer to question 4 times $40.
6. The answer to question 4 divided by 5.

Conclusions:
1. Answers will vary, depending on the community.
2. Yes. Industrial trash was not included.

10–2. MINERAL RESOURCES

BACKGROUND: Some rocks called ores are valuable because they contain a usable mineral. Three ways mineral deposits are produced are discussed. Conservation and recycling are ways of increasing the mineral supply for growing populations. The student pans for "gold" in the Investigation.

MOTIVATION: Have samples of some common ores such as hematite (for iron), bauxite (for aluminum), galena (for lead), sphalerite (for zinc), malachite (for copper), and argentite (for silver) available.

INVESTIGATION

Teaching Hints: If possible, do this Investigation out of doors where a source of water is available. If done indoors, have several students do the panning and then give each group a pan to pick out the gold. The least mess would be for you to do the panning. It is especially interesting if some iron pyrites is available, since it looks and feels just like real gold.

Iron pyrites is available from mineral supply houses. Another approach would be to paint lead shot with gold paint, and to have the students determine how much it would be worth in today's market if it were real gold.

Conclusions:

1. The source is near the stream, before point G.
2. Its density is greater than sand and rock.
3. Increase the sample size used and the pan size. Other suggestions are possible.

10–3. WATER RESOURCES

BACKGROUND: Water is one of the most important natural resources. Although nearly all the water that the earth started with is still on earth today, it is not necessarily in the right place at the right time. We are polluting the available freshwater sources rapidly. This section deals with the water sources, pollution of water, and local water budgets.

MOTIVATION: Fill a container with water and ask students to trace it back to its source. Ask the students to identify parts of the water system that are susceptible to pollution.

INVESTIGATION

Teaching Hints: The water budget of Savannah, Georgia, is studied with Table 10–1, p. 284. The Savannah budget is compared to that of Yuma, Arizona.

Tall plastic or paper drinking cups can be cut to a height of 10 cm for the two containers. Small disposable aluminum pans are recommended for the overflow. If time permits, the Yuma budget could be studied as well. You could have some students do Savannah while others are doing Yuma. Then compare results.

Answers to Questions:

1. 70 mm.
2. 20 mm.
3. 50 mm.

4. They are the same.
5. The container overflowed.
6. No.
7. Feb., Mar., Nov., Dec.
8. Yuma has a much smaller rain income, never has a monthly surplus, and has its greatest income in winter instead of summer.
9. It is necessary to bring in water all year long to balance the budget for Yuma.

Conclusions:

1. Yuma, Arizona. Yuma usually has a monthly deficit.
2. No. Water may have been stored from a surplus.

10–4. NONRENEWABLE ENERGY RESOURCES

BACKGROUND: A comparison is made of past and present sources of energy. Today's fossil fuels are classified as nonrenewable. The possibility of using nuclear fuel is brought out. A comparison of fission and fusion is made.

MOTIVATION: It seems that the amount of energy used by individuals must be cut back. Use the Table on p. 293 to discuss the ways a family could save money and conserve fuel. Not long ago the cost of a kWh was less than a penny. Today it is greater than 10 cents. Most people cannot cut back enough to keep the same size fuel bill.

INVESTIGATION

Teaching Hints: Data is given for the fuel demand and the corresponding population in the United States for 1950 through 1980. See p. T64.

You may wish to plot a few points on each graph on the chalkboard and assist in doing the calculation for the answer to question 3. Information can be obtained from U.S. Committee on Energy Awareness, P.O. Box 37012, Washington, D.C. 20013.

Answers to Questions:

1. About 1967.
2. 3,800 billion barrels.
3. 19,000 barrels per person.
4. About 240 million.
5. About 6,800 billion barrels.
6. About 28,300 barrels per person.
7. In 1990 there will be about 1.5 times the demand that existed in 1967.
8. About 33,000 barrels per person. (Answers will vary.)

Conclusions:

1. The growth per person of fuel demand is greater than the population growth.
2. Answers will vary.

10–5. RENEWABLE ENERGY RESOURCES

BACKGROUND: The alternatives to fossil fuels for energy are discussed. Solar, geothermal, energy from moving water and wind, and biomass conversion are discussed.

MOTIVATION: Discuss problems related to depending on nonrenewable fuels, such as pollution, availability, and political factors affecting supplies. Demonstrate the energy of the sun with a solar cell or radiometer. Wind energy could be demonstrated with a pinwheel.

INVESTIGATION

Teaching Hints: The student makes a model solar collector and compares the temperatures.

Have the students work in groups of three. The light source should be at least a 60-watt incandescent lamp. If possible, use direct sunlight.

Safety: Place a flag of tape around the top of the outside thermometer so that it cannot roll off the table.

Answers to Questions:

1. The thermometer in the box.
2. The thermometer in the box.

Conclusions:

1. The thermometer inside the box absorbed more heat. Black paper absorbed more heat and the plastic cover held the heat in.
2. The thermometer outside the box lost more heat. There was no way of holding the heat.
3. Solar collectors are painted black inside to absorb the heat better. The transparent cover allows light in but keeps the heat from leaving.

TECHNOLOGY

Teaching Hints: Research in Geothermal energy is underway in Cambourne, England. For more information, see *New Scientist*, October 18, 1984, page 32.

CHAPTER REVIEW

Vocabulary:

1. natural resource 2. pollution 3. ore 4. Conservation 5. mineral deposit 6. recycling 7. thermal pollution 8. watershed 9. fossil fuel 10. Petroleum 11. Hydrocarbons 12. nuclear energy 13. Fission 14. Fusion 15. solar cell 16. geothermal power

Answers to Questions:

1. Answers will vary. Forests are a natural resource. Many useful products are made from trees.

2. A renewable resource, such as a forest, can be recovered and used again. A nonrenewable resource, such as a fossil fuel, requires a very long time to form and cannot be replaced.
3. Lack of food to feed the increasing population and the pollution that results from increased population are two problems.
4. Pollution must be prevented to keep from harming the environment to the degree that it cannot recover.
5. A mineral deposit requires a very long time to form, and thus it is considered nonrenewable.
6. If the watershed loses its plant cover, the water will drain too rapidly, causing flooding, filling of reservoirs with mud and reducing the available water, and polluting the streams with debris.
7. Coal as an energy source pollutes the air with excess sulfur dioxide and carbon dioxide; damages the environment during mining operations; and endangers the lives of miners.
8. Fusion is the building of more complex nuclei from simpler nuclei. Fission is the reverse.
9. For billions of years there has been more radiant energy available from the sun than has been needed. This is not expected to change. Therefore, solar energy is considered not limited and renewable.
10. Solar energy would be nonpolluting, with no limit to the availability, whereas fossil fuels are limited and nonrenewable. Installations for solar energy are quite expensive now, whereas fossil fuel plants are relatively low in cost.
11. Petroleum and natural gas are found together and they both are made up of hydrocarbons.

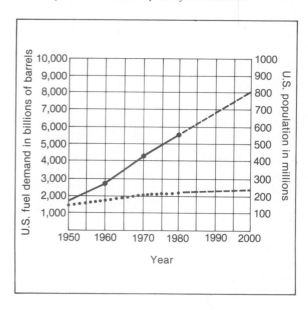

CHAPTER 11 AIR

SECTION/ TIME NEEDED	MATERIALS (30 STUDENTS)	BASIC SCIENCE SKILLS	PACING GUIDE
11-1. Composition of the Earth's Atmosphere (1-2 periods)	10 graduated cylinders 200 mL limewater 10 beakers 10 straws 10 candles 10 cardboard squares matches 10 flasks 10 stirring rods 5 L vinegar blue litmus paper	Classifying (p. 312) Compare and Contrast (p. 315)	Slow Learners: Students should learn the names of the different gases in the air. Discuss sources of air pollution in your locality. Above Average: Discuss the relative amounts and properties of each of the gases in the air. Mainstreaming: Have students make a diagram or poster of Fig. 11-1, p. 310.
11-2. Atmospheric Pressure and Temperature	10 large containers 10 small beakers 10 straws water	Sequencing (p. 319) Cause and Effect (p. 322)	Slow Learners: Have students make a chart of the layers of the atmosphere shown in Fig. 11-13 (p. 320). Place in the troposphere—Mt. Everest and clouds; stratosphere—jets; mesosphere—meteors; thermosphere—satellites. Above Average: Discuss the reasons given on pp. 318-319 for the changing temperatures in the atmosphere. Mainstreaming: Have students organize vocabulary under "air pressure" and "layers."

SUPPLEMENTAL MATERIALS

Audiovisual

Air Expansion by Heat, EBE, No. S-80742, color filmloop.
Something In The Air, Caterpillar film, color (28 min).
The Noise Boom. NBC film, color (26 min).

Print

Baum, R. M. "Stratosphere Chemistry." *Chemical and Engineering News*, September 13, 1982.

Chameides, W. L., and D. D. Douglas. "Chemistry in the Troposphere." *Chemical and Engineering News*, October 4, 1982.
Riehl, H. *Introduction to the Atmosphere*. New York: McGraw-Hill, 1978.
Schaefer, V. J. *A Field Guide To The Atmosphere*. Boston: Houghton Mifflin, 1981.

Computer

Weather. Educational Activities. APL II Series; TRS-80; COM 64, Pet.

CHAPTER OVERVIEW

Chapter 11 is the first of three chapters concerned with the atmosphere and the hydrosphere. The importance of air as an essential ingredient for life and as an integral part of the energy cycle is emphasized.

The composition of the atmosphere, the nitrogen and carbon dioxide cycles, and pollution are discussed first. Atmospheric pressure and temperature and their relation to layers of the atmosphere are then considered.

11–1. COMPOSITION OF THE ATMOSPHERE

BACKGROUND: We need to have a constant supply of air as an essential part of the living process. The composition of the air is discussed together with the action of the nitrogen and carbon dioxide cycles. Emphasize the importance of the atmosphere in producing an environment supportive of life. The natural cycles tend to maintain the present composition of the atmosphere. However, these natural processes may be altered by human activities. Air pollutants are considered as foreign substances in the air.

MOTIVATION: Have the students read the motivation on page 309; then discuss the need for food, water, and air. Display a container of ice water on which moisture has condensed. Most students will be aware of the water vapor (gas) in the air. Ask for suggestions for methods of demonstrating the presence of other gases in the air.

Hint: Avoid detailed discussion of the role of water in atmospheric processes. This topic is developed later.

INVESTIGATION

Teaching Hints: In this Investigation, students test for gases in the air that can form acids and acid rain.

A 250-mL conical flask and a 150-mL beaker are recommended. Glass tubing could be used in place of the rubber tubing. Common book or stick matches are recommended. Use white vinegar. Purified water from the grocery store may be used in place of distilled water.

The limewater is a saturated solution of calcium hydroxide. An easy way to make limewater is to dissolve calcium hydroxide in a beaker of distilled (purified) water. It dissolves rapidly until the water becomes saturated. When white, undissolved calcium hydroxide begins to persist in the water, filter out the undissolved calcium hydroxide. Always use freshly prepared limewater and pretest it by doing procedure A.

Carbonic acid (H_2CO_3) is the weak acid found in carbonated beverages. The sulfur dioxide produces

sulfurous acid (H_2SO_3), which is a stronger acid. Some sulfur dioxide in air may be oxidized to sulfur trioxide, which produces sulfuric acid (H_2SO_4), a very strong acid. Nitrogen oxides from burning fuels also produce acids with moisture. It is the mixture of all of these that make acid rain.

Safety: If you feel uncomfortable having the students do the carbon dioxide test, you may wish to demonstrate it or have one of your trusted students demonstrate it.

Answers to Questions:
1. The limewater turns cloudy.
2. The limewater turns cloudy.
3. Carbon dioxide.
4. Answers will vary. Usually called a disagreeable, choking odor.
5. The litmus turns pink (red).
6. The litmus turns pink (red).

Conclusions:
1. The carbon dioxide and sulfur dioxide dissolved in the moisture, producing an acid that turned the litmus red.
2. As a water drop falls through air containing large amounts of carbon dioxide and sulfur dioxide, it dissolves some of the gases and becomes a drop of acid.

11–2. ATMOSPHERIC PRESSURE AND TEMPERATURE

BACKGROUND: The pressure of air is introduced as the weight of the air above the point of observation. Methods of measuring air pressure are described. The atmosphere has distinct layers that are defined by temperature and composition.

MOTIVATION: Demonstrate atmospheric pressure in the following way. Obtain an empty, rectangular one-gallon can, such as those containing duplicator fluid or paint thinner. Rinse the can well with water, then put about 50 mL of water in it. With the cap removed, heat the can until the water boils and vapor is seen escaping. Turn off the heat and immediately replace the cap or insert a rubber stopper in the opening that fits tightly. As the can cools, condensation of the water vapor and cooling of the air inside will lower the pressure within the can. Atmospheric pressure outside will then slowly collapse the can as it cools. For quicker results, place the can under cold water from a faucet.

Have barometers available for students to inspect. You could discuss how an aneroid barometer could be used as an altimeter.

INVESTIGATION

Teaching Hints: Students investigate the effect of air pressure by observing the water in an inverted beaker whose open end is under water.

A small fish tank is recommended. However, any container deep enough for the size beaker being used will do. The soda straw with the flexible section works best. As with any activity that uses large amounts of water, be prepared for spills.

Some students may be surprised that the water remains in the beaker even above the water in the container. You may wish to comment that this would be true even if the beaker were over 30 feet tall! Since mercury is 13.6 times heavier than water and it is held to a height of 76 cm by atmospheric pressure, a column of water would be held to a height of 1,033.6 cm or 33.9 feet!

Answers to Questions:

1. The water stays in the beaker.
2. The level in the beaker is higher.
3. The water flows out of the beaker.
4. They are the same level.
5. The air stays inside the beaker.
6. The level in the beaker is lower.
7. The air flows out of the beaker.
8. They are the same level.

Conclusions:

1. Since there is no air pressure inside the beaker, the atmospheric pressure pushing down on the surface of the water in the container keeps the water up in the beaker. The straw allows the pressure above the water in the beaker to equal atmospheric pressure and the water level is lowered.
2. The water doesn't enter the beaker because it is already filled with air. The air pressure inside the beaker was allowed to become equal to that outside the beaker, allowing the water to rise to the same level inside the beaker as outside.
3. The water in the bottle will stay at about the same level that it was before the stopper was removed.

COMPUTER

Teaching Hints: You will need some advance planning to run this program since the necessary data consists of temperature and barometer readings for 14 days. Students can collect actual readings, or they may collect 14 days of readings from a national newspaper, making note of regional differences across the country.

The 14-day period during which you and your students collect weather data may not provide readings that allow for clear comparisons or contrast. If that is the case, you can use the opportunity to discuss the need for sampling over longer periods of time in order to increase the accuracy of predictions.

The concept of probability may be brought up in the context of PROGRAM FORECASTER and the collection of data. Students have usually heard weather forecasters speak in terms of the likelihood of weather patterns. You can discuss the bases for this likelihood or probability.

CHAPTER REVIEW

Vocabulary:

1. thermosphere 2. ionosphere 3. barometer
4. stratosphere 5. troposphere 6. mesosphere
7. pollutant 8. photosynthesis

Answers to Questions:

1. Nitrogen and oxygen make up 99% of the air by volume.
2. The water vapor varies from almost nothing over the deserts, to as much as 3% over the tropical oceans.
3. Sulfur and nitrogen oxides combine with raindrops to produce acid rain. Acid rain damages the forests and life in lakes.
4. The carbon dioxide cycle is the process by which carbon dioxide is taken from the air, and returned.
5. Carbon monoxide and hydrocarbons from auto exhaust make smog.
6. The accumulation of carbon monoxide, a poisonous gas, can endanger life.
7. Pollutants do not stay in the lower atmosphere very long, but there are no weather changes in the upper atmosphere to remove pollutants, so they may remain and build up over a long period.
8. The aneroid barometer, because it is small, does not contain a liquid, and is easier to use.
9. Half of the weight of the air is found in the troposphere.
10. The thermosphere has the highest temperatures.
11. Your ears would pop when the trapped, higher pressure air in them is released to balance the lower outer pressure.
12. The temperature and pressure decrease as you go higher.
13. The temperature would be 7 degrees Celsius.
14. Ozone is found in the upper stratosphere and lower mesosphere. It is important because it filters out most of the dangerous ultraviolet rays.
15. The ionosphere, which reflects radio signals, has changed to a higher position. This allows radio signals from farther away to be reflected to the radio.

CHAPTER 12 ENERGY AND WATER IN AIR

SECTION/ TIME NEEDED	MATERIALS (30 STUDENTS)	BASIC SCIENCE SKILLS	PACING GUIDE
12-1. Atmospheric Energy (1-2 periods)	Skill-Building Activity: 15 clinical thermometers 15 laboratory thermometers Investigation: 30 lab thermometers paper 20 small containers 2 L soil water electric lamp	Cause and Effect (pp. 330–338) Predicting (p. 337) Measuring (p. 340)	Slow Learners: Show and discuss examples of transferring heat: radiation—light bulb, conduction—spoon in hot water, convection—put some food coloring or a potassium permanganate crystal down the side of a beaker. Heat this side of the beaker slowly. Above Average: Discuss the result of the Coriolis effect on aircraft flying north and south. Mainstreaming: Outline information in this section.
12-2. Water Enters the Air (1-2 periods)	30 thermometers 45 pieces cotton cloth string water 1 pt glycerine 1 pt alcohol 15 index cards	Cause and Effect (pp. 342, 346) Measuring (pp. 343–344)	Slow Learners: Discuss common examples of evaporation. Demonstrate condensation by adding ice to a metal container filled with water. Above Average: Have students calculate the relative humidity. A relative humidity table is found in Chapter 12 of the Teacher's Resource Book. Mainstreaming: Have students read summary and study vocabulary before studying this section.

SUPPLEMENTAL MATERIALS

Audiovisual

Air Pressure and Winds, Edwin Shapiro Co., captions, filmstrip.
Global Circulation Model I, Thorne, No. GC 400, color filmloop.
Global Circulation Model II, Thorpe, No. GC 40, color filmloop.
Water Vapor in the Atmosphere, Edwin Shapiro Co., captions, filmstrip.
What Makes the Wind Blow? EBE, 2249, color film.

Print

Bernard, H. W., Jr. The Greenhouse Effect. Cambridge, MA: Ballinger, 1980.
Gribbin, J. R. Future Weather and the Greenhouse Effect. New York: Delacourt Press, 1982.
McKean, K. "Hothouse Earth." Discover, Dec. 1983.
Patton, A. R. Solar Energy for Tomorrow's World. New York: Messner, 1980.

Computer

Water and Weather Series. Focus Media, Inc. COM 64, Pet; APL II Series; TRS-80, Models III, 4

CHAPTER OVERVIEW

Energy from the sun heats the earth and atmosphere. The greenhouse effect traps some of the heat.

Unequal heating of the air causes pressure and temperature changes, which cause the air to move. In addition, water in the atmosphere is studied. Absolute and relative humidity are studied and quantified.

12-1. ATMOSPHERIC ENERGY

BACKGROUND: Solar energy received by the earth and its effect on the atmosphere are described. The greenhouse effect is emphasized. Uneven heating of the atmosphere produces wind patterns. The changes produced by the Coriolis effect are also introduced.

It may be possible to relate local wind patterns to the land, mountain, and sea breezes described in the text. The nearest office of the Weather Service or general aviation pilots may be able to provide information about local winds.

MOTIVATION: Compare the effect of radiation from an infrared lamp and light from an ordinary incandescent lamp on the skin and on thermometers at a distance of about 1 meter. Students should not look at the lamp for a long period of time. A photographer's light meter might be used to help distinguish between visible light and invisible radiant energy.

Create unequal heating of the atmosphere, in terms of high- and low-pressure regions. Attach a 50-cm piece of string to the center of a meter stick so it balances. Tie the meter stick to some support. Open two small paper bags and tape them, open end down, to the ends of the meter stick so it still balances. Place a small lighted candle a short distance below the opening of one of the bags. Warm air rising into the bag should cause it to rise visibly. Discuss what happens.

SKILL-BUILDING ACTIVITY

Teaching Hints: Label thermometers in pairs, one A, the other B; one clinical, the other laboratory. Do not label all the clinical ones as B; mix them up.

Have other thermometers, with different temperature ranges, for viewing if available. Refer to the Fahrenheit-Celsius scale in Fig. 1-18, p. 19, to show the relationship between the two scales.

Safety: Show how to shake down a clinical thermometer. Have everyone do it with you. Caution on the possibility of striking the thermometer against something. Discuss why you do not shake a laboratory thermometer. Note that thermometers are fragile.

Answers to Questions:
1. Answer depends on the labels assigned.
2. Usually 94 to 108 degrees.
3. Usually 0.1 degree.
4. Usually −10 to 110 degrees.
5. Usually 1 degree.
6. Answers will vary. It may be kept in its holder or have something attached to it to keep it from rolling.
7. It is too easy to break the bulb end.
8. Usually 35 to 37 degrees Celsius. 37 degrees Celsius is normal body temperature.

9. Usually 94.0 to 96.0 degrees. Normal body temperature is 98.6 degrees Fahrenheit, but that high a reading is unlikely.

Conclusions:
1. Descriptions will vary. Precautions should include: avoid rolling of thermometer, do not use it as a stirring rod, avoid hard knocks of the bulb.
2. The laboratory thermometer reads the temperature where it is located. The clinical thermometer holds the last and highest temperature until it is reset.

INVESTIGATION

Teaching Hints: Baby food jars or 6 cm petri dishes can be used. Make certain the soil or dark sand is dry. Provide 250-mL beakers for dispensing water. A gooseneck lamp with a 60-watt bulb works best. If a ring and ringstand are available, you may wish to use them to support the thermometers.

The thermometers should be just barely below the soil or water surface. Using 6 cm petri dishes, in 2 minutes the changes were: water, from 25 to 27 degrees Celsius; river sand, from 25 to 34 degrees; air, from 26 to 32 degrees.

Safety: Thermometers are fragile, and the glass cuts. Wash hands after handling soil. Wear eye protection. Label all containers for disposal of materials.

Answers to Questions:
1. Usually the soil; air is a close second.
2. The water is slowest.
3. Usually the soil.
4. The water.

Conclusions:
1. Soil, air, water.
2. Soil, air, water.
3. Soil absorbs heat faster and cools faster than air or water. Water heats slowly and cools slowly.

12-2. WATER ENTERS THE AIR

BACKGROUND: Energy released during condensation is the driving force of weather changes. This section deals with humidity of the atmosphere and the conditions that affect it.

A solid changing directly to a gas is called sublimation. Dry ice and mothballs are examples. Ice sublimes in dry air below 0°C. The reverse process, in which water vapor condenses directly as a solid, may also occur. Frost is formed by this process.

MOTIVATION: Demonstrate the cooling effect of evaporation. Apply a small amount of water to the back of one hand of a student. To the other hand, apply an equal amount of rubbing alcohol, not revealing which liquid is applied to each. Ask the student to describe the sensations. Discuss the different amounts of energy needed to cause each liquid to evaporate.

Have students demonstrate how they would show water vapor is present in the atmosphere.

INVESTIGATION

Teaching Hints: If students have had little experience with using thermometers, do the Skill-Building Activity, p. 340, before doing the Investigation.

Match thermometers in pairs that show the same temperature. Provide the cotton cloth in small squares. Small baby food bottles are good containers for the liquids. Isopropanol (rubbing alcohol) and glycerine may be purchased at any drugstore. Provide a common waste container for the used cloth and string.

Measure the room air temperature; you need to know it for grading answers.

Safety: Thermometers break and the glass cuts.

Answers to Questions:
1. Answers will vary.
2. The dry bulb is the control.
3. Each gave a cooling effect.
4. The alcohol.
5. The glycerine.

Conclusions:
1. Evaporation has a cooling effect.
2. The water evaporates from your skin and swimsuit, causing a cooling effect.
3. The alcohol has a greater cooling effect than water.

CHAPTER REVIEW

Vocabulary:
1. radiant energy 2. greenhouse effect 3. convection 4. convection cell 5. Coriolis effect 6. humidity 7. dew point 8. relative humidity 9. condensation

Answers to Questions:
1. About 33% of the solar energy is reflected back into space and about 20% is absorbed by the atmosphere, allowing only 47% to reach the surface.
2. The ultraviolet rays that cause sunburn can pass through the clouds.
3. The land and water change the radiant energy into longer waves. The carbon dioxide and water in the atmosphere block the outgoing waves, causing the atmosphere to become warmer.
4. Convection cells form in the atmosphere, and the resulting winds spread the energy over the earth.
5. The doldrums are located at the equator. It is the area where heated air is rising.
6. Land and sea breezes are caused by the unequal heating and cooling of land and water.
7. A monsoon responds to the differences in summer and winter temperatures. High summer temperatures cause convection cells that move air from the sea over the land. In the winter, the air moves from over the land to the sea.
8. In the Northern Hemisphere, the Coriolis effect bends winds to the right; in the Southern Hemisphere it bends winds to the left.
9. The "polar easterlies" are the surface winds of a convection cell of cold air moving toward the equator, rising where it meets the westerlies, the air turning northward at high altitudes, cooling and returning to the surface at the poles to complete the circle.
10. You feel cool in a wet bathing suit because the water evaporates from the suit, taking energy away with it.
11. The water molecules at the surface absorb radiant energy, making them move faster than the rest of the molecules. These fast-moving molecules move away from the liquid entering the air as vapor.
12. Humidity is the amount of water vapor in the air. Relative humidity is a comparison of the amount of water vapor in the air to the amount of water vapor the air can hold at a given temperature.
13. Dew is the liquid that condenses from saturated air. Frost is the solid that condenses from saturated air when the air temperature is below freezing.
14. Fog forms from air saturated with water vapor when tiny drops of liquid water condense around dust particles in the air.
15. Both fog and clouds form in the same way. They differ in that a fog forms close to the surface, whereas clouds form higher in the air.
16. Water cools and warms more slowly than the land surrounding it. This in turn keeps the air warmer at night and cooler during the day.
17. (a) The air can hold about 18 g/m³. (b) The relative humidity is therefore 9/18 or 50%.

CHAPTER 13 WEATHER

SECTION/ TIME NEEDED	MATERIALS (30 STUDENTS)	BASIC SCIENCE SKILLS	PACING GUIDE
13-1. Water Leaves the Air (1-2 periods)	15 shiny cans 15 ice cubes 15 stirring rods 15 thermometers water	Cause and Effect (pp. 351, 357) Observing (p. 357)	Slow Learners: Have students observe clouds. Use Fig. 13-9 (p. 355) to identify them. Above Average: Calculate cloud heights for different days, using the method shown in the Investigation on p. 357. Use a sling psychrometer for more accuracy. Call local weather bureau to verify your results. Mainstreaming: Have students make a chart classifying the clouds in Fig. 13-5 (p. 355) as low, middle, or high clouds.
13-2. Air Masses (1-2 periods)	30 sheets tracing paper pencils	Cause and Effect (p. 360) Classifying (pp. 362-363)	Slow Learners: Discuss examples of what happens to air in closed spaces such as caves, cellars, and cars. Above Average: Use a weather map to show locations and movements of the different air masses and to predict weather. Mainstreaming: Use a weather map to show air masses and fronts.
13-3. Weather Changes (1-2 periods)	30 sheets tracing paper pencils	Sequencing (p. 366) Cause and Effect (p. 368)	Slow Learners: Bring in newspaper articles on tornadoes and hurricanes. Above Average: Using a weather map showing movement of air masses, discuss why tornadoes and hurricanes occur frequently in certain areas. Mainstreaming: Show a filmstrip with taped captions on storms, tornadoes, and hurricanes.
13-4. Predicting Weather (1-2 periods)	paper and pencil	Measuring (p. 373) Sequencing and Predicting (p. 377)	Slow Learners: Have students keep a daily record of local weather. Above Average: Cut out the weather map from a newspaper for a week. Discuss any weather patterns and make forecasts based on the maps. Mainstreaming: Have students read weather maps, using the legend in the map or the appendix.

SUPPLEMENTAL MATERIALS

Audiovisual

Beyond Disaster, Florida Dept. of Commerce, color film (28 1/2 min).

Condensation and Precipitation, Edwin Shapiro Co., captions filmstrip.

Meteorology From Space, NASA, 22 frames, audiotape, color film.

Storms, The Restless Atmosphere, EBE, color film (22 min).

Story of a Storm, EBE, No. S-80741, color film.

The Cold Front, BFA, Super 8, color film.

The Origins of Weather, EBE, color film (13 min).

The Warm Front, BFA, Super 8, color film.

The Weather Watchers, NASA, HQ-290, color film.

Tornado, Thorne, No. GC 413, color filmloops.

Understanding Weather and Climate, Jubbard, No. WFS-763, set of 6, color filmstrips.

Print

Aylesworth, T. G. *Storm Alert: Understanding Weather Disasters*. New York: Messner, 1980.

Compton, G. *What Does a Meteorologist Do?* New York: Dodd, Mead, 1981.

Dunlop, S. and F. Wilson. *The Larousse Guide to Weather Forecasting*. New York: Larousse, 1982.

Field, F. *Dr. Frank Field's Weather Book*. New York: Putnam's, 1981.

Hardy, R. and others. *The Weather Book*. Boston: Little, Brown, 1982.

Computer

Forecast. CBS Software, COM 64, APL II Series; TRS-80, Models III, 4.

Home Automatic Weather Station. Vaisala COM 64, VIC 20.

Weather Fronts. Cambridge Development Laboratory, APL II Series.

Weather Command. Educational Audio Visual, Inc., APL II Plus with 48K.

CHAPTER OVERVIEW

Several aspects of weather are studied in this chapter, such as water in the atmosphere and cloud formation, precipitation, air masses and the weather they cause, cyclonic and local storms. Weather prediction is also discussed.

It will probably be easy to get students involved in simple studies of weather. They have all seen weather maps in newspapers and probably wondered what they were all about. They also have watched television to get a weather prediction. While studying this chapter, students will find that they can make simple forecasts if they are given complete information. This necessary information is not usually available in newspapers.

13–1. WATER LEAVES THE AIR

BACKGROUND: In order for water vapor to condense from the atmosphere, air must be cooled and some solid surface must be present. When droplets of liquid water are produced, they are usually small enough to remain suspended to form clouds. Precipitation occurs when cloud droplets grow too large to remain suspended.

Condensation of water vapor from the air is an energy-releasing process. Thus cloud formation always provides a supply of energy that may be very destructive when released in violent storms.

MOTIVATION: Demonstrate how a cloud is made. You will need a one-gallon glass jug. Saturate the air in the jug by adding water and shaking. Pour the water out. Add condensation nuclei by using the smoke from an extinguished match. Lower the temperature to its dew point. This can be done by placing your mouth on the mouth of the jug, making a tight seal. Increase the air pressure in the jug by blowing into the jar while keeping an air tight seal. Remove your mouth quickly. This causes a sudden drop in temperature that should produce a cloud. Discuss the factors needed for the formation of a cloud. You can show the importance of condensation nuclei by repeating the demonstration, but leaving out the smoke particles.

Identify any local cloud types in terms of the simple classification scheme described in the text. Many localities have characteristic fog patterns that can usually be accounted for in terms of the local topography.

INVESTIGATION

Teaching Hints: For showing dew points, shiny aluminum cans can be obtained from supply houses, or soup cans with the paper label removed will work satisfactorily for one year. A couple of ice cubes per can should be sufficient. Students tend to allow too much dew to form, which takes a long time to remove. The ice should be removed as soon as the very first dew forms.

Safety: DO NOT ALLOW STUDENTS TO STIR WITH THER-MOMETERS. Thermometers should be removed while stirring. If glass stirring rods are not available, plastic soda straws work well.

Answers to Questions:

1. Answer depends on conditions.
2. Answer depends on conditions.
3. Answer depends on conditions.
4. Answer depends on conditions.
5. Answer depends on conditions.

Conclusions:

1. Surfaces would become wet with dew.
2. It makes the predicted height lower.

13–2. AIR MASSES

BACKGROUND: Development of air masses is discussed in some detail in order to account for their temperature and humidity characteristics. General movement of air masses across the continental United States is described along with weather characteristics produced. Weather fronts are identified as a feature of the boundaries of air masses.

Most local weather extremes, such as droughts and floods, can usually be explained in terms of changes in the usual movement of air masses.

MOTIVATION: Ask students to compare temperature and humidity of air trapped in mini-air masses. Compare the air trapped in the classroom with the air outside. A similar comparison can be made with large enclosed spaces inside buildings, such as shopping malls and sports arenas.

Discuss the origins, movements, and effects of the particular kinds of air masses that have the most noticeable effect on local weather and climate. Students might be asked to predict the local effect of specific changes in movement of air masses.

INVESTIGATION

Teaching Hint: A map for this Investigation is included as a master in the Teacher Resource Book, p. 13-5.

Answers to Questions:

1. Yes.
2. 13°, 19°, 21°, 23°.
3. 11°C.
4. 24°C.
5. Students might name southern Illinois or Indiana, or central Kentucky.
6. Students might name eastern Wyoming, western Nebraska, or northern Colorado.
7. Along the line separating the warm air mass and the cold air mass.
8. Cirrus clouds changing to stratus, gentle rain, and rising temperatures. Yes.

Conclusions:

1. It is necessary to have a large number of temperatures reported by stations across the map. Then, by joining stations having the same temperatures, lines can be formed.
2. A cold air mass has lower temperatures in the center, whereas a warm air mass has higher temperatures in the center.

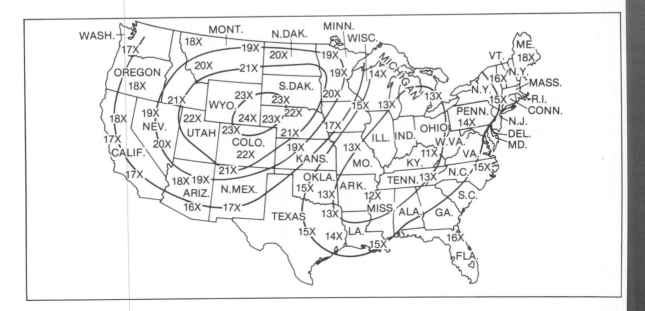

13-3. WEATHER CHANGES

BACKGROUND: This lesson deals with large-scale weather systems as well as more localized storms. Most weather changes in the continental United States result from the movement of cyclonic systems generated along the polar front. Development of these cyclones is treated in some detail. Thunderstorms and tornadoes are described as primarily local weather events.

MOTIVATION: Bring in newspaper or magazine articles on tornadoes and hurricanes, local if possible, to discuss.

SKILL-BUILDING ACTIVITY

Teaching Hint: Fig. 13–24 is provided in the Teacher Resource Book (p. 13–7).

Answers to Questions:

1. The path was northwesterly from the Atlantic, north of South America. It curved east of the coast of Florida and hit the coast about halfway between Savannah and Norfolk. It then headed south along the coast to West Palm Beach, where it turned out to the ocean in a northeasterly direction.
2. It reversed its path at the coastline between Savannah and Norfolk on day 5.
3. Savannah, Jacksonville, West Palm Beach, and Miami.
4. Over the sea off the southeastern coast.
5. On day 4, warnings to Savannah and Norfolk. On day 5, warnings to Savannah, Jacksonville, West Palm Beach, and Miami.

6. The path was westerly past San Juan, where it turned southwest along the southern coast of Cuba, turning northwest past Havana, then north until it hit the coast of Florida near Tallahassee.
7. Hurricane Wanda traveled faster.
8. Havana or Tallahassee.
9. On day 5, Tampa, Tallahassee, Mobile, and New Orleans.
10. Well inland past Tallahassee.
11. It will lose its strength and die out over land.
12. In the Northern Hemisphere, the winds blow counterclockwise around a hurricane.
13. The other source of energy is the energy released when water vapor condenses into liquid water.

Conclusions:

1. The hurricanes come in from the Atlantic Ocean, moving westward. They may turn northward, just to the east of Florida, then moving up the eastern coast, or just west of Florida, then moving inland from the coast of the Gulf.
2. They may send out warnings for people to guard against destruction of life and property.

13-4. PREDICTING WEATHER

BACKGROUND: The forecasting of weather is described in terms of the observations made and the instruments used for those observations. Details of the most common instruments are given. This lesson also discusses the information usually found on a weather map and shows how a forecast is made using it.

MOTIVATION: Display some of the weather-observing instruments you may have available. Ask students for

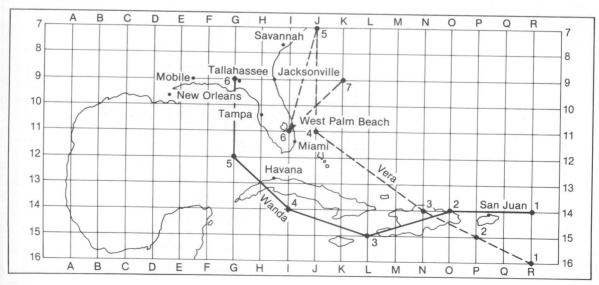

predictions made by TV weather personalities the previous evening. Compare the forecast with the actual weather. Discuss the probable source of the information used by the TV forecaster. (Probably the local weather station.)

SKILL-BUILDING ACTIVITY

Teaching Hint: It would be an excellent idea to bring in a comparable current newspaper weather map and have it xeroxed for individual student copies.

Answers to Questions:

1. Chicago, Detroit, and Washington.
2. Cold and warm fronts and low pressure are shown.
3. 24 hours.
4. About 800 km.
5. East.
6. Southeast.
7. Chicago and Detroit.
8. Great Falls, Reno, and Nashville.
9. Great Falls, Reno, Washington, and Nashville.
10. Great Falls, Reno, Washington, and Nashville will change from cloudy and rain to clear and cool.

Conclusion:

1. At least two weather maps are needed showing the movement of weather fronts and pressure systems.

TECHNOLOGY

Teaching Hints: For more information, write or call: NOAA/Satellite Data Services Division, Room 100, World Weather Building, Washington, DC 20233; 301-763-8111.

A magazine containing up-to-date material on this subject is *Weatherwise*, published by the American Meteorological Society, 4000 Albemarle Street NW, Washington, DC 20016; 202-362-6445.

COMPUTER

Teaching Hint: Ask your students to draw conclusions *on the basis of the data*. They may be tempted to go back to the text and repeat relationships as described verbatim. In order to draw conclusions, they will find it necessary to run a number of trials.

CHAPTER REVIEW

Vocabulary:

1. True 2. False, air mass 3. True 4. False, anticyclone 5. False, front 6. True 7. True 8. False, cyclone 9. False, cold front.

Answers to Questions:

1. Air may be lifted by rising over a mountain, by currents caused by uneven heating of the earth, or by warm air moving up over cold air.
2. Clouds may not form if solid surfaces are not present.
3. Stratus are low clouds, altostratus are 3–7 km in altitude, and cirrostratus are above 7 km.
4. Nimbostratus and cumulonimbus are rain clouds.
5. In warm areas cloud droplets grow and finally fall as rain. In cold areas the upper part of clouds are ice crystals, which grow by cloud droplets freezing on them until they drop as snow.
6. Both air masses have cool and moist air. The ones from the northern Pacific Ocean bring much snow and rain to the Pacific Northwest. They then move on to give mild, fair weather to the eastern United States. The ones from the North Atlantic invade the New England states, bringing summer rains and winter snows.
7. Warm, moist air masses come in from the South Atlantic, bringing warm and stormy weather to the central and eastern states. Dry, hot air masses from northern Mexico move northward to give hot, dry summers to the southwestern states.
8. Weather is a description of the conditions of the atmosphere at a particular place for a short period of time. Climate is a description of the weather conditions for a particular place over a long period of time.
9. Fronts are the result of warm and cold air masses meeting. The warm air rises over the cold air, producing condensation. The motion determines how severe the disturbance will be.
10. In a cyclone there is both warm and cool moist air. The warm air moves upward while cool air moves into its place in a clockwise motion. In an anticyclone, the air is cool and dry. The cool air moves downward and spreads outward and counterclockwise from the center.
11. Both storms are the result of circular moving winds originating from low pressure due to the upward movement of air. Both require condensing moisture to continue. A hurricane covers a huge area compared to the tornado and forms in the tropics over water, whereas the tornado usually forms over land as a result of thunderstorm activity.
12. Tornadoes are made from thunderstorm activity.
13. The wind vane gives the direction of the wind. The anemometer measures the speed of the wind.
14. Weather Stations use a rain gauge that allows 0.01 in. of rain to collect in a bucket. It then tips over to spill the collected water. This causes 0.01 in. rain to be recorded and at the same time another bucket tips into place to collect another 0.01 in.
15. A weather forecast tells where the fronts, lows, and highs will be located. Special forecasts give warnings of floods, tornadoes, large thunderstorms, and hurricanes. Some give frost warnings and pollution forecasts.

SECTION/ TIME NEEDED	MATERIALS (30 STUDENTS)	BASIC SCIENCE SKILLS	PACING GUIDE
14-1. The Earth Hidden Beneath the Sea (1–2 periods)	30 sheets graph paper pencils	Classifying (p. 388, 391) Problem Solving (p. 389)	Slow Learners: Have blindfolded students use a line and weight to measure the various heights of books placed on the floor. Compare this to the method described on p. 389. Above Average: Compare the ocean floor features and land features. Mainstreaming: Have students read summary and vocabulary before teaching this section.
14-2. Sea Floor Sediments (1–2 periods)	15 straws 15 razor blades 15 felt tip pens sea floor model (made from clay layers)	Classifying (p. 399) Inferring (p. 400) Sequencing (p. 400)	Slow Learners: Demonstrate how sediments form on the ocean floor. Put equal amounts of top-soil, gravel, chalk dust, and sand in a jar and shake. Place jar down and discuss what happens in the jar. Above Average: Have students read the history in diagrams of core samples. For example, a diagram could show from top to bottom, soil, volcanic dust, land plants, volcanic dust. Mainstreaming: Teacher should read or use tapes of key paragraphs for students to use as they read.

SUPPLEMENTAL MATERIALS

Audiovisual

Formation of Sedimentary Strata, EBE, No. S-80752, color filmloop.
Ocean Bottoms, BFA, Super 8, color.
Oceans and Their Histories, Creative Visuals, MH642327-2, color filmstrip.
Oceanography: A Voyage to Discovery, UEVA, color film (20 min).

Sediment Deposition, EBE, No. S-80743, color filmloop.
Sedimentation, Gordon Flesch, P85-0149/IS2, Super 8, color filmloop.

Print

Lambert, D. *The Oceans*. New York: Warwick, 1980.
Mann, C. "The Weather Tamers." *Science Digest*, December 1983.
Time-Life Editors. *Floor Of The Sea: Planet Earth Series*. Chicago: Time-Life Books, 1985.

CHAPTER OVERVIEW

This chapter deals with the ocean floor and its exploration. It covers sea floor features and how they might have been formed. A more detailed study of sea floor sediments and their sources follows. Sea floor sediments, caused by land erosion, build up and become elevated to mountains; then erosion starts again.

14–1. THE EARTH HIDDEN BENEATH THE SEA

BACKGROUND: This section deals with the sea floor and its principal features. The topography was revealed by sonar depth measurements. Submersible vehicles now allow direct observation of small areas of the deep sea floor. Plate tectonics came largely from the discovery of the mid-ocean ridge.

MOTIVATION: Use a globe or map to show the part of the earth's surface covered by water (about 70%).

INVESTIGATION

Teaching Hints: Echo-sounding data is used to determine the depth of the ocean and to prepare a graph that represents a profile map. After the profile map is made, the student concludes by labeling the ocean floor features found on it.

The student must use the scale for the axes as given in Table 14–2 in order for the ocean floor features to be obvious. The vertical axis should be numbered from top to bottom. "0" should be at the top. A student graph is in the Teacher Resource Book, p. 14-5.

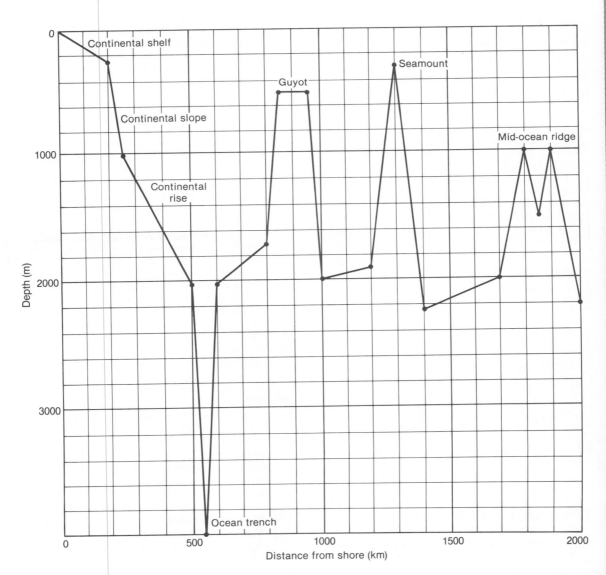

Answers to Questions:
1. The ocean bottom is deeper.
2. The ocean floor has many peaks and valleys.

Conclusions:
1. A sound is made by an echo sounder. The sound travels through the water and strikes the ocean bottom. The sound is then reflected back to a receiver on the ship. The time for the round trip is measured, multiplied by the speed of sound in water, and then divided by two.
2. See graph on page T77.

14–2. SEA FLOOR SEDIMENTS

BACKGROUND: This section discusses sea floor sediments, solids that enter the oceans.

MOTIVATION: Fill a 5 gallon pail with layers of humus, sand, clay, etc. and insert a pipe or core sampler.

SKILL-BUILDING ACTIVITY

Teaching Hints: Use single-edge razor blades or wood scalpels. You will need a felt pen marker if you use plastic soda straws. Fill a small (10 cm) aluminum pie pan with layers of colored clay or Play-Doh. Make some layers thicker than others. Make a list of the colors and thicknesses in the order you put them in. Students are asked about this order. Play-Doh should be kept moist on top.

Answers to Questions:
1. Answer depends on your sample.
2. Answer will describe the top layer.
3. Answer will describe the bottom layer.
4. Answer depends on your sample.
5. Answer depends on your sample.
6. The particles would be smaller in the deep ocean.
7. The top sample is tan, the bottom gray.
8. No. The samples differ greatly.
9. Any four of the following: remains of living things, stream-carried particles, windblown dust, volcanic dust, small pieces of meteorites.

Conclusions:
1. As materials enter the ocean, the heavier particles settle first, lighter next, and so on, forming layers.
2. The youngest is on top, oldest on the bottom. A layer is always younger than the one below it.
3. The sample could have come from a continental shelf, continental slope, or continental rise.
4. The larger rock pieces probably came from the continental shelf.

CHAPTER REVIEW

Vocabulary:
1. trench 2. continental slope 3. red clay 4. ooze 5. seamount 6. continental shelf 7. guyot 8. submarine canyon 9. abyssal plain 10. continental rise 11. coral

Answers to Questions:
1. Pacific, Atlantic, Indian, Arctic, Antarctic.
2. There is only one global ocean because all the oceans are connected to one other.
3. 1510 m/s $\times$ 3.2 s = 4,832 m; 4,832 m $\div$ 2 = 2,416 m.
4. Continental shelf, continental slope, continental rise, abyssal plain, mid-ocean ridge.
5. As a volcano erupts under the sea, it builds a tall mound of material. The mound of material is called a seamount. If the seamount grows tall enough to be above the ocean surface, it becomes an island. If the waves erode the top of the island flat and it sinks below the surface of the ocean, it becomes a guyot.
6. The two main sources are rivers, which bring rock fragments to the sea, and animals and plants in the oceans, which leave their remains as a sediment.
7. Ooze is the very fine sediment made of the remains of sea plants and animals. It is found where animal and plant life are abundant.
8. Hydrothermal vents are hot water springs found along the sides of the mid-ocean ridge. They are like oases because they supply minerals, energy, and food to support living things in the dark ocean.
9. Oil is mined and most of the world's fish are caught along the continental shelves.
10. A coral reef grows around a volcanic island. The island sinks and is slowly worn away. As the island disappears, the reef grows upward until it finally remains as a doughnut-shaped atoll.
11. The movement of crustal plates pushes the sea floor layers of sediments into the continents.
12. Sediments moving down the continental slopes erode the slopes to make canyons.
13. The sediments on the continental shelves are mainly the larger pieces of rock fragments brought to the sea by the rivers. The abyssal plains are made of the fine fragments carried to the oceans.
14. Red clay sediment is found over most of the deep sea floor. It is made up of the smallest particles carried to the oceans by rivers, windblown dust, volcanic dust, and tiny meteorite particles.
15. Sea floor sediments contain clues to the conditions on the earth in the past, such as the types of life and climate that existed and volcanoes that erupted.

SECTION/ TIME NEEDED	MATERIALS (30 STUDENTS)	BASIC SCIENCE SKILLS	PACING GUIDE
15–1. Sea Water (1–3 periods)	Investigation: 15 straws 15 rulers masking tape 1 lb salt water 1 lb modeling clay 15 tall containers (milk cartons) Skill-Building Activity: 15 ring stands 15 rings 30 styrofoam cups 15 thermometers 15 test tubes 15 rocks 15 graduated cylinders 15 beakers string 15 Bunsen burners 15 wire screens 15 balances 15 test tube holders	Cause and Effect (p. 410) Compare and Contrast (pp. 411, 415)	Slow Learners: Have students prepare artificial sea water by adding 35 g (5 level teaspoons) NaCl to 965 grams of water. Above Average: Have students test artificial sea water for NaCl using silver nitrate solution. Distill and test clear water obtained for NaCl. Mainstreaming: Have students classify the properties of sea water under composition, temperature, and pressure.
15–2. Resources from the Sea (1–2 periods)	15 ring stands 15 rings 15 wire screens 15 test tubes 15 Bunsen burners 15 flasks 15 beakers 15 clamps 15 graduated cylinders 15 pieces, glass tubing colored salt water (artificial sea water)	Cause and Effect (p. 417) Compare and Contrast (p. 424)	Slow Learners: Have students make a list of types of food from the sea. Above Average: Have students give oral or do written reports on protecting the ocean resources as discussed on p. 423. Mainstreaming: Have students classify the resources from the sea under the headings food, mineral, and energy.

SUPPLEMENTAL MATERIALS

Audiovisual

New Water for a Thirsty World, Sterling, color (22 min).
Understanding Oceanography, Hubbard, No. OFS-726, set of 6, color filmstrips.

Print

National Geographic Society. *Romance of the Sea.* Washington, DC: National Geographic Society, 1981.

Computer

Ecology Simulations-II. Creative Computing, COM Pet, APL II Series.

CHAPTER OVERVIEW

The physical composition, temperatures, and pressures of the sea are introduced. Its role as a life support system is explained. The importance of the sea as a source of food and of other resources is dealt with in this chapter.

15–1. SEA WATER

BACKGROUND: This section describes the composition of sea water and accounts for differences that exist in different parts of the sea.

The total amount of dissolved solids in the sea is enormous. It is estimated that if these solids were removed from the sea water, they would cover all the earth's land surfaces to a depth of 50 meters.

MOTIVATION: Trace a drop of water from the ocean to vapor, to condensation, to rain, to lakes and streams, to useful purposes, and back to the ocean. It has been estimated that water spends about 98 percent of its time in the ocean.

Ask students to suggest ways to distinguish sea salt from pure sodium chloride in the form of table salt. You should emphasize that the bulk of the dissolved substances in sea water is made up of sodium and chloride ions.

INVESTIGATION

The students make a hydrometer, which is capable of telling the difference between simulated sea water and tap water. In the process, the students learn that objects float higher in sea water than in fresh water.

Teaching Hints: Modeling clay should be soft for best results. Do not use Play-Doh, since it is water soluble. A liter (quart) container, such as a milk carton, must be used for the tall container in order for the ocean water solution to be approximately 3.5% in part E. One level teaspoon of regular table salt weighs approximately 7 g. Since accuracy is not critical, you could have the students measure 5 teaspoons of salt for the sea water. Provide a container for the disposal of the salt water. This salt water can be used again in the next Investigation or in the Investigation in Chapter 16, Section 1.

Answers to Questions:
1. Readings will vary.
2. Readings will vary, but the reading should be higher than the reading in question 1.
3. In the ocean.

Conclusions:
1. The dissolved salts make objects float higher in water.
2. In the Great Salt Lake.

SKILL-BUILDING ACTIVITY

Teaching Hint: Have the students use the scientific skill of comparing and contrasting in the study of heat absorption of water and rock. The conclusions require the student to explain why land cools faster at night and heats faster in the daytime than a body of water.

Answers to Questions:

The data for the table in the activity will vary from group to group but will show that, for the same temperature, the same size rock sample as the water sample will change the temperature of water less.

Conclusions:
1. The temperature change of the water in cup A was less than in cup B.
2. Since the data indicated that the rock heated water less than an equal amount of water, rock apparently requires less energy to change its temperature than water and thus will heat faster than water, given the same amount of sunlight.

15–2. RESOURCES FROM THE SEA

BACKGROUND: The importance of the sea as a source of food, minerals, fossil fuels, and water increases as the world's growing population strains the productivity of the land. The concept of the food chain in the ocean can be made clearer by comparing it to the food chain involved in the production of a common food such as meat.

The political problems related to the peaceful development of marine resources are a continuous area of negotiation and discussion between nations.

MOTIVATION: Ask the class where they could find an ounce of gold. One answer would be in 20,000 tons of sea water. However, they would have to extract the ounce of gold dissolved in it. Also, some method would have to be found to dispose of the 7,000 tons of other solids present.

INVESTIGATION

Artificial seawater is distilled using a simple air-cooled apparatus. The original sea water and the distillate are compared to assure that the salt did not carry over with the distillate.

Teaching Hints: Use 35 grams of table salt in 1 liter of water for the sea water. Add enough food coloring to give a definite color. The color will not distill into the test tube. The length of the glass tubing will depend on your apparatus. It should be long enough to have the beaker away from the heat of the burner and on either the table or a firm support. For better results, have the students add an ice cube or two to the beaker. Bend the glass ahead of time and insert the short end into the stopper. Use glycerine or oil to lubricate the tubing when inserting it into the stopper. If boiling chips or beads are available, you may want to have the students add them to their flasks during the distillation process. This will help prevent the boiling over of the salt water into the test tube.

Answers to Questions:
1. It has lost its color and taste.
2. Pure water.
3. Answers willl vary from a cloudy liquid to a bluish-white solid.
4. The salt and coloring.
5. It is less.
6. Some of the water escaped as steam.
7. Yes.
8. It tastes salty.

Conclusions:
1. A solution is boiled in a flask that has a glass tube leading from it. As the liquid boils away, the water vapor cools down in the glass tube, turning back into a liquid. The liquid is collected in a container.

2. The sea water tasted salty and had a color, whereas the water in the test tube was colorless and had no salty taste.

CHAPTER REVIEW

Vocabulary:
1. upwelling 2. krill 3. pack ice 4. salinity
5. thermocline 6. plankton 7. zooplankton
8. nodules 9. phytoplankton

Answers to Questions:
1. Sea water has about 300 times the salinity of drinking water.
2. The salinity is 56 (g of dissolved salts).
3. Chloride ions, 55.2%; sodium ions, 30.4%.
4. The salinity of sea water is greater at the equator than it is near the poles.
5. The sea is salty because of the ions that are being added to the water at the mid-ocean ridge, volcanic activity, and because of the dissolved materials from the rivers that empty into the sea.
6. The ocean surface layer is warm, lighted by sunlight, with abundant life and oxygen supply, whereas the deep ocean water is cold, dark, and has little oxygen and life.
7. The evaporation of water from the tropical seas adds moisture and heat energy to the atmosphere.
8. The oceans can provide more than twice the food because they occupy more than twice as much surface as the land.
9. Plankton is the beginning of the food chain in the ocean.
10. The best areas for fishing are where upwelling occurs.
11. Upwelling provides the minerals for abundant plankton growth.
12. Gold is not taken from the sea because the processs costs more than gold is worth.
13. The nodules contain manganese, iron, copper, zinc, silver, gold, and lead.
14. Energy may be taken from the sea as petroleum and by the motion of tides, winds, and waves.
15. The biggest problem is deciding who owns the ocean resources.

SECTION/ TIME NEEDED	MATERIALS (30 STUDENTS)	BASIC SCIENCE SKILLS	PACING GUIDE
16–1. Underwater Currents (1–2 periods)	30 beakers, 250-mL 15 balances sodium chloride (salt) food coloring water	Cause and Effect (pp. 429–431)	Slow Learners: Place food color or a crystal of potassium permanganate down one side of a beaker of water. Place a low flame directly below this side. Above Average: Have students research upwellings and their importance for sea life and food. Mainstreaming: Have students read the summary and study the margin vocabulary.
16–2. Surface Currents (1–2 periods)	15 sheets carbon paper 15 pcs cardboard 15 sheets paper 15 ball bearings (1 cm or larger) 15 pins pencils	Cause and Effect (pp. 434–435)	Slow Learners: Use a wave tank to show surface waves. Above Average: Obtain from travel agencies brochures describing beaches, e.g., of the Bahamas, Florida, or California. Mainstreaming: Have students draw the different surface currents on a globe or world map.
16–3. Waves (1–2 periods)	15 pans 15 straws food coloring water source	Observing (p. 441)	Slow Learners: Demonstrate a wave using a rope or a Slinky. Above Average: Have students compare different waves by their height, wavelength, frequency. Mainstreaming: Use overheads to show the parts of a wave.
16–4. Giant Waves (1–2 periods)	30 rulers 30 pcs. graph paper pencils	Predicting (p. 448) Cause and Effect (p. 449)	Slow Learners: To show how a tsunami forms, have students fill a rectangular pan half-full of water. Tilt one end of the pan slightly and tap the deep-water end. Above Average: Predict the times of tides from a position of the moon, earth, and sun. Mainstreaming: Have students make diagrams similar to the figures in this section.

SUPPLEMENTAL MATERIALS

Audiovisual

Growth of Sea Waves, BFA, Super 8, color filmloop.
How Level Is Sea Level? EBE, No. 2993, color (13 min).
Ocean Currents, SP, T825, 2 color/sound filmstrips.
Ocean Tides, EBE, No. 1497, color (14 min).
Orbital Motion Beneath Waves, BFA, Super 8, color.
Origin of Water Waves, EBE, No. S-80834, color filmloop.
Surface Currents on the Ocean, BFA, Super 8, color.
The Moon and the Tides, BFA, Super 8, color filmloop.
Turbidity Currents in the Sea, BFA, Super 8, color filmloop.
Waves on Water, EBE, No. 2253, color (16 min).

Print

Hsu, K. J. *The Mediterranean Was a Desert: A Voyage of the Glomar Challenger.* Princeton, NJ: Princeton University Press, 1983.

Computer

Shore Features. Cambridge Development Laboratory, APL II Series.

CHAPTER OVERVIEW

In this chapter the oceans are treated as dynamic, moving, ever-changing bodies of water. The sun is the major cause of ocean currents. The sun causes ocean currents directly by heating the water unevenly and producing density differences, and indirectly by causing the winds. How salinity and temperature differences cause currents are also discussed.

The patterns and paths of the surface currents are the subjects of section 2. Section 3 is concerned with waves. The topics discussed include the source of most waves, the way waves develop and move, and how waves behave on entering shallow water. *Wavelength, period, crest, trough, fetch, swell, breaker,* and *rip current* are described. Section 4 deals with the way tsunamis (sometimes improperly called "tidal waves") are caused, how they change, how they travel, and the effect they have on shore. The relation of tides to the sun and moon is discussed. The possible use of tides as an energy source is given some attention.

16–1. UNDERWATER CURRENTS

BACKGROUND: Near the sea surface, currents are caused almost entirely by winds. Deep currents are independent of wind action since they are usually caused by changes in density of the sea water or by movements of sediments down the continental slopes.

The speed of deep currents is connected to their cause. Density currents are generally slow, and in some cases are hardly noticeable, as in the slow motion of cold, bottom water away from the poles. Other density currents may be fairly swift, as in the case of the Gibraltar current.

Turbidity currents can achieve very high velocities for short periods of time.

MOTIVATION: Ask students to imagine that they have been set adrift in the middle of the ocean, on a raft without a sail. Would the raft remain stationary? Most will say that surface currents will carry the raft in some direction. Now ask the class to imagine being submerged in a submarine without power. Would the submarine move? Leave the question open until students have studied this section and have an awareness of the existence of deep ocean currents.

INVESTIGATION

Students make a mixture of two different-density solutions that do not immediately mix to study some of the phenomena that are related to density currents.

Teaching Hints: Small jars or small, clear plastic containers could be substituted for the beakers. Use dark food colors. Use the salt solution from previous Investigations if it has been saved. Balances are not needed if you supply teaspoons for measuring the salt. A liter (quart) milk carton could be used for making the salt water. In step D, the waters will tend to mix if students pour too rapidly, or if they do not wait until all the currents are gone in the tap water. With care, step F can be repeated a number of times to study the internal waves.

Internal waves are common in the oceans. They are thought to be responsible for the destruction of the submarine *USS Thresher*. Studies are still being done to understand this ocean phenomena better.

Answers to Questions:
1. The colored water stays separate on the bottom.
2. The colored water has a wave on it which travels back and forth between the layers of water.
3. Yes.
4. The internal waves are smaller, have a similar shape, and move slower than the surface waves.

Conclusions:
1. Saltwater.
2. Internal waves are formed between layers of unequal density when they move.

16-2. SURFACE CURRENTS

BACKGROUND: In this section, the surface currents of the North Atlantic are described in some detail to illustrate the general circulation in most ocean areas. The general circulation in the other ocean areas is also described. The Coriolis effect causes surface currents to be deflected from the direction of the wind causing the current. This deflection varies from about 15 degrees in shallow water to about 45 degrees in deep water. It is this deflection that is largely responsible for the basic circular motion of surface currents in most of the world's oceans.

MOTIVATION: Have students try to answer the question in the motivation for this section, p. 434.

INVESTIGATION

A model of the Coriolis effect is studied to see the direction in which motion of objects is affected in the Northern Hemisphere and in the Southern Hemisphere. This motion is then related to the circular motion of global ocean currents.

Teaching Hints: A large ball bearing works best. Small ones may not mark through the carbon paper. A piece of soft wood, ceiling tile, etc. would make an excellent substitute for the cardboard. A piece of masking tape placed where the hole is made through the paper will give more strength and minimize tearing as the papers are rotated. The Coriolis effect may also be demonstrated by the following: Chalk a slate globe. Spin the globe slowly in a counterclockwise direction as seen from above. Use an eyedropper to drop water on the earth. Move the eye dropper north and south. The path the drops follow shows the Coriolis effect. Also to illustrate the effect, a Lazy Susan tray can be turned while a straight line is drawn from the center to the edge.

Remind the students that the Coriolis effect works on anything that moves any distance on the earth in a northerly or southerly direction. It is an effect due only to the rotation of the earth and otherwise would not be present. It is often expressed as one proof that the earth does rotate on its axis.

Answers to Questions:
1. It is a straight line.
2. The track curves to the right.
3. The track curves to the right.
4. The track curves to the left.

Conclusions:
1. The rotating paper caused track no. 2 to be curved.
2. The paper was rotated in opposite directions for the tracks marked 2 and 3.
3. In the North Atlantic, the direction of rotation of the earth causes currents to curve to the right as

they move. In the South Atlantic, the direction of rotation of the earth causes currents to curve to the left as they move.

16-3. WAVES

BACKGROUND: Waves in the sea are responses to the addition of energy to the water. Ordinary waves seen on the sea surface are the result of energy transferred from the winds. The form and behavior of these waves is understood in terms of the circular motions of the individual water particles.

Waves tumble and break when their height exceeds one-seventh of their wavelength. Thus, very large waves can break in the ocean although this is likely to happen only when hurricane-force winds blow over a large expanse of sea.

MOTIVATION: Demonstrate waves generated on a rope or spring. Special slinkys and coil springs are available from science supply houses. Tie one end of the rope or spring to a fixed object (or have someone hold it still). Shake the other end to generate a pulse (wave).

INVESTIGATION

Using water in a shallow pan, the student produces waves by wind. A study of the waves produced is related to the fetch, how hard the wind blows, and the wavelength of the waves produced.

Teaching Hints: Any shallow dish or pan about 20 cm across will do. Dark food coloring is easiest to observe. No ruler was given to measure the wavelength. The idea is to get the student to relate the wavelength to the size of the container. You may tell them the diameter of the shallow pan.

Answers to Questions:
1. Small irregularly shaped and sized waves are made moving in the direction of the wind.
2. The fetch is the distance over which the air from the straw is blowing.
3. It sinks and begins to spread out.
4. Answers will vary. It usually moves in the water in the direction of the waves.
5. The food coloring represents the water below the surface.
6. The waves.
7. The wave returns from the side, moves back to the center, and passes on toward the other side.
8. The crest. It became larger.
9. The wavelength is equal to the diameter of the shallow container.

Conclusions:
1. By pushing on the surface of the water, the wind piles up some water, which moves off as a wave.

2. The constant movement of waves in a given direction will cause a surface current in that direction.
3. The energy came from the wind.

16–4. GIANT WAVES

BACKGROUND: Waves produced by winds have wavelengths that are relatively short. Energy added to sea water in two other ways produces waves that have very long wavelengths. One way is a sudden movement of the sea floor as in an earthquake. The other way is the result of the gravitational effect of the sun and the moon.

MOTIVATION: Demonstrate the production of waves in a large plastic pan partly filled with water. Ask students how a wave could be generated other than by blowing on the water. When someone suggests that hitting the container would cause a wave, show that it really does.

SKILL-BUILDING ACTIVITY

A table of data concerning water heights at different times in Boston Harbor is used for making inferences concerning the cause of tides.

Teaching Hints: You may wish to discuss the definition for "infer" by reading from a dictionary. There is little difference in science between "infer" and "conclude." Both are derived from facts or evidence through logical thinking. If you live on the seacoast, you may wish to obtain the tide listings from the newspaper and discuss them with the class. Be prepared to explain any discrepancies from the usual 6-hour cycle between high and low tides.

Answers to Questions:
1. 12 hours. 12 hours.
2. There is a connection between the moon and high tide.
3. The moon's gravity may be pulling the water up, making the tidal bulge.
4. High tide.
5. It would decrease the attraction.

Conclusion:
1. The attraction of the moon and the water on the moon side of the earth and the lack of attraction of the moon and water on the opposite side of the earth.

CHAPTER REVIEW

Vocabulary:
1. FALSE. *tide* 2. TRUE 3. FALSE. *swell* 4. FALSE.

breaker 5. FALSE. *fetch* 6. TRUE 7. FALSE. *current* 8. TRUE

Answers to Questions:
1. The sun's energy causes ice to melt, adding fresh water to the sea, thus making the sea water less dense. The sun's energy causes evaporation of fresh water from the sea, making the sea water more dense.
2. Cold water currents are most likely found near the sea floor. Warm water currents are found close to the sea surface.
3. Dense water flows out of the Mediterranean Sea near the bottom at the Strait of Gibraltar while the less dense ocean water moves in at the top. This current is caused by the increased density of Mediterranean sea water due to evaporation.
4. Turbidity currents carve out submarine canyons in the continental slopes. They also move large amounts of sediment from the continental margin to the deep sea floor.
5. The prevailing winds set the currents in motion. The Coriolis effect causes the currents to curve in a huge circular pattern.
6. The Gulf Stream is a warm current of water flowing out of the Gulf of Mexico between Florida and Cuba and up the east coast of the United States.
7. The Gulf Stream is caused by the North Equatorial Current. It splits into the North Atlantic Drift and the Canary Current in the North Atlantic.
8. A thick fog results from the collision of the Labrador Current and the Gulf Stream.
9. Surface currents may carry warm or cold water near shorelines, causing a different pattern of winds to develop and making a change in the climate of the immediate area.
10. The motion is an up-and-down circular motion.
11. The speed and wavelength decrease as the wave height increases until the wave tumbles forward.
12. In the past tsunamis have resulted in the deaths of thousands of people and much destruction of property.
13. In the open sea they have a wavelength of about 200 km and a wave height usually less than one meter. At the shoreline the height increases so much as to cause flooding.
14. The major cause of tides is the moon's gravity.
15. There are usually two high tides and two low tides in one 24-hour period. If a high tide occurs first, a low tide will follow in about 6 hours. Then a high tide will return in about 6 more hours, followed by a low tide in 6 hours.

CHAPTER 17 THE SUN

SECTION/ TIME NEEDED	MATERIALS (30 STUDENTS)	BASIC SCIENCE SKILLS	PACING GUIDE
17-1. Structure of the Sun (1-2 periods)	paper pencil	Cause and Effect (pp. 462-463)	Slow Learners: Have students use two index cards to view the sun. Put a large pinhole in one card. Hold the card with the pinhole above the other card so that the image of the sun will fall on the bottom card. Above Average: Have students do a library report on how sunspots can affect the earth. Mainstreaming: Show a sound filmstrip on the sun. Have students make a diagram of the sun similar to Fig. 17-1 (p. 460).
17-2. The Sun's Energy (1-2 periods)	15 showcase bulbs with sockets and cords 15 different gratings pencils paper	Compare and Contrast (pp. 468, 469)	Slow Learners: Use a prism to break up sunlight or white light into its colors. Above Average: Have students do a library report on fusion. Mainstreaming: Read the text to students, or use tapes of key paragraphs while students read along.

SUPPLEMENTAL MATERIALS

Audiovisual

Energy, Harnessing the Sun, Sterling, color film (25 min).
New View From Space, NASA, HQ214, color film (28 min).
Solar Flares, BFA, Super 8 color filmloop.
Solar Prominences, BFA, Super 8 color filmloop.

Print

Buckley, S. *Sun Up to Sun Down*. New York: McGraw-Hill, 1979.
Hecht, J. *Laser: Super Tool Of The 1980's*. Boston: Houghton Mifflin, 1984.
Smithsonian editor. *Fire of Life, The Smithsonian Book of the Sun*. New York: Norton, 1981.

Computer

Solar System Astronomy. Cambridge Development Laboratory, APL II Series.

CHAPTER OVERVIEW

The study of our sun in this chapter begins with the structure of the sun, giving some attention to sunspot activity. The connection between solar winds and auroras on the earth is also included. Some details are given on the way solar energy is generated, how it escapes, and the forms in which it reaches the earth.

17-1. STRUCTURE OF THE SUN

BACKGROUND: The nature of the sun allows us to explain and understand its full significance. The sun's diameter is 110 times the earth's. Its mass is 330,000 times that of the earth. The density is about one-fourth that of the earth, which suggests that gases exist on the sun. The sun has layers of different sizes and temperatures.

Disturbances on the sun affect the earth. The cycle of sunspot activity seems to affect plant growth. Auroras are the result of particles sent out by the sun interacting with the earth's magnetic field and gases in the earth's atmosphere.

MOTIVATION: Discuss the introductory paragraph concerning auroras with the class. Ask if anyone has seen "northern lights." Have them describe their experiences. Compare theirs with the description in the text and discuss differences. Do not try to explain why auroras occur until students have had a chance to read the section.

SKILL-BUILDING ACTIVITY

The Skill-Building Activity uses problem-solving skills to answer the question "Does the sun rotate?" By using sunspot positions, the student is able to determine that the sun does rotate and can derive a suitable period of rotation.

Teaching Hints: Most sunspots change or entirely disappear before they make one rotation. Some larger ones have been observed to travel around with the sun several times. Observations of these sunspots give an average of about 27 days for sunspots near the sun's equator to make one rotation. Sunspots closer to the poles require an average of about 33 days to make one rotation. In this activity, the fraction of rotation is close to one-fifth. This would give a 30-day period of rotation, which is very close since the sunspots appear to be about halfway between the equator and pole.

Answers to Questions:
1. You could watch permanent markings move.
2. Most students will conclude that the earth rotates, carrying the feature around once every 24 hours.
3. They move together.
4. Two days.
5. 28 days.
6. 28 days.

Conclusions:
1. The motion of sunspots shows that the sun's surface moves as though the sun rotates.
2. Particular sunspots move a given distance together in a certain number of days. An estimate can then be made as to how many days it would take for the sunspots to move completely around the sun one time.
3. That the sunspots move as a result of the rotation of the sun on its axis.

17-2. THE SUN'S ENERGY

BACKGROUND: The sun is composed of 74% hydrogen and 25% helium. The other 1% represents all the other elements in small amounts. The huge mass of the sun causes core temperatures that strip electrons from atoms, leaving only nuclei. Collision of nuclei at the core temperature causes fusion. Einstein explained the tremendous energy released during the fusion process as being accompanied by a slight loss in mass. Hydrogen fusion will continue for at least 5 billion more years.

Nuclear energy is produced only in the sun's core. Nuclear energy changes to radiant energy by absorption in the sun's outer layers. Radiant energy moves through empty space at 300,000 km/sec and takes about 8 minutes to reach the earth. Our senses detect infrared waves through the skin and visible light by the eyes. Ultraviolet waves cause the skin to burn or tan. The sun sends out all kinds of radiant energy. 90% of it is in the form of visible and infrared waves. Gases in the upper atmosphere absorb most of the other 10%, which consists of dangerous ultraviolet, gamma, and X-rays.

MOTIVATION: Most students will have some familiarity with sunlight. Discuss with them the idea that sunlight is made up of the colors of the rainbow. You could display the spectrum, using either a diffraction grating or a glass prism, and note the colors of the rainbow. Ask if there are other colors that we cannot see but may be able to feel. Point out that infrared light is detected by certain cells in the skin as heat. The Investigation on p. 474 will give the students more first-hand experience with the visible spectrum. You may wish to discuss some of the industries built around the production and detection of electromagnetic radiation (TV, radio, etc.).

INVESTIGATION

By use of a diffraction grating, the student studies the visible spectrum. By referring to Figure 17–12 on p. 472, the wavelengths of the shortest and longest wavelengths of visible light are estimated.

Teaching Hint: One single, unfrosted showcase bulb that is lighted and placed so that all the students can see it will work for the whole class. Inexpensive diffraction grating can be obtained and mounted between microscope slides. It can also be bought mounted. Label the top of each after determining that the sprectrum is spread out on the right and left sides as you look at the light.

Safety: The students will not be looking at the sun in this activity, but you should still caution them that to do so would cause injury to eyes.

Answers to Questions:
1. Violet, blue, green, yellow, orange, red.
2. The rainbow.

3. About halfway between 10^{-7} and 10^{-6} m.
4. Violet.
5. The colors repeat several more times.
6. Red.

Conclusions:

1. Violet, blue, green, yellow, orange, red.
2. The longest is about 10^{-6}, the shortest is about halfway between 10^{-6} and 10^{-7}.
3. Ultraviolet is just outside the short wavelength side of visible light.
4. Infrared is just outside the longest wavelength of visible light.

TECHNOLOGY

Teaching Hint: A magazine that will be sure to publish any information on changes in the progress of the Hubble Space Telescope is *Sky and Telescope*, published by Sky Publishing Corp., 49 Bay State Road, Cambridge, MA 02138; (617) 864-7360.

COMPUTER

Chapter 17   Shadow Moves
Teaching Hint: Examples of REM statements would be:
   145 REM Enter latitude position
   175 REM Enter length of stick
   295 REM Enter Data statements

CHAPTER REVIEW

Vocabulary:
1. corona 2. sunspot 3. radiant energy
4. prominence 5. photosphere 6. chromosphere
7. solar flare 8. aurora

Answers to Questions:

1. The sun's diameter is 110 times the earth's, its mass is 333,000 times that of earth, and its density is only one-fourth that of the earth.
2. The matter in the sun is mostly a mixture of bare atomic nuclei and free electrons called plasma.
3. The photosphere is the layer of the sun that produces visible light.
4. The corona is the outermost layer of the sun. It surrounds the sun like a halo. It has no outer boundary; some of its material escapes the sun, forming the solar wind.
5. Looking directly at the sun can cause permanent damage to the eyes.
6. There seems to be a similar cycle in the pattern of tree rings to that found in the cycle of sunspots.
7. Two shields protect us from the solar wind, the atmosphere and the magnetic field of the earth.
8. The solar storm may cause auroras and also interfere with radio communications.
9. If the sun made its energy from burning coal, it would last only 6,000 years. The sun must be as old as the earth; therefore, it would have burned up long ago.
10. The fusion process combines hydrogen nuclei to produce helium nuclei. In the process, some mass is converted to energy.
11. The equation says that a very small amount of matter can disappear and be replaced by a large amount of energy.
12. Radiant energy is the result of the gases in the photosphere absorbing energy from the sun's interior and converting it to radiant energy.
13. Radio, television, and radar waves are commonly made on earth.
14. We can see the visible light and feel the infrared as heat.

SECTION/ TIME NEEDED	MATERIALS (30 STUDENTS)	BASIC SCIENCE SKILLS	PACING GUIDE
18-1. The Inner Planets (1-2 periods)	30 rulers 60 straight pins 30 8 × 10 cardboard string paper pencils	Compare and Contrast (p. 481)	Slow Learners: Show filmstrip on the inner planets. Above Average: Discuss how the conditions on Venus could occur on the earth. Mainstreaming: Have students identify the inner planets from photos.
18-2. Outer Planets (1-2 periods)	30 rulers 30 pencil compasses 30 sheets unlined paper pencils	Compare and Contrast (pp. 489, 492)	Slow Learners: Show filmstrip on the outer planets. Above Average: Discuss the pros and cons for spending money on exploring the solar system. Mainstreaming: Have students make a poster of the solar system.
18-3. Smaller Members of the Solar System (1-2 periods)	30 rulers 60 straight pins 30 8 × 10 cardboard string paper pencils	Classification (p. 498) Cause and Effect (p. 498)	Slow Learners: Use newspaper or magazine articles on Halley's comet. Above Average: Have students research and report on the Cort Cloud. Mainstreaming: Show a sound filmstrip on asteroids, meteors, and comets.
18-4. The Moon (1-2 periods)	pencils paper	Compare and Contrast (pp. 502, 503)	Slow Learners: Use a filmstrip on the moon. Above Average: Compare the conditions on the moon with those on the earth. Mainstreaming: Using photos of the moon's surface, have students make a list of the conditions on the moon.
18-5. The Earth and Moon (1-2 periods)	15 balls, half-painted pencils paper	Cause and Effect (pp. 509, 510)	Slow Learners: Use a light source (sun), small ball (moon), large ball (earth) to demonstrate phases and eclipses. Above Average: Have students research superstitions on phases, e.g., full, harvest, hunter's, planter's moon. Mainstreaming: Have students make diagrams showing how phases and eclipses occur.

SUPPLEMENTAL MATERIALS

Audiovisual

Mars—Is There Life? NASA (HQ263), film (14 1/2 min).
Mars and Jupiter, BFA, Super 8 color filmloop.
Our Solar System, NASA (HQ234), film (4 1/2 min).
Scale Model of Solar System, BFA, Super 8 color.
Solar System, Edwin Shapiro Co., captions, filmstrip.

Print

Cooper, H. F. S., Jr. *The Search for Life on Mars: Evolution of an Idea.* New York: Holt, Rinehart and Winston, 1980.
Gore, Rick. "The Planets: Between Fire and Ice." *National Geographic,* January 1985.

Henbest, N. *The New Astronomy.* New York: Cambridge University Press, 1983.
Hoyt, W. G. *Planets X And Pluto.* Tucson, AZ: University of Arizona Press, 1980.
Kopal, Z. *The Realm of the Terrestrial Planets.* New York: Wiley, 1979.
Menzel, D. H. and J. M. Pasachoff. *A Field Guide to the Stars and Planets.* Boston: Houghton Mifflin, 1983.
Muirden, J. *Astronomy Handbook.* New York: Arco, 1982.

Computer

Introduction to Science Series. Focus Media, Inc., APL II Series.
Planetarium Computer. Focus Media, Inc., APL II Series.
Solar System: The Discovery of Pluto. Tandy, TRS Color Computer.
Titan Experience. Muse Software, APL II Series.

CHAPTER OVERVIEW

The planets, asteroids, comets, and meteors as well as the moon and its relation to the earth are the subjects of this chapter. The inner planets are treated separately from the outer planets as are the smaller members of the solar system. An expanded section on the earth-moon system is included.

18–1. THE INNER PLANETS

BACKGROUND: Modern space explorations have changed some previous ideas about the solar system and added vastly to what is known. About 99% of the mass of the solar system is contained in the sun. Billions of smaller bodies orbit the sun.

Conditions on the planets of the solar system vary widely, depending on their closeness to the sun and size. Similarities exist between the inner planets, which are different from the outer planets.

Mercury is difficult to see since it is too near the sun. Information about it is gathered from flybys of spacecraft. It has no atmosphere and extreme temperatures ($-170°C$ to $+430°C$).

Venus has a thick cloud layer that keeps the surface from our direct view. The atmosphere is mostly carbon dioxide. Surface temperatures of $475°C$ occur.

Earth is also an inner planet. The atmosphere is mostly nitrogen. 70% of its surface is covered by water. The earth's surface temperature averages about $15°C$.

Mars has received more attention than the other inner planets. The red color of Mars is caused by soil that is rusted. Mars has been studied since Galileo first saw the "canals" on its surface.

MOTIVATION: Discuss answers to "Why find out what the other planets are really like?" 1. It is the inquisitive nature of people, and one reason that they investigated the New World in 1492. 2. Refinement of space technology gives many spinoffs such as inexpensive small computers and some new medicines. 3. Technology exists to explore all the planets with pilotless probes.

INVESTIGATION

The students draw scale models of the orbits of the inner planets. They then compare and contrast the orbits while describing them.

Teaching Hints: Kite string is excellent to use here. Have scissors available for each ball of string. Soft pine wood pieces about the size of the paper work best for anchoring the pins. The Investigation in section 3 involves this same method for drawing the orbits of some asteroids and a comet.

Conclusion:
1. Answers may vary. Most will describe the orbits as being circular, although they are in fact an ellipse. A circle cannot have more than one center.

18–2. OUTER PLANETS

BACKGROUND: The planets from Jupiter outward to Pluto are exceedingly far from earth. The Voyager spacecraft have given us close-up pictures of Jupiter and Saturn. The giant planets, Jupiter, Saturn, Uranus, and Neptune are much larger than the inner planets, often have many satellites.

Jupiter is about 2 1/2 times the mass of all the other

planets put together. It is eleven times the diameter of earth but has a density that is only one-fourth that of the earth. It gives off more heat than it receives, has a magnetic field, takes 12 years to orbit the sun, rotates every 10 hours, rotates faster at the equator than at the poles (suggesting it is not solid), is 95% hydrogen and helium. Jupiter has a temperature of −130°C at the top of its clouds. Jupiter moons have received a lot of attention lately. Galileo discovered the four largest, which can be seen with a low-power (20×) telescope. Some of the moons have large areas of ice.

All other outer planets, except Pluto, are similar to Jupiter. Saturn has many rings whereas one faint one is found around Jupiter. Saturn has 17 satellites in addition to rings. Saturn orbits the sun about every 29 years. Its rotation rate is about 11 hours. Uranus and Neptune are similar and are about 3 1/2 times the diameter of the earth, appear to have solid cores covered with ice and liquid hydrogen. Uranus has 5 moons. It is considered to be an "oddball" planet. It spins on its axis nearly level with its orbital plane. Neptune has a mass 17 times that of the earth. It has one moon. Pluto may consist entirely of ice. Pluto's orbit is tilted out of the pattern of the other planets and is now inside the orbit of Neptune. It may have, at one time, been a satellite of Neptune. Pluto has one satellite.

MOTIVATION: Discuss the implications of reliance on instruments for making observations in space as well as on earth to explore the ocean depths.

INVESTIGATION

The students draw models of the size of each planet in the solar system and, using the same scale, get an idea of the distances planets are from the sun.

Teaching Hints: You may wish to furnish pieces of cardboard the size of the unlined notebook paper to assist in drawing the planet model circles. Using adding machine paper and a scale of 1 cm = 10,000,000 km, students could make a position model of the solar system about 18 feet in length. They could then get an idea of the vast space found in the region of the outer planets.

Answers to Questions:
1. 1400 cm. (14 m)
2. Jupiter, Saturn, Uranus, and Neptune.
3. Mercury, Venus, Earth, Mars, and Pluto.
4. No.
5. 1,400,000 cm. (1400 km)

Conclusion:
1. In order to make a reasonable size model for distance and a big enough size for the planets, two different scales must be used.

18–3. SMALLER MEMBERS OF THE SOLAR SYSTEM

BACKGROUND: Asteroids occur mostly in a zone between Mars and Jupiter. It has been suggested that they are the remnants of a planet that never formed. Some asteroids out of this region actually cross the earth's orbit. They range in size from a few kilometers to about 1,000 km in diameter.

Meteors are very small and most are less than 1 gram in mass. They enter the earth's atmosphere and burn up at altitudes near 100 km. Those that reach the ground are called meteorites.

Comets are icy bodies that orbit the sun, moving near and through the outer edges of the solar system.

MOTIVATION: The UFO controversy is one of the few earth phenomena that has defied explanation through centuries. Display photos of UFO's and discuss how these solar system objects are explained. (More than 90% are explained as natural phenomena.) How will we ever know what UFO's are?

INVESTIGATION

Students draw the orbits of some asteroids and a comet. These paths are then compared and contrasted to the paths made by the inner planets from the Investigation done in section 1.

Teaching Hints: If the Investigation in section 1 was not done, you will want to do it before doing this one. The comet path should be drawn on the opposite side or on another sheet so the student will not get the wrong impression that the scales are the same. The solar objects were chosen to give a variety of asteroid orbits. You may wish to provide information for more orbits, see the section on physical data for the planets, their satellites, and some asteroids in a current handbook of physics and chemistry for this information.

By 1975, nearly 10,000 artificial objects had been placed in orbit in a narrow band about the earth. About 40% of these still orbit the earth, plus all those that have been sent up since 1975. Thus, there are about three times as many earth satellites in this narrow belt about the earth as the known asteroids in the 200 million-kilometer-wide asteroid belt.

An Apollo class asteroid, designated 1976AA, has a period of 347 days, perihelion of 118 million km, an aphelion of 171 million km, and will pass close to earth in 1996. The earth's average distance from the sun is 150 million km. It has been calculated that this asteroid has 3 out of 4 chances of striking the earth within the next 24 million years. If it did, it would make a crater about 20 miles across.

Conclusions:

1. The typical asteroid.
2. Apollo.
3. Hildago.
4. The path of Halley's comet is much flatter, passes closer to the sun, goes farther from the sun, and is very much longer than the path of a typical asteroid.

18–4. THE MOON

BACKGROUND: In the past few decades, explorers have landed on the moon and finally provided the basis for a scientific understanding of the moon's true nature. This section deals with the modern scientific view of the moon.

The mass of the moon is much less than that of the earth. Its diameter is also much smaller, about one-fourth that of the earth. This means that gravity on the moon is less than on earth, about one-sixth as great. This low gravity is the main reason for the lack of atmosphere on the moon. The moon and the earth probably have very similar histories. Any information about the moon may give a clue about the earth's past.

MOTIVATION: A quarter and a paper disk from a paper punch represent the relative sizes of the earth and the moon. Tape them to the chalkboard and ask the class what they think the objects are supposed to represent.

INVESTIGATION

A speculative exercise is done on listing materials that are important to survival on the moon's surface. This is an exercise originated by NASA.

Teaching Hints: You may wish to have students work in groups of three as a crew. The crew could work together on the list. Crews could be named and a competition for the best survival score encouraged. Whether students work in crews or individually, have them place their scores on the chalkboard. Rank them and discuss any questions.

Provide students with the following astronaut list either by duplication or writing on the chalkboard at the proper time. Reasons for rank are those provided by NASA.

1. Two 50-kg tanks of oxygen (for breathing).
2. 20 L of water (replenish body loss).
3. Star chart (principal means of finding direction).
4. Four packages of food concentrate (daily food requirements).
5. Solar-powered radio (signaling and receiving).
6. 20 m nylon rope (climbing, securing packs, etc.).
7. First-aid kit (injury or sickness).
8. Large piece of insulating fabric (shelter from sun).

9. Three signal flares (location marker when in sight of base ship).
10. Two 45-caliber pistols (useful as propulsion devices).
11. Case of dehydrated milk (nutrition source).
12. Portable heating unit (useful only on dark side of moon).
13. Flashlight (useful only on dark side of moon).
14. Magnetic compass (useless since moon has no magnetic poles).
15. Box of matches (little or no use on moon).

Answers to Questions:

1. See 1, 2, and 3 of the previous list.
2. See 13, 14, and 15 of the previous list.
3. Answers will vary.
4. Answers will vary.
5. Answers will vary.
6. Answers will vary.

Conclusions:

1. The moon lacks oxygen, water, and food.
2. In order to survive on the moon, you would need a supply of oxygen, water, and food.

18–5. THE EARTH AND MOON

BACKGROUND: It is the changing appearance of the moon and its monthly schedule of motions across the sky that is most impressive to ordinary observers on the earth. The explanation for these observations lies in the relative motions of the earth, moon, and sun. These motions are the topic of this section.

Various calendars that have been used by people through the ages are almost always related to the cycles of the moon. The confusing features of the present calendar may need to be cleared up.

Eclipses of the sun and moon are discussed in detail.

MOTIVATION: Using a basketball to represent the earth and a tennis ball for the moon, draw a "sun" on the board. While moving the moon around the earth, ask what phase the moon would be in at various locations. Or ask what the moon would look like at various places.

INVESTIGATION

Teaching Hints: The drawing of the sun should be large enough and located where it can be seen from all parts of the room. For the moon models, 7.5-cm (3-inch) Styrofoam spheres work excellently. The first students to use them could darken one-half of the sphere by using a felt pen or other suitable black ink source. Any ball or sphere will work, but Styrofoam doesn't bounce!

While the students have everything set up, you can convince them that the moon rotates once for each revolution around the earth. Have them hold the moon model so the white side is always toward them. Remind them that we know that we always see the same side of the moon. Use the sun as a fixed reference point, then have them turn one quarter of a revolution at a time and note that the moon rotates one-quarter of a turn with respect to the sun. When the moon has revolved around the earth one time, it will have also rotated one time on its axis.

Answers to Questions:

1. Full moon.

2.

3. Quarter moon (last quarter).

4.

5. New moon

6.

7. Crescent moon.

8.

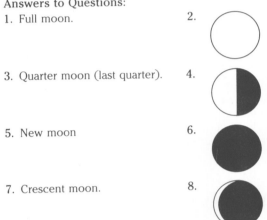

Conclusion:

Drawing should look similar to Figure 18–31 (p. 518) with the phases properly labeled.

TECHNOLOGY

Teaching Hints: Hypotheses concerning the possible extinction of certain life forms during the history of the earth have been part of the recent popular science literature and are more likely than more scholarly subjects to have reached student notice. As a project, students could be encouraged to follow up this research, reviewing current science magazines. Results of their research can provide material for discussion about the values and perspectives of different scientific source materials.

COMPUTER

Teaching Hints: Have several groups of students compose the definitions. Each set of definitions could be entered as a separate program. (Give each a different name.) Students should run the program containing other students' definitions. This provides more interest as well as a check on the accuracy and clarity of student communications.

CHAPTER REVIEW

Vocabulary:

1. asteroid 2. Moon phase 3. maria 4. comet
5. eclipse 6. new moon 7. meteor 8. meteorite
9. gibbous 10. sister theory 11. ray 12. ellipse

Answers to Questions:

1. The sun is the controlling body of the solar system because of its huge mass and gravity.
2. The group of planets near the sun are composed mainly of rocky material that could withstand the very hot conditions found early in the solar system's history. The outer planets, being much farther from the sun, formed from the gases and ices that could exist in the very cold conditions.
3. Mercury is the smallest inner planet. The earth is the largest inner planet.
4. Venus, Earth, and Mars have atmospheres. Atmospheres cause weathering and erosion of surface features that does not occur on planets with no atmosphere.
5. The extra energy comes from shrinking.
6. The outer cloud layer is made up mostly of hydrogen and helium gas separated into bands. There are also small amounts of water, ammonia, methane, and other compounds from the lower layers, which give the cloud cover color. Toward the interior the gases become thicker, blending into an ocean made mainly of liquid hydrogen. Near the center, the liquid hydrogen becomes similar to a metal. Jupiter probably has a small rocky core at its center.
7. Io, Europa, Ganymede, and Callisto.
8. The rings are divided into three large bands, each of which is made of many smaller rings and particles.
9. Uranus is tipped over on its side with its north pole pointing toward the sun.
10. The asteroid belt is located between the orbits of Mars and Jupiter.
11. Comets are made of ices, rocks, and dust that come from a cloud surrounding the solar system.
12. One kind of meteorite is made of rock; the other kind is made of iron mixed with nickel.
13. Mars' moons are probably two captured asteroids.
14. With a pair of binoculars, you can see rugged mountain ranges, wide level plains, and hundreds of large and small craters on the moon.
15. The fine dust came from the breakup of the surface by meteorites.
16. The sister theory suggests that the moon and earth are twin planets formed at the same time.
17. We see the same side of the moon because the moon completes one orbit and it rotates once in the same time period.
18. A partial eclipse of the sun occurs when the outer part of the moon's shadow hits the earth.

CHAPTER 19 STARS AND GALAXIES

SECTION/ TIME NEEDED	MATERIALS (30 STUDENTS)	BASIC SCIENCE SKILLS	PACING GUIDE
19-1. Observing Stars (1–2 periods)	1 lb modeling clay 15 rulers pencils chalk chalkboard	Observing (pp. 521, 526)	Slow Learners: Have students study different constellation legends. Above Average: Show students how to use a star finder. Mainstreaming: Outline the ways distances to stars are measured.
19-2. The Life of a Star (1–2 periods)	15 rulers 15 batteries (size D) 15 1.5-V flashlight bulbs 15 light sockets 15 diffraction gratings 15 steel wire, no. 30, uncoated, 3 m long masking tape black paper	Classifying (p. 530)	Slow Learners: Using colored construction paper, have students make models of the different kinds and sizes of stars. Above Average: Have students research unusual stars, e.g., neutron stars, black holes. Mainstreaming: Have students make a poster of the different types of stars, similar to Fig. 19–11.
19-3. Galaxies (1–2 periods)	none	Comparing and Contrasting (p. 540)	Slow Learners: Discuss the time needed to travel between stars and galaxies, at the speed of light. Above Average: Have students study the size and origin of the universe. Mainstreaming: Show a sound filmstrip on galaxies.

SUPPLEMENTAL MATERIALS

Audiovisual

Astronomy, Millikan, set of 6, sound duplicating masters, color filmstrip.

How Many Stars? MIS color film (11 min).

The Stars, Creative Visuals, MH 396020, set of 6 color filmstrips.

The Universe, Creative Visuals, MH 396027, set of 6 color filmstrips.

The Universe, Edwin Shapiro Co., captions, filmstrip.

Mystery Of Stonehenge, McGraw-Hill, color (57 min).

Print

Bartusiak, M. "The Planet Hunters." Science Digest, January 1984.

Clark, D.H. Superstars. New York: McGraw-Hill, 1984.

Durham, F. and R. D. Purrington. Frames of the Universe. New York: Columbia University Press, 1983.

Gold, M. "The Cosmos Through Infrared Eyes." Science '84, March 1984.

Gore, Rick. "The Once and Future Universe." National Geographic, June 1983.

Hoyle, F. The Intelligent Universe. New York: Holt, Rinehart and Winston, 1984.

Moore, P. Travelers In Space and Time. New York: Doubleday, 1984.

Overbye, D. "Cosmic Winter." Discover, May 1984.

Provenzo, E. F., Jr. and R. D. Purrington. Rediscovering Astronomy. La Jolla, CA: Oaktree, 1980.

Reeves, H. Atoms of Silence, An Exploration of Cosmic Evolution. Cambridge, MA: MIT Press, 1984.

Sullivan, W. Black Holes: The Edge of Space, the End of Time. Garden City, NY: Doubleday, 1979.

Tauber, G. E. Man's View Of The Universe. New York: Crown, 1979.

Computer

Argos Expedition. CBS Software, COM 64.
Astronomy Keyword. Focus Media, APL II Series; TRS-80, Models III, 4.
Astronomy: Stars for All Seasons. Educational Activities, APL II Series; TRS-80, Models I, III, 4; COM 64.
Project Space Station. HesWare, APL II Series; COM 64; IBM PC, PC Jr.
Stellar Astronomy. Cambridge Development Laboratory, APL II Series.

CHAPTER OVERVIEW

A complete understanding of the nature of the earth must include its role not only as a member of the solar system but also as a member of the universe. Starting from the view of the night sky as seen from the earth, the student finds out how the stars are mapped, what makes up a galaxy, and how the distances to stars and galaxies are measured. Observation of brightness and calculation of surface temperature lead to a pattern that suggests a certain life history that applies throughout the universe.

19–1. OBSERVING STARS

BACKGROUND: There is a lot of evidence that people have studied the heavens since ancient times. Ancient stone monuments show sun positions. Ancient viewers gave names to stars and groups of stars. Astrologers through the ages have explained the past, the present, and the future by numbers related to sky patterns. Ancient navigators determined positions by star location.

MOTIVATION: Organize a star-viewing party. A couple of telescopes and an interest in doing it are the starting place. A hobby could result. Use a current issue of *Sky and Telescope* to determine what is visible.

INVESTIGATION

Using their eyes, students perform a number of experiments that show how parallax is used in judging distances. The optimum conditions for using parallax are derived, and its limitations are determined.

Teaching Hints: Place vertical marks, near the top of the chalkboard, at 10 cm intervals for at least 2 meters across the board. Number these consecutively. Make a smaller mark midway between each of the numbered marks. If the chalkboard is not properly positioned to be seen by all, the marks could be placed on adding machine tape and attached to the wall.

The distance to the moon was first measured by parallax in 1752. Two observations were made, one in Germany and the other 8,050 km away in South Africa. The distance to the moon was determined to be 386,400 km.

Answers to Questions:
1. The pencil shifted to the left and no longer lined up with the object on the wall.
2. The same thing happened as in question 1.

Conclusions:
1. There was less parallax farther away from the pencil.
2. Move to the left or right. If the object does not appear to move against the background, it must be far away.
3. You must compare the position of a nearby object with an object in the distant background from two different locations.

19–2. THE LIFE OF A STAR

BACKGROUND: How do we get information to answer questions like: "If our sun is using up its mass and converting it to energy, will it eventually die out?" Starlight gives information about stars. Brightness can be used to determine distance and the rate at which energy is released from a star. The spectrum of a star gives information about the composition of the star.

MOTIVATION: Demonstrate that color is related to how hot a star is by lighting a 100-watt bulb using a rheostat to allow only a small amount of current at first then a larger amount of current. Ask one or two students to feel the heat given off when the filament is glowing red, then orange, then white. The hotter, the whiter. The same principle applies to stars.

SKILL-BUILDING ACTIVITY

The science skill of inferring using experimental data is practiced in this Activity. The student uses the analysis of light coming from a flashlight bulb to infer how star color indicates a star's temperature.

Teaching Hints: All the materials can be purchased from a local hardware store except the miniature sockets for the bulb. These are available from science supply houses and may be available from a local hobby shop. If you prefer, an uncoated, incandescent 110-V bulb could be used with an A.C. variable-control wall switch in place of the individual units described. Four of these units placed around the room could be used with larger groups.

The following bar graphs illustrate roughly what happens as the brightness of the bulb is varied.

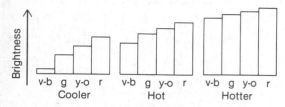

Encourage the students to locate the constellation Orion and see the colors of Betelgeuse, top left corner of Orion, and Rigel, bottom left corner of Orion. Capella is between the Orion constellation and Polaris.

Answers to Questions:
1. Near the end closest to the light bulb.
2. Yes. The blue, green colors.
3. The red remains.
4. Rigel.
5. Capella.

Conclusions:
1. Rigel.
2. It appears yellow because this is its brightest color.

19-3. GALAXIES

BACKGROUND: There are many billions of stars in our galaxy. We are in only one of many billions of galaxies.

MOTIVATION: To give an idea of how large one billion is, have students figure how long it takes to say a typical number larger than 100 million (like 293,467,962) and then figure the time needed to count by ones up to 1 billion. The answer is about one hundred years if you say one number every three seconds without resting or taking time out for food or sleep. Now the idea of 100 billion stars should have some meaning.

Discuss the likelihood that in some other star system there is a planet like the earth that has life on it.

TECHNOLOGY

Teaching Hint: For a more complete account of space lab's accomplishments, see *newscientist*, August 23, 1984. (For back issues, write: Dee Knapp, IPC magazines Ltd., 205 East 42nd Street, New York, NY 10017; (212) 867-2080.

COMPUTER

Teaching Hint: While the computer will perform almost any calculation more quickly than a person with pen and pencil can, it is most efficiently used for large quantities of data. When programming and entry time are considered, students will probably find that the computer is not always the best tool for the job.

CHAPTER REVIEW

Vocabulary:
1. Polaris 2. constellation 3. quasar 4. light-year
5. parallax 6. absolute magnitude 7. black hole
8. Ursa Major 9. supernova 10. main sequence
11. Milky Way 12. galaxy

Answers to Questions:
1. Most constellations were named after animals, ancient gods, or heroes.
2. Locate the Big Dipper constellation. Use the two stars that form the side of the bowl as pointers. Go in a line five times longer than the distance between the pointers to find the North Star.
3. The North Star belongs to Ursa Minor.
4. Visible light, radio waves, and infrared.
5. A telescope must be placed in space to receive all the kinds of radiant energy.
6. From starlight, the scientists know the kinds of atoms in the star and the temperature of the star.
7. Planets are seen to move among the constant star patterns when viewed from night to night.
8. The brightness method is used to measure distances to far away stars.
9. The temperature of all the red stars is similar; however, the absolute brightness of a red giant is much greater than that of a regular size red star.
10. A star begins as a cloud of gas and dust. Gravity forces act to shrink the cloud. As it shrinks, it heats up. One or more gravity centers develop enough heat to start the fusion of hydrogen. Each center of gravity then becomes a star.
11. After using up nearly all its hydrogen, the shrinking star becomes hot enough to cause helium nuclei to join, making heavier elements. This results in the release of larger amounts of heat, causing the star to expand into a red giant.
12. After a star becomes a supernova, it collapses to become a neutron star.
13. We see the faint band of light because of the many stars that shine as we look toward the middle of the Milky Way galaxy.
14. In the Andromeda constellation there is a fuzzy patch of light. With a telescope, it looks like a pinwheel of stars in a spiral shape.
15. The red-shift seems to show that the galaxies that are the farthest away are moving away at the highest speeds. If this is true, then the universe is expanding.

H · O · L · T

EARTH SCIENCE

WILLIAM L. RAMSEY

LUCRETIA A. GABRIEL

JAMES F. McGUIRK

CLIFFORD R. PHILLIPS

FRANK M. WATENPAUGH

HOLT, RINEHART AND WINSTON, PUBLISHERS
New York • Toronto • Mexico City • London • Sydney • Tokyo

THE AUTHORS

William L. Ramsey
Former Head of the Science Department
Helix High School
La Mesa, California

Clifford R. Phillips
Former Head of the Science Department
Monte Vista High School
Spring Valley, California

Lucretia A. Gabriel
Science Consultant
Guilderland Central Schools
Guilderland, New York

Frank M. Watenpaugh
Former Science Department Chairman
Helix High School
La Mesa, California

James F. McGuirk
Head of the Science Department
South High Community School
Worcester, Massachusetts

About the Cover:

The photograph on the cover shows mesas and buttes in Monument Valley. The mesas and buttes are traces of what was once a broad plateau. They rise more than 300 meters above the valley floor. Monument Valley is located on the Utah–Arizona border on the Navajo Indian Reservation.

Photo Credits appear on page 563.
Cover photograph by Shelly Grossman/Woodfin Camp & Assoc.

Art Credits:

Andres Acosta, Robin Brickman, Ebet Dudley, Leslie Dunlap, General Cartography Inc., Jean Helmer, John Jones, Kathie Kelleher, Network Graphics, James Watling, Craig Zuckerman

ACKNOWLEDGMENTS

Teacher Consultants

Pat Courtade
Gulf Comprehensive High School
New Port Richey, Florida

Sandy Meyer
Caledonia Junior/Senior High School
Caledonia, Minnesota

Julia Harlan
Wangenheim Junior High School
San Diego, California

William A. Moore Jr.
Director of Science Exploration
   and Experimentation Center
Junior High School 240
Brooklyn, New York

Larry Nilson
Wayland Junior High School
Wayland, Massachusetts

J. Peter O'Neil
Waunakee Junior High School
Waunakee, Wisconsin

Bob Whitney
Dover Plains Junior/Senior High School
Dover Plains, New York

Content Critics

Dr. John E. Callahan
Professor of Geology
Appalachian State University
Boone, North Carolina

Dr. William Imperatore
Professor of Meteorology
Appalachian State University
Boone, North Carolina

Dr. Frank McKinney
Professor of Oceanography
Appalachian State University
Boone, North Carolina

Dr. Thomas Rokoske
Professor of Astronomy
Appalachian State University
Boone, North Carolina

Safety Consultant

Franklin D. Kizer
Executive Secretary
Council of State Science Supervisors
Lancaster, Virginia

Computer features written by

Anthony V. Sorrentino, D.Ed.
District Coordinator of Computer
   Instruction
Monroe-Woodbury School District
Central Valley, New York

Computer Consultant

Nicholas Paschenko, M.Ed.
Computer Coordinator
Englewood Cliffs School System
Englewood Cliffs, New Jersey

Readability Consultant

Jane Kita Cooke
Assistant Professor of Education
College of New Rochelle
New Rochelle, New York

This book is about your home. Every city, state, or country where you might live is located somewhere on the planet Earth. The earth is not only the planet on which you live. It also provides you with air, water, food, and most of the other materials you need to live. As you study this book, you will learn how scientists have come to understand much about the way the earth works. This scientific knowledge will then help you to understand and appreciate your home planet.

How to Use This Book

HOLT EARTH SCIENCE consists of 19 chapters organized into five units. In the first chapter, What Is Science?, you will learn about science and some of the tools scientists use to solve problems. The two units following Chapter 1 are concerned with geology, the study of the solid part of the earth. The third unit deals with meteorology, the study of the layer of gases that surrounds the earth. In unit four, oceanography, the study of the oceans and the earth's surface below the oceans, is explored. In the fifth and final unit, astronomy, the study of the universe beyond the earth, is presented. Each unit is divided into chapters. Chapter Goals at the beginning of each chapter tell you what you will learn in that chapter. At the end of the chapter, a Chapter Review made up of Vocabulary, Questions, Applying Science, and a Bibliography will help you to remember what you learned in that chapter and to explore some applications and extensions of that knowledge. Each chapter is further divided into sections that begin with a list of section objectives. You will find many interesting and helpful diagrams and photographs in each section. Science words that may be new to you are printed in boldface type. Their definitions are printed alongside in the page margin. Each section ends with a brief Summary of the important ideas developed in the section. Questions at the end of the section will help you to test if you have accomplished the objectives for that section. Most sections are followed by an Investigation or a Skill-Building Activity. These Investigations and Activities will help you to develop good laboratory techniques and to practice basic science skills. Three features of special interest called Careers, Technology, and ¡Compute! are found throughout the book. These features will tell you about careers in earth science, about the ways technology affects our lives, and about the ways computers are used in science.

Safety

In your study of earth science, you will learn many fascinating things about the natural world. However, the study of science can also involve many potential dangers. Science classrooms and laboratories contain equipment and chemicals that can be dangerous if not handled properly. You should always follow the directions and cautions in the book when doing an Investigation. In addition, your teacher will explain to you the precautions and rules that must be followed to insure the safety of you and your classmates. Your responsibility is to follow these rules carefully and to demonstrate a positive attitude toward laboratory safety.

Work Habits

- Never work alone in a science laboratory.
- Never eat, drink, or store food in a science laboratory.
- Wash hands before and after work and after spill cleanups.
- Restrain loose clothing, long hair, and dangling jewelry.
- Never leave heat sources unattended (e.g. gas burners, hot plates, etc.).
- Do not store chemicals and/or apparatus on lab bench, and keep lab shelves organized.
- Never place glassware or chemicals (in bottles, beakers/flasks, wash bottles, etc.) near the edges of a lab bench.
- Read all lab procedures before doing laboratory work.
- Analyze accidents to prevent repeat performances.
- Protection should be provided for the lab worker and lab partner.
- Do not mix chemicals in the sink drain.
- Have actions pre-planned in case of an emergency (e.g., what devices should be turned off, which escape route to use, and how to use a fire extinguisher).

Safety Wear

- Approved eye or face protection should be worn continuously.
- Gloves should be worn which will resist penetration by the chemical being handled and which have been checked for pinholes, tears, or rips.
- Wear a laboratory coat or apron to protect skin and clothing from chemicals.
- Footwear should cover feet completely; no open-toe shoes.

CONTENTS

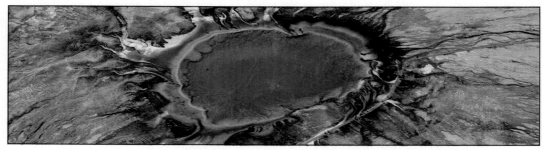

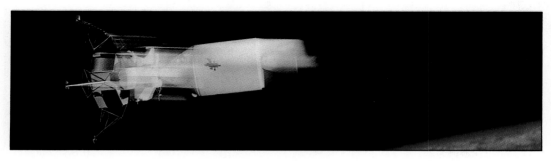

This volcanologist is wearing protective clothing and a mask to shield him from the intense heat of the lava. He is using a steel pipe in order to take a sample of the molten lava.

WHAT IS SCIENCE?

CHAPTER GOALS

1. Describe the branches of earth science and their importance in our lives.
2. Explain how scientists find answers to questions.
3. Explain why making careful observations and measurements is important in science.

1–1. What Is Earth Science?

At the end of this section you will be able to:
- ☐ Explain how scientific study of the earth began.
- ☐ Describe four main branches of earth science.
- ☐ Explain why earth science is important.

Job: Collect some lava. Study the lava to find the source of the melted rock. Such a job is one type of work an earth scientist does. The picture on the left page shows an earth scientist taking a sample of lava. Lava is the melted rock that comes from volcanoes. By studying these samples, earth scientists can learn much about volcanoes.

THE SCIENCE OF THE EARTH

A volcano can be one of nature's most frightening shows. In a few minutes a volcanic eruption can throw out millions of kilograms of rocks and deadly clouds of hot gases. Whole mountains or islands can be blasted away. There are records of cities that have been destroyed by volcanoes. About 1450 B.C. a volcanic eruption took place on an island near Greece. In one day, most of the island and its people disappeared. This event may have been the basis of the story of the lost land of Atlantis.

In Mexico in 1943, a farmer discovered a crack about 25 meters long in his cornfield. The crack had opened during the night. Red-hot stones and ash blew out of the crack. By the next morning, a volcano ten meters high had appeared. Within a year, the volcano grew to a height of over 300 meters. Ash from the volcano covered much of the countryside. The farmer lost his cornfield. In the years that followed, thousands of people lost their homes.

Fig. 1-1 The final moments of some victims in Pompeii. Their bodies left openings in the hardened volcanic material. The openings were then filled with plaster to make these molds.

Observations Any information that we gather by using our senses.

Reinforcement: Ask students to describe some superstitions. Students should explain the difference between superstitions and science, which is based on facts.

Perhaps the best-known volcanic disaster took place in A.D. 79. It was the loss of the Roman city of Pompeii (Pom-**pay**). The city, with at least 2,000 people, was buried by volcanic material. See Fig. 1-1. Volcanic material has kept the city just as it was before being buried.

Stories have been made up to explain volcanoes. For example, the Romans believed that volcanoes were caused by a god. They thought the Roman god Vulcan had huge ovens under the ground. Volcanoes (named after Vulcan) were the chimneys of those huge hidden ovens. Scientific study of volcanoes began when people started to make careful **observations** (ob-sur-**vay**-shuns) of the earth. *Observations* are any information that we gather by using our senses. For hundreds of years, people had lived in many places where old volcanoes quietly rested. They did not know that the mountains and fields were once the scene of fire and explosion. Toward the end of the 18th century scientists studying the earth began to find that volcanoes were not unusual geographic features. They found rocks believed to have formed from lava. Some mountains were found to be old volcanoes. Scientists began to see that much of the earth's surface was shaped by volcanoes. See Fig. 1-2. For the first time, scientists became aware that the earth's surface could change. If volcanoes could build mountains, then the shape of the earth's surface must not always have been the same. By observing the changes in the earth's surface, scientists believed they could find out how the earth works. This was the beginning of the scientific study of the earth.

*The mountain was formed by a volcano.

Fig. 1-2 How do you think this mountain was formed?

THE EARTH SCIENCES

Earth is the planet we live on, and earth science is the study of our planet. Earth science can be divided into four branches. One branch is called **geology.** *Geology* is the study of the solid part of the earth. Scientists called geologists study the matter and the structure of the planet. They do this to find out how the earth formed and how it changes. Some geologists study volcanoes because volcanoes change the surface of the earth.

The earth is surrounded by a layer of gases that make up its atmosphere. **Meteorology** is the branch of earth science that studies the atmosphere. The scientists who work in the field of *meteorology* are called meteorologists. They also observe weather and study climates. See Fig. 1–3.

Oceans cover most of the earth's surface. **Oceanography** is the branch of science that studies oceans. *Oceanography* also includes the study of the earth's surface below the oceans. The scientists who make these studies are called oceanographers, or *marine scientists*. See Fig. 1–4.

Word Study: Geology: *geo + logos* (Gr.), earth description.

Geology The branch of earth science that studies the solid part of the earth.

Meteorology The branch of earth science that studies the atmosphere.

Oceanography The branch of earth science that studies the oceans and the ocean floor.

Word Study: Oceanography: *okeanos + graphein.* (Gr.), to write about the ocean.

Enrichment: In many cases, the understanding of geology requires an understanding of meteorology.

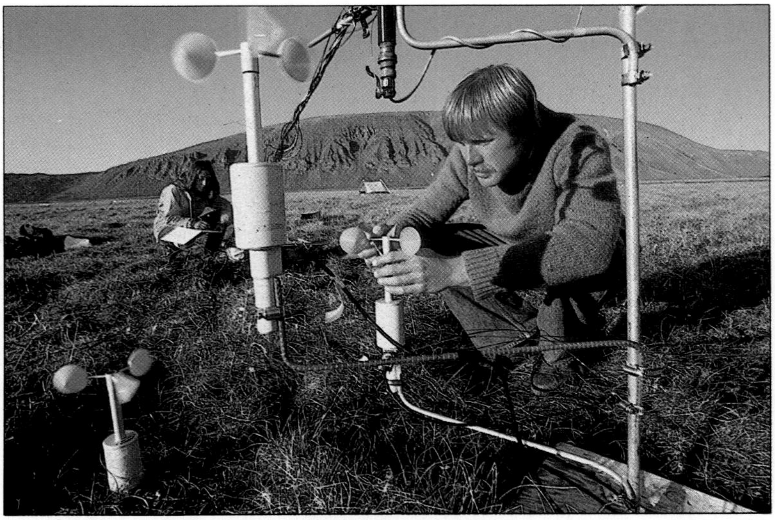

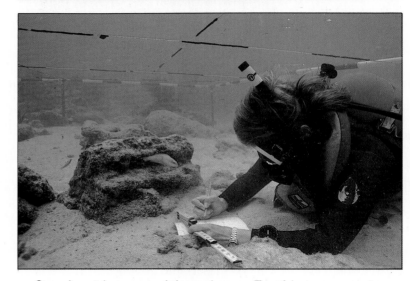

Fig. 1-3 Meteorologists study the weather.

Fig. 1-4 Oceanographers study the oceans.

Astronomy The branch of earth science that studies the universe.

Our planet is a part of the universe. For this reason, **astronomy** is a part of earth science. *Astronomy* is the study of the universe beyond the earth. Astronomers study other planets, the sun, and the stars. See Fig. 1-5. In this book, you will learn about geology, meteorology, oceanography, and astronomy. Each will be studied in a different unit.

WHY EARTH SCIENCE IS IMPORTANT

Our planet is always being changed by powerful forces. Almost every day the news tells you about volcanoes, earthquakes, storms, floods, and drought. At some time, these forces

Fig. 1-5 Astronomers study the universe.

will touch the life of every person. What causes earthquakes and weather? Why do volcanoes suddenly erupt? Finding the answers to questions like these can save lives and property.

Understanding the earth also can improve the way we live. Exploring the land and oceans will help find oil, coal, and metals. Almost all of the things we use every day come from the earth. Human-made wastes cause pollution of air and water. We must understand the atmosphere and oceans to control pollution and its effects. From space, we can see the earth as never before. Already pictures of the earth from space have helped to locate valuable deposits of metals and sources of pollution.

SUMMARY

Scientific study of the earth began with observations. The four main branches of earth science are geology, meteorology, oceanography, and astronomy. Scientific study of the earth can improve our lives on this planet.

QUESTIONS

Use complete sentences to write your answers.

1. Explain how observations of volcanoes led to scientific study of the earth.
2. Describe the four main branches of earth science.
3. Give one example of why the study of earth science could be important to you.

INVESTIGATION

See teacher's commentary for teaching hints, safety, and answers to questions.

OBSERVATIONS

PURPOSE: To make observations of a very common event. Then to ask a question that requires more facts to be gathered through experiment in order to arrive at a conclusion.

MATERIALS:

7 index cards   2 sheets of paper
stapler

PROCEDURE:

A. Staple two index cards together using one staple near the center of the cards. Staple four other index cards together in the same way.

B. Compare the weight of one card to the two stapled together.

   1. Which is heavier?

C. Hold one card in one hand and the two stapled cards in the other. With your hands at waist height, drop them both at the same instant. Note which hits the floor first. Do this several times.

   2. Which falls through the air faster?

D. Pick up the two-card pile in one hand and the four-card pile in the other.

   3. Which is heavier?

   4. Which pile do you think will fall through the air faster?

E. Holding the cards waist high, drop them at the same instant. Note which pile strikes the floor first. Do this several more times.

   5. Which pile fell through the air faster?

   6. Was your answer for question 4 above correct?

F. At this point you could conclude that weight determines how fast an object falls through air. A scientist would ask, "Is the speed at which an object falls through the air determined only by its weight?" Another experiment is needed to answer this question. Crumple one sheet of paper into a ball.

   7. Which is heavier, the ball or the flat sheet?

   8. On the evidence you have at this time, which should fall through air faster?

G. Hold the two pieces of paper waist high and drop them at the same instant. Do this several times.

   9. Do objects having the same weight always fall at the same speed?

   10. What explanation can you give for this result?

H. Fold the flat sheet of paper into quarters. Drop the folded paper from waist height and observe its speed.

   11. What do you observe about the folded paper's speed compared to the flat sheet?

CONCLUSIONS:

1. What two factors affect the speed at which objects fall through the air?

2. What do you think would happen to the speed of falling objects if air were not present?

3. What must you do to be certain of your answer to question 2?

1–2. Scientific Thinking

At the end of this section you will be able to:

- ☐ Explain how asking questions can lead to scientific thinking.
- ☐ List and explain the steps in scientific problem solving.
- ☐ Distinguish between a *hypothesis* and *scientific theory*.

It was a place for camping, hunting, and fishing. There were green forests, meadows, and lakes. This was the area around Mount St. Helens in the state of Washington. At 8:32 A.M. on May 18, 1980, the mountain blew its top. About 250 billion kilograms of rock and dust were blasted out by the eruption. The forests were turned into a landscape that looked like the surface of the moon. See Fig. 1–6.

Fig. 1-6 (left) Mount St. Helens before it erupted. (right) Mount St. Helens after the eruption.

ASKING QUESTIONS

Since that May morning in 1980, scientists have been studying Mount St. Helens. They are searching for answers to questions about the eruption. One important question is: Does a volcano give warning signs that it is going to erupt? Scientific thinking begins with such a question. Finding an answer could help to save many lives in the future.

Fig. 1-7 *Oil geologists use the scientific method to find oil.*

SCIENTIFIC PROBLEM SOLVING

Research is a way scientists look for answers to questions. Scientists who want to find a way to tell when a volcano will erupt must gather as much information as possible about volcanoes. Scientific research is often done by following a number of steps. The steps are called the **scientific method.** The *scientific method* is a guide to scientific problem solving. It is not a set of instructions that will always result in an answer. See Fig. 1-7. Applying the scientific method, research on the eruption of Mount St. Helens might take the following steps.

1. State the problem. Sometimes hot liquid rock called lava forces its way to the surface of a volcano. When this happens, it is called a volcanic eruption. An erupting volcano also gives off gases. Sometimes the lava flows out quietly. Other times there are explosions. The lava flow and explosions can damage property and injure or kill people in the area.

The first step of the scientific method is to identify and state the problem so that it can be researched. This is not always easy. The problem with Mount St. Helens is to find a way of telling when an eruption will happen. In science, telling what is likely to happen is called predicting. The problem is then stated in the form of a question. The question could be: "Can the eruptions of Mount St. Helens be predicted?" The question must clearly state the problem. Then the research can begin.

2. Gather information. Research usually begins by gathering all information that is already known. For example, scientists had records that showed earthquakes happening before the eruption of Mount St. Helens. These records came from instruments placed on the volcano. See Fig. 1-8. The records showed that earthquakes began around Mount St. Helens several months before it erupted. Most of the earthquakes were small. Sometimes as many as fifteen small earthquakes were recorded in one hour.

3. Form a **hypothesis.** A *hypothesis* (hie-**poth**-uh-sus) is a possible answer to a question. The question to be answered here is, "Can the eruptions of Mount St. Helens be predicted?" A hypothesis often comes from first seeing some kind of pattern in observations. For example, the earthquake records seem to show that groups of small earthquakes come before eruptions of Mount St. Helens. Thus a hypothesis could be stated as: Mount St. Helens is likely to erupt shortly after many small earthquakes are observed

4. Test the hypothesis. Scientists think of a hypothesis as only one of many possible answers to a question. They do not consider a hypothesis to be correct until it has been tested. A hypothesis is tested by doing one or more **experiments.** An *experiment* is a setup that enables observations to be made. Ideally, *experiments* are set up so that they cannot be disturbed. In this way, scientists can be sure that the results of the experiment are correct. These experiments are called *controlled* experiments. In these experiments, scientists are able to *control* anything that may affect the experiment. Therefore, many controlled experiments are done in laboratories. For example, suppose light affected an experiment. In the laboratory scientists could block out the light. However, many experiments in earth science cannot be set up under controlled conditions. For example, earthquakes and volcanoes cannot be made to happen. So earthquakes and volcanoes must be tested by a different kind of experiment. In this kind of experiment, scientists make as many observations as they can. From these observations, they then try to find a way of telling when an

Fig. 1-8 A seismograph on Mount St. Helens. It records earthquakes.

Hypothesis A statement that explains a pattern in observations.

Experiment A setup to test a particular hypothesis.

Enrichment: You might want to tell students that the factors which are controlled and manipulated in an experiment are called *variables.*

Reinforcement: Hold up various objects and have students make observations. Point out which are good observations (facts). Students often express biases, using terms that do not have the same meaning for everyone.

eruption will take place. For these scientists, the volcano has become their laboratory.

5. **Accept or change the hypothesis.** Experiments will show if the hypothesis is correct. Since 1980, scientists have used earthquakes as early warnings of volcanic eruptions. They have been able to predict many Mount St. Helens eruptions. Some were predicted up to three weeks before they happened. Often other experiments will show how the hypothesis can be improved. For example, there were other kinds of observations made before the Mount St. Helens eruption. Scientists observed that parts of the mountain moved out slightly. These movements were measured by instruments placed on the mountain. See Fig. 1-9. Scientists also observed that certain gases were given off just before the eruption. These other ob-

Fig. 1-9 A laser range finder is used to measure the movements of parts of Mount St. Helens. Measurements can be made to an accuracy of about one part per million.

servations, along with earthquake activity, can be used to make better predictions. When scientists find that their predictions are correct, they then can accept the hypothesis as correct.

SCIENTIFIC THEORIES

The hypothesis that used earthquakes as early warnings of volcanic eruptions was tested many times and found to be correct. Other observations were also used to help predict volcanic eruptions. Very often several or many hypotheses will lead to the development of a **scientific theory.** A *scientific theory* is a general explanation made up of many hypotheses, not just one. A scientific theory that only predicts eruptions on Mount St. Helens is not complete. It must answer a wider range of questions. For example, can the theory predict eruptions on other volcanoes? Why do earthquakes and volcanoes happen in the same places? A good scientific theory is able to answer these questions. However, even good theories may change as new observations are made.

Some scientific theories have been tested many times and found to be correct. A theory that has been found to be correct every time it is tested is called a *scientific law.* An example of a scientific law is the law of gravity. The force of gravity on the earth obeys this scientific law. Even bodies in space, such as the sun and planets, obey this law.

Scientific theory A general explanation that includes many hypotheses that have been tested and found to be correct.

Enrichment: Scientific laws often require centuries to develop. The beginnings of the law of gravity can be traced back to antiquity. Aristotle, Galileo, Kepler, Newton, and finally Einstein have all provided a part of the understanding of this law of nature.

1. The question, "Does a volcano give signals that it is going to explode?" has led to much scientific research about Mt. St. Helens in search of the answer.
2. The steps used as a guide to scientific thinking are: (a) state the problem, (b) gather information, (c) form a hypothesis to explain the observations, (d) experiment to test the hypothesis, (e) use experimental results to accept or change the hypothesis.
3. A scientific theory combines a number of related hypotheses that have been tested many times and found to be correct.

SUMMARY

Scientific thinking begins with asking questions. The scientific method is a guide that scientists use to find answers. Scientific research may lead to a possible answer or hypothesis. After a hypothesis has been tested and found to be correct, it may become part of a scientific theory. When a theory has been found to be correct, it becomes a scientific law.

QUESTIONS

Use complete sentences to write your answers.

1. Give an example of how asking a question can lead to scientific thinking.
2. List the steps that can be used as a guide to scientific problem solving.
3. Explain the difference between a scientific theory and a hypothesis.

INVESTIGATION

See teacher's commentary for teaching hints, safety, and answers to questions.

AN EXPERIMENT

PURPOSE: Make a hypothesis and carry out an experiment to check the hypothesis.

MATERIALS:

2 thermometers (Celsius)

2 pieces of dry cloth

1 medicine dropper

PROCEDURE:

A. Copy the table shown below.

Condition	Temperature (°C)	
	First Reading	Second Reading
Bulbs Bare		
Bulbs Covered		

B. Place both thermometers side by side on the table. Take care that they will not roll off the table. Place the bulbs of the thermometers 2 cm apart. See Fig. 1–10.

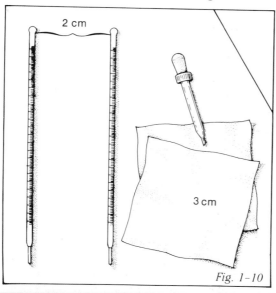

Fig. 1–10

C. Allow the thermometers to remain undisturbed for two minutes. Then read both thermometers and record their temperatures in your table. The reading on the two thermometers may differ by a degree or two, even though they are at the same temperature.

D. Wrap a piece of dry cloth around the bulb of each thermometer. Put them back in the same place as before. Again, after two minutes, read each thermometer and record in your data table.
1. Was there a change in the reading for either thermometer?
2. What effect would you predict there would be if you wet the cloth on one of the thermometers with water?

E. The answer you gave for question 2 is called a "hypothesis." A hypothesis needs to be tested. What you do in testing the hypothesis is called an "experiment."
3. Design an experiment that will test your hypothesis.

F. Carry out your experiment. Record your results in the data table. Record the temperature of both thermometers. Describe the condition of each in the data table.

CONCLUSIONS:
1. Using complete sentences, describe your hypothesis.
2. Using the results of the experiment, tell whether or not your hypothesis was correct.
3. Does the scientific method always give a correct hypothesis? Explain.

1-3. Measuring

At the end of this section you will be able to:

☐ Explain why measurements are important in making scientific observations.

☐ Name and use the SI units of measurement.

☐ Explain why measurements are likely to contain error.

Look at Fig. 1-11. What would happen to the ball held in each dragon's mouth during an earthquake? Even a small earthquake could cause a ball to fall into the mouth of the toad below. This instrument was used by the Chinese around A.D. 130 to detect earthquakes. The strength of the earthquake could also be measured. It was measured by the number of balls dropped.

MEASUREMENTS IN SCIENCE

An earthquake takes place very quickly. It leaves no record except in the damage that is done. However, the ancient Chinese could make a record of earthquakes using the instrument shown in Fig. 1-11.

Fig. 1-11 The ancient Chinese used this instrument to detect and measure earthquakes.

Enrichment: In 1866 the United States Congress adopted the metric system as the official system of measurement. Since then the yard, foot, and inch have all been defined as a given fraction of the meter. In 1960 many revisions were made, including the change in name to SI.

Fig. 1-12 Earthquakes from all over the United States are recorded by these instruments at the National Earthquake Information Service in Golden, Colorado.

Like the ancient Chinese, modern scientists also try to make observations that can be kept as a record. This is a reason for scientific observations often being made in the form of **measurements.** A *measurement* is an observation that can be made by counting something. Measuring may be as simple as counting the number of balls falling from the mouths of the dragons. However, modern earthquake measurements are not so simple. Instruments now used to record earthquakes measure their strength by traces on paper held by turning drums. See Fig. 1-12. Scientists make as many observations as possible in the form of measurements. A measurement is written as a number value. Then it can be understood because numbers have the same meaning for everyone.

All measurements are made by using a number and a unit. The number tells how many. The unit tells what is being counted. For example, you might tell someone your height in feet and inches. Feet and inches are units commonly used in the United States for measuring length. However, people in other countries would have a problem understanding feet and inches. Units of measurement used in almost all parts of the world are based on the metric system. The modern form of the metric system is called the International System of Units, which is usually written as SI. SI stands for the French name *le Système International.* Common metric units are the same as SI units.

Scientists use SI units to make measurements. One reason is that SI units are easy to use. This is because they are based on the number ten. All SI units can be changed into larger or smaller units by using the number ten. For example, the kilometer is equal to 1,000 (10 × 10 × 10) meters. One meter is equal to 100 (10 × 10) centimeters. Changing units such as feet into larger or smaller units is not so simple. To change feet into inches, you must first know that 1 foot equals 12 inches. To change feet into yards, you must know that there are 3 feet in 1 yard. With SI units, you only must remember that the size of all units can be changed by using the number ten.

MAKING MEASUREMENTS

Measurements are usually made with the use of tools or instruments. Instruments can improve our observations. For example, telescopes are instruments that help us to see deep into

Fig. 1-13 Radiotelescopes have detected some of the most distant objects in the universe.

space. See Fig. 1-13. Microscopes help us to see objects that are too small to be seen with just our eyes. Measurements also make observations more useful. To say something is large is not enough. Measurements are needed to show how large.

1. Length or distance. The basic SI unit used to measure length or distance is the **meter** (m). A *meter* is slightly longer than a yard. Lengths or distances that are much longer than one meter are measured in larger units. The unit most often used to measure distances much longer than one meter is the *kilometer* (km). A kilometer is equal to 1,000 meters. "Kilo-" means 1,000.

Units smaller than a meter are used when measuring lengths shorter than a meter. A unit equal to 1/100 (0.01) of a meter is called a *centimeter* (cm). A unit smaller than a centimeter is called a *millimeter* (mm). A millimeter is equal to 1/1,000 (0.001) meter.

A basic unit such as a meter can be changed into a larger or smaller unit by using a prefix. A prefix is always put in front of another word. For example, "kilo-" and "milli-" are prefixes added to the word "meter." Each of these prefixes tells how much larger or smaller the basic unit has been made. Table 1-1 shows the prefixes that can be used with all basic SI units. Lengths can be measured in SI units using a metric ruler. Metric rulers are often marked in centimeters, which are numbered.

Meter The basic SI unit of length. A meter is equal to 39.4 inches.

Enrichment: The Golden Gate Bridge is a little more than 1 km long.

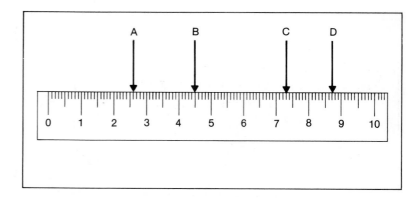

*The readings are A, 2.6 cm; B, 4.5 cm; C, 7.3 cm; D, 8.7 cm.

*Fig. 1-14 The numbered divisions on this metric ruler are centimeters. What are the readings at A, B, C, and D in centimeters?**

Each centimeter is divided into millimeters, which are not numbered. See Fig. 1-14. A common metric ruler is one meter in length and is called a meter stick.

Prefix	Meaning	Example
Milli-	1/1,000 (.001)	Millimeter
Centi-	1/100 (.01)	Centimeter
Deci-	1/10 (.1)	Decimeter
Deca-	10 times	Decameter
Hecto-	100 times	Hectometer
Kilo-	1,000 times	Kilometer

Table 1-1

2. Mass. Mass is the amount of matter an object has. Measuring mass means finding the amount of matter in an object. Measuring how much an object weighs is one way to find its mass. The weight of an object usually depends on its mass. Thus mass and weight are often used as if they were the same. However, they are not the same. The weight of an object can change. This is because weight also depends on the object's distance from the earth. Thus someone in space, far from the earth, is nearly weightless. But the mass or amount of matter in the person's body cannot change. The weight of an object remains the same only when measured on the earth.

Kilogram An SI unit of mass equal to 1,000 grams.

The SI unit for mass is the **kilogram** (kg). One *kilogram* of mass weighs about 2.2 pounds. A smaller SI unit of mass is a gram (g). One gram is 1/1,000 (0.001) kg. A nickel coin weighs about 5 g. There are 28.35 g in 1 ounce.

Mass can be measured by using a balance. See Fig. 1-15. The balance works like a seesaw. The mass of the object on the left pan is balanced by moving the sliding weights, or riders, on the

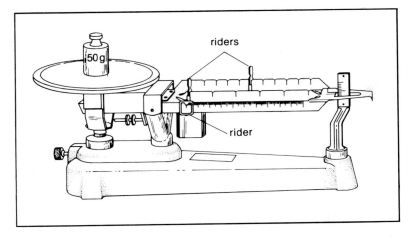

Fig. 1-15 A balance is used to measure mass.

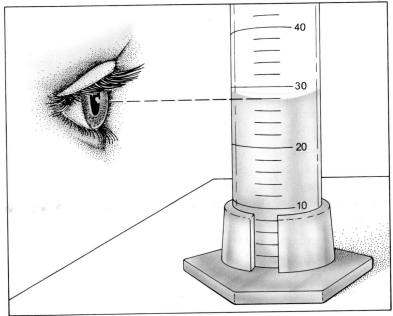

Fig. 1-16 To read a graduated cylinder, hold up the cylinder so that the surface of the liquid is level with your eye. Read the mark closest to the bottom of the curve. Numbered marks are ml or cm. What is the reading shown?

right. Each rider gives a reading. The readings are then added to get the mass of the object.

3. Volume. All materials take up space. Rocks, water, and even air take up space. Liquids usually are measured by their volume. The volume of a liquid is the amount of space the liquid takes up. The volume of a liquid can be measured with a graduated cylinder. See Fig. 1-16. The liquid is poured into the cylinder. Holding the cylinder with the liquid surface at eye level, read the mark closest to the surface. Since the liquid surface is curved, use the mark closest to the bottom of that curve.

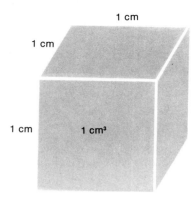

1 cm

1 cm

1 cm

1 cm³

Fig. 1–17 A cube that measures 1 cm on each side has a volume of 1 cm³. A sugar cube has a volume of about 1 cm³.

The unit often used when measuring the volume of liquids is the **liter** (L). A *liter* is a little more than a quart. Small volumes are measured in milliliters (mL). One mL is 1/1,000 (0.001) liter. One mL is equal in volume to a cube that is 1 cm on each side. Such a cube has a volume of one cubic cm. This is written as 1 cm³. See Fig. 1–17. Thus, 1 mL and 1 cm³ are equal and are used in the same way. The table below shows the relationship between the units commonly used for measuring liquids.

Unit	Symbol	Value
liter	L	1,000 mL
milliliter	mL	0.001 L

The volumes of solids can also be measured. However, the volumes of solids are measured in cm³. When a solid has a *regular* shape, such as a cube, its volume can be measured by multiplying the lengths of its three sides. For example, the cube shown in Fig. 1–17 has a volume of 1 cm × 1 cm × 1 cm or 1 cm³. If a solid has an *irregular* shape, such a common rock, its volume can be measured by finding out how much water it can displace. One way this can be done with a small object is to fill a graduated cylinder partially with water. Read the volume. Then lower the solid into the water. The solid will cause the water level to rise. The difference between the new water level and the original level is equal to the volume of the solid.

4. Temperature. The most widely used temperature scale in the world is the **Celsius** (C) scale. Most scientific measurements of temperature are based on the *Celsius* scale. The freezing temperature of water on the Celsius scale is 0°C. The boiling temperature of water is 100°C. You are probably used to temperatures being given in the Fahrenheit scale. The Celsius and Fahrenheit scales are shown in Fig. 1–18.

5. Density. Sometimes measurements are made by using two kinds of units. For example, the measurement of **density** uses units of mass and volume. *Density* tells how much mass there is in a certain volume of material. The units for density are grams per cubic centimeter (g/cm³). For example, the density of lead is 11.3 g/cm³. This means that 1 cm³ of lead has a mass of 11.3 g. Aluminum has a density of 2.7 g/cm³. This means that 1 cm³ of aluminum has a mass of 2.7 g. Thus a piece of lead has a much larger mass than a piece of aluminum with the same volume. Therefore, lead is said to be much more dense than aluminum.

Liter A unit of volume. A liter is equal to 1.05 quarts.

Celsius The temperature scale used most in science. It is based on the freezing point of water, 0°C, and the boiling point of water, 100°C.

Density The amount of mass in a given volume of a substance; density = mass/volume.

Enrichment: Density is an example of an *operational definition*. This is a definition based on experimental observation and including only that which has been or can possibly be subjected to such experimental operations. See the Investigation on p. 108.

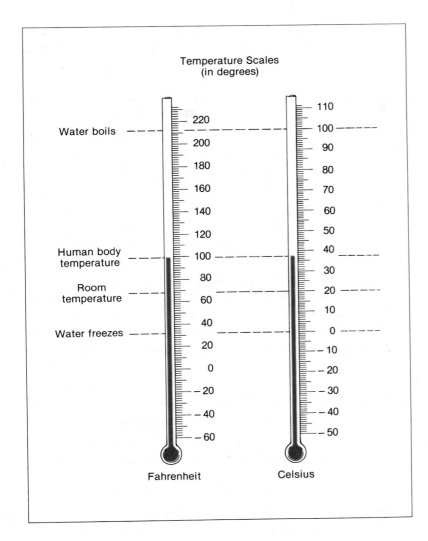

Fig. 1-18 A comparison of the Celsius and Fahrenheit temperature scales. You can use this graph to change from one scale to the other. Find the temperature on one scale. Then use a ruler to move across to the other scale to find the equivalent temperature.

ESTIMATING

Each time a measuring instrument is used, you will need to estimate the final answer. For example, suppose that you want to measure the length of this book. The edge of the book will not be exactly even with the smallest mark on the ruler. You probably will choose the mark that is closest to the edge of the book. The smallest mark on most metric rulers measures millimeters. Therefore, your measurement could easily be off by almost as much as 1 mm. See Fig. 1–19. There is no way you can remove this error. No measurement using instruments can be exact. This is because all measurements made by instruments need some estimating.

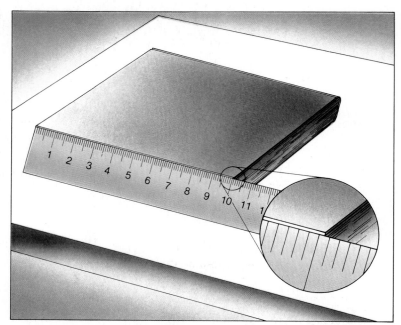

Fig. 1-19 When measuring length using a metric ruler, you will have to estimate to the nearest mm.

Scientists know that their measurements contain some error. They try to keep the errors as small as possible. One method of keeping the error small is to make the same measurement many times. Some errors make measurements too large. Other errors make measurements too small. These errors will tend to balance each other. The measurements are then averaged into a final measurement. Averaging is one way scientists try to make measurement error as small as possible.

SUMMARY

Measurements are important because they make observations more useful and have the same meaning for everyone. The basic SI units are the meter, liter, and kilogram. All measurements made by instruments need some estimating.

QUESTIONS

Use complete sentences to write your answers.
1. Why are measurements used in science?
2. Describe each of the SI units of measurement for (a) length, (b) volume, and (c) mass. Compare each to a common measure.
3. Describe two metric units of length smaller than a meter.
4. Explain what causes measurements to contain error.

1. Measurements are easily kept as a permanent record, are expressed with numbers that mean the same for everyone, and are easily communicated.
2. The metric unit for (a) length is the meter, which is slightly more than a yard; (b) volume is the liter, which is slightly more than a quart; and (c) mass is the gram, a five-cent coin is about five grams.
3. The metric unit of length, the meter, is subdivided into 100 equal parts called centimeters. The centimeter is further subdivided into 10 equal parts called millimeters.
4. Each time a measuring instrument is used, you have to estimate the final answer.

See teacher's commentary for teaching hints, safety, and answers to questions.

MEASURING WITH THE SI SYSTEM

PURPOSE: To use the SI system in making measurements of length, mass, and volume.

MATERIALS:

ruler   graduated cylinder
10 index cards   rubber stopper
balance

PROCEDURE:

A. Copy the following table. Record all your measurements in your copy of the table.

Measurement	Index Card	Rubber Stopper
Length (cm)		————
Width (cm)		————
Thickness (cm)		————
Mass (g)		
Volume (cm³)		

B. Use the metric side of the ruler to measure the length and width of one index card. Record your results.

C. Measure the thickness of one index card to the nearest 0.1 cm.
    1. What is your measurement?

D. Use your ruler to measure the thickness of ten index cards that are held tightly together. Find the thickness of one card by dividing by ten. Record this in your table to the nearest hundredth of a cm.

E. Find the volume of one index card (multiply its length × width × thickness). Use the thickness obtained in step D. Record this.

F. Determine the mass of one index card. Record in the table.

G. Determine the mass of the stopper and record in your table.

H. Find the volume of the rubber stopper by doing the following:

    2. Fill the graduated cylinder about half-full with water. Note and record the volume of water. Remember to read to the bottom of the curved surface of the water.

    3. Carefully lower the stopper into the cylinder until it is completely underwater. Note and record the new volume of water.

    4. The difference in the two volumes is the volume of the solid. The measurements on the graduated cylinder are in milliliters (mL). Each 1.0 mL is equal to 1.0 cm³. Record the volume of the irregular solid in your table to the nearest cm³.

CONCLUSIONS:

1. Name and give the symbols for the units of length, mass, and volume of the SI system used in this activity.

2. Explain why ten index cards were used for the measurements made in steps C and E.

3. Why do you think the volume of the stopper was measured using a graduated cylinder rather than a ruler?

4. If you were given a cube and a small ball, explain how you would measure the volume of each.

SKILL-BUILDING ACTIVITY

USING A BALANCE

PURPOSE: A balance is an instrument used in determining the mass of an object. You need to know how to use the balance properly to measure mass accurately.

MATERIALS:

balance   objects to weigh

PROCEDURE:

A. A balance must always be checked with no mass on it. It should be placed on a level surface and adjusted until the indicator shows that it is level. When the pointer moves equally on both sides of the center of the scale, it is in balance. See Fig. 1–20. An equal-arm balance, like the one in Fig. 1–20, has right and left sides that are the same size and shape. If your balance is an equal-arm balance, find the mass of each of two different objects by doing the following:

   a. Place the object of unknown mass on the left pan of the balance.

   b. Place known masses on the right pan until the pointer again moves equally on both sides of the center of the scale.

   c. If your balance has a numbered scale and a rider on that scale, you can use the rider to help balance the unknown mass.

   d. The unknown mass is equal to the total of the known masses plus the rider reading.

B. Another kind of balance is the triple-beam balance shown in Fig. 1–15. It has one pan and three movable riders. If your balance

is a triple-beam balance, weigh two objects by doing the following:

   e. Place the object of unknown mass on the pan.

   f. Move the riders until the pointer moves equally on both sides of the center of the scale.

   g. The unknown mass is equal to the total of the readings of the riders. Always report mass to the smallest unit on the scales. For example: back rider = 10 g, middle rider = 100 g, front rider = 8.6 g; the mass is 118.6 g.

CONCLUSIONS:

1. Describe how you would use an equal-arm balance to find the mass of an object.

2. Describe how you would use a triple-beam balance to find the mass of an object.

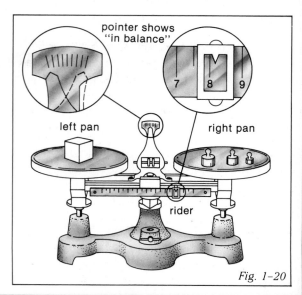

Fig. 1–20

T E C H N O L O G Y

INTRODUCTION

Welcome to the future, a world created by scientific research and development, a world of questions and possibilities. Science is much more than knowledge of facts that have already been discovered. Science is an activity for answering questions and seeking new knowledge. It is also an attempt to explain what new facts mean and how these facts can be used. Science is also the foundation upon which new technology is built. In the Special Features sections, we have tried to inform you of current research and technological developments that have actually extended the subject matter of science or changed the nature of what you must learn in order to live and work in modern society. The Special Features sections go beyond textbook material to encourage you to explore, on your own, events in the scientific world— events that are constantly changing and that often raise more questions than they answer. For example, you will read about weather satellites helping to save crops and lives, about new technology creating another gold rush in California, about computerized images from heat-sensing satellites charting warm ocean currents, and of the possibility of harnessing the wind energy to supply a share of our electrical energy.

The earth and planetary sciences are undergoing amazing changes because of the information available through new technologies. The new technologies, the result of the application of scientific thought and research, have also created a society of different jobs and careers. To find work in

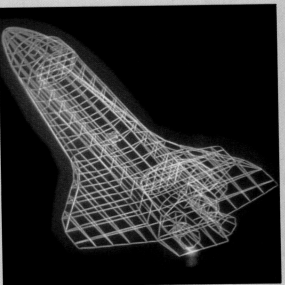

today's job market, you have to understand the changes brought about by the addition of new technologies and the elimination of old ones. For example, typographers used to set books into print using physical type —that is, actual letters engraved on metal. They arranged the letters line by line for each page of print. The pages of this book were not printed that way. Instead, the words were entered into a word processor, stored on a floppy disk, and became type when a compositor programmed a computer to translate the words on the floppy disk onto printed pages. Even the owl in the computer section was not drawn by a human artist; it was made by a computer. Changes such as these mean learning skills different from those that people had to learn even as recently as ten years ago. Computers are becoming a fact of daily life, and learning to use them is essential. The section titled Compute! will help you to learn science through computers and to learn more about computers while applying information in your text to the programs. That section includes points of interest for both the student who wants an introduction to computers and one who is looking for new software to try.

In each Special Features section, you will find words that may not be completely familiar to you, such as *rotor, oscillation,* and *sensor.* As you learn to use these words comfortably, your understanding will increase. Be patient with yourself; this is only the beginning of a fascinating journey into a future of many possibilities.

CHAPTER REVIEW

VOCABULARY

On a separate piece of paper, match the number of each sentence with the term that best completes it. Use each term only once.

observation	scientific theory	liter (L)	kilogram (kg)
experiment	scientific method	meteorology	Celsius
hypothesis	measurement	oceanography	astronomy
geology	meter (m)	density	

1. The branch of science dealing with the universe beyond the earth is _____.
2. A scientist who studies volcanoes works in the branch of earth science called _____.
3. A(n) _____ is made by counting something.
4. Hypotheses are tested by planning a(n) _____.
5. A(n) _____ is something you know from seeing, hearing, or feeling.
6. The basic unit of length in the metric system is called the _____.
7. A(n) _____ is a possible answer to a question.
8. The _____ is a metric unit of volume.
9. When a hypothesis is found to be correct, it becomes part of a(n) _____.
10. The _____ is the SI unit of mass.
11 The temperature scale widely used in science is the _____ scale.
12. _____ is the study of the oceans and the ocean floor.
13. A guide used by scientists in solving problems is called the _____.
14. One who studies weather works in the field of _____.
15. _____ is the amount of mass in a given volume of a substance.

QUESTIONS

Give brief but complete answers to each of the following questions. Unless otherwise indicated, use complete sentences to write your answers.

1. What is earth science?
2. How did the scientific study of the earth begin?
3. Why are volcanoes a part of geology?
4. Why is astronomy a branch of earth science?
5. How can understanding earth science save lives?

6. Describe how scientific thinking starts.

7. In your own words, describe the five steps that may be used as a guide to scientific problem solving.

8. When does a scientific hypothesis become a part of a scientific theory?

9. When does a scientific theory become a scientific law? Give an example of a scientific law.

10. How are experiments usually done in earth science?

11. Why are controlled experiments considered ideal experiments?

12. Why do scientists use SI units for measuring?

13. What is the purpose of making the same measurement many times?

14. Convert: (a) 80 cm to meters; (b) 30 cm to millimeters; (c) 12 mm to centimeters.

15. Convert: (a) 512 g to kilograms; (b) 6.12 L to milliliters; (c) 450 ml to liters.

APPLYING SCIENCE

1. Write a job description for (a) a geologist, (b) an astronomer, (c) an oceanographer, and (d) a meteorologist.

2. Write at least six observations of your classroom. Then compare your list with the lists of your classmates. (a) Were your observations written in the same or in a different way? (b) Did any of the observations use numbers? Were these observations easier to understand? Why?

3. Collect labels, containers, etc., that show SI units. Show your collection to the class.

BIBLIOGRAPHY

American Metric Council. *The Metric System Day by Day.* Washington, D.C.: American National Metric Council, 1977.

Campbell, Norman Robert. *What is Science?* New York: Dover.

Davies, Paul. *The Edge of Infinity.* New York: Simon & Schuster, 1982.

Gore, Rick. "Eyes of Science." *National Geographic,* March 1978.

National Bureau of Standards. *The International System of Units (SI).* Special Publication 330. Washington, D.C.: U.S. Dept. of Commerce, 1981.

Zubrowski, Bernie. *Messing Around with Drinking Straw Construction.* Boston: Little, Brown, 1981.

This Landsat photo from an orbiting spacecraft shows the beauty and variety of the earth's surface. Urban areas (red), forests (green), marshes (blue) are 3 of 38 identifiable areas.

A UNIQUE PLANET

CHAPTER GOALS

1. Develop a hypothesis that explains present observations of the solar system and how it began.
2. Explain how information about the inside of the earth is obtained.
3. Develop models to represent the earth's size, shape, and surface.
4. Describe the location of a place on the earth.
5. Describe the earth's motions using observations of the sun and stars.

2-1. Origin of the Earth

At the end of this section you will be able to:

- ☐ Use present observations of the solar system to explain how it began.
- ☐ Explain the collision and dust cloud hypotheses.
- ☐ Describe why the dust cloud hypothesis is used to explain how the solar system formed.

The photo at the left is a view of the surface of the earth. It was taken by *Landsat*, a spacecraft above the earth. Looking down on the earth from space has given scientists a new view of the planet we live on.

THE EARTH AND PLANETS

Any hypothesis that explains how the earth formed must also explain certain observations. First, the earth is one of nine planets and other bodies that move around the sun. The earth, planets, sun, and smaller bodies are called the **solar system.** Each planet in the *solar system* follows a path or *orbit* around the sun. A hypothesis that explains how the earth formed should explain why the earth is part of the solar system. Second, the hypothesis should explain why the orbits of the planets fall on nearly the same plane. The planets move as if they were on a flat surface such as the top of a desk. Third, the hypothesis also should explain why the planets are different. Four of the planets closest to the sun are like the earth. They are made

Solar system The sun with its nine planets and smaller bodies that move around it.

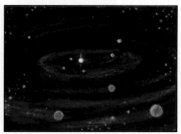

Fig. 2-1 Formation of the solar system: (a) a slowly rotating gas-and-dust cloud begins to shrink; (b) rotation accelerates, forming a disk with matter collecting at the center; (c) shrinking continues as the sun forms, with rings of material circling the new star; (d) the planets grow from the rings, following paths around the sun.

mostly of heavy solids. However, four of the planets far from the sun are made mostly of light gases.

THE COLLISION HYPOTHESIS

In science a hypothesis is often changed. Sometimes a better hypothesis is developed. The new hypothesis might be able to explain observations better. One hypothesis tried to explain the origin of the earth and other planets by saying that the planets formed when another star passed very close to the sun. The passing star then tore large pieces of matter away from the sun. These pieces of the sun cooled and formed bodies that became the planets. This hypothesis has at least two problems. Material drawn out of the sun would be so hot that it would expand instead of contract. Thus solid planets could not form. The hypothesis also says that the sun and another star would have to come very close. Observations show that stars are spread very far apart in space. The chance of such a near collision is very small. These problems, along with others, led scientists to search for a better hypothesis.

THE DUST CLOUD HYPOTHESIS

The hypothesis accepted by most scientists today is the dust cloud hypothesis. See Fig. 2-1. It says that the solar system came from a huge cloud of dust and gas in space. Such clouds can be seen with telescopes. The cloud of dust and gas was much larger than the solar system is today. This hypothesis says that the solar system started to form when this cloud of gas and dust began to shrink and spin slowly. The explosion of a nearby star may have caused the cloud to shrink. As it became smaller, it turned faster. The same thing happens to spinning skaters. When they pull their arms in, they spin faster. See Fig. 2-2.

As the cloud turned faster, it also became flatter. The shape began to look like a giant wheel or disk. This became the plane of the solar system. At the same time, most of the matter was gathering into a body around the center of the disk. As this body grew, it became hotter. Thus a hot new star, the sun, was formed. As the disk of gas and dust was still turning around the sun, smaller bodies were also beginning to grow. These smaller bodies were too small to have enough heat to become stars. They became the planets and smaller bodies of the solar system. Thus the sun came to have a system of large and small

Fig. 2-2 *When a skater draws her arms closer to her body, she spins faster.*

bodies around it. They moved in orbits on the plane of the disk from which they grew. Fig. 2–1 shows how the solar system formed from a cloud of gas and dust.

After the planets formed, they began to change. In the hot inner part of the young solar system, the heat from the sun caused the planets to lose most of the lighter gases. The heavier solids remained. The planets far from the sun were in the cold outer parts of the solar system. They were able to hold the lighter gases. This is one reason why the planets closest to the sun are different from the planets far from the sun.

At the present time, the dust cloud hypothesis is able to explain many observations. Future observations may show this hypothesis to be correct. Or the hypothesis may be changed as new observations are made.

SUMMARY

The collision and dust cloud hypotheses try to explain many observations of the solar system. The dust cloud hypothesis is accepted by most scientists today. It says that the solar system formed from a spinning cloud of dust and gas.

QUESTIONS

Use complete sentences to write your answers.

1. What observations must a hypothesis explain for how the earth formed?
2. Using the collision hypothesis, explain how the earth and other planets formed.
3. Why is the collision hypothesis no longer accepted by most scientists?
4. Using the dust cloud hypothesis explain how the solar system formed.

Enrichment: Astronomers are using radiotelescopes to peer into the interior of our galaxy in the hope that they will be able to see stars or solar systems being born.

1. The facts the hypothesis must account for are: (a) The earth is part of a large family of planets. (b) The planets' orbits fall in the same plane. (c) The planets differ from one another.
2. The collision hypothesis asserts that the sun originally had no system of bodies circling it. Another star passed very close. The passing star tore large chunks of matter from the sun. These pieces of the sun cooled and finally formed solid bodies held by the sun in orbits.
3. The hypothesis had two serious problems. First, the chunks torn from the sun would spread out, not cool down and become solid. Second, observations show that the stars presently are spread so far apart that it is very unlikely that near collisions could have occurred.
4. Fig. 2–3 on page 31 shows the main steps in the dust cloud hypothesis. Students should describe these steps in their own words.

See teacher's commentary for teaching hints, safety, and answers to questions.

A MODEL OF THE DUST CLOUD HYPOTHESIS

PURPOSE: To make and study a model of the dust cloud hypothesis.

MATERIALS:

celery seeds   round pan or dish
small spoon   container for water

Fig. 2–3

PROCEDURE:

A. Place the pan on a flat surface. Fill it about half-full of water. Obtain a half-teaspoon of celery seeds. Sprinkle the seeds over the surface of the water as evenly as possible. See Fig. 2–3. Have your partner observe what happens as this is done. Observe only the seeds that float on the water.

1. What was the very first thing that happened to the floating seeds?

2. What happened to the floating seeds after a few minutes?

3. Which part of the dust cloud hypothesis would this show?

B. Observe what happens to the floating seeds for a few more minutes.

4. What do you observe happening to the seeds?

5. Which part of the dust cloud hypothesis would this show?

C. Observe the seeds until there is no longer any movement. The forces that are causing this reaction allow you to see, in a very short period of time, what would take billions of years for gravity to achieve on the scale of our solar system.

6. Describe the pattern of seeds.

7. What do you think the pattern will be like in a few hours?

D. Pour out the water and seeds into the container given by the teacher.

E. Once again, place the pan on a flat surface. Then fill the pan about half-full of water. Obtain another half-teaspoon of celery seeds. Use your finger to stir the water in the pan to make it move in a circle. Remove your finger and then, immediately but slowly, sprinkle the celery seeds over the moving water surface.

8. In the very beginning, what happens to the floating seeds?

F. Watch the seeds until all the motion has stopped.

9. Describe the final pattern of seeds after motion has stopped.

CONCLUSIONS:

1. In what way does this model act like the dust cloud hypothesis?

2. In what ways does this model differ from the dust cloud hypothesis?

3. How is the second model a better one for the dust cloud hypothesis?

2–2. Birth of a Planet

When you finish this section you will be able to:

- ☐ Explain how scientists study the inside of the earth.
- ☐ Describe the inside of the earth using the terms "core," "mantle," and "crust."
- ☐ Explain how the earth might have formed continents, oceans, and an atmosphere.

The moon is the earth's closest neighbor in space. Using the moon as a laboratory has helped scientists learn about the earth's early history. For example, a rock taken from the moon looks like many rocks you can pick up on the earth. See Fig. 2–4. But moon rocks tell a different story. Almost all rocks on the earth's surface have been changed many times during the earth's history. However, moon rocks have hardly changed since the time the earth began. Unlike earth rocks, moon rocks give scientists information about the early solar system. Since the earth formed as a part of the solar system, the moon's story is also about the earth's beginning. From what we know about the moon, we can make a model of the early earth.

Fig. 2-4 An astronaut uses a special rake to collect rock chips from the surface of the moon.

EXPLORING THE INSIDE OF THE EARTH

Somehow, in its early history, the inside of the earth must have formed layers. Earthquake waves show the inside of the earth to be made up of layers. You probably have felt vibrations caused by a passing train or truck. The vibrations pass through the ground as a kind of wave. Earthquakes also produce waves. Large earthquakes can produce some waves that pass through the earth. When those waves return to the earth's surface, they carry information about the materials inside the earth. This is similar to the way X-rays can give information about the inside of your body. It has helped scientists to "see" the inside of the earth.

THE EARTH'S LAYERS

Crust The thin, solid outer layer of the earth.

Enrichment: The deepest well drilled goes down about 8 km.

Earthquake waves show that the inside of the earth is made up of three main layers. Covering the outside of the earth is a thin, solid layer called the **crust.** The thickness of the *crust* differs from place to place. The parts of the crust that make up the sea floor are from 4 to 7 km thick. Under the continents, the crust is about 35 km in thickness. Beneath some mountains, the crust is as much as 70 km thick. Since we can study the crust directly, we know more about it than we do about the other layers. However, the crust makes up only 1 percent of the earth's volume.

In the early 1900's a Yugoslavian scientist named A. Mohorovicic observed that earthquake waves changed their speed. They changed their speed a short distance beneath the earth's surface. He believed that this was because the waves had entered a kind of rock different from the crust. Below the crust, another layer of rock had been found. This layer was called the

Mantle The layer of earth below the crust having a thickness of about 2,870 km.

mantle. The boundary between the crust and the *mantle* is now called the *Moho*, after Mohorovicic. The mantle is a very thick layer of about 2,870 km. It makes up about 80 percent of the earth's volume and almost two-thirds of the earth's mass.

Core The center layer of the earth, made up of an inner and an outer part, having a radius of about 3,500 km.

Below the mantle, deep in the earth's center, is a very heavy **core.** The *core* has a radius of about 3,500 km. The outer core is liquid. The outer core's thickness is about 2,000 km. The inner core is solid. The inner core's radius is about 1,500 km. The core makes up 19 percent of the earth's volume and nearly one-third of its mass. Only rocks made of iron and nickel are heavy enough to compare to the mass of the core. Most of the

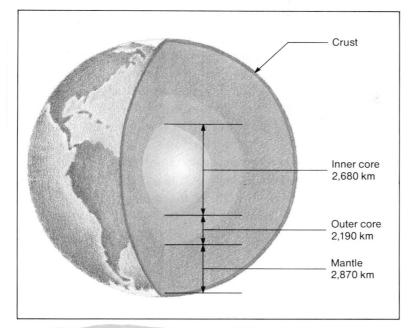

Crust

Inner core
2,680 km

Outer core
2,190 km

Mantle
2,870 km

Fig. 2–5 The interior of the earth is made up of layers, as shown in this diagram. The thickness of the crust is thinner than can be shown correctly in a diagram this size.

core is thought to be made up of these two metals. Fig. 2–5 shows the layers of the earth.

People who drill into the earth for minerals and oil find that the rocks deep in the crust are hotter than those on the surface. The deeper they drill into the earth, the higher the temperature. Measurements in the upper part of the earth's crust show that the temperature goes up about 30°C for each kilometer in depth. Scientists believe that the temperature in the mantle increases to about 3,000°C. Thus scientists find the earth to be made up of three main layers and a very hot interior. How can this be explained?

THE EARTH'S BEGINNING

Moon rocks show that the moon and planets of our solar system formed about 4.5 billion years ago. The moon and earth share much of their early history. Most moon rocks are similar to rocks found around volcanoes on the earth. This means that the moon was once very hot. During that time, the inside of the moon formed layers. This may be similar to the way the earth formed its layers. Planets, such as the earth, that are made up of layers are said to be **differentiated.** Scientists are not completely sure about how the earth became *differentiated* because the process took place so long ago. But layers and heat in

Differentiation The process by which planets became separated into layers.

the earth show that the earth once was very hot. Because of the heat, the entire earth was made up of melted materials. Then heavy substances such as iron sank and formed the core. Lighter materials floated and became the mantle and crust. In time, the young planet cooled. The crust and much of the mantle became solid. However, some of the heat still remains inside the earth.

Among the materials that formed was water. This water was locked up in certain minerals. As the earth began to cool, hot, melted rock poured out from the inside of the earth. Continents probably formed from the lightest rocks that cooled. The light rocks formed solid islands that floated on the still-soft crust. Water vapor escaped and turned into liquid water. In time, enough water gathered on the surface to form the world's first oceans. At the same time that water vapor was escaping from the earth's interior, other gases were also given off. The lightest of these gases escaped into space. Others were held by the earth's gravity. Thus a thin covering of gases remained over the surface to form the earth's first atmosphere.

Scientists believe that about four billion years ago the earth ended its development. All of the forces still working today to change the planet were set into motion. Today's earth was shaped by these forces.

SUMMARY

Most of what scientists know about the inside of the earth is based on earthquake waves. The earth is divided into three main layers—the core, mantle, and crust. The earth's continents, oceans, and atmosphere were formed because of heating and melting.

QUESTIONS

Use complete sentences to write your answers.

1. How do earthquakes give information about the inside of the earth?
2. Describe each layer of the earth.
3. Describe how the earth might have formed layers.
4. Explain how the continents, oceans, and atmosphere formed on the earth.

1. Scientists obtain information about the earth's interior by studying the behavior of earthquake waves as they pass through the earth.
2. The crust is a thin, solid layer averaging about 35 km in thickness under the land. Beneath the crust is the mantle, which extends to a depth of almost 2,900 km. Deep in the earth's center is a very dense core with a radius of 3,740 km. It is made up of a solid inner core surrounded by a liquid outer core.
2. The earth may have grown from a core of iron. In time it gathered lighter materials, which became the mantle and crust. Heat from collisions, gravity, and radioactivity caused the earth to become molten. As new material was added to the growing earth, the heavy materials sank inward while the lighter materials moved upward to join the mantle or crust.
4. The water originally was locked up in minerals. As the earth began to cool, molten rock containing these materials poured out onto the surface of the earth, releasing the water as vapor. The water vapor condensed, eventually forming the earth's oceans. At the same time, the gases that went to make up the earth's first atmosphere also were released.

See teacher's commentary for teaching hints, safety, and answers to questions.

THE ORIGIN OF A LAYERED EARTH

PURPOSE: To use observations from two experiments to show how layers were formed in the earth.

MATERIALS:

250-mL beaker   light small rock
soil sample   water
heavy small rock

PROCEDURE:

A. Fill the beaker about half-full of water. Then obtain 60 mL of the soil sample. Add the soil sample to the water in the beaker and stir. Allow the mixture to sit undisturbed for about five minutes.

　1. What evidence of layering do you find in the beaker?

B. Look closely at the material that has settled to the bottom.

　2. Compare the size of the particles at the bottom with those found at the top.

C. Look closely at the top of the water in the beaker.

　3. Describe any particles floating on the surface.

　4. How does the size of these particles compare with those that sank to the bottom?

D. Pick up the two rock samples one at a time.

　5. Which rock, the heavy one or the light one, would you predict to fall faster in the water?

E. Test your prediction by experiment. Use the same container of water and soil from before. Drop both rocks at the same time from just above the surface of the water. Your partner will observe which hits bottom first and what happens when they strike bottom.

　6. Was your prediction correct?

F. The table below shows the average density of the layers of the earth.

Layer	Density (g/cm³)
Crust	2.8
Mantle	4.5
Core	10.7

　1. What is the density of the crust?

　2. The density of the earth is 5.5 g/cm³. Compare the density of the crust with the density of the earth.

　3. How do you explain the difference?

　4. Explain how scientists were able to determine the density of the mantle and the core from knowing the density of the crust and the earth.

CONCLUSIONS:

　1. If the earth was melted at one time, would this have helped layers to form? Explain your answer.

　2. If the earth was melted, would light material or heavy material end up in the center as a core?

　3. If there are layers inside the earth, would the materials of which they are made be similar?

　4. Why are sources of heavy metals most often found in those areas where volcanoes once were active?

2-3. Models of the Earth

When you finish this section you will be able to:

☐ Show how the size and shape of the earth can be found.
☐ Describe how maps are made from globes.
☐ Read information from maps.

In 1492, when Columbus sailed to find the Far East, some of the crew nearly forced him to turn back from his voyage. At that time, many believed the world was nearly flat. Some crew members were afraid that the ships would come to the edge of the world and fall off. People living on the earth cannot see that the earth's surface is curved. However, you have seen pictures taken from space that clearly show the earth as round. See Fig. 2-6. But how could you prove the earth to be round from your own direct observations?

Fig. 2-6 The earth as seen from outer space.

SIZE AND SHAPE OF THE EARTH

The ancient Greeks knew that the earth was round. About 300 B.C. Aristotle gave two reasons for believing in a round earth. He observed that when the earth's shadow was cast on the moon, it had a curved shape. Also, people moving north from Greece reported that stars moved higher in the sky. The stars seem to move lower in the sky as the people moved south. Aristotle said that this could only be explained if the earth was round. See Fig. 2-7.

Skills: Aristotle's reasoning is a good example of the use of cause and effect.

| North Star as seen at equator | As seen at Chicago | As seen at North Pole |

The ancient Greeks also were able to measure the size of the earth. A Greek named Eratosthenes (air-uh-**tahs**-thuh-nees) made this measurement about two thousand years ago. He observed that on June 22, at noon, the sun's rays reached the bottom of a deep well in Syene, a city in Egypt. He also observed that the buildings did not cast shadows. This meant that the sun was directly above Syene. On the same date at noon, Eratosthenes knew that in Alexandria, a city to the north, buildings did cast shadows. He asked how buildings in Alexandria could cast shadows and at the same moment buildings in Syene cast no shadows? The only possible answer was that the earth's surface was curved. See Fig. 2–8. The sun's rays did not strike all places on the earth's curved surface at the same angle.

Fig. 2-7 Evidence that the earth is round can be seen in the way that the North Star is seen higher in the sky when moving north or lower when moving south.

Fig. 2-8 (a) On a flat surface, shadows cast by two different buildings of the same size would be the same length. (b) On a curved surface, shadows cast by two different buildings of the same size would be different.

Eratosthenes also saw that he could measure the distance around the earth by using the shadows. When the sun was not directly overhead, shadows would be cast. The sunlight would be striking everything at an angle. Eratosthenes was able to measure the angle of the sun's rays in Alexandria. He found the angle to be 7°. A complete circle is 360°. Seven degrees is about one-fiftieth of a complete circle. So the distance between Alexandria and Syene was one-fiftieth of the distance around the earth. See Fig. 2–9. Eratosthenes also knew that the distance between Alexandria and Syene was about 800 km. So 800 km was one-fiftieth of the distance around the earth. He then multiplied 800 km by 50. His result was 40,000 km for the distance around the earth. This was very close to the modern measurement of 39,989 km.

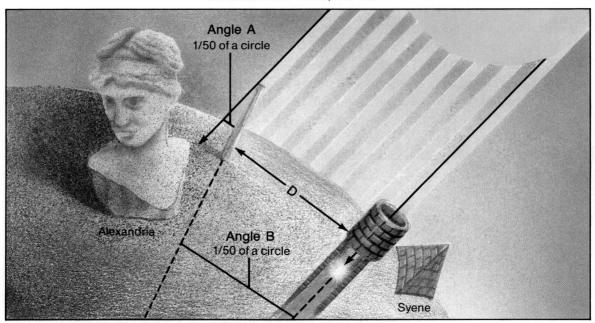

Angle A
1/50 of a circle

Alexandria

Angle B
1/50 of a circle

D

Syene

Fig. 2-9 Eratosthenes' method depended on measuring angle A. It was equal to angle B, which had the same relationship to a full circle that distance D did to a full circle.

Modern measurements have also shown that the earth is a little larger around the equator than around the poles. This means that it is not perfectly round. However, the earth is so close to being exactly round that its shape is said to be round.

MODELS OF THE EARTH

When we want to study the large features of the earth's surface, such as continents and oceans, some kind of earth model must be used. An example of an earth model is a globe. See Fig.

2–10. In many ways, a globe is the best model of the earth. Like the earth, the globe is round. But globes are often not easy to use. They are usually too small to show the earth's surface with much detail. A globe also cannot be carried easily. Because of this, maps often are used as models of the earth. Maps can show the earth's surface on a flat sheet of paper.

The earth's surface shown on a flat sheet of paper is always out of shape. Imagine trying to flatten a large piece of an orange peel. The curved shape of the peel will stretch and tear. In the same way, the curved surface of the earth, when shown on a flat surface, is stretched out of its true shape.

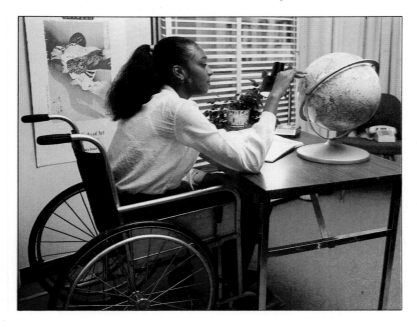

Fig. 2-10 A globe is a model of the earth.

MAKING MAPS

Making the curved surface of the earth into a flat map is done by using a **map projection.** There are different ways of projecting the curved surface of the earth onto a flat surface. No one way is perfect. Some projections distort the size of the continents, others distort the shape. You can imagine how a *map projection* is made by thinking of a globe made of glass. If you put a light inside the glass globe, the globe's surface could be projected on a piece of paper held next to the globe. One of the most common kinds of map projections is called a Mercator projection. It is made as if a sheet of paper were wrapped

Map projection The curved surface of the earth shown on the flat surface of a map.

around a globe, making a cylinder. See Fig. 2-11. This is the kind of map projection most often used in textbooks. Only the area of the earth's surface near the equator is shown accurately on this kind of map. Regions near the poles appear much larger than their true sizes. See Fig. 2-12.

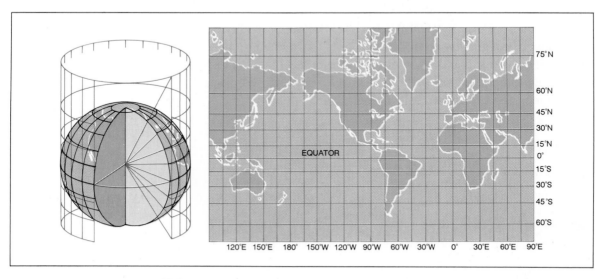

Fig. 2-11 (above) A Mercator projection shows only the areas of the earth's surface near the earth's equator accurately.

Fig. 2-12 (right) On the left is Greenland as it appears on maps using a Mercator projection. On the right is Greenland in its true size in relation to other masses.

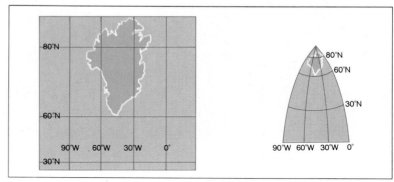

Another type of map projection uses a cone instead of a cylinder. This is called a conic projection. See Fig. 2-13. Maps showing small parts of the earth's surface, such as road maps, use this kind of map projection. Suppose that a flat sheet of paper is held so that it touches a globe at a single point such as the North Pole. A map then can be projected up onto the paper. This kind of map is called a polar projection. See Fig. 2-14. A polar projection only shows the correct size of regions near the poles. As you can see, all maps are not the same. Each shows only part of the earth accurately and other parts out of shape.

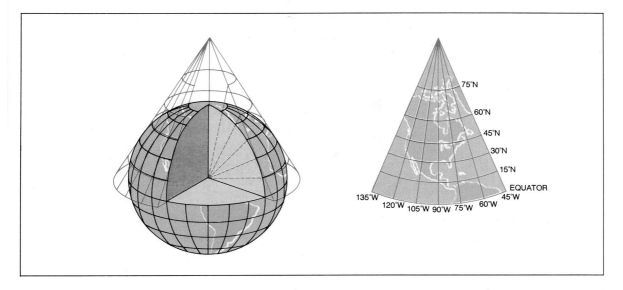

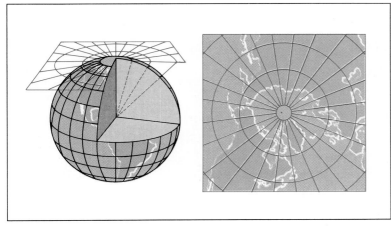

Fig. 2-13 (above) A map projection based on a cone. Maps made this way can show the true size of small areas of the earth's surface.

Fig. 2-14 (left) A polar projection shows the regions near the poles without distortion.

READING MAPS

A map can be used to provide different kinds of information about the earth's surface. For example, a *political map* shows boundaries between countries or states. These are human divisions of the earth's surface. Cities and towns are also shown. *Relief maps* show the land surface by using different colors for mountains and valleys. Political maps and relief maps are often combined. Weather can also be shown on maps.

Suppose that you want to use a map to guide you on a trip. The first thing you need to know is the direction you will be going. Common types of maps are made so that north is at the top and east is at the right when the map is held upright.

Suggestion: Hand a road map to a student. Watch and then comment on the automatic arranging of the map in a manner to put north at the top, etc.

You can also use a map to measure how far you will be going. This can be done by using the **scale** of the map. All maps have a *scale*. The scale shows the relationship between the distance on a map, or the size of a model, and the actual measurement that it represents. For example, a model airplane might have a scale of 1 to 1,000. This means that the model would be one-thousandth the size of the real airplane. Scales on maps show a relationship between distances. For example, 1 cm on a map might represent 1 km on the ground. The scale of the map may be shown as a bar graph. Or the scale may be written as 1/10,000. The numerator, 1, gives the map distance. The denominator represents the actual distance on the earth's surface. See Fig. 2–15. A map scale may also be written as 1:10,000. The first number is the distance on the map. The second number is the actual distance on the earth's surface. Thus, 1 cm on the map would be equal to 10,000 cm, or 100 m, on the ground. Most maps show the scale in a lower corner of the sheet.

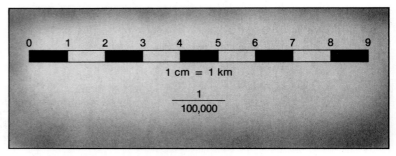

Fig. 2–15 The scale on a map can be shown by a graph (top) or as a fraction (bottom). On this map scale, 1 cm equals 1 km in actual distance.

Another kind of information found in the lower corners of maps are symbols. Symbols are used to show features such as roads, buildings, and railroad tracks. Some symbols look like the features they represent. To show what these and other symbols mean, maps have a **legend.** A *legend* explains the meaning of each symbol used on a map.

TOPOGRAPHIC MAPS

One type of map that is very useful in earth science is called a **topographic map.** See Fig. 2–16. This is because *topographic maps* show the surface features of the earth. This is done by using **contour lines.** Lines on a map that connect points having the same elevation, or height, are called *contour lines*. Contour means shape. Contour lines are used to show the shape of the land. Each contour line is labeled with the elevation of all the

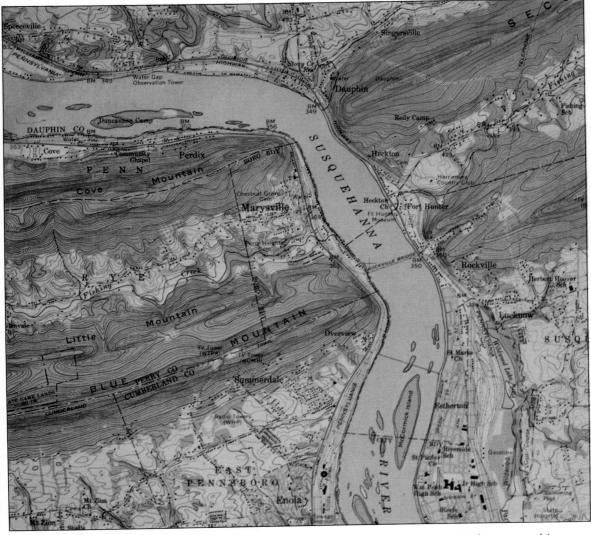

Fig. 2–16 A topographic map shows the surface features of the earth.

points found on the line. Fig. 2–17 shows how contour lines are drawn to show mountains and valleys. The difference in elevation between the contour lines is called the *contour interval*. The contour interval on different topographic maps is not always the same. The contour interval is chosen according to the size of the map and the elevations that must be shown. For example, the contour interval shown in Fig. 2–17 is 100 m. A smaller contour interval would cause the lines to be closer together and harder to see. Other symbols commonly used on topographic maps are shown in Fig. 2–18.

A UNIQUE PLANET

Fig. 2-17 (right) Contour lines show the changes in the elevation on the earth's surface.

Fig. 2-18 (below) Common symbols found on a topographic map.

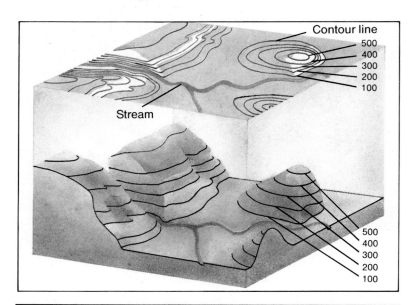

Symbol	Color	Meaning	Symbol	Color	Meaning
	Black	Buildings		Blue	Stream
				Blue	Stream that does not flow constantly
	Black	school and Church		Blue	Lake or pond
	Red or black	Road or highway		Brown	Depression
	Black	Railroad		Blue or green	Marsh or swamp

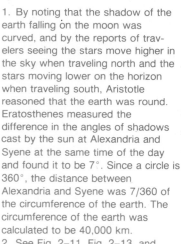

1. By noting that the shadow of the earth falling on the moon was curved, and by the reports of travelers seeing the stars move higher in the sky when traveling north and the stars moving lower on the horizon when traveling south, Aristotle reasoned that the earth was round. Eratosthenes measured the difference in the angles of shadows cast by the sun at Alexandria and Syene at the same time of the day and found it to be 7°. Since a circle is 360°, the distance between Alexandria and Syene was 7/360 of the circumference of the earth. The circumference of the earth was calculated to be 40,000 km.

2. See Fig. 2–11, Fig. 2–13, and Fig. 2–14. The students should describe in their own words the manner in which these projections are shown to be made.

3. A political map shows boundaries between countries or states. It also shows cities and states. A relief map shows what the land's surface is like. A map can also show weather. Answers might also include distance between places and special features such as bridges and roads.

4. The topographic map is used in earth science. Its main feature is the use of contour lines giving elevations of land showing valleys and hills.

SUMMARY

The earth's size and shape can be found from observations. Maps are made by projecting the curved surface of the globe onto a flat surface. Maps give us information about the earth's surface.

QUESTIONS

Use complete sentences to write your answers.

1. Explain how the ancient Greeks were able to tell the earth's size and shape.
2. Name three kinds of map projections.
3. List three kinds of information that maps provide.
4. Describe the type of map that is very useful in earth science.

INVESTIGATION

See teacher's commentary for teaching hints, safety, and answers to questions.

CONTOUR MAPS

PURPOSE: To draw a contour map.

MATERIALS:

4 pipe cleaners	masking tape
soup plate	grease pencil
mixing bowl	metric ruler
plastic cup	2 sheets of paper

PROCEDURE:

A. Turn the plate, bowl, and cup upside down on a flat surface. Place the bowl on top of the plate and the cup on top of the bowl to form a "hill." Tape the three objects together. See Fig. 2–19.

B. Tape two sheets of paper together. Place the objects on the paper. Draw guide lines on the paper and on the plate so that you can align them when you remove the plate. Draw a line around the edge of the plate.

C. Hold the ruler straight up on the table. At the 5-cm mark, use the grease pencil to mark a line on the bowl. Make this measurement at several places around the bowl.

D. Twist the ends of the pipe cleaners to form one long piece. Wrap it around the bowl at the 5-cm mark. Have a partner hold the pipe cleaner while you remove the taped objects. Place the pipe cleaner on the paper and trace the circle it makes.

E. Do the same for elevations of 10 cm and 15 cm.
1. How many circles have you drawn?
2. What are each of their elevations measured in centimeters? Label them.
3. If 1 cm = 2 m on your model, what contour interval would that represent?

F. Use the ruler to approximate the highest point of the hill.
4. What is the highest elevation in meters?
5. If there were a school at that elevation, how would you show it? Draw it in.
6. Many cups have an indentation on the bottom. How would you show this on your contour map?
7. How would you show a marsh between 10 m and 20 m?

CONCLUSIONS:
1. How could you change your model to show that one side is steeper than the other?
2. If you did that, what would the contour lines look like?
3. What would the contour lines look like if a stream ran down the hill?

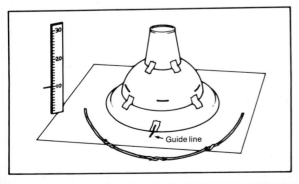

Fig. 2–19

Fig. 2-20 The location of any place on earth can be found. What two pieces of information are given here?*

*Distance and direction.

Latitude The distance in degrees north or south of the equator.

2-4. Position on the Earth's Surface

At the end of this section you will be able to:

☐ Describe a location by using latitude and longitude.

☐ Find a place on the earth's surface by using latitude and longitude.

If someone wants to find where you live, you can give the person an address. Your address probably is made up of two things. It has a street name and a number. In the same way, the "address" of any place on the earth's surface can be found. See Fig. 2-20. Even a place in the middle of the ocean can be found if it can be described as the point at which two lines cross.

LATITUDE

The equator is an imaginary line that divides the earth into two equal parts. It is drawn around the globe midway between the North Pole and South Pole. One of the kinds of lines used to find places is based on the equator.

Suppose that you want to find an island in one of the world's oceans. First you must know if the island is in the northern or southern part of the earth. Once you know which part, you need to know how far north or south of the equator the island is. This can be done by using other lines running parallel to the equator. Each of these lines is called a parallel of **latitude**. A parallel of *latitude* is measured by using the equator as a starting point of 0°. Then it is possible to find how many degrees north or south of the equator the parallel of latitude is. Such a description is called the latitude. Latitude tells the distance in degrees north or south of the equator. For example, the latitude of the North Pole is 90° north or 90°N. In the same way, the latitude of the South Pole is 90°S. A place halfway between the North Pole and the equator would have a latitude of 45°N. See Fig. 2-21. Every place on the earth's surface can be described by a latitude. Can you see why the latitude of a place can never be greater than 90°?

Latitude can be found by using the North Star. One way to do this is by measuring how far the North Star is above the horizon at night. The latitude can be found by measuring the number of degrees the North Star is above the horizon.

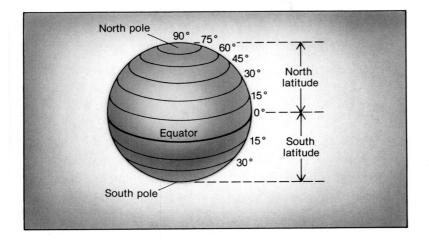

Fig. 2-21 *Find the parallel of latitude that would pass through a place halfway between the equator and the North Pole.*

*45°N.

LONGITUDE

The latitude of a place only describes how far north or south of the equator a place is. But you also need to know its east–west location. Another set of lines, called *meridians,* is used to show east–west locations. These lines are drawn from pole to pole. The prime meridian is used as the starting point of 0°. It runs through Greenwich (**Gren**-ich), England. The location of a place east or west of the prime meridian is called its **longitude.** *Longitude* shows the distance east or west of the prime meridian in degrees. For example, a place might have a longitude of 30° west or 30°W. It would be found on the meridian that is 30° west of the prime meridian. See Fig. 2–22. The east and west longitude lines meet at the 180° meridian on the other side of the earth.

Longitude The distance in degrees east or west of the prime meridian.

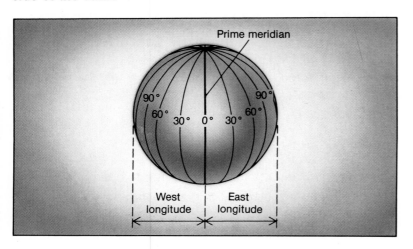

Fig. 2-22 *Longitude measures the distance east or west of the prime meridian.*

FINDING LOCATION BY LATITUDE AND LONGITUDE

By using both latitude and longitude, any place on the earth's surface can be found. For example, Washington, D.C., has a latitude of 39°N and a longitude of 77°W. Therefore, Washington can be found on the parallel of latitude that is 39° north of the equator. Washington is also on the meridian that is 77° west of the prime meridian. Washington, D.C., is found where these two lines meet. See Fig. 2–23. Thus the latitude and longitude of a place are like an address, showing where it is on the earth.

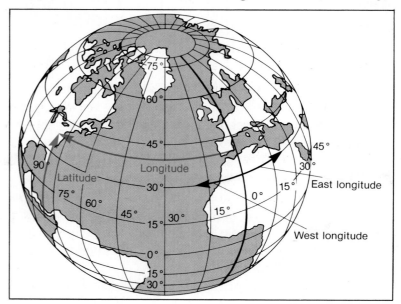

Fig. 2-23 Washington, D.C., is found at 39°N and 77°W.

SUMMARY

To find a place on the earth's surface, lines of latitude and longitude are used. Latitude is measured from the equator and longitude is measured from the prime meridian. Both latitude and longitude can be measured by using the position of the North Star and the sun.

QUESTIONS

Use complete sentences to write your answers.

1. Describe how you would find an island located near 15°S and 45°E.
2. What state is found at 60°N and 150°W?

1. The latitude tells that the island is 15° south of the equator; the longitude tells that it is 45° east of the prime meridian. The island's location is the point at which these two lines meet.

2. You would know that it is located 60° north of the equator and 150° west of the prime meridian. The location is on the south coast of Alaska.

See teacher's commentary for teaching hints, safety, and answers to questions.

READING LATITUDE AND LONGITUDE

PURPOSE: To find places on a map using latitude and longitude.

MATERIALS:
pencil   paper

PROCEDURE:

A. Study the map of the world in Fig. 2–24. Answer the following questions.

1. At about what latitude and longitude is your school?

2. At about what latitude and longitude is the capital of the United States?

3. At about what latitude and longitude is the center of Africa?

4. At about what latitude and longitude is the United Kingdom (England)?

5. What longitude does not seem to pass through any land masses?

6. What continent is found at 25°S and 135°E?

7. What state of the United States is found at 20°N and 155°W?

8. What country is found at 40°N and 140°E?

9. What country is found at 25°N and 105°W?

CONCLUSIONS:

1. Can the location of a city be found by using only its latitude or only its longitude? Why?

2. Explain how the location of any place on the earth's surface could be described.

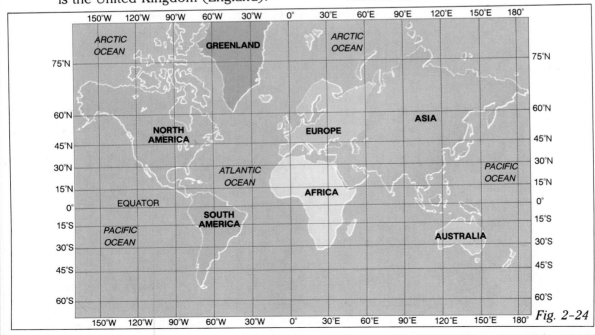

Fig. 2–24

2-5. Motions of the Earth

At the end of this section you will be able to:

- ☐ Explain the effects of the earth's rotation.
- ☐ Describe the earth's motion around the sun.
- ☐ Explain the relationship between the tilt of the earth's axis and the seasons.

Fig. 2-25 This photo shows the sun's position in the sky at the same time of day on different dates throughout the year. The figure-eight shape is caused by the sun's changing position in the sky.

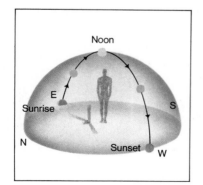

Fig. 2-26 The motion of the sun across the sky each day can be shown by the moving shadow cast by a stick.

How do you think the picture of the sun in Fig. 2-25 was made? The picture was made by someone moving at a speed of 106,000 km/h in space. The person was traveling on the moving earth. So you and everything on the earth are moving at 106,000 km/h. In this section you will study evidence that you are a rider on a moving earth.

THE EARTH'S ROTATION

Each day the sun appears to rise in the morning, cross the sky, and set in the evening. It seems to move across the sky from east to west, as shown in Fig. 2-26. There are two ways to explain the movements of the sun. First, it could be said that the sun moves around the earth each day. This idea explains the observations of the sun. This idea was believed to be true for many centuries. A Greek named Ptolemy reasoned that the earth could not be moving. If the earth moved, he said, a bird on a tree branch would be swept off. The idea that the sun moved around the earth was accepted for 1,500 years.

But in the 1500's a Polish scientist, Copernicus (koe-**pur**-nih-kus), proposed that the sun does not move around the earth. Why, then, does the sun appear to move across the sky? The correct explanation is that the earth is spinning like a bicycle wheel. This spinning motion is called *rotation*. Just as a wheel turns on an axle, the earth turns on its **axis**. The earth's *axis* is an imaginary line running through the center of the earth from pole to pole. The direction of earth's rotation is west to east. For example, if you put your finger on a globe and push toward the east direction on the globe, you will cause it to rotate west to east, the same way as the earth.

Axis The imaginary line through the center of the earth on which the earth rotates.

Enrichment: On earth, a person's motion in space is a combination of the following: (a) earth's rotation; (b) earth revolving about sun; (c) sun and system revolving about galaxy center; (d) movement of galaxy away from center of universe.

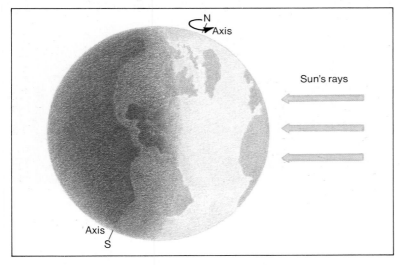

Fig. 2-27 The rotating earth produces day and night.

For everyone on earth, the most observable effects of the earth's rotation are day and night. As the earth rotates, every place passes through sunlight and darkness. See Fig. 2-27. A shadow made by a stick during a sunny day can be used to follow the path of the sun across the sky as the earth rotates. The shadow stick could be made into a sundial and used to tell time. When the sun is highest in the sky, the shadow's position on the ground could be marked twelve noon. Numbers could be put on each side of this mark to show daylight hours before and after noon. You would have made a kind of sundial. See Fig. 2-28. However, since the sun moves, people a short distance east or west of each other would have different times on the sundial. This would happen when people set their clocks

Fig. 2-28 A sundial can be used to measure the passage of time during the day.

for noon when the sun was highest in the sky. Because the sun moves from east to west, the sundials in the west would show noon later than those in the east. For example, Philadelphia is a short distance west of New York City. The sun is overhead in Philadelphia about four minutes later than in New York. Cities near each other would have different times. Thus the use of sundials is not the best way to tell time.

To prevent this problem, the earth has been divided into 24 equal north–south strips, or time zones. Each time zone covers 15° of longitude. The boundaries between time zones have been made so that they do not pass through cities. Within each zone, all places have the same time. This is called **standard time.** All clocks in a time zone are set to the *standard time* for that time zone. See Fig. 2-29. When you travel east and into the next time zone, you must set your watch ahead one hour. When traveling west and into the next time zone, you must set your watch back one hour.

Standard time The time based on one meridian and used for a given time zone.

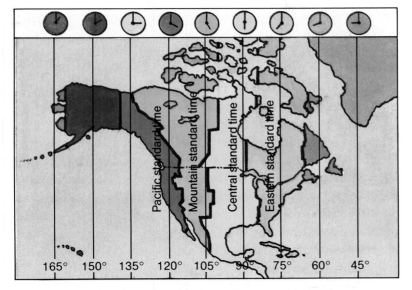

Fig. 2-29 This map shows time zones for North America and Hawaii. The boundaries between time zones are arranged so that all of any state is in the same zone. Clocks in Hawaii and most of Alaska are set two hours earlier than Pacific standard time.

If you traveled all the way around the world, you would cross 24 time zones. Each time zone differs from the one next to it by one hour. This means that a trip around the world would cause you to gain or lose 24 hours, or one day. To adjust for this, an international date line has been placed between two of these zones. See Fig. 2-30. It was placed through the Pacific Ocean at the 180° meridian in order to avoid crossing any land masses. When crossing the international date line going west, the day

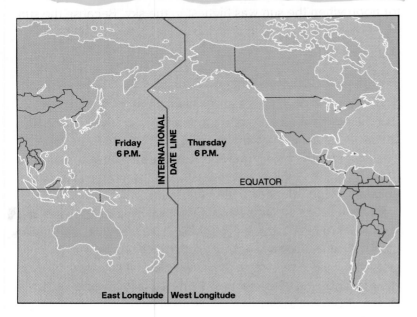

Fig. 2-30 The international date line is located in the Pacific Ocean to avoid crossing land where people live.

changes to the next day. When going east, the day changes to the day before. For example, when it is Sunday west of the international date line, it is Saturday east of it.

MOTION AROUND THE SUN

Each planet follows a curved path, or orbit, around the sun. There is a gravitational attraction between each planet and the sun. This attraction keeps the planet in its orbit around the sun. The orbit of the earth is not a perfect circle. The orbit is an oval shape called an *ellipse* (eh-**lips**). See Fig. 2–31. The sun is not located at the center of the ellipse but is slightly off to one side. As a result, the earth, as it moves in its orbit, is not always the same distance from the sun. During the year, the distance from the earth to the sun changes. The average distance between the earth and the sun is 150 million kilometers.

Fig. 2-31 The earth is not always the same distance from the sun because its orbit is elliptical. The stars seen in the summer sky are different from those seen in the winter sky.

Observation of the stars at different times of the year shows that the earth follows an orbit around the sun. If you look at the sky on a clear night, you will see stars and some patterns of stars. Six months later, you will see different stars and different patterns. As the earth moves in its orbit around the sun, different stars come into view. See Fig. 2–31.

If you think of a planet as moving in its orbit on a plane, its axis will not be straight up, or at 90° to that plane. The earth's axis is tilted 23½° to the plane of its orbit. As the earth moves around the sun, the North Pole will be tilted toward the sun during part of the earth's orbit. The axis will be tilted away from the sun during another part of the orbit.

SEASONS

The tilt of the earth's axis and the earth's movement in its orbit around the sun cause the seasons. In the Northern Hemisphere, the North Pole is tilted most toward the sun about June 22. This is the first day of the summer season. This day is called the **summer solstice** (**sol**-stus). The word "solstice" means "sun stops." This is because the sun's path on this day will reach its highest position in the sky for the year. The sun's rays strike the earth's surface more directly on the *summer solstice* than on any other day of the year. See Fig. 2-32b. Thus the Northern Hemisphere receives more solar energy about June 22 than at any other time of the year. This day also has the longest time of light and shortest time of darkness in the year. Within the area of the North Pole, there are 24 hours of daylight. See Fig. 2-33, position A. In some time zones, clocks are set ahead one hour because of the longer days. This system is called *daylight-saving time.* More daylight and the large amount of solar energy cause the warmer weather in the summer months.

Summer solstice The time of the year, about June 22, when the earth's axis is tilted most toward the sun.

Enrichment: Research of a number of ancient stone monuments and megaliths show that many were used as calendars. They were erected to show events such as the summer solstice, winter solstice, etc.

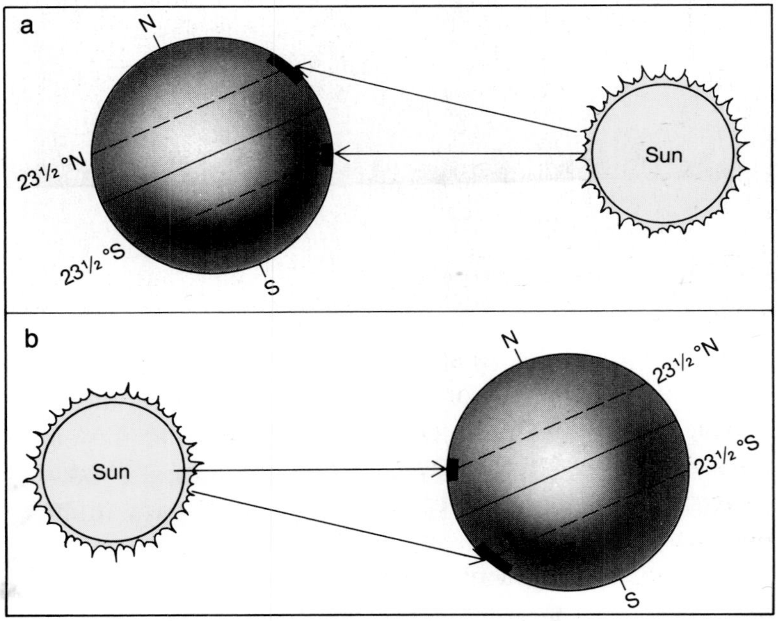

Fig. 2-32 The sun's rays deliver more energy when they strike directly than when they strike at an angle.

As the earth moves in its orbit, the sun's path across the sky each day becomes lower. The days become shorter and the nights longer. Three months after the summer solstice, about September 22, the length of day becomes equal to the length of night. The Northern Hemisphere is tilted neither toward nor away from the sun. See Fig. 2–33, position B. This is the first day of autumn, or fall. This day is called the **autumnal equinox** (**ee**-kwuh-noks). *Equinox* means "equal night."

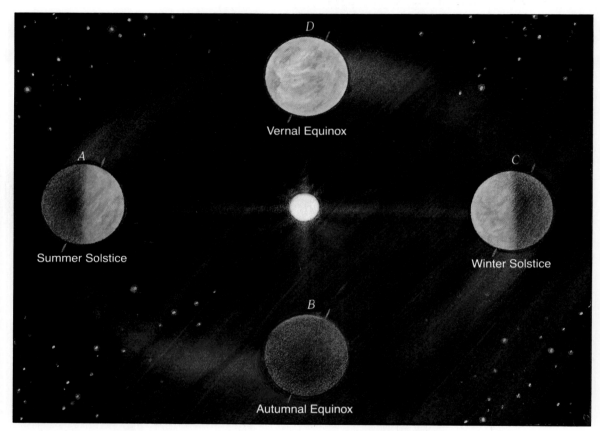

Vernal Equinox

A

Summer Solstice

C

Winter Solstice

B

Autumnal Equinox

Fig. 2–33 The tilt of the earth's axis remains almost the same over very long periods of time. Position A: summer solstice; Position B: autumnal equinox; Position C: winter solstice; Position D: vernal equinox.

During the autumn, or fall season, the path of the sun across the sky continues to lower as each day passes. As the earth moves in its orbit, the days become shorter and the nights longer. Three months after the autumnal equinox, about December 22, the Northern Hemisphere is tilted away from the sun. See Fig. 2–33, position C. This is the first day of winter and is called the **winter solstice**. The sun's path has reached its lowest position in the sky for the year. The sun's rays make their smallest angle with the earth's surface. The Northern Hemisphere receives the smallest amount of solar energy of the year and has the shortest day and longest night. In the area of the North Pole, there are 24 hours of darkness. The small amount of solar energy and the short days cause the cool winter weather during the winter season.

After the *winter solstice*, the days begin to grow longer and the nights shorter. The sun's path in the sky moves higher as each day passes. Three months after the winter solstice, about March 21, the length of the day becomes equal to the length of the night. The earth is tilted neither toward the sun nor away from it. See Fig. 2–33, position D. This day is called the **vernal equinox.** This is the first day of spring.

After the *vernal equinox*, the earth moves for another three months toward its summer solstice position. The days grow longer and the nights shorter. The days are now longer than the nights. The sun's path across the sky moves higher each day. In June the summer solstice will occur again. Then the cycle of seasons will begin again.

It takes about one year to complete the cycle of seasons. A year is defined as the length of time it takes the earth to complete one revolution in its orbit. One way of measuring the year would be to measure the amount of time between two vernal equinoxes. However, the time between two vernal equinoxes is about 20 minutes shorter than one year. This is due to the effect of the moon's gravity on the earth. It causes the earth's axis to move. In 13,000 years, the earth's axis will point toward the sun on about December 22. In other words, summer in the Northern Hemisphere will take place in December. After 13,000 more years, the earth's axis will return to its present position.

SUMMARY

Day and night are the most observable effects of the earth's rotation. The earth revolves around the sun in an elliptical orbit. The tilt of the earth's axis and the earth's revolution cause the seasons.

QUESTIONS

Use complete sentences to write your answers.
1. What causes day and night?
2. Describe the earth's path around the sun.
3. Why do we need standard time zones and the international date line?
4. Describe the tilt of the earth's axis during each season.

Vernal equinox The time of the year, about March 22, when the earth's axis is tilted neither toward nor away from the sun.

Enrichment: This tilting motion is called precession. Precession causes the earth's axis to move in a slow circle, completing one turn approximately every 26,000 years. This motion is similar to the wobbling of a spinning top as it slows down.

1. The earth's rotation on its axis once every 24 hours causes the sun to shine on one side of the earth at a time.
2. The path is the shape of an ellipse with the sun a little off center. This results in the earth varying its distance from the sun during its orbit.
3. Standard time zones are used to avoid confusion in telling time as you travel from one place to another. Since there are 24 time zones as one goes completely around the earth, it is necessary to have a place, called the international date line, where the time changes from one day to the next.
4. In the Northern Hemisphere, the axis is tilted most toward the sun at the summer solstice. The axis is tilted neither toward nor away from the sun at the fall equinox. The axis is tilted farthest away from the sun at the winter solstice. Once again, at the vernal equinox, the axis is tilted neither toward nor away from the sun.

INVESTIGATION

STAR MOVEMENT

PURPOSE: Why do different stars appear in the sky during different times of the night and the year?

MATERIALS:
pencil   paper

PROCEDURE:

A. The only stars that are visible at midnight are those that are above the horizon, as you can see in Fig. 2–34.

    1. What letters in Fig. 2–34 show stars visible at midnight?

    2. What letters in Fig. 2–34 show stars not visible at midnight?

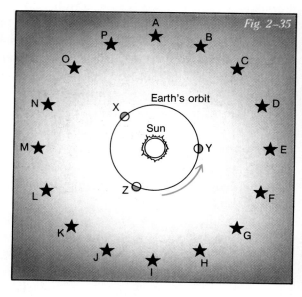

Fig. 2–35

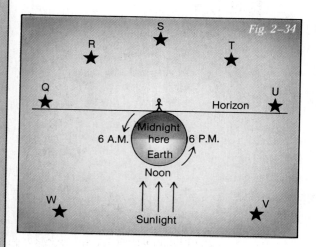

Fig. 2–34

B. The earth moves around the sun. It takes the earth one year to complete its orbit around the sun. Fig. 2–35 shows the earth at three different positions on its year-long orbit around the sun. They are shown in the diagram as X, Y, and Z. Stars are shown in positions A to P.

C. Look carefully at Fig. 2–35. At earth position X, put a piece of paper on the diagram so that the edge of the paper represents the horizon at midnight.

    3. Which stars will be visible (above the horizon) at midnight when the earth is in position X?

D. Do the same thing for earth positions Y and Z.

    4. Which stars will be visible at midnight when the earth is in position Y?

    5. Which stars will be visible at midnight when the earth is in position Z?

CONCLUSIONS:

1. Why aren't all stars seen at night?

2. Explain why different stars are visible at midnight at different times of the year.

3. Explain why certain stars, such as the North Star, can be seen all year round.

CAREERS IN SCIENCE

CARTOGRAPHER

Since ancient times, maps have shown where places are and, often, how to reach them. Today, cartographers continue to produce a variety of maps. Road maps guide us on our trips. Topographic maps show the earth's land features. Resource maps show the distribution of our planet's natural resources. Star maps describe the universe.

Cartographers must present data accurately and according to scale. Their tools include measuring devices, rich colors, and even computers and images sent from satellites. In fact, some cartographers are involved in mapping what we know of other planets.

To work in the field, cartographers need a college degree and must have completed courses in geography, economics, drawing, design, and computer science. Jobs are available in publishing companies, the government, and universities. For further information, write: Association of American Geographers, 1710 16 Street N.W., Washington, DC 20009.

SCIENCE EDITOR

Science editors work on textbooks, magazines, and journals; some even develop computer software. A science editor supervised the way this book is designed and the information that appears in it. An editor reviews the work written by the author and decides if anything needs to be changed, or added, or omitted. He or she must make sure that the information contained in the article or book is accurate. Science editors also work with the artists and photographers who provide illustrations for the written material, which is called copy.

To become a science editor, you need a college degree and strong writing skills. A knowledge of science and math is essential, and teaching experience is helpful. Jobs are available with a variety of publishing companies. For further information, write: National Association of Science Writers, P.O. Box 294, Greenlawn, NY 11740, or Association of Earth Science Editors, c/o American Geological Institute, 5205 Leesburg Pike, Falls Church, VA 22041.

See teacher's commentary for answers.

VOCABULARY

On a separate piece of paper, match the number of each sentence with the term that best completes it. Use each term only once.

axis	crust	longitude	winter solstice	scale
contour line	equinox	orbit	standard time	mantle
core	latitude	solar system	summer solstice	topographic map

1. The sun, planets, and smaller bodies orbiting the sun are called the _____.
2. The _____ is the thin, solid layer covering the outside of the earth.
3. The _____ is a very thick layer just under the crust of the earth.
4. The very dense center of the earth is called the _____.
5. A(n) _____ shows the features of a small part of earth's surface.
6. Points having the same elevation are shown by a(n) _____ on some maps.
7. The _____ of a location is the distance north or south of the equator in degrees.
8. The distance east or west of the prime meridian is called the _____.
9. A(n) _____ shows size relationship between a map or model and actual size.
10. The _____ is an imaginary line on which the earth rotates.
11. The time all clocks are set to within a given time zone is called the _____.
12. The _____ is the day when the sun follows its highest path across the sky.
13. The day when the sun follows its lowest path across the sky is the _____.
14. The curved path a planet makes around the sun is its _____.
15. A time when day and night are of equal length is called a(n) _____.

QUESTIONS

Give brief but complete answers to each of the following questions. Unless otherwise indicated, use complete sentences to write your answers.

1. Explain why the planets closest to the sun are different from the planets farthest from the sun.
2. If the planets formed from the same material as the sun, why didn't the planets become stars?
3. How can rocks found on the moon give scientists information about how the earth began?

4. What is the Moho?

5. What evidence is there that the earth was once melted?

6. Compare the seasons in the Southern Hemisphere with the seasons in the Northern Hemisphere.

7. Why do the boundaries of the time zones go around cities?

8. What would be the effect of the earth's axis being tilted more than 23½°? Less than 23½°?

9. Why are the summer months warmer than the winter months in the Northern Hemisphere?

10. Why was the international date line drawn in the ocean?

11. Where will the earth's axis be pointing toward on about June 22 in approximately 13,000 years?

APPLYING SCIENCE

1. Suppose you go to the moon. The first thing you see is some footprints in the dusty surface. Write a hypothesis to explain the footprints. What changes, if any, will you make in your hypothesis if you then see vehicle tracks? Following the vehicle tracks, you discover the remains of a moon-lander. What changes in your hypothesis will you make? Now will you know the true origin of the footprints?

2. From a topographic map of your area, find the elevation of your home, a friend's house across town, and your school. Also find out what is the highest elevation in or near your town. Your local Department of Engineering and Development will have a topographic map of your area if you can't find one in the library.

3. What problems would exist if there were just one time zone for the entire earth?

BIBLIOGRAPHY

Jespersen, James, and Jane Fitz-Randolph. *From Sundials to Atomic Clocks*. New York: Dover, 1982.

Owen, Tobias. "The Evolution of Titan's Atmosphere." *The Planetary Report*, November/December 1983.

Pampe, W. R., ed. *Maps and Geological Publications of the United States: A Layman's Guide*. Falls Church, VA: American Geological Institute, 1978.

Waldrop, M. Mitchell. "Before the Beginning." *Science 84*, 1984.

Bent rock layers are the work of great forces inside the earth. These once flat layers of rock were squeezed together by moving continents. When continents collide, they act like a giant vise.

PLATE TECTONICS

CHAPTER GOALS

1. Give evidence for the theories of continental drift and sea floor spreading.
2. Explain how continental drift and sea floor spreading led to the theory of plate tectonics.
3. Use the theory of plate tectonics to explain how the plates in the earth's crust move.
4. Explain how some mountains, island arcs, and sea floor trenches may form at the boundaries of crustal plates.
5. Using the theory of plate tectonics, describe how continental landmasses become larger.

3-1. Moving Continents

At the end of this section you will be able to:

- ☐ Explain the theory of continental drift.
- ☐ Use the theory of *plate tectonics* to explain how plates in the earth's crust move.
- ☐ Relate *convection* to movements of large parts of the earth's crust.

Once there were oceans in the midwestern United States, and part of Alaska was a tropical island. Ice used to cover much of the Sahara desert, while swamps existed in Antarctica. These are pieces of a giant puzzle. Put together, the pieces can show how the planet we live on works.

DRIFTING CONTINENTS

For a long time, scientists looked at the many surprising discoveries on the continents of the earth. They tried in different ways to explain each of the discoveries. Look at the pieces of the jigsaw puzzle in Fig. 3-1. It is hard to see what the whole picture looks like when the pieces are apart. The same is true for seeing how the earth works. For a long time, scientists did not see their discoveries as pieces of the same puzzle. Thus they could not see how the different parts of the earth's picture could be put together.

Fig. 3-1 It is difficult to see the whole picture when pieces of the jigsaw puzzle are apart.

Have you noticed that a world map looks a little like a jigsaw puzzle? For example, the continents of South America and Africa seem to fit together like pieces of a puzzle. See Fig. 3-2. Could these continents once have been connected?

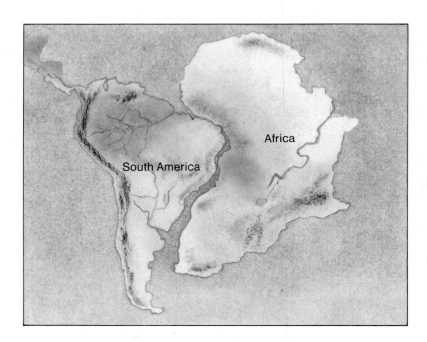

Fig. 3-2 The fit of South America and Africa led to the theory of continental drift.

The idea that the earth's continents were once a large, single landmass that moved apart is called the theory of *continental drift*. It was proposed as a scientific theory in 1912 by a German scientist named Alfred Wegener. Wegener called this landmass Pangaea, which means "all lands." Wegener pointed to the fit of South America and Africa, as well as other evidence. Remains of very similar plants and animals were found in Africa, South America, Australia, India, and Madagascar. See Fig. 3-3. He asked how plants and animals could be so similar if the lands were always separated by such wide oceans. Wegener also pointed to evidence of glaciers in warm lands, such as the continent of Australia. He reasoned that such continents once must have been in colder places on the earth and moved to where they are now. However, Wegener's theory of continental drift was not accepted by most scientists. They found it hard to believe that giant continents could push their way through the solid crust of the ocean floor. They said it would be as impossible as ships being able to move through a frozen sea. Some scientists tried to explain the similar plants and animals in another way. They said that the continents were once joined by a large landmass that was now below the ocean. There were still missing pieces to the puzzle. Thus the theory of continental drift did not become a part of the scientific view of the earth at that time.

Fig. 3-3 *The theory that the lands were once joined together can explain why the fossils of these ancient reptiles have been found across different continents.*

SEA FLOOR SPREADING

Finding out how the continents move through the ocean floor was not easy. However, modern scientific instruments have been able to map the sea floor. One of the most surprising discoveries on the sea floor was a chain of mountains. These

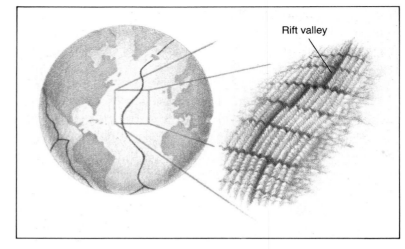

Rift valley

Fig. 3–4 The colored line on the globe shows the location of the mid-ocean ridge system. The enlargement shows the surface features of the mid-ocean ridge.

Mid-ocean ridge A chain of underwater mountains found on the floor of the oceans.

Rift valley A deep, narrow valley formed where the earth's crust separates.

mountains, almost completely covered by the oceans, were named the **mid-ocean ridge.** The *mid-ocean ridge* is the largest system of mountains in the world. It circles the earth like the endless seam of a baseball. See Fig. 3–4.

One of the best-known parts of this giant chain of underwater mountains is the mid-Atlantic ridge. In Fig. 3–4, you can see how those mountains would look if the water in the Atlantic Ocean could be drained away. Then the mid-Atlantic ridge would be seen as a wide range of mountains running down the middle of the floor of the Atlantic Ocean. In certain places, the ridge has been pushed out of line, forming other mountain ranges. Running through the middle of the ridge is a deep, narrow valley called a **rift valley.** A *rift valley* forms when the ocean crust separates along the mid-ocean ridge.

As scientists explored the mid-ocean ridge, they made two important discoveries. First, the layer of mud that covered most of the sea floor was missing from the mid-ocean ridge. Instead, rock samples taken from these mountains showed that the rocks were formed by volcanoes. *Pillow lava,* a kind of rock formed when molten lava is cooled quickly by water, was common. See Fig. 3–5. Second, scientists found that the center of the mid-ocean ridge was much warmer than any other part of the sea floor. Therefore, the center of the mid-ocean ridge must be a place where hot material from inside the earth is rising toward the surface. A new theory, called *sea floor spreading,* was proposed to explain these discoveries. The theory of sea floor spreading says that the mid-ocean ridge is a crack in the

Fig. 3–5 Pillow lava is common along the mid-ocean ridge.

earth's crust. Molten rock from the inside of the earth constantly rises from the center of this crack. Cooled by the sea water, the lava becomes solid. It is then pushed outward and away from the ridges by new molten rock. Thus new sea floor is always being formed. The new sea floor then spreads out everywhere along both sides of the mid-ocean ridge. See Fig. 3–6. Evidence for sea floor spreading is found in the different ages of parts of the sea floor. The youngest rocks are always found along the sides of the mid-ocean ridge. The oldest rocks are found farthest from the ridge.

Suggestion: Sea floor spreading can be demonstrated as follows: (a) staple 2 sheets of paper together along one end; (b) push two desk tops together; (c) push the open ends of the paper up through the crack between the desk tops until some of the paper falls to each side, mimicking the mid-ocean ridge; (d) slowly push the paper up through the crack between the desks to mimic sea floor spreading.

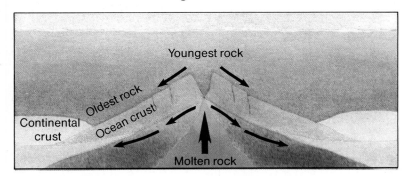

Youngest rock
Oldest rock
Ocean crust
Continental crust
Molten rock

Fig. 3–6 Molten lava rises in the rift valley along the mid-ocean ridge. New sea floor is created and is moved outward along each side of the ridge.

THE THEORY OF PLATE TECTONICS

Some scientists saw sea floor spreading as the beginning of a new theory to explain how continents move. They proposed that the crust is broken up into large sections, or plates. These plates grow at the mid-ocean ridge and move away at a speed of 1 to 20 centimeters per year. Continents are carried along with some of the moving plates, the way logs frozen into blocks of ice are carried along a river. Continents are usually parts of larger plates, but they could also make up entire plates. Thus continents did not have to push their way through the solid crust of the earth to move. They could move as sections of the earth's crust moved. The new theory used the ideas of sea floor spreading and continental drift. This theory has been given the name **plate tectonics** (tectonics comes from the Greek word "tekton," meaning "builder").

Plate tectonics The theory that the earth's crust is made up of moving plates.

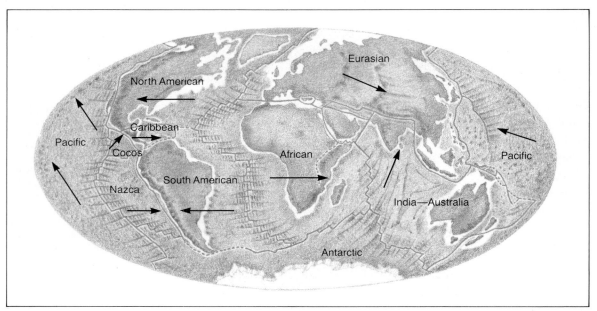

Fig. 3-7 The boundaries of the crustal plates are shown on this global map. Arrows show the direction each plate is moving.

The modern theory of *plate tectonics* says that the earth's crust is divided into seven major plates of different sizes. The largest plates have the same names as the continents they carry. See Fig. 3-7. As you can see, the size and shape of the continents are not exactly the same as the plates that carry them. Unlike the sea floor, which is always being formed along the mid-ocean ridge, the rocks that make up the continents are much older. They are also lighter than the rocks that make up most of the crustal plates. Thus the continents ride along on top of the moving crustal plates. The theory of plate tectonics has changed the way scientists think about the earth.

WHAT MOVES THE CRUSTAL PLATES?

Powerful forces are needed to move the gigantic crustal plates. Scientists believe that the forces start inside the earth.

Suggestion: A convection cell can be demonstrated by the following: (a) place a beaker of water on a tripod or stand; (b) add one crystal of potassium permanganate to the water at one side of the beaker; (c) slowly heat the beaker directly below the crystal.

CHAPTER 3

The forces are caused by the heat inside the earth. They are similar to the forces produced when cold air is added to hot air. You probably have seen what happens to cold air when a freezer is opened. Since cold air is denser than hot air, it sinks. At the same time, the less dense warm air rises. A current is formed that moves and mixes the air. Movement of material caused by a difference in temperature is called **convection.** Scientists believe that there may be very slow-moving *convection* currents set up in the mantle as heat moves out toward the crust. See Fig. 3–8.

Sea floor spreading could be caused by material being carried up by convection currents. The slow currents in the mantle cause material to be added to the crust along the mid-ocean ridge. At other places in the crust, currents in the mantle could cause one plate to be pulled down beneath the edge of another. This returns crustal material back to the inside of the earth.

Scientists do not have a complete answer to the question of what forces move the crustal plates. The question cannot be answered until there is a better understanding of what lies beneath the moving plates that make up the earth's surface.

SUMMARY

The theory of continental drift said that the earth's continents were once a single landmass. The discoveries of the mid-ocean ridge and sea floor spreading were used as evidence to form the theory of plate tectonics. This theory says that the earth's crust is divided into separate moving plates. Powerful currents in the mantle may be the driving force causing the plates to move.

QUESTIONS

Use complete sentences to write your answers.
1. List three observations Alfred Wegener used as evidence to support the theory of continental drift.
2. Explain what is happening along the mid-ocean ridge.
3. Describe how continents move, using the theory of plate tectonics.
4. How do convection currents help explain the movement of the crustal plates?

Convection Movement of material caused by a difference in temperature.

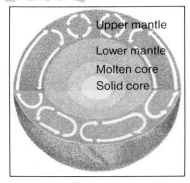

Upper mantle
Lower mantle
Molten core
Solid core

Fig. 3–8 Convection currents in the mantle are believed to cause the crustal plates to move.

Skills: The theory of plate tectonics can be used to reinforce the skill of drawing conclusions from facts.

1. Alfred Wegener used the following to support his theory of continental drift: (a) the fit of the continents of South America and Africa; (b) remains of similar plants and animals found in lands separated by wide oceans; (c) evidence of glaciers in warm lands, such as the continent of Australia.
2. New sea floor is always being created. The new sea floor then spreads out everywhere along both sides of the mid-ocean ridge.
3. Continents are carried along with some of the moving plates, the way logs frozen into blocks of ice are carried on a river.
4. Convection currents deep in the earth move molten material upward at the mid-ocean ridge, causing the new crust formed to push the plates apart. At other places, crustal plate edges are plunging downward into the earth's interior.

3–2. Action at the Plate Edges

At the end of this section you will be able to:

- ☐ Compare three kinds of boundaries between the crustal plates.
- ☐ Describe how some mountains, island arcs, and sea floor trenches may be formed.
- ☐ Explain how continents grow.

Walk westward across the valley shown in Fig. 3–9 and you would leave northern California for what was once part of Mexico. Travel to southern Florida and you would be in a land that was once part of Africa. Or move eastward into Connecticut or Massachusetts. You would be leaving the continental United States for what was once part of Europe. Places such as these exist all over the earth.

PLATE BOUNDARIES

The earth's surface is cracked into many crustal plates. As these cracks open, crustal plates move. Each crustal plate is a strong, rigid piece of the crust with an average thickness of about 100 kilometers. The crustal plates "float" on the layer of the earth called the mantle. Studying the way earthquake waves move through the mantle shows that the mantle's upper layer is partly molten. Thus crustal plates are able to slide over the soft upper part. This hot, soft upper part of the mantle is called the **asthenosphere.** The *asthenosphere* goes down about 250 kilometers into the mantle.

Since the crustal plates are solid, they move mostly along their edges, or boundaries. Boundaries are always marked by a kind of fault, or crack in the crust where plates meet. There are three kinds of plate boundaries. Each kind of boundary depends on how the separate plates move.

One kind of plate boundary is formed along the mid-ocean ridge. There the plates are being pushed apart as new crust is added. See Fig. 3–10 (a). If the plates move apart in one place, they must come together in another place. Thus not all boundaries between the plates are places where they are moving apart. Some plate boundaries are formed where two or three plates are coming together.

Fig. 3–9 A portion of the San Andreas fault in California.

Asthenosphere The partly molten upper layer of the mantle.

Word Study: Asthenosphere: *asthenos* + *sphaira* (Gr.), weak sphere.

Enrichment: Crustal plate movement has been measured by reflectors left on the moon by Apollo 11, 14, 15, and 16. Laser beams sent from the earth to the moon and reflected back show a displacement when movement occurs.

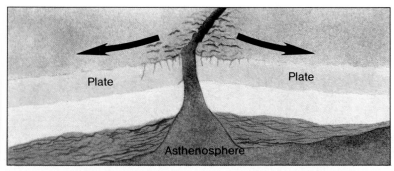

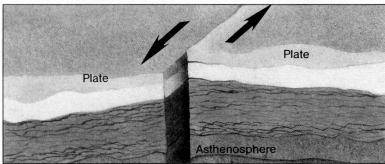

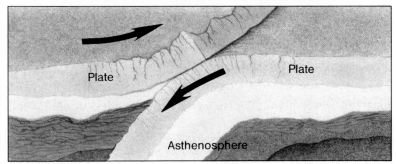

A second kind of plate boundary is formed where the plates slide past one another. The motion of each plate is parallel to the plate boundary. See Fig. 3-10 (b). A third kind of plate boundary is formed where two plates move toward each other. The edge of the heavier plate is usually pushed down beneath the lighter plate. See Fig. 3-10 (c). This process is called **subduction.** When *subduction* takes place, one of the plates is forced down into the mantle. There the plate is destroyed by melting as its material becomes part of the mantle. As material is removed from the plate by subduction, new material is added to the plate along the mid-ocean ridge. Equal amounts of material are removed and added. Otherwise, the size of the earth would be changing all the time.

Subduction Where one plate is forced beneath another plate.

Enrichment: Valuable metal deposits are found in plate boundaries.

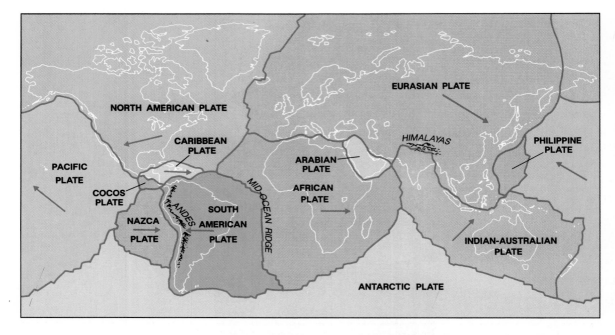

Fig. 3–11 Some of the
world's largest mountain
ranges occur along plate
boundaries.

Reinforcement: To help students
visualize plates, use a wax pencil or
water soluble marker to draw plate
boundaries on a globe.

COLLISIONS THAT CHANGE THE CRUST

When you look at Fig. 3–11, you can see that the plate carrying the South American continent is moving westward. The
Nazca plate on the Pacific Ocean floor, west of South America,
is moving eastward toward the continent. These two plates
meet along the western edge of South America. This collision
of plates seems to be crumpling up the western edge of South
America and forming the Andes (**an**-deez) Mountains at the
boundaries of the crustal plates. Fig. 3–12 shows subduction
occurs as South America rides up over the Nazca plate. It is believed that the melting material below the mantle is being
added to the Andes Mountains in two ways. The rubbing of the
plates adds material to the base, while some of the melted material rises to the surface and forms volcanoes.

*Fig. 3–12 The Andes moun-
tains in South America were
formed when the Nazca plate
was pushed below the South
American plate.*

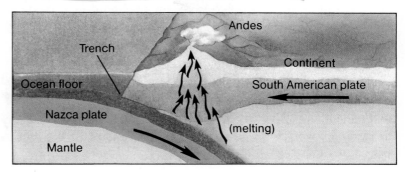

The rock that makes up the continents is lighter than the rock of the plates on which the continents are carried. Thus the continents "float" on the plates like gigantic rafts. When continents meet, they do not sink into the mantle. See Fig. 3–13. Instead, the two continents come together and crumple into mountain ranges. The two continents then become joined into a single landmass. For example, India became attached to Asia in this way. India was once a separate continent, carried on a northward-moving plate. The Indian plate collided with the Asian plate. The continents carried on the plates were crumpled and folded when this occurred. The Himalayas (him-ah-**lay**-ahs), a very thick layer of mountains, were formed. See Fig. 3–14. The Alps in Europe were also formed in a similar way. They were created when parts of the African plate collided with the Eurasian plate. Millions of years from now the Mediterranean Sea may disappear as Africa meets the Eurasian plate.

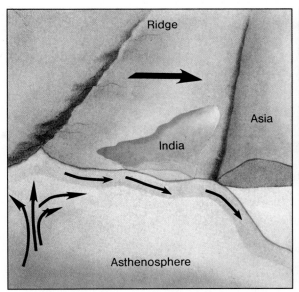

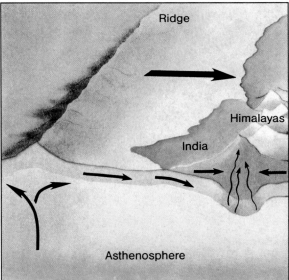

If you look at the world map in Fig. 3–7, a line of islands can be seen along the western side of the Pacific Ocean. These islands are caused by two ocean plates meeting. Scientists believe that the Pacific plate is slowly being pushed under Asia. As it sinks, it is melted by the high temperatures of the mantle. The light, melted material is forced up, forming strings of volcanoes. Because of the slow movements of these plates, the volcanoes are formed offshore under the water. The lava piles up

Fig. 3–13 India "crashed" into Asia about 45 million years ago, producing the Himalayas.

Fig. 3-14 *The Himalayas.*

until the volcanoes rise above the water and form islands. See Fig. 3-15. The volcanoes form a curved shape, called an *island arc.* The reason for the curved shape of island arcs can be seen by pushing down on one side of a tennis ball. The Aleutian Islands and Japan are examples of island arcs. As the plates continue to move, the island arcs may eventually get crushed against continents.

In some places in the ocean, one crustal plate plunges below another to form a *trench*. Some of the deepest parts of the ocean are at these places. Trenches are long, narrow valleys in the ocean floor. Their depth is usually greater than 6,000 meters. Most of the world's trenches are located near island arcs in the Pacific Ocean, as shown in Fig. 3-16.

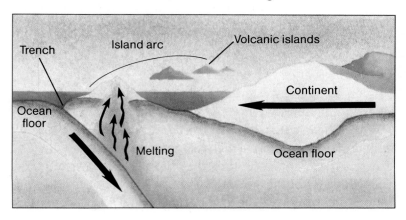

Fig. 3-15 *Volcanic islands and sea floor trenches can be created when one crustal plate is pushed beneath another.*

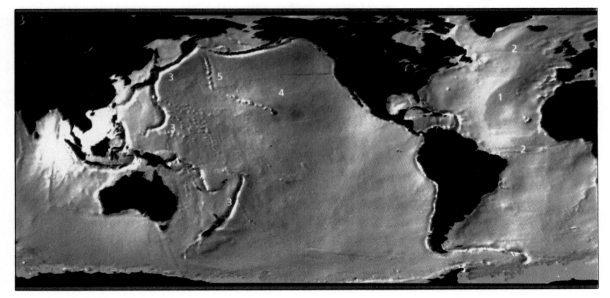

GROWTH OF CONTINENTS

If you tried to measure the age of the North American continent, you would find that it isn't all one age. You would find that the continent is oldest at its center and becomes younger toward its coasts. Thus the outer parts of the continent must have been formed after the center formed. It seems as if the continent grew. Many scientists believe that continents do grow. Throughout the world, small pieces of continental crust are scattered over the ocean floor. These submerged *micro-continents* are usually several hundred kilometers across. They are believed to be parts of continents that somehow became separated. A part of the ocean floor, carrying a micro-continent, may collide with a continental plate. If the ocean floor is subducted, the small piece of continental crust can become attached to the edge of another continent. A continent can then grow by adding small pieces to its edges. Thus the original continents were much smaller than they are today. Much of the west coast of North America and almost all of Alaska seem to have been added in this way. See Fig. 3–17.

Continents also may have grown when large island arcs or pieces of sea floor crust were pushed against and over the continents, forming a "thin skin" over the older continental rock. There is evidence of a thin layer of rock in the Appalachians. The crust may have wrinkled to form these mountains.

Fig. 3–16 Many of the world's largest trenches on the sea floor are shown in this photo taken from space.

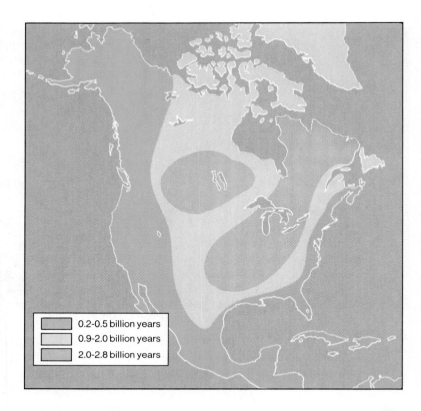

0.2-0.5 billion years
0.9-2.0 billion years
2.0-2.8 billion years

Fig. 3-17 The edges of the North American continent are younger in age than its center.

1. One boundary is represented by the mid-ocean ridge, where the plates are being pushed apart; a second kind of boundary occurs when two plates collide head on, causing one of the plates to ride up over the other; the third kind of boundary results when two plates slide past one another.
2. Colliding crustal plates produce all three. When subduction of one causes the land on the other to rise, or when continental plates crumple against one another, mountain ranges are formed. When two plates collide under the ocean, one may plunge down under the other, forming a deep ocean trench. When two plates meet, subduction of one may form molten lava that rises to produce island arcs.
3. Continents grow by the adding on of one or more of the many micro-continents found scattered on the floor of the oceans during the collision of an ocean plate with a continental plate.

SUMMARY

Along their boundaries, crustal plates may be moving apart, past each other, or together. The world's great mountain ranges, volcanoes near these mountains, deep ocean trenches, and island arcs, are caused by plates coming together. As a plate moves, some material along its edge may be destroyed. In other places, new material is added to the plate's edge. Continents may grow when small pieces of continental crust are added to their edges.

QUESTIONS

Use complete sentences to write your answers.

1. Describe the three kinds of boundaries that exist between crustal plates.
2. How does plate tectonics explain the formation of mountain ranges, trenches, and island arcs
3. Explain how small pieces of continental crust may be added to a continent.

TECHNOLOGY

SEISMIC REFLECTION PROFILING

Scientists today consider the study of the earth's crust to be one of the great frontiers of modern earth science and a key to understanding the nature and evolution of our planet. For a long time, earth scientists concentrated their study on the uppermost surface of the earth, relying on general theories to explain the crustal layers beneath it. The most common theory held was that, under a thin layer of sedimentary rock at the surface, lay a thick band of granite with a thick layer of volcanic rock below that.

Today, new techniques, such as deep drilling and measurement of tremors that pass through the earth, have allowed scientists to explore the geology of the thick crust of the continents. The deepest hole drilled for scientific purposes is more than 10,900 meters (about 11 kilometers) beneath the surface, and that record will soon be broken. As a deep hole is drilled, a long cylinder, or core, of rock is cut and brought to the surface for scientists to study.

The COCORP Project, based at Cornell University, has used a second technique, called seismic reflection profiling, to explore deep beneath the earth's surface. The principle behind seismic reflection profiling is similar to that used in echo-sounding from a boat. Two crews using truck-mounted vibrators are needed for this technique: one to send tremors, or seismic waves, through the earth and another to receive and measure them. Crews move the source and receiver slowly over the earth's surface like a giant caterpillar. Huge quantities of data are recorded and must be processed by a computer. The final product is very detailed information about the structure and other physical properties of the rock along the path taken by the source and receiver.

Many deep holes and seismic measurements are needed to build a reliable base of information, and the cost of these information-gathering techniques has often limited scientists. Nonetheless, based on the information now available, many scientists have revised their theories. They found that, while it is true that the crust is generally divided into zones or bands of different types of rocks, considerable variation in the bands exists.

Earth scientists have long believed that the rocks of the crust, and thus the continents themselves, are shaped by the up-and-down movement or shifting of rocks over time. The variation in the bands of rock uncovered by these new techniques has strengthened the evidence for that theory. It also has provided evidence that makes many believe the masses of rock shift not only up and down but sideways as well.

The next few hundred years will be an exciting time for earth scientists, as the whole crust—not just its surface—is explored and charted. The results will be of practical use in the search for petroleum and the prediction of earthquakes. This exploration will also give earth scientists the tools to learn still more about how continents were formed and will change over time.

INVESTIGATION

See teacher's commentary for teaching hints, safety, and answers to questions.

MAKING A CRUSTAL PLATE MODEL

PURPOSE: To assemble a model of the seven major crustal plates, identify each by name, and locate some earth features related to crustal plate boundaries.

MATERIALS:

scissors
glue

3 color pencils
tracing paper

PROCEDURE:

A. Fig. 3–18, page 81, shows the pieces of a jigsaw puzzle. Each piece represents a map of one crustal plate. Carefully trace the edge of each piece onto a plain sheet of paper. Next, trace the continent outlines and the numbers shown.

B. Cut out the pieces you have drawn. Be careful not to cut off the numbers.

C. Assemble the plates as you would a jigsaw puzzle. The pieces should fit very closely. Glue the assembled map onto a piece of paper. You will use this map in several more activities, so save it and treat it with care.

D. The crustal plates are named according to continents or other major features on them. The names of the seven plates you traced are: African, Antarctic, Australian-Indian, North American, Pacific, South American, and Eurasian. On your map the Pacific plate is in two pieces. Outline each crustal plate on your new map by tracing around it.

E. Label the plates on your assembled map with their correct names.

1. Which plate does not bear the name of a continent?

2. List the names of the plates that bear the name of a single continent.

3. List the names of the plates that bear the names of more than one country or continent.

F. Fig. 3–4 on page 68 shows the location of the mid-ocean ridge. Draw the ridge on your map, using the color pencil listed in your key.

G. Refer to Fig. 3–11 on page 74, which shows some mountain ranges that are related to crustal plate boundaries. Make marks similar to those found in Fig. 3–11 on your crustal plate map, using the color listed in the key.

4. List the names of the mountain ranges shown on your map.

5. Explain how these are thought to be due to crustal plate activity.

H. See Fig. 3–16 on page 77, which shows where some major ocean trenches are. Using the proper color from your key, mark the locations of those shown.

CONCLUSIONS

1. List the names of the seven major crustal plates. Describe why each was so named.

2. Name three earth features that are found along crustal plate boundaries.

3. Explain how the mid-ocean ridge is related to crustal plate activity.

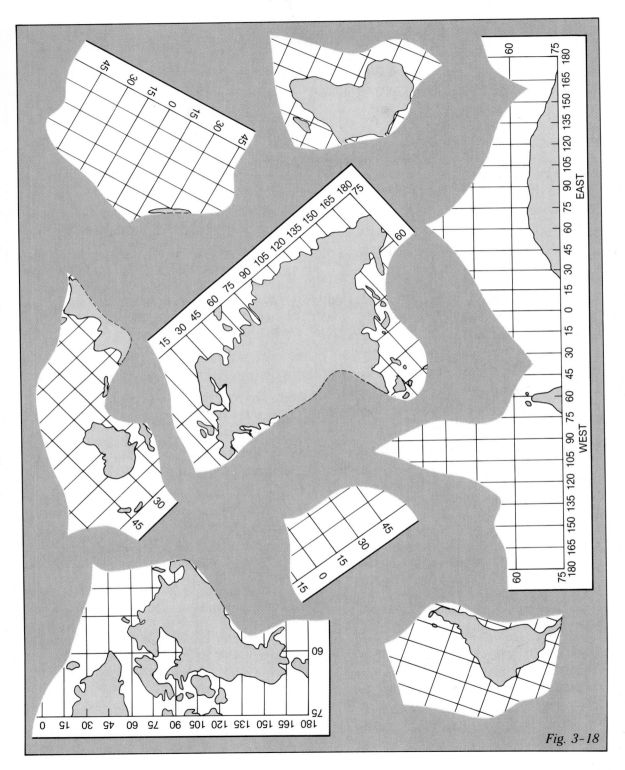

Fig. 3-18

¡COMPUTE!

SCIENCE INPUT

The study of crustal plates has produced many changes in the way earth scientists think about the beginnings of the earth's history. The theory of plate tectonics, or shifting plates, has been helpful in explaining how ridges and other landforms have been created. In addition, knowing the direction of plate shifting is useful in dating certain features of the earth.

Before you can use a theory, however, it is important to be familiar with all its parts. Understanding plate tectonics requires that you learn many new words and concepts.

COMPUTER INPUT

The computer can be used as a learning tool in a number of ways. One way is to create a game. Instead of asking straightforward questions, a game format can permit some hinting. In Program Plate Game, a game format is used. Your answer will not be input as a complete word, but as separate letters until the entire correct word is printed on the screen. There is, in this way, a continual cycle of feedback.

Feedback is information about your answers; it helps you to learn more efficiently. Your teachers' comments and grades on your assignments are feedback. They give you information you can use to judge the quality of your work. The computer's feedback helps in this way, too. It acts as a private tutor and has the advantage of giving you an immediate response.

In Program Plate Game you will use the computer to learn the words and concepts of plate tectonics explained in this chapter. The computer will give you feedback and, at the end of the program, will tell you how many times you have given the wrong answers.

WHAT TO DO

Make a data chart that lists brief definitions for the new words and concepts in this chapter. A sample chart appears below. Make your definitions short and clear.

Enter Program Plate Game using the information from your data chart to write data statements 311 to 319. Data statement 310 is printed in the program as an example. The program gives you the definition and asks you to enter the word or concept being defined, letter by letter. If the letter you choose is in the word, it will be printed on the screen in its proper place. If the letter is not in the word, it will count as a wrong choice.

SAMPLE DATA CHART

Term	Definition
Core	Center part of the earth
Mantle	Next to crust / largest layer
Crust	Thin solid outer layer of the earth
Mid-ocean ridge	Underwater mountains

PROGRAM NOTES

If you changed the nature of the data statements, this program could be applied in a variety of ways. You could use words and concepts from other chapters, or from other subjects. For example, you could change the word to the name of an historical figure, and change the definition to an event or achievement associated with that person. Then the program would print out the achievement and you would need to supply the person's name.

TECTONIC TUTOR: PROGRAM PLATE GAME

GLOSSARY

HARDWARE   The machinery itself, including the CPU (central processing unit), keyboard, and monitor (screen).

SOFTWARE   The programs used in the computer. Usually, these are available on a floppy disk or cassette tape.

REM   Short for *remark,* a command in BASIC, written for the user, not the computer. REM statements help the user understand the program, but they are not read by the computer.

PROGRAM*

```
100   REM PLATE GAME
110   FOR X = 1 TO 20: PRINT : NEXT
120   READ W$,H$: IF W$ = "0" THEN
      END
130   T$ = "":Q = 0
140   FOR X = 1 TO LEN (W$):T$ = T$ +
      "-":NEXT
150   PRINT : PRINT : PRINT : PRINT H$"-
      "T$: PRINT : INPUT "PICK A
      LETTER —> ";L$
160   G$ = T$:T$ = "":Q = Q + 1:Q1 =
      0
170   FOR X = 1 TO LEN (W$)
180   IF MID$ (W$,X,1) = L$ AND Q1 = 0
      THEN Q = Q - 1:Q1 = 1
190   IF MID$ (W$,X,1) = L$ THEN T$ =
      T$ + L$: GOTO 210
200   T$ = T$ + MID$ (G$,X,1)
210   NEXT
220   FOR X = 1 TO LEN (W$)
230   IF MID$ (T$,X,1) = "-" THEN GOTO
      150
240   NEXT
250   PRINT W$;"- ";Q;" WRONG
      LETTER(S)": PRINT : PRINT
260   FOR X = 1 TO 15: PRINT : NEXT :
      GOTO 120
300   REM DATA STATEMENTS
310   DATA CORE, CENTER PART OF THE
      EARTH
311   DATA 0,0
320   END
```

BITS OF INFORMATION

Feedback is an important function of computer learning systems. It is a term that was first applied to both computers and humans by Norbert Wiener, a pioneer in computer electronics. He explained feedback as a process of communication involving a response to an original statement or input that helps a person or machine make a necessary adjustment. Wiener's most well-known book is called Cybernetics, *which is a term used to define the science of control and communication.*

Because they can manipulate data quickly, computers can give feedback almost instantaneously, making many pieces of modern technology self-adjusting. For example, as the space shuttle is operating, computers on the spaceship and on earth are constantly monitoring the performance of the many pieces of equipment on board. If a problem is spotted by the computer, many adjustments are made immediately and automatically to assure the success of these space missions.

* All the programs in this book are written for the Apple IIc.

CHAPTER REVIEW

VOCABULARY

On a separate piece of paper, match each term with the number of the description that best explains it. Use each term only once.

plate tectonics   island arcs   asthenosphere   rift valley
trenches   micro-continents   sea floor spreading   subduction
convection   pillow lava   continental drift   mid-ocean ridge

1. Describes the mid-ocean ridge as a place where hot material is rising.
2. A valley with steep sides.
3. The Aleutian Islands and Japan are examples of.
4. Differences in temperature causing movement in material.
5. The idea that the earth's continents are moving.
6. A chain of underwater mountains.
7. The theory that the earth's crust is made of moving plates.
8. Where one plate is forced beneath another.
9. A kind of rock formed when lava is cooled by water.
10. Long, narrow depressions in the ocean floor.
11. Pieces of continental crust scattered over the ocean floor.
12. The partly molten upper layer of the mantle.

QUESTIONS

Give brief but complete answers to each of the following questions. Unless otherwise indicated, use complete sentences to write your answers.

1. Why was the idea of continental drift not accepted by most scientists?
2. Describe the theory of continental drift.
3. What two important discoveries about the mid-ocean ridge led to the sea floor spreading theory?
4. How does sea floor spreading explain why the youngest rocks are found near the mid-ocean ridge?
5. Use the theory of plate tectonics to explain how the continents move.
6. What causes convection currents in the earth's mantle?
7. What did scientists find out about the ages of rocks on the sea floor?

     CHAPTER 3

8. Where are crustal plates moving apart?

9. Give an example of where two plates are sliding past one another.

10. Give an example of where crustal plates are coming together.

11. Name some mountain ranges that have been formed by plate collisions.

12. What is an ocean trench and where are most of them found?

13. Give two examples of island arcs.

14. Why do ocean plates sink under continental plates when they meet?

15. Why do continents crumple when pushed together?

APPLYING SCIENCE

1. A crustal plate moves an average of 5 cm a year. How far will the crustal plate move in (a) 5 years? (b) 50 years? (c) 1,000 years? How does this explain why it is difficult to see that landmasses move?

2. According to the theory of plate tectonics, will California ever fall into the Pacific Ocean? Explain.

3. If the sea floor is spreading at the mid-ocean ridge, does this mean that the earth is getting larger? Explain.

4. Why don't continents fit together exactly, the way a jigsaw puzzle does?

5. Using Fig. 3–7 on page 70, describe where your home will be on the earth millions of years from now.

BIBLIOGRAPHY

Alexander, George. "Going, Going, Gone." *Science 81*, May 1981.

Bingham, Roger. "Explorers of the Earth Within." *Science 80,* September/October 1980.

Harrington, John W. *Dance of the Continents: Adventures with Rocks and Time.* Boston: J. P. Tarcher (Houghton Mifflin), 1983.

Miller, Russel. *Continents in Collision* (Planet Earth Series). Chicago: Time-Life Books, 1984.

Siever, Richard. *The Dynamic Earth.* San Francisco: W.H. Freeman, 1983.

Time-Life Editors. *Floor of the Sea* (Planet Earth Series). Chicago: Time-Life Books, 1984.

Life begins after only a few months on a hardened lava flow. Much of the earth's crust was formed from sheets of lava. Every day there is new crust added to the earth.

BUILDING THE CRUST

CHAPTER GOALS

1. Identify the ways that the earth's crust responds to strain caused by movement of the crustal plates.
2. Describe how the different types of volcanoes form.
3. Explain why earthquakes, volcanoes, and mountains occur in the same areas.
4. Describe the geologic features produced by magma.

4-1. Earthquakes

At the end of this section you will be able to:

- ☐ Explain why most earthquakes occur.
- ☐ Define the term "fault" and describe three ways that rocks can move along faults.
- ☐ Name and describe three kinds of earthquake waves.
- ☐ Describe safety procedures to follow in case of an earthquake.

On the afternoon of Good Friday, March 27, 1964, most of the people of Anchorage, Alaska, were at home. Schools were closed and offices let out early. The streets were quiet and peaceful. Suddenly, at 5:36 P.M., the ground began to shake. The people of Anchorage, however, were not worried. They were used to the ground shaking. But this would be the beginning of one of the largest earthquakes ever to take place in this century.

For the thousands of people who lived in Alaska, that day would be remembered for the widespread destruction that took place. Thousands of homes and most of the industry were destroyed. For scientists, however, this was an opportunity to study the forces at work inside the earth. They made many observations and measurements of the ground. From this information they began to understand what happens when an earthquake strikes. But even more important was that scientists began to understand what causes an earthquake and where earthquakes are most likely to take place. With this new information, scientists hope to be able to predict future earthquakes. In this way, lives may be saved.

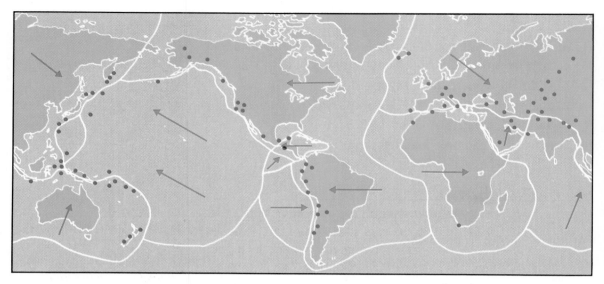

Fig. 4-1 The colored dots on the map show the location of large earthquakes during recent years.

WHAT IS AN EARTHQUAKE?

Fig. 4-1 shows the places where large earthquakes took place during a six-year period. This map shows that most serious earthquakes occur where the crustal plates come together. The boundaries between the plates are actually huge cracks in the crust. As the plates slowly move, their edges grind together. The plates move against each other along these cracks in crustal rocks. A place where the rock has moved on one or both sides of a crack is called a **fault.**

The rocks along a *fault* may move in three ways. First, the rocks may be pulled apart. This can happen where the plates are separating, as they are along the mid-ocean ridge. Then the rocks along one side of the fault slide down against the rock on the other side. This is called a normal fault. See Fig. 4-2 (a). Second, the rocks may be pushed together, as they are when two plates meet head-on. One side of the fault usually moves up. This is called a reverse fault. See Fig. 4-2 (b). A third kind of fault takes place when the blocks of rock on each side move in opposite directions. This is called a strike-slip fault. This kind of fault occurs where two plates are sliding past each other. See Fig. 4-2 (c).

Faults are also found away from the edges of crustal plates. Many scientists believe, however, that almost all faults are caused by moving plates.

Fault The result of movement of rock along either side of a crack in the earth's crust.

Word Study: Fault: *fallere* (L.), to deceive.

Enrichment: The strongest earthquake that has occurred within historical times in the United States took place in 1812 along the Mississippi River near New Madrid, Missouri. This is an example of tectonic activity within the interior of a crustal plate.

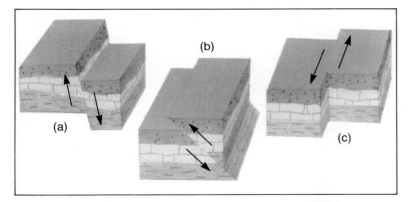

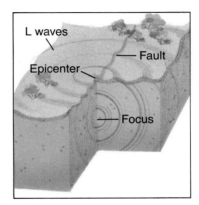

The plates that make up the earth's crust are in slow, constant motion. As they move, the rocks that make up the plates bend, stretch, or get squeezed. When rocks can no longer change shape, they break at their weakest point. These cracks, or fractures, are also faults.

Almost all large earthquakes take place when rocks suddenly snap along faults. If the rocks along a fault were constantly moving, their surfaces would soon be worn smooth. The rocks would slide gently like a well-oiled drawer. However, the rock surfaces are jagged. The sides of a fault area usually are locked together like a sticky drawer. This causes stress. When the stress becomes too great, the rocks suddenly snap and slip along the fault. This movement causes a sudden release of the built-up stress, which causes an earthquake.

The point beneath the earth's surface where the rocks move along the fault is called the **focus** See Fig. 4–3. Directly above the *focus* on the surface is the **epicenter.** The *epicenter* is the place where the earthquake is usually felt most strongly.

EARTHQUAKE WAVES

Sudden movement along a fault causes vibrations in the form of waves. These waves move out from the focus in all directions. The waves are felt on the surface as an earthquake. Earthquake waves are called **seismic (size-mik) waves.** *Seismic waves* are a form of energy. They can travel through solid rock, loose sand, and gravel. Seismic waves are felt most often on loose, water-soaked earth. During an earthquake this loose earth shakes like jelly in a bowl. In general, the buildings that are damaged most during earthquakes are those built on loose rock. See Fig. 4–4.

Fig. 4-2 (above left) *(a) Movement along a fault may be caused by plates moving vertically; (b) two plates can be pushed together along a fault where one plate moves up relative to the other; (c) plates can slide past each other.*

Fig. 4-3 (above right) *The focus is the exact place where movement occurs along a fault. The epicenter is the place on the surface above the focus.*

Focus The point of origin of an earthquake.

Epicenter The point on the earth's surface directly above the focus of an earthquake.

Seismic waves The vibrations caused by movement along a fault.

Fig. 4-4 This damage occurred when a large block of loose sand, gravel, and clay slid down toward the shore.

There are three kinds of seismic waves, *primary* (P) *waves*, *secondary* (S) *waves*, and *surface* (L) *waves*. Primary waves are caused by back-and-forth vibrations in the rock. They travel through the earth faster than the other waves. Secondary waves are caused by the up-and-down motions of the rock. They do not travel as fast as the primary waves. Surface waves cause the surface of the land to move in much the same way as ripples move on water. They cause most of the destruction on land. The three types of seismic waves are shown in Fig. 4–5.

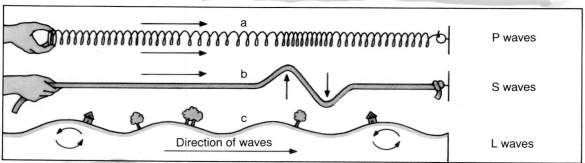

a — P waves

b — S waves

c — Direction of waves — L waves

Fig. 4-5 (a) Primary waves are like the back and forth wave moving along a stretched spring; (b) secondary waves resemble the up and down wave moving along a rope; (c) surface waves are like waves on water.

Seismic waves travel out from an earthquake focus and pass through the earth. Each kind of seismic wave moves in its own way through different parts of the earth. For example, secondary waves cannot travel through liquids. The fact that secondary waves do not travel through the earth's core helped scientists to understand that the earth's core is not completely solid. With this information, scientists can study the inside of the earth.

MEASURING EARTHQUAKES

Seismic waves can be recorded by an instrument called a **seismograph** (**size**-muh-graf). Fig. 4-6 shows how one kind of *seismograph* works. In 1935, Charles Richter (**rik**-tur) used seismograph records to measure the energy given off by an earthquake. The *Richter scale* uses a scale of numbers to show the amount of energy released by an earthquake. Each number on the scale represents ten times as much energy as the next lower number. For example, an earthquake measuring 5 on the Richter scale would be ten times larger than one with a measurement of 4, while an earthquake measuring 6 would be a hundred times larger than one measuring 4. Almost all earthquakes are very small. That is, they have Richter measurements of 4 or less. Large earthquakes with values of 8 or more usually take place only once every five to ten years. However, any earthquake with a rating of 6 or more on the Richter scale can cause damage to buildings. Fig. 4-7 shows where the largest earthquakes have taken place in the continental United States.

Fig. 4-6 (below) *When seismic waves travel through the earth, the bedrock vibrates, causing the drum to vibrate as it turns. The pen records the vibrations on the drum.*

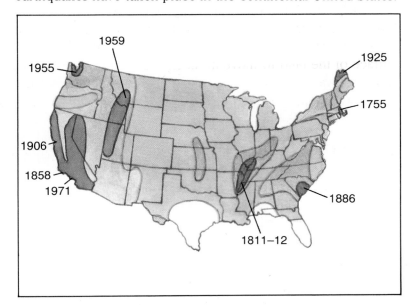

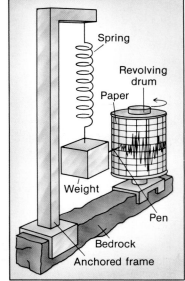

Fig. 4-7 (left) *The darkest color shows the areas where the largest earthquakes have struck in the past. Dates mark the location of very large earthquakes.*

Large earthquakes release more energy than hundreds of thousands of small ones together. This means that small earthquakes do not release large amounts of energy a little at a time. The total energy released by small earthquakes is not enough to prevent the buildup of stress in the rocks. Large, dangerous earthquakes give off most of the stored energy.

EARTHQUAKE SAFETY

It is estimated that throughout history 74 million people have lost their lives in earthquakes. See Fig. 4–8. Some places in the world have many more earthquakes than other places. However, no place is completely safe from earthquake damage. Some faults are known to be the source of earthquakes. For example, the city of San Francisco is located very near a fault that caused a major earthquake in 1906. Buildings in cities near such faults should be built in ways that will help to prevent them from falling during an earthquake. Dams should not be built near active faults. Failure of a dam during an earthquake can cause a very sudden and dangerous flood. Nuclear power plants should also be built in places away from faults. An earthquake could cause the release of radioactive materials from these plants.

Fig. 4-8 Earthquakes can cause damage to homes and property.

Earthquakes usually take place without any warning. If one takes place where you are, the actual shaking probably will last less than a minute. The entire earthquake may be over before you have time to protect yourself. But if there is time, a few simple rules can be followed to reduce the danger. Many schools and buildings already have plans in case of an earthquake.

Move into an open area to get away from buildings that might collapse. However, if you are in a large building, stay inside. Do not run outside where you might be struck by falling objects. Try to stand in a doorway. Get under a desk or table that will protect you from falling parts of the building. Stay away from chimneys, brick walls, and large glass areas. These are likely to fall during an earthquake.

CHAPTER 4

After the earthquake, help any injured persons. Stay away from fallen electric wires or power poles. Do not move around the neighborhood to check the damage. You probably will be getting in the way of emergency crews. Pay attention to instructions you may be given by officials such as police officers and firefighters.

We cannot stop the movement of the great crustal plates. However, scientists soon may be able to predict when and where an earthquake will happen. They have found that the earth often gives signals before an earthquake. For example, the speed of primary waves seems to slow down before an earthquake. This change in speed can be observed from a small earthquake or explosion. Measurement of the change in speed may make it possible to predict larger earthquakes. Research also has shown that certain gases are released into well water just before an earthquake. The presence of these gases in well water could be a signal of a coming earthquake. Also, the earth bulges in places where major earthquakes have occurred. Animals seem to become very nervous and excited a short time before an earthquake takes place. Research is being carried on to try to find other signs that usually occur before an earthquake happens.

Experiments are now under way to try to measure the buildup of stress along faults. In the future, earthquake predictions may become as common as today's advance warning of bad weather. Being able to predict earthquakes will make it possible to prevent some loss of life and property.

SUMMARY

Earthquakes occur when rocks suddenly move along a fault. Earthquakes send out P, S, and L waves, which can be recorded on a seismograph. Knowing what to do during and after an earthquake can decrease the chance of injury.

QUESTIONS

Use complete sentences to write your answers.
1. What is a fault?
2. Describe three ways that rocks may move along faults.
3. Describe the three kinds of seismic waves.
4. Explain what you should do during and after an earthquake.

Enrichment: Warning signs that have been studied include velocity change of earthquake waves, ground swellings, changes in electrical conductivity of rocks, variations in the geomagnetic field, and appearance of radon gas in wells. Although all have been observed to change prior to some earthquakes, none of these have proved to be reliable predictors.

1. A fault is a place where the rock on either or both sides of a crack has moved.
2. One kind of motion found along a fault is when the rocks on each side are pulling away from one another. Two other motions are the result of colliding crustal plates. One causes an upward motion of the rocks on one side of the fault. The other case is when the rocks on the two sides of the fault travel in opposite directions.
3. Primary waves are produced by the back and forth vibrations of the rock. They travel through the earth faster than the other waves. Secondary waves are caused by the up and down vibrations of the rock. They cannot travel through the liquid parts of the earth. Surface waves cause the surface of the land to move in much the same way as ripples on water.
4. During an earthquake, you should (a) move into an open area if possible; (b) if in a large building, stay there; (c) inside a building, stand in a doorway or get under something for protection; (d) stay away from chimneys, brick walls, and large glass areas. After an earthquake, (a) help injured persons, (b) check for fires, (c) look for leaking gas and turn off valves if possible, (d) stay away from fallen electric wires or power poles, (e) do not get in the way of emergency crews, and (f) pay attention to instructions given by officials.

See teacher's commentary for teaching hints, safety, and answers to questions.

PLOTTING EARTHQUAKES

PURPOSE: To plot a number of earthquake positions and to look for a pattern in these positions.

MATERIALS:

crustal plate map   paper
pencil

PROCEDURE:

A. The following table lists 15 locations, in the degrees of latitute and longitude, where earthquakes have occurred. This list is only a small sample of all the earthquakes that occur each year.

Earthquake	Latitude	Longitude
1	60°N	152°W
2	45°N	125°W
3	35°N	35°E
4	30°N	115°W
5	30°N	60°E
6	20°N	75°W
7	50°N	158°E
8	40°N	145°E
9	15°N	100°E
10	15°N	105°W
11	10°S	105°E
12	5°S	150°E
13	0°	80°W
14	25°S	75°W
15	50°S	75°W

B. On the crustal plate map you made in the Investigation section on page 80, locate the positions where each earthquake occurred and place its number at that position. For example, earthquake number 1 occurred on the coast of southern Alaska at 60°N latitude and 152°W longitude. Place a small number 1 at this location.

C. After locating all 15 earthquakes, look at your map to answer the following questions. Use complete sentences to write your answers.
   1. Describe where the earthquakes are located.
   2. Is there a relationship between the earthquake positions and the shape of the continents?
   3. Describe how the locations of the earthquakes are related to crustal plate boundaries.

CONCLUSIONS:
   1. Describe the pattern of the locations of the 15 earthquakes you plotted.
   2. Copy and complete the following hypothesis: The pattern found by plotting a number of representative earthquake locations on the world map suggests that many future earthquakes may be expected to occur _____.
   3. From the information on your map, predict if an earthquake will take place where you live.
   4. Which is easier to predict, where or when earthquakes will happen? How can earthquake predictions be improved?

4–2. Volcanoes

At the end of this section you will be able to:

- Explain what causes volcanoes.
- Point out the locations on the earth's surface where most volcanoes are found.
- Describe how volcanoes erupt.

On August 24, A.D. 79, the scene in Pompeii was peaceful and normal. Though the city was located near Mount Vesuvius, a large volcano, the people were not worried. The volcano had not erupted for hundreds of years. Suddenly, Mount Vesuvius erupted with an explosion ten times more powerful than the eruption of Mount St. Helens. Volcanic ash buried the city and its people. Scientists believe that there are 33 sites in the western states, Alaska, and Hawaii where an eruption could take place at any time.

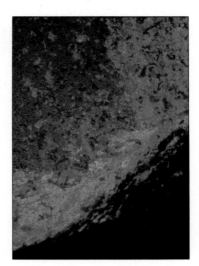

Fig. 4-9 Lava is magma that appears on the earth's surface.

MAGMA AND LAVA

In certain places in the earth's mantle there are pockets of red-hot, molten rock called **magma.** About 95 to 320 km beneath the surface is a zone of rock where seismic waves are slowed. The rock in this region is very close to its melting point. *Magma* is formed in this zone. For example, the action around the plate edges can cause deep pockets of magma to be formed.

Magma is always red-hot. Its temperature is usually between 500°C and 1,200°C. The liquid magma is less dense than the solid rock around it. Being less dense, magma tends to work its way up toward the surface like a hot-air balloon rising in the air. Sometimes the magma cools before reaching the surface. It then hardens into solid rock while still beneath the surface. Often magma does reach the surface as a liquid. When magma comes out onto the earth's surface, it is called **lava** (**lahv**-uh). See Fig. 4-9. A **volcano** begins to form at the place where *lava* appears on the surface.

THE RING OF FIRE

The world's most active *volcanoes* on land are found mainly in a narrow belt that circles the Pacific Ocean. This chain of volcanoes is often called the "Ring of Fire." See Fig. 4–10. Other

Magma Molten rock beneath the earth's surface.

Lava Magma that has flowed out onto the earth's surface.

Volcano A place where lava reaches the surface and builds a cone or other surface feature.

Word Study: Volcano: *Volcanus* (L.), god of fire.
Lava: *labes* (L.), to fall.

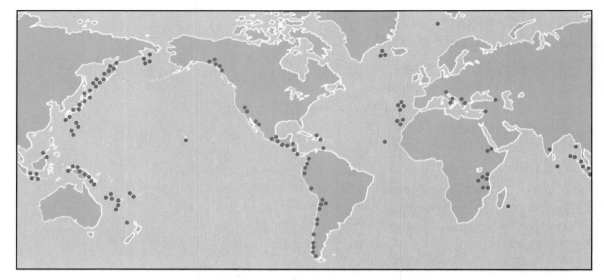

Fig. 4-10 The colored dots on the map show the locations of many of the earth's volcanoes.

Reinforcement: The density of melted rock is always less than the solid rock from which it came. The rock in the subducting plate rises as it is melted for the same reason that a cork bobs to the surface in water. The rising lava probably creates its own fractures that allow it to reach the surface.

Fig. 4-11 Many volcanoes are formed from the collision of two plates that make up a part of the sea floor.

centers of volcanic activity are located along the mid-ocean ridge and in the Mediterranean region. These areas are the active parts of the earth's crust along the boundaries of the crustal plates. These active zones are also the location of most of the world's earthquakes.

Volcanoes are found in greatest numbers along two kinds of plate boundaries. First are the zones of subduction. Here, one plate is sinking beneath the other and producing magma. For example, volcanoes such as Mount St. Helens in the Pacific Northwest are caused by subduction. See Fig. 4-11. There the Pacific plate is sinking beneath the North American plate.

Another example of volcanoes formed by subduction is on the other side of the Pacific near Japan. There the Pacific plate is being pushed beneath the Asian plate. See Fig. 4-11. The re-

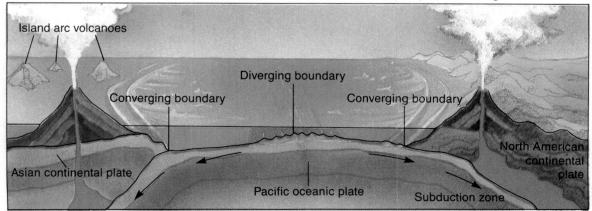

Island arc volcanoes

Diverging boundary

Converging boundary

Converging boundary

North American continental plate

Asian continental plate

Pacific oceanic plate

Subduction zone

sulting volcanoes form a chain of island arcs. These are the volcanic islands of the Aleutians, Japan, and the southeastern Pacific.

A second zone of volcanic activity is found along the boundaries where plates move apart. Along the mid-ocean ridge, for example, constant eruptions take place beneath the sea. In the North Atlantic, part of the mid-Atlantic ridge rises above the water. This is the island of Iceland. The volcanoes found in Iceland are caused by magma rising where plates are spreading apart. See Fig. 4-12. The volcanoes found in East Africa are also caused by plate spreading. There the continental plate is being pulled in two. A place where a continent is breaking apart is called a *rift*. Scientists believe that rifts in continents may become oceans millions of years from now.

Most volcanoes are located along plate boundaries. However, some volcanoes, like the Hawaiian Islands, are found in the middle of plates. Scientists believe that these volcanoes are caused by a huge column of rising magma. This hot spot remains in place while the moving plate is carried over it. A chain of volcanoes can result. The Hawaiian Islands are part of such a chain stretching out across the Pacific Ocean. See Fig. 4-13.

Compare and contrast: The lava from a subducting plate makes its own conduits to the surface. Along the mid-ocean ridge the spreading of plates fractures the solid crust and allows the molten material below to escape. The pressure of the water prevents the escape of gases from the lava that might cause violent eruption. Thus volcanic activity along the deep mid-ocean ridge is not violent.

Enrichment: A large body of magma lies beneath much of the Yellowstone Park region. A similar body of magma is also believed to be beneath the Mammoth region in California.

BUILDING THE CRUST

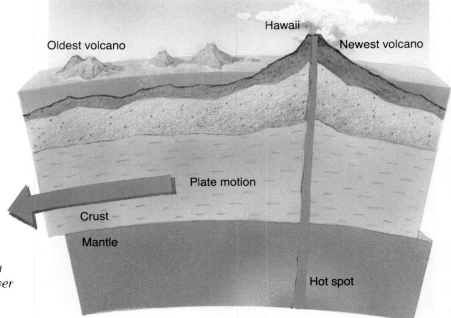

Oldest volcano · Hawaii · Newest volcano · Plate motion · Crust · Mantle · Hot spot

Fig. 4–13 The Hawaiian Islands were formed when the Pacific plate moved over a hot spot.

VOLCANIC ERUPTIONS

The opening through which magma reaches the surface is called a *vent*. Around a vent there is often a bowl-shaped feature called a *crater*. A single volcano often has several vents through which lava reaches the surface.

Volcanoes are not all alike. They differ in the way they erupt, or throw out materials. Also, magma is not always the same. It may be heavy, dark, and thick or it may be light and thin. Some magma contains steam and other gases that bubble up like the gas in a bottle of soda pop.

How a volcano erupts depends mainly on the kind of magma and what is inside the vent. Gentle eruptions usually happen when the vent is not blocked and the magma is thin. Then the lava flows out in a steady stream. If the vent is sealed by old, hardened lava or other materials, an explosion can occur. Trapped gases build up pressure in the magma. An explosion can be set off suddenly by any change. For example, Mount St. Helens erupted when part of the mountain broke away along a fault. See Fig. 4–14. An eruption along the side and near the top of a vent, as with Mount St. Helens, sends the blast to the side as well as upward. Many volcanoes may erupt quietly at one time and with explosions at another time.

Fig. 4–14 (opposite page) (a) A landslide, caused by an earthquake, begins; (b) large amounts of rock and ice slide down the side of the mountain; (c) steam and volcanic ash are released; (d) Mt. St. Helens explodes, giving off rock and ash.

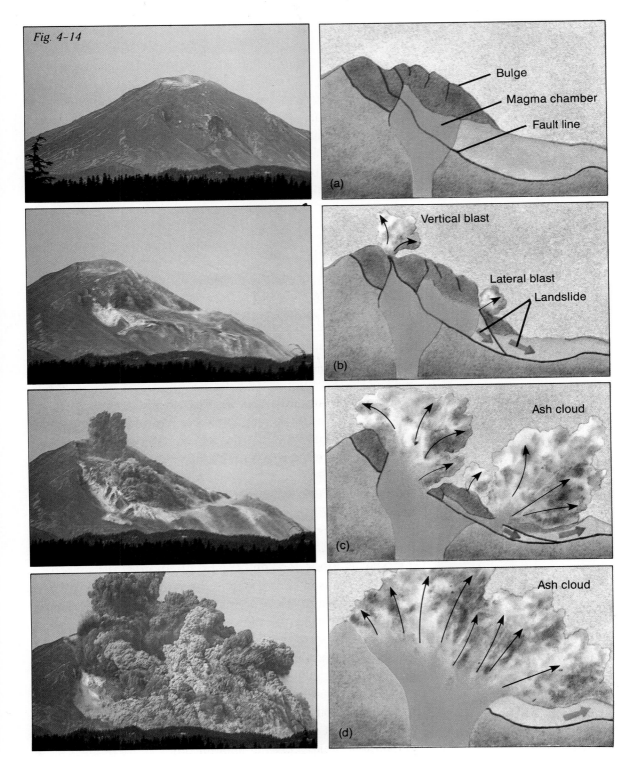

Fig. 4-14

(a) Bulge
Magma chamber
Fault line

(b) Vertical blast
Lateral blast
Landslide

(c) Ash cloud

(d) Ash cloud

When a volcano erupts, living things are destroyed by the high temperatures. The temperature of lava is about 1,000°C. Living things also may be affected by poisonous gases given off by the volcano. The volcano also gives off clouds of ash which can travel around the world. The ash can block out part of the sunlight and cause a cooling of the atmosphere. Ash deposits can clog air filters in cars, cause power failures, and make breathing difficult. But volcanoes can provide benefits as well. When they erupt, valuable materials are brought to the earth's surface. Volcanic deposits can contain metals, such as gold, silver, iron, and copper. Lava rocks supply materials for fertile farmland.

Today, some scientists are looking for ways to tap the heat given off by erupting volcanoes. Currently, some countries, such as Iceland, use heat from magma in the earth. This heat is called *geothermal energy*. Geothermal energy supplies much of Iceland's heat and hot water.

An eruption can end in three ways. First, the supply of gas in the magma may drop. Then there is not enough pressure to force the magma out of the vent. Second, the magma chamber may be emptied enough to bring an end to the eruption. Third, the vent may become blocked when magma hardens.

During explosive eruptions, lava is blown out. Particles of lava harden as they fly through the air. The smallest of these particles are called *volcanic dust* and *ash*. Larger particles called *cinders,* which are several centimeters across, are thrown out as well. There may also be large pieces, up to one meter across. These are called *bombs*. Bombs are often soft enough to become shaped like loaves of bread as they fly through the air. See Fig. 4–15.

There are three kinds of shapes for volcanoes. The shape of a volcano is caused by the type of eruption. First, the ash and cinders may build up around a vent. The result is a tall, narrow *cinder cone* like the one shown in Fig. 4–16 (a). Second, a volcano may begin with an explosion followed by quiet lava flows. Eruptions of this kind usually build up a large, cone-shaped mountain. This is called a *strato,* or *composite, volcano.* The mountain consists of alternating layers of lava and pieces of rocks thrown out during the violent periods. See Fig. 4–16 (b). A third kind of volcano forms when there is almost never an explosive eruption. It forms a mountain built up of layer upon

Fig. 4–15 Volcanic bombs are large pieces of hardened lava thrown out by volcanoes. They often make a whistling sound like a falling bomb as they fall.

layer of lava. This is called a *shield cone*. It has a broad, flat shape with gently sloping sides. See Fig. 4–16 (c). The Hawaiian Islands are a group of shield cones built up over a hot spot on the Pacific Ocean floor.

Fig. 4-16 (a) A cinder cone, (b) a composite volcano, (c) a shield volcano.

SUMMARY

Magma is present in certain places in the crust and upper mantle. Magma may reach the surface to form a volcano. Volcanoes form mainly along plate boundaries. The type of eruption determines the kind of volcanic cone that is formed.

QUESTIONS

Use complete sentences to write your answers.

1. How does a volcano begin?
2. Where are most of the active volcanoes found?
3. What two kinds of crustal plate boundaries produce the greatest numbers of volcanoes?
4. Describe three ways volcanoes may behave and the kind of volcanic cone formed in each case.

1. A volcano begins when magma breaks out onto the surface as lava.
2. Most active volcanoes are found in a narrow belt that circles the Pacific Ocean, along the mid-ocean ridge, and the Mediterranean Sea.
3. Most volcanoes are produced in boundaries that have subduction and boundaries where crustal plates are moving apart, as in the mid-ocean ridge.
4. Explosive volcanoes form a tall, narrow cinder cone. Composite volcanoes may first explode and then have quiet lava flows, usually building up a large, cone-shaped mountain. Volcanoes that rarely explode build up mountains that have broad, flat, gently sloping sides called shield cones.

INVESTIGATION

PLOTTING VOLCANOES

PURPOSE: To plot the positions of a number of volcanoes.

MATERIALS:
world map
paper
color pencil

PROCEDURE:

A. The following table shows 17 locations, in degrees of latitude and longitude, where volcanoes have formed.

Volcano	Latitude	Longitude
1	60°N	150°W
2	45°N	120°W
3	20°N	105°W
4	0°	75°W
5	65°N	15°W
6	40°N	30°W
7	17°N	25°W
8	45°N	15°E
9	30°N	60°E
10	55°N	160°E
11	40°N	145°E
12	5°S	155°E
13	10°S	120°E
14	5°S	105°E
15	15°S	60°E
16	30°S	70°W
17	55°S	25°W

B. Use a color pencil to plot each volcano position on the map of the world. For example, plot the position for volcano number 1 by going across the 60°N latitude line to where it crosses the 150°W longitude line. Write a small 1 at this point. This point should be on the southern coast of Alaska. Continue to plot the remaining volcanoes. These volcanoes are only some of the many that have formed in these regions.

C. Look at your finished map and answer the following questions.

1. Which ocean has volcanoes that stretch to the north and south near its midpoint?

2. Which ocean has a ring of volcanoes on the land surrounding it?

3. Is there a general pattern in the location of volcanoes?

D. Look at the map on which you plotted the earthquakes. Compare the volcano locations to the earthquake locations.

4. How are they related?

5. What is the relationship between the location of volcanoes and crustal plate boundaries?

CONCLUSIONS:

1. In general, where on earth are volcanoes most likely to occur?

2. Explain why volcanoes and earthquakes may be caused by the same forces.

3. Earthquakes very often happen before volcanoes erupt. Explain how eruptions of volcanoes may be predicted using earthquakes.

4-3. Building the Land

At the end of this section you will be able to:

- ◼ Explain two ways that the earth's crust may be strained
- ◼ Describe three ways in which parts of the crust may be raised to form mountains.
- ◼ Compare the way that plateaus form with the way that mountains form.

Seashells can be found at the tops of some mountains. Old beaches can be found hundreds of meters above sea level. See Fig. 4–17. Could these locations once have been underwater and then risen to their present height?

Fig. 4-17 How can you tell that this area was once covered by water?

*The ripples in rock show that this area was once underwater.

FORCES THAT CHANGE THE CRUST

The rocks that make up the earth's crust are always being put under stress and strain. Forces in the earth cause rocks to be squeezed together, melted, pulled apart, or twisted. Since rocks do not change shape easily, the rocks become strained. It is as if giant hands are always working to push, pull, and bend a rigid crust. When the solid rock of the crust is moved, it is called **diastrophism.** Huge forces must act on the rocks to produce *diastrophism.* You already have seen how the crust is made up of great moving plates. Much of the force needed to bend and twist the rocks in the plates comes from the moving plates. As they grind together, they create forces that are felt all through the rocks.

Diastrophism Movement of solid rock in the crust.

Word Study: Diastrophism: *diastrophe* (Gr.), to twist.

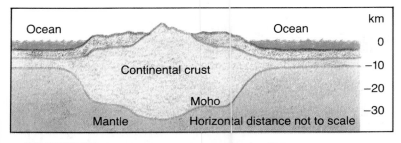

Fig. 4-18 The crust is thin under the ocean floor, thicker under the continents, and thickest under mountains.

The rock that makes up the mantle is more dense than the rock that makes up the crust. Because of the difference in density, the continents and sea floor are floating on the mantle. Suppose you drop a piece of wood in water. Part of the wood will sink into the water. The wood will then float because it is less dense than the water. The same is true for the crust floating on the mantle. The crust floating on the mantle is like different-sized logs floating in water. Large, thick logs float higher in the water than small logs. Mountains, being the thicker parts of the crust, float higher than other crustal parts. They also have roots that reach deeper into the mantle. See Fig. 4-18. Each part of the crust is in a state of balance as it rides on the mantle. This state of balance between different parts of the lower-density crust floating on the higher-density mantle is called **isostasy** (ie-**sahs**-tuh-see).

Isostasy The state of balance between different parts of the lightweight crust as it floats on the dense mantle.

Word Study: Isostasy: *isos + stasia* (Gr.), equal stand.

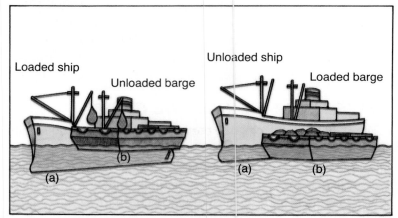

Fig. 4-19 Ships floating in water act like the crust floating on the mantle.

Observing what happens to floating ships shows how *isostasy* works. See Fig. 4-19. Cargo is moved from ship A to barge B. Ship A rises in the water and barge B sinks. The crust, floating on the mantle, acts in a similar way. For example, a heavy rock can fall from a mountain to the land below. The weight of the rock moves from one area of the crust to another. Because of

Fig. 4-20 Folds are caused
when rocks are put under
strain for a long period of time.

isostasy, the area that loses the weight rises. The area that gains the material sinks. Thus the rise and fall of different parts of the crust create strain in the surrounding rocks.

The rise and fall of parts of the crust take place very slowly. Strain builds up in the rocks over a long period of time. Sometimes the rocks crack. If there is movement along either side of such a crack, it is called a fault. If there is no movement along such a crack, it is called a **joint**.

Forces that squeeze rocks together do not always cause joints or faults. Instead, a slow, steady pressure can make some rocks bend without breaking. When rocks bend in this way, it is called **folding.** *Folds* in rocks may be small wrinkles. They also can be large enough to form mountains. See Fig. 4-20.

Joint A crack in solid rock with no movement along the sides of the fracture.

Folding Bending of rocks under steady pressure without breaking.

SHAPING THE LAND

Mountains are caused by diastrophism. There are three main ways that a part of the crust can be raised enough to form mountains. First, large blocks of rock can be separated by faults. Then the blocks may be tilted, like a row of books that have fallen over. Mountains made in this way are called **fault-block mountains.** The Grand Tetons in Wyoming are examples of *fault-block mountains*. See Fig. 4-21 (a).

A second type of mountain range is formed by giant folds in the rocks. The Appalachian Mountains, for example, contain the remains of many great folds. During their long history the Appalachians have been worn down. Today we see long, parallel ridges and valleys, which were formed when the rock layers

Fault-block mountains Mountains formed when large blocks of rock are tilted over.

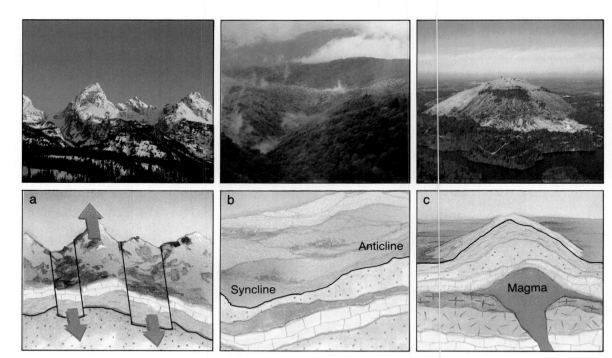

a

b
Anticline
Syncline

c
Magma

Anticline Rock layers folded upward.

Syncline Rock layers folded downward.

Word Study: Anticline: *anti* (against) + *klinein* (to lean) (Gr.), lean against. Syncline: *syn* (with) (Gr.) + *klinein*, lean with.

Dome mountains Mountains formed when a part of the crust is lifted by rising magma.

Plateau Large raised area of level land.

were bent into folds. See Fig. 4–21(b). The raised part of the fold that forms the ridges is called an **anticline (ant**-ih-kline). The ridges in folded mountains like the Appalachians are *anticlines.* The lowered section of a fold is called a **syncline** (**sin**-kline). *Synclines* form the valleys between the ridges. Look at Fig. 4–21b. again.

A third type of mountain is made when a large body of magma rises close to the surface. The crust is pushed up and lifted by the rising magma. The part of the crust that is raised makes **dome mountains.** The Black Hills of South Dakota are *dome mountains.* See Fig. 4–21(c).

PLATEAUS

Scattered all over the earth are very large regions of land that are lifted above the level of the surrounding crust. A large region of raised land is called a **plateau** (pla-**toe**). *Plateaus* can be caused by the same forces that build mountains. In the case of plateaus, large areas of the crust are bent upward. The Colorado Plateau, where the Grand Canyon is found, was formed this way. The Colorado River cut down into the land as the plateau was slowly raised. This formed the Grand Canyon. See Fig. 4–22 (a). Plateaus also can be formed when lava pours out and

covers a large part of the land surface. The Columbia Plateau in the state of Washington is a lava plateau. See Fig. 4–22 (b).

While large areas of land are being uplifted into plateaus, some small areas are sinking. Death Valley in California is an example of such a place. The large block that makes up the floor of Death Valley is slowly tilting as one end sinks. See Fig. 4–22 (c).

SUMMARY

The solid rock is not as quiet as it seems. Huge forces are at work to strain the rock. Sometimes the rock moves to form mountains. Three types of mountains are folded, faulted, and domed. Parts of the crust can be lifted to form plateaus.

QUESTIONS

Use complete sentences to write your answers.

1. Describe two ways the crust may be strained.
2. Compare how folded, faulted, and domed mountains are formed.
3. Describe two ways in which a plateau may be formed. Give an example of each.

Fig. 4–21 (opposite page) *(a) Large blocks of crust are pushed up to make fault-block mountains; (b) layers of rock bend to form folded mountains; (c) magma pushing up the crust forms dome mountains.*

Fig. 4–22 (above) *(a) The Grand Canyon, a plateau made by a river cutting into slowly rising land; (b) the Columbia Plateau, formed from lava; (c) Death Valley, formed when an area of land sank.*

1. The crust may be strained by crustal plate movements and density differences of crust and mantle.
2. Folded mountains result from slow bending without breaking the crust. Faulted mountains result from a slab riding over another or tilting of blocks in a fault. In dome mountains, magma has pushed up the crust.
3. Plateaus may be formed by forces lifting the land along faults, as in the Colorado Plateau, or by lava pouring over a large surface, as in the Columbia Plateau.

INVESTIGATION

See teacher's commentary for teaching hints, safety, and answers to questions.

DENSITY AND ISOSTASY

PURPOSE: To find the density of some common rocks and relate these to isostasy.

MATERIALS:

shale rock   balance
granite rock   graduated cylinder
basalt rock   water

PROCEDURE:

A. Copy the following table:

Rock	Mass (g)	Volume (cm³)	Density (g/cm³)
shale			
granite			
basalt			

B. Obtain samples of shale, granite, and basalt. Make certain each rock will fit into your graduated cylinder. Measure the mass of each rock and record the information in your data table.

C. Using the graduated cylinder and water, find the volume of each rock. Remember to tilt the graduated cylinder and add the rock slowly in order to avoid breaking the bottom of the cylinder. Record your results in the data table.

D. Find the density of each rock sample by dividing the mass by the volume. Record these measurements in the data table.
   1. Which rock has the greatest density?
   2. Which rock has the least density?

E. The three most common rocks found on the earth's surface are shale, granite, and basalt. Find the average density of your three rocks by adding their densities and dividing the total by three.

   3. What is the average density? This average is very close to the average density of all the rocks on earth's surface.

   4. Which rock has nearly this density?

F. Scientists have found the density of the earth by dividing the earth's mass by its volume. This gives an average density for the earth of 5.5 g/cm³.

   5. How does the average density of the three rocks compare to the average density of the earth?

   6. In isostasy, the rocks of earth's crust are said to float on the mantle. Which of the rocks you measured would float highest if resting on the mantle?

   7. Which rock would float deepest in the mantle?

CONCLUSIONS:

1. Which has the higher density, the entire earth or the rocks found on the earth's surface?

2. Explain how the results of this activity support the idea that the earth has a lower-density crust floating on a higher-density mantle.

3. Explain how isostasy causes an uneven surface of the crust.

4. Explain why the largest mountains sink deeper into the crust.

5. Explain why mountains rise as sediment is worn away from the mountain's surface.

CAREERS IN SCIENCE

GEOLOGICAL TECHNICIAN

Geological technicians assist geologists and spend much of their time working outdoors. A technician's tools may be as simple as a hammer and chisel or they may involve heavy equipment. Technicians assisting geological researchers take samples of the earth's crust and determine which rocks are contained there. Other technicians are involved in studying earthquakes and their causes, and must be familiar with some of the computerized equipment in the field. Another group of geological technicians is engaged in the search for deposits of oil and natural gas, which are vital to our country's energy future. All of these technicians carry out the work designed by a geologist, who usually heads the team.

Most technicians have completed at least a two-year program at a technical school. Jobs are available with mining and petroleum companies and the government. For further information, write: American Geological Institute, 5205 Leesburg Pike, Falls Church, VA 22041.

SURVEYOR

Surveying is a job for someone who likes to work outdoors. Surveyors measure the shape and contour of the land and its waterways, and they help to set its boundaries. With this information, detailed maps and charts can be prepared. These are vitally important for engineers who are planning a highway, a pipeline, a mining project, or some other type of construction, such as a shopping center or a railroad.

Surveyors usually work in groups called "field parties." For example, one person may operate the theodolite (a device to measure angles), while another holds a rod used to measure the height and rise of the land at a particular location. High school courses in math, physics, and mechanical drawing provide good preparation for this kind of work. On-the-job training is available, but advancement usually requires college courses leading to a degree. For more information, write: American Congress on Surveying and Mapping, 210 Little Falls Street, Falls Church, VA 22046.

CHAPTER REVIEW

VOCABULARY

On a separate piece of paper, match each term with the number of the statement that best explains it. Use each term only once.

seismic wave   fault   lava   plateau   focus
diastrophism   volcano   seismograph   epicenter   magma

1. A place where the rock on either or both sides of a crack has moved.
2. The location along a fault where movement causes an earthquake.
3. The location on the surface directly above the focus of an earthquake.
4. Records seismic waves.
5. Vibrations caused by movement along a fault.
6. Molten rock below the earth's surface.
7. Place where lava is produced.
8. Magma that has reached the earth's surface.
9. The movement of the solid rock of the crust.
10. A large area of raised land.

QUESTIONS

Give brief but complete answers to each of the following questions. Unless otherwise indicated, use complete sentences to write your answers.

1. How are earthquakes related to crustal plates?
2. Describe three ways that rock may move along a fault.
3. What is the difference between the focus and the epicenter of an earthquake?
4. Describe the three kinds of earthquake waves. Explain how they differ.
5. What is the Richter scale? Give an example of how it is used.
6. Tell what you should do if an earthquake occurs when (a) you are outside a building or (b) you are inside a building.
7. What is the difference between magma and lava?
8. Where are earthquakes and volcanoes most likely to occur?
9. Describe two sources of strain which cause the rock of the earth's crust to become bent and twisted.
10. Compare and contrast the structure of cones produced by explosive, composite, and nonexplosive volcanoes.

11. What is the difference in the way the Aleutian Islands formed and the way Iceland formed?

12. What is the difference between a joint and a fault?

13. How much more energy is released by an earthquake that measures 7 on the Richter scale than one that measures 4?

14. Why are the ridges of the Appalachian Mountains called anticlines and the valleys called synclines?

15. How could you tell the difference between a folded mountain and a fault-block mountain by their layers of rock?

APPLYING SCIENCE

1. Imagine that a large earthquake occurred in your town or city. (a) List some of the dangers that could occur. (b) Would hospitals and police and fire departments be able to handle such an emergency? Explain your answer. (c) Describe an earthquake emergency plan that would reduce injuries to people and damage to property.

2. Erupting volcanoes can both damage and benefit humans. Make a list of the damages and a list of the benefits. Do you think erupting volcanoes have been harmful or beneficial to humans? Explain your answer.

3. The largest mountain systems in the United States, such as the Appalachians and Rockies, are both folded and faulted. Explain how a mountain range could develop both folds and faults.

4. Plot the locations of earthquakes and volcanic eruptions during the school year on the crustal plate map used in the activity section on page 80. Use a color pencil different from the one used in the activity.

BIBLIOGRAPHY

Courtillot, V., and G. E. Vink. "How Continents Break Up." *Scientific American*, July 1983.

Decker, Robert and Barbara Decker. *Volcanoes*. San Francisco: Freeman, 1981.

Francis, Peter, and Stephen Self. "The Eruption of Krakatoa." *Scientific American*, November 1983.

Kimball, Virginia. *Earthquake Ready*. Culver City, CA: Peace Press, 1981.

Walker, Bryce. *Earthquake* (Planet Earth Series). Chicago: Time-Life Books, 1982.

WEATHERING AND EROSION

CHAPTER GOALS

1. Explain how the rocks on the earth's surface are always being changed.
2. Understand how water changes the earth's surface.
3. Describe how water occurs beneath the earth's surface.

5-1. *Weathering and Erosion*

At the end of this section you will be able to:

- ☐ Name the two processes that work together to wear down the earth's surface.
- ☐ Describe the two processes that break rock into smaller pieces.
- ☐ Explain how soil is formed.
- ☐ List three ways in which gravity can cause rocks and soil to move.

The earth's surface is always changing. Sometimes the changes are sudden, like the earth movement that destroyed the highway shown in Fig. 5-1 on page 114. More often, the changes are slow and take place in small steps.

ROCKS WEAR AWAY

No matter where you look on the surface of the earth's continents, you see the results of a constant battle. The battle is between two kinds of forces that cause the land to be pushed up and then worn down again. On one side of the battle are the forces beneath the surface. These powerful forces cause the crust to be faulted, folded, and lifted. On the other side of the battle are the processes of **weathering** and **erosion** (ih-**roe**-zhun). The term *weathering* means all the processes that break rock into small pieces. You have seen examples of weathering at work on sidewalks. A new sidewalk is smooth and free of cracks. Later the surface becomes rough and cracked. The top layer is worn off. Large cracks appear if roots from nearby trees grow underneath the sidewalk. These processes

A natural arch begins when erosion along cracks in some rock formations creates rock walls. Sometimes slabs of rock have soft layers that are easily worn away. Eventually, a window is worn through the rock wall. In time, the window enlarges and an arch is formed.

Word Study: Erosion: *erodere* (L.), to gnaw off.

Weathering All the processes that break rock into smaller pieces.

Erosion All the processes that cause pieces of weathered rock to be carried away.

Fig. 5-1 Landslides can be caused when heavy rains or earthquakes loosen material on a steep slope.

are similar to the way rocks weather at the earth's surface. See Fig. 5-2. Once rock has been broken up by weathering, small pieces can be moved by gravity, water, ice, or wind. Everything that happens to cause pieces of rock to be carried away is called *erosion*.

Weathering and erosion, along with folding and faulting, shape the mountains, hills, plateaus, and plains that make up the earth's surface. If there were no weathering and erosion, the earth would look very different. It might appear as rough as the surface of the moon. On the other hand, weathering and erosion alone could flatten mountains on the continents within 25 million years. The battle between the forces that build the

Fig. 5-2 The solid granite rock on this hillside is being reduced to fragments by the processes of weathering.

earth's crust and the forces that wear it down goes on all the time.

HOW WEATHERING HAPPENS

Rock material is usually formed below the surface of the crust. It is not exposed to the weather. As long as it remains buried, no weathering takes place. Only rocks that are formed on the surface, or are uncovered, are exposed to the weather. One of the main forces in weather is the water in the air. In general, weathering happens where the solid part of the earth (the *lithosphere*) comes into contact with the liquid water of the earth (the *hydrosphere*) or air (the atmosphere). These three parts of the earth meet in zones called **interfaces**. Weathering takes place in *interfaces*.

Physical and **chemical weathering** are two processes of weathering. In *physical weathering*, rocks are broken into smaller pieces. In *chemical weathering*, substances in the rock are changed into different ones. Weathering is often a combination of both physical and chemical breakdown.

PHYSICAL WEATHERING

1. Frost. The effect of *frost* action on a rock is an example of physical weathering. Frost action happens when water enters the cracks in rock. If the temperature drops below 0°C, the trapped water freezes. Water expands when it freezes. Freezing water produces a very large force as it expands. You may have seen water pipes that broke when the water inside froze. When water freezes inside a crack in a rock, the crack is made larger. The water in the crack freezing, melting, and freezing again repeatedly can break the rock apart. Frost action is an important agent of physical weathering in cold climates.

2. Exfoliation. Another kind of physical weathering begins with rock that is buried deep underground. A rock deep beneath the surface is under great pressure from the weight of all the rock above. This buried rock expands when it is exposed by erosion. Its outer layers may be loosened and peel off like the layers of an onion. This is one example of a process called **exfoliation**. See Fig. 5–3. Weathering by *exfoliation* causes rocks to take on a rounded shape as the layers peel off. Large rounded knobs of rock can be formed by exfoliation. See Fig. 5–4.

Word Study: Interface: *inter* (between) + *facere* (to make) (L.), make between.

Interface A zone where any of the three parts of the earth (the lithosphere, hydrosphere, and atmosphere) come together.

Physical weathering The ways rock breaks up into smaller pieces without any change in the materials in the rock.

Chemical weathering The ways rock breaks down by changing some of the materials in the rock.

Word Study: Exfoliation: *exfoliatus* (L.), stripped of leaves.

Exfoliation The peeling off of scales or flakes of rock as a result of weathering.

Fig. 5-3 How was this rock weathered?

Reinforcement: Ask students to identify evidences of weathering that they may have observed in exposed bricks, pavement, sidewalks, and stone, such as marble, used on the outside of buildings. Try to describe the type of weathering that might have been responsible for each example given.

3. Heating. Heating also can cause rock to break apart. Solid materials, such as the substances that make up rock, expand when they are heated. Most kinds of rock are made up of many different substances called *minerals.* A rock can be identified by the minerals it contains. The different minerals in rocks may expand by different amounts when the rock is heated. This can cause the rock to crack. Over a long period of time, the heating during the day and cooling at night can make the outer layers of the rock break into small pieces.

4. Plants. Plants can be agents of physical weathering. Plant roots work their way through small cracks in the rock. As the roots grow, they expand and break apart the rock.

Fig. 5-4 Half Dome in Yosemite National Park is an example of a large dome of rock produced by exfoliation.

CHEMICAL WEATHERING

1. Water. When *chemical weathering* takes place, the minerals in the rock are changed. Water is one agent of chemical weathering. Some materials in rock can be dissolved in water. These dissolved minerals are picked up by water running over rocks in streams or from rain. The rock is broken up by loss of some of its material to the water. Some minerals that make up rock are able to absorb water the way a sponge does. This absorbed water increases the size of those parts and strains the rock. The weakened rock then is more easily broken apart.

Fig. 5–5 Rust is caused by oxygen and water acting on iron.

2. Oxygen. The air contains the gas *oxygen*. Oxygen is another agent of chemical weathering. A small amount of oxygen from the air can join with some minerals in rock when they are wet. The minerals are then changed into new substances. The combining of minerals or other materials with oxygen is called *oxidation*. You have seen pieces of rusting iron. See Fig. 5–5. Oxygen can cause the oxidation of wet iron to form rust. Rocks often contain iron that can rust just like an old chain or car bumper. The red, yellow, and orange colors often seen in weathered rock result from a similar process.

3. Carbon dioxide. The air also contains *carbon dioxide* gas. Carbon dioxide joins with water to form a weak acid called carbonic acid. This acid is able to dissolve some of the minerals in rock.

4. Plants. Chemical weathering can also be caused by the acids from some plants. The simple plant lichen (**like**-en) is an example. Lichens grow on the surface of rock. See Fig. 5–6. An acid is produced that dissolves some of the materials in the

Fig. 5–6 Lichens produce acids that break down rock surfaces.

rock. Lichen is often called a "pioneer" plant because it is one of the first plants to grow in a rocky area.

SOIL

Word Study:
Humus: *humus* (L.), earth.

A gardener may think of soil as material in which to grow plants. A farmer knows that crops could not be grown easily without soil. The earth scientist is interested in soil as a substance formed by the physical and chemical weathering of rock and the decay of living matter. Looking at a handful of soil will show that it is made up of particles. These particles could be the remains of a mountain. The particles can be divided into two groups. One group is made up of **organic** material. The other group is made up of pieces of weathered rock. *Organic* means coming from living things. Most of the organic material in soil comes from decaying plants. Some organic material also comes from animal remains. Many living things too small to be seen also are present in soil. In time, all the organic material is changed into **humus** by bacteria. *Humus* is a very important part of the soil. It is one of the main sources of the food and minerals needed for the growth of new plants. Without a supply of the materials supplied by humus, soil is not fertile. Soils differ in the amount of humus and other organic matter they contain. For example, the soil in swamps may be made almost entirely of organic matter. In deserts, on the other hand, the soil usually has only a small amount of humus.

Soil also contains pieces of weathered rock. These particles are not all the same size. The larger pieces of rock are called gravel. The smaller pieces, from grain size up to two millimeters, are called sand. The smallest pieces are called silt and clay. These particles are as fine as flour or dust.

Making soil is a very slow process. Weathering must first break down the solid rock. As time passes, and weathering continues, soil begins to form separate layers. The layers in soil are called **horizons** (huh-**rie**-zuns). Each *horizon* has its own properties and is different from the other layers. Humus in the top layer gives it a dark color. This dark-colored top layer is called the *A horizon*. The A horizon is also called topsoil. It is the best layer in which to grow plants.

In young soil there are only two horizons. These two layers are the A horizon and the *C horizon*. The C horizon is made of partly weathered rock.

Organic Coming from any living thing.

Humus Material in soil that is from the remains of plants and animals.

Horizon Any separate layer of soil.

If erosion does not remove the topsoil, a third layer will develop in time. Water moving through the A horizon dissolves some of the minerals. The finest soil particles, the size of silt or clay, are carried through the soil by the water. These grains and dissolved minerals are collected in a layer called the *B horizon*, or subsoil. It may take about 10,000 years for this layer to form. The B horizon is a hard soil. It is difficult to plow or dig. Only large, strong plant roots can enter it. Water can not pass quickly through the subsoil.

A soil with well-developed A, B, and C horizons is called *mature*. The steps in the development of a mature soil are shown in Fig. 5-7. Mature soils are not formed in dry regions. The moisture needed for the chemical weathering processes is absent in dry areas.

Reinforcement: Soil in cold or dry climates is usually thin and poorly developed. A warm, humid climate produces a thick layer of soil. This is a result of the speed of chemical weathering in each kind of climate. Ask students to relate the kind of soil found in a region to the climate, and explain why the difference exists.

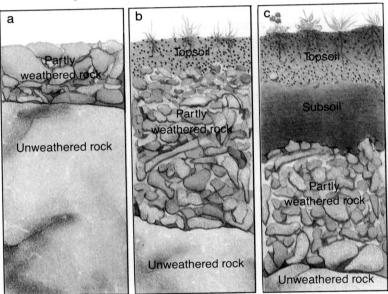

Fig. 5-7 (a) unweathered rock; (b) immature soil has an A horizon and a C horizon; (c) mature soil has an A horizon, a B horizon, and a C horizon. The photo shows mature soil.

Plants growing in soil remove some of the minerals. If these minerals are not replaced, the soil becomes less able to support the growth of new plants. If the plants are not removed, they will return the minerals to the soil when they decay. Soil that is used to grow crops must have the minerals replaced in the form of fertilizers.

EROSION CAUSED BY GRAVITY

The effects of weathering are best seen where the ground is not level. Broken rock pieces can fall from a cliff or slide down

Word Study: Talus: *talutium* (L.), a slope.

Fig. 5–8 A pile of talus has collected at the bottom.

Talus A pile of broken rock that gathers at the bottom of a cliff or steep slope.

Landslide A sudden movement of large amounts of loose material down a slope.

Creep A slow downhill movement of loose rock or soil.

a slope. A pile of broken material at the bottom of a cliff or steep slope is called **talus**. Fig. 5–8 shows *talus* on the side of a canyon wall. Gravity is an agent of erosion because it pulls pieces of rock down to the bottom of a slope.

Gravity can also move large amounts of loose, weathered material. The sudden movement of large amounts of loose rock or soil is called a **landslide**. *Landslides* may be made up of part of a slope or the side of a mountain. Some landslides are very dangerous. These are the ones in which a large amount of rock material suddenly tumbles down a mountain slope. Such a landslide can take place where rock layers are tilted. If a weak layer beneath the surface gives way, the upper layers rush down the slope. Millions of kilograms of rock and loose material may suddenly move. Earthquakes often can be the cause of landslides. Landslides all over the world have cost many lives. Entire towns and villages have been wiped out. Sudden floods have been caused when large landslides block rivers or raise the level of lakes so that water spills over dams.

The movement of weathered rock by gravity is not always sudden. Sometimes the movement is only about one to two centimeters a year. Slow downhill movement of loose rock or soil is called **creep**. It is often hard to tell that *creep* is taking place. There is usually no change in the way the slope looks. However, careful observation of trees, telephone poles, and fence posts can show evidence of creep. See Fig. 5–9. Creep is the most common kind of erosion caused by gravity. It is the

Fig. 5-9 (left) *Soil creep.*

Fig. 5-10 (right) *Slump usually leaves a curved surface.*

Slump The sudden downward movement of a block of rock or soil.

main way weathered rock moves downhill. Sometimes a part of a slope may slide down more or less in one piece. This is called a **slump**. A *slump* usually takes place where the slope of the land has been changed due to erosion. For example, the banks of streams often slump into the eroded stream channel. See Fig. 5-10. Slumping is also seen along cliffs at the shore where waves have cut into the base of the cliff.

SUMMARY

The earth's land surfaces are the result of two opposing forces. Movements of the crust raise the land to form features such as mountains. At the same time, the agents of weathering and erosion work to wear down the land features. Rocks are made into smaller pieces by physical and chemical weathering. Soil is formed from pieces of weathered rock and the remains of plants and animals. Gravity causes rocks to move.

QUESTIONS

Use complete sentences to write your answers.
1. Describe the two processes of weathering and erosion.
2. Name two kinds of weathering and explain how they are different.
3. Explain how the three layers of soil are formed.
4. What causes slumping, landslides, and soil creep?

1. Both are processes that wear away the land. Weathering refers to the processes that break the materials down, whereas erosion refers to the processes that move the materials once they are broken down.
2. Chemical weathering and physical weathering are the two kinds of weathering. They differ in that chemical weathering is the dissolving of rock whereas physical weathering is the breaking of the rock into smaller pieces by tumbling down a hillside, etc.
3. The top, or A, horizon is formed by the weathering of rock and the accumulation of plant and animal materials. The bottom, or C, horizon is formed of partly weathered rock. The middle, or B, horizon is formed by the water draining down through the top layer and carrying with it the small particles and dissolved minerals.
4. Slumping, landslides, and soil creep are all caused by gravity.

See teacher's commentary for teaching hints, safety, and answers to questions.

WEATHERING

PURPOSE: To investigate some examples of chemical and physical weathering.

MATERIALS:
test tube
test tube rack
limestone rock
soda water
glass container

PROCEDURE:

A. Half-fill a test tube with soda water. Rest it carefully in the test tube rack.

B. Drop a small piece of limestone into the soda water.
   1. Describe any reaction you see.

C. Without moving or shaking the container, look closely at the limestone in the soda water.
   2. What forms on the rock itself?

D. Gently shake the container so that the rock moves.
   3. What happens to the bubbles on the rock?

E. Soda water is a solution of carbon dioxide gas in water, making the water acidic. Rain also dissolves some carbon dioxide gas from the air.
   4. Would you expect rain to have a similar effect on limestone rock?

F. Limestone is dissolved by an acid. In so doing, carbon dioxide gas is made. Look again at the limestone in the soda water. Do not shake the container.
   5. Are there gas bubbles rising from the surface of the rock?

G. Let the soda water and limestone sit for a few minutes. Look again at the limestone rock.
   6. What evidence do you see that the rock is dissolving?
   7. What kind of weathering is shown by the limestone in soda water?

H. Half-fill the glass container with water. Place a small piece of limestone in the water. Shake the rock around in the water until no more air bubbles rise from it.

I. Remove the rock from the water and place it in the waste container.

J. Look at the bottom of the water in the container.
   8. Are there any pieces of rock in the bottom of the container?
   9. What kind of weathering is shown here?

CONCLUSIONS:

1. Which was an example of: (a) chemical weathering? (b) physical weathering? Explain what happened in each example.

2. Compare how weathering took place in your investigation to how weathering occurs in nature.

3. What would be the effect of shaking the rock in soda water as compared to plain water?

4. Is it possible for rocks in nature to undergo both chemical and physical weathering? Explain your answer.

5-2. Water on the Land

At the end of this section you will be able to:

☐ Describe the parts of the *water cycle*.

☐ Name the parts of the earth's *water budget*.

☐ List three ways in which running water erodes the land.

☐ Compare the features of a young river and an old river.

A flood can make a peaceful river or stream become a killer. See Fig. 5-11. Swiftly moving water can tear houses from the ground. While floods destroy property and take lives, they are a natural part of the constant movement of the earth's water.

Fig. 5-11 Most floods are caused by heavy rainfall.

THE WATER CYCLE

About 97 percent of the earth's water supply is in the oceans. Smaller amounts, about 2 percent, are in glaciers and ice, below the ground, in lakes or rivers, or in the air. See Fig. 5-12.

The earth's water moves from one storage place to another. Water that was once in the Pacific Ocean may now be in your faucet. It may also be in the clouds in the sky. Water moves from oceans to air and land and back to the oceans again. This movement occurs all over the earth and is called the **water cycle.** A cycle is something that has no beginning or end. The *water cycle* consists of steps that are repeated over and over.

Water cycle The processes by which water leaves the oceans, is spread through the atmosphere, falls over the land, and runs back to the sea.

WEATHERING AND EROSION

Word Study: Evaporate: *e* (away) + *vapor* (steam) (L.), steam away. Precipitation: *praecipitatus* (L.), thrown out. Transpiration: *trans* (across) + *spirare* (breathe) (L.), to breathe across.

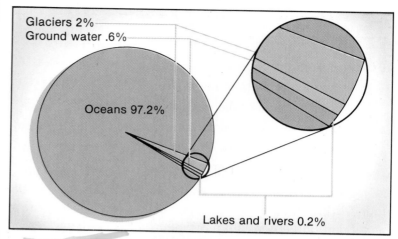

Glaciers 2%
Ground water .6%

Oceans 97.2%

Lakes and rivers 0.2%

Fig. 5–12 This graph shows where the earth's water supply is found.

Evaporation The process of water changing from a liquid to a gas at normal temperatures.

Precipitation Water that falls from clouds, usually as rain or snow.

Ground water The water that soaks into the ground, filling the openings between rocks and soil particles.

Transpiration The process by which plants take up water through their roots and release water vapor from their leaves.

The first part of the water cycle happens when sunlight falls on a body of water. Heat from sunlight causes some of the water to **evaporate** (ih-**vap**-uh-rate). *Evaporation* causes liquid water in the sea to change into a gas called water vapor. The water vapor then enters the atmosphere. Winds spread the water vapor through the atmosphere.

The next part of the water cycle causes the water to fall out of the atmosphere. Water vapor in the air becomes rain or snow and falls as **precipitation** (prih-sip-uh-**tay**-shun). Large amounts of *precipitation* fall on the land. Several things can happen to this water that falls on the ground. It can sink into the ground or it can form rivers or streams. Rivers and streams move quickly over the land surface. The water in rivers and streams flows back into the sea within a few weeks. If the water did not return to the sea, evaporation would cause the sea level to drop by one meter each year.

Rainwater that soaks into the ground becomes **ground water.** Some *ground water* is returned to the air by plants. This process is called **transpiration** (trans-puh-**ray**-shun). Plants take in ground water through their roots. They then return water vapor to the air through openings in their leaves. Only a small amount of ground water is taken up by plant roots and *transpired* back into the air. Instead, most of the water sinks deep into the crust. It fills the spaces between cracks and other openings in the rock. Almost all ground water becomes part of an underground water system. The water in this system moves very slowly. As it seeks lower levels, it eventually flows back into the ocean. The complete water cycle is shown in Fig. 5–13.

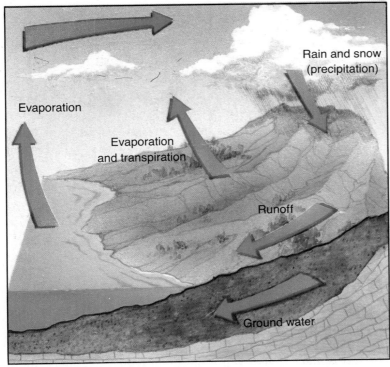

Rain and snow
(precipitation)

Evaporation

Evaporation
and transpiration

Runoff

Ground water

Fig. 5-13 The parts of the water cycle are shown by the arrows of this diagram.

A single drop of water may spend a few days in the air, or 50,000 years frozen in the ice of a glacier. It may be stored in a lake for a hundred years, or it can be deep underground for thousands of years. Given enough time, a drop of water will complete the cycle.

THE EARTH'S WATER SUPPLY

The earth is on a *water budget*. A budget is a record of income and outgo. Precipitation such as rain or snow is the income of the water budget. Evaporation and transpiration are the outgo. At any time, an area may have more water coming in than going out. Some water enters the ground. But ice melting in the spring or a heavy rainstorm can add extra water to an area. Thus the water budget may be out of balance for a while. The opposite may also happen. In the summer, evaporation and transpiration may increase. The weather may be very dry and the water supply may drop. One place may have too little or too much water. However, the water budget for the entire earth is balanced. The amount of water entering the air is equal to the amount of water falling as precipitation.

Reinforcement: Discuss the source of the local water supply. Find out what water treatment, if any, is used.

Modern civilization depends on a ready supply of water. All living things need water. For example, a person can live only a few days without water to drink. Water also serves in many other ways to support our lives. Crops that supply food are affected by the amount of water available. The oceans, lakes, and rivers of the world are used as highways to transport goods. Most of the great cities of the world have grown around river valleys or seaports. The cities, towns, industries, and farms have kept growing until many have reached the limits of their supply of water.

In the United States, each person uses about 7,000 L of water each day. This amount of water could fill a bathtub more than 100 times. Only about 400 L are actually needed for each person. The remaining 6,600 L are used for farm irrigation and industry. This water is "borrowed" from the water cycle. It must be collected, stored, and sent out to homes, fields, and factories. Only a tenth of the precipitation falling on the United States is used. The rest returns to the oceans. Altogether, the water cycle provides more water than is needed. However, the water is not always in the right place at the right time. For example, the western United States has 60 percent of the land. But it receives only 25 percent of the total water falling as precipitation. Many areas have dry periods, such as California in 1976–1977 and Florida in 1981.

We cannot change the water budget in a particular place. We must be careful how our valuable water is used. *Water conservation* must become a part of our lives. All the ways of saving water and making it available for use are included in water conservation. The way water conservation can be carried out is discussed in Chapter 10.

RUNNING WATER

Suppose the photograph shown in Fig. 5–14 could have been made about nine million years ago. It would not show the exposed layers of rock. Scientists believe that the San Juan River carved this great valley. Another example of erosion by running water is the Grand Canyon, which is about 1.6 km deep and 14 to 29 km wide. It shows the power of running water to cause erosion on the earth's surface.

Water moving over the earth's surface moves soil, rock, and even boulders. It can shape the land. Erosion by running water

begins with raindrops that come down on exposed soil. Raindrops break up lumps of soil and stir up the soil. Some rain sinks into the soil. Some moves over the surface. Water that moves over the surface of the land is called **runoff**.

Runoff may come from melting snow or ice as well as from rain. Runoff begins as a shallow layer, or sheet, of water flowing downhill. Pebbles and rocks on the surface quickly break the sheet into tiny streams. These tiny streams form shallow **channels.** *Channels* meet and are widened and deepened by the water flowing through them. Small streams come together forming larger ones. A group of streams collects runoff from a large area. All the streams from an area flow into one large stream or river to form a river system. See Fig. 5–15. Streams that flow into rivers are called **tributaries** (**trib**-yuh-ter-ees). The large area that sends runoff into a river is called the *drainage basin,* or *watershed*.

EROSION BY STREAMS

Running water erodes the land surface in three ways. One way is when soil and rock materials dissolve in water. Over many years, running water can dissolve away some of the rock. But water itself is not able to change the rocks very much. Most erosion by streams is done in a second way. This is when particles of rock are carried along by the water. The rocks that are small enough to be picked up by the water wear away the channel, just like sandpaper wears away wood. A third way is when pieces of rock too large to be carried by the water are

Fig. 5–14 (left) *The San Juan River has created Gooseneck Canyon, Utah.*

Fig. 5–15 (right) *All the streams in an area join to form larger ones.*

Runoff Water running downhill over the land surfaces.

Channel The path of a stream of water.

Tributary A stream that flows into a river.

Word Study: Channel: *canalis* (L.), pipe or canal.
Tributary: *tributum* (L.), to contribute or pay.

Fig. 5-16 Contour plowing follows the shape of the land surface to help prevent rapid runoff with heavy erosion.

rolled and bounced along the channel. They chip and wear away the rock of the channel. Thus streams are able to move loose materials and, at the same time, cut down into solid rock.

Erosion takes place rapidly where the soil is loosely packed or where the slope is steep. Heavy rainfall washes away soil more rapidly than light showers do. This is because the soil cannot absorb all of the rain as it falls. A rainstorm may cause a huge flow of mud to move down a slope or valley suddenly. These sudden mud flows may destroy houses or other buildings in their paths.

Careless use of the land also speeds the rate of erosion. Plants protect the soil from washing away. When the natural plant cover is removed, the soil is exposed to runoff and is eroded. It takes a long time to replace the lost soil. Nature needs about 1,000 years to make 2.5 cm of soil. Grass and trees planted along roadways and overpasses help hold the soil in place. Farmers use contour plowing to reduce erosion in their fields. See Fig. 5-16.

THE LIFE HISTORY OF A RIVER

River systems are formed where streams of running water come together. The bottom of a river channel is called the *bed*. Its sides are called the *banks*. The place at which the river flows into a larger body of water is called the *mouth*. Rivers are often described as young or old. These terms refer to the amount of erosion of the river channel, not to the age of the river. The amount of erosion can be seen by the shape of the channel.

1. Young rivers. A young river is still cutting through the channel. Thus there is a large difference in elevation between its watershed area and its mouth. As a result, the water flows swiftly through the channel. The soil and rock, carried or rolled by the water, scrape the bed. Large amounts of rock and soil from the bed are carried off. The channel is cut deeper. Erosion of the river bed gives most young rivers a V-shaped valley. See Fig. 5–17. Because the bed is rocky, a young river is likely to have waterfalls and rapids.

2. Mature rivers. As a river reaches maturity, the cutting action becomes very slow, or stops completely. The difference in height between its source and mouth becomes less. The water flows through the channel more slowly. Erosion speeds up along the river banks. The river bed widens. Waterfalls and rapids are smoothed over. The valley of the river channel becomes wider and may take on a U-shape.

3. Old rivers. Finally, the slope of the river channel becomes so small that the water moves slowly. The slow-moving water is able to carry only the finest material. A bend in an older river may become a wide curve. This is because the water always moves faster along the outer bank of a bend. The faster-moving water has more energy to erode rock and soil. Water flows slower along the inner bank of a bend and tends to drop its load of soil and rock. See Fig. 5–18. These two actions cause

Fig. 5–17 How can you tell if this river is young or old?

*The river is young because of the V-shaped valley, rock bed, and rapids.

Enrichment: Many rivers pass through several cycles of aging. An old river can be rejuvenated by uplift that will cause it again to enter a youthful stage.

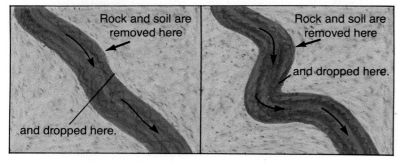

Rock and soil are removed here

Rock and soil are removed here

and dropped here.

and dropped here.

Fig. 5–18 The actions that cause the bend in a river to become wider are shown in this diagram.

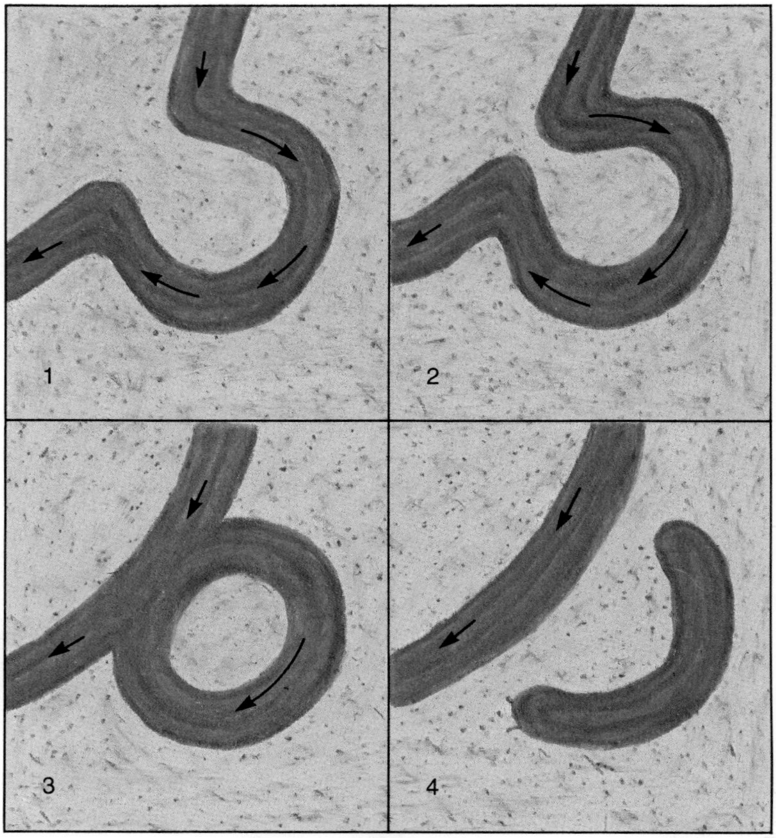

Fig. 5-19 Steps in the formation of an oxbow lake are: (1) a meander develops, (2) erosion cuts across the base of the meander, (3) the river bypasses the old meander, (4) a separate curved lake is left.

Meander A wide curve in the channel of an old river.

Word Study: Meander: *maiandros* (Gr.), (the name of) a winding river.

the curve to grow into more of a loop. In time, the bend can become a horseshoe-shaped loop called a **meander** (mee-**an**-dur). See Fig. 5-14. Further erosion may cut the base of the *meander*. An *oxbow lake* is formed in this way. See Fig. 5-19.

The river system may grow until it is large enough to drain the average runoff from its watershed. When the amount of runoff is greater than usual, the river is not large enough to handle the extra water. A flood results. Heavy rain or fast-melting snow in the spring can cause floods. Most rivers have a small flood once in a while. Major floods occur less often. Unfortunately, these floods often cause disasters. Lives may be lost. Buildings, roads, and crops can be destroyed.

Flood water deposits soil and rock on both sides of a river channel. Repeated flooding builds up layer on layer of this river soil. Such an area is called a *flood plain*. Older rivers often have wide, flat flood plains. The soil on a flood plain is fertile and makes good farm land.

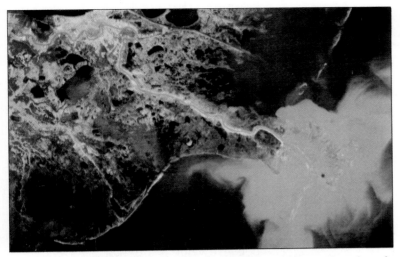

Rivers empty into a lake or the sea. Much of the soil and rock material carried by the river is dropped at the mouth. A deposit of these particles is built up in the larger body of water, causing a **delta** to be formed at the mouth of the river. The *delta* formed by the Mississippi River reaches far into the Gulf of Mexico. See Fig. 5-20.

Delta A deposit formed at the mouth of a river or stream.

Word Study: Delta: *delta*, Δ (Gr.), the fourth letter of the Greek alphabet, which has the shape of a triangle.

SUMMARY

Almost all water on the earth's surface takes part in an endless cycle. Water moves from the oceans to the atmosphere, falls on the land, and runs back to the sea. Every area has a water budget. It shows the income from precipitation and the outgo through evaporation and transpiration. While running across the land, water erodes mainly by carrying or rolling rock and soil. All runoff water enters a river system. A river ages as the speed of water moving through its channel becomes slower.

QUESTIONS

Use complete sentences to write your answers.

1. Diagram and describe the processes in the water cycle.
2. Explain how the earth's water budget at a particular time and place could be out of balance.
3. Give an example of three ways in which running water erodes the land.
4. Compare the features of a young river with the features of an old river.

1. The diagram should resemble Fig. 5-13, page 125. The description should match the processes represented by the arrows in Fig. 5-13.
2. On the local scale, the rate of evaporation may exceed the rate of precipitation, or the rate of precipitation may exceed the rate of evaporation at a given time.
3. Running water may move soil, rock, or boulders. The rocks that are carried along by the running water scrape away at the sides and bottom of the channel, eroding the channel to a larger size. In places where the soil is loose and the slope is steep, the running water may cause a mud slide.
4. A young river has a steeper slope, fast-running water, usually rapids and waterfalls, and a V-shaped river valley. An old river has a more level slope and slow-running water. Bends in the river become wide curves, some curves become meanders, some meanders may become oxbow lakes, and the river valley takes on a U shape.

INVESTIGATION

See teacher's commentary for teaching hints, safety, and answers to questions.

WATER BUDGET

PURPOSE: To graph and study a typical water budget for Minneapolis, Minnesota.

MATERIALS:
graph paper
2 different color pencils

PROCEDURE:
A. Copy the following table.

E. Shade in the area with the second color where the evaporation is greater than the precipitation. Mark this area "deficit."
4. Which months show a deficit?

F. Using your table, find the surplus or deficit for each month. Do this by subtracting the evaporation from the surplus. Record this in your table in the horizontal row headed "P–E." A positive value shows a surplus, a negative value is a deficit.

MONTHLY PRECIPITATION AND EVAPORATION (in mm) FOR MINNEAPOLIS, MINNESOTA, DURING A ONE-YEAR PERIOD													
	Jan.	Feb.	Mar.	Apr.	May	June	July	Aug.	Sept.	Oct.	Nov.	Dec.	Total
Precip.	20	20	35	50	85	105	90	85	75	50	35	25	
Evap.	0	0	0	35	90	125	150	125	80	40	0	0	
P–E													

B. Plot a line graph of the monthly precipitation for Minneapolis. Make two axes. The vertical axis will represent the precipitation. The horizontal axis will represent the months. Use a color pencil to draw the line.
1. Which month shows the greatest amount of precipitation?
C. On the same axes, plot a line showing the evaporation using a different color pencil or a dashed line.
2. Which month shows the greatest amount of evaporation?
D. Shade in the area with one color where the precipitation is greater than the evaporation. Mark this area "surplus."
3. Which months show a surplus?

G. Fill in the last column of your table by finding the total precipitation and the total evaporation. Subtract the total evaporation from the total precipitation.
5. How much is the surplus or deficit for the year?

CONCLUSIONS:
1. Under what conditions does a surplus occur?
2. How does a deficit occur?
3. At the end of the year, how can you tell if there was a surplus or deficit?
4. Why does the amount of evaporation during the months change?

5-3. Water Beneath the Land

At the end of this section you will be able to:

- ☐ Explain how water enters the ground.
- ☐ Relate the water table to springs and wells.
- ☐ Describe how limestone *caverns* are formed.

One day in 1981, in Winter Park, Florida, the hole shown in Fig. 5-21 suddenly appeared. Within a few hours it had swallowed buildings, cars, and most of a swimming pool. Could this also happen where you live? You will find out how to answer that question in this section.

Fig. 5-21 Homes in Winter Park, Florida, fell into a huge sinkhole when the ground collapsed.

HOW DOES WATER ENTER THE GROUND?

Three things can happen to water that falls on the ground. Some of it may be returned directly to the atmosphere by evaporation. Some of the water may form runoff. Most of the time, some also sinks into the ground. Any water that sinks below the surface is called *ground water*. Water can sink into the soil because there are open spaces between the soil particles. The ground water fills these spaces. The rock below the soil may contain many joints and cracks. Water fills these empty spaces too. Water also enters most rock that seems to be solid. This is because almost all rocks are made up of grains of different minerals. This causes spaces between the different grains in most solid rock. Water can fill these tiny spaces. The spaces are called *pores*. The amount of empty space in soil or rock is called its **porosity** (pour-**ross**-ih-tee). The *porosity* of some kinds of rock is 25 percent. This means that one-quarter of the rock is empty space! All rock and soil has a certain porosity. The porosity controls how much water rock or soil can hold. See Fig. 5–22.

Porosity The amount of open spaces or pores in soil or rock.

Word Study: Porosity: *poros* (Gr.), passage.

Reinforcement: Compare the number of people standing in a hallway to porosity. A few people would be like high porosity, whereas many persons crowded together would resemble low porosity. Permeability can then be related to how fast a person might be able to move down the hallway under each condition.

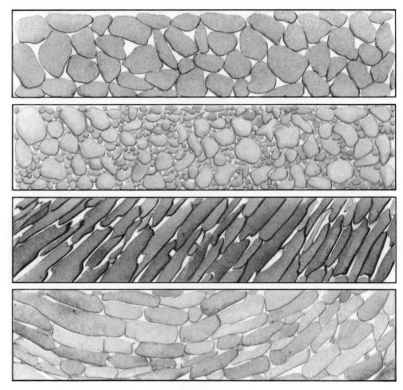

Fig. 5–22 There are open spaces or pores in soil or rock.

Permeability A measure of the speed with which water can move between the pore spaces in a rock.

Another important property of soil and rock is **permeability** (per-me-uh-**bill**-ih-tee). *Permeability* describes how rapidly water can move between the pore spaces or cracks. In rock with many large pore spaces (high porosity) or with many cracks, water can usually move quickly. Therefore, the rock has high permeability. On the other hand, water cannot move quickly in rock that has very small pore spaces or is without cracks. Therefore, this type of rock has low permeability.

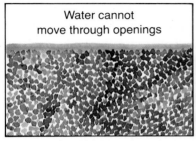

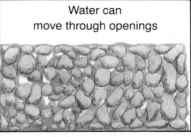

Fig. 5-23 Soils made up of fine particles have low permeability.

Soils, such as clay, with a large amount of very small particles have low permeability. The pore spaces between the tiny clay particles are very small. See Fig. 5-23. Once such small pore spaces are filled with water, the water is held in place. This means that more water cannot enter. Layers of clay or rock with low porosity stop the movement of ground water.

How far down into the crust can ground water go? The answer to this question is not known. It is known that water passes through joints, cracks, and pores in rock below the soil. There is probably a level, several kilometers deep, where the water can sink no further. Rock there is most likely under pressures great enough to close up the cracks and pore spaces.

THE WATER TABLE

Each time it rains some water sinks into the soil. The water moves downward through the soil and rock until it reaches a level where the pore spaces, cracks, and joints are already filled with water. The top of this level is called the **water table.** In general, the *water table* reflects the shape of the land surface. It rises under hills and falls beneath low places. In some places the land surface falls below the water table. Lakes, swamps, and rivers are found at these places. See Fig. 5-24.

Word Study: Permeability: *permeabilis* (L.), passable.

Water table The level below the ground surface where all open spaces are filled with water.

Fig. 5-24 In some low places the surface of the land falls below the water table.

The water table in a region may stay at the same level for a time. However, changes in the water budget may cause it to rise or fall. For example, heavy rains can cause the water table to rise. During dry periods, the depth of the water table usually falls.

Springs are found where the water table comes to the surface without a low place where the water can flow. Many springs are found on hillsides. Most springs disappear when the water table drops during a dry period. One kind of spring flows all the time. This kind of spring is supplied by ground water sitting on top of a layer of low porosity. The temperature of ground water is normally near the average temperature of the region. Thus spring water feels cool in summer and warm in winter. In some places, however, ground water is heated before it comes to the surface in a spring. Such places are called *hot springs*. The water of hot springs is heated by a body of magma or by hot gases escaping from magma. The water in hot springs often has a large amount of dissolved minerals. Some of these minerals may be deposited around the hot spring. See Fig. 5–25. *Geysers* are hot springs that shoot hot water and steam high into the air. Steam created underground forces the water out in a geyser. See Fig. 5–26.

Digging to a level below the water table creates a *well*. Most modern wells are made by drilling a hole to a depth below the water table. A pipe is usually put in to keep the well open.

* The minerals come from the hot water inside the earth.

Fig. 5–25 Where do the minerals around the hot spring come from?

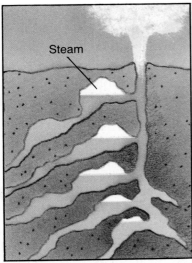

Enrichment: Well and spring water from underground is often thought to be pure. It may be very clear; however, it may also contain bacteria and other substances that make it harmful to drink. Thus it should be tested before being used as drinking water.

Fig. 5-26 A geyser is produced by steam pressure building up when water boils in underground chambers.

Openings in the bottom of the pipe allow water to flow into the pipe. Then the water must be pumped or lifted to the surface. A well provides water as long as the water table does not drop. Often a well will exhaust the water from the part of the water table close by. In time, the water flows back and the well again has water. If the well is drilled into rocks with low permeability, the water will flow into the well very slowly. Removing the water too fast from such a well will cause it to dry up until more water is able to flow in.

The water in some wells may flow out of the top of the well. These wells are called *artesian* (are-**tee**-zhun) wells. In artesian wells, the water comes from a porous rock layer. This layer is sandwiched between two nonporous layers. The porous layer is called the *aquifer* (**ak**-wuh-fer). The aquifer slants downward from both the place where it picks up water and from the surface. This downhill movement of water in an aquifer gives the pressure needed to force the water from an artesian well. See Fig. 5-27.

THE WORK OF GROUND WATER

Mention a cave and almost everyone thinks of a dark and mysterious place. There is a mystery about caves, but it has nothing to do with dark secrets or hidden treasures. The real mystery is how a large cave can form in the rock beneath the surface. The answer to this mystery can be discovered by tracing the flow of ground water.

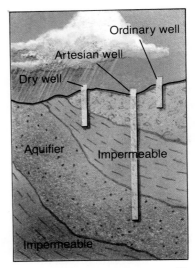

Fig. 5-27 In an artesian well, pressure is created by the downhill flow of water. This causes the water to rise to the surface.

In some areas, thick layers of limestone are found. This rock lies beneath a thin covering of soil. Rain and ground water have a special effect on limestone. As the rain falls, it dissolves some carbon dioxide gas from the air. The rain becomes a very weak acid. Some types of limestone are easily dissolved by this acid. Water that contains dissolved limestone is called *hard water*. Hard water does not make lather with soap. As the acid ground water enters cracks in the limestone, some of the cracks are widened. With each new rainfall, more limestone is dissolved. The cracks widen further. In time, the ground water dissolves away a group of connected underground passages. A **cavern** (**kav**-ern) is formed. See Fig. 5–28. *Caverns* may run underground for several kilometers.

Word Study: Cavern: *caverna* (L.), cave.

Cavern Any series of underground passages made when water dissolves limestone.

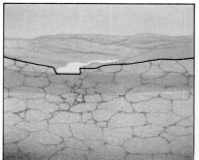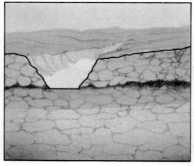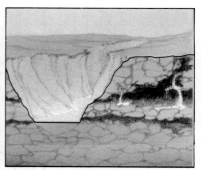

Fig. 5–28 A cavern is made when ground water dissolves limestone. The cavern first grows when the water table drops. As the river erodes the valley floor, the water table continues to drop until the cavern becomes empty.

There are many strange rock formations inside caverns. See Fig. 5–29. From the roof, long icicle-shaped stones hang down. These are called *stalactites* (stuh-**lack**-tights). The mounds of stone that seem to be growing up from the floor are called *stalagmites* (stuh-**lag**-mights). Stalactites and stalagmites feel cool and damp to the touch. They are formed when the ground water drips from the ceiling. The limestone, once dissolved in the water, is deposited. As each drop of water falls, it leaves behind a tiny amount of limestone. Limestone is also deposited on the floor where the drop lands. The stalactites slowly grow longer and the stalagmites grow taller. They may join as one solid stone column reaching from the cavern floor to ceiling.

Sometimes the roof of a cavern falls in. When this happens, a deep, round hole in the ground is formed. Caverns that have fallen in are called *sinkholes*. Fig. 5–21 shows a sinkhole. Sinkholes are likely to occur only in areas where there are also caverns. The sinkhole often fills with water to become a pond. In some regions where there are large deposits of limestone, there

Fig. 5-29 Unusual rock formations are found in caverns.

are many sinkholes. There are few surface streams in these regions because most runoff flows into the sinkholes. Surface water flowing into caverns can form underground streams. These streams, or small rivers, usually flow through the cavern until they reach an outlet such as a spring or river. The only place that ground water moves in underground streams is in caverns. Otherwise, ground water is found almost everywhere beneath the surface. It is spread through the pore spaces and cracks in soil and rock.

SUMMARY

Some of the water that falls on the earth's land surface sinks beneath the surface. The water moves through cracks or small openings in the soil and rock. Below the surface there are layers where all the spaces in the rock are filled with water. When the ground surface falls below this water-filled layer, the water flows out onto the surface, forming springs and artesian wells. Water moving through openings in certain rocks may dissolve the rock, creating caverns.

QUESTIONS

Use complete sentences to write your answers.

1. What two soil properties control the amount of water that sinks into the soil?
2. What relationship exists between the water table and a well?
3. How are springs produced?
4. Describe the conditions necessary to form a cavern.

SKILL-BUILDING ACTIVITY

COMPARING AND CONTRASTING THE PROPERTIES OF PERMEABILITY AND POROSITY OF SOIL SAMPLES

PURPOSE: To examine three soil samples, comparing their properties to find how they affect soil porosity and permeability.

MATERIALS:

3 beakers	100-mL graduated
gravel	cylinder
sand	water
soil	labeling tape

PROCEDURE:

A. Fill one beaker three-fourths full of gravel. Shake it a little so that the rocks settle. Label this beaker A.

B. Fill a second beaker to the same level with sand. Shake this beaker so that the sand settles. Label this beaker B.

C. Fill the third beaker to the same level with soil. Shake this beaker to settle the soil. Label this beaker C.

D. Look closely at the three beakers.
   1. In which beaker does the amount of space between particles seem to be the greatest?
   2. In which beaker does the amount of space between particles seem to be the smallest?
   3. Which material do you think will hold the most water in the spaces between its particles?
   4. Which material do you think water will flow through most easily?

E. Fill a graduated cylinder with 100 mL of water. Into beaker A, gently pour just enough water to cover the top of the particles. The idea is to fill the spaces between the particles of the material with water. Observe how quickly the material becomes soaked.
   5. Subtract the amount of water in the graduated cylinder from 100 mL to find how much water was needed to fill beaker A to this level.

F. Repeat step E, using beaker B.
   6. How much water was needed to fill beaker B to the top of the particles?

G. Repeat step E, using beaker C.
   7. How much water was needed to fill beaker C to the top of the particles?
   8. Which beaker held the most water in the material? The amount of water it held is a measure of its porosity.
   9. Which beaker allowed the water to soak through the material the fastest? This is a measure of its permeability.

CONCLUSIONS:
   1. Compare the porosity of the three soil samples.
   2. How does the permeability of the soil samples differ?
   3. What is necessary in order for a soil sample to have a great amount of porosity?
   4. In what way is the porosity of a substance related to its permeability?

CAREERS IN SCIENCE

HYDROLOGIST

WASTEWATER-TREATMENT-PLANT OPERATOR

Hydrologists are scientists who study the earth's water supply. They probe beneath the surface to find out where underground aquifers and springs are located. (An aquifer is a layer of rock, gravel, or sand from which water can be obtained.) These underground sources can be especially important in dry areas that may be facing a water shortage.

Hydrologists also study surface waters. Some, for example, are involved with efforts to prevent flooding along our rivers. Others study irrigation methods and the effects of soil erosion on farmland. While hydrologists spend part of their time outdoors, they also conduct research in laboratories and in libraries.

To become a hydrologist, a college degree is required, with courses in geology, math, and computer science. Jobs are available in private industry and government agencies. For further information, write: American Geophysical Union, 1909 K Street N.W., Washington, DC 20006.

Keeping the country's lakes and rivers free of pollution is extremely important. A potential source of pollution is sewage from homes and work sites. This sewage and wastewater must first pass through treatment plants where it can be cleaned before entering our waterways. The machines in these plants are run by wastewater-treatment-plant operators. They monitor the antipollution devices and take water samples to make sure the water is clean. If breakdowns occur in pipes or valves, operators have the know-how to fix them. They also apply chemicals to water to purify it.

Operators learn through on-the-job training, and programs in wastewater technology are also available in two-year colleges. Employment opportunities in this field are expected to increase. Operators are employed in private industry and in community treatment plants. For more information, write: Water Pollution Control Federation, 2626 Pennsylvania Avenue N.W., Washington, DC 20037.

¡COMPUTE!

SCIENCE INPUT

Many landform features and events are created by the processes of geologic weathering and erosion. Slump, creep, landslides, caverns, and springs—landforms discussed in Chapters 5 and 6—all have been shaped through the forces of nature such as wind, waves, running water, glaciers, rain, and earthquakes. Studying how erosion occurs is important. In the practical sense, if the processes are understood, then damage may be prevented. Erosion is also a factor in the aging of landforms. Therefore, if you were a scientist of earth history, you would need to understand erosion to be able to estimate how old a landform is.

COMPUTER INPUT

Learning a new field requires memorizing a great many new terms and concepts. A modern computer can be used to make old-fashioned learning drills somewhat easier and even fun. The computer can be programmed to review material over and over again without ever getting tired and without ever getting angry at you for giving the wrong answers. The computer is a friendly machine when it comes to this type of activity.

Sometimes a computer, a computer language, or a word processor is described as being "user friendly." Usually this means that the terms for commands and responses, the labels on the keyboard, or the words and symbols used in operating the machine are more like everyday conversational words than like a new technical language.

Look at the words in Program Erosion. While you may not be familiar with the way the symbols are being used, the words and numbers are often used in the same way as in sentences or mathematical equations. These programs have been written in BASIC, a computer language that is not very technical and quite "user friendly."

WHAT TO DO

You will be programming your computer to ask you to make an association between a landform (a dune, for example) or a process (a landslide, for example) and the type of erosion that caused it. The computer will tell you whether or not you are right, and give you a score by telling you what percent of your answers are correct.

First, you must collect and record data on erosion for your data statements. In one column, make a list of all the landforms and processes discussed in Chapters 5 and 6. Next to each item write the type of erosion that caused it. When there is more than one kind of erosion at work, list each, separated by a slash (/). The program requires that you remember each part of your definition, or your answer will be considered incorrect. A sample data chart appears below. (The program will hold 100 data statements.)

After your data chart is complete, enter Program Erosion into your computer. Be careful to type your data statements as shown in the example statement 301, and to end with statement 401 as shown. Save the program on disk or tape and then run it.

GLOSSARY

DIM   Short for *dimension*. This instruction tells the computer how many spaces are needed in that part of its memory reserved for information or data. In this program, dimensions are given for information that will be used in the quiz questions. See line 110 below. There can be 100 questions and 100 answers.

PRINT   Instruction to the computer to print on the computer screen. There is a different instruction that starts the printer.

PROGRAM

```
100   REM EROSION DRILL AND
      PRACTICE
110   DIM Q$(100), A$(100)
115   REM READ IN QUESTIONS,
      ANSWERS
120   FOR X = 1 TO 100
130   READ Q$(X), A$(X)
140   IF Q$(X) = "0" THEN GOTO 160
150   NEXT
160   PRINT: PRINT
165   REM COMPUTER PICKS RANDOM
      QUESTION
170   N = INT(RND(1) * (X - 1) + 1)
180   PRINT "WHAT TYPE OF EROSION?"
190   PRINT Q$(N);: INPUT "-----> ";R$
200   PRINT
210   IF R$ = "F" THEN END
220   QC = QC + 1
225   REM COMPARE RESPONSE WITH
      ANSWER
230   IF R$ = A$(N) THEN PRINT
      "CORRECT" : C = C + 1: GOTO 250
235   REM ANSWER TO INCORRECT
      RESPONSE
240   PRINT "CORRECT ANSWER ---> ";
      A$(N)
245   REM SCORING
250   PRINT "OUT OF     "; QC; "   YOU
      WERE CORRECT ON =     ";
      C:PRINT "SCORE =     "; INT (C/QC
      * 100 + .5); "%"
260   GOTO 160
300   REM DATA STATEMENTS FOLLOW
301   DATA DUNES, WINDS
401   DATA 0,0
999   END
```

PROGRAM NOTES

This program can be used to help learn any kind of association. For example, your data chart could have listed football or baseball players and the league and club in which they play.

A computer that contains a large database can seem to "know" a lot. However, the computer only knows what has been put into it. If you put in the wrong definition, the wrong definition will come out. You also have to be quite precise. For example, if you enter "Rain/earthquakes" as agents of erosion for "landslide," and then, when the program asks for an answer, you type in "Earthquakes/rain," the computer will not recognize your answer as correct.

EROSION DATA CHART

Landform or Process	Agent of Erosion
Dunes	Winds
Sea cave	Waves
Mudflow	Rain
Landslide	Rain/earthquakes

BITS OF INFORMATION

Computer games are probably one of the most popular uses for personal computers. Complete with colorful graphics, there are adventure games, war games, mazes, and chess or checker games. For those who are absorbed with videogames and other uses of computers, there never seems to be enough software. You can exchange and buy software through the Young People's Logo Association (YPLA). YPLA lists programs in a number of computer languages. For membership information, write to: YPLA, 1208 Hillside Drive, Richardson, TX 75801.

CHAPTER REVIEW

See teacher's commentary for answers.

VOCABULARY

On a separate piece of paper, match the number of each sentence with the term that best completes it. Use each term only once.

humus	meander	creep	precipitation
water cycle	delta	landslide	weathering
evaporation	chemical	interface	erosion
talus	slump	physical	organic

1. _____ is the breaking of rock into small particles.
2. The processes which cause rock particles to be carried away are called _____.
3. A(n) _____ is a zone where any of the three parts of the earth come together.
4. In _____ weathering, rock is broken up without any change in the material of the rock.
5. _____ weathering acts to change the materials of the rock.
6. The _____ is the material in soil that comes from the decay of plants and animals.
7. The pile of broken material at the bottom of a cliff is called _____.
8. A(n) _____ is the sudden movement of large amounts of loose rock or soil.
9. The slow downhill movement of loose rock or soil is called _____.
10. Downhill movement of part of a slope in one piece is called _____.
11. The _____ is the process by which the water leaves the oceans, is spread through the atmosphere, falls over the land, and runs back to the oceans.
12. _____ is the changing of a liquid into a gas at normal temperature.
13. A(n) _____ is a deposit formed at the mouth of a river.
14. The wide curve of an old river is known as a(n) _____.
15. Falling rain or snow is called _____.
16. _____ means "coming from living things."

QUESTIONS

Give brief but complete answers to each of the following questions. Unless otherwise indicated, use complete sentences to write your answers.

1. List four ways in which rocks can undergo physical weathering.
2. Name some agents of chemical weathering.

3. Give examples of how plants cause physical and chemical weathering.
4. Describe how the earth's rock surface becomes mature soil.
5. Why is gravity an agent of erosion?
6. What three things can happen to precipitation?
7. When does a water surplus occur?
8. Trace how runoff becomes a river.
9. What type of stream is most effective in cutting through bedrock?
10. Name the parts of a stream.
11. What is the relationship between soil porosity and permeability?
12. Why are lakes, swamps, and rivers found below the water table?
13. Compare the formation of a spring that disappears during a dry period to the formation of a spring that flows all the time.
14. Explain why ground water can produce a cavern.
15. Describe how each of the following form in caverns: (a) stalactites, (b) stalagmites, and (c) columns.

APPLYING SCIENCE

1. Look around your neighborhood for examples of weathering and erosion. List the examples you find in table form. Identify each example as either physical weathering, chemical weathering, or erosion. In the case of erosion, describe how the erosion occurred.
2. Visit a recent cut in a hillside or an excavation site. Take pictures of the soil layers that have been exposed. Determine whether the soil at the site is young or mature. Discuss your results and share the pictures with your class.

BIBLIOGRAPHY

Fodor, R. V. *Angry Waters: Floods and Their Control.* New York: Dodd, Mead, 1980.

Gardner, Robert. *Water, The Life-Sustaining Resource.* New York: Messner, 1982.

Gold, Michael. "Who Pulled the Plug on Lake Peigneur?" *Science '81,* November 1981.

Gunston, Bill. *Water.* Morristown, NJ: Silver Burdett, 1982.

Kitfield, James. "The Greening of Germany." *Omni,* April 1983.

Nothing on the earth's surface can escape the effects of erosion. When ice, wind, and waves move over land, they pick up and move all kinds of material from the earth's surface.

ICE, WIND, AND WAVES

CHAPTER GOALS

1. Explain how ice, wind, and waves erode the land.
2. Give examples of ice, wind, and wave erosion.
3. Describe the deposits left by ice, wind, and waves.
4. Describe the areas most affected by ice, wind, and waves.

6-1. Glacial Ice

Enrichment: The photo on the left page is of Glacier Bay, Alaska. In regions such as Alaska, Greenland, and Antarctica, the snow line is at or close to sea level. Many glaciers reach the sea. As they extend into the water, blocks of ice break off and become icebergs in a process called calving.

At the end of this section you will be able to:

- ☐ Define *valley glaciers* and *continental glaciers*.
- ☐ Explain how glaciers are able to cause erosion.
- ☐ Compare two types of deposits left by glaciers.
- ☐ Identify two types of theories that explain how *ice ages* may have been caused.

The greatest rivers on the earth are made of solid ice. Like huge plows, they move and shape large parts of the earth's surface. In some places huge amounts of ice reach the shore and flow out into the ocean. These slowly moving masses of ice are an important force in changing the surface of the earth.

GLACIERS

Most snow that falls on land melts to form runoff. The water returns to the sea. There are places on the earth, however, where it is too cold for all the snow to melt. As the snow falls, year after year, it piles up. The snowflakes are squeezed together and become ice. When the weight becomes great enough, the body of ice and snow begins to move. This large body of moving ice and snow is called a **glacier.**

Word Study: Glacier: *glacies* (L.), ice.

Glacier A large body of moving ice and snow.

VALLEY GLACIERS

One kind of *glacier* on the earth today is a **valley glacier.** *Valley glaciers* are found in mountainous areas all over the world. See Fig. 6-1. They look like rivers of dirty ice filling mountain valleys. Often several valley glaciers join to form a large valley glacier. Valley glaciers are able to move because of the pull of gravity. Their great weight causes them to slide downhill. A very thin film of liquid water is formed under the

Valley glacier A river of ice in a mountain valley.

Fig. 6–1 (left) *A valley glacier.*

Fig. 6–2 (right) *The ice near the surface of a glacier breaks to form crevasses.*

glacier from melting ice. Heat caused by the friction of ice scraping against the mountainside usually causes this melting. The liquid water allows the ice to slide more easily over the rock. Glaciers also move because parts of the ice are under pressure for a long time. Each small crystal of ice is able to slip a tiny amount. Think of what would happen to a tall stack of cards if each card slipped a little. The sum of all the individual movements would make the whole stack move. In the same way, the ice in a glacier can creep slowly. As the glacier moves over uneven ground, the ice near its surface breaks and forms cracks that run across the glacier. Cracks also may appear near the edges when one part of the glacier moves faster than other parts. These great cracks are called *crevasses* (kruh-**vass**-ez). See Fig. 6–2. Since the ice near the surface is not under pressure, it does not flow. Instead, the ice cracks.

Some valley glaciers move down their valleys at the speed of about one meter each day. They are fed from snow that builds up on their upper ends. As they move down the valleys, their lower ends melt. This combination of downward movement and melting usually makes it seem as if the glaciers are not moving at all. Sometimes a glacier appears to retreat because its lower end melts faster than it moves forward. Other times the glacier may advance when it moves forward faster than its lower end melts.

CONTINENTAL GLACIERS

The other kind of glacier is a **continental glacier.** *Continental glaciers* are huge sheets of ice that may cover thousands

Continental glacier A sheet or mass of ice that covers a large area of land.

Fig. 6-3 A continental glacier.

of square kilometers. Two such glaciers, called icecaps, cover most of Antarctica and Greenland today. See Fig. 6-3. In places, the Antarctic icecap is over 4,500 meters thick. It covers great mountain ranges running across the continent. Ninety percent of the world's ice is in the Antarctic icecap. Antarctica also contains 75 percent of the world's freshwater.

Icecaps are shaped like wide domes. They are able to move because of their great weight. See Fig. 6-4. The weight of the ice causes outward movement in all directions from the thick center. Edges of icecaps break off when they reach the sea. These huge pieces of ice float away as *icebergs*. Valley glaciers that reach the ocean also form icebergs.

THE WORK OF GLACIERS

Glaciers are responsible for the appearance of much of the earth's surface at high altitudes and latitudes. Sharp mountain peaks, wide valleys, level plains, lakes, and rivers have been made by glaciers. The movement of glaciers can erode and shape the land. Rocks buried in the ice at the bottom and sides of the glacier are like the teeth of a file. As glaciers move, the rocks scrape and dig at the land. Solid rock can be polished and often shows deep scratches where moving glaciers have left their marks. See Fig. 6-5 on page 150.

Glaciers that form in mountains change the shape of the mountains. The changes begin at the top of the valley where snow and ice collect. Frost action breaks rock from the valley walls. Rock is picked up from the valley floor as the glacier

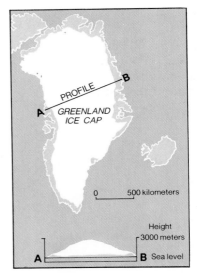

Fig. 6-4 The Greenland icecap is shaped like a dome. See profile A–B. Its weight at the center causes the ice to flow toward the shore.

Fig. 6-5 *What evidence is found on these rocks to show that they were once covered by a glacier?**

* The surface of the rock is scratched and polished.

moves. The upper end of the valley is changed into a rounded, steep-walled basin called a *cirque* (**surk**). Many cirques can form close together in mountains. They cause the ridges between them to become very sharp and jagged. These ridges are called *arêtes* (uh-**retts**). Very steep peaks, called *horns*, may also be formed. As the glacier grows, it moves down the valley. The floor and walls of the valley are worn away. In time, the original V-shape of the valley is changed into a U-shape. The steps by which the glaciers cause mountains to become rugged are shown in Fig. 6-6. Unlike valley glaciers, which form only in mountains, continental glaciers cover large areas of the land surface except for the highest mountains. Continental glaciers also change the land differently from the way valley glaciers do. Instead of reshaping mountains, the moving icecap grinds down the land to a nearly level surface.

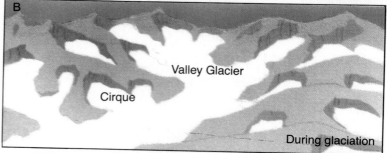

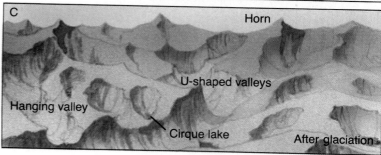

Fig. 6-6 (a) *Mountains before they are changed by glaciers.* (b) *Glaciers growing in mountain valleys.* (c) *Appearance of mountains after the glaciers have disappeared.*

GLACIAL DEPOSITS

A moving glacier can pick up rock material. When the ice melts, all of this material is deposited. There are two kinds of deposits left by glaciers. One kind consists of material deposited directly from the melting ice. This is called **till.** *Till* is made up of unsorted or broken rock of all sizes mixed together, from fine particles to boulders. Layers or ridges of till are called **moraines.** Ridges of till formed along the sides of glaciers are called *lateral moraines.* Lateral means "side." Till that is deposited in front of the glacier is called *terminal moraine.* Terminal means "end."

A second kind of deposit is made when glacial ice melts and forms streams flowing out from the glacier. These are called outwash deposits. Streams flowing from the glacier's front drop sand and gravel, depositing the largest particles first, then the smaller particles. Unlike till, which consists of unsorted materials, outwash deposits are sorted. That is, outwash deposits are formed in separate layers of sand, gravel, and finer particles. If the streams flow out onto level ground, the material is deposited in wide, even sheets that form an outwash plain. Large blocks of ice may break off the glacier and become buried in material washed out of the melting glacier. When these blocks melt, a large water-filled hole called a *kettle lake* is left. See Fig. 6-7. Many of the lakes and ponds of Michigan, Wisconsin, and Minnesota were formed this way.

Till Material deposited directly from a glacier.

Moraine A layer of till.

Word Study: Till: a Scottish word used to describe the rocky soil of Scotland; moraine: *morena* (Fr.), a heap of earth.

Enrichment: Cape Cod in Massachusetts is a terminal moraine deposit.

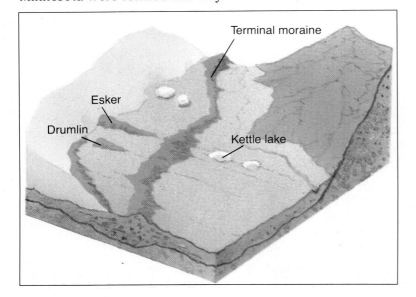

Fig. 6-7 The retreat of a glacier leaves a variety of glacial deposits.

Other deposits left by glaciers are called *drumlins* and *eskers*. Drumlins are long, low hills made of till. They range in height from about 15 to 60 meters. One side of the hill is usually steep and faces the direction the glacier moved. The Battle of Bunker Hill in the Revolutionary War was fought on a drumlin. Eskers are deposits made by streams flowing under the ice. Fig. 6-7 shows the kinds of deposits that can be made by glaciers.

ICE AGES

Glaciers covered much of the earth's surface ten thousand years ago. What is now New York City was then covered by a layer of ice about one kilometer thick. See Fig. 6-8. At that time, the sea level was much lower than it is today. This was partly a result of much of the earth's water supply being trapped in the ice sheets. The huge glaciers of that time eroded much of North America. They left features that now tell us where they were. For example, the Great Lakes were formed by glacial action. These huge lakes fill basins that were made when the advancing glaciers made river valleys wider and deeper. Water from the melting glaciers was trapped by dams built up from moraines. Other evidence of glacial erosion is seen in the plains of midwestern North America and all of the New England states. The climate of the earth has warmed since then. The last of the great ice sheets disappeared from North America about ten thousand years ago. Now glaciers are found only in high mountains and near the poles.

Scientists believe that several periods of glaciation have taken place during the earth's long history. Each cool period that allowed much of the earth's water to be frozen into ice was

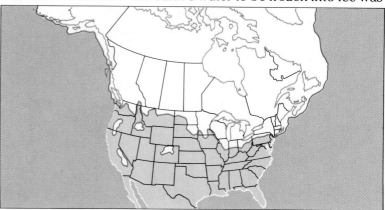

Fig. 6-8 About ten thousand years ago, glaciers covered many parts of North America.

an *ice age*. Each ice age was followed by a warm period. There is evidence that there have been at least four, and maybe as many as ten, ice ages during the last million years. During an ice age, glaciers grow and join together. In time, ice sheets several kilometers thick are formed. Because of their weight, these ice-caps spread out in all directions.

There is no one scientific theory that explains why the earth has had ice ages. Scientists look for reasons why the earth's climate should become cool.

1. **Events on the earth.** One type of theory uses events on the earth itself as reasons for cooling. For example, one theory is based on the drift of the continents. The positions of continents could prevent the movement of warm ocean waters toward the poles. This could cause regions near the equator to become very warm. Closer to the poles, the climate would be much cooler. Another theory is based on volcanic eruptions. An increase in volcanic activity might throw large amounts of dust into the atmosphere. This could prevent some of the sun's rays from reaching the earth's surface. The earth would then become cooler as it received less heat from the sun.

2. **Events outside the earth.** Another type of theory that explains the ice ages is based on reasons outside the earth. One theory suggested that at times the sun may produce less heat. The sun does go through short periods when it gives off slightly less heat. Whether this takes place over long periods of time is not known. A theory that is gaining support from many scientists is based on changes in the earth's orbit and the tilt of its axis. The elliptical orbit of the earth around the sun changes slightly. At times, this could have put the earth a little farther from the sun, producing cooler summers. In addition, changes in the tilt of the earth's axis can cause cooler summers. For example, the tilt of the earth's axis today is what causes the earth to have seasons. The combined changes in the earth's orbit and the tilt of its axis could have lowered the earth's temperature. This could have caused some snow to remain through the summer. If this occurred during many summers, snow would continue to collect until glaciers formed. Changes in the tilt of the earth's axis and orbit have their largest effect on the amount of heat received from the sun in the Northern Hemisphere. Those kinds of changes do seem to have taken place around the same time that the most recent ice ages have occurred.

Enrichment: Scientists believe that there is a difference of only 2.3°C in the ocean temperature and 6.5°C on land between an ice age and the present.

Enrichment: This theory is called the Milankovitch theory, after a Yugoslav astronomer.

In the future, scientists may be able to forecast when the earth will enter another ice age or a warm period. Presently, there are 25 million cubic kilometers of ice in the earth's glaciers. The melting of that ice would raise the sea level by about 65 meters. Many large cities of the world such as New York, London, and Tokyo would be flooded. This is a practical reason for trying to gain a scientific understanding of the reasons for the ice ages.

It is certain that the earth's climate will change. Even now the climate is changing. There is not yet enough scientific evidence to determine if the earth will enter another ice age. The world may be warming as a result of the burning of large amounts of fuels such as coal, oil, and natural gas. This has released large amounts of carbon dioxide into the atmosphere. Carbon dioxide can trap the heat the earth receives from the sun. This is similar to the way glass in a greenhouse traps heat from the sun. The trapped heat could melt the glaciers. Some scientists believe that this could raise the sea level enough to flood coastal cities such as New York.

SUMMARY

Large amounts of snow and ice may build up in mountain valleys and over areas near the poles. These masses of snow and ice that move become glaciers. Materials carried in the moving ice erode the land and are deposited when the glacier melts. The earth passes through periods called ice ages when its climate becomes cooler and huge glaciers form.

QUESTIONS

Use complete sentences to write your answers.

1. What are the differences between a valley glacier and a continental glacier?
2. Explain how the movement of a glacier can erode and shape the land.
3. Give an example of a land feature formed by till and one formed by outwash deposits. Tell how they differ.
4. Describe the two types of theories used to explain why ice ages occur.

1. Valley glaciers look like dirty rivers of ice filling mountain valleys. Continental glaciers are huge sheets of ice covering very large areas of land. A valley glacier moves in one direction, downhill, while a continental glacier moves out in all directions from the center of its dome shape. Valley glaciers cause mountains to have sharp peaks and the valleys to be ground to a U shape. Continental glaciers smooth over the land causing it to become more level.
2. Glaciers pick up rocks, which become embedded in the ice. These rocks act like the teeth of a file, grinding against whatever the glacier moves over.
3. Drumlins and eskers are made by till. Outwash plains are made by outwash deposits. Till is made of unsorted rocks of all sizes. Outwash deposits are sorted, or are in separate layers of sand, gravel, and finer particles.
4. The two general theories are (1) that events on the earth itself, such as the screening of sunlight from the earth by volcanic dust, could cause the cooling; and (2) that events outside the earth, such as slight changes in the earth's orbit and axis, could cool the earth.

SKILL-BUILDING ACTIVITY

See teacher's commentary for teaching hints, safety, and answers to questions.

EROSION BY RUNNING WATER: A STUDY IN CAUSE AND EFFECT

PURPOSE: To study the factors that determine the rate of erosion and amount of erosion caused by running water.

MATERIALS

stream table	soil
water source	sand
overflow container	gravel

PROCEDURE:

A. Water running off melting ice from a glacier or snow pack is one of the major causes of erosion.

   1. What conditions do you think affect the action of running water on soil, sand, and gravel?

B. One condition to study would be the effect of the slope of the land on erosion. Set up the stream table as shown in Fig. 6-9. Using a mixture of soil, sand, and gravel, let the water flow at the same rate for a given amount of time. Stop the water and observe the different-size particles that are moved by the running water.

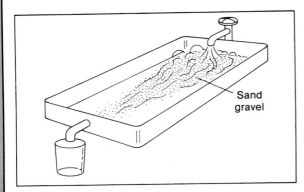

Fig. 6-9 A stream table.

   2. What size particles were moved by the water?

C. Increase the slope. This can be done by placing one book under the beginning of the stream table. If the speed of the water does not change, add one book at a time until you notice a change in speed.

   3. What effect did increasing the slope have on the speed of the flowing water?

   4. What effect does this have on the different-size particles moved?

D. Remove the books so that the stream table returns to its original height. Slowly increase the volume of water flowing. Observe the order in which the different-size particles move. Do this until particles of all sizes are moving.

   5. As the volume of water was increased, in what order did the different-size particles move?

CONCLUSIONS:

   1. What effect would increasing the slope of a stream have on the amount of stream erosion?

   2. What effect would increasing the volume of water flowing have on the amount of stream erosion?

   3. What factors could change the slope or volume of water of a stream?

   4. Imagine you were planning a house near a stream. From your observations in this Activity, how would your plans differ (a) if the stream were fast-moving? (b) slow-moving?

6-2. Wind and Waves

At the end of this section you will be able to:

☐ Explain how the wind can cause erosion as well as build deposits.

☐ Describe two types of wind deposits.

☐ Identify two ways in which waves wear away the land.

☐ Describe shoreline features caused by wave erosion.

Have you ever felt a strong wind? See Fig. 6–10. Sometimes the wind can blow you over unless you lean against its force or hold on to something. It is the energy in the wind that you feel. And it is this energy that is able to cause erosion and build deposits on the earth's surface.

Fig. 6-10 Some places in the world can have very strong winds.

WIND EROSION

Wind is moving air. Like running water, wind can move pieces of rock. But wind moves only small particles, such as dust and sand, through the air. Wind can only move large pieces of rock by rolling or bouncing them along the ground. Wind causes erosion only where many small rock particles are exposed. For example, in dry climates such as in deserts, the growth of plants that would cover the soil is prevented. Therefore, the wind is able to move the soil particles across the bare ground. Beaches, dry flood plains, and plowed ground are also exposed to wind erosion. Strong winds in these places carry away the sand and soil.

Besides moving small particles, wind has another effect. Particles of sand usually have sharp edges. Sand blown by the wind can **abrade** (uh-**brade**), or wear away, anything in its path. The ability of moving sand to *abrade* sometimes is used to clean the outside of stone or brick buildings. Particles of sand are blown against the dirty surfaces. As the sand is blown against the surfaces, the dirt is worn away. See Fig. 6-11 *(left)*. Wind can cause erosion of solid rock in the same way. Wind erosion does not have any effect on things higher than about one meter above the ground. The sand particles that shape rock by abrasion are not lifted very high by wind action. This means that most of the strange large rock formations often found in deserts were not carved by wind. Wind action only polishes and cleans structures near the ground. See Fig. 6-11 *(middle)*.

The ability of wind to carry dust or move sand depends on its speed. A strong wind is able to move large amounts of loose material. Most deserts have big areas of bare ground where all the sand and soil particles have been blown away. Only large particles such as gravel and stones are left behind. See Fig. 6-11 *(right)*. In dry regions, a strong wind may cause huge dust storms. During a large dust storm, a cubic kilometer of air may carry as much as 900,000 kilograms of dust. A storm covering a large area is able to move many millions of kilograms of dust. During periods of drought, the growth of plants and farm crops is prevented. Wind then becomes an important agent of soil erosion in farming areas.

Abrade To wear away.

Word Study: Abrade: *abradere* (L.), to scrape away.

Enrichment: An oasis may form where winds erode the surface to a depth where water is present.

Fig. 6-11 (left) Sand is being blasted against this building to abrade and clean the surface. (middle) Wind causes unusual rock formations in deserts. (right) Everything except the heavier rocks has been blown away.

WIND DEPOSITS

Windbreak A barrier that causes wind to move more slowly.

A barrier that causes the wind to lose speed is called a **windbreak.** For example, a bush or a large rock can cause the wind to slow down. When the wind slows down, it drops part of its load of sand. A mound of sand often is deposited around a *windbreak* such as a bush or a large rock. Once the sand deposit is started, it acts to block the wind further. In time, the mound grows and becomes a **dune.** *Dunes* are commonly found just inland from beaches. Wind carries sand from the beach and drops it around nearby plants or other windbreaks. Deserts contain much dry, loose material that is easily moved by the wind. Thus in many deserts there are dunes covering large areas.

Dune A deposit of wind-carried materials, usually sand.

Word Study: Dune: *dun* (Old English), hill.

Dunes can take many shapes. But all dunes have two things in common. Every dune has a gentle slope, called the *windward* side, which faces the wind. On the other side of the dune, away from the wind, the slope is much steeper. This is called the *leeward* side. Sand is blown up over the top of the dune. It then drops down on the other side. See Fig. 6–12. This causes the dune to move slowly in the direction the wind blows. Dunes can move over roads, railroad tracks, farmlands, and even buildings! Plants and fences are often used as windbreaks to stop the movement of dunes.

Fig. 6–12 (left) *A dune has a gentle slope facing the wind and a steeper side away from the wind.*

Not all the particles carried by the wind are deposited as dunes. Smaller and lighter particles are carried farther than the dune-building sand. This light dust eventually comes to rest. There is probably no place on the earth that does not have some wind-carried dust on it. In some places, the dust covers the land like a blanket. See Fig. 6–13. Thick layers of this wind-

Fig. 6–13 (right) *A deposit of loess.*

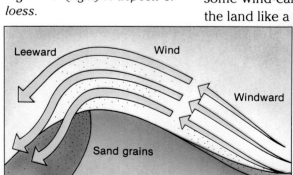

Leeward   Wind

Windward

Sand grains

blown dust are called **loess** (**less**). Deposits of *loess* are usually tan in color. They can be up to 30 meters thick, covering hills and valleys. Loess deposits are soft and easily eroded. A large area in northern China is covered with loess. The Chinese deposits came from the deserts of Asia. In the United States, loess is found along the eastern side of the Mississippi Valley. Deposits are also found in eastern parts of the Pacific Northwest. These deposits probably came from the beds of lakes and streams that dried after the glaciers retreated.

Loess A deposit of windblown dust.

WAVE EROSION

Wind is also able to move water and cause erosion. Winds blowing across oceans or lakes create waves. Waves travel across the water surface toward the **shoreline.** A *shoreline* is the place where land meets water. When waves reach a shoreline, they pound against it. During storms, large waves hit with great force. See Fig. 6–14. This constant pounding of the waves erodes the rocks of the shoreline. Waves break up rocks in two ways. First, waves force water and air into cracks in the rocks. This causes pressure to build up in the cracks. When the pressure is great enough, the cracks become larger. Large blocks of rock are broken up and moved by waves in this way. Second, pieces of broken rock cause additional erosion. This occurs when the rock pieces are smashed against the shore by the waves. Hitting the shore, they wear away more rock. As they are thrown around by the waves, these rock pieces also grind and rub against each other. This wears down the pieces. They become smaller and more round over a period of time.

Shoreline The place where the land meets water.

Enrichment: Rich farmland in Nebraska, Kansas, Iowa, Illinois, and Missouri is made up of loess deposits.

Fig. 6-14 Waves pound against the rocks along the shoreline.

Fig. 6–15 A magnificent sea cliff.

Sea cliffs The steep faces of rocks eroded by waves.

Sea cave A hollow space formed by waves at the base of a sea cliff.

Terrace A platform beneath the water at the base of a sea cliff.

Sea stack A column of rock that is left offshore as sea cliffs are eroded.

The effect of waves depends on the force of the waves and the type of rock that makes up the shoreline. Some rocks can be broken down at a faster rate than others. Fig. 6–15 shows a **sea cliff.** A *sea cliff* is a steep face of rock eroded by waves. The waves come straight into the shore and continually wear away the base of the sea cliff. The rocks at the top of the cliff eventually tumble down. Then they are ground up by wave action. Waves also cause **sea caves** to be formed. *Sea caves* occur when waves hollow out soft rock in a sea cliff. As the sea cliff is worn back by this process, a platform is formed at the base of the cliff. The platform is called a **terrace.** *Terraces* often slow down the erosion of the cliff. If the terrace is wide, waves lose most of their force before they reach the base of the cliff.

Unequal erosion also produces **sea stacks.** *Sea stacks* are columns of rock that stand a short distance offshore. They are formed when the softer rock is worn away. See Fig. 6–16.

Fig. 6–16 Sea stacks along a shoreline.

Fig. 6-17 (left) *A sandy beach.* (right) *A pebble beach.*

SHORE DEPOSITS

Rock particles produced by the grinding action of waves are usually pushed toward and along the shore. Many parts of the shoreline become covered by this ground-up material. These areas are called **beaches.** *Beaches* may be covered by fine sand or larger particles such as pebbles. See Fig. 6-17 (*left* and *right*). The kind of material found on beaches reflects the makeup of the coastline. Beaches also are made of materials carried to the shore by rivers and streams. The sand on a beach may be light or dark in color depending on its source. Much of the white sand on the coast of Florida came from erosion of the southern Appalachian Mountains.

In the summer, waves that reach the beach are usually small. They carry small sand particles to the shore. The wide sandy beaches formed during the summer months can be changed during the winter months. Winter storms bring bigger waves to the beach. Large waves often carry sand away and deposit it offshore to form long underwater *sand bars*. See Fig. 6-18.

Beaches Parts of the shoreline covered with sand or small rocks.

Word Study: Beach: *bece* (Old English), brook.

Fig. 6-18 A sandbar is formed offshore from material carried away from the shore.

Fig. 6-19 *Sand is moved along a beach in shifting waves.*

Drift of sand along beach ⟶

Incoming waves

Path of typical grain of sand

Longshore current Movement of water that is parallel with the shoreline.
Spit A long, narrow deposit of sand formed where a shoreline changes direction.

Fig. 6-20 (left) *Sand is deposited where a shoreline changes direction.* (right) *Toward the bottom of the photo, a spit has formed from sand moved alongside the beach.*

Waves almost always reach the beach at a slight angle. As a result, some of the water from the breaking wave moves parallel with the beach. Most of the water, however, flows straight back down into the surf. As a result of these two motions, sand grains are moved along a beach in a back-and-forth direction as shown in Fig. 6-19. Each day, this wave action can move sand and small beach pebbles many meters along the shore.

Waves striking the shore at an angle cause some water to move parallel to the shore. This motion is called a **longshore current**. *Longshore currents* also move large amounts of sand along the shoreline. This load of sand is often deposited when the shoreline changes direction. See Fig. 6-20 *(left)*. Sand is deposited at the point where the shoreline curves. This deposit is called a **spit**. *Spits* are long sand bars connected to the shoreline. They point in the direction that the longshore current moves. See Fig. 6-20 *(right)*.

The shape of a shoreline can also be caused by changes in the level of the sea. The shoreline, with its sea cliffs and terraces, seems to have been caused by a drop in sea level. It also

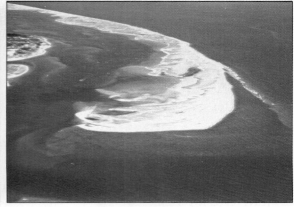

could have been caused by a lifting of the land. When the sea level drops or the land is lifted, new areas of shore are exposed to wave erosion. This causes many sea cliffs to be formed. On the other hand, the sea level may rise or the land may sink. Such a rise in sea level creates an irregular shoreline because stream valleys are flooded. The flooding of stream valleys can form bays and inlets along the coast.

Many things can change the shape and features of a shoreline. Careful study is needed to form some general conclusions about how a particular coastline was formed.

PROTECTING AND PRESERVING THE SHORES

Waves and currents make any shoreline a place of constant change. These changes are important because almost two-thirds of the world's population lives along the coasts. The coastal land is used for many human activities including seaports, fishing, and recreation. But many of the ways people use the shore can cause problems. For example, houses and other buildings can be destroyed if they are built too close to the water. See Fig. 6-21. There is no way to prevent the shoreline from changing. However, a scientific understanding of the way waves create beaches and other features will make it possible to predict and help control harmful changes.

SUMMARY

Wind can cause erosion in a dry region by picking up small particles of sand. These moving sand particles may strike against and wear away solid rock. Slowing down of the wind causes deposits to build up. Thick layers of windblown dust are found in some places. Waves wear away the land along the shore. Many features are formed along a shoreline by the work of waves and currents. The shape of a shoreline usually depends on the effect of the changing sea level.

QUESTIONS

Use complete sentences to write your answers.

1. How does wind cause erosion?
2. Describe two kinds of deposits caused by wind.
3. Explain how waves can erode a rocky shoreline.
4. Name and describe at least five shoreline features formed by wave erosion.

Fig. 6-21 Houses and other buildings are easily destroyed near the shoreline during a storm.

1. Wind can move small particles of rock and dust a great distance. Wind causes erosion where small rock particles are exposed.
2. Winds may deposit sand in mounds called dunes, near beaches and at the base of mountains. Winds may also carry fine dust many miles to form thick layers of fine material called loess.
3. Waves force water and air into cracks in the rocks, enlarging the cracks and finally breaking the rock. The resulting pieces of rock are smashed against the rocky shore and abrade more rock.
4. (1) Sea cliffs are steep faces of rock eroded by waves; (2) terraces are underwater platforms extending out from a sea cliff; (3) sea caves are hollowed out places in a sea cliff; (4) sea stacks are lone columns of rock that stand alone a short distance offshore; (5) beaches are parts of the shoreline covered by eroded particles.

INVESTIGATION

See teacher's commentary for teaching hints, safety, and answers to questions.

WINDBLOWN DEPOSITS

PURPOSE: To study the nature of windblown deposits.

MATERIALS

newspaper	spoon
bottle of dry sand	goggles
box lid	water

PROCEDURE:

A. Spread out the newspaper to cover the desk top. Place the box lid near the center of the paper.

B. Remove the lid from the bottle of sand and place it in the middle of the box lid. Pour one spoonful of sand into the bottle lid. Your setup should look like Fig. 6–22.

Fig. 6–22 Your setup should look like this.

C. Put on goggles or other protective eye wear. With your face about 15 centimeters from the lid, blow gently down onto the sand. Increase the strength of your wind until sand is being thrown from the lid. Continue blowing like this for five to ten seconds.

D. Examine the material on the paper by rubbing your finger over it. Do the same thing to the material trapped in the box lid and the material left in the bottle lid. If no material is found on the newspaper, blow a little harder for five to ten seconds and retest.
  1. Where are the finest grains found?
  2. How did this process separate the sand into many different-sized particles?
  3. Wind carries very fine particles for many miles before it slows down enough to settle and make a layer of dust. A layer of this material is called by what name?

E. Dump the material in the box lid and bottle lid into the waste container provided. DO NOT PLACE SAND IN SINK.

F. Repeat step B. Sprinkle this pile of sand with water until the surface is completely wet.

G. Repeat steps C and D.
  4. What effect does wetting the sand have on the amount of material blown away?

H. Discard the wet sand in the container provided. NO SAND IN THE SINK.

CONCLUSIONS:

1. Describe the kind of material most likely to be blown some distance by wind.

2. Look at Fig. 6–11 *(right)* on page 157. Explain why large areas of desert have this appearance.

TECHNOLOGY

WIND FARMS

Sometimes a "new" technology is an old idea updated. That's what has happened to the windmill, a technology probably associated most closely with Holland. Versions of the windmill, however, have been in use in the United States for a long time, particularly in the rural areas of the country. They have been used primarily to pump water in irrigation systems, but they have also provided electricity. In fact, in the 1930's wind-generated electricity from the "windcharger" was common throughout the Midwest before utility lines covered the country. Now "wind farms" are being developed to "harvest the wind" in order to produce electricity on a large scale. The four giant wooden blades of the old storybook windmills have been replaced by a wide variety of designs. When used to generate electricity, they are called wind turbines. If they have moving blades that rotate around a stationary cylinder, you may see them referred to as rotors.

Many of the United States' wind farms are concentrated in the Altamonte Pass in California, shown in the photo. There the wind commonly blows for 15 to 20 days in a row without stopping, dies down for a day, and then starts all over again. This pattern is repeated from mid-March to mid-September. At one location in the Altamonte Pass, 100 wind turbines generate electricity day in and day out. Not far away, there are 50 more wind turbines designed and built by another manufacturer. At still a third location are the first five of several hundred wind turbines planned by a third developer. There are a few other sites of wind farms in Montana, Wyoming, and New Hampshire. Individual wind turbines are used to produce electricity on a smaller scale in both rural and urban areas.

Each type of turbine is slightly different in design. Ranging in height from 12 meters to 22 meters, they usually have two or three fiberglass blades, which may be 15 to 21 meters in diameter. Several of the newer "windmills" have particularly unusual designs. The Darrieus "egg beater" rotor has two or more curved aluminum blades that are attached, at the bottom and top of the blade, to a central power shaft. The Department of Agriculture has experimented with the Darrieus rotor as a water pumper to irrigate farms in Bushland, Texas. Engineers are also testing models of wind turbines with no moving parts at all, a design which creates far fewer mechanical problems.

A wind turbine is designed to be capable of generating a certain amount of electricity, but the exact amount it actually produces depends on the wind.

What if there isn't any wind? These latest wind turbine projects solve that problem. They are connected to an electrical system that stores the extra electrical energy produced when there is a great deal of wind. They can then be used when the air is calm and there is little or no wind. The biggest fault of windpower has always been that you couldn't depend on it as a constant source of energy. Backup energy sources have always been needed. The possibility now exists that wind could become a safe and reliable source of energy. This hope is shared in other countries where wind energy is being developed.

CHAPTER REVIEW

See teacher's commentary for answers.

VOCABULARY

On a separate piece of paper, write TRUE next to the number of each statement that is true. Next to the number of each false statement, write FALSE, then make the statement true by writing the correct term in place of the underlined incorrect term.

1. A <u>longshore current</u> is a river of ice in a mountain valley.
2. Rock material deposited directly from a glacier is known as <u>till</u>.
3. A <u>terminal moraine</u> is a ridge of till.
4. Sand blown by wind can <u>abrade</u> anything in its path.
5. A barrier that causes wind to move more slowly is called a <u>spit</u>.
6. A place where water meets land is a <u>windbreak</u>.
7. A steep face of rock eroded by waves is known as a <u>beach</u>.
8. The underwater platform at the base of a sea cliff is called a <u>terrace</u>.
9. A <u>sea cave</u> is a hollowed-out space found at the base of a sea cliff.
10. A lone column of rock that stands a short distance offshore is known as a <u>sea cliff</u>.
11. A <u>sea stack</u> is a part of the shoreline covered by sand or small rocks.
12. The narrow deposit of sand formed where a shoreline changes direction is a <u>dune</u>.

QUESTIONS

Give brief but complete answers to each of the following questions. Unless otherwise indicated, use complete sentences to write your answers.

1. What conditions are necessary to cause a glacier to form?
2. Where on earth are continental glaciers found today?
3. Explain how glaciers are able to move.
4. Compare the two types of theories that explain the ice ages.
5. What is an ice age? When did the last one occur on earth?
6. Why is it important to understand what causes ice ages?
7. Where does wind erosion occur most often? Why?
8. Explain why wind erosion cannot account for most of the strange large rock formations found in the desert that appear to have been eroded.

9. How do plants help prevent erosion by wind?

10. What causes longshore currents?

11. How are sea caves, terraces, stacks, and sea cliffs related?

12. Contrast the kind of shoreline features that result from a lowering of the sea level with those that result from the raising of the sea level.

13. How can you tell if an area was eroded by (a) ice, (b) wind, or (c) waves?

14. Describe the deposits left by (a) ice, (b) wind, and (c) waves?

15. What types of areas are affected most by (a) ice, (b) wind, and (c) waves?

APPLYING SCIENCE

1. Find examples of erosion by ice, wind, or waves near where you live. Describe how the erosion is taking place. Explain what is being done about it.

2. Using an electric fan, some fine sand, a shallow box, and some paper to catch any escaping sand, set up the conditions necessary to study the movement of sand by wind. Study the use of windbreaks, the forming of dunes, the movement of dunes, etc. Take pictures or write a report of your results.

3. Refer to Fig. 6–20 (right) on page 162. Make a series of drawings to show what will happen in the future if the longshore current continues.

4. Make a poster for display in the classroom that shows the important features made by a valley glacier. Show and label such features as a cirque, arêtes, horns, U shaped valley, and the outwash plain formed by the glacier. Make the poster as drawings, cut-outs, or with pictures showing each of the features. Obtain permission before removing pictures from magazines, etc.

BIBLIOGRAPHY

Chasan, James Peter. "Columbia Glacier in Retreat." *Smithsonian*, January 1983.

Gallant, Roy. "Wind, Sand, and Space." *Science 81*, November 1981.

Radlauer, Ruth. *Grand Teton National Park*. Chicago: Children's Press, 1980.

Simon, C. "Columbia Glacier's Last Stand." *Science News*, January 1984.

Time-Life Editors. *Glacier* (Planet Earth Series). Chicago: Time-Life Books, 1984.

Time-Life Editors. *Ice Ages* (Planet Earth Series). Chicago: Time-Life Books, 1984.

Most rocks can tell us something about the earth's past. Some rocks, like the one below, tell us that there were once fish living in the place where the rocks were found.

EARTH'S HISTORY

CHAPTER GOALS

1. Describe the principle of uniformitarianism and explain how it is used in the study of rock layers.
2. Discuss how fossils are formed and the kinds of information that can be obtained from studying them.
3. Describe two ways the relative age of rock layers can be found.
4. Explain how radioactivity can be used to find the age of rocks and fossils.
5. Name the four main parts of the earth's history and explain why the earth's history is divided into parts.

7-1. The Rock Record

At the end of this section you will be able to:

- Explain how the idea that "the present is the key to the past" helps to explain the earth's history.
- Compare the ages of different rock layers.
- List four kinds of information that may be learned from an *unconformity*.

The Grand Canyon is like a book that holds a part of the earth's history. The pages of the book are the layers of rock seen on the canyon walls. In the same way as printing on a book's pages carries information, the rocks contain the story of how they were formed. In this section, you will begin to understand how scientists can read some of the history recorded in the rocks.

TIME AND CHANGE

Almost all scientists believe that the earth was formed about 4.7 billion years ago. It is hard for anyone to imagine such a long period of time. The following comparison may help you to appreciate how long 4.7 billion years is. Suppose that an imaginary writer started to write a history when the earth was formed. This writer is very lazy and only types one letter every ten years. On each page there are 7,200 letters. One thousand pages make up one volume of the long history. Thus it takes the

Fig. 7-1 The rock layers of the Grand Canyon are like pages in a history book; each layer tells another story.

writer 72 million years to type each volume. During the 4.7 billion years since the earth was formed, the writer would have typed 65 volumes. The story would now be on page 278 of the 66th volume. See Fig. 7-2. The entire history of the United States would be recorded in the last line. Columbus' discovery of America would be found in the line above. About ten lines above would describe the first Egyptian pyramids. Cave paintings made 20,000 years ago would be on the same page. Most scientists believe that humans have been present on the earth for only a short part of its long history.

Each person can see only a very small part of the earth's long history. During a person's lifetime, it is hard to see that the earth's surface is always changing. If your grandfather climbed

Fig. 7-2 The "history of the earth" written at the rate of one letter every ten years and one volume every 72 million years. The formation of the oldest rocks known would begin in volume 14!

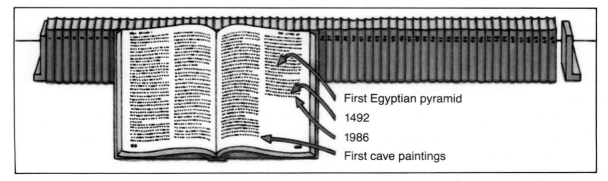

First Egyptian pyramid

1492

1986

First cave paintings

a mountain, you can climb that same mountain. But millions of years from now that mountain will probably be gone. In North America, erosion removes an average of six centimeters of the land every 1,000 years. During the billions of years of earth's history, even the tallest mountains have been eroded away. They have been replaced by other mountains raised by the forces that bend the earth's crust.

The changes on the earth's surface have been studied for many centuries. Writings about earthquakes date back to the Greeks more than 2,300 years ago. Aristotle believed that earthquakes occurred when air was absorbed by the ground. He said that when the air was heated by fires inside the earth, the ground exploded.

Then during the seventeenth and eighteenth centuries, many geologists believed that the surface features of the earth were formed by sudden events called *catastrophes*. For example, features such as mountains and canyons were explained to have been caused by sudden earthquakes. This idea said that the features on the earth were formed over a very short period of time. After these features were formed, they remained as we see them today. Thus the earth did not have to be very old.

But in 1795, a geologist named James Hutton proposed a different idea. He thought that the present is the key to the past. This idea is called the principle of **uniformitarianism.** The principle of *uniformitarianism* states that processes, such as mountain building, weathering, and erosion, take place very slowly and over long periods of time. It says that the geological processes now operating on the earth were active in the past.

Geologists believe the processes that are working today have always existed. The forces that build mountains today were at work millions of years ago. The river now running in a canyon began wearing away the land thousands of years ago. Thus, by studying the present, we can learn about the processes in the past. For example, look at Fig. 7–3. According to the principle of uniformitarianism, the ripples in the ancient sandstone *(top)* were made by the same process that made the new ripples in the sand *(bottom)*.

We know the earth must be very old because its features take a long time to form or wear away. For example, rocks have been found with fossils of sea animals that lived more than 10 million years ago. Some of these rocks are now in mountains

Uniformitarianism The principle that the same processes acting on the earth today also acted on the earth in the past.

Fig. 7–3 (top) *The ripples in ancient sandstone were formed by the same process, wind, that made the ripples in the sand* (bottom).

Word Study: Uniformitarianism: *uni + formis* (L.), one form.

Enrichment: The theory that the earth's landscape was shaped by rapid world-wide upheavals is called *catastrophism.* It was the commonly held theory during the 17th and 18th centuries.

2,000 meters above sea level. These mountains have been building at a rate of only .2 millimeters per year for ten million years. Yet that is a short time compared to the billions of years the earth has existed. The evidence shows many cycles of mountain building and erosion. But the earth's features seem to change little in a human life span. Major changes can only be seen over spans of thousands or millions of years. Thus, the features we can see today are like a giant cover spread over the long history written in the rocks below. Beneath our modern geography is the record of ancient formations, many of which existed before human history began. Most modern scientists now believe that uniformitarianism is a useful guide to understanding the past. The forces that changed the earth in the past are thought to be similar to those causing changes today.

HOW ROCKS CAN TELL A STORY

Reading the rock record is like reading a story written in a mysterious code. The principle of uniformitarianism is one clue to the code. There are other clues.

In a group of rock layers, the bottom layer is generally the oldest. The top layer is the youngest. As an example, you can make a pile of books by placing one book on top of another. The book that was put down first will be in the pile for the longest time. It forms the oldest layer in the pile, and it is also the one at the bottom. This is an example of the **law of superposition.** The *law of superposition* is believed to work when rock layers are formed. Each new layer is laid over the one under it. Therefore, each layer is younger than the one beneath it. In the Grand Canyon, for example, the upper layers are seen near the top. Older layers are near the bottom of the canyon where the Colorado River is still eroding the land.

There are certain cases where the law of superposition does not apply. Forces are always creating folds and faults in the crust. Sometimes rock layers are bent or broken so much that old layers may be pushed on top of younger ones. As you can see in Fig. 7-4, it is sometimes difficult to tell how layers may have been arranged before they were folded or faulted.

Comparing the positions of rock layers can tell only which layer is older or younger. Older or younger refers to **relative age.** *Relative age* does not tell how many years have passed since a rock layer was formed. It only shows that one layer is older than another. Another principle can be used to tell the

Law of superposition The principle that, in a group of rock layers, the top layer is generally the youngest and the bottom layer the oldest.

Relative age A method of telling if one thing is older than another. It does not give the exact age.

*It is difficult to tell which rock layers are the oldest where they are overturned.

Fig. 7-4 Can you tell which of these rock layers are the oldest from their position?

Crosscutting The principle that faults or magma flows cutting through other rocks must be younger than the rocks they cut.

relative age of rocks. It is called the principle of **crosscutting** relationships. For example, *crosscutting* can occur with a fault or a flow of magma within layers of rock. A fault is always younger than the layers it cuts across. A flow of magma is always younger than the rock layers around it.

ROCKS RECORD EARTH MOVEMENTS

Fig. 7–5 shows the arrangement of rock layers in a part of the Grand Canyon. You can see that the lower rock layers are tilted compared to those above. Layers of rock most often form in a level position. They generally lie flat with the other layers, like pages in a book. The lower layers in the photograph must have been tilted after they were formed. But how can the level layers above be explained? A reasonable explanation would be that the lower layers were formed, then tilted by crustal movements. Over a long period of time, erosion leveled the surface of the layers. Conditions then changed and the erosion stopped. New layers were deposited on the old eroded surface. The eroded surface forms a boundary between the old and new layers. Such a boundary is called an **unconformity.** An *unconformity* separates younger layers from older layers that were exposed to erosion. Unconformities may also exist between layers that have not been tilted. An unconformity usually shows that rock layers are missing. These are the layers that were lost to erosion. These missing layers are like pages torn from a book. They are gaps in the complete story that once existed in the rock record. With more information, these gaps may someday be closed.

Unconformity A boundary separating younger rock layers from older layers that were exposed to erosion.

Fig. 7–6 shows four steps in the formation of an unconformity. First, layers are formed underwater. See Fig. 7–6 (a). Level layers such as these are made when material settles to the bottom of a body of water. Second, the land is lifted and tilted after the lower layers are formed. See Fig. 7–6 (b). Third, erosion levels these tilted layers. See Fig. 7–6 (c). Last, the eroded surface is submerged under the sea. Erosion stops and new deposits are formed. See Fig. 7–6 (d). Level rock layers on top are formed.

Unconformities give us clues about the history of a region. First, they help us to place events in order. For example, the rock layers below the unconformity must have been tilted before they were eroded. This is because all the layers have been

Fig. 7–5 The boundary separating the tilted layers from the level layers is called an unconformity.

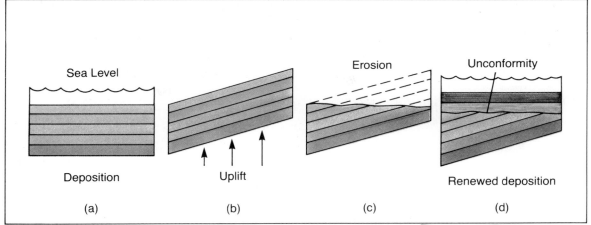

Sea Level

Deposition

(a)

Uplift

(b)

Erosion

(c)

Unconformity

Renewed deposition

(d)

eroded. Otherwise, only the top layer would be eroded. Unconformities also tell us something about the location of the dry land and the sea in the past. The rocks below the unconformity were formed as level layers underwater. Their surfaces could be exposed to erosion only if they were raised above sea level. The rock layers above the unconformity were formed when the eroded surface submerged again. Thus unconformities are useful clues to the relative ages and history of layered rocks.

Fig. 7-6 These diagrams show the steps in the formation of an unconformity.

SUMMARY

Layers of rocks in the earth's crust contain a record of the earth's history. The same processes that shape the crust today also operated when those rock layers were formed. The law of superposition, the law of crosscutting, and the idea of relative age allow us to understand events in the past. Sometimes an unconformity is found in the rock record. Such a gap can also provide clues to the history of a part of the crust.

QUESTIONS

Use complete sentences to write your answers.

1. Use the principle of uniformitarianism to explain ripples found in some rock.
2. You are looking at the layers of rock made visible by a highway cut through a hill. Which layer is the oldest? Which layer is the youngest? Describe one case in which you cannot be sure of your answers.
3. List in proper order the four events that may have occurred where an unconformity is found.

1. Wind caused ripples to form in sand. Then the sand hardened into sandstone.
2. The oldest layer is the bottom layer. The youngest layer is the top layer. If the layers are bent or folded with breaks, some of the older layers could be pushed up over younger layers.
3. Erosion of nearby higher land caused the formation of level layers, with the older ones at the bottom and the younger ones at the top. The land was then lifted and possibly tilted after the layers had formed. Erosion leveled the raised layers. Eroson stopped and new layers were formed.

INVESTIGATION

INTERPRETING ROCK LAYERS

PURPOSE: To use a diagram of a section of the Grand Canyon for reviewing some of its history.

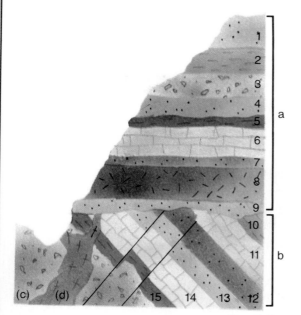

Fig. 7–7

MATERIALS:
pencil   paper

PROCEDURE:

A. Study Fig. 7–7 and write the answers to the following questions:

  1. Which numbered layer is the oldest of the horizontal layers in group a?

  2. Which numbered layer is the oldest of the tilted layers in group b?

  3. Which is the older, group a or b?

  4. Was the tilted layer of group b level at one time in history?

  5. Did the movement along the fault in the group b layers take place before or after layer 9 was laid down?

  6. Which layer of rock was the first to form after the tilting of group b?

  7. Which layers of rock must have been laid down before the magma, labeled d, pushed its way in?

B. An unconformity represents a time during which erosion was taking place at a particular location. Material was being carried away to some other location. The layer being eroded lost its flat appearance. When new layers form over this uneven surface, the bottom of the first new layer is also uneven. Look at Fig. 7–7 again to answer the following questions.

  8. Which two layers have uneven bottoms and thus show periods of erosion? Identify the layer that shows the oldest unconformity.

C. Layer 3 and layer 7 are deposits of limestone. Layer 3 contains the remains of ancient sea life. Layer 7 contains no such remains and is more fine grained and dense than layer 3. Answer the following questions using this information.

  9. What can you say about the conditions that existed at the time layer 3 was laid down? What reasons can you give for layer 7 not having ancient remains?

CONCLUSION:

  1. This activity reviewed four clues about interpreting rock layers. List the four in the order they were used.

7–2. The Record of Past Life

At the end of this section you will be able to:

☐ Define *fossil.*

☐ Describe six ways fossils can be formed.

☐ Point out several kinds of information that can be obtained from the study of fossils.

Fig. 7–8 tells a story. It shows human footprints made about 6,000 years ago. The rock in which the footprints are found was made from volcanic ash. Scientists believe that the footprints were made by a person running from an erupting volcano in Hawaii. What information do you think can be gained about this person from the footprints?

Fig. 7–8 Human footprints about 6,000 years old.

EVIDENCE OF ANCIENT LIFE

What could you learn by studying a person's footprint left in soft ground? Measuring the depth of the print could give an idea of the person's weight. The area covered by the impression gives a clue about the person's height and general size. Many of the kinds of plants and animals that have lived on the earth no longer exist. Scientists know about those ancient life forms only by their remains left in rocks. It is like describing a person from a footprint. Both remains and impressions are records of what happened in the past. See Fig. 7–9 (*top*).

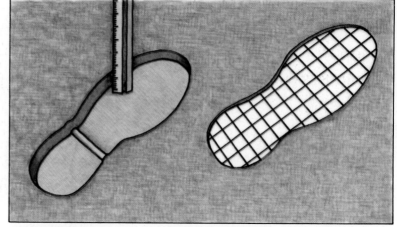

Fig. 7-9 (top) *Measurements are made of footprints in order to hypothesize about the animal that made them.* (bottom) *What can you tell about the animal that made this footprint?**

Fossil Any evidence of, or the remains of, a plant or animal that lived in the past.

*Because of the large size and depth of the footprint, the animal was probably very large.

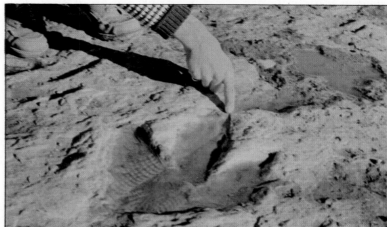

Any naturally preserved part, trace, or entire remains of a plant or animal that lived in the past is called a **fossil.** A *fossil* can be a piece of wood, a tooth, or a bone that was part of a living thing. A footprint or an impression also can be a fossil. Although a footprint is not a part of an animal, it is evidence that the animal once lived. See Fig. 7-9 (*bottom*). A footprint is an example of a *trace fossil.* Any mark left by an animal walking, crawling, or burrowing is called a trace fossil.

HOW FOSSILS FORM

A fossil also can be a part of a living thing. Usually the remains of plants and animals are destroyed by the process of decay. Of all the different plants and animals that have lived, very few have become fossils. However, sometimes decay is prevented from occurring. For example, the remains of plants

and animals may contain hard parts, which are preserved better than the soft parts. Also, the plant or animal remains may be protected very quickly.

1. Ice. In a few cases, fossils have been protected by ice and preserved by freezing. Some frozen woolly mammoths have been found in the frozen areas of Siberia. See Fig. 7–10 (*left*). Woolly mammoths were ancestors of today's elephants. They lived 8,000 to 20,000 years ago during the last ice age. Unlike elephants, they were very hairy. See Fig. 7–10 (*right*).

2. Amber. Fossils of ancient insects are found in hardened tree sap *(amber)*. The insects became stuck in the sap. Later the sap became as hard as stone. Amber containing whole, perfectly preserved insects can now be found. See Fig. 7–11.

Fig. 7–10 (left) *This young woolly mammoth was preserved in the permafrost in Siberia. It was discovered in 1977.* (right) *The woolly mammoth lived during the last ice age.*

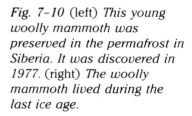

Enrichment: Layered rocks called *stromatolites,* formed by the growth of successive layers of bacteria, are believed to be the oldest fossils. Some stromatolites are thought to be 3.5 billion years old.

Fig. 7–11 The remains of this insect are implanted in a piece of amber.

3. Tar. In the center of the city of Los Angeles there are famous pools of tar. They are known as the La Brea Tar Pits. This sticky tar was formed by petroleum oozing out onto the surface. See Fig. 7-12. Tigers, wolves, horses, and other animals became trapped in the tar about 15,000 years ago. Many sank into the tar when they probably mistook the tar for a pool of water. Others attacked the trapped animals and were trapped themselves. The bones and teeth of these animals have been dug up from the ancient tar pits. These remains are very well preserved.

Fig. 7-12 The La Brea Tar Pits hold the remains of many prehistoric animals that were accidentally trapped in them.

4. Burial. Most organisms are not preserved as well as those in ice, amber, or tar. This is because decay begins quickly in most places when the remains are exposed to air. If the remains are covered by sand or soil, the decay process is slowed. Bones, teeth, shells, and other hard parts do not decay as fast as soft parts. When an animal or plant falls in a place where it becomes covered, its hard parts may be preserved. The remains of animals living in the sea and other bodies of water often sink to the bottom. There they are covered by mud and often become fossils. Most rocks that contain fossils were formed from the material on the bottom of bodies of water. Another way that fossils form is when ash falling from volcanoes has covered whole forests. These fossil forests are later found with the trees still in place.

5. Replacement by minerals. Although plant and animal remains may be covered over, decay can still continue. Other processes can also begin. The water that passes through the soil carries dissolved minerals. These minerals may slowly enter the remains of the dead animal or plant. The minerals take the place of some of the original material. The bones and shells become stronger with the addition of hard mineral material. When all of the original material has been washed away the fossil becomes completely mineral. Such fossils are said to be **petrified.** Fig. 7–13 shows *petrified* trees. These trees were once part of a forest. Most of the trees' living material was replaced by minerals.

6. Molds and casts. Some plant and animal remains are not replaced by minerals. They are completely dissolved. For example, a clam shell might become buried in soft mud. The mud would harden into rock. The shell would slowly dissolve and leave an impression of the shell in the rock. This type of impression is called a *mold.* Shellfish, fish, and plants leave molds in many rocks. A mold sometimes becomes filled with mud or mineral material. This results in the formation of a *cast.* The cast has the original shape of the living thing. See Fig. 7–14. Many kinds of soft-bodied plants and animals, such as jellyfish, leave fossils only in the form of molds or casts.

Fig. 7–13 A part of a petrified tree.

Petrify To replace the material in the remains of a living thing with hardened mineral matter.

Word Study: Petrify: *petra* (L.), rock or stone.

Suggestion: Some of your students may have fossil collections at home. Ask them to bring their collections to school and display them for the rest of the class.

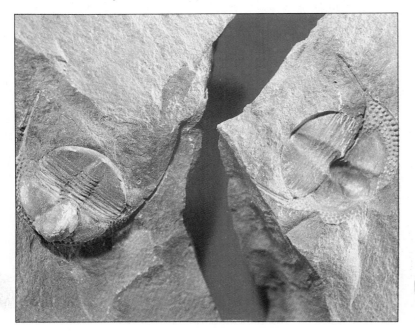

Fig. 7–14 (left) A cast and (right) a mold fossil. The cast does not contain any part of the original animal. It was formed inside a mold fossil.

Fig. 7-15 The dinosaur,
Tyrannosaurus Rex.

WHAT CAN BE LEARNED FROM FOSSILS?

The study of fossils can provide us with important knowledge about the earth's past. Dinosaur bones, for example, can give scientists much information about the animal, such as its size and shape. The shape of the teeth gives clues about the kind of food the animal ate. Marks on bones show where muscles were attached. The shapes of the bones indicate how the different parts of the animal's body might have been used. See Fig. 7-15. Trace fossils such as footprints were made while the animals were still alive. These prints are records of where the animal lived. See Fig. 7-16.

Fossil plants provide clues about the conditions that existed in ancient times. We know, for example, that swamps are found only in warm climates. Ancient swamps, buried long ago, have developed into beds of coal. Today, coal is found in Antarctica. This suggests that the climate of this now-frozen land was once very warm. Also, fossil seashells are found on Antarctic mountain tops. This suggests that this land was once underwater.

Fig. 7-16 Fossil footprints of the same kind of dinosaur squatting (left) and walking (right). No direct fossils of this dinosaur have been found. The drawing was made by using the trace fossils to estimate the animal's size and general appearance.

One important theory coming from the study of fossils is the story of changing life. Many plants and animals that were common at one time are no longer in existence. Their stories are found in the layers of rock. A climb up the walls of the Grand Canyon is like a trip through the earth's history. The oldest layers at the bottom of the canyon contain no fossils. The next layers contain fossils of simple water plants. Climbing higher, layers with fossils of simple animals are found. Fish fossils are found in higher layers. Still farther up, fossils of land animals can be found. See Fig. 7-17.

Fig. 7-17 This shows the different forms of life in different geological times as they are represented by the fossil remains in the rock layers of the Grand Canyon.

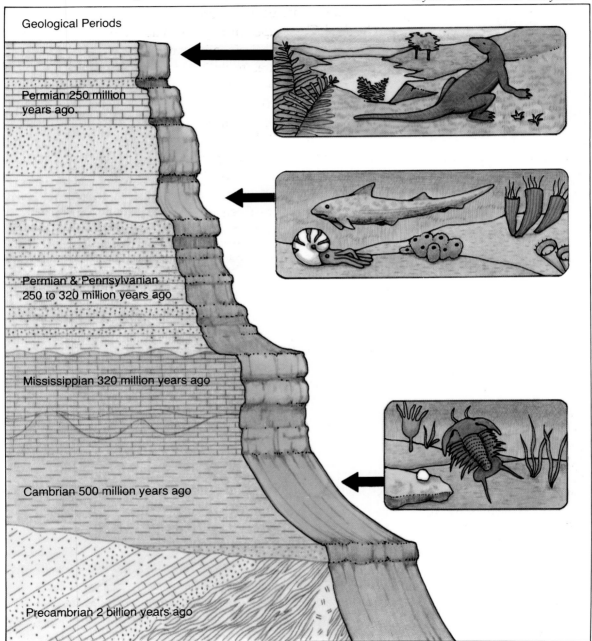

Geological Periods

Permian 250 million years ago.

Permian & Pennsylvanian 250 to 320 million years ago

Mississippian 320 million years ago

Cambrian 500 million years ago

Precambrian 2 billion years ago

Fossils found in the Grand Canyon and in many other places have been studied. Based on the fossil evidence, most scientists agree with the following theories:

1. Life has probably existed on earth for at least three billion years.

2. Simple forms of life appeared first. Higher forms of life developed from simple forms.

3. Some forms of life no longer exist.

The fossil record is difficult to read. There is no evidence from fossils showing how the different forms of life may have developed. There are many missing parts in the fossil record. One scientist estimates that only 1 percent of all existing fossils have been discovered. If and when these missing parts are discovered, other theories may develop to explain the earth's history of life. Much work remains before the story of the earth and its life forms is nearly complete.

SUMMARY

Ancient forms of life have left evidence of their existence in the form of fossils. In a few cases, a living thing may be completely preserved. More often, the remains are changed in some way. A part or all of the original material may be replaced with mineral matter. Some fossils consist only of the impressions left by the remains or the activities of the living things. The study of fossils suggests that life has developed slowly during the earth's history, from simple forms to more complex forms. Fossils provide information about past environments and have helped us to learn about the earth's history.

QUESTIONS

Use complete sentences to write your answers.

1. What is a fossil?
2. Describe the ways fossils may be formed.
3. How does a fossil become petrified?
4. List some of the information that can be obtained from the study of dinosaur fossils.
5. Coal is found in Antarctica, and fossil seashells are found in Antarctic mountains. What does this tell us about Antarctica's past?

1. A fossil is any naturally preserved part, trace, or entire remains of a plant or animal that lived in the past.
2. A mark could be left by an animal walking, crawling, or burrowing. A part of a living thing that has not been completely decayed may be left. Examples would be the frozen remains of woolly mammoths, insects found in hardened tree sap, and animal bodies found in tar pits. Bones, teeth, and shells that decay very slowly may be left in rock layers. As decay occurs, minerals may be deposited in the place of the decaying parts, forming a petrified fossil. An impression of a once-living plant or animal may form a mold. The filling of a mold by minerals may form a cast in the shape and size of the living thing.
3. A fossil becomes petrified by the replacing of slowly decaying parts of an animal or plant with the minerals found in the water of the area.
4. The shape of dinosaur teeth can give clues about the kind of food the animal ate. Marks on bones show where muscles were attached. The shape of the bones indicates how the different parts of the animal's body might have been used. Footprints made while the animals were still alive are records of the actual behavior of the animals.
5. Coal was made from plants that lived around swamps in a warm climate; therefore, Antarctica must have had these conditions at one time. Sea shells found in the mountains indicate that the area was at one time submerged under the sea.

INVESTIGATION

MOLD AND CAST FOSSILS

PURPOSE: To make some molds and casts of common objects.

MATERIALS:

modeling clay
object for mold
plaster of paris

mixing container
mixing stick
water

PROCEDURE:

A. Roll the modeling clay into a smooth, flat layer.

B. Press a leaf, shell, coin, key, or similar object into the clay. Make sure there is a definite impression in the clay. Remove the object.

    1. How is the mold similar to the original object?

    2. How is the mold different from the original object?

C. Using additional clay, build a low wall that measures two or three millimeters high around the impression.

D. Using the mixing container, mix enough plaster of paris to fill the impression to the top of the wall. Mix enough water with the plaster of paris so that it looks like whipped cream.

E. Fill the mold to the top of the wall. You will have to wait about 30 minutes for the mold to harden. If time is not available, label your container with your name and set the mold aside to harden overnight.

F. After the mold has hardened, carefully pull off the clay.

G. Study the cast you have made.

    3. How is the cast like the original object?

4. How is the cast different from the original object?

5. How much of the cast would you need in order to identify the object that made it?

H. Exchange casts with another student. See if you can identify the object the other student's cast represents.

    6. What is the object?

I. Look back at Fig. 7–9 (*bottom*), page 178. Use it to answer the following questions.

    7. What kind of trace fossil is this?

    8. What kinds of information does this fossil provide?

CONCLUSIONS:

    1. Explain the difference between a cast and a mold.

    2. Which types of things about the original object cannot be identified from a cast?

    3. Identify the kind of fossil in Fig. 7–18 below. Explain how the fossil probably formed.

Fig. 7–18

7-3. Age of Rocks

At the end of this section you will be able to:

☐ Explain how scientists can determine the relative ages of rocks.

☐ Describe how the age of rocks can be measured using *radioactive decay*.

☐ Name three *radioactive* materials used to find the ages of rocks and fossils.

How could you describe the ages of the people in the photograph shown in Fig. 7–19? Some could be described as teenagers. Others could be called adults or elderly. Another way is to measure their ages in years. A certain person might be 45 years old. Scientists express the ages of rocks in similar ways.

*They could be described as young, middle-aged, and old, or by their exact ages.

*Fig. 7–19 How might you describe the ages of the people in this photo?**

RELATIVE AGES OF ROCKS

Earlier in this chapter, you learned that the *relative age* of a rock layer could be found. This is done by studying the position of rock beds along with the presence of other features such as hardened magma and faults. Another way to compare the ages of rock layers is to look at the rock material itself.

Any rock layer that is found over a large area and can be identified easily is called a **key bed.** A layer made from volcanic material can be a *key bed*. A volcanic eruption may throw a large amount of ash into the air. The ash may settle over a

Key bed A rock layer that is easily recognized and found over a large area.

CHAPTER 7

large area, forming a single layer. For example, the eruption of Mount St. Helens spread a layer of ash over hundreds of thousands of square kilometers. In time, the ash will be buried and form a rock layer. Scientists studying the Mount St. Helens eruption have learned much about how to recognize rock layers formed by volcanoes.

Other events can create key beds. For example, a change in sea level produces layers that may serve as key beds. See Fig. 7-20. This layer can be traced among the other beds of rock. All layers below the key bed are older than the layers above it. Wherever the key bed is found, it can be used to find the relative ages of rock layers above and below. See Fig. 7-21.

Fig. 7-20 Can you see a layer that might be used as a key bed?

*The horizontal layer of sandstone can be used as a key bed.

Enrichment: Key beds are often the result of a period of uniform deposition in a body of water that results in a sedimentary rock layer covering a wide area.

*The dark brown layer is the only layer that appears in all three sections of rock layers. Therefore, it could be used as a key bed.

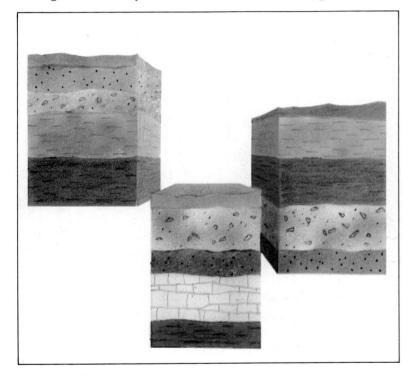

Fig. 7-21 Which of the rock layers could be used as a key bed?

INDEX FOSSILS

Fossils can be used to find the more accurate age of rocks. Such a fossil is called an **index fossil.** An *index fossil* is the remains of a plant or animal that lived during only a small part of the earth's history. Any rock containing an index fossil must have formed during the time the fossil lived. Thus the age of the rock can be determined from the period when the fossil was alive. Also, an index fossil must come from a form of life that

Index fossil The remains of a plant or animal that lived during only a small part of the earth's history.

lived in many parts of the world. An example of one kind of index fossil is the *trilobite*. See Fig. 7-22. The trilobite was a shell animal. It is believed to have lived about 500 to 600 million years ago. There were over a thousand kinds of trilobites living in oceans all over the world. Most were only a few centimeters long. A few hundred million years ago, all trilobites disappeared from the earth. Their shells are now commonly found as fossil molds and casts in certain rock layers all over the world. Trilobites lived for a short part of the earth's history and were very plentiful. Thus they make good index fossils. Any rocks that contain trilobite fossils probably formed between 500 and 600 million years ago. There are many other kinds of index fossils. Each kind comes from a certain part of the earth's history. The discovery of any kind of index fossil in a rock layer helps to find the age of that rock.

Fig. 7-22 Trilobites are good index fossils.

Suppose that the same kind of index fossil is found in rock layers at different places. Then the rocks in the layers must be about the same age. For example, trilobites are found in rocks on one side of the Grand Canyon. The same fossils are also found in rocks on the other side of the canyon. This means that the layers in each side of the canyon must have been formed at the same time. The layer was cut through, forming a canyon.

The same thing must be true for rock layers in all places on the earth containing a certain index fossil. No matter where those rocks were formed, they must be nearly the same age. For example, certain rocks in Europe contain the same kind of trilobites as found in the Grand Canyon. Thus the European rocks must be the same age as the Grand Canyon rocks. Scien-

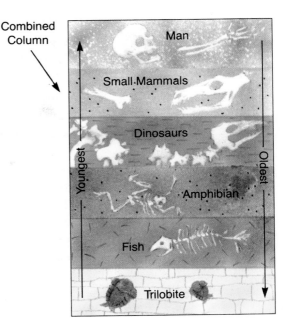

(a)

(b)

(c)

Combined Column

Man

Small Mammals

Dinosaurs

Amphibian

Fish

Trilobite

Youngest

Oldest

tists have used fossils to compare the ages of the rock at locations all over the world. They have combined all these observations to make a **geologic column.** Fig. 7-23 shows how a *geologic column* can be made. A geologic column has been made for the different rock layers found all over the world. The geologic column represents rock layers everywhere on the earth's surface. They are in the same order in which they were laid down. It is like an imaginary Grand Canyon where rock layers all over the earth could be seen at the same time. Scientists can compare any particular rock layer to its position in the geologic column. The relative age of any rock can then be found.

FINDING THE EXACT AGE OF ROCKS

Key beds can be used to tell the relative age of rocks. Index fossils show that rocks were formed at a certain period of time. But scientists also can measure the ages of many rocks more exactly. They do this with the help of **radioactive** substances. A *radioactive* substance is one that changes over a period of time. All matter is made up of tiny units called atoms. A radioactive substance has atoms that change into other atoms. This is called *radioactive decay*. A radioactive atom decays into another kind of atom at a regular, steady rate. Thus radioactive decay can be used to measure the ages of certain rocks.

Fig. 7-23 (left) *Layers having the same color and fossils are the same geologic age.* (right) *The diagram shows a geologic column with each layer in order by its age.*

Geologic column An arrangement showing rock layers in the order in which they were formed.

Enrichment: Using the law of superposition, have students explain how the geologic column was derived in Fig. 7-23.

Radioactive Describes a substance that changes over a period of time into a completely new substance.

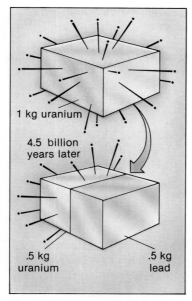

1 kg uranium

4.5 billion years later

.5 kg uranium

.5 kg lead

Fig. 7–24 After 4.5 billion years, half of all the uranium atoms will have changed into lead.

1. Uranium. Uranium is a radioactive substance found in many rocks. It slowly decays into lead. Measurements show that it takes 4.5 billion years for one-half of the uranium in a rock to change into lead. See Fig. 7–24. This length of time is called a **half-life.** Because of this very long *half-life*, radioactive uranium is used to find the age of very old rocks. Scientists carefully measure the amount of uranium and lead in a rock. Using these measurements, the length of time that decay has been going on in the rock can be found. For example, equal amounts of uranium and lead in a rock would show that one-half of the original amount of uranium had changed into lead. Such a rock would be about 4.5 billion years old. One half-life would have passed since the rock was formed. A rock that old has not yet been found. The oldest rock known today is about 4 billion years old. It was found in western Greenland.

2. Potassium. Radioactive potassium is another substance found in many rocks. It too is used to find the age of rocks. Potassium decays into argon. The half-life of radioactive potassium is 1.3 billion years. Radioactive potassium has a shorter half-life than uranium. Thus it can be used to measure the age of younger rocks.

3. Carbon 14. A scientist can find the age of once-living materials such as a fossil shell or a piece of wood. All living things absorb from the atmosphere a radioactive form of carbon called carbon 14. Carbon 14 has a half-life of about 5,800 years. As long as the plant or animal is still alive, the amount of carbon 14 in its cells remains the same. This is because the decaying carbon 14 is constantly being replaced. When the plant or animal dies, the decaying carbon 14 is no longer replaced. The amount of carbon 14 in its cells starts to decrease. The fossil's age is found by measuring the amount of carbon 14 in the fossil. That amount is compared to the amount found in a modern living thing. For example, suppose the amount of carbon 14 in an ancient piece of wood is measured. It is found to be one-half the amount of carbon 14 found in wood from a tree still living. This would mean that one half-life of carbon 14 had passed since the old wood stopped taking in the radioactive carbon. See Fig. 7–25 (a) and (b). Thus the wood would be said to have an age equal to one half-life of carbon 14, or about 5,800 years. Can you tell the approximate ages of the wood shown in Fig. 7–25 (c) and (d)?

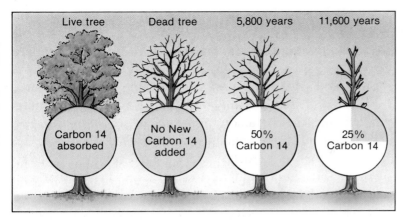

Fig. 7-25 *The amount of radioactive carbon in a piece of wood can be used to find its age.*

Carbon 14 is useful for finding the age of young fossils that are between 1,000 and 75,000 years old. Carbon 14 dating has provided a way of dating fairly recent events in the earth's history. For example, the age of fossil wood in Europe and North America has been measured. The results show that these areas were covered by ice as recently as 11,000 years ago.

SUMMARY

The relative ages of rocks can be found by using key beds. Index fossils can be useful in finding the period in the earth's history when the rock was formed. Measurement of the amounts of radioactive materials in rocks can be used to find the number of years since the rocks were formed. Radioactivity can also be used to measure the ages of the remains of once-living things.

QUESTIONS

Use complete sentences to write your answers.
1. Describe two ways that key beds may be formed.
2. Explain why index fossils are used to help find the age of a rock layer.
3. Why can radioactive decay be used to measure the age of rocks?
4. Explain how uranium can be used to tell the age of very old rocks.
5. Name a radioactive substance that can be used to tell the age of young rocks. Give its half-life and tell what substance it changes into.

1. Key beds may be formed by a volcano spreading dust over a very large area that in later years becomes a rock layer. A change in sea level may cause layers of rock to form over a large area producing key beds.
2. Any rock containing an index fossil must have formed during the period of time the animal lived.
3. Radioactive decay occurs at a regular, steady rate.
4. Uranium decays into lead. In 4.5 billion years one-half of the uranium will decay into lead. By knowing the ratio of uranium to lead in a given rock, its age can be calculated.
5. Radioactive potassium can be used to determine the age of young rocks. Its half-life is 1.3 billion years, and it changes into argon.

SKILL-BUILDING ACTIVITY

RECORDING DATA

PURPOSE: To gather and record data using a model of radioactive decay and then to determine half-life based on the data obtained.

MATERIALS:

50 sugar cubes   pencil
shoe box   paper
felt-tip pen   graph paper
clock or watch

PROCEDURE:

A. You will carry out instructions which will give data similar to radioactive decay. If possible, data should always be recorded in a table as the data is collected. Copy the table below. You may work in teams to perform steps B through F.

B. With a felt-tip pen, mark one side of each cube with an "X."

Shake Number	Number with Marked Side	Number Remaining
1		
2		
3		
Etc.		

C. Carefully place the 50 marked cubes into a shoebox. One member of the team should act as timekeeper. You will have two minutes to do steps D, E, and F. Work quickly and carefully. Read the directions before you begin.

D. Gently shake the shoebox so that the cubes roll around inside.

E. Open the box and remove all cubes that show their marked sides up. Count them. Do not return them to the box. Replace the lid on the box.

F. Record the number of cubes you removed. Find the number of cubes remaining and record.

G. Repeat steps D, E, and F nine more times. Add lines to your data table to record the data.

H. The half-life is the time it takes for one-half of a radioactive substance to change into a new substance. In this model, the cubes in the box represent the radioactive substance. The ones you removed represent the new substance.

   1. How many shakes were needed before half of the cubes changed to a new substance?

   2. If you followed the two-minute time schedule, how much time did it take for half the cubes to change?

   3. What is the half-life for the sugar cubes in this case?

CONCLUSIONS:

   1. Suppose the half-life of a radioactive substance is 5,000 years. After 10,000 years, how many atoms out of each 100 present at the start will have changed to a new substance?

   2. Plot your data on graph paper. The horizontal axis will represent the shake number. The vertical axis will represent the number of cubes that showed their marked sides up. Describe the line plotted.

7-4. An Earth Calendar

At the end of this section you will be able to:

☐ Name four main *eras* of the earth's history.

☐ Describe the conditions that existed during each era of the earth's history.

☐ Compare the length of time occupied by each main division of the earth's past.

You can make a kind of calendar for your life by using important events. This calendar would mark off the time you live into separate divisions. One part of the calendar might be from the time you were born until you started to walk and talk. Another period might begin at the time you started going to school. Each change in the way you live could divide your life into parts. The record left in rocks shows that the earth also has a history that can be divided into a kind of calendar. One difference between the two calendars would be that, unlike your life history up to now, most of the earth's history is still unknown.

A TIME SCALE FOR THE EARTH'S HISTORY

Scientists believe that the oldest rocks found to date on the earth's surface have an age of about 3.8 billion years. This conclusion has been reached by measuring radioactive materials in many rocks.

The earth's history, as recorded in the rocks, begins that long ago. However, the earth itself must be older. The radioactive clocks in rocks begin ticking when the rock hardens from magma. Most scientists believe that the earth came into existence about 4.7 billion years ago. There must have been a period of about a billion years when the newly formed earth was developing the crust we find today. The calendar of the earth's history began when the solid crust formed.

What kind of picture of the earth's history comes from the study of rocks and their fossils? The record is not complete. Scientists do not agree on all the details. However, one fact seems clear. There have been great disturbances in the crust along with sudden changes in the forms of life on the earth. It is the evidence of these turning points that provides the reasons for dividing the earth's history into parts.

Enrichment: The original geologic calendar was entirely in terms of relative ages. In recent years, as methods of measuring geologic time in absolute units have been developed, the use of absolute dates has been added.

1. Eras. Scientists divide the history of the earth into long time periods called **eras.** The *era* in which we are now living began about 65 million years ago. It is called the *Cenozoic* (sen-uh-**zoe**-ik) *Era* (recent life). The era before the Cenozoic was the *Mesozoic* (mess-uh-**zoe**-ik) *Era* (middle life). It lasted about 160 million years. Before the Mesozoic was the *Paleozoic* (pay-lee-uh-**zoe**-ik) *Era*. The length of the Paleozoic Era was about 345 million years. As you see, the divisions that make the earth calendar are not of equal length.

The time covered by these three eras may seem long. However, they cover only 13 percent of the earth's history. The time before the Paleozoic makes up the remaining 87 percent of the earth's total history. The record of all the events before the Paleozoic is very difficult to read. Almost all fossils are found in Paleozoic and younger rocks. Since very few fossils are found before the Paleozoic Era, scientists cannot divide the time before the Paleozoic into separate eras. Thus it is simply lumped together into one very long time space called *Precambrian* (pree-**kam**-bree-un) time. See Fig. 7–26.

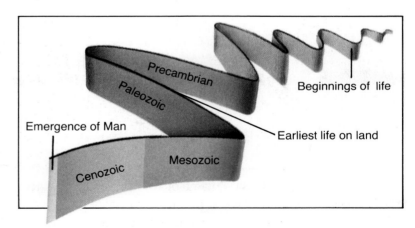

Fig. 7–26 In this chart, the length of Precambrian time is compared to the geologic eras.

Just as the yearly calendar we use is divided into months, weeks, and days, the earth calendar is divided into smaller and smaller parts. A very long division is an era. Study of the fossils found in the rocks of each era shows that the life forms changed during each era. For example, during the Mesozoic Era, dinosaurs appeared. They developed into many different forms. Some of the dinosaurs became large meat eaters. Others developed into huge plant eaters that lived in shallow lakes. Some even were able to fly. See Fig. 7–27.

Fig. 7-27 *Many different kinds of dinosaurs lived during the Mesozoic era.*

2. Periods. Study of fossil remains of life forms such as dinosaurs allows scientists to divide the eras into smaller divisions called *periods*. The Paleozoic Era has been divided into seven periods. The Mesozoic Era has been divided into three periods. The Cenozoic Era has two periods.

3. Epochs. Careful study of the fossils within each period also shows changes. Thus periods are further divided into smaller divisions called *epochs*. However, only the seven epochs that make up the two periods in the Cenozoic Era have been named. Epochs that make up periods in other eras have not been named. Table 7-1 lists the eras, periods, and epochs of the earth's history.

Era	Period	Epoch	Length in Years	How Long Ago It Ended (years)
Precambrian				570 million
Total Length of Precambrian Time			4 billion	570 million
Paleozoic	Cambrian		70 million	500 million
	Ordovician		70 million	430 million
	Silurian		35 million	395 million
	Devonian		50 million	345 million
	Mississippian		25 million	320 million
	Pennsylvanian		40 million	280 million
	Permian		55 million	225 million
Total Length of Paleozoic Era			345 million	225 million
Mesozoic	Triassic		35 million	190 million
	Jurassic		54 million	136 million
	Cretaceous		71 million	65 million
Total Length of Mesozoic Era			160 million	65 million
Cenozoic	Tertiary	Paleocene	12 million	15 million
		Eocene	16 million	37 million
		Oligocene	11 million	26 million
		Miocene	14 million	12 million
		Pliocene	10 million	2 million
	Total Length of Tertiary Period		63 million	2 million
	Quaternary	Pleistocene	2 million	_____
		Present Time	_____	_____
Total Length of Cenozoic Era			65 million	_____

Table 7–1

PRECAMBRIAN TIME

According to the most widely accepted scientific theory, the solid earth was formed about 4.7 billion years ago. In time, an atmosphere and then oceans were produced. Most scientists agree that life probably began in the ancient seas. It is likely that at least one billion years passed before the first signs of life appeared in the oceans. At first there were probably only tiny bits of living matter. From these came the great variety of plants and animals that have lived on the earth. The actual record of life during Precambrian time is scarce. Rocks formed during this long part of the earth's history contain almost no fossils. This is probably because most very ancient forms of life did not have the hard parts that commonly form fossils. There is some evidence of life in Precambrian seas, such as impressions of soft-bodied animals. But these impressions are very rare. Microscopic examination of some Precambrian rocks shows traces of a simple kind of life. As Precambrian time came to an end, the stage was set for the appearance of many living forms.

THE PALEOZOIC ERA

The beginning of the Paleozoic Era, about 600 million years ago, marked the start of a time filled with important events. The earth's surface was very different from what it is today. The moving crustal plates caused continents to wander all over the face of the earth. At the beginning of the Paleozoic Era, there seems to have been one large and five smaller continents. They were widely separated from each other. All were located near the earth's equator. There was no land near the North or South poles. Later in the Paleozic Era, these continents moved together. By the end of the Paleozoic Era, all the separate continents had joined to form a single supercontinent called *Pangaea* (pan-**gee**-uh). An enormous single ocean covered the remainder of the earth's surface. See Fig. 7–28.

There were many changes in living things during this era. During the early Paleozoic Era, the level of the sea was high. As a result, shallow seas covered much of the land surfaces. These warm, shallow seas became filled with **invertebrates.** These are animals that do not have backbones. However, many had outside shells or plates. Some of the *invertebrates* living in the

Invertebrates Animals without backbones.

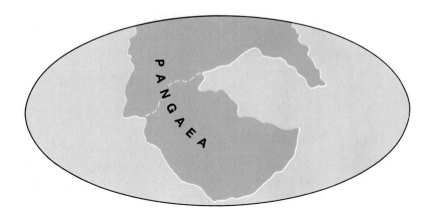

Fig. 7-28 During the Paleozoic era, separate continents came together to form the supercontinent called Pangaea.

Vertebrates Animals that have backbones.

Word Study: Vertebrate; *vertebra* (L.), a joint.

early Paleozoic seas were trilobites, worms, sponges, snails, and starfish. As time went on, the first fish appeared in the Paleozoic seas. These ancient fish were the first of the **vertebrates.** *Vertebrates* are animals with backbones. Insects also first appeared during this time.

Forests of giant ferns and similar plants grew in the swampy areas that covered much of the land. See Fig. 7-29. Most of the world's supply of coal comes from these Paleozoic forests. The growth of land plants also provided shelter and food for the animals that were coming out of the sea, such as insects, scorpions, and land snails. Late Paleozoic rocks also have fossils of land vertebrates. *Amphibians* such as frogs and their relatives appeared along with the first reptiles.

Toward the end of the Paleozoic Era, the continents began to join and form larger landmasses. The collisions formed mountains. Thus the total area of the land grew smaller, just as a rug covers less floor space if it is crumpled up. The total size of the oceans grew as the continents took up less area. Since the same amount of water covered a larger area, the level of the sea fell. The shallow seas that covered much of the continents drained back into the oceans. The continental climates became drier. Living things on land had to change to fit the new conditions. These changes marked the end of the Paleozoic Era.

Fig. 7-29 The life in a Paleozoic swamp may have looked like this.

Club Moss

Horsetail

Icthyostega

Tree Fern

Seed Fern

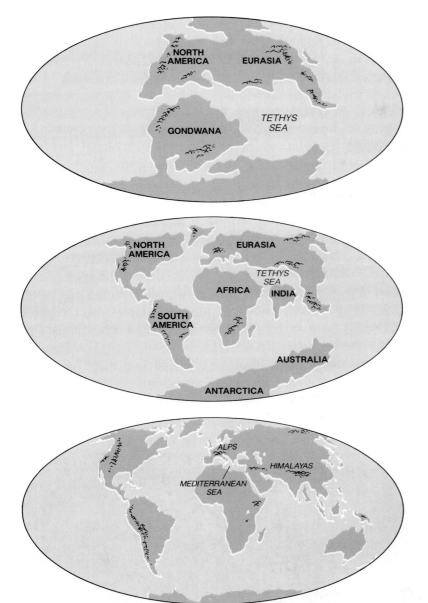

Fig. 7–30 During the Mesozoic era the present continents were formed. (a) First, the supercontinent Pangaea separated into North America and Eurasia and Gondwana. (b) Then, Gondwana divided into South America and Africa, India, Australia, and Antarctica (c) India moved north to collide with Eurasia.

THE MESOZOIC ERA

The Mesozoic Era began with the breakup of Pangaea. It first separated into three parts. See Fig. 7–30 (a). Later, during the Mesozoic Era, they separated into the continents as we know them today. See Fig. 7–30 (b) and (c). The changes in the earth's crust probably caused changes in the life forms. A rise in sea level may have been caused by sea floor spreading. When large amounts of magma pour out from the mid-ocean ridge, the sea floor is lifted. This would cause the level of water in the ocean to be higher. As the sea level rose, nearly half the land areas became covered with shallow seas and marshes. These conditions were perfect for reptile life forms. The Mesozoic Era became the Age of Reptiles. Dinosaurs, turtles, crocodiles, lizards, and snakes roamed the thick forests that covered much of the land.

The most common of these animals were the dinosaurs. These giant reptiles were some of the largest animals ever to live. See Fig. 7–31. But the dinosaurs, along with many other Mesozoic reptiles, were doomed. The moving crustal plates that opened the Mesozoic Era had slowed. The level of the seas dropped. The climate became cooler. The warm, wet condition that allowed the reptiles to thrive came to an end. The fossil record shows that the dinosaurs disappeared in a surprisingly short time. There is evidence in the rock layers that the dinosaurs may have disappeared because a comet or other large body collided with the earth. Such a collision would have thrown a huge amount of dust into the atmosphere. This would have caused the climate over the entire earth to become much cooler. Dinosaurs and other reptiles may have been wiped out because they could not change enough to meet the new conditions. The Mesozoic Era ended with the disappearance of the dinosaurs and many of their relatives.

Fig. 7–31 The outstanding form of life during the Mesozoic era was the dinosaur, of which there were many varieties.

Fig. 7–32 Some mammals that existed in the late Cenozoic era.

THE CENOZOIC ERA

The Cenozoic Era began about 65 million years ago. During this time the earth's surface has developed into the form we see today. The shallow seas that covered much of the land during the past are gone. However, the crustal plates continue to move. Their grinding collisions raise mountains and cause volcanic activity. The climates of the modern world range between the chill of the poles and the heat of the tropics. It is an era of great variety and change.

Mammals are one form of life that has been successful in living with the variety of climates in the Cenozoic Era. If the Mesozoic Era can be called the Age of Reptiles, then the Cenozoic Era can be called the Age of Mammals. Most of the ancestors of modern mammals appeared during the late Mesozoic. See Fig. 7–32. During the Cenozoic Era, many more kinds of mammals developed. They include hunting animals such as cats and wolves. There are also plant eaters such as horses and cattle. Animals able to run fast, such as deer and rabbits, have all developed to fit the conditions of the Cenozoic Era. Some mammals, such as whales and porpoises, have even gone back to the sea.

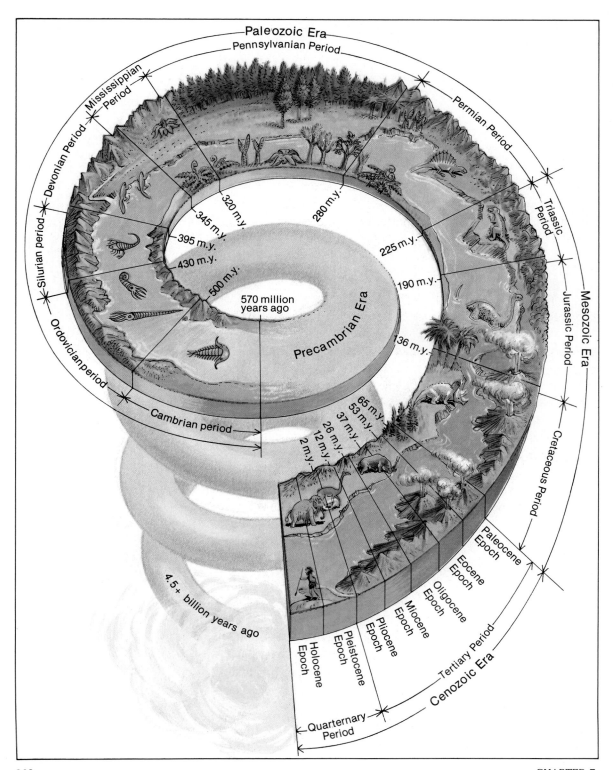

Paleozoic Era

Pennsylvanian Period

Mississippian Period

Devonian Period

Silurian period

Ordovician period

Cambrian period

Permian Period

Triassic Period

Jurassic Period

Cretaceous Period

Mesozoic Era

Paleocene Epoch

Eocene Epoch

Oligocene Epoch

Miocene Epoch

Pliocene Epoch

Pleistocene Epoch

Holocene Epoch

Tertiary Period

Quarternary Period

Cenozoic Era

Precambrian Era

570 million years ago

4.5 + billion years ago

320 m.y.

345 m.y.

395 m.y.

430 m.y.

500 m.y.

280 m.y.

225 m.y.

190 m.y.

136 m.y.

65 m.y.

53 m.y.

37 m.y.

26 m.y.

12 m.y.

2 m.y.

There is reason to believe that the changes marking the Cenozoic Era will continue into the future. The modern world is only one stage in the earth's history. See Fig. 7–33. There is evidence that the Atlantic Ocean will become larger. It has been growing wider during the past 180 million years. Within the next 100 million years, a huge new mountain range will probably appear. It will grow along the Atlantic coast of North America. On the other side of North America, much of southern California and Baja California may become a small continental island. Australia will probably move north. There may be a collision between Australia and Japan or the eastern part of Asia. The resulting combination of the Indian, Antarctic, and Pacific oceans would result in the world's largest ocean.

Fig. 7–33 (opposite page) *This diagram shows the earth's history divided into the eras, periods, and epochs that most scientists use to describe the earth's past.*

SUMMARY

The earth's history can be divided into parts to make an earth calendar. Each major part of the earth calendar is separated from the next by some worldwide change. The record left in the rocks shows that there have been four main divisions, or eras, in the earth's past. Each era shows changes within its time space that allows it to be divided into smaller parts. Each era of the earth's history has been marked by certain conditions that favored the development of different life forms.

QUESTIONS

1. Name the era with which each of the following is identified: (a) Age of Reptiles, (b) Age of Mammals, (c) 87 percent of earth's history, (d) coal and oil formation, (e) vertebrates, and (f) invertebrates.

1. (a) Mesozoic, (b) Cenozoic, (c) Precambrian, (d) Paleozoic, (e) Paleozoic, (f) Paleozoic.

2. Make a chart like the following. List the four main eras of the earth's history in the first column and then complete the other columns.

Era	Conditions Existing	Organisms Found	Time Era Lasted
1. Precambrian	cooling earth	simple sea life	4 billion yr.
2. Paleozoic	one supercontinent, sheets of ice	invertebrates, fish, plants	375 million yr.
3. Mesozoic	separate continents	reptile life, dinosaurs	160 million yr.
4. Cenozoic	changes to present conditions	mammals	70 million yr.

INVESTIGATION

See teacher's commentary for teaching hints, safety, and answers to questions.

GEOLOGIC TIME SCALE

PURPOSE: To prepare a model of the geologic time scale and relate to it some of the events that occurred during earth's history.

MATERIALS:

adding machine tape   paper
meter stick   pencil

PROCEDURE:

A. Copy the following table on a separate piece of paper.

Era	Length (years)	Length on Tape (cm)
Precambrian	4 billion	400
Paleozoic	345 million	34.5
Mesozoic	160 million	16.0
Cenozoic	65 million	6.5

B. The left column in the table above shows the main divisions of the geologic time scale. Use the table on page 196 to fill in your table with the remaining number of years for all four eras.

C. Measure and cut off a strip of adding machine tape five meters long.

D. One meter's length of tape will represent one billion years. One billion is the equivalent of 1,000 million.

   1. How many years would one millimeter represent if one meter represents one billion years?

   2. How many years would one centimeter represent on this scale?

E. Change the number of years of each era to the number of centimeters that will represent it. Use the scale of one meter equals one billion years. Record these values in your table. Note that the first one is done for you.

   3. Which time period is the longest?

   4. Which time period is the shortest?

F. Draw a line across the tape very near one end. Label it "Begin Precambrian." Measure four meters from this line and draw another line. Label it "End Precambrian." Starting at this line, measure the distance on the tape that will represent the Paleozoic Era and draw a line. Label the beginning and end of this era. In the same manner, draw in the lines to represent the remaining two eras. Next, refer again to the table on page 196 to answer the following question.

   5. In what era did the following exist or begin to exist? (a) invertebrates, (b) dinosaurs, (c) mammals.

G. For each of the other eras, mark on your tape information about the types of life that existed.

   6. Which of the life forms in question 5 lived on the earth for (a) the longest period of time? (b) the shortest period of time?

CONCLUSION:

1. Compare the length of Precambrian time to the lengths of the different eras.

2. Give one way in which the geologic time scale is divided.

CAREERS IN SCIENCE

ARCHAEOLOGICAL ASSISTANT

Much of what we know about the past is due to archaeologists and their assistants. At a "dig," as an investigation site is called, they uncover the remains of earlier civilizations. Archaeological assistants dig through the rocks and soil. There they might find hunting tools, pottery, or cooking utensils used by the people who lived there. They might even discover part of a skeleton. Assistants also take part in marking and classifying these items for study. These artifacts, as they are called, will provide a picture of how people lived centuries ago.

Each year, digs are under way throughout the world. Students in high school or college may apply to participate in them. Courses in social studies, history, and geography provide a particularly useful background. It is also important to be able to do careful and detailed work. For more information, write: Archaeological Institute of America, 53 Park Place, New York, NY 10007, or contact your local museum or the anthropology department of a university near you.

GEOCHRONOLOGIST

How old is a newly discovered animal fossil? When was a particular mountain formed? Such questions can be answered by geochronologists. These scientists try to determine when events occurred in the earth's past by studying earth's formations and fossil remains. Geochronologists rely on a variety of dating methods. One of these involves the radioactive element carbon 14. Carbon is present in all living things, but the amount and type of carbon depends on how long it has been there. Because early humans used obsidian in many of their implements, this substance is useful for dating too. With this evidence, geochronologists try to piece together the earth's history and build a picture of what it might have looked like.

To work as a geochronologist, you will need at least a college degree, with courses in geology, chemistry, and paleontology (the study of fossils) including experience in field work. For more information, write: American Geological Institute, 5205 Leesburg Pike, Falls Church, VA 22041.

¡COMPUTE!

SCIENCE INPUT

Scientists estimate the dates of events in the earth's history in a number of ways. For example, they may study the types of rocks and sediment found in different layers of the earth, or they may look at evidence of plant and animal fossils in different layers. These events are dated relative to one another. In other words, a very specific time usually cannot be given. Instead, one event is said to come before or after another. In this way, the geological or biological events are arranged in an orderly fashion. Another way to state this is to say that they are placed in a sequence. Two sequences are listed below: the first is a sequence of geological eras; the second is a sequence for the evolution of animals on earth.

SEQUENCES

A	B
Geologic Eras	**Animal Evolution**
Precambrian	Bacteria
Paleozoic	Tribolites
Mesozoic	Dinosaurs
Cenozoic	Mammals

COMPUTER INPUT

The term "sequencing" refers to the logical arrangement, or order, of instructions placed in the computer. The computer follows each instruction in the numbered order, from lowest to highest. The programmer, through his or her understanding of (1) the purposes of the program and (2) the most efficient ways of using the computer's capacities, designs the order of instructions.

WHAT TO DO

Using data from this book, you will program the computer to print out several sequences. You will also learn how to add data into an existing program in order that it may be read in proper sequence. Written below is Program Order, which is designed to list the changes in continent formation. Enter the program and run it. Then, enter it again, using different numbers for each statement. For example, try the following two sets of numbers: (A) 204, 205, 206, 207, and (B) 100, 150, 160, 170.

Did either of these changes alter the order in which the sequence was printed out?

Suppose you want to add another event, the breakup of the supercontinent, to the process. In the program the statement would be written:

PRINT "BREAKUP OF SUPERCONTINENT"

But where would it go in the sequence? Which numbering system, example (A) or (B), would allow room to add a number? Give the statement a number and add it to the program. Run it. How did it change the order of events printed on the screen, or monitor?

Now, you can write your own sequencing program. On a separate piece of paper, rearrange the list of scrambled geologic time periods in proper sequence. Then, using Program Order as a model, write a program that will print the events on your screen correctly. By using multiples of ten for line numbers, leave space in the sequence for remark (REM) statements. Divide the periods by using the era in your rem statements.

To write your program, rearrange the following according to the information in Table 7–1, page 196.

Cretaceous Period
Precambrian Period

SEQUENCING: PROGRAM ORDER

Ordovician Period
Quaternary Period
Jurassic Period
Permian Period
Cambrian Period
Tertiary Period
Triassic Period
Pennsylvanian Period
Devonian Period
Mississippian Period
Silurian Period

Enter and run your new program. Each time you enter a program, remember first to save it on tape or disk.

GLOSSARY

RUN	The command in BASIC that starts the program you have entered.
PROGRAM	A list of step-by-step instructions that a computer reads and follows; also called software.
STORE	To save information in the computer's memory for future use.

PROGRAM

```
100   REM CHANGING CONTINENTS IN
      ORDER
110   PRINT "ONE SUPERCONTINENT"
120   PRINT "NORTHWARD DRIFTING OF
      SUPERCONTINENT"
130   PRINT "PRESENT CONTINENT
      SHAPES"
140   END
```

PROGRAM NOTES

The statements in programs are usually numbered in a way that will allow for additions without disturbing the sequence. However, instructions that should follow one after the other may be given consecutive numbers so that the necessary steps are not interrupted. If you would like to look at a number of programs to study their sequencing and numbering, check your library or local newsstand for one of the many computer magazines that are written for young adults. They contain a wide variety of programs.

BITS OF INFORMATION

Are you thinking of buying a computer? First, read about computers and do some window-shopping. You want a computer that will fit the family budget as well as perform the kinds of functions you require. You might also want one that is compatible with your school computer so you can practice at home what you learn in school. For a 50-page illustrated guide to home computers, write to: "How to Buy a Home Computer," Electronics Industries Association, P.O. Box 19100, Washington, DC 20036. Enclose a stamped, self-addressed envelope that is 6" × 9" or larger, with 54 cents postage.

CHAPTER REVIEW

VOCABULARY

On a separate piece of paper, match the number of each blank with the term that best completes each statement. Use each term only once.

eras	geologic column	relative age	uniformitarianism
fossils	invertebrates	key beds	vertebrates
half-life	petrified	superposition	
index fossils	radioactive decay	unconformity	

To understand the processes of the past, scientists use the principle of __1__, which states that they are the same as the processes acting in the present. By studying the positions of rock layers, and applying the law of __2__, the __3__ of the rock layer can be found. An eroded surface that sometimes can be seen between two rock layers is called a(n) __4__.

Remains of plants and animals, called __5__, can be found in rocks. One example is when the remains are replaced by minerals in __6__ wood.

Plant or animal remains that can be used to tell the age of rocks are called __7__. They can also be used to compare rock layers all over the world, particularly rock layers called __8__ that are easily recognized over a large area. The different rock layers can be combined to form the __9__. The most accurate method used today to measure the age of rocks is __10__. One example is uranium, which has a(n) __11__ of 4.5 billion years.

The earth's history can be divided into four main __12__. During the early Paleozoic Era, there were many animals without backbones. They are called __13__. As time went on, animals with backbones, called __14__, appeared.

QUESTIONS

Give brief but complete answers to each of the following questions. Unless otherwise indicated, use complete sentences to write your answers.

1. Why are older rock layers sometimes found on top of younger rock layers?
2. Give an example of a formation that demonstrates uniformitarianism.
3. Compare the fossils found in recent rock layers with the fossils found in old rock layers.
4. In what way can a fault or flow of magma in rock layers be used to tell the age of the rock layers?

5. What conditions are necessary for the preservation of a fossil?
6. Describe five ways in which ancient animals and plants can become fossils.
7. Why do only a few plants and animals become fossils?
8. Give three examples of how trace fossils are formed.
9. What can each of the following tell about the age of rock layers? (a) A key bed, (b) an index fossil, (c) uranium, (d) radioactive potassium, and (e) carbon 14.
10. What is the geologic column?
11. Why can't carbon 14 be used to tell the age of all rocks?
12. What evidence is used to divide the earth's history into eras?
13. What special events mark the beginning of each of the four eras?
14. Why did reptiles flourish during the Mesozoic Era?
15. Compare the general climate of the Mesozoic and Cenozoic eras.

APPLYING SCIENCE

1. Describe how a leaf falling from a tree could become a fossil.
2. Write a list of all the living things in your home. From this list, make another list of those living things that probably could become fossils. Answer the following: (a) Is your fossil list smaller than the list of living things in your home? Why? (b) How is your fossil list similar to the fossil list of living things one million years ago?

BIBLIOGRAPHY

Burgess, Robert F. *Man: 12,000 Years Under the Sea: A Story of Underwater Archaeology.* New York: Dodd, Mead, 1980.

Harrington, John W. *Dance of the Continents: Adventures with Rocks and Time.* Boston: J.P. Tarcher (Houghton Mifflin), 1983.

Mannetti, William. *Dinosaurs in Your Backyard.* New York: Atheneum, 1982.

Mossman, David, and William A. S. Sarjeant. "The Footprints of Extinct Animals." *Scientific American*, January 1983.

Stuart, Gene S. *Secrets From the Past.* Washington, D. C.: National Geographic Society, 1979.

Weitzman, David. *Traces of the Past: A Field Guide to Industrial Archaeology.* New York: Scribner's, 1980.

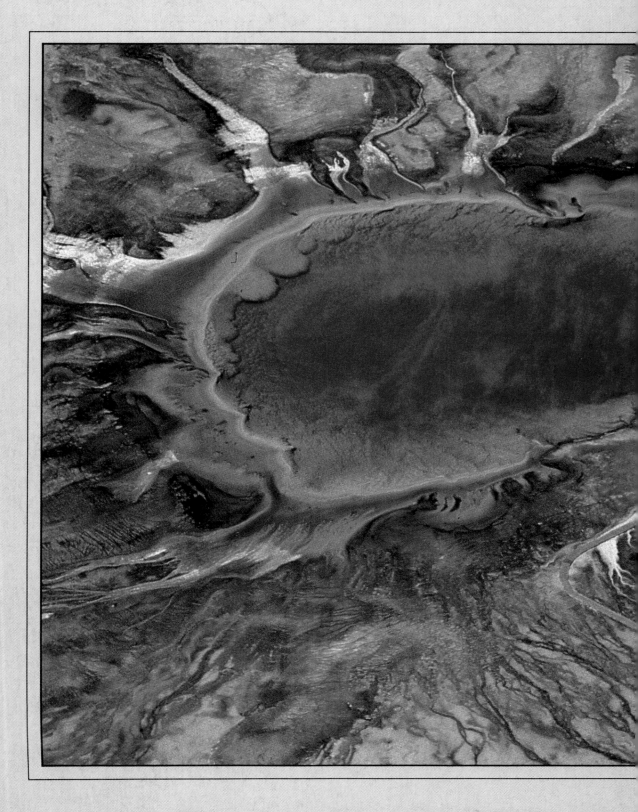

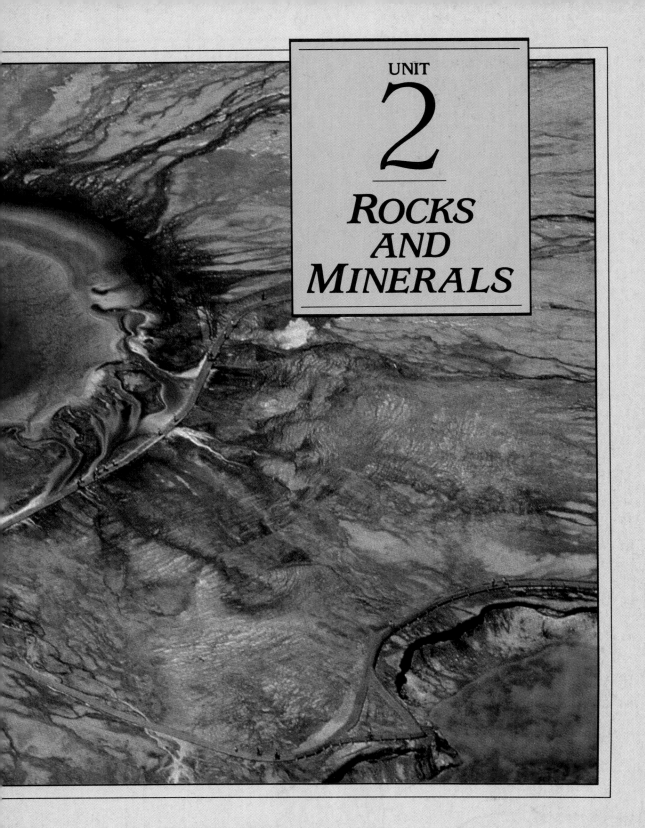

UNIT

2

ROCKS
AND
MINERALS

There are materials in the earth's crust with properties that are the same all around the world. These materials are called minerals. Sulfur is one of about 2,000 minerals found on the earth.

MINERALS

CHAPTER GOALS

1. Describe an atom using the words *nucleus*, proton, neutron, and electron.
2. Explain how atoms may differ in structure and how these differences account for their ability to form compounds.
3. Describe three main divisions of the earth.
4. Explain the difference between a rock and a mineral.
5. Relate atoms, *elements*, *compounds*, and *minerals*.
6. List and describe several properties useful in identifying minerals.

8–1. Atoms, Elements, and Compounds

At the end of this section you will be able to:

☐ Describe how *atomic particles* make up an atom.

☐ Explain how one kind of atom is different from another.

☐ Show how atoms are able to join together to make a chemical *compound*.

The pieces of a jigsaw puzzle can fit together in only one way to make a complete picture. Atoms also can join together only in certain ways to make chemical compounds. The mineral quartz, for example, is a compound formed when silicon and oxygen atoms fit together in a particular way, like pieces in a puzzle. The joining together of atoms produces the many different kinds of materials that form the crust. In this section you will see why atoms behave as they do.

Word Study: Proton: *protos* (Gr.), first.
Neutron: *neutralis* (L.), of neuter gender.
Electron: *electrum* (L.), amber.

INSIDE THE ATOM

A single atom is so small that it is invisible even under the most powerful microscope. Yet atoms are made up of even smaller particles called **atomic particles.** Although *atomic particles* cannot be seen, scientists have been able to learn some important facts about them. For example, the particles with the greatest mass are *protons* (**proe**-tahnz) and *neutrons* (**noo**-trahns). Protons and neutrons are about the same size.

Atomic particles Small particles that make up atoms. Protons, neutrons, and electrons are three kinds of atomic particles.

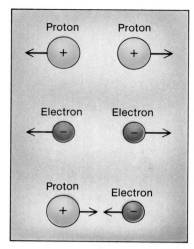

Fig. 8-1 *Two protons or two electrons carry like charges that cause them to repel each other. A proton and an electron carry opposite charges and attract each other.*

Nucleus The central part of an atom that contains the protons and neutrons.

Reinforcement: Static electrical charges can be demonstrated by the following: (a) A comb rubbed with a piece of cloth will attract bits of paper; (b) an inflated toy balloon rubbed with a cloth will cling to a wall.

Enrichment: Atoms of the same element that contain different numbers of neutrons are called isotopes. For example, there are three isotopes of hydrogen containing zero, one, and two neutrons.

One of the smallest particles is the *electron* (eh-**lek**-trahn). An electron is about two-thousandths (0.002) of the mass of a proton. These particles—protons, neutrons, and electrons are the parts that make up atoms.

Protons and electrons have electrical charges. Protons have a positive electrical charge. Electrons have a negative electrical charge. They behave the same way as magnets. If you have ever used magnets, you probably have seen what happens when the ends of two magnets are brought together. They may be pushed apart or pulled together. The forces that are active between protons and electrons are shown in Fig. 8-1. Atomic particles that carry like electrical charges push each other apart. On the other hand, different charged particles, such as a proton and an electron, pull together, or attract. The attraction between protons and electrons helps to explain how atoms are arranged. Protons and neutrons are found only in the center of the atom. Neutrons have no electrical charge; they are neutral. The protons and neutrons in the center of an atom make up the **nucleus** (**noo**-klee-us). Because the *nucleus* always contains protons, it has a positive electrical charge. Electrons are found around the nucleus. They move at a very high speed. In fact, they move so fast that they can make millions of complete trips around a nucleus in one second. The attraction between the positive nucleus and the negative electrons keeps the electrons around the nucleus.

CLIMBING THE ATOMIC LADDER

Table 8-1 lists 20 kinds of atoms, beginning with hydrogen. Hydrogen is the smallest atom. Look carefully at the number of protons and electrons as you go down this list. The table shows that there is always the same number of protons and electrons in an atom. Now look at the number of neutrons. As you can see, the number of neutrons can be equal to, greater than, or less than the number of protons or electrons.

The study of all known atoms, from smallest to largest, is like a trip up a ladder. At each step a proton and an electron are added. Look again at Table 8-1. Neutrons also may be added, but not always one at a time. At the top of the ladder is the largest atom, with more than 100 protons, more than 100 electrons, and more than 100 neutrons.

Name	Protons	Neutrons	Electrons	Symbol	Electron Shells 1 2 3 4			
Hydrogen	1	0	1	H	1			
Helium	2	2	2	He	2			
Lithium	3	4	3	Li	2	1		
Beryllium	4	5	4	Be	2	2		
Boron	5	6	5	B	2	3		
Carbon	6	6	6	C	2	4		
Nitrogen	7	7	7	N	2	5		
Oxygen	8	8	8	O	2	6		
Fluorine	9	10	9	F	2	7		
Neon	10	10	10	Ne	2	8		
Sodium	11	12	11	Na	2	8	1	
Magnesium	12	12	12	Mg	2	8	2	
Aluminum	13	14	13	Al	2	8	3	
Silicon	14	14	14	Si	2	8	4	
Phosphorus	15	16	15	P	2	8	5	
Sulfur	16	16	16	S	2	8	6	
Chlorine	17	18	17	Cl	2	8	7	
Argon	18	22	18	Ar	2	8	8	
Potassium	19	20	19	K	2	8	8	1
Calcium	20	20	20	Ca	2	8	8	2

Table 8–1

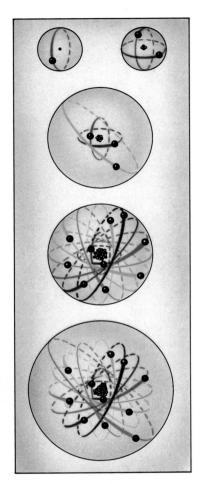

Fig. 8–2 Electrons move around an atomic nucleus within shells. Each kind of atom shown in the diagram has a different number of electrons and shells. Using Table 8–1, identify each kind of atom.

Electron shell An area where a certain number of electrons move at a definite distance from the nucleus.

Element A substance made up of only one kind of atom.

Also listed in the table are letter symbols for each kind of atom. These symbols are a short way of writing the names of the atoms. For example, the symbol for a hydrogen atom is H. The symbol for a helium atom is He, and for lithium, Li. Sometimes the symbol comes from Latin. For example, the symbol for iron is Fe. Fe comes from the Latin name for iron, *ferrum*.

Electrons in atoms move in a path, or an **electron shell.** Each *electron shell* is an area that is a definite distance from the nucleus. See Fig. 8–2. Electrons move within these shells. Each shell is a certain size and can hold a definite number of electrons. Electron shells for some atoms are shown in Table 8–1.

ELEMENTS

Each different kind of atom is said to belong to a different chemical **element.** An *element* is a substance that cannot be broken down further into any other substance. An atom is the smallest complete part of an element. An element contains

only one kind of atom. For example, the element silicon is
made up of only silicon atoms. The element oxygen is made up
of only oxygen atoms. Elements are the building blocks of all
kinds of matter. There are more than a hundred different kinds
of chemical elements known to exist. However, only eight elements make up most of the earth's crust. Oxygen makes up almost half of the weight of the earth's crust. The next most
common element is silicon. Oxygen, silicon, and the six other
elements shown in Table 8–2 make up about 99 percent of the
weight of the earth's crust.

Element	Percent by Weight in Crust
Oxygen	46.60
Silicon	27.72
Aluminum	8.13
Iron	5.00
Calcium	3.63
Sodium	2.83
Potassium	2.59
Magnesium	2.09
Other elements	1.41

Table 8–2

HOW ATOMS COMBINE

1. Gaining and losing electrons. Very few atoms are found
alone in nature. Usually two or more different atoms combine
to form a chemical **compound**. A *compound* is a substance
made up of two or more kinds of atoms combined together.
Atoms combine because of their electron shells. Each shell can
hold a definite number of electrons. For example, the first shell
can hold only two electrons. The second shell can hold eight
electrons. If an atom has an outer electron shell that is only
partly filled, it is able to become part of a compound.

Look at chlorine in Table 8–1 on page 215. As you can see, a
chlorine atom has seven electrons in its outer shell. This third
shell can hold a total of eight electrons. Thus chlorine needs
one electron to fill its outer shell. Now look at sodium in Table
8–1. It has only one electron in its outer shell. When one sodium atom transfers its one outer electron to a chlorine atom,
sodium chloride, or table salt, is formed. This gives both atoms
complete outer electron shells. However, the transfer of one
electron changes the sodium and chlorine atoms. Each no

CHAPTER 8

longer has an equal number of negatively charged electrons and positively charged protons. The loss of one electron in sodium causes it to have one more proton than electron. This extra proton causes the sodium atom to be positively charged. The chlorine atom that gains the electron now has one more negative electron. The chlorine atom becomes negatively charged. An atom that takes on a positive or negative electrical charge caused by the gain or loss of an electron is called an **ion.** A sodium atom becomes a positive *ion* by the loss of an electron. A chlorine atom becomes a negative ion by gaining an electron. Since opposite electrical charges attract, the sodium ion and chloride ion join together to form sodium chloride. In the same way, many compounds are made up of ions held together by electrical charges.

Word Study: Ion: *ion* (Gr.), to go.

Ion An atom that has become electrically charged as a result of the loss or gain of electrons.

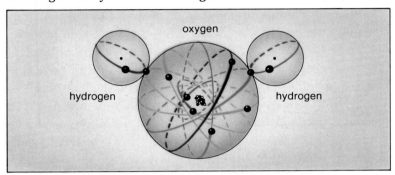

Fig. 8-3 Two hydrogen atoms and one oxygen atom can share their electrons and join together.

2. Sharing electrons. Not all atoms combine by gaining or losing electrons. Another way that atoms may combine is by sharing electrons. Look at the number of electrons in the hydrogen and oxygen atoms in Table 8-1. Oxygen has only six electrons in its outer shell. It can accept two more electrons to complete its outer shell. Two hydrogen atoms can give those two electrons to an oxygen atom. Each hydrogen atom can share its one electron with the oxygen atom. The oxygen atom then has a full electron shell of eight. In turn, the oxygen atom can share its outer electrons with the hydrogen atoms. One electron is shared with each hydrogen atom. Each hydrogen atom then has a full electron shell. See Fig. 8-3. When two hydrogen atoms share electrons combined with one oxygen atom, a **molecule** (**mol**-ih-kyool) of water forms. A *molecule* is the smallest part of a substance that still has the properties of the substance. The water molecule is represented by the chemical formula H_2O. A formula tells how many of each kind of

Word Study: Molecule: *molecula* (L.), small mass.

Word Study: Compound: *componere* (L.), to put more.

Molecule The smallest part of a substance with the properties of that substance.

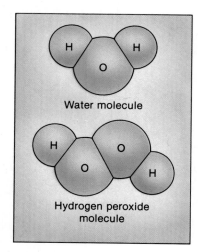

Water molecule

Hydrogen peroxide
molecule

*Fig. 8-4 A water molecule
compared to a molecule of
hydrogen peroxide.*

Enrichment: Hydrogen peroxide used
as an antiseptic is usually a 2%
solution.

1. (a) Electron; located in shells
about the nucleus; has a negative
charge. (b) Proton; located in the nucleus; has a positive charge.
(c) Neutron; located in the nucleus;
has no electrical charge.
2. The next larger atom will have
one more electron and one more
proton. The number of neutrons may
remain the same or be more or less
than before.
3. In sodium chloride, the sodium
atom gives one electron in its outer
shell to the chlorine atom, giving the
chlorine atom a negative charge and
the sodium atom a positive charge.
In water, an oxygen atom shares
two different electrons in its outer
shell with two different hydrogen
atoms.

atom are in each molecule. The 2 in the formula means that
there are two atoms of hydrogen for every atom of oxygen.

All the molecules of any one substance are alike. They all
have the same properties. A molecule of a compound such as
water, for example, always has the same kind and the same
number of each type of atom. A water molecule always contains two hydrogen atoms combined with one oxygen atom.
The properties of water come from the properties of these molecules. Every chemical compound has its own set of properties
because it always contains the same number and kind of atoms.
Molecules of different substances are not alike. For example, a
molecule of water is different from a molecule of sugar. They
differ in the number and kinds of atoms.

Atoms also may combine in different ways. For example, two
compounds can be formed from hydrogen and oxygen atoms.
The two compounds are water, H_2O, and hydrogen peroxide,
H_2O_2. In water, there is one oxygen atom in each molecule. In
hydrogen peroxide, there are two oxygen atoms in each molecule. See Fig. 8–4. Although water and hydrogen peroxide are
made of hydrogen and oxygen atoms, the number of oxygen
atoms is different. Thus their properties are different. You can
drink water, whereas pure hydrogen peroxide would cause
burns in your mouth and throat.

SUMMARY

Everything in the earth is made up of atoms. Atoms, in turn, are
made up of smaller atomic particles. Each atom is made up of a
nucleus with electrons moving around it. Each kind of atom
contains a different number of protons, electrons, and neutrons. If their electron shells are incomplete, atoms can join together to form chemical compounds. Atoms can combine by
gaining, losing, or sharing electrons.

QUESTIONS

Use complete sentences to write your answers.
1. List the particles found in the atom and give the location as
   well as the charge of each.
2. Describe the difference in the number of electrons, protons,
   and neutrons between one atom and the next larger atom.
3. Give an example of each of the two ways atoms may combine to make compounds.

CHAPTER 8

See teacher's commentary for teaching hints, safety, and answers to questions.

MODELS FOR ELEMENTS AND COMPOUNDS

PURPOSE: To make models of elements and compounds using paper clips as atoms.

MATERIALS:
9 paper clips (3 each of 3 different sizes)

PROCEDURE:

A. Separate the paper clips into piles according to size.
   1. If each size represents one type of atom, how many different elements would these represent?

B. In order to be seen, real atoms must combine in different ways. You will join paper clips together one way only, end to end, to make chains. Join together three paper clips of the same size.
   2. Which does this model represent, an element or a compound?

C. Use three of the remaining paper clips to form a chain that represents a compound.
   3. How does this model differ from the one made in B above?

D. Take apart your element and compound models and return the clips to their separate piles.

E. Using only two paper clips at a time, make as many different chain models that represent compounds as you can. Make a sketch of each model before you take it apart.
   4. How many different models could you make with just two clips?

F. Using only three paper clips at a time, make as many different models that represent compounds as you can. Remember, compounds are made when two or more kinds of atoms combine. You will need to take apart models to make new ones. Make a sketch of each model before you take it apart.
   5. How many different three-clip chain models did you make from the three sizes?

G. Using only four paper clips at a time, make as many different models that represent compounds as you can. Make a sketch of each model before you take it apart to make others. Use "s" for small, "m" for medium, and "l" for large to sketch your models.
   6. How many different models did you make with four clips?

CONCLUSIONS:

1. What effect did increasing the number of different atoms have on how many compounds could be formed?

2. There are a number of atoms that can combine with more than two other atoms each. What effect would this have on the number of compounds possible?

3. There are about 90 different kinds of natural elements. Most of these combine with other elements in several different ways. Explain why these elements could combine to make the millions of compounds known to exist.

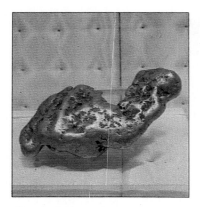

Fig. 8-5 A gold nugget found in a stream.

8-2. Earth Materials

At the end of this section you will be able to:
- ☐ Describe the *lithosphere* of the earth.
- ☐ Compare *minerals* and rocks.
- ☐ Name a mineral that is a compound and list the elements found in it.

Fig. 8-5 shows a gold nugget. Once a part of a larger rock, the nugget was separated from its parent rock by erosion. Its lucky finder picked it from a deposit made by the running water that once carried it. A lump of gold stands out from the other pieces of rock. It has its own special properties that make it easy to identify. Thus you would have no trouble picking a gold nugget from a handful of rocks.

MINERALS AND ROCKS

The materials that make up the earth are often divided into three parts. The **atmosphere** (**at**-muhs-fear) is the layer of gases that surround the earth. All the bodies of water on the earth make up the **hydrosphere** (**hie**-druh-sfear). The outer layer of solid material that makes up the earth is called the **lithosphere** (**lith**-uh-sfear). Each of these three divisions is made up of materials that can be observed and described. The materials of the *atmosphere* and *hydrosphere* will be studied in other sections. The upper part of the *lithosphere*, commonly called the crust, will be studied in this section.

The solid continents and sea floors make up the earth's crust. Materials in the crust can be compared to materials in a kitchen. Some things are the same in all kitchens. Salt and ordinary sugar are good examples. Each substance has its own properties, such as taste and color. Even if you had different brands of table salt, you could not tell one from another by its taste or color. Since they are all the same substance, they have the same properties. Similarly, there are materials in the crust with properties that are the same no matter where they are found. These substances are called **minerals** (**min**-uh-rulz). A *mineral* is a solid, naturally forming substance that is found in the earth. Each mineral has its own definite set of properties.

Atmosphere The layer of gases that make up the outside of the earth.

Hydrosphere All the bodies of water on the earth's surface.

Lithosphere The earth's outer layer averaging about 100 km in thickness.

Word Study: Atmosphere: *atmos* (Gr.), vapor.
Hydrosphere: *hydro* (L.), water.
Lithosphere: *lithos* (Gr.), stone.

Word Study: Mineral: *minera* (L.), mine.

Mineral A solid, naturally forming substance that is found in the earth and that always has the same properties.

Just as salt and sugar can be identified by their properties of taste, minerals can be identified by their properties.

On the other hand, some things found in a kitchen are not always the same substance. Soups or salads, for example, can be different combinations of foods. These combinations are not always the same. In the same way, different parts of the earth's crust are not always the same. If you pick up a piece of rock at one place, it may be made of materials that are different from a rock taken from another place. These two pieces of rock would not be the same because they are probably made up of different combinations of minerals. *Rock* is a solid part of the crust that is made up of one or more minerals. Most rocks contain more than one kind of mineral. See Fig. 8–6.

A mineral compares to a rock as a tree compares to a forest. A forest is made up of separate trees. Likewise, a rock is made up of separate mineral grains. Just as trees come in different shapes and sizes, the mineral grains in a rock are also different in size and shape. Most forests contain different kinds of trees

Enrichment: There are over 2,000 minerals in the earth's crust. However, only about 12 minerals make up 99 percent of the crust. These are called the rock-forming minerals.

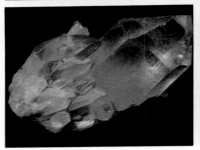
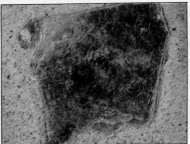
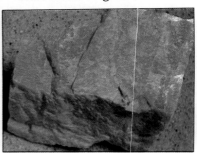

Fig. 8–6 (top) *Granite is a rock made up of various minerals.* (bottom) *Three minerals found in granite.*

just as rocks usually contain different minerals. However, in some forests there may be only one kind of tree. Some rocks also can be made up almost entirely of grains of a single type of mineral. Forests may be alike and be put into the same group even though they are never exactly alike. Rocks may also be alike and be put into the same group. But even rocks in the same group are never exactly the same.

MINERALS: A CLOSER LOOK

Suppose that you could take a piece of rock into a laboratory to find out exactly what it is made of. If you looked at a thin slice of the rock under a special microscope, it would be possible to see the mineral grains that make up the rock. See Fig. 8–7. But what would you see if you looked at a single mineral grain under a microscope with no limit to its power? The super microscope would show that the mineral grain was made up of atoms. It would take more than one million of those atoms to reach across the period at the end of this sentence.

Fig. 8-7 Individual mineral grains of gypsum magnified 30×.

Fig. 8–8 A sample of quartz crystals.

If you could see the atoms in the mineral, you would find that they are not all alike. For example, suppose that you were examining the mineral shown in Fig. 8–8. It is a common mineral called *quartz*. You would see that it contains two kinds of atoms, silicon and oxygen. Most minerals are made up of at least two kinds of atoms. That is, minerals are usually made up of two or more chemical elements. However, a few elements, such as gold, may be found in the pure form. A gold nugget is a lump of a single kind of mineral. It always has the same properties. Thus gold is easy to identify when found mixed with pieces of rock.

Each of the separate atoms in a mineral substance must be joined to one another. Otherwise the mineral grain would not exist as a solid piece of matter. In the mineral quartz, for example, the silicon and oxygen atoms must be joined together. The way in which these two kinds of atoms combine gives quartz special properties. Thus quartz can be identified as a separate mineral, just as gold can be identified by its properties. Quartz and almost all other kinds of minerals are examples of mineral compounds.

The atoms of a compound are not simply mixed like peanut butter and jelly in a sandwich. The atoms combine in such a way that a new substance, a compound, is formed. All mineral

Enrichment: Many minerals contain silicon and oxygen as their principal elements. All of these minerals are called silicates. Feldspar and mica are examples of silicate.

compounds, such as quartz, always contain the same kinds of atoms joined in the same way. See Fig. 8–9. Thus a mineral can always be identified by the special properties that come from the arrangement of its atoms.

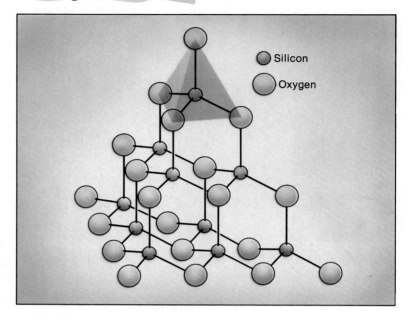

Silicon

Oxygen

Fig. 8–9 A silicon atom and oxygen atoms are joined in this pattern. How does the number of oxygen atoms and the number of silicon atoms compare?*

*There are two oxygen atoms for each silicon atom.

SUMMARY

The part of the lithosphere nearest the surface, the crust, is made of rock. If you looked closely at a piece of rock, you would probably see that it is made up of separate grains. Each of these individual grains is likely to be a mineral. Each mineral has its own set of properties. Most minerals are made up of two or more kinds of atoms. Atoms combine to produce chemical compounds. Each compound has its own properties because it is always made up of the same kinds of atoms.

QUESTIONS

Use complete sentences to write your answers.
1. Name and describe the three main divisions of the earth's materials.
2. What is the difference between a rock and a mineral?
3. List in order the three most abundant elements found in the minerals of the earth's crust.
4. What two elements combine to form the mineral quartz?

1. (a) The lithosphere is the layer of solid material near the earth's surface, commonly called the crust. (b) The atmosphere is the layer of gases that surrounds the earth. (c) The hydrosphere is made up of all the bodies of water on the earth.
2. A mineral is made of one kind of substance. Rocks are made up of several kinds of minerals in various amounts.
3. Oxygen, silicon, aluminum.
4. Silicon and oxygen are the elements that combine to form quartz.

INVESTIGATION

See teacher's commentary for teaching hints, safety, and answers to questions.

MODELS OF MOLECULES

PURPOSE: To make models showing how atoms combine.

MATERIALS:
set of 12 marked
 cards

PROCEDURE:

A. The cards you will use represent atoms of hydrogen, oxygen, carbon, and nitrogen. Cards may be joined together only by placing the mark on the edge of one card next to a similar mark on another card. Select two cards marked "H," and one marked "O," and place them separately in front of you. To understand why the marks must be on the edges of the cards, answer the following questions.

1. How many electrons does an atom of hydrogen have? (See Table 8–1, page 215.)

2. The first electron shell of any atom is filled when it has two electrons. How many more electrons would a hydrogen atom need to fill its shell?

3. Look at the H card. How many marks does it have on its edges? How many electrons does an atom of oxygen have in its second shell?

4. The second shell of any atom is not filled until it has eight electrons. How many more electrons does an atom of oxygen need to fill its second shell? Look at the O card. How many marks does it have on its edges?

B. Join the three cards to make a model of a water molecule.

5. What does the "2" in H_2O stand for?

C. Using two hydrogen (H) cards and two oxygen (O) cards, make a model of an H_2O_2 molecule. This is a model of hydrogen peroxide.

6. Compare your model to Fig. 8–4, page 218. How are they alike?

D. Use the N card, which stands for nitrogen, and attach as many hydrogen cards to it as will fit.

7. How many H cards fit on one N card?

E. Use the C card, which stands for carbon, and attach as many hydrogen cards to it as will fit.

8. How many H's fit on one C? Writing the C first, then the H, write the symbols that would represent this molecule similar to the way that H_2O represents the water molecule.

F. Remove one H card from the C card. Attach in its place another C card.

9. How many H cards are needed to satisfy the marks on this C card?

CONCLUSIONS:

1. Look at Table 8–1 and explain why there are four places marked on the edges of the C card for joining to other atoms.

2. Write the symbols for the nitrogen–hydrogen molecule you made in D.

3. Name and give the symbols for two kinds of molecules that contain only hydrogen and oxygen atoms.

Fig. 8-10 The British crown jewels contain many valuable minerals.

8-3. Mineral Properties

At the end of this section you will be able to:

☐ Describe how to identify a mineral such as salt.

☐ List and describe four properties of minerals that can be observed with the eye or hand alone.

☐ List and describe four properties of minerals that can be observed by using simple tools.

The large red jewel in the British crown shown in Fig. 8-10 is called the Black Prince's ruby. For centuries it has been part of the treasure that makes up the British crown jewels. Now it is known that this is not a ruby. It is a mineral called red spinel. Spinel is a form of the mineral corundum. It is much less valuable than a ruby. All minerals, such as spinel and ruby, have individual properties that can be used to identify them.

ORDERLY ATOMS

Like rubies and spinels, common table salt is also found as a mineral. When salt is found in its natural state in the earth, it is called *halite* (**hal**-ite). Halite is the mineral name for salt. Like most minerals, halite is a chemical compound. It is made up of the two elements sodium and chlorine. How can halite be identified? Tasting is not always a good method. Many mineral compounds are poisonous. You should never taste any substance that is not known to be a food. Also, many compounds other than halite have a salty taste. However, halite does have one characteristic that makes it easy to identify.

You can see this characteristic if you look at grains of ordinary salt with a magnifying glass. Many of the salt grains will be seen as tiny cubes. See Fig. 8-11. Each cube is a salt **crystal** (**krist**-'l). *Crystals* are formed because of the orderly way in which atoms or ions are arranged. For example, the compound sodium chloride, or salt, is made up of sodium and chloride ions. These ions can fit together only in the pattern shown in Fig. 8-12 (a). Thus salt and its mineral form, halite, are seen as crystals with the shape of a cube. See Fig. 8-12 (b).

Each mineral is made up of crystals with a definite shape. Fig. 8-8 on page 223 shows the crystal shape of the common mineral quartz. Usually the crystal shape of a mineral is not easily

Crystal A mineral form with a definite shape that results from the arrangement of its atoms or similar units.

Fig. 8-11 Many of the grains of salt from a salt shaker appear as cubes.

seen. Large and perfectly formed mineral crystals are not commonly found. This is because most of the time the crystals grow in a limited space. They are crowded together in the mineral. This results in mineral grains made up of very small crystals. The shapes of the individual crystals are not visible to the eye. However, the arrangement of the atoms or similar units in very small crystals can be found by using X-rays. For example, X-ray examination of the "ruby" in the British crown revealed the crystal pattern of spinel.

Word Study: Crystal: *krystallos* (Gr.), ice.

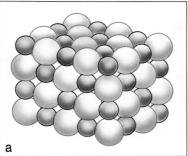

Fig. 8-12 (a) The shape of a crystal of halite is the result of the orderly arrangement of the sodium and chloride ions. (b) Pieces of the mineral halite.

OBSERVING THE PROPERTIES OF MINERALS

Since the crystal shape of a mineral usually is not easy to observe, other properties must be used. Most minerals have other properties that can be observed with the eyes and hands alone or with simple tools. Some of those properties are listed below.

1. Color. Minerals come in many colors. Quartz, for example, may be colorless. It can look clear. But it also can be milky white, brown, or pink. See Fig. 8–13 (left). Ruby and red spinel are found in different shades of red. As you see, the same mineral, such as quartz, may have different colors. Or different minerals, such as ruby and spinel, may have the same color. For

Suggestion: The growth of crystals can be demonstrated by preparing a saturated salt solution. Suspend a weighted string in the solution and leave it over a period of several days or weeks. Crystals will usually form on the string as the solution evaporates. Slow evaporation will produce larger crystals. Canning or pickling salt will produce a solution free of cloudiness.

Enrichment: A newly discovered mineral is named by its discoverer.

Fig. 8-13 (left) *Quartz is found in many colors due to the dissolved mineral impurities.* (right) *The mineral pyrite has a metallic luster.*

Suggestion: Students should be given as much direct contact as possible with mineral samples to become familiar with their properties. If possible, use minerals that are known to occur locally.

Enrichment: Color often depends on the presence of small amounts of different elements in a mineral. For example, ruby and sapphire are both forms of corundum. Ruby is colored by traces of chromium, while sapphire contains traces of iron, or titanium, or both, giving it colors of yellow, green, pink, purple, and blue.

Suggestion: Minerals with streaks that are different from their colors include feldspar, pyrite, and quartz. Pyrite is particularly effective to demonstrate the color of a streak.

these reasons, color by itself is not the best way to identify minerals. Color is helpful only when used with other properties.

2. Luster. The way light is reflected from the surface of a mineral is called luster. Some minerals reflect light in the same way that the shiny surface of metal does. See Fig. 8–13 *(right)*. Such minerals are said to have a metallic luster. Other minerals have a luster like glass, wax, or pearl. Some minerals have dull surfaces that hardly reflect light at all.

3. Feel. Rubbing your finger over the surface of a mineral can help you to identify it. Some minerals feel powdery. Some feel rough like sandpaper. Pencil lead is made from the mineral called graphite. This mineral is said to have a greasy texture.

4. Density. How heavy a piece of a mineral feels is determined by its density. You can tell the difference between salt and iron from their densities. A cupful of salt is lighter than a cupful of iron. In the same way, you can compare the weights of equal size pieces of different minerals. A piece of gold, for example, weighs eight times as much as the same size piece of halite. The density of gold is much greater than that of halite.

The four mineral properties described above can be identified by the use of eyes and hands alone. Four additional properties can be revealed by using simple tools. These tests are:

5. Streak. A powder mark may be left when a mineral is rubbed against a rough surface. This mark is called a *streak*. Streak color helps to identify a mineral. The color of the streak can be seen best on a white surface. The back of a bathroom tile can be used as a plate for making streak marks. The color of

Fig. 8-14 Making marks on a streak plate can help identify a mineral.

the streak is not always the same as the color of the mineral used to make it. See Fig. 8-14.

6. Fracture or cleavage. How can you quickly tell the difference between a piece of glass and a piece of clear plastic? One way would be to break each piece. Glass breaks in a completely different way than plastic. The difference is easy to see. Many kinds of minerals break in ways that help to identify them. Most minerals break unevenly. An uneven break is called a *fracture*. Fig. 8-15 shows a curved fracture. Not all minerals fracture. Some split along smooth surfaces. A smooth break is called *cleavage*. Minerals cleave in one, two, or three directions. See Fig. 8-15. Some minerals show their cleavage only slightly. The broken surface of a mineral sample must be examined carefully to discover how it breaks. Cleavage should not be confused with crystal shape. A mineral that shows cleavage will break in such a way that the broken pieces have the same general shape as the large original piece. A crystal, on the other hand, breaks into pieces that do not look like the original crystal.

Suggestion: Layers of mica can be cleaved with a knife blade. A sample of calcite can be cleaved by tapping a knife blade gently while held firmly on the sample. (Wear goggles.)

Suggestion: Students may be interested in collecting and displaying samples of minerals that can be found locally. Ask any students with mineral or gem collections at home to show them to the rest of the class.

Fig. 8-15 (left) *This sample of obsidian fractures when broken.* (right) *Iceland Spar, a type of calcite, exhibits cleavage.*

HARDNESS SCALE		
Hardness Number	Simple Test	Mineral
1	Fingernail scratches it easily.	Talc
2	Fingernail barely scratches it.	Gypsum
3	Copper penny just scratches it.	Calcite
4	Steel knife scratches it easily.	Fluorite
5	Steel knife barely scratches it.	Apatite
6	Scratches glass easily.	Feldspar
7	Scratches steel and glass easily.	Quartz
8	Scratches all other common minerals.	Topaz
9	Scratches topaz.	Corundum
10	Hardest mineral.	Diamond

Table 8-3

7. **Hardness.** Minerals differ in hardness. Hardness is a measure of a mineral's resistance to being scratched. Hardness is one of the best ways of identifying minerals. If mineral A can scratch mineral B, A is harder than B. If A can scratch B and B can scratch A, they both have the same hardness. Scientists have developed a scale to test the hardness of minerals. This scale is shown in Table 8-3. As you can see, it ranks minerals from one to ten. Ten is the hardest. The hardest mineral is diamond. Number one on the scale is the softest mineral. A mineral with this hardness, called talc, is found in talcum powder. A mineral with a high hardness number can scratch any mineral with a lower number. The hardness of an unknown mineral is tested by finding out which minerals on the scale it can scratch and which it cannot scratch. For example, gypsum is harder than talc or a fingernail. But gypsum is softer than calcite or a copper penny. Thus gypsum has a hardness between two and three on this scale.

Some minerals, such as diamond and quartz, are hard enough to be cut and polished into **gems.** To be considered a valuable *gem*, a mineral substance must also have an attractive appearance and be rare. See Fig. 8-16.

8. **Magnetism.** A few minerals are attracted to magnets. Perhaps you have put a magnet into a pile of sand. If so, you found that some small mineral particles, called magnetite, cling to the magnet. These magnetic properties are useful in helping to identify such minerals.

Suggestion: A common mineral test not mentioned in the text is the use of diluted hydrochloric acid solution to detect calcite. A sample of calcite or a rock containing calcite will give carbon dioxide bubbles when the acid is dropped on it. This is a commonly used test since calcite is an abundant mineral.

Gem An attractive or rare mineral substance that is hard enough to be cut and polished.

Fig. 8-16 Attractive, rare minerals, when cut and polished, can be very valuable.

9. Special properties. There are some properties that are shown by only a few minerals. For example, some minerals give off visible light when exposed to ultraviolet light or X-rays. This property is called *fluorescence*. Other minerals give off light when they are heated. Properties such as these are useful for identifying only a small number of the many different kinds of minerals.

SUMMARY

Atoms or other units that make up minerals are arranged in patterns that give each mineral a crystal form. Other properties of minerals can be observed with simple tests. Color, luster, feel, and general density can be determined without using anything other than eyes and hands. To test for the properties of streak, fracture or cleavage, hardness, and magnetism, simple tools must be used.

QUESTIONS

Use complete sentences to write your answers.

1. How can you tell that the mineral halite and table salt are the same substance?
2. List four mineral properties that can be tested using only the eyes and hands.
3. Describe four tests of mineral properties that require simple tools.
4. Why is the diamond called a gem?

1. By comparing the shape of the crystals of the halite and table salt and noting that they both are the shape of cubes.
2. Color, luster, texture, and density.
3. (a) Rub the mineral against a rough, white surface to see the color of the streak left on the surface. (b) Break or separate the crystal to see if it fractures or cleaves. (c) Test its hardness by seeing what it will scratch or what scratches it. (d) Test its magnetic property to see if it is attracted by a magnet.
4. A diamond is rather rare and has an attractive appearance when cut and polished.

SKILL-BUILDING ACTIVITY

SEQUENCING: MAKING AND USING A MINERAL-HARDNESS SCALE

PURPOSE: To make a sequence, or scale, of hardness for use with minerals.

MATERIALS:
chalk
penny
mineral sample
pencil

PROCEDURE:

A. Scratch the wood of a pencil with your fingernail.

   1. Did your fingernail scratch the wood or did the wood make a scratch on your fingernail?

   2. Which is harder, pencil wood or your fingernail?

B. Perform the same test to see which is harder, your fingernail or chalk.

   3. Did your fingernail scratch the chalk? (A mark that rubs off is not a scratch.)

   4. Which is harder, fingernail or chalk?

C. Look at Table 8–3, page 230.

   5. Name two minerals that are not as hard as your fingernail and give the hardness number for each.

D. Perform a test to see which is harder, your fingernail or pencil lead.

   6. Which is harder, pencil lead or your fingernail?

E. Perform a similar test to see if the wood of a pencil is harder than a piece of chalk.

   7. Which is harder, pencil wood or chalk?

F. Perform tests to find out which is harder, chalk or pencil lead. Do the same for pencil wood and pencil lead.

   8. Which is harder, chalk or pencil lead?

   9. Which is harder, pencil wood or pencil lead?

G. Perform a test to see if your fingernail is harder than a copper penny.

   10. Which is harder, fingernail or copper penny?

   11. List in order of hardness the five objects you tested, with the softest first and the hardest last. This is a sequence of objects according to their hardness.

H. Obtain a mineral sample from your teacher and test it for hardness, using only your fingernail and a copper penny.

   12. Where does the hardness of this sample fit in the sequence you made for the answer to item 11?

CONCLUSIONS:

1. Explain how a mineral-hardness test is performed, using an example.

2. Refer to Table 8–3 on page 230. What was the hardness of the mineral you tested in part H?

3. You are given two gems. Both look like diamonds but one is an imitation. Use what you have learned from this activity to tell how you could identify the real diamond.

A NEW GOLD RUSH: CLASH WITH THE ENVIRONMENT?

Not every new technological advance is welcomed with open arms. Some are considered a mixed blessing and raise as much anxiety as interest. For example, take the new "gold rush" planned to begin in 1985 in the Mother Lode area in Jamestown, California. (*Mother Lode* refers to the main vein of a deposit.) The last operating mines, the survivors of the gold rush of 1849, were closed in 1942. Gold mining, however, has continued in the United States in one

form or other since those early gold rush days. Prospectors still pan for gold dust in streams or dig in the surface of rocks for "placer" gold, which consists of nuggets and grains of gold. While most individuals do not strike it rich, many still find gold mining profitable and exciting.

But the newest gold rush is not on a small scale. The Jamestown project is large-scale. It will operate 24 hours a day, using new technology to separate the gold from other minerals and the old technology of strip mining to pull the ore, a mixture of gold and quartz, from the hills. Instead of shafts being dug, mountains will be taken apart and millions of tons of rock will be moved. Conveyor belts miles long will carry ore directly from open pits to mills. There, the ore containing the gold will be crushed. The ore will then be washed in a mixture of water, pine oil, and other chemicals. This mixture is made to foam so that the gold and iron pyrites (fool's gold) will stick to the bubbles and be skimmed off the top. The iron pyrites is then separated from the gold by mixing it with a compound called thiourea. Thiourea is found in the urine of cattle and other animals. The iron

pyrites–thiourea byproduct can be used as fertilizer, in glassmaking, or to produce sulfuric acid. This method is said to be environmentally safe. Since thiourea can be considered a fertilizer, a spill would not harm the environment.

Previous methods of recovering gold were borrowed from the techniques used in copper mining. In that method, the finely ground ore was mixed with cyanide in large open pits. The gold concentrate that resulted was then melted and formed into ingots. Environmentalists were concerned that the cyanide technique could create a potential for disaster because the cyanide might leak into and poison the ground water. The thiourea method was chosen instead.

Another environmental concern with the new project is the noise that will be generated by the strip-mining activities operating 24 hours a day, seven days a week. To protect the environment, the mining companies intend to pay the county to monitor both the noise and ground water pollution. The mining companies will also replant some areas they've mined and flood the mine pits to create reservoirs.

Newer techniques may be required if this 25-year project develops. The problems of strip mining in this region of California may grow. The population in this region is expected to increase at least five times during this period. Strip mines and housing developments do not usually become good neighbors. It will be interesting to follow this project in years to come to see how industrial and environmental interests can be managed.

¡COMPUTE!

SCIENCE INPUT

Geologists and geological engineers use their knowledge of the physical properties of a mineral to distinguish it from other minerals. Knowing a mineral's characteristics is essential to using that mineral. For example, if you are drilling through a mountain, you need to know how hard the rock will be and how it is likely to react when drilled or blown away with explosives. If you are building with rock, you have to be aware of the degree of softness of your material, or else the structure may fall. Knowing the properties of minerals is necessary for both their identification and use.

COMPUTER INPUT

It is possible to use the computer as an encyclopedia. Data (information) can be saved in the computer's memory and then recalled (brought back at a later date). One advantage of the computer is that you can manipulate the information in a computer more easily than the information in a book. You can choose to recall only part of the information, or you can rearrange it. These capacities exist because of the way in which computers store information. Computers do not actually store words and numbers as we normally use them. The computer works on the basis of a *binary number system*. *Binary* means "having 2 parts." A computer cannot actually read. It can only tell whether or not magnetic impulses are present. (Electrical flows create a magnetic field.) When the flow is on, a magnetic field is present; when the flow is off, the magnetic field is absent. These different states are represented by 1 (present) or 0 (absent). In other words, they are represented by a binary number system. This kind of number system allows the computer to rearrange the data more quickly than if it were using words and conventional numbers.

WHAT TO DO

On a separate piece of paper, make a chart of minerals and their properties. The presence or absence of a mineral property will be coded into a system of 0's and 1's. Use this book and other earth science reference books to research the properties of the different minerals.

In Program Search, the computer will help you to identify an unknown mineral by its properties. To provide the data for statements 301 to 400, code the characteristics of as many minerals as you wish, according to the chart code. Each mineral code should have 7 digits. For example, calcite would be 0000010. It has no streak color (0) and therefore belongs to streak color group 000. The mineral itself is a light color (0), has cleavage (1), and is not harder than 5 (0). The mineral codes will be your database. A database is a fund of information stored in the computer's memory. When you run the program, it will ask you to describe, according to the binary number system, the characteristics of the mineral to be identified. After you have input the information, the computer will review its memory and, if information about that mineral has been entered into the database, the computer will print out its name.

CHART CODE

1. Does the mineral have a streak color?
   1 = Yes; 0 = No
2. To which color group does the streak belong?
   000 = no streak color
   100 = black/grey
   010 = red/brown
   001 = some other color
3. Is the mineral itself a dark color (black, green, red, blue, or brown)?   1 = Yes; 0 = No
4. Does the mineral have cleavage?
   1 = Yes; 0 = No

5. On the hardness scale, is the mineral harder than 5?
   1 = Yes; 0 = No

PROGRAM

```
100   REM SEARCH
110   DIM M$(100),C$(100)
115   REM REPLACE THIS STATEMENT
      WITH CLEAR SCREEN STATEMENT
120   FOR X = 1 TO 100
130   READ M$(X),C$(X)
140   IF M$(X) = "NO" THEN GOTO 160
150   NEXT
160   M = X - 1
170   PRINT : INPUT "STREAK COLOR
      (1 = YES/0 = NO)? ";C1$
175   C2$ = "000": IF C1$ = "1" THEN
      PRINT : PRINT "100 = BLACK/
      GREY": PRINT "010 = RED/
      BROWN": PRINT "001 = OTHER":
      INPUT "WHICH COLOR? (100, 010,
      001)? ";C2$
180   PRINT : INPUT "DARK COLOR =
      BLACK, RED, GREEN, BROWN, BLUE
      (1 = YES/0 = NO)?";C3$
190   PRINT: INPUT "CLEAVAGE (1 = YES/
      0 = NO)?";C4$
200   PRINT: INPUT "HARDER THAN 5
      (1 = YES/0 = NO)?";C5$
210   M$ = C1$ + C2$ + C3$ + C4$ +
      C5$
230   FOR X = 1 TO M
240   IF M$ = C$(X) THEN PRINT M$(X)
250   NEXT
260   GOTO 170
300   REM PLACE DATA STATEMENTS
      BELOW
301   DATA CALCITE, 0000010
400   END
```

GLOSSARY

BIT   Short for *binary digit*. A bit is a piece of computer-coded information. The coding process is done by the computer hardware, not by the program. The hardware is designed to translate the data input into binary numbers, which can be stored electronically in the computer's memory.

BYTE A collection of eight bits. The power of a computer's memory is often expressed as 8K or 16K. Each K represents 1,024 bytes of information. The more bytes in the memory, the greater its capacity.

PROGRAM NOTES

The longer the program, the more chances there are for error. Enter the program carefully. Besides typing it into the computer without errors, you must also research the data carefully. If the information in your data chart is incorrect, then, even if your program runs, you will be learning misinformation.

BITS OF INFORMATION

There is a computer program called "Prospector," which can analyze geological field data and identify different kinds of deposits. In 1982 Prospector, through the analysis of information provided by geologists, identified a $100 thousand mineral deposit in an area that had been mined for years with little success. Prospector's program was designed to imitate the thought processes of 20 top geologists.

CHAPTER REVIEW

VOCABULARY

On a separate piece of paper, match each term with the number of the statement that best explains it. Use each term only once.

atmosphere   lithosphere   compound   electron shell
atomic particles   nucleus   element
crystal   mineral   ion
hydrosphere   molecule   gem

1. The part of an atom that contains protons.
2. The earth's crust.
3. The layer of gases that surrounds the earth.
4. A substance found in the earth that always has the same properties.
5. An area a definite distance from the nucleus where a number of electrons move.
6. An electrically charged atom.
7. An attractive mineral that is hard enough to be cut and polished.
8. Electrons, protons, and neutrons.
9. A mineral form having a definite shape.
10. A substance made up of only one kind of atom.
11. A substance composed of two or more kinds of atoms joined together.
12. All the bodies of water on the earth.
13. Two or more atoms joined together.

QUESTIONS

Give brief but complete answers to each of the following questions. Unless otherwise indicated, use complete sentences to write your answers.

1. Explain the relationship between (a) atoms and elements, and (b) compounds and minerals.
2. Explain the sentence, "A mineral compares to a rock as a tree compares to a forest."
3. Name one mineral and tell why it is a mineral and not a rock.
4. Name three atomic particles and tell where they are found in the atom.
5. Why does an electron attract a proton and repel another electron?
6. How does an oxygen atom differ from a carbon atom?
7. An atom has 45 protons. (a) In what part of the atom are they found? (b) How

many electrons would this atom have? (c) In what part of the atom are the electrons found?

8. Why does it take two atoms of hydrogen to join with one atom of oxygen to make a water molecule?

9. What is the difference in the way the atoms join to make a water molecule and the way the atoms join to make sodium chloride?

10. Give an example of how the same two kinds of atoms may combine in different ways.

11. Why is it not a good idea to taste a mineral to help you identify it?

12. Compare the test for streak with the test for hardness of minerals.

13. Compare fracture and cleavage of minerals.

14. List several minerals that can be scratched by feldspar. Why does feldspar scratch these minerals?

APPLYING SCIENCE

1. Find several minerals. Use a mineral identification key from a high school earth science or geology text to identify the minerals. Write a report of your results.

2. Grow crystals of some common compounds and bring in the crystals to show to the class. See *Crystals and Crystal Growing,* by Alan Holden and Phyllis Morrison, for information on procedures.

3. Alloys are mixtures of metallic elements. Find out what elements are in the following alloys: (a) stainless steel, (b) bronze, (c) sterling silver, (d) brass, (e) 14-K gold, (f) alnico.

BIBLIOGRAPHY

Dietrich, R. V. *Stones, Their Collection, Identification, and Uses.* San Francisco: Freeman, 1980.

Holden, Alan, and Phyllis Morrison. *Crystals and Crystal Growing.* Cambridge, MA: MIT Press, 1981.

MacFall, Russell P. *Rock Hunter's Guide: How to Find and Identify Collectible Rocks.* New York: T. Y. Crowell, 1980.

O'Donoghue, Michael. *Minerals and Gemstones.* New York: Van Nostrand Reinhold, 1982.

The rocks of the earth come in many different colors, shapes and sizes. But each rock tells a story of how it was formed. In this chapter, you will learn to read some of this story.

ROCKS

CHAPTER GOALS

1. Explain how geologists classify rocks.
2. Discuss two ways that igneous rocks are formed, giving an example of a rock formed in each way.
3. Describe three kinds of sediments and how they form rocks, and name a rock formed from each.
4. Explain why sedimentary rocks are an important tool in the study of earth's history.
5. Tell how heat and pressure may change existing rock layers into new rock.
6. Describe the steps in the natural recycling of rocks on earth.

9–1. Magma and Igneous Rocks

At the end of this section you will be able to:

- ☐ Name the major groups of rocks and briefly describe how each is formed.
- ☐ Define *igneous* rock.
- ☐ Classify igneous rock into two major types.

You may know people who can't resist carrying home a rock or two every time they go on an outing in the country. Perhaps you have a rock or two in your room right now. Why collect rocks? Often it is simply a "different" or "colorful" rock that ends up in one's pocket. But John Young and Charlie Duke, shown in Fig. 9–1, page 240, were sent to the moon to gather rocks for a different reason. By gathering and studying rocks from a number of locations, scientists hope to gain a better understanding of the moon's history. In this section you will learn how rocks reveal their history.

Reinforcement: Display common examples or pictures of igneous, sedimentary, and metamorphic rocks.

THE THREE ROCK GROUPS

A **rock** is a piece of the earth's crust. *Rocks* are usually made up of two or more minerals mixed together. They come in all sizes, shapes, and colors. Some rocks are denser than others; some are harder than others. There seems to be a large variety of rocks found in the earth's crust. Careful study shows that

Rock A piece of the earth that usually is made up of two or more minerals mixed together.

Fig. 9-1 Studying rocks can tell us how they formed. Here, Apollo 16 astronauts are collecting samples of moon rocks.

Igneous rock Rock formed as molten material cools.

Word Study: Igneous: *igneus* (L.), fire.

Word Study: Metamorphic: *meta + morphe* (Gr.), change form.

Word Study: Extrusive: *ex + trudere* (L.), to thrust out.

there are some important differences among rocks. These differences result from the way rocks are formed.

Geologists classify rock into three basic groups: *igneous* (**ig**-nee-us) *rock*, *sedimentary* (sed-ih-**ment**-uh-ree) *rock*, and *metamorphic* (met-uh-**mor**-fik) *rock*. **Igneous rock** is formed when molten material cools. Most sedimentary rock is created when small pieces of rock are cemented or squeezed together. Metamorphic rock is igneous or sedimentary rock that has been changed by great heat and/or pressure. In this section, you will learn about *igneous rock*.

IGNEOUS ROCK

Rock formed from molten material is called igneous rock. The word "igneous" means "coming from fire." Molten material reaching the earth's surface is called lava. Molten material beneath the earth's surface is magma. Igneous rock may be formed from either lava or magma.

1. Extrusive rock. Rock formed by lava coming to the surface is called *extrusive* (ek-**stroo**-siv) *igneous rock*. Extrusive means "pushed out." Lava that pours out onto the earth's surface begins to cool and harden as it comes into contact with the air. See Fig. 9-2. The appearance of igneous rock depends

Fig. 9-2 Igneous rock forms as lava hardens.

largely on how fast the lava cools. When lava cools very quickly, there is little time for crystals to grow in the hardening rock. The rock formed has a glasslike texture and is called *volcanic glass*. The rock called *obsidian* is an example of volcanic glass. See Fig. 9-3. Obsidian was used to make arrowheads because its broken edges are as sharp as broken glass. However, lava usually cools slowly enough for small crystals to form in the rock. Thus most of the igneous rock that is formed from lava cooling at or near the earth's surface has small crystals. It is classified as fine textured. It does not look or break like glass.

Fig. 9-3 This deposit of obsidian (volcanic glass) formed when lava cooled very rapidly.

Volcanic lava usually contains gases. If there is a small amount of gas, the lava pours out quietly. If there is a large amount of gas, the lava may come out with an explosion. Molten lava may be blown high into the air. Tiny pieces of hardened lava fall back to the ground as volcanic ash. See Fig. 9–4. As volcanic ash piles up on the ground, the pieces are squeezed together to form rock called *tuff*. Pumice (**puhm**-iss) is formed when gas bubbles are trapped in hardened lava. Pumice looks like a hard sponge and can float on water.

Not all lava comes to the surface of the crust as a result of volcanic eruptions. It can also come to the surface through many fractures. Then the lava spreads out in great sheets. Such an outpouring may cover the existing landscape and form a lava plateau.

Huge amounts of lava also reach the surface beneath the sea along the mid-ocean ridge. The cooling of this lava creates new sea floor. This slowly pushes apart the great crustal plates. The ocean floor is primarily made up of the igneous rock *basalt*, a dark-colored, fine-grained rock.

2. Intrusive rock. What about the huge amount of magma that never flows out onto the surface? This magma can also cool into rock. Igneous rock formed beneath the surface of the earth is called *intrusive* (in-**troo**-siv) *igneous rock*. Intrusive means "forced in." Fig. 9–5 shows intrusive rock called shiprock. Lava in the throat of a volcano first hardened. Then erosion wore away the mountain around the body of hardened lava, exposing it. Intrusive rocks cool slowly. There is time for large crystals to grow. Intrusive rocks are classified as coarse textured.

Suggestion: Show a sample of granite and rhyolite (preferably the same color). Granite is coarse grained, rhyolite is fine grained. Both contain the same minerals.

Fig. 9-5 Shiprock in New Mexico.

Dikes Bodies of igneous rock formed underground in cracks that cut across layers of existing rock.

Sills Bodies of igneous rock formed underground in cracks that run parallel to layers of existing rock.

There are two types of igneous rock bodies formed deep in the earth. **Dikes** and **sills** are examples of one type. They form when magma is forced into cracks in rock and hardens. See Fig. 9-6. If erosion wears away the rock, *dikes* and *sills* are exposed at the surface. Dikes cut across the layers of rock. Sills run parallel to the layers.

Enrichment: Continents are primarily granite. Granite often is used in building construction.

Fig. 9-6 A dike formed by intrusive igneous rock.

Batholiths Very large bodies of igneous rock that are formed underground.

A second type of intrusive rock forms when very large bodies of magma harden within the crust. These are called **batholiths** (**bath**-uh-liths). *Batholiths* may stretch for hundreds of miles under the surface. Batholiths usually form the cores of mountain ranges. The upward movement of magma in the crust seems to be one of the steps in mountain building. The Sierra Nevada range is the result of a great batholith. See Fig. 9-7. A part of the batholith may be exposed as the mountains erode.

Suggestion: Show a large piece of granite. Point to the large crystals of minerals it contains.

Fig. 9-7 The Sierra Nevada batholith.

In general, intrusive igneous rock cools very slowly. Magma in batholiths probably takes many thousands of years to cool. Therefore, rock formed this way usually has large crystals. Granite, a kind of rock with large mineral crystals, is an example of the kind of rock found in batholiths.

SUMMARY

Every piece of rock has a story to tell. The origin of a rock is an important part of its history. One of the best ways to sort different kinds of rock is to classify them according to how they are formed. Igneous rock is formed directly from molten material. The cooling of the molten material can take place on the surface or beneath the ground. The size of the mineral crystals shows how fast the rock has cooled.

1. Igneous rock is formed from molten material. Sedimentary rock is formed when small pieces of rock or minerals are cemented or squeezed together. Metamorphic rock is rock changed by heat and/or pressure.
2. Igneous means "coming from fire."
3. (a) Extrusive igneous rock, (b) intrusive igneous rock, (c) extrusive igneous rock, (d) extrusive igneous rock, (e) intrusive igneous rock.

QUESTIONS

Use complete sentences to write your answers.

1. List the three major groups of rocks and describe how each is formed.
2. What does the word "igneous" mean?
3. Classify the following as either extrusive igneous rock or intrusive igneous rock: (a) obsidian, (b) a dike, (c) tuff, (d) pumice, (e) granite.

SKILL-BUILDING ACTIVITY

See teacher's commentary for teaching hints, safety, and answers to questions.

THE LABORATORY BURNER

PURPOSE: To become familiar with the operation of the laboratory burner.

MATERIALS:

burner   goggles
matches

PROCEDURE:

A. There are several types of laboratory gas burners. All are alike in the way they work. The Bunsen burner is the most common kind. Look at Fig. 9–8. Locate on your burner the labeled parts of the burner.

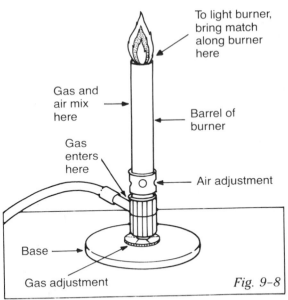

Fig. 9–8

B. Connect your burner to the gas source. CAUTION: WEAR GOGGLES ANYTIME YOU ARE LIGHTING AND USING THE LABORATORY BURNER.

C. To light a burner, bring a lighted match up alongside the top of the barrel, then turn on the gas source. If the lighted match is held directly over the barrel, the escaping gas may blow out the match. Fig. 9–9 shows what the flame should look like.

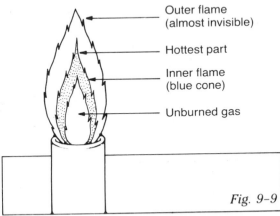

Fig. 9–9

D. If the flame is large and bright yellow, there is too much gas. Use the air adjustment to allow more air to enter until the flame is blue.

E. A gap between the flame and the top of the barrel means that too much air is being admitted. Use the air adjustment to decrease the amount of air entering at the base.

F. If the flame seems to be burning inside the barrel, turn off the gas. Cut down on the amount of air and light the burner again. CAUTION: THE BARREL MAY BE HOT.

CONCLUSIONS:

1. Make a drawing of your burner and label all parts.

2. Describe how to light the burner with a match and adjust it to have a blue flame.

INVESTIGATION

See teacher's commentary for teaching hints, safety, and answers to questions.

CRYSTAL FORMATION IN IGNEOUS ROCK

PURPOSE: To show how the rate of cooling affects crystal size.

MATERIALS:

test tube	test tube holder
stearic acid	index card
goggles	small plastic spoon
Bunsen burner	metric ruler

PROCEDURE:

A. Fill the test tube about one-third full of stearic acid. Because molten rock (lava) is extremely hot, you will use stearic acid to represent it. Stearic acid melts at about 80°C and forms crystals upon cooling. CAUTION: USE GOGGLES WHILE LIGHTING BURNER OR HEATING ANYTHING IN A TEST TUBE. ALWAYS POINT THE OPEN END OF THE TEST TUBE AWAY FROM YOURSELF AND OTHERS.

B. Light and adjust the burner to give a blue flame about six centimeters high.

C. Attach the test tube holder to the test tube near its open end.

D. Heat the test tube until the stearic acid is completely melted. Hold the test tube about four centimeters above the flame. Move the test tube around in the heat to allow all parts of the stearic acid to be heated. The test tube will blacken if the burner is not properly adjusted. If this happens, stop heating and wipe the test tube clean with a dry paper towel. Then adjust your burner to give a pale blue flame.

E. When the stearic acid is completely molten, pour about half of it onto an index card. Pour the remaining molten material into the small plastic spoon. Prop up the handle of the spoon so that the molten material will not spill.

1. Which sample is cooling faster?

F. Look closely at the sample on the index card while the other sample continues to cool.

2. Describe what you see.

G. Locate one of the longest crystals on the card. Measure its length, using a metric ruler.

3. What is its length in millimeters?

H. After the stearic acid in the spoon has hardened, look closely at its crystals.

4. Describe the shape of the crystals.

5. In general, how do these crystals compare in size to the ones made on the index card?

6. What is the length of the largest crystal you can find in the spoon sample?

7. How does the rate at which stearic acid cools affect its crystal size?

CONCLUSION:

1. Based on your observations in this activity, how do you think the rate at which lava cools affects crystal size in igneous rock?

2. What size crystals would you expect to see in lava that has poured out across the earth's surface and hardened?

3. What size crystals would most likely occur when magma cools and hardens inside the vent of a volcano?

9-2. Sediments

At the end of this section you will be able to:

- ☐ Define *sediment.*
- ☐ Describe three ways that sediments are formed.
- ☐ Name one rock formed by each of the three kinds of sediments.
- ☐ Explain how rocks can give information about the earth's past.

The coal you see being mined in Fig. 9-10 was formed from plant material laid down long ago in ancient swamps. Dead plant material builds up in swamps because the lack of oxygen in the water prevents decay. The organic sediment has been pressed between layers of rocks until it has become a rock layer itself. The exact cause of the formation of the original sediments may never be known. However, one thing is certain. In the future, coal most likely will continue to be a major source of energy. It has been estimated that, within the continental United States, the earth's crust holds more energy in coal than the total known earth reserves of oil. We need only to learn how to use this source with minimum damage to our environment.

Fig. 9-10 Coal mining in the United States.

FORMATION OF SEDIMENT

Scientists believe there is evidence showing that about four billion years ago most of the earth's surface was covered with molten rock. The first rock of the crust was igneous rock formed as this molten material cooled. Igneous rock formed the surface of the young earth. Since its formation, the igneous rock on all the continents has been exposed to the agents of erosion.

The chief agent of erosion is running water. Water flowing over the surface can pick up small pieces of rock. Also, small amounts of many of the minerals in rock dissolve in the water. The small rock pieces and dissolved minerals carried by running water are called a *load.* The flooding stream in Fig. 9–11 is carrying a large load. Every stream, from the smallest trickle to the biggest river, carries a load. For example, near its mouth the Mississippi River moves a load of about 15,000 kilograms each second.

The amount of material carried by a stream depends on how fast the water is moving. Increasing the speed of the water in a stream increases the size of the load it can carry. When a stream flows into a larger body of water, such as a river, lake, or ocean, it slows down. When it slows down, the stream can carry less material. Part of the load falls to the bottom. The material that settles at the bottom of the water is called **sediment** (**sed**-ih-ment). *Sediment* is deposited where streams of running water slow down.

Sediment Any substance that settles at the bottom of water.

Fig. 9–11 This flooding stream is carrying a large amount of sediment.

The rock pieces in sediments are not all the same size. The smallest rock fragments make up what is commonly called *mud*. Mud is the most abundant of all sediments. It is a mixture of clay and silt particles. Larger pieces of broken rock make up *sand*. One-fourth of all sediment is sand. Coarse pieces of gravel are also found in sediments. Gravel sediments are abundant in some places, but usually make up only a small portion.

Sediments are usually deposited in layers when they settle at the bottom of a body of water. Fig. 9–12 shows an example of such layers. Generally, the larger rock particles are the first to fall to the bottom. Gravel settles before sand, and sand before silt and clay. Because the different particles settle at different rates, layers are formed.

Enrichment: Clay comes from the weathering of feldspar.

Fig. 9-12 Sedimentary rock showing its many layers.

SEDIMENTARY ROCK

Most sediments collect very slowly. Many thousands of years may be needed to build a layer of sediment that is only a few centimeters thick. Each layer will grow in thickness until conditions change. For example, a layer of mud might settle in quiet water that is not moving. When the water begins to move, the layer of mud stops growing. When the water becomes quiet again, a new mud deposit is started.

Layers of sediment usually become covered by other layers. Particles are squeezed together by the weight of the layers above. Sometimes, minerals such as quartz or calcite fill the space between the particles. When this happens the sediments are cemented together. The rock made when a layer of sediment becomes solid is called a **sedimentary rock.**

Sedimentary rock A kind of rock made when a layer of sediment becomes solid.

TYPES OF SEDIMENTARY ROCK

1. Cementing and compaction. The most common of all *sedimentary rocks* is *shale*. See Fig. 9-13 *(top left)*. Shale is formed when clay particles are compacted, or squeezed together. When grains of sand become cemented together, *sandstone* is formed. See Fig. 9-13 *(top right)*. Gravel and large, round particles become cemented together to make *conglomerates* (kun-**glom**-uh-ruts). See Fig. 9-13 *(bottom)*.

Fig. 9-13 (top left) *Shale;* (top right) *sandstone;* (bottom) *conglomerate.*

2. Chemical. The second kind of sedimentary rock forms from *chemical sediments*. A body of water may contain a large amount of dissolved minerals. Chemical sediments form when the water evaporates and the dissolved mineral matter remains as a deposit. For example, if the water supply to a lake is shut off, the water in the lake eventually will disappear. As the water evaporates, the lake becomes a salt lake. When the lake is completely dry, a deposit of minerals remains. *Gypsum* (**jip**-sum) and *rock salt* are formed in this way. See Fig. 9-14.

3. Organic. The third type of sediment is called *organic* (or-**gan**-ik) *sediment*. The word "organic" means "life." Organic sediment is made from the remains of living things. Clams, oysters, snails, coral, and some types of fish are among the animals that live in salt water. Each of these has a hard shell or a skeleton. When these animals die, their soft body parts decay. Their hard parts, however, fall to the bottom of the sea. Layers of such sediment become a sedimentary rock called *limestone*. Limestone also may contain mud and sand. See Fig. 9-15. Another organic sediment was formed from plants that lived in swamps millions of years ago. When these plants died, their remains eventually formed coal. Coquina (koh-**kee**-nuh) is also a sedimentary rock that formed from shells that hardened together.

Reinforcement: Show the students a piece of chalk. Chalk is a variety of limestone.

Reinforcement: Use acid test on chalk, limestone, and seashells to show similarity.

Enrichment: Limestone may also form as a chemical precipitation.

Enrichment: Peat is the first stage of change from plant remains to coal. As more sediment accumulates, more pressure changes peat to lignite (brown coal). Continued pressure and heat produce bituminous (soft coal) and finally anthracite (hard coal).

Fig. 9-15 Limestone is formed from organic sediments.

Fig. 9-16 This sandstone was once an ancient desert. The layers of sediment are clearly seen.

HISTORY FROM SEDIMENTS

Sedimentary rocks are an important tool in the study of the earth's history. They are a record of the conditions in the earth's past. Coal is a sedimentary rock that comes from the remains of plants that lived in swamps. The coal found in Antarctica shows that the climate of that land was different in the past.

Sedimentary rocks containing molds of seashells tell of seas once spread over the present continents. Sedimentary rocks containing glacial deposits tell of former glaciers. Sandstone gives evidence of ancient deserts. See Fig. 9–16. Sedimentary rocks also tell about past life. The fossil remains of many plants and animals are preserved in the sedimentary layers.

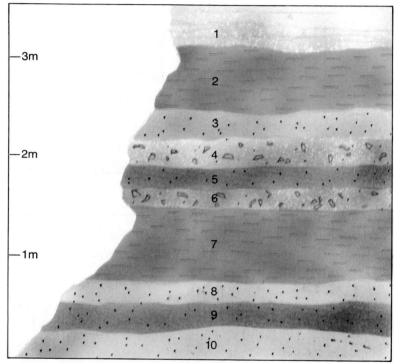

Fig. 9–17 Profile of a section of land. Layer 1: gypsum; layers 2,7: shale; layers 3,5,8,9,10: sandstone; layers 4,6: conglomerate.

One of the first things a student in geology learns to do is to study drawings that represent profiles of measured sections of land. Fig. 9–17 shows such a drawing. The layers of sedimentary rock differ in thickness and composition. Often samples of the rock from each layer are taken back to the laboratory to be studied. One property that is usually checked is the grain size. The grain size gives information about the movement of the water that deposited the sediment. For example, large grains would show the water to have been moving rapidly. Only the largest grains would have settled. The numbers in Fig. 9–18 give the grain size of the samples in order, from the finest grains found in shale samples to the coarsest grains found in conglomerates. Also shown is a graph of the grain size of each sample alongside the layer in which the sample was found.

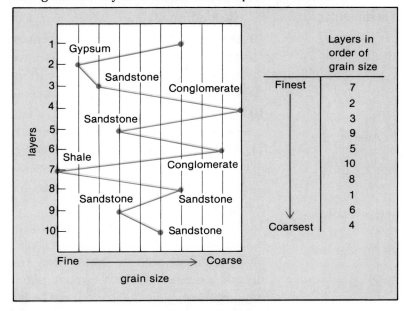

Fig. 9–18 Grain sizes of the samples.

SUMMARY

The earth's surface is no longer completely covered with a layer of igneous rock. There are processes at work on the surface that reduce igneous rock to sediments. Sediments are moved from one place and deposited in another. Layers of sediments can become changed into many different kinds of sedimentary rocks. These rocks contain a record of much of the earth's history.

QUESTIONS

Use complete sentences to write your answers.

1. What is meant by "sediment"?
2. Describe three ways that sediments may form and give an example of a rock made from each kind.
3. Give three examples of what sedimentary rocks can tell us about earth's history.

1. A sediment is any substance that settles at the bottom of water.
2. (a) Rock material settles at the bottom of water to produce sandstone or conglomerate. (b) Minerals are left behind when water evaporates. Gypsum is a rock made from such a sediment. (c) Organic material may settle at the bottom of water to produce coral, limestone, or coal.
3. They tell that the climate of Antarctica was different; that seas once spread over the continents; that glaciers advanced and retreated; etc.

INVESTIGATION

See teacher's commentary for teaching hints, safety, and answers to questions.

STUDYING SEDIMENTARY ROCK

PURPOSE: To study a profile of sedimentary rock layers.

MATERIALS:
paper   pencil

PROCEDURE:

A. Look at Fig. 9–17, page 252. Answer the following questions.

   1. What kind of rock is found at the 1-meter level?

   2. What kind of rock is found just above the 2-meter level?

B. Study Fig. 9–17 and Fig. 9–18. Answer the following questions.

   3. What would cause the shale found at the 1-meter level to be slightly finer grained than the shale located near the top of the 2.5-meter level?

   4. Which one of the sandstone layers has the largest grain size?

   5. Which kind of rock layer shown is a chemical sediment?

   6. Which rock layer was originally deposited as a very fine silt?

C. Much more can be learned about the history of a section of land. Remember that larger grains of sediment require faster-running water to carry them. Study Fig. 9–18 and refer to Fig. 9–17 to answer the following questions.

   7. What kind of rock was formed by sediments transported by the fastest-moving water?

   8. How many times during the history of this section did the water show such a swift current?

   9. As the grain size of the sedimentary rock became finer, what must have happened to the speed of the current of water that delivered the sediment?

  10. Which kind of rock layer was laid down when the water was very calm?

  11. Calm water with little current is typical of large, shallow inland seas. How many times did the water become this calm?

  12. How do you explain the layer of gypsum just above the last layer of shale?

D. Shallow inland seas or lakes often last decades, during which time many different forms of life may spring up in the water. When water no longer drains into the low area, the sea will evaporate.

  13. What kind of sediment would show that the water had completely evaporated from an inland sea or lake?

  14. How many times did this happen for the profile shown?

  15. What reason can you give that sea life may not have existed in this inland sea or lake?

CONCLUSIONS:

   1. Starting with the oldest rock layer, list its number and name the sediment from which it formed.

   2. Give the conditions that were present to cause each of the sediments.

9-3. Changed Rocks

At the end of this section you will be able to:

- ☐ Define *metamorphic*.
- ☐ Name two agents that produce metamorphic rock and explain how each may occur.
- ☐ Describe the *rock cycle*.

Skill: The study of the formation of metamorphic rock reinforces cause and effect relationships.

Can you identify the lump of material shown in the middle of the photograph in Fig. 9-19? It is the remains of a car. If you look carefully, you can see that the two headlights have been pushed together. They are now located next to each other. Applying pressure to the car certainly changed it! Rock in the earth's crust can be changed in a similar way. In this section you will learn about the ways in which rock is changed.

Fig. 9-19 Minerals in rocks can be crushed together the way these cars are crushed.

METAMORPHIC ROCK

Look again at the remains of one of the wrecked cars shown in Fig. 9-19. Is there anything in the lump of twisted metal that tells you what kind of car it was? Now look at Fig. 9-20. The rock, like the car, has been squeezed by great pressure. The rock was once a sedimentary rock. The sediments have been flattened and pressed tightly together. The original sediments can still be seen. The rock is a **metamorphic rock.** Metamorphic means "changed in form." Great pressure has changed a sedimentary rock into a *metamorphic rock*.

Enrichment: Metamorphic rocks are generally found on continents.

Metamorphic rock Rock that is changed by the action of heat or pressure without melting.

ROCKS

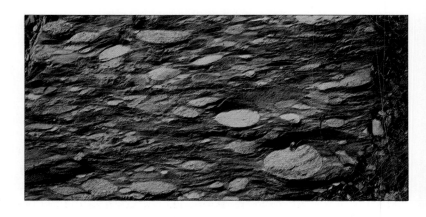

Fig. 9-20 Great pressure has squeezed these rocks together.

Pluton Any large, intruded rock mass formed from magma.

MAKING METAMORPHIC ROCK

Both igneous and sedimentary rocks can become metamorphic rocks. This can occur when heat, or pressure, or both act on the rock. Many igneous rocks and all sedimentary rocks are formed on or near the earth's surface. Then how are they put under pressure great enough to change them? There are several ways this can be done. For example, rocks on the surface may be covered by layers of sediment. The weight of these layers might provide the needed pressure. Movements of the crust also twist, flatten, and stretch the mineral grains that make up the rock.

Heat is another agent that acts to change rocks. The interior of the earth is hot. Rock buried under several kilometers of earth can be affected by heat as well as by pressure. Magma may push into existing rock layers, causing great pressures as well as bringing heat with it. The magma, after giving up its heat, hardens into rock and is known as a **pluton** (**ploo**-tahn). Rock under high pressure can be heated to a high temperature without melting. When this happens, the rock can change. The minerals in that rock may grow into large crystals. New minerals may form as atoms rearrange. Sometimes gases trapped in the rock are released. New gases or liquids from surrounding rocks may mix with the rock material. Rocks changed in this fashion are often hard to trace to their original form.

Metamorphic rock is often found near mountains. If a mountain is the result of a *pluton,* hot magma comes in contact with rock material. During this kind of mountain building, rock changes result from both heat and pressure. Such changes formed most of the world's important deposits of ore minerals.

Folding and faulting often occur with mountain building. These processes twist and squeeze existing rock material. Metamorphic rock formed at the roots of mountains is exposed only after millions of years of erosion.

Quartzite (**kwort**-zite) is the metamorphic rock made from sandstone. Sand grains in quartzite are fused together more firmly than they are in sandstone. Quartzite is harder than sandstone. *Slate* is a metamorphic rock made from clay or shale. Slate can be found as black, red, green, or blue rock. *Marble* is another metamorphic rock that can vary in color. Marble can be made from limestone. The color of marble depends on the minerals mixed in with the limestone. *Gneiss* (**nice**) is a metamorphic rock with a streaked or banded appearance. See Fig. 9–21. Gneiss usually has the same minerals as granite. However, the changes that produce gneiss are so great that it is hard to determine the original rock material. Without the bands, most gneiss looks like granite. Many different rocks probably can form into gneiss. In the same way, fine-grained rocks can be formed into the metamorphic rock called *schist* (**shist**).

Fig. 9–21 (top left) *Quartzite;* (top right) *slate;* (bottom left) *marble;* (bottom right) *gneiss.*

THE ROCK CYCLE

The formation of metamorphic rock is one step in an endless process that has been going on for billions of years. The process is called the **rock cycle.** This process consists of the mixing and reusing of the earth's rock material. For example, hot magma exists beneath the earth's surface. When magma is forced to the surface, it is called lava. Lava cools and hardens, forming extrusive igneous rock. Agents of erosion act on this rock. The magma in the crust may also harden below the surface, forming dikes, sills, or batholiths. These intrusive igneous bodies are brought to the surface by erosion or crustal movement. Agents of erosion act on these bodies as soon as they reach the surface.

Erosion results in sediments. Sedimentary rock is formed from deposits of sediments. This rock becomes buried by other layers. Sedimentary rock may remain buried, or may be brought to the surface, where it is eroded.

Any rock that is buried below the surface can be changed by heat and pressure into metamorphic rock. The new metamorphic rock eventually is brought to the surface. Erosion then acts on this form of rock.

Scientists believe that a large portion of the *rock cycle* is taking place at the edges of the crustal plates. Magma is coming to the surface along the mid-ocean ridge and near trench areas. In the trenches, old rock is being forced down into the mantle. In the mantle, it is melted and mixed with other rock material. This material may then make its way back to the crust as new lava or magma. Such recycling has been going on since the earth was formed. Fig. 9–22 represents the rock cycle.

Rocks can take shortcuts through the rock cycle. For example, igneous rock can be changed into metamorphic rock. It does not have to weather first. Exposed sedimentary rock may weather and become part of new sedimentary rock.

The rocks of the earth's crust have gone through the rock cycle many times. This is the reason that rocks near the earth's surface never have ages greater than 3 billion years, while the earth is probably 4.5 to 5 billion years old. All original rocks of the earth's crust have been changed by the rock cycle since they were first formed. The age of rocks in the crust shows only how long they were formed as part of the rock cycle.

Rock cycle The endless process by which rocks are formed, destroyed, and formed again in the earth's crust.

Skill: Study of the rock cycle reinforces sequencing skills.

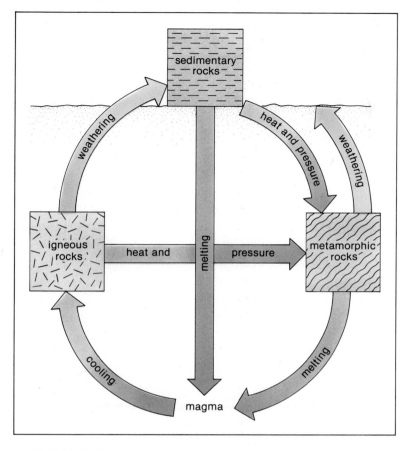

Fig. 9-22 The rock cycle.

SUMMARY

Metamorphic rock is formed by the action of heat and pressure on existing rock. Slate, gneiss, marble, and schist are some common metamorphic rocks. The formation of metamorphic rock is one step in the rock cycle. The rock cycle is the name given to the process that recycles the rock materials of the earth.

QUESTIONS

Use complete sentences to write your answers.

1. What does "metamorphic" mean?
2. Describe two ways that metamorphic rock may be formed.
3. Name two common metamorphic rocks and the original rock from which each was made.
4. Draw a diagram of the rock cycle.

1. Metamorphic means "changed in form."
2. Metamorphic rock may be formed by great pressure squeezing the original rock particles into a new form. Metamorphic rock also may be made by heat great enough to cause a new form of rock.
3. Quartzite, formed from sandstone; slate, formed from clay or shale; etc.
4. See Fig. 9-22.

INVESTIGATION

See teacher's commentary for teaching hints, safety, and answers to questions.

THE FORMATION OF METAMORPHIC ROCK

PURPOSE: To use clay to demonstrate one way that rocks are changed.

MATERIALS:

modeling clay, two 1-cm cubes (different colors)

bread, two 1-cm squares

waxed paper

eyedropper

water

PROCEDURE:

A. Form about ten small spheres from each piece of clay. These represent pieces of sediment. Each color represents a different mineral.

B. Put the spheres together loosely in a pile. Call this the original form of the sediment.

C. Press the pile of clay spheres with the palm of your hand, but not hard enough to remove all the space between the spheres.

   1. Which would the pile represent, sedimentary or metamorphic rock?

D. Take apart a small part of the pile and inspect the spheres.

   2. Describe the spheres now.

   3. Compare the space between particles in this flattened model with the space in the original form.

E. Press the remaining pile of spheres until their shapes are completely destroyed.

   4. Compare the appearance of the clay now with the form it was in before.

F. Look at Fig. 9–20, page 256.

   5. What evidence do you see that this rock has been under great pressure?

G. Scientists often work with models and try to discover things in an indirect way. For example, to test the effect that great pressure may have on the amount of water a rock can hold, one might do the following experiment.

H. Place a 1-cm square of bread on waxed paper. Using an eyedropper, release water one drop at a time onto the bread. Count the number of drops the bread can soak up before water starts coming from the lower edges onto the waxed paper.

   6. How many drops did you count?

I. Flatten a second 1-cm square of bread until you have pushed most of the air from it. This reduces the space between the particles of bread, just as pressure does in rock.

J. Using the procedure in step H, count the number of drops this piece of bread holds.

   7. How many drops did it hold?

CONCLUSIONS:

   1. What effect does pressure have on the grain size and shape of particles in rock?

   2. What evidence would you look for to tell whether the rock you have is the sedimentary rock, sandstone, or its metamorphic form of quartzite?

   3. Which would hold more water, sedimentary rock or its metamorphic form?

CAREERS IN SCIENCE

METALLURGICAL ENGINEER

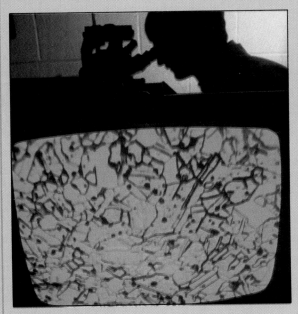

Metallurgical engineers are concerned with practical ways of changing raw metals into useful products. Some of these engineers study methods of removing metal from ores such as iron ores. Other engineers are interested in how these metals can be changed into finished products. These products include space exploration devices, medical equipment, and power plant machinery. Metallurgical engineers study the properties of metals and develop ways of casting metal and shaping it. These processes are used in the industrial production of steel and other metals, as well as in casting operations used by sculptors of metal.

Metallurgical engineers have at least a bachelor's degree, and many have advanced training. They are employed by mining companies and in industries that manufacture machinery, electrical parts, and other types of equipment. For more information, write: Engineers' Council for Professional Development, 345 East 47 Street, New York, NY 10017.

STONEMASON

Do you like the feel of a piece of marble or limestone in your hands? Do you enjoy building things? Stonemasons work with different types of stone, assisting in the construction of large buildings such as offices, and hotels, and churches. They also work on smaller jobs such as floors, fireplaces, and window sills. The artistry in the face of a building is often the work of the stonemason. To make sure the stones fit together, masons must measure them accurately. Then they must carefully position the stones, using mortar to hold them together. Masons must be concerned with the safety and stability, as well as the appearance, of their work.

Individuals who want to become masons usually start as apprentices. In the apprenticeship program, they learn skills such as sketching and how to read blueprints. Jobs for stonemasons are available with construction companies. For further information, write: Associated General Contractors of America, Inc., 1957 E Street N.W., Washington, DC 20006, or visit a construction site in your own area.

SCIENCE INPUT

Rocks, as they form and decay, go through constant changes. The specific changes depend on the environment in which the rocks are located. This ongoing effect of changing from one rock type to another is called the rock cycle. Understanding the nature of this cycle is useful in determining the geologic history of an area. Environmental conditions act on the features of the land to create their varying forms.

COMPUTER INPUT

Computer programmers draw an outline of their program using a system of symbols, lines, and terms called a flowchart. A flowchart provides a shorter, faster way of communicating than sentences. Most important, however, it lets both the programmer and other users of the software keep track of the steps of the program. A flowchart outlines a series of paths and processes that provide the instructions for the program.

The flowchart in this chapter represents the rock cycle. It outlines the various paths of rock originations and rock decay, providing a short review of some of the material you learned in this chapter. The program has been written on the basis of the flowchart. Examine the chart very carefully and try to answer the questions *before* entering and running the program. To make it easier to study, draw a larger version of the flowchart. The glossary will help you to keep track of what each step means.

FLOWCHART QUESTIONS

1. What type of rock is formed from molten rock? What types are formed from sediments? What form does partially melted rock take?
2. The future of a rock is greatly affected by whether or not it lies on the surface. How does that factor influence the rock's future?
3. The flowchart has no end or stop point. Why?

WHAT TO DO

Using your interpretation of the flowchart, try to answer the flowchart questions.

Enter Program Flowchart into your computer. When you run it, it will ask you questions about different types of rock. Choose one of the alternatives. Compare each screen step of the program with the flowchart. Compare the answers you gave to the flowchart questions with the answers the computer gives. Run the program several times to help you review the steps in the rock cycle.

GLOSSARY

 Start or stop.

 Shows the direction the program will follow.

 A single step in the program.

Input or Output.

 A decision point: A question must be answered, or two things are compared. "Y" means yes and "N" means no.

PROGRAM

```
100   REM FLOWCHART
110   REM REPLACE THIS STATEMENT
      WITH CLEAR SCREEN STATEMENT
120   PRINT "HOT MOLTEN ROCK"
130   PRINT "COOLS AND FORMS
      IGNEOUS ROCK"
140   PRINT "DIASTROPHISM OCCURS"
150   INPUT "IS THE ROCK AT THE
      SURFACE (Y/N)? ";A$
155   PRINT
160   IF A$ = "N" THEN GOTO 210
165   IF A$ <> "Y" THEN GOTO 150
170   PRINT "WEATHER AND EROSION
      OCCUR"
180   PRINT "SEDIMENTS FORM"
190   PRINT "SEDIMENTARY ROCK
      FORMS"
200   GOTO 140
210   PRINT "HEAT AND PRESSURE ARE
      APPLIED"
220   INPUT "DOES THE ROCK MELT (Y/
      N)? ";A$
225   PRINT
230   IF A$ = "Y" THEN GOTO 120
235   IF A$ <> "N" THEN GOTO 220
240   PRINT "THE ROCK IS CHANGED"
250   PRINT "METAMORPHIC ROCK IS
      FORMED"
260   GOTO 140
```

PROGRAM NOTES

A clear screen statement is an instruction to the computer to remove all data from the TV screen or monitor. Not all computers use the same commands. Consult your user's manual for the clear screen statement appropriate to your computer.

FLOWCHART

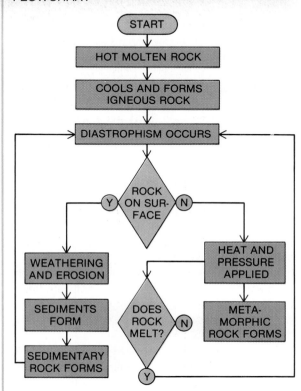

BITS OF INFORMATION

Can you imagine a computer small enough to sit on your lap, a computer that would let you write letters, or make notes, and even do programming? Such computers are available and can weigh as little as four kilograms. They are not yet a replacement for regular computers. They have small screens and can handle only a small range of software. But they certainly can be helpful for the many people who spend a great deal of time traveling and wish to work en route.

CHAPTER REVIEW

VOCABULARY

On a separate piece of paper, match each term with the number of the phrase that best explains it. Use each term only once.

rock   sedimentary rock   rock cycle   pluton
igneous rock   batholiths   sills
sediment   metamorphic rock   dikes

1. Underground bodies of igneous rock in cracks crossing layers of existing rock.
2. Any large intruded rock mass formed from magma.
3. A piece of the earth made up of two or more minerals.
4. May make up the cores of mountain ranges.
5. The endless process by which rocks are formed, destroyed, and formed again.
6. Rock made when a layer of sediment becomes solid.
7. Rock that has been changed by heat or pressure.
8. Material that settles at the bottom of water.
9. Rocks formed when molten material cools.
10. Bodies of igneous rock formed underground that run parallel to layers of existing rock.

QUESTIONS

Give brief but complete answers to each of the following questions. Unless otherwise indicated, use complete sentences to write your answers.

1. What is an igneous rock?
2. Name the two groups of igneous rock and tell how they differ in the way they form.
3. List three rocks formed from lava.
4. How can you tell whether an igneous rock has cooled rapidly or slowly?
5. Name and describe three kinds of igneous rock structures formed underground.
6. What is the most abundant of all sediments? Name and describe two other sediments from which rock is formed.
7. How does a sediment become sedimentary rock?
8. Name two common sedimentary rocks formed from rock fragments.

9. What is chemical sediment? Name one rock formed from chemical sediment.
10. How did coal form?
11. What is organic sediment?
12. How does heat change rock?
13. What two agents cause metamorphic rock?
14. How do igneous and metamorphic rocks differ in the way they form?
15. Does a rock cycle occur on the moon? Explain your answer.

APPLYING SCIENCE

1. Organize a rock-hunting party. This can be an actual outing to gather rocks or a hunt through magazines and other pictorial sources to locate pictures of rocks. Prepare a display of your results, grouping the minerals and rocks according to whether they are igneous, sedimentary, or metamorphic. Label as many as possible with their scientific names. Use a mineral identification key.
2. Make a list of the rocks and minerals that were used to construct your school building.
3. Make a poster-size drawing of the rock cycle. Attach to it, in the proper places, examples of igneous, sedimentary, and metamorphic rock.
4. Many public buildings are made of stone. Locate as many as you can in your area. Determine what stone each is made of, whether the stone is igneous, sedimentary, or metamorphic. Inside the buildings there may be other types of stone for floors, stairways, or walls. Be sure to list all types you find. Find out if any of the stones were taken from the local area.

BIBLIOGRAPHY

Bates, Robert L., and Julia A. Jackson. *Our Modern Stone Age*. Los Altos, CA: Kaufman, 1981.

Dietrich, R. V. *Stones: Their Collection, Identification, and Uses*. San Francisco: Freeman, 1980.

MacFall, Russell P. *Rock Hunter's Guide; How to Find and Identify Collectible Rocks*. New York: Crowell, 1980.

Radlauer, Ruth. *Grand Teton National Park*. Chicago: Children's Press, 1980.

CHAPTER

10

Everything we use, wear, eat, or drink comes from the earth. The earth has always been able to supply us with building materials, clothing, food, and water. Will this always be true?

EARTH RESOURCES

CHAPTER GOALS

1. Tell what is meant by a natural resource and explain why some are renewable and some are nonrenewable.
2. Discuss several important problems having to do with population growth.
3. Explain what a local water budget is and the importance of protecting the watersheds as well as preventing pollution of air and water resources.
4. Tell the difference between a rock and an ore.
5. Describe how mineral deposits form and why it is necessary to conserve our mineral resources.
6. Discuss our dependence on fossil fuels for energy.
7. Explain the role of nuclear energy and other alternative energy resources in our future needs for energy.

Enrichment: Water is a very important resource. An average of 7,000 L of water per day (two million liters per year) is used by each person. The sources of fresh water are rivers, lakes, and underground water. These sources are constantly being replenished by rainfall. The chapter opener shows water running down Weeping Rock, in Zion National Park, Utah.

10-1. People and Resources

At the end of this section you will be able to:

☐ Define *natural resources*.
☐ Explain how the growth of the earth's population affects the food supply.
☐ Describe three kinds of pollution.

You are now living on a kind of space station. It is called the planet Earth. A space station like the one shown in Fig. 10-1, page 268, must supply all the needs of its passengers. Similarly, the earth is the source of the materials needed to support its population. See Fig. 10-2, page 269.

NATURAL RESOURCES

The earth has everything people need to live. It gives us food and the materials needed to make clothing and shelter. The fuels that furnish energy come from the earth's crust. Any material from the earth that can be used in some way by people is called a **natural resource**. *Natural resources* are made into various products. For example, this book was made from many natural resources. The paper came from forests. Chemicals

Natural resource Any of the substances from the earth that can be used in some way.

Fig. 10-1 In the future, the space shuttle will bring supplies to a space station like this. Planet Earth is also like a large orbiting space station.

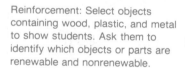

Reinforcement: Select objects containing wood, plastic, and metal to show students. Ask them to identify which objects or parts are renewable and nonrenewable.

from rock and air helped to change the wood into paper. Minerals supplied the metals used for printing the book. The energy used for printing came mostly from coal or oil. It is easy to forget that the raw, natural materials that are used for our basic needs come from the earth. The rows of food in a supermarket do not look like the fields in which the food grew. We do not think of cotton fields or herds of sheep when we shop for jeans and sweaters. Air in buildings is heated and cooled. Often, there is little thought of the fuels needed to make the artificial environments in which we spend much of our time. It is important to look at the natural environment to remember how natural resources are used in everyday life.

There are two kinds of natural resources. Some resources, such as forests, wildlife, and water, are *renewable*. A renewable resource can be replaced in a short time after being used. Other earth resources are *nonrenewable*. These resources are not replaced or are replaced very slowly. For example, it may take millions of years to replace the supplies of minerals and fossil fuels, such as oil, taken from the earth's crust. Thus nonrenewable resources must be used carefully.

THE EARTH'S POPULATION AND FOOD SUPPLY

Are there enough resources to feed the world's people? The world's population is growing rapidly. In 1700, there were about 600 million people on earth. One hundred years later, the population had increased to 1 billion. At that rate, the population of the earth doubled every 150 years. However, by 1980 the population was 4.5 billion. See Fig. 10-2. The earth's population is now growing at a rate that will cause it to double every

33 years. If the population continues to grow at the present rate, in 66 years there will be four additional people for each person alive today. This rapid growth has been brought about largely by better control of disease and a more dependable supply of food.

Thousands of years ago, our ancestors learned to use the land to grow their food. They used muscle power and simple tools. Those farming methods changed very little from ancient times until about 150 years ago. Then the machines that were developed in the Industrial Revolution were first used for agriculture. The machines made it possible for a single person to farm many acres of land. See Fig. 10-3. Farmers also began using artificial fertilizers. This greatly increased the amount of food they could grow. They also began using chemicals to control pests that damaged crops. In countries such as the United States, the average farmer today produces enough food for 30 people. But improved farming methods have not reached all parts of the world. For example, an average farmer in India produces only enough food for several people. One way of feeding the world's growing population is by using better farming methods in all parts of the world.

The amount of land that can be used for growing food is limited. About 60 square meters of farmland are needed to grow food for one person. This is an area about the size of a tennis court. It takes about 10 percent of the earth's land to feed the nearly five billion people now on the earth. Much of the land contains mountains, deserts, and land not useful for farming. A

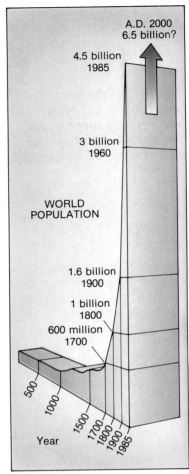

Fig. 10-2 As this graph shows, the earth's population has more than doubled during the last century.

Enrichment: It takes about 4.5 kilograms of grain to produce 0.5 kilograms of beef.

Enrichment: Mention that vegetables and grains can now be grown in tanks that contain special solutions of chemicals dissolved in water. This process is known as hydroponics.

Fig. 10-3 Farm machines have increased our ability to grow food.

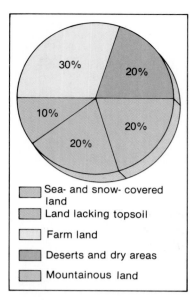

30%

20%

10%

20%

20%

☐ Sea- and snow- covered
   land
☐ Land lacking topsoil
☐ Farm land
☐ Deserts and dry areas
☐ Mountainous land

Fig. 10-4 This chart shows how much of the earth's surface is utilized for farming.

Enrichment: Since 325 B.C. farmers have used powdered fish to feed livestock.

Pollution The adding of harmful substances to the environment.

total of 30 percent of the earth's surface can be used for farming. See Fig. 10-4. However, many of these areas have climates or other conditions that require new farming methods to make the land productive. Scientists have been able to develop new kinds of crops that grow faster, resist disease, and have more food value. For example, rice is one of the most commonly grown foods in the world. But rice is difficult to grow, has diseases, and often produces a grain that lacks essential food values. By using improved crops, along with the best methods of farming of all available land, the earth could probably provide food for a population of 40 to 50 billion people.

What about resources from the sea? Oceans cover more than twice as much of the earth's surface as the land does. Plants grow as well in the sea as they do on the land. Therefore, the sea produces about twice as many plants as the land. But few people eat plants from the sea. Most of the seafood that people eat is fish. These fish make up only a small part of the food from the sea. But the plants from the sea that might be used for food are too small to be farmed. Increasing the food supply from the oceans would require a special kind of farming called "ocean farming." Ocean farming could be done by fencing off and fertilizing large areas of the sea.

The use of better farming methods on more of the land may also make it possible to feed the world's population for the next century. However, if the population continues to double every 33 years, there will be a time when enough food cannot be grown. Then population growth will stop, mainly because people will starve. As the world's population continues to grow, better farming methods will be needed to continue to feed the world's people.

PEOPLE AND THE ENVIRONMENT

Lack of food is not the only problem that is caused by increased population. One of the most serious problems is the pollution of our most valuable natural resources—air, water, and soil. The adding of harmful substances to the earth's environment is called **pollution.** There are three major kinds of *pollution:* air pollution, water pollution, and soil pollution.

1. Air. Automobiles, power plants, factories, the burning of fuels for heating, and burning of trash are major causes of air pollution. Most coal and petroleum contain sulfur. When the

Enrichment: There are many state laws regulating the burning of trash, fuels for heating, and car emissions. The EPA regulates many of these activities.

Fig. 10-5 The burning of fuels releases gases that cause air pollution.

Enrichment: Partly burned gasoline contributes to the formation of smog.

coal or oil is burned, deadly sulfur dioxide gas is put into the air. See Fig. 10-5. Breathing air containing sulfur dioxide can cause health problems. Sulfur dioxide also causes *acid rain*. Rain and snow mix with sulfur dioxide to form an acid. The acid rain and snow cause damage to resources such as forests and lakes. Air pollution also comes from such things as bits of rubber that spin off automobile tires and gases given off in car exhaust.

2. Water. Water can be polluted in three ways. One of the most dangerous kinds of water pollution occurs when untreated sewage is dumped into streams, lakes, or the ocean. This waste may add disease-producing bacteria and viruses to the water. Sewage can also pollute water in another way. Chemicals in the sewage can help plants such as algae to grow. Algae float on the surface of ponds and streams. When the algae and other plants decay, they use up the oxygen from the water. Other plants and animals no longer can live in the water because of the lack of oxygen. Water can also be polluted by poisonous chemicals, such as those used on crops to control pests. These chemicals can be washed off the land and enter the water supply. These poisons affect the living things in the water. They also are harmful to the people who eat the food from the polluted waters.

Enrichment: DDT, an effective insecticide, is no longer used because its stability allowed it to accumulate to dangerous levels in the soil. DDT washed into rivers, where fish became contaminated.

3. Soil. A third form of pollution is caused by the buildup of harmful substances in the soil. Some soil pollution comes from the air. Most, however, is caused by the chemicals used to control pests and plant diseases. See Fig. 10-6. Small amounts of chemicals, such as *ethylene dibromide* (EDB) have been found

Fig. 10-6 Chemicals used to control pests and diseases on crops can remain in the soil.

in water wells in Florida. EDB has been found to cause cancer in laboratory animals. Many poisonous chemicals in containers are buried in landfills. Some of these chemicals have leaked into ground water. Ground water supplies drinking water for more than half of the population of the United States.

Chemicals are not the only pollutants. The trash you throw away adds to pollution. Trash is only one part of the pollution problem that you have some control over. Burning trash may pollute the atmosphere. Dumping it without careful planning may pollute the land around you. Unless pollution is controlled, our natural resources will be destroyed. It may take many years before they are renewed.

SUMMARY

Like a spacecraft, the earth must supply everything needed by the world's population. However, this planet may be in danger of becoming overcrowded. Even if the earth is able to feed a large population, adequate food will do little good if the environment is polluted to dangerous levels.

QUESTIONS

Use complete sentences to write your answers.
1. What is a natural resource?
2. What is a renewable resource? Give an example.
3. Why are minerals called nonrenewable resources?
4. What is the main problem with population growth?
5. List three major kinds of pollution and give an example of each kind.

1. Any material from the earth that can be used in some way by people is called a natural resource.
2. A renewable resource can be replaced after being used. An example is a forest.
3. Minerals cannot be replaced or are replaced very slowly.
4. The main problem is supplying food for the increased population.
5. Air pollution—caused by burning coal and oil, which produce sulfur dioxide gas. Water pollution—caused by the dumping of untreated sewage into streams, rivers, lakes, or oceans. Soil pollution—caused by the use of insecticides, such as EDB, that wash into the ground water supply.

272

See teacher's commentary for teaching hints, safety, and answers to questions.

TRASH AND POLLUTION

PURPOSE: To determine the amount of trash produced in your town based on the measured amount of trash you make in a day.

MATERIALS:
pencil   paper

PROCEDURE:

A. Make a list of all the things you and your family throw in the trash each day. Include lunch bags, food storage bags, candy wrappers, cans, bottles, notebook paper, paper towels, napkins, etc.

B. Find the weight, in kilograms, of each item on your list. If you must guess, use the table below for some rough weights.

Examples	Weight (kg)
bottle, soft drink	.500 kg
can, soft drink, soda	.023 kg
lunch bag	.008 kg
candy wrapper	.006 kg
notebook paper (ea.)	.005 kg
newspaper	.500 kg
milk carton	.020 kg
napkin	.002 kg
plastic fork or spoon	.004 kg

C. Find the total weight, in kilograms, of the things you throw out each day.
    1. What is the total weight?

D. Individually, your waste may not seem like much. But you live in a community of many people. To get an idea of the amount of waste your community produces each day, multiply the amount of your waste by the number of people in the community. Your teacher will tell you the population of your community.

    2. What is the total weight, in kilograms, of the trash thrown out daily by your community?

    3. What groups of people may throw out more trash per day than students? What groups would produce smaller amounts of trash?

    4. A metric ton is 1,000 kilograms. How many metric tons of trash are produced by the people in your community each year?

E. In some communities it costs $40 to get rid of each metric ton of trash. Multiply the number of metric tons of trash made by your community by $40.

    5. How much would it cost your community to rid itself of the trash at $40 per metric ton?

F. A trash truck can haul about five metric tons a day to the disposal area. Divide the total number of metric tons of trash by five to find the number of trash trucks needed each day.

    6. How many trash trucks are needed each day?

CONCLUSIONS:

1. How many metric tons of trash were produced in your entire community daily?

2. Do you think it is likely that your community produces more trash than this? Explain your answer.

Fig. 10-7 Aluminum cans can be recycled.

Ore A rock that contains useful minerals and can be mined at a profit.

Mineral deposit A body of ore.

10–2. Mineral Resources

At the end of this section you will be able to:

☐ Compare ordinary rocks with *ores*.

☐ List three ways that mineral deposits may be formed.

☐ Using an example, show the need to conserve and *recycle* mineral resources.

Used aluminum cans are valuable. You can be paid for aluminum cans turned in to collection centers or aluminum recycling plants. See Fig. 10–7. However, no one will pay you for old "tin" cans. The "tin" cans are made up mostly of iron. Why are aluminum cans valuable but not cans made of iron? The answer to this question can be found by learning more about the way we find and use the earth's mineral resources.

WHAT IS A MINERAL DEPOSIT?

Some minerals are useful because they contain certain chemical elements that are valuable. For example, a mineral that contains the element copper might be useful. The copper metal could be separated from the mineral and used to make electric wires. Other minerals are useful just as they are found. Limestone, for example, is used to make cement. Any kind of rock that is useful because it contains a valuable mineral is called an **ore.** To be called an *ore,* the rock must be able to be mined at a profit.

Some chemical elements, such as aluminum, are found in many different minerals. Aluminum is such a common part of minerals that it is likely to be found in almost any rock you may pick up. On the other hand, some elements, such as copper, are much less common in minerals. Thus ordinary rocks are likely to contain only very small amounts of this element.

FORMATION OF MINERAL DEPOSITS

Rocks in certain places in the earth's crust may contain much larger amounts of a particular element or mineral than are ordinarily found. A body of rock that contains a large amount of an element or a mineral is called a **mineral deposit.** A *mineral deposit* is formed when changes in the earth's crust cause minerals to separate from their parent rock. The minerals are first separated. Then they are deposited within another body of rock, forming an ore. Most important mineral deposits are formed in one of three ways.

1. Dissolving. First, minerals can become separated from rocks by dissolving in hot water. The hot water may come from ground water that is heated by bodies of magma below the surface. Some hot water may also come from the magma itself. As the hot water moves through the rocks, it dissolves minerals. This hot mineral mixture then enters cracks in cooler rocks. As the water cools, the dissolved minerals settle out as solid material. This may form a mineral vein, or streak, running through rocks. See Fig. 10-8. Ores of copper, gold, silver, lead, and tin are often found in veins.

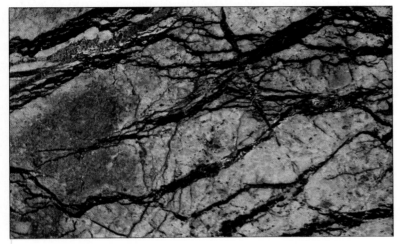

Fig. 10-8 This rock has numerous mineral veins of copper running through it.

In the past few years, scientists have found that the mid-ocean ridge is also a place where mineral deposits are forming from heated water. Research submarines exploring some parts of the mid-ocean ridge have found hot springs on the sea floor. Cold sea water enters through fractures. It is then heated by the magma rising beneath the ridge. The heated water dissolves minerals, then rises again to the sea floor. See Fig. 10-9, page 276. Hot water bursts through the sea floor at many vents near the rift in the center of the ridge. As the water cools, it deposits solid minerals in structures like chimneys. Many of the chimneys pour out a stream of black mineral-rich water. See Fig. 10-10, page 276. The mineral deposits built up along the mid-ocean ridge are carried away by the spreading sea floor. Scientists believe that many mineral deposits now found on land originally were formed along the mid-ocean ridge.

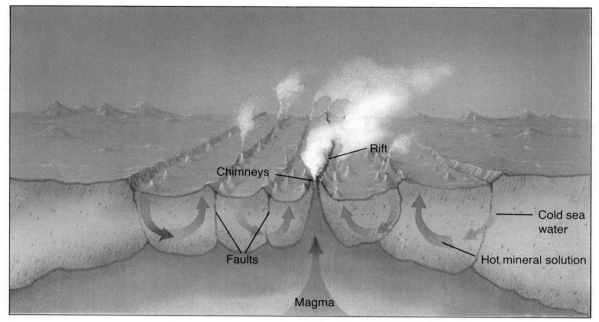

Labels in figure: Chimneys, Rift, Cold sea water, Faults, Hot mineral solution, Magma

Fig. 10-9 Cold sea water is heated by magma along the mid-ocean ridge. The heated sea water carrying dissolved minerals rises back to the sea floor as many hot springs.

Fig. 10-10 The hot, mineral-rich water is moving rapidly upward through this "chimney" in the mid-ocean ridge.

2. Weathering. A second way that ore deposits are formed is by weathering of rock. Minerals that make up the ore deposit are separated from the rock by the weathering process. For example, aluminum is found most often in minerals from which it cannot be separated profitably. Weathering of the aluminum-bearing minerals breaks them down into a form from which the aluminum can be taken. Deposits of aluminum ore are found in warm, humid regions. There the weathering processes have been most effective. Weathering may separate the denser minerals, such as gold, from their parent rocks. The denser mineral grains are moved by streams and often settle where the current is weak. Deposits built up by stream action in this way are called *placers*.

3. Cooling magma. Mineral deposits may also be formed in a third way. A body of magma may be pushed into the crust and cooled. Within the cooling magma, some minerals may separate and then sink toward the bottom. A layer of ore that is rich in some important metals can be formed in this way. The ores of chromium and nickel metals are examples of this. They are usually found within a mass of igneous rock formed from cooled magma. Hot magma may also change the rock it invades. Some mineral deposits may be caused by these changes around the edges of a body of magma.

MINERALS AND THE FUTURE

Mineral deposits form very slowly. It has taken millions of years to form most of the ores that are now in the earth's crust. For this reason, minerals are a nonrenewable resource. As the world's population grows, the need for mineral resources also increases. In some countries, such as the United States, the need for minerals grows faster than the number of people. Each year, the amount of mineral resources needed by each person becomes greater. This has been caused by a demand for more new kinds of products and the use of minerals to make agricultural fertilizers. The need for mineral resources in other parts of the world is also growing each year. How can the world's need for minerals be met?

One way to increase the supply of minerals is by finding new deposits. There are still large areas of the earth's continents that have not been explored fully for mineral deposits. There are also large supplies of certain minerals on the sea floor. The prospect for finding new mineral deposits is good. However, many deposits will be expensive to mine. For example, the minerals on the sea floor are found mostly in deep water. It will be expensive to bring these minerals to the surface and then to the land.

The need for minerals can also be met by **conservation.** Controlling the way any natural resource is used is called *conservation.* One method of conservation of mineral resources is **recycling.** *Recycling* is the process by which a mineral is recovered and used over and over again. A large volume of metals, among them aluminum, is now recycled. One hundred used cans can make ninety new ones. Recycled aluminum also saves energy. It takes about one-fifth as much energy to melt used aluminum as it does to separate the metal from its ore. Iron is also recycled. About one-third of the iron used today comes from the recycling of larger items, such as wrecked cars. At this time, it does not save money to collect and recycle small amounts of iron, such as "tin" cans. At some time in the future, the need for conservation of all mineral resources may make it necessary to recycle everything made of iron.

Metals are not the only valuable resources that can be recycled. For each metric ton of old newspaper that is recycled, about 17 trees are saved from being cut down to make new paper. Trash becomes a valuable resource if the used metals,

Conservation Saving the natural resources by controlling how they are used.

Recycling The process of recovering a natural resource and using it again.

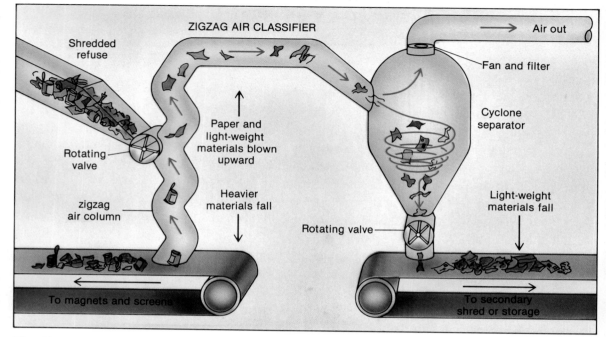

Fig. 10–11 *This kind of system can be used to separate trash before recycling.*

1. When the rock contains a useful mineral in sufficient quantity to be mined at a profit, it becomes an ore.
2. Water heated by magma may dissolve minerals from the underground rock and then travel up into cracks of cooler rock, depositing its minerals in veins. A second way is by the weathering of rock, with the minerals being carried down streams to be deposited in slower water. A third way is for magma to be pushed up into the crust, where it slowly cools, allowing the minerals to separate and sink to the bottom, forming a layer of minerals.
3. We may run low on some minerals in the future because (1) minerals are a nonrenewable resource and (2) the demand for minerals increases faster than the increase in population.
4. The mineral resources may be increased by exploring for new sources, and the existing resources can be extended by conservation.

paper, and other materials can be separated and recycled. Fig. 10–11 shows a system that can be used to separate the useful materials in trash. Recycling plays an important part in the conservation of all natural resources.

SUMMARY

Some rocks contain enough useful minerals to be taken from the earth at a profit. Such rocks make up ores that are found as mineral deposits in the earth's crust. Mineral deposits form in the crust as a result of the action of hot water, magma, and weathering. Since mineral resources are nonrenewable, future needs can be met only by finding new deposits or by conservation.

QUESTIONS

Use complete sentences to write your answers.
1. What makes a rock an ore?
2. Describe three ways mineral deposits may form.
3. Give two reasons why we may run low on some minerals in the future.
4. How can the world's need for minerals be met?

INVESTIGATION

See teacher's commentary for teaching hints, safety, and answers to questions.

PANNING FOR GOLD

PURPOSE: To separate a mineral from other materials.

MATERIALS:
shallow dish or pan
container of water
large container for
 overflow water
sand, gravel, fool's
 gold mixture

PROCEDURE:

A. Place a handful of the mixture in the shallow pan. You are to separate the parts of the mixture by using the technique called panning. If your classroom lacks the necessary facilities, the panning may be done as a demonstration. Fool's gold, the mineral iron pyrites, resembles real gold so closely that, during the California gold rush, many mistook it for the real thing. Your main goal is to recover as much of this mineral from the mixture as possible.

B. Fill the shallow pan nearly full of water. Hold the pan with both hands and, using a circular motion, gently rock the contents back and forth. DO THIS OVER THE OVERFLOW WATER CONTAINER IN CASE OF SPILLS.

C. When the water becomes cloudy, gently and carefully tip the pan and allow the cloudy water to run off into the overflow container.

D. Set the pan down and pick out any large stones and set them aside.

E. Fill the pan with water and repeat the motion as in step B. This time, tilt the pan forward slightly so that the water and sand slowly flow over the side and fall into the overflow container. Continue this until most of the sand has been removed from the pan. Give the pan a quick backward motion to spread out the remaining material.

F. Pick out any of the gold-colored pieces you see. Continue until you think you have recovered all the mineral.

G. Look at Fig. 10–12, which represents a stream system. By panning, a prospector found placer gold at points A, C, E, and G. He found no gold at points B, D, and F.

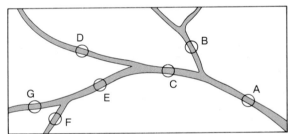
Fig. 10–12

CONCLUSIONS:

1. In Fig. 10–12, where would the source of the gold most likely be found?

2. Why did the fool's gold stay in the pan?

3. What could be done to increase the amount of mineral recovered each time you panned?

10-3. Water Resources

At the end of this section you will be able to:

☐ Identify three of the ways in which water can become polluted.

☐ Compare the amount of water used by households, industry, and agriculture.

☐ Explain the importance of local water budgets and *watersheds*.

Name something that you use every day and that costs about 15¢ a metric ton delivered to your house. You are correct if you answered that water fits this description. Turn on a faucet and out comes a stream of clean, running water. Will there always be enough? The answer is yes, if two problems can be solved. One problem comes from pollution of the water supply. The other problem is that nature does not always provide the water where it is needed.

PROTECTING THE WATER SUPPLY

Many millions of years ago the earth had more than nine million billion metric tons of water. Almost all of that water is still available today. Water is a renewable resource. Water is used, but it is not used up. Almost every drop is returned to the natural water cycle. In this cycle, water moves to the sea, evaporates, and falls as rain to the earth again. Water is made pure as it moves through the water cycle. In this way, nature gives us pure water.

Since 1900, the amount of water used in the United States has doubled every 25 years. People now use more water than ever before. See Fig. 10-13. If the population continues to grow at the present rate, three times as much water will be needed in the year 2000 as is needed now. There is plenty of water to meet those future needs. About 12 trillion liters of rain fall on the United States each day. We use only about one-tenth of it. However, none of that water can be used unless it is protected from pollution.

1. Sewage. One kind of pollution comes from untreated sewage wastes. See Fig. 10-14. Raw sewage in the water supply carries germs that cause diseases. However, the water flowing

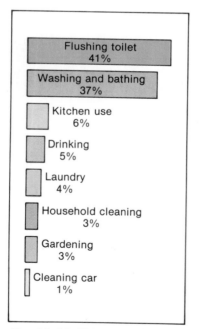

Fig. 10-13 This graph shows how water is used in the average household.

- Flushing toilet 41%
- Washing and bathing 37%
- Kitchen use 6%
- Drinking 5%
- Laundry 4%
- Household cleaning 3%
- Gardening 3%
- Cleaning car 1%

Fig. 10-14 Raw sewage is being flushed into the water. This can cause serious pollution problems.

in streams can break up and destroy the dangerous wastes. Some of the wastes are dissolved and some settle to the bottom. The moving water also mixes with the air. The wastes in streams combine with oxygen. The dissolved oxygen in the water acts on the wastes, making the sewage harmless. But in an overloaded stream, waste materials use up all of the oxygen dissolved in the water. The stream cannot purify itself until more oxygen is taken in from the atmosphere. Increased amounts of waste material in the water prevent the natural action from catching up. Thus the stream becomes polluted. Pollution from sewage can be prevented in two ways. One way is to control the amount of wastes dumped into the water. A second way puts the raw sewage through a treatment plant. A sewage treatment plant destroys the germs. Many treatment plants also remove much of the material that would use the oxygen in the water.

2. Chemicals. A second kind of pollution is caused by dumping chemicals that cause the growth of too many plants. The oxygen supply of a stream or lake can be upset. One such chemical is called *phosphate* (**fahs**-fate). Phosphates are found in some detergents, cleaning materials, and fertilizers. They cause small forms of plant life, such as algae, in the water to grow rapidly. When these plants decay, they use the dissolved oxygen in the water. Many of the living things in that body of water are killed. This upsets the natural balance of life in the water. To maintain a pure water supply, materials that damage the natural balance of life cannot be allowed to enter streams and lakes.

Suggestion: You may want to review the water cycle found in Chapter 5, page 123.

Enrichment: The numbers on bags of lawn fertilizer indicate the percentages of nitrogen, phosphorous, and potassium found in it. For example, 6-10-4, a general purpose fertilizer, means 6% nitrogen, 10% phosphorous, and 4% potassium.

Fig. 10-15 Wastes from industry may be one of the major sources of chemical pollution.

Chemicals that come from industrial wastes may be the most serious pollutants over a long period of time. This kind of pollution can be caused by chemicals directly entering the water supply. See Fig. 10-15. The chemicals used in agriculture can also be a source of pollution. Many of these chemicals wash down through the soil or run off in streams. See Fig. 10-16. Some chemicals enter the ground water supply and cause pollution in water taken from wells. Streams may carry the chemicals over a wide area. The polluted water then becomes part of the ground water or collects in lakes and reservoirs. Most of the water now used in homes contains very small amounts of chemicals from industry and agriculture. Over long periods of time, the effect of these chemicals on human health and other forms of life is not known.

3. Oil spills. A very serious kind of pollution comes from oil spills in the oceans. The oil can be released by accidents to tankers or offshore oil wells. In past years, much of the oil pollution came from tankers that cleaned their oil-storage tanks with sea water. Now there are international laws to prevent such pollution. However, accidental oil spills will continue to be a problem. Oil pollution in the sea can kill much of the natural life. The floating oil spreads out. Much of it finally sinks to the sea floor. Some may reach the shore, where beaches are ruined and the marine life often is killed. Spilled oil is very hard to remove from the sea surface or the land along the shore. The solution to the problem can be found only in measures that will prevent this dangerous form of pollution from happening.

Fig. 10-16 Chemicals used as fertilizers and for pest control in agriculture can enter the ground water supply.

4. Thermal pollution. Another way in which water can become polluted is through its use as a cooling agent. Power plants, for example, produce a large amount of waste heat. Many power stations use water from rivers to remove this heat. Water used to cool power plants becomes heated. This heated water is sent back to nearby rivers, lakes, or oceans. The heated water may affect the natural balance of plant and animal life where it is released. The release of heat into streams, other bodies of water, or the atmosphere is called **thermal pollution.** *Thermal pollution* can kill the eggs or young of animals that live in water, such as fish. The increase in temperature also reduces the oxygen in the water. This can change the normal life cycle of plants and animals in the water.

Enrichment: Europe has many high-temperature incinerators that change dangerous wastes into carbon dioxide, water, and other gases.

Thermal pollution Release of heat into streams, lakes, oceans, or the atmosphere.

THE WATER SUPPLY

About 97 percent of the earth's water is found in the oceans. Over 2 percent of the remaining water is "locked up" in the polar icecaps and glaciers. The supply of water available for human use must come from the 1 percent that remains. This water moves between the sea and the land.

It may seem to you that the important uses of water are the household uses. Your life would certainly be different without a supply of water in your house. However, households use only a small part of the total water supply. Industry uses about three times more water each day than is used in all the households in the country. In order to make an automobile, for example, more than 100,000 liters of water are needed. In order to make one liter of gasoline, nearly 200 liters of water are needed.

Almost all of the water used in factories and homes can be cleaned and made pure. It can then be stored and used again before it is returned to the natural water cycle. Some water cannot be returned to the water cycle easily, such as water used for crops. In the United States, more water is used for farming than for all other uses combined. Water needed by crops is brought to the fields from deep wells or distant sources. The water used for farming may be taken up by plants. The rest runs over the surface, evaporates, or seeps into the ground. This water cannot be used by people until it makes its way through the water cycle. The water will continue to move through the cycle until it reaches a large body of water or becomes part of the ground water.

LOCAL WATER SUPPLIES

The water cycle describes how moisture moves between bodies of water, the air, and the land. It is also important to know how much water there is at a given location. A *local water budget* is the record of moisture income, storage, and outgo at a given location. The local water budget will tell you whether there is a lack of water or an oversupply of water at any particular time. Surplus water is stored in the soil, as part of the ground water, or becomes runoff. This runoff may be stored in lakes or reservoirs. Any water that cannot be stored flows away in streams to the ocean. Tables 10–1 and 10–2 give the amount of moisture income and outgo each month for a typical year in Savannah, Georgia, and Yuma, Arizona. Compare these two tables. Which place has the greatest yearly surplus of water?*

* Savannah, Georgia

WATER BUDGET FOR SAVANNAH, GEORGIA												
	Jan.	Feb.	Mar.	Apr.	May	June	July	Aug.	Sept.	Oct.	Nov.	Dec.
Income (mm) (rain)	70	80	88	73	80	135	175	180	150	75	50	70
Outgo (mm) (evaporation, use)	20	25	40	75	125	165	175	155	120	75	28	18
Monthly surplus (mm)	50	55	48	0	0	0	0	25	30	0	22	52
Monthly deficit (mm)	0	0	0	2	45	30	0	0	0	0	0	0

Table 10–1

WATER BUDGET FOR YUMA, ARIZONA												
	Jan.	Feb.	Mar.	Apr.	May	June	July	Aug.	Sept.	Oct.	Nov.	Dec.
Income (mm) (rain)	17	14	10	4	2	0	4	12	6	4	6	15
Outgo (mm) (evaporation, use)	18	19	20	90	140	190	210	200	165	100	30	15
Monthly surplus (mm)	0	0	0	0	0	0	0	0	0	0	0	0
Monthly deficit (mm)	1	5	10	86	138	190	206	188	159	96	24	0

Table 10–2

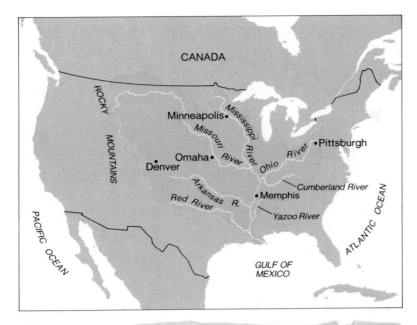

Fig. 10-17 The watershed of the Mississippi.

A local water supply is affected by its **watersheds.** All of the land area that is drained into a stream or river is a *watershed.* Every square meter of land is part of a watershed. Each stream has its own watershed. A watershed may be as small as a single field or as large as several states. The watershed of the Mississippi River, for example, includes almost half of the United States. See Fig. 10-17.

Watershed The land area that drains into a stream or river.

The flow of water in streams is determined mostly by its watershed. A watershed made up of land without plants will not soak up water. Muddy water can run off rapidly and cause floods. Since there are no plants to hold the soil together, soil is washed into the streams and reservoirs. This reduces the amount of water that can be stored. On the other hand, a watershed that is covered with plants will absorb most of the water falling on it. Floods and droughts are less common where watershed areas are covered with forests or other plants.

A watershed can be damaged if the protective plant covering is removed. This may happen because of poor farming methods, fires, or the cutting of trees for lumber. See Fig. 10-18. Today many watersheds are destroyed by the building of paved lots, such as shopping malls. If there is to be a steady flow of water into streams and rivers, all watersheds must be protected. Damaged watersheds must be repaired by replacing the plants.

EARTH RESOURCES

Fig. 10-18 Forest fires are a common cause of damage to watersheds.

1. Water is a renewable resource because in the water cycle it is purified when water evaporates back into the air.
2. The water supply can be polluted by raw sewage, chemicals, and oil spills.
3. Industrial and agricultural chemicals drain into the streams and ground water. As the stream runs across the land, the chemicals soak into the soil and get into our water wells, reservoirs, lakes, etc. These water supplies are not treated for the chemicals, so they find their way into our homes.
4. The heat kills the eggs of water animals, as well as some animals and plants that live in the water, and decreases the oxygen supply of the water.
5. The amount used in industry is three times that used in all the homes, while the amount of water used by agriculture is more than industrial and home use combined.
6. The watersheds must be protected to control the flow of water into the streams in order to prevent floods and droughts.

SUMMARY

The earth provides a water supply that is able to renew itself. However, the water supply must be protected from pollution in order for us to continue to have enough water to meet our needs. Most of the fresh water supply is used by agriculture and industry. Although there is plenty of water, it is not always found where needed. Many regions must plan carefully in order to have sufficient water to meet their needs. Destruction of watersheds can affect the water supply of streams and rivers.

QUESTIONS

Use complete sentences to write your answers.
1. Explain why water is a renewable resource.
2. Name three ways in which water can become polluted.
3. Describe the manner in which industrial and agricultural chemicals get into our home water supply.
4. Why is thermal pollution a problem?
5. Compare the amount of water used in industry and agriculture with that used in the home.
6. Why is it important to protect the watersheds of streams?

INVESTIGATION

See teacher's commentary for teaching hints, safety, and answers to questions.

LOCAL WATER BUDGETS

PURPOSE: To study the local water budgets for two different cities.

MATERIALS:
meter stick
two 10-cm tall
  containers
small pan
container of water
  (water source)

PROCEDURE:

A. Look at Table 10–1, page 284. It shows the water budget for Savannah, Georgia.

   1. What is the water income for January?

   2. What is the water outgo for January?

B. Set one of the 10-cm tall containers in the pan. This will represent Savannah's water supply. Fill the container to a depth of 70 mm (7 cm). This represents the 70-mm water income for the month of January.

C. Now remove 20 mm of water from the container. This amount of water, lost by use of plants, evaporation, and other ways, is the outgo for January.

   3. What is the depth of the water left in the container? This water represents the water stored in the soil, lakes, reservoirs, etc.

   4. How does this compare with the monthly surplus for January shown in Table 10–1?

D. Fill the second container to 80 mm (8 cm) for the income shown for February. Remove 25 mm of the water for outgo. The remaining 55 mm are the surplus for February. Pour this into the first container with the January surplus.

   5. Describe what happened.

E. Any water that overflows into the pan represents water that cannot be stored. It runs off to another place. The water left in the container represents stored water in the water budget for future use. Pour the water in the pan back into the water source.

F. Continue this for the remaining months. Notice that in April, May, and June, the outgo is greater than the income. You will need to pour some water out of the first container to meet this outgo.

   6. Was there any month during the year that the surplus ran out?

   7. In what months were the lakes and reservoirs full?

G. Look at Table 10–2. It shows the water budget for Yuma, Arizona.

   8. Describe three ways that Yuma's water budget differs from Savannah's.

   9. Why do you think millions of dollars have been spent to route water from the Colorado River to Yuma?

CONCLUSIONS:

   1. In which city would you expect to find a desertlike environment? Why?

   2. If the water budget shows a deficit for a given month, does that mean there is no water available for use? Explain your answer.

10-4. Nonrenewable Energy Resources

At the end of this section you will be able to:

- List the *fossil fuels* and describe how each is formed.
- Describe how the supply of fossil fuels will change in the future.
- Compare the processes of *fusion* and *fission* as sources of *nuclear energy*.

Think about a world in which most work had to be done by using energy supplied by human or animal muscles. See Fig. 10-19. How would your life be different? What happens when the electric power is shut off? What would happen if the gasoline supply were cut off? Without a steady supply of energy, there would be many changes in the way you live.

Fig. 10-19 Without fossil fuels for tractors, we would have to rely on horse power to plow fields. Think of other ways your life style would change without fossil fuels.

FUELS FROM THE EARTH

Two hundred years ago, muscle power and power from burning wood were the major sources of energy. Today, such sources supply only a tiny part of the world's energy needs. The Industrial Revolution brought machines that use large amounts of energy. Most of this energy has come from the burning of **fossil fuels.** These fuels have been formed from the decay of plant and animal fossils in the earth's crust. The *fossil fuels* are coal, petroleum, and natural gas.

Fossil fuel Mineral fuel found in the earth's crust.

1. Coal. The energy in coal came from the sun. Plants living in swamps long ago stored solar energy. The swamps became buried because of changes in the crust. Then the plant material was slowly changed to coal deep beneath the surface. Thus the huge deposits of coal that exist today were once living plants.

There are several forms of coal. Each form is a stage in a process by which plant materials are changed by increasing pressure and heat. The process changes plants into pure carbon. The first stage is *peat*. The plant material is only slightly changed in this stage. Peat looks like rotted wood and is found on or near the surface of the ground. It is used as a fuel in some parts of the world. The second stage in forming coal is *lignite* (**lig**-nite), often called "brown coal." Lignite is brownish black, catches fire easily, and burns with a smoky flame. Deposits of lignite are found beneath the surface. The third stage in the making of coal is *bituminous* (bih-**too**-mih-nus), or soft coal. It is dark brown or black in color and breaks easily into blocks. Bituminous coal burns with a smoky flame. It is the most abundant kind of coal. More heat and pressure on a coal deposit will change bituminous into a much harder kind of coal called *anthracite* (**an**-thruh-site). This is a hard, brittle, black form of coal that is almost pure carbon. It burns with a blue, nearly smokeless flame and produces great heat. Anthracite is a kind of metamorphic rock. The other forms of coal are sedimentary rock. The different forms of coal are shown in Fig. 10-20.

Enrichment: It takes about 45 meters of plant material to make a 1-meter layer of coal.

Enrichment: The Soviet Union has about 65% and the United States about 27% of the world's coal deposits.

Enrichment: The United States has more energy in coal deposits than the known petroleum energy reserves of the entire world.

Suggestion: As a demonstration, heat pieces of coal in a test tube fitted with a one-hole stopper containing a short glass tube. With a match, ignite the gases produced (mostly methane). Smoke and the odor of compounds of sulfur will also be noticeable.

Fig. 10-20 Stages in coal formation.

Peat   Lignite   Bituminous   Anthracite

Fig. 10-21 Strip-mining for coal is safer, but it destroys the landscape and the natural habitat for wildlife.

Petroleum A fossil fuel that is usually a liquid, that comes from the decay of ancient life in warm, shallow seas.

Enrichment: During World War II, the Germans converted coal to a liquid fuel to power their war machines.

Hydrocarbons The principal substances found in petroleum. They contain only hydrogen and carbon.

Enrichment: Carbon dioxide will increase the greenhouse effect. This is the ability of the atmosphere to change short-wavelength light into longer-wavelength heat waves which cannot escape back through the atmosphere without being absorbed. The result is a warming of climate. An effect could be the melting of the icecaps with subsequent rise of the sea level.

There are serious problems in using coal as a major source of energy. For example, much coal contains large amounts of sulfur. When the coal is burned, the sulfur produces sulfur dioxide gas. This gas causes a serious air pollution problem. Sulfur dioxide gas also is a cause of acid rain. Power plants that use coal must install special equipment to prevent the release of large amounts of sulfur dioxide from their smokestacks. Carbon dioxide gas, which is also released by burning coal, may build up in the atmosphere. This carbon dioxide may cause the earth's climate to change. Many scientists have predicted a warming of the earth.

Mining coal also creates serious problems. Coal is obtained from many deposits by strip mining. See Fig. 10-21. The damage caused to land by strip mining must be repaired to preserve the natural environment. Another problem with coal is the danger to the miners who work in underground coal mines. Lives are lost because of accidents. Diseases can be caused by breathing coal dust. However, these problems are not likely to prevent large amounts of coal from being used as a main source of energy. In the future, coal may be used as a liquid fuel. It then can help to replace **petroleum** (puh-**troe**-lee-um). Coal can also be used to make a gas that can be used in place of natural gas.

2. *Petroleum* is a liquid that comes from the decay of ancient life forms in warm, shallow seas. The remains of these plants and animals settled on the sea floor and were covered by sediments. After about 100 million years, this material was changed to substances made of hydrogen and carbon. These substances are called **hydrocarbons.** Petroleum is a mixture of many different *hydrocarbons*. Petroleum can be separated into different products, such as gasoline, kerosene, and motor oil. See Fig. 10-22 *(top).* Some of the hydrocarbons in petroleum are used to make plastics, rubber, soaps, and some medicines.

Petroleum was formed from the mud on the sea bottom. The mud was later changed into the sedimentary rock called shale. Pressure usually forced the petroleum out of the shale. The petroleum then moved into rock with more open spaces, such as sandstone. Petroleum may also have collected where layers of sandstone and limestone were joined. Water may also have been trapped in the sedimentary rock. The oil, being less dense

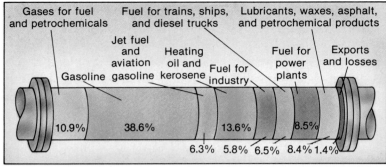

Gases for fuel and petrochemicals

Fuel for trains, ships, and diesel trucks

Lubricants, waxes, asphalt, and petrochemical products

Jet fuel and aviation gasoline

Heating oil and kerosene

Fuel for industry

Fuel for power plants

Exports and losses

Gasoline

10.9% 38.6% 13.6% 8.5%

6.3% 5.8% 6.5% 8.4% 1.4%

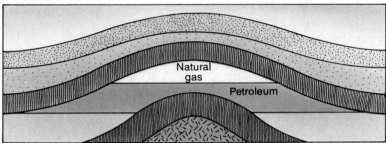

Natural gas

Petroleum

Fig. 10-22 (top) *Refining separates petroleum into these products.* (bottom) *The petroleum is trapped beneath a dome of non-permeable rock forming an oil pool.*

than water, floated upward. The oil stopped rising when it met a layer of rock that it could not flow through. The oil collected below this layer and formed an *oil pool*. Fig. 10-22 *(bottom)* shows one of the most common arrangements of rock layers that contain oil pools.

3. Natural gas is also a hydrocarbon. Petroleum and natural gas are usually found together. Natural gas is often trapped with the petroleum beneath the solid rock covering an oil pool. See Fig. 10-22.

FOSSILS FUELS IN THE FUTURE

The supplies of fossil fuels in the earth's crust are limited. It has taken millions of years to make them. Thus they are nonrenewable. When the supplies are used up, there will be no quick way to replace them. There are probably greater amounts of coal than of petroleum remaining within the earth. Coal comes from sediments formed in the shallow seas that once covered large parts of the continents. Because the continents are very old, enough time has passed for large deposits of coal to be formed. Petroleum, on the other hand, is found in sediments laid down along the edges of continents. As the crustal plates moved and collided, these sediments were destroyed. Petroleum and natural gas deposits are younger than coal and are less common. It is not known how much oil and gas are left.

Fig. 10-23 A piece of oil shale that has been cut and polished to show the brown streaks that yield oil.

Most of the petroleum and gas deposits still to be found in the United States probably are found in the shallow waters just off some parts of the coasts. There is a huge amount of petroleum found in a kind of rock called *oil shale*. See Fig. 10-23. Large deposits of oil shale are found in parts of Colorado and Utah. However, the petroleum in oil shale is not liquid. The rock must be mined and then heated to release the oil. This is a difficult and expensive process. Many problems must still be solved before oil shale can become a source for our petroleum needs.

Most of the energy that is used at the present time comes from the burning of fossil fuels, which are nonrenewable. Industry is the biggest energy user. Making steel, for example, uses large amounts of energy. Forms of transportation, such as automobiles, also use a large share of the energy supply. Industry and transportation together use about 60 percent of the energy produced in this country each year. The rest is used by homes and businesses.

The total amount of energy used in the United States has grown rapidly during recent years. If energy continues to be used at the present rate, the amount needed will double every 20 years. This increase in the need for energy is caused by the growing population. Also, the amount of energy used by each person is increasing every generation. You use twice as much energy as your parents did when they were your age. Table 10-3 shows how much energy is used by home appliances each year.

For the past hundred years, the growing need for energy has been met by using fossil fuels. However, fossil fuels cannot be renewed in a short period of time. This has created an energy crisis. At the present rate of use, the supply of petroleum will run out. It probably will be used up within the next 50 to 100 years. The supply of coal will last much longer. In fact, at the present rate of use, coal reserves will last about 200 years. However, burning coal can cause serious damage to the environment. In the future, renewable sources of energy may be able to replace the fossil fuels. These energy sources will not run out because they will constantly be replaced. Until this happens, however, the only answer to the present energy crisis may be careful conservation of the present energy supply.

ESTIMATED ANNUAL ENERGY CONSUMPTION OF ELECTRIC APPLIANCES UNDER NORMAL USE			
Appliance	Estimate of Energy Used Each Year* (kW-h)**	Appliance	Estimate of Energy Used Each Year* (kW-h)**
Blender	15	Washing machine (nonautomatic)	76
Broiler	100	Water heater	4,219
Carving knife	8	Air cleaner	216
Coffee maker	106	Air conditioner (room)	860*
Deep fryer	83	Electric blanket	147
Dishwasher	363	Dehumidifier	377
Egg cooker	14	Fan (attic)	291
Frying pan	186	Fan (circulating)	43
Hot plate	90	Fan (rollaway)	138
Mixer	13	Fan (window)	170
Microwave oven	190	Heater (portable)	176
Range		Heating pad	10
with oven	1,175	Humidifier	163
with selfcleaning oven	1,205	Hair dryer	14
Roaster	205	Heat lamp (infrared)	13
Sandwich grill	33	Shaver	1.8
Toaster	39	Sun lamp	16
Trash compactor	50	Toothbrush	0.5
Waffle iron	22	Radio	86
Waste disposer	30	Radio/Record Player	109
Freezer (15 cu ft)	1,195	Television	
Freezer (frostless 15 cu ft)	1,761	Black & white tube type	350
Refrigerator (12 cu ft)	728	solid state	120
Refrigerator (frostless 12 cu ft)	1,217	Color tube type	660
Refrigerator/Freezer (14 cu ft)	1,137	solid state	440
(frostless 14 cu ft)	1,829	Clock	17
Clothes dryer	993	Floor polisher	15
Iron (hand)	144	Sewing machine	11
Washing machine (automatic)	103	Vacuum cleaner	46

* Based on 1,000 hours of operation per year. This figure will vary widely depending on area and specific size of unit.
SOURCE: Edison Electric Institute
** kW-h = kilowatt-hour

Table 10–3

Nuclear energy Energy that is produced by changes in the nuclei of atoms.

Fission The splitting of the nuclei of certain kinds of atoms.

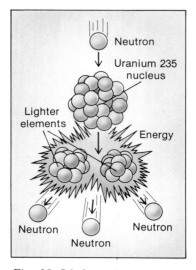

Fig. 10-24 A neutron can strike certain kinds of atomic nuclei causing fission with a release of energy.

Nuclear energy has already begun replacing fossil fuels. *Nuclear energy* comes from changes in the *nuclei* (**noo**-clee-ie), or cores, of certain atoms.

There are two ways to make energy from the nuclei of atoms. One method is called **fission** (**fish**-un). In *fission*, the nuclei of atoms are split into smaller nuclei by neutrons. See Fig. 10-24. When fission takes place, a large amount of heat is given off. The heat can be used to produce steam to run electric generators. See Fig. 10-25. Uranium is the fuel most often used in these power plants. A piece of uranium the size of a marble can produce the same amount of heat as 100 metric tons of coal. But there is not very much uranium in the earth. Uranium could be used up within the next 50 years.

Many people believe that nuclear plants are not safe. One problem is the radioactive wastes that come from the power plants. These wastes must be removed from time to time. They are stored until they lose their radioactivity. The wastes can remain dangerously radioactive for hundreds, even thousands, of years. A place must be found to put these dangerous substances where they cannot escape into the environment. Some people believe that nuclear power plants are unsafe because of the way they are cooled. Plants are cooled by water. If the cooling system were to stop working, the plant would overheat, and could even melt. Then the radioactivity would escape. However, the people who build and operate nuclear plants believe that the problems can be solved. All nuclear plants use safety systems, and research is reducing the risks.

Fig. 10-25 This power plant produces electricity by using the heat from nuclear fission.

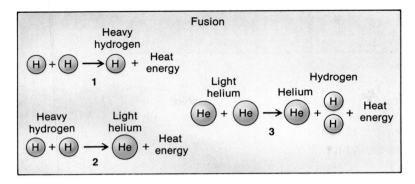

Fusion

Heavy hydrogen

$H + H \longrightarrow H + $ Heat energy

1

Heavy hydrogen   Light helium

$H + H \longrightarrow He + $ Heat energy

2

Light helium   Helium   Hydrogen

$He + He \longrightarrow He + \begin{array}{c} H \\ H \end{array} + $ Heat energy

3

Fig. 10-26 Fusion is joining together atomic nuclei, causing energy to be released.

A second method of producing nuclear energy may solve some of the problems caused by nuclear plants. This method is called **fusion.** *Fusion* (**fyoo**-zhun) is the joining of atomic nuclei to make new and larger nuclei. See Fig. 10-26. When fusion of nuclei takes place, very large amounts of heat are released. This heat may be used to run electric generators. But nuclear fusion happens only at very high temperature and pressure. It is hard to make machines that can withstand these temperatures and pressures. Scientists are trying to find ways to use nuclear fusion as a source of energy. It would solve many of the world's energy problems. The fuel needed for nuclear fusion is a form of hydrogen that can be taken from sea water. At some time in the future, the sea may become an energy resource that will never run out.

Fusion The joining together of nuclei of small atoms to produce larger atoms with the release of energy.

Enrichment: Fusion is the source of the sun's energy.

SUMMARY

During the past century, people have been using more and more fossil fuels, such as oil and coal. These fuels take millions of years to form. The supply of fossil fuels will be running out unless we conserve energy. Nuclear energy may replace fossil fuels in the future.

QUESTIONS

Use complete sentences to write your answers.

1. Name and describe the four forms of coal.
2. Name two other fossil fuels and tell how they formed.
3. Explain why there will be a short supply of fossil fuels in the future.
4. Describe two ways to produce energy from the nuclei of atoms.

1. Peat is a form of coal that represents the earliest stage in the development of coal. It is found on the surface and resembles rotted wood. Lignite, or brown coal, is located just under the surface. Bituminous coal is dark brown to black in color and breaks easily into blocks. Additional heat and pressure forms coal that is nearly pure carbon and is called anthracite. Anthracite is a kind of metamorphic rock.
2. Two other fossil fuels are petroleum, formed from the remains of plants and animals that lived in the seas, and natural gas, formed as part of petroleum deposits.
3. Petroleum will be used up in 50 to 100 years.
4. Energy can be produced from the nuclei of atoms by the process of fission or fusion. Fission is the splitting of complex atoms to form simpler atoms and release energy, while fusion is the building of more complex atoms from simpler ones, with a release of energy.

SKILL-BUILDING ACTIVITY

USING GRAPHS TO STUDY ENERGY USED IN THE UNITED STATES

PURPOSE: To prepare graphs for the study of energy used in the United States during the last half of the 20th century.

MATERIALS:

graph paper   pencil

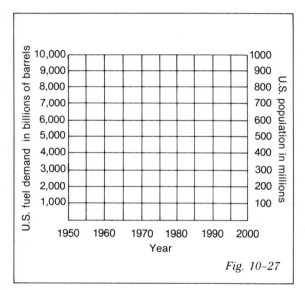

Fig. 10–27

PROCEDURE:

A. On graph paper, draw axes like those in Fig. 10–27 and plot the values shown below. Use a solid line for Fuel Demand/Year and a dotted line for Population/Year.

Year	Fuel Demand (billions barrels)	Population (millions)
1950	1,800	150
1960	2,800	180
1970	4,300	210
1980	5,600	220

B. Use your graphs to answer the following questions.
   1. In what year was the population of the United States 200 million?
   2. In that year, what was the United States' fuel demand?

C. One way to see national energy used is to note how much energy is required to support one person. To do this, divide your answer to question 2 by 200 million.
   3. What was the answer?

D. Using a dashed line, extend the solid line (fuel demand) and the dotted line (population) to the year 2000. Keep the shape of each graph the same. Then answer the following questions.
   4. What will the population of the United States be in 1990?
   5. What will the United States' fuel demand be in 1990?
   6. What will the fuel demand per person be for the year 1990?
   7. How does this compare to your answer to question 3?

E. The future population and demand for fuel involves many things not included in these graphs. The effects of alternate energy sources, conservation, and future weather conditions are not known.
   8. What will the United States' fuel demand per person be in the year 2000?

CONCLUSIONS:
   1. Explain the relationship between population growth and fuel demand.
   2. What must be done so that the future fuel demand can be met?

10–5. Renewable Energy Resources

At the end of this section you will be able to:

- ☐ List four ways that solar energy can be used.
- ☐ Explain how *geothermal* energy can be used.
- ☐ Describe how moving water, wind, and plants can provide energy.

Sunlight contains energy. However, this energy must be gathered and changed into a useful form such as electricity. The mirrors shown in Fig. 10–28 are helping to change sunlight into electricity.

Fig. 10–28 Solar mirrors can be used to turn solar energy into electricity.

SOLAR ENERGY

When sunlight falls on your skin, the warmth you feel is part of the *solar energy* that the earth receives from the sun. Each year the sun sends 500 times more energy to the United States than we use. The entire energy needs of the country could be met by using only a small part of that solar energy. However, solar energy must be gathered and changed into forms that can be used.

1. Passive. A house can be built in a way that allows it to gather solar energy. A solar-energy system that doesn't have any motors, pumps, or other moving parts, is called *passive*. "Passive" means "not active." A passive solar house has windows facing the sun. Sunlight enters the house through the windows and heats materials inside the house, such as stone or

Reinforcement: This is similar to a car collecting heat in the sun. The greenhouse effect occurs.

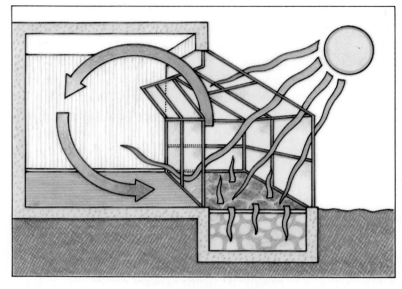

Fig. 10-29 During the daytime, sunlight enters through the windows and heats the air and floor. The warm air rises while the cool air sinks. At night, the warm floor provides heat.

Fig. 10-30 (left) This drawing illustrates how this house uses solar energy directly to heat water. The water is heated as it passes through solar collectors on the roof. (right) Even older homes can have solar collectors installed.

brick. See Fig. 10-29. The warmed materials then heat the living space. The materials can also store some of the heat for the night.

2. Collectors. Solar collectors are usually boxes with glass or plastic tops. They are placed on roofs facing the sun. Inside the boxes are tubes, which are usually filled with water and antifreeze. However, another liquid or air may also be used. The tubes and the inside of the box are painted black to absorb more heat. After the water is heated, it is moved to a storage tank where it is ready to be used. See Fig. 10-30. One problem with solar heating is the lack of sunlight on cloudy days.

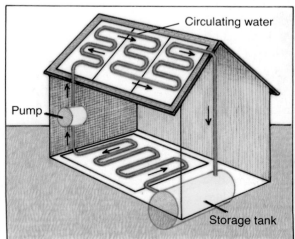

Circulating water

Pump

Storage tank

3. Solar cells. Sunlight can be changed into electricity by the use of **solar cells.** Each *solar cell* produces only a small amount of electricity. Therefore, a large number of cells are needed to create a useful amount of electricity. While solar energy itself is free, solar cells are presently very expensive to make. For this reason, it would be costly to use enough solar cells to provide the electricity needed in a house. Their use now is mostly in spacecraft and other places where no other source of electricity can be used.

Solar cell A device that is used to change sunlight directly into electricity.

Enrichment: Solar cells were first used in space in 1958 for the *Vanguard* satellite.

4. Concentrators. Sunlight, if concentrated enough, can be used to boil water and produce steam. The steam can then be used to turn generators to produce electricity. Solar power plants have been built that use mirrors to reflect sunlight from a large area onto a smaller surface. See Fig. 10–31. This is similar to using a magnifying glass to concentrate sunlight.

Fig. 10–31 A central, solar power station uses mirrors to reflect sunlight onto steam generators.

GEOTHERMAL ENERGY

In some places it is possible to use **geothermal** (jee-oe-**thur**-mul) **power.** *Geothermal power* comes from heat within the earth's crust. The heat causes ground water to become hot enough to make steam. One source of geothermal power uses this steam when it is released through wells. Only a few of

Geothermal power Energy obtained from heat in the earth's crust.

Word Study: Geothermal: *gaia* + *thermos* (Gr.), earth heat.

the earth's locations can use geothermal energy. Some places have been found where bodies of magma are close to the surface. Another way to use the earth's heat is by drilling deep wells at these places to produce very hot water. The hot water can then be used as heat or to generate electricity.

WATER, WIND, AND PLANT POWER

We usually think of solar energy as just sunlight. However, solar energy also exists in other forms: It causes water to evaporate and fall as rain; by heating the atmosphere, it powers the wind; and sunlight is stored by plants as they grow. Water, wind, and plants are indirect and renewable sources of solar energy.

Hydroelectric power Electricity produced by water moving downhill.

1. Water power. Water power, or **hydroelectric** (hie-droe-eh-**lek**-trik) **power,** has been used for many years to produce electricity. See Fig. 10–32. *Hydroelectric power* comes from water that is flowing downward. As the water flows down, it turns a machine that makes electricity. The energy of falling water really comes from solar power. Heat from the sun causes water to evaporate, mostly from the sea. When the water falls as rain and runs back to the sea, stored solar energy is released. Hydroelectric power is constantly renewed as the water cycle goes on. The world still has large amounts of water-power resources that are not being used.

Enrichment: In Solana County, CA, is a windmill 107 meters high when the 61-meter blade is upright. It is expected to provide energy for 1,100 homes. As of June, 1985 California accounted for 95% of the commercial wind energy produced in the United States and 75% of the world's total. Denmark is in second place.

Enrichment: Windmills were believed to be used to grind grain in 1000 B.C.

Fig. 10–32 Water flowing through this dam turns machines that produce electricity.

2. Wind. Wind also comes from solar power. Wind is caused by the heating of air. Like water power, wind is renewable. The energy of wind is never used up. There is a large amount of energy in wind. Some of this energy can be used to produce electricity by turning a windmill. See Fig. 10-33. Using even a small part of the total amount of energy in winds would meet the world's power needs. Windmills seem to work best on plains, the tops of mountains, and along some coastal areas. Small windmills have been in use in the United States since about 1920. These were mainly on farms. However, the wind cannot always be used as a source of power. There are not many places where the wind blows steadily enough to run a power station made up of windmills.

3. Biomass. Plants are like solar collectors. They absorb energy in sunlight. Leaves, wood, and other parts of plants are made when the plants use solar energy. Thus wood and other plant materials contain stored solar energy. This source of energy is called **biomass.** The plant materials that make up *biomass* can be changed into a liquid fuel. For example, plant material that contains sugar or starch can be made into alcohol. The alcohol can be burned as a motor fuel or mixed with gasoline to make *gasohol.* An acre of corn can produce about 1,045 liters of ethanol. But using farms to produce fuel also uses up land that could be used for growing food.

Fig. 10-33 Wind energy produces electricity by turning this modern windmill.

Biomass   Plant materials that are used to produce energy.

1. A house can be used as a solar collector by having windows facing the sun to let solar energy in. Solar collectors on roofs of houses use panels to heat tubes filled with water. The water is pumped to a storage tank where the heat is stored for use. Solar cells convert sunlight into electricity. Concentrators use mirrors to reflect sunlight onto a smaller surface. This can be used to change water to steam to turn generators that produce electricity.
2. Geothermal energy can be used to make steam, which can be released through wells to run electric generators. Geothermal energy also produces hot water, which can be used to heat homes.
3. Hydroelectric power comes from water moving downward. It can be used to turn generators to make electricity. Wind can be used to turn windmills, which generate electricity. Biomass is plant material that can be burned for heat or converted into other fuels such as alcohol.

SUMMARY

As the supplies of fossil fuels are used up, renewable energy sources must help to meet our energy needs. The most important renewable source of energy is the sun. Solar energy can be used directly. It also can be used in the form of moving water, wind, or plant materials. The earth's heat can be used as a source of energy in certain locations.

QUESTIONS

Use complete sentences to write your answers.

1. Describe four ways that solar energy can be used.
2. Give two ways in which geothermal energy can be used.
3. Explain how energy can be provided by hydroelectric power, wind, and biomass.

See teacher's commentary for teaching hints, safety, and answers to questions.

MAKING A SOLAR COLLECTOR

PURPOSE: To make a solar collector and find out how it absorbs light to make heat.

MATERIALS:

box	2 thermometers
black paper	transparent wrap
tape	light source

PROCEDURE:

A. Make a copy of the following table:

Condition	Inside Thermometer	Outside Thermometer
After 5 min. with light		
After 5 min. without light		

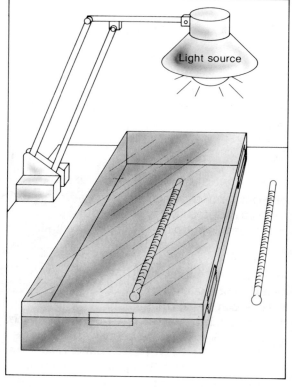

Light source

B. Line the inside of the box with black paper.

C. Tape one thermometer inside the box. Be sure that the numbers on the thermometer are showing.

D. Cover the open side of the box with transparent wrap. Tape the edges to the sides of the box.

E. Place the box and the other thermometer under a light source as shown in Fig. 10–34. After five minutes, read both thermometers. Record the temperatures in your table.

   1. Which thermometer had the higher temperature?

F. Remove the light source. After five minutes, read both thermometers. Record the temperatures in your table.

   2. Which thermometer had the higher temperature?

CONCLUSIONS:

1. Which thermometer absorbed more heat when placed under the light source? Why?

2. Which thermometer lost more heat when the light was removed? Why?

3. From the results of this activity, explain why solar collectors have a transparent cover and are painted black inside.

TECHNOLOGY

CONVERTING GEOTHERMAL ENERGY TO ELECTRICITY

Heat energy from the earth may be free for the taking. What sounds like a dream may fast be becoming an economic reality. People have used the earth's heat, called geothermal energy, for cooking, recreational activities, and for heating homes since before the time of the Romans. Increasingly sophisticated technology has allowed us to tap heat sources deeper and deeper into the earth. In Reykjavík, the capital of Iceland, hot water from wells deep in the earth has been piped into homes, factories, and schools since 1930. Some of these wells are over 2 kilometers deep, and the temperature of the water from them is almost 225°C. Reykjavík is located over a gigantic reservoir of hot water, and can use geothermal energy directly. However, if geothermal energy is to be used more generally, it must be converted to a form of energy that can be transmitted from the source of heat to the place where its use is required. That form of energy is electricity.

The world's largest geothermal electricity project is located at Geysers, California, this facility produces enough electricity to meet the requirements of San Francisco. Geysers is unusual in that the water it uses is brought to the surface as steam. This steam can be used directly to turn turbines to generate electricity. More commonly, hot water comes to the surface under pressure, and energy must be added to change it to steam. This hot water usually contains dissolved minerals, which present one of the major problems to be solved if geothermal energy is to be used economically. Some water contains up to 35 percent dissolved minerals. When pressure is reduced and the water turns to steam, these minerals settle. This settling can reduce the diameter of a 7.6-centimeter pipe to as little as 1.3 centimeters in 100 hours and destroy parts made of normal metals.

A number of different approaches to converting geothermal energy to electricity are being tested today. At one facility in California, water at 210°C is pumped to the surface. As the pressure on the water is reduced, some of it "flash boils," or turns to steam. In this way, about 20 percent of the liquid is used to turn turbines. The rest of the water is pumped back into the ground. Another approach is the binary-cycle (two cycle) method. In this method, the hot water is used to heat another substance, which has a lower boiling point. This second liquid vaporizes and drives a turbine. With this method, scientists hope to eliminate the problem of dissolved minerals and to allow the use of lower temperature water.

With the potential radiation problems of nuclear plants, the pollution problems surrounding plants that burn coal, and the high cost of gas and oil, alternative energy sources are always being sought. Geothermal generation of electricity could be an important technological advance for the western United States and other parts of the world. (The photo shows a geothermal plant that is now in operation in Italy.) Geothermal energy, like wind energy, can provide a safe and clean source of fuel to bolster depleting reserves of fossil fuels.

CHAPTER REVIEW

See teacher's commentary for answers.

VOCABULARY

On a separate piece of paper, match the number of the sentence with the term that best completes it. Use each term only once.

fossil fuel	watershed	conservation	fission
geothermal power	thermal pollution	mineral deposit	fusion
solar cell	pollution	natural resource	petroleum
nuclear energy	recycling	ore	hydrocarbons

1. A(n) _____ is any substance from the earth that can be used in some way.
2. The adding of harmful substances to the environment is called _____.
3. A kind of rock that contains useful minerals is called a(n) _____.
4. _____ is the saving of natural resources by controlling how they are used.
5. A(n) _____ is a body of ore.
6. The process of recovering a natural resource and using it again is _____.
7. When water is used as a cooling agent, _____ may occur.
8. All the land drained by a stream or river is called a(n) _____.
9. A mineral fuel found in the earth's crust is _____.
10. _____ is usually a liquid fossil fuel.
11. _____ are the principal substances found in petroleum.
12. The kind of energy produced from the nuclei of atoms is called _____.
13. _____ is the splitting of the nuclei of certain kinds of atoms.
14. _____ is the joining together of small atoms to make larger atoms.
15. A(n) _____ changes sunlight directly into electricity.
16. Energy obtained from heat in the earth's crust is called _____.

QUESTIONS

Give brief but complete answers to each of the following questions. Unless otherwise indicated, use complete sentences to write your answers.

1. Name a natural resource and tell why it is one.
2. What is the difference between a renewable resource and a nonrenewable resource? Give an example of each kind.
3. Describe two problems resulting from the earth's population growth.
4. Explain why it is important to prevent pollution.

5. Why are mineral deposits considered to be nonrenewable?

6. How can a damaged watershed cause floods, pollution, and a reduction of water supply?

7. List three problems of using coal as an energy source.

8. What is the difference between fusion and fission?

9. Why is solar energy a renewable energy source?

10. Compare using solar energy to using fossil fuels for energy.

11. How are petroleum and natural gas related?

APPLYING SCIENCE

1. You already know that aluminum cans and newspapers are recycled. Discuss recycling with a number of people to find out what other materials are being recycled. List at least five other materials being recycled and describe how they are being recycled.

2. Organize a trip to the facilities of your local water supplier. Find out where the water comes from and what is done to purify it for your use. Also ask, what plans have been made for any future population growth. Find out if the water supplier has a plan to prevent pollution of the water supply and how you and your classmates might help in the plan.

3. Write to the U.S. Committee for Energy Awareness, P.O. Box 37012, Washington, DC 20013, or to your local member of Congress, and ask to be sent the most recent update on "Energy Projections to the Year 2000," which is prepared from time to time by the U.S. Department of Energy. They will also send you the most recent information on alternate energy sources. Take the material to school and discuss the information with your teacher or class.

BIBLIOGRAPHY

Branley, Franklyn M. *Water for the World*. New York: Crowell, 1982.

Eldridge, Frank R. *Wind Machines*. New York: Van Nostrand Reinhold, 1980.

Findley, Rowe. "Our National Forests, Problems in Paradise." *National Geographic*, September 1982.

Hellman, Hal. "The Day New York Runs Out of Water." *Science 81*, May 1981.

Kraft, Betsy Haney. *Oil and Natural Gas*. New York: Watts, 1982.

Roberts, Leslie. "Reactors In Orbit." *Science 83*, December 1983.

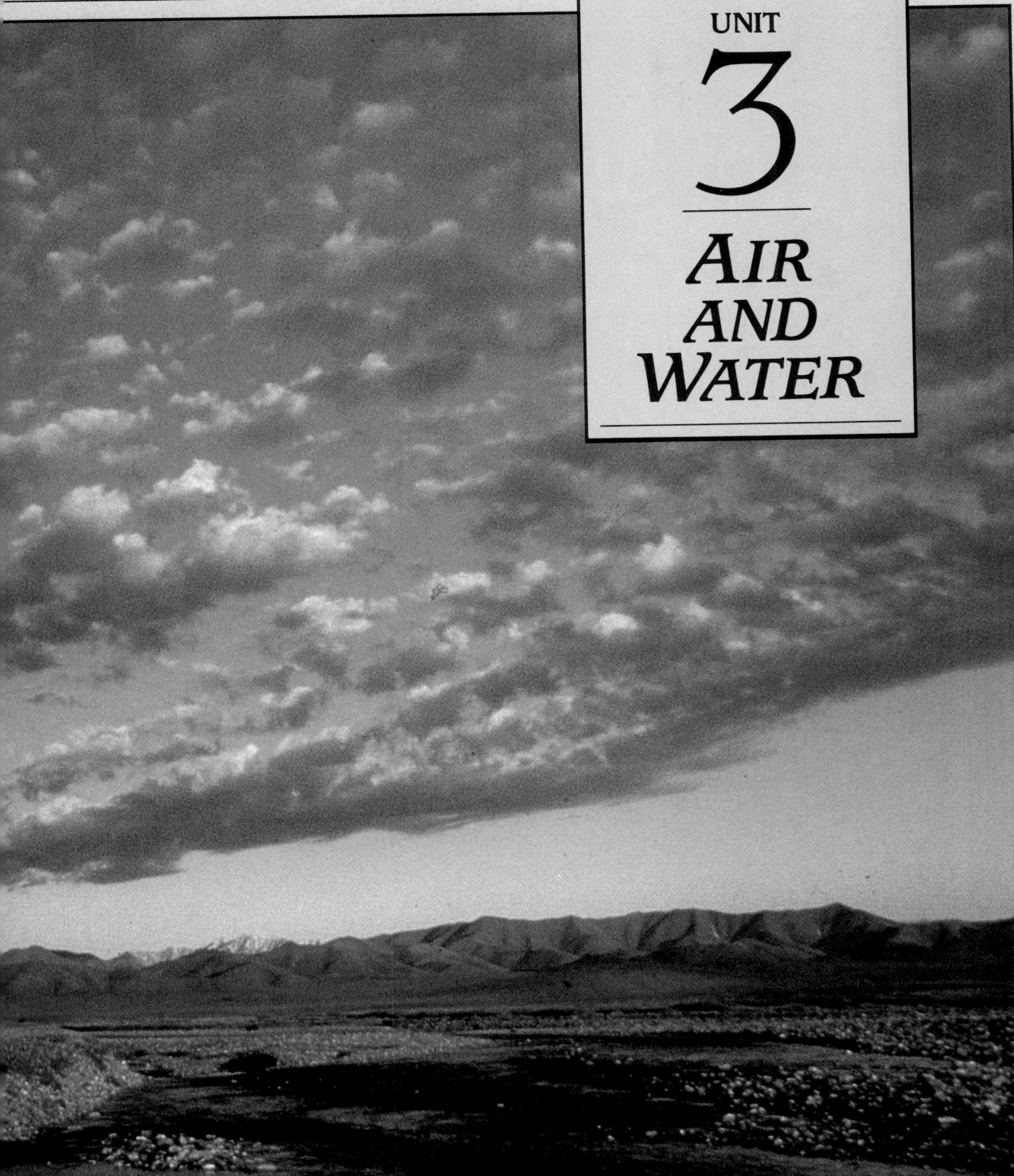

UNIT

3

AIR AND WATER

Our planet is surrounded by a layer of gases called the atmosphere. Without these gases, there would be no clouds or colorful sunsets. The earth's surface would be as barren as the moon.

AIR

CHAPTER GOALS

1. Name the main gases that make up the atmosphere.
2. Describe air pressure and how it is measured.
3. Explain how the air in your area may become polluted.
4. Compare the four layers of the atmosphere.

11–1. Composition of the Earth's Atmosphere

At the end of this section you will be able to:

- ☐ List the five most abundant gases that are found in the atmosphere.
- ☐ Describe two kinds of air pollutants.
- ☐ Explain how the most common man-made air pollutants are produced.

You could live about five weeks without food and about five days without water. However, you could live only five minutes without air. For this reason alone, the atmosphere might be called the most important of all earth resources.

Word Study: Atmosphere: *atmos* + *sphaira* (Gr.), vapor sphere.

GASES IN THE AIR

The air that you take into your lungs with each breath is not a single gas. It is a mixture of many gases. The amount of some of the gases changes from time to time. However, the five gases described below make up 99.99 percent of the total volume of the atmosphere. Air also usually contains small amounts of dust and other solid particles.

1. Nitrogen is the most plentiful of all the gases in air. It makes up about 78 percent of the volume of air. See Fig. 11–1. You do not use the nitrogen in the air. The nitrogen taken in with each breath is normally exhaled. Most plants, on the other hand, use nitrogen. However, plants cannot take the nitrogen directly from air. Some bacteria, called nitrogen-fixing bacteria, are able to take nitrogen from the air and change it into forms that plants can use. These bacteria live in the soil and in the roots of some plants, such as clover. Lightning makes nitrogen

Enrichment: Gravity holds the earth's atmosphere. Small bodies such as Mercury and the moon have too little gravity to hold an atmosphere.

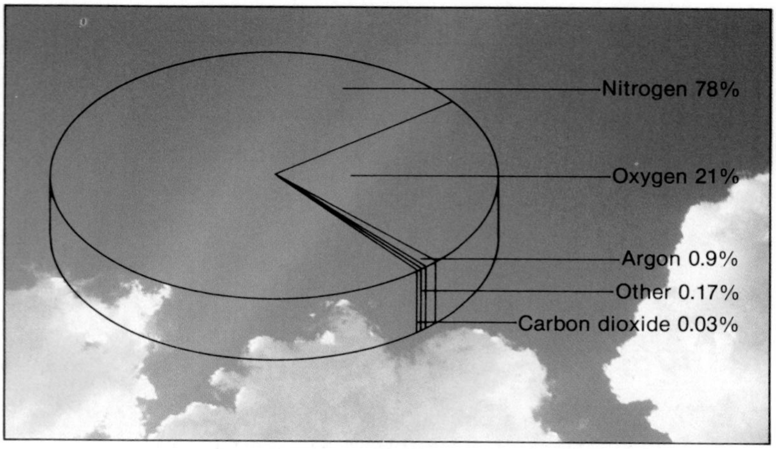

![Pie chart showing composition of air: Nitrogen 78%, Oxygen 21%, Argon 0.9%, Other 0.17%, Carbon dioxide 0.03%]

Nitrogen 78%

Oxygen 21%

Argon 0.9%
Other 0.17%
Carbon dioxide 0.03%

Fig. 11-1 Nitrogen and oxygen make up most of the volume of air.

(The total of 100.1% is caused by rounding off.)

Suggestion: To demonstrate the percentage of oxygen in air, do the following: Wedge a small wad of steel wool in the closed end of a test tube, and invert the test tube in a beaker of water; in several days the formation of rust will be apparent and the water will have risen about one-fifth the distance into the test tube.

compounds in the air. These compounds can be used by plants, too. The compounds are carried by rain into the soil and absorbed through the roots of the plants. Animals that eat plants get the nitrogen they need in this way. The waste products of animals and dead plants break down and release nitrogen into the soil. Other bacteria cause nitrogen to be returned to the atmosphere. The process by which nitrogen is taken from the air and then returned to the air is called the *nitrogen cycle*. See Fig. 11-2. The nitrogen cycle returns as much nitrogen to the air as is removed. This keeps the amount of nitrogen in the air at a constant level.

2. Oxygen is the second most plentiful gas in air. It makes up about 21 percent of the air by volume. Animals remove some of the oxygen from the air as they breathe. Oxygen is also removed by the decay of plant and animal matter and by the

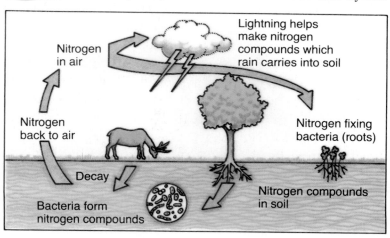

Fig. 11-2 The nitrogen cycle.

weathering of rocks. At the same time, green plants give off oxygen through a process called *photosynthesis* (foh-toe-**sin**-the-sis). Photosynthesis replaces the oxygen that is used. In this way, the amount of oxygen in the air stays the same.

3. Argon is a gas found in very small amounts in air. It makes up 0.9 percent of the air. Argon does not combine chemically with other substances. The atmosphere also contains tiny amounts of neon, helium, and other rare gases that are similar to argon.

4. Carbon dioxide makes up only 0.03 percent of the air, but it is an important gas. Carbon dioxide is used by plants during photosynthesis. The carbon dioxide is taken in through the plant leaves. Green plants use carbon dioxide to make their own food. Oxygen is released as a byproduct as plants carry on photosynthesis. For animals, carbon dioxide is a waste product. Animals give off carbon dioxide as they obtain energy from their food. The process by which carbon dioxide is taken from the air, then returned, is called the *carbon dioxide cycle*. See Fig. 11–3. The amount of carbon dioxide used by plants seems to be equal to the amount released by animals and other natural processes. However, this balance is upset by the burning of fuels, such as coal and oil. Burning these fuels dumps billions of metric tons of carbon dioxide into the atmosphere each year. As a result, the amount of carbon dioxide in the atmosphere is increasing slightly each year.

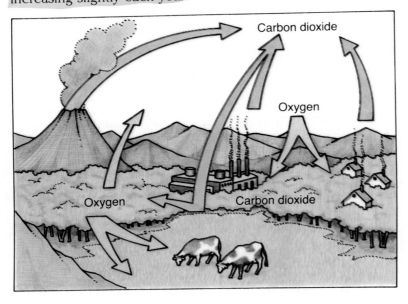

Fig. 11–3 The carbon dioxide cycle.

5. Water vapor. One of the most important gases in the atmosphere is *water vapor*. Unlike the first four gases described, the amount of water vapor in air varies greatly. The amount of water vapor in extremely cold air or in desert regions is very small. On the other hand, the warm air over warm oceans contains large amounts of water vapor. The amount of water vapor in air may be as much as 3 percent by volume. But most of the time it makes up less than 1 percent of the volume of air.

AIR POLLUTANTS

Every year millions of metric tons of substances not normally found in the air are dumped into the atmosphere. These substances in the atmosphere are *air pollutants*. There are two kinds of air pollutants: *solids* and *gases*. Some kinds of solid pollutants come from natural sources. These are dust and microscopic living things such as bacteria. Volcanoes are also a source of dust pollution in the atmosphere. Some volcanoes throw dust particles high into the atmosphere. It may take several years for the smallest of these volcanic dust particles to return to earth.

People, mainly in cities, also add large amounts of solid particles to the air. Smoke from the burning of fuels such as coal gives off particles of carbon and other solids, which enter the air. Burning wood in home heating stoves has caused a smoke pollution problem in some cities. Automobiles also give off small, solid particles of carbon. Automobiles that burn leaded gasoline give off tiny particles of lead. Industries may give off dust from operations that involve cutting, grinding, and crushing materials. One very harmful form of solid pollution is asbestos. See Fig. 11-4. Asbestos is a mineral used in making many things. It is found, for example, in some insulating materials and in some automobile parts. Breathing air that is polluted with small particles of asbestos is known to cause lung cancer.

CAUSES OF AIR POLLUTION

1. Automobiles. Gases cause more air pollution than solids do. The automobile is one of the major sources of pollutant gases. One of the most harmful gases given off is *carbon monoxide*. This very poisonous gas may build up in the air near heavy traffic or in any closed space where a motor is running. Automobiles also give off *nitrogen oxides* and *hydrocarbons*.

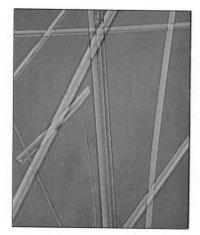

Fig. 11-4 Asbestos was used as a fire retardant in buildings until it was deemed unhealthful.

Warm air

Cooler air

Fig. 11-5 (top) *Smog is often formed in cool air that has been trapped beneath a warmer layer.* (bottom) *A photo of smog over a city where cool air from the ocean is trapped by warm air from inland. The surrounding mountains also help to hold the smog over the city.*

These two kinds of gases together can produce smog. Smog can cause serious health problems and damage rubber and other materials. See Fig. 11-5.

2. Coal and oil. Not all pollutant gases come from automobiles. Burning coal and oil in power plants and industries can produce *sulfur dioxide*, along with nitrogen oxides. Sulfur dioxide is dangerous to breathe. It also combines with water to produce an acid. This acid damages metal, and even some kinds of stone. Sulfur dioxide and nitrogen oxides dissolve in raindrops to make acid rain. Acid rain may be causing serious harm to forests and to the life in lakes.

Most pollutants do not stay in the atmosphere for a long time. Many solid particles fall out of the air or are washed out by rain and snow within a few weeks. Most gases are broken down or absorbed by moisture within several months. However, pollutants that are carried to the upper parts of the atmosphere may remain there for up to three years. For example, a powerful volcanic eruption may throw dust to a great height. The volcanic dust will be trapped in the upper atmosphere. There are no weather changes at those high altitudes to help clean the air. The dust may affect the weather as it is carried all over the world.

Serious air pollution problems are usually caused by people in cities or other areas with large populations. Laws have been passed to help control the sources of air pollution in such areas. For example, automobile makers are required by law to provide special converters in all automobile exhaust systems. The converter keeps carbon monoxide and hydrocarbon gases from being given off. Automobile makers also must make engines that give off smaller amounts of pollutants than earlier engines. Other laws control power plants and industries that give off pollutants. These laws have already helped greatly in reducing air pollution almost everywhere. More laws and public information programs that show the benefits of clean air are still needed. Polluted air cannot be purified and reused as water can. But pollutants put into the air can be reduced.

SUMMARY

The atmosphere is made up of a layer of gases that surrounds the solid earth. This layer is made up of a mixture of gases. The most abundant gases are nitrogen and oxygen, along with small amounts of carbon dioxide, water vapor, and some rare gases. Air pollutants can be solids or gases. Human-made gases cause the most serious kinds of air pollution.

QUESTIONS

Use complete sentences to write your answers.

1. Name the five most abundant gases in the atmosphere. List them in order and describe the amount of each that is found in the air.
2. List the sources of solid air pollutants.
3. Name four gases that are common air pollutants. Explain how each is produced.

1. Nitrogen: 78%; oxygen: 21%; argon: almost 1%; carbon dioxide: .03%; water vapor: varies from almost none to as high as 3%.
2. Solids can come from volcanoes and the burning of fuels, such as coal, wood, and gasoline.
3. Carbon monoxide, nitrogen oxides, and hydrocarbons are produced by automobile exhaust. Sulfur dioxide is produced by the burning of coal and oil in power plants and industry.

INVESTIGATION

ACIDS FROM THE AIR

PURPOSE: To test for gases in the air that form acids.

MATERIALS:

graduated cylinder	matches
limewater	flask
beaker	stirring rod
straw	vinegar
candle	blue litmus paper
pie tin	

PROCEDURE:

A. Pour 10 mL of the limewater into the beaker. CAUTION: LIMEWATER WILL DAMAGE SKIN IF LEFT ON IT. RINSE OFF WITH WATER. Use the straw to blow your breath through the limewater solution until a change occurs in the limewater. Do NOT get limewater in your mouth.

   1. What happens to the limewater?

Fig. 11–6

B. The result in procedure A is the test for carbon dioxide. Carbon dioxide enters the air in large amounts by the combustion of fuels. Attach the candle to the pie tin by using a few drops of molten wax from the burning candle. Place the flask upside down over the burning candle. See Fig. 11–6. Allow the candle to go out. Then place the flask upright and add 10 mL of limewater. Swirl it around to mix with the gases in the flask.

   2. Describe what happens to the limewater.

   3. What gas must have been in the flask?

C. Carbon dioxide dissolves in water to produce carbonic acid. Another gas found in air as a result of the combustion of fuel oil, gasoline, and coal is sulfur dioxide. It, too, forms an acid when dissolved in water. Sulfur dioxide is also made by burning match heads. It has a noticeable odor. Light a match and notice the odor that comes from the burning match head.

   4. Describe the odor.

D. Using a stirring rod, place a drop of vinegar on a piece of blue litmus paper.

   5. What happens to the litmus color?

E. The result in procedure D is the test for an acid. Both carbon dioxide and sulfur dioxide are given off when a match burns. These gases must dissolve in water to become an acid. Moisten a piece of blue litmus paper with distilled water. Light a match and hold the moistened blue litmus above the flame of the match until you see a color change.

   6. What happens to the litmus color?

CONCLUSIONS:

   1. Why did the moist litmus change color when held over a burning match?

   2. Explain how drops of water falling through the air may produce acid rain.

11–2. Atmospheric Pressure and Temperature

At the end of this section you will be able to:
- ☐ Explain how air pressure is measured.
- ☐ Predict how pressure and temperature will change as you move upward from the earth's surface.
- ☐ Name four layers of the atmosphere beginning at the earth's surface.

Fig. 11–7 (right) *Hang gliders depend on the upward motion of warm air.*
Fig. 11–8 (below) *The column of air pushing down on you weighs more than 1,000 kg!*

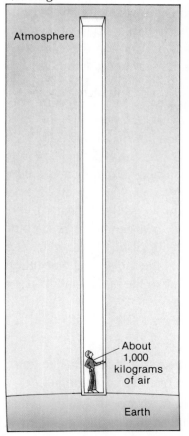

Atmosphere

About 1,000 kilograms of air

Earth

Air cannot be seen. Only its effects can be seen. The hang glider shown in Fig. 11–7 has just soared off a cliff. A column of warm air lifts the glider and keeps it in the air. How can you tell that you live at the bottom of an ocean of air?

MEASURING AIR PRESSURE

As you walk around, you are carrying almost 1,000 kilograms of air on your head and shoulders. See Fig. 11–8. The weight of this air pushing down causes air pressure. You do not feel this pressure because there is equal pressure inside your body. Sometimes you can see the effect of air pressure. For example, air pressure makes it possible for you to drink through a straw. The liquid moves up the straw because you remove the air from inside the straw by breathing in. This causes the air pressure inside the straw to drop. The air pressure on the liquid then pushes it up the straw. See Fig. 11–9.

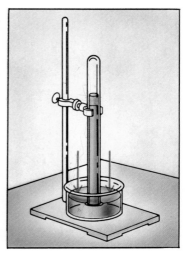

Fig. 11–9 When you drink through a straw, you are causing a pressure change.

Fig. 11–10 A simple mercury barometer.

Air pressure can be measured by using the weight of air. The air pushes a liquid up into a tube from which the air has been removed. Atmospheric pressure is measured by an instrument called a **barometer** (buh-**rom**-uh-tur). One kind is a mercury *barometer,* shown in Fig. 11–10. A glass tube about one meter long is closed at one end. It is then filled with mercury. Mercury is used because it is a very dense liquid. The tube is then placed, open end down, into a dish of mercury. The mercury runs out of the tube until its weight is equal to the weight of air on the mercury in the dish. In other words, the mercury is pushed up into the tube by the air pressure.

At sea level, the mercury runs out of the tube until the height of the mercury left in the tube is equal to 760 millimeters. For this reason, the air pressure at sea level is said to equal 760 millimeters of mercury. If air pressure increases, the height of mercury in the tube will rise. If air pressure decreases, the height of the mercury in the tube will fall. A measurement of the height of mercury in a barometer is used to express air pressure.

Mercury barometers are not easy to move from one place to another. A more useful kind of barometer is an *aneroid* (**an**-uh-roid) barometer. It does not use a liquid. Aneroid means "without liquid." An aneroid barometer uses a round, sealed metal can. One side of the can squeezes in or springs out as the air pressure changes. This movement causes the pointer on a dial to move. The air pressure can then be read. For example, on an average day at sea level, the dial might give a reading of 760 millimeters or 29.92 inches.

Barometer An instrument that is used to measure atmospheric pressure.

Suggestion: Demonstrate the air pressure as follows: Place 50 mL of water in a 1-gallon metal container having a narrow opening (duplicator fluid container, etc.). Heat container with top off until steam is emitted for at least 30 seconds, then turn off heat, quickly place cap on or push a solid rubber stopper snugly into the opening, and allow to cool. Container collapses under the atmospheric pressure.

Enrichment: An altimeter is a special kind of aneroid barometer used in aircraft.

Fig. 11-11 An aneroid barometer shows air pressure on a dial.

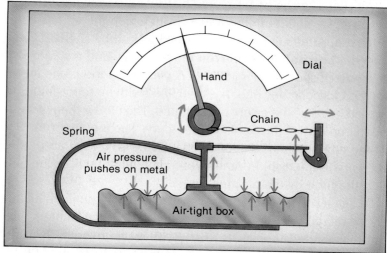

Fig. 11-12 This diagram shows how an aneroid barometer works.

Some aneroid barometers, called *barographs*, keep a continuous record of air pressure. The recordings made by this kind of barometer appear as a line on paper held on a revolving drum. The line traced on the paper shows how the air pressure changes over a period of time.

AIR PRESSURE AND TEMPERATURE

When you take a trip to the mountains, you may notice that your ears pop. This happens because of a slight drop in the air pressure. If you were at the top of a mountain five kilometers high, about half of all the atmosphere would lie below and half above. Thus on the mountain there is less air above you than there is at sea level. This causes the air inside your ear to have a higher pressure than the air outside. When pressure inside and outside of your ears becomes equal, a popping sound is heard.

In addition to lower air pressure than at sea level, the temperature on the mountain top is colder. At an altitude of five kilometers, the temperature averages 30°C colder than at sea level. Pressure and temperature continue to drop at higher altitudes. At a height of about 11 kilometers, the pressure drops to 25 percent of that at sea level. At the same altitude, the temperature is about −55°C. This altitude marks the boundary between two layers of the atmosphere.

LAYERS OF THE ATMOSPHERE

1. Troposphere. Temperature changes at certain altitudes divide the atmosphere into layers. Closest to the earth is the **troposphere** (**troe**-puh-sfear), the densest layer of the atmosphere. This is the layer in which you live. Almost all weather takes place in the *troposphere*. Gravity causes the gases of the atmosphere to be pulled toward the earth's surface. This causes the troposphere to have about half of the total weight of the atmosphere. In the troposphere, the temperature drops an average of 6.5°C for each kilometer of increasing altitude. At a height of about ten kilometers the temperature stops dropping. See Fig. 11-13, page 320. This marks the upper boundary of the troposphere. The upper boundary of the troposphere is higher over the equator than at the poles.

2. Stratosphere. Above the troposphere, at an altitude of about 20 kilometers, is the clear, cold layer called the **stratosphere** (**strat**-uh-sfear). It extends upward to an altitude of about 50 kilometers. Temperatures in the lower *stratosphere* are very cold. However, above the altitude of 20 kilometers, the temperature begins to rise again. At the height of about 50 kilometers, the temperature rises to about 0°C. A small amount of a special form of oxygen, called *ozone,* is the reason for the temperature rise. This small amount of ozone is very important. Ozone is able to filter out the sun's most dangerous ultraviolet rays. If the full strength of those rays were to reach the earth's surface, they would harm living things.

3. Mesosphere. The next layer above the stratosphere is called the **mesosphere** (**mes**-uh-sfear). See Fig. 11-13. It is the coldest layer of the atmosphere. The *mesosphere* extends

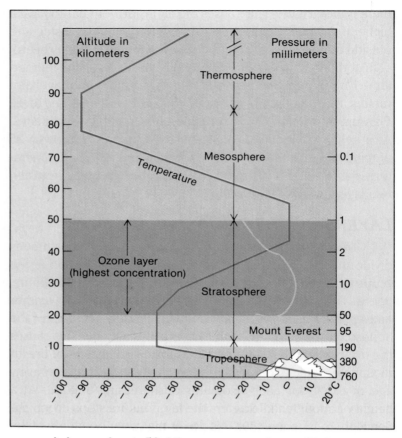

Fig. 11–13 Changes in temperature mark the boundaries between layers of the atmosphere.

Thermosphere The layer of the atmosphere above the mesosphere, extending from about 85 km to about 600 km, where temperature increases with increasing altitude.

Ionosphere A region of the upper atmosphere that contains many electrically charged particles.

upward from about 50 kilometers to about 85 kilometers, where the temperature decreases to about −90°C.

4. Thermosphere. Above the mesosphere is the **thermosphere** (**thur**-muh-sfear). The *thermosphere* reaches to an altitude of about 600 kilometers, with the temperature going up to 2,000°C. Air in the thermosphere is very thin. It is only one ten-millionth (0.0000001) as dense as air at sea level. Beyond the thermosphere, the earth's atmosphere blends gradually into the very thin gases found in deep space.

Within the thermosphere is a region called the **ionosphere** (ie-**on**-uh-sfear). It begins at a height of about 80 kilometers and goes up to about 400 kilometers. The *ionosphere* also makes up the upper part of the mesosphere. This region contains electrically charged particles called *ions*. Ions are atoms or molecules that are charged when electrons are gained or lost. This happens in the ionosphere when solar energy is absorbed by the gases. The ionosphere plays an important part in

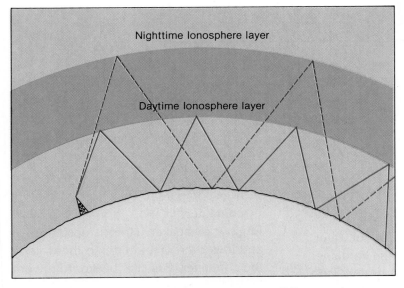

Fig. 11-14 Radio waves are reflected from the ionosphere.

the way radio waves can be sent between different places on the earth. The ionosphere can reflect many kinds of radio waves back toward the earth, just as a mirror reflects light. This allows many kinds of radio waves to travel long distances by bouncing back and forth between the earth and the ionosphere. See Fig. 11-14. Radio signals often cover greater distances at night than during the day. This happens when the ionosphere is at a higher altitude at night.

SUMMARY

Air pressure is caused by the weight of the atmospheric gases. Air pressure is measured by barometers. Moving upward through the atmosphere, pressure decreases as the temperature changes. The atmosphere is divided into layers because of these temperature changes.

QUESTIONS

Use complete sentences to write your answers.

1. Explain how two kinds of barometers work for measuring air pressure.
2. Describe how pressure and temperature change as you move upward in the lower atmosphere.
3. List the four layers of the atmosphere starting at the earth's surface and going upward.

1. A mercury barometer uses the weight of air to push mercury up a tube from which the air has been removed. An aneroid barometer uses a round, sealed metal can. One side of the can moves in or out as the air pressure changes.
2. Both the temperature and pressure become lower as you move upward in the lower atmosphere.
3. Troposphere, stratosphere, mesosphere, and thermosphere.

INVESTIGATION

See teacher's commentary for teaching hints, safety, and answers to questions.

ATMOSPHERIC PRESSURE

PURPOSE: To show the effect of air pressure.

MATERIALS:

large container   water
small beaker   drinking straw

PROCEDURE:

A. Fill the large container nearly full of water. Lay the beaker on its side in the container and let it fill completely with water. Turn the beaker upside down underwater. Carefully lift the beaker until its open end is just below the surface of the water.

   1. What happens to the water in the beaker?

   2. How does the water level in the beaker compare to the water level in the container?

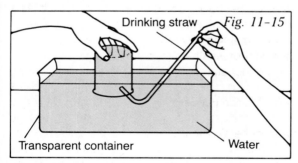

Drinking straw   *Fig. 11–15*

Transparent container   Water

B. Hold the beaker as you did above with its open end just underwater. Put your finger over one end of the straw. Keep your finger over the outside end of the straw as you place the other end into the water in the beaker. See Fig. 11–15. Release your finger and observe what happens. (If nothing happens to the water in the

beaker, it is because some water got into the straw. Remove the straw, shake it dry, and try again.)

   3. What happens to the water inside the beaker?

   4. How does the water level in the beaker now compare to the water level in the container?

C. Empty the beaker. Hold it upside down and lower the open end into the water in the container. Push the beaker to the bottom of the container.

   5. What happens to the air inside the beaker?

   6. How does the water level in the beaker compare to that in the container?

D. Put your finger over the outside end of the straw. Hold the beaker as you did for procedure C and insert the other end of the straw into the beaker. Remove your finger from the outside end of the straw.

   7. What happens to the air inside the beaker?

   8. How does the water level in the beaker now compare to the water level in the container?

CONCLUSIONS:

   1. How does air pressure explain your observations in procedures A and B?

   2. How does air pressure explain your observations in procedures C and D?

   3. A stoppered bottle half-full of water is set upside down in the water. What happens when the stopper is removed with the mouth of the bottle underwater?

CAREERS IN SCIENCE

AIR ANALYST

Air analysts monitor the pollution levels in the atmosphere. They take air samples and analyze them for pollutants. For example, analysis may show that pollution levels in a city are too high. Analysts may also be asked to take samples of the air in industrial worksites and other locations such as school buildings, where the health of large numbers of persons might be affected by pollutants in the air. Devices are used that collect dust so it can be examined for particles of lead, rock, coal, or asbestos. After the samples are analyzed, reports of the findings are issued. If the pollution level is unsafe, analysts suggest methods that can be used for reducing pollution.

For a position as an air analyst, you would need a college degree, with courses in chemistry, meteorology, and mathematics. For more information, write: Air Pollution Control Association, P.O. Box 2861, Pittsburgh, PA 15230. Your local branch of the Environmental Protection Agency is also a good source of information.

METEOROLOGICAL TECHNICIAN

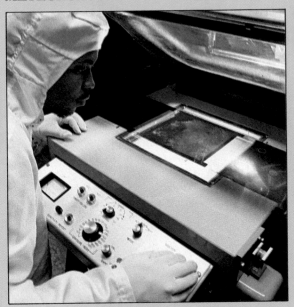

There's an old saying that everyone talks about the weather but no one does anything about it. As a meteorological technician, you would be able to "do something" by working with meteorologists in the field of weather forecasting. Technicians work at weather stations where they make observations and record data important to forecasting, such as precipitation (levels of rainfall), wind speeds, cloud cover, and barometric pressure (the pressure of the atmosphere). They may also be involved in receiving information transmitted from weather satellites about wind patterns and ocean currents. This information is sent to the National Oceanic and Atmospheric Administration's computer center where national weather maps are prepared. Technicians are also involved in weather research projects and in developing instruments.

Specialized training is offered at community colleges and technical institutes. For more information, write: American Meteorological Society, 45 Beacon Street, Boston, MA 02108.

¡COMPUTE!

SCIENCE INPUT

Some weather factors are valuable predictors of weather and some are not. Two commonly considered weather variables are temperature, measured with a thermometer, and air pressure, measured with a barometer. Before the use of weather satellites, which give more up-to-the-minute information for forecasters, the thermometer and barometer, along with past experience, were the chief tools for weather prediction. In this exercise you will use these "old-fashioned" tools to gather data. The modern computer will then be used to analyze the data.

COMPUTER INPUT

The computer's ability to manipulate and rearrange information is especially important. When data are collected, great quantities of information can be provided. If the information is not organized according to some system, none of it will make sense.

For example, imagine someone asking you to collect data about the students in your class. First you would have to know what kinds of questions you intended to answer with the data. Otherwise, you wouldn't know what kinds of information to collect. Once you collected the data, you would then have to arrange it or it would seem like a hodgepodge of information. If you sort the data and provide order, however, patterns become clear and it becomes easier to draw conclusions.

A computer can manipulate data easily. It can let the facts tell a story from many points of view. In this exercise, you will sort weather data to see if any patterns emerge that can help predict weather in the future.

WHAT TO DO

First, you must collect the data for the computer to sort. For 14 days, at approximately the same time each day, record the temperature (degrees Celsius), the air pressure (millibars), and the present state of the weather (fair, rain, possible rain). Make a chart of that data. A sample data chart is provided for your reference.

Enter Program Forecaster into your computer. Enter your data in statements 401 to 415. Use the form given in the example shown in statement 401. When you run the program, it will ask you how you would like to sort the data, by temperature or air pressure. Try both methods. After you've analyzed the data, decide which instrument was more useful for predicting the weather: the thermometer or the barometer.

GLOSSARY

HIGH LEVEL LANGUAGE	A programming language that uses simple terms from everyday vocabulary for its commands.
BASIC, LOGO, PASCAL, FORTRAN	Examples of high level languages using the alphanumeric system.
LOW LEVEL LANGUAGE	Machine language. It uses electronic data, not words, to execute the program.
BINARY NUMBER SYSTEM	Example of low level machine language.

PROGRAM

```
100  REM FORECASTER
110  DIM T$(15), AP$(15), SW$(15)
120  FOR X = 1 TO 15
130  READ T$(X),AP$(X),SW$(X)
132  H = X
135  IF T$(X) = "0" THEN H = X - 1:
     GOTO 150
140  NEXT
150  PRINT: PRINT "1) SORT
     TEMPERATURE & STATE OF
     WEATHER"
160  PRINT "2) SORT AIR PRESSURE &
     STATE OF WEATHER"
170  PRINT: INPUT "    WHICH CHOICE
     (1 OR 2)? ";C$
180  C = VAL(C$)
190  IF C < 1 OR C > 2 THEN GOTO 150
200  K = 0
210  FOR X = 1 TO H - 1
220  IF C = 2 THEN GOTO 240
230  IF VAL(T$(X)) < = VAL(T$(X + 1))
     THEN GOTO 290
235  GOTO 250
240  IF VAL (AP$(X)) < = VAL (AP$(X +
     1)) THEN GOTO 290
250  T$ = T$(X):T$(X) = T$(X + 1): T$(X
     + 1) = T$
260  AP$ = AP$(X):AP$(X) = AP$(X + 1):
     AP$(X + 1) = AP$
270  SW$ = SW$(X):SW$(X) = SW$(X +
     1): SW$(X + 1) = SW$
280  K = K + 1
290  NEXT
300  IF K > 0 THEN GOTO 200
310  PRINT "DAY"; TAB( 8)"TEMP."; TAB
     (15)"PRES."; TAB( 25)"STATE OF
     WEATHER"
320  FOR X = 1 TO H
330  PRINT X;
332  IF C = 1 THEN PRINT TAB( 8)T$(X);
335  IF C = 2 THEN PRINT TAB
     (15)AP$(X);
337  PRINT TAB(25)SW$(X)
340  NEXT
350  GOTO 150
400  REM PLACE DATA STATEMENTS
     HERE
401  DATA 55, 1023, FAIR
402  DATA 45, 1010, RAIN
500  DATA 0,0,0
999  END
```

SAMPLE DATA CHART

Day	Temperature (Degrees C)	Air Pressure (Millibars)	State of Weather
#1	45°	1010	rain
#2	55°	1023	fair

BITS OF INFORMATION

Would you like to set up computer communication between your school and one in another neighborhood? It's possible if both locations have computers and modems. A modem (short for modulator-demodulator) changes the electronic impulses inside your computer into vibrations that a telephone can understand and back again into electronic impulses to the other computer.

CHAPTER REVIEW

See teacher's commentary for answers.

VOCABULARY

On a separate piece of paper, match each term with the number of the statement that best explains it. Use each term only once.

barometer   mesosphere   thermosphere   photosynthesis
ionosphere   stratosphere   troposphere   pollutant

1. The outermost layer of the atmosphere.
2. A region of the upper part of the atmosphere containing many electrically charged particles.
3. An instrument used to measure atmospheric pressure.
4. The atmospheric layer just above the troposphere.
5. The layer of the atmosphere closest to the earth's surface.
6. The upper atmospheric layer that filters the sun's ultraviolet rays.
7. A substance not normally found in air.
8. The process by which plants release oxygen.

QUESTIONS

Give brief but complete answers to each of the following questions. Unless otherwise indicated, use complete sentences to write your answers.

1. What two gases make up 99% of the air by volume?
2. Why is it difficult to list the percentage of water vapor found in air?
3. How is air pollution causing serious damage to forests and to the life in lakes?
4. What is meant by the *carbon dioxide cycle?*
5. Which pollutants from auto exhausts produce smog?
6. Explain why you should never run an auto engine in a closed garage.
7. Pollutants that get into the upper atmosphere may become a much greater hazard than those in the lower atmosphere. Why is this?
8. Which would you rather take with you on a trip to the mountains, a mercury barometer or an aneroid barometer? Explain your choice.
9. Which layer of the atmosphere contains about half the total weight of air?
10. In which atmospheric layer does the highest temperature exist?
11. Why do your ears pop when you go from a lower to a higher level?

12. Describe what happens to the temperature and pressure as you go higher and higher into the troposphere.

13. Predict what the temperature would be at an altitude of two kilometers above a location on the ground where the temperature is 20 degrees Celsius.

14. In which atmospheric layers is ozone found? Why is it important?

15. Just after dark you notice that the radio station you are listening to, which is some distance from your town, begins to fade away, and you start receiving a station even farther away. Why is this happening?

APPLYING SCIENCE

1. Write down the location of a distant radio station that you can receive at night. Make a diagram showing the location of the station and your location. Show on the diagram how the radio waves were able to reach your location at night and not during the day.

2. Locate an aneroid barometer. On a weekend day, make readings on it every two hours from the time you get up in the morning until you go to bed that night. Note the kind of weather at the time of each reading. Make a graph of the readings and try to account for any pattern you find in the data.

3. If you were to climb Mount Everest, an 8,845-meter-high mountain, what changes in the atmosphere would you expect? What materials would you bring with you so that you could better survive these changes?

4. The automobile as a producer of air pollutants has been a target for state and federal governments. Find out what kinds of gases are being regulated as auto exhaust. A good source of information is a local automobile dealer. Most would be willing to allow you to find out how their autos rate against the state and federal regulations, if you explain that you are doing a class project.

BIBLIOGRAPHY

Allen, Oliver. *Atmosphere* (Planet Earth Series). Chicago: Time-Life Books, 1983.

Franklin, Deborah. "From Bust to Dust." *Science News*, June 1984.

Gedzelman, Stanley D. *The Science and Wonder of the Atmosphere*. New York: Wiley, 1980.

Golden, Frederick. "Lady in a Cage." *Discover*, July 1984.

Every day energy from the sun affects our lives. During the day, we can see the sun's light. On sunny days we can feel its warmth. In the summer many of us make use of its tanning rays.

ENERGY AND WATER IN AIR

CHAPTER GOALS

1. Describe how the sun's energy reaches the earth and what its effect is on the earth's atmosphere.
2. Explain the role of convection in producing winds.
3. Describe the system by which air moves between the equator and the poles.
4. Explain how water enters and leaves the atmosphere.

12–1. Atmospheric Energy

At the end of this section you will be able to:

- ☐ Name three kinds of radiant energy.
- ☐ Explain how the shape of the earth and the presence of water affect the heating of the earth.
- ☐ Explain what is meant by *convection cells* and give several examples.
- ☐ Name and describe the wind belts of the earth.

Have you ever seen the sun at midnight? The left-hand page shows how the sun looks at midnight near the North Pole at certain times during the year. The colored rings might represent all the forms of energy given off by the sun. Some of that energy is trapped by the atmosphere.

ENERGY FROM THE SUN

The sun releases a great amount of energy. About two-billionths (0.000000002) of that energy is received by the earth. Even this tiny part of the sun's energy is a very large amount. It is 13,000 times greater than all the energy used by everyone on earth.

If the earth had no atmosphere, its surface would be like the surface of the moon. The days would be very hot and the nights very cold. The atmosphere prevents this from happening. During the day, the atmosphere reflects some energy back into space. At night, it holds some of the sun's energy from the day. Thus the atmosphere is able to prevent the earth from getting too hot during the day and too cold at night.

Fig. 12-1 *Moving air, which gets its energy from the sun, keeps this sailboard moving.*

Fig. 12-2 *The atmosphere allows less than half of the sun's energy to reach the earth's surface.*

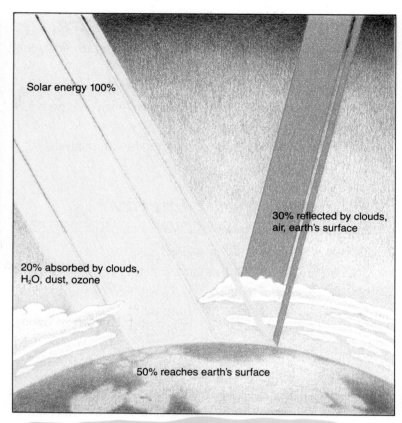

Solar energy 100%

30% reflected by clouds, air, earth's surface

20% absorbed by clouds, H₂O, dust, ozone

50% reaches earth's surface

About 30 percent of the solar energy that reaches the earth is reflected back into space. Clouds are the main reflectors of solar energy. About 20 percent of the solar energy falling on the earth is absorbed directly by the air, and only about 50 percent reaches the surface. See Fig. 12–2.

ATMOSPHERIC EFFECTS

Radiant energy A form of energy that can travel through space.

The sun's energy reaches the earth as **radiant** (**rade**-ee-unt) **energy.** *Radiant energy* is energy that can travel through space. *Light* is one of several kinds of radiant energy. It is the only kind that can be seen. When sunlight falls on your skin, you feel warmth. This is caused by a kind of radiant energy called *infrared* (in-fruh-**red**) *energy.* Sunburn is caused by another kind of radiant energy, *ultraviolet* (ul-truh-**vie**-uh-let) *energy.* Unlike visible light, infrared and ultraviolet energy cannot be seen. Fig. 12–3 shows the kinds of radiant energy given off by the sun.

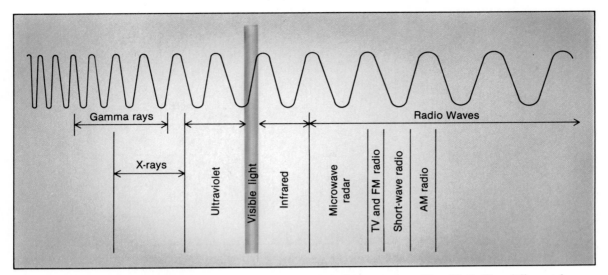

Fig. 12-3 *The different forms of radiant energy are shown on this chart. We can see only the narrow band of visible light.*

Clouds block infrared rays, along with most of the light of the sun. Thus the air feels cooler when a cloud covers the sun. Most of the sun's ultraviolet energy is absorbed by a layer of ozone gas in the upper atmosphere. The small amount of ultraviolet rays that get through this layer can cause sunburn. However, ultraviolet rays pass through clouds. Therefore, you can get sunburned on a cloudy day.

Gases in the air catch rays of blue light from the sun. While other colors pass through, most of the rays of blue light are spread out in all directions. These blue rays are what make the sky look blue. Above the atmosphere, space appears black since there are no gases to catch the light. When the atmosphere contains large particles of dust or water vapor, the other colors of visible light are spread out. This causes the sky to look less blue and more milky, or white. The blueness of the sky shows how free it is of dust and other particles.

Have you ever noticed how hot a car with its windows closed becomes when it is left in the sun? The atmosphere is heated in much the same way as the air in the car. The inside of the car gets hot when radiant energy passes through the car windows. The energy is absorbed by the interior of the car. The energy is then given off again, but it has been changed into longer waves. The glass in the car windows will not allow the longer-wave radiant energy to escape. Energy becomes trapped inside the car, causing the materials and the air inside the car to get hotter and hotter.

Reinforcement: Use a prism to show that blue light is bent the most.

Fig. 12–4 (top) *The greenhouse effect occurs in a greenhouse because the glass windows trap the air.* (bottom) *The same thing happens to a city, but the warmer air is trapped by the atmosphere.*

Greenhouse effect The name given to the method by which the atmosphere traps energy from the sun.

The land and water of the earth are also warmed by absorbing radiant energy. This shorter-wave energy is changed by the land and water, and sent back into the air as longer waves. The longer waves cannot pass out of the atmosphere. This energy is blocked and absorbed by the carbon dioxide and water vapor in the air. The trapped energy heats the air. This heating is called the **greenhouse effect.** This term is used because the atmosphere acts like the glass in a greenhouse, or the windows of a car. The atmosphere allows radiant energy to enter, but prevents it from leaving. The *greenhouse effect* heats the atmosphere. See Fig. 12–4. By adding more carbon dioxide to the atmosphere, more heat will be trapped. The earth already may be getting warmer. The burning of fossil fuels has already put large amounts of carbon dioxide into the air.

HEATING THE EARTH'S SURFACE

1. Water and land. Large amounts of heat energy can be absorbed by water without changing the temperature of the water very much. You could confirm this by heating water in a metal pot. The metal of the pot and its handle will soon become too hot to touch, while the water inside is only warm. However, once the water is heated and removed from the heat source, the pan will cool quickly, while the water remains warm for a long period of time. Because water heats slowly, air near oceans and lakes stays cool during the day. The air in these areas also remains warm at night because water cools slowly. But the land is like the metal pot. When the sun is out, land heats quickly. At night, land quickly loses the heat it absorbed during the day. Thus the air around land areas is warm during the day and cool at night.

2. The shape of the earth. The heating of the earth's surface also is affected by the round shape of the earth. If the earth were flat, all areas would receive the same amount of energy from the sun. But because the earth's surface is curved, polar regions receive less energy than the regions near the equator. See Fig. 12–5. Regions around the equator receive direct rays from the sun, while polar regions receive slanting rays. Slanting rays spread the energy over a larger area than direct rays do. The direct rays falling near the equator have more energy than slanting rays. Near the equator, the air gets hotter than it does at the poles. Without some way to transfer heat from the equator to the poles, the equator would become hotter and the poles would become colder.

Fig. 12–5 (left) *Areas near the equator, which receive direct sunlight, have little temperature change year-round.* (below) *Areas near the poles receive slanting sun rays and have large temperature changes throughout the year.*

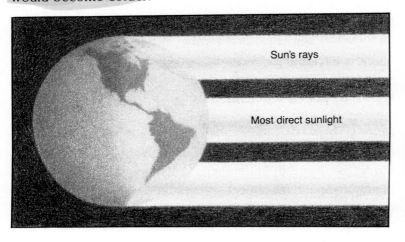

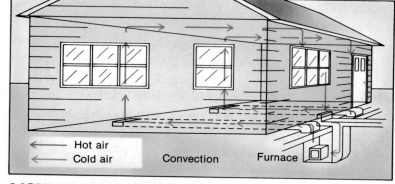

Fig. 12-6 Many homes are heated by convection currents.

Reinforcement: Place a balloon over the mouth of a flask. Heat. As air expands, so does the balloon. Air takes up more space and therefore is less dense.

MOVING THE ATMOSPHERE

Air moves when it is not heated evenly. When air is heated, it becomes less dense and lighter. The surrounding cooler air, which is more dense and heavier, will force the warmer air upward. Cool air always moves toward warm air because of this difference in density. The movement of air because of the difference in temperature and density is called **convection.**

Convection Movement of air caused by differences in its temperature.

Many heating systems in houses use *convection.* In this system, heated air usually comes into a room near the floor. This warm air is pushed up and moves toward the ceiling. Cool air moves downward toward the floor, where it is mixed with warm air. See Fig. 12-6. The air keeps moving until the air in the room is the same temperature everywhere.

Skills: Study of convection reinforces cause and effect.

Convection also causes air to move between the earth's poles and the equator. Air near the poles is cooled, becoming dense and heavy. This cold, heavy air settles downward. Air near the equator is warmed, causing it to become less dense and lighter. The warmer air rises and streams from the equator toward the poles. Cold air from the polar regions moves toward the equator. See Fig. 12-7. Heat is moved from areas of high temperature to areas of low temperature. Thus the sun's energy is spread more evenly over the earth. This has made it possible for life to exist on much of the land surface of the earth.

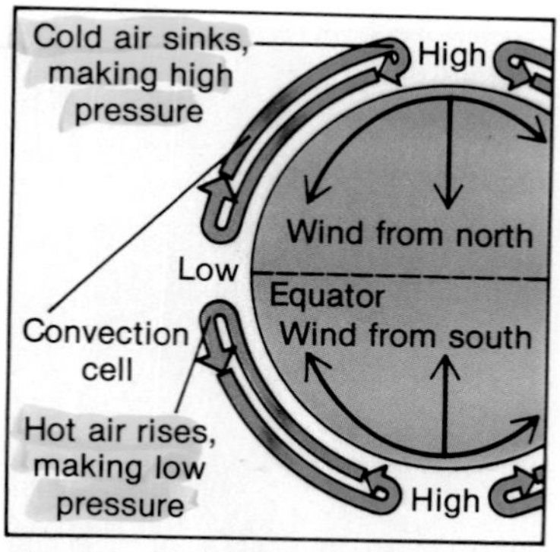

Fig. 12-7 Cold air moves toward the equator and warm air moves toward the poles.

MOVING AIR

When air moves horizontally, it is called *wind.* Wind almost always is caused by differences in temperature in the atmosphere. Temperature affects air pressure. Warm air has a lower air pressure than cool air. As cool air moves into an area of low pressure, the warm air is lifted. As the warm air rises, it is

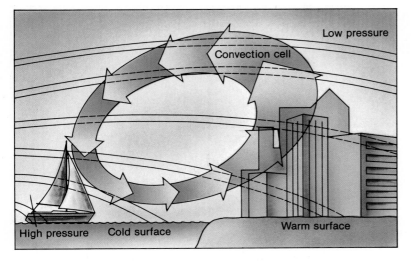

Enrichment: The fastest winds ever recorded at the surface of the earth moved at a speed of 372 km/h. They blew across Mount Washington, New Hampshire, in 1934.

Fig. 12-8 Wind moves air toward a low-pressure area, forming a convection cell.

cooled. It sinks back toward the ground and is warmed again. The complete circle of moving air is called a **convection cell.** Movement of air in *convection cells* causes winds along the ground to flow from areas of high pressure to areas of low pressure. See Fig. 12-8.

Convection cell A complete circle of moving air caused by temperature differences.

1. Mountain and valley breezes. An example of a convection cell is found on mountains. During the night, the air around mountain tops cools more rapidly than the air in the valleys. The cooled air has a higher air pressure than the warmer air. The cooled air moves down the mountain slopes, forming what is called a *mountain breeze*. During the day, air near the mountain tops heats more rapidly than air in the valleys. Because of this uneven heating, the direction of the airflow is reversed. A *valley breeze* is formed. See Fig. 12-9.

Reinforcement: Have students look outside. Ask, "How can you tell that the wind is blowing?" Or ask the same question about Fig. 12-1.

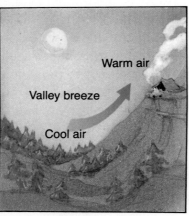

* Mountain breezes form when air cools around mountaintops at night. The cooler air then moves down the mountain slope. Valley breezes form when air near mountaintops heats more rapidly than the air in the valley. As the warmer air rises, cool air from the valley moves up the mountain slope.

Fig. 12-9 (left) *How do mountain breezes form?* (right) *How do valley breezes form?**

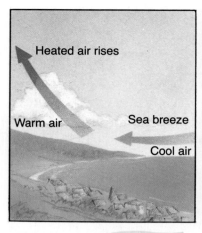 

Fig. 12-10 (left) *How do sea breezes form?* (right) *How do land breezes form?**

* Sea breezes form during the day when air above the land is heated more quickly than the air above the water. This creates a low pressure above the land. The higher pressure cool air then moves from the water to the land. Land breezes form at night when air above the land cools more quickly than the air above the water. This creates a higher pressure above the land. The cooler air then moves from the land to the water.

2. Land and sea breezes. Convection cells are also formed during the day and night in coastal areas. During the day, the land heats faster than the water. The air above the land becomes warmer than the air above the water. An area of low pressure is created above the land. Cool air moves from over the water toward the land. This creates a *sea breeze* in the late afternoon. At night, the land and the air above it cool more rapidly than water. This causes the air movement to be reversed. Air moves from land to water, forming a *land breeze*. See Fig. 12-10. Sea and land breezes occur almost every day along coastlines in warm climates.

3. Monsoons. A *monsoon* is a giant land and sea breeze. A monsoon affects large parts of entire continents. Instead of following the temperature changes of day and night, the monsoon is caused by the differences in summer and winter temperatures. High summer temperatures cause air from the cooler sea to move to the warmer land. In the winter, the air moves from over the cooler land to the warmer sea. Monsoon winds are found all over the world. One of the best known is the monsoon wind that affects India. During the summer months, warm, moist air sweeps in from the Indian Ocean. Much of the food needed for the huge Indian population depends on the rains brought by the summer monsoon.

GLOBAL WINDS

While many winds are caused by local conditions, there is a pattern of winds that covers the entire earth. It is caused by the uneven heating of the earth. The belt of hot air around the equator is a low-pressure area. Caps of cold air at the poles are high-pressure areas. If the earth did not turn, one giant convection cell would carry cold air from the poles to the equator. Another giant convection cell would carry warm air from the equator to the poles. But the earth is turning. It turns once every 24 hours. Because of this movement, places on the equator move at a speed of over 1,600 kilometers an hour. A place very near one of the poles, however, is hardly moving!

Fig. 12–11 shows what happens because of this difference in speed. Anything moving over the earth follows a curving path caused by the earth's rotation. This is called the **Coriolis** (kore-ee-**oe**-lus) **effect,** after the person who first described it. The *Coriolis effect* makes a moving object in the Northern Hemisphere drift to the right when facing in the direction it is moving. In the Southern Hemisphere, a moving object drifts to the left when facing in the direction it is moving. The Coriolis effect is seen most when something moves a great distance. Air moves in a curved path between the equator and the poles.

Coriolis effect Bending of the paths of moving objects as a result of the earth's rotation.

Reinforcement: Show Coriolis effect by drawing a chalk line on a slowly spinning globe from one of the poles to the equator.

North Pole
Straight path of ball
Equator
Earth not spinning
Curved path of ball
Earth spinning

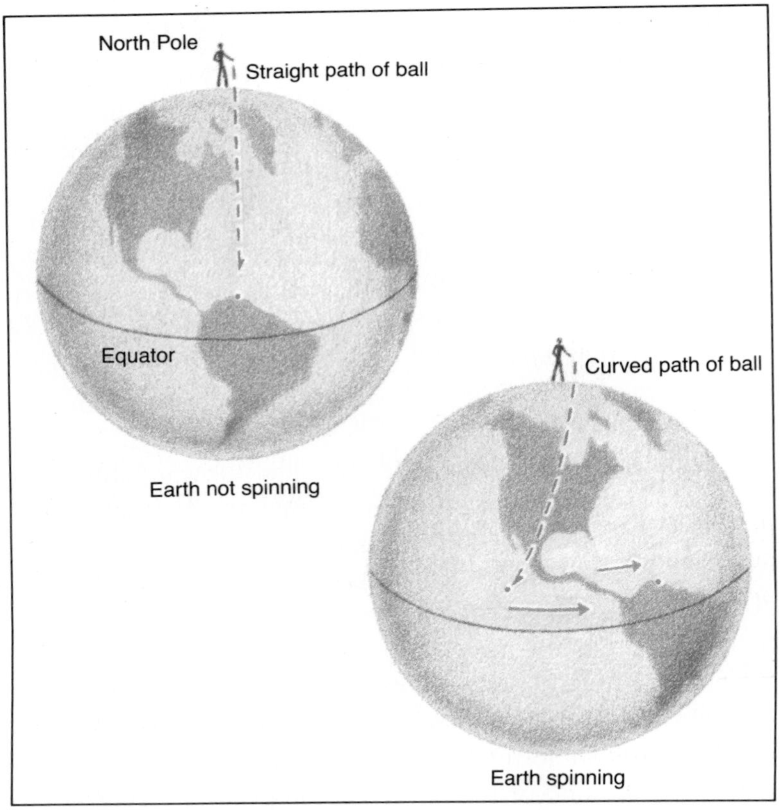

Fig. 12–11 The Coriolis effect is noticeable over great distances.

Near the equator is a belt of air that is strongly heated by the sun's nearly vertical rays. This heated air rises, thus creating a belt of low pressure called the *doldrums* (**dole**-drumz). The air in this belt moves mostly upward. There are only weak winds near the earth's surface. This region got its name, doldrums, from the days when sailing ships were stalled there for lack of wind. The rising air currents from the doldrums turn north and south toward the poles as they become upper-level winds. This movement, or circulation, of air between the equator and the poles, combined with the effects of the earth's rotation, produces a series of *wind belts* in each hemisphere.

1. Trade winds. In the upper parts of the troposphere, the air is cooled as it moves toward the poles. About one-fourth of the way to the poles, some of the air sinks back to the surface. At the surface, the sinking air divides into two parts. In the Northern Hemisphere, one part moves toward the North Pole. The other part moves back toward the equator and creates the warm, steady belt of *trade winds*. The Coriolis effect bends these winds so that they appear to come from a northeasterly direction. The trade winds blow steadily and gently, completing a giant convection cell that began in the doldrums.

2. Westerlies. The story is different, however, for the air that moves from the area where the air sinks toward the poles. These are bent, by the Coriolis effect, to the right in the Northern Hemisphere and to the left in the Southern Hemisphere. Since this makes the winds appear to come from the west, they are called the *westerlies*. Notice that winds are named according to the direction from which they come. The westerlies make up the second of the earth's wind belts. Unlike the trade winds, the westerlies are often strong winds that were once feared by crews of sailing ships.

3. Polar easterlies. The westerlies blow toward the poles until they meet very cold air moving toward the equator from the poles. This band of cold air moving away from the poles is called the *polar easterlies*. The polar easterlies make up the last of the earth's wind belts. The complete pattern of wind belts for the entire earth is shown in Fig. 12–12.

The wind belts describe an overall picture of the movement of air in the earth's atmosphere. At any particular place, the wind patterns are formed from many things. There are local effects, such as land or sea breezes, along with the general air-

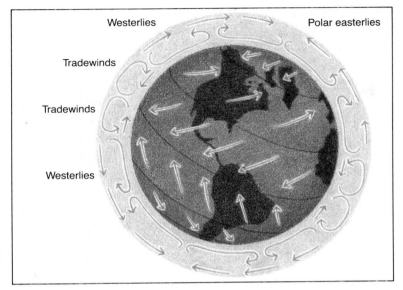

Westerlies   Polar easterlies

Tradewinds

Tradewinds

Westerlies

Fig. 12-12 The earth's wind belts.

flow of the wind belts. Winds do not always flow at the same speeds or along the same paths.

SUMMARY

Sunlight delivers a very large amount of energy to the earth. Without the atmosphere, this energy would quickly escape back into space. But some gases in the atmosphere trap the solar energy that has reached the earth's surface. The unequal heating of the atmosphere causes winds to blow from high-pressure areas to low-pressure areas. Rotation of the earth causes the wind path to bend. Each hemisphere has a group of major wind belts that combine to move air between the warm equator and the cold poles.

QUESTIONS

Use complete sentences to write your answers.

1. List three kinds of radiant energy that come from the sun. Identify which one can be seen, which one is felt as heat, and which one causes sunburn.
2. What does the shape of the earth have to do with the air near the equator becoming hotter than the air at the poles?
3. Give three examples of winds occurring in places where the earth's surface is not heated equally.
4. Name and describe the wind belts of the earth.

ENERGY AND WATER IN AIR

1. Light, the kind we see; infrared, the kind we feel as heat; ultraviolet, the kind that causes sunburn.
2. Because of its round shape, the sun's rays falling on the earth are more direct at the equator than at the poles.
3. Mountain and valley breezes, land and sea breezes, and monsoons.
4. (a) trade winds: winds made by the air descending about a fourth of the distance from the equator to the poles. The winds blow from the southeast and northeast toward the equator.
   (b) westerlies: winds which blow toward the poles from the west and start from the same area as the trade winds.
   (c) polar easterlies: a band of cold air moving away from the poles.

SKILL-BUILDING ACTIVITY

See teacher's commentary for teaching hints, safety, and answers to questions.

USING A THERMOMETER

PURPOSE: To develop skill in the use and care of thermometers.

MATERIALS:
laboratory thermometer
clinical thermometer

PROCEDURE:

A. There are many types of glass thermometers. You will look at two that are in common use—the laboratory (Celsius) thermometer and the clinical thermometer. The clinical thermometer is used for taking body temperature, is small in size, has a narrow range, and usually has a Fahrenheit (F) scale.
   1. Which of your thermometers is the clinical thermometer?
   2. What are the lowest and highest temperatures shown on it?
   3. What part of a degree is each smallest division on the scale?

B. Look at the laboratory thermometer.
   4. What are the lowest and highest temperatures shown on it?
   5. What part of a degree is each smallest division on the scale?

C. Regardless of the kind of glass thermometer you work with, all must be handled with great care because they break easily. Clinical thermometers are usually triangular in shape and do not tend to roll. Look at the laboratory thermometer.
   6. Describe what you may do to keep it from rolling off the table.

D. In order for heat to affect the liquid in the thermometer, the bulb holding the liquid must be as thin as possible. This makes it the weakest part of the thermometer. Look at the bulbs of both thermometers.
   7. Why should you never use a thermometer as a stirring rod?

E. Make a gentle fist around the bulb of the laboratory thermometer. Hold it this way for at least three minutes. BEWARE, THERMOMETERS BREAK EASILY.
   8. Look at the smallest division on the scale. What is the temperature shown? The units are as important as the number! A reading is incomplete without the unit. (Use °C with the laboratory thermometer and °F with the clinical thermometer.)

F. Another difference in the clinical thermometer is that it contains a wiggly constriction in the liquid column near the bulb. This keeps the liquid column from returning automatically to lower readings. Before making a reading, it is necessary to shake the liquid down by a few quick movements of the wrist. Your teacher will show you this motion. Repeat the activity described in procedure E, using the clinical thermometer.
   9. What is the reading on the scale?

CONCLUSIONS:
   1. Describe the laboratory thermometer and give several reasons why care is needed when it is used.
   2. How does the laboratory thermometer differ from the clinical thermometer?

See teacher's commentary for teaching hints, safety, and answers to questions.

ABSORBING RADIANT ENERGY

PURPOSE: To study the rate at which soil, water, and air absorb radiant energy.

MATERIALS:

sheet of paper water
2 small containers 3 thermometers
soil electric lamp

PROCEDURE:

A. Place the sheet of paper on a table and arrange the two small containers side by side near the middle of the paper. Fill one container nearly full with soil and the other nearly full with water. Put a thermometer in the soil so that the bulb is just barely below the surface of the soil. Place another thermometer in the water so the bulb is just barely below the water surface. Lay the third thermometer on the paper near the edges of the two containers.

B. Copy the following table. You will record all your measurements in your table.

	Beginning Temp. °C	Final Temp. °C	Temp. Difference °C
Soil			
Water			
Air			

C. Read each of the thermometers and record the results in your table in the column labeled "Beginning Temp. °C."

D. Arrange the lamp so that it is about 15 centimeters above the table and will shine equally on the thermometers in the soil, water, and air.

E. Turn on the lamp and let it shine on the three thermometers for two minutes.

F. Turn off the lamp and immediately read the three thermometers. Read the one in air first, the one in soil second, and the one in water last. Record your results in the column labeled "Final Temperature °C." Leave the thermometers undisturbed until later.

G. Find the difference between the beginning and final temperatures for each of the three thermometers. Record this information in your table in the column labeled "Temperature Difference °C."

1. Which substance—air, water, or soil—absorbed the radiant energy from the lamp with the greatest temperature change?

2. Which absorbed the least?

H. After a few minutes, look again at the three thermometers.

3. Which has cooled the fastest?

4. Which has cooled the slowest?

CONCLUSIONS:

1. List the three substances studied in this investigation in the order in which they changed temperature when absorbing radiant energy. List the one that showed the the greatest change first.

2. List the three substances studied in the order in which they cooled. List the one that cooled fastest first, the slowest last.

3. Explain your answers to conclusions 1 and 2.

12–2. Water Enters the Air

Fig. 12-13 Water that is exposed to air evaporates.

At the end of this section you will be able to:
- ☐ Explain why you feel cool when water evaporates from your skin.
- ☐ Define *humidity, absolute humidity,* and *relative humidity,* and explain how they are related.
- ☐ Describe how dew, fog, and frost form.

It is not difficult to see that water evaporates. Wet clothes dry when hung outside; the streets dry after a rain; any trapped water exposed to the air slowly disappears. See Fig. 12-13. If you have ever felt cold when the water evaporates from a wet bathing suit, you have made an important discovery about the evaporation of water. This discovery will help to explain why evaporated water, or water vapor, is a small but very important part of the atmosphere.

HOW WATER ENTERS THE AIR

The chill you feel from a wet bathing suit or when you step out of a shower demonstrates an important fact about the evaporation of water. Heat energy is absorbed when water evaporates. The chill you feel when water evaporates from your skin is caused by the loss of heat from your body. The heat energy your body loses goes into the air with the water vapor.

Water vapor that enters the atmosphere from the surface of the sea uses energy from the sun. Radiant energy causes some water molecules at the sea's surface to move faster than the rest of the water molecules. These fast-moving water molecules leave the liquid. They become water vapor in the air. See Fig. 12-14. About one-third of all the solar energy received at the earth's surface is spent evaporating sea water.

Around the world, millions of tons of water are evaporated every minute. Most of the water vapor in the atmosphere comes from the sea. But lakes, rivers, moist soil, plants, and animals also release water vapor into the air. Industrial plants, cooling towers, and a number of other human activities are important sources of water vapor in certain areas.

Air moves the water vapor once evaporation has taken place. The winds of the atmosphere act like a giant transport system

Reinforcement: Wet a towel. Ask students to list the ways of speeding up the drying time.

that moves huge amounts of the water vapor. The winds carry more water than the largest streams found on the surface of the earth.

MEASURING MOISTURE IN THE AIR

The moisture in the air is called the **humidity** (hew-**mid**-ih-tee). Air does not always have the same *humidity*. The amount of moisture in the air may be anywhere from 0 to about 4 percent of the volume of the air. The amount of water contained in a volume of air can be measured. This measurement is called the *absolute humidity*. For example, if one cubic meter (1 m³) of the air contained ten grams of water, the absolute humidity of the air would be expressed as ten grams per cubic meter (10 g/m³).

Air at a given temperature never holds more than a certain amount of water vapor. For example, when the air in your classroom is 25°C, it can hold no more than 22 grams of water in every cubic meter. Air that contains all the water vapor that it can hold at that temperature is said to be *saturated*. Thus, if each cubic meter in your classroom contained 22 grams of water at a room temperature of 25°C, the air would be said to be saturated. The warm air near the earth's equator can hold about ten times as much water vapor as the cold polar air.

Air is not usually saturated. Most of the time, the air contains less water vapor than it can hold. One measurement that often is used to describe the amount of water vapor in the air compares how much water vapor the air contains to how much the air can hold at its temperature. The measurement is called the **relative humidity.** *Relative humidity* is always given as a percentage. For example, one cubic meter of air at 25°C can hold 22 grams of water. If that volume of air is holding only 11 grams of water, its relative humidity is .50 or 50 percent (11 g/m³ divided by 22 g/m³). A relative humidity of 50 percent means that the air contains half of the water vapor it can hold at that temperature. The relative humidity of the air has a strong effect on how comfortable you feel. When the relative humidity is high, moisture does not evaporate quickly from the skin. This makes you feel sticky and warmer.

As air is cooled, a temperature is reached at which the amount of water vapor in the air is equal to the amount of moisture it can hold. Suppose, for example, that the cubic

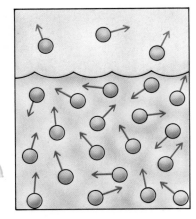

Fig. 12-14 Fast-moving molecules move out of a liquid.

Humidity The amount of water vapor in the air.

Relative humidity The amount of water vapor in the air compared to the amount of water vapor the air can hold at that temperature.

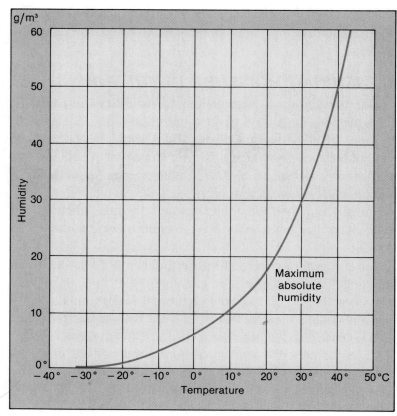

g/m³

*About 30 g/m³ and about 17 g/m³.

Fig. 12-15 According to this graph, how much water is needed to saturate the air at 30° C? at 20° C? *

Reinforcement: To demonstrate condensation, a beaker of cold water placed over a low Bunsen burner flame becomes wet on the outside. Large amounts of water vapor are made by the combustion of fuels.

Dew point The temperature at which air becomes saturated with water vapor.

Condensation The process by which a gas is changed into a liquid.

Enrichment: The condensation of a gas directly to a solid, or the evaporation of a solid directly to a gas, is called sublimation.

meter of air at 25°C with 11 g/m³ of water is cooled. The graph in Fig. 12–15 shows that when air reaches a temperature of about 12°C, it will be able to hold only 11 g/m³. The air will become saturated. Its relative humidity will become 100 percent.

CONDENSATION

The cooling of air will often cause it to become saturated with water vapor. The temperature at which the air becomes saturated is called its **dew point.** When air reaches its *dew point*, its relative humidity is 100 percent. Cooling to a temperature below the dew point causes the water vapor to begin to change back into a liquid. The process by which a gas is changed into a liquid is called **condensation.**

The water that collects on the outside of a glass of ice water is an example of *condensation.* The air touching the cold glass is chilled below its dew point. Water that has condensed on the cold surface is *dew.* On cool nights with no wind, dew may form on cool surfaces of automobiles, rocks, and plants. When

the dew point is below 0°C, water vapor forms a solid instead of a liquid, called *frost*. Frost forms when the temperature is below freezing. See Fig. 12–16.

In some cases, the cooling of air below its dew point causes small droplets of water or ice crystals to condense on dust in the air. When this occurs near the earth's surface, it is called a *fog*. If it occurs higher in the air, it is called a *cloud*. Fog also forms when air is blown across a cold surface. This kind of fog often occurs when warm, moist air is blown from bodies of water across cold land surfaces.

Fig. 12-16 (left) *How does frost form?* (right) *How does dew form?**

* Frost forms when water vapor freezes. Dew is formed when water vapor condenses on a surface.

Enrichment: "Contrails" made by high-flying aircraft are caused by the condensation of water from the combustion of the fuel; hence the name.

Reinforcement: Ask students to explain why they can see their breath on a cold day but not on a warm day.

Enrichment: Demonstration for fog formation: Place some hot water in the bottom of a milk bottle or similar transparent container. Place an ice cube in the mouth of the bottle; a fog will soon form.

1. When water evaporates from your skin, your skin feels cooler.
2. Humidity refers to the presence of water vapor; absolute humidity refers to the amount of water vapor in a cubic meter of air; relative humidity compares the absolute humidity to the amount of water in the air when it is saturated.
3. Dew, fog, and frost all form when the temperature of the air is lowered to or below its dew point.

SUMMARY

Energy is absorbed when liquid water evaporates. The water vapor carries this energy with it into the air. Most of the moisture in the air comes from the sea. The amount of moisture that can be held by air is greater as the temperature rises. When air is holding all the water vapor it can, the air is saturated. Lowering the temperature of air will reduce its ability to hold water vapor. Water may condense from cooled air.

QUESTIONS

Use complete sentences to write your answers.

1. How could you demonstrate that evaporation of water requires energy?
2. Explain the relationship between humidity, absolute humidity, and relative humidity.
3. Explain how dew, fog, and frost are similar.

INVESTIGATION

See teacher's commentary for teaching hints, safety, and answers to questions.

TEMPERATURE CHANGES DUE TO EVAPORATION

PURPOSE: To find out how the temperature changes when liquids evaporate.

MATERIALS:

2 thermometers	container with
cotton cloth	glycerine
(3 pieces)	container with
string	alcohol
container with	index card
water	paper towels

PROCEDURE:

A. Allow both thermometers to lie on the table for five minutes.

B. Copy the following table.

Thermometer	Temp. Before Fanning	Temp. After Fanning
Dry bulb Bulb with water		
Dry bulb Bulb with glycerine		
Dry bulb Bulb with alcohol		

C. Read one of the thermometers. This is the temperature of the air.

1. What is the temperature of the air?

D. Wrap a small piece of cotton cloth around the bulb of one of the thermometers. Tie with string. Dip the cloth in water until completely wet.

E. Hold both thermometers up and fan them gently with an index card for exactly three minutes.

F. Read the thermometers, wet bulb first, immediately after fanning them. Record the temperatures in your table.

G. Remove the cloth from the thermometer and wipe the thermometer dry. Repeat steps D through F, using alcohol in place of water.

H. Remove the cloth from the thermometer and wipe dry. Repeat steps D through F. This time use glycerine instead of water.

I. Remove the cloth from the thermometer. Place the pieces of cloth and string in the containers provided.

2. What is the reason for using a dry bulb in each case above?

3. In each case, did evaporation cause a cooling effect or did it cause a heating effect?

4. Which liquid gave the greatest difference in temperature when it evaporated from the surface of the thermometer?

5. Which liquid gave the least difference in temperature when it evaporated from the surface of the thermometer?

CONCLUSIONS:

1. What effect on temperature does evaporation of a liquid have?

2. Explain why you feel cold when you get out of a swimming pool into the windy air.

3. Why is using alcohol to cool a fever better than using water?

CAREERS IN SCIENCE

CLIMATOLOGIST

Climatologists study weather conditions of the past and try to make forecasts for the future. To make such forecasts, they analyze records showing past humidity (percent of water in the air), temperatures, precipitation (levels of rainfall), and other elements.

Today, some climatologists are involved in studying the "greenhouse effect"—the gradual warming of the earth's atmosphere—as well as the effects of warm ocean currents and volcanic eruptions on changing weather patterns.

Climatologists have college degrees, with courses in meteorology, physics, and advanced mathematics. These scientists are employed by the National Oceanic and Atmospheric Administration (NOAA), by organizations such as the hurricane center in Florida, and by private industry. They may also act as consultants to radio and television weather broadcasting stations. For further information, write: American Meteorological Society, 45 Beacon Street, Boston, MA 02108.

FORESTRY TECHNICIAN

You may have seen news broadcasts showing forestry technicians at work fighting forest fires. Those firefighters are only one type of forest management technician assisting foresters in the job of maintaining the nation's forests. Some technicians help to oversee the proper cutting of trees, as well as the replanting of timber in areas where trees have been removed. Others help to plan what kinds of trees will be planted. Forestry technicians are also involved in building roads and cutting trails through forests for hikers and campers. In addition, the increasing use of our national parks and forests for recreation has created the need for persons with skills in forestry, conservation, and public relations.

The field is very competitive, and openings are limited. Formal training, which is available at community colleges or technical schools, will help you to obtain a job. For further information, write: The U.S. Department of Agriculture, Forest Service/ Human Resource Program, P.O. Box 2417, Washington, DC 20013.

CHAPTER REVIEW

VOCABULARY

On a separate piece of paper, match the number of each blank with the term that best completes each statement. Use each term only once.

condensation Coriolis effect humidity
convection   dew point   radiant energy
convection cell greenhouse effect relative humidity

Energy that the earth receives from the sun is called ___1___. The energy then becomes trapped by the earth's atmosphere in a process called the ___2___. When the earth's surface is heated unequally, air moves by the process of ___3___. Sometimes air will move in a complete circle called a(n) ___4___. The ___5___ will cause the moving air to change direction.

Warm air will absorb water vapor called ___6___. If the temperature is lowered to its ___7___, the air will become saturated. At that temperature, the ___8___ will become 100 percent. Some of the water vapor may turn back into liquid water in a process called ___9___.

QUESTIONS

Give brief but complete answers to each of the following questions. Unless otherwise indicated, use complete sentences to write your answers.

1. What happens to the amount of solar energy reaching earth as it enters and passes through the atmosphere?
2. Explain why you can get a sunburn on a cloudy day.
3. How does the atmosphere trap radiant energy?
4. How does the atmosphere spread solar energy over the earth?
5. Where are the doldrums located and what causes them?
6. What causes land and sea breezes?
7. Describe a monsoon as a convection cell.
8. Compare the Coriolis effect in the Northern Hemisphere with the Coriolis effect in the Southern Hemisphere.
9. Describe one of the three major wind belts as a very large convection cell.
10. Why do you feel cool in a wet bathing suit?

11. Explain how water evaporates from the seas.

12. What is the difference between humidity and relative humidity?

13. How are dew and frost related?

14. Describe how fog forms.

15. Compare fog with a cloud.

16. Why does the air near oceans and lakes stay cool during the day and warm during the night?

17. A sample of room air has a temperature of 20°C and holds 9 g/m³ of water. (a) Use the table in Fig. 12–16 to find out how much moisture the air can hold at this temperature. (b) Calculate the relative humidity of the sample.

APPLYING SCIENCE

1. Compare the placement of air conditioners near the ceiling with the placement of heaters, which are usually near the floor.

2. Why does fog form after clear nights more often than after cloudy nights?

3. Explain how the Coriolis effect would affect a flying airplane.

4. It is common practice, in winter, for the weather service to give the "chill factor" for cold temperatures. The chill factor is the temperature we feel due to the effect of wind. In the summer, the weather service often will give a temperature that includes the effect of humidity when the temperature goes above 32°C (90°F). Why would high humidity make us feel hotter at higher temperatures? Contact your local Weather Service Office to find out how this kind of temperature is determined.

BIBLIOGRAPHY

Allen, Oliver. *Atmosphere* (Planet Earth Series). Chicago: Time-Life Books, 1983.

Moses, L., and John Tomikel. *Basic Meteorology*. California, PA: Allegheny Press, 1981.

Reck, Ruth. "The Albedo Effect." *Science 81*, July/August, 1981.

Schaefer, Vincent J., and John A. Day. *A Field Guide to the Atmosphere*. Boston: Houghton Mifflin, 1983.

Lightning lights up the sky during a thunderstorm. Lightning can strike the earth with 100 million volts of electricity and with temperatures five times hotter than the sun's surface.

WEATHER

CHAPTER GOALS

1. Explain how clouds form and precipitation occurs.
2. Relate the movement of air masses to weather and climate.
3. Explain how fronts and severe storms are produced.
4. Explain how precipitation forms.

13-1. Water Leaves the Air

At the end of this section you will be able to:

- ☐ Name the conditions necessary for clouds to form.
- ☐ Describe three types of clouds.
- ☐ List four kinds of precipitation.

Some mornings when you go outside, you walk into a cloud. A fog has settled in during the night. The fog may seem to come from nowhere. But it is a visible part of the cycle in which water enters and leaves the air.

HOW CLOUDS FORM

Fog is a kind of cloud that forms at ground level. Fog, like all other clouds, is made of drops of water so small that they can float in the air. For example, water vapor enters the air when you take a hot shower. When the warm, moist air strikes a cool surface, the water vapor leaves the air and changes back to a liquid. This process is called *condensation*. A bathroom mirror becomes coated with a film of condensed water. Two conditions are necessary for water vapor in air to condense. First, the air must be cooled. At lower temperatures, air holds less moisture. Fog often forms at night because the temperature of air near the ground drops.

A second condition necessary for water vapor in the air to condense is the presence of some kind of solid surface. For example, the solid surface of a mirror provides a place on which water vapor can condense. Water condenses on a surface such as a mirror because the mirror is cooler than the warm, moist air in the bathroom. Water molecules that make up water vapor tend to stick to solid surfaces. More water is then attracted until

Enrichment: Fog tends to form on "C" nights: clear, calm, cool.

Fig. 13-1 During a shower, water vapor condenses on the cooler surface of the mirror.

Fig. 13-2 At the dew point, warmer, moisture-filled air will also condense.

Enrichment: If all the water in the atmosphere were condensed, it would cover the earth with about 2.5 cm (1 inch) of rain.

drops form. See Fig. 13-1. Without a surface on which the molecules can land, water vapor usually will not condense.

In order for water vapor to condense from the air, the same conditions must exist in nature as in a bathroom. The first step is the lowering of the temperature of the air to its *dew point.* The dew point is the temperature at which the water vapor in air begins to condense. For example, air can be cooled to its dew point when it touches a cool surface. For this reason, dew forms at night on surfaces such as grass. Air is also cooled when it mixes with colder air. See Fig. 13-2. This happens when you open the door of a freezer. Air is also cooled when it moves upward in the atmosphere. As the air rises, it expands because the atmospheric pressure decreases. The air cools as its molecules move farther apart as a result of the drop in pressure.

The second step needed for water vapor to condense is to have a solid surface available. Just as dew condenses on grass, water vapor high in the atmosphere can condense on solid particles in the air. The air is filled with tiny solid particles. There are salt particles from the sea and dust from soil and rocks. These solid particles are carried all over the earth by winds. Smoke also adds particles to the air. All of these kinds of solid particles provide places for liquid water to collect when water vapor from the air condenses.

When air is lifted and cooled, its water vapor may condense and form clouds. This can happen in several ways. For example, when air moves over mountains, it is lifted. Clouds are often found around mountaintops. See Fig. 13-3 *(left).* Air also is lifted when dense, cold air pushes beneath warmer air. As the

lighter, warm air rises, clouds are formed. Also, when air is heated, the warm air rises. The rising air then cools and forms clouds. See Fig. 13-3 *(right)*. Thus, in order for clouds to form, the air temperature must first drop to its dew point. Then tiny solid particles must be present to provide a surface on which the water vapor can condense.

Fig. 13-3 (left) *A cloud may form when air is cooled as it is lifted over a mountain.* (right)*Clouds may also form when warm air rises and then is cooled at higher altitudes.*

KINDS OF CLOUDS

Most clouds are formed high above the earth's surface. As the rising air is cooled, tiny droplets of water gather around small solid particles in the cooling air. These cloud droplets are so small that they float in the air. The slightest rising air currents will keep these tiny droplets suspended. As more water vapor condenses around the droplets, they become larger and larger. Each cloud droplet reflects some of the sunlight falling on it. It is this reflected light that gives clouds their familiar white color.

Clouds come in many shapes and sizes. Observing clouds can provide clues to changes in the atmosphere and the weather to come. A system of names, based on the height and appearance of clouds, is used in *meteorology* (meet-ee-uh-**rol**-uh-jee). Meteorology is the study of the atmosphere.

Often the sky is a mixture of different clouds that are carried in several directions by winds at various altitudes. Careful observation of cloud shapes shows that there are three separate kinds of clouds. Layers or flat patches are called *stratus* (**strat**-us) *clouds* (spread out). See Fig. 13-4 *(left)*. Rising mounds that develop individually, often with tops shaped like

Fig. 13-4 (left) *Stratus clouds*. (middle) *Cumulus clouds*. (right) *Cirrus clouds*.

cauliflower, are called *cumulus* (**kyoo**-myoo-lus) *clouds* (piled up). See Fig. 13-4 *(middle)*. Feathery clouds made of ice crystals formed at high altitudes are called *cirrus* (**sear**-us) *clouds* (curled). See Fig. 13-4 *(right)*. Sometimes a cloud has two of these basic shapes at the same time. Then the cloud is identified as a combination of the two. For example, a layer of clouds with a piled-up surface is called *stratocumulus*.

The system of naming clouds also takes their height into account. Low clouds are found at an altitude below two kilometers. Middle clouds are found between two and seven kilometers. These clouds are always named by adding *alto* before the basic name. Thus a stratus cloud at an altitude of three kilometers is called altostratus. Above seven kilometers are high clouds that have *cirro* added before the basic name. For example, a stratus cloud at that altitude is called cirrostratus. Cumulus clouds have vertical development, since they may grow upward to high altitudes.

Word Study: Nimbus: *nimbus* (L.), rain cloud.

Rain or snow may come from stratus or cumulus clouds. In either case, the term *nimbus* is used with the basic name to form the names nimbostratus or cumulonimbus. Fig. 13-5 shows the way clouds are named according to their shapes and altitudes.

Enrichment: At any one time, about 50% of the earth's surface may be covered with clouds, but only 6% would be receiving precipitation.

PRECIPITATION

Any liquid or solid water that falls to the earth's surface is called precipitation. Precipitation begins with the cooling and condensing of water vapor. A cloud produces precipitation when its droplets or ice crystals become large enough to fall as

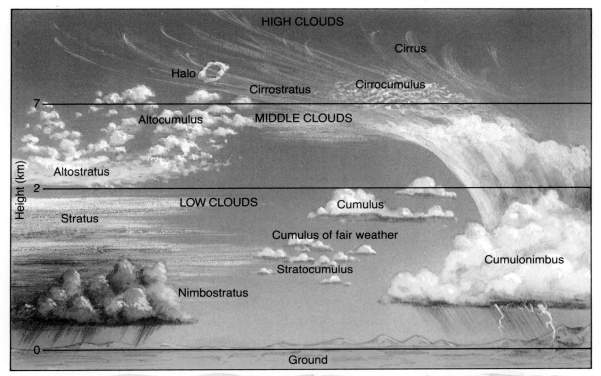

HIGH CLOUDS

Cirrus

Halo

Cirrostratus   Cirrocumulus

7 —

Altocumulus   MIDDLE CLOUDS

Height (km)

Altostratus

2 —

LOW CLOUDS   Cumulus

Stratus

Cumulus of fair weather

Cumulonimbus

Stratocumulus

Nimbostratus

0 —

Ground

Fig. 13-5 Classification of clouds according to height and form.

rain or snow. There are four kinds of precipitation: rain, snow, sleet, and hail. The form in which precipitation reaches the earth's surface depends on the amount of water vapor in the air, the air temperature, and air currents.

1. Rain. Droplets of water remain suspended in the air as long as they are small. In some clouds, droplets become too large to remain suspended. These larger droplets begin to fall. As they move downward, they collide with other droplets. As a result of the collisions, they combine with other drops and grow even larger. These drops can grow to be a hundred or a thousand times larger than the original cloud droplets. Many of these drops become heavy enough to fall as raindrops. See Fig. 13-16. Some scientists think that electrical charges may attract cloud droplets to one another in the formation of raindrops. Rain probably forms this way in warm areas.

2. Snow. In cooler parts of the earth, precipitation can be formed in a different way. Most clouds in these cooler regions have ice crystals in their upper parts. Water from the air condenses on the cool surfaces of the ice crystals and freezes. The process continues until the ice crystals become snowflakes

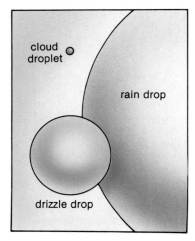

cloud droplet

rain drop

drizzle drop

Fig. 13-6 Raindrops are large drops. The sizes here are exaggerated.

Fig. 13-7 Hail is precipitation in the form of lumps of ice.

large enough to fall. Usually, the air that is below the clouds is warmer than in the clouds. This causes the snowflakes to melt and fall as rain to the earth. However, if the air is cold all the way to the ground, snow falls.

3. Sleet. Sometimes snowflakes strike a layer of warm air and melt into raindrops. They may pass through a cold layer, refreeze and become frozen rain. Precipitation that reaches the ground in the form of frozen rain is called *sleet*.

4. Hail. Hail usually begins as a frozen raindrop. As the frozen raindrop falls, it may be blown up by strong, rising air currents. There more water droplets collect on it and freeze. The larger frozen droplet falls again. Strong currents of air may carry the frozen water droplet up again, where more water freezes on it. Each time the droplet is lifted, a new layer of ice is added, increasing its size. Finally, it falls to the ground as a hailstone. Hail usually falls as small ice particles, but may be as large as walnuts, or even baseballs. See Fig. 13-7.

Knowledge of the role played by ice crystals in the forming of rain has led to the first scientific attempts to make rain. The method is called *cloud seeding*. In cloud seeding, substances that form ice are put into clouds. Two substances commonly used are dry ice and *silver iodide* (**eye**-uh-died). Dry ice is solid carbon dioxide. Its low temperature causes ice crystals to form from the water vapor in the air. These ice crystals then produce snow or rain. Crystals of silver iodide also act as seeds on which ice crystals can grow.

SUMMARY

Clouds form when certain conditions are found in the atmosphere. Precipitation occurs when cloud droplets or ice crystals are too large to remain suspended in the air. Clouds are seeded by adding substances that cause ice crystals to form inside clouds.

QUESTIONS

Use complete sentences to write your answers.

1. Give an example of how clouds might form.
2. Name three kinds of clouds, based on shape, and describe the appearance of each kind.
3. Describe four kinds of precipitation.

INVESTIGATION

See teacher's commentary for teaching hints, safety, and answers to questions.

DEW POINT

PURPOSE: To find the dew point of the classroom air and to predict the height at which clouds form.

MATERIALS:

shiny can	water	stirring rod
thermometer	ice	

PROCEDURE:

A. Measure the air temperature. Hold the thermometer in the air for two minutes (do not touch the bulb of the thermometer). Read the temperature.

   1. What is the temperature of the air?

B. Fill a clean, dry can about half full with water. Do not breathe on it. If moisture forms on the can, refill it with slightly warmer water and wipe the outside dry.

C. Add a piece of ice to the water. Put in the thermometer. Stir with a stirring rod. DO NOT USE YOUR THERMOMETER AS A STIRRING ROD. Be careful not to breathe on the can. Look carefully at the can to see if dew or moisture is collecting on the surface. It may help to run your finger over the surface to feel the dew. See Fig. 13–8. When dew forms, remove the ice and measure the water temperature.

   2. At what temperature did dew first appear on the can?

   3. Compare the air temperature to the dew point.

D. Clouds usually do not form near the ground because the temperature is above

Fig. 13–8

the dew point. As the height increases, the air temperature decreases. When the temperature equals the dew point, clouds can form. To find this temperature, subtract the dew point from the air temperature.

   4. What is the temperature difference?

E. The air temperature decreases 1°C for every 100 m above the ground. Find the height where clouds will form by multiplying the temperature difference by 100 m.

   5. At what altitude is the temperature equal to or just below the dew point?

CONCLUSIONS:

1. Describe what would happen if the temperature in the classroom were lowered to the temperature you found as the dew point.

2. To predict the height at which clouds would form, it is assumed that the dew point remains the same, but it actually decreases slightly with increasing height. How does this affect your prediction for the cloud height?

13–2. Air Masses

At the end of this section you will be able to:

☐ Define *air mass* and *front*.
☐ Describe the four air masses that affect North America.
☐ Describe three major climate zones.

Have you ever gone into a cave or a cellar on a hot day? In those places, the air is cool and damp. It feels much different from the air outside. Air trapped inside a closed space, such as a cave or cellar, takes on the temperature and humidity characteristics of that place. What happens when parts of the earth's atmosphere remain stationary, like air held in a cave?

Fig. 13–9 Air trapped inside a closed space, such as a cave, takes on the temperature and humidity of that place. That is why caves are usually cool and damp.

HOW AIR MASSES ARE MADE

When air remains still for a time over one part of the earth's surface, it takes on the properties of that region. For example, air that sits over warm ocean waters becomes warm and moist. On the other hand, air that stays over the cold regions near the poles becomes very cold. Because low temperatures prevent air from holding much moisture, the air there is also dry. A large body of air that takes on the temperature and humidity of a region is called an **air mass.** An *air mass* usually covers millions of square kilometers. It reaches from the surface of the earth to a height of three to six kilometers. At any particular altitude within this huge body of air, the temperature and humidity are the same or nearly the same. Once an air mass begins to move, it moves as a body.

Air mass A large body of air that has taken on the temperature and humidity of a part of the earth's surface.

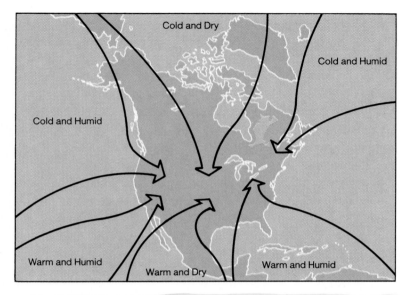

Cold and Dry

Cold and Humid

Cold and Humid

Warm and Humid

Warm and Dry

Warm and Humid

Fig. 13-10 *The source areas shown here affect weather in North America.*

Air masses can only form at places where the earth's surface is almost the same over large areas and the winds are not strong. In this way, the air's temperature and moisture levels become similar to those of the area below it. The oceans near the equator and the regions near the poles are places where air masses can form. Areas on the earth's surface where air masses are formed are called *source regions*. There are two general source regions for the air masses that affect North America. In these source regions, four kinds of air masses are formed. See Fig. 13-10.

1. Cold and dry polar air. Northern Canada and Alaska are the source regions for cold, dry polar air. During the winter, these polar air masses bring very cold weather as they are carried by the prevailing westerlies across the eastern part of the country.

2. Cool and moist air. Cool, moist air masses are formed in the northern Pacific Ocean. These masses bring rain and snow to the Pacific Northwest. As the air masses move across the country and pass over the mountains, they usually lose most of their moisture. These now-drier air masses bring mild, fair weather to the eastern United States. Moist polar air masses also form in the northern Atlantic Ocean. They usually drift toward Europe. However, these cold, moist air masses sometimes invade the New England states, bringing rain in the summer and heavy snow in the winter.

Enrichment: "Cold" or "warm" refers to the temperature relative to the ground. A "cold" air mass is colder than the ground beneath it.

3. Warm and moist air. Air masses coming from the south are mostly warm and moist. The ones that have the greatest effect on the weather in North America are formed over the Gulf of Mexico, the Caribbean Sea, and parts of the Atlantic Ocean. These air masses move north, bringing warm and often stormy weather to the central and eastern United States. Warm, moist air from the south may also move into the desert regions of Arizona and southern California during the summer, causing heavy thunderstorms.

4. Hot and dry air. Hot, dry air masses also form during the summer in northern Mexico and the southwestern part of the United States. Northward movement of these air masses is responsible for much of the hot, dry summer weather of the American Southwest.

FRONTS

As air masses move, they meet each other. See Fig. 13–11. The air in these masses does not mix easily. This is mainly because of the differences in density between the two bodies of air. Cold air has a greater density than warm air. Mixing air with different densities is like trying to mix oil with water. Because they don't mix easily, the boundary between the two air masses tends to be sharp and distinct. Boundaries separating air masses are called **fronts.** The temperature, pressure, and wind direction of the air are generally different on opposite sides of the *front*.

Most fronts form when one air mass moves into a region already occupied by another air mass. If a cold air mass moves

Front The boundary separating two air masses.

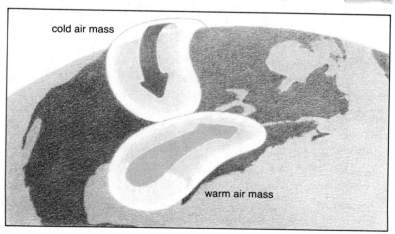

Fig. 13–11 *When air masses meet, they do not mix.*

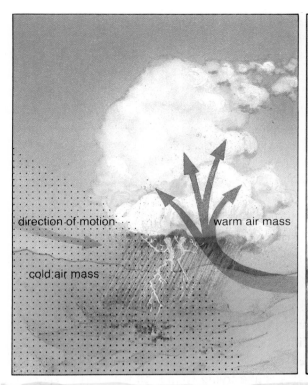

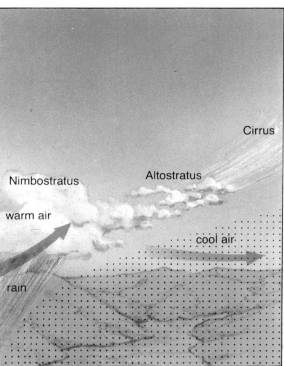

into an area occupied by warmer air, *a cold front* is created. The cold air does not mix with the warmer air. Instead, the colder, denser air pushes under the warmer air like a wedge. See Fig. 13–12 *(left)*. The lifting of the warm air usually causes cumulus clouds to form along the cold front. The clouds along a cold front usually bring heavy rain. Cold fronts move quickly because the dense, cold air easily pushes the warmer air. The stormy weather that comes with a cold front usually passes quickly. The air mass behind a cold front brings cool, dry weather.

When a warm air mass pushes into an area occupied by cold air, a *warm front* is formed. Because it is less dense, the warm air rises up over the colder air. See Fig. 13–12 *(right)*. As the warm air is lifted, cirrus clouds are formed high in the sky. As time passes, the cloud cover becomes lower and thicker, until a solid sheet of stratus clouds covers the sky. A long period of gentle rain may occur, followed by slow clearing and rising temperatures. Warm fronts do not move as quickly as cold fronts. Thus the weather changes that follow a warm front are not as noticeable as the changes following a cold front.

Fig. 13–12 (left) *A cold air mass moving in on a warm air mass, forming a cold front.* (right) *A warm air mass moving in on a cold air mass, forming a warm front.*

Sometimes air masses move parallel to the front that separates them. Then the front does not move and is called a *stationary front*. A stationary front may be either a warm or a cold front.

Air masses can be located by using the temperature readings of many areas. When the readings are recorded on a map, lines can be drawn through all the points of equal temperatures. Such lines are called **isotherms** (**ie**-suh-thurms).

On the map shown in Fig. 13–13, the *isotherm* through all the points of equal temperature (15°C) represents an air mass.

Isotherm A line on a map joining locations that record the same temperature.

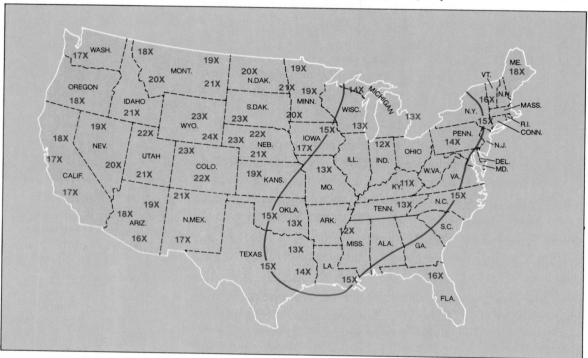

Fig. 13–13 *A line called an isotherm is drawn on this map to connect all locations having a temperature of 15° C.*

Climate The average weather at a particular place.

CLIMATES

Moving air masses and fronts are the main causes of day to day weather changes. **Climate,** however, is the average weather at a particular place over a long period of time. The *climate* of an area is determined by the type of air masses that are found over its location. There are three major climate zones. See Fig. 13–14.

1. Tropical. The climate of regions near the equator is caused by warm air masses. The average temperature during the year is 18°C and above. This climate zone is called *tropical*.

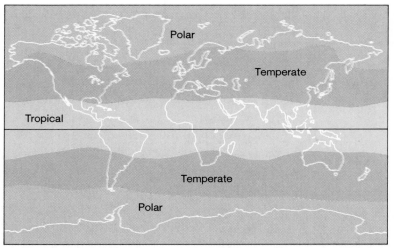

Reinforcement: Ask students to explain why warm fronts aren't as noticeable as cold fronts. Warm air, being less dense than cold air, does not push cold air as easily as cold air pushes warm air.

Fig. 13-14 The three main climate regions are caused by movements of warm and cool air masses.

1. An air mass is a large body of air that takes on the temperature and humidity of a region. It is made by air standing still for a period of time.
2. Northern Canada and Alaska form cold, dry polar air that affects the eastern part of the continent. Cool, moist air masses form in the northern Pacific Ocean, affecting the Northwest, and in the northern Atlantic Ocean, affecting the Northeast. Warm, moist air masses form over the Gulf of Mexico, the Caribbean Sea, and parts of the Atlantic Ocean. They affect the central and eastern parts of the United States. Hot, dry air masses form during the summer in northern Mexico and the southwestern United States, affecting the midwestern United States.
3. A front is the boundary between a cold air mass and a warm air mass. It can be located by the differences in temperature, pressure, and wind directions on opposite sides of the front.
4. Tropical climates are located near the equator and are the result of warm air masses. Polar climates are the result of air masses formed near the poles and have an average temperature of less than 10°C for the year. Temperate climates are caused by both warm and cold air masses in the middle latitudes. The weather is very different day to day.

2. Polar. *Polar* climates are caused by air masses formed near the poles. Average yearly temperature for regions with a polar climate is 10°C or less. The summers are short, and the winters are long and harsh.

3. Temperate. Between the equator and the poles are regions with *temperate* climates. This kind of climate is affected by both warm and cold air masses. The weather in the temperate climatic regions can change from day to day and from one part of the year to another. Most of the United States is found in this kind of climatic region.

SUMMARY

Air masses are produced when air remains over one area for a long period of time. Boundaries between different air masses can be identified because different air masses do not mix easily. These boundaries, or fronts, bring certain kinds of weather as they pass over a place. The kinds of air masses that usually move over a region determine its climate.

QUESTIONS

Use complete sentences to write your answers.
1. What is an air mass? How is an air mass usually made?
2. Describe the four air masses that affect North America. Indicate where each forms.
3. What is a front? How is a front usually located?
4. Describe the three major climate zones.

INVESTIGATION

LOCATING AIR MASSES

PURPOSE: To use temperatures for locating cold and warm air masses.

MATERIALS:

tracing paper   pencil

PROCEDURE:

A. Trace the map in Fig. 13–13, page 362, and write the temperatures and X's as shown.

B. Study the solid line that has been drawn through all stations (X's) that have a temperature of 15°C. Notice that the line sometimes goes between stations of 14°C and 16°C. This is because there is a 15°C temperature about halfway between those readings even though there is no reporting station at that point.

     1. If temperatures were available from several stations in Canada near the Great Lakes, do you think the 15°C line would form a closed loop?

C. On your map, find a station where the temperature is 13°C. Starting at that station, draw a light line to show where 13°C locations would be. Remember, if there is no 13°C station nearby, you will need to draw halfway between the 12°C and 14°C stations. Continue the line until it has gone through all the stations and all locations that are at 13°C. Your line should form a closed loop. Smooth it out and darken it similarly to the 15°C line. This is the 13°C line. Check with your teacher to be sure it is drawn accurately.

D. Draw lines for 17°C, 19°C, 21°C, and 23°C. Not all will be closed loops. Only the odd number temperature lines are to be drawn.

     2. Which of the lines are closed loops?

E. Look at your completed map.

     3. What is the lowest temperature found in the smallest closed loop in the eastern half of the map?

     4. What is the highest temperature found in the smallest closed loop in the western half of the map?

F. The closed loops show where the air masses are located. Cold air masses have isotherms that decrease in temperature as you move toward the center. Look at your completed map.

     5. Where is the center of the cold air mass? (Name the region of a state near the center.)

G. Warm air masses have isotherms that increase in temperature as you move toward the center.

     6. Where is the center of the warm air mass? (Name the region of a state near the center.)

     7. Where would you expect a front to occur for the conditions given on this map?

     8. For a warm front, what weather might you expect to find along the front? Would the weather change severely after the front passed?

CONCLUSIONS:

1. Explain how air masses on a map can be located.

2. How can you tell where the center of a cold or warm air mass is found?

13–3. Weather Changes

At the end of this section you will be able to:

☐ Trace the steps in the development of a *cyclone*.

☐ Describe an *anticyclone*.

☐ Explain how a *hurricane* forms.

☐ Describe the conditions that cause thunderstorms and tornadoes.

Weather changes are caused by moving air masses. Whenever you see the weather change from sunshine to rain, or from hot to cool, it is because a new air mass has taken the place of an old air mass. Sometimes the weather change can be very sudden and violent, as in the case of a tornado. Tornado winds can cause damage like that shown in Fig. 13–15. They cause more deaths annually in the United States than any weather disturbance except lightning. Like all weather changes, tornadoes begin with the formation of an air mass.

Fig. 13–15 Tornado winds caused this damage.

WEATHER FRONTS

During World War II, high-flying bombers were met by a river of fast-moving winds of about 120 to 240 kilometers per hour. These winds form the *jet stream*. See Fig. 13–16. The jet stream occurs at an altitude about 10 to 15 kilometers above the surface.

The jet stream is formed along the boundary of a cold and a warm air mass called the **polar front**. The huge mass of cold air

Enrichment: The jet stream usually has a slight "S" curve in its path from west to east. Unusual weather accompanies deviations of this path. Meteorologists are studying the relationship of the jet stream and weather.

Polar front The boundary where cold polar air meets warm air.

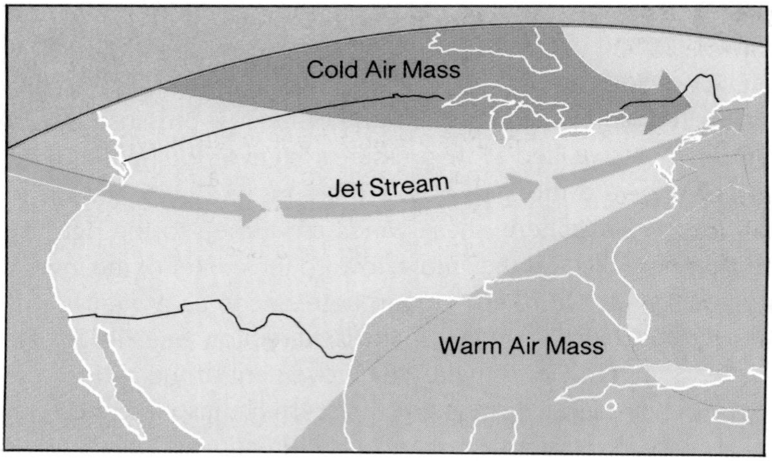

Cold Air Mass

Jet Stream

Warm Air Mass

Fig. 13–16 Jet streams are strong upper-level winds formed when cold, polar air meets warmer air.

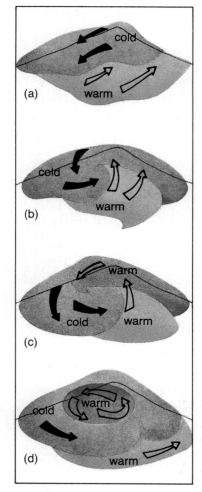

Fig. 13–17 The formation of a cyclone.

Cyclone A kind of storm that is formed along the polar front.

forms around the North Pole. This cap of dense air eventually begins to push slowly southward. The earth's rotation then causes the cold air mass to move west, forming the wind called the *polar easterlies*. At the same time, a warm air mass moves toward the poles and to the east, forming the winds called the *westerlies*. The polar easterlies and the westerlies meet and move past each other, like traffic on opposite sides of a highway, forming the *polar front*. At high altitudes near the polar front, a band of very strong winds that moves from west to east forms the jet stream. Pilots in aircraft traveling eastward often use the jet stream to save fuel. They can thus increase their speed by as much as 160 kilometers per hour. Traveling west, they follow a path that avoids the jet stream or they fly below it.

Often the smooth flow of the winds past each other at the polar front is disturbed. A disturbance, or interference, in the flow can be caused by many things. For example, interference can be caused by mountains or by the differences in temperature between land and sea surfaces. Any interference can produce a small wave, or bulge, in the polar front. See Fig. 13–17. At the bulge, the warm air pushes into the cold air, making a warm front. The cold air moves around and behind the warm air, forming a cold front. The faster-moving cold front soon catches up with the warm front. The warm, moist air is squeezed upward, leaving only cold air near the ground.

CYCLONES

Lifting of the warm, moist air almost always causes a storm center with dark clouds and heavy precipitation. A storm center formed in this way is called a **cyclone** (**sie**-klone). The air in a *cyclone* tends to have an upward movement. This results in an area of low pressure. Because of the low-pressure area, a cyclone is also called a low-pressure system or *low*. Winds tend to blow toward the area of low pressure in a cyclone. In the Northern Hemisphere, these winds are turned to the right by the Coriolis effect as they move toward the center of the low. A cyclone in the Northern Hemisphere becomes a giant swirl, slowly twisting in a counterclockwise direction. See Fig. 13–18. In the Southern Hemisphere, the movement of the air is in the opposite direction. Lows in the Southern Hemisphere are giant swirls, slowly twisting in a clockwise direction.

A mass of cool, dry air may also develop a spinning motion. The greater density and higher pressure of cool air causes an outward flow of air. Winds created by this air movement are given a clockwise twist by the spinning motion of the earth or the Coriolis effect. The result is a large, slowly spinning body of air called an **anticyclone** (an-tee-**sie**-klone). In the Northern Hemisphere, *anticyclones* turn in a clockwise direction. Their motion is counterclockwise in the Southern Hemisphere. Because they are formed by cool air that is usually not very moist, anticyclones often bring fair, dry weather. Anticyclones are also called high pressure systems or *highs*.

HURRICANES

Hurricanes (**hur**-ih-kanes) are cyclonic storms formed over warm ocean water near the equator. *Hurricanes* are usually smaller, but more intense, than cyclones formed near the polar front. Hurricanes that affect North America are produced in the late summer and early fall. During that time of the year, conditions in the atmosphere cause rapid upward motion of very warm, moist air over the warm part of the ocean. As the warm air rises, its moisture condenses, releasing huge amounts of energy. As more moist air is drawn into the developing storm, a great column of rapidly rising air is created. The rotation of the earth twists the rising column. See Fig. 13–19.

Fig. 13–18 *A cyclone seen from an orbiting spaceship.*

Anticyclone A large mass of air spinning out of a high-pressure area.

Hurricane A small but intense cyclonic storm formed over warm parts of the sea.

Word Study: Cyclone: *kykloun* (Gr.), revolving.

Suggestion: Remind students of the Coriolis effect through discussion.

Enrichment: Hurricanes are also called typhoons (Asia), cyclones (India), and willy-willies (Australia).

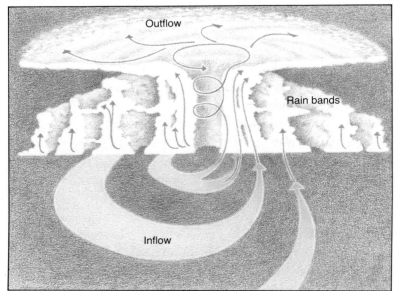

Outflow

Rain bands

Inflow

Fig. 13–19 *Formation of a hurricane.*

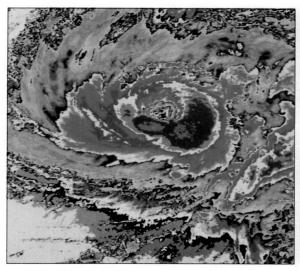

Fig. 13-20 (left) *Different colors show the different temperatures of this hurricane.* (right) *Satellite photograph of hurricane Alicia. Clouds can be seen to circle the "eye."*

When they are fully developed, hurricanes in the Northern Hemisphere have bands of clouds spinning in a counterclockwise direction around a center of calm air. See Fig. 13-20. The center of a hurricane is called the *eye*. Winds circling the eye may reach speeds greater than 100 kilometers an hour. As hurricanes come near land, winds fling destructive waves onto the shore. Much of the damage caused by hurricanes occurs when they strike the shore with their huge waves. As the storm moves inland, heavy rain causes flooding. Hurricanes usually die out as they move over land or cooler water as their supply of warm, moist air is cut off.

THUNDERSTORMS

Fig. 13-21 A thunderhead.

Most people are familiar with thunderstorms. These violent storms are produced when warm, moist air is lifted very rapidly. Towering clouds develop quickly. These clouds belong to the type called cumulonimbus. They are often called *thunderheads*. See Fig. 13-21. Thunderstorms often accompany advancing cold fronts. The cold air wedges under the warm air and lifts it. Cold fronts are sometimes marked by a continuous line of thunderstorms and strong winds. Such a line is called a *squall* (**skwawl**) *line*. See Fig. 13-22 *(left)*. Local conditions can give rise to thunderstorms. For example, warm, moist air blown up the side of a mountain, or heated by an area of warm ground, may form small thunderstorms.

CHAPTER 13

Fig. 13-22 (left) *A squall line is made of many thunderstorms.* (right) *Lightning.*

Lightning occurs in a thunderstorm because the ice crystals and water droplets in the cloud become electrically charged. Electricity builds up in the thundercloud until it discharges as a lightning flash. Most often, the lightning jumps within the clouds, but it may pass between the clouds and the earth. See Fig. 13-22 *(right)*. The passage of lightning through the air causes heating. The heating expands the air and creates a tremendous sound wave that we hear as thunder.

Very heavy rain can fall during a thunderstorm. Thunderstorms also produce precipitation in the form of *hail*.

TORNADOES

The most destructive of all weather disturbances are *tornadoes* (tor-**nade**-oez). Tornadoes are small, but extremely violent storms that form along cold fronts. A tornado usually drops from a thunderstorm as a thin, white, funnel-shaped cloud. Scientists are not really sure how a tornado forms. But they do know the kinds of weather conditions that are likely to produce tornadoes. Most tornadoes occur in the spring and in the central and southeastern U.S. Everything needed for a tornado-producing thunderstorm is present. The conditions are most likely to occur along the polar front where cool, dry air and warm, moist air meet. The cool, dry air forms over Canada and moves south. The Rocky Mountains cause the air to move in a southeasterly path. At the same time, warm, moist air from the Gulf of Mexico moves northward, where it meets the cool, dry air. The two air masses collide, mostly in the central U.S. The difference in temperature on each side of the front is usually very large, especially in the spring. This condition creates the strong thunderstorms that produce tornadoes.

Fig. 13-23 The funnel cloud of a tornado.

1. A disturbance on the polar front; a wave develops; a warm front and a cold front form; warm air rises over cold air; cold air joins with warm air above it; the center becomes a low-pressure area; heavy clouds and precipitation results.
2. An anticyclone results from cold air forming a high-pressure region; a cyclone results from warm air rising over a cold air mass, forming a low-pressure area. The winds of an anticyclone move outward in a clockwise direction; the winds of a cyclone move inward with a counterclockwise motion. Anticyclones bring dry, cool, clear weather, while cyclones bring unsettled, stormy weather.
3. A hurricane is smaller, more intense, and forms over the ocean and near the equator.
4. Warm, moist air is lifted rapidly. Towering clouds develop quickly. Thunder, lightning, and heavy rains with winds often occur.
5. Tornadoes usually drop out of thunderstorms as a thin, white or dark, funnel-shaped cloud.

Up to 1,000 tornadoes hit the United States each year. Most of these take place in a belt from Texas to Michigan, called "Tornado Alley." Tornadoes vary in size. Some are small, with diameters of tens of meters. Others can be hundreds of meters in diameter. Tornadoes usually last only a few minutes, but are very destructive. Their winds, sometimes moving as fast as 800 kilometers an hour, can destroy almost everything in their paths. See Fig. 13-23. Today, scientists are doing experiments in order to understand better the cause of tornadoes.

SUMMARY

A very large, spinning weather system can result when cold air from the poles meets warmer air. Moist air that is made to rise rapidly can create thunderstorms. Hurricanes affecting North America usually come from the warm parts of the air over the oceans near the equator. The collision of warm and cool air sometimes provides the conditions that are necessary to make tornadoes.

QUESTIONS

Use complete sentences to write your answers.
1. List the steps in the development of a cyclone.
2. Compare an anticyclone to a cyclone.
3. How does a hurricane differ from a cyclone?
4. Trace the development of a thunderstorm.
5. How are tornadoes related to thunderstorms?

SKILL-BUILDING ACTIVITY

See teacher's commentary for teaching hints, safety, and answers to questions.

PREDICTING THE PATHS OF HURRICANES

PURPOSE: To use the National Weather Service method of tracking hurricanes to predict their paths and possible danger zones.

MATERIALS:
tracing paper   pencil

PROCEDURE:

A. Trace the map in Fig. 13–24 on the next page.

B. The table below gives the position of two hurricanes for each day of a hurricane watch. On your map plot the position of hurricane Vera for each day. For example, on Day 1 this hurricane was where the line labeled J meets the line labeled 23. Put a small dot at this point.

HURRICANE VERA		HURRICANE WANDA	
Day	Position	Day	Position
1	R–16	1	R–14
2	P–15	2	O–14
3	N–14	3	L–15
4	J–11	4	I–14
5	J–7	5	G–12
6	I–11	6	G–9
7	K–9		

C. Label each point with the number of the day. Draw a dotted line that joins the locations. Label it "Vera."

  1. What is hurricane Vera's path?

  2. Where and on what day does Vera reverse its direction?

3. Which cities shown on the map would have been damaged most by Vera?

4. Where do you think hurricane Vera will be near on Day 8?

5. On what days and to what cities shown would you have issued hurricane warnings for hurricane Vera?

D. Plot hurricane Wanda. Use a dot for each point, number each one, and connect them with a solid line. Label it "Wanda."

  6. Describe the path of Wanda.

  7. Which hurricane traveled faster?

  8. What other place would be in danger from hurricane Wanda?

  9. On what day and to what cities shown would you issue hurricane warnings?

  10. Where do you think hurricane Wanda will be on Day 7?

  11. What will happen to the hurricane as it continues inland?

  12. What is the direction of the winds around a hurricane?

  13. The sun provides some energy for hurricanes, but what is another source of their energy?

CONCLUSIONS:

  1. What general path do hurricanes take as they form off the southeastern coast of the United States?

  2. What valuable service can be given by those who plot the paths of hurricanes?

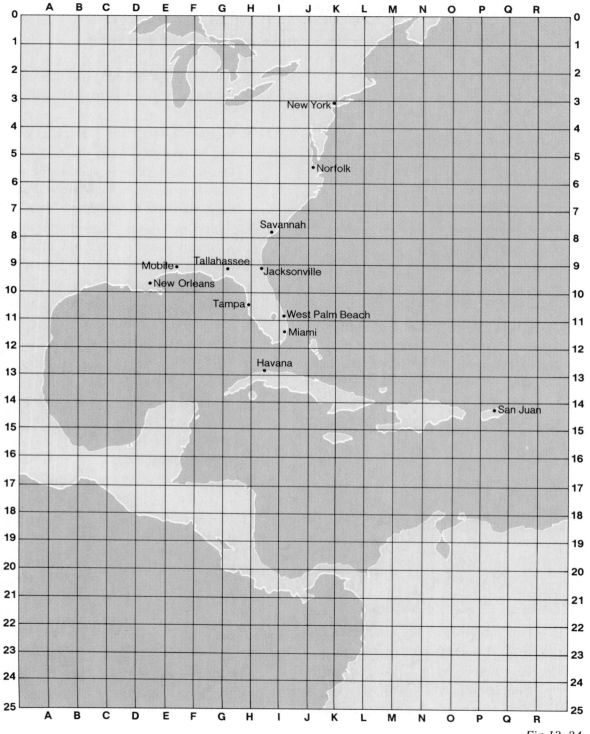

Fig. 13-24

372

CHAPTER 13

13–4. Predicting Weather

At the end of this section you will be able to:

- ☐ Name three weather instruments.
- ☐ Explain how a weather forecast is made.

On a cold winter morning, a television weather announcer received a phone call from an angry viewer: "Please come out and shovel your 'partly cloudy' off my driveway!" The evening before, the reporter had predicted no snow. That weather prediction was wrong. Twenty-five centimeters of snow had fallen. Why is weather prediction so difficult? In this section you will learn how a weather forecast is made.

WEATHER INSTRUMENTS

Weather prediction begins with measurements of the atmosphere. In Chapters 11 and 12 you learned about measuring temperature, relative humidity, and air pressure. Other measurements must also be made. For example, the direction and speed of the wind is an important measurement. Wind direction is measured by a *wind vane*. See Fig. 13–25. A wind vane is turned by the wind so that it points into the wind. Winds are named for the direction from which they come. A wind blowing from the south to north, for example, is called a south wind. The speed of the wind is measured by an **anemometer** (an-uh-**mom**-uh-tur). See Fig. 13–25. An *anemometer*

Word Study: Anemometer: *anemos* + *metros* (Gr.), wind measure.

Anemometer An instrument used to measure wind speed.

Fig. 13–25 Identify the wind vane and the anemometer shown in this photograph.

Description of Wind	Wind Speed (km/h)	Description
Calm	0.0–0.7	Still; smoke rises vertically
Light air	0.8–5.4	Smoke drifts, but wind vane remains still
Light breeze	5.5–12.0	Wind felt on face; leaves rustle; wind vane moves
Gentle breeze	12.1–19.5	Leaves and small twigs move constantly, flags extended
Moderate breeze	20.0–28.5	Raises dust; moves twigs and thin branches
Fresh breeze	29.0–38.5	Small trees in leaf begin to sway
Strong wind	39.0–50.0	Large branches move; wires whistle; umbrella difficult to control
Stiff wind	50.5–61.5	Whole trees sway; somewhat difficult to walk
Fresh gale	62.0–74.5	Breaks twigs from trees; walking difficult
Strong gale	75.0–83.0	Blows off roof shingles
Whole gale	83.5–102.5	Trees uprooted; much structural damage
Storm	103.0–117.5	Widespread damage
Hurricane	118.0 and above	Extreme destruction

Table 13–1

usually has a small cup at the end of each of three long arms. The cups catch the wind, spinning the arms of the instrument. The speed at which the arms turn gives a wind-speed measurement. The wind speed is usually given in meters per second or miles per hour. Wind speed can also be described by the terms shown in Table 13–1. A wind vane and an anemometer are often used together, as shown in Fig. 13–25, page 373.

Precipitation is measured by an instrument called a *rain gauge*. A rain gauge can be any container with straight sides, such as a jar. The depth of rain that falls into the container is measured with a ruler. Official weather stations use a rain gauge that allows 0.01 inch of rain to collect in a small bucket. In the United States rain is measured in inches. In other countries the measurements are in millimeters. When the bucket is filled, it tips over to spill the collected water and causes 0.01 inch of rain to be recorded. At the same time, another bucket tips into place and begins to collect another 0.01 inch of rain. This method allows a measurement of how fast the rain is falling as well as the amount. Solid precipitation, such as snow, is melted in order to record its amount. The depth of snow is also measured. Radar can also show approximately how much precipitation is falling over a very large area.

Not all measurements of the conditions in the atmosphere are made at the earth's surface. Balloons are sent up carrying instruments that send back their readings by radio. See Fig. 13-26. Temperature, humidity, and pressure are measured as the balloon rises. Those measurements are transmitted back to a ground station. The position of the balloon is also tracked to give measurements of upper level winds.

Weather satellites give information about weather that cannot be obtained in any other way. See Fig. 13-27. For example, satellites are able to provide pictures of clouds covering large parts of the earth's surface. Daytime pictures of clouds are made with television cameras using ordinary light. At night, the clouds are photographed by cameras that record the heat given off by the clouds. Positions of clouds can show low-pressure

Fig. 13-26 A weather balloon.

Enrichment: The first weather balloon was sent up in 1785 by Dr. John Jeffries. He measured the temperature, humidity, and pressure at 3,000 meters.

Fig. 13-27 A weather satellite.

Enrichment: There are over 8,000 weather stations around the world making observations and recording data four times each day.

systems and fronts. Some satellites also carry instruments that can measure air temperature and humidity at different levels above the earth's surface. Thus satellites can provide information about atmospheric conditions in regions where there are few weather stations.

MAKING WEATHER FORECASTS

Each day the National Weather Service, a government agency, receives a great deal of weather information. Daily, it gets thousands of reports from weather stations, ships and aircraft, radar stations, satellites, and balloon measurements of the upper atmosphere. This information is used to prepare charts and maps that show present weather conditions. The kind of weather map that you have seen might look like the one in Fig. 13–28. Symbols show areas of low and high pressure, fronts, and precipitation. Weather maps may also show wind direction and speed, atmospheric pressure, and clouds.

The first step in preparing a weather forecast is to study the most recent weather maps. From those maps, the meteorologist making the forecast can get a general picture of the present

Fig. 13–28 Weather maps use symbols to show the weather.

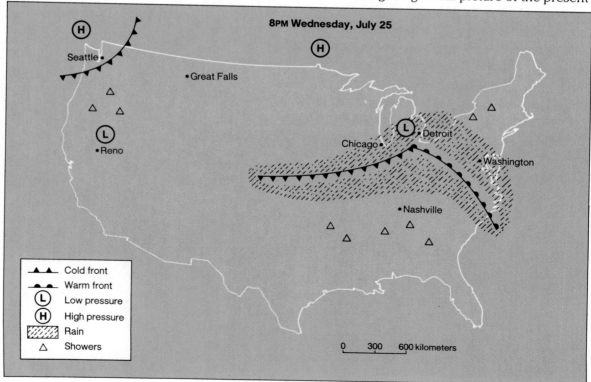

weather conditions. The second step is to compare the present map with others made 24 hours before. Maps showing weather over the entire Northern Hemisphere may be used. This gives a general picture of the existing weather systems, along with their direction of movement and speed. Since satellite photographs show the actual weather, they can be used to check the maps. Reports from radar stations also give information about the location of precipitation. The third step is to use computers to prepare maps that show how weather conditions can be expected to change within the next 24 hours. By putting all of this information together, the meteorologist makes a forecast. The forecaster must decide where the fronts, lows, and highs will move. Usually the forecast is made by a local weather office, using information supplied by the National Weather Service. Special forecasts are also prepared to give warnings of floods, tornadoes, large thunderstorms, and hurricanes. Some weather offices also provide frost warnings for farmers and air pollution forecasts.

A forecast is usually most accurate for the next 6 to 24 hours following the forecast. A forecast made for a period of 2 or 3 days is usually fairly accurate. However, the accuracy becomes less with each passing day. Beyond 5 days, the forecast is usually not dependable.

SUMMARY

Weather prediction depends on observing the conditions in as much of the atmosphere as possible. This information is used to prepare weather maps that show the main weather characteristics at a particular time. Forecasts are made by predicting how the weather map may look as time goes on.

QUESTIONS

Use complete sentences to write your answers.

1. List three measurements made in the atmosphere that are used to predict weather changes.
2. What measurements are made by weather balloons?
3. What information about the weather can be obtained by weather satellites?
4. Describe the three steps involved in making an accurate weather forecast.

1. Temperature, relative humidity, air pressure, wind direction, and wind speed are used to predict weather changes.
2. Temperature, humidity, pressure, and wind speed and direction in the upper atmosphere are measurements made by weather balloons.
3. Weather satellites can show cloud cover over large parts of the earth, give positions of lows and fronts, and some can give temperature and humidity at different levels above the earth's surface.
4. The first step is to study the most recent weather maps. The next step is to compare the present map with others made during the previous 24-hour period. Finally, computers are used to prepare maps that show how weather conditions can be expected to change over the next 24 hours.

SKILL-BUILDING ACTIVITY

COMPARING AND CONTRASTING

PURPOSE: To use information on a weather map to predict weather.

PROCEDURE:

A. Look at the weather map in Fig. 13–28, page 376. Use the legend to answer the following questions.

   1. Which cities shown have overcast skies and rain?

   2. What kinds of fronts and pressure systems are shown near those cities?

B. Look at the weather map in Fig. 13–29 below. Compare and contrast this weather map with the one in Fig. 13–28 to answer the following questions.

   3. How many hours elapsed between the time each of the weather maps was made?

   4. Using the scale on the map, how far did the cold front move along a line from Seattle to Great Falls?

   5. In what direction did the low pressure near Detroit move?

   6. In what direction did the cold front near Chicago move?

   7. Which cities on the map had the weather change from cloudy and rainy to clear?

   8. Which cities on the map had the weather change from clear to cloudy with showers or rain?

   9. Which cities on the map should have the greatest change in weather in the next 24 hours?

   10. Describe this change in weather.

CONCLUSIONS:

   1. What information is needed in order to forecast the weather?

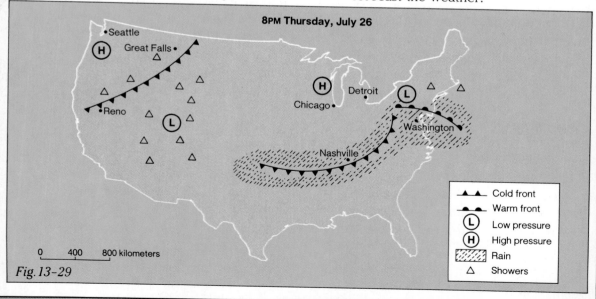

8PM Thursday, July 26

Fig. 13–29

Legend:
- ▲▲▲ Cold front
- ●—● Warm front
- Ⓛ Low pressure
- Ⓗ High pressure
- ▨ Rain
- △ Showers

0   400   800 kilometers

TECHNOLOGY

WINDSAT JOINS THE WEATHER SATELLITES

At the present time, data collected by several different kinds of weather satellites help meteorologists make their forecasts. NOAA–7 and NOAA–8 are in orbit over the poles. Each spacecraft, called a Television and Infrared Observation Satellite (TIROS), covers the whole earth twice a day. Using television devices and infrared (heat) sensors, they send back information to help predict the global movement of storms. Two other satellites, called Geostationary Operational Environmental Satellites (GOES), are in circular orbits about 35,800 kilometers above the equator. GOES are different from TIROS in that they maintain their orbits from approximately the same points near the equator. This makes it easier to track them from earth and to collect data. Receivers can be directed to one point, whereas TIROS must be tracked as they move in their polar orbits. Both types of satellites carry instruments to monitor the sun's activity, to collect and relay information radioed from remote parts of the world, and to send satellite images and National Weather Service maps to both amateur and professional users.

These satellites have already made important contributions. They have, for example, increased the accuracy of weather prediction, especially through observations of weather patterns. Winds over the continental United States blow mostly from the west; therefore, observations of weather patterns over the Pacific Ocean are important to forecasters. GOES satellites, with their heat sensors, can make observations on the progress of freezing

temperatures. Warnings are then issued to fruit growers in Florida and southeast Texas so that measures can be taken to prevent or limit damages to crops. Weather satellites are now helping to save lives as well. The World Weather watch was established in 1968. The United States was the first to commit spacecraft to this program of global weather observations, which, in recent years, has launched a new service—sea search and rescues. NOAA–8—equipped with instruments from France and Canada—and two Russian polar orbiters can detect distress signals from anywhere on the globe and relay them to rescue forces via local ground stations provided by many other countries. More than 185 lives have been saved since the program began in June 1982.

Even greater benefits may be gained from *Windsat,* a proposed new weather satellite. An artist's conception of *Windsat* can be seen in the illustration provided by NASA. *Windsat* was designed by RCA Commercial Communications Laboratories. Its function will be to produce maps of the earth's winds using infrared lasers to detect wind patterns. Information about winds is essential if long-range weather forecasts are to become more accurate.

Windsat will generate two laser pulses per second into the atmosphere. These beams will bounce off pollen and dust that are carried by winds and then return to the satellite. Sensors on board will then compute the direction and speed of the wind.

Windsat will, no doubt, be a more-than-welcome addition to the satellite family.

SCIENCE INPUT

In this chapter you learned how clouds were formed. Because of the need for rain, it is critically important to understand the conditions necessary for clouds to exist. Perhaps with more research about cloud development, techniques other than cloud seeding will be found that will enable us to use the water vapor in the air for agricultural purposes whenever the need exists. In this exercise, you will use your knowledge of the relationship between the altitude of rising air, air temperature, and dew point. (You may want to review Chapter 11, which discusses atmospheric pressure, and Chapter 12, which explains dew point.)

COMPUTER INPUT

As you have already seen in previous chapters, computer programs can be written to produce quizlike questions. Sometimes the questions, or data for the questions, are entered by the user, as was done in Program Plate Game and Program Erosion. Other times, different questions are created by using the computer's capacity to generate random numbers. When something is random it has no particular order or reason. A random number is one that has just as much of a chance to be chosen as any other. In a lottery, great efforts are made to have the numbers picked at random so that every ticket has an equal chance of being chosen. In a computer game, random numbers are sometimes used to create constantly changing situations that cannot be controlled by the player. In Program Cloud Formation, the computer's ability to generate random numbers is used to set up different atmospheric conditions. You must then decide if clouds will be formed or air will rise under each set of conditions.

In this exercise, the computer is used in three ways: (1) to produce a number of different situations; (2) to create a series of questions that test your knowledge of how clouds form; and (3) to give you feedback.

WHAT TO DO

After reviewing the necessary chapters, enter Program Cloud Formation into your computer. Save it on disk or tape and run it. The program may be used to practice independently or to test your knowledge in group competition. The program can best be used to produce data that you can then study to gain a better understanding of the relationship between all the variables the program contains. On a separate piece of paper, record the data produced.

PROGRAM NOTES

There are two kinds of variables: dependent and independent. An independent variable is one that remains constant in the particular problem you are considering. A dependent variable changes according to the character of the independent variable. In this program, cloud formation and rising air

CLOUD FORMATION CHART

altitude of rising air	air temperature	dew point	Does cloud form?	Does air rise?

WEATHER: PROGRAM CLOUD FORMATION

are the dependent variables. Altitude, air temperature, and dew point are independent variables. You are given information about the three independent variables and asked to predict changes in the two dependent variables.

PROGRAM

```
100   REM CLOUD FORMATION
110   FOR X = 1 TO 10: PRINT : NEXT
120   FOR X = 1 TO 10: PRINT : NEXT
130   T = INT ( RND (1) * 30 + 1)
135   D = INT ( RND (1) * 30 + 1): IF D >
      T THEN D = T
140   T1 = INT ( RND (1) * 30 + 1)
142   E = INT ( RND (1) * 10 + 1)
145   PRINT "ALTITUDE OF RISING AIR"
150   PRINT "KM        AIR
      TEMP    TEMP    DP"
170   PRINT E; TAB( 10)T1; TAB( 20)T;
      TAB( 27)D
173   PRINT
175   INPUT "DOES CLOUD FORM (Y/N)?
      ";A$
176   IF A$ = "Y" AND D = T THEN
      PRINT "CORRECT!!"
177   IF A$ = "N" AND D = T THEN
      PRINT "INCORRECT"
178   IF A$ = "Y" AND D < > T THEN
      PRINT "INCORRECT"
179   IF A$ = "N" AND D < > T THEN
      PRINT "CORRECT!!"
180   IF D = T THEN PRINT "TEMP = DP
      - CLOUD FORMS"
185   IF D < > T THEN PRINT "TEMP <>
      DP - CLOUD DOES NOT FORM"
190   INPUT "DOES AIR RISE (Y/N) ? ";A$
191   IF A$ = "Y" AND T1 < T THEN
      PRINT "CORRECT!!"
192   IF A$ = "Y" AND T1 > = T THEN
      PRINT "INCORRECT"
193   IF A$ = "N" AND T1 < T THEN
      PRINT "INCORRECT"
194   IF A$ = "N" AND T1 > = T THEN
      PRINT "CORRECT!!"
200   IF T1 < T THEN PRINT "TEMP1 <
      TEMP2 - AIR RISES"
210   IF T1 > = T THEN PRINT "TEMP1
      > = TEMP 2 - AIR DOES NOT RISE"
220   PRINT
240   GOTO 120
```

GLOSSARY

RND	A command in BASIC that instructs the computer to generate a random number.
VARIABLE	Any factor or thing or number that can change. In this program, altitude, air temperature, dew point, rising air, and cloud formation are variables.

BITS OF INFORMATION

A program that creates models of different kinds of situations is called a simulation. Most often, these programs come with graphic displays; that is, images (pictures) that look like the real situations appear on the screen. An important area where computer images are used is the field of flight simulation for pilot training. The trainee is placed in an exact copy of the cockpit of an aircraft. Every dial, switch and knob present in a real cockpit are there to give the pilot-in-training the feeling of being in an actual plane. Images made by a computer portray what the pilot would see if he or she were flying. Thus, without endangering any lives, the person in training can gain a great deal of flight experience.

CHAPTER REVIEW

VOCABULARY

On a separate piece of paper, write TRUE next to the number of each statement that is true. Next to the number of each false statement, write FALSE, then make the statement true by writing the correct term in place of the underlined incorrect term.

1. Meteorology is the study of the atmosphere.
2. An anticyclone is a large body of air that has taken on the temperature and humidity of a region of the earth.
3. A small, intense cyclonic storm formed over warm parts of the oceans is called a hurricane.
4. A cyclone is a large mass of air spinning out of a high-pressure area.
5. The boundary separating two air masses is called a polar front.
6. An isotherm is a line on a map joining all the locations that record the same temperature.
7. The climate is the average weather at a particular place.
8. A tornado is a kind of storm formed along the polar front.
9. The boundary where cold polar air meets warm air is called a warm front.

QUESTIONS

Give brief but complete answers to each of the following questions. Unless otherwise indicated, use complete sentences to write your answers.

1. Describe several ways in which air is lifted and cooled to form clouds.
2. Explain why just cooling moist air may not cause clouds to form.
3. What is the main difference between stratus, altostratus, and cirrostratus clouds?
4. In what way are nimbostratus clouds related to cumulonimbus clouds?
5. Compare the way precipitation forms in warm areas with the way it forms in cold areas.
6. Describe the air masses that form in the northern Pacific Ocean and northern Atlantic Ocean and how they affect the weather in the United States.
7. Compare the two air masses that come from the south and the effect they have on the weather of North America.
8. What is the difference between weather and climate?
9. Explain how fronts and severe storms are produced.

10. Compare the kind of air and its movement in a cyclone to the kind of air and its movement in an anticyclone.

11. Compare and contrast a hurricane and a tornado.

12. How are tornadoes related to thunderstorms?

13. What do a wind vane and an anemometer tell about wind?

14. Describe how a weather station rain gauge works.

15. What kinds of information are given by the National Weather Service in a weather forecast?

APPLYING SCIENCE

1. Use the information about air masses that are common to your area and the general wind patterns to explain why your region has the kind of weather that it does during the summer and during the winter. Write a report describing the reasons for your explanation.

2. Make daily observations of the atmosphere for a week. Note, by name, the kinds of clouds present, whether there is rain, smog, or fog associated with the clouds, and any other weather events that might have occurred during the week. Report your results to the class.

BIBLIOGRAPHY

Canby, Thomas Y. "El Nino's Ill Wind." *National Geographic*, February 1984.

Gedzelman, Stanley D. *The Science and Wonder of the Atmosphere*. New York: Wiley, 1980.

Linn, Alan. "The Earth Spins, So We Have the Coriolis Effect." *Smithsonian*, February 1983.

Moses, L., and John Tomikel. *Basic Meteorology*. California, PA: Allegheny Press, 1981.

Moyer, Robin. "The Great Wind." *Science 80*, November 1980.

Olesky, Walter. *Nature Gone Wild*. New York: Messner, 1982.

Olson, Steve. "Computing Climate." *Science 82*, May 1982.

Stommel, Henry and Elizabeth. *Volcano Weather: The Story of 1816, the Year Without a Summer*. Newport, RI: Seven Seas Press, 1983.

CHAPTER

14

Instruments have become vital to studying the sea. However, nothing can replace the first-hand knowledge that comes from direct observation of conditions beneath the sea.

THE OCEANS

CHAPTER GOALS

1. Name and describe the main features of the sea floor.
2. Compare the deep sea floor with the sea floor near the edges of the continents.
3. Identify the kinds of sediments found on the sea floor and describe how they are formed.
4. Describe how a record of the earth's history can be found in sea floor sediments.

14–1. The Earth Hidden Beneath the Sea

At the end of this section you will be able to:

☐ Name four ocean basins and describe how they were formed.

☐ Explain how the sea floor is explored by scientists.

☐ List three groups of features found on the ocean bottom.

☐ Describe the sea floor near the edges of the continents.

A world ocean covers seven-tenths of the earth's surface. While people speak of separate oceans and seas, it is actually one great ocean. The water in the ocean is not a very large part of the earth. If the planet were the size of an orange, the ocean would be only about as thick as a sheet of paper. However, without its thin covering of water the earth's surface would be very different. The weather would be different: The poles would be much colder and the tropics much hotter, and most scientists believe that there would be no life.

THE OCEAN BASINS

Suppose that the surface of the earth had no continents. Then the water in the ocean would cover the entire earth to a depth of thousands of meters. But the continents rise above sea level, leaving the water to fill the lower parts between them. The continents do not divide the ocean into parts. Instead, the ocean separates the land masses into huge islands. See

Enrichment: The ocean is a foreign environment for humans. In order to study the oceans, special equipment and skills are needed. A basic requirement for survival is a continuous supply of oxygen. But another serious problem is increased pressure during descent. Pressure nearly doubles for every 10 meters in depth. At a depth of 1000 meters, the pressure is 100 times that of the surface. For depths below 50 meters, specially trained divers use scuba (self-contained underwater breathing apparatus) equipment which supplies air at the correct pressure up to about one hour. Below 100 meters, it is necessary to enclose the diver completely in a protective suit.

Enrichment: The total area of the earth is about 510 million km². The ocean covers about 360 million km² or 71 percent.

Enrichment: The term "seven seas" refers to the areas of the ocean known before the 15th century.

Fig. 14-1 From space the earth's surface can be seen to be mostly covered by water.

Fig. 14-1. The water-filled regions surrounded by continents and other raised parts of the crust make up the *ocean basins*.

There are three large and two smaller ocean basins. Each of them is called a separate ocean. The largest are the Pacific, Atlantic, and Indian oceans. The Arctic Ocean, at the north end of the globe, is smaller. The southern parts of the Atlantic, Pacific, and Indian oceans that surround the continent of Antarctica are often called the Antarctic, or Southern Ocean. However, the Antarctic Ocean does not occupy a separate ocean basin like the four other oceans.

The Pacific Ocean is the largest and deepest of all oceans. It covers more than one-third of the earth's surface. It is so large that all of the earth's lands could fit into the Pacific Ocean. Smaller water-covered parts of the earth are called *seas*. A sea may be a part of an ocean, such as the part of the Atlantic called the Caribbean. Seas may also be separate from the ocean, as the Mediterranean is. Since all oceans and seas are connected with each other, there is actually only one global ocean.

Ocean basins are formed when the moving crustal plates cause continents to break apart. The oceans and seas of today fill basins that began to be formed when the ancient supercontinent broke apart about 200 million years ago. Before the breakup of the supercontinent, there was only a single super-ocean. Since the continents are still moving and breaking apart, there will be different ocean basins millions of years from now.

EXPLORING THE SEA

People have lived by the ocean and sailed on it for thousands of years. However, what lay beneath the water was a mystery for a long time. Only in the past hundred years has the part of the earth's surface covered by water become known to scientists. Most of the knowledge about the ocean basins has been discovered within the past 30 to 40 years.

During early years of sea exploration, there was no accurate way of determining what the ocean bottom looked like. Explorers tried to find out by lowering a weight on a line or wire until it hit bottom. See Fig. 14–2 (*left*). But this was not an accurate method. Ships often drifted, causing the weight to drop down at an angle. Sometimes the line or wire was not long enough to reach bottom. Also, these measurements were made in only a few places. From these incomplete and often inaccurate measurements, maps of the ocean bottom were drawn. These early maps showed the sea floor as an almost level plain.

During World War I, a way to locate submerged submarines was found. This was done by using sound waves. The method is called *echo sounding*. In this method, sound waves sent from a ship through the water bounced off solid objects and returned to the ship. See Fig. 14–2 (*right*). The bouncing effect of sound waves is similar to the way a ball bounces. The farther away the object that the ball strikes, the longer it takes the ball to return. This principle also applies to sound waves. The time it takes for the sound waves to return to the ship shows how far away from the ship a submerged submarine can be found.

Enrichment: The science of oceanography began with the voyage of the H.M.S. Challenger from 1872 to 1876. This was the first attempt to make a comprehensive study of the global ocean.

Reinforcement: Soundings were first used by ships entering or leaving harbors or moving near the shore to avoid grounding.

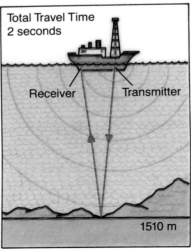

Fig. 14-2 (left) *Measuring the depth of the ocean by sounding was often inaccurate.* (right) *Echo sounding uses sound signals that are bounced off the sea floor.*

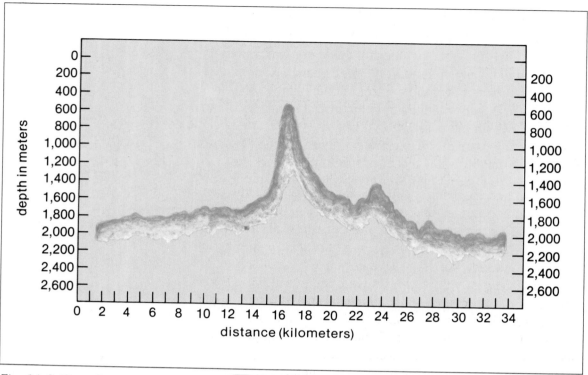

Fig. 14-3 The echo sounder produces a picture of the shape of the sea floor. The tracing shows hills and valleys much higher or lower than they actually are.

Reinforcement: Ultrasonic sound waves are used for echo sounding to distinguish them from the audible sounds made by the ship. The chart made by most echo sounders exaggerates the vertical scale by about ten times.

Oceanographers (oe-shuh-**nog**-ruh-furs), scientists who study the sea, soon realized that sound waves could be used to explore the ocean floor. The *echo sounder* was developed. The echo sounder sends out sound signals that bounce off the ocean bottom. For example, the time it takes for the sound signal to travel from the ship to the bottom and back again is measured to be two seconds. The amount of time it takes for the sound signal to reach the bottom can be found by dividing this measurement by two, which gives one second. By multiplying the speed of sound waves in water, 1,510 meters per second, by the time measurement of one second, the depth of the ocean at a given point is found to be 1,510 meters. See Fig. 14-2 (*right*), page 389. The differences in depth from point to point, recorded by the echo sounder, create a picture of what the sea floor looks like. See Fig. 14-3. Much of the sea floor has now been explored with echo sounders.

FEATURES OF THE SEA FLOOR

The pictures show that earth's surface beneath the water has hills, mountains, steep cliffs, and smooth plains. Oceanographers have divided these forms into three groups: the continental margins, the mid-ocean ridge, and the ocean basin floor. Many of these features are now seen to be related to the forces that shape all parts of the crust.

1. Continental margins. The areas between the sea floor and the continents are marked by the *continental margins*. Most continents are part of the same crustal plate as the nearby sea floor. Thus the continental margin usually is not the edge of a crustal plate. Instead, it is the part of the ocean where the continent begins to rise above the level of the sea. A continental margin is made up of three parts: the continental shelf, the continental slope, and the continental rise. The first of these parts is a region of shallow water called the **continental shelf.** See Fig. 14–4. The *continental shelf* is a part of the continent that has been flooded. Most parts of a continental shelf are nearly flat and slope gently downward toward the ocean basin floor. The depth of the ocean over a continental shelf is usually about 120 meters. The width of a continental shelf can vary from narrow to wide. It averages about 75 kilometers. During the last ice age, large amounts of water were locked up in glaciers. This caused the sea level to drop by 90 to 120 meters and exposed large areas of the continental shelves. Evidence of the land animals and plants that lived on these once exposed parts can now be found on the sea floor of the continental shelves.

Continental shelf The flat submerged edge of a continent.

Enrichment: Sunlight can penetrate seawater to a depth of about 100 meters. Thus the continental shelves are the only regions of the sea floor where light reaches the bottom and there is an abundance of life.

Fig. 14–4 The shape of the sea floor in the North Atlantic. The differences in elevation are not as large as shown in this diagram.

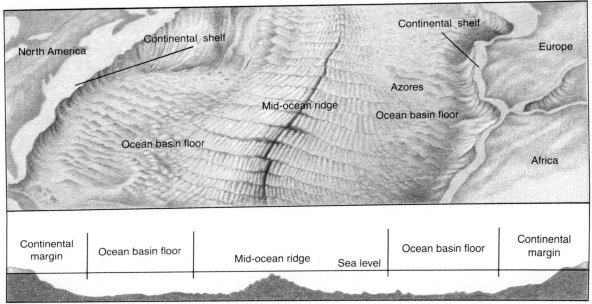

North America
Continental shelf
Continental shelf
Europe
Azores
Mid-ocean ridge
Ocean basin floor
Ocean basin floor
Africa

Continental margin	Ocean basin floor	Mid-ocean ridge	Sea level	Ocean basin floor	Continental margin

Submarine canyon Any deep valley cut into a continental shelf.

Reinforcement: Submarine canyons are the principal avenues by which sediments from the continental margins reach the deep sea floor. The powerful currents created by sudden movement of material down a submarine canyon is called a *turbidity current.*

Continental slope The sloping part of the sea floor that marks the boundary between the sea floor and continental shelf.

Fig. 14-5 The continental slope and the continental rise separate the continental shelves from the ocean basin floor.

Deep canyons called **submarine canyons** have been cut in many places along the continental shelves. See Fig. 14–5. Some of these canyons were cut by rivers when the sea level was lower during the glacial periods. Other *submarine canyons* seem to have been formed by underwater landslides. These landslides occur when sediments that have built up on the continental shelves suddenly break loose and slide down toward the ocean basin floor. The moving sediments erode the continental shelves in the same way that rivers erode the land.

While the continental shelves make up less than 10 percent of the sea floor, they have been studied in detail by oceanographers. This is not only because they are close to land, but also because valuable resources are found there. The continental shelves are rich in a large variety of life. Most of the world's supply of fish is caught there. Large amounts of petroleum and natural gas are found in the rock layers that make up the continental shelves. About 25 percent of all oil and gas now produced comes from wells drilled from ocean platforms on the continental shelves. This percentage will increase in the future as other supplies are used up.

The water depth increases rapidly where a continental shelf ends. This marks the beginning of the **continental slope.** See Fig. 14–4. The *continental slope* is the boundary between the

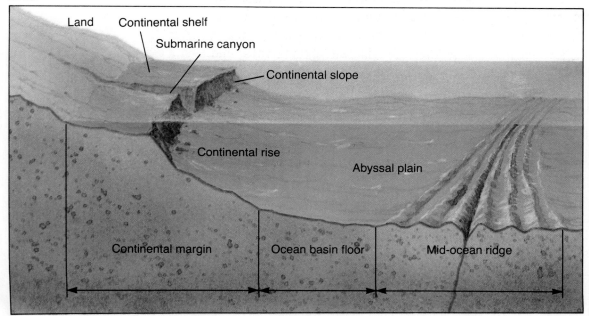

continental crust and the oceanic crust that makes up the ocean basins. If the bottom of a pie pan were the ocean basin floor, the sides of the pan would be the continental slopes.

Beyond the continental slope is the third part of the continental margin. This area, where the slope becomes less steep, is the **continental rise.** See Fig. 14-5. The *continental rise* may reach out for hundreds of kilometers until it finally blends with the ocean basin floor.

2. Mid-ocean ridge. An important feature found on the ocean floor is the *mid-ocean ridge* system. See Fig. 14-5. The mid-ocean ridge system is a continuous chain of wide mountains rising two to four kilometers above the ocean floor. The ridge is crossed by many faults. According to the theory of plate tectonics, the mid-ocean ridge system is where new crust is formed. Hot liquid magma from below the crust pushes up and out near the center of the ridge. As the magma hardens, new sea floor is created. The newly made crust cools and shrinks and settles to a lower elevation as it moves away from both sides of the ridge. This causes the sea floor to slope away from the sides of the mid-ocean ridge.

Parts of the mid-ocean ridge have been seen and photographed. Scientists are able to descend for short periods of time in specially built submarines called deep submersibles. However, this exploration is limited to small areas. These limited explorations have revealed places along the mid-ocean ridge where hot water pours out. These *hydrothermal vents*, or warm-water sea floor springs, provide energy, minerals, and food to support smaller sea organisms. In the total darkness of the sea floor these fountains of hot water are like oases in the desert. See Fig. 10-10, p. 276.

3. Ocean basin floor. On either side of the mid-ocean ridge is the ocean basin floor. See Fig. 14-5. On the deep ocean floor are very flat areas called **abyssal** (uh-**bis**-ul) **plains.** See Fig. 14-5. *Abyssal* means deep. The *abyssal plains* are formed by deposits of sediments. These sediments cover most of the hills and valleys that were made as new crust was produced along the mid-ocean ridge. Abyssal plains are the most level areas on the earth. They are flatter than any plains on the continents. The abyssal plains make up more than 60 percent of the ocean basin floor. Close to the continents, the abyssal plains contain sediments that were carried off the continental shelves and

Continental rise A gently sloping area at the base of the continental slope.

Reinforcement: The continental rise is made up of sediments that accumulate at the base of the continental slope.

Abyssal plain A flat, wide deposit of sediments that covers the deeper parts of the ocean basin floor.

Enrichment: Abyssal plains are widespread where there are no trenches along the continental margins that trap sediments moving down the continental slopes.

down the submarine canyons. Farther out to sea, the sediments come from different sources.

Not all of the ocean basin floor is flat. Scattered over the entire sea floor are thousands of submerged volcanoes called **seamounts.** See Fig. 14–6. Some are in groups, while others stand alone. A large *seamount* can rise three to four kilometers above the sea floor. Many volcanic islands, such as Hawaii and Tahiti, are the exposed tops of huge seamounts that have grown from the sea floor.

Some submerged seamounts have flat tops. These are called **guyots** (gee-**yos**). See Fig. 14–6. A *guyot* appears to have been an island at one time. As the sea floor spread away from the mid-ocean ridge and sank, causing the guyot to submerge, wave erosion flattened its top.

The deepest parts of the sea floor occur where one crustal plate collides with another. As one plate moves beneath the other, an ocean **trench** is created. *Trenches* are found along the edges of ocean basins, particularly in the Pacific. See Fig. 14–6. The deepest of these trenches reach almost 11 kilometers below the sea surface.

Seamount A submerged volcano rising from the ocean floor.

Guyot A submerged volcano whose top has been flattened by wave erosion before it sank below the surface.

Trench A deep valley on the ocean floor found along the edges of some ocean plates.

Enrichment: Mount Everest, the tallest mountain on land, has an elevation of 8,845 meters above sea level. The seamount that forms the island of Hawaii has an elevation of 9,150 meters from the sea floor, making it the tallest mountain on the earth.

Word Study: Guyots were named after the first professor of geology at Princeton University.

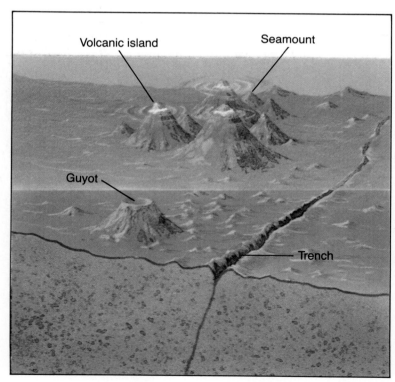

Fig. 14-6 The ocean basin floor has many unusual characteristics.

CORAL REEFS AND ATOLLS

Large numbers of small animals called **coral** (**kor**-uhl) may live in warm, shallow sea water. Each of the *coral* animals has a chalky skeleton, which is usually connected to its neighbors. The coral skeletons make a framework that also contains the remains of other small animals and plants. In time, a large structure called a *coral reef* is formed. See Fig. 14-7. Some coral reefs grow out from an island and remain attached to it. The islands that make up the Florida Keys have this kind of coral reef. A *barrier reef* is separated from land and usually lies some distance from the shore. Part of the east coast of Australia is protected by the Great Barrier Reef. Another kind of coral reef, called an *atoll* (**ay**-tawl), is circular and often not connected to any visible land. An atoll is formed in stages. First it grows as a submerged reef around a volcanic island. The island slowly sinks as the sea floor moves away from the mid-ocean ridge. As the island disappears, the reef grows upward until it finally remains as a doughnut-shaped atoll.

Coral A small animal that lives in great numbers in the warm, shallow parts of the sea.

Fig. 14-7 Coral is an animal that, in partnership with other small plants and animals, builds large reefs in shallow parts of the ocean.

SUMMARY

The earth's surface is divided into elevated continents and submerged ocean basins. Scientific exploration of the part of the earth's surface that is covered by water shows it to be made up of three kinds of features. The continental margins make up the shallow parts of the sea that border the land. The mid-ocean ridge occupies the central regions of the sea floor. The ocean basin floor makes up the remaining parts of the sea floor. Small coral animals build the foundations for reefs found in warm, shallow parts of the ocean.

QUESTIONS

Use complete sentences to write your answers.

1. Name the present ocean basins and explain how they were formed.
2. Explain how an echo sounder is used to measure ocean depths.
3. List three kinds of features found on the ocean floor.
4. Why are continental shelves so important?

1. Presently, there are the Pacific, Atlantic, Indian, and Arctic ocean basins. They were formed when the moving crustal plates caused the supercontinent to break apart and move the continents to their present locations.
2. An echo sounder sends a sound into the ocean. The time it takes for the echo to return is used to measure the depth of the ocean.
3. Any three of the following: mid-ocean ridge, hydrothermal vents, seamounts, guyots, coral reefs, islands, trenches.
4. Answers will vary. Continental shelves contain valuable resources. For example, most of the world's supply of fish is caught on the continental shelves.

INVESTIGATION

See teacher's commentary for teaching hints, safety, and answers to questions.

MAPPING THE OCEAN FLOOR

PURPOSE: To make a map of a section of the ocean floor.

MATERIALS:
graph paper
pencil

Speed of Sound(m/s)	Time (s)	Distance (m)	Depth (m)
1,510	2	3,020	1,510
1,510	3	4,530	2,265
1,510	4	6,040	3,020

Table 14–1

PROCEDURE:

A. The depth of the ocean at any place can be found by using a special instrument called an echo sounder. A sound is made by the echo sounder and the sound then travels through the water. When the sound strikes the ocean bottom, it is reflected back to a receiver on the ship. The time taken for the round trip is measured. In part B of this lab, you will see how the depth of the ocean bottom is calculated.

B. In Table 14–1, the average speed of sound in sea water and the time taken for a round trip is given. The distance the sound travels can be found by multiplying the speed of sound and the travel time (column 1 × column 2). Do this for each of the travel times given. The distance the sound travels is the distance from the ship to the ocean bottom and back. To find the depth of the ocean bottom, divide the distance traveled (column 3) by two.

   1. What does a longer travel time for the sound tell you about the bottom?

C. Draw a graph with the distance from shore on the horizontal axis and the depth of water on the vertical axis. Number the horizontal scale from 0 to 2,000. Beginning with 0 at the top, number the vertical scale from 0 to 4,000. Plot the points for the pairs of numbers in Table 14–2. Then connect the points.

Distance (km)	Depth (m)	Distance (km)	Depth (m)
0	0	1,000	2,000
200	249	1,200	1,903
250	1,004	1,300	249
500	2,001	1,400	2,250
550	3,903	1,700	2,000
600	2,001	1,800	1,004
800	1,736	1,850	1,500
850	491	1900	1,019
950	491	2,000	2,197

Table 14–2

   2. Describe the shape of the ocean floor.

CONCLUSIONS:

   1. Explain how the depth of the ocean floor is measured by an echo sounder.

   2. Find and label the following features on your graph: guyot, mid-ocean ridge, continental shelf, continental slope, continental rise, seamount, and ocean trench.

14-2. Sea Floor Sediments

At the end of this section you will be able to:

- ☐ Identify two sources of sediments in the sea.
- ☐ Describe the kinds of sediments found in different parts of the sea floor.
- ☐ Explain how part of the earth's history is recorded in sea floor sediments.

A strange kind of rain occurs in the sea. It is a gentle rain of solid particles that slowly fall through the water and settle on the bottom This slow, but steady, solid rain causes almost all the sea floor to become covered with a layer of sediments.

Fig. 14-8 The visible load of this river can be seen when it empties into the sea.

SEDIMENTS ON THE CONTINENTAL SHELVES

Most rivers empty into the sea. Carried along with the water of the rivers are the products of rock weathering. Part of the material carried by the rivers is made up of small pieces of rock that can be seen. These particles are called the *visible* load. See Fig. 14-8. They consist of very fine particles of silt and clay as well as larger fragments of rock that are rolled or bounced along the stream bed. Some of the material carried by the stream is dissolved in the water. Dissolved material makes up the *invisible load*. Dissolved materials brought to the sea by rivers help to make the sea "salty." Both the visible load and the

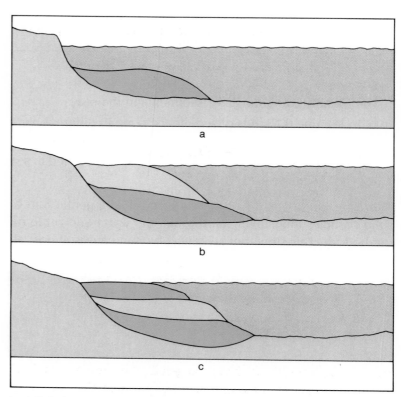

Fig. 14-9 Thick layers of sediments can build up on the continental shelf.

invisible load end up as sediments on the sea floor. These sediments from rivers are a major source of sediments in the sea.

Much of the visible load of a river is deposited close to the shore when the river empties into the sea. Large particles are dropped near the mouths of rivers. Waves and currents caused by waves may carry much of this material along the shore. The material is later deposited, forming sandy beaches. Sometimes, huge underwater, fan-shaped deposits are also formed where rivers empty into deep water. These deposits are like deltas hidden beneath the water. Finer particles are carried farther out and deposited on the continental shelves. Almost all of the sediments found on the continental shelves come from the land. Some continental shelves have very thick layers of sediments that have been deposited there.

As the edge of a continent moves away from the mid-ocean ridge, it tends to sink. Thus the continental shelf may also slowly sink at the same time as sediments are being deposited. This allows many layers of sediments to be built up, as shown in Fig. 14-9. Sediment layers many kilometers in thickness can be built up. Some layers of sediments formed this way are now found as part of the rock within today's continents. Apparently, these sediments were laid down in ancient times along the edges of continents that existed at that time. The movement of crustal plates pushed these thick layers of sediments into the bodies of the continents. They are found there today, bent and broken, as parts of mountains.

TYPES OF DEEP SEA SEDIMENTS

The sea floor is almost completely covered by sediment. Part of the sediment is carried down from the continental margins. The rest has slowly settled down to the bottom from the water above. Sea floor sediments are grouped according to where they came from.

1. Remains of living things. There is a type of sediment formed from the remains of living things. Near the equator and along the west coasts of continents, the conditions in the sea favor the abundant growth of living things. The most common forms of life in these waters are floating plants and animals that are so small that they can be seen only with a microscope. See Fig. 14–10. When these living things die, their remains sink to the bottom. These remains make up most of the sediment found on the sea floor in the deep parts of the ocean far from the continents. The sediment formed from remains of living things is called **ooze.** Beneath the ocean areas where sea life is abundant, a thick layer of *ooze* covers the entire sea floor.

Ooze A kind of sea floor sediment formed from remains of living things.

Word Study: Ooze: *wose* (Old English), mire or slime.

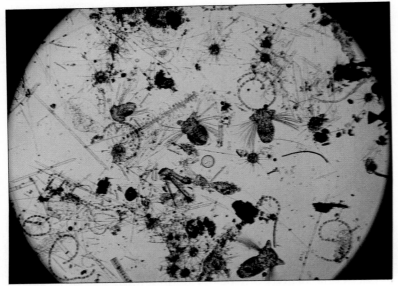

Fig. 14–10 Many parts of the ocean contain large amounts of microscopic plants and animals.

Red clay A kind of sediment found on the floor of most of the open sea.

Enrichment: The deep-sea drilling ship *Glomar Challenger* has greatly expanded scientific knowledge of the sea floor during the past two decades.

2. Red clay. Most of the sea does not support enough life to produce a heavy layer of ooze on the bottom. Most of the open sea floor is covered with a sediment called **red clay.** However, this sediment is usually not red and does not contain much of the kinds of minerals that make up ordinary clay found on land. Instead, it is made up of the smallest particles carried by rivers into the sea. *Red clay* also contains windblown dust, volcanic dust, and even tiny particles from meteorites. Red clay is deposited very slowly. It may take 1,000 years for one millimeter of red clay to settle. Oozes, on the other hand, can be deposited at the faster rate of one centimeter every 1,000 years. On the continental shelves, sediments build up at the rate of one meter every 1,000 years.

Until the spreading of the sea floor from the mid-ocean ridge was discovered, scientists were puzzled about the lack of very thick sediments on the deep sea floor. During the hundreds of millions of years of earth's history, the deep sea floor should have built up a very thick layer of sediments. However, the exploration of the ocean showed that much of the sediment layer was missing. Now scientists understand that the layer was removed by moving crustal plates on the sea floor. The sea floor and the sediments carried on it were destroyed where the crustal plates collided. Thus no part of the ocean bottom is old enough to have built up a very thick layer of sediment.

A RECORD OF ANCIENT CLIMATES

Scientists can obtain samples of sea floor sediments for study. One of the most common methods uses a sharp-edged pipe that is dropped to the sea floor from a ship. When it hits the bottom, the pipe is driven into the sediments. Then it is brought back to the surface and a long cylinder of sediments, called a *core*, is taken from it. See Fig. 14–11. The layers of sediments can then be seen in the core.

Because the sediments build up over millions of years, they contain clues to the conditions in the ocean in the past. For example, the remains of certain animals and plants that were known to be alive at a particular time tell the age of a sediment layer. Because some of the animals and plants lived only in warm water or cold water, the temperature changes at the sea surface in the past can be found. In this way, the sea floor sediments contain a record of the earth's most recent ice ages.

Fig. 14-11 A core sample of the sea floor reveals the layers of sediments.

Even changes that take place on land are recorded in the ocean sediments. For example, dust from volcanoes sometimes reaches the ocean bottom. Sediments carried to the oceans from streams and rivers tell us what the soil on the earth's surface was like in the past.

SUMMARY

Almost all the weathered rock material that is carried into the sea by rivers remains there for a long time. Most of this material becomes part of the sediments that cover the sea floor. In some parts of the ocean, most sediments are the remains of living things. Heavy deposits of sediments are laid down on the sea floor near the edges of the continents. A fairly accurate record of the past temperature conditions on the earth can be found in sea floor sediments.

QUESTIONS

Use complete sentences to write your answers.

1. What two kinds of materials are brought to the sea by rivers?
2. Describe the various kinds of materials found on the sea floor at the following locations: (a) continental shelves; (b) deep seas where life is abundant; (c) deep seas where life is not abundant.
3. How do scientists obtain samples of sea floor sediments?
4. What kind of information is obtained from sea floor sediment samples?

1. Rivers carry solid particles such as silt, clay, and larger pieces of rock, as well as dissolved materials.
2. (a) The continental shelves contain the erosion products of rock such as silt, sand, and rock fragments. (b) The deep sea floor has a layer of ooze where life is abundant. Ooze is the remains of the living animals and plants. (c) In the deep sea where life is not abundant, the floor has a layer of red clay, which is made mostly of the finest material brought to sea by the rivers.
3. Samples of the sea floor are taken by dropping a heavy, sharp-edged pipe to the bottom, where it is driven into the sediments and traps material inside it.
4. They contain clues to the conditions in the oceans in the past.

SKILL-BUILDING ACTIVITY

See teacher's commentary for teaching hints, safety, and answers to questions.

SEQUENCING AND A MODEL OF A CORE SAMPLE

PURPOSE: To make a model of a core sample and use the idea of sequence in the study of it.

MATERIALS:

straw   marking pen
sea floor model   razor blade

PROCEDURE:

A. Push the straw straight down into the clay model of sediments of the ocean floor that your teacher has prepared. Be sure to push it all the way to the bottom.

B. Remove the straw. Using the marking pen, mark the straw to indicate the top of the soil sample.

C. Using the razor blade, carefully slice the straw and clay core lengthwise to expose the core sample. Leave the core sample in the straw to protect it. The layers in the sample represent layers of sediments.

D. Make a drawing of the core sample on your paper. Label the top and bottom of the sample. Shade in the different layers and label each according to its color.

E. Use your drawing and the model core sample to answer the following questions.

1. How many sedimentary layers are there in the core sample?

2. According to the sequence of the layers, which layer in the core sample was deposited last?

3. In the sequence of layers, which layer in the sample was deposited first?

4. Which layer is the thickest? If the rate at which materials enter the ocean has been nearly the same, what does the thickness of this layer tell about the relative time it took for it to form?

5. Starting with the oldest, describe each layer in the order in which the layers were deposited.

6. If your core sample was taken in the deep part of the ocean, how would you expect the particle sizes to compare with a sample made near shore?

D. Look at Fig. 14–11 on page 401.

7. What differences do you see between the two core samples being studied?

8. Could these core samples be from the same area of the sea floor? Explain.

9. Name four materials you might find in a core sample.

CONCLUSIONS:

1. Explain why a sea floor core sample is made up of a sequence of layers.

2. Describe how to tell the relative age of sediment layers in a sequence of them found in a core sample.

3. In a sequence of sediment layers found in a core sample, an unusually thick layer is found. What explanation can you give for it being so thick?

4. A core sample is made up of several layers of sand, with the top layer having some slightly larger rock pieces. From what part of the sea floor could this sample have come?

TECHNOLOGY

FORECASTING EL NIÑO

Today, satellites with computerized devices that send back images of ocean currents and weather patterns are helping us to understand how closely affected we are by events that seem far away.

Every few years, air pressure drops over the southeastern Pacific. The trade winds that usually blow across the Pacific toward Asia slow down or disappear. The warmer water that these trade winds normally hold back begins to move toward South America. (Since these warm waters occur off Peru around Christmas, Peruvian fishermen named this effect El Niño, meaning "the child.") In 1982–1983, this warming extended to one-third of the Pacific Ocean. In some places the ocean temperature rose 11 degrees. The warmer water heated the atmosphere above it. The speed and position of the jet stream was affected. It was pushed farther south. Unusual weather resulted. There were tornadoes in Los Angeles and 20 centimeters of snow in Atlanta in one day in April. Record-breaking rains and heavy flooding took place in Ecuador, northern Peru, California, Louisiana, and Cuba. The worst droughts in history were recorded in 1983. Fish moved to colder waters, causing the fishing industry to lose tremendous amounts of money. Because of the decrease in fish and other marine life, birds and other animals did not reproduce.

What was the cause of these natural disasters? The answer to this question lies in a series of complicated patterns of high- and low-pressure systems, ocean currents and countercurrents, and jet

streams called El Niño and the Southern Oscillation (ENSO).

Many meteorologists and oceanographers believe El Niño, the warming of the water, is either caused by or is closely related to the Southern Oscillation, a giant rocking of tropical air masses over the Pacific. This rocking causes pressure systems to seesaw back and forth. When there is low pressure over the southeastern Pacific, there is high pressure over Australia, which usually brings dry weather. This occurred in 1983, when Australia suffered a great drought, the results of which can be seen in the photo.

Scientists are carefully analyzing the data collected during the 1982–1983 El Niño. Because of the worldwide effects that both El Niño and the Southern Oscillation have on weather climate, these scientists hope to be able to predict the return of El Niño. Mathematical models are being prepared that will allow the use of computers to study the effects of weather changes and food production. Already in use are computerized images transmitted from satellites. These photos are made with scanners sensitive to concentrations of chlorophyll. High levels of chlorophyll indicate the presence of plant life, which is an indication that the water temperature has risen. The goal of this research is to help us prepare for the changes in weather, climate, and ecological balance that the currents of El Niño and the winds of the Southern Oscillation bring. An improved understanding of these currents and winds may help to reduce the damage they can cause.

CHAPTER REVIEW

VOCABULARY

On a separate piece of paper, match each term with the number of the statement that best explains it. Use each term only once.

continental shelf	continental rise	guyot	ooze
submarine canyon	abyssal plain	trench	red clay
continental slope	seamount	coral	

1. A deep valley in the ocean floor.
2. The sloping part of the sea floor that separates the sea floor and continents.
3. The kind of sediment found on the floor of most of the open sea.
4. A kind of sea floor sediment made from living things.
5. A submerged volcano rising from the sea floor.
6. The flat, submerged edge of a continent.
7. A submerged volcano whose top has been flattened.
8. A deep valley cut into the continental shelf.
9. The flat, wide deposit of sediment that covers the deep ocean floor.
10. A gently sloping area at the base of the continental slope.
11. Small animals that live in great numbers in the warm, shallow seas.

QUESTIONS

Give brief but complete answers to the following questions. Unless otherwise indicated, use complete sentences to write your answers.

1. Name five common ocean basins.
2. Explain why there is really only one global ocean.
3. It takes 3.2 seconds for a sound to go to the sea floor and return. Sound travels 1,510 meters per second in the sea. How deep is the ocean at this location?
4. Starting with the continent coastline, list the following ocean floor features in the order they would be found as you moved toward the deep ocean: (a) mid-ocean ridge; (b) abyssal plain; (c) continental shelf; (d) continental rise; (e) continental slope.
5. Describe how each of the following is the result of a volcano in the ocean: (a) guyot; (b) seamount; (c) island.
6. What are the two main sources of sediment in the sea?

7. What is "ooze," and where is it found?

8. What are "hydrothermal vents," and why are they like oases in the desert?

9. What part of the ocean is mined for oil? What other resource is found there?

10. Describe how an atoll forms.

11. What finally happens to a layer of sea floor sediment?

12. How are submarine canyons made?

13. Compare the sediments found on continental shelves with that found on the abyssal plains.

14. Where is the sediment called "red clay" found, and of what is it composed?

15. What do sea floor sediments tell about the earth's history?

APPLIED SCIENCE

1. Using the information from your study of sea floor sediments, explain why a large puddle of muddy water leaves a layer of very fine sediments near the middle of it when it evaporates.

2. Write a paper discussing why there are so many coral reefs in the middle of the Pacific Ocean near the equator. You may want to look up additional information in the library.

3. Prepare a poster to show to the class the kinds of studies an oceanographer might do. Use the information presented in this chapter and any other information you may find in the library.

4. Make a poster-size drawing, to show the kinds of features and materials found on the ocean floor. Label all parts. Be sure to include the following: continental shelf, continental slope, submarine canyon, trench, guyot, seamount, abyssal plain, mid-ocean ridge, coral, ooze, red clay. Draw in some of the resources found in the sea and on the sea floor.

BIBLIOGRAPHY

Britton, Peter. "Cobalt Crusts: Deep Sea Deposits." *Science News*, October 30, 1982.

Britton, Peter. "Deep Sea Mining." *Popular Science*, July 1981.

Mathews, Samuel W. "New World of the Oceans." *National Geographic*, December 1981.

Time-Life Editors. *Restless Oceans* (Planet Earth Series). Chicago: Time-Life Books, 1984.

Great demands are placed on the ocean's supply of fish. Entire schools of fish are being taken every hour by factory ships. Will the sea continue to provide food in the future?*

WATER IN THE OCEAN

CHAPTER GOALS

1. Describe the properties of sea water.
2. Explain how the sun's heat affects the sea.
3. Identify and describe the resources that can be obtained from the sea.

* Improved fishing methods will probably not be able to provide enough food in the future. Fishing, however modern, is nothing more than hunting. Fishing to feed the world is similar to trying to feed the world by hunting wildlife. Sea farming, like land farming, will be needed.

15–1. Sea Water

At the end of this section you will be able to:

☐ Describe the composition of sea water.

☐ Explain how the sea became salty.

☐ Compare the temperature in the upper and lower parts of the ocean.

There is an old folk legend that explains why the sea is salty. According to this tale, somewhere on the bottom of the sea there is a magical salt mill. This mill is supposed to be grinding away all the time and pouring salt into the water of the ocean. In this section you will find out why this old story is partly true.

SEA WATER

A glass of sea water taken from almost any place in the open sea looks like a glass of ordinary drinking water. But the salty taste of sea water would quickly show that they are not the same. If you boiled the sea water until it all evaporated, white crystals would be left. A sample of sea water would leave behind an amount of solid material equal to about 3.5 percent of weight of the sea water. In other words, 1,000 grams of sea water contains about 965 grams of water and 35 grams of dissolved salts. See Fig. 15–1 (a), page 408. The number of grams of dissolved salts in 1,000 grams of water is called the **salinity** of the water. The *salinity* of sea water is 35, which is about 300 times the salinity of most ordinary drinking water.

The two most abundant dissolved substances in sea water are sodium ions and chloride ions. Ions are atoms or groups of atoms that have lost or gained electrons. Many substances are made of ions. For example, sodium ions and chloride ions

Enrichment: If all the water could be removed from the ocean, a layer of salts about 60 meters thick would cover the entire ocean floor.

Word Study: Salinity: *sal* (L.), salt.

Salinity The number of grams of dissolved salts in 1,000 grams of sea water.

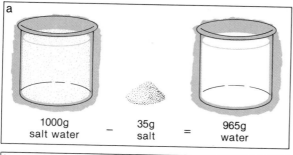

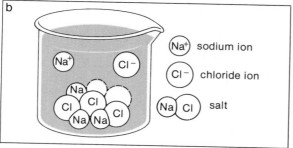

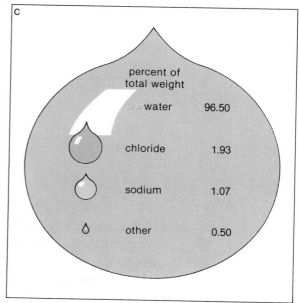

Fig. 15-1 (a) Sea water contains 3.5 percent dissolved salts by weight. (b) Salts such as sodium chloride separate into ions in sea water. (c) How does the weight of chloride ions compare to the weight of sodium ions?*

*Chloride ions weigh more.

make up ordinary table salt. Ordinary salt is the chemical compound called sodium chloride. When salt dissolves in water, it breaks up into sodium ions and chloride ions. See Fig. 15-1 (b). Thus when we refer to dissolved salt in the ocean, we are really referring to the dissolved ions coming from the salt.

Sea water is not the same as table salt dissolved in tap water. See Fig. 15-1 (c). This shows the percent by weight of sodium and chloride ions in sea water. You can see that there are also other ions present. Fig. 15-1 (c) shows that sea water contains more than sodium and chloride ions. Thus, sea water is more than just table salt dissolved in water.

There are other kinds of ions found in sea water. There are sulfate ions, magnesium ions, calcium ions, and potassium ions. See Fig. 15-2. Other substances, such as gold, are also present in very small amounts. Oceanographers believe that all of the chemical elements found in the earth's crust are present in sea water, mostly in the form of ions.

The salinity of water taken from different places in the ocean is not always the same. Water near the equator has a higher salinity than water closer to the poles. One cause of this difference is the warmer temperatures. This is caused by a higher rate of evaporation near the equator. Water that evaporates leaves the salts behind. Thus the water that remains after evaporation has occurred becomes more salty.

CHAPTER 15

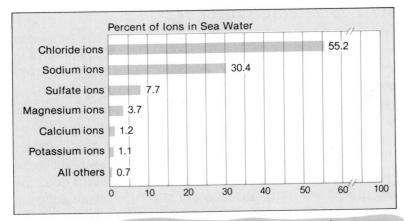

Percent of Ions in Sea Water

Chloride ions	55.2
Sodium ions	30.4
Sulfate ions	7.7
Magnesium ions	3.7
Calcium ions	1.2
Potassium ions	1.1
All others	0.7

0   10   20   30   40   50   60   100

Fig. 15-2 This bar graph shows the most abundant kinds of ions in sea water.

Although the salinity of sea water may change, the proportion of the salts in sea water does not change. The salts in sea water are always present in the same proportion. The reason for this is shown in Fig. 15-3. This figure shows two glasses of water. When a spoonful of salt is dissolved in one, the water in that glass then has a fixed number of sodium and chloride ions. Adding water does not change the number of each ion. In the same way, adding or removing water from the ocean does not change the proportion of dissolved salts in sea water.

Sea water also contains large amounts of dissolved gases. These gases come from the atmosphere. The temperature of the water affects the amount of gases that will dissolve. Warm water is less able to hold dissolved gases than cold water.

The amount of oxygen dissolved in sea water is also affected by living things. Usually the amount of dissolved oxygen in the ocean is highest in the upper water layers. This is because plants living in these sunny upper layers make oxygen. In the darker lower depths, where there are fewer plants, more oxygen is used up by living and decaying things than is produced.

Enrichment: Some elements, such as vanadium, are present in sea water in such small amounts that they can be detected only because plants and animals concentrate them in their tissues.

Fig. 15-3 The mixing of A and B does not change the proportion of dissolved materials.

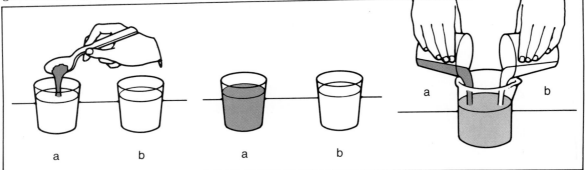

WHY IS THE SEA SALTY?

The scientists who first studied the sea believed that its salts came from the land. They thought that the rivers that flowed into the sea carried salts that came from rocks. When the water evaporated from the ocean, these dissolved materials would remain in the sea. However, this theory could not explain why some ions were found in sea water in large amounts. One example was the amount of chloride ions. It was much greater than the amount that could have come from the weathering of rocks. Scientists had to look for other ways that salts could be added to the sea. The discovery of sea floor spreading along the mid-ocean ridge helped to solve the mystery. When magma flows up into the ridge from the inside of the earth, chloride ions are given off. Thus scientists now believe that the volcanic activity is an important source of the materials dissolved in the sea. The magical "salt mill" of the old folk tale is actually the volcanic activity that creates new sea floor. Volcanic eruptions on land also produce materials that are washed into the sea by streams. Since the water in the ocean originally came from volcanic activity that took place during the earth's early history, sea water has always been salty. Since then, more salts have been added. Some salts have been added by rock weathering on the land. Also, many of the salts have come from rocks and sediments on the sea floor. Presently the makeup of sea water is not changing. Ions are slowly being removed by living things and the formation of sediments. But these are being replaced at about the same rate. Thus the makeup of sea water is not changing very much.

Reinforcement: An example of a substance whose concentration is in equilibrium is calcium. Plants and animals remove calcium from sea water to build shells and skeletons at very nearly the same rate at which it is added.

Fig. 15-4 A change in temperature with depth occurs in most areas of the oceans.

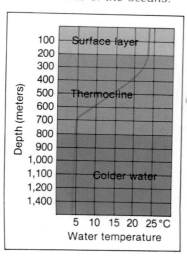

TEMPERATURE

The temperature of ocean water changes with depth. The ocean can be divided into three layers according to their temperatures: surface layer, thermocline, and a cold, deep layer.

1. Surface layer. The sun's rays falling on the sea surface heat the water. However, the rays do not reach more than a few meters into the water. Only the water very close to the surface can be heated. Waves cause the warm water to be mixed with cooler water from below to form a *surface layer*. Within the surface layer the temperature of the water remains the same. In most parts of the sea, the surface layer has the same temperature down to about 200 meters.

2. Thermocline. Below the surface layer, the temperature begins to drop rapidly. This layer of decreasing temperature is called the **thermocline.** The *thermocline* is the boundary between the warm surface water and a layer of cold deep water.

3. Cold, deep layer. In this layer, temperatures fall only a few more degrees. At depths greater than about 1,500 meters, the water temperature is usually less than 4°C. See Fig. 15-4. Since the world's oceans average about 4,000 meters in depth, most sea water is less than 4°C. Thus, most sea water is very close to freezing.

Near the poles, the sea receives little heat from the sun. Thus the sea surface in polar waters does not usually have a warm surface layer. Parts of the ocean near the poles are covered by a floating ice layer called *pack ice.* See Fig. 15-5 and Fig. 15-6. Because of its dissolved salts, sea water will not freeze at 0°C, as fresh water does. It must reach a temperature of −2°C in order to freeze. As sea water begins to freeze, the dissolved salts are not included in the ice that is formed. Thus pack ice is not salty. At the same time, the unfrozen water near the pack ice becomes more salty than normal. Usually, the layer of pack ice is no more than five meters thick because the ice protects the water below from freezing.

The ocean acts like a storage tank for heat on the earth's surface. There is more heat stored in the upper three meters of the

Thermocline A zone of rapid temperature change beneath the sea surface.

Word Study: Thermocline: *therme* (heat) + *klinein* (to lean) (Gr.), gradual temperature change.

Enrichment: Below the thermocline in the open sea the water temperature is uniformly between 0°C and 4°C.

Reinforcement: Pack ice is always frozen directly on the sea surface. Icebergs are formed from pieces of glaciers that have reached the sea. Floating blocks of pack ice are not more than five meters in thickness and do not represent a hazard to ships.

Fig. 15-5 Pack ice floating on the sea surface near the North Pole.

WATER IN THE OCEAN

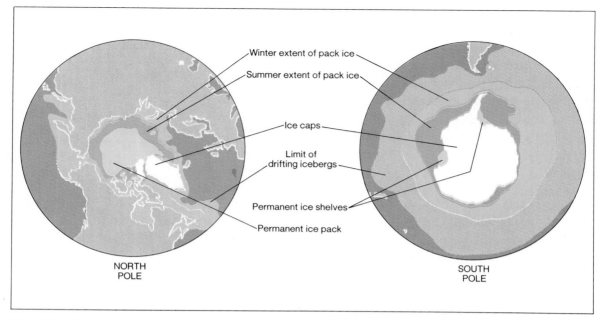

Fig. 15-6 Pack ice on the sea surface in polar regions is shown on these maps.

Image labels: Winter extent of pack ice; Summer extent of pack ice; Ice caps; Limit of drifting icebergs; Permanent ice shelves; Permanent ice pack; NORTH POLE; SOUTH POLE

sea than in the entire atmosphere. Water also absorbs and releases heat much slower than air does. For this reason, the sea affects climates all over the world. In the winter, the water does not cool as quickly as the air. This causes the water to be warmer than the air. Thus, water has a warming effect on the atmosphere in the winter. In the summer, the water does not heat as quickly as the air. This causes the water to be cooler than the air. Because of this, the ocean tends to cool the atmosphere in the summer.

The oceans absorb nearly 75 percent of the sun's energy that reaches the earth. Scientists have been looking for ways to use this energy. One system uses a liquid such as ammonia, which turns into vapor at a low temperature. The warm ocean water would be used to vaporize the ammonia. The ammonia vapor would then drive a turbine to make electricity. A cable might be used to carry the electricity to the shore.

The warmest surface waters are found near the equator. There the sun's rays strike the earth most directly. In those tropical parts of the sea, surface temperatures of 21°C to 26°C are common. Those high temperatures cause evaporation of sea water into the atmosphere. The evaporation from tropical seas has two important effects on the atmosphere. First, the air becomes very moist. The warm, moist air masses formed over the oceans near the equator have a strong influence on the world's weather and climates. The second effect of the evaporation of water from the sea is the addition of heat to the atmosphere. The water vapor leaving the sea surface removes heat energy from the sea and adds it to the atmosphere. It is this energy that is released by hurricanes and other storms that form over the warm parts of the ocean.

PRESSURE

When you dive deep in a swimming pool, you will often feel an increase in pressure on your ears. The weight of the water above you causes this increase in pressure. In the same way, the weight of water in the ocean causes the pressure to become greater with increasing depth. If you were diving in the ocean, at a depth of 10 meters you would feel a pressure that was about double that at the surface. Because of the increase in pressure, divers can go down only about 100 meters with scuba equipment and about 300 meters using diving suits. In the deepest parts of the sea floor, the pressure is more than 1,000 times greater than the pressure at the surface.

Reinforcement: Pressure in the sea is measured in units called *bars*. One bar is nearly equal to ordinary atmospheric pressure. Sea water pressure increases by one bar for each 10 m of depth.

SUMMARY

Sea water contains many dissolved substances. These materials have become dissolved in the water as a result of volcanic activity and the weathering of rock on the land. Sea water also contains dissolved gases from the atmosphere and living things. Most sea water is very cold. Only the upper layers of the ocean are warmed by the sun. The weight of water causes water pressure to increase with depth.

QUESTIONS

Use complete sentences to write your answers.

1. What is meant by the salinity of water?
2. Why is sodium chloride dissolved in water not the same as sea water?
3. List, in order of abundance, the six most abundant substances found dissolved in sea water.
4. Describe the temperature of each of the three layers of sea water as depth increases.

1. Salinity is the number of grams of dissolved salt in 1,000 grams of water.
2. Sea water contains other substances besides sodium chloride.
3. Chloride ions, sodium ions, sulfate ions, magnesium ions, calcium ions, potassium ions.
4. The surface layer down to about 200 meters is warm; the thermocline shows a rapid decrease in temperature; in the cold, deep layer below 1,500 meters, the water is usually cold, about 4°C.

INVESTIGATION

See teacher's commentary for teaching hints, safety, and answers to questions.

THE EFFECT OF SALINITY ON FLOATING OBJECTS

PURPOSE: To find out what effect salts dissolved in water have on floating objects.

MATERIALS:

modeling clay	tall 1-L container
plastic straw	water
masking tape	sodium chloride
ruler	pencil

PROCEDURE:

A. Make a ball of modeling clay about two centimeters in diameter (about the size of a marble). Push the straw into the ball of clay far enough so that it sticks firmly, making a watertight seal.

B. Cut a piece of masking tape 8 centimeters long and 1.5 centimeters wide. On the nonsticky side, draw a pencil line down the center the full length of the tape. Make a scale on the line by placing marks across it at one-millimeter distances. Make the first mark and each fifth mark longer than the others. Number the longer marks 0, 5, 10, 15, and so on.

C. Attach the scale to the straw so the end marked "0" is very near the open end of the straw. See Fig. 15–7.

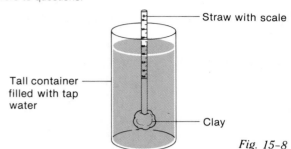

Straw with scale

Tall container filled with tap water

Clay

Fig. 15–8

D. Fill the tall container nearly full with tap water. Carefully place the straw in the container, clay end down. See Fig. 15–8. The straw should float in an upright position, with part of the scale sticking out of the water. If necessary, add or remove clay to make it float.
   1. What mark does the water come to on the scale? (It does not have to be zero.)

E. Remove the straw from the water. Make the water similar to sea water by dissolving 35 grams (5 level teaspoons) of sodium chloride (salt) in the water.

F. Place the straw in the salt water. Take a reading of where the water comes to on the scale.
   2. What is the reading?
   3. Would an ocean-going vessel float higher in a freshwater river or in the ocean?

CONCLUSIONS:

1. What effect do salts dissolved in water have on floating objects?

2. The Great Salt Lake in Utah has about 25 percent dissolved salts in it due to many years of evaporation. Would an object float higher in the Great Salt Lake or in the ocean?

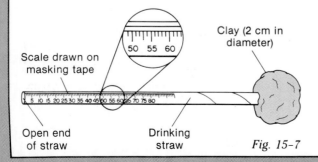

Scale drawn on masking tape

Clay (2 cm in diameter)

Open end of straw

Drinking straw

Fig. 15–7

See teacher's commentary for teaching hints, safety, and answers to questions.

THE HEAT ABSORPTION OF WATER AND ROCK

PURPOSE: To use the ideas of comparing and contrasting in a study of heat absorption.

MATERIALS:

beaker	graduated cylinder
water	2 styrofoam cups
wire screen	test tube
ring stand, ring	string
Bunsen burner	thermometer
balance	test tube holder
rock sample	

PROCEDURE:

A. Copy the following table.

Mass of rock	
Mass of water in test tube	
Beginning temp. of water, cup A	
Final temp. of water, cup A	
Temp. change due to rock	
Beginning temp. of water, cup B	
Final temp. of water, cup B	
Temp. change due to water	

B. Fill the beaker half-full of water and place it on the wire screen on the ring stand over a Bunsen burner. Heat to boiling. Go on to steps C and D while water is heating.

C. Use the balance to determine the mass of the rock. Record in your data table.

D. Using a graduated cylinder, measure an amount of water equal to the mass of the rock (volume of 1 mL of water equals mass of 1 g of water). Transfer this water to a test tube. Place the test tube of water in the water being heated.

E. After the beaker of water has come to a boil, adjust the burner flame so that it keeps the water barely boiling. Use the string to hang the rock in the boiling water. Make sure that it does not rest on the bottom of the beaker. After about five minutes, the rock and the test tube of water will be at the same temperature.

F. Using the graduated cylinder, pour the same amount of cold water into each of two styrofoam cups. The cups should be about one-third full. Label one cup A, the other B.

G. Measure the temperature of the water in cup A. Record in your table. Hold the thermometer in cup A. Quickly transfer the rock from the boiling water to the cup. Wait until the temperature stops rising. Record this final temperature of cup A.

H. Measure the temperature of the water in cup B. Record this in your table. Hold the thermometer in cup B. Using the test tube holder, transfer the test tube of water from the boiling water to cup B by pouring the test tube of water directly into cup B. Wait until the temperature stops rising. Record this final temperature of cup B.

CONCLUSIONS:

1. Compare the temperature change of the water in cup A with that in cup B.

2. Explain why land, which is mostly rock, cools faster at night and heats faster in the daytime than a body of water.

15–2. Resources from the Sea

At the end of this section you will be able to:

☐ Explain why certain parts of the sea produce more fish than others.

☐ Describe two resources other than food that can be obtained from the sea.

☐ Identify the major problems that prevent the orderly development of ocean resources.

Fig. 15-9 Much of the food we eat comes from the sea.

Look at some of the food that can be found in a supermarket. See Fig. 15–9. How much of the food that you eat comes from the sea? Today, many fishing boats sweep across the sea, taking every living thing. Will the sea continue to provide food in the future?

FOOD FROM THE SEA

On the earth's surface, the sea covers more than twice the area covered by land. The amount of plant growth on each square meter of the surfaces of the land and sea is about the same. Therefore, the sea is as able as the land is to produce food. Then why does less than 10 percent of the world's food supply come from the sea?

Part of the answer is found in the way we collect food from the sea. Almost all plant life in the sea is found in **plankton.** *Plankton* are plants and animals that float near the sea surface. Most are so small that they can be seen only with a microscope. The microscopic plants in plankton are called *phytoplankton.* They are found mostly in the upper few meters of the sea, where there is enough light to supply the energy needed for growth. Tiny animals, called *zooplankton*, then feed on the phytoplankton. Both kinds of plankton are eaten by larger animals. These, in turn, become food for larger fish and other marine animals. Thus phytoplankton are the beginning of a *food chain* that supports the animal life in the sea. If plankton could be harvested easily, the sea could provide a much larger amount of food. However, no practical way to harvest plankton has been developed. It is the other end of the food chain that supplies the harvest from the sea.

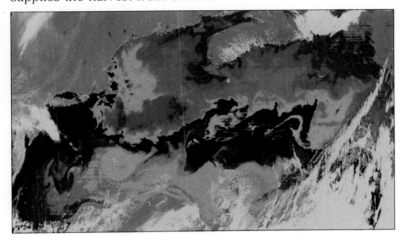

Today the food taken from the sea is mostly fish. Fish are easily caught in nets and are found in large numbers in some parts of the ocean. Most of the fish are caught in the parts of the ocean over the continental shelves. The water over the continental shelves is rich in the phytoplankton that are the beginning of the food chain that supports the fish. See Fig. 15–10. Some areas near the coasts are natural fish farms, partly because of **upwelling** of sea water that occurs there. *Upwelling* is the rising to the surface of cold water from below. This carries minerals needed for plant growth to the surface. Upwelling also occurs in some parts of the open ocean. See Fig. 15–11.

Plankton Small plants and animals that float near the surface of the sea.

Word Study: Plankton: *planktos* (Gr.), to wander or drift.
Phyto: *phyton* (Gr.), plant.
Zoo: *zoos* (Gr.), animal.

Enrichment: Phytoplankton supply about 80 percent of the oxygen present in the earth's atmosphere.

Enrichment: An unusual food chain exists around the thermal vents found on the sea floor near the mid-ocean ridge. Bacteria thrive on the dissolved substances in the hot water and become the primary part of a food chain that supports a community of animals around the vents.

Fig. 15–10 This photograph taken from a satellite shows the sea surface off the northeast coast of the United States. Cape Cod is visible in the upper middle. Red and orange colors show a large amount of phytoplankton, whereas blue indicates water that is low in life.

Enrichment: Almost all of the world's fishing grounds are areas of upwelling. About 90 percent of the catch is fish, with the remaining 10 percent consisting of shellfish and crustaceans.

Upwelling The rise of water to the sea surface.

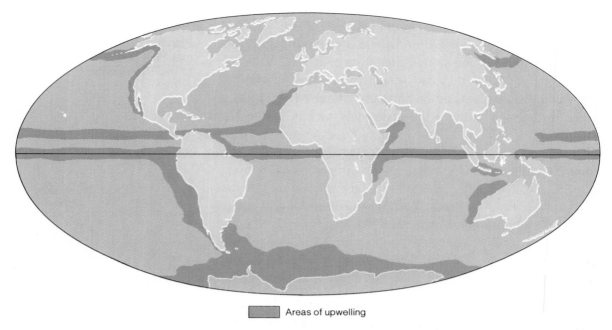

Areas of upwelling

Fig. 15-11 Parts of the ocean where upwelling occurs are shown on this map. Fish are usually most abundant in these regions.

This situation is commonly found off the west coast of South America, the California coast, and the areas of the equator.

Today we are catching nearly one-third of all the fish large enough to be taken in nets. There is a limit to the number of fish that can be taken without destroying the source of supply. Increasing the catch might cause the fish population to drop. If the number of many kinds of fish in the sea is reduced to a low level, we may destroy a large part of the entire fish population. Some kinds of fish that were once plentiful, such as cod, are now much less abundant. Some fish populations that have nearly disappeared from the oceans are now being protected from fishing crews by laws. Scientists are trying to find all of the conditions in the sea that affect the fish populations. Then they will know how much fish can be harvested without affecting the number produced. Using that information would help to insure that there would always be a supply of fish. Animals that are lower in the food chain could also be added to the catch. For example, a possible new food source is *krill*. This is a small, shrimplike animal that is found in large numbers in the cold waters of the Southern Hemisphere. See Fig. 15–12. Krill can be used directly for food or as part of other foods. Other ocean animals, such as krill, that are not now used in large amounts may become an important new food source.

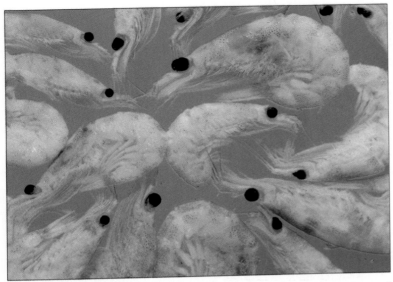

Fig. 15-12 Krill is a small animal found in huge numbers in some parts of the ocean.

There may also be another way to increase the amount of food we get from the sea. Farming methods can be used in the sea, just as they have been used on land. Making sea farms, however, is not a simple task. Large parts of the oceans would have to be fenced. The water would need to be fertilized. Huge floating tanks might have to be built to hold mineral-rich water brought up from the bottom. Until such ocean farms are in operation, most food from the sea will come from fishing. See Fig. 15–13.

Enrichment: In some parts of the world, such as Japan, seaweed is an important source of food.

Fig. 15-13 Fishing is the main way that food is obtained from the sea.

Fig. 15-14 Salt is the most abundant mineral taken from the sea.

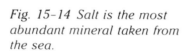

Nodules Lumps of sediment found on the sea floor that contain valuable minerals.

MINERALS FROM THE SEA

The sea contains huge amounts of dissolved minerals. Although there are large amounts, most of these minerals are dissolved in a vast amount of water. Thus, it is very expensive to take these minerals from sea water. For example, it is possible to take a valuable mineral, such as gold, from sea water. But the process costs more than the gold is worth. At the present time, salt, bromine, and magnesium are the only minerals taken from the sea. See Fig. 15-14.

In the future, another way to obtain minerals from the sea may be used. In some places, large areas of the deep ocean floor are covered with lumps of minerals called **nodules** (**nod**-yoolz). The *nodules* are about the size of potatoes or eggs. See Fig. 15-15. Scientists do not know all the details about how nodules are formed. However, it is known that they grow very slowly. It may take a million years to form one of the larger nodules. They often contain large amounts of manga-

Fig. 15-15 These nodules were taken from the deep sea floor.

nese or iron. They also contain smaller amounts of copper, zinc, silver, gold, and lead. Thus the nodules are a vast reserve of these metal ores. Taking the nodules from the sea bed will require special mining methods. The nodules might be swept up by a machine that works like a giant vacuum cleaner. They also might be gathered by automated submarine vehicles that would travel between the sea floor and a ship. Nodules may be an important source of some minerals in the future.

Sea water is also a resource. In some parts of the world, fresh water supplies are so small that water is taken from the sea. The dissolved salts can be removed by either of two methods. The water can be heated and evaporated, leaving the salts behind. The vapor is then cooled and changed back into fresh water. A second method uses a plastic film that allows pure water, but not dissolved salts, to pass through it. Fresh water supplied by these methods is expensive. The cost is too high for the water

Enrichment: In the 1950's, Japanese people eating fish from a part of the Bay of Tokyo were afflicted by a mysterious disease. It was discovered that the cause was mercury pollution of the water from a paper manufacturing plant. Fish at the top of the food chain contained high concentrations of the mercury released into the water.

Fig. 15-16 Desalination plants take the salts out of the sea water.

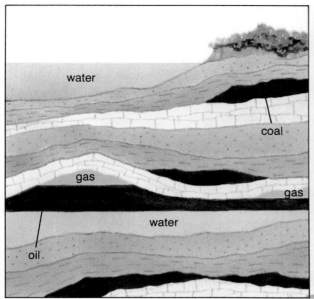

Fig. 15–17 (left) *Layering beneath the ocean showing the oil pools.* (right) *Offshore oil rigs drill into the ocean floor.*

to be used for irrigation. Now there are about 1,500 plants in the dry parts of the world that produce fresh water from the sea for human and industrial use. See Fig. 15–16, page 421.

ENERGY FROM THE SEA

The most important source of energy from the sea is petroleum. A large part of the world's reserves of petroleum and natural gas lie beneath the sea. Most of these offshore deposits are found along the continental shelves. It is difficult to reach these offshore deposits. It is also more expensive to drill for oil or gas on the sea floor than it is on land. Offshore drilling has become common because supplies of gas and oil are becoming harder to find on land. The oil and gas from the sea will have an even greater importance in the future. See Fig. 15–17.

The ocean is also filled with another source of energy. Most of this energy comes from the sun. Heat supplied by the sun is stored in the water and also causes the winds that produce waves and currents. Some of the sun's energy is also captured by the plants that grow in the sea. Motion of the water caused by the tides also can be an energy source. However, all of this energy is spread through the entire ocean. It is very expensive to build the equipment that can take large amounts of this energy from the water. The waves striking about 100 kilometers of shoreline have enough energy to supply electric power to a million homes. But it would be very difficult and expensive to build the equipment necessary to do this. It is likely that for many years the most important source of energy from the sea will continue to be the fuels taken from the sea floor.

PROTECTING THE OCEAN RESOURCES

The food, minerals, and the fuels in the sea are more difficult to obtain than those found on land. However, the difficulty in obtaining the resources is not the biggest problem. The real problem is deciding who owns the sea's resources. Countries cannot agree on questions about using the resources from the sea. How far out to sea do the borders of a country extend? Do nations have the right to take fish without limit in the open sea? Who has the right to collect minerals from the deep sea floor? Do countries without a coastline have any rights to ocean resources? Until there is agreement among nations there cannot be full use of the sea's resources.

Pollution of the sea is also a problem that must be solved among nations. Chemicals and other wastes that are dumped into the sea cause pollution that can affect the entire ocean. Only if all nations work together will the resources of the sea be protected.

In the last 20 years, the amount of fish taken from the oceans has increased three times. This can be attributed to advances in technology such as echo sounders to locate schools of fish. In many areas off coasts, the supply of fish has been wiped out. Unless "overfishing" is stopped, it is conceivable that in the next decade, there will be no healthy fishing stocks left in the oceans.

Reinforcement: The most common sources of pollution are oil spills, toxic wastes and insecticides.

SUMMARY

Food taken from the sea is mostly in the form of fish. The ability of the sea to provide food can be increased by using other forms of life and by creating sea farms. Most of the ocean's mineral wealth lies on or beneath the sea floor. The ocean is an energy resource. The development, use, and preservation of ocean resources depends on cooperation among nations.

QUESTIONS

Use complete sentences to write your answers.

1. Why are most of the ocean's fish found over the continental shelves?
2. Name three minerals presently being taken from the sea.
3. What is the most important energy resource that is taken from the sea?
4. List four questions that must be answered before much of the ocean's resources can be used.

1. There are abundant food supplies for fish in the sea over the continental shelves.
2. Salt, bromine, and magnesium are taken from the sea.
3. Petroleum is the most important energy resource taken from the sea.
4. (a) How far out to sea does a country's border extend? (b) Do nations have a right to take fish from the sea without a limit? (c) Who has a right to take minerals from the deep sea floor? (d) Do countries without a coastline have any rights to ocean resources?

See teacher's commentary for teaching hints, safety, and answers to questions.

CHANGING SEA WATER TO FRESHWATER

PURPOSE: To use the process of distillation for changing sea water to freshwater.

MATERIALS:

graduated cylinder	wire screen
colored salt water (artificial sea water)	Bunsen burner
	clamp
flask	beaker
ring stand	glass tubing
ring	test tube
tap water	goggles

PROCEDURE:

A. Using a graduated cylinder, measure 25 mL of the colored salt water and place it in the flask. Place a drop of the salt water on your tongue. The salt content should be apparent. CAUTION: NEVER TASTE ANY SUBSTANCE IN THE LABORATORY WITHOUT PERMISSION.

B. Set up the equipment as in Fig. 15–18.

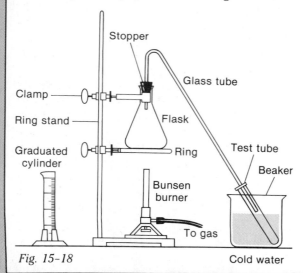

Fig. 15–18

C. CAUTION: WEAR GOGGLES FOR THIS PROCEDURE. *Gently* boil the salt water until there is almost none left in the flask. If steam escapes from the test tube, slow the heating. This process is called distillation. When the first signs of dryness or solid material appear, stop heating the liquid. NEVER BOIL A LIQUID TO COMPLETE DRYNESS IN A FLASK. Turn off the burner. Be careful of the hot apparatus. Let it cool.
  1. How does the liquid in the test tube differ from the original liquid?
  2. What do you think the liquid is?
D. When the apparatus has cooled, remove the flask. NEVER PLACE WATER IN HOT GLASSWARE.
  3. Describe the material left in the flask.
  4. What do you think is left in the flask?
E. Measure the amount of liquid collected in the test tube.
  5. How does the amount in the test tube compare with what was in the flask?
  6. Account for any difference found in your answer to question 5.
F. Add some tap water to the cooled flask. Shake the water around in the flask.
  7. Does the material in the flask dissolve when tap water is added?
  8. Taste the solution that results and describe the taste.

CONCLUSIONS:

  1. Describe how salt can be removed from sea water by distillation.
  2. Compare the properties of the sea water used in this investigation to the water obtained in the test tube.

CAREERS IN SCIENCE

COAST GUARD ENLISTED PERSONNEL

Each year, hundreds of lives are saved by the efforts of the U.S. Coast Guard. The Coast Guard, a branch of the armed services, conducts search and rescue missions along the coastline and on inland waterways, using both ships and helicopters. It is also the Coast Guard's duty to patrol the nation's coastlines and protect them from attack.

To join the Coast Guard, you must be at least 18 (17 with parents' consent), but no older than 26. You must also be in good physical condition and have a high school diploma. The term of enlistment includes four years of active service and two years in the reserves. You may move frequently or stay at one duty station.

Jobs in the Coast Guard vary from a seaman or seaman's apprentice, who might do labor on board ship or in the shipyard, to trained personnel such as electricians, mechanics, communication specialists, and radar technicians. Training is provided by the Coast Guard. For further information, write: Commandant, United States Coast Guard, Washington, DC 20590.

TECHNICAL WRITER

Technical writers translate difficult scientific concepts into language that most people can easily understand. A technical writer may write a magazine article on the newest underwater diving gear. She or he may prepare manuals on how to operate computers, or write advertising copy to describe new medicines. Technical writers may also prepare audio-visual aids on everything from the latest satellite to the most recent discovery in cancer research. There is a wide range of available work.

To become a technical writer, a college degree and strong writing skills are generally required. Competition is high, and potential employers usually look for people with a good educational background. Jobs are available in publishing companies and in high-tech companies, in advertising firms and with scientific organizations that produce publications. Many writers also work independently on a free-lance basis. For more information, write: Society for Technical Communication, Inc., 1010 Vermont Avenue N.W., Suite 421, Washington, DC 20005.

CHAPTER REVIEW

VOCABULARY

On a separate piece of paper, match each term with the number of the statement that best explains it. Use each term only once.

salinity   pack ice   phytoplankton   upwelling   nodules
thermocline   plankton   zooplankton   krill

1. The rise of bottom water to the sea surface.
2. A small shrimplike animal that is often found in the cold waters of the Southern Hemisphere.
3. The floating layer of ice on the ocean near the poles.
4. The number of grams of dissolved salts in 1,000 grams of sea water.
5. A region of rapid temperature change beneath the sea surface.
6. Very small plants and animals that float near the sea surface.
7. Tiny animals that feed on phytoplankton.
8. Lumps of sediment found on the sea floor that contain valuable minerals.
9. Very small plants found in the upper few meters of the sea.

QUESTIONS

Give brief but complete answers to each of the following questions. Unless otherwise indicated, use complete sentences to write your answers.

1. How does the salinity of sea water compare to that of drinking water?
2. What is the salinity of a 1,000-gram sample of water if only 944 grams of it is pure water?
3. Give the names and the amounts, in percentages, of the two most abundant substances found in sea water.
4. Compare the salinity of surface sea water at the equator with that near the poles.
5. Why is the sea salty?
6. Compare the conditions found in the ocean surface layer with those found in the deep sea.
7. What two important effects on the atmosphere result from the evaporation of water from the tropical seas?
8. Why can the oceans produce twice the food that land provides?

9. What is the beginning of the food chain for life in the ocean?

10. Where are the best places for fishing located?

11. Why is upwelling necessary for abundant life in the ocean?

12. The ocean contains a large amount of gold and other minerals, but gold is not taken from sea water. Why is this?

13. What minerals do nodules contain that make them so valuable?

14. Describe the ways energy may be taken from the sea.

15. What is the biggest problem with taking resources from the ocean?

APPLYING SCIENCE

1. Determine the salinity of a sample of drinking water by doing the following: (a) Weigh a clean, dry 500-mL beaker. (b) Measure into the beaker 250 grams of water. (c) Boil until nearly dry. CAUTION: NEVER BOIL UNTIL COMPLETELY DRY! (d) Allow beaker to sit, away from heat, until completely dry. (e) Weigh beaker. The difference in original weight and final weight of beaker is the amount of dissolved material in the water for .25 kilogram of water. (f) Calculate the salinity.

2. Make up a sample of artificial sea water by dissolving 35 grams of rock salt in 965 grams of water. Try to get drinkable water from the salt water by freezing it.

3. In 1984 much of the krill usually present in the Antarctic waters disappeared. Find out why this happened and its effect on the life in the ocean.

4. Use a biology book to find out what plankton look like. Write a list of the ocean food chain in which plankton are involved.

5. Find out the details of recent developments or uses of sea resources such as: sea farming; sea mining; energy from the sea; seaweed products; etc. Report to the class on your findings.

BIBLIOGRAPHY

Britton, Peter. "Cobalt Crusts: Deep Sea Deposits." *Science News*, October 30, 1982.

Britton, Peter. "Deep Sea Mining." *Popular Science*, July 1981.

Matthews, Samuel W. "New World of the Oceans." *National Geographic*, December 1981.

Time-Life Editors. *Restless Oceans* (Planet Earth Series). Chicago: Time-Life Books, 1984.

Gentle ripples show the motion of the sea. Through its motion, the saltiest waters become less salty, and the coldest waters warmer. Eventually, all water will move around the world.

THE MOTIONS OF THE SEA

CHAPTER GOALS

1. Explain the different ways that underwater currents can form in the oceans.
2. Describe the movement of the surface currents in the oceans and the factors that cause them.
3. Explain how surface currents can affect climates.
4. Describe the parts of a wave, and explain how a wave forms on the sea surface.
5. Explain what causes tides and describe the normal pattern of these giant waves.

16-1. Underwater Currents

At the end of this section you will be able to:

- ☐ Explain how differences in the temperature and salinity of sea water produce ocean currents.
- ☐ Describe how *turbidity currents* can be formed.
- ☐ Describe the path followed by deep currents of the Atlantic and Pacific oceans.

Much of the world's population depends on food from the sea. Fishing boat captains find that they can usually depend on great catches at certain places. Studies of the ocean currents show that deep water rises in these areas. This supplies the food needed for sea life. Knowing where the ocean currents rise, the oceanographer can now guide sea captains to places that are full of sea life.

Word Study: Current: *currens* (L.), running.

TEMPERATURE

When the water in the sea moves in a certain direction, a **current** is formed. One type of *current* is formed when water is heated unequally by the sun's energy. For example, water near the equator is heated. This lowers the density of the water.

Water near the poles is cooled. This makes the water more dense. The cold water sinks and moves toward the warmer water near the equator. For example, the cold water in the Arctic and Antarctic sinks to the ocean floor. Then, the cold

Current Water moving in a particular direction.

Skills: Study of currents reinforces cause and effect.

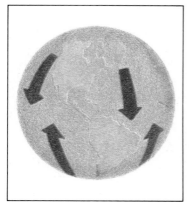

Fig. 16-1 Deep currents move generally from the poles toward the equator.

Fig. 16-2 The salinity of the sea water can be increased by freezing or evaporation.

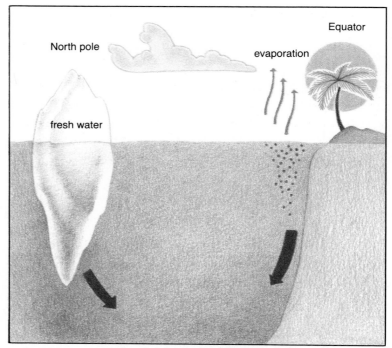

water moves very slowly along the bottom of the ocean toward the equator. Some of the cold water mixes with the water above and moves back toward its starting place. The rest of this very cold water slowly moves through the deep oceans away from the polar regions where it was formed. See Fig. 16-1. Most of this icy water flows deep in the ocean for many years before finally making its way back to the surface.

SALINITY

Currents are also formed by changes in the salinity, or amount of dissolved salts in sea water. When salt dissolves in water, the density of the water increases and the water sinks. One way salinity is increased is when water freezes. When ice is formed in the sea, only the freshwater freezes. The salt in the sea water is forced out and goes into the water that does not freeze. Therefore, water surrounding newly formed ice has more salt, or a higher salinity. Another way salinity can increase is when water evaporates at the surface. See Fig. 16-2. In both cases, the surface water sinks because it has a higher density than the water under it.

Fig. 16-3 shows a current caused by differences in salinity. It is found at the entrance to the Mediterranean Sea at the Strait of

Gibraltar (ji-**brawl**-tar). The dry, warm climate of the Mediterranean causes evaporation from the surface of the water. This causes the salinity of the water to be much greater in the Mediterranean than it is in the Atlantic Ocean. The dense water from the Mediterranean sinks below the surface and flows westward past Gibraltar into the Atlantic. At the same time, the lighter Atlantic water flows eastward, past Gibraltar, on top of the heavy outgoing flow. During World War II, some submarines slipped past the defense at Gibraltar by shutting off their engines and riding this density current.

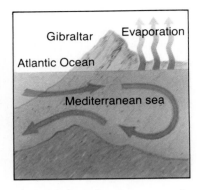

Fig. 16-3 A density current is found at the entrance of the Mediterranean Sea because of differences in salinity.

TURBIDITY CURRENTS

Another type of current caused by changes in water density is called a **turbidity** (tur-**bid**-ih-tee) **current.** *Turbidity currents are strong, downslope movements of muddy water. They form from the sand and mud that build up on the continental slope.* Sometimes they suddenly fall under their own weight and mix with water. Since the muddy water is more dense than the surrounding water, it rushes down the slope, causing a deep current. Turbidity currents may also be started by severe storms, earthquakes, or floods. Submarine canyons in the continental slope are probably carved out by turbidity currents where rivers carrying sediments flow into the ocean. These currents also move large amounts of sediment from the shallow continental margins to the deeper sea floor.

Turbidity current A current that is caused by the increased density of water as a result of mixing with sediments.

Enrichment: The boundaries between water of different densities in the sea are similar to the boundaries between cold and warm air masses. Just as it does in the atmosphere, turbulence occurs at these boundaries. Submarines have encountered violent internal waves at such boundaries.

DEEP SEA CIRCULATION

Currents that are very deep in the sea are hard to study. Thus they are not understood as well as the currents near the surface. However, the main currents in the deepest parts of the Atlantic, Pacific, and Indian oceans have been charted.

Along the floors of these oceans, there is a northward flow of cold, dense water that comes from the Antarctic region. This current is formed when the cold water freezes. The freezing action leaves extra salt in the unfrozen water. This causes the water around the continent of Antarctica to be very dense. This water sinks and then moves northward along the bottoms of the three oceans that surround Antarctica. The speed of this icy cold bottom current is quite slow. It may take at least several

Word Study: Turbidity: *turbidus* (L.), disturbed.

Enrichment: Deep ocean currents may be traced by use of special floats that remain submerged at a selected depth while sending out a sound signal.

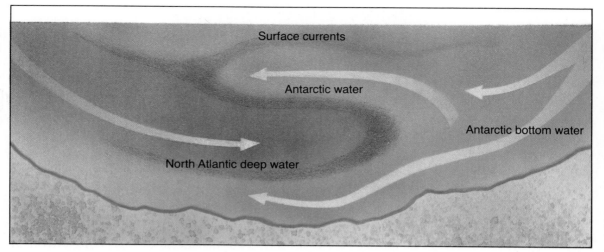

Surface currents

Antarctic water

Antarctic bottom water

North Atlantic deep water

Fig. 16–4 Deep currents in the Atlantic result from water sinking near Antartica in the south and Greenland in the north.

Enrichment: The slow-moving deep current away from the polar regions is called *polar creep.*

1. The cold water in the Arctic and Antarctic sinks to the ocean floor and moves toward the equator. Warm water around the equator rises to the top and moves toward the poles to replace the cold sinking water.
2. As water evaporates or is turned into ice, the salinity of the remaining water increases. This makes the water more dense. This dense water sinks and moves away from the area where it is produced. Less dense water moves in to take its place.
3. Turbidity currents form when sand and mud tumble down an undersea canyon under their own weight. This may be caused by severe storms, earthquakes, or floods.
4. It is at the poles that the greatest changes in density are occurring, due to very cold temperatures and the formation of ice. That causes the water to be very dense and to sink. The water moves away from the poles at the bottom of the oceans.

hundred years for this water to travel from Antarctica to the northern Atlantic and northern Pacific oceans.

In the Atlantic, there is a deep current that moves south. This current is caused by the cold water that sinks near Greenland. This water near Greenland is less dense than the water that comes from the Antarctic. These two currents flow in opposite directions, meeting a little north of the equator. The less dense, North Atlantic water rides up over the heavier, Antarctic water. See Fig. 16–4. However, neither of these currents is affected by the surface movement of ocean water.

SUMMARY

Deep currents are caused by differences in the temperature or salinity of the water. Turbidity currents are caused by down-slope movements of sand and mud. All of the world's oceans have slow currents of near-freezing water moving close to the bottom, away from the poles and toward the equator.

QUESTIONS

Use complete sentences to write your answers.

1. Give an example of how differences in temperature cause ocean currents.
2. Explain how changes in salinity cause ocean currents.
3. Explain how turbidity currents form.
4. Why do the deep currents in the Atlantic and Pacific oceans begin at the poles?

INVESTIGATION

See teacher's commentary for teaching hints, safety, and answers to questions.

DENSITY CURRENTS

PURPOSE: To see what happens when waters of different densities are put together.

MATERIALS:

liter container   2 250-mL beakers
water   food coloring
sodium chloride   balance
  (salt)

PROCEDURE:

A. Prepare a salt water solution by dissolving 35 grams (5 level teaspoons) of sodium chloride in 1 liter of water.

B. Fill one of the beakers half-full with tap water. Allow this to sit undisturbed for several minutes. Go on to step C.

C. Fill the other beaker half-full with the salt water. Add four or five drops of food coloring to the salt water and mix well.

D. Slowly pour one-third of the colored water down the side of the beaker of tap water. Make sure the two solutions do not mix when poured together. Do not move the beaker of tap water. See Fig. 16–5. If the colored water mixes with the tap water, pour out the mixture and repeat steps B, C, and D.

E. Let the mixture sit a minute. Just as with air masses, water masses of different densities do not tend to mix.

  1. What do you observe?

F. Without lifting it from the table, slowly tilt the beaker of colored water and tap the beaker on one side. Hold it this way until the colored water settles in a new position. Then *gently* place the beaker back in its original position. Watch the colored water for a few minutes.

  2. What do you observe?

G. Repeat step F. This time watch carefully for waves at the border between the salt water and the tap water. These are called *internal waves*.

  3. Do the waves bounce off the side of the beaker?

H. Again repeat step F. Watch the waves on top of the tap water as well as the internal waves. It may be more difficult at this time since the colored water may be mixing into the tap water. If that happens, you may need to prepare another sample.

  4. Compare the size, shape, and speed of the internal waves with those of the surface waves.

CONCLUSIONS:

  1. Which is denser, tap water or salt water?

  2. Explain how internal waves are formed.

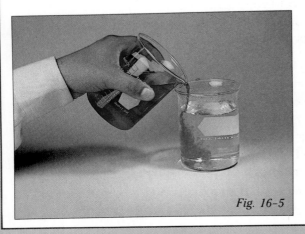

Fig. 16–5

16-2. Surface Currents

At the end of this section you will be able to:

- ☐ Describe how the continents and the Coriolis effect change the motion of surface currents in the oceans.
- ☐ Describe the general pattern of surface currents in the oceans.
- ☐ Name and describe the paths of the main surface currents of the North and South Atlantic.

The beaches in southern California are much like the beaches in southern Florida. See Fig. 16-6. The air temperature is warm all year and the sun shines much of the time. However, if you entered the water, you would immediately feel one big difference. The water at the beaches in California is cold, with temperatures rarely reaching higher than 21°C. In contrast, the water at the beaches in Florida is warm, with temperatures rarely lower than 21°C. How can the water off the coasts of two sunny, warm regions be so different? *

* While air temperature affects the water temperature, other factors are present. Cold water currents must rise from deep water along the coast of California, whereas warm water currents flow to the coast of Florida.

Fig. 16-6 Southern California beaches have water temperatures colder than southern Florida.

MOVING WATER

Water near the surface of the sea is moved mainly by winds. We know this because of the winds over the northern Indian Ocean. There the winds move in opposite directions in summer and winter. When the winds reverse, so do the surface currents. However, there are other factors that influence water motion near the ocean's surface. For example, the continents act as barriers, blocking the free movement of the ocean water. Also, the earth's rotation influences the motion of the sea water. The Coriolis effect causes the currents in the Northern Hemisphere to turn to the right from the direction of water flow. Thus surface currents do not necessarily move in the same direction as winds. In the Southern Hemisphere, the Coriolis effect causes a twist to the left from the direction of the water flow. In both hemispheres, this causes the water at and near the surface to form huge circles of moving water. In the Northern Hemisphere, the direction of this movement is clockwise. In the Southern Hemisphere, the motion is counterclockwise. See Fig. 16-7. There are also many smaller circles in each ocean.

Winds also may cause *upwelling*, or the rising of water from the deep layers of the ocean to the surface. This happens when wind moves warmer surface water away from the shore. As the warm surface water is moved away, it is replaced by cold water from below. This slow upward flow of cold water creates the area of cold water off the shore of California.

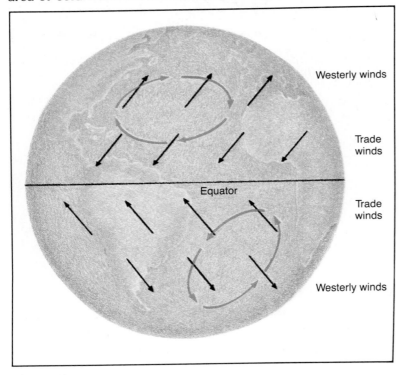

Westerly winds

Trade winds

Equator

Trade winds

Westerly winds

Fig. 16-7 The Coriolis effect and the action of wind produce huge circles of moving water.

CURRENTS OF THE NORTH ATLANTIC

1. **The Gulf Stream.** Around 1770, Benjamin Franklin learned that some ships had crossed the Atlantic from England to America in about two weeks less than the expected time. The captains of these ships described a swift, eastward current flowing in the North Atlantic. By sailing south to avoid this current, they made a faster voyage. Franklin measured the water temperature and the speed of the current during several trips between America and England. In 1779, he published the first scientific chart of an ocean current. This current is known as the Gulf Stream.

The Gulf Stream begins with waters moved by the trade winds. These winds blow from the northeast and southeast near the equator. Because of the Coriolis effect, the surface water moves westward in the Atlantic just north of the equator. This flow is called the North Equatorial (ee-kwuh-**tor**-ee-ul) Current. The North Equatorial Current pushes water against the coast of Central America into the Gulf of Mexico. This causes water to spill out of the Gulf of Mexico between Florida and Cuba, forming the Gulf Stream. See Fig. 16–8. In the Atlantic Ocean, the Gulf Stream flows around the Florida coast, bringing warm water to its beaches. The result is that the water temperature rarely drops below 21°C throughout the year. The Gulf

Fig. 16–8 This satellite photograph of the southeastern United States coast clearly shows the Gulf Stream.

436

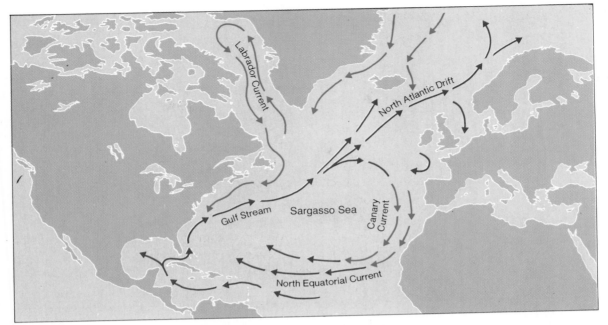

Fig. 16-9 Currents in the
North Atlantic. Warm currents
are shown in red.

Stream moves northward along the eastern coast of the United States and then turns east toward Europe. The Gulf Stream moves huge amounts of water. It sweeps along more than four billion tons of water each second. This is more than a thousand times the amount of water moved by the Mississippi River.

2. Labrador Current. Near Greenland, the Gulf Stream meets the cold, south-flowing Labrador (**lab**-ruh-dor) Current. The collision of the warm Gulf Stream and the cold Labrador Current causes the ocean in that area almost always to be covered by a thick fog.

3. North Atlantic Drift. Beyond Greenland, the path of the Gulf Stream is not as easily identified. Part of the current moves on toward northern Europe, forming the North Atlantic Drift. A slow-moving current is called a "drift." Another branch of the Gulf Stream turns south, forming the Canary Current. The Canary Current moves toward the equator and begins to complete the huge circle of currents that move around the North Atlantic.

4. North Equatorial Current. Off the coast of Africa, the Canary Current joins the North Equatorial Current to complete the circle formed by the North Atlantic currents. See Fig. 16–9. The center of the North Atlantic has no major currents. In an area called the Sargasso Sea, large amounts of seaweed are found because no major current sweeps it away.

Enrichment: The Sargasso Sea gets its name from sargassum, a kind of floating seaweed that covers much of the surface in that area. At one time, sailors believed that the seaweed could trap their ships.

THE MOTIONS OF THE SEA

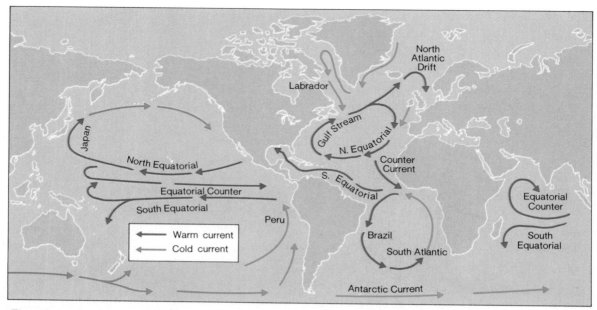

Fig. 16–10 The principal ocean currents of the world.

CURRENTS OF THE SOUTHERN OCEANS

The currents in the North Pacific move in a huge clockwise circle, very much the same as currents do in the North Atlantic. However, in the southern parts of the Atlantic and Pacific oceans, the currents flow in a counterclockwise direction. The Antarctic Current follows an unusual path. It completely circles Antarctica; no landmass blocks its movement.

Currents in the Indian Ocean are controlled by two factors. One factor is the South Equatorial Current, which divides as it moves toward Africa. One part turns to the north and the other part moves to the south. The southern branch turns back toward Australia. Near Australia the current breaks up, forming many small currents that move among the South Pacific islands. The other factor affects the northern area of the Indian Ocean. There the currents follow the strong monsoon winds. During different seasons of the year, the currents change their directions. The principal ocean currents of the world are shown in Fig. 16–10.

OCEAN CURRENTS AND CLIMATE

Ocean currents, such as the Gulf Stream and the North Atlantic Drift, have a strong influence on temperature and climate. Wind blowing across these currents and onto land can raise the land's average temperature. For example, northwestern Europe has a higher average temperature than other areas at the same latitude. This occurs because of the steady westerly winds. These winds blow over the North Atlantic Drift toward northwestern Europe.

Ocean currents also help transfer heat from areas near the equator to the poles. Otherwise the regions near the equator would become warmer, while the poles would become colder.

Scientists believe that changes in ocean currents also may cause weather patterns to change. For example, a warm current of equatorial water appears about once every ten years along the west coast of South America. It is called "El Niño," which means "the child," because it usually appears around Christmas. Fish that normally inhabit the waters along the coast of Peru disappear because they cannot live in the warm water. In 1982, a very strong El Niño current developed, causing the surface water in the eastern Pacific to become very warm. Unusual winter storms and a very hot summer in North America resulted when the atmosphere was heated by the warm water in the Pacific.

SUMMARY

Although winds are the cause of surface currents in the sea, other influences such as land barriers and the Coriolis effect change the direction the water moves. Major surface currents in the northern and southern parts of the Atlantic and Pacific oceans form huge circles. Other oceans have their own pattern of currents. Surface currents in the sea are an important influence on the atmosphere, and therefore affect the climates of the continents.

QUESTIONS

Use complete sentences to write your answers.

1. Explain how ocean currents are affected by land and the Coriolis effect.
2. Describe the general pattern of surface currents in the North and South Atlantic.
3. Name the five main currents of the North Atlantic and briefly describe each.
4. Name two main currents of the South Atlantic and briefly describe each.

1. Landmasses cause currents to be deflected along their shores. The Coriolis effect causes the currents in the Northern Hemisphere to be deflected to the right, while in the Southern Hemisphere, currents are deflected to the left. This causes very large circular currents in the oceans.

2. The general pattern of surface currents in both the North and South Atlantic is that of a large circle, with the currents going clockwise in the North Atlantic and counterclockwise in the South Atlantic.

3. (1) Gulf Stream: a warm current flowing north along the east coast of the United States and then turning east toward Europe. (2) North Atlantic Drift: a warm current; a branch of the Gulf Stream that continues toward northern Europe. (3) Canary Current: a cool current; the branch of the Gulf Stream that turns south toward Africa. (4) North Equatorial Current: a warm current that moves westward from Africa toward North America. (5) Labrador Current: a cold current that flows southward between North America and Greenland.

4. (1) Antarctic Current: a current that completely circles Antarctica. (2) South Equatorial Current: a current that moves toward the African coast, where it divides into two parts.

See teacher's commentary for teaching hints, safety, and answers to questions.

THE CORIOLIS EFFECT

PURPOSE: To demonstrate how the Coriolis effect occurs.

MATERIALS:

sheet of paper	ball bearing (1 cm
cardboard	or larger)
carbon paper	pencil
pin	

PROCEDURE:

A. On a level table, place the sheet of paper on the cardboard and put the carbon paper, carbon side down, on the paper. Then stick a pin in the center of the carbon paper to hold the paper, carbon paper, and cardboard in place.

B. Starting in one corner of the paper, roll the ball bearing across the carbon paper to the other corner. Lift the edge of the carbon paper and label the track No. 1. Draw an arrow on the track showing the direction of the roll.
    1. Describe the track made by the ball as it rolled over the carbon paper.

C. The ball bearing represents moving water. The paper and carbon represent the earth. When viewed from the North Pole, the earth turns in a counterclockwise direction. Hold a corner of the paper and carbon with one hand and the cardboard with the other hand. Keep the cardboard still while you turn the paper and carbon counterclockwise slowly around the pin. While you are doing this, have your partner carefully roll the ball bearing across the carbon paper from one corner of the paper to the other.

D. Lift a corner and label the track No. 2. Place an arrow on the track to show the direction the ball bearing moved.
    2. Describe the track made by the ball as it rolled over the carbon paper.

E. When viewed from the South Pole, the earth turns in a clockwise direction. While you carefully turn the papers in a clockwise direction, have your partner slowly roll the ball bearing across the carbon paper again. Label this track No. 3. Indicate the direction the ball bearing moved with an arrow.

F. Remove the pin and carbon. Inspect the tracks.
    3. Describe the track the ball made when the papers were turned in a counterclockwise direction (track No. 2).
    4. Describe the track the ball made when the papers were turned in a clockwise direction (track No. 3).

CONCLUSIONS:

1. What caused the difference between tracks 1 and 2?
2. What caused the difference between the tracks marked 2 and 3?
3. Look at Fig. 16–10 on page 438. Use the information from this investigation to explain why the large circles of currents in the North Atlantic and South Atlantic curve in the directions shown.

16-3. Waves

At the end of this section you will be able to:

- ☐ Describe a wave in terms of its height, wavelength, and period.
- ☐ Explain how waves are formed.
- ☐ Describe the motion of water in waves.

Suggestion: If available, a ripple tank can be used to illustrate waves. A light bulb placed below the tank will allow the entire class to view the waves reflected on the ceiling. Pulses and periodic waves can be shown.

If you have ever seen surfing, or been in the ocean, you have seen or felt the power of the waves. See Fig. 16–11. Many people can feel the up-and-down motion of waves while in a boat. Anyone who has been caught by surprise and thrown over by a breaking wave knows that waves carry energy. Where do waves get their energy?*

* Most waves get their energy from wind.

Fig. 16–11 Large waves are every surfer's wish. Where do these waves come from?*

* Large waves come from the wind blowing on the water. The longer the wind blows on a wave, the larger the wave.

WHAT IS A WAVE?

Wind produces the familiar waves that are almost always present on the sea. Even a gentle breeze blowing across the surface of still water causes waves. Air never moves in smooth streams, but has a certain amount of whirling motion. The whirling motion of moving air makes small dents in the water's surface. This whirling action also raises a part of the water. The wind then pushes on the raised part of the water's surface and a wave is created. As long as the wind continues to blow, the wave will move and grow larger.

If you toss a stone into a quiet pool of water, waves are created. The falling stone carries energy into the water. Anything

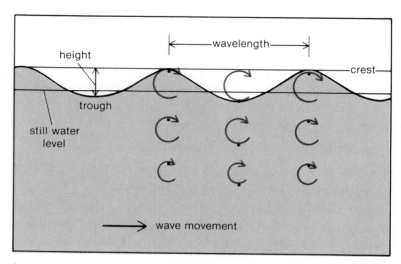

Fig. 16-12 The parts of a wave.

Fetch The distance over which the wind blows steadily, producing a group of waves.

that adds energy to water can make waves. All waves have two major parts. The raised parts are called *crests*. Between the crests are low areas called *troughs* (**trawfs**). Waves can be described by measurements of their crests and troughs. For example, the *wave height* is the distance between the lowest part of a trough and the highest part of the next crest. The distance between two neighboring crests, or two neighboring troughs, is called the *wavelength.* See Fig. 16–12. The time it takes for two crests to pass a given point is the *wave period.*

When the wind begins to blow over calm water, patches of ripples appear. These small waves are called "cat's paws." If the wind continues at the same speed, the little waves will grow. The shape and size of the waves that are produced depend on three things: the speed of the wind, how long the wind blows, and the **fetch.** *Fetch* is the distance over which the wind blows steadily. Strong winds blowing for many hours over a long fetch produce waves with the greatest heights, wavelengths, periods, and speeds. However, waves can only grow to heights equal to about one-seventh of their wavelengths. Waves that build up higher than that break and tumble over. Thus waves that have short wavelengths cannot get very big. Really large waves can only be formed on the open sea, where strong winds blow across fetches of thousands of kilometers.

The wind usually makes waves of many sizes. Wind-caused waves tend to move in the direction of the wind. The larger waves grow even larger since they are better able to take energy from the wind. These large waves move away from the

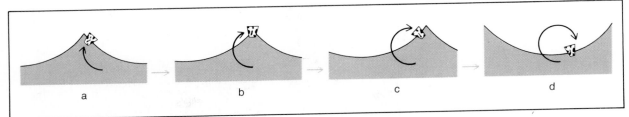

place where they were formed and travel great distances across the ocean. Long, rolling waves moving across the open sea are called **swells**. If you watch the waves that reach the shore, it is possible to make good guesses about their origins. The long *swells* that roll in slowly were probably formed by storms some distance away. Small, choppy waves are usually young. They were probably made by local winds.

Fig. 16–13 Water particles move in circles as the wave passes.

Swells Long, regular waves that move great distances across the oceans.

MOTION OF WATER IN WAVES

A swell formed in the middle of an ocean does not move water from the middle of the ocean along with it. Water does not move over great distances with the waves. It is the energy in the waves that moves. You can see this for yourself. Watch the action of a floating cork as a wave passes. The cork is lifted up, carried forward, but then back again. The cork moves in a circle. See Fig. 16–13. The cork moves ahead only slightly as the wave passes. Water particles also move through circles as a wave passes. The movement of the wave across the water is the result of these circular motions. Below the surface, water also moves in circles as the waves go by. However, the size of these circles gets smaller with greater depth. See Fig. 16–14. Submarines traveling at depths greater than 100 meters escape most of the circular movement of wave action.

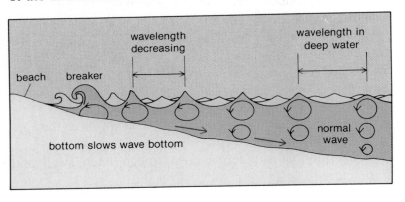

Fig. 16–14 Breakers form when the lower parts of a wave hit the sand on the shore.

Enrichment: Waves with a height of about 15 meters are commonly produced during severe cyclonic storms. Waves generated during hurricanes can be as high as 20 to 35 meters.

Breaker A wave whose top has curled forward and crashed down against the shore.

1. Winds blowing across the water.
2. Height: The distance from trough to crest in a wave. Wavelength: The distance from a crest to the crest of the next wave. Period: The time it takes for two successive crests or troughs to pass a point.
3. As waves approach the shore, they slow down, the bottom slowing more than the top. The top finally leans too far forward and tumbles over to form a breaker.
4. Crest: The part of a wave that is raised to form a hill. Trough: A low area between crests in waves. Fetch: The distance over which the wind blows steadily to form a set of waves. Swell: Long, rolling waves moving across the open sea. Rip current: Water piled on shore by waves suddenly returns to the sea.

When waves come close to the shore, they enter shallow water. Near the shore, the circling water particles in the waves begin to rub against the sea floor. The bottoms of the waves are slowed. As the bottoms of the waves slow, the tops may lean too far forward and tumble over in foamy crashes. See Fig. 16–15. Such waves are called **breakers.** Usually water thrown up on the shore by *breakers* runs back as a weak current below the surface. Sometimes the water piles up on the shore and then suddenly rushes back toward the sea in strong, swift *rip currents.* Rip currents are dangerous to swimmers. A person caught in such currents can be carried out to deep water.

SUMMARY

Sea waves are usually caused by the wind blowing across the water. The size of a wave is determined by the strength and duration of the wind and the fetch. Waves can be described by measurements such as height, wavelength, and period. Water particles move in circular paths as a wave passes. When waves reach shallow water near the shore they may form breakers.

QUESTIONS

Use complete sentences to write your answers.
1. What causes most surface waves?
2. Describe the meaning of the terms height, wavelength, and period as used in discussing waves.
3. Briefly explain the changes that occur in waves as they approach shore.
4. Define the following terms: crest, trough, fetch, swell, and rip current.

INVESTIGATION

MAKING WAVES

PURPOSE: To produce and study waves made by wind.

MATERIALS:

pan	drinking straw
water	food coloring

PROCEDURE:

A. Fill the pan with water to a depth of about one centimeter. Let the water sit for a few minutes.

B. Aim the straw so that when you blow through it gently, the air will go across the surface of the water from the center of the pan. See Fig. 16–16.

Fig. 16–16

C. Blow through the straw and observe the waves.
  1. Describe the waves produced.
  2. What is the *fetch* in this case?

D. While disturbing the water as little as possible, place a drop of food coloring on the surface of the water near the center of the pan.

3. What happens to the food coloring?

E. Aim the straw at the center of the pan and blow through it gently. As the waves move on the surface, observe what happens to the food coloring.
  4. What happens to the food coloring?
  5. What do you think the food coloring represents?
  6. Which move faster, the waves or the food coloring?

F. Now blow one puff of air down from directly above the center of the pan. Observe what happens when the wave hits the side of the pan.
  7. What did you observe?

G. Increase the force at which you blow the puff of air.
  8. What is the name of the part of the wave you are watching as it moves? How did the wave change?

H. Time the puffs of air so that as the wave returns to the center of the pan, you produce another wave. In this manner only one wave at a time is seen as it goes to the side and reflects back to the center of the pan.
  9. Describe the wavelength of this wave. Remember, the wavelength is the distance the wave traveled by the time another is generated.

CONCLUSIONS:

1. How are waves produced by wind?
2. How are the currents made?
3. Where did the energy come from to make the waves and currents?

16–4. Giant Waves

At the end of this section you will be able to:

- ☐ Explain how a *tsunami* is formed.
- ☐ Describe the cycle of high and low tides each day.
- ☐ Explain the cause of tides.

Skills: Study of tsunamis and tides reinforces comparing and contrasting as well as cause and effect.

The destruction shown in Fig. 16–17 was brought about by a few huge waves on May 26, 1983. Not far from this location, a school bus had just brought 43 school children to the beach for a picnic. All the children and their bus were swept out to sea. In 1896, on the same Japanese island, a few similar waves swept away 27,000 people and 10,000 houses. In this section you will learn how such huge waves are caused.

Enrichment: Tsunamis are also called seismic sea waves because of their origin. The term "tidal wave" has been dropped in order to avoid confusion with the tides.

Fig. 16–17 Tsunami destruction can be severe.

KILLER WAVES

Imagine what would happen in a swimming pool if the bottom were suddenly pushed up or dropped down. All the water in the pool would move. Something similar happens when there is a sudden change on the sea floor. During an earthquake, an underwater volcanic eruption, or a mudslide, a large part of the sea floor may rise or fall. Such disturbances on the sea floor cause a very large wave, called a **tsunami** (soo-**nahm**-ee). *Tsunami* is Japanese for "harbor wave." The people of Japan have had many experiences with these waves.

Tsunami A giant wave caused by a sudden disturbance on the sea floor.

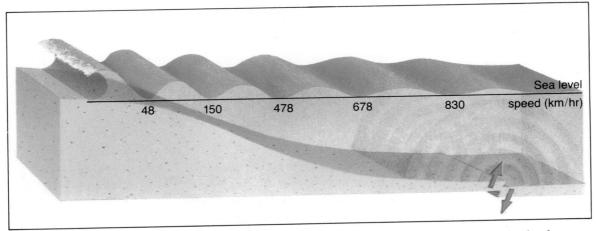

48	150	478	678	830

Sea level
speed (km/hr)

Fig. 16-18 The birth of a tsunami.

Tsunamis are very long waves. In the open sea, they often have wavelengths of about 200 kilometers. But heights of tsunami waves are usually less than one meter. A ship in the open ocean would hardly notice tsunamis passing by. However, tsunamis move at speeds up to 960 kilometers an hour. When these giant waves approach land, their heights can increase to 30 meters. See Fig. 16-18. When tsunamis hit a coastline, serious flooding can occur. Coastal towns can be heavily damaged. Tsunamis occur in groups of five to ten waves and about 5 to 20 minutes apart. Thus people should not return to the shore until the last of the waves has hit.

A warning system has been set up for the Pacific Ocean, where tsunamis are most common. When an earthquake that might cause a tsunami occurs, people living in coastal regions are warned by radio. Although tsunamis move at high speeds, earthquake waves travel much faster. Thus, with tsunamis, people usually can be given several hours' warning to move away from the shore to higher ground.

TIDES

Most people along coasts do not experience the long waves called tsunamis. There is, however, a long wave with which they are familiar. This large wave is called a **tide.** It makes the sea level rise and fall regularly in most places. Observation of *tides* shows that they are very long, regular waves. The crest of the wave brings a high tide. The trough of the wave brings a low tide. The period of the wave is about twelve hours. High and low tides are about six hours apart.

Tide A regular rise and fall of the water level along the shore.

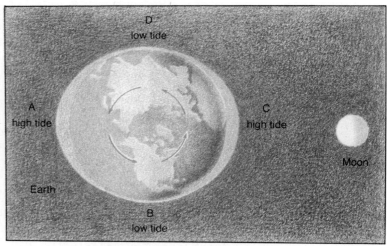

Fig. 16-19 There are two high tides and two low tides each day.

What causes such huge waves in the sea? The answer to this question involves the moon and the sun. The earth, moon, and sun all attract each other with a gravitational force. The moon's gravitational force has a greater effect on tides than does the sun's. However, when the sun's gravitational force is added to the moon's gravitational force, the tides are even higher. Because water in the sea is free to move, the moon's gravity pulls sea water into a bulge toward the moon. A similar bulge is formed on the side of the earth opposite to the moon. On that side of the earth, the moon's gravitational force on the water is much less than on the solid earth. This causes the solid earth to be pulled more than the water. Thus the water seems to rise as a tidal bulge. This means that the sea always has two bulges. One is facing toward the moon and the other is farthest from the moon.

While the earth is turning, the bulges stay lined up with the moon. In Fig. 16-19, points A and C are at high tide. As the earth rotates, points D and B will be at high tide about six hours later. When will points A and C be at high tide again? If you were measuring the tides at one point, you would observe two high tides and two low tides about every twenty-four hours. There are about six hours between each high and low tide.

The sun's gravitational force has less effect on the tides than the moon has because the sun is much farther away. However, when both the moon and sun are lined up on the same side of the earth, their gravitational pull is combined. Then unusually high tides, called *spring tides*, occur. See Fig. 16-20 (a). Spring

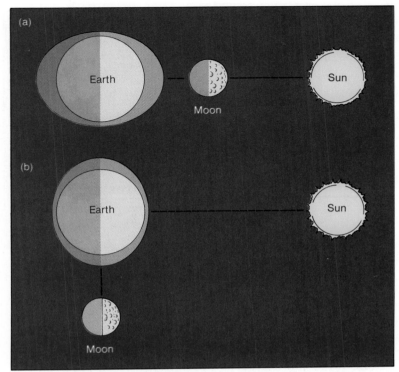

Fig. 16-20 (a) Spring tides occur when the moon, sun, and earth line up. (b) Neap tides occur when the sun and the moon are at right angles to each other.

tides occur twice each month. They do not refer to the spring season of the year. When the moon and sun are at right angles to each other, unusually low tides occur. These are called *neap tides*. See Fig. 16-20 (b). Neap tides also occur twice each month.

TIDAL EFFECTS

Not all parts of the sea are affected in the same way by the tides. The shapes of the ocean basin and the sea floor make the water respond in different ways. Each part of the sea has its own pattern of tides. For example, some places on the shore of the Gulf of Mexico have only one high and one low tide each day. In some places, the changes in water level from high to low tide are very large. The water level in the Bay of Fundy in Canada changes about ten meters. This is about the same as the height of a three-story building. See Fig. 16-21, page 450. Along straight coastlines and in the open sea, the water level does not change more than 50 centimeters. The behavior of tides at a place can only be predicted from observations made over a long period of time.

Fig. 16-21 (top) *Low tide and* (bottom) *high tide in the Bay of Fundy.*

Tidal bore A large wave that leads a tide upriver.

A single wave, called a **tidal bore,** may be found leading a tide up a river. See Fig. 16-22. In order for such waves to form, two conditions must exist. First, a large tide difference, from six to ten meters, must occur in the river's tidal basin. Second, the mouth of the river must be broad and have a gradual, sloping bottom. On the Severn River in England, the *tidal bore* is so great that surfers may ride it upstream for miles.

At places where tidal changes are large, tidal power may one day be used as a source of energy. To use tidal power, a dam would have to be built to trap water at high tide. The water

Fig. 16-22 A tidal bore.

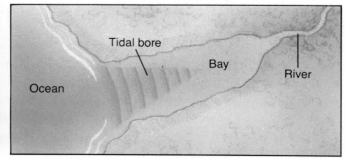

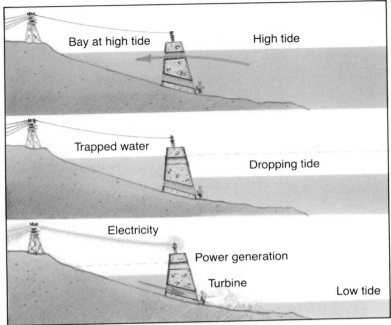

Labels on figure:

Bay at high tide — High tide

Trapped water — Dropping tide

Electricity — Power generation — Turbine — Low tide

Fig. 16–23 The principle of a tidal dam.

would then be released during low tide. See Fig. 16–23. The flowing water would be used to generate electricity. An experimental tidal power plant has been constructed at the mouth of the Rance River in France.

SUMMARY

Giant waves are produced by large earth movements on the sea floor. Caused by earthquakes or volcanic eruptions, they can be dangerous to people living on the coast. In most places, the sea level changes in a regular way every day because of tides. Tides are caused mainly by the effect of the moon's gravity on the earth and its ocean waters. Tidal effects are different from place to place.

QUESTIONS

Use complete sentences to write your answers.

1. Describe several causes of tsunamis.
2. Explain why there are usually two high tides and two low tides each day.
3. The moon passed overhead on the seacoast at 9 P.M. Explain why you would expect the highest tide of the day to occur at about this time.
4. How can the behavior of tides be predicted at a place?

1. A tsunami may be caused by an underwater earthquake, an underwater volcanic explosion, or a large underwater mudslide.

2. There is a tidal bulge on the side of the earth near the moon and at the same time on the side farthest from the moon. Between each tidal bulge, there is a tidal trough. Since the earth rotates on its axis once each day, the tidal bulges and troughs are seen to move past a given point once each day.

3. The moon's attraction for the water is greatest when directly overhead and this causes the highest tide.

4. The behavior of tides can be predicted by making observations over a period of time.

SKILL-BUILDING ACTIVITY

See teacher's commentary for teaching hints, safety, and answers to questions.

AN INFERENCE FROM THE TIDES

PURPOSE: To use water height data and other information for making an inference about the cause of tides. An inference is a conclusion drawn from observations.

MATERIALS:
graph paper   ruler
pencil

PROCEDURE:

A. A measuring tape attached to a pier piling in Boston Harbor was used to make readings of the water level. Table 16–1 shows those readings. Make a graph of the data. Label the vertical axis "Height of Water (m)" and the horizontal axis "Time (hr)."

B. The highest readings are *high tides*. The lowest readings are *low tides*. Look at the graph.

    1. How many hours apart are the high tides? How many hours apart are the low tides?

C. On January 1, the moon had passed overhead a little before the 9 P.M. measurement was taken. On December 31, the moon had passed overhead a little before 8 P.M. The 8 P.M. reading was the highest for the day.

    2. What conclusion (inference) would this suggest?

D. On Jan. 2, at 10 P.M. the moon had passed overhead a little before the measurement was made. The 10 P.M. reading was the highest for January 2.

    3. What role could the moon's gravity play here?

E. Look at the graph again. Around 9 A.M. in Boston, the moon would be on the opposite side of the earth.

    4. Is there a high tide or a low tide in Boston at that time?

    5. Of all the water in the oceans, the water at Boston, and north and south of Boston, would be the farthest from the moon at 9 A.M. Would this increase or decrease the moon's attraction?

CONCLUSIONS:

1. What causes tides?

2. Why are there two high tides in Boston in one day?

HEIGHT OF WATER IN BOSTON HARBOR ON JANUARY 1												
Time	1 A.M.	2 A.M.	3 A.M.	4 A.M.	5 A.M.	6 A.M.	7 A.M.	8 A.M.	9 A.M.	10 A.M.	11 A.M.	12 noon
Height of water (meters)	2.0	1.7	1.6	1.9	2.4	2.9	3.4	3.8	4.0	3.7	3.3	2.7
Time	1 P.M.	2 P.M.	3 P.M.	4 P.M.	5 P.M.	6 P.M.	7 P.M.	8 P.M.	9 P.M.	10 P.M.	11 P.M.	12 midnight
Height of water (meters)	2.2	1.6	1.2	1.3	1.6	2.1	2.6	3.4	4.6	3.5	3.2	2.8

Table 16–1

CAREERS IN SCIENCE

MARINE TECHNICIAN

Marine technicians can enjoy fascinating careers that are involved with almost every aspect of the oceans. Some technicians work with oceanographers charting the ocean floor. Others work with marine biologists who are developing new sources of food from the sea. Still others participate in the undersea efforts to recover ships and their sunken treasures.

Technicians may be found on board ships or on stationary platforms in the sea. They may work near shipyards and docks or in the laboratories of a marine industry. Technicians are the people who follow through on the plans, check details, and help gather data for a particular project.

To become a technician, you'll need a two-year college degree. Courses in science, math, and drafting are helpful. Jobs are available in private industry and the government. For more information, write: American Society of Limnology and Oceanography, I.S.T. Building, University of Michigan, Ann Arbor, MI 48109.

MARINE GEOLOGIST

The world of the marine geologist is the vast ocean floor. Geologists gather samples of the sediment that lies on the ocean bottom. Then they perform experiments on these samples to find out what they contain. Marine geologists may also explore the ocean floor for oil and mineral deposits. The quality of these deposits is analyzed, and it is the marine geologist who must decide if they can be brought to the surface for human use. A person in this field might also be involved in studying the underwater crust of the earth to help understand our planet's history. Most geological research involves the use of sophisticated technologies and equipment.

To enter the field of marine geology, you must have a bachelor's degree. For high level positions, an advanced degree is also necessary. Jobs are available with oil companies, universities, and the government. For more information, write: American Geological Institute, 5202 Leesburg Pike, Falls Church, VA 22041.

CHAPTER REVIEW

VOCABULARY

On a separate piece of paper, write TRUE next to the number of each statement that is true. Next to the number of each false statement, write FALSE, then make the statement true by writing the correct term in place of the underlined incorrect term.

1. A current is a regular rise and fall of the water level in the sea.
2. A turbidity current is a current caused by the increased density of water that has mixed with sediments.
3. A fetch is a long, regular wave that moves a great distance across the ocean.
4. A swell is a wave whose top has curled forward and crashed down against the shore.
5. The tide is the distance over which wind blows steadily, producing waves.
6. A tidal bore is a large wave that leads a tide upriver.
7. Any water moving in a particular direction is called a breaker.
8. A giant wave caused by a disturbance on the sea floor is called a tsunami.

QUESTIONS

Give brief but complete answers to each of the following questions. Unless otherwise indicated, use complete sentences to write your answers.

1. Give an example of how the sun's energy causes an increase in the density of ocean water and an example of how it causes a decrease in the density of ocean water.
2. Where are the currents of cold water most likely to be found in the ocean? Where are the currents made up of warm water found?
3. Describe a well-known density current found in the Mediterranean Sea. What causes this current?
4. What effect do turbidity currents have on the ocean floor?
5. Why is the general pattern of surface ocean currents circular in the North and South Atlantic?
6. What is the Gulf Stream?
7. Name the current that causes the Gulf Stream. Which two currents does it split into in the North Atlantic?
8. What common effect is caused when the Labrador Current meets the Gulf Stream?

9. Explain how surface currents can affect climates.

10. What is the pattern of motion for water particles near the surface as a wave passes?

11. Describe what happens to the speed, wave height, and wavelength of a wave as it approaches the shore and forms a breaker.

12. Why are tsunamis called "killer waves"?

13. How is a tsunami in the open sea different from one at the shoreline?

14. What is the major cause of tides on earth?

15. Describe a usual cycle of tides during a one-day period. In your answer, include the time periods between high and low tides.

APPLYING SCIENCE

1. What would be the effect on climates near the poles and the equator if the oceans did not circulate?

2. Explain the difference in temperature of the water at the beaches in California and Florida.

3. Explain how the El Niño may affect the air temperature on the west coast of the United States.

4. Refer to Fig. 12–12, page 339. Compare the global wind patterns with the ocean currents shown in this chapter. Make a drawing to show the wind patterns and the ocean currents on a map similar to the map shown in Fig. 16–10 on page 438.

BIBLIOGRAPHY

Britton, Peter. "Waves Challenging Giant Rigs." *Popular Science*, January 1981.

Frances, Peter, and Stephen Self. "The Eruption of Krakatau." *Scientific American*, November 1983.

McKean, Ken, and Mayo Mohs. "Tracking the Killer Waves." *Discover*, August 1983.

Matthews, Samuel W. "New World of the Oceans." *National Geographic*, December 1981.

National Geographic Editors. "The Promise of Ocean Energy." *Special Report on Energy*, February 1981.

Time-Life Editors. "Restless Oceans" (Planet Earth Series). Chicago: Time-Life Books, 1984.

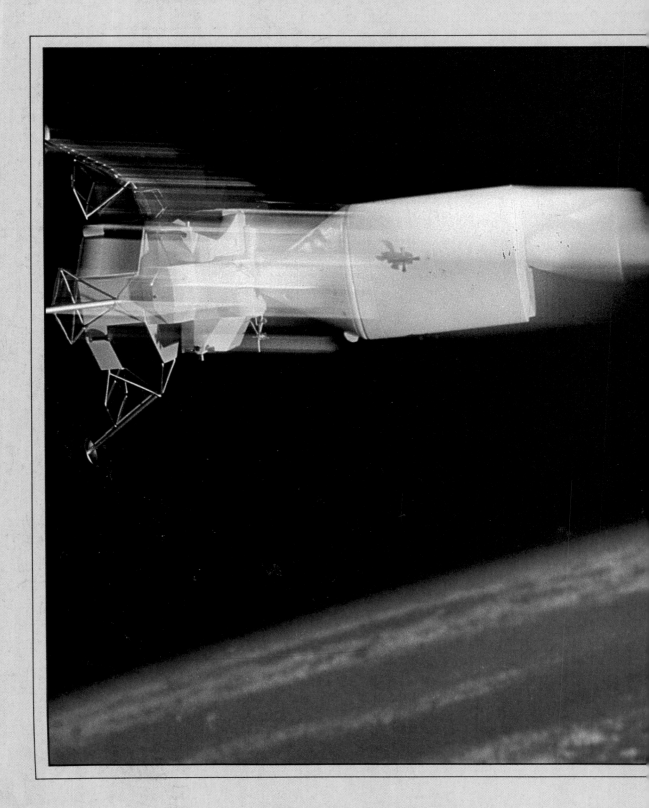

UNIT 5

EARTH AND SPACE

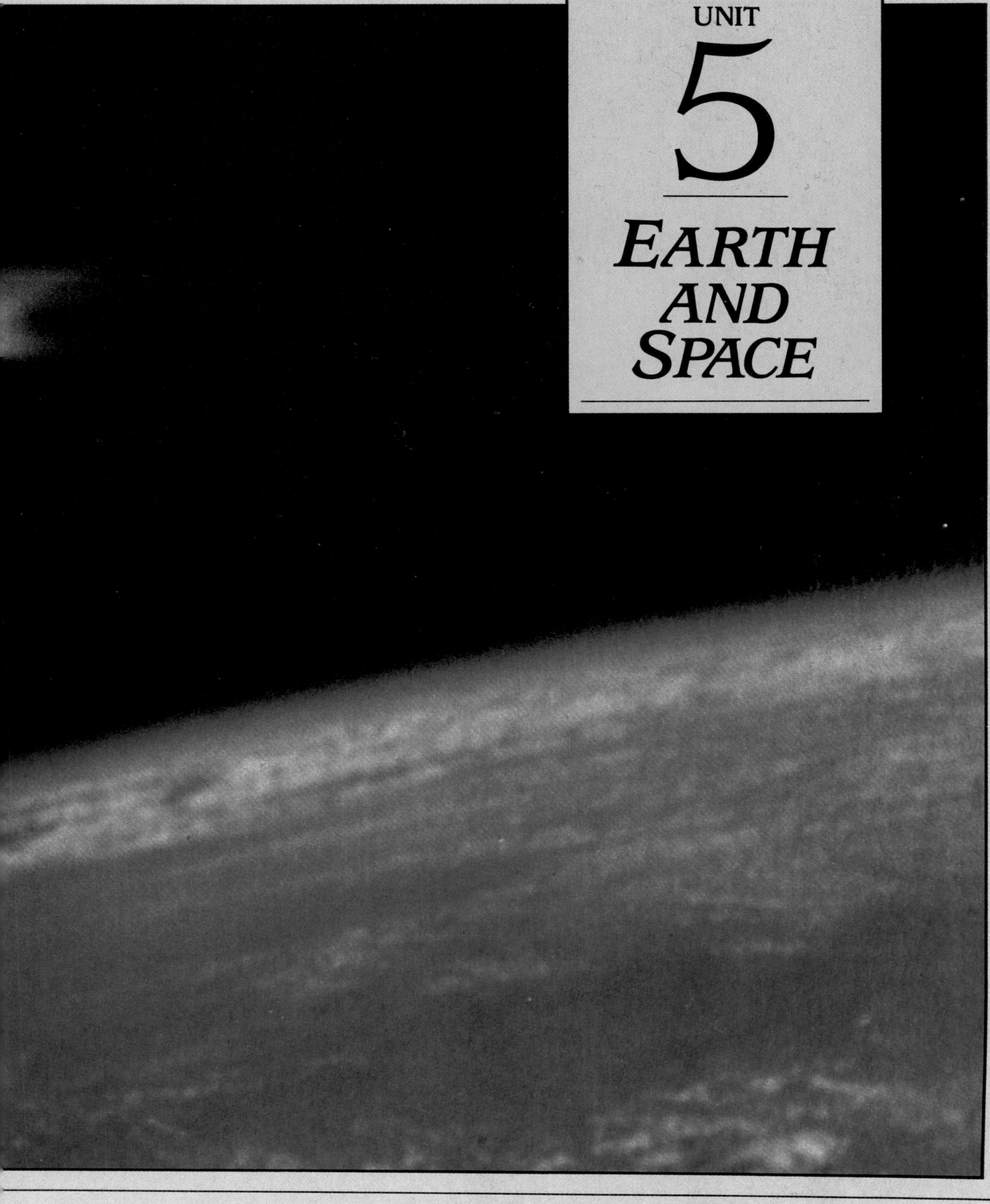

A total solar eclipse occurs when the earth passes through the moon's shadow. In any given place, a total eclipse of the sun is likely to be seen only once in several hundred years.

THE SUN

CHAPTER GOALS

1. Describe the structure of the sun.
2. Explain how the sun changes.
3. Identify the source of the sun's energy.
4. Describe the form in which solar energy is received by the earth

Enrichment: Before the year 2000, only one total solar eclipse will be seen in the entire United States. This will be visible from many parts of the country on July 11, 1991.

17–1. *Structure of the Sun*

At the end of this section you will be able to:

- ☐ Name and describe the layers of the sun.
- ☐ Describe sunspots.
- ☐ Explain how disturbances on the sun may affect the earth.

On a dark, clear night in the far north, the sky begins to glow. Green and red patches of light dance across the sky. These flowing curtains of light move in strange shapes as they expand and shrink. For a while, it seems as if the sky is on fire. Then the lights fade away, leaving a faint reddish streak that also gradually fades away. These beautiful "northern and southern lights" are often seen at night near the north and south polar regions. Would you think of the sun as their cause? In this section you will see how these lights are caused when particles from the sun reach all the way to the earth.

LAYERS OF THE SUN

The sun is a huge ball of very hot gas. It would take more than a million earths to make up the same volume as the sun. The sun's density is low because it is made up of gases.

The sun is made up of several layers that blend into one another. Because all the layers are gases, there are no sharp dividing lines between them. Because the sun is so hot, the gases in the sun are mostly a mixture of bare atomic nuclei and free electrons. This matter is called *plasma*. On the earth, plasmas are found only where the temperatures are very high, such as in the flame of a welding torch.

Enrichment: If the sun were the size of a beach ball, the earth would be the size of a pea about five meters away.

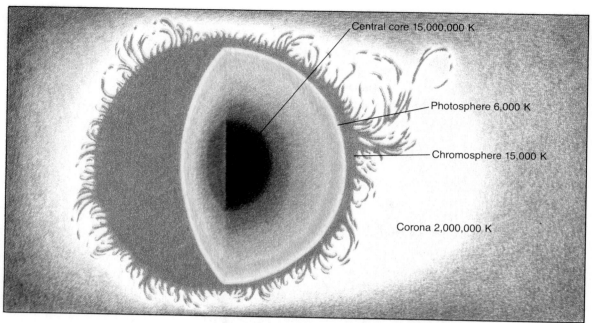

Central core 15,000,000 K

Photosphere 6,000 K

Chromosphere 15,000 K

Corona 2,000,000 K

Fig. 17–1 The parts of the sun.

The sun can be divided into four layers: the core, photosphere, chromosphere, and corona. See Fig. 17–1.

1. **Core.** The changes that produce the sun's huge supply of energy take place in its center, or core. The temperature in the core has never been measured because it is so deep within the sun. Scientists believe that the temperature near the sun's center is at least 15,000,000 degrees on the Kelvin scale. Energy coming from this central power plant is carried outward until it reaches the sun's surface layer.

2. **Photosphere.** Energy from the sun's interior reaches the surface. There it causes a thin layer of gases to give off a brilliant light. This glowing layer of gas, called the **photosphere** (**fote**-oh-sfeer), produces the visible light that comes from the sun. The *photosphere* is the part of the sun that we see. CAUTION: DO NOT LOOK DIRECTLY AT THE SUN AT ANY TIME. EVEN A BRIEF LOOK CAN PERMANENTLY DAMAGE YOUR EYES. The photosphere layer has a temperature of about 6,000 degrees on the Kelvin scale.

3. **Chromosphere.** Just above the photosphere are two layers that make up the sun's atmosphere. First is a layer, called the **chromosphere** (**kroe**-moh-sfeer), that gives off a faint red light. This light cannot be seen against the bright background of the photosphere.

Reinforcement: Astronomers measure temperatures with the Kelvin (K) scale. Remind students that 0 K = −273°C, or K = °C +273°

Photosphere A thin outer layer of gas in the sun, which produces visible light.

Word Study: Photosphere: *phot* + *sphaira* (Gr.), light sphere.

Word Study: Chromosphere: *chroma* (Gr.), color; Corona: *corona* (L.), crown.

Chromosphere The layer in the middle of the sun's atmosphere, which gives off a faint red light.

4. Corona. The *chromosphere* blends into a much less dense layer of faintly glowing gas. This layer, called the **corona** (kuh-**roe**-nuh), surrounds the sun like a halo. There is no outer boundary for the *corona*. Some of the sun's material escapes completely and streams out to, and past, the earth. This stream of particles from the sun's corona makes up the *solar wind*. The parts of the sun are shown in Fig. 17–1, page 460.

Corona The outermost layer of the sun.

Enrichment: The yellow-white color of the sun is an indication of its surface temperature. Hotter stars have a blue-white color. Cooler stars have a red color.

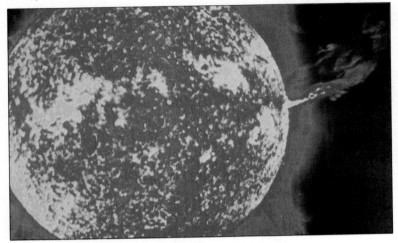

Fig. 17–2 *The surface of the sun has a rough appearance.*

SUNSPOTS

The gases that make up the sun are always moving. Energy pouring from the sun's core causes storms in the hot gases. Thus the surface of the sun appears to be rough because of these motions. When seen through a special telescope, the visible surface of the sun resembles the peel of an orange. See Fig. 17–2. The most easily seen storms on the sun's outer layers are **sunspots.** *Sunspots* are huge storms that extend from the surface of the sun into its interior. These are areas that appear dark when seen against the bright background of the photosphere. Sunspots form when a part of the sun's surface becomes cooler. This is caused by the blocking of the hot gases that constantly rise from below. The temperature inside a sunspot is about 2,000 degrees on the Kelvin scale cooler than the surrounding surface. This lower temperature makes it appear darker than the bright photosphere. A large sunspot may be five times as large as the earth's diameter. Some large groups of sunspots last for several weeks. Sometimes the same sunspots are seen for months. Small sunspots usually last for a few days to a week.

Enrichment: Sunspots were recorded by Chinese astronomers as early as 200 B.C. In the western world, Galileo first observed sunspots in 1613.

Sunspots Areas on the sun's surface that appear dark.

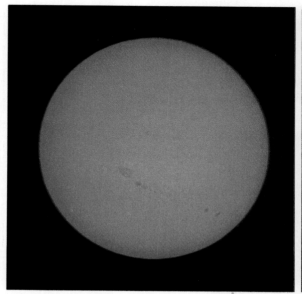

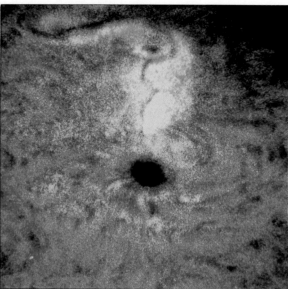

Fig. 17–3 (left) *Sunspots can usually be seen on the sun's surface.* (right) *Prominences are great clouds of glowing gas.*

Enrichment: Places closest to the sun's equator take 25 1/3 days to make one rotation. Areas halfway between the equator and poles take 28 days, whereas points near the poles take about 33 days. The average rotation period is 27 days.

Prominence A huge cloud of gases that shoot out from areas of sunspots.

Compare and contrast: Ask students to contrast the sun's rotation with the earth's. Two differences should be mentioned: (a) The sun's period is longer at the poles but shorter near the equator. (b) All points on the earth's surface complete one rotation each 24 hours, whereas places on the sun have different periods.

Most sunspots are seen in groups close to the sun's equator. See Fig. 17–3 *(left)*. As the sun rotates, the sunspots appear to move across the face of the sun. Sunspots circle around the sun and return to their original position in about 27 days. Thus the sun appears to rotate on its axis about once every 27 days. However, because the sun is made of gas, not all of it turns at the same speed. Places near the sun's equator move faster than places near the poles. Separate spots and groups of spots may change their shapes as they rotate with the sun.

Huge clouds of bright gases, called **prominences** (**prom-ih-nen-ses**), shoot out from the sun's surface in the areas of sunspots. A *prominence* may extend thousands of kilometers above the sun's surface. A prominence can be seen in Fig. 17–3 *(right)*. Prominences often form a huge arch between pairs of sunspots. Sometimes giant fountains of bright gas suddenly shoot up from the sun's surface. These are solar flares. They usually appear as explosions near sunspots.

DO SUNSPOTS AFFECT THE EARTH'S WEATHER?

Observations of sunspots over many years show a general pattern when they appear. See Fig. 17–4. The number of sunspots increases every few years. After reaching a peak, the number of sunspots decreases. Then the cycle begins again. There is some evidence that the sunspot cycle may affect the

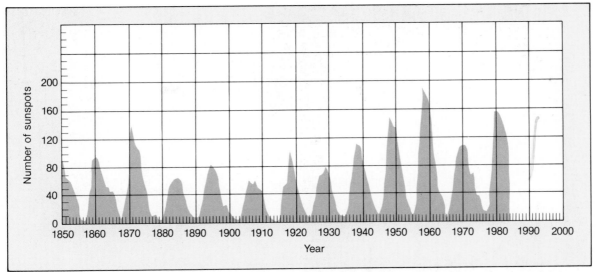

Fig. 17-4 The number of sunspots seen each year since 1850 is shown on the graph.

Enrichment: Sunspot activity reaches a maximum every 11.2 years on the average. The last maximum was in 1979. The next maximum should be in 1990.

earth's environment. For example, the study of many tree rings shows a pattern very much like the sunspot cycle. During sunspot cycles, the space between the rings changes from the normal pattern. The distance between tree rings shows how fast the tree has grown each year. See Fig. 17-5. Each ring represents one year of growth. During years when there is abundant rain, the tree grows rapidly. The distances between rings become larger. In dry years, the tree grows less. The distances between rings are smaller. Thus the distances between the tree rings are a record of wet and dry periods. This could mean that sunspots have some effect on the earth's weather by changing the amount of rain that falls in some places.

Fig. 17-5 During sunspot cycles, the space between the rings, indicating tree growth, varies from the normal pattern.

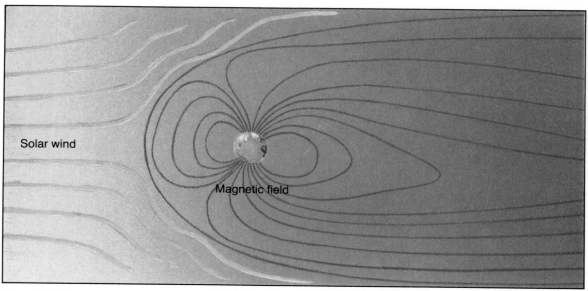

Fig. 17-6 The earth's magnetic field causes the solar wind to flow around the earth.

Enrichment: Spaceships have been proposed with sails that would use the solar wind for power. The pressure of the solar wind on a Mylar sail of 929 square meters would cause a spaceship to reach the speed of the *Apollo* spaceships (40,000 kilometers an hour) in about six weeks.

Enrichment: The solar wind extends out to Saturn, a distance of 1,448 million kilometers from the sun. If the solar wind could be considered as an extension of the sun's corona, then the earth is orbiting within the solar atmosphere. If the earth's atmosphere were as large, it would reach to the moon.

SOLAR WIND

The space between the planets and the sun is not empty. Solar flares may cause gases from the sun's corona to be thrown out into space. These gases are made of electrically charged particles, and they move out through the solar system as solar wind. See Fig. 17-6. The solar wind moves past the earth at speeds of 300 to 700 kilometers per second. This is much faster than the speed of a moving bullet. The solar wind could be a danger to living things. However, we have two shields that protect us, the atmosphere and a magnetic field. The earth's atmosphere is able to absorb most of the solar wind that reaches the earth. Also important is a magnetic field that surrounds the earth far out into space.

The earth acts like a giant magnet. This can be shown by holding a bar-shaped magnet so that it can turn freely. It will slowly turn and come to rest pointing north and south. A needle in a compass acts in the same way. The magnetic effect may be caused by slow movements of hot materials within the core and mantle. Like all magnets, the earth has two opposite magnetic poles, called north and south. However, these are not the same places as the geographic North and South poles, which are on the earth's axis. But compass needles point generally north and south. This is because the magnetic poles are near the geographic North and South poles. See Fig. 17-7.

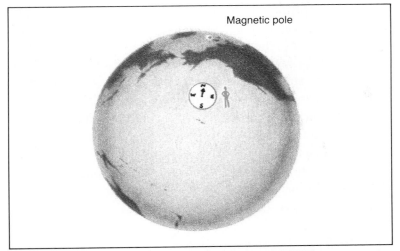

Magnetic pole

Fig. 17-7 The latitude of the North Pole is 90° N. The magnetic North Pole is at 76.2° N and 101° W. When electrically charged particles come near the earth, they are guided to the earth's magnetic poles. The particles strike the gases in our atmosphere, causing auroras.

The earth's magnetism pushes the solar wind to either side as it reaches the earth. The solar wind flows around the earth's magnetic field, like a river flowing around an island, as you can see in Fig. 17-6. Some of the charged particles pass through the earth's magnetic field and become trapped. When these charged particles strike the gases in the earth's atmosphere, light is given off. These greenish-white and blue lights are seen as **auroras,** or northern and southern lights. See Fig. 17-8. *Auroras* are most commonly seen near the poles. This is because the particles are guided by the magnetic field toward the North and South poles. Sometimes, they are seen in places closer to the equator. This appears to be caused when solar flares form on the sun's surface.

Enrichment: Topographic maps give magnetic declination or variation, which is the number of degrees east or west that a compass differs from true north.

Auroras The northern and the southern lights produced when charged particles from the sun are trapped in the earth's magnetic field.

Enrichment: The colors of the auroras are an indication of the energy of the solar particles striking the molecules of gases in the atmosphere to produce light. Red indicates low-energy particles, whereas high-energy particles produce green light. The auroras usually are seen as curtains of light because the incoming particles follow the magnetic lines of force, which are nearly vertical near the poles.

Fig. 17-8 The northern lights as seen in Alaska.

Solar flares appear quickly, like the explosion of a pool of gasoline. Prominences may suddenly rise upward from the location of the flare. Then the flare quickly fades away. However, the burst of energy from the solar flare may blow out a part of the sun's atmosphere into space. This causes a sudden increase in the solar wind. Several days after the solar flare is observed, its effects may be felt when the strong solar wind reaches the earth. The increased solar wind can cause a solar storm in the earth's atmosphere.

A solar storm can effect the earth's atmosphere in three ways. First, x-rays from the solar wind interfere with high frequency radio waves. This can cause radio blackouts and scramble signals from satellites. A second effect is caused by protons and electrons in the solar wind. These particles may push through the earth's magnetic field and produce auroras. If the solar wind is very powerful, a third effect takes the form of sudden increases in electrical power in power lines. This can produce interruption in the supply of electricity in some areas.

SUMMARY

The sun is a glowing sphere of gas made up of four layers. The energy pouring from the sun's core causes storms in the hot gases. The storms that occur on the sun's outer layers are seen as sunspots, which appear as dark areas on the bright surface. The number of sunspots increases and decreases in cycles every few years. Sunspots may have some influence on the earth's weather and thus have a possible effect on the growth of plants. A solar wind moves out from the sun through the solar system. The earth's magnetic field causes the solar wind to flow around the earth. Sometimes part of the solar wind is trapped by the earth's magnetic field and produces auroras and other effects in the earth's atmosphere.

QUESTIONS

Use complete sentences to write your answers.
1. Name and describe the four layers of the sun.
2. Describe sunspots and how they form on the sun.
3. Name two other disturbances on the sun's surface that appear related to sunspots.
4. What effect does the solar wind have on the earth?

1. The core produces the sun's huge supply of energy. The photosphere produces the light you see. The chromosphere is above the photosphere and gives off a faint red light. The corona is a layer of glowing gas that looks like a halo. The corona reaches far out into space.
2. Sunspots are huge storms on the surface of the sun. They develop when a part of the sun's surface becomes cooler due to the blocking of the hot gases that rise from within the sun.
3. Solar prominences and solar flares are two other disturbances that are related to sunspots.
4. The solar wind causes auroras and the interruption of some radio communications.

SKILL-BUILDING ACTIVITY

See teacher's commentary for teaching hints, safety, and answers to questions.

DOES THE SUN ROTATE?

PURPOSE: To use problem solving skills in finding the answer to the question, "Does the sun rotate?"

MATERIALS:

paper   pencil

PROCEDURE:

A. We must look at the sun from our position in space and tell whether or not it rotates. Often the scientist uses another example as a model to help find an answer to a problem.
   1. If you looked at earth from space, how could you tell that the earth rotates?
   2. Suppose you noticed that a certain feature on the earth's surface returned to view every 24 hours. What would you conclude?

B. The sun's surface is so hot that everything there exists as a gas. Usually such a surface would give no marks to watch for evidence of rotation. There are certain times, however, when gigantic sunspots can be seen to move across the sun's surface. Look at Fig. 17–9, in which the bottom photo was taken two days after the top photo.
   3. Do the sunspots appear to move together or independently?
   4. The movement shown by the photos is ¼ the distance around the sun. How long did this movement take?
   5. How many days would it take the sunspots to move around the sun?

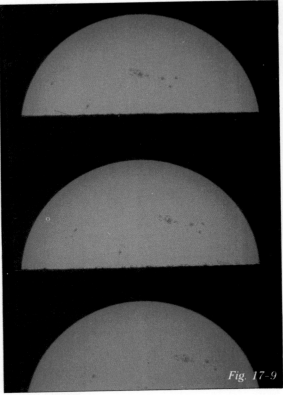

Fig. 17–9

6. Assume the sunspots do not move on their own but are carried around by the rotation of the sun. How many days does it take for the sun to rotate once?

CONCLUSIONS:
   1. How can you tell that the sun rotates?
   2. Explain how you determined the length of time it takes the sun to make one complete rotation.
   3. What assumption had to be made about sunspots in order to answer question 6?

17–2. The Sun's Energy

At the end of this section you will be able to:
- ☐ Identify the source of the sun's energy.
- ☐ Describe the energy that leaves the sun.
- ☐ Name the kinds of *radiant energy* received by the earth.

Each second, the sun gives off an amount of energy equal to 200 billion hydrogen bombs. How the solar furnace works remained a mystery for a very long time. However, today scientists not only understand how the sun works, but can even copy the way it makes energy. See Fig. 17–10.

Fig. 17–10 An H-bomb uses the same source of energy as the sun.

WHAT MAKES THE SUN SHINE?

Think about how much energy the sun gives off. Each second, the sun produces as much energy as the United States would use in three million years. And it has been giving off about this same amount of energy for at least the last four to five billion years. What can be the source of this marvelous power?

The most simple hypothesis would be that the sun gets its energy from fire. In other words, the sun might be hot and bright because something is being burned. However, this would be very unlikely. For example, suppose that the sun could be made of coal. Astronomers know that the sun has a mass of 1.8×10^{30} kilograms. An amount of coal equal to the mass of the sun would last only about 6,000 years. Clearly, the source of the sun's energy cannot be anything like ordinary burning.

Then, in 1835, a German scientist named Hermann von-Helmholtz developed a theory that, for a while, seemed to work. Helmholtz said that the sun's great mass caused its outer layers to be pulled in by gravity. Thus the sun was constantly shrinking. The shrinking caused the gases inside the sun to be squeezed and become heated. The shrinking of the sun would then cause it to get very hot and to shine. Helmholtz found that the sun would need to shrink only about 50 meters each year in order to produce its energy. This is a very small part of the sun's diameter of 1.4 million kilometers. However, there was one problem with this theory. Even when shrinking so slowly, the sun must have been very much larger in the past. Only 25 million years ago it would have been so big that the earth would have been inside the sun. Thus the earth could not have been older than about 25 million years. However, geologists who measure the rate of decay of radioactive elements in rocks have gathered evidence that shows the earth to be at least 4.5 billion years old. Since the earth cannot be older than the sun, this means that the sun must have been shining for at least several billion years. By the end of the 18th century, most astronomers realized that the age of the sun according to the Helmholtz's theory could not be correct. Another explanation was needed.

A clue to the mystery was provided by Albert Einstein in 1905. Einstein said that matter could be changed into energy. This principle is expressed by the famous equation: $E = mc^2$, where (E) is energy, (m) is mass, or the amount of matter changed, and (c) is the speed of light. The equation states that a small mass can be turned into a large amount of energy. Thus, the sun could produce great amounts of energy by losing some of its mass. Each second the sun changes more than 600 million tons of hydrogen nuclei into helium nuclei, with a loss of mass. Other kinds of nuclear fusion reactions also help supply the sun's energy. For example, the nuclei of carbon, nitrogen, and oxygen can be involved in a complicated reaction that also produces helium. Because the sun is so huge, it would take 100 trillion trillion years to use up its entire mass. Now scientists could build a theory to explain how the sun produces energy. Further scientific observations provided the details needed to make a complete theory.

THE SOLAR FURNACE

The sun is made up of an enormous amount of gases. Hydrogen gas makes up 74 percent of the sun's mass. Helium makes up 25 percent. The remaining 1 percent is probably made up of small amounts of all the other known chemical elements. All the matter in the sun presses down on its center with a huge force. This creates a pressure within the sun's core that is nearly one billion times the air pressure at the earth's surface. The gases near the sun's center are squeezed together so strongly that they are changed.

The high temperatures in the core cause the atoms in the gases to lose their electrons. Only the nuclei are left. The very high pressures and temperatures cause some of these atomic nuclei to join together. This process is called *fusion*. In the sun, fusion joins four hydrogen nuclei into a helium nucleus plus two hydrogen nuclei. The helium and hydrogen that are produced have a little less mass than all the four hydrogen nuclei had originally. See Fig. 17–11. The small amount of mass that disappears is changed into energy. Smaller amounts of other kinds of atomic nuclei are also changed. These nuclei also produce helium along with energy.

Each second, the sun changes about 4.2 billion kilograms of hydrogen into energy. There is enough hydrogen in the sun to supply energy at the present rate for at least five billion years.

Scientists now believe that the puzzle of the sun's energy has been solved. The answer is in the constant fusion of atomic nuclei, within the sun's core, that changes mass into energy.

Scientists have duplicated reactions similar to those that occur on the sun. One place is in the hydrogen or H-bomb. In order to reach the high temperatures needed to start the fusion reaction, an atomic bomb is used first. Another place where scientists are duplicating fusion is in fusion reactors. Here, scientists are attempting to release energy in controlled amounts. The major problem has been finding a way to contain the very high temperatures needed for fusion. At present, fusion reactors have been able to contain the high temperatures for only thousandths of a second. Someday scientists hope to find a way for producing energy from a fusion reactor continuously. This would be similar to the reactions that cause the sun to shine. The development of such a method would supply much of humankind's future need for energy.

Enrichment: The process of changing hydrogen to helium in the sun is called the proton-proton cycle. In the first step, two protons combine to form a deuterium nucleus containing one proton and one neutron. A positron and neutrino are given off. This changes one of the protons into a neutron. In the second step, helium-3 is formed with a photon being given off.

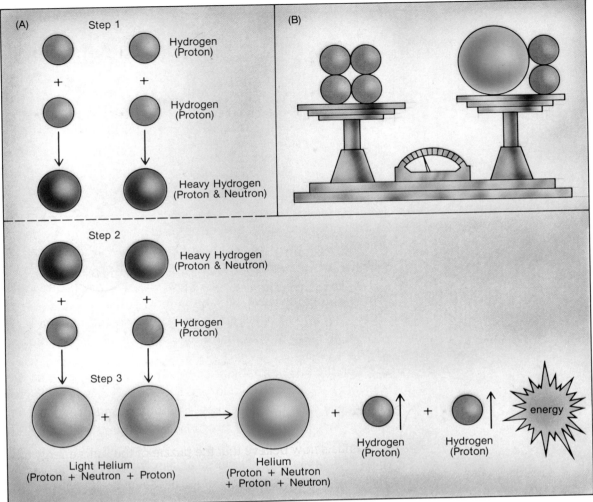

Step 1

Hydrogen (Proton)

+

Hydrogen (Proton)

Heavy Hydrogen (Proton & Neutron)

(B)

Step 2

Heavy Hydrogen (Proton & Neutron)

+

Hydrogen (Proton)

Step 3

Light Helium (Proton + Neutron + Proton)

Helium (Proton + Neutron + Proton + Neutron)

+ Hydrogen (Proton) + Hydrogen (Proton) energy

SUNLIGHT

The processes that supply the sun's energy take place deep inside the sun. Here the temperatures are high enough for nuclear changes to occur. Energy produced inside the sun then makes its way out toward the surface. The gases near the surface absorb this energy and become hot. This causes the gases in the sun's outer layers to give off *radiant energy*. Radiant energy is energy that travels through space in the form of waves. All of the many forms of radiant energy can move through space at the very high speed of 300,000 kilometers per second. Radiant energy can travel from the sun to the earth in about eight minutes.

Fig. 17–11 (A) The three steps by which hydrogen nuclei combine in the sun to form helium. Step 1: Four hydrogen nuclei form two heavy hydrogen nuclei. Step 2: Each heavy hydrogen nucleus combines with another hydrogen nucleus to form a light helium. Step 3: Two light helium nuclei form a normal helium nucleus and two hydrogen nuclei. (B) At the end, the total mass is less than the total mass at the beginning.

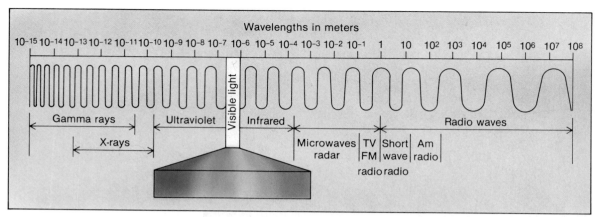

Fig. 17-12 The spectrum of radiant energy.

There are also other sources of radiant energy on the earth. For example, radio, television, and radar waves are all forms of radiant energy. The different kinds of radiant energy are shown in Fig. 17-12. The longest waves are radio, television, and radar waves. Infrared, visible light, and ultraviolet waves are shorter. Very short wavelengths are found in X-rays and gamma rays. Notice that visible light makes up only a small part of radiant energy.

Of all the kinds of radiant energy that make up sunlight, we are able to sense only two forms. The eye sees a very small part of the spectrum as visible light. The other form of radiant energy in sunlight that can be sensed is infrared waves. These waves are felt on the skin as heat. Ultraviolet waves in sunlight can cause sunburn and tanning of skin. Too much exposure to ultraviolet waves can also seriously damage your skin. Prolonged exposure to ultraviolet waves over a period of years has also been linked with skin cancer.

Most of the radiant energy leaving the sun is in the form of visible light and infrared rays. Together they make up more than 90 percent of the energy sent out by the sun. Almost all of the remaining 10 percent is made up of ultraviolet rays, X-rays, and gamma rays. Fortunately for all the living things on earth, not every kind of radiant energy given off by the sun reaches the earth's surface in large amounts. The gases in the upper parts of the earth's atmosphere absorb X-rays, gamma rays, and most of the ultraviolet rays. For example, ozone, which is a form of oxygen, is responsible for blocking out most of the ultraviolet rays. For this reason, the small amount of ozone gas in the atmosphere is very important. See Fig. 17-13.

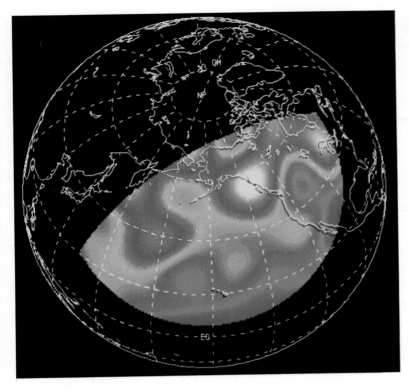

Fig. 17–13 Ozone gas on earth filters out some of the sun's harmful radiation.

SUMMARY

The secret of the sun's energy lies deep within its core. The high temperature and pressure cause separate hydrogen nuclei to join together. In this process, a small amount of the mass of the hydrogen is changed into energy. Energy from the core of the sun moves out into space. By using its hydrogen as fuel, the sun will be able to produce energy for billions of years. The earth receives this radiant energy from the sun, mostly in the form of visible light and infrared waves.

QUESTIONS

Use complete sentences to write your answers.

1. What is the source of the sun's energy?
2. What is radiant energy?
3. Name and give the percentages of the kinds of radiant energy sent out by the sun.
4. Which kinds of radiant energy are absorbed by the earth's atmosphere?

1. The source of the sun's energy is the constant fusion of atomic nuclei within the sun's core, which changes mass into energy.
2. Radiant energy is the kind of energy that travels through space as waves.
3. Ninety percent of the radiant energy leaving the sun is in the form of visible and infrared waves. Most of the remaining 10 percent is made up of ultraviolet, X-rays, and gamma rays.
4. The earth's atmosphere absorbs the X-rays, gamma rays, and most of the ultraviolet rays.

See teacher's commentary for teaching hints, safety, and answers to questions.

SEPARATING LIGHT

PURPOSE: To separate light into its various colors.

MATERIALS:

diffraction grating   pencil
light bulb   paper

PROCEDURE:

A. A diffraction (dif-**rack**-shun) grating is a piece of plastic having thousands of parallel lines molded into its surface. The lines can easily be damaged by fingerprints and scratches. Hold the diffraction grating to your eye and look through it at the light bulb. See Fig. 17–14. CAUTION: NEVER LOOK AT THE SUN. YOU MAY DAMAGE YOUR EYES. You should be able to see colors spread out to the right or left. If not, turn the grating a quarter of a turn so that the colors are on the right and left sides.
1. List the colors you see.
2. Where in nature have you seen this set of colors before?

Fig. 17–14

B. The colors you see are the same colors in the same order that you would see if you looked at a rainbow. (DO NOT LOOK AT THE SUN!) This is the visible light spectrum. Look at Fig. 17–12 on page 472.
3. Estimate the shortest wavelength given for the visible light spectrum.

C. The shortest wavelength of visible light will be the color nearest the light bulb. Use the grating to look at the light bulb.
4. Which color is the one with the shortest wavelength?
5. As you look farther away from the bulb, past the full spectrum, what do you see?

D. The longest wavelength of visible light will be the color in the spectrum farthest from the light bulb. Use the grating to look at the light bulb.
6. Which color is the one with the longest wavelength?

CONCLUSIONS:
1. List the colors of the visible light spectrum, starting with the color of the shortest wavelength.
2. What is the shortest wavelength of light that can be seen? What is the longest wavelength of light that can be seen?
3. Name the kind of radiant energy that is just outside of the visible spectrum on the short wavelength end.
4. Name the kind of radiant energy that is just outside the visible spectrum on the long wavelength end.

TECHNOLOGY

EXPLORING SPACE: THE HUBBLE SPACE TELESCOPE

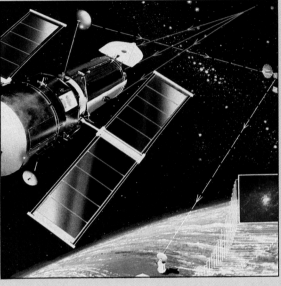

Called both the most important astronomical happening for the 20th century and perhaps the most difficult and demanding project ever undertaken by NASA, the Hubble space telescope, is now scheduled to be launched in 1986. The telescope was appropriately named after the American astronomer Edwin P. Hubble (1889–1953). Hubble's research demonstrated the existence of nebulae outside our galaxy. His work, resulting in what is now called Hubble's Law, led to an estimate of the size of the universe. He is credited with having begun the study of the universe beyond our galaxy. The Hubble space telescope will continue and extend this research. We will see what we only thought to exist before.

This massive instrument, which weighs 11,339 kilograms and is 13.4 meters long and 4.3 meters in diameter, will be placed in an orbit 515 kilometers above the earth. From that point, the telescope will be looking at objects that are 14 billion light years away. It will actually be looking back in time at stars just forming! With the information sent back by this instrument, scientists hope to determine how our galaxy was formed. The images from space will be collected with the mirror of the telescope, approximately 2.4 meters in diameter. Light will be fed into five instruments, including two new cameras—one a wide-field planetary camera, the other a faint-object camera. The wide-field camera is expected to be the most used instrument. Each instrument and camera has been designed especially for the space telescope. Each component must be tested thoroughly to make certain that it will withstand the force and vibrations of lift-off as well as the low temperature of space. Since the objects being viewed are so far away from the telescope, it will take about 24 hours to collect a single complete image. For that 24 hours, there can be no movement of the device, or the image will be distorted.

The Hubble space telescope is expected to operate for fifteen years. To make this possible, the space telescope has been designed so that the instruments and other parts can be repaired or replaced by astronauts, or be brought back to earth by a shuttle for major repairs. The telescope will be visited every five years for this purpose.

When finally launched and placed in orbit, the Hubble telescope is certain to have an effect on the way we look at the universe. Many astronomers believe that the most important discoveries will be unexpected, just as the discovery of the moons of Jupiter was unexpected by Galileo when he first aimed a telescope at the planets. We know it will give us information about 27th and 28th magnitude stars and the formation of galaxies. Scientists also expect to learn more about quasars and black holes. And astrophysicists may find data that could verify one of their GUTs, Grand Unified Theories. GUTs are theories which try to explain the various particles of matter and their behaviors with one overall hypothesis. What other questions do you think they might be able to answer by seeing into the outer reaches of space?

SCIENCE CONCEPT

The sun is vital to our existence. It is essential to the process of photosynthesis, which provides oxygen and food. It provides heat and, of course, it provides light. Because of the earth's revolutions, the sun's light and heat affect different parts of the globe differently. As you learned in the beginning of this text, the earth's tilt causes a change in the direction of the sun's perpendicular rays. This process creates the change in seasons. Although you can't see or feel these motions, you can observe them in changes in shadows. At any given time, a shadow's length will differ as you move north or south. It will also differ from day to day at the same location. Studying data on the length of shadows in different places at different times gives us information about the earth's movement around the sun.

COMPUTER INPUT

If a cycle of events occurs on a regular basis, it can be written in a mathematical form. Once information is in that form, it can be entered into a computer. The computer is designed to perform many calculations quickly. This is especially helpful when you need to make the same calculation over and over, changing just one or two parts of the mathematical formula each time. You could program all the factors into the computer and come out with the figures in a matter of seconds. In this case, we will program the computer to produce data about different shadow lengths based on the mathematical relationships between the earth and the sun.

WHAT TO DO

Before entering Program Shadow Moves, record your answers to the following questions on a separate piece of paper.

A. You are standing at 23.5 degrees north latitude, looking at an upright stick ten centimeters in length.
1. On what day is the stick's shadow the longest? On what day is it the shortest?
2. In what months does the shadow get longer? In what months does it get shorter?
B. Would your answers to the preceding questions be different if you were standing at 40 degrees north latitude or at 60 degrees north latitude?

Enter Program Shadow Moves. The computer will ask you for the latitude of your location and the length of the stick. (This program will perform the correct mathematical calculations for latitudes between 23.5 and 60 degrees north of the equator.) When these are entered, the computer will calculate the length of the shadow at that latitude on the 21st of each month. Each time you run the program, record the data. (See the sample data chart.) Compare the computer's data output with your responses to the questions above. Record the data for at least five different latitudes. Plot your data in graphs to see the changes in shadow length more easily. Try to draw conclusions about the relationship between the different variables involved. For instance, if you keep stick length and time (month and day) the same, what happens to shadow length as you go north? If you keep location (latitude) and stick length the same, what happens as you go from winter to summer?

SAMPLE DATA CHART

Latitude	Month	Shadow Length
23.5° N	Jan.	
	Feb.	
	Mar.	
	Etc.	

GLOSSARY

GOTO A command in BASIC that instructs the computer to return to a previous statement or to go to another part of the program and follow the directions there.

PERIPHERALS Computer accessories. A peripheral might be a printer, a joystick, the TV screen, or the keyboard. Everything except the actual electronic circuits are peripherals.

PROGRAM

```
100   REM SHADOW MOVES
110   DIM A(12), M$(12)
120   FOR X = 1 TO 12
130   READ M$(X), A(X)
140   NEXT
150   INPUT "WHAT IS THE LATITUDE
      (23.5 — 66)? ";LA$
160   LA = VAL(LA$)
170   IF LA < 23.5 OR LA > 66 THEN
      GOTO 150
180   INPUT "HOW LONG IS THE STICK?
      ";ST
190   PRINT "MONTH"; TAB(15)"SHADOW
      LENGTH"
200   FOR X = 1 TO 12
210   A = 90 — (LA + A(X))
220   AD = TAN (ABS (A / 57.3))
230   PRINT M$(X); TAB(15) INT((ST / AD)
      * 10) / 10
240   NEXT
250   PRINT
260   INPUT "REPEAT Y/N" A$
270   IF A$ = "Y" GOTO 150
275   END
300   DATA  JAN 21, 15.9, FEB 21, 7.8,
      MAR 21, 0
310   DATA  APR 21, —7.8, MAY 21,
      —15.9, JUN 21, —23.5
320   DATA  JUL 21, —15.9, AUG 21, —7.8,
      SEP 21, 0
330   DATA  OCT 21, 7.8, NOV 21, 15.9,
      DEC 21, 23.5
```

PROGRAM NOTES

REM statements, written into the program, are useful since they alert the user to the purposes of each section. In Program Shadow Moves, there is only one REM statement. Several others could be added to clarify the steps. Can you write REM statements to enter before lines 150, 179, and 300 that would let the user know what kind of step is coming up next?

BITS OF INFORMATION

Computer learning takes many forms. For instance, a company in Palo Alto, California, has designed a system called Dial-A-Drill. Subscribers to the system call, using a pushbutton phone, and a computerized voice asks questions in math or spelling. Answers are given by pressing the buttons on the phone. Dial-A-Drill keeps track of your right and wrong answers.

CHAPTER REVIEW

VOCABULARY REVIEW

On a separate piece of paper, match the number of the statement with the term that best completes it. Use each term only once.

photosphere   corona   aurora   solar flare
chromosphere   sunspot   prominence   radiant energy

1. The _____ is the outermost layer of the sun.
2. A dark, cooler area on the sun's surface is called a(n) _____.
3. The kind of energy given off by the sun that travels through space as waves is called _____.
4. A large arc of glowing gas seen on the sun's surface is known as a solar _____.
5. The _____ is the layer of the sun that we can see.
6. The layer of the sun that gives off a reddish glow is called the _____.
7. A(n) _____ may cause an interference of radio communications on earth.
8. The northern or southern lights are called a(n) _____.

QUESTIONS

Give brief but complete answers to the following questions. Unless otherwise indicated, use complete sentences to write your answers.

1. Compare the size, mass, and density of the sun with that of the earth.
2. What kind of matter is found in the sun?
3. Which layer of the sun produces the visible light?
4. Describe the corona of the sun.
5. Why should you never look directly at the sun?
6. What relationship exists between tree rings and sunspots?
7. How are we protected from the dangers of the solar wind?
8. In what ways does a solar storm affect the earth?
9. Explain why the sun could not be making its energy by a common burning process such as burning of coal.
10. Describe the fusion process by which the sun makes its energy.
11. Of what importance is the equation $E = mc^2$ in the way the sun makes its energy?
12. What causes the sun to give off radiant energy?
13. What kinds of radiant energy are commonly made on earth?

14. Identify the kinds of radiant energy, found in the spectrum shown in Fig. 17–12, page 472, that we can either see or feel.

APPLYING SCIENCE

1. Cut a two-centimeter-square hole in a piece of cardboard. Cover the hole by taping a piece of aluminum foil over it. With a sharp needle, punch a neat hole in the aluminum near its center. Use this device to cast an image of the sun onto a piece of white paper. CAUTION: NEVER LOOK DIRECTLY AT THE SUN. EYE DAMAGE MAY RESULT! Shading the paper by placing it in a cardboard box will make the image show up better. Adjust the distance between pinhole and paper screen to give as large an image as possible. Locate any visible sunspots and make a drawing of them. Watch the sunspots for several days and note any changes. Calculate the period of rotation of the sun based on your observations.

2. One kilogram of wood will supply about 12 million (1.2×10^7) joules of energy if burned. A joule is a unit of energy. Using $E = mc^2$, calculate the amount of energy given off if the one kilogram were converted completely into energy ($c = 3 \times 10^8$ m/sec, 300 million m/s). Compare this to the amount of evergy given off when one kilogram of wood is burned.

3. In direct sunlight, use a fine spray of water from a hose nozzle to produce a rainbow. Move the spray with respect to the sun to make the best rainbow. Make a drawing to show the location of the nozzle, sun, and rainbow. Relate the arrangement of the rainbow colors to the spectrum shown in Fig. 17–12.

4. The equation "speed = wavelength × frequency" relates these three properties of waves. Find out the frequency (Hz/s) of your favorite radio station and use the equation to calculate the wavelength of the radio waves. Use the speed of radiant energy given in the text. Your answer will be in kilometers. Convert it to meters and relate this to the wavelengths given in the spectrum of Fig. 17–12.

BIBLIOGRAPHY

French, Bevin M., and S. P. Maran. *A Meeting with the Universe*. Pasadena, CA: Planetary Society, 1984.

Kidder, Tracy. "A Blemished Sun?" *Science 81*, July/August 1981.

Lampton, Christopher. *The Sun*. New York: Franklin Watts, 1982.

Willamson, Samuel J., and Herman Z. Cummins. *Light and Color in Nature and Art*. New York: Wiley, 1983.

The moon is seen as it enters the earth's shadow (right) and as it leaves the shadow. The next lunar eclipse visible in parts of North America will occur August 16, 1989.

SATELLITES OF THE SUN

CHAPTER GOALS

1. Compare the characteristics of the inner planets with those of the outer planets.
2. List and describe the smaller members of the solar system.
3. Describe the moon's features and its history.
4. Explain the motions of the earth–moon system as it revolves around the sun.

18-1. The Inner Planets

At the end of this section you will be able to:
- ☐ Explain why the solar system contains two general kinds of planets.
- ☐ Compare the sizes of the three planets closest to the earth.
- ☐ Compare the conditions on the surfaces of Mercury, Venus, Earth, and Mars.

Suppose that observers from outer space have been watching the earth since it was formed. For more than four billion years, they would not have seen anything leave the earth. Then suddenly, starting about thirty years ago, they would have noticed that the people of the earth began to send off tiny spacecraft. At first, the spacecraft only flew around the earth. Then they landed on the moon. Later, the observers would have seen other spacecraft fly to the earth's neighboring planets. The observers might conclude that the people of the earth had begun a new age of exploring.

THE SOLAR SYSTEM

The solar system is made up of the sun and billions of smaller bodies. Each of these bodies follows an orbit around the sun. The orbits are not perfect circles. Instead, the orbits have the shape of an **ellipse** (eh-**lips**). An *ellipse* is like a circle that has been flattened into an oval shape. An egg has the shape of an ellipse. However, many of the orbits, such as the earth's, are only slightly elliptical.

Reinforcement: The difference between a circle and an ellipse can be emphasized by pointing out that all points on a circle are equidistant from its center, whereas an ellipse has two focal points.

Ellipse The oval shape of the orbits followed by members of the solar system.

Fig. 18-1 The planets compared with each other and the sun.

Reinforcement: Have students make up a sentence using the first letter of the names of the planets in order of distance from the sun. This will help lower-level students remember the names and order of the planets.

Reinforcement: Draw a circle 139 cm in diameter on the board to represent the sun. Next to it, draw a circle 1 cm in diameter to represent the earth.

The bodies of the solar system move in orbits. They are held by the force of gravity from the huge mass of the sun. In fact, the sun makes up more than 99 percent of the solar system's mass. The sun is the controlling body of the solar system. The nine planets are the best known and, in most cases, the largest other bodies of the solar system. The planets can be separated into two major groups, the inner and outer planets. The inner group of planets includes Mercury, Venus, Earth, and Mars. These planets are the smaller ones, consisting of solid rock with metal cores. The four outer planets include Jupiter, Saturn, Uranus, and Neptune. These are giant planets. They are made of ices and gases, with small solid cores. See Fig. 18-1. Pluto is the farthest planet from the sun. It is very small and seems to be different from all the other planets.

Scientists believe that the way the solar system was formed explains the differences between the inner and outer planets. The inner planets are made up mostly of solid rock and metals. They were formed close to the sun's heat and had their lighter materials driven off. Materials such as stone and metals were not easily driven off. On the other hand, the outer planets are made up mostly of ices and gases. They were formed farthest from the sun's heat and held on to the lighter materials. Also, the large masses and strong gravity of the outer planets help to hold the lighter gases. The materials that formed the ices and gases now make up most of the outer planets. Since the earth is one of the small inner planets, its history must be similar to that of its closest neighbors. Thus the spacecraft that explore these planets give scientists a better understanding of the earth. We will take a closer look at each of the planets that formed in the same way that the earth did.

MERCURY

Mercury (**mur**-kyuh-ree) is slightly bigger than the moon but much smaller than the earth. Mercury is not easily seen in the night sky because it is so close to the sun. However, just after the sun sets at night, Mercury may be seen near the horizon. At dawn, it rises shortly before the sun comes into view. The earth's atmosphere makes it difficult to see Mercury this low in the sky. During the day, the brightness of the sun makes observations difficult. However, spacecraft have flown near Mercury, sending back pictures and information.

The surface of Mercury is covered with craters. See Fig. 18-2. When the solar system was very young, it was filled with smaller bodies left over from the time the planets formed. Many of these bodies struck the planets, leaving craters as scars. On Mercury, these craters have remained. Mercury has no atmosphere, and thus there is no weather to cause erosion. The photographs of Mercury's surface also show many giant curving cliffs that extend hundreds of kilometers, cutting across the craters. Some of the cliffs are as much as three kilometers high. They were caused by the wrinkling of Mercury's surface when the planet was cooling and shrinking. Early in its history, Mercury also went through a period of volcanic activity. This has produced large lava plains that cover about one-quarter of its surface.

Mercury rotates very slowly on its axis. A day on Mercury is equal in length to 59 earth days. Mercury's long days, combined with its closeness to the sun and lack of an atmosphere, cause its surface to become very hot. Temperatures soar to around 400°C during the long period of daylight. On the dark side, the temperature drops to about −180°C.

Mercury seems to be a planet that has hardly changed for the past four billion years. It may show how the surface of the earth could have looked in its early history.

Fig. 18-2 The surface of Mercury photographed from a spacecraft.

Enrichment: Objects low in the sky must be seen through more air than objects high in the sky. This is the reason that sunsets and sunrises appear reddish. The light from the setting and rising sun travels through enough air to scatter almost all light except red.

Enrichment: Due to its long period of rotation and short period of revolution, there are 1.5 Mercury days in every Mercury year, and 6.2 Mercury days in every earth year.

VENUS

Venus (**vee**-nus) comes closer to the earth than any other planet. Venus is covered by a thick layer of clouds. See Fig. 18–3 *(left)*. The clouds cause it to reflect sunlight. Thus Venus is seen as a brilliant object in the night sky. It is seen just after sunset or before sunrise. Venus spins very slowly on its axis. Venus rotates in a direction opposite to the direction it moves in its orbit. It takes 243 earth days for Venus to complete one entire rotation. Since it takes 225 days for Venus to orbit the sun, there is less than one Venus day in a Venus year.

For a long time Venus was called earth's twin. Both planets are similar in size, mass, and density. However, the atmospheres of the two planets are very different. The atmosphere of Venus is made up almost entirely of carbon dioxide. It is 90 times denser than the earth's atmosphere. This causes the atmospheric pressure on the surface of Venus to be much greater than on the earth. It is about the same as the pressure in our oceans at a depth of about one kilometer. This very heavy atmosphere traps heat because of the greenhouse effect. Short-wave radiation passes through the atmosphere and is absorbed by the surface. When the radiation is given off by the surface, it leaves as long-wave radiation, which cannot pass out of the atmosphere. The atmosphere also spreads the heat evenly over the surface of the planet. The temperature difference between the sunlight and dark sides of the planet or between its equator and its poles is only a few degrees. Temperatures on the surface

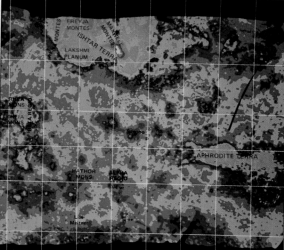

Fig. 18–3 (left) *A close-up photograph of Venus taken from a spacecraft shows its heavy layer of clouds.* (right) *A radar map of the surface of Venus. Light-colored areas are highlands.*

are about 475°C. This is about as hot as the temperature used to grill a steak outdoors.

Until spacecraft were able to visit Venus, little was known about this planet because of its cloud cover. Now crewless space vehicles have examined its surface by radar and landed on its surface. It was discovered that the surface of Venus had high mountains, valleys and canyons. See Fig. 18–3 *(right)*. Many of these features were the result of volcanic activity that probably also created the heavy atmosphere of Venus. There was no water found on Venus.

Enrichment: In October 1975 two Soviet spacecraft, called *Venera 9* and *10,* landed on Venus. They sent back the first photos of the surface of Venus.

In the past, Venus may have had much less carbon dioxide in its atmosphere. With conditions like that, its surface temperature would have been more like that of the earth. However, once the carbon dioxide starts to build up in the atmosphere, the trapped heat causes more carbon dioxide to be released. This, in turn, causes more heat to be trapped. Eventually, a thick cloud layer covers a very hot surface. Small changes in a planet's atmosphere can set off a chain of events that may cause a great change in the climate. The conditions on Venus should make us very careful about releasing large amounts of carbon dioxide into the earth's atmosphere.

Enrichment: The earthlike planets are called "terrestrial planets."

MARS

Mars is a smaller planet than the earth. It is about half as large in diameter, with only about one-tenth (0.1) the mass of earth. It probably does not have a heavy core like the earth. Mars turns once on its axis in a little over 24 hours. A visitor from earth would feel very much at home with the length of day and night on Mars. Its axis is tilted at nearly the same angle as the earth's. Thus, Mars also has four seasons. However, it takes Mars nearly two earth years to complete its orbit around the sun. Thus each of its seasons is six months long.

At the surface, the atmospheric pressure on Mars is only about two-hundredths (0.02) that of the earth's atmosphere. The Martian atmosphere is made up mostly of carbon dioxide, with about 3 percent nitrogen. But it is very thin, and therefore cannot trap much heat from the sun. As a result, the surface of Mars is always very cold. Even during the summer, the temperature near the Martian equator probably does not get above 0°C. Like the earth, Mars has polar icecaps of frozen water. Nighttime temperatures near the poles approach −200°C.

Fig. 18-4 The largest volcano on Mars is called Olympus Mons.

Many forces helped to shape the surface of Mars. There are many craters that were formed during its early history. Large areas of the planet have been flooded with lava, forming lava plains. Huge volcanic cones are found on Mars' surface. The largest is about 600 kilometers across and 25 kilometers high. See Fig. 18-4.

Scientists were surprised to find evidence that Mars once had running water. Photographs of the Martian surface show many channels and other features that could only have been made by running water. See Fig. 18-5. However, the surface of Mars is now dry and the atmosphere contains no water vapor. Where has the water gone? Much of it is locked up in the polar icecaps. There may also be a thick layer of frozen water beneath the planet's surface. There also seems to be frozen water in the soil particles just below the surface.

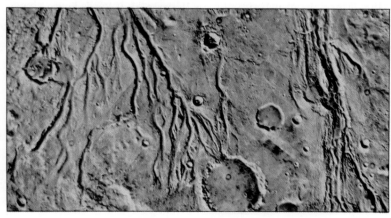

Fig. 18-5 An example of the channels on the surface of Mars that were made by running water.

Windblown material has also played a part in forming the Martian surface. Huge dust storms sweep across the planet during its spring season. The winds reach much higher speeds on Mars than on the earth. Thus the materials carried by the wind cause much more erosion of the rock than on the earth. Large deposits of wind-carried material are very common all over Mars.

In 1976, two United States spacecraft landed on Mars' surface. Those spacecraft carried small laboratories that tested the soil for life. The experiments, however, revealed no sure sign of any kind of life.

Many kinds of evidence show that the climate of Mars was once much different than it is today. At one time, millions or billions of years ago, this planet must have had a warmer climate, running rivers, and a denser atmosphere. It must have been a very different kind of planet from the freeze-dried Mars we are now observing. What caused this change on Mars is not known. Perhaps the same events that caused the earth to pass through its cold periods have also affected Mars. Further exploration of Mars may lead to a better understanding of the causes of all planetary climates.

SUMMARY

The solar system contains two kinds of planets, the solid, earthlike planets, and planets that are very large and are made mostly of ice and gas. The planets closest to the sun—Mercury, Venus, Earth, and Mars—are called the inner planets. Mercury lacks an atmosphere and is covered with craters. Venus has very high temperatures caused by its heavy covering of clouds. Mars now has a cold, dry surface with many kinds of features.

QUESTIONS

Use complete sentences to write your answers.

1. Why are the inner planets so different from the outer planets of the solar system?
2. Compare the sizes of Mercury and Venus to Earth.
3. List conditions for Mercury, Venus, Earth, and Mars. Include atmosphere, surface features, temperature (°C), length of day (Earth days).

1. When the planets formed, the inner planets, being so close to the hot sun, could only be made of rock, while the outer planets were far enough away and so cold that they could retain their ices and gases.
2. Mercury is much smaller, about the size of the moon, whereas Venus is almost the same size as the earth.
3. MERCURY: has no atmosphere; its surface is made up of craters, cliffs, and lava plains; temperature ranges from −180°C to 400°C; 59 earth days. VENUS: atmosphere dense clouds of mostly CO_2; surface made up of mountains, valleys, canyons; temperature 475°C; 243 earth days. EARTH: atmosphere light, N_2 and O_2; surface made up of oceans, continents, icecaps; temperature ranges from −100°C to 100°C; 1 earth day. MARS: atmosphere very thin, CO_2, N_2; surface made up of craters, lava plains, icecaps; temperature ranges from 0°C to −200°C; 1 earth day.

INVESTIGATION

See teacher's commentary for teaching hints, safety, and answers to questions.

PLANET ORBIT MODELS

PURPOSE: To draw scale models of the inner planets' orbits.

MATERIALS:

plain paper	2 straight pins
cardboard	string
pencil	ruler

PROCEDURE:

A. Place a sheet of paper over the cardboard. Draw a small circle near the center of the paper. Through the center of the circle, press a pin through the cardboard to anchor it well. Label the circle SUN.

B. Look at Fig. 18–6 (a). You will use the method shown in the figure to draw the orbits of Mars, Earth, Venus, and Mercury, in that order, on the same sheet of paper.

Planet	Distance Between pins (cm)	Distance Around loop (cm)
Mars	1.7	19.9
Earth	0.2	12.1
Venus	0.1	8.8
Mercury	1.0	6.5

scale: 1.0 cm = 25 million km

The following table shows how far apart to set the second pin from the sun pin, and how large a loop to make.

C. When making the orbits, do not move the pin labeled SUN. Move only the second pin the proper distance each time. Move it in a straight line so the holes it makes are in a line.

D. To make the string loop, cut a piece of string several centimeters longer than the distance around the loop. Bring the ends of the string together and tie a knot one-half the distance shown in the table. For example, for Mars, cut a piece of string about 22 centimeters long. Bring the ends together and tie a knot at 9.9 centimeters. Check the knot as you tighten it. See Fig. 18–6 (b).

E. Keep the string tight as you move the pencil while making the orbit. You may wish to practice once or twice on a separate sheet of paper. Draw the orbits carefully. Label each orbit with its planet's name as soon as you draw it.

CONCLUSION:

1. Describe the shapes of the orbits of the inner planets.

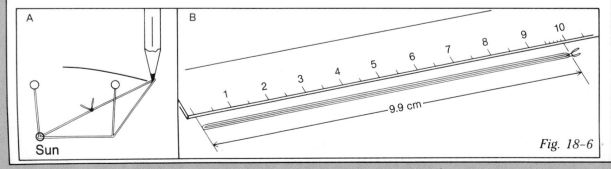

A
Sun

B

9.9 cm

Fig. 18–6

18–2. Outer Planets

At the end of this section you will be able to:

☐ Describe the main features of Jupiter, Saturn, Uranus, and Neptune.

☐ Compare Jupiter's moons with the inner planets.

☐ Explain how Pluto is different from the other planets.

Five hundred years ago, frail sailing ships set out on exploratory voyages. Their voyages changed the world. Similarly, in modern times, spacecraft have been sent out to explore the outer parts of the solar system. See Fig. 18–7. What these spacecraft have discovered has been as much of a surprise as the discovery of the New World was in the 15th century.

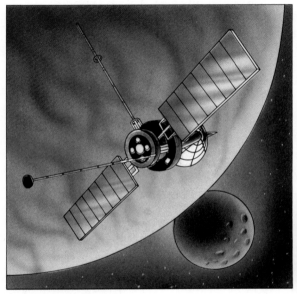 

Fig. 18–7 (left) *An unmanned spacecraft.* (right) *A sixteenth century sailing ship.*

JUPITER

Jupiter (**jou**-pih-ter) was the greatest and strongest among the ancient Roman gods. It also is the giant among the planets. Jupiter has almost 2.5 times the mass of all the other planets put together. Its diameter is 11 times the diameter of the earth. Although it would take more than 1,000 earths to equal Jupiter's volume, Jupiter has only 318 times the earth's mass. This is because the density of Jupiter is less than one-third that of the earth. It is also more than five times farther from the sun than the earth and therefore receives far less heat. Strangely, Jupiter gives off more heat than it receives from the sun. Some of this heat is left over from the time the planet was formed. More heat is also produced by the planet's slow shrinking that squeezes and heats the inside of Jupiter. Like the earth, Jupiter is also surrounded by a magnetic field. Charged particles from the sun are trapped in this magnetic field. This belt of trapped particles makes up a danger zone around Jupiter. Spacecraft approaching the planet must protect their instruments from the effects of these particles.

The surface of Jupiter is covered with a cloud layer that is separated into colored bands. See Fig. 18–8. The clouds are made up of about 88 percent hydrogen and 11 percent helium gases. There are also small amounts of water, ammonia, methane, and other compounds. The hydrogen and helium, being transparent, are not visible. The color in the clouds comes from different kinds of colored compounds that are found in lower layers of the cloud covering. Jupiter probably does not have a solid surface. Inside, the gases that make up its atmosphere simply become thicker and denser. Finally the dense atmosphere blends into an ocean made up mainly of liquid hydrogen. Near the center of Jupiter, the pressure is so great that the liquid hydrogen becomes similar to a metal. It is this liquid metallic hydrogen within Jupiter that causes its powerful magnetic field. Jupiter probably has a small rocky core at its center.

Fig. 18–8 The upper cloud layer of Jupiter is made up of colored bands.

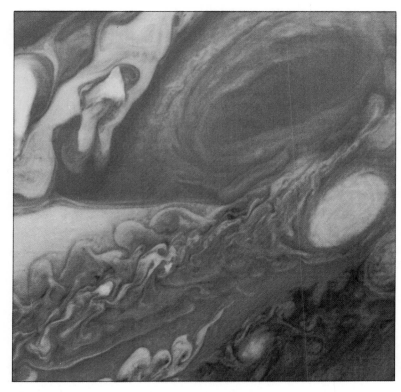

Fig. 18-9 Winds cause the swirls and currents seen in the colored bands on Jupiter.

It takes this giant planet nearly 12 earth years to complete its orbit around the sun. But Jupiter rotates very rapidly. One day on Jupiter is less than ten hours long. The rapid rotation of Jupiter causes its outer cloud layers to separate into colored bands. Heat coming from the inside causes the clouds to have high and low pressure regions just as in the earth's atmosphere. Jupiter's rapid rotation causes these highs and lows to wrap completely around the planet. Strong winds flow between the pressure belts, causing swirls and twisting motions in the gases. See Fig. 18-9. Storms similar to hurricanes develop in the atmosphere from time to time and appear as colored spots. One such storm is called the Great Red Spot. It has been observed in the same place on Jupiter for over three centuries.

Enrichment: The Great Red Spot has been observed on Jupiter for 300 years. It is believed to be a kind of permanent storm in the upper atmosphere of the planet.

THE MOONS OF JUPITER

In 1610, using a telescope he had constructed, the Italian scientist Galileo discovered that Jupiter had four large moons. These four satellites are called the Galilean moons of Jupiter. They are named, in order of increasing distance from the planet, *Io* (**eye**-oh), *Europa* (yoo-**roh**-puh), *Ganymede* (**gan**-ih-meed), and *Callisto* (kuh-**lis**-to). Each of these satellites is larger than the planet Mercury. See Fig. 18–10. Each moon is quite different from the huge planet it orbits. Io, Jupiter's closest moon, has a bright yellow-to-orange-colored surface that, surprisingly, is free of craters. It has active volcanoes whose lava flows may have covered craters and given it a fairly smooth surface. The next satellite, when moving away from Jupiter, is Europa. It is about the size of the earth's moon. Europa has a light yellow-colored surface that is covered by dark cracks. There are also large deposits of ice on Europa's surface. The two outermost satellites, Ganymede and Callisto, are similar in appearance. Their surfaces are covered with a mixture of rock and ice marked by many craters. Jupiter has at least 12 other satellites that are much smaller than the four Galilean moons. Around Jupiter, there is also a faint ring that is made up of very small particles.

The Galilean satellites of Jupiter share many characteristics with the earthlike inner planets. Future exploration of these moons may provide better understanding of the working of planets like the earth.

Fig. 18–10 Photographs of the Galilean moons of Jupiter. Clockwise, beginning at the bottom left: Io, Europa, Ganymede, and Callisto.

Fig. 18–11 The colored bands in Saturn's upper cloud layers can be seen in this photograph.

SATURN

Saturn (**sat**-urn) is the next planet beyond Jupiter. It is the least dense of all the planets. Its density is less than that of water. If there could be a large enough ocean, Saturn would float in it. This means that Saturn, like Jupiter, is rich with the light element hydrogen. Also, like Jupiter, Saturn is believed to have a small, rocky core surrounded by a thick layer of liquid hydrogen. It produces more heat than it receives from the sun. Saturn rotates very rapidly, once every 10 hours and 40 minutes. This causes its outer cloud layers to separate into bands. Photographs of Saturn taken from spacecraft show bands of pale yellow, golden brown, and reddish brown colors as shown in Fig. 18–11.

Saturn's most observable feature is its great system of rings. See Fig. 18–12. Saturn's rings are made up of many small pieces of ice or substances coated with ice. The size of the chunks of ice material ranges from a few centimeters across to several meters. The rings are divided into three large bands, each of which is made of many smaller rings. These are similar to the rings on a phonograph record.

URANUS AND NEPTUNE

If you were on Uranus (**yoor**-uh-nus) or Neptune (**nep**-toon), the sun would look so far away that it would be hard for you to tell it from other stars. Thus both Uranus and Neptune are very cold. Each has a diameter about four times that of the

Enrichment: One of Saturn's moons, Titan, is the second largest satellite (after Ganymede) in the solar system. Titan is larger than Mercury and about three-quarters the size of Mars.

Fig. 18–12 Saturn's rings are hundreds of separate ringlets, like grooves in a record album.

Uranus (top) and its satellite Miranda (bottom).

Enrichment: In 1979 Neptune became the most distant planet. Once every 248 years, Pluto's orbit brings it inside the orbit of Neptune for a period of 20 years. Thus Neptune will be the farthest planet until 1999.

1. Jupiter, Saturn, Neptune, and Uranus are made mostly of ices and gases. Jupiter has a large red spot in its atmosphere. Saturn has many rings. Uranus and Neptune have faint bands.

2. Jupiter's moons are Io, Europa, Ganymede, and Callisto. Io has a bright, yellow-to-orange-colored surface. Europa has a light yellow-colored surface covered by dark cracks. Ganymede and Calisto are covered with a mixture of rock and ice marked by many craters.

3. Pluto differs from the other outer planets in that it is the smallest planet in the solar system, has a tilted, more elliptical orbit, and seems to be made up of frozen methane and other ices.

earth. Each has a small, solid core surrounded by a thick layer of ice and liquid hydrogen. The planets have a pale greenish or blue-green color with only very faint bands like Jupiter and Saturn. Uranus and Neptune take about 17 hours to complete one rotation. Uranus rotates in an unusual way because it is "tipped over." Its north pole points almost directly toward the sun, rather than up and down as with the other planets. Uranus has five large satellites and ten smaller ones. It also has at least eleven rings similar to those of Saturn. Close-up photos from the Voyager 2 spacecraft have shown that Uranus' rings are made of dark boulder-size rocks.

PLUTO

There are reasons to believe that Pluto did not form the same way as any other planet. First, Pluto is a very small planet. Its diameter seems to be about equal to the distance between Los Angeles and Washington, D.C. This makes Pluto the smallest planet. Second, Pluto's orbit is different from the orbits of the other planets. While the orbits of other planets lie in nearly the same plane, Pluto follows a strange, tilted orbit that is the most elliptical of all the planetary orbits. Third, Pluto has a very low density and seems to be made up of frozen methane and other ices. Such a composition is unlike that of any other planet. Answers to many of the questions about Pluto must await some future time when spacecraft can finally reach the very edge of the solar system.

SUMMARY

The outer planets are made up mostly of liquids, ices, and gases. Jupiter, Saturn, Uranus, and Neptune have a small, rocky core surrounded by a thick layer of liquid hydrogen, ice, and clouds. Pluto is different from the other outer planets and is believed to have formed differently from them.

QUESTIONS

Use complete sentences to write your answers.
1. Compare the main features of Jupiter, Saturn, Neptune, and Uranus.
2. Name and describe the four largest moons of Jupiter.
3. How does Pluto differ from the other outer planets?

INVESTIGATION

See teacher's commentary for teaching hints, safety, and answers to questions.

A SCALE MODEL OF THE SOLAR SYSTEM

PURPOSE: To draw models of the size of each planet in the solar system and, using the same scale, to get an idea of the distances planets are from the sun.

MATERIALS:

unlined paper	pencil compass
pencil	ruler

PROCEDURE:

A. It is difficult to get a good idea of the relative sizes of the sun and its system of planets. Since they are all very large, we will use circles as scale models to represent them. The scale that we will use is 1 cm = 1,000 km.

    1. The sun has a diameter of 1,400,000 km. Using the scale, what diameter circle would represent the sun?

B. Copy the table below. Complete the third column by finding the scale diameter of each planet.

Name	Diameter (km)	Scaled Diameter (cm)
Mercury	5,100	5.1
Venus	13,000	13
Earth	13,000	13
Mars	6,900	6.9
Jupiter	140,000	140
Saturn	120,000	120
Uranus	53,000	53
Neptune	50,000	50
Pluto	6,000 (?)	6.0

C. On one side of an unlined sheet of paper draw a circle to scale for each planet, using a pencil compass. The circles may overlap. Label each circle with the name of the planet it represents.

    2. Name four large planets that are similar in size.

    3. Name five small planets that are similar in size.

D. Copy the table below. Complete the third column by finding the scale distance each planet is from the sun. Use the same scale as for planet size.

Name	Distance from Sun (km)	Scale Distance from Sun (cm)
Mercury	58,000,000	58,000
Venus	110,000,000	110,000
Earth	150,000,000	150,000
Mars	230,000,000	230,000
Jupiter	780,000,000	780,000
Saturn	1,400,000,000	1,400,000
Uranus	2,900,000,000	2,900,000
Neptune	4,500,000,000	4,500,000
Pluto	5,900,000,000	5,900,000

E. Look at your scale distances and answer the following questions.

    4. Is it possible to draw a scale distance model of the solar system inside the classroom using this scale?

    5. What is the scale distance to Saturn in kilometers?

CONCLUSION:

1. Why do commercial models of the solar system use one scale for the sizes of the sun and planets and a different scale for the distances of the planets?

18-3. Smaller Members of the Solar System

At the end of this section you will be able to:

- ☐ Explain how a belt of *asteroids* may have been formed as part of the solar system.
- ☐ Compare the properties of *meteors* and *comets* with other smaller members of the solar system.
- ☐ Describe two ways in which moons may have formed around planets.

On June 30, 1908, a mysterious explosion occured in Siberia. People asleep in their tents 80 kilometers away were blown up into the air. In a nearby town, windows were broken. The explosion was detected as far away as England. The explosion was caused by an unknown object from space. The solar system contains many bodies that float between the planets.

ASTEROIDS

Between the orbits of Mars and Jupiter is a wide region of the solar system that does not contain a planet. In this zone are found a huge number of lumps and boulders. These objects are are called **asteroids.** See Fig. 18-13. The total number of *as-*

Asteroid A smaller body in the solar system, usually found in an orbit between Mars and Jupiter.

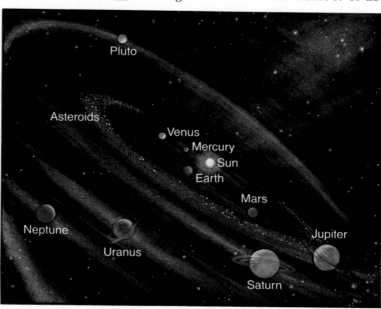

Fig. 18-13 The belt of asteroids lies between the orbits of Mars and Jupiter. The orbits of the planets are not shown with the correct size in this diagram.

teroids is not known. Only the larger ones reflect enough sunlight to be seen through telescopes. About 2,000 asteroids have been discovered. An asteroid is said to have been discovered when it has been photographed several times and its orbit is determined. See Fig. 18-14. Thousands more have appeared in photographs, but no one has spent the time necessary to find their orbits. About 230 asteroids are known to have diameters greater than 100 kilometers. The largest, called *Ceres* (**seereez**), has a diameter of about 1,000 kilometers. There are probably thousands more with diameters of a few kilometers or less. All the asteroids put together probably would not equal the mass of a small planet. Ceres alone makes up about 30 percent of the total mass of all the known asteroids.

At one time, scientists thought that the asteroids might be the wreckage of a small planet. This planet was supposed to have been torn apart by Jupiter's tremendous force of gravity. However, scientists now know that there is not enough material in the asteroids to make a planet. They now think that most of the asteroids were once part of about 50 large bodies that formed along with the planets. But Jupiter's gravity caused them to start crashing into each other. These collisions broke them into the pieces that are found today. Only a few of the original larger bodies, such as Ceres, remain. Many smaller asteroids seem to be bits of matter left over as larger bodies grew in the young solar system. Asteroids probably existed in other parts of the solar system. But then the planets acted like giant vacuum cleaners that swept up these small objects. The large distance between Mars and Jupiter allowed the leftover material to remain as the belt of asteroids.

Most of the asteroids follow orbits that keep them in the region between Mars and Jupiter. However, sometimes an asteroid passes close enough to Mars or Jupiter to be pulled into a new orbit. This causes some asteroids to follow orbits that carry them to other parts of the solar system. For example, 28 asteroids follow paths that cause them to cross the orbit of the earth as they move around the sun. Some of these can come close to the earth and could collide with this planet. However, such collisions with asteroids are very rare. It is believed that it was an asteroid that caused the mysterious explosion in Siberia. The only very recent crater known to exist on the earth is found in Arizona. See Fig. 18-15.

Fig. 18-14 Asteroids appear as streaks in telescopic photographs.

Fig. 18-15 This crater in Arizona was made when a meteorite struck the earth about 30,000 years ago.

METEORS

While collision with large objects is not common, many of the very small objects in the solar system often collide with the earth. Most of these objects are very small pieces of rock. They usually have a mass of less than one gram. When they reach the earth's atmosphere, friction causes most of them to get so hot that they burn up at altitudes near 100 kilometers. The bright streaks they create in the sky are called **meteors** (**meet**-ee-orz), or "shooting stars." Some *meteors* look like huge balls of fire. Sometimes one of the objects is so large that parts of it may reach the earth's surface. These chunks of material that reach the earth's surface are called *meteorites*. They usually can be identified by marks on their surfaces. These marks are caused by the heat produced by friction as they pass through the atmosphere. About 90 percent of the meteorites found are made of stone similar to ordinary rocks on the earth. The others are made mostly of iron mixed with nickel. Most meteorites seem to be asteroids that have wandered out of their usual place in the solar system. Some meteorites found on the earth have not changed since the solar system formed. Thus scientists are able to learn much about the early history of the solar system from their studies of meteorites.

COMETS

Scientists believe that the entire solar system is surrounded by a distant cloud. This cloud is made up of what have been called "dirty snowballs." Chunks of ice, rock, and dust slowly move around the sun at distances so great that it takes the sun's light more than a year to travel there. This cloud was probably formed along with the rest of the solar system about five billion years ago. Because it is so far away, we would not even know this cloud was there. However, sometimes the gravity of a passing star pulls one of the clumps of ice and rock from the cloud. It may begin moving toward the sun. The icy body then becomes a **comet.** Each *comet* follows a different path as it moves in toward the sun.

Comets often are affected by the gravity of Jupiter and Saturn. For example, a comet can be pulled by Jupiter's gravity into an orbit that takes it close to the sun. When this happens, the comet may become a member of the group of comets that are seen from the earth at different times.

Meteor A bright streak of light in the sky, which is caused by rock material burning up in the atmosphere.

Enrichment: A meteoroid about the size of a grain of sand will look as bright as the brightest star as it vaporizes in the upper atmosphere. A meteoroid the size of a golf ball would look as bright as the full moon.

Enrichment: Comets discovered today are first named for their discoverer. Later they are assigned scientific names in order of their passage around the sun. For example, Comet, 1984, 1.

Comet An icy body that comes from the edge of the solar system and moves toward the sun.

A comet that follows an orbit bringing it near the sun can become one of the most unusual objects we ever see in the sky. When a comet passes the orbit of Mars, it may begin to grow a long tail. The sun's heat and the solar wind cause material to glow and stream away from the comet. See Fig. 18–16. A comet's tail always points away from the sun, like long hair streaming away from the wind. See Fig. 18–17. Some comets develop two tails, whereas others may hardly have a visible tail.

Some comets can be seen from the earth at regular times. For example, Halley's comet has been seen about every 76 years for many centuries. However, each time a comet makes a passage around the sun it loses some of its materials. Thus a comet usually becomes smaller and dimmer after each appearance until it finally disappears. A comet may also break into smaller parts that continue to follow the original orbit. Meteor showers are seen when these pieces collide with the earth's

Fig. 18-16 Solar wind and heat are responsible for the long tail that is characteristic of many comets.

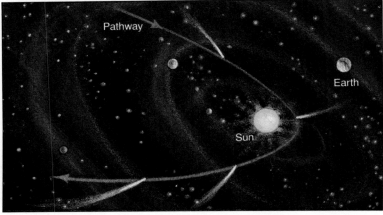

Fig. 18-17 The tail of a comet always points away from the sun.

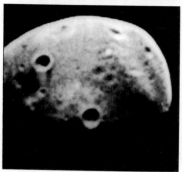

Fig. 18–18 Mars has two tiny, irregularly shaped moons. (top) Phobos, the larger, inner moon. (bottom) Deimos, the smaller, outer moon.

1. In the region between Mars and Jupiter, about 50 small, rocky bodies formed at the same time as the other planets. Due to Jupiter's gravity, these have collided and most have broken up into thousands of smaller pieces.
2. Comets are made of ice, rock, and dust. They come from a distant cloud that surrounds the solar system. Meteors are made of stone or iron mixed with nickel. Most seem to have been asteroids at one time.
3. Moons may have formed around a planet at the same time as the planet formed, or a moon may have been captured by the planet's gravity long after the planet formed.

atmosphere. At times during the year, as many as 50 to 100 meteors are seen in an hour. Such meteor showers occur when the earth moves through one of the groups of bodies that were once part of comets. Fairly heavy meteor showers can be seen each year near the dates of May 4 to 6, August 10 to 14, and October 19 to 23. Most of these meteor showers are heaviest after midnight.

MOONS

Other kinds of bodies found in the solar system are moons. Planets may come to have one or more satellites, or moons, in two ways. First, a planet's moons can be formed along with the planet. For example, the large moons of Jupiter seem to have been formed like a miniature solar system around the planet as the planet itself formed. The inner moons have rocky interiors and the outer moons have icy interiors. Saturn also has a similar system of moons, with at least 17 satellites. The largest moon of Saturn, called *Titan* (**tie**-tun), is the only satellite in our solar system to have an atmosphere.

A second way that a planet may come to have moons is by the capture of large asteroids. For example, the two satellites of Mars, called *Phobos* (**foh**-bohs) and *Deimos* (**dee**-mohs), are asteroids that are held by Mars' gravity. See Fig. 18–18.

SUMMARY

The most numerous of the smaller bodies in the solar system are asteroids, which are found in a region between the orbits of Mars and Jupiter. Icy bodies from the most distant parts of the solar system may become visible as comets when they begin to follow orbits that bring them close to the sun. Meteors are objects that sometimes collide with the earth but are usually burned up in the atmosphere. A planet's moons may be formed along with the planet or captured at a later time.

QUESTIONS

Use complete sentences to write your answers.

1. What explanation is given by scientists for the presence of the asteroid belt?
2. In what ways do comets differ from meteors?
3. Describe two ways moons may have formed around planets.

See teacher's commentary for teaching hints, safety, and answers to questions.

THE PATHS OF ASTEROIDS AND COMETS

PURPOSE: To draw the paths of several asteroids and a comet, and to compare these with the paths of planets from the Investigation in Section 18–1.

MATERIALS:

plain paper	pencil
cardboard	string
2 straight pins	ruler

PROCEDURE:

A. The procedure you will use is the same as in the Investigation of Section 18–1. Place a sheet of paper over the cardboard and stick one pin near the center of the paper, anchoring it firmly. Draw a small circle around the pin and label it SUN. You will not move this pin when making the asteroid orbits.

B. Use the method shown in Fig. 18–19 to draw the orbits of the asteroids on one side of your paper and the orbit of the comet on the other side. The comets have a much larger path that goes from very near the sun to the outer boundary of the solar system. The comet path cannot be drawn on the same scale as the asteroids.

C. Cut one piece of string at a time, making each several centimeters longer than the distance around the loop given in the table below. Bring the ends of the string together and tie a knot one-half the distance shown in the table. For example, for the typical asteroid, the distance around the loop is 6.4 centimeters. Cut a piece about 9 centimeters long, and tie the ends together, making a loop 3.2 centimeters long. See Fig. 18–6(b) on page 488.

Object	Distance Between Pins (cm)	Distance Around Loop (cm)
Typical asteroid	0.6	6.4
Apollo	2.5	7.0
Hildago	11.0	28.0
Halley's Comet	10.4	21.2

CONCLUSIONS:

1. Which orbit is most like the orbits of the planets drawn in the Investigation of Section 18–1?

2. Most asteroids follow a typical path. Which asteroid has a path that is much closer to the sun than that of the typical asteroid?

3. Which asteroid has a path that takes it much farther from the sun than the typical asteroid?

4. Compare the path of Halley's comet to the path of a typical asteroid.

Sun   Fig. 18–19

18–4. The Moon

At the end of this section you will be able to:

☐ Compare the surface of the moon with the surface of the earth.

☐ Give a brief description of the moon's history.

☐ Describe three theories that may account for the origin of the moon.

One accomplishment of the space program has been travel to the moon. See Fig. 18–20. Some scientists predict that, by the early 21st century, humans will be able to live on the moon. Could you be one of the first "natives" on the moon?*

* The cost of a program that would build colonies on the moon is an estimated 84 billion dollars. However, this is not much more than the cost of the Apollo program to land men on the moon, which cost 73 billion dollars.

Fig. 18-20 The panel instruments of an Apollo spacecraft with the moon in the background.

THE SURFACE OF THE MOON

The moon is the earth's nearest neighbor in space. It is so close that you can easily see the main features of the moon's surface. With a pair of binoculars, you can see rugged mountain ranges, wide level plains, and hundreds of large and small craters. These features stand out because there is no atmosphere to hide them. However, it was not until astronauts landed on the moon, beginning in 1969, that the actual conditions of earth's closest neighbor became known.

In some ways, the moon resembles the earth. Both have a crust and a mantle. The moon may also have a core like the earth, although it is probably much smaller. The inside of the moon may also be hot like that of the earth.

However, the astronauts who visited the moon were more impressed by the differences between the surface of the moon and that of the earth than by their similarities. First, there is no atmosphere on the moon. But instruments have been able to measure small amounts of the gases argon, neon, and helium. Also, there is no water on the moon. No sign of life has been found on the moon. It appears that the only living things ever to have been on the moon were the astronauts from earth. During the day the temperature on the moon's surface rises to above 100°C. During the night the temperature drops to below −100°C. The moon has no atmosphere to shield it from the sun's heat or help hold the heat during the night.

The moon is smaller than the earth. Its mass is also much less than the earth's mass. Because of this, the force of gravity felt on its surface is only one-sixth that felt on earth. Thus the astronauts on the moon's surface were able to jump higher and move by leaps and bounds, as if they had springs on their feet. They fell in slow motion and landed softly.

The surface features of the moon have not been changed by wind and water. Many of the features that can be seen today on the moon have been there since its early history. The moon is not free from all change, however. The lack of an atmosphere allows the surface to be hit constantly by small meteorites and solar wind particles. This breaks up some of the rocks on the surface. On the moon, the small pieces of rock are not washed away or blown away as they would be on earth. As a result, a fine dust covers almost all the moon's surface. See Fig. 18–21. The astronauts were surprised to find that the moon dust is a dark gray color. It looks tan from the earth.

HISTORY OF THE MOON

The rocks brought back from the moon are surprisingly old. All these moon rocks are more than three billion years old. However, almost all the rock on the earth's surface is younger than that. Changes that take place on the earth produce young rock, which covers the older rock. On the moon, there seem to have been no major changes for three billion years.

Evidence from the moon rocks seems to support the following outline of the moon's history. The moon came into being at the same time as the earth. This was about 4.6 billion years ago. At first, its outer layers were hot and molten. Then the heavy

Fig. 18–21 Moon dust is stirred up by this astronaut on the moon.

Fig. 18-22 (left) *Features on the moon's surface can be seen with the eye alone. Tyco can be seen as the large crater near the bottom.* (right) *Craters cover much of the moon's surface.*

Maria Lava plains that are seen as dark areas on the moon's surface.

Rays Light-colored streaks that lead out from some of the craters on the moon.

materials settled downward. The lighter materials rose to the surface. A crust formed as the outer parts cooled and hardened. During its earliest years, the moon had many volcanoes. However, the volcanoes stopped erupting about three billion years ago. Also, during this early period, the moon was struck many times by huge chunks of material that were flying through space. These collisions tore great basins in the moon's surface. Volcanic eruptions later filled these low places with lava. These lava-filled basins now appear as dark-colored plains on the moon's surface. Early astronomers thought that these dark areas were seas and called them **maria** (**mar**-ee-ah), which is Latin for "seas." Most of the rest of the moon's surface is made up of light-colored highlands. The highlands are the remains of the moon's original crust, which were not covered by the lava flow. The dark *maria* and the light highlands are easily seen when you look at the full moon. See Fig. 18-22 *(left).*

The highlands of the moon are marked by thousands of craters. See Fig. 18-22 *(right).* These craters are the scars left by meteorites of many different sizes that crashed into the moon's surface throughout its history. A few of them are several hundred kilometers in diameter.

Some craters have streaks of light-colored material leading out from their edges. These streaks are called **rays.** The *rays* consist of material that was thrown out of the crater when it was formed. One large crater, named Tycho (**tie**-ko), has a large system of rays. Tycho is easily seen with the aid of binoculars when the moon is full. See Fig. 18-22 *(left).*

WHERE DID THE MOON COME FROM?

Three theories have been put forward to explain the origin of the moon.

1. Daughter theory. The *daughter theory* suggests that the moon was once part of the earth. It says that the rotation of the earth and the gravity of the sun produced a bulge in the earth. Then, long ago, a part of the earth broke away. It flew into space, and became the moon.

2. Sister planet theory. The *sister planet theory* says that the earth and moon are twin planets that formed at the same time.

3. Capture theory. The third theory is called the *capture theory*. It suggests that the moon was formed in some other part of the solar system and was "captured" by the earth's gravity.

It is unlikely that the moon was once part of the earth, as the daughter theory suggests. The theory is unlikely because there are too many differences in the chemical makeup of the two bodies. The capture theory does not agree with the principles of gravity. According to these principles, if the earth captured the moon, the orbit of the moon would be different than it is today. The sister theory seems the most likely to be true. However, not all scientists agree with this theory. The information gained from exploring the moon has not settled the question about the moon's origin.

SUMMARY

The surface of the moon is very different from the earth. This is because the moon does not have an atmosphere. The rocks of the moon have changed very little since its early history. They show that the moon's surface was once hot and molten, and covered with volcanoes. It is believed that the moon formed along with the earth.

QUESTIONS

Use complete sentences to write your answers.

1. Compare the conditions on the moon's surface with the conditions found on the earth's surface.
2. Outline the moon's history as supported by a study of its rocks.
3. Describe three theories that answer the question, "Where did the moon come from?"

Enrichment: The far, or back, side of the moon, that is never visible, is not as smooth as the side we see. Almost half of the visible side is covered by maria, whereas the far side has many more craters. The near side probably experienced more volcanism in the moon's early history because of the greater tidal forces caused by the earth.

1. Answers may vary but should make the following points: There is only one-sixth the gravity on the moon's surface compared to the earth; the moon's surface features, mostly mountains, craters, and lava plains, have not been eroded by wind and water; the moon has no atmosphere; the moon's surface is mostly covered with a layer of dust, while the earth's is covered mostly with water; and the rocks are old compared to earth rocks.
2. The outline of the moon's history is as follows: (a) It came into being at the same time as earth. (b) In the beginning, its outer layers were hot and molten, allowing the heavier materials to settle downward and the lighter materials to rise. (c) A crust formed as the outer materials cooled and became hard. (d) The moon had many volcanoes, which stopped erupting about three billion years ago. (e) During the time when volcanoes were active, the moon was struck many times by material from space, causing great basins that were filled with lava from volcanoes.
3. The daughter theory suggests the moon was once a part of the earth that broke away early in the earth's history and became the moon. The sister planet theory suggests that the earth and moon are twin planets that came into being at the same time. The capture theory suggests that the moon came into being at some other part of the solar system and was later captured by the earth's gravity.

INVESTIGATION

See teacher's commentary for teaching hints, safety, and answers to questions.

SURVIVAL ON THE MOON

PURPOSE: To find out what is important for survival on the moon.

MATERIALS:

paper   pencil

PROCEDURE:

A. You and two members of your crew are returning to the base ship on the sunlit side of the moon after carrying out a 72-hour exploration trip. Your small rocket craft has crash-landed about 300 kilometers from the base ship. You and the crew need to reach the base ship. In addition to your spacesuits, your crew was able to remove the following items from the rocket craft:

```
 4  packages of food concentrate
20  m nylon rope
 1  portable heating unit
 1  magnetic compass
 1  box of matches
 1  first-aid kit
 2  50-kg tanks of oxygen
20  L of water
 1  star chart
 1  case of dehydrated milk
 1  solar-powered radio set
 3  signal flares
 1  large piece of insulating fabric
 1  flashlight
 2  45-caliber pistols, loaded.
```

B. Using what you know about the moon, rate each item in the above list according to how important it would be in getting you back to the base ship. List the most important first, the least important last.

Number them 1 through 15. Answer the following:

1. Which three items were the most important? Explain.
2. Which items would be useless? Explain your answer.

C. Compare your list with the one supplied by your teacher. Astronauts would list the items in this order.

D. To score your list against the astronauts' list, do the following:

3. Beside each item on your list, place the number that represents the difference between your ranking and the astronauts' ranking. For example if you listed oxygen first, you would write 0 in front of oxygen on your list. If you had listed it third, then you would write 2, and so on.
4. After placing a score beside each item on your list, add up the individual scores to get a total. Compare your score with those of other students.
5. What is your total score?
6. The lower your total score, the closer you came to surviving the return trip to the base ship. How did your chance of surviving compare to other students' chances?

CONCLUSIONS:

1. What does the moon lack that humans need for survival?
2. What materials would you need to survive on the moon?

18-5. The Earth and Moon

At the end of this section you will be able to:

- ☐ Describe the motions of the earth–moon system.
- ☐ Relate each *phase* of the moon to a position in its orbit around the earth.
- ☐ Explain how *eclipses* of the sun and moon occur.

High above the earth's surface the space shuttle orbiter releases a space telescope. See Fig. 18–23. The telescope will remain in an orbit around the earth while the space shuttle returns to earth. From its position high above the earth's atmosphere, the space telescope can see deep into space. The space telescope is one example of many different kinds of human-made satellites that have been developed to orbit the earth.

Fig. 18-23 *The space shuttle is a useful tool for releasing and repairing satellites and for conducting experiments in space.*

THE EARTH–MOON SYSTEM

A **satellite** is a body that moves in an orbit around a much larger body. For example, the space telescope is a *satellite* of the earth because its mass is very small compared to the mass of the earth. Is the moon also a satellite of the earth? The moon is often called the earth's natural satellite. This is because the mass of the earth is so different from that of the moon that the moon could be called a satellite of the earth. However, rather

Satellite A body that moves in an orbit around a larger body.

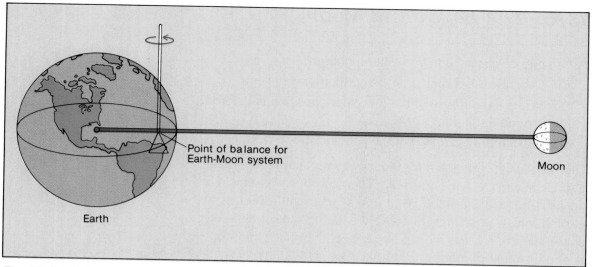

Fig. 18-24 If the earth and
the moon were connected, the
balance point of the system
would be within the earth.

than the moon being a satellite of the earth, it and the earth
make up a system that is like a pair of planets moving around
the sun together. If the earth and moon were connected by a
solid bar, the system would balance at a place inside the earth.
See Fig. 18-24. Instead of the earth moving in a smooth orbit
with the moon circling the earth, the earth and moon together
orbit the sun. At the same time, each turns around the balance

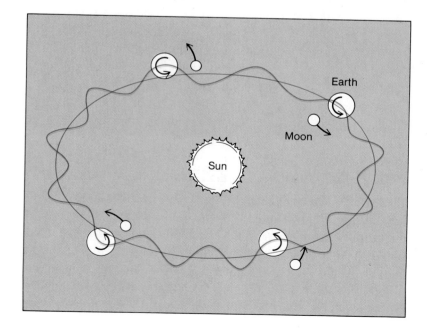

Fig. 18-25 The earth and the
moon move around the sun
as a single system.

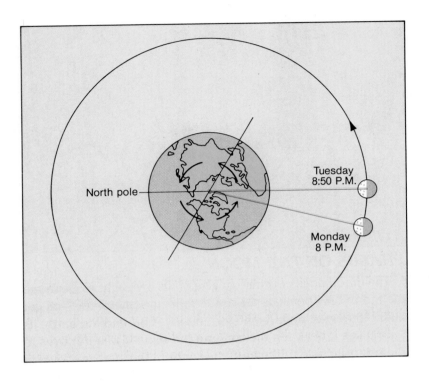

North pole

Tuesday
8:50 P.M.

Monday
8 P.M.

Fig. 18-26 The moon moves ahead in its orbit each day, so a place on the earth must rotate about 50 minutes longer each day to bring the moon back into view.

point. See Fig. 18–25. However, when the moon is seen from the earth, it looks as if the moon is circling the earth. For this reason, the moon is usually said to move in an orbit around the earth.

The moon makes one complete trip around the earth every 27 1/3 days. The moon also turns on its axis once every 27 1/3 days as it moves around the earth. As a result, the same side of the moon always faces the earth.

We can see the moon appear on the eastern horizon every day. This is because the earth's eastward rotation carries us around until the moon can be seen. The moon appears to rise in the east, then move across the sky, and set in the west. If the moon did move around the earth, it would always rise and set at the same time of day. However, the moon does move around the earth in an eastward direction. Thus the earth must turn a little farther each day to "catch up" with the moon and bring it back into view again. It takes about 50 minutes each day for the earth to rotate the extra amount needed to "catch up" with the moon. This means that the moon rises about 50 minutes later each day. See Fig. 18–26.

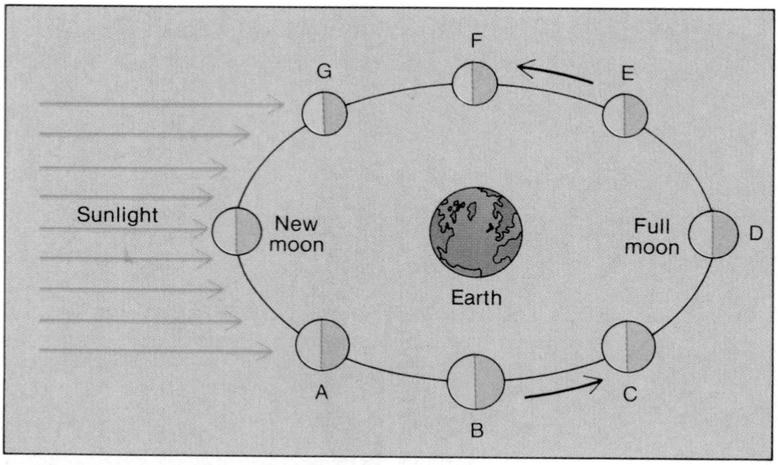

Fig. 18-27 The moon moves constantly between the sun and the earth, then away from the sun.

PHASES OF THE MOON

The moon does not shine or give off its own light. We can see the moon only because the moon reflects some of the sunlight falling on its surface. As the moon moves around the earth, its path takes it between the sun and the earth. Then it moves to the opposite side of the earth, away from the sun, as shown in Fig. 18-27. The position of the moon, the earth, and the sun determines how the moon looks to us as we see it from the earth. The different amounts of the lighted and dark parts of the moon seen from the earth are called the **moon's phases** (**faze**-uhz). See Fig. 18-28.

When the moon is between the earth and the sun, its dark side is facing the earth. This *moon phase* is called the *new moon*. If you were standing on the dark side of the moon during the new moon phase, you would see the sunlit side of the earth. As the moon moves in its orbit, a small part of the lighted side can be seen from the earth. This is the *crescent* (**kres**-ent) phase. When the moon has moved one-quarter of the way through its orbit, one-half of its bright side is seen from the earth. This phase is called the *first quarter*. After the first quarter, comes the football-shaped *gibbous* (**gib**-us) phase. After this phase, the amount of sunlit surface increases until all of the sunlit side of the moon can be seen from the earth. The fully-lit side of the moon is called the *full moon*. As the moon continues its orbit, the sunlit part of the moon begins to move out of view. The phases following full moon are gibbous, last quarter, crescent, and then another new moon. It takes 29 1/2 days to go through one complete cycle of moon phases. For

Moon phase The changing appearance of the moon as seen from the earth.

Fig. 18–28 The phases of the moon (left to right and top to bottom): *crescent, quarter, gibbous, full.*

example, from one full moon to the next is 29 1/2 days. However, the time it takes the moon to make one revolution around the earth is 27 1/3 days. The reason is that the earth also moves around the sun. Thus when the moon makes one revolution around the earth in 27 1/3 days, the earth has moved farther around the sun. The moon must move in its orbit about two more days or 29 1/2 days to return to its original phase.

THE MOON AND THE CALENDAR

In the past, people used phases of the moon to measure the passage of time. The calendar month originally came from the 29 1/2-day cycle of moon phases. A year, or the time it takes for the earth to make one complete orbit of the sun, is about 365 1/4 days. But, in a 365-1/4 day year, there are more than twelve 29 1/2 day cycles of moon phases. Since 365 1/4 cannot be divided evenly by 29 1/2, Western societies do not use a calendar based on the phases of the moon. Western calendars use a system of 12 months that add up to 365 days. To make up the extra quarter day of each year, a whole day is added every fourth year. In this way, we keep our calendar in step with the earth's movement around the sun.

Enrichment: The full moon is nine times brighter than the quarter-moon. During the full moon, sunlight is reflected more directly toward the earth.

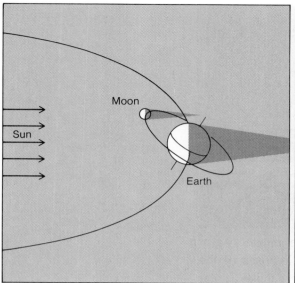

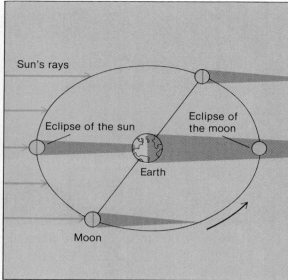

Fig. 18–29 (left) *The moon's tilting orbit makes its shadow usually miss the earth. The orbit is tilted less than is shown.* (right) *The moon's shadow causes a solar eclipse. The earth's shadow causes a lunar eclipse.*

Eclipse The earth or the moon passing through the shadow of the other.

Fig. 18–30 *A solar eclipse.*

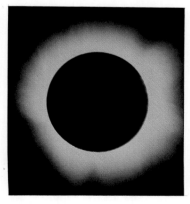

ECLIPSES

Both the earth and the moon cast long shadows in space. At the new moon, the moon is between the earth and the sun. You might wonder why the shadow from the moon does not fall on the earth every new moon phase. The answer is found in the way the moon's orbit is slightly tilted in relation to the earth's orbit around the sun. See Fig. 18–29 *(left)*. Usually, the moon crosses too high or low between the earth and the sun for its shadow to fall on the earth. However, as the earth and the moon move together around the sun, the moon sometimes crosses directly between the sun and the earth. See Fig. 18–29 *(right)*. When this happens, an **eclipse** of the sun occurs. Only the tip of the moon's shadow falls on the earth's surface. Thus only a small part of the world is able to see a total *eclipse* of the sun. See Fig. 18–30. The shadow moves so quickly that the eclipse lasts only a few minutes at any one place. In the outer part of the shadow, where the light is not completely blocked out, only a part of the sun is seen to be covered. This is called a *partial eclipse*. Because the moon follows an elliptical orbit around the earth, it is not always the same distance away. During some eclipses, the moon is too far from the earth for its shadow to reach the earth's surface. This causes an *annular eclipse*. Annular means "ring." During an annular eclipse, the outer edges of the sun are still visible, forming a ring.

Fig. 18-31 The beginning of
an eclipse of the moon.

Word Study: Eclipse: *eclipsis* (L.), a
failing.

An eclipse of the moon occurs when the moon passes through the earth's shadow. See Fig. 18-31. Eclipses of the moon are as rare as eclipses of the sun. However, unlike a solar eclipse, an eclipse of the moon is seen by everyone on the dark side of the earth.

SUMMARY

Both the earth and the moon move around a balance point located below the earth's surface. Together they make up the earth–moon system. As the moon orbits the earth, its appearance changes as it passes through its phases. Sometimes eclipses are produced when the shadow of the moon falls on the earth or when the moon moves through the earth's shadow.

QUESTIONS

Use complete sentences to write your answers.

1. Starting with a moonrise, describe the moon's movement in the sky. Explain why the moon will rise 50 minutes later each day.
2. List the eight phases of the moon, in the order in which they occur, starting with the new moon.
3. During which phase is the moon between the sun and the earth?
4. During which phase is the earth between the sun and the moon?
5. What causes an eclipse of the sun?
6. What causes an eclipse of the moon?

1. The moon rises in the east, moves westward across the sky, and sets in the west. Since the earth rotates in the same direction as the moon travels in its orbit, the earth must rotate farther the next time to make the moon come into view in the east as a moonrise.
2. New moon, crescent phase, first quarter, gibbous, full moon, gibbous, last quarter, crescent phase.
3. The moon is between the sun and the earth at the new moon phase.
4. The earth is between the sun and the moon during the full moon phase.
5. The moon crosses directly between the earth and the sun, blocking out some of the sun's light.
6. The earth's shadow must fall on the moon as the earth passes between the sun and the moon.

INVESTIGATION

See teacher's commentary for teaching hints, safety, and answers to questions.

THE PHASES OF THE MOON

PURPOSE: To demonstrate the phases of the moon using a model.

MATERIALS:

ball, one-half painted black	pencil
chalkboard	paper

PROCEDURE:

A. The ball represents the moon; your head, the earth; and a drawing on the chalkboard, the sun. Stand by your desk with your back to the sun.

B. Turn the unpainted side of the ball toward the sun. See Fig. 18–32 *(left)*. Hold it at eye level in front of you.

   1. What phase of the moon does this represent?

   2. On a sheet of paper, draw a circle. Make your circle look like the moon model as you see it.

Fig. 18–32

C. Turn your body clockwise one-quarter of a turn (90 degrees). See Fig. 18–32 *(right)*. Again hold the moon model at eye level with the light side toward the sun.

3. What phase of the moon is this?

4. On your paper, make a drawing of the moon model as you see it. With your pencil, shade in the part that is dark.

D. Turn another quarter clockwise. You should now be facing the sun. Hold the moon at eye level with the light side toward the sun.

   5. What phase of the moon does this represent?

   6. On your paper, draw a circle and shade it in to look like the model as you see it.

E. Turn one-eighth turn counterclockwise. Again turn the light half of the moon toward the sun.

   7. What phase of the moon is this?

   8. Make a drawing of what the moon model looks like.

F. Return to the position in step D. If you have the time, try continuing this same process until you have returned to your original position. Try to find a position that represents the crescent phase that occurs just after new moon and compare it to the drawing you made for the answer to question 8. Find positions that demonstrate the gibbous moon phases.

CONCLUSION:

1. Make a drawing similar to Fig. 18–27 on page 510 showing the positions of the moon, earth, and sun that you demonstrated in this activity. Label each moon position with its proper phase.

SAVING PLANET EARTH

Suppose a comet were on a collision course with earth. Could anything be done? At a conference held by NASA in 1981, scientists and weapons experts discussed this possibility and the methods to be used to save the planet.

Is this a fantasy space game? No, but new findings, especially those whose meaning is unclear, present new questions. In this case, careful study, combining the work of paleontologists and geochronologists, has sug-

gested that mass extinctions have occurred about every 26 to 30 million years. A deposit of clay, rich in an element usually seen only in meteorites, was found worldwide. This layer separates rocks that contain dinosaur fossils from rocks that do not.

What could have happened during that period to eliminate the dinosaur? Whatever it was also produced the rich layer of clay and occurred every 26 to 30 million years. Some scientists believe it takes a period of approximately 26 million years for our solar system to travel through the crowded central portion of the Milky Way. Perhaps, scientists hypothesize, an asteroid crashed into the earth at that point. The tremendous explosion that resulted could have thrown so much dust and so many fragments of both the asteroid and earth into the atmosphere that sunlight could not reach the earth. The earth became cold, plants died, and the dinosaurs starved or froze to death. Eventually, this dust settled out of the atmosphere, forming the worldwide layer of clay. Once the sun could get through the atmosphere, life continued. If this happened

once, perhaps it could happen again. And if the earth were to be bombarded with comets, the atmosphere would again be filled with dust and other particles. Sunlight would be blocked and the earth would cool. Would some life forms again disappear? At the 1981 NASA conference, it was agreed that this need not happen, not if the new technology of space were combined with the peaceful use of rocketry. With enough forewarning, a missile could be launched to meet the oncoming asteroid at a distance far removed from earth. In fact, weeks before the orbit of the comet could send it crashing into the earth, a computer-directed missile could be fired toward the comet. Last-minute adjustments could be made from earth to keep the missile on target. Scientists and weapons experts agree that it would not be necessary to use nuclear weapons nor to hit the asteroid directly. The explosive force itself behind the comet would actually be enough to push it into another orbit.

This scene may never be played out. It is possible that data from further research will disprove the theories concerning the mass extinctions of certain plants and animals. Part of planning for the future, however, is anticipating as many possibilities as are imaginable. In this case, the combined research efforts of scientists interested in the earth's past and those interested in its future could help prevent a possible disaster. In any case their further research will seek a greater understanding of the earth's relationship to other bodies in space.

¡COMPUTE!

SCIENCE INPUT

Our knowledge of the solar system changes as the technology of space telescopes and space travel becomes more and more advanced. In the earliest times, astronomical theories depended on observations from earth. Today, computerized images of the planets have been sent from space vehicles launched precisely for research purposes. Through all these scientific and engineering efforts, we have learned a great deal about the characteristics of these satellites of the sun and the entire solar system. Scientists hope that further research will push us past the limits of our own galaxy where we may gain an understanding of the structure of the universe. Program Satellite will help you to review the data presented in your textbook chapters about our solar system's planets.

COMPUTER INPUT

A computer can perform a number of functions, that is, a number of tasks, as you have already seen. Program Planets uses the computer's ability to select random numbers to design a four-question quiz for you (see Chapter 13 Compute! for the definition of random). Program Planet will ask questions, tell you when you're correct, and at the end of the quiz, tell you your final score. Other programs in the Compute! section have used the computer's ability for mathematical calculations. More complicated programming might involve computer graphics, to create images or graphs or very sophisticated mathematical computations.

Not every computer can perform all possible computer functions. What the computer can do depends on its power, or memory capacity. When talking about computers, users may refer to an 8K RAM or a 16K RAM machine. This information describes the amount of storage a computer has.

(See Chapter 8 Compute! for a definition of "bits" and "bytes.") 8K means that the particular computer can manipulate and store approximately 8,000 bytes of information. Different computer functions require different amounts of storage capacity. Therefore, if you were buying software, it would be important to know how much storage it needed. A program needing 16K RAM will not run on your 8K RAM computer.

WHAT TO DO

On a separate piece of paper, make a chart similar to the one below, listing all terms and their definitions. The definitions should be short and simple. Study the chart before entering the program. When you run the program, it will produce a quiz of four questions using the characteristics of the planets you input. Lines 220 to 320 are for data statements. Your terms and definitions are data. Be sure to type the statement in the correct form. An example is given in line 220.

The program will tell you the correct answer if your answer is wrong. After four questions it will give you a score and tell you how many questions out of four you've answered correctly. Run the program at least five times, recording the terms you've missed so you know the ones you need to study.

DATA CHART

Sun

Satellite	Characteristic
Mercury	closest to the sun
Jupiter	has 6 satellites
Venus	covered by a thick layer of clouds
Mars	believed to have been like earth
etc.	. . .

GLOSSARY

RAM   Random Access Memory. That part of the computer's memory in which data can be input, stored, changed, and deleted.

ROM   Read Only Memory. That part of the computer's memory that may not be changed or erased.

PROGRAM

```
100   REM PLANETS
110   DIM T$(10), D$(10)
120   FOR X = 1 TO 10: READ
      T$(X),D$(X): NEXT X:K=0
130   FOR Q = 1 TO 4: LET N = INT (10 *
      RND (1) + 1)
140   PRINT : PRINT "WHICH SUN
      SATELLITE HAS THIS
      CHARACTERISTIC?"
150   PRINT "--> ";D$(N): INPUT "-->
      ";A$
160   IF A$ = T$(N) THEN PRINT "YOU'RE
      CORRECT!"
170   IF A$ = T$(N) THEN K = K + 1
180   IF A$ < > T$(N) THEN PRINT "THE
      CORRECT ANSWER IS:  ";T$(N)
190   FOR T = 1 TO 2000: NEXT T: NEXT
      Q
200   PRINT "--> END OF QUIZ"
210   PRINT "--> ";K;" OF 4 ANSWERS
      WERE CORRECT"
220   DATA   MERCURY, CLOSEST TO
      THE SUN
310   DATA
320   END
```

PROGRAM NOTES

This program is written to be used with ten data statements. New words and definitions can, however, replace the old ones at any time. The program can, in that way, be used for more than ten planet characteristics or for any other set of words and their definitions. If you wanted to add more data statements, you would change line 110, which tells how many spaces of memory are being saved for data. You would also change line 140, of course, if you had changed the subject of your terms and definitions.

BITS OF INFORMATION

Space exploration depends heavily on the use of computers. The technology of space is so complicated that, without the tremendous capacity of the computer to do many kinds of calculations at the same time and quickly, it would be impossible to organize a space mission. Computers also permit two-way input into space vehicles. On a space mission it is not only the astronauts who pilot the spacecraft. The scientific and engineering ground crew can also assist through communication with the spacecraft's computer. This provides an additional safety factor.

Satellites also are in two-way communication with earth. They transmit data about conditions on earth and in the atmosphere. NASA has developed a suitcase-size, computerized rescue kit based on satellite communications. If, for example, you were lost at sea, you would open the kit and transmit a signal to a nearby communications satellite. The satellite would then transmit your signal to a station on land.

CHAPTER REVIEW

VOCABULARY

On a separate piece of paper, match the number of each statement with the term that best completes it. Use each term only once.

meteor	gibbous	meteorite	new moon
ray	eclipse	maria	asteroid
comet	sister theory	moon phase	ellipse

1. A smaller body of the solar system found in orbit between Mars and Jupiter is called a(n) _____.
2. _____ is the changing appearance of the moon as seen from the earth.
3. Lava plains on the moon are known as _____.
4. A(n) _____ is an icy body that comes from the edge of the solar system.
5. A(n) _____ results when the earth or the moon passes through the shadow of the other.
6. The _____ phase occurs when the moon is between the sun and the earth.
7. A(n) _____ is a bright streak of light in the sky.
8. A natural object that survives its plunge through the earth's atmosphere is known as a(n) _____.
9. The _____ moon phase occurs just before or after the full moon.
10. The _____ is an explanation of the origin of the moon that seems most likely true.
11. A light-colored streak leading out from a crater on the moon is called a(n) _____.
12. A(n) _____ is the name for the shape of the path followed by a solar system member.

QUESTIONS

Give brief but complete answers to each of the following questions. Unless otherwise indicated, use complete sentences to write your answers.

1. Why is the sun called the controlling body of the solar system?
2. Compare the two general kinds of planets in the solar system and how they formed.
3. Which one of the inner planets is the smallest?
4. Which of the inner planets have an atmosphere? What effect has this had on their surfaces that is not true of planets that have no atmosphere?

5. What is the source of the extra energy that Jupiter gives off that it doesn't receive from the sun?
6. Describe the makeup of Jupiter, starting with the outer cloud layer and going in to the center of the planet.
7. List the four Galilean moons of Jupiter in order from innermost to outermost.
8. Describe the rings of Saturn.
9. What unusual feature does Uranus have?
10. Where is the asteroid belt located?
11. What are comets made of and where do they come from?
12. Describe two kinds of meteorites that reach the earth's surface.
13. What explanation is given for the way in which Mars got its moons?
14. What surface features can you see on the moon with a pair of binoculars?
15. Why is a large part of the moon's surface covered by a fine dust?
16. Describe the theory that seems to best explain how the moon was made.
17. Why do we always see the same side of the moon?
18. What is a partial eclipse of the sun?

APPLYING SCIENCE

1. Using a pair of binoculars, or a small telescope, locate Jupiter in the night sky. Look for the four Galilean moons. The binoculars will need to be held steady against a fence post or similar stable object to keep it from vibrating. Make a drawing of the location of the moons of Jupiter that you can see.
2. Using a pair of binoculars, or a small telescope, look at the moon. Make a drawing of the part of the moon that is visible. Label any of the surface features that you can identify.

BIBLIOGRAPHY

Branley, Franklyn M. *Halley's Comet 1986*. New York: Dutton, 1983.
Chapman, Clark R. *Planets of Rock and Ice*. Pasadena, CA: Planetary Society, 1984.
Gore, Rick. "The Planets, Between Fire and Ice." *National Geographic*, January, 1985.
Kowal, Charles. "The Chiron Mystery." *Omni*, June 1983.
Miller and Hartman. *The Grand Tour*. New York: Workman, 1981.
Time-Life Editors. *Solar System* (Planet Earth Series). Time-Life Books, 1983.

Stars looking like bright streaks circle the North Star above two observatories. Observatories are used to gather and read many different kinds of information carried in starlight.

STARS AND GALAXIES

CHAPTER GOALS

1. Describe a way to locate stars seen in the night sky.
2. List ways that the distance between the earth and a star can be measured.
3. Identify the main source of scientific information about stars.
4. Trace the steps in the life history of a typical star.

19-1. Observing Stars

At the end of this section you will be able to:

- ☐ Explain one way *constellations* can be useful.
- ☐ Describe ways of gaining information about stars.
- ☐ List two ways of finding the distance between the earth and a star.

There are more stars in space than all the drops of water in the oceans or all the grains of sand on the beaches of the earth. Except for the sun, the stars are also so distant from the earth that even the largest and most powerful telescopes show them only as pinpoints of light. Since the stars are so far away, it might seem difficult to learn about them. But information about the stars is carried in their light. By reading this information, scientists have learned what stars are made of, how large they are, and how they are moving in space.

STAR PATTERNS

Long ago people thought that stars were lights on a huge dome that surrounded the earth. However, we now know that each star is actually a sun.

On a clear, dark night, about 2,000 stars can be seen. With a little practice, you can begin to see that there are patterns among the stars. You can begin to recognize the pictures that people have used for thousands of years to identify these patterns. According to this old way of studying the stars, the particular star patterns can be seen as forming the outlines of people,

Fig. 19-1 This is a photograph of the constellation
Orion. Can you see the figure
of a hunter?

Constellation A group of stars
in a particular pattern.

Word Study: Constellation: *com +
stella* (L.), with star.

animals, or familiar objects. A group of stars in such an arrangement is called a **constellation** (kon-stuh-**lay**-shun). Most *constellations* are named after animals or ancient gods and
heroes. Often it is hard to see how a group of stars makes an
outline of something. See Figs. 19-1 and 19-2. At times, it may
be easier to recognize a constellation as a baseball diamond
rather than a horse with wings. However, no matter how you
identify them, it is useful to be familiar with some of the con-

Fig. 19-2 Orion the hunter is
depicted here with the
prominent stars shown. The
three bright stars forming the
belt are easily found in the
winter sky.

522

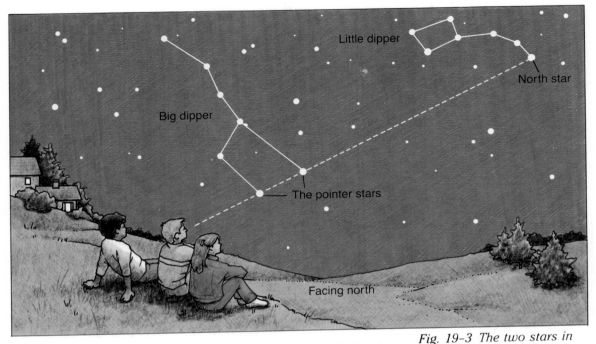

Fig. 19–3 The two stars in the bowl of the Big Dipper point to Polaris, the North Star.

stellations. They can be used as guides, or maps, to find certain stars in the sky. Once you become familiar with a few constellations, it is easy to find other constellations or individual stars.

The group of stars known as the Big Dipper is an example of how to use constellations as guides. The Big Dipper is part of a constellation called *Ursa Major*, or Big Bear. This constellation is easily seen in the northern sky all during the year in the Northern Hemisphere. Two stars forming the side of the bowl of the Big Dipper can be used as pointers to find the North Star. A line drawn through these pointer stars and going about five times the distance between them leads to the star called *Polaris* (poh-**lar**-us), or North Star. See Fig. 19–3. Polaris is the star at the end of the handle of the Little Dipper, which is part of the constellation called *Ursa Minor*, or Little Bear. A camera aimed at Polaris with the shutter left open for several hours will produce a photograph similar to Fig. 19–4. All stars except Polaris will show as a circular streak on the film. Polaris is almost directly above the North Pole. Thus all the other stars seem to circle around it as the earth rotates on its axis.

When we look at the sky, we see the stars in constellations as if they were located on a kind of ceiling over the earth. For example, the stars in the Big Dipper all seem to be at the same

Fig. 19–4 The stars seem to circle in the sky as a result of the earth's rotation.

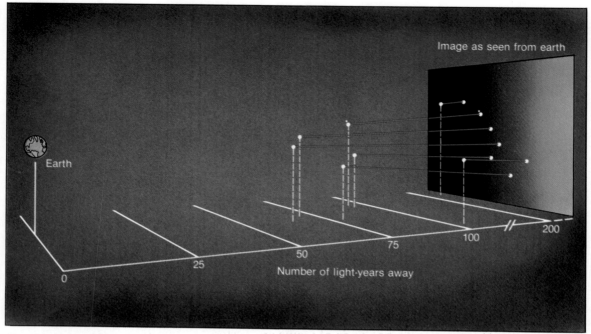

Image as seen from earth

Earth

25   50   75   100   200

0

Number of light-years away

Fig. 19-5 The stars that make up the Big Dipper are at different distances from the earth.

distance from the earth. This is so because of their great distance. But the stars are actually widely scattered at different distances from us. See Fig. 19-5.

You may see little connection between the appearances of the constellations and their names. This is because the names were invented by people in a different society thousands of years ago.

STARLIGHT

Every star is able to send some kind of message to the earth. The message is in the form of radiant energy, which can travel through space. For example, when you look at a star, you can see some of its radiant energy in the form of visible light. Stars also give off invisible forms of radiant energy. There are radio waves, X-rays, ultraviolet, infrared, and gamma rays. *Astronomers* study the stars by capturing some of their radiant energy with telescopes.

Some kinds of telescopes can be used on the earth's surface. The most familiar are the telescopes that use lenses or special mirrors. These telescopes collect and magnify the visible light from stars. Radio telescopes capture and study radio waves that reach the earth's surface. Infrared rays can also be detected by ground-based telescopes. However, X-rays, ultraviolet rays, and gamma rays from stars are mostly blocked by the earth's atmosphere. Thus the telescopes that capture these forms of energy must do their observing from satellites or space vehicles high above the atmosphere.

All of the kinds of radiant energy sent out by stars carry information about the stars' atoms. The light from a star can be compared to music that is made up of sounds produced by different instruments. Each instrument adds its own sound to the music. In the same way, the different atoms in a star add a characteristic part to the energy given off by the star. Scientists study starlight by first collecting it with a telescope. Then a special instrument on the telescope, called a *spectroscope,* separates the light into its characteristic parts. These parts are then recorded on photographic film or in computers for later study.

By studying the records made by spectroscopes, astronomers are able to learn about what kinds of atoms are in stars. All scientific knowledge about stars comes from the messages that can be read in starlight.

HOW FAR AWAY IS A STAR?

When looking at the night sky, you can see that not all stars have the same brightness. Actually, some of the brightest objects seen are planets. Venus, Mars, Jupiter, and Saturn are easily seen without a telescope. Planets can be sorted out from stars by observing their positions from night to night. Since stars are much farther away from the earth than the planets, stars appear to keep the same position among other stars. However, planets are much closer. Thus they appear to move against the background of stars.

Some close stars actually do seem to move very slightly when observed every six months. As the earth moves in its orbit, these stars seem to shift their positions in relation to the more distant stars. This is similar to the way a nearby tree appears to move against the distant background when viewed from a moving car. See Fig. 19–6. Trees that are farther away appear to move less against the distant background. This apparent motion can be used to give an idea of how far away the trees are.

Reinforcement: Planets can be distinguished from stars by their lack of "twinkling." This is because the stars, being much farther away, are point sources of light. Planets are close enough to be seen as discs of light.

Enrichment: The star closest to the sun is Alpha Centauri. It is 4.3 light-years away. It is actually a triple star system with one star, called Proxima Centauri, slightly closer than the other two. It would take a spacecraft 850,000 years to reach Alpha Centauri at the same average speed traveled by the *Apollo* spacecraft on their trips to the moon.

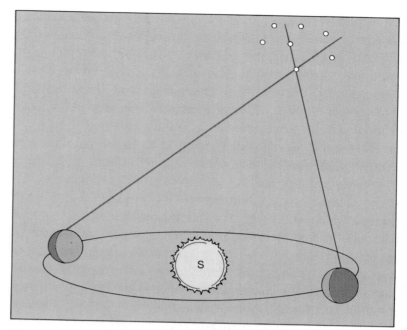

Fig. 19–6 *Close stars seem to shift their position when seen from opposite sides of the earth's orbit.*

This apparent motion is called **parallax** (**par**-uh-laks). By using *parallax*, the distances to nearby stars can be measured.

At first astronomers used kilometers to measure how far from earth objects in space were. But as they began to measure the distances to stars, the numbers became too large to use easily. For example, the distance of the closest star to the earth, other than the sun, is 4×10^{13} (40 trillion) kilometers. So astronomers decided to use another unit, the **light-year,** to measure large distances. A *light-year* is the distance light travels in one year. Light travels about 300,000 kilometers per second. To figure out how far light would travel in one year, you have to find out how many seconds there are in one year. You can do this by multiplying 60 (the number of seconds in one minute) by 60 (the number of minutes in one hour) by 24 (the number of hours in one day) by 365 (the number of days in one year). This gives you the number of seconds in one year. The number of seconds in one year can then be multiplied by the speed of light to give more than 9×10^{12}, or 9 trillion, kilometers. Thus it is easier to say that the nearest star is 4 light-years away than to write out the number of kilometers.

Scientists use other methods to measure the distances to stars that are far away. One method makes use of the brightness of the star. Have you ever judged the distance to a street light by

526

how bright it looked? If you have, you were estimating distance using brightness. The actual amount of light a star sends out is called the **absolute magnitude** (**ab**-suh-loot **mag**-nih-tood). Stars have different *absolute magnitudes*. Some are 10,000 times brighter than the sun. Others are only ten-thousandths (0.0001) as bright as the sun. Once the absolute magnitude of a star is known, its distance can be found by comparing its absolute magnitude to its brightness as actually seen from the earth. You might judge the distance to a street light by making the same kind of comparison.

Astronomers have different ways of finding the absolute magnitudes of stars. One of the best ways makes use of a kind of star whose brightness changes during periods of one day or up to many days. The length of time it takes for these stars to change their brightness is related to their absolute magnitudes. Once the absolute magnitude of these stars is found, the distance to them, and others close to them, can be found. This is done by comparing their brightness, as actually seen from the earth, to their absolute magnitudes. For example, a star that has a high absolute magnitude may appear dim. This means that the star is probably very far away.

SUMMARY

Anyone who observes the night sky notices that there are patterns in the arrangement of the stars. These patterns are now used as guides to locate the positions of stars. Detailed information about stars comes mainly from the energy that they send out into space. Observing closer stars from different points in the earth's orbit can show their distance from the earth. Stars farther away show a change in brightness that can be used to determine their distance from earth.

QUESTIONS

Use complete sentences to write your answers.

1. Why is knowing the location of some constellations useful?
2. How is information gained about stars?
3. What various kinds of scientific information can we get from starlight?
4. List two methods used to measure the distance from earth to stars. Which one can be used only for close stars?

Absolute magnitude A measure of the actual brightness of a star.

Word Study: Magnitude: *magnitudo* (L.), greatness.

1. Knowing the location of a constellation helps to locate and identify individual stars.
2. By studying the radiant energy given off by stars, information is gained.
3. The kinds of atoms that make up a star and the temperature of the star can be learned from starlight.
4. Parallax and star brightness. Parallax can be used only for close stars.

INVESTIGATION

THE PARALLAX EFFECT

PURPOSE: To find out how the parallax effect helps us judge distance.

MATERIALS:

pencil	modeling clay
chalk	ruler
chalkboard	

PROCEDURE:

A. Hold your pencil at arms length with the eraser end up. Close your right eye and, using only your left eye, line up the pencil with some object on the far wall of the room. Without moving the pencil or your head, open your right eye and close your left eye.

   1. Describe what happened to the position of the pencil compared to the object on the far wall.

B. What happened in step A is called the "parallax effect." It is the basis of our ability to tell distances with just our eyes. Hold the pencil again as in step A. While keeping your right eye closed, move your head to the right. Do not move the pencil.

   2. In what way is the effect like that in step A?

C. Copy the following table.

Distance of Pencil from Eye (cm)	First Eye (Number)	Second Eye (Number)
30		
60		
90		

D. Draw 10 vertical lines on the chalkboard 10 cm apart. From left to right, number the lines from 1 to 10.

E. To see the parallax effect, one must observe a near object (pencil) against a distant background (chalkboard) by looking at it from two different locations (each eye). Place the pencil, pointed end up, on the table between you and the lines drawn on the chalkboard or wall. Anchor the pencil in clay so it does not move. Hold your head level, 30 centimeters from the pencil. Using one eye only, sight the point of the pencil so it lines up with one of the numbered lines on the chalkboard or wall. Then, without moving your head, open the other eye and close the first eye. Note the number with which the point of the pencil now lines up. Estimate fractions between numbers. Record these two numbers in your table.

F. Move your head 60 centimeters from the pencil and repeat step E. Record the numbers in your table.

G. Move your head 90 centimeters from the pencil and repeat step E. Record the numbers in your table.

CONCLUSIONS:

1. How did moving farther from the pencil affect the parallax?
2. Our eyes judge distances to objects automatically until the objects are far from us. How could we use parallax to judge distances that are too far for just our eyes to judge? (Hint: See step B.)
3. What conditions are necessary to be able to judge distance by parallax?

19-2. The Life of a Star

At the end of this section you will be able to:

- ☐ Explain why stars have different colors.
- ☐ Relate the color of a star to its absolute magnitude and surface temperature.
- ☐ List the stages in the life history of a typical star.

Circling high above the atmosphere, the space telescope will have a much clearer view of the universe than telescopes on the ground. See Fig. 19-7. Outside the atmosphere, the space telescope will probe seven times deeper into space than before.

Fig. 19-7 *The orbiting space telescope can see objects ten times more clearly and fifty times farther away than telescopes on the ground.*

STAR COLORS

The messages carried by the light from stars have shown scientists that there are many kinds of stars. Some stars are larger than the earth's orbit around the sun. Others are smaller than the earth. There are stars made of gas thinner than air and others that are harder than diamond. One kind of star repeatedly blows up like a balloon, then shrinks, and blows up again, in an endless cycle. One difference among stars is their color. They appear at night in a wide variety of colors. This is because the color of starlight is related to the temperature on the star's surface. For example, you have probably seen how an iron poker held in a fire changes color as it gets hot. As a piece of iron is heated, it first becomes red. After further heating, the

Reinforcement: Compare the study of stars from the ground to the study of birds from the bottom of a lake. The atmosphere distorts and absorbs light waves very much as water does. From space, the space telescope can be used to see clear images of stars.

iron glows with a yellow light. Finally, it becomes a brilliant blue-white. In the same way, the color of a star shows its surface temperature.

The coolest stars have surface temperatures of about 3,000 K or higher. Astronomers have discovered that there is also a relationship between the star's surface temperature and its brightness, or absolute magnitude. Generally, the higher the temperature of the star, the higher the absolute magnitude.

TYPES OF STARS

When stars are plotted on a chart according to their surface temperature and absolute magnitude, they fall into three main groups. See Fig. 19–8. Most stars fall into a narrow band on the chart. The band begins with hot stars at the upper left and ends with cool stars at the lower right. Since most known stars fall into this group, they are called the *main sequence* (**see**-kwens) stars. A second group of stars appears at the upper right of the chart. These are bright stars even though their red color shows low surface temperatures. Their brightness results from their great size. The stars in this group are called *red giants*. One red giant is so large that, if it were placed at the center of the solar system, its surface would reach past Mars. A third group of stars appears at the bottom of the chart. These are called *white*

Enrichment: The chart is called an H–R diagram after Einar Hertzsprung and Henry Russell, who studied the relationship between absolute magnitude and star temperature.

Fig. 19-8 Stars are grouped in this chart according to their surface temperature and absolute magnitude.

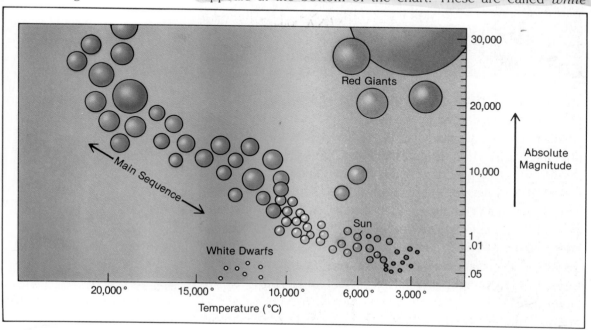

dwarfs. They are very hot and produce white light. But they are very small and therefore not very bright. Some white dwarfs are only the size of the earth.

LIFE OF STARS

Most stars have life histories that are measured in billions of years. The groups of stars shown on the chart in Fig. 19–8 have helped astronomers piece together the life history of stars. The stars in each group are believed to represent different stages in the life history of typical stars. A star begins as a cloud of gas and dust. Such clouds are common in the space between stars. Many such clouds have been observed. See Fig. 19–9. The force of gravity pulls the clouds of dust and gas together. As the center of the cloud shrinks, it becomes hot. Finally, the temperature rises high enough to begin the process of fusion. In fusion, hydrogen nuclei are brought together to make helium nuclei, giving off energy in the process. The star begins the first stage of life as its interior furnace begins to supply energy. The heat produced prevents the star from shrinking any more. The star becomes a main sequence star. It shines with a brilliant blue-white light.

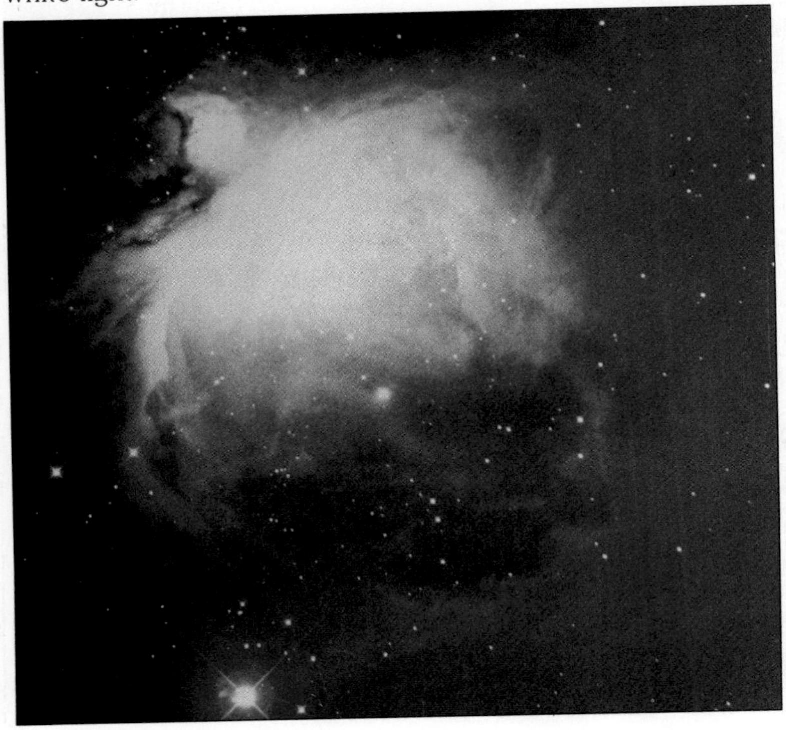

Enrichment: White dwarfs have densities greater than that of any known terrestrial material. A white dwarf the size of the earth would have a density nearly the same as the sun. Some white dwarfs may be made of crystalline carbon, the same as diamond.

Fig. 19–9 A cloud of gas and dust seen in the constellation Orion.

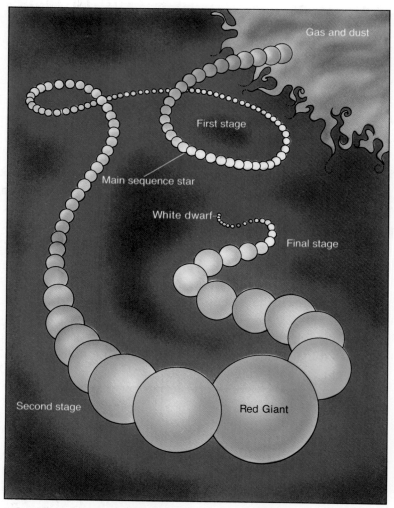

Fig. 19-10 *Stages in the life history of a typical star.*

In this stage the young star has a large supply of hydrogen. See Fig. 19-10. For a long period of time, it shines with a steady light. The star will be in this stage longer than in any other stage in its life. Thus, most known stars are in the main sequence stage. The sun, for example, is a main sequence star.

Eventually, a star will have used up nearly all of its hydrogen. Then its gravity starts to make it shrink. This makes the interior even hotter. It becomes possible for the helium nuclei to join to make heavier elements. These reactions give a new burst of energy to the star. It begins the second stage in its life history. See Fig. 19-10. The star expands to giant size. But its large size causes the surface temperature to drop. It becomes one of the red giants seen on the chart in Fig. 19-8.

When the helium fuel is also used up, the star shrinks again. Now the star enters its third and final stage. See Fig. 19–10. Its smaller size allows more energy to reach its surface, and it shines with a white light. A white dwarf seen on the chart in Fig. 19–8, page 530, has been formed. White dwarfs are very old stars that have largely used up their supply of nuclear fuel. They slowly grow colder until they burn out. They finally die as black dwarfs, like the ashes from a fire.

The larger the size of a star at birth, the faster it goes through the three stages. This is because the force of gravity is much greater. Gravity causes the star to shrink faster. For example, the star Sirius (**seer**-ee-us) is 2.5 times the mass of the sun. It is going through its life stages about 15 times faster than the sun. The sun probably has about five billion years to go before it will become a red giant. At that time it will expand so much that the solar system will be destroyed.

UNUSUAL STARS

Very large stars have a different life story. Stars having more than about four times the mass of the sun use up their hydrogen fuel very quickly. These larger stars explode at the time when a smaller star would only be through half its life. The explosion can become as bright as 10,000 suns. Such a flare-up in the sky is called a **supernova** (soo-pur-**noe**-vuh). Among the stars seen from the earth, a *supernova* is observed about once every 140 years. These stars collapse after the explosion. The great force crushes the atoms, leaving only neutrons. The result is a *neutron star*. A neutron star is about 15 kilometers in diameter and is very dense. One spoonful of its matter would weigh more than 9×10^7 million metric tons on the earth. If the earth were squeezed down to the same density, it would fit into a baseball park.

Neutron stars are also known as *pulsars*. At first, scientists thought that pulsars were an alien radio source. They give off rapid pulses or bursts of radio waves, light, and X-rays. A neutron star can spin on its axis as many as thirty times each second. A source of energy in its surface sends out a stream of radiation in a pattern similar to that of the rotating light in a lighthouse. When the earth is in the path of one of these streams of radiation, radio telescopes can pick up the signals as rapid pulses.

Enrichment: A supernova observed in A.D. 1054 was brilliant enough to be seen during the day. A hot, expanding cloud of gas known as the Crab Nebula now remains as a remnant of that supernova.

Supernova A large star that suddenly explodes.

Astronomers believe that some very large stars may collapse with such force that they pass beyond the stage of being a neutron star. They become *black holes*. A black hole is an object so small and dense that its gravity will allow nothing, not even light to escape from it. Because no energy of any kind can come from a black hole, there is no way to observe it directly. If you tried to take a flash picture of a black hole the light would only disappear down the hole. However. astronomers believe that they may be able to locate some black holes by the effect they have on nearby objects. Black holes will swallow anything that comes near. For example, a black hole close to a star would strip matter off its neighbor. The material lost from the star would then give off energy such as X-rays as it disappeared down the black hole. Powerful X-ray sources are believed to be the result of black holes taking in matter from nearby stars.

One of the most likely possibilities for a black hole is Cygnus X–1. It was the first X-ray source discovered in the constellation Cygnus. Cygnus X–1 is about six to eight times the mass of the sun and about 8,000 light-years from the earth. Scientists continue to search for other possible black holes.

SUMMARY

The color of a star is determined by its surface temperature. Red stars are coolest, while blue stars are the hottest. Sorting out stars according to their color and brightness is the key to learning all about their life histories. Most stars move through several stages, during which they change in color and size. They end up as small, black dwarfs. Large stars have a different fate. They may end up as a very small dense object with unusual properties.

QUESTIONS

Use complete sentences to write your answers.
1. What determines the color of a star?
2. What two properties of stars help us learn their relative age?
3. Starting right after a star is formed from gas and dust, list three changes it goes through before it becomes a black dwarf.

See teacher's commentary for teaching hints, safety, and answers to questions.

STAR COLORS AND TEMPERATURES

PURPOSE: To use an experiment to infer why a star's color shows its temperature.

MATERIALS:

masking tape	miniature socket
plastic ruler (30 cm)	flashlight battery
No. 30 uncoated	(size D)
steel wire (3 m)	black paper
1.5 V flashlight bulb	grating

PROCEDURE:

A. Cover each edge of the plastic ruler with masking tape. Wind about 3 meters of the wire around the ruler so the turns do not touch. Tape each end of the coil in place, leaving about five centimeters free on one end. Put the bulb in the miniature socket.

B. Tape one socket wire to the free end of the coil. Tape the other wire to the base of the flashlight battery. Put the setup on the black paper. See Fig. 19–11 *(top)*.

C. Like stars, electric light bulbs radiate a continuous spectrum. You will control the amount of current flow and the temperature of the bulb by touching the center post of the battery to different places on the coil of wire. See Fig. 19–11. Try this now.

   1. When does the bulb glow brightest?

D. From the brightest light, in about five moves find where the bulb just glows. At the same time, your partner will watch the light through a grating that breaks up the spectrum of light. To block stray light, cup your hand around your eye.

   2. Are all the color bands visible when the light is brightest? Which colors seem to fade most?

   3. Looking closely at the bulb, describe the spectrum when the bulb is dim.

E. Betelgeuse (**bet**-'l-jooz), Rigel (**rye**-jul), and Capella (kuh-**pell**-uh) are three stars in the winter constellation Orion (oh-**rye**-un). They appear bright red, bluish, and yellow-white, respectively, as shown in the bar graphs. See Fig. 19–11 *(bottom)*.

   4. Which star gives the brightest violet-blue (v-b) color band?

   5. Which star has a yellow-orange band brighter than its other colors?

CONCLUSIONS:

   1. Based on the information you obtained from the light-bulb experiment, which star in step E is the hottest star?

   2. Why is Capella yellow-white?

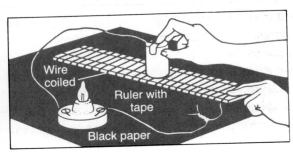

Wire coiled

Ruler with tape

Black paper

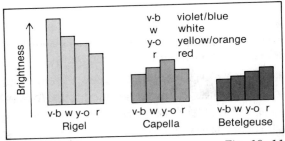

v-b	violet/blue
w	white
y-o	yellow/orange
r	red

v-b w y-o r
Rigel

v-b w y-o r
Capella

v-b w y-o r
Betelgeuse

Brightness

Fig. 19–11

19-3. Galaxies

At the end of this section you will be able to:

☐ Explain what a galaxy is.

☐ Describe the structure of the Milky Way galaxy.

☐ Give three theories for the origin of the universe.

On a clear moonless night you can see a bright, thin cloud stretching across the sky. See Fig. 19–12. This cloud of light actually consists of billions of stars. It is so large that it would take light 100,000 years to travel from one edge to the other.

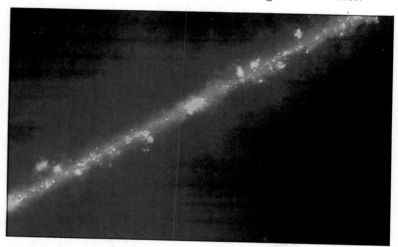

Fig. 19–12 The Milky Way as seen by the Infrared Astronomy Satellite (IRAS).

Milky Way A faint band of light seen spread across the night sky, produced by the stars in our galaxy.

Galaxy A huge system of stars held together by gravity.

Word Study: Galaxy: *gala* (Gr.), milk.

THE MILKY WAY

The faint band of light we see on any clear, dark night is called the **Milky Way.** With binoculars or a small telescope, the *Milky Way* can be seen to be made up of thousands of faint stars. The combined light of these stars makes the Milky Way appear to be a continuous streak across the sky. Why are so many stars seen in one part of the sky?

The sun and all other stars seen from the earth are part of a huge system of stars called a **galaxy.** If the *galaxy* were the size of North America, the stars would be like tiny specs, smaller than a thousandth of a centimeter, and 150 meters apart. The galaxy to which the sun belongs contains more than 100 billion stars. The stars are held in a galaxy by their gravity, which tends to pull the stars toward each other. The galaxy to which the sun belongs is usually called the Milky Way galaxy.

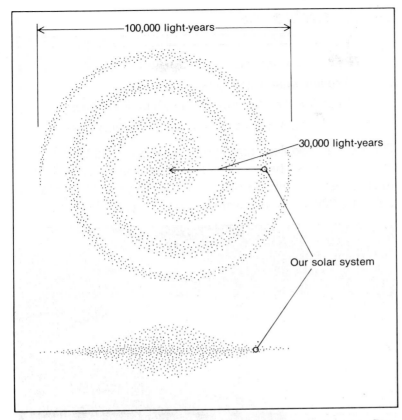

100,000 light-years

30,000 light-years

Our solar system

Fig. 19-13 Our solar system is near the edge of our galaxy, the Milky Way.

The galaxy is shaped like a disk with a bulge at its center. The solar system is located closer to the edge than to the center of the galaxy. Thus, looking toward the galaxy's center, we see many more stars than when looking elsewhere. See Fig. 19-13.

OUR GALAXY

On a very clear, dark night, if you look at the constellation of Andromeda, it is possible to see a small, fuzzy patch of light among the stars. A telescope shows that this is a galaxy about 2.2 million light-years away. See Fig. 19-14, page 538. It has a spiral shape that makes it look like a pinwheel. Scientists think that the Milky Way galaxy would also look like this if we could see it from far away.

Our galaxy is more than 100,000 light-years from edge to edge. It has a thick center that is about 10,000 light-years across. Surrounding the center is a flat disk that is divided into spiral arms. The sun is located in one of these spiral arms about 30,000 light-years from the center, as can be seen in Fig. 19-13.

Enrichment: The galaxy has rotated only 20 times since the solar system was formed.

Fig. 19-14 The Andromeda Galaxy.

The Milky Way galaxy is made up of about 200 billion stars. Most of these stars are located in the flat disk made up of the spiral arms. To us, the stars appear to be crowded together in the sky. However, this is a result of our looking at the stars many thousands of light-years into space. This makes the stars appear to be stacked together. Actually, the stars are very far apart. To get some idea of relative distances, imagine how far from each other a few baseballs would be if you scattered them across the entire United States. However, if seen from the moon, the baseballs would appear to be grouped together. There are 60 known stars within 17 light-years of the sun. The closest is 4.3 light-years away. Between the stars there are large amounts of gas and dust.

All the stars in the Milky Way galaxy revolve around its center, in the same way that the planets revolve around the sun. The sun takes about 230 million years to make one revolution around the galaxy.

THE UNIVERSE

Telescopes that look deep into space beyond our own galaxy show that the universe contains billions of other galaxies. Just as stars are seen to differ from each other, not all galaxies are alike. In addition to the spiral shape of the Milky Way and Andromeda galaxies, some galaxies are shaped like footballs or like spheres. Other galaxies have no general shape. All galaxies that can be observed have one thing in common. They are all moving away from each other at a very high speed. Scientists can tell that the galaxies are moving apart because of the information that can be read from their light. There is more red than there would be if the galaxies were not moving or were moving toward the earth. This is called the *red shift* of the starlight. The red shift in the light from the galaxies is caused when the light waves are spread out as the source moves away. You may have heard the same effect with sound waves that come from a moving source. For example, the sound of a siren from a moving car changes as the car moves past you. The sound waves are crowded together as the car approaches then spread apart as the car moves away. In the same way, light waves from a distant galaxy can show that it is moving away.

The red shift seems to show that the galaxies that are farthest away are moving at the highest speeds. If this is true, then the entire universe must be expanding. To get a clearer sense of this, think of the galaxies as dots painted on a balloon. The balloon expands as it is blown up. The dots are seen to move away from each other. In the same way, if the universe is expanding, then all of the galaxies will be seen to be moving away from each other.

One of the most puzzling of all objects seen by astronomers is the **quasar** (**kway**-zahr). *Quasars* seem to be young galaxies that are still forming. The amount of energy quasars produce is much greater than the energy given off by ordinary galaxies. Thus quasars are bright enough to be seen at great distances. Many quasars seem to be 10 to 15 billion light-years away. The light from quasars shows a very large red shift, indicating that they are moving away at high speeds.

HOW DID THE UNIVERSE BEGIN?

The idea that the universe is expanding has led scientists to the *Big Bang* theory. This theory says that at one time the entire universe was concentrated in a small volume. Then, about 15 to 20 billion years ago, there was a tremendous explosion that caused the universe to begin expanding. About one million years after the Big Bang, huge clouds of hydrogen and helium began to form. The clouds were the beginning of the galaxies. In time, stars started to grow in the galaxies. Within about 10 billion years after the Big Bang, all of the galaxies that are now seen had formed. Today, we see the results of that Big Bang as a rapidly expanding universe with galaxies moving away from us. Scientists have also found that all space is filled with weak radio waves. These radio waves are believed to be left by the great explosion that was the Big Bang.

Other theories have been developed to explain the universe. For example, the *steady-state* theory says that new matter is always being created in space. As the expanding universe carries the old stars away, new ones are created to replace them. Thus, according to the steady-state theory, the universe is in a state of balance, with no beginning or end. Another theory is the *oscillating universe* theory. *Oscillating* means "to swing back and forth." The oscillating universe theory says that gravity will finally stop the outward movement of the galaxies caused by the Big Bang. Then they will move back together, forming a hot mass, until another Big Bang sends them flying out again. This is repeated over and over, so the universe will continue to expand and shrink forever.

At the present time, most astronomers believe that the steady-state and oscillating universe theories cannot be proven. There is evidence that the universe started with an explosion and has continued to expand since that time. The way the universe may have begun and how it might end is a question that scientists still are trying to answer.

SUMMARY

Like most stars, the sun is part of a galaxy. The galaxy to which the sun belongs is shaped something like a saucer, with spiral arms reaching out from a central bulge. We see evidence of our galaxy in the form of a band of stars crossing the sky. The universe contains billions of other galaxies that are apparently moving apart as the universe expands.

QUESTIONS

Use complete sentences to write your answers.
1. What is a galaxy?
2. What causes the Milky Way we see in the night sky?
3. Describe the size and shape of the Milky Way galaxy.
4. Explain three theories for the origin of the universe.

1. A galaxy is a huge system of stars.
2. The Milky Way is caused by the light coming from thousands of faint stars seen as we look from the edge in toward the center of our galaxy.
3. The Milky Way is more than 100,000 light-years from edge to edge. It has a thick center that is about 10,000 light-years across. Surrounding the center is a flat disk that is divided into spiral arms.
4. The Big Bang theory says that the entire universe was at one time concentrated in a small volume. Then a tremendous explosion caused the universe to expand. The clouds of gas formed the galaxies. The steady-state theory says that, as old stars are carried away, new stars are created to replace them. The oscillating universe theory says that gravity will cause the universe to stop expanding and move back together. Then another Big Bang would occur.

TECHNOLOGY

SPACELAB

Spacelab 1 flew for ten days in November 1983. In that time, it produced information to fill 280,000 volumes of an encyclopedia. It will probably take ten years to analyze all the data. *Spacelab 1* was an important first because it not only took scientific experiments into space but also took the scientists with them. It was significant too as an international research effort joining the forces of the European Space Agency and NASA.

The specialists on *Spacelab 1* carried out 72 experiments designed by scientists in 11 countries. These included investigations into how the biological clocks of living things are affected by zero gravity; how zero gravity affects the human immunological system; and how space sickness is caused. *Spacelab 1* carried the Imaging Spectrometric Observatory (ISO), a set of unique instruments for studying the atmosphere.

Spacelab produced a new kind of scientist—one who is competent to perform experiments in biology, physics, chemistry, and astronomy in a very limited period of time. *Spacelab* produced a new space traveler, the payload specialist, who is not one of the crew. The mission also proved that people in space were extremely valuable. As equipment broke down, the scientists in space could repair it. A camera designed to produce extremely clear photographs of 2,590-square-kilometer regions of the earth jammed. This could have led to a major disappointment, since up to that time only 40 percent of the earth had been mapped well enough for resource planning. What *Spacelab* could do in weeks would normally take from 10 to 20 years of expensive photographing from planes. What happened? The camera was repaired on board and produced 1,000 important photographs.

Why are *Spacelab* experiments so important? Space can provide conditions that do not normally occur on earth, making some scientists and manufacturers wonder if certain things that would be impossible to make on earth might be built or developed in space. For example, one experiment was designed to test the possibility of manufacturing single, perfect crystals of silicon in the weightlessness of space. These crystals could be used to make faster and less expensive computer chips. On earth, under the influence of gravity, melting silicon is affected by currents of heat. The particles do not always cool evenly, producing impurities. Videotapes from *Spacelab* showed no evidence of heat currents on board. Styrofoam spheres, more uniform than any on earth, were produced. More research is needed to know whether mineral crystals produced in space are so free of defects that it is worth the tremendous cost of manufacturing in orbit. Orbiting space stations may someday produce synthetic minerals that must now be mined on earth. *Spacelab* is scheduled for four more flights before 1988. The scientific community awaits these opportunities for new experiments. Each of these flights offers another chance for increased international cooperation in scientific research.

¡COMPUTE!

SCIENCE INPUT

We can measure the distance to some nearby stars if we know the parallax angle. You learned in this chapter that the parallax is the apparent shift of position of a star when seen from different points in the earth's orbit. The parallax angle of a star is the angle it makes across the sky as seen from one of the points in the earth's orbit.

Using what is called the method of trigonometric parallax, we can calculate the distance to nearby stars. The formula is:

$$\text{DISTANCE TO STARS (light years)} = \frac{3.26}{\text{PARALLAX ANGLE (seconds)}}$$

Parallax angles used to be measured with photographs taken six months apart. Now new devices called measuring engines enable scientists to measure stars five times more distant than could be determined previously.

COMPUTER INPUT

The computer's ability to do mathematics quickly and accurately is a great help in doing long calculations. This ability represents the immense progress made in mathematics because of new technology. One of the first tools used to assist in the solution of mathematical problems was the abacus. The slide rule, invented in 1630, was a wonderful tool for mathematicians, engineers, and scientists because it was small, portable, and required only the application of human energy. Attempts to develop mechanical calculators, powered by hand, began in the 17th century. The first practical device for calculation, however, wasn't invented until 1889. It only did multiplication. The electronic revolution brought us the hand calculator, which gave us speed as well as convenience in mathematical work. The full range of the computer's abilities, though, can make the hand calculator seem like an ancient tool.

In this exercise, the speed of a computer's mathematical capabilities will be matched against your own and those of a hand calculator.

WHAT TO DO

You will need paper and pencil, a watch or clock with a secondhand, hand calculator, and a computer. On a separate piece of paper, copy the data chart below.

Using the formula given above and the data for the parallax angle of some nearby stars, calculate with pencil and paper the distance from the earth to those stars. Do the same calculations using a hand calculator. Record the time it took to do each of these tasks.

Enter Program Parallax. (Remember to check your manual for the correct command for a clear screen statement.) Save the program on tape or disk and then run it. Record the time it took for the computer to do the calculations.

DATA CHART

Star's Name	Parallax Angle	Distance
Sirius	.379	
Rigel	.005	
Alpha Centauri	.760	
Altair	.198	
Barnard's Star	.544	
Procyon	.294	
Wolf 359	.402	
Beta Cassiopeia	.073	

GLOSSARY

DISK DRIVE   That section of the hardware that "reads" the program stored on disk. Some software is in cassette form, not on disk. In that case a drive is not necessary.

PIXEL   A picture element. A pixel is a point or tiny box on the screen. The screen is composed of vertical and horizontal pixels.

RESOLUTION   The number of pixels on a TV screen or monitor. High resolution means there are a large number of pixels. Low resolution means there are fewer. The higher the resolution, the sharper the image or display.

PROGRAM

```
100   REM PARALLAX
105   REM REPLACE THIS STATEMENT
      WITH CLEAR SCREEN STATEMENT
110   INPUT "WHAT IS THE PARALLAX
      ANGLE? ";P$
120   P = VAL(P$)
130   IF P = 0 THEN GOTO 110
135   REM FORMULA FOR PARALLAX
140   D = 3.26/P
150   PRINT "STAR DISTANCE = ";D;"
      LIGHT YEARS"
160   END
```

PROGRAM NOTES

When calculating your time, be sure to record the time it took to enter the program and to correct any errors. After comparing calculation times among the three methods, add the entry and correction time to the computer's time. How does that change the differences among the three methods? Do you think a computer is always the most efficient method for mathematical calculation? In what situations might you choose to use a hand calculator instead of a computer?

BITS OF INFORMATION

You don't have to be a seasoned programmer to become involved in the computer world. Many young adults have even won computer contests without programming. Two 17-year-old students from Harbor Spring, Michigan, won a five-day, all-expense trip to Washington, D.C., and $500 in cash. Their idea, which they carried out, was to give senior citizens computer lessons. Another 11-year-old from Dubuque, Iowa, won $4,000 from a food company for her idea for a nutrition program. If you are interested in computer contests, keep watch for news through your school computer club, computer magazines, computer retail stores, and manufacturers' advertisements.

Your local library may have a computer newsboard. If it doesn't, you might ask the librarian about the possibility of adding such a bulletin board, to announce computer events that are open to people in your community or state.

CHAPTER REVIEW

VOCABULARY

On a separate piece of paper, match each term with the number of the phrase that best explains it. Use each term only once.

absolute magnitude	black hole	main sequence	supernova
light-year	Ursa Major	Milky Way	constellation
parallax	quasar	Polaris	galaxy

1. Commonly called the North Star.
2. Any particular pattern of stars.
3. A small, starlike object, very far from the earth, that produces huge amounts of energy.
4. A unit used in measuring distances to stars.
5. The shift in position of a star when seen from different parts of the earth's orbit.
6. A measure of the actual brightness of a star.
7. Thought to be the result of the collapse of a very large star.
8. Used to help locate the North Star.
9. A large star that suddenly explodes.
10. The group to which most stars belong.
11. A faint band of light running across the sky.
12. A huge system of stars.

QUESTIONS

Give brief but complete answers to each of the following questions. Unless otherwise indicated, use complete sentences to write your answers.
1. How did most of the constellations get their names?
2. Describe how you would locate the North Star.
3. To what constellation does the North Star belong?
4. List three kinds of radiant energy that telescopes on the earth can receive.
5. Where must telescopes be placed to receive all of the different kinds of radiant energy given off by stars?
6. Describe two kinds of scientific information astronomers get from starlight.
7. How can you tell whether an object is a star or a planet?
8. What method is used to measure the distance from earth to faraway stars?

9. Why can't a red star be identified as a red giant, considering only its surface temperature?
10. Describe how stars are born, according to scientific theory.
11. According to theory, how does a star become a red giant?
12. How is a supernova related to a neutron star?
13. Why do we see the Milky Way as a faint band of light running across the sky?
14. Locate and describe a galaxy that astronomers believe looks like the Milky Way galaxy.
15. What evidence do scientists have for believing that the universe is presently expanding?

APPLYING SCIENCE

1. Observe the night sky and locate the North Star. Extend a slightly bent line from the pointer stars of the Big Dipper beyond the North Star until you see a constellation of five stars that form a *W*. This constellation is Cassiopeia. Draw a sketch to show the relative positions of the Big Dipper, North Star, and Cassiopeia.
2. Look in a newspaper, an almanac, or the current issue of *Sky and Telescope* to find which planets are visible. Locate any one of the planets in the night sky. View it for several nights at the same time and explain how you know it to be a planet.
3. Explain why the moon appears to move with you as you ride along in a car, while the trees, street lights, buildings, etc., all move past you.

BIBLIOGRAPHY

Couper, Heather, and Terence Murtach. *Heavens Above: A Beginner's Guide to the Universe*. New York: Watts, 1981.

Gold, Michael. "The Cosmos Through Infrared Eyes." *Science 84*, March 1984.

Kunitzen, Paul. "How We Got Our Arabic Star Names." *Sky and Telescope*, January 1983.

Menzel, Donald H., and Hay M. Pasachoff. *A Field Guide to the Stars and Planets*. Boston: Houghton Mifflin, 1983.

National Geographic Editors. *Our Universe*. Washington, DC: National Geographic Society, 1980.

Whitney, Charles. *Whitney's Star Finder*. New York: Knopf, 1981.

GLOSSARY

A

Abrade To wear away.

Absolute magnitude A measure of the actual brightness of a star.

Abyssal plain A flat, wide deposit of sediments that covers the deeper parts of the ocean basin floor.

Air mass A large body of air that has taken on the temperature and humidity of a part of the earth's surface.

Anemometer An instrument used to measure wind speed.

Anticline Rock layers folded upward.

Anticyclone A large mass of air spinning out of a high pressure area.

Asteroid A smaller body in the solar system, usually found in an orbit between Mars and Jupiter.

Asthenosphere The partly molten upper layer of the mantle.

Astronomy The branch of earth science that studies the universe.

Atmosphere The layer of gases that make up the outside of the earth.

Atomic particles Small particles that make up atoms. Protons, neutrons, and electrons are three kinds of atomic particles.

Auroras The northern and southern lights produced when charged particles from the sun are trapped in the earth's magnetic field.

Autumnal equinox The time of the year, about September 22, when the earth's axis is tilted neither toward nor away from the sun.

Axis The imaginary line through the center of the earth on which the earth rotates.

B

Barometer An instrument used to measure atmospheric pressure.

Batholiths Very large bodies of igneous rock that are formed underground.

Beaches Parts of the shoreline covered with sand or small rocks.

Biomass Plant materials that are used to produce energy.

Breaker A wave whose top has curled forward and crashed down against the shore.

C

Cavern A series of underground passages made when water dissolves limestone.

Celsius The temperature scale used most in science. It is based on the freezing point of water, 0°C, and the boiling point of water, 100°C.

Channel The path of a stream of water.

Chemical weathering The ways rock breaks down by changing some of the materials in the rock.

Chromosphere The middle layer in the sun's atmosphere which gives off a faint red light.

Climate The average weather at a particular place.

Comet An icy body that comes from the edge of the solar system.

Compound A substance made up of two or more kinds of atoms joined together.

Condensation The process by which a gas is changed into a liquid.

Conservation Saving the natural resources by controlling how they are used.

Constellation A group of stars in a particular pattern.

Continental glacier A sheet or mass of ice that covers a large area of land.

Continental rise A gently sloping area at the base of the continental slope.

Continental shelf The flat submerged edge of a continent.

Continental slope The sloping part of the sea floor that marks the boundary between the sea floor and continental shelf.

Contour line A line drawn on a topographic map joining points having the same elevation.

Convection The movement of material caused by differences in its temperature.

Convection Movement of air caused by temperature differences.

Convection cell A complete circle of moving air caused by temperature differences.

Coral A small animal that lives in great numbers in the warm shallow parts of the sea.

Core The center layer of the earth made up of an inner and an outer part having a radius of about 3,500 km.

Coriolis effect Bending of the paths of moving objects as a result of the earth's rotation.

Corona The outermost layer of the sun.

Creep A slow downhill movement of loose rock or soil.

Crosscutting The principle that says faults or magma flows cutting through other rocks must be younger than the rocks they cut.

Crust The thin, solid outer layer of the earth.

Crystal A mineral form with a definite shape that results from the arrangement of its atoms or similar units.

Current Water moving in a particular direction.

Cyclone A kind of storm that is formed along the polar front.

D

Delta A deposit formed at the mouth of a river or stream.

Density The amount of mass in a given volume of a substance.

Dew point The temperature at which air becomes saturated with water vapor.

Diastrophism Movement of solid rock in the crust.

Differentiation The process by which planets became separated into layers.

Dike Body of igneous rock formed underground in cracks that cut across layers of existing rock.

Dome mountains Mountains formed when a part of the crust is lifted by rising magma.

Dune A deposit of wind-carried materials, usually sand.

E

Eclipse The earth or moon passing through the shadow of the other.

Electron shell An area where a certain number of electrons move at a definite distance from the nucleus.

Element A substance made up of only one kind of atom.

Ellipse The oval shape of the orbits followed by members of the solar system.

Epicenter The point on the earth's surface directly above the focus of an earthquake.

Era A time period in the earth's history. Each era is separated by major changes in the crust and changes in living things.

Erosion All the processes that cause pieces of weathered rock to be carried away.

Evaporation The process of water changing from a liquid to a gas at normal temperatures.

Exfoliation The peeling off of scales or flakes or rock as a result of weathering.

Experiment A setup to test a particular hypothesis.

F

Fault The result of movement of rock along either side of a crack in the earth's crust.

Fault-block mountains Mountains formed when large blocks of rock are tilted over.

Fetch The distance over which the wind blows steadily producing a group of waves.

Fission The splitting of the nuclei of certain kinds of atoms.

Focus The point of origin of an earthquake.

Folding Bending of rocks under steady pressure without breaking.

Fossil Any evidence of or the remains of a plant or animal that lived in the past.

Fossil fuel Mineral fuel found in the earth's crust.

Front The boundary separating two air masses.

Fusion The joining together of nuclei of small atoms to produce larger atoms with the release of energy.

G

Galaxy A huge system of stars held together by gravity.

Gem An attractive or rare mineral substance that is hard enough to be cut and polished.

Geologic column An arrangement showing rock layers in the order in which they were formed.

Geology The branch of earth science that studies the solid part of the earth.

Geothermal power Energy obtained from heat in the earth's crust.

Glacier A large body of moving ice and snow.

Greenhouse effect The name given to the method by which the atmosphere traps energy from the sun.

Ground water The water that soaks into the ground, filling openings between rocks and soil particles.

Guyot A submerged volcano whose top has been flattened by wave erosion before it sank below the surface.

H

Half-life The length of time taken for half a given amount of a radioactive substance to change into a new substance.

Horizon Any separate layer of soil.

Humidity The amount of water vapor in the air.

Humus Material in soil that comes from remains of plants and animals.

Hurricane A small but intense cyclonic storm formed over warm parts of the sea.

Hydrocarbons The principal substances found in petroleum. They contain only hydrogen and carbon.

Hydroelectric power Electricity produced when water moves downhill.

Hydrosphere All the bodies of water on the earth's surface.

Hypothesis A statement that explains a pattern in observations.

I

Igneous rock Rock formed when molten material cools.

Index fossils The remains of a plant or animal that lived during only a small part of the earth's history.

Interface A zone where any of the three parts of the earth (the lithosphere, hydrosphere, and atmosphere) come together.

Invertebrates Animals without backbones.

Ion An atom that has become electrically charged as a result of the loss or gain of electrons.

Ionosphere A region of the upper atmosphere containing many electrically charged particles.

Isostasy The state of balance between different parts of the lightweight crust as it floats on the dense mantle.

Isotherm A line on a map joining locations that record the same temperature.

J

Joint A crack in solid rock with no movement along the sides of the fracture.

K

Key bed A rock layer that is easily recognized and found over a large area.

Kilogram An SI unit of mass equal to 1,000 grams.

L

Landslide A sudden movement of large amounts of loose material down a slope.

Latitude The distance in degrees north or south of the equator.

Lava Magma that has flowed out onto the earth's surface.

Law of superposition The principle that, in a group of rock layers, the top layer is generally the youngest and the bottom layer the oldest.

Legend Information on a map that explains the meaning of each symbol used on a map.

Light-year The distance light travels in one year.

Liter An SI unit of volume. A liter is equal to 1.05 quarts.

Lithosphere The earth's outer layer averaging about 100 km in thickness.

Loess A deposit of windblown dust.

Longitude The distance in degrees east or west of the prime meridian.

Longshore current Movement of water parallel with the shoreline.

M

Magma Molten rock beneath the earth's surface.

Mantle The layer of earth below the crust having a thickness of about 2,870 km.

Map projection The curved surface of the earth shown on the flat surface of a map.

Maria Lava plains that are seen as dark areas on the moon's surface.

Meander A wide curve in the channel of an old river.

Measurement An observation made by counting something.

Mesosphere The atmospheric layer above the stratosphere extending from about 50 km to 85 km, where temperature decreases with increasing altitude.

Metamorphic rock Rock that has changed by the action of heat or pressure without melting.

Meteor A bright streak of light in the sky caused by rock material burning up in the atmosphere.

Meteorology The branch of earth science that studies the atmosphere.

Meter The basic SI unit of length. A meter is equal to 39.4 inches.

Mid-ocean ridge A chain of underwater mountains found on the floor of the oceans.

Milky Way A faint band of light seen spread

across the night sky produced by the stars in our galaxy.

Mineral A solid, naturally forming substance that is found in the earth and that always has the same properties.

Mineral deposit A body of ore.

Molecule The smallest part of a substance with the properties of that substance.

Moon phase The changing appearance of the moon as seen from the earth.

Moraine A layer of till.

N

Natural resource Any of the substances from the earth that can be used in some way.

Nodules Lumps of sediment found on the sea floor that contain valuable minerals.

Nuclear energy Energy that is produced by change in the nuclei of atoms.

Nucleus The central part of an atom that contains the protons and neutrons.

O

Observations Any information that we gather by using our senses.

Oceanography The branch of earth science that studies the oceans and the ocean floor.

Ooze A kind of sea floor sediment formed from remains of living things.

Ore A rock that contains useful minerals and can be mined at a profit.

Organic Coming from any living thing.

P

Parallax The apparent shift of position of a star when seen from different parts of the earth's orbit.

Permeability A measure of the speed with which water can move between the pore spaces in a rock.

Petrify To replace the material in the remains of a living thing with hardened mineral matter.

Petroleum A fossil fuel that is usually a liquid, that comes from the decay of ancient life in warm, shallow seas.

Photosphere A thin outer layer of gas in the sun which produces visible light.

Physical weathering The ways rock break into smaller pieces without any change in the materials in the rock.

Plankton Small plants and animals that float near the surface of the sea.

Plate tectonics The theory that the earth's crust is made up of moving plates.

Plateau Large raised areas of level land.

Pluton Any large intruded rock mass formed from magma.

Polar front The boundary where cold polar air meets warm air.

Pollution The adding of harmful substances to the environment.

Porosity The amount of the open spaces or pores in soil or rock.

Precipitation Water that falls from clouds, usually as rain or snow.

Prominence A huge cloud of gases that shoot out from areas of sunspots.

Q

Quasar A small object very far away which produces a huge amount of energy.

R

Radiant energy A form of energy that can travel through space.

Radioactive Describes a substance that changes over a period of time into a completely new substance.

Rays Light-colored streaks leading out from some craters on the moon.

Recycling The process of recovering a natural resource and using it again.

Red clay A kind of sediment found on the floor of most of the open sea.

Relative age A method of telling if one thing is older than another. It does not give the exact age.

Relative humidity The amount of water vapor in the air compared to the amount of water vapor the air could hold at that temperature.

Rift valley A deep, narrow valley formed where the earth's crust separates.

Rock A piece of the earth that is usually made up of two or more minerals mixed together.

Rock cycle The endless process by which rocks are formed, destroyed, and formed again in the earth's crust.

Runoff Water running downhill over the land surfaces.

S

Salinity The number of grams of dissolved salts in 100 g of sea water.

Satellite A body that moves in an orbit around a larger body.

Scale The relationship between the distance on a map and the actual distance on the earth's surface.

Scientific method A guide scientists use in solving problems.

Scientific theory A general explanation that includes many hypotheses that have been tested and found to be correct.

Sea cave A hollow space formed by waves at the base of a sea cliff.

Sea cliffs The steep faces of rocks eroded by waves.

Sea stack A column of rock left offshore as sea cliffs are eroded.

Seamount A submerged volcano rising from the ocean floor.

Sediment Any substance that settles out of water.

Sedimentary rock A kind of rock made when a layer of sediment becomes solid.

Seismic waves The vibrations caused by movement along a fault.

Seismograph An instrument that records seismic waves.

Shoreline The place where the land meets water.

Sill Body of igneous rock formed underground in cracks that run parallel with layers of existing rock.

Slump The sudden downward movement of a block of rock or soil.

Solar cell A device that is used to change sunlight directly into electricity.

Solar system The sun with its nine planets and smaller bodies that move around it.

Spit A long, narrow deposit of sand formed where a shoreline changes direction.

Standard Time The time based on one meridian and used for a time zone.

Stratosphere The atmospheric layer above the troposphere, between about 20 km and 50 km, showing a temperature increase with increasing altitude.

Subduction Where one plate is forced beneath another plate.

Submarine canyon A deep valley cut into a continental shelf.

Summer solstice The time of the year, about June 22, when the earth's axis is tilted most toward the sun.

Sunspots Areas on the sun's surface that appear dark.

Supernova A large star that suddenly explodes.

Swells Long, regular waves that move great distances across the oceans.

Syncline Rock layers folded downward.

T

Talus A pile of broken rock that gathers at the bottom of a cliff or steep slope.

Terrace A platform beneath the water at the base of a sea cliff.

Thermal pollution Release of heat into streams, lakes, oceans, or the atmosphere.

Thermocline A zone of rapid temperature change beneath the sea surface.

Thermosphere The layer of the atmosphere above the mesosphere, extending from about 85 km to about 600 km, where temperature increases with increasing altitude.

Tidal bore A large wave that leads a tide upriver.

Tide A regular rise and fall of water level in the sea.

Till Material deposited directly from a glacier.

Topographic map A map that shows the shape of the land surface.

Transpiration The process by which plants take up water through their roots and release water vapor from their leaves.

Trench A deep valley on the ocean floor found along the edges of some ocean plates.

Tributary A stream that flows into a river.

Troposphere The most dense layer of the atmosphere. It lies closest to the earth's surface.

Tsunami A giant wave caused by a sudden disturbance on the sea floor.

Turbidity current A current that is caused by the increased density of water as a result of mixing with sediments.

U

Unconformity A boundary separating younger rock layers from older layers that were exposed to erosion.

Uniformitarianism The principle that the same processes acting on the earth today also acted on the earth in the past.

Upwelling The rise of water to the sea surface.

V

Valley glacier A river of ice in a mountain valley.

Vernal equinox The time of the year, about March 22, when the earth's axis is tilted neither toward nor away from the sun.

Vertebrates Animals that have backbones.

Volcano A place where lava reaches the surface and builds a cone or other surface feature.

W

Water cycle The processes by which water leaves the oceans, is spread through the atmosphere, falls over the land, and runs back to the sea.

Water table The level below the ground surface where all open spaces are filled with water.

Watershed The land area that drains into a stream or river.

Weathering All the processes that break rock into smaller pieces.

Windbreak A barrier that causes wind to move more slowly.

Winter solstice The time of the year, about December 22, when the earth's axis is tilted most away from the sun.

APPENDIX A: THE SI (METRIC) SYSTEM

METRIC SYSTEM PREFIXES

Greater than 1:	Less than 1:
kilo (k) = 1,000 hecto (h) = 100 deka (da) = 10	deci (d) = 0.1 centi (c) = 0.01 milli (m) = 0.001 micro (μ) = 0.000 001

METRIC–ENGLISH EQUIVALENTS

1 km = 0.621 miles
1 m = 39.37 inches
1 cm = 0.394 inches
1 mm = 0.0394 inches
1 kg = 2.2046 pounds
1 g = 0.035 ounce
1 L = 1.06 quart
1 mL = 0.00106 quart

COMMONLY USED UNITS

	Unit	Can be compared to
Length	kilometer (km) 1000 m	The distance covered by walking for twelve minutes.
	meter (m)	The height of a door knob.
	centimeter (cm) 0.01 m	The thickness of a bread slice.
	millimeter (mm) 0.001 m	The thickness of a dime.
Mass	kilogram (kg) 1000 g	The mass of this book.
	gram (g) 0.001 kg	The mass of a paper clip.
Volume	liter (L) 1000 mL	Four times the volume of one cup.
	milliliter (mL) 0.001 L	The volume occupied by two nickels.

APPENDIX B: CHARACTERISTICS OF PLANETS

Planet	Average Distance From Sun (millions of km)	Time Taken to Go Once Around Orbit	Period of Rotation	Mass (Earth = 1)	Number of Satellites
Mercury	57.9	88.0 days	59 days	0.055	0
Venus	108.2	225.0 days	243 days	0.815	0
Earth	149.6	365.25 days	24 hrs	1.0	1
Mars	227.9	687.0 days	24 hrs 37 min	0.108	2
Jupiter	778.3	11.9 yrs	9 hrs 54 min	318.0	16
Saturn	1427.0	29.46 yrs	10 hrs 40 min	95.2	17
Uranus	2870.0	84.0 yrs	17 hrs	15.0	15
Neptune	504.0	165.0 yrs	17 hrs 54 min	17.0	2
Pluto	5900.0	248.0 yrs	6 days 9 hrs	0.002	1

APPENDIX C: WEATHER MAP SYMBOLS

A weather map is made by placing symbols and numbers at locations on a map. The symbols and numbers represent the weather observed at those locations. A set of lines (isobars) connects the places that have the same atmospheric pressure. Heavier lines with solid triangles or half-circles on them show fronts. The symbols for fronts are:

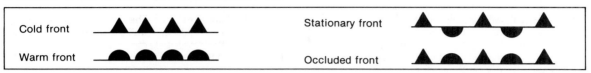

Observations taken at individual locations are plotted on a small circle on the map where the readings were taken. These circles along with symbols describing the weather conditions are called *station models*. Some symbols used to make station models are:

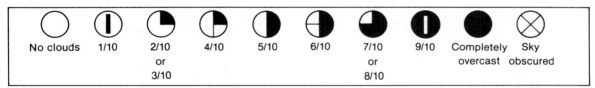

1. Cloud cover

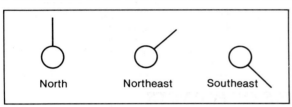

2. Wind direction

3. Wind speed

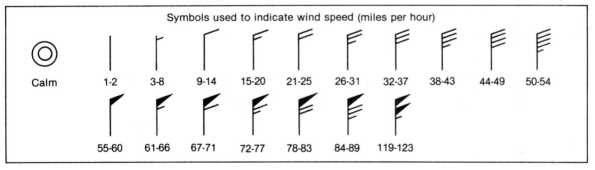

For example, a basic station model might be:
In addition to the symbols shown above, station models also include other numbers and symbols.
An example of a complete station model is shown on right.

Wind from northwest at 15 m.p.h., sky one-half covered with clouds.

*mb = millibars

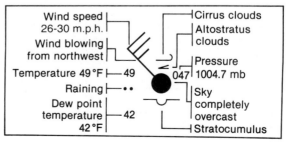

APPENDIX D: PROPERTIES OF SOME MINERALS COMMONLY FOUND IN ROCKS

LIGHT-COLORED MINERALS ABUNDANT IN IGNEOUS ROCKS				
Mineral	Hardness	Cleavage or Fracture	Color	Streak
Quartz	7	Fractures with curving lines on broken surface	Colorless or smoky gray to pink	None
Feldspar	6	Cleavage in two directions, nearly 90°	Usually white to pink	None
Mica (Muscovite)	2–2.5	Cleavage in one direction; splits into thin sheets	Colorless to slightly brown or greenish	None
DARK-COLORED MINERALS ABUNDANT IN IGNEOUS ROCKS				
Mineral	Hardness	Cleavage or Fracture	Color	Streak
Mica (Biotite)	2–2.5	Cleavage in one direction; splits into thin sheets	Dark brown to black	None
Chlorite	1–2.5	Cleavage in one direction; may show uneven fracture	Pale green to brown, rose and yellow	Light green
Hornblende	5–6	Cleavage two ways, at 56° and 124°	Black to greenish-black	Gray-green
Augite	5–6	Cleavage two ways at about 90°	Dark green to black	Gray-green
Olivine	6.5–7	Uneven fracture with curved lines on broken surface	Olive green to brown	None
LIGHT-COLORED MINERALS COMMONLY FOUND IN SEDIMENTARY ROCKS				
Mineral	Hardness	Cleavage or Fracture	Color	Streak
Calcite	2–3	Cleavage in three directions at greater than 90°	Colorless	None
Dolomite	3.5–4	Cleavage in three directions at greater than 90°	White or various colors	White
Halite	2.5	Three cleavages at 90°	Colorless to light gray	None
Gypsum	2	Good cleavage in one direction	Colorless to gray-white	None

DARK-COLORED MINERALS COMMONLY FOUND IN SEDIMENTARY ROCKS				
Mineral	Hardness	Cleavage or Fracture	Color	Streak
Hematite	5.5–6.5	Uneven fracture	Reddish-brown to black	Brown-red
Magnetite	6	Irregular fracture	Black; metallic luster (attracted to a magnet)	Black
Limonite	1–2.5	Earthy fracture	Yellow-brown or dark brown to black	Yellow-brown
LIGHT-COLORED MINERALS COMMONLY FOUND IN METAMORPHIC ROCKS				
Mineral	Hardness	Cleavage or Fracture	Color	Streak
Chrysotile	2.5–4	Splintery fracture; may separate into fibers (asbestos)	Greenish-gray	None
Talc	1	Cleavage in one direction making thin flakes	White to gray; greasy feel	White
DARK-COLORED MINERALS COMMONLY FOUND IN METAMORPHIC ROCKS				
Mineral	Hardness	Cleavage or Fracture	Color	Streak
Corundum	9	Irregular fracture	Usually brown	None
Epidote	6–7	Cleavage in two directions, distinct in one direction poor in the second; fracture uneven	Green to dark brown or black	None
Garnet	6.5–7.5	Irregular fracture	Red and brown	None
MINERALS COMMONLY FOUND IN VEINS IN ROCKS				
Mineral	Hardness	Cleavage or Fracture	Color	Streak
Pyrite	6–6.5	Uneven fracture	Brass yellow	Black
Galena	2.5	Cleavage in three directions at 90°	Silver-gray	Black
Sphalerite	3.5–4	Cleavage in six directions at 60°	Usually dark brown to black	Yellow-brown
Chalcopyrite	3.5–4	Uneven fracture	Yellow-brown metallic luster	Green-black

INDEX

sounding of depth, *389, 390, 391;* temperature of, *410*–413, 439
ocean basins, 387–388; floor of, *391, 392, 393, 394;* mapping, 396; sediments on, 393–394, 397–401
ocean currents, 428–452
ocean water, desalination of, *421*
oceanic trenches of world, *76, 77,* 394
oceanographers, 3, *4,* 390, 408
oil, 249, 268, 311; pools, 291, *422;* shale, 290, *292;* spills, 282
ooze, 399–400
orbit, of earth, 29, *56, 508;* of planets, 29, 482, 494
ore, 274
ore deposits, 256
organic material, 118
organic sediment, 247, 251
Orion, *522, 531*
oscillating universe theory, 540
outwash deposits, 151
oxbow lake, *130*
oxidation, 117
oxygen, 217; in atmosphere, *310*–311; in earth's crust, *216;* in minerals, 223, *224;* and pollution, 271, 281; in sea water, 409; in weathering, 117
ozone, 319, 472–*473*

P

Pacific Ocean, 388
Pacific plate, 96
pack ice, *411, 412*
Paleozoic Era, 194, 195, *196, 197*–198
Pangaea, 67, *197, 199*
parallax, *526, 528*
parallel of latitude, 48–*49,* 50
peat, 289
periods, in earth's history, 195, *196, 202*
permeability, 134–*135,* 137, 140
petrified tree, 181
petrify, 181
petroleum, 288, 290, 291, 292, 392, 422; as pollutant, 270
phases of moon, 508–509, *510,* 514
phosphates, as pollutants, 281
photosphere, **460**
photosynthesis, 311

physical weathering, 115–116
phytoplankton, *417*
pillow lava, *69*
placers, 276
planet(s), 481, *482*–495, *525;* birth of, 33, 35,
plankton, **417**
plants, as agents of weathering, 116, *117*–118; and transpiration, 124
plasma, in sun, 459
plate, crustal, *70*–71
plate boundaries, 72–*74*
plate tectonics, 64–80; theory of, 69–**70,** 393
plateaus, 106–107
Pluto, 482, 494
pluton, **256**
polar easterlies, 338
polar front, **365,** 366
polar projection map, 42–*43*
Polaris, *39, 48, 523*
political map, 43
pollution, **270,** 273; air, 270–271, 312–314; chemical, 281, *282;* soil, 270, 271–*272;* water, 270, 280–*281, 282,* 423
Pompeii, *2*
population of earth, *267,* 268–*269,* 270
pores, in soil, *134,* 135, 139
porosity, **134,** 140
potassium, radioactive, 190
Precambrian Era, *194, 196*–197
precipitation, **124,** 125, 354–*356;* 374
pressure, water, 413
prime meridian, 49–50
prominences, **462**
protons, 213–*214*
Ptolemy, 52–53
pumice, 242
pyrite, *228*

Q

quartz, *223,* 226, 227, *228,* 249
quartzite, *257*
quasars, 539

R

radar, weather, 377
radiant energy, **330,** *331*–332, 341, 342, 471, *472,* 473, 524–525

radio waves, *321,* 472
radioactive decay, 189, 192
radioactive potassium, 190
radioactive substance, **189,** 193
radioactive uranium, 190
radioactive wastes, 294
radiotelescopes, *15*
rain, formation of, 354, *355; See also* precipitation
rain gauge, 374, 375
rainfall, 280
rainstorm, and erosion, 128
rays, moon, **504;** of radiant energy, 472
recycling, *274,* **277,** 278
red clay, **400**
red giants, 530, 532, 533
red shift, 539
reefs, 395
relative age, 173
relative humidity, **343,** *344,* 373
relief maps, 43
Reptiles, Age of, 200
resources, from oceans, 416–423; *See also* natural resources
Richter, Charles, and earthquake energy, 91
Richter Scale, 91
rift, 97
rift valley, **68**
"Ring of Fire," 95–97
rip currents, 444
river(s), age of, 129; sediments from, 398
river systems, 129, 130
rock(s), **239**–263; classes of, 239–240; exfoliation of, 115–*116;* finding age of, *171,* 186–191, 193, 196–197; frost action on, 115; grain size, 252, *253;* igneous, 239, **240,** *241*–244, 246, 248; key beds in, **186**–*187,* 189; metamorphic, 240, **255**–256, 257; minerals in, 116, 221–222; permeability of, 134–*135,* 137; porosity of, **134,** 140; relative age of, 186, 189; sedimentary, 240; weathering of, 397, 410
rock cycle, 258–*259*
rock layers, coal as, 247; fossils in, 182, 188, 196; interpretation of, *168, 171, 172, 173, 174, 175,* 176

PHOTO CREDITS

Chapter 1: p. xii, United States Geological Survey; p. 2 (tl) G. Gualco/U.P./ Bruce Coleman; (tr) Walter Rawlings/ Robert Harding Assoc.; p. 3 Nicholas DeVore III/ Bruce Coleman; p. 4 (tr) Stephen J. Krasemann/ D.K.R. Photo; (cr) Bill Curtsinger/ Photo Researchers; p. 5 Dan McCoy/ Rainbow; p. 7 (cl) James Sugar/ Black Star; (cr) John Marshall; p. 8 John de Visser/ Black Star; p. 9 Ralph Perry/ Black Star; p. 10 Ralph Perry/ Black Star; p. 13 Mark Tuschman/ Discover Magazine; p. 14 Eric Kroll/ Taurus Photos; p. 15 James Sugar/ Black Star.

UNIT 1: pp. 26–27 Peter Turner/ Image Bank

Chapter 2: p. 28 NASA/ Photo Researchers; p. 31 (tl) Focus on Sports; (tr) Focus on Sports; p. 32 HRW Photo by Richard Haynes; p. 33 NASA; p. 38 NASA Science Photo Library Int'l/ Taurus Photos; p. 41 HRW Photo by Russell Dian; p. 45 HRW Photo by Russell Dian; p. 48 David Falconer/ Bruce Coleman; p. 52 (tr) Dennis di Cicco/ Sky & Telescope Magazine; p. 54 Robert Harding Picture Library; p. 61 (tl) Charles Henneghien/ Bruce Coleman; (tr) HRW Photo by Richard Haynes.

Chapter 3: p. 64 S. J. Krasemann/ Peter Arnold; p. 66 (tr) HRW Photo by Richard Haynes; p. 69 (tl) Woods Hole Oceanographic Institution; p. 72 Meffert/Stern/ Black Star; p. 76 (t) Galen Rowell/ Peter Arnold; p. 77 NASA/ Jet Propulsion Laboratory; p. 79 Department of Geological Sciences/ Cornell University.

Chapter 4: p. 86 GOLT/ Photo Researchers; p. 90 (t) Marshall Lockman/ Black Star; p. 92 Gene Daniels/ Black Star; p. 95 Ken Sakamoto/ Black Star; p. 97 Snead/ Bruce Coleman; p. 99 (tc) Gary Rosenquist/ Earth Images; (bc) Gary Rosenquist/ Earth Images; (bl) Gary Rosenquist/ Earth Images; p. 100 Peter Arnold/ Peter Arnold, Inc.; p. 101 (tl) W.E. Ruth/ Bruce Coleman; (tc) Robert Harding Picture Library; (tr) United States Geological Survey; p. 103 Robert P. Carr/ Bruce Coleman; p. 105 W.H. Hodge/ Peter Arnold; p. 106 (tl) Manuel Rodriguez; (tc) Michael Gallagher/ Bruce Coleman; (tr) Russ Kinne/ Photo Researchers; p. 107 (tl) Manuel Rodriguez; (tc) Grant Heilman; (tr) Bob & Clara Calhoun/ Bruce Coleman; p. 109 (tl) Grafton M. Smith/ Image Bank; (tr) E. R. Degginger/ Bruce Coleman.

Chapter 5: p. 112 Manuel Rodriguez; p. 114 (t) Bart Bartholomew/ Black Star; (b) William L. Ramsey; p. 116 (t) Joy Spurr/ Bruce Coleman; (b) Manuel Rodriguez; p. 117 (cl) Michael Markew/ Bruce Coleman; (br) Stephen J. Krasemann/ D.R.K.; p. 119 U.S.D.A./ Soil Conservation; p. 120 Manuel Rodriguez; p. 121 (tl) J. Callahan; (tr) Jacques Janeoux/ Peter Arnold; p. 123 (c) Jack W. Pykinga/ Bruce Coleman; p. 127 (tl) Manuel Rodriguez; (tr) Stephen J. Krasemann/Peter Arnold; p. 128 Grant Heilman; p. 129 (tr) Stephen Fuller/ Peter Arnold; p. 131 NASA; p. 133 Orlana Sentinel Star/ Black Star; p. 136 Peter Arnold; p. 137 (tr) Harvey Lloyd/ Peter Arnold; p. 139 Adolahe/ Southern Light; p. 141 (tl) Paul Shambroom/ Photo Researchers; (tr) Michael Heron/ Woodfin Camp.

Chapter 6: p. 146 Tom Bean/ Tom Stack; p. 148 (tl) Steve Coombs/ Photo Researchers; (tr) Manuel Rodriguez; p. 149 (tl) George Holton/ Photo Researchers; p. 150 (tl) Leland Brown/ Photo Researchers; p. 156 Sabine Weiss/ Photo Researchers; p. 157 (bl) Robert P. Carr/ Bruce Coleman; (bc) Lillian N. Bolstad/ Peter Arnold; (br) Sund/ Image Bank; p. 158 (br) Grant Heil-

man; p. 159 Loraine Wilson/ Robert Harding Picture Library; p. 160 (tr) Baron Wolman/ Woodfin Camp & Assoc.; (br) Robert Frerck/ Woodfin Camp & Assoc.; p. 161 (tl) Hank Morgan/ Rainbow; (tr) HRW Photo by Ken Lax; p. 162 (bl) Jack Fields/ Photo Researchers; (br) John S. Shelton; p. 163 (tr) Bruce Roberts/ Photo Researchers; p. 164 HRW Photo by Richard Haynes; p. 165 David Madison/ Bruce Coleman.

Chapter 7: p. 168 David Stone/ Rainbow; p. 170 Robert Harding Picture Library; p. 172 (tr) Manuel Rodriguez; (cr) Manuel Rodriguez; p. 173 (bl) G.R. Roberts/ Photo Researchers; p. 174 William Ramsey; p. 177 John Reader/ National Geographic Society; p. 178 (cr) William Ramsey; p. 179 (tl) Tass/ Sovfoto; (bl) George Roos/ Peter Arnold; p. 180 Photo Works/ Photo Researchers; p. 181 (tr) Stephen J. Krasemann/ D.R.K. Photo; (bl) Runk/ Schoenberger/ Grant Heilman/ N. Museum/ Franklin & Marshall College; p. 182 (tl) Manuel Rodriguez; p. 185 J. Fennell/ Bruce Coleman; p. 186 HRW Photo by Ken Karp; p. 187 (tr) Grant Heilman; p. 188 Runk/Schoenberger/ Grant Heilman; p. 205 (tl) John Curtis/ Taurus Photos; (tr) Walter H. Hodge/ Peter Arnold.

UNIT 2: pp. 210–211, Michael Freeman/ Bruce Coleman

Chapter 8: p. 212 Krafft-Explorer/ Photo Researchers; p. 220 Novosti/ Image Bank; p. 221 (cl) Photo Researchers; (bl) Karen Tompkins/ Tom Stack; (bc) Karen Tomkins/ Tom Stack; (br) Ft. Worth Museum of Science & History/ Susan Gibler/ Tom Stack; p. 222 Alfred Pasieka/ Taurus Photos; p. 223 Runk/ Schoenberger/ Grant Heilman/ Franklin & Marshall College; p. 226 Granger; p. 227 (tl) Bob Gossington/ Tom Stack; (cr) Manuel Rodriguez; p. 228 (tl) M. Liacos; (tr) B.M. Shaub; p. 229 (t) HRW Photo by Richard Haynes; (bl) J. Fennell/ Bruce Coleman; (br) Adrian Davies/ Bruce Coleman; p. 231 Larry Dale Gordon/ Image Bank; p. 233 Gabe Palmer/ Image Bank.

Chapter 9: p. 238 Alan Pitcairn/ Grant Heilman; p. 240 NASA/ Black Star; p. 241 (t) John Running; (b) E.R. Degginger/ Bruce Coleman; p. 242 Fred Ihrt/ Image Bank; p. 243 (t) Shelly Grossman/ Woodfin Camp; (c) Robert P. Carr/ Bruce Coleman; p. 244 Bob & Clara Calhoun/ Bruce Coleman; p. 247 Michael G. Borum/ Image Bank; p. 248 Keith Gunner/ Bruce Coleman; p. 249 Keith & Donna Dannen/ Photo Researchers; p. 250 (tl) Leo Touchet/ Woodfin Camp; (tr) Walter Rawlings/ Robert Harding Picture Library; (c) Jack Baker/ Image Bank; p. 251 (t) Bob McKeever/ Tom Stack; (b) David W. Hamilton/ Image Bank; p. 252 (tl) William Felger/ Grant Heilman; p. 255 Nicholas de Vore III/ Bruce Coleman; p. 256 Dan McCoy/ Rainbow; p. 257 (cl) David Krasnor/ Photo Researchers; (cr) Stephen J. Krasemann/ Photo Researchers; (bl) Art Kane/ Image Bank; (br) Budd Titlow/Naturegraphics/ Tom Stack; p. 261 (tl) Dan McCoy/ Rainbow; (tr) Dan McCoy/ Rainbow.

Chapter 10: p. 266 Bob & Clara Calhoun/ Bruce Coleman; p. 269 (b) Nicholas DeVore III/ Bruce Coleman; p. 271 Nicholas DeVore III/ Bruce Coleman; p. 272 Nicholas DeVore III/ Bruce Coleman; p. 274 Jeff Foott/ Bruce Coleman; p. 275 Dr. George Gerster/ Photo Researchers; p. 276 (bl) Woods Hole Oceanographic Institute; p. 281 Mark Sherman/ Bruce Coleman; p. 282 (tl) Mark Sherman/ Bruce Coleman; p. 286 Katherine S. Thomas/ Bruce Coleman; p. 288 Wendell Metzen/ Bruce Coleman; p. 289 (cl) Horst Schafer/ Peter Arnold; (cr) Paola Kock/ Photo Researchers; (bl) Robert Whitney; (br) Robert Whitney; p. 290 Marc Solomon/ Image Bank; p. 292 Dennis Hogan/ Tom Stack; p. 294 (br) Ron Sherman/ Bruce Coleman; p. 297 Peter Arnold; p. 298 (br) Jonathan T. Wright/ Bruce Coleman; p. 299 George Gerster/ Photo Researchers; p. 300 Terence Moore/ Woodfin Camp; p. 301 George Gerster/ Photo Researchers; p. 303 Ph. Charliat/ Photo Researchers.

UNIT 3: pp. 306–307 Stephen J. Krasemann/ D.R.K. Photo

Chapter 11: p. 308 Wards Science/Science Source/ Photo Researchers; p. 310 (t) William Thompson/ Earth Images; p. 312 Alfred Pasieka/ Taurus Photos; p. 313 (cl) Robert P. Carr/ Bruce Coleman; p. 315 HRW Photo by Richard Haynes; p. 316 J. Serafin/ Peter Arnold; p. 317 (tl) HRW Photo by Richard Haynes; p. 318 (tr) Jules Bucher/ Photo Researchers; p. 323 (tl) Michal Heron/ Woodfin Camp; (tr) NASA/ Science Source/ Photo Researchers.

Chapter 12: p. 328 Bjorn Bolstad/ Photo Researchers; p. 330 (tl) Hanson Carroll/ Peter Arnold; p. 333 (br) Bill Evans; p. 342 Norman Owen Tomalin/ Bruce Coleman; p. 345 (tl) Thomas W. Putney/ Delaware Stock Photos; (tr) Dan McCoy/Rainbow; p. 347 (tl) L.L.T. Rhodes/ Taurus Photos; p. 347 (tr) Eric Kroll/ Taurus Photos.

Chapter 13: p. 350 John Deeks/ Photo Researchers; p. 352 (tl) HRW Photo by Russell Dian; (tr) Jonathan Wright/ Bruce Coleman; p. 353 (tl) Dave Baird/ Tom Stack; (tr) Russ Kinne/ Photo Researchers; p. 354 (tl) Tom Stack/ Tom Stack & Assoc.; (tc) Hans Namuth/ Photo Researchers; (tr) Dale Jorgenson/ Tom Stack; p. 356 Norman Owen Tomalin/ Bruce Coleman; p. 357 HRW Photo by Richard Haynes; p. 358 Richard Weiss/ Peter Arnold; p. 365 (tr) Brian Parker/ Tom Stack; p. 367 (tr) Shostal; p. 368 (t) NOAA; (bl) Doug Millar/ Photo Researchers; p. 369 (tl) Henry Lansford/ Photo Researchers; (tr) Spencer Swanger/ Tom Stack; p. 370 Frederic Lewis; p. 373 Joel Gordon/ D.P.I.; p. 375 (tl) Joe Munroe/ Photo Researchers; (bl) Wards Science/ Science Source/ Photo Researchers; p. 379 NOAA.

UNIT 4: pp. 384–385 Guido Alberto Rossi/ Image Bank

Chapter 14: p. 386 Bob Evans/ Peter Arnold; p. 388 Wards Science/ Science Source/ Photo Researchers; p. 395 Peter Arnold; p. 397 Vivian M. Peevers/ Peter Arnold; p. 399 Runk Schoenberger/ Grant Heilman; p. 401 Dan McCoy/ Rainbow; p. 403 Sullivan & Rogers/ Bruce Coleman.

Chapter 15: David Doubilet; p. 411 E.R. Degginger/ Bruce Coleman; p. 416 D. Brewster/ Bruce Coleman; p. 417 NASA; p.

419 (t) Tom McHugh/ Steinhart Aquarium/ Photo Researchers; (b) Frank W. Lane/ Bruce Coleman; p. 420 W.H./ Image Bank; p. 421 (t) George Whiteley/ Photo Researchers; (b) Dan McCoy/ Rainbow; p. 422 (tr) Shell Oil Company; p. 425 (tl) Christina Thompson/ Woodfin Camp; (tr) Jim Pickerell.

Chapter 16: p. 428 Bob Evans/ Peter Arnold; p. 433 HRW Photo by Richard Haynes; p. 434 Larry Mulvehill/ Photo Researchers; p. 436 NOAA; p. 441 John Bryson/ From RG/ Photo Researchers; p. 444 Ron Church/ Photo Researchers; p. 445 HRW Photo by Richard Haynes; p. 446 Kyodo Photo Services; p. 450 (tr & cr) Clyde H. Smith/ Peter Arnold; (bl) Weston Kemp; p. 453 (tl) Eric Kroll/ Taurus Photos; (tr) Jack Fields/ Photo Researchers.

UNIT 5: pp. 456–457 Rainbow

Chapter 17: p. 458 Dan McCoy/ Rainbow; p. 461 NASA/ Science Source/ Photo Researchers; p. 462 (l) Ward's Scientific/ Photo Researchers; (r) G.C. Fuller/ S.P.L./ Photo Researchers; p. 463 (bl) Michael P. Gadomski/ Bruce Coleman; p. 465 (bl) Jack Finch/S.P.L./ Photo Researchers; p. 467 Ward's Science/ Science Source/ Photo Researchers; p. 468 U.S. Navy/ S.P.L./ Photo Researchers; p. 473 NASA/ Science Source/ Photo Researchers; p. 474 HRW Photo by Richard Haynes; p. 475 NASA/ Robert Harding Picture Library.

Chapter 18: p. 480 R. Royer/ Science Photo Library/ Photo Researchers; p. 483 Astron Society of the Pacific/ Science Source/ Photo Researchers; p. 484 (l) NASA; (r) NASA; p. 486 (t) NASA; (b) Astron Society of the Pacific/ Science Source/ Photo Researchers; p. 490 NASA; p. 491 NASA; p. 492 NASA; p. 493 (tl) NASA; (br) NASA/ A.M.N.H.; p. 497 (tr) Lowell Observatory; (br) Litton Industries; p. 499 (cl) NASA/ A.M.N.H.; p. 500 (tl & bl) NASA/ National Space Science Data Center; p. 502 Rainbow; p. 503 NASA; p. 504 (tl) Dennis di Cicco; (tr) NASA; p. 507 NASA; p. 511 (tl) Allen Green/ Photo Researchers; (tr) Jules Bacher/ Photo Researchers; (cl) Dennis di Cicco; (cr) Russ Kinne/ Photo Researchers; (bl) Jules Bacher/ Photo Researchers; (bc) Wards Science/ Science Source/ Photo Researchers; (br) Russ Kinne/ Photo Researchers; p. 512 (bl) Tersch Observatories; p. 513 Wards Science/ Science Source/ Photo Researchers; p. 514 HRW Photos by Richard Haynes; p. 515 Clyde H. Smith/ Peter Arnold.

Chapter 19: p. 520 James Sugar/ Black Star; p. 522 (tr) Wards Science/ Science Source/ Photo Researchers; p. 523 (br) Bill Belknap/ Photo Researchers; p. 531 Ronald Royer/ Photo Researchers; p. 536 NASA; p. 538 Hale Observatories/ Science Source/ Photo Researchers; p. 541 NASA.

CALIFORNIA

Leland F. Cooley

STEIN AND DAY/*Publishers*/New York

This edition first published in 1984.
Copyright © 1973, 1984 by Leland Frederick Cooley
All rights reserved, Stein and Day, Incorporated
Printed in the United States of America
STEIN AND DAY/*Publishers*
Scarborough House
Briarcliff Manor, N.Y. 10510

Library of Congress Cataloging in Publication Data

Cooley, Leland Frederick.
    California.

    1. California—History—Fiction.   I. Title.
PS3553.0564C3   1984   813'.54   84-40237
ISBN 0-8128-2987-5

For Danny
and Allison
and "Gwennie"

"We *californios* never knew what riches we possessed until some outsider with clearer vision landed on our hospitable shores or scaled our icy sierras or staggered across our burning deserts to take them from us—"

Anita Maria Dolores Chard Lewis
on her eightieth birthday,
El Rancho de las Flores, Proberta,
California, December 25, 1918.

CONTENTS

CALIFORNIA

RELATIONSHIP OF PRINCIPALS

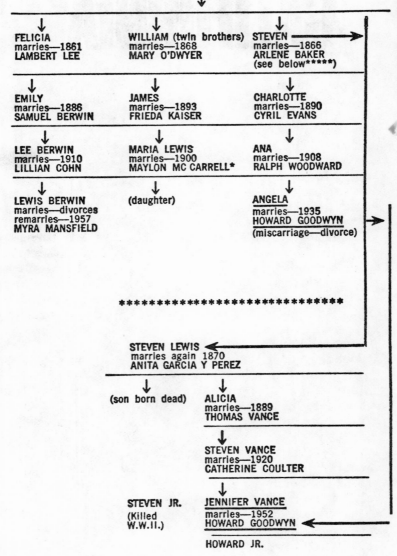

JOHN LEWIS
marries—1842
MARIA ROBLES

FELICIA
marries—1861
LAMBERT LEE

WILLIAM (twin brothers)
marries—1868
MARY O'DWYER

STEVEN
marries—1866
ARLENE BAKER
(see below****)

EMILY
marries—1886
SAMUEL BERWIN

JAMES
marries—1893
FRIEDA KAISER

CHARLOTTE
marries—1890
CYRIL EVANS

LEE BERWIN
marries—1910
LILLIAN COHN

MARIA LEWIS
marries—1900
MAYLON MC CARRELL*

ANA
marries—1908
RALPH WOODWARD

LEWIS BERWIN
marries—divorces
remarries—1957
MYRA MANSFIELD

(daughter)

ANGELA
marries—1935
HOWARD GOODWYN
(miscarriage—divorce)

****

STEVEN LEWIS
marries again 1870
ANITA GARCIA Y PEREZ

(son born dead)

ALICIA
marries—1889
THOMAS VANCE

STEVEN VANCE
marries—1920
CATHERINE COULTER

STEVEN JR.
(Killed
W.W.II.)

JENNIFER VANCE
marries—1952
HOWARD GOODWYN

HOWARD JR.

Pro Bono Publico

Howard D. Goodwyn, forcibly struck by a forty-year-old memory, stood half lost in reflection on the steps of the capitol building in Sacramento as California's Governor, Ronald Reagan, began to introduce him to the audience.

The day was a "valley scorcher" and Howdy longed to get back to his huge southern California ranch and the cooling southwest ocean breeze.

Both Howdy and the governor were outsiders who had come to California as young men. Both had left indelible imprints on their adopted state. Neither had done it the easy way.

Reagan had started as a sports announcer and Howdy, a half-starved cowboy singer from Guymon, Oklahoma, had bummed his way to Hollywood to find a job. Now they were standing on the capitol steps, each preparing in his own way to take another long step up the achievement ladder. Howdy Goodwyn smiled inwardly at the unlikely bedfellowships arranged by the political machine and by single-minded self-interest.

He couldn't speak for Reagan, but he could honestly tell himself that his ambition, the driving force that had made him one of the richest men in the West, was not now entirely selfish. He would give back something, in some way best for him, to the state that had given him everything. As tough as the going had been during those early depression years, California, incredibly Los Angeles itself, had sensitized him to a new kind of beauty. The twilight nights spent on the porch of the Franklin Avenue boarding house watching the ugly urban sprawl transformed by nightfall into a jeweled sea were etched in his memory. They had touched something in him and reawakened the same soaring elation he had felt back home sprawled beneath a tree to stare up and beyond the stars to a world of unreal aspirations. They had seemed unreal in Guymon but not in

15

Hollywood. That was thirty-odd years ago. Today Ronald Reagan and Howard Goodwyn were one shallow granite step apart and the governor was saying:

"At the risk of implying that there's no business like show business if you want to succeed in California politics, I want to say that *because* of, not in spite of, this man's intimate knowledge of people and his responsiveness to their needs, I know of no other man in the state who is better qualified to occupy the office I hope soon to vacate, *if* he can be persuaded, than Howard D. Goodwyn. Ladies and gentlemen, I'll resist the obvious and urge you, instead, to say, 'Hi' to Howdy."

The enthusiastic rattle of applause from the modest gathering of press representatives, party faithful, and curious passersby brought Howdy back to the present. Ducking his head in a subtle imitation of the late Will Rogers, a deliberate mannerism that over the years had served him well, Howdy smiled and waited. When it was quiet, he spoke.

"Governor Reagan, friends, I think it was President Truman who said, 'If you can't stand the heat, stay out the kitchen.' I'd like to change that a bit and say, 'If you can't stand the heat, stay out of —'" When the expected ripple of laughter started, Howdy broke off in apparent consternation and shook his head. "I was about to stay, 'If you can't stand the heat, stay out of Sacramento,' but—"

The laughter he had deliberately triggered broke over him and the governor's was loudest of all. After a moment, he signaled for silence. "That would sure be misunderstood, wouldn't it. What I really wanted to say was, 'Come on down to my end of the state and I'll show you some real heat.' I mean the heat that comes from hot issues, the heat that a lot of us generate under our collars when we see what's happening to California because a lot of other people still seem to feel that it's the best place in the world to live."

An older woman in the audience turned to her husband and said in a stage whisper, "He hasn't changed all that much. He still gives you confidence, just like he did on the radio when he sold all those things."

The man snorted derisively. "What he sold was snake oil and that's what he's still selling, only in a fancier wrapper. He's a typical native con man. They come natural to this state."

16

The woman caught the quick gleam of amusement in Howdy's eyes and lowered her own in embarrassment as he began to speak again.

"I'm real glad for this chance to tell you about some of the things the governor and I have been discussing this afternoon because they are things that affect us all."

Holding up his left hand, Howdy prepared to tick off some points, starting with his strong, stubborn thumb. "The new people who come to California have to find a place to live. So we build suburbs for them and new towns. The trouble is that in the past few years we've been building housing on our best agricultural land because that's where developers find it cheapest to build. Every year subdividers, most of them from out of state, take up to a hundred thousand acres of prime growing land out of production and bury it under slabs, driveways, and streets.

"Then we build more freeways to serve the new communities and new shopping centers and every mile of concrete and asphalt kills enough fertile soil to feed five thousand people one or more basic staples."

In the growing crowd an elderly man turned to a stranger next to him. "He's dead right. Ever since the Mexicans, California's best land has been fair game for outsiders. Damned state never has learned to protect itself."

A reporter for a Northern California newspaper chain nodded in agreement and added sotto voce: "But let us not forget while he's preaching the native gospel that Goodwyn's as much an outsider as the rest."

"Forty years ago, maybe. But he married into an old Mexican family that came up to Monterey in the late seventeen hundreds. Some of that rubbed off on him. He's got an honest feel for land."

The newsman laughed dryly. "And he's got one hell of a feel for the money land makes him, too."

Howdy's last statement had caused a murmur in the crowd and a number of people had begun taking notes as he continued.

"If I were to spell it all out, we'd melt in our own juice. But I would like to add another fact or so that will point up our need to do some very careful planning real soon."

"For instance, we have about one hundred million acres in the state of California. About one half of our land is in

17

national parks, state parks, and national and state forest preserves. A little more than one fifth of our state is given over to deserts and high mountains. But because twenty million people, about one out of every ten persons in the United States, lives in California now, ten percent of the remaining land on which we must live and work and grow our food is polluted."

He paused for effect. "I said, 'polluted.' One way or another we've polluted it."

The charge, nobody doubted that it was a fact, caused a new outburst of exclamations. Howdy went on.

"And if we don't do something constructive, starting right now, in thirty years there won't be any usable prime land left in California. And that's not only a critical prospect for us, but for the rest of the U.S.A. too, because we grow one fourth of all this country's table food and forty-three percent of its fruits and nuts—"

A wag on the fringe of the crowd called out, "Wrong! We got one hundred percent of the country's nuts here." Howdy grinned at the governor and joined in the laughter.

"The point is," he continued, "I'll run out of fingers before I've listed a tenth of the dangerous problems that face our state. I don't pretend to know the answers to all of them. But I know a good place to start. Because we believe people can live in communities built on productive land without spoiling it, we've set aside five thousand acres of our Los Nidos Ranch for a new, ecologically sound living concept that we call Micropolis Two Thousand. We believe that if we don't learn to do that, and *muy pronto,* your grandchildren and mine will be part of a new bunch of Forty-Niners who'll be digging up this state, turning over asphalt and concrete looking for the most precious resource California has ever had: good, rich land that will raise food twelve months of the year."

Governor Reagan nodded approval and led a round of applause. The reporter nudged the elderly man. "While he's helping California, let us also not forget that he'll be making a profit that would turn the old Comstock and Railroad robbers green with envy."

His companion regarded him tolerantly. "One thing about being young, son—you can get over it. Meantime, don't be so sanctimonious. The founder of your paper made his and so did young Bill Hearst. Ever been to San

18

Simeon?" The reporter did not respond and the old man's seamed face rearranged itself into a dry smile. "Profit's the name of the game, son. If you get your money by giving folks a fair shake, there's nothing wrong with it. And Goodwyn's always given this state a fair shake so far as I can see."

The response was cut off as Howdy began his concluding remarks and the band from Sacramento's John A. Sutter High School buttoned up loosened uniform collars preparatory to unlimbering their instruments.

Except for the inevitable scattering of obvious skeptics, Howdy saw friendly, or at least interested, faces. Reassured that he still retained his "edge," he started his concluding remarks.

"In the two hours I've spent with the governor this afternoon I was impressed again by the extent and the depth of his knowledge of California's problems. He did me the honor of thinking that I understand those problems well enough to be useful in public office."

Howdy lowered his head slightly and succeeded in looking a trifle more humble. "I'm not sure about that. But one thing I am sure of is that I'm just as concerned about our problems as Governor Reagan and I will pledge the very best I can do to help get our state back on the track. At this point in time that's not a political promise because I haven't said I would be a candidate for any office. But if I should come to believe, and the people believe along with me, that I can best serve the state in office then I'll make that an honest political promise, too." Raising a hand, Howdy smiled sincerely at his audience. "Thank you for standing and listening."

Governor Reagan, concealing the disappointment he felt that Howard Goodwyn had not committed himself more directly, nodded to the band leader as he urged the crowd to prologed applause. A moment later the strident brass strains of *California, Here I Come* drowned out the ovation.

Well back in the crowd, an affluent-looking man in his middle years addressed his younger companion.

"California, here we go. If that smooth-talking, humble-pie shit kicker runs for governor and gets in, he'll drain the rest of our northern water into his southern California swimming pools and the only cash left in the state

19

will be the millions the son-of-a-bitch is stashing in his pockets. He's got to be stopped."

The outburst produced a smile. "That, my friend, is going to be about as easy as bottling a fart in a whirlwind."

Book I

The John Lewis Story

1 SOAKED THROUGH and struggling to control a chill, John Lewis watched from his hiding place in the dense madrone grove as the woodcutters' ox-drawn *carreta* lumbered by. The loosely lashed load of gnarled oak limbs tilted dangerously as the primitive wooden wheels slithered in and out of the soggy ruts that angled down the slope.

John had negotiated the icy half-mile swim to shore with nothing more alarming than a brush with an impertinent Pacific sea lion and an unexpected encounter with an otter that had surfaced to poke an inquisitive nose into his face, let out a startled squeak, and submerge again in the pre-dawn darkness.

Still ahead were a host of uncertainties. As hazardous as any was the prospect of having to spend a night in the densely forested mountains ringing Monterey Bay with no weapon but his sheath knife. From tales told aboard ship by older hands who had sailed in the California hide trade, John knew of the giant grizzlies that hunted there and of the packs of timber wolves whose slashing teeth, it was said, could reduce a man to a quivering mass of gory ribbon in minutes. Willingly, he would risk these beasts to reach his destination, the renegade camp at Branciforte, even though he knew well that he could not count on asylum.

Isaac Graham and Will Chard usually welcomed able-bodied men to their private militia of trappers, deserters, and adventurers, but from what he had heard, John wondered if they would welcome one as young as he.

If it turned out that Graham or Chard did not like the cut of his jib, there was a strong likelihood that they would turn him over to the Mexican governor, Juan Alvarado, or to Monterey's military commandant, Colonel José Castro, in return for the sizeable cash bounty the

23

master of the *Reliant* was certain to post in the custom-house and in the waterfront *cantinas*.

So the decision was simple. There was no alternative but to find Graham and his men and make a place for himself, for, once accepted, there was security of a sort among the feared *renegados*.

"Decent women leave the streets when Graham and his companions ride down the calle principal," they said. "And only the strongest men dare to stand downwind of them, so dreadful is the stink of their greasy buckskins and their aguardiente-laden breaths."

When the *carreta* disappeared into the timber again, John Lewis appraised his own condition and smiled grimly. "I may not be old enough or rough enough for them," he told himself, "but by heaven I'll be high enough by the time I get there!"

Shouldering his pack of spare clothing, John eased out of the thicket and made his way to the summit of the ridge. Dawn was breaking as he stopped to catch his breath. Below him the settlement of Monterey, capital and sole port of entry to Alta California, lay snugged darkly in the curving embrace of the low hills that enclosed the lower end of the bay. To the northwest, John could see a party of fast-moving horsemen crossing the barren sand dunes, heading toward the Pájaro River road leading inland to San Jose. Beyond, he could make out—lying at anchor in the bay—the familiar silhouette of his own vessel, the brig *Reliant*.

The sight of the ship's masts and spars, dead black against the lightening sky, produced a curious turmoil in him. Suddenly he found himself thinking of the letter to his sister in Brooklyn that he had entrusted to Ned. He wondered if the third mate had believed the story they had cooked up, or whether Captain Nathaniel Burke, a hard master to deceive, had put Ned in irons while he devised yet another of his cruel punishments for complicity in a shipmate's desertion.

Alice would be eighteen now. The thought of her started a quick flood of memories. Four years earlier his mother had died of consumption. A year later an undiagnosed illness had taken his father. Within a week, creditors had swarmed over the family's small general store in Middletown, Connecticut, and picked it clean.

Distant relatives in New York had taken Alice; but he

24

had been allowed to keep a tiny back room in the family house until he had finished his schooling on condition that he work for his keep. The tight-fisted new owners, determined to make the utmost of their bargain, had set him to doing the most menial work. When the flinty-faced mistress added the revolting task of emptying the family night jars, he stayed on just long enough to secure his diploma from the Middlesex Academy. Then, with his poor belongings and a book or two gathered into a makeshift bundle, he confronted the woman, charged her with her shortcomings as a "Christian lady," and set out for Boston. There he had met Ned James, then a stranger also in dire circumstances, and the two signed aboard the *Reliant* as apprentice seamen.

Annoyed at himself for wasting precious minutes in useless reverie, John reshouldered his oilskin pack and began the steep descent down the back side of the first range. If he could manage the unrealistic pace of four miles an hour through the rough country he had chosen as the best route to elude possible pursuers, it would still take another ten hours to reach his doubtful haven.

The lower limb of the sun had just begun to flatten against the Pacific horizon off to his left as he reached the high ground overlooking the south side of the small Pájaro River. Concealing himself, he studied the far bank. There was little traffic on the river road; but upstream to his right he could see tell-tale clouds of dust and he knew that each one concealed a party of riders.

Certain that it would be foolhardy to attempt the crossing and the half-mile run to cover on the far side, he settled down to wait for nightfall. When the last crimson thread of sunset had narrowed to nothingness, he gathered himself and moved stealthily down to the water's edge. There, poised on the tip of a gravel bar that jutted nearly across the narrow waters, he swung his pack pendulum-like several times and let it go. It arced across the open water and landed heavily in the moist sand at the foot of the opposite bank. A moment later he cleared the current, retrieved the pack, and began the cautious climb out of the stream bed.

Halfway up the bank a sound startled him. Straining to hear, he crouched beneath some overhanging brush, scarcely breathing. After a time he relaxed. The valley was full of wandering long-horned Mexican range cattle.

25

One of them probably had dislodged a small cobble while coming down to water. Taking a firm new grip on the pack, he scrambled to the top of the bank, peered along the road once more, then forced himself upright and set off at a dead run for the cover of a distant brush-lined draw.

Where the horsemen came from John Lewis would never know. All at once they came bearing down on him from four directions—two from the rear flanks and two from the front. He never saw the hissing rawhide *reatas* that snared his feet and arms and jerked him viciously to the ground.

Not until he was bound hand and foot and the toe of a soft Spanish boot pushed him roughly face-up did he see his captors. The one who stood over him was the only man he could really see. He was young and slender and he was smiling.

"*Buenas tardes, Señor Lewis! Créamos mucho gusto en verle!* We are sorry to detain you but it is dangerous for sailors to make a *pasada* in these mountains. Since we have a tradition of hospitality in Alta California we have come to escort you back to your ship."

Still gasping from the exertion of the run and the shock of the ambush, John struggled wildly. The men watched, laughing, and allowed him to continue, knowing that the rawhide bonds were biting deeper with each convulsion. No more than a minute elapsed before the strands across his throat left John croaking for help. Finally, when he was half suffocated, the leader knelt down. While two companions pinned his legs, the *reata* around John's neck was loosened. Once more he was rolled over, this time on his face, while the bonds were rearranged to pinion his hands behind him.

On a signal, one of the men released the rawhide looped around his ankles and he was jerked to his feet and boosted up behind the saddle of the leader's horse. His captor mounted as did the other three, and the party set off downriver toward the bay. They had ridden in silence until the man in the saddle turned to John. "*Dispensame, Señor Lewis. Estoy remiso! Me llamo Gregorio Robles—a sus ordenes.* Not always do we make such extreme precautions for the safety of our guest." He chuckled and said something in the rapid musical California dialect. The remark amused his companions.

The horses started across a shallow ford. In midstream Goyo Robles leaned forward and spurred his mount. The animal responded with a leap that sent John tumbling backward over its rump into the water. When he came to the surface, gasping, he was hauled to the opposite bank by the laughing men and remounted. Helpless to do more than try to balance with his hands tied behind him, John clenched his teeth and prepared himself for the next indignity. But none came.

Goyo led them down a steep bank to the bottom of a dry wash. The pace quickened then and an hour later the first lights of Monterey were visible. John had spent less time in Monterey than in the other trading ports along the Alta California coast. It was Mexico's most northern department and reputedly the most picturesque of the pueblos. Situated at the southerly end of the great curving crescent of the bay, it had no more than two score buildings. These were built along three broad packed-earth streets that ran from the base of the foothills down to the narrow, kelp-strewn strip of sandy beach. Behind the pueblo, dwarfing its adobe buildings, were the grass-carpeted hills with their scattered stands of oak and evergreens.

Finally, in darkness, the horsemen halted before a one-story, whitewashed adobe building whose sides and back were enclosed by a high mud-block wall. Goyo and two of his companions helped John down and half dragged him inside.

Doña Gertrudis Robles and her sixteen-year-old daughter, Maria, scurried from their chairs in alarm as Goyo and his men pulled John over to a crudely made rawhide chair and bound him to it.

In response to his mother's demanding "Que pasa?," Goyo straightened and turned to her. "My dear mother, my dear sister, we have an unexpected guest who will stay with us until morning. And because we value his presence so highly, we will make certain he does not leave us for fear of imposing on our well known *californio* hospitality. Permit me to present an exceptional yanqui *marinero*, Señor Juan Lewis." Goyo indicated his captive with an elaborately ceremonious bow.

John's sullen glare moved from the kindly-faced, patrician older woman, in whose eyes he thought he detected a look of sympathy, and fixed on the girl. Despite the chill that had begun to make him shake and the cruel bite of

27

the rawhide lashings, he found himself held by her beauty. Even though her head was lowered, he could tell from the curving sweep of the long black lashes that her eyes were uptilted and enormous, widely set in a delicate heart-shaped face. Like her mother, she was small-boned and finely made.

Goyo watched his prisoner with malevolent amusement. "So, my beautiful sister makes you forget your discomfort?"

Without warning he stepped forward and slashed John across the face with the backs of his fingers. His hand drew back for a second blow as Maria cried out in wordless protest and Doña Gertrudis called her son's name sharply.

Goyo stayed the blow and resorted to soft, mocking laughter. "Perhaps I do you an injustice, Señor Lewis. Perhaps in your present state of discomfort you do not have the evil thoughts of most yanqui pigs who look at my very desirable sister."

Then he brightened. "Well, it is nothing, this small slap in the face, or so it will seem to you when you are returned to your ship for punishment."

Doña Gertrudis moved to her son's side. "Why is this necessary? What has this man done?"

"This is necessary if we wish to collect the very rich reward that is offered for our captive. He is worth twice as much as any other yanqui sailor who has deserted. He is as strong as a bull. And he is smart, this one. He does not yet have twenty years and he is permitted to bargain for hides for his company. Impressive. No?"

Spreading his arms dramatically, Goyo sighed. "And because we *californios* are compassionate we shall see him safely back to his ship and collect not one but *two hundred dollars* as a reward that will be used to drive other yanqui pigs from Alta California." He turned menacingly toward John. "We have been fools. We have trusted you here as our brothers did in Texas. We will not make that mistake even if our governor does not agree and Mexico City does not care."

For a long moment, Goyo Robles stood poised over his captive, a hated yanqui: Then he addressed the women, "Mama, Maria, go prepare some food for our guest. He hoped to partake of Señor Graham's hospitality tonight, but we have spared him that cruel fate." A final check of

28

the bonds satisfied Goyo and after a short, whispered conversation, he left with both men and the women.

John looked around the room. To his left was a crudely made fireplace. On the rough-hewn mantel two home-made candles burned on either side of an ornate painting of the Virgin Mary holding the infant Jesus. The floor was hard-packed earth, the walls whitewashed mud plaster, and the only furniture in the room was the table and half a dozen crudely made straight-back chairs.

Now that he was alone, John thought about his decision to desert. Once again he told himself it had been the only endurable course open to him. Even though his return to the ship now would put his survival in doubt. The *Reliant* was at the mercy of a violent young master. If a sailor survived the floggings Captain Nathaniel Burke administered with near-insane relish, he would forfeit all pay, all liberty, and all but one meal a day until the ship returned to home port in Boston. There the offender would face charges of criminal violation of the Ship's Articles.

John's shipmate, Ned James, had, up to the last, tried to stop him. As John's hour to escape neared, he had whispered, his voice tight with apprehension, "Are you sure you want to go ahead with this folly, John? You know what will happen if they catch you."

"I know what will happen if I stay aboard. Maggoty meals. Porridge for another year. Walking barefoot over stony beaches balancing a hundred pounds of hides on my head. No. I've got nothing to go back to now except my sister, who is well taken care of by relatives. There may be nothing on this coast either but I have to see for myself. Scottish and English merchants are doing well here. Why not a Yankee?"

Deep in thought, John did not see Maria as she timidly reentered the room. She had been sent back by Goyo and told to watch the prisoner. When John looked up, startled, she cast her eyes down and seated herself at the table. She hoped she was concealing the excitement that had persisted from the moment she had seen the young yanqui. She had seen many foreign sailors while she and her mother pursued errands along the calle principal but none had disturbed her the way this strapping American had. Even at first glance there had been a quality of *gentileza* in his face and in the intelligent dark blue eyes that seemed to conceal little of what he was feeling. Maria was certain

29

she had seen admiration in his eyes when he first caught sight of her. But there had been a flash of something else there, too, and it had started a strange restlessness in her.

John, watching the play of candlelight on the girl's face, thought again that in all his wanderings he had never seen one so lovely. Suddenly he became aware of a smarting sensation on his cheek. Inclining his head, he tried to raise his shoulder high enough to brush the flesh against his shirt. Instantly, Maria was at his side, bending close.

"Your cheek is scraped, señor." Using the tips of her fingers, she brushed gently at the grains of sand clinging to the flesh. Maria's manner was carefully impersonal but John could not rid himself of the feeling that the girl's light touch was also a subtle caress. The idea excited him momentarily, then the impossibility of it all moved him to wry amusement. "You've lost touch with reality, mate," he told himself. "Here you are trussed up like a hog for slaughter and you are entertaining romantic fantasies about the sister of a man who would cheerfully slit your throat if it were not for the bounty on your bloody, stupid head."

Maria was using a lace-bordered linen handkerchief to complete her ministrations. From the handkerchief, in fact from all of her clothing, emanated the clean scent of lavender and as she hovered over John, a new emotion was added to his turmoil. Like an aura, the scent of lavender had always seemed to surround his mother and his sister. Now, the haunting odor triggered a sudden and particularly poignant new onset of homesickness.

Maria straightened to inspect her work. "There, señor, you are fortunate. The skin is not badly broken."

As she moved to return the handkerchief to its place in the low-scooped neckline of her cotton blouse, her eye found another area that needed attention. Leaning so close that John's innards knotted with unfulfilled need, she pursed her lips in concentration and brushed away more sand particles trapped in the tangle of deep brown curls near his jawline.

Maria was so engrossed that she did not see her mother's knowing smile as she paused in the doorway bearing a large earthenware platter of beans, tortillas, and a cup of wine.

"It will be necessary for us to feed ... our guest, Maria," she said as she handed her daughter the wine cup.

30

It took all of John Lewis' will power to keep from wolfing down the beans in their white flour tortillas. Laced with fiery jalapena sauce, they could make a man forget his hunger quickly and beg for more to drink.

When Maria placed the wine cup to his lips, he grimaced involuntarily at the sour taste of the Champlain trade wine from New York. He could not understand why the *californios* did not make their own from the grapes brought to the territory by the padres almost a half century earlier.

Goyo returned before the feeding was finished. Leaning against the table, he watched the scene and made no attempt to conceal his amusement. From time to time he addressed mocking remarks to the women in the soft, rapid *californio* dialect that John found almost unintelligible despite his diligent study of Spanish grammar aboard ship.

At the meal's end John's tremors had not subsided. He needed something hot, preferably tea, and decided to ask for it. Instantly, Goyo was all grave concern. "But, my dear friend, the fire in the *cocina* has been dead for hours. We had no idea we should be asked to entertain so late." He reached out and felt the blanket. "We shall give you another blanket, immediately."

"Why don't you let me get into dry clothes? I have some in my oilskins."

Goyo studied the bundle that had been tossed in the corner beside the front door. For an instant John thought he would agree, but Goyo turned back, smiling warily. "I think another blanket is better. If you will forgive me, I do not trust you with your hands free."

"Then untie my feet and let me get into dry drawers and trousers. I won't try anything. You can cut off my shirt if you wish. Then I shall be glad to have a second blanket."

The slender *californio* laughed. "If I free your legs you will make like an angry mule. No?"

John struggled to hold his temper. "No! Would I be fool enough to try to run to Branciforte with your chair lashed to my back like a turtle in a shell? I will not try to escape. You have my word."

Goyo's smile turned to a thin smirk. "The word of a gentleman, no doubt?"

"The word of an honest sailor!"

31

Instantly Goyo's expression turned to delight. "How fortunate! Not only have we saved a life but we have spared that rarest of all men, an honest sailor."

John turned appealing eyes to the women, but it was obvious that they lacked the courage to intercede. For a time he sat silently, glowering at the *californio*. Deliberately then he allowed the tremors to increase, using caution so they did not exaggerate. A bead of moisture from his nostril formed on the end of his nose. Lowering his head, he tried to wipe it against his shirt front. Before Goyo could stop her, Maria darted forward and touched it away with a cloth. Her brother's quick annoyance turned as quickly to amusement. *"Madre de Dios!* This is a big baby! First his nose, and then what next I wonder?"

Humiliated and angry, John jerked suddenly and nearly upset the chair. Goyo's hand shot to the haft of the vicious-looking knife stuck in the side of his right legging.

John sneered at him. "You would use that on a trussed-up man, wouldn't you?"

"I would welcome the excuse, Señor Lewis!"

"I'll give it to you, the first chance I get, Robles. Meantime, if you think the captain is going to pay you two hundred dollars for a sailor with the chills and fever, think again. Since there are no honest sailors, what makes you think you'll find an honest master? You've been to the customhouse. You should know better."

When Robles made no immediate response, John knew that his words had struck home.

"Bueno!" He moved to the table and shoved the lamp aside. "Maria. Bring the bundle here." Both mother and daughter went for it and hoisted it to the tabletop. Goyo worked the clove hitch free and the oilskins sagged open revealing the meager necessities John had been able to take with him. Robles snorted contemptuously and turned his face away: it was not the clothing he found offensive but the odor of whale oil that had begun to emanate from the impregnated wet weather gear.

The girl pulled out a pair of woolen underdrawers, and his white dress ducks. Goyo took them from her and dismissed both women from the room.

When they were alone, he walked around to the back of the chair to examine the bonds. Satisfied, he began loosening the raw-hide loops around John's ankles. "If you had known how to sit on a horse this would not be necessary.

32

Now, because you are so clumsy, I must be not only your benefactor but your *camerero.*" He glanced up, smirking. "And also I must kneel at your feet."

The wet straps on the heavy leather shoes troubled him. He worked at them impatiently. "No wonder yanqui feet stink!" Then he unfastened the broad leather belt that secured the tar-smeared work trousers. *"Levántese sus asentaderas!"*

John strained to keep his backside off the chair long enough for his captor to slip both trousers and underdrawers free. Then he sat again while the wet cloth was worked from his legs. Extremely uneasy, he waited naked from the waist down as Goyo, on his feet again, stood back and studied him with wonder. "Incredible! On such a big yanqui one would expect more." He indicated the genitals trapped in a tangle of damp curls.

"Well, no matter." The *californio* extended his arms in a gesture of resignation. "But our Spanish women have much more to be grateful for than they know."

Angry enough to kill, John endured the mockery and swore that one day he would even the score. Robles reached for the dry clothing, then reconsidered. "I have a better idea. Without them you will be just as warm and less willing to leave us. *Como no?*" He walked to the door and paused to say, "I hope you are sincerely grateful for our hospitality, señor."

The smile that widened John Lewis' mouth made Goyo Robles all the more anxious to rid himself of his prisoner and collect the reward at the earliest moment. "I shall never forget it, Robles. And I shall see that you are properly thanked—if it takes me the rest of my life."

Goyo left and returned shortly carrying a second blanket. Contemptuously, he flung it around Lewis' shoulders. "If it is any satisfaction to you, señor, neither of us will sleep tonight."

With his boot, he pushed a chair away from the table and settled himself on it. "I am going to sit with you, my friend. At first light we will dress you and deliver you to the longboat at the customhouse. If you are uncomfortable now I suggest you think of the discomfort that awaits you at the captain's hands. It will help you to appreciate your stay with us."

While Robles was removing his spurs and preparing to

33

rest his feet on the table, John surreptitiously tested the painful braided rawhide bonds.

In the dimly lit room, listening to the night sounds that reached him through the two small unglazed windows, John suffered some remorse. He did not regret his attempt to escape—only his poor timing. Had he waited until the night before the *Reliant* was due to sail for San Pedro again he doubted that Nathaniel Burke would have lingered more than an hour or two before hoisting anchor to slip out on the tail of the ebbing tide. He had been too eager for his freedom. Now there would be no second chance.

On the bay the ship's bell tolled twice, marking one o'clock in the morning. From the hills came the eerie yelping of coyotes and overhead the stillness was disturbed from time to time by the raucous cry of a night bird.

Tipped back in his chair, Goyo Robles struggled against fatigue and watched his prisoner through eyes grown heavy-lidded. John was grateful when the *californio* chose to fight sleep with conversation.

"Juan Lewis."

"Yes?"

"Have you not wondered in what manner we found you?"

"I have."

"It was not difficult, my friend."

"How so?"

"At first you did not make mistakes. We knew you would come ashore below the old fort."

John was surprised. "How would you know that? The only man who knew I was leaving the ship and where I was going would die before he would tell."

Robles laughed easily. "When we heard you were missing we knew you would swim to the creek. They all do. It is the only safe way. So we started the dogs there but you stayed in the water so you left no scent."

"And then?"

"And then you made a mistake, up near the top. You stepped into fresh ox dung. We followed you for two or three leagues until we were certain of your destination, then we took a shorter route and waited."

"How did you know I would go to Branciforte?"

"Because every sailor on the California coast knows

34

that Isaac Graham and Will Chard will take them in. Nobody else would dare."

Goyo leaned forward. "But I must compliment you, Señor Lewis. You were much faster than we reckoned. For a sailor you move well on land. You did not keep us waiting too long at the river. For that we thank you."

John smiled bitterly in the near darkness. Half aloud, he recited the sailors' maxim.

"A man's a fool and more o' the same
To bet his gold on another's game."

Goyo glanced over curiously. "What have you said?"

"Nothing. I was remembering a short sermon—too late."

The *californio* made a show of being pleased. "Ah! You are a religious man."

John laughed mirthlessly. "Only when I'm in trouble."

The retort amused Goyo. "*Verdad, amigo!* Except for the padres and the women the same may be said of us all."

They fell silent then, listening to the night sounds. Faintly, they heard the *Reliant's* bell sound one-thirty. First light would not begin to gray the sky behind the Monterey hills until past five. There would be four more hours to endure.

John pondered his possibilities. He weighed the chances of surviving Captain Nathaniel Burke's punishment against those of convincing Robles that he had no personal ambitions beyond visiting the northern valley before seeking another ship to Yerba Buena. He knew Burke would flog him. He had seen the ship's master at work with the cat o' nine tails. No seaman had died from the floggings but neither had any been completely whole again.

It would do little good to protest to Robles that any other ship would be preferable to the *Reliant*. As unrelenting a disciplinarian as Nathaniel Burke was known to be, he was still considered more humane than many masters.

Somehow John managed to doze. The muffled stomping of horses in the adobe-walled compound in back of the house roused him just as the *Reliant's* bell struck the couplets that marked three o'clock.

A sharp sound nearby made him start. Goyo Robles stood in the doorway, silhouetted against the rectangle of

35

lightening sky. He was listening intently. Some yards away a party of horsemen, apparently riding down the calle principal toward the customhouse, clattered by noisily. When the sound diminished, Goyo turned and took his spurs from the table.

"Another half hour and we go, yes?"

John shrugged. "Do I have any choice?"

Robles sat down and began to secure the ornate spurs to his boots. "Señor Lewis, you have none." He strapped the second spur in place and rose. "Maria will watch you while I awaken my friends. Then we will dress you and escort you safely back to your ship. That prospect may please me more than it pleases you, my friend. But that is how it will be." He stepped into the adjoining room and a few minutes later John heard voices.

The blanket that had been draped across his upper legs had begun to slip. In vain he tried to maneuver it back into place. The best he could do was catch it between his knees to keep it from slipping off completely. By twisting his shoulder and working the upper blanket with his chin he was able to divert a corner to his lap. He had barely managed to cover himself when Goyo came back followed by Maria. Goyo indicated John with a nod. "For an uncivilized yanqui he is very docile, but you watch him until I get the others."

The girl took one of the chairs at the table and sat with eyes lowered and hands folded in her lap.

"Maria?"

Startled, she looked up, then lowered her head again. "Si, señor?"

"Do you speak any English, Maria?"

"Poquito, señor. Pero, prefiero hablar español."

John laughed ruefully. "I would prefer to speak Spanish, too. But I do not know many words. Will you speak in English with me?"

Maria Robles lifted her head slowly and looked at him for a moment. "I will try if you wish, señor."

"I do wish. Indeed I do. I have not spoken with a lady for three months. May I ask how many years you have?"

"I have sixteen years, señor."

John shifted and winced as the bonds dug deeper into his wrists. "I have a sister. She has seventeen years and I think she is as beautiful as you."

36

The girl's luminous dark eyes brightened with interest. "Where does she live, señor?"

"In the United States of America, in a city called New York."

Maria Robles' lips formed a silent "Oh." Briefly, they were without words. Then she looked up again. "New York is a place very distant, is it not?"

Suddenly it seemed to John that New York and home and family and old friends were all on another planet. He nodded wearily. "It is very distant, *muy distante. Muy lejos, sí.*"

The words surprised the girl. "*Señor! Usted habla muy bien!*"

He bowed awkwardly. "*Gracias, señorita.* I do not speak well. But I try. I must speak more but I have no opportunity. And now I will have none at all because," he nodded toward the harbor, "your brother is going to see to it that I am returned to the ship in irons."

Maria Robles lowered her head unhappily. "My brother would have all foreigners deported. He is afraid of yanquis, especially because of Texas. So is Castro. They try to make Governor Alvarado think in the same manner."

John alerted. "Doesn't he agree?"

Maria Robles shook her head. "Once he did. But no longer, I think."

In his eagerness to grasp at straws John forced the chair forward. The motion caused the blanket to slip from his shoulders. Frantically, he managed to clamp a fold between his knees. When he leaned over in a futile attempt to secure a corner in his teeth the girl jumped up. Her face averted, she caught the displaced edges and redraped them over his legs and shoulders. In other circumstances John would have been embarrassed, but between the lines of her aimless conversation he had detected a faint note of hope.

"Miss Maria."

"Yes, señor?"

"If Governor Alvarado knew that I want nothing more than to become a good citizen here," he indicated the town with a nod, "like Mr. Larkin—do you think he might let me stay?"

"He knows you are here, señor. My brother told him yesterday. He said nothing. He would be sympathetic but he does not want to go against his friends until all Alta

37

California is a sovereign Mexican state. If you were an important man and your government might be offended, that would be different."

John finished the thought. "But a common sailor is another thing. True?"

"Si, señor. For my brother, that is true."

He was considering how best to ask her to plead his case with the governor when Goyo returned with his three companions. "So, my friend, we go now."

He dismissed Maria, who retreated to the adjoining room. Some minutes later, flanked by two men on either side, John was led to the door. Out of the corner of his eye he saw Maria Robles reappear and half turned to thank her. Instantly, Goyo spun him away roughly. "I said we go—in the direction of the harbor—not my sister!" They forced him through the door then and Goyo, carrying a loaded pistol with a second one stuck beneath his sash, walked behind as they made their way down the calle principal toward the customhouse.

In Middletown or in Boston the distance would have measured no more than several long blocks. To his right were the walled gardens at the rear of a row of dwellings that faced on the next street. He knew that one of them was Governor Alvarado's private residence and that others belonged to leading native families and to trading agents. On his left was the imposing unfinished residence being constructed for Thomas Larkin. Several of the buildings had long pole lean-tos set from the eaves to the ground on their weather sides to keep the driving rain from eroding the soft adobe bricks.

Off to the southeast, just growing visible in the graying dawn, he could make out the tower and belfry of the presidio chapel, the most impressive structure in Monterey.

Some yards from the customhouse John heard loud voices speaking in English. The sounds were coming from one of the several cantinas, the miserable adobes that housed the bars patronized by sailors and by the "young women of good family" who would devote themselves to the itinerant customers "for a price."

When they were opposite the hitching posts, Goyo raised a hand. "*Espérame aqui!*" Quietly, he moved closer to the horses tethered there and examined their brands. A moment later, visibly upset, he returned and began speaking rapidly to his companions. John caught enough to

38

learn that the animals belonged to Isaac Graham and Will Chard and their men. For an instant he thought of shouting for help but reason told him that even if he were heard the commotion would probably only compound his trouble.

Robles and his men conferred briefly and John found himself being half dragged toward the customhouse.

A pre-dawn visit from Isaac Graham had prevented the clerk, Ignacio Torres, from dispatching an oarsman to summon the *Reliant*'s longboat. A former trapper turned distiller, Graham was in the midst of making his usual sub rosa arrangements to ship casks of illegal whiskey up the coast when Goyo Robles and his men burst into the room, forcing John ahead of them. Graham whirled, his hand on the long knife stuck in his belt. "Hyar now, Goyo. You're a mite early throwing white boys off the shore, ain't you?"

Robles, taken by surprise, pretended to ignore the question. He steered John to a heavy chair against the opposite wall. Curious, Graham sauntered over for a look at the captive. After a brief inspection he turned back to Robles. "I axed you, Goyo, but I didn't hear you do me no courtesy by answering."

The *californio* completed an unnecessary testing of the rawhide bonds, then he looked defiantly at the one-time Tennessee trapper, whose eyes gleamed with malevolent humor.

"If we had been wise, Señor Graham, we would have begun these deportations much sooner!"

Isaac Graham's large nose seemed to thrust like a threatening weapon from the bramble of his dirty red beard, but he said mildly, "The p'int's took and noted, amigo. I'm storing it against the day you and Bautista'll be calling on our rifles again." He brushed past Goyo and walked to John. "What be your name, son?"

"Lewis. John Lewis, sir."

"How old be you?"

"Nineteen, sir. Coming on twenty."

"What did you ship as?"

"Common seaman, sir. And apprentice supercargo."

Graham's overhanging brow lifted. "I take that to mean you're book smart then?" Before John could respond, the trapper turned with amazing speed. "Goyo! You and your boys stay put!" He glanced over at the unhappy clerk who had half risen from his desk. "You too, Ignacio. I'll take it

39

unfriendly if you're in such a sweat to get your bounty ye cain't let a friend among ye get caught up on news from his homeland."

Unmistakably reluctant to challenge the mountain man, Goyo and his men stood as they were. Graham turned back to John. "Why are ye making tracks? Steal? Let blood?"

"No sir. I've been a year and a half on this coast buying and droghing hides. Mostly south. Now they want to keep me against the articles. I see no prospect of getting back to Boston for another year. I thought I'd like to see the country. I was trying to get to your place when they caught me."

An expression of great delight transformed Isaac Graham's face. His high-pitched nasal voice soared. "Ye was coming visiting? How nice! I sure am touched. It ain't every day we get friendly callers." He moved closer and laid a heavy hand on John's shoulder. "Burn my britches! My eyes might get to tearing from happiness," he turned slowly toward Goyo Robles, "if it wasn't for the dishospitable nature of some."

He squeezed John's shoulder reassuringly. "Goyo, you and your friends wouldn't do me and mine out of a chance to visit with a blood brother from the United States, now would ye?"

Goyo's dark eyes shifted from the sailor to his companions and to Torres, from whom he knew he would get no help. His reply was caught in a throat strictured by rage. Robles knew he must act or lose his prisoner. His pistol might take care of Graham but the trapper's lightning-quick throwing knife would take care of him. In either event the shot would bring Will Chard and the others from the cantina in seconds. Because of Goyo's open antagonism toward *extranjeros* neither Castro nor Alvarado would be likely to take his word that Graham had precipitated the trouble. He forced the words out. "This man is a fugitive. We have an obligation to return him to his captain."

Isaac Graham's eyes widened innocently. "Well, now. On whose law are ye standing?"

No formal extradition agreement existed and Goyo Robles knew it. Had such a law been adopted Graham himself would have been expelled from Alta California. A foreigner caught entering Alta California by land or sea

40

might be expelled on any improvised pretext and by any means, at the option of the local authorities. Goyo answered stiffly, "It is in the interests of our commerce to see that ships trading with us are not deserted by their best hands. Masters will pass us for other ports if they know we give asylum to their deserters. You should know, Señor Graham, that more sailors desert here in Monterey than in all other ports."

"That be the truth, Goyo. That be the truth." Graham turned away casually, and before they realized what he was up to he had unsheathed his big knife and slashed the bonds that secured John Lewis' hands.

Smiling amiably, he replaced the knife. "How much is on his head, Goyo? Fifty? Hundred dollars?"

Without waiting for an answer Graham stepped between John and the *californio*, who had moved toward him. "You stand stock where you are Goyo, because you're too pretty to spill your juice." Over his shoulder he addressed John. "Son, you hightail to the cantina and fetch a man named Will Chard."

Massaging his wrists, John bolted for the door. Less than a minute later he flung aside the cantina's beaded entrance drape and called Chard's name. A tall, clean-shaven, neatly dressed young man with long copper-brown hair rose from the table. "I'm Chard, sailor. What do you want?"

"Mr. Graham wants you at the customhouse right away. He's got trouble."

The words were scarcely out before Will Chard barked an order and John found himself being swept through the door by a dozen charging men. Seconds later their horses passed him at a dead run. John was only halfway down the path when he saw the men pull up their mounts, leap off, and disappear into the building. John ran up out of breath and Chard and another man met him in the customhouse doorway. Chard barked, "This is Bill Dockey. You go with him."

A grimy frontiersman dressed in fringed buckskins grasped John's arm and pushed him to one of the mounts grazing quietly between its trailing reins. "Get on, boy!"

John pulled his arm free of the shorter man's strong fingers. "My gear's in there. Everything I own."

Bill Dockey retrieved the reins and bunched them on

41

the animal's neck. "Don't fret. You'll get them. Now get on, boy. Get on!"

Awkwardly John mounted the animal. By the time he was up and had maneuvered his square-toed seaman's boots into the stirrups, Dockey was beside him. "Can you ride a horse—fast?"

John nodded uncertainly. So far as he knew, he had done nothing to encourage the animal, but an instant later the tough little mount was following Dockey's larger horse in a flat run along the low sandy embankment that bordered the edge of the bay.

Ten miles north, around the gentle arc of coastline and not far from the mouth of the Pájaro River, Bill Dockey turned inland to the bank of a narrow brush-lined creek. Hidden from view, they rested and watered their mounts. Soon they were on their way again.

They crossed the Camino Real well west of the point where Goyo Robles and his men had set up their successful ambush late the night before. Another two hours brought them to the outskirts of the pueblo of Branciforte.

A well-worn trail led up the north slope of a small canyon. They followed it until they came to a grove of redwood trees, and finally emerged at a high meadow rounded by taller redwoods. Scattered around its periphery were a dozen or more log houses. When they rode into view, men began to gather. Dockey turned to John. "You're new. They'll be asking for news. If you want to fit cozy here, you tell it to them."

"But I've been away from the states for over a year. I don't have any news."

Dockey grinned. "Then make up some." He started to ride on ahead but checked his mount and looked back. "If you do make it up, boy, make it interesting so I can enjoy it too!"

Dockey rode to the largest of the cabins and dismounted as Isaac Graham's men encircled them curiously. Not a word was spoken until John had come down stiffly from his animal. Then his guide addressed the crowd. "This here's a sailor name of—" He frowned and turned, questioning. "Name of what?"

"My name is John Lewis—out of Boston."

Dockey nodded. "Name of John Lewis, out of Boston. Mr. Lewis is arranging to miss his ship and Isaac'd like us to help him in that noble work."

Several men guffawed. John identified them easily as sailors who had "gone over the side." At least he would start with that in common.

Dockey, a short, wiry, hollow-cheeked man whose plains dialect betrayed a trace of cockney, indicated the doorway. "This is Cap'n Graham's headquarters and meeting place. He wants you to wait here until he gets back. We'll boil up some coffee. Go on in."

The cabin consisted of one room made of small, straight redwood boles. The peaked roof was supported by hand-hewn rafters. The furniture was crude and sparse, a table and a scattering of chairs and stools. Most small articles were suspended from pegs driven into the log walls. An area about eight by twenty feet at the far end of the cabin was partitioned off by canvas draped over a line. A wooden bunk was visible through the half-opened hanging. A makeshift rack above the bunk held a collection of long rifles. There were no proper windows and what light there was entered through a series of high slit-ports. Obviously, this cabin was intended to serve as the defense center in case of attack, since the ports commanded four sides of the structure.

Dockey pointed to a covered container and said to John, "If you want a chaw of dried beef, it's in that crock."

John smiled gratefully. "I could use something. Thanks." He helped himself to a foot-long strip of mahogony-hued dried beef and worried an end off between his teeth.

Dockey watched him struggling. "It ain't very filling but it'll keep you so busy you won't notice being hungry."

Lewis maneuvered a stubborn wad to one cheek and nodded. "It beats the salt horse aboard ship."

"Salt horse?"

The man turned from the stove to give him an incredulous look. "That ain't even human. What were you on—a Frenchman?"

Lewis shook his head. "I was on the *Reliant*, a Yankee brig. I don't know what the salt meat was. We didn't dare ask."

Dockey turned away, shaking his head. "I don't reckon you did."

They were silent for a time until John broached the

43

matter that was much on his mind. "Do you think Mister Graham will let me stay here until the *Reliant* sails?"

Dockey shrugged. "That's up to him. We elected him cap'n. If he likes you, that'll be right with us—unless you got some quirk that don't fit in."

"I'm peaceable. I do my work and mind my own business. I got on with my mates aboard ship."

The statement earned him a skeptical look. "Is that why you jumped?"

"No, I jumped because I found out the master was going to leave me on the coast again, curing and droghing hides. I spent six months sleeping in a hide shed in San Diego. I'm not of a mind to do that any more."

"What are you of a mind to do, boy?"

John answered honestly. "Don't know. But if they take me back aboard ship it'll be dead, in canvas."

Dockey chuckled. "It ain't the first time we heard that talk."

John decided that his best chance to persuade Graham that he could fit would lie in finding out as much as he could about this strange little isolated community of men with no apparent means of support. "I don't mean I don't like work. I'll be glad to work at anything you men do if Mr. Graham will let me stay here for a while—at least until the *Reliant* is on her way to San Diego again."

Dockey continued fussing at the stove. John decided to press a little more. "I could fit in, if you men do any of the usual kinds of work."

Dockey indicated the wooden water bucket. "Fetch me a dipper."

John complied and Dockey held it poised over the big coffee pot. "If we did the usual kind of work, boy, we wouldn't be here."

Will Chard and his men rode up an hour before sunset. Isaac Graham saw John Lewis standing in the doorway and called to a man named Joe Majors. "Give the lad his duds."

Majors released the rawhide saddle ties and let the tarpaulin drop to the ground. John recovered it and started to carry it inside.

Graham watched, amused. "Hyar now! Leave it outside. I don't aim to take no sailor boys under my robe."

Majors started to guffaw but a sharp look from Chard

sobered him. Graham fumbled beneath his beard to loosen the laces at the neck of his deerskin shirt. "I think I need a hair of the dawg."

Chard followed Graham inside but John, embarrassed, waited outside the door. A moment later, Graham called to him. "Come on in, lad. If you're a jug-sucker at such a tender age, you're welcome to a little fortify-cashun of the liver."

Graham smeared his bewhiskered mouth with the back of his hand while Chard took his pull at the demijöhn. The powerful *aquardiente de trigo* seemed to have no more effect than plain water on either men.

Chard gave the neck a token swipe with his palm and offered the jug to John. Raw liquor was not what he wanted but he took the jug, tipped it expertly over the crook of his arm, and deliberately metered out and swallowed a large draught. As the rough spirits seared his gullet, he thanked God he'd learned to hold his own in the cantinas in San Diego and San Pedro. Almost nonchalantly he returned the jug to the table.

Graham eyed him as he reached for the jug. "Ye look like a man, ye talk up like one—and ye drink like one. What else can ye do like a man, son?"

"Pretty much whatever's needed." The statement was made with no trace of boastfulness.

"From your looks I've a mind to believe you. How about ye, Will?"

Chard finished his leisurely appraisal of the young six-footer. "I'm inclined to accept that without further proof."

Graham laughed. "I expect Henry Mueller will be asking him to prove up though."

John understood the implication immediately. He would have his measure taken and the leader would make no decision about allowing him to stay until he proved himself. He didn't have to wait long.

Later that evening when the men were gathered around the fire there was a clamor for late news from the States. John had been trying to remember everything he had heard during the past year. As he sifted through the rumors and gossip he labeled each account for what he believed it to be:

"We spoke the barkentine, *Jessie Dawkins,* out of Charleston. Her master came aboard and told Captain Burke that a fire was burning the whole city down. He

45

said that by the time he cleared the roads over a thousand buildings had burned to the ground and only God Almighty knew the end to it. He told us it was no use putting in for anything. Not biscuits or water.

"There was a copy of the *New Orleans Picayune*, a new journal, that the *Dawkins'* master traded for our Boston journals. It had an account of a steamer named the *Moselle*. She was a stern-wheeler on the Ohio River near Cincinnati. She blew her boiler and steamed alive over a hundred souls."

A voice from the fringe spoke up. "What's happening in Washington?"

John Lewis shook his head. "We didn't hear of any late happenings. The first thing President Van Buren did was to sign the bill making the Michigan territory into a state. There are twenty-six regular states now."

Henry Mueller, an oaken barrel of a man in his late thirties, spoke up for the first time. John was surprised at the incongruity of the high-pitched voice issuing from the solid body. "I don't know about you so'jers, but I want this here smart-ass sailor boy to tell us more lies. A thousand buildings burning to the ground in Charleston and a hundred bodies boiled in Cincinnati. Them's lies. But I think I'd rather hear made-up news than the true happenings anyway—especially if they're made up by an artistic liar."

Will Chard started to rise but Isaac Graham rested a hand lightly on his knee. The man named Majors was the first to respond. His manner was facetious. "Why, Henry, ye ain't a bit choosy about your words, are ye?"

Mueller bowed with mock humility. "I admit I'm short of learning, Joe." Suddenly his manner changed. "But I reckon I got me enough words to fix the proper name to a liar when I see one."

A nervous stir ran through the group. John understood that any attempt to defend the truth of his statements would be taken as a sign of weakness. Besides, the truth was not at issue. He rose slowly from his place and forced himself to speak in an amiable tone. "I expect you'll own that I have a few words too?"

Mueller edged forward on the stump. "I grant ye that. In fact, no honest man would have need for so many words as ye been spewing tonight."

Still pretending amiability, John said, "And so in your

46

ox-dumb head you think that any man who can outtalk you is a liar! Any schoolboy reciting his lesson would be marked a liar by you too because you don't have enough light to understand even that."

Nervous laughter broke out among the men.

Henry Mueller had come to his feet upon hearing the words, "ox-dumb." The young sailor was not behaving as he had predicted. Instead of offering a defense, part apology and part outrage, Lewis was giving as well as he was taking. Isaac Graham, beaming behind the snarled brush of beard, rapped Chard's thigh with the back of his hand. "Might be we fotched up with a good one."

The second-in-command nodded without taking his eyes off the adversaries who were moving closer. "You'll know soon enough. But I want Mueller's knife."

Mueller usually enjoyed these encounters. But the word "dumb" had triggered hurt and anger whose roots went deep into the past. "No boy or man never called me 'dumb' and lived to do it twice . . . and if such there ever be . . . it won't be a scurvy bugger boy!"

"MUELLER!"

Will Chard was on his feet. "Throw that knife over here!"

Mueller glanced down at the buckhorn haft showing above his deerskin legging and seemed to hesitate. Chard's voice rang out again. "Right now, Mister!"

The move was so fast John scarcely saw it. The heavy blade grazed him and buried itself two inches in the log behind Chard, who ignored the quivering weapon as he addressed Mueller again. "I think it would be well to remember two things, Henry: you started this—and the captain and I intend to see it done fair."

Mueller's charge came before Chard had finished his warning. It was the man's favorite opening maneuver with those who did not know him. The flying, battering ram head-butt in the solar plexus usually ended the fight before it started. John was spared the full impact by instinctively angling away but Mueller's hooked left arm snagged his body and they both went down. Mueller's thick fingers caught at John's neckband and ripped it away. Immediately he was on his feet again, charging. John had only seconds to ward off the attack with a chopping blow that caught the older man on the side of the head. It did no damage but it bought John time. He began circling to get

47

the fire at his back. Mueller watched for a moment then winked at the others. "Now here's a smart bugger-boy. Going to get the fire behind him so I can't see clear and I'll jump in it, like as not." He laughed and deliberately turned his back on Lewis to address the men. "Gentlemen. It's natural to the working of a liar's head to be figgering all the time. And its natural to a coward, to boot. You seen him weaseling around? I say he's no part of a man. I say he's a yellow fancy dancer—no sailor boy. What say ye?"

Mueller knew his men. He knew they would not reply but he also knew that if John attempted a surprise attack from behind it would instantly be mirrored in their eyes though none would call out a warning. At this game they were impartial spectators. Long since, they had judged Mueller as a good fighting man. But the sailor was a new sort. Like Chard, he was articulate and smart. Mueller was trying to goad Lewis into a fight on Mueller's own terms.

As the trapper continued addressing his audience, John Lewis moved several steps closer. "Mueller.'

The man whirled with mock surprise and indignation written on his beefy face. "Mueller? Mueller, did ye call me? *Mister* Mueller. That's what you call me. And lucky you are that I don't ask ye to kneel and kiss my hand!"

John managed a convincing show of equanimity. He knew it would be fatal to allow the older man's obvious tactics to stampede him. "Very well then, *Mister* Mueller. I want to say that if you can fight as well as you can jaw then I admit I'm whipped. And if it's all the same to you, I'd just as soon kiss your hand. It's a sight cleaner than your ugly face."

The mock surprise changed to disbelief as a loud guffaw rose from the tensed audience. For an instant Henry Mueller's mouth hung slack. Then it ovaled and a wordless bellow trumpeted from it as he charged. A step from closing he feinted to his left. When John counter-moved to avoid the charge, Mueller, as agile as a cat now, reversed his angle and the two collided and fell, John on the bottom.

A shout rang out that neither man heard. John, struggling beneath two hundred pounds of hard muscle and bone, fought frantically to keep the clawing fingers from tearing out his eyes. Suddenly he found one of his legs free. He drove it with all of his strength into his assailant's crotch. The blow produced an agonized bellow of rage but Mueller did not falter. Twice more John drove his knee into the older man's scrotum until he forced him to twist to one side for protection. The move freed his other leg. He lifted it high and crashed the heel of his heavy sailor's boot down into the small of Mueller's back. The kidney blow brought little more than a grunt. More blows followed until Mueller squirmed still farther away to avoid the hammering boot. Instantly, John twisted free. Mueller tried to grasp his thick hair but missed as John lurched backward in a half-somersault, rolled up on his knees, and shot to his feet.

There was no waiting now and John knew it. The first sledge-fisted blow struck and split Mueller's cheekbone while he was still scrambling to his feet. A second blow caught him on his Adam's apple and a squared boot-toe sank deep into his solar plexus. John attacked now with the intent to kill. If a fair fight in this company meant nothing barred short of clubs, knives, and guns, then he intended to use every advantage he possessed. He aimed another kick and somehow Mueller managed to grab his leg and they both went down again. He freed himself with a piston-like drive of his other boot, ripping a broad strip of skin from the back of Mueller's hand. Both men scrambled to their feet again, sucking for air in long, hollow gulps. For a moment they stood eyeing one another and John was aware for the first time of the strange silence that had fallen over the men around him.

He heard Mueller taunting him again in the high, raspy, nasal twang. "If you aim to play smart bear to my *dumb ox*, bugger boy, then before you do I'd call to mind that a sight more bears then bulls get killed in the pens—and ye ain't even a weaned cub yet!" He spat blood. "Killing you will be easy!"

Smiling, Will Chard turned to Graham. "Henry's not building his argument on the facts."

The captain nodded. "He ain't fighting his usual fight either."

What was obvious to Graham was also obvious to Mueller's companions. The man was behaving strangely. It had something to do with Lewis' first words—something to do with being called a "dumb ox," for those were the words that Mueller had just repeated and there had been unmistakable hatred in his voice as he said them.

Mueller was moving again, circling, advancing in a slow-closing spiral. John stood his ground, turning with him, until he found himself being crowded into close quarters again. He felt certain, after assessing the damaging impact of the two blows to Mueller's head, that he could knock the man senseless if he could get within range without being trapped in the crushing embrace. Another charge would come at any moment now. Unless he could change his own strategy the result could be indecisive. As he continued to maneuver to keep Mueller in front of him, John was remembering the frequent fights aboard the *Reliant*. One stood out. It had been a grudge fight between a Swede and a Basque. It had ended when the Basque resorted to a tactic common to his people. The surprise attack had sent the big Swede crashing head first into a hatch coaming. Three days later he was dead of a skull fracture. At the time John had thought the tactic unfair but his shipmates had laughed at him.

Mueller was circling faster now, maneuvering to get his own back to the fire, to position himself so that his charge would carry his opponent backward into the circle of spectators where a close friend or two might give him some covert help in the confusion.

Snarling vile epithets, goading, attempting to confuse, he circled closer and closer. Realizing that time was running out, John made his decision. Apparently confused, he stopped and glanced around almost as though he were seeking an avenue of escape. Mueller charged. When he was a short step away John sprang upward, gathered his knees close, then rammed both legs outward like two great uncoiled springs. The boots, held close together, struck Mueller's breast bone. The blow expelled the wind from his deep-barreled chest in an explosive, asthmatic whoosh, and his thickset body catapulted backward, seemed to leave the ground, and crashed into Isaac Graham and Will Chard. John heard neither the shout of protest that went up from the men nor Graham's surprised outcry. Before Mueller's body had come to rest John leaped at him again and brought his one hundred eighty-four pounds down, feet first, onto the man's belly. A stream of sour puke geysered from the gaping mouth. All caution gone now, John had only one purpose, to kill the man who would kill him if he did not. He threw himself on the sprawled body and began driving mauled fists into Mueller's jaws with a fury that brought more cries of protest from the onlookers. Swinging madly, John scarcely felt the hands pulling. When they began to haul him away he shook them off and continued the murderous punishment. They seized him again and finally, exhausted and struggling for air, he allowed himself to be pulled free and pinioned.

Henry Mueller strangled on the clotted, half-chewed beef he had wolfed down little more than an hour before the fight. They tried to save him by forcing fingers down his throat—and smooth twigs—but his face mottled and turned blue, and his eyes, bulged wide with terror and appeal, glazed and after a minute or so the battered hulk of a body ceased its convulsing and lay still.

Fifty men, inured to violence and death, stood looking down in shocked disbelief. None was more stunned than Isaac Graham, for it was not the way death should come to a man such as Mueller.

Ever since the late twenties Graham had watched Henry Mueller drink and fight, always with roaring good humor. Once he had seen him douse a slumbering drunk's long hair with raw whiskey and set it afire. The man had been marked for life but that was no more than the usual prank played on the unwary. Graham had agreed with the others:

"It was a thing to even up someday, but it wasn't something to hold a grudge over." But this was different. This lad was different.

Chard, sensing the uneven temper of Graham and the men as they stood silently looking down at the still warm corpse, moved unobtrusively beside Bill Dockey. "Take the lad to my cabin and be sly about it. Stay there with him." As Dockey started to move away he stopped him. "And better load up a couple of my pistols and see he knows how to use them."

Unnoticed, the little mountain man eased John Lewis back into the darkness. An hour passed before Will Chard returned to the cabin. "You didn't kill him, lad. Graham knows that and so do the men. But they don't like the way you fight."

Swollen, torn, and dirty, and still struggling with his smoldering rage, John sat on the edge of the bunk pressing a wad of rag soaked in *aquardiente* against the abrasions on his cheek where Mueller's fingers had clawed the flesh. "I would have killed him. He was trying to kill me."

Chard nodded. "I believe that. For some reason I don't fathom, your words unsettled him mightily."

John removed the compress and examined the dark red stains. "He's properly settled now."

Chard laughed full out. "You're not much for remorse, are you?"

"I feel none for men who are animals."

Chard looked surprised. "How about that. You're a right high-minded young man, if you mean what you say." He studied Lewis for a while. "How old are you?"

"I'll be twenty in a month." Turning to Dockey, Chard indicated the door. "Bill—fetch Lewis' things from Isaac's cabin as fast as you can without making a show of it." He turned back to John. "This may not be the best place for you for a while. When is the *Reliant* posted to sail?"

"On the morning tide, unless Captain Burke lays over hoping to get me returned."

"He might. For a good hand. No master's ever in a hurry on this coast unless he's outrunning weather. And he knows you're here and who you're with. Robles will see to that."

Chard walked to the cabin door and appraised the sky. "You'd better ride tonight. I'll send Dockey with you to Natividad. You stay there at the distillery."

John glanced at the compress and set it aside. "What if he's no happier about me than the others?"

"Dockey's my man, and he's fair. Also, he's felt Henry Mueller's weight on him a time or two. He won't be doing much mourning, believe me!" He smiled cryptically. "And, my friend, neither will I."

John was plainly surprised. "What about Graham? I mean if you aren't—uh—"

Chard cut him off. "If we don't see eye to eye, then why am I cahooting with him?"

John nodded and Chard looked off, reflectively. "We are useful to each other, for the time being. Our men guarantee us a certain security. I find it useful to take advantage of same—until I'm in a position to go into the lawful trade. He eyed John narrowly. "If you are wondering about my concern for you, lad, it is simply this—all things being equal, I may find you useful also. Does that interest you?"

52

The answer was immediate and direct. "That depends on what you have in mind and whether it suits my purpose."

Chard regarded John with amusement. "Tell me, lad. Aboard ship were you ever trussed up and flogged?"

"No. I did my work and minded my business."

"And I take it you weren't as quick to retort with Captain Burke as you are with me."

John grew wary. "I can't see that crawling is going to get me where I want to go."

Still amused, Chard moved closer to inspect the wounds. "We'll find out where you want to go later. Do you speak Spanish?"

"Some."

"Good. Then you'll understand the Californio saying, *'Por el canto se conoce el pájaro.'* If it's true that a bird is known by its song, then I'd say I may have caught myself a young eagle!"

Dockey came in with the bundled tarpaulin. Chard issued instructions about horses and the three moved quietly to the corrals at the edge of the woods. When John and Dockey were mounted, Chard indicated a low pass through the hills. "Go to Alviso's and stay there until I can ride in. I want to judge the temper here before we add any replacements to our company."

John glanced at Dockey. "Just set me a course. I can ride alone. I don't want to bring trouble down on anybody. The men will blame him for helping me."

Secretly pleased, Chard dismissed the idea. "They'll blame me. Bill's my man. They'll not fault him for following orders. Alviso and his partner, Butron, are friendly. They leased us the land for our distillery. They like our money and our whiskey better than Alvarado's home rule. Now make tracks, lad. Pronto!"

Goyo Robles was outraged. Tokoya, an Indian and a former menial with Will Chard's band who had been banished because of his affinity for the powerful *aguardiente*, had spread the rumor that Graham and Chard were plotting to take over all of Alta California.

Goyo was determined to get to the bottom of the rumor. He told Maria, "The Indian's rumor must be taken seriously. He hasn't enough intelligence to invent tales. Perhaps, at last, I can persuade Alvarado to send Castro and his soldiers against Graham. They must be driven out of Alta California."

Maria pulled at her brother's sleeve. "You dream, Goyo. If Juan's own uncle, Mariano Vallejo, dares to say publicly that he believes the foreigners are necessary to insure our independence from Mexico, then you are stupid to believe you can force Juan Alvarado to act. All you wish to do is avenge your honor. You are angry because Lewis was taken from you. Why were you not angry when other American sailors, who were not half so agreeable, strutted like arrogant cocks on our streets, accosted the women, and rode off with Graham?"

Goyo glared at her. "So you found this one agreeable? As agreeable as he found you, without a doubt!" He turned away. "All the more reason to rid this land of foreigners, and with the help of God I am going to do that before it is too late!"

Crude by established standards, the distillery at Natividad was a master work of improvisation. Copper sheeting salvaged from wrecked vessels had been worked into large pots, set over brick fireboxes. The condensing coils spiraled down into the pots from large domed covers and were cooled by an ingenious system of wooden drip troughs. The alcohol dripped from the condensing coils into holding casks.

At peak production the distillery could produce three hundred gallons every seven days. Harry Morrison ran the still. He was a compact bandy-legged Scotsman in his mid-thirties. He had been a British Navy pharmacist who had deserted after a fatal fight in the Vieux Carré in New Orleans. After he had shown John Lewis the workings of the still, he beamed at the sailor's compliments.

"Aye! But the great problem is, Lewis, that good whiskey wants its color from ageing in char and not from the tincturing and tampering wi' nature we do here. Graham calls this 'bourbon' for sentimental reasons. But a mon that knows good whiskey would never call it more than fuel for a spirit lamp. Personally, the sweetening wi' molasses suits it more for pouring on Johnny cakes than for serious drinking. I wouldna downrate our liquor to foreigners but I miss the smoky tang of peat in it."

After the meal, John, Bill Dockey, and Morrison sat talking over a jug of *aguardiente*. John was forced to con-

54

clude that Graham's own private stock was no improvement over the run-of-the-still spirit that Morrison was serving. The Scotsman watched closely as his guest downed the first swallow with no visible reaction. "Aye, laddie, ye drink well! Either you've got a martyr's self-possession or a leather gullet!"

Still hurting from the fight, John Lewis drank so well that he had no trouble sleeping on the hard floor. Neither was he bothered by the ubiquitous fleas that infested most of the California coast. Resigned to their constant attacks, the tolerant *californios* had even composed several serenades to them. One of the songs listed the searching out and killing of the vermin as one of the services a bride and groom perform for each other.

Will Chard arrived at Natividad late the following afternoon and watched John help Morrison and Dockey set new batches of mash. Lewis handled the *quintales* of wheat as though they weighed half their hundredweight. When Morrison indicated the newcomer and nodded covert approval, Chard called John aside. "The *Reliant* cleared for San Pedro yesterday, so you are safe enough on that score unless Nathaniel Burke has made some arrangement to have you caught and sent south aboard another vessel."

Relieved, John sought further reassurance. "What about your men?"

Chard pursed his lips. "I may not sense their temper as well as Graham but I think you'll be all right, though it would be well to stay away from the camp for a time."

John nodded. "Then I'll be on my way up the valley. I'll stop at Yerba Buena for a while and look around."

"Why? Doesn't this place suit you?"

"It may be I don't suit this place."

"You suit Harry," Chard said emphatically. "Why don't you stay here and learn? I told you two days ago that I had plans that you might fit into."

John shrugged. "I have no plans. But if yours are to my liking then I'll be glad to learn the ropes, though I won't promise to make whiskey a life work."

Will Chard laughed. "If you'd settle for that, lad, I'd not settle for you! Now go about your work and I'll see that you're snugged in here and left alone—but for Harry's nagging. You'll find he is under the delusion that we should be making ambrosia for the gods."

55

From then on, John saw no one but Morrison and the Indians who did menial labor around the Natividad distillery. The Scotsman found his new hand a willing worker and a quick learner, who became so adept that the delicate malting process was entrusted to him.

Three weeks later, on a morning in mid-May, Isaac Graham and Will Chard arrived at the distillery. Their arrival made John uneasy. They greeted John curtly but they were not unfriendly; both seemed preoccupied. He found out that Graham and Chard were under pressure to in-increase the output of the stills. They didn't spell it out but it was apparent that the customs officials in Monterey were demanding a larger "under the table" share for not enforcing the law prohibiting the distillation of hard liquors.

Morrison responded quickly. "I've got enough copper here for another pot but we'll be needing more for the condenser. Aye! But the bonniest sight that could greet my eyes would be a brick kiln so we could fire our own and make a decent high temperature charcoal oven." The Scotsman's burr thickened with the urgency of his appeal. "Thot's what we're needing for the filtering, mon. Give me that and I'll make ye a whiskey so good they'll change the law and make it legal!"

Graham wouldn't promise the material for the kiln. "We be needing a sawmill first, one driven by water power. I've great plans for making lumber. It be profitable too—with no silent partners at the customhouse."

It was obvious to John that Isaac Graham was purposely ignoring him. He tried to fathom the reason, and concluded that he was at Natividad under Will Chard's protection and that Graham had probably agreed to the arrangement, with some reluctance.

He was surprised when Graham called to him as he was about to go to his bed. "Hyar, boy!" Graham sauntered over with thumbs hooked in his belt. "Harry tells me your brain be as strong as your back. That's what we want— men that work fair, eat fair—and fight fair."

He cocked his head speculatively while his tongue probed a segment of upper gum almost barren of molars. Dislodged particles of food spattered on the dry earth. The stench of rotting teeth made Lewis wince. "Ye'll pass muster on the first two points, but the boys be a mite uncertain on the other." The lice in his beard occupied him

56

briefly as, with folded arms, he studied John. "I be uncertain myself. But I don't suppose leaping like a flying mule's worse than gouging or hair lifting—in the sight of the Lord. Killing one good man for another just ain't sensible. But I'd rather seen ye stomp a savage than my old friend Henry. But it wasn't a savage testing ye, was it? So, stay on hyar and Will will tell ye when the cool spell sets in over the hill. Harry's measure of ye weighs mighty on your side—a point to remember if ye get itchy."

Graham turned away abruptly and went to fix his bed in the private quarters in back of the distillery. The fetid aura surrounding him lingered.

Wakeful in his bedroll, John pondered the strange monologue. What Graham appeared to be saying was simple; he and the men did not approve of his fighting tactics but the practical need to replace a good man now dead with a good live one guaranteed a sort of uneasy amnesty. Meanwhile Harry Morrison was to be his custodian if not quite his jailor. Now that the threat of return to the *Reliant* had been removed, Lewis wondered what Isaac Graham's reaction would be if he deserted from his present company.

"He'd probably see it as an excuse to hunt me down and 'lift my hair,' as he calls scalping," John expressed the thought half aloud, recalling the grisly souvenirs that hung from Graham's belt. Not all of the scalps were Indian.

Harry Morrison was pleased to have an intelligent companion who was a good listener. Morrison spoke freely of his youth, his service in the Royal Navy, the fatal fight in the Vieux Carré that had been forced on him and his subsequent flight. He often mentioned frustration at not being able to make a good product at the distillery and of his desire to see the production of whiskey legalized.

After several more weeks of working together, Morrison surprised him. "Ye'll be havin' to make some decisions soon, won't ye, laddie?"

Wary, John tried to parry the first question of a personal nature that the man had made. "Such as?"

"Whether or not ye'll be wanting to return to the camp or stay on here wi' me?" The little Scotsman's transparent attempt to disguise his secret hope pleased John.

He smiled humorlessly. "Not much of a decision to make. Not in my circumstances."

57

"How so, laddie?"

"As I see it, I'm little better off now than I was aboard ship. If the air doesn't clear at the camp I'll be trading flogging for fighting."

Morrison's eyes narrowed shrewdly. "You'd do a lot worse than to stay on here wi' me."

John looked dubious. "I'm not cut out to be a whiskey baron. I think I lean toward trade."

"Trade's a profitable venture, laddie, if ye can get your boot in the door in Monterey and the proper connection with a Boston mercantile house. But I wud'na think those who monopolize it will look wi' favor on more competition."

"Who are the men who control the trade in Monterey?"

"Captain Cooper, David Spence, and lately, Tom Larkin—for the most part. And the Mexican officials who meet wi' them in the lee of the longboat to collect their 'tithe,' so to speak."

The revelation did not surprise John. He knew that Larkin had come to Monterey seven years earlier as a clerk and accountant for Cooper. Now the dour, thirty-seven-year-old Yankee from Charlestown, Massachusetts, had become rich enough to outdo his benefactor. He was building the most pretentious residence on the Calle Principal, a two-story house of adobe brick, redwood, and oak with a spacious veranda, upper balcony, and wrought-iron-barred windows. The *Reliant* had delivered much of Larkin's elegant furniture from New England's best craftsmen, along with fine glassware and silver. Larkin had checked each bill of lading and quibbled over the slightest irregularity. Captain Burke had cursed all the way from the landing to the anchorage.

In reply to questioning about Larkin, Morrison said to ask Chard. "He's been trying for years to get his boot in the door. He knows them all—and all that's worth knowing about them."

"How much do you know about Chard—and Graham?"

The question caught Harry Morrison by surprise. "More than I'm likely to tell ye, laddie."

John smiled at the rebuff and did not press. However, several minutes later Morrison surprised him in turn by reopening the subject. "What do ye know of Chard and Graham yourself?"

"Nothing—except what I can see."

"And what do ye think ye see, laddie?"

John pondered for a moment. "A mountain man who stinks like a backwater slough—who pretends to be a captain and leader—and a possible leader who's too clean and smart to be the mountain man he pretends to be."

A burst of laughter exploded from Morrison. "By God, Lewis, ye've a pair of sharp eyes in that head of yours." For a moment he seemed to be debating with himself. Finally, he said reflectively, "Well, laddie, what I do know can't hurt me, if ye'll use a little discretion."

John nodded.

"I'll tell ye aboot Chard first. I met him in Los Angeles, a place only a desperate sailor would find himself. He had come from New York in thirty-two and set out a vineyard. He had a very poor and very small brandy still which I was able to improve. But between the pressure from the Yankee merchants and the squeeze demanded by the officials, we were soon out of business. In early thirty-six I caught a bark to Monterey and Chard went to Santa Barbara, where he met Graham. Graham had been a hunter and trapper in Kentucky and drifted to New Mexico and then Santa Barbara. Chard was near starving, so when Graham told him some tall tales about Monterey, he threw in wi' the roughest crew this side of Botony Bay and came to Monterey. That's where we met again. I couldn't find an ear among the merchants, so Chard spoke wi' Graham." He swept the room with an arm. "And the result is to be seen hereaboot."

An obvious paradox troubled John. Will Chard was well-mannered, educated, impressive looking, strongly built, and had all the attributes of a leader. Why then, he asked Morrison, had he not been able to make a place for himself among lesser men.

"I'd say perhaps it's a lack in him. A lack of self-esteem. He's ten times the mon Graham is—but Graham believes himself to be ten times the mon he is. Will's capable of running things if he's given the chance." Morrison raised a finger. "I think 'given' is the key word. If Chard learns to take his opportunities, then he'll come into his own." He shook his head sadly. "And meantime, laddie, who knows what's behind the faults in any of us who hide out here on the California coast?"

Will Chard did not return to the distillery until the last

59

week in June. A replacement for John Lewis rode in with him. Chard evaded John's questions about plans. "You'll have to trust me, lad. But I give you my word as a—" He interrupted himself to laugh softly. "I give you my word that I'll lead you into no danger. If it reassures you at all, let me tell you that I have a deep selfish interest in seeing you safe."

Harry Morrison watched them pack up and mount. John had never seen him so crestfallen. The little Scotsman ignored the awkward attempt at thanks and a farewell and addressed Chard instead. "I dinna ken the reasoning that deprives me of the best apprentice I've had—not to mention intelligent company. Also, it's a rare thing to find a sailor thot drinks well. I'll be missing the lad."

Will Chard chuckled. "It's a rarer thing to see you sentimental!"

Morrison bristled. "Aye! So rare it is ye'll never see it. It's for practical reasons I'll be missing the laddie." He sniffed disdainfully. "I'm as sentimental as a hangman!" Without further comment he turned away abruptly and reentered the still house.

For the first few miles Will Chard seemed to be leading them in the general direction of the Pájaro River valley. But when he turned southwestward toward the Salinas River, Lewis grew puzzled, then concerned. That course would take them to Monterey, a place he still deemed unsafe.

Chard, sensing his anxiety, sought to reassure him. "Do not worry about Monterey, lad. The mission we are about to undertake is worth a great deal more to those in power there than the price on your head. And it could put a great deal more money than that in your pocket, and eventually in mine."

John reined up. "I don't mean to head into any trouble, not with my eyes closed," he said.

Chard laughed. "You are a hard-mouthed young mustang, aren't you, lad? I thought by now you'd come to trust me."

"I'm grateful to you for getting me out of a mess. And I'm disposed to trust you more than most."

"Thank you." Chard pursed his lips thoughtfully. "I did not want to let Morrison or that rabble at the distillery know we were heading for Monterey. Morrison's more

60

than dependable. He'd fight for you, lad—that much impression you've made on him. But if nobody knows what we're up to, then nobody can slip. So let's ride on and I'll tell you what I have in mind. If you don't want to go along with it, you're free to go back to the distillery, or to the camp . . ." He waved a hand. "Or to wherever you wish to go. Agreed?"

Lewis hesitated, then nodded. "Agreed."

They rode in silence side by side, reining their animals to a slow canter through rippling, satiny green seas of native oats and stretches of shimmering yellow wild mustard. The clustered blossoms reached to the swinging skirts of their saddles. In front of them clouds of blackbirds whirled upward and wheeled away, darkly iridescent in the sun. Shoulder deep and scarcely visible in the cover on all sides, deer and herds of valley antelope were feeding. Overhead, the sky's clear, azure dome arced away to the horizon. It must be, John thought, one of the most bountiful and beauty-blessed lands on earth. More than anything else he wanted to be free to stay on and explore it.

Watching his companion covertly, Chard smiled to himself. He could guess that the young New Englander was coming under the same spell that Alta California had cast on him when he had ridden north from arid southern California to Santa Barbara and Monterey with Isaac Graham's party. He had been entranced by the oak-shaded open land, by shadowy canyons, and the crystal streamlets cascading from the redwood-clad slopes.

Lewis broke the silence. "I don't think I would go back to Graham's camp."

Secretly relieved, Chard responded cautiously. "I think you'd find, as I have, that the future there is limited." He stood in the stirrups to point off to the northeast. "This valley is rich. But there are others that more than match it. The Sacramento, the San Joaquin, the Sonoma, the Napa."

John looked at him with new interest. "Have you seen them?"

"Only the Sonoma and the Napa. But I have talked with Colonel Vallejo at his Rancho Petaluma. He has ridden up the Sacramento as far as some old Indian villages. I would trust his judgment. The colonel has an eye for beauty as well as for money!" It was time now, Chard judged, to set the hook. "The plan I have in mind could take you to the Sacramento valley and to the others too, if it interests you."

John grew wary again.

Chard, sensing his unasked question, said, "I am not at liberty to tell you the exact plan we have in mind but I can tell you about the men who will underwrite it."

"What would I be asked to do—beside go sightseeing?"

"Do what smart travelers do—ask intelligent questions and keep a sharp eye out for important detail."

Unaccountably a passage from Numbers came to John's mind: ". . . and Moses sent them to spy out the land of Canaan and see the land, what it is, and the people that dwelleth therein, whether they be strong or weak, few or many . . ."

He said to Chard, "I don't know that I am suited by nature to be a spy."

Chard tried to conceal his surprise. "If you don't like that word, lad, I'm sure we can find another. They call mass murderers 'generals' and mortal gods 'sea captains' and 'masters.' Would 'seeker after the truth' be more to your liking?"

John's response was swift. "Perhaps, if it's bestowed with full knowledge of the futility of the task."

If you speak such profundities in the presence of Captain Cooper and Tom Larkin, you may give them reason to be wary of you, lad. Except for Spence, they are self-learned and suspect glibness in men."

So it was out, Lewis thought. It was something that would serve the principal merchants in Monterey. Under their protection he would probably be safe. But whatever the mission, it would also serve Chard's ends. If he couldn't learn the purpose behind the job he was being offered, he could at least try to find out about the men he would serve. His first question produced a cautious response.

"Taking bad things first—against Larkin is his lack of humor and his penuriousness. In favor is his quick mind, his orderliness, and what some would call his scrupulous honesty—although I cannot always tell precisely where sharp trading and larceny overlap. He is ambitious." Chard nearly added "and ruthless." He wondered again how much of Larkin's apparent restraint was due to the man's unusual obligation to Captain J. B. R. Cooper. That was important. Lewis would learn of it in due time, but not now.

62

"Tom Larkin is a family man, originally from Massachusetts. He puts little store in frivolity. Most of his time is spent improving his business and his house." Chard pondered briefly. "There isn't a great deal more I can tell you about him." He would have preferred to say, "that I am at liberty to tell you."

"Cooper is from Alderney Island—a former English sea captain—naturalized Mexican now, and baptized Catholic, as is Spence. Both are married to Mexican ladies. Cooper was master of the *Rover*. A brig, I believe she was. He traded in the Pacific islands and along the coast of South America. To quote him, he tired of the "inhuman demands of the sea" and became a trader, merchant, retailer, and rancher instead. He's prospered by being all things at once. He has a son they call "John the Baptist" because he was christened John Baptist Henry. John Cooper—a scholarly sort—is away in the Sandwich Islands much of the time. William Hartnell tutored him earlier— some years before he founded his College of San Jose. John Cooper is the elder statesman among the Anglo-Californians, I'd say. Though he and David Spence arrived only a year apart, Spence coming after, from Scotland."

Lewis knew that Spence was a trader, too. Chard supplied some detail. "I'm less sure about Spence's background, lad. But I've been told that he was the California agent for Begg and Company in Glasgow. I believe he supervised the salting down of beef for ship's stores in Monterey and Santa Barbara before he decided to go into trading for himself so he could stay on the coast." Chard glanced at John meaningfully. "This land seems to lay a hand on men, doesn't it, lad?"

It was dusk when Will Chard and John Lewis picked their way out of the canyon and rode onto the upper edge of the sloping, brush-dotted mesa overlooking Monterey and the bay. Chard guided them past outlying adobes to Captain Cooper's residence.

Cooper, a ruddy-faced man with iron-gray hair, greeted them cordially. John saw that his left hand had been crippled.

He introduced them to his wife, Doña Encarnacion, who showed them to adjoining bedrooms.

The house, one of the largest in Monterey, was invitingly furnished with excellent pieces that had been brought to California around the Horn a decade earlier. In many ways the furnishings resembled those in John's boyhood home in Middletown and awakened painful nostalgia in him.

During the evening meal the conversation touched on many topics, none of which related to the purpose of the visit. At the meal's end Doña Encarnacion excused herself and Captain Cooper got a bottle of Chilean brandy from the sideboard. He was about to serve it when Thomas Larkin and David Spence arrived. They acknowledged the introduction with friendly reserve and some minutes later, around the fireplace, with brandies, John had an opportunity to study the merchant princes of Monterey.

Larkin, tall, prim, and in his late thirties, had wispy receding hair that framed a broad, high forehead and merged untidily into scruffy sideburns. His blue eyes were bland and watery as the result of a head cold. His mouth was compressed into a permanent expression of mild disapproval.

The Scotsman, David Spence, seemed kindlier and more open. His blue-gray eyes were caught at the outer corners in fine webs of wrinkles that could be harbingers of a ready smile, and his mouth was pleasant above a determined chin. The squared face and well-shaped head created the impression of a mature and capable man.

Absorbed by his observations, John Lewis was caught off guard when Cooper addressed him. "Mister, if knowing you're in the presence of a former ship's master makes you a bit stiff, perhaps you'll slack up when I tell you that technically you and I are in the same boat. We're both deserters, you know."

Larkin smiled thinly and David Spence laughed and reached over to rest a hand on the knee of Lewis' worn wool winter blues. "That should reassure you, lad. It means that under the correct circumstances, and with the proper opportunity, a criminal can redeem himself and become a useful citizen in Alta California."

Captain Cooper chuckled at John's consternation. "Port captains are blunter than judges, Lewis. What are your plans?"

John glanced from Will Chard to Thomas Larkin, whose face remained impassive. "I have no plans, sir, except not to get put back aboard a Burgess ship in violation of the articles."

Cooper nodded reassuringly. "We know about that.

You've no need to worry. It could be that you'll find an opportunity to go inland for a time. Would that appeal to you?"

"That depends, sir. All things considered, I'm a bit puzzled by the kindness I'm being shown."

"That may be less than altruism, son. You're a presentable young man and, judging from what Will tells us, capable. It may not be kindness we're offering you. It may simply be the lesser of two evils."

Thomas Larkin dabbed at his nose with a pocket cloth and leaned forward. He did not feel well and his voice was raspy and without humor. "Your presence here, Mr. Lewis, means that you are capable of independent action. Life here is filled with uncertainties. If you should take it into your head to act upon your own, it would be fair to say that these uncertainties would become far more certain."

David Spence glanced at Larkin, frowning, then turned to John. "By now, lad, you've had ample time to figure that out for yourself. Mr. Larkin simply means that if you move outside the protection of friendly associates who wish you well, your future state could not be guaranteed. This is an open land. A very rich land. It attracts men of varied character and if you are curious about why we are talking to you, I'll be plain about it. We made our proposal to Will Chard. He feels, and on second thought we agree, that he is too well known to do the job So it is not that we have singled you out of many for our favors, it is simply that you seem to possess the qualifications and you are here, now, when we have a need for someone."

Cooper removed his pipe and pointed the glistening stem at John. "Every man-jack who has ever managed to make something of himself, did so because he was the right man in the right place at the right time. What Mr. Spence is suggesting is that—due to no particular perception of your own—you may be such a man."

Cooper reached for a sheaf of papers on the table beside him. "You'll be easier if we get to the point, lad. The charter brig *Clementine* anchored here three days ago. She's out of Honolulu for Yerba Buena and other coastal ports with general merchandise. But she's got some suspicious passengers aboard. Her master was my guest two nights ago, so my information is reliable. He's carrying an odd young fellow who calls himself Captain Johann Au-

65

gustus Sutter. He is a German-born Swiss who claims to have been an officer in Charles the Tenth's personal guard in Paris. He's lugging an uncommon packet of introduction letters from leading merchants and politicians in the United States, from the commander of the North American department of Russian Alaska, at Sitka, and from James Douglas of the Hudson Bay Company." He indicated David Spence. "We have seen these credentials. They are a fact. Already he's presented them to Governor Alvarado. The governor was impressed. He asked Sutter to dine with him. Sutter was refused permission to land in Yerba Buena with a company of five white men and eight *kanakas*. He claims he wants to explore the headwaters of the Bay of San Francisco and the rivers flowing to it. His purpose is to set up an outpost at the head of the navigable waters."

Recognizing the threat such an establishment might pose for these men who had created their own secure monopoly on the California coast, John blurted the obvious question. "An outpost? Who for?"

All but Larkin smiled. Cooper leaned forward, pipestem jabbing again. "That's what you could be finding out for us."

John looked from one to the other, still questioning. "What about Governor Alvarado? Would he give permission to Sutter?"

Spence nodded emphatically. "Aye! That he will."

Open disbelief shone in John's dark blue eyes. "That could be like the master inviting the crew to start the seacocks!"

Cooper nodded emphatically. "Precisely."

"But why in the name of common sense would the governor do that?"

"Alvarado could have two reasons," Cooper said. "First, he's much influenced by Castro and Robles, who fear Isaac Graham and his—uh—men." He indicated Will Chard. "Castro has seen the power of Graham's army and as military commander of the Department of the North, that's not calculated to make Castro rest easy. There's a saying that one Yankee musket is worth ten Mexican cannon. Secondly, the governor may well want someone he can trust to watch the back door. Mexico City wants no more outsiders here but they won't give Alvarado enough troops to man the necessary garrisons. A man like Sutter

66

could be a useful pair of eyes. Alvarado will bribe him to be loyal—probably with promises of land, promises he may not keep any better than he kept his promises to Isaac Graham."

"But Sutter himself is an 'outsider.' The letters would not change that."

David Spence replied, "Citizenship would be part of Alvarado's price for permission to set up the post. I met Sutter yesterday. He was adroit in avoiding direct answers to questions. The man is fluent in French. We are not unmindful that France is very much interested in California. And so are others.

"Sutter could be useful to Alvarado. If he's the agent of a foreign power, those of us who seek to protect this country's best interests should know about it as soon as possible." A suggestion of a twinkle shone in his eyes. "We need to have the most accurate possible appraisal of this Sutter. That is best done by someone he does not suspect of an ulterior motive." He paused for emphasis. "Now, that about states our position. I hope it also suggests the importance of the mission."

John sensed the merchants' anxiety. After a silence, Thomas Larkin addressed him. "Mr. Lewis, we are not unmindful of the uses of incentive. We do not wish to put a price on your freedom and while you are in our employ we will see that you are properly outfitted, given pocket money, and we will deposit to your credit, in Captain Cooper's firm, two dollars per day." He looked at John questioningly.

John held back no longer. "Sir, if you gentlemen feel you can trust me, I would find the arrangement quite acceptable."

Their relief was almost audible. Cooper lumbered out of his chair, offering more brandy, speaking as he refilled John's glass first. "You'll stay close by here now, John. As soon as we know for certain that Sutter has prevailed, we'll make arrangements to get you to Yerba Buena ahead of the *Clementine*." He nodded in the direction of the bay where the bark lay at anchor. "She'll make slow headway beating to windward. A good horse will get you there well ahead of her."

David Spence spoke next. "It will not be wise for Sutter to see you in our company." He removed several silver coins from his waistcoat pocket and handed them to John.

67

"Two leagues north, off the San Jose road, I have an empty house. Will knows it. You stay there until we send for you. I recommend that you and Will ride out tonight."

Within twenty-four hours John Augustus Sutter was informing all who would listen that he was now recruiting loyal California citizens to man a frontier colony for the governor. In poor Spanish, broken English, and in fluent French and German, he boasted that he would found a settlement to rival Monterey and that after he had received his own citizenship, he would be given a huge land grant generous portions of which would be transferred to all *californio* citizens who would come north to settle and work.

Sutter lingered only four days in Monterey after attempting without much success to arrange credit and to sign on men. Disappointed, he boarded the *Clementine* and sailed for Yerba Buena on the morning tide.

Just two hours later, in his home, Cooper was giving instructions to John Lewis. "Yerba Buena's a very small place, John. You'll be conspicuous. This letter identifies you as an agent-at-large for J. B. R. Cooper Company. Use it only if the Mexican authorities question you. As far as Sutter is concerned, you're just another merchant seamen who's jumped ship. Sutter needs men, and should leap at the chance to get you. But don't seem too eager. Stay backed off a bit. Keep your advantage."

Will Chard chuckled, "Tell him you're an expert distiller, that you were apprenticed to one of the best whiskey distillers from Scotland. You'll be indispensable."

Cooper pointed to the clothing and supplies piled on a table. "We've got you rigged out proper, John. How soon can you be on your way?"

"If you set me a course, I'm ready now."

"We've done that—with notes on the friendly places to stop for food and a fresh mount and a bed, too." He handed Lewis a map. "Will is going to ride part way with you."

In less than an hour John and Will Chard were riding north. Chard indicated the low hills ahead of them. "From San Juan keep to the flat country east of the mountains until you get to the south end of the bay and to the old Spanish Trail he's marked on the map. The streams are low now. You'll have no trouble."

Chard reined up. "I'll leave you here. Remember—keep Cooper posted. Send your messages aboard a coaster as he directed. It's customary for them to carry letters. If you're questioned too much, use the letter of authorization, but remember, the connection must be kept a secret unless you are taken by authorities. If they ask you about citizenship, say you are waiting out your year's probation."

In just sixteen hours of riding, with only the briefest stops for food and less than four hours of sleep, John rode to the hills overlooking Yerba Buena on the Bay of San Francisco. There he hobbled his mount and made a temporary camp.

From the top of a rise he could see the anchorage. No vessel as large as the brig was swinging on its cable. If the *Clementine* had worked her way north as far as the Farallones, she would wait until morning before attempting the dangerous passage through the narrows. Resisting the temptation to go into the settlement, John returned to the little camp and did what he could to make himself comfortable. Wearily, he gathered a load of fallen oak limbs, arranged them as the *vaqueros* did to make a "long fire" that would last the night, and settled into his covers.

For twelve hours he slept, scarcely stirring. When he finally awakened, the sun was a half hour above the green-splashed, yellow ochre hills on the east side of the bay. From the leather *talego*—the large saddlebag filled with provisions—he took the makings of a meager breakfast. Then he caught his mount and began a cautious exploration, taking care to keep away from El Camino Real, the old trail bearing the extravagant title, the King's Highway. To the north and west of it he recognized the twin peaks called Los Pechos because of their marked resemblance to a pair of young, upthrust breasts. Setting himself a course to the south of them he began a long probe through the wooded ravines in the direction of the ocean. It was past midday by the time he had penetrated the densely grown canyons and gained the summit of the protective ridge. He waited, hoping to catch sight of the *Clementine*'s sail. After several hours he remounted and rode back to his camp. Sometime during the third day, while he was exploring the countryside south and west of Yerba Buena, the *Clementine* made her way into port. In the late afternoon Lewis saw

69

her, riding at anchor. Relieved that his wait was over and excited in the spite of his continued admonitions to himself to be calm, he broke camp the following morning and rode into Yerba Buena.

2 "BUT, MY DEAR JOHANN, I insist you can accomplish the same things much closer to Yerba Buena—and do them faster and better."

Jean Vioget, fellow Swiss, addressed John Augustus Sutter as the two picked their way down the grassy trail leading to the waterfront.

Sutter, reflecting, shaped his drooping handlebar mustache and smiled benignly. "True. And I will also be under constant surveillance of everybody including your *Oberbefehlshaber* Vallejo! What I wish to accomplish must be done without interference."

Vioget studied Sutter's short, wide back. If Sutter had been German instead of Swiss, he would have chosen "arrogant" to describe him. As it was, stubborn Teutonic determination was softened a bit by inborn Swiss diplomacy. "Johann, for whom do you undertake this outpost?"

Sutter stopped in his tracks. "For *me*. Only for me. I have seen what is wrong with these frontier communities. They lack organization, discipline. I shall establish the model for all such trading posts. It will exceed even Russia's Fort Ross."

"But that's bound to awaken anxiety in General Vallejo. He is already concerned about the *schabig armeekorps* of Graham and Chard at Branciforte. He fears all military forces except his own."

"Vallejo will not fear that which he will not see," Sutter said and turned abruptly toward the shed that housed the trading activities of Spear, Hinckley and Leese.

Immediately on landing, Sutter had begun negotiations with Nathan Spear and William Hinckley for two charter schooners to carry himself and the crew he hoped to hire to the location of his outpost. With characteristic impatience Sutter wished his plans to spring into being, fullblown. "Jean, do you know why God created the universe in only six days? Because he had no one to obstruct him!"

The pair descended the path to a point just above the muddy high tide line then turned north to the crude but serviceable warehouse.

Spear and Hinckley and their young bay captain Bill Davis were waiting for Sutter at the warehouse. The two small schooners Sutter wanted to buy were some yards offshore. An hour or so later, with negotiations completed, largely on Sutter's promises and good faith, they were rowed to the moored craft for inspection. Sutter was delighted. His limited English and Spanish almost failed him. Standing in the bow of the *Isabella* like a conquering admiral, he lapsed into such an excited torrent of German that even poor Vioget had trouble following. Struggling to control his impatience, Sutter turned to the two merchant partners. "I will need many things! Many things. Both these vessels will be filled to the gunwales when I sail this coming week."

Hinckley and Spear exchanged startled glances, then looked to Davis. Spear voiced their concern. "Even if we can fill your complete list, Captain Sutter, it will take until the end of next week before Mr. Davis can have these vessels ready for such a long voyage."

Sutter closed his eyes. "The end of the week is out of the question! It is already midsummer. Every day is precious. The stupidity of your local authorities that required me to waste weeks and travel to the governor himself has already cost me dearly." He shook his head with finality. "Impossible! I must sail on the morning of the fourth day or make other arrangements!"

Spear, Hinckley and Leese had a virtual monopoly on bay schooners that made other arrangements patently impossible. But they were not inclined to argue with Sutter's ridiculous request, since he offered the possibility of an expanding volume of business.

Sutter and Vioget left the warehouse together and walked to the rickety structure of embedded poles that was the embarcadero's only wharf. They stopped abruptly at the sight of a young man, obviously not a native, seated on an upturned rowboat. The stranger was staring dejectedly out across the bay. Sutter's questioning look brought a quick shrug from Vioget. Frowning, Sutter walked to within a yard before the young man noticed him and

roused from his reverie. Sutter bowed and smiled warmly.
"Good day! Good day!"

"Good day, sir." The tone was flat.

Sutter extended his hand. "Allow me to introduce myself, Captain John Augustus Sutter." He turned. "And this
is my good friend and countryman, Jean Jacques Vioget, a
native of Yerba Buena."

John rose slowly, steadying the bundle of belongings
on the upturned keel with his left hand. "Pleased to know
you gentlemen. My name is Lewis. John Lewis."

The men shook hands, Sutter with great vigor. With his
usual lack of inhibition, he flung an arm in the direction of
the two schooners and announced grandly, "I am admiring
my new merchant fleet. Only this morning I have acquired
these two splendid vessels. In four days, God willing, my
company and I will set out to explore the great unknown
river wilderness," he drew himself up proudly, "under a
special commission from Governor Alvarado at Monterey."

The young man's obvious lack of interest puzzled and
annoyed him. "You have a touch of fever, perhaps?"

As anxious as he was about this first encounter with
Sutter, John was barely able to conceal his amusement. He
shook his head. "I'm in round good health." Casually, he
reseated himself on the upturned boat.

Somewhat hesitantly, Jean Vioget said, "I do not recall
seeing you here before, M'sieur Lewis. Have you come
with the *Clementine?*"

Sutter exploded. "Of course he has not! I came on the
Clementine. I know every man, rat, and roach aboard
her!"

Vioget tried again. "You are from where then, M'sieur
Lewis?"

John allowed a second of silence to go by before he said
slowly, "Originally, I came from New England."

Sutter's hands fluttered impatiently. "So? Where? What
city?"

"Lately, Boston."

Sutter was growing plainly annoyed. "Lately? How did
you arrive?"

"I arrived in a Boston merchantman, sir, and on this
coast lately could mean a year, even two, as you perhaps
know."

Sutter's china blue eyes widened. "You are a fugitive,

73

then? Is it not foolish to sit here and risk capture? Or are you so desperate for food that you don't care?"

John rose wearily. "I am full up with food and injustice. And I hold that no man is a fool who escapes unjust persecution."

Sutter looked nervously from Lewis to Vioget and back. "Yes, yes." Catching himself, he cleared his throat. "I mean, no indeed!"

Jean Vioget broke in. "You must excuse our curiosity, M'sieur Lewis. A stranger in this far off land is occasion enough for questions. We hope for news. Forgive us, but do you know anybody in Yerba Buena?"

"No. I arrived this morning—overland."

"From Monterey?"

"From San Juan Bautista."

Lewis nodded and Vioget was encouraged to press more questions. "But you've been in Monterey, San Pedro, and Los Angeles?"

"Not Los Angeles. San Pedro and San Diego."

Sutter interrupted. "Ah, yes. I was correct. You are a merchant sailor who is—shall we say—at liberty. Am I correct?"

John's tanned face turned grim. "You are correct. I'd rather carry a price on my head than more California hides."

Sutter's shrewd little eyes darted from one to the other. He was remembering his own bankruptcy and the threat of a debtors' prison in Switzerland. "By God, Lewis, someday I shall tell you why we are brothers!" Radiating sympathy, he sat down. "Do you know anything about these people, Lewis?"

"The *californios*?"

'Yes."

"Some are good and some are bad."

"Have any of them offered to help you?"

John's answer was a dejected shrug.

Indignant, Sutter shot to his feet. "God in Heaven! I should have known better than to ask. Barbarians! *Verdammt* barbarians! But what can you expect from a people who invented the Inquisition?" He braced his palms on his bulging hips and declared, "*I* offer you help, my friend. And I will pay you to accept." Leaning close, he scrutinized John's face. "Personally, I will pay. So you hear? Per-

74

sonally, from my own purse. Tell me, are you a good seaman, my friend?"

"Able-bodied, sir. At ease aloft, below, or on the quarter deck."

Sutter pointed to the smaller of the two schooners at anchor. "Could you be a reliable master of such a vessel as that one?"

John was astonished at the man's impulsiveness but when he looked at the open-decked schooner, he could barely keep a straight face. The craft was scarcely larger than the *Reliant's* longboat. "A fore and aft rig is no problem, sir, if it has better than a rag of sail, a straight keel, and a stout tiller."

"By God, Jean, in this country, I cannot make mistakes. Everything favors my plans." Sutter turned to John. "Lewis, do I interest you?"

There was no need to worry now. John knew that the more reluctant he seemed the more information he would get. He left the question unanswered until the Swiss nudged his leg. "Do I?"

"Sir, you're going off exploring rivers. I'm a deep-water man, not an oysterman. I appreciate your interest but—"

Sutter popped up again, fairly bouncing on the balls of his feet. "Do you know what I offer you, Lewis? Immunity from the law! A position as captain of the *Nicholas*. *Captain* Lewis. Captain John Lewis. Very distinguished. You will command your own ship, under me, of course. With Vioget here." Turning, he pointed to the larger schooner on whose transom was painted the name *Isabella*. "Captain Davis will command her and together, Lewis, we will be the first party to explore the great inland rivers. And when we find and claim our place, we will build our settlement—and my great fort."

In spite of himself Lewis' eyes reflected surprise. "Your fort?"

Sutter hastened to cover himself. "An outpost. Only an outpost. In English these distinctions confuse me. I use the word loosely. Our mission is peaceful. We will keep the great unexplored inland from falling into foreign hands. In a year I will be a Mexican citizen and in another year land will be granted to me by the Governor. As a loyal Mexican it will be my duty to see that the flag of my adopted country flies over this domain."

Sutter puffed with importance. "And those with me will

75

also be made citizens and I will have the power to grant land."

"How long would I be asked to sign?" John asked.

It was not a question Sutter had anticipated. He sputtered. "You do not sign. No! With Johann Augustus Sutter your *word* is good. And remember, Mr. Lewis, you will be paid in *dollars*. Twenty each month and food and quarters. That is because you are an officer, a captain."

John put a hand on his belongings, rather uncertainly, and said, "Well, I thank you for your proposal, Captain Sutter. It's interesting, but I'm not sure." He nodded toward the hillside. "I think I'd better find me a place to sleep and think on it."

Vioget, as eager now as Sutter, sprang to his feet. "You have no place to sleep? Only the woods?"

"I've slept very well in the woods for weeks now."

Resolutely, Vioget reached for John's duffle. "Tonight you sleep in my house. And you eat there, too. It is not Versailles but it has a good roof. You will not sleep on the ground like an animal."

John allowed the duffle to be taken from him. Then, flanked by the two men, he walked up the path to Vioget's cabin.

Early in November the Russian schooner *Constantine* put into Monterey for water and some emergency supplies. The master delivered to Captain John Cooper a packet of letters. Within the hour, David Spence, Thomas Larkin, and Will Chard were listening as he read aloud:

New Helvetia,
October 3, 1839

Captain J. B. R. Cooper
Monterey, Alta California.

Sir:

Finding the wherewithal to write this was almost as difficult as finding a means to send it. First, I want to advise you that Captain Sutter is writing to your firm asking for credit to acquire provisions to sustain our small company here at New Helvetia.

Captain Sutter is a man not easily fathomed. He is

76

given to much talk that might seem boasting, but he may be one of those *rara avis* whose deeds match his own accounts of same. Except for a confrontation here with over one hundred savages in full regalia and armed with spears and bows and arrows, an emergency which the Captain brought us through peacefully by reason of his courage and diplomacy (he lays claim to a special relationship with all native savages as a consequence of his experiences on the Great Plains) we have suffered no untoward hardship, beyond discomfort from the heat and from stinging insects and a certain poisonous oak-like bush that resembles the infectious ivy of the East.

The Mexican flag has flown at full staff from the moment we stepped ashore here on the banks of the Rio de Los Americanos. It is raised at sun-up each day with full military ceremony, including drum and bugle, and is lowered in like manner each evening. Even though we arrived here with two fully loaded trade schooners and a pinnace, we are woefully short of supplies. On the voyage up the Bay of San Francisco we put in at the embarcadero of Don Ignacio Martinez at Carquinez where Captain Sutter arranged with that gentleman to supply us with some livestock, seed, and certain equipment not available at Yerba Buena. Two savages were sent from here to show Don Ignacio's *vaqueros* the road but they have not been heard from. As a consequence we have all but exhausted our supplies from Hinckley and Spear and rely more and more on wild game. Fortunately the area abounds in fish, waterfowl, elk, deer, antelope, and bear.

The savages hunt for themselves; their women, a filthy lot, gather acorns, roots, and certain insects including maggots which they prize as food. They also thrive on a cooked swill made of grains which they slop into their mouths with their hands from a long hog trough.

The captain, by rewarding them with various trinkets and by generous gifts to their chief, has induced these natives to make the mud bricks for the construction of a fort, which will be the first proper building in New Helvetia.

Captain Sutter has two small brass cannon. He has

made a demand upon Hinckley and Spear for a half dozen more of larger bore for the defense of the fort. (You will also find a requisition for same in his message to you.) I believe I enjoy the captain's confidence now, but am wary of questioning his intended use of the ordinance. He seems eager to reassure all that the cannon are only to discourage an uprising of savages. To my mind the argument does not brace square when I remember that he is incredibly proud of his power of persuasion with the savages. I believe the cannon are intended more to impress Governor Alvarado of his intention of keeping his bargain. Still, it is too early to fathom Sutter's ulterior purpose. My first resistance to him has lessened somewhat and I am persuaded to judge him sincere and just in his dealings with us.

We are now four whites, ten Kanakas, and three-score savages. The Indians have come from as far north as Rio de Las Plumas and the Rio Jesu Maria to join this settlement. It was at Rio de Las Plumas that the crews of the charter schooners threatened mutiny unless we turned back. Otherwise, we might well have established this fort deeper in the wilderness for the waters of the Buena Ventura or Rio Sacramento promise to remain navigable for some leagues.

The valley is one of great beauty. Even in autumn the ground is covered with tall green graze beneath great, spreading oak trees. Their groves extend from the distant coastal mountains in the west to the first evergreens in the Sierra foothills in the east. The Indian chief, Narcisco, tells me that stands of oak and other trees line the rivers for many days journeying to the north.

Captain Sutter believes the black loam along the rivers and the redlands higher up will produce prodigious amounts of any crop suited to the climate. Hereabouts, the cottonwood trees along the river are garlanded to fifty feet with arm-thick vines festooned with great purple clusters of delicious wild grapes. I do not doubt this land's fecundity. Proof of it may be seen on every hand. Not only is the valley literally infested with wild game for the table, but also with those of value for their furs. Lately, the skies are

78

filled at dawn and at dusk with myriad migrating waterfowl. Wedges of geese and duck move across the sky during most of the day. Every foray, for whatever purpose, is certain to set off a feathered explosion of crested quail and raise clouds of wild pigeons and dove. The streams and upper tidewaters teem with a dozen varieties of fish. To this New Englander, accustomed to hard scratch, this is a veritable land of Goshen.

Given the opportunity, I would explore this region far more thoroughly than the captain is now likely to.

If this report does not illuminate as much as you had hoped, I ask your indulgence. Rest assured that I shall enter in my journal all events of possible significance and convey same to you, Sir, at reasonable intervals.

> Your obedient servant,
> John Lewis

Captain Cooper laid the letter aside. The men appeared to be occupied with their own thoughts. Cooper said slowly, "I incline to think we've chosen our man well. Sutter's argument for cannon 'doesn't brace square' the lad says here. I'll wager Lewis stays on until he finds out which way the wind blows."

Will Chard said, "A settlement means new settlers and a fort means new soldiers. In the end I wonder if Alvarado will find this bargain to his liking?"

Cooper grunted and rose stiffly. "He will not! Jed Smith and Joe Walker opened the back door over the Sierra six years ago. Something tells me Sutter is going to make sure the latchstring is out." He knocked the soggy heel from his pipe into the fireplace.

There was some desultory speculation about the wisdom of extending credit to Sutter.

Thomas Larkin argued for caution. "Sufficient credit to my mind means enough for food, livestock, seed, and some building material. It does not include cannon, powder, and metal for grape shot and balls. Lewis' point is well taken. Why cannon to defend Sutter against natives he boasts will peaceably do his bidding? Against whom, then? Industrious men who wish to settle there?" He regarded them sourly. "As unlikely as the possibility that he'd have occasion to defend himself against an invading army." He clasped one hand in the other as though to deter it from the temptation of dispensing money unwisely. "I will agree only to the minimum risk until he shows whose colors he flies."

Chard was the last to leave Cooper's house. As he left, Cooper said, "You've done us a service, Will. I do not think I speak too hastily in the lad's favor. He has a good fist for a quill too! Come to think about it, after reading the lad's view of the valley, I may have been hasty in trading my grant on the American River for the Bodega holdings."

Chard was surprised. "I didn't know there were grants so far north."

"There are. Mine was called El Rancho del Rio Ojotska. Governor Echeandía had a notion it would be useful to have citizens holding land there. When I married my good missus, the land was gifted to us. I've never set foot on it. But now I think this fellow, Sutter, has."

The reference to Doña Encarnacion Vallejo Cooper gave Will Chard the opportunity to inquire about something that had been puzzling him. "I have respected your wishes in this matter, but may I ask why you did not want Lewis to know that General Mariano Vallejo is your brother-in-law and that Governor Alvarado is Señora Cooper's nephew?"

Cooper returned the questioning smile. "And also that Tom Larkin is my half-brother? We could not reasonably expect the lad to understand such an apparent conflict of interest. If I had tried explaining he might have got wary. I doubt if his ironbound New England schoolboy mind could understand that in doing the opposite of what the *californio* officials want, we are actually preserving their long-term interests. He'll find out in time. But by then he'll know enough about California to understand."

Chard mounted without responding and Cooper moved to drop the gate poles. "How will you be occupying your time until spring?"

The seemingly innocent question made Will Chard smile inwardly. "For want of a choice, I'll be at my present pursuits—in the hope that Lewis' findings will benefit us all."

Blandly, Cooper returned Chard's probing gaze. "Aye. A sound course, I'd think."

On the long ride back to Graham's camp in the woods near Branciforte, Will Chard struggled to chart a clear path into the future. He was disappointed that the merchants had not really expressed their gratitude for his own part in providing them with good eyes and ears.

He wondered about the possibility of joining Sutter. If his settlement was successful, it could mean excellent trading possibilities along the river. A chance to grow. Then he remembered that Spear, Hinckley and Leese would monopolize any new bay and river trade—and with Vallejo's blessings. Jacob Leese had also married one of Mariano Vallejo's sisters. That made him Cooper's brother-in-law and, like Cooper, he was also Governor Alvarado's uncle by marriage.

He laughed bitterly. His plans to become a merchant trader by worming his way into the chosen group were both foolish and hopeless. Their flunky, really. Damn the groups and their stranglehold monopoly. It had to be broken and there was only one way to do it. California must be opened up to new settlers.

Chard considered the elements he must deal with. Their complicity was overwhelming. First there was Alvarado, subject to Castro's and Robles' influence and subject to the government in Mexico City—all determined not to let any more foreigners in. He smiled, however, as he thought of the weakness of the governor's plan to install Sutter and his settlement as protectors against infiltrators in the north. Any man who had explored the inland country knew the futility of such a plan. Chard wondered, too, how Vallejo would take the news that Sutter, not yet naturalized, had been commissioned by his nephew to establish an outpost fort.

Vallejo might well be the key to the situation. He was the wealthiest and most influential of the native *californios*. He could trade heavily on the fact of his huge ranch and powerful military garrison in the Sonoma Valley. He openly favored immigration now as the only sensible way to establish a strong and independent Alta California, one strong enough to bargain on favorable terms with any of the three foreign powers who were unquestionably interested in California.

But it was sheer stupidity to believe that Alta California would ever stand united against a common enemy or welcome a friendly power for the common good. If Alvarado

81

cracked under the strain, then Mexico City would appoint another weak man, another Gutiérrez, and that would mean another revolt in the south by Pico and his supporters. Pio Pico would never stop trying to bring the capital and the customhouse to Los Angeles. If that happened it would mean economic disaster to Monterey. To forestall that the merchants in the north would prefer a takeover by a responsible foreign power—annexation, something that Vallejo would also welcome.

The merchants in the south had built huge trading empires, too. They would have to defend them and they would fight unless the tapering off of the hide trade continued. Open immigration and annexation would be the only way left to build their markets and that had to mean new merchants in new markets. The vision excited Chard and hardened his resolve to find a way in.

John Lewis kept his journal faithfully. Through the fall and winter of 1839 he recorded the progress at New Helvetia.

December 3, 1839

The first Indian trapping party set out this morning. Captain Sutter believes fur will be prime by the time they arrive on the northern streams. The Indians assure him that many beaver will be found on the Rio De Los Americanos and the Rio de las Plumas. The Captain hopes to receive two thousand pelts worth four dollars each. The trash fur will be used for garments and covers here.

Señor Martinez finally sent about one fifth of the supplies the Captain ordered. He also released our two mission Indians. He had impressed them without pay for the past four months. The Captain's good humor and optimism are remarkable. We are sorely in need of everything but he refuses to be downcast.

February 10, 1840

Two deserters from the Hudson's Bay Company trapping parties arrived last night. One is a Frenchman who speaks several Indian tongues and may be useful. The other is from the Ohio Valley. They are not the most promising specimens but we welcomed them. We are now six whites and only six Kanakas. Four of the Sandwich Islanders returned to Yerba Buena with Captain Davis on the *Isabella* and will not return. Two of them were women and will be missed. The Captain believes we will lose the others if this cold continues but nonetheless he plans to petition the Sandwich Island King for more immigrants. His audacity may turn the trick.

Tomorrow we set out to study the pattern of high water in these parts. We are in danger of being washed away by spring floods if we remain here. Captain Sutter will study the situation and choose a higher location for his fort and the permanent settlement of New Helvetia.

March 4, 1840

Trouble! Several weeks ago the Captain banished four Indians of trapping party No. 1. They had returned from the streams empty-handed. The Frenchman, Octave Custot, who has joined us from Martinez rancho, learned the truth from the trapper, Matignon. The Indians had done well with beaver but had met a party of Hudson's Bay men north of the Rio Buena Ventura. The whites had given them whiskey in exchange for our entire catch. Sutter sent them packing north to their Korusi village.

Last night the Captain's Sandwich Island bulldog set up a terrible commotion. We sprang from our beds to find a party of our own Indians preparing to assassinate Captain Sutter. The dog, of a sturdy breed brought to the tropical islands by the New England missionaries, had fastened himself on the hapless savage leader and ripped him badly. The clamor sent the others fleeing.

Captain Sutter brought the wounded savage to his room, attended to him, then forced him to name his accomplices. Most had fled but we apprehended a dozen and fetched them back. Matignon advised the Captain to execute them forthwith as an example. But Captain Sutter refuses to deal with them as criminals. He seems to count them merely unruly children and disciplines accordingly. He did promise to deal

83

severely with any further attempts. (I agree with Matignon. From this night on I shall not sleep easily.)

March 10, 1840

Our troubles worsen!

There has been much unrest among the Indian workers here despite Captain Sutter's Christian forbearance. We have discovered since the incident at the Captain's door last week that most of the Indians have been concealing weapons in preparation for a general uprising. Also, they have been stealing and slaughtering our precious livestock.

Two days ago we discovered that many had deserted New Helvetia. Our scouts found they had withdrawn twenty miles southeast to a position just beyond the Rio Consumnes. Thoroughly angered at last, Captain Sutter mounted an armed party and took after them.

We found them at dawn this morning and killed six of their number including their chief. Many more were wounded. After an hour they threw down their weapons and begged for mercy. The Captain was forced to restrain Matignon, Custot, Harrow, and myself for, though none of our party suffered so much as a scratch, we felt that unless a proper example be made of them we should never know peace.

I must count Captain Sutter a courageous man but a foolish one. Again he has promised the savages forgiveness if they will return and work. He also promised to improve their lot and to give them more glass beads and trinkets. We shall see if this is sufficient. I suspect it will not be since he is teaching them the rewards of blackmail and revolution.

Meanwhile, I shall watch this experiment from behind a loaded rifle and a well-sharpened knife.

In mid-June Captain Sutter sent John Lewis to Monterey with a letter for Governor Alvarado. The letter stated that Sutter would be arriving in the capital toward the end of August to receive his Mexican citizenship. There had been no time for John to advise Will Chard or Captain Cooper of his trip. When he arrived at the Cooper home minutes after sundown the Captain's greeting was

84

cordial enough but Lewis sensed that the man was disturbed. "By God, John—come in! You're welcome, of course. But I'm not sure you'll be happy with the news I have for you."

Cooper helped him unsaddle and they turned the animal free in the corral.

Once inside the house, Cooper wasted no time on preliminaries. "Alvarado's shipped Will Chard and Isaac Graham off to prison in Mexico, at San Blas. About two score of Graham's men were sent with them. Castro put them aboard, in chains, by God, like animals!"

Dumbfounded, questions began tumbling from John's lips.

Cooper held up a restraining hand. "Lad, we think we know who's behind this but we can't prove it. The whole lot are charged with treason. Alvarado issued the warrant and Castro and his soldiers ambushed them in their cabins."

"Is Will all right?"

"He's unhurt. But it's a miracle. Two of the men were badly shot up and cut. That bastard, Castro, hamstrung Naile like an ox. He'll never walk proper again, if he lives."

"What about Harry Morrison? Did Castro go to Natividad?"

"Aye. They caught men there too, but not Morrison, so far as I know."

"Thank God for that!" But John was not reassured. Harry Morrison would never have deserted his men at the distillery, and it seemed unlikely that he would not be there.

Cooper continued his indignant recital. "They threw the lot of them into one little room at the old barracks. No food. No water. Tom Larkin and I protested and we managed to get a little to them. It's been five weeks now that Alvarado's refused to see us. He's refused to listen to anyone, even to Mariano Vallejo, who came down to protest."

Still incredulous, John struggled to curb his mounting anger. "I know those men! They have no minds for anything political. It would be contrary to Graham's plans to try such a thing. And you know Chard."

Captain Cooper made a placating gesture. "Aye. We do indeed. We've made the strongest representations, particularly for Chard. But Alvarado's still convinced that Graham was up to something. There was talk of a fake horse race that would be the excuse to bring all of his men into Monterey to bet on a new stallion Graham's said to have bought. The way Castro had it, the men would wait for the proper moment then pounce on the troops, kidnap Alvarado, and take Monterey."

"Take it for whom, for God's sake?"

Cooper shrugged. "That's another question that's not been answered, lad. But it's been suggested that the United States was behind it, as it doubtless is in Texas."

"That's bilge! I don't know about matters in Texas. But I'd stake my life that Graham and Chard are no part of any such plot here. I rode in with a letter from Sutter to the governor. When I deliver it, I'll speak for them."

Cooper took John by the arm and steered him to a chair. "You'll not do that, lad. Unless you want to risk chains and a trip to San Blas yourself. And it strikes me you were not very smart to be carrying the post for Sutter either, not to Monterey. It might be well for you to remember that it was the three of us who suggested you join Sutter. Not Alvarado."

John knew he was hearing the truth. Suddenly he felt foolish. In his effort to do an effective job for the merchants he had come to think of himself as their man, enjoying their immunity. Now, with the new turn of events, he realized how little immunity he really had.

Cooper held out a hand. "You'd best give me the letter. I'll get it to the governor. A schooner arrived from Yerba Buena this morning. As port captain I'd naturally receive the mail packets. I'll see that Alvarado gets the letter this evening, at his home."

A bit uncertainly, John removed the folded, wax-sealed paper from his pocket and handed it over. Cooper tucked it in his jacket. "Good. I'll tell Doña Encarnacion that you'll have supper with us. You'll stay here tonight. And I wouldn't show myself around the town too much."

Cooper left the room and returned shortly with two glasses of sherry. He handed one to John. "You'll be wanting to know what's being done for Will. Everything possible. The British consul in Tepic is looking into the matter now—acting for the United States too. David Spence got him onto it. Larkin's written to the Department of State in Washington and it's my opinion that Larkin will go to

Mexico City himself, if necessary. Also, we've forwarded some funds to the consul for the men's use, informally, of course."

Unable to believe that the governor could turn on the man who had marched with him, who had been responsible for sending the Mexican governor, Gutiérrez, into exile and insuring home rule for the *californios*, John began a bitter indictment of Alvarado.

Cooper cut him off. "It's treachery, right enough. But you're laying it at the wrong door. The root of it goes back to Castro's jealousy and his fear of Graham's riflemen. Goyo Robles and his friend Antonio Osio have cleverly encouraged it. But the perfidious act that brought about the sneak raid properly should be laid at the door of the church."

"The *church?*"

"Aye. Or at least at the feet of Father Varga."

Bewildered, John gestured helplessly. "But the church keeps away from politics now."

Cooper agreed. "True. True. It began with Goyo Robles. He was told that Tokoya, an Indian who had been one of Graham's men, heard Graham and Chard plotting to take over all of Alta California. Goyo wasted no time persuading Father Varga to seek out the Indian and administer the oath of truth to him." Cooper spread his hands. "Who is not to believe a priest? Anyway, it was sufficient to make Alvarado take action." He added slyly, "We have it on very good authority. Señorita Maria Robles, no less."

John was half out of the chair. "Robles' own sister? *Why?*"

"Because, lad, it seems she doesn't agree with her brother in all things." John thought he detected a suggestion of a twinkle in the older man's eyes. "I'm told by my good wife that Señorita Maria feels some Yankees should be excepted—particularly a young seaman from the *Reliant* who she thought was staying with Graham."

John hoped that he was managing to conceal his excitement. "Perhaps I'll have a chance to thank the lady for her concern."

Cooper nodded gravely. "Later perhaps."

Returning to safer ground, John asked, "Alvarado would take the testimony of an Indian, even under oath, against men who helped him to power?"

87

"Lad," Cooper's tone was sympathetic, "the whole affair is preposterous. But it's a fact. All we can hope for now is the triumph of justice, and soon." A vagrant thought caused him to chuckle. "As a matter of fact, some justice has been done already. The officials at San Blas also interned Castro and two of his officers and sent them to Mexico City. We do not know the reason for sure. The speculation is that Castro was so eager to get rid of his prisoners that he hustled them off without providing a bill of particulars or formal charges against them. The Mexican authorities would not look kindly on anyone dumping forty or fifty unwanted revolutionaries on them. The central government probably feels that since Alvarado makes so much of being a home-rule governor he should have taken the responsibility for sentencing and deporting them. And I quite agree. A fair trial here, without dealing with the practical matter of an illegal distillery, would probably have found the men innocent. My guess is, neither Alvarado nor Castro cared to risk that."

Thoughtfully, John resumed sipping his sherry. Cooper was right. It would be folly to attempt to intercede with Alvarado for Chard. It would be good to see Maria Robles again, to thank her for her concern. But that probably was impossible. The most important thing now was to find some way to rid himself both of the onus of fugitivism and of his known association with Graham and Chard. Until that was done there would be no security for him, even though the merchants might vouch for him privately.

"Sir."

"Yes, John?"

"In time, would you see any way that I might be able to move freely around Monterey? I'm not a fugitive. I was within my legal rights. I'm tired of skulking in the night. I intend nothing but good for California. I find it has great beauty and I'd truly like to stay on the coast without continual fear for my freedom."

As he spoke, John watched Cooper's expression change from mild interest to serious concern. Finally he looked up. "Yes, lad. There's a way. To tell you the truth, I hadn't thought of it when we sent you to keep an eye on Sutter." He nodded in the direction of an adjoining room. "Do you know the contents of that letter?"

"Yes, sir. I wrote it for the Captain."

"It is confidential?"

John considered briefly. "No, sir. It is mainly to advise the governor that the Captain will come to Monterey in August to receive his citizenship and to apply for land."

"Is that all?"

A slow grin warmed John's face. "He did add a postscript, sir, asking the governor to persuade you and Mr. Larkin, also Mr. Spence, to extend him more credit."

Cooper snorted. "By God, I could have guessed that!" He put his glass aside. "All right, John. Wait until Sutter's a citizen, then get him to vouch for you and countersign your formal application for citizenship. You'll get it if Sutter convinces Alvarado that you're indispensable to both of them."

Captain Johann Augustus Sutter, soon to be Don Juan Augusto Sutter, citizen of the Republic of Mexico, entered Monterey on the afternoon of a perfect August day at the head of a cavalcade consisting of an armed guard of honor of which John Lewis was the leader, two *vaqueros*, and some thirty horses.

The lookout posted by Governor Alvarado spied the party and spurred to a gallop to bring word to the fort. Then the hills around Monterey echoed with the booming of the old cannon. It was the signal the capital had been waiting for: Monterey had its excuse for a fandango.

Somewhat uncertain about his situation, John was careful to allow the gathering crowd to screen him from the governor. Some weeks before, Sutter had sent a letter urging the governor to grant immunity and to give citizenship to John Lewis, "this intelligent and courageous young man who has become my good right arm and without whom I should experience great difficulty in hastening the establishment of this fort so necessary to the defense of the interior."

John's anxiety had mounted until two weeks ago when Sutter received a note from Alvarado saying: "The young man may accompany you to Monterey on this happy occasion, certain that he travels under my protection."

Sutter's cavalcade arrived at the customhouse and the crowd pressed in to hear the governor's official speech of welcome. On foot, holding his own mount, John watched the people. He estimated there were over a hundred. In addition to the festively dressed Montereyans, the Indians, and *mestizos*, there were the crewmen from several of the ships lying at anchor in the bay. Once he thought he

89

caught a glimpse of Goyo Robles and Antonio Osio down in front and he craned to see if Maria had come with her brother. He did not see Captain Cooper sidling unobtrusively through the crowd. "Welcome, lad. Why do you stay back?"

Startled, he turned to find the merchant offering his hand. He took it, somewhat surprised, for it was the first time Cooper had made the conventional gesture.

"I'm not much for ceremonies, sir."

Cooper moved closer, speaking softly. "Sutter is to stay at the Alvarado home. I could not be certain you would ride in. But on the chance, we've prepared a room for you. We'll expect you when this nonsense is finished." He waved in the direction of the speaker. "There's a fandango tonight. Tomorrow there'll be games and a bull and bear fight at the old presidio. At some point we'll make time for a little personal visit. We're anxious to know how things proceed." He looked at John closely. "Sutter has spoken for you, eh?"

"More properly you would say that I spoke for myself—on paper—and Captain Sutter signed."

Pleased at the forthrightness of the reply, Cooper said, "I'll wager that was the best you've ever written." He gave John's shoulder a pat. "If things work out, you'll find Spence, Larkin, and myself ready to speak for you, too, John." He moved away as unobtrusively as he had come.

Twilight had descended by the time the ceremonies granting Sutter citizenship were over. John waited with Captain Cooper for the arrival of the Spences and the Larkins, who would join them for dinner.

"We'll not be able to talk as freely as we'd like, lad," Cooper said. "Not that the women aren't dependable—if we threaten them judiciously—but if we bore them with business, they'll even the score and bore us with fashion. So tell me now, briefly, what you think Sutter's game is."

"It's too early to tell. There are two more trappers now. Nine of us in all. But we don't have enough of anything—not even good straw for bricks. None of the supplies the governor promised have arrived. If Sutter has anything in mind beyond setting up his fort and getting enough cattle and seed for crops, he's not shown his hand yet."

"If he got the items he needs for building and planting, do you think he'd make profitable use of them?"

90

Lewis nodded. "I think he'd try."

"Are you suggesting that we give him much more credit?"

"No sir. But I'm suggesting that unless he gets some credit soon, I'll be out of a job."

The fandango was held in front of the two-story government house overlooking the harbor. An area nearly fifty *varas* square had been broomed clean by Indian women and sprinkled down the previous night and again in the late afternoon. Surrounding the dancing area were dozens of pitch flares tied in place on high stakes. Strung between them on heavy cord were brightly colored pennants.

Dressed in their most extravagant finery, Governor Alvarado, Secretary of State Casarin, and their wives and the honored guest, Johann Augustus Sutter, emerged from the government house. The shabbily uniformed Monterey Military Band played earnestly, making up with determination what they lacked in talent. The official party watched as most of Monterey's men and women danced the *californio* variations of the *paso doble*, the *vals*, the *jota*, the *son*, and the bouncy *cortejo de las pulgas*, a favorite with the more daring young couples, whose pantomimed pinching gestures imitating the courtship of the fleas scandalized the older women.

Because he was not familiar with the dances, John watched from the sidelines. Most of his attention was directed to Maria Robles. Among a score of overly ornamented unmarried girls and young wives, most of whom were dressed in expensive silk gowns with daringly low cut bodices, Maria's beauty stood out. Her slender body moved with regal grace and then with fiery abandon through the intricate variations of the dances. She paired with Goyo to dance the controversial *cortejo*. Her delicately suggestive responses to the overtures of the male flea brought laughter and applause from all, including the governor.

Grudgingly, John acknowledged the grace of the *californio* men, who were equally at home in a saddle or on the *pista de danze*. Dressed in flaring pantaloons and fine cotton shirts with elaborately decorated black velvet vests, Goyo, Antonio, and a half dozen other young men

91

deserved the admiration they were receiving from the female spectators. As he watched them, John recalled a comment Harry Morrison had made during a discussion of the trouble with Goyo. "Riding and dancing and plotting and gambling and fornicating. That's all they're good for, laddie! But ye'll find that knavery's their greatest accomplishment."

The dancing lasted until past midnight. When it was over, John made his leisurely way up the gentle slope to the Cooper adobe. Weary from the long ride to Monterey and the festivities, he slept until mid-morning, when Captain Cooper rapped on the door. "If you've turned out, John, there are some gentlemen here for a talk."

In the sitting room he was greeted by Larkin and Spence, who presented him to Don José Robles. Don José smiled warmly and extended his hand. "I have waited for the opportunity to apologize for my son's questionable hospitality. I was visiting Los Robledos. Otherwise, my friend, you would have had an entirely different welcome."

Surprised, John found himself at a loss for a graceful response. Don José changed the subject. "Señor Lewis, I'm afraid we are about to impose on you a second time." John Cooper indicated a chair.

"I hate to put you to work on an empty stomach, lad, but your employer, Mr. Sutter, has made a proposal to Don José and I took the liberty of suggesting that he speak to you about it."

"Señor Lewis, your Captain Sutter proposes that I sell him fifty quintales of wheat and barley on credit at nearly double the market price. I am anxious of course to do anything to assist the Captain in a project so favored by our governor. But the credit is the question. I am not a man of great means and the loss of that income could be of great consequence to me and my family." He hesitated. "I intend no disservice to Señor Sutter but we are all agreed that his credit is, well, perhaps not soundly established yet."

Had Don José not made a point of stating that an error in judgment would jeopardize his family as well, John would have leaned over backward to justify the extension of credit. But now the problem confronting John was how to answer fairly for all. Before he could frame his answer, Larkin broke in, "I think it's obvious, Don José, that our friend is troubled by some doubts. So I'll make an alternate proposal. I'll pay you market price, in cash, and take upon myself the risk of extending credit to Mr. Sutter on the terms he proposed to you. Since the governor has blessed the undertaking at New Helvetia I do not feel the risk unseemly. The man is already modestly entered on our books."

John saw Cooper's eyes widen with surprise, then crinkle with amusement. Little wonder, he thought, that Larkin, Cooper's half brother, had come to Monterey as his clerk and had managed in so short a time to become the leading merchant in the north.

With Graham's distillery all but shut down, Sutter would have a market for his spirits. Whiskey required grain. John wondered about Harry Morrison. In all likelihood he was no longer at Natividad. If Morrison was safe and could be induced to set up Sutter's still, there would be some justification for extending credit to make *aguardiente*. If the output reached Monterey, Larkin would market it. The local officials, for a price, had closed their eyes to Graham's illegal traffic. Surely, he thought, they would be stone blind to Larkin's. As it was, the merchants were able to land dozens of kegs of good Scotch whisky through the surf at night while the collector of customs conveniently found engagements elsewhere.

Larkin interrupted John's musings. "In order to insure my risk, Mr. Lewis, I would be willing to sell Captain Sutter the grain at a modest bonus if he will contract his entire output, beyond his household needs, to my company. Do you think that is possible and practical?"

"It could be very fair for all, sir. And it could be practical if we can be assured of the try pots and copper for a larger still."

David Spence, silent until now, said, "We can assure Captain Sutter these items." He turned to the others. "Can we not, gentlemen?"

Cooper nodded. "Aye. No problem there."

Lewis wondered at the good grace with which they had apparently taken their outmaneuvering by the Yankee merchant until he decided that so far as the Monterey merchants were concerned, "One hand washes the other."

After the men left, John Cooper rested a hand paternally on John's shoulder. "A good start was made for ev-

93

erybody this morning. I am glad you did not endorse Sutter's credit too enthusiastically. That will count for you here, too." He gave the shoulder a thump. "Now get some food, son."

3 SHORTLY AFTER MIDDAY, John strolled slowly toward the arena where the bloody bull and bear fights, the favorite diversion of the *californios,* would take place. Deliberately, he had allowed Maria and her mother Doña Gertrudis, to move ahead of him. When they passed, Maria had turned, pretending a casual interest in the crowd. He made no effort to acknowledge recognition but determined to find a seat near her. The night before, her gleaming black hair had been pulled into a large bun worn low on her neck. This afternoon it was divided into two glossy braids into which had been woven bright strands of colored silk. Her elaborately embroidered blouse was gathered tightly around her tiny waist beneath a bright green sash. Her skirt was ankle-length, black and decorated with a bold design of yellow, red, and blue flowers and large green leaves. She was, John thought, extraordinarily beautiful. The risk with Goyo notwithstanding, he wanted more than anything to know her better. The memory of those hours when she had acted as his reluctant jailer had stayed with him through the months.

There were times when the need for a woman turned him restless and finally tormented him until he, or nature, had relieved the suffering. But during none of those times of faceless fantasy had the girl been Maria. "A mon's balls be a bloody burden to carry alone too long, laddie." Harry Morrison had observed once during a restless period at Natividad. "But it's nae good to risk the pox with a verminous savage." Sailors and frontiersmen as well relieved their sex urgencies as they relieved their bladders and bowels and that was the end of it until they could get ashore and find a woman. John had found tolerable women in South America and in the ports along the Central American coast and he had washed his parts with raw brandy and endured the frightful sting that could keep him

95

clean. But always the experiences had left him feeling anxious and somehow unsatisfied.

Now, as he neared the arena, a voice called his name. He turned to see Captain Cooper and Doña Encarnacion. "You'll sit with us, John." Grateful, he fell into step, hoping the Robles family would be seated nearby.

At the grandstand he helped the captain get Doña Encarnacion to a seat that commanded a view of the entire enclosure. John liked the kind, buxom, dark-eyed woman who, though she spoke little to him, made him feel welcome.

Finally seated fashionably but not much more securely than the common spectators opposite them, John looked around and discovered that they were two tiers above and a few feet to the right of the Robles party. When Captain Cooper and Don José exchanged greetings, Don José signaled up and called out softly, *"Muchas gracias, señor."*

Maria turned with studied nonchalance, then averted her eyes as John smiled and replied, *"De nada, señor."* He saw the girl turn to her father and question him. Don José's answer was interrupted by the release of the first bull. He was a handsome animal—a brown and white Mexican longhorn, heavy-necked and wild, one of the largest, most ferocious-looking range bulls John had seen. A cry of anticipation echoed from the stands as the animal hurtled from the *toril* and came to a braced-leg halt just beyond mid-ring. On the far side of the arena, Goyo Robles and Antonio Osio mounted their stallions. Moments later they entered the ring, holding *reatas* looped and ready.

In perfect accord, Robles spurred his animal and reined it to the left to distract the bull while Osio made ready to lasso a hind leg. The entire maneuver was executed so quickly that John found it difficult to follow.

The instant the bull began its charge, Osio's *reata* streaked out and snared a hind leg. The move required incredible timing. The hoof was clear of the ground for only a fraction of a second. To prevent his companion's mount from taking the full straight-line shock, Goyo Robles swerved his own mount to divert the charging animal into a turn. As he did so the yardwide spread of twisted horns grazed the stallion's flank. When Osio's braced mount brought the bull to a jarring halt, Goyo's *reata* leaped from his hand and the rawhide loop dropped over the deadly horns. While Osio's superbly trained animal kept the strain, the rider dismounted, rushed to the bull, seized its tail and began to twist it. Shouts of encouragement went up from the stands as the bull, caught front and rear, thrashed helplessly and began to fall. A final twist brought the animal crashing to the ground. At the same instant a helper, on foot, raced out dragging a long length of rigging chain. While Osio maintained the painful pressure on the bull's tail the links were secured to the animal's right rear leg.

On a signal from Goyo, two more horsemen entered the ring, their mounts plunging and straining to drag a heavy iron-strapped cage fitted with hardwood runners. A wave of excitement swept over those close enough to catch a glimpse of the bear inside the cage. John estimated the cage to be ten feet long by four feet wide and nearly as high. He was amazed that the animal appeared to fill it. Don José half rose, then turned to call out to Cooper. "See him! *Madre de Dios!* He is the largest I have seen in many years. Goyo and Antonio caught him near Gavilan. They say it took ten men to cage him. I believe it." He craned for a better view. "Goyo thinks he weighs six hundred kilos."

At roughly two and one fifth pounds to a kilo, John computed the animal's weight at better than thirteen hundred pounds—by any measure a monster.

While one of the riders dismounted, the helper dragged the free end of the chain to the cage door. The horseman removed a heavy pair of blacksmith's tongs from a saddle thong and waited until the kneeling helper had pried up the slotted cage door sufficiently for the tool to insert beneath it. After several attempts that enraged the grizzly the tongs were clamped around a forefoot and the stump-thick member was forced out far enough to secure the chain.

As the moment of the fight drew near, John found his excitement mounting. Because of the chains that linked them, both animals were at a disadvantage. In a natural confrontation, the battle might well cover a league or more. It would be a different matter in the arena. The bear, finding himself bested as often happened, might succeed in freeing itself and turn on the spectators in its panicked attempt to escape.

John had heard stories of the first bull and bear fight in

97

California. There had been no arena. The rawhide tethers had snapped and the retreating animal had clawed and slashed its way to freedom, killing or maiming a dozen people.

He wished now that he could see Maria's face in order to watch her reaction to the bloody encounter. He knew that his own sister quite probably would be sickened at the first sight of violence.

A hush fell over the crowd and John could see that the helper was preparing to raise the door. Goyo and Antonio were waiting to slip their *reatas* free. The other horsemen had fastened their *reatas* to the rear of the cage to pull it clear of the arena.

Osio increased the twist strain on the bull's tail as Robles' *reata* slackened. A flick of his wrist carried the loop free. Retrieving the braided rawhide as he moved, he urged his mount back in order to bait the bull into a charge as soon as it regained its feet. Then, in a series of moves executed so rapidly that they seemed to be one, Osio released the tail, freed his *reata,* and leaped into the saddle.

For a long moment the bull lay still, its rage-reddened eyes rolling wildly. Then, with the sudden realization that it had been freed, it convulsed, gave a great heave, and came up on all fours. Frothing, it charged straight at Goyo's mount.

A great roar went up as the force of the rush jerked the outraged grizzly from its cage. Fascinated by the raw brutality of the encounter, John forgot Maria for the moment. The grizzly, a deceptive monster who appeared awkward to the uninitiated as he was revealed in all his enormity, scrambled to his feet, snapped frantically at the chain around his foreleg, then saw the bull. Taking his natural enemy for his true antagonist, he sprang with incredible agility and landed full upon the bull's back. The chain thrashed crazily as both animals crashed to the ground. The bull, surprised from behind, struggled to turn its horns on the unseen adversary. A bucking heave sent the bear sprawling. In an instant it was on its feet again. Faced off, each sought the instinctive advantage needed for the kill. The bull wanted its horns in the bear's soft belly and the bear wanted its claws and teeth in the bull's thick throat. The bull charged first. The bear rose to meet it and the impact shook the ground and brought a wild shout from the spectators.

98

While the bear was still erect, attempting to get its forelegs in position for a crushing embrace, the bull's head dropped and hooked up viciously. Another shout from those nearest the action brought the audience to its feet. The great curving horn had gouged deep into the bear's belly. Driving home his advantage, the bull continued to hook and each new thrust brought an outraged protest from the grizzly.

Suddenly the bear reared back and fell free. Blood pulsed in a thick stream from the gore wound. Men shouted and women and children screamed, calling for the kill.

Dazed, the bull stood with lowered head. On its feet again, the grizzly rushed in low and struck a clubbing blow at the bull's heavy cheekbone. The great head and horns snapped to one side. Quicker than any spectator's eye the bear's jaws drove in under the bull's muzzle and clamped high up on the throat. A vicious twist brought both animals tumbling to the ground. The grizzly's hind legs doubled and drove out straight again. Four-inch claws slashed into the bull's underbelly and laid it open down to the pendulous testicles. Burgeoning entrails, opalescent in the sun, bulged from the ragged rip. An agonized bawling filled the arena.

Through the screaming crowd, John caught glimpses of Maria Robles. She was watching, engrossed, her fingers pressed against her mouth as though to stifle an outcry. Another shout went up as the grizzly's claws ripped the exposed intestines and spilled their steaming feces on the ground.

In the arena again, Goyo Robles and Antonio Osio maneuvered their mounts into position to separate the animals. Seconds later, the stallions straining at the *reatas* pulled the combatants apart. The intestines that were still caught in the hooks of the bear's claws trailed from the open belly cavity in gleaming ribbons, tautened elastically and snapped apart as the bull began to convulse.

The helper loosened the chain around the bear's leg and secured a second *reata* to it. The free end was carried into the cage and threaded through the back bars. Osio therew two hitches around the saddle horn and dragged the maddened grizzly back into its cage. John, still watching Maria's reactions, turned back to the ring in time to see Goyo's knife slash across the bull's throat and plunge hilt-deep to the jugular vein. A font of blood spurted from the gash, soaking his arm to the elbow in crimson gore. Laughing, he pulled the knotted handkerchief from his head and wiped at it.

While wagers were being settled around the arena, the men at the pens examined the bear and decided it could fight again. The second contest was brief. An old range bull, no longer agile, did not survive the second charge and in less than fifteen minutes, Goyo and his helpers dragged its still twitching carcass into the street where, amid swarms of flies and a pack of wild street dogs, the first victim was being crudely butchered by the Indians.

The crowd, well pleased by the first fight but disappointed by the second, grumbled its way out of the stands. John helped Captain Cooper assist Doña Encarnacion down from the seats. The Robles family reached the ground first but were caught in the crowd making its way out. Without meaning to, Lewis found himself within inches of Maria. For the first time he saw her close up in full light. During the brief moment that their eyes met, he found himself trembling. Without warning, the idea of returning to New Helvetia with Sutter became unthinkable. Somehow he had to find a way to stay in Monterey.

His mind was whirling and he was scarcely aware that Don José was presenting Doña Gertrudis and Maria and her two cousins, Dolores and Secundino, or that Goyo had ridden up and was watching them from the end of the stands. John acknowledged the introductions awkwardly.

Don José was obliquely apologetic about the quality of the fights. "The second bull, that for certain would have finished the bear, broke its leg. The old one was all the men could find nearby at the last minute. We are sorry to perform so badly for our guest of honor." He glanced toward John Sutter, who was with the governor, talking animatedly. "But one day, here in Monterey, we shall have *corridas* as fine as those of Mexico City, Señor Lewis. Then you will not see just a contest between brutes. You will see man ennobled by the ultimate discipline," he turned toward Goyo, who was dismounting, "the triumph of reason over force. A great matador becomes almost a god when he also triumphs over man's fear of death. It is an ancient rite, not understood by all." He turned back and looked at John questioningly. "Have you ever seen such a *corrida de toros?*"

John shook his head. "No, sir. But I think I would enjoy one. I admire skill."

Out of the corner of his eye he saw Goyo pushing through the crowd to join them.

Captain Cooper cleared his throat and turned hastily to greet Goyo. "The bear was magnificent! Will he fight again?"

Goyo nodded curtly and made no effort to conceal his displeasure at John Lewis' presence. "He will fight again, captain, and he will fight well." His dark, narrow face broke into a malevovent grin. "He is a native bear. To him range bulls are intruders."

The gold flecks in Maria Robles' brown eyes turned to little points of fire as she glared mutely at her brother. She took her father's arm. "We are detaining Captain Cooper, papa." She urged him forward. "Besides, we should arrange to visit in a more comfortable place."

Goyo smirked at his sister. "Why not at our house? The last time Señor Lewis was there he found it very difficult to leave." The impertinence brought a gasp from Maria. Goyo pushed past John and went toward Governor Alvarado's party.

Goyo's mother, Doña Gertrudis, a slender, once-pretty woman in her mid-forties, looked after her son unhappily.

Don José said hastily, "I know Goyo. He feels he did badly at the morning games. And he blamed himself for the young bull's broken leg. He wanted things to go stupendously well today," he glanced toward the governor, "for his excellency."

A half hour of daylight remained but it was gloomy in the Coopers' big adobe. John helped his host light the lamps and accepted a sherry. Cooper, seated opposite, seemed preoccupied. Several minutes passed in silence before the captain set his drink aside. "This business with Goyo is not necessarily personal, lad. In time he'll come to accept you. Having you taken away from him by Graham was bad enough. But having you back here under Alvarado's protection is rubbing salt in his wound." He frowned. "And I don't suppose you've noticed, but his sister's disposition to be friendly hasn't helped him either."

John pretended indifference. "She seemed polite."

Cooper looked up, amused. "Aye, lad. That she can be. They're well raised. Don José's one of the few true gentlemen among the *californios*. The daughter's got a lot of fire like Goyo though. Ask the suitors who pester her with serenades. She can torment them unmercifully."

John suffered a quick stab of misgiving. Certainly he had seen evidence of her spirit at the fandango. But during his captivity and also at the bull and bear fight, he thought he had sensed another quality—an underlying disapproval of cruelty. At the arena she had averted her eyes during the worst of it and had not seemed to revel in the kill as most of the others had.

Cooper interrupted his thoughts. "When do you think Sutter will return to New Helvetia?"

"Within the week. Vioget is overseeing things, but the Captain doesn't trust him to handle the savages."

"I wouldn't trust *anybody* to handle them! Not even the *padres* any more since Governor Echeandía tied their hands with the secularization order nine years ago. Left to themselves, Indians will never work. If Sutter can get a day's work out of them, I doff my cap to him."

Lewis knew he was right. "Captain Sutter will never build what he wants at New Helvetia unless he can get good men to settle there."

"I wouldn't debate that point, lad," Cooper replied. "John, go on back with Sutter for a while longer. We'll ease his credit a wee bit and watch him. By the way, is he paying you as he promised?"

"Not in cash. I have no need for money up there. He's keeping it in his strongbox."

"You keep it in *your* strongbox, lad! Then you'll know it's there."

John felt foolish. "The captain's got the only one worth locking. I have no reason to mistrust him—so far." And he wondered then about the money the three merchants were crediting to him and whether or not he had better worry about that, too.

The formal ceremony investing Johann Augustus Sutter with citizenship in the Republic of Mexico was held in private in the office of Secretary of State, Jimeno Casarin. As soon as the proclamation was posted, Sutter began a drive to sign up men for New Helvetia. He bought dozens of drinks in the cantinas. Indirectly, he urged merchant sailors to desert. After two days of hard recruiting the

102

best he was able to manage were six Spanish-speaking former mission Indians.

Late in the afternoon, John walked the dejected new citizen to the governor's home, where he had remained as a guest. A few yards from the entry, John stopped deliberately. For forty-eight hours he had done little but think of some pretext on which he could remain in Monterey. On the previous day he had met Maria on the Calle Principal. She had been with Don José and Doña Gertrudis. She had gone out of her way to speak to him and invited him to attend a *partida del campo*. He had declined with obvious reluctance, explaining that he was obliged to return to New Helvetia. There had been no mistaking the girl's disappointment. Casually, she had let him know that Goyo had gone to Branciforte for a few days. Don José, apparently still anxious to make amends for his son's behavior, had urged John to persuade Sutter to stay over and join them. John had been sorely tempted, but he decided against it because he was afraid of revealing his real purpose.

After a restless night, his strategy had come clear. Now bracing himself, he said to Sutter, "Captain, I haven't spoken before because I did not want to risk disappointing you at a time like this. But I believe it is possible that Captain Cooper and the others will extend more credit."

Sutter's china-blue eyes rounded. "*Wunderbar!* I shall write a list immediately! I shall present it tomorrow."

"Excuse me, captain, but I don't think that would be wise."

"No, why not? Why not?"

"I have taken advantage of my position as a guest in Captain Cooper's home to plead our case. But I have had to do it with great care. The merchants here do not have much confidence in your plan."

Sutter began to puff indignantly and John hurried on. "Only because they lack your imagination, captain, but if you will let me stay on here for a while, I think I know how we can win them to your way of thinking."

Mollified, Sutter was eager again. "*Ja? Ja?* How? Tell me—how?" When he was excited or angered, Sutter sprinkled his heavily accented English with German words.

"I am certain I can convince them of the importance of

103

your work at New Helvetia," John said as calmly as he could. "They ask me many questions. If I—"

Sutter's thick arms flung outward. "You have no more important work than this, Lewis. Stay. By all means, as long as necessary!"

A troubled expression clouded John's face. Sutter saw it. "What is wrong? Something is wrong, Lewis?"

"Sir, I am here in Monterey under your protection. When you leave I will have nobody to turn to—in case—"

Sutter dismissed his anxiety. "You have the governor's word. That is the best protection."

"But was it offered beyond this visit?"

"But of course it was! There was no time limit!"

Still unconvinced, John glanced toward the house. "All the same, sir, I'd feel better, I'd have a freer mind to work for you, if you could speak to him again—to make certain."

Sutter gestured expansively. *"Gewiss!* Certainly! I'll speak to him immediately." He stepped closer, smiling conspiratorily. "It is a magnificent plan, Lewis. Alvarado will be happy to agree. If my credit comes from the merchants, that will be even less he will have to pay from the treasury. It would be good business for him to agree. I shall speak to him immediately."

Sutter started to move away, then stopped abruptly. "These people are much friendlier after I asked to become a citizen." His voice dropped to a breathy whisper. "Have you ever considered this idea?"

Quite honestly, John replied that he had. He did not say that the thought had come to him at the same time he had been overwhelmed by Maria Robles.

"Good, Lewis. Good. When I ask for the governor's continuing protection for you I will be very clever. I will also hint that your great admiration for him inspires you to seek citizenship." He tapped John's shirtfront for emphasis. "And I shall tell him how important you are to his plans in New Helvetia—and how smart you are," he drew away a little, "and how much cleverness and courage awakened love can give a man." John felt his face drain and his guts knot and prayed that Sutter hadn't noticed. "Alvarado himself found such courage, my young friend, because of his love for California. And you and I have it too—because we share that love with him." He clapped

John on the shoulder. "Don't you worry!" He moved away. *"Buenas tardes, amigo!"*

That night, after dinner, John Cooper and David Spence were both aware of John's preoccupation. Puzzled, Cooper broached the matter. "You look like you've started a seam, lad. What's troubling you?"

It was the question John Lewis hoped to elicit and the one he feared. The outcome would probably depend on how successfully he had concealed his interest in Maria Robles. He said slowly, "Something unexpected has come up, sir. Captain Sutter wants me to stay in Monterey for a while."

The two merchants exchanged anxious glances. "What for, lad?"

"He does not feel that he's had time enough to impress upon the governor the importance of getting all of the supplies he needs and enough workers to do the job in time."

Cooper made a wordless sound of annoyance and fussed with his pipe. David Spence shook his head. "He'll never get the workers here in Monterey, even with Graham gone. As for the supplies, we're prepared to let him have modest amounts."

Lewis looked from one to the other, concerned. "I know for a fact that you and Mr. Larkin could not supply all he needs. Captain Sutter feels—"

Cooper exploded. "Just what in hell is this Sutter a captain of? I wish somebody'd explain that to me! If he was a captain in Charles the Tenth's Palace Guard I'll wager he went over to Louis Philippe's side after the first shot. As a drummer boy. Captain indeed!"

Spence laughed. "A good point. But never mind his title. I suspect it's like much else about him—subject to question." He turned back to John. "You were saying, 'Captain Sutter feels—'?"

"He feels that his best chance of getting the fort and the shops built in time to be effective is for me to convince the governor to appropriate a good sum for it and order some conscript workers north to get things done. He will suggest to the governor that I might make a good citizen."

Both men were startled by an idea that had not occurred to either of them. Spence said firmly, "Sutter must not

105

suspect that you're our man. You'll have to stay here for a while, but that may not be bad in the long term. You may be able to make Alvarado see the urgency of Sutter's need, particularly if you emphasize the trappers who keep coming in illegally. If Sutter's credit is guaranteed by the state we would be happy to supply everything he needs."

Cooper was pleased. Spence's plan was perfect: they could have the business without the risk. But Lewis must not be allowed to meet with Alvarado until they had properly instructed him. Quickly, he addressed Spence. "David, I'm thinking—it might be wise if you spoke to Alvarado about prolonging Lewis' visit and possible citizenship before our illustrious new citizen does."

Spence rose immediately. "I agree, John." He moved toward the door. "Since I was on my way there on other matters, I'll speak privately with the governor this evening."

The Robles adobe, overlooking Monterey Bay, stood high on the wooded mesa at the upper end of the Calle Principal. Most of the year the family occupied the rambling, single-story structure in preference to the larger, more isolated *casa grande* at El Rancho de los Robledos on the inland side of the hills. Don José customarily spent ten days of each month there overseeing the cattle raising and the dry-farming activities. During these absences, Doña Gertrudis occupied herself with one of the finest gardens in Monterey. She had begun with seeds and cuttings brought from the missions at Carmel and Santa Clara, and in two decades had managed to grow grapes, oranges, lemons, a wide variety of vegetables, and a flower garden that was the envy of every woman on the north coast.

Maria found her there picking tiny, fiery *pimientas* for a hot sauce. Mechanically, she took the basket from her mother. Doña Gertrudis picked quickly for several minutes, then straightened, curious about her daughter's unusual silence. "Do you feel well, Maria?"

The girl shrugged listlessly. "Yes, *mamá*, I am well, thank you."

Doña Gertrudis frowned and peered at her more closely. "Then why do you look so," she fluttered her hands, "so miserable?"

"Because I am."

"Oh?" A knowing smile softened her thin, dark face.

106

"Does your unhappiness have anything to do with this?" She tapped the bosom of the long gray cotton apron that protected her dress and saw the quick little flash of surprise in her daughter's eyes. "Yes?"

"It is not my heart that is troubled, Mama. It is my conscience. I think Juan Lewis could have stayed a few more days. I think he did not want to because of Goyo. I am sure he believes we all feel the same way about foreigners."

Doña Gertrudis dropped a last handful of peppers into the out-held basket and slipped an arm around Maria's waist. "Your father has made it very clear that we are friendly. And I know at least six young men who would die joyfully if you would look at them the way you looked at Señor Lewis."

Maria pulled away to confront her. "I have looked at him only with sympathy."

Doña Gertrudis regarded the girl gravely and clucked. "What a pity. Truly a pity. Then you have not seen that this young yanqui giant is very attractive, that he is always neat and clean, even in his shabby clothing, that his hair is very dark brown—like Juan Alvarado's—and always trimmed," she reached out and turned Maria to her gently, "and that his eyes are the shadowy blue of the sea on a cloudy day."

Maria twisted away again and thrust the basket at her mother. Indignantly she braced her fists on the out-thrusting curves of her small hips. ". . . and that they always look wary—like an animal that is being hunted? Indeed, Mama, I have seen that look ever since Goyo, Antonio, and the others hunted him down like a bear, tormented him, and dragged him into this house. They forced him to the camp of that horrible Isaac Graham. And now they force him to hide in the wilderness with that ridiculous little *pisaverde,* Sutter. Juan Lewis wishes to stay in California and make a new life. He has risked much to try."

Doña Gertrudis lifted her hand soothingly. "And he will, dear little daughter, if that is his true desire. Have you forgotten how Goyo stormed all over the house shouting 'traitor!' after Juan Alvarado warned him that Señor Lewis was under his protection. That is a start."

Violently, Maria thrust her head forward. "Some day my idiot brother will call Alvarado a traitor to his face. Then he will be sent off to San Blas, too."

Doña Gertrudis gasped. "Watch your mouth, Maria. It is almost as though you wished it."

"I do, *mamá*! I do!" She faced her mother defiantly, then turned and fled to the house.

For a time Doña Gertrudis stood, smiling thoughtfully. She was certain now that her seventeen-year-old daughter, so much like herself at the same age, had fallen in love. It did not trouble her that the man was a yanqui. Among the girls in Monterey, Maria was outstanding just as Encarnacion Vallejo had been when John Cooper had chosen her, and Adelita Estrada when she had caught the eye of the much respected David Spence. Secretly, Doña Gertrudis had always hoped that Maria would fall in love with one of the foreigners. The Alta California girls who had married them in Monterey, in Santa Barbara, in Los Angeles, and in San Diego were the grand señoras now, the ones with position, the ones with the great homes and the fine things. But more important, they would be secure when the Americans came—as surely they would. Don José, who had spent long hours in futile argument with Goyo—who seemed determined to blind himself to the truth—had said they would.

As she moved toward the house, Doña Gertrudis decided to ask her husband to speak for John Lewis.

Spence returned to Cooper's home the next morning. Together they counseled John on his approaching audience with the governor. Spence was not optimistic.

"The governor is aware of you but I feel he neither likes nor dislikes you, lad. And the news that you would be interested in applying for citizenship did not exactly put him into a Highland Fling. He is under much pressure from Castro and Goyo. As I see it, you do not bring any special skill that could be useful to the governor at the moment and you can't count on much advantage from our sponsorship. So we must provide you with as much advantage as we can by giving you the essentials of Alvarado's background."

Spence rose and paced slowly. "Alvarado conceived the idea of an independent Alta California, but other, stronger, men, won it for him, too easily, I think. He heads the government because he is the only man capable of discharging the duties of office, except his uncle, Mariano Vallejo, who has other more practical ambitions."

"Alvarado was left fatherless at an early age. His mother, Josepha, was Vallejo's eldest sister. She was a great favorite of the royalist governor, Pablo Vicente Solá, who had sponsored the political career of her father, José Francisco Alvarado. Solá brought books to Monterey—against the wishes of the church—and personally tutored young Juan. Those books are still here. They are the best in California—the ones the priests didn't burn. You'll find them in Alvarado's library. They are his only true love. He is a scholar by instinct and an accountant and admininstrator by necessity. He is rather better at keeping accounts than at keeping civil order, but he does remarkably well with what he has to work with. He listens to Castro because in the larger decisions he lacks confidence in his own judgment. Castro, as military commander of the department, and Robles and Osio, exploit that lack of confidence whenever they can. They are directly responsible for the Graham affair." Spence frowned thoughtfully. "Alvarado is self-educated, but well-educated, I'd say. He's a realist with figures but an idealist with all else. He's a philosopher, not a politician."

The frown gave way to a compassionate expression that surprised John. "And above all, he's a very lonely man who was too long under the influence of doting women. Had his own father lived, or had Solá stayed in Alta California, he might have become the most effective statesman the *californios* could produce."

Cooper pulled a large silver watch from his waistcoat pocket and opened the case. "It's time to be on our way. Unlike most *californios*, the governor prizes punctuality." John rose somewhat reluctantly. He was disappointed. The information had been too general.

"Can you tell me his favorite books?"

A pleased smile lit Spence's angular face. "Well, I doubt it would be the Holy Bible. He was nearly excommunicated some years back for reading Fenelon's *Telemaque*. The bishop let him off with an official excoriation and he's had little to do with the church since. But I know he's read Cervantes and Gil Blas. *The Life of Cicero* is a favorite, I'm told, and Goldsmith's *Greece*. He admits to reading all twenty volumes of Buffon's *Natural History* and he admires Rousseau. I know he had a good translation of *Julie*. There are others, I'm sure, but I have seen these in his chambers."

John was familiar only with Cervantes and Rousseau and with some of Cicero's writings. There was little common ground there. The Bible would have been much safer.

Alvarado received them cordially and showed them into a small sitting room beyond his own austere office. A *mozo* brought a fine dry Spanish sherry and John sipped and listened to the polite small talk at which David Spence excelled. The Scotsman's Spanish was excellent. Spence had urged John to converse in Spanish, however haltingly. "Do it as evidence of your sincere intention to adopt their customs." Twice before, John recalled, Spence had referred to the Alta Californians in the third person. It was evidence that neither merchant truly regarded himself as more than a technical citizen.

The governor's large expressive eyes, mirroring each fleeting mood, commanded John's attention. He was so deeply absorbed studying Alvarado that he did not realize immediately he was being addressed.

"... and we are always pleased when a traveler finds our land so pleasant that he wishes to remain with us. We thank you for such a sincere compliment. But may I ask, Señor Lewis, why one who comes from so favored a land as New England would wish to leave it and forfeit his United States Citizenship for a land as primitive as Alta California?"

Aware that David Spence was watching intently, tacitly warning him, John said, "Your Excellency, New England is a favored land for those who are in a favored position. As one who had no mother or father or close family other than a younger sister who lives in New York with charitable friends, I found my position difficult."

The governor's manner changed from polite interest to quick sympathy. He waited for John to continue. "I found that without sympathetic friends in important positions most doors were closed. So I chose the career of a sailor."

The governor smiled disarmingly. "That we know, Señor Lewis. We also know that your ship's master holds you in the highest esteem and places a premium on your services."

John's stomach muscles knotted. "Your Excellency, I am grateful to the master for the opportunity and for his confidence. I kept my part of the contract, but he was not prepared to do the same. When I give my word, it is given. I expect others to do likewise."

As he had from the beginning, David Spence continued to help John over the more difficult phrases. The governor smiled openly at the translation of the last part of the statement. "A commendable quality, my young friend."

Somewhat reassured, John thought "So far, so good. *Estoy con un pie en la casa*. I have one foot in the house."

"If it were possible to live where you desire, what work would you give your life to?"

John had no answer ready; he could not place himself in competition with his benefactors. Slowly he said, "First, Your Excellency, if I am permitted to remain in California I should like to travel more. I would hope to discover what my new home most urgently requires. Then I would try to learn to do that work as well as possible." John could almost hear Cooper and Spence breathe, "Well done, Lad!"

Abruptly, Alvarado changed the tenor of his questioning. "Señor Lewis, tell me please, what is your opinion of our friend, Señor Sutter, and his works at New Helvetia?"

It was not a moment for equivocation but John needed time. He smiled apologetically, and indicated to David Spence that he had not fully understood the question.

The Scotsman undertook a deliberate and unnecessarily long translation. John nodded and leaned forward earnestly to address the governor. "I cannot say, Your Excellency, that I am privileged to call Captain Sutter my close friend. But I have found him to be energetic and courageous. I think he is a good teacher for the Indians and he maintains military discipline and demands proper respect for the Mexican flag."

"Would you say that his first concern now is to build his fort and make it self-sufficient?"

"I think so, Your Excellency."

"Do you believe he can stop unlawful immigration from the north and east?"

Lewis recognized another dangerous question. "It is my opinion, Your Excellency, that Captain Sutter will be able to protect the frontiers better when more men can be spared from the building and planting."

Alvarado steepled his fingers. "Do you think, Señor

111

Lewis, that the captain, as you call him, will need a garrison of professional soldiers?"

Without hesitation, Lewis said, "I am certain of it, Your Excellency, particularly if there is trouble with the Indians from the north. But a garrison would not be practical unless the fort can sustain it, or unless supplies and arms can be sent from outside sources."

"Sources such as Monterey?"

Lewis spread his hands. "I was not suggesting, Your Excellency, I was expressing an inexpert opinion. I am not an experienced observer in such matters."

"Perhaps that is an advantage." The governor straightened to indicate that the interview was about to end. Spence and Cooper eased forward on their chairs. Alvarado remained seated and addressed the two merchants. "I am obliged to you for this opportunity to visit with Señor Lewis. I take your willingness to sponsor him as evidence of your high regard for him. I should think your confidence in this young man is well placed." He rose and shook hands with Cooper first, then with Spence. Then he turned to John. "We shall consider your wishes carefully, my young friend. I must tell you that in addition to these gentlemen you have the support of Don Juan Sutter, of Doña Encarnacion, also of Don José and Doña Gertrudis Robles." He moved toward the door. "But I think with at least one other in the Robles family, it is still necessary to do some winning."

After the meeting, John suffered a curious feeling of remorse. He was uneasy about the devious tactics he had employed and considered himself a poor finagler. Just before sundown, he climbed the hill back of Monterey. Below him, overlooking the great curving bay, was the capital of this new land that could become his home. Monterey's most imposing houses lined the west side of the principal street. Nearest, and directly below him, was the Robles house. Somewhere in it was Maria.

Monterey, as pueblo and town, was scarcely a half century old. In the mild coastal climate there was no need to build against sub-zero winters or blustery, hurricane-threatened autumns. John had felt the intense heat of the Sacramento Valley. But it was dry—garments could be washed and worn again within the hour. And he had felt the cold and broken skim ice along the edge of the river. The worst of California that he had seen, except for the

112

great desert, was said by some to be kinder than the best of New England. And perhaps it was. But there were golden days in New England, too, days for adventuring along the streams, autumn days when the trees caught fire overnight with such color that their beauty made a body ache. There were crystal days when a glistening mantle of snow softened the land.

"Juan Lewis, Citizen of the Republic of Mexico." He spoke the words aloud, then repeated them. But the alien sound lingered. He thought of Maria and said, *"Don Juan y su esposa, Doña Maria."*

Restless, he rose and walked along the grassy contour of the hill until the twin belfries of the Presidio Chapel were directly below him. Two weeks had passed since he had last written to his sister in Brooklyn. He recalled, with a wry smile, the closing paragraph of the letter, now eastward bound aboard the Medford-built ship, *California:*

> ... and so, my dear sister, there seems little to tie me to the old home and much to attach me to the new. I have been singularly fortunate in the making of connections here, not because I am extraordinary but because of the need here for any able-bodied man who is half literate. California will not be built by the timid, the stable, or the 'salt of the earth.' It will be built by adventurers who left nothing behind and for whom the excitement of an uncertain future is reward enough. Should the reward I find turn out to have been worth the seeking, I shall offer to share it with you in this new land, if such be your wish.

Three weeks after the meeting, John received a clue that he didn't recognize about the probable nature of the governor's decision. He had had a chance meeting in the afternoon with Maria and her mother. Afterward, Doña Encarnacion Cooper, listening as John talked with her husband, questioned him. "Do you remember exactly how Maria said it?"

"Oh, yes. She said, 'Father tells us that Governor Alvarado is well disposed toward your request for citizenship. He may grant it if he is convinced that you are willing to meet all of the usual qualifications.' She made a point of emphasizing the word 'all.' For some reason it seemed to amuse her."

113

Doña Encarnacion's plump brown face broke into a knowing smile. Cooper studied her suspiciously. "I know that look, my love. Let us hear the devious female logic that prompts it."

She turned to John. "What do you say most occupies the minds of men?"

His reply came without hesitation, "Business."

Still suspicious, Captain Cooper tried to anticipate his wife. "Aye, that would be true of responsible men. But what are you driving toward?"

Ignoring him, Doña Encarnacion countered with a second question. "And what do you think most occupies the mind of women?"

John frowned. "Well—their children, I suppose, and their homes."

The woman laughed sympathetically and shook her head. "No, no, my young man. Women's minds are most occupied by men. A married woman's mind is most occupied by *her* man. He is the beginning of everything else for her. She may deny it, but it is the truth."

Thoroughly confused, John turned from one to the other. Cooper still seemed bewildered. "Don't ask me, lad. I've been married to her for nearly a score of years and there is still much about her I'll never fathom," he extended his arms helplessly, "simply because she's a woman!"

Doña Encarnacion sniffed. "Simply because *you* are a man!" She lowered her needlework. "Maria understood what the governor meant when he said '*all* of the *usual* qualifications.' And she was amused because she knew that you would not understand until it had been stuck to the end of your nose. I think, Juan Lewis, that the governor chose a very clever and willing messenger."

A soft whistle escaped from Captain Cooper's pursed lips as an understanding smile spread over his weathered face. "Ah, yes. A year of probation and a promise of loyalty to country, to *family,* and to church." He braced his hands atop his knees. "By God, lad, I believe you've got your answer all right!"

That night, in bed, John tossed restlessly as he realized that the governor was attaching binding strings to his official consent. It was strange, he thought, even though he had admitted to himself his love for Maria Robles, Alvarado's almost mandatory condition that he embrace a *californio* woman and probably the Roman Catholic Church rankled him. He did want Maria. Everything about her appealed to him. In any female company she would be judged outstanding. From the beginning he had sensed in her a capacity for passion. When she gave herself the giving would be unrestrained. The thought of being with her had excited him to the point of distress every night since the fiesta. Suddenly annoyed, John threw off the Mexican blanket, rose, dressed quietly, and left the house.

The night was bright and clear. As he walked alone down the Calle Principal he heard the reedy wheezing of a concertina coming from one of the waterfront cantinas. On an impulse he started to enter the place for an *aguardiente*. As he reached the door, a figure moved stealthily from between the buildings. He braced for trouble. Then he recognized the *mestizo* woman known as *La Mofeta*, the Skunk, who sold the services of her temporarily adopted Indian daughters to visiting sailors. On several occasions he had taken one of her older half-breed girls to the beach. Afterward, he had cleaned himself painfully but effectively with strong brandy. But the rumor was current now that the girls had all been exposed to the Spanish pox. Some weeks earlier, sailors aboard an English bark had murdered and thrown overboard one of their shipmates who had caught the infection and threatened to spread it through the fo'cas'le. Lurid accounts of how the deadly Monterey Bay sharks had torn the body and each other in their greed to devour it were still heard on the waterfront.

The woman ignored his rebuff and called out. A heavily shawled girl emerged from the shadows and stood demurely, with lowered head. The woman pulled the girl in front of her. *"Nueva, señor. Limpia. Muy limpia. Por cinco reales, señor. Solamente cinco."*

Lewis peered closer and saw that the girl was indeed a new addition. If she was not as clean as the woman promised at least she was made bearable by a generous splashing of cheap perfume. It had not been his intention to look for a girl, but he paid and took her to a dune a quarter of a mile along the beach. When he tried to enter her she cried out pitifully and began sobbing. Lewis pulled away. "What's the matter? *Qué pasa?*"

The girl continued to sob, apparently unable to speak.

115

He shook her. *"Dígame. Qué pasa?"* Still there was no response but the sobbing. Puzzled, frustrated, and awkwardly hobbled by his trousers, he was propped on an elbow beside her. Impulsively, he pulled the shawl and loose fitting blouse away from her body. In the dim light he could make out the half-developed breasts of a child.

"Nina! Dígame! Cuantos años tiene? Dígame—pronto!"

The girl's reply was almost inaudible. *"Doce, señor . . ."*

Lewis gasped. *"Ay! Doce? Verdad?"*

Between sobs the girl managed to nod. *"Si, señor. Doce . . ."*

A wave of disgust engulfed Lewis. He struggled upright and secured his trousers. Then he pulled the sobbing girl to her feet and urged her gently toward the lights of the town. A block from the cantina he left her, still weeping, and returned to the Cooper adobe. Behind the house he relieved himself joylessly. It was nearing dawn when he finally fell asleep.

The first opportunity to spend time with Maria Robles came early in November. Doña Gertrudis invited the Coopers, the Spences, and the Larkins to share a wild boar that Goyo and Antonio Osio had killed in the hills behind El Rancho de los Robledos. While the animal was being roasted on a *barbacoa* constructed of green poles in the walled rear yard, John and Maria, grateful for the warmth in the chill night, stood back from the pit watching and talking.

"You have never eaten *jabalí asado,* Juan Lewis?"

"No, señorita. Only salt pork." His distasteful grimace made her laugh. "Goyo says it is the best of all roast meats."

John glanced toward the house. "I had the feeling when I arrived that Goyo would have enjoyed roast Yankee more!"

Maria said reproachfully, "You must understand, Juan Lewis, that my brother is not being personal. When Goyo understands that you wish to become a loyal Mexican citizen, he will be your friend."

"I would not depend on that, señorita."

"It is the truth. But let us talk of more pleasant things. Of course, you will stay in Monterey now. What will you do?"

"If the governor asks me to do the impossible I shall try

116

to find a berth on a ship sailing back to the United States."

He could feel disappointment settle heavily on Maria. "But, Juan Lewis, the governor likes you—very much. He would not ask you to do something impossible in order to become a citizen."

A curious perversity seized John. He nodded in the direction of the distant presidio chapel. "I'm told that I might have to convert and become a Catholic."

"But a man must have *some* church, Juan Lewis. The Catholic Church is the only one in Alta California. Señor Cooper and Señor Spence belonged to other churches when they came here. But they are Catholic now. Talk to them. They will tell you that it is a very beautiful religion. The padres would not expect you to believe everything— in the beginning."

Even though her growing anxiety pleased John, he still felt perverse. "But they took *californio* wives. They had to become Catholics in order to marry them. Is that not the truth?"

"But they loved their women. To please them they converted willingly, Juan Lewis. It is a small thing to do for a woman you love. Is *that* not the truth?"

"It is no small thing to take vows. But I have no need yet to worry about changing my faith, do I? Many things come before that. I must work. Then I must try to find some girl who could love me . . ."

Maria Robles drew herself up indignantly. "Juan Lewis, if you walk around with your eyes—" Before she could finish, the rear door of the adobe opened and Goyo Robles was silhouetted in the opening. Ignoring them, he approached with knife in hand and sank the long point into the boar's heat-swollen belly. Instantly, a stream of clear fat spluttered into the coals as Goyo examined the roast. Maria moved to his side. "I am very hungry, Goyo, will it be ready soon?"

Goyo pretended amazement. "My dear sister! What a surprise! Suddenly now, you have an appetite for pigs." He laughed derisively and strode away. Maria could feel John's anger as her brother swaggered past him. Silently she cursed Goyo and his *nationalistas*. The moment had been lost now and she cursed the big yanqui beside her who was not only blind, she told herself bitterly, but deaf and dumb, too.

117

The meal was the best John had ever eaten in Alta California. The roast wild boar was far more delicate than he had imagined it could be. Urged on by Maria and her mother, he ate shamelessly.

Goyo watched silently and spoke hardly at all. He and Antonio Osio, a strongly built, swarthy young man with unruly straight black hair, seemed to be listening and appraising everything that was being said.

The conversation was dominated by Thomas Larkin, who described his inland journey to Vallejo's Rancho Petaluma. Afterward the subject turned to Sutter. Goyo then concentrated intently, his narrow, sharp-featured face mirroring mistrust.

Larkin turned to Doña Encarnacion. "Your brother, Don Mariano was most hospitable. I spent almost one week at his *casa grande* and learned much from him. He met Sutter and Mr. Vioget in the Sacramento Valley surveying the land Sutter wishes to apply for. A considerable amount of land. He wants two very large grants reaching from the American River on the south to the Feather River on the north. Don Mariano said that Sutter has added several good French Canadians to his staff. Three of them are trappers. Also, he's hired that Morrison fellow who escaped from Natividad when Castro arrested Graham and his men."

A vast sense of relief came over John. The little Scotsman was safe. His attention returned to Larkin, who seemed to be speaking in the hope that his words would reach other ears. "Don Mariano's conclusion was that Captain Sutter is welcoming dangerous foreigners to the fort. Apparently he welcomes like brothers all who come, especially the French-speaking trappers. It seems to Don Mariano to be a curious attitude for one who was commissioned by the Secretary of State as the law-enforcement agent for the interior." He turned again to Doña Encarnacion. "Your brother hopes that His Excellency's trust has not been misplaced."

Absorbed as he was in the merchant's account, and in Goyo's reactions, John was still aware of Maria. There was no real need for it but she had taken over the supervision of the serving table, ordering the Indian girls to bring more of some dishes and remove the remains of others. Several times he saw Doña direct an approving glance at her daughter.

Maria whispered her orders a bit imperiously, John thought, pretending to ignore the guests but making it impossible for them to ignore her.

Shortly before ten o'clock, John took advantage of a lull to go to the *retrete*. When he returned he found the guests on their feet preparing to depart. As he stood admiring a tall, ornately carved Spanish chest, Maria appeared beside him.

"It is very beautiful, yes?"

John smoothed his palm gently over the dark oak. Its age-mellowed patina was warmed by the reflected glow from the fire. "It was not made here, I imagine."

"No, Juan Lewis. It is from Burgos in Spain. Governor Borica brought it to Monterey. He gave it to San Carlos Borromeo. My grandfather received it as a gift from the padres for his service to the mission."

Leaning closer, John let his fingers trace the carved door panels depicting four saints at their devotions. Maria watched him, smiling. "You love nice things, Juan Lewis."

He nodded. "We had some fine pieces in my home when I was a child. I remember them." He straightened and glanced around the room again. "You have many fine things here." He was referring particularly to the silver candelabra, to a large leather-bound trunk strapped with hammered brass, and to a long, beautifully made refectory table. Above it, in an oval mahogany frame, was a large portrait of a bemedaled, distinguished-appearing gentleman dressed in the court fashion of a half century earlier. The blue eyes and long copper hair arrested John.

"My uncle, Estevan Robles. He was the brother of my father's father. Charles the Third made him a special emissary to Mexico City." Maria clasped her hands beneath her chin and gazed up at the painting. "I used to think he was the most beautiful man!" She turned, frowning critically, and studied John's hair. "In the bright sun I think your hair is almost the same color. But your eyes are darker." She clucked sadly. "Too bad!" Lewis resisted an impulse to take hold of her regal little neck and shake her playfully.

"And my nose is thicker and my hands are larger, and my feet too probably, and in one thousand years I could never be a fine gentleman."

"Oh, but you are, Juan Lewis! Even your rough clothes ..." She broke off in confusion. "I mean, you cannot hide

119

what you are." She pointed to the picture. "If a good artist painted you, you would see that I am right."

John laughed. "If I paid him enough."

Doña Encarnacion interrupted to say good night. He excused himself and went to thank Maria's parents. For his host's sake, he was determined to say the conventional minimum to Goyo but when he turned to look for him, he discovered that he and Osio had gone.

At the door he said his formal thanks to Maria. She acknowledged them with an exaggerated curtsy. "Our house is your house, señor. We hope you will return soon. By the way, Juan Lewis, since you are learning Spanish, do you know the word, 'ciego'?"

He thought a moment. "I don't believe so, Why?"

Smiling mischievously, Maria curtsied again and disappeared. On the way home, he asked Doña Encarnacion what the word meant.

"*Ciega?* In English, Juan, you would say 'blind.'"

4. THROUGH THE WINTER and well into the spring of 1841, John, working as a clerk in Captain John Cooper's establishment, received packets of letters from Sutter. The tenor was always the same, extravagantly optimistic predictions for the future and petulant often angry, complaints that not enough of his basic requirements were being forwarded. There were increasingly urgent appeals to John to speak on his behalf. The most recent letter—John suspected Harry Morrison had written it for Sutter—told of a major invasion of Hudson's Bay Company trappers along the upper Sacramento River and its tributaries. Because of the consequences, Lewis showed the letter to Cooper. The merchant studied it carefully. "It's time to see if we can't move the governor to keep some promises. I'm still not sure what that poor fool is up to, but whatever it is he's trying to do it all himself. Until he gets some dependable whites up there, he's not going to establish anything that'll stand. Alvarado's going to have to dip into the treasury. Meantime, we can ship him enough to tide him over."

John refolded the letter. "What is the best way to handle the governor, sir?"

"The best way, lad, is to make him feel that his plans are being threatened. There are two conditions under which a man fights hardest—when he's scared for his life and when he wants something real bad. We've got a good straight haul on him both ways. I'll be stopping by there in a few minutes. We'll see what happens."

Shortly after Monterey's siesta was over, Captain Cooper returned to the adobe office. John looked up from an inventory sheet. "Lad, the governor's suddenly taken a notion that he'd like to chat with you in about fifteen minutes. Will you be ready?"

Surprised that an appointment had been managed so

121

quickly, John said, "I'll be there. Is there anything I should know?"

"Nothing, lad. Among other things, I happened to mention to His Excellency that you'd received some disturbing news from the north. For some reason, he got an urge to know about it. I told him he'd best hear it from you."

Alvarado's preliminary greetings were brief but cordial. He came to the point of the meeting quickly and said to John, "I am told there is great difficulty at New Helvetia. The government is fortunate to have you here in Monterey since you know our esteemed Señor Sutter's problems from personal experience," John caught the fleeting smile, "and quite possibly you would be less apt to exaggerate them. No? And, I understand he has written you."

"Captain Sutter feels a great responsibility—"

Alvarado lifted a hand to interrupt. "Perhaps Señor Sutter will also be a captain in Alta California, some day. Forgive me. Continue, please."

"*Señor* Sutter feels a great responsibility is resting on him. As Your Excellency's representative he fears he will earn your disfavor because he is unable to secure the simple things he needs to establish his fort and keep out the foreigners."

"Oh? But I have been told that Señor Sutter was able to get all he needs from the merchants."

"He has received less than half of the things he has tried to purchase. He lacks the money to pay."

Alvarado reached for a letter. "I have received letters, too, my friend. Sometimes I think our Señor Sutter has more trouble with our language than with our Indians. And I think he tries to frighten me into more assistance. He tells me he has reports of many groups of Canadian and American trappers on the waters above the Rio de las Plumas and he lacks sufficient strength to drive them out. Do you believe that is true?"

"Yes, Excellency. It is true. He has no dependable force."

The governor appeared to be lost in thought. After several minutes he looked up. "We are grateful to you, my friend, for telling us so many things. Sometimes it is difficult for an administrator to obtain correct information. Very little of the advice a governor receives is free of selfish interest."

John stirred uneasily, wondering how to interpret the remark. The governor went on. "If Señor Sutter is so short of good help, then he must have committed you to a very important task to spare you?"

John's uneasiness increased. He sensed it was no time for deviousness. "Señor Sutter feels it is very important for me to stay in Monterey to assist in getting supplies. It is his intention that I return to New Helvetia."

Alvarado's dark eyes saddened. "What a pity. I am sure there will be sadness in many of our houses. It occurs to me that your presence in New Helvetia could be useful to many people." Pondering, he returned to his chair. "For instance, if I were a merchant who had extended Señor Sutter much credit, I might find it useful to know how my materials were being used."

John's uneasiness verged on panic now. Obviously, the governor suspected he was working with the merchants.

Alvarado said slowly, as though to ensure complete understanding, "It could also be useful to the government to have loyal and experienced eyes watching our distant interests." He leaned forward. "Is it possible that you will be able to return soon?"

Although the room was chilly, John's hands were clammy. The extent of the duplicity he was involved in overwhelmed him. He struggled to keep from saying aloud, "First, the merchants sent me to spy. Then I schemed to come back and stay back. Now *you* ask me to spy." He longed to shout, to demand an end to his dual, now triple role. He wished to God now that he had not met Maria Robles and fallen in love with her. Without that he could find the courage to say what he wanted to say, then try to find a ship out of this accursed land, so beautiful on the outside and so rotten in its internal affairs—a land ruled by men filled with deceit, treachery, and selfishness, and Sutter was no exception. But he did love Maria. And it was no time to do what he had done so many times, let anger rule him and uproot him. "Excellency, it is possible for me to return to New Helvetia tomorrow. But I am in the captain's service." He made no attempt to correct Sutter's self-assumed title. "I have not accomplished what he expects. If I return now, I will have failed him."

"You will not have failed." The governor indicated the paper on which he had scratched notes some minutes ear-

123

lier. "I have decided Don Juan Sutter needs practical assistance. The government will give him two thousand dollars in credit to be spent with our merchants and you, my friend, will be given an excellent reason to return. More than that, Don Juan Lewis, when you have enough information of importance to return to Monterey you will write me a personal letter and I shall demand your immediate presence here, to receive your citizenship."

John's sense of relief was so vast that the governor laughed. "I understand well, my friend! It is difficult to wait the outcome of critical things."

Early the following morning he intercepted Don José Robles, Doña Gertrudis, and Maria on their way to mass at the presidio chapel. Casually, he let them know that he would be returning to New Helvetia soon. Their disappointment was unmistakable. Maria appeared stricken until he added that the governor had indicated full citizenship for him, possibly within a few months. He accepted their congratulations gracefully and resumed his pretended errand to the customhouse.

During Monterey's two-hour siesta John had taken to climbing the hills back of the capital. It helped him think and gain perspective on his problems. Seated on a sun-warmed sandstone outcropping, he started as he saw Maria leave the Robles house almost furtively and hurry up the hill toward him. There was no doubt that she was taking pains to keep well concealed.

For the first few minutes he was able to follow her bright red and yellow wool shawl as it flashed between the tree boles and clumps of underbrush. And then he lost her.

Goyo, working in the compound, saw her too. It was not unusual for Maria to wander in search of wild onions and water cress. But he saw that she was hurrying now, half-running. Curious, he put a bridle on one of the horses and mounted. He circled to the west of the house so she wouldn't suspect that he was following. When he was certain that she could not see him, he turned southward and spurred his horse viciously into an uphill gallop.

John began working his way down the hill to intercept Maria. He caught sight of her, less than fifty yards below him, still climbing rapidly and obviously tiring. Moving toward her, he called her name. She froze when she heard

124

his voice, a hand clapped over her mouth. He stepped into the open then and she cried out with relief. A moment later he found himself holding her as she braced against his body, supporting herself by clutching at the cloth of his upper sleeves.

"Oh, Juan Lewis, you frightened me!" The words came in one explosive breath. She took in another great gulp of air. "I saw you over there—I see you every day—but I did not see you leave."

Self-conscious, John freed her hands and urged her gently down onto a half-buried boulder. "Rest, Maria, get your breath first, then tell me what the trouble is." Annoyed with herself, she shook her head violently. *"Madre de Dios,* Juan Lewis! I ran too quickly. I am no longer a young girl, you know."

John burst out laughing. "You are an ancient hag, Maria. Even a blind man knows that!" He grasped her shoulder and shook her playfully. "Now tell me, what is the difficulty?"

Maria turned to him appealingly. "Oh, Juan Lewis, everything is wrong, everything!"

"Everything is *not* wrong. What are you talking about? Tell me!"

Seated close enough to feel the warmth radiating from her body, John looked down at her pale, oval face, at the luminous gold-flecked brown eyes, and the beauty of her made him ache.

"Juan Lewis—what is wrong is that you are leaving. It is not the proper time, because—because—"

"Because what? Why?"

Maria's face contorted with anguish. "Because you are blind and I am shameless."

"I am not blind, Maria."

Seized by a sudden fury she began beating on his chest with clenched fists. "Then you are cruel, Juan Lewis, because you force me to say it! You force me to say 'I love you!' Everybody in Monterey knows that I love you. Everybody but you!"

From his hiding place a hundred yards away, Goyo, choked with rage, watched his sister go into John Lewis' arms. He could feel the storm of their passion. He saw John rise quickly, lift her as though she were a child, stand her on the rock, and take her in a huge, hunched embrace. Sickened, Goyo watched Maria as she kissed the

125

big American wantonly, urging herself against him until he was forced to thrust her away. Moments later, they were together again, down on the steep grassy slope. Maddened with hatred, Goyo Robles leaped down the far side of the hill toward his tethered mount.

Several hours later, at the Robles house, Goyo confronted Maria. His smile frightened her. "Well, my little maiden—or should I say, my little 'woman'—did you have an interesting afternoon?"

Maria eyed him suspiciously and averted her eyes. "It was nothing unusual."

Goyo regarded her with mock amazement. "Nothing unusual?"

Maria cried out in terror as she was snatched to her feet and struck in the face. "Nothing unusual?" He struck her again and his palm left a broad reddened welt on her right cheek. "It is nothing unusual for a rutting whore to throw herself at a man and grovel around on the ground with him?" He struck her again and continued to strike with his open hand until she sagged and slipped from his grasp to the hard earthen floor. The cotton blouse that had ripped off in his hand he flung down on her naked torso and he jabbed the pointed toe of his riding boot into the flesh of her firm, outthrust breasts. "You like the ground, you dirty little whore! Stay there! Even *La Mofeta* would not offer your body for sale. That is how depraved you are. For a *californio* lover you do not even have a promising smile. But for a yanqui criminal you have everything and you beg him to take it."

Sobbing, Maria tried to protest. He abused her with his boot again and spat on her. "Weep, my little whore of a sister. Weep here and soon you will weep at the grave of your yanqui, too—if you can find it."

Caught up in turmoil, John excused himself after supper with the Coopers and went for a walk. The feel, the smell, and the volcanic passion of Maria still overwhelmed him. He cursed the caution that had made him keep his feeling to himself for so long. He wished that he had declared himself first.

His thought strayed again to the complexities of California politics, the jealous maneuverings that set brother against brother, California against her parent Mexico, Los

Angeles against Monterey, and Monterey against any group that coveted the customhouse. As darkness settled he turned inland to the San Jose road. Several times he heard rather than saw riders. Some were nearby. Finally, aware of the seeping night chill, he angled southwestward until he came to a meandering stream.

Midway across a narrow log footbridge, he heard the start of hoofs. Before he fully realized his predicament, he was confronted by a horseman spurring his animal to reckless speed. The bridge, scarcely four feet wide and without handrails, offered no possibility for escape other than over the side into the unknown darkness. John's warning shout caused the horse to shy when it was upon him. Instinctively his hand shot up to grab the bridle. As his fingers closed around the cheek-strap, the animal reared and its hindquarters slipped off the logs. An instant later, horse, rider, and intended victim plunged into the brackish water.

When he touched the muddy bottom John pushed to the surface and found the water came only to his armpits. The animal, screaming in terror, was thrashing on its side, struggling to keep its head above the water. Seizing the coiled halter rope, John pulled up until the horse gained its forelegs. From the darkness beyond the animal he heard a choking cry. Then silence. Its footing secure again, the big animal lunged at the slippery bank, driving its powerful hind legs wildly. John, caught between the narrow banks, had no chance to maneuver past it to help the man. After another series of futile lunges the horse wheeled and floundered off downstream in the darkness. Instantly, John began calling out as he searched for the rider. There was no sound. He made his way downstream until his leg struck a submerged object. Seconds later his fingers clutched at loose clothing and he pulled an unconscious form to the surface. Holding the man's head above water with his left hand, he groped along the bank with his right hand, seeking a place to climb out. Twice he tried to hoist himself up by the thick overhanging grass but the roots let go. Ten minutes of treacherous going followed. Several times John and his unconscious charge slipped below the surface. Finally he found that by driving his sheath knife into the slimy incline he could pull himself and the unconscious man up by degrees.

He gained the grassy bank and collapsed face down. He

127

lay there until his great heaving breaths began to subside. Then he realized that his hands were sticky and warm with fresh blood. He felt the rider's face and discovered that the temple and cheek had been laid open to the bone. Propping himself up, he grasped the clothing and turned the limp body over on its back. As he leaned close to listen for signs of breathing, he discovered that the man he had struggled to save was Goyo Robles.

Suddenly John understood the meaning of the hoofbeats in the darkness nearby and a murderous rage welled up in him. Grasping Goyo's long, coarse hair, he raised his knife intending to plunge its blade deep into the *californio's* throat. For a long moment the point trembled within inches of the bloody flesh. Then with an anguished curse he flung Goyo's head aside. An instant later the sound of voices reached him and he heard the muted thudding of hoofbeats moving nearer. John wiped his hands on the grass and replaced his knife. Then, running quietly, he followed the stream until he came to its small, thin-veined delta at the beach. There he turned west toward the town. An hour later, dry and fortified by strong Chilean brandy, he told Captain Cooper and David Spence what had happened.

Cooper, white around the mouth, glowered toward the bay. "That murderous bastard! I've a good notion to drag him up in front of Alvarado myself. He's a danger to us all!" He turned back to John, "You're sure he was alive?"

"He was breathing. I couldn't see but I think he was kicked unconscious by the horse."

Cooper snorted. "It would have been Divine Providence if the horse had killed him. God knows he's killed enough of them!"

David Spence, who had been listening with grave concern, rose from his chair. "Time enough to tell Alvarado about it later. I think I'll pay a little visit to the Robles adobe in my official capacity. If he was hurt as badly as it sounds, perhaps Providence will spare us much more of him anyway."

The *alcalde* departed and returned within the hour. "He's alive. Osio and the others found him. He's not very talkative at the moment but I gather his horse shied at something on the log bridge and they both went over the side. Osio says they found him about a hundred yards downstream. He thinks Goyo must have been dragged and worked loose when the animal leaped up the bank. Doña Gertrudis and Maria are going to make a novena to the Holy Mother for saving him."

Spence tried unsuccessfully to refrain from smiling and Cooper cursed half aloud. "He'll need nine novenas before I'm through with him! One way or another I'm going to keelhaul that rat!"

John drained the last of his second brandy and set the glass aside. "I'd like to attend to that myself, captain. I don't want anybody to go to the governor for me."

Cooper grumbled. "Good enough. How do you intend to handle it?"

The fact was that John had no idea how he would confront Goyo Robles. But he knew he would. "I'm not sure yet, sir. But you can wager that I will."

Ten days passed before John's opportunity came. He had managed to see Maria briefly each day. She seemed strangely subdued but through her John was able to follow the course of Goyo's surprisingly fast recovery. When he learned that Goyo would be visiting Juan Alvarado the following afternoon at five, he determined to confront him before the meeting.

During the morning he worked with Cooper on a list of basic requirements that he felt Sutter would need. As the afternoon wore on John found frequent excuses to go to the door. A few minutes before five he saw Goyo, his face heavily bandaged, emerge from the Robles house and walk slowly down the hill toward the adobe capitol. Ordinarily a *californio* would mount his horse to cross the street. John suspected that Goyo had been told not to ride until the flesh had joined properly. Quickly, John crossed the calle. Then he hurried to Cooper's house and entered through the back door. Seconds later, he left by the front doorway and waited, partially hidden behind a stand of tall willows that grew beside the railed footbridge leading to the governor's offices.

Goyo did not see John in time to avoid him. Smirking, he bowed slightly. "So. We meet on another bridge."

John blocked his way. "I did not know who I was saving, or you would be dead now. But if you or any of your friends ever make another move against me, I swear by our Lord Jesus Christ, I will kill you without mercy."

There was no fear in Goyo's fiery black eyes as he stared back at the yanqui towering over him. Inwardly he

129

cursed the bad turn of fortune that had spared John. In his mind's eye he could see the yanqui and his sister again, lost in their passion on the grassy slope. At some place, at some time, he would be the one who killed. He eased back warily as John stepped closer. "Do you understand me, Goyo? Do you understand what I have said?"

The bandage distorted Goyo's attempted smile and made him wince. "I understand, Juan Lewis, yanqui defiler of virgins. And I swear also to the Holy Mother that I shall have the pleasure of seeing my sister weep at your grave."

Before John could respond, Robles slipped past him and strode across the bridge.

John was able to prolong his stay in the capital for another ten days. The last meeting with Maria took place in the untended garden of the presidio chapel after vespers. Doña Gertrudis had accompanied her daughter and had deliberately prolonged her devotions to allow them time together.

In the winter darkness Maria clung to John with her cheek pressed hard against his coarse woolen shirtfront. "Oh, Juan—I love you so much, and I will be like poor old Concepcion Arguello waiting for her Resanov. You will go away and I will never know what happened to you. I will wait for years and then I shall have to join an order!"

John chuckled mirthlessly. "And if I stay here your brother and his cutthroats will see to it that nobody else knows what happened to me also."

"Goyo will change in time, Juan Lewis. He does not know about love. He has never loved because he does not want the responsibility. He is filled with the political madness. He plays games and pretends he is doing the work of a patriot. But he will accept you when we are married. He will find someone else to suspect then. I am certain."

John slipped a finger under Maria's chin and lifted her face. "You are dreaming, dear little girl. Goyo would turn on me again just as he turned on Graham and Chard. He is treacherous. He would even turn on Alvarado."

Her eyes blazing, Maria pulled away. She had not spoken of the beating or the loathing for her brother that had followed his unfounded accusations, for when she learned that she had been seen a great guilt assailed her. She knew

130

that if John had persisted she would have allowed willingly all that he asked. She despised Goyo now, but for some reason she did not understand she felt compelled to defend him. "Goyo and Juan Alvarado are blood brothers."

"So were Cain and Abel. I will never trust your brother, Maria. Don't ask me to. To him I will always be an outsider whose hair and skin are the wrong color, who goes to the wrong church, if he goes at all."

To soften his words, John took her hands and put them around him. He kissed her tenderly and then with such fierceness that she protested and freed herself.

"Please, Juan Lewis! You will not be gone forever. You will return soon, I know. I will wait and I will write letters and when you return you can finish your instruction and we can be married here with my father's blessing and with the governor's too. His Excellency likes you, Juan Lewis."

"He doesn't really know me."

"He knows you through your friends."

John laughed bitterly. "He believes the last man who talks to him. Your brother makes certain *he* is that one!

Again Maria protested, "No! That it not true. The governor listens to Goyo, yes. But if he believed those lies he would not be trusting you in New Helvitia. There are times when Juan Alvardo has very much a mind of his own."

Doña Gertrudis, her knees sorely abused by the overlong devotions, appeared in the doorway of the chapel. Maria kissed John hastily and together they walked toward the Robles adobe. At the footbridge near the government house, John left them. Doña Gertrudis took his big hand in both of hers. "God will go with you, Juan Lewis. And He will keep you safe. We shall pray for you— Maria and I."

It was still dark when John, leading two pack animals heavily laden with personal clothing and some Indian trade goods for Sutter, rode out of Monterey along the San Jose road.

The trip to New Helvetia took four days. The second night was spent at the Peralta rancho on the east side of the Bay of San Francisco. The third night John spent at Doctor John Marsh's Rancho de los Meganos. He had heard about the penurious misanthrope but he was not quite prepared for the man. Apparently humorless and far from hospitable by *californio* standards, the gaunt Marsh was said to have graduated from Harvard University with a degree in the liberal arts. He had arrived in Los Angeles in 1836 and was granted a license to practice medicine solely on his own contention that he was a skilled physician and surgeon. It was generally conceded that Marsh had treated successfully most of those who had come to him. Even on so short an acquaintance John found that he could believe the stories that Marsh had stocked his rancho by demanding that usurious medical fees be paid in livestock, seed, and equipment.

The suspicious Marsh ordered a straw bed to be fixed for John in the attic. He was to share the cramped quarters with two *mestizo vaqueros* who customarily slept there and acted as the owner's bodyguards. He had no doubt that Marsh had ordered them to keep a watchful eye on him as well.

When John climbed the ladder to stow his things he discovered that the walls below the overhanging eaves were perforated with powder-blackened loopholes. At supper that evening the doctor told him that the upper floor was often used for defense against bandits, marauding Indians, and renegade white trappers who judged the remote rancho to be an easy target. Marsh bitterly blamed Captain Sutter for attracting dangerous rabble with his compulsive hospitality.

The parlor was overrun with books. Among the rows of classics in English, Greek, and Latin Lewis found several impressive, well-worn medical volumes. There was no doubt that John Marsh, bona fide doctor or not, possessed an even more comprehensive library than Governor Alvarado. After supper John surprised his host by losing himself for an hour in a copy of Pliny's *Letters* printed in Latin. Marsh tested him by quoting long passages which John translated easily. The doctor seemed to mellow a bit then, but John knew that the eccentric *ranchero* was not one most men could ever warm to.

Marsh was quietly doing everything he could to encourage American immigrants to come West. He had been among those picked up in Castro's surprise raid and accused of plotting to take over Alta California for the United States. Cooper had told John that the doctor was released later after pleas from both Larkin and Vallejo. There was little doubt, John felt, that Marsh represented a far greater danger to Alvarado and the Mexican government than the illiterate mountain man, Graham.

Captain Sutter welcomed John Lewis as a prodigal son. Harry Morrison managed to conceal his own elation until he and John found a brief moment alone. "Oh, my God, laddie! What a relief it's going to be to have a two-sided conversation for a change." They had only minutes together before Sutter, jubilant over the news of his new credit, fired the small brass cannon to announce a celebration.

An old ox, well past its work prime, was killed and butchered to make a special feast for the Indians. In the Captain's quarters venison and several varieties of wildfowl and game birds were served for the dozen white workers along with vastly improved *aguardiente* from Harry Morrison's crude still.

At his request John shared a hut with the little Scotsman. They talked late and John questioned him about the raid that had taken Graham and Chard and the others and how he had managed to evade Castro's soldiers. "No credit to me, John, I was over bargaining wi' Butron for some malting barley. I took a demijohn of wally spirits and we talked and drank too late. So I decided not to ride back until early morning. Just before sun up, one of the half-breeds I had cutting wood for the stills come ass-spanking across the fields to tell me the soldiers were going to hunt me doon. Butron hid me in a bug-infested straw pile for two days. When it was safe again I rode east to the Consumnes River country and lived off the land until the damned savages got a wee bit too clubby. I did keep them calmed doon by cooking up a stinking brew of roots and berries mixed wi' a bit of abacadabra. When they finally caught the notion that it wasna' going to make them into eagles like I promised I reckoned it was past time to go calling on my dear friend Sutter."

John laughed sympathetically. "You've had me worried."

Morrison snorted and resettled himself in his blankets. "Nae half as much as I've worried myself, laddie. But this place'll do nicely until I can sniff the wind." He braced himself up on one elbow. "I'm afeared to ask, but what's happened to my still at Natividad?"

John did not have the heart to tell the little Scotsman

133

that he had not troubled to find out. "If I know anything about Alvarado he's got somebody running it at a profit to himself." John twisted on the straw pallet and changed the subject. "What do you think of the Captain by now?"

"Well, John, me young friend—if he gets one tenth of the things done that he's promising, then he'll have nae trouble getting himself crowned emperor of California. The only trouble is, like the bean stew he feeds us three times a week, I dinna ken how much is substance and how much is wind."

The first four months of 1841 passed quickly at New Helvetia. The merchants, certain that their bills would be paid directly from the treasury, forwarded to Sutter what was needed for the still and the blacksmith shop, and work had been started on the main building of the fort. John estimated the protected area would finally measure about five acres of hilltop. Sutter's plans called for walls fifteen feet high and three feet thick. Workrooms and shops were located around the inner walls and one of the bastions would house the prison. John noted in the journal he was keeping for Alvarado that Sutter intended to defend the walls with twenty cannon. He wondered what sort of invasion Sutter expected. The capital of Alta California itself had only eleven cannon, as defense against invasion by foreign fleets.

Early in May, Captain Sutter summoned John to a conference. Jean Vioget had gone down river to Yerba Buena to determine survey points on the proposed land grants. Sutter needed assurance that Alvarado would look with favor on the request for land grants. He planned to send John to Monterey with maps and persuasive arguments.

Sutter pointed to the crudely drawn map spread before him. "We do not ask for much land. Less than fifty thousand acres. In two pieces. Nine thousand acres here at the confluence of the American River and the Sacramento, and forty thousand acres here on both sides of the Feather River."

John studied the drawing. "Why do you not ask for this land?" He pointed to the great area of rich valley lying between the two parcels.

Sutter's shrewd little pale blue eyes twinkled. "Because, John, never ask for too much! This is a modest request. Alvarado will know it is the truth when I say it will be easy to defend the state land between my two grants. We will have an armed outpost on the Yuba river and another on the Feather. With the fort here, intruders would be caught in the pincers. It is excellent military strategy." Sutter smiled conspiratorially. "And very good political strategy. When we have the two original grants, then we shall wait for the proper moment to ask for the land in between."

He spread his arms expansively. "And then, Johann Sutter will be the master of one hundred fifty thousand acres, the largest grant in Alta California. My fort will be the heart of a great center of commerce—the richest in the west." He pressed a cautioning finger against his lips. "But we will not speak about that to Alvarado. Take the map. Go tomorrow. Early. And you will see that Sutter does not forget to reward those who work well for him."

John Lewis did his work well. On June 18, 1841, Secretary of State Casarin confirmed to Don Juan Augusto Sutter eleven leagues of land in two parcels as described in the survey done for Sutter by Jean Vioget.

In Monterey again, Sutter was dumbfounded and then secretly elated to find that Alvarado had included in the terms of the grant a proviso that he in turn must grant to bona fide heads of families sufficient land to establish small ranchos for the purpose of guaranteeing stable manpower to protect the interior. The provision astounded Sutter because it seemed to be wholly illogical, a contradiction of all that Alvarado and Castro had espoused. Surely by now, Sutter thought, they understood that he would not be able to recruit native *californio* families to the interior. His most glowing predictions and most generous promises had failed to attract more than a score of undesirables. It must be obvious to the governor and to his advisers that any such settlement would have to be undertaken by the very immigrants from the north and east that the government seemed to hold in deadly fear. It was incredible. But now the success of New Helvetia was assured. He would be free to openly encourage settlers to establish families to protect the very land they would be overrunning. They would be his dependents economically; as Special Minister of the Interior he would hold the power of life and death over them. Sutter could hardly believe his own good fortune.

Two days later, shortly after daybreak, most of Monterey turned out to see the newly landed citizen on his way. From the moment of his arrival, Sutter again had started to recruit additional help for New Helvetia. He had combed every cantina in the capital, inviting, coaxing, and even indirectly threatening new deserters. This time he had managed to commit a dozen drifters most of whom the officials were openly relieved to see gone. Unmolested by responsible authorities in Castro's absence, these were men who roamed the capital at will, drinking, fighting, and stealing.

Captain Cooper saw the men assembled and turned to Sutter, his mouth agape. "God's cheek, man! How will you ever manage such a crew of rascals in the wilderness?"

Supremely confident, Sutter laughed. "It is very simple, my friend. I will allow them nothing stronger to drink than water."

One of the recruits was a big Negro who called himself Tennessee. Strong and amiable, he had been an expert blacksmith before he fled the slave compound and escaped to California aboard a trading vessel. Sutter learned that the man was a skilled craftsman and outdid himself at persuasion. He offered Tennessee a generous salary and promised him that he would be set down in "the great history books" as the first "darky" ever to hold a position of high responsibility in Alta California.

Diminishing his pleasure somewhat was the disappointment Sutter felt upon learning that John Lewis was to be detained in Monterey at the governor's request. But his mood changed immediately when John hinted that citizenship might be forthcoming. "Good! Very good, Lewis! When you say your oath, tell the governor that you wish to take up land near me. I will have Vioget measure off three leagues on the Sacramento north of the Feather River. We will work the land together. I will supply everything and we will share."

John professed to be pleased and Sutter, at the head of his motley *caballada,* marched grandly off to New Helvetia filled with plans more extravagant than ever.

Full citizenship in the Republic of Mexico was conferred upon John Lewis on the tenth of July. Following the ceremony in the governor's office, attended by the merchants and their families and by Don José and Doña Gertrudis, the head of the Robles family announced the betrothal of his daughter Maria to John Lewis.

For all but Goyo Robles it was an occasion for rejoicing. He managed, at his father's request, to endure the ceremony but refused to congratulate the couple.

Five days later, Monterey was plunged into despair by the arrival of the schooner *Bolina* from Mazatlán. Aboard were Isaac Graham, William Chard, and a dozen others of the company. All had been acquitted of treason, fully pardoned with promises of indemnity, fitted out with new clothing and arms, and returned to Alta California at the expense of the Republic of Mexico.

Stricken by the central government's affront to his authority, Governor Juan Alvarado retired to his inland rancho and left the government in the hands of the secretary of state and to *alcalde* David Spence.

The following day a dispatch arrived from Sutter addressed to Captain Cooper. It advised that a plainsman who had crossed the sierra reported a wagon party of two hundred men, women, and children rolling westward from the United States by way of South Pass. John, again in Cooper's employ, read the dispatch and was amused that Sutter could hardly conceal his jubilation. Sutter also demanded a great many additional supplies. John's computation showed that the order would exceed the governor's guaranteed credit by more than a thousand dollars. Skeptically, Cooper examined the list. "It's more of the same, John. We'll not be sending so much as one keg of nails until we've got our funds from the treasury. And with Alvarado away, no doubt nursing his injuries on a bottle, I'm more than a little apprehensive about receiving payment for what we've sent already."

"Suppose Alvarado should refuse to pay?"

"In that case, lad, we'll do as we've done in the past. We'll simply arrange to have our next cargo or two landed here free of customs duty." He smiled slyly over the steel-rimmed magnifying spectacles he had put on to read the dispatch. "As a matter of fact, if we've nothing more than profit in mind, I should think David and Tom would prefer such a settlement."

That afternoon John wrote a long letter to Sutter explaining Cooper's inability to extend more credit and urg-

137

ing him to do everything possible to help Harry Morrison increase the size of the still:

> —for there is no faster source of cash revenue than that of whiskey. The Natividad distillery continues to produce, but only very small amounts. Even with Isaac Graham back it is doubtful that it will ever produce the quantity and quality of spirit realized under Morrison's direction. Most particularly since Will Chard wishes to forsake that activity and disassociate himself with Graham. So a ready and large market exists and established merchants here will take the product in fair exchange for credit.

John said nothing about the misgivings aroused by Alvarado's absence or about Graham's swaggering public boasting that he would find a way to settle his score with "Bautista."

John and Maria saw Will Chard only briefly before he left for Branciforte. Though somewhat thinner than when he had been taken away to San Blas, he looked thoroughly fit. Maria found him an imposing figure. While Chard was still angered at the injustice of his arrest, he was not as bitter as Graham. Chard surprised them by revealing that Thomas Larkin, working quietly behind the scenes, had been a great influence in securing their release. Larkin, without fanfare, had gone to Mexico to intercede. Chard told them that Larkin, deploring the Mexican's inhuman treatment of their prisoners, had managed to convince the government that Alvarado and Castro had acted without cause. No written charges had been filed and no proof offered. John wondered whether or not Spence and Cooper knew of Larkin's intervention. He suspected not. Chard's statements reinforced John's conviction that Thomas Larkin was playing for far higher stakes than just a preeminent position among the traders on the California coast. His mysterious behavior in this matter, John felt, lent credence to the persistent rumor that he sent confidential reports to high officials in the United States government.

Intrigue on all sides infested Monterey. Committed now to assume the responsibilities of the head of a family, John began to feel a deepening sense of anxiety. The realities of the situation pointed to violent change in Alta California.

First, there was Alvarado's curious turnabout in the matter of settlers in the interior followed by Graham's release and the governor's irresponsible disappearance; Castro's continued absence; the presence of large groups of respectable family immigrants from the United States, not just mountain men and deserters who in the past had formed the majority of immigrants to Alta California. All of this was added to rumors that the United States would soon be formally at war with Mexico over Texas.

John was certain his best potential lay in becoming a merchant. But anything in Monterey would be limited by the tolerance of the established merchants, who were wary of anyone not working for them. Moreover, any upheaval in the government would be most acutely felt at Monterey. In Yerba Buena and Santa Barbara there were fewer merchants and fewer opportunities to trade.

Only one new merchant had managed to gain a foothold in Yerba Buena. William Leidesdorff had established a warehouse and forced Spear, Hinckley and Jacob Leese to accept him. For reasons he wasn't sure he understood, John found himself under a growing compulsion to travel to the Bay of San Francisco to visit Leidesdorff. Leidesdorff's reputation for charm and hospitality had already reached Monterey. Colonel Vallejo confessed to his sister, Doña Encarnacion, that his four attractive young daughters had been vying shamelessly to attract the newcomer's attention. Apparently, Leidesdorff's appeal was extraordinary.

On August sixteenth, Governor Juan Alvarado, appearing ill and much thinner, returned to Monterey to take over the reins of government. He had been brought back hurriedly by a report from Colonel Vallejo that an armed United States Navy flotilla had dropped anchor in the Bay of San Francisco two days earlier. Commander of the expedition, Commodore Charles Wilkes, immediately dispatched a party of sixty men in six whaleboats and a launch under the command of a Lieutenant Ringgold to explore and take soundings along the inland waterways and up the Sacramento River. Alvarado, commander of the capital's ineffective forces in Castro's absence, attempted to ready the Mexican Fort for a token defense. Next he called a meeting of Monterey's able-bodied men to form a militia.

139

Maria, clinging to John's arm as he got ready to attend the meeting, urged him to stand with the governor. "He needs help now, Juan, and he would rather die than turn to Graham again."

John laughed and slipped an arm gently around her. "I was one of Graham's cutthroats once. Do you really think they're so dangerous?"

"Do not laugh, Juan Lewis! They are dangerous now. There are wagers being made that Graham will try to kill Alvarado before the summer is finished."

John snorted. "All Graham wants now is his thirty-six thousand dollars indemnity. He will do nothing to risk that, I promise you." He brushed her hair with his lips. "And I promise you that I'll stand with the governor."

Speaking from the porch at Government House, Alvarado addressed the fifty-odd men. He had placed the capital's permanent garrison of thirty soldiers under the command of an arrogant non-commissioned officer named Chavez. As they listened, nothing about the plan reassured John Lewis or the other Anglo-Americans present. But in the end they pledged to defend Monterey. Muskets were distributed, most of which were rusty and bore French arsenal imprints dated during the Napoleonic campaigns. By common consent the men volunteered to use their own weapons and to provide their own powder and shot.

Later at a meeting with Spence, Hartnell, and John, Captain Cooper, acting as senior militia leader, summed up the situation. "It's no secret that the French, the British, and the Yankees all have warships on Pacific stations in case Mexico gives them an excuse to take over. But it will not be done this way. Save your powder for ducks and geese, my friends. The expedition is harmless and an accurate survey of the rivers emptying into the bay could be very useful information in the right hands. I'd a deal rather have those Yankee tars up there surveying the waterways than have them down here surveying our women!"

John listened and speculated about asking a leading question. Larkin had not been present at Alvarado's meeting. Rachael Larkin had sent word to the governor that her husband had been confined to his bed since the evening before but John had seen Larkin at work in his office earlier that day. Perhaps, if he could phrase it properly, the question would reveal something of the thoughts of the other three. "Captain, the French and the British have had warships visit our coast much more often than the Americans have. Why do you suppose they have not undertaken such a survey?"

Half-smiling, Cooper answered, "John, lad—being mindful that I'm not blessed with occult vision—I'd hazard a guess it's because the French and the British have not had such far-sighted *citizens* in Alta California."

So there it was. A key piece to the puzzle. Given a choice, John had no doubt that both Cooper and Spence would favor the Union Jack flying over Alta California. He wondered if Cooper's oblique reply had been a tacit admission that the British had been outmaneuvered and that accurate inland water charts in American hands would be considered now to be "in the right hands."

Early in September, the unpredictable Sutter committed a serious political affront, this time an internal one. Governor Alvarado summoned John to his office and told him that the Russian commander at Fort Ross had sailed the schooner *Constantine* up the Sacramento River to New Helvetia for an unofficial visit with Sutter. Alvarado asked John to go to New Helvetia to check on Sutter's "progress in fortifying the fort." John understood that he was also to report on Sutter's relationship with the Russians.

Sutter greeted John like an affectionate father and presented him to Commander Rotscheff. The commander spoke no English and Sutter translated the amenities.

During dinner John tried to follow the animated conversation in French and managed to catch a few of the words that shared common roots with Spanish. From time to time, Sutter would provide a sketchy explanation. John gathered that the discussion related to Fort Ross and the Russian installations on Bodega Bay.

After the meal, Sutter indicated tactfully that he and Rotscheff wanted to talk in private. Lewis excused himself and returned to Harry Morrison's enlarged adobe hut.

Morrison showed him the distillery and Lewis saw that once again the Scotsman had managed a wonder of improvisation. The still was not as large as the one at Natividad but it was more efficient. Lewis found the whiskey raw but palatable. More to his liking was the brandy Morrison was distilling from wild grape wine. Later, in the hut where they could not be overheard, the Scotsman recounted the events of the last several months. They were largely a repetition of Sutter's endless promises and disappointing performance. Little was new. Martinez and Peralta had sent *majordomos* to try to collect unpaid debts. Doctor Marsh appeared in person to threaten. Sutter had not been able to mollify him and after his departure he had cursed the doctor and called him an unfeeling, money-mad devil. Vallejo would extend no credit. It was becoming apparent to Lewis that the Swiss who had talked himself in so readily was now in the process of talking himself out. The realization made it all the more imperative that he find out just what the Russian's visit portended. The potential possibilities were alarming.

Four days later, Rotscheff sailed aboard the *Constantine* for the Bay of San Francisco to reprovision before sailing north to the Russian outpost at Fort Ross.

John stood at the landing with Sutter as the vessel cast off. It was impossible not to feel Sutter's repressed excitement. "Such a good visit, Lewis. Such a good visit!" He rubbed his hands. "Very soon, my young friend, they will know that Johann Sutter will not be stopped by usurers like Marsh or by doubters like Martinez or Peralta and Vallejo. They will see!"

John and Sutter waited to ride back to the fort until the vessel disappeared around the bend. John debated the wisdom of prying. Sutter might welcome a safe audience for his boasting since he was extremely pleased with himself.

They rode in silence for some yards, Sutter astride his favorite mount, a leggy, ridge-backed mule that he called Katy. John rode beside him on a skittish, half-broken mustang. When he felt enough time had passed, John said, "The Russian commander seemed very pleased with things here."

"Ja! Ja! And why not? It was a very profitable visit for everybody. No?" The question was a deliberate lead. John felt Sutter studying him craftily and decided to wait and see if his apparent lack of real curiosity would egg the captain on. Minutes passed before Sutter eased his mule closer. "Lewis—you are one of the few people I trust. Always when I ask you to do something you do it very well

142

for me. So now, maybe, I ask you to do something more. Ja?"

John smiled modestly. "I try to do what I can."

"Ja! I believe that, Lewis. So now I believe you can keep a tight mouth. Ja?"

"When it's best to be silent I can manage that, most of the time. But I prefer not to be told things if having such knowledge makes it dangerous for me and for others."

Sutter laughed gleefully and addressed the mule. "Did you hear what he says, Katy? He doesn't want to hear secrets if they are dangerous. But we can trust this young man, Katy. That is why we tell him that we have just purchased from the Russians all of Fort Ross and everything that is in it—livestock, tools, cannon, muskets, pistols, powder, shot—everything." Twisting in the saddle, he swept a strong, stubby arm back toward the river. "Including the *Constantine*—and I have bought it all, over a hundred thousand dollars in value, Lewis, for only thirty thousand dollars. Would you like to hear how I managed this?"

John's answer was a dumbfounded stare that made Sutter laugh. "Good. In my office I will show you. And when the time is proper I will reveal the purchase and you will see that Johann Augustus Sutter will very soon be the most powerful man in Alta California," he shook a stubby finger at John in good-natured warning, "and you will rejoice that you did not call me a fool and a *poseur* like some of the others!"

Alone with Sutter in his headquarters, John Lewis listened spellbound to Sutter's story. Sutter's ultimate purpose was to establish a personal duchy strong enough to dominate all of Alta California. If he could induce Isaac Graham, who bore a just grudge against the government, to join him, then Castro's pitiful army and Vallejo's irregulars would fall before the first charge. Suddenly John's fantasies seemed frighteningly real. Sutter could make any kind of a bargain he wished with the Russians, even have himself declared supreme commander of Pacific North America. Backed by his European military experience and by trained Imperial forces sent by the Czar, after he defeated the Mexican forces, Sutter could turn northward and easily defeat the token British garrisons along the Columbia. He could consolidate the entire Pacific Coast under the Double Eagle from Baja California to Russian Alaska.

143

The prospect made John's flesh creep. He tried to envision the sometimes ludicrous Sutter wielding such power and he wondered if Napoleon himself, in the beginning, had not seemed as ludicrous for the role of world conquerer as the fleshy, balding, mustachioed man recounting with childish glee how he had single-handedly outsmarted the Russians and the rest of Alta California.

Later that evening, John made careful notes on the details of the transaction. They included the promise of the incredibly small down payment of two thousand dollars in cash, bringing the total obligation to thirty-two thousand dollars.

The autumn of 1841 was a crucial one for Alta California and for John Lewis. He was spared the necessity for breaking his promise to keep Sutter's secret when word came to Monterey that Sutter had boarded the Russian vessel *Helena* in the bay of San Francisco and entered into formal contract with the Russian government.

Colonel Mariano Vallejo had been negotiating for the Russian holdings at Fort Ross and Bodega ever since Rotscheff's initial overtures. He was livid on hearing the news. He raced to the capital and arrived shortly after Castro's unexpected return from Mexico City. One of Santa Anna's high military commanders, General Manuel Micheltorena, had secured his release. Unlike Vallejo, Castro retreated into bitterness and displayed little interest in the threat that Sutter now posed.

John was present to confirm details as Vallejo stormed around his nephew's office and Castro slumped dispiritedly while Alvarado studied the figures Lewis had laid before him. Near hysteria, Vallejo broke out again. "Mother of God! I repeat. We are finished! My dear nephew—you must realize now what a traitor you have welcomed into our midst. By supporting this man Sutter you have delivered Alta California into the hands of its destroyer. You must realize," he leaned across the desk and rapped the paper viciously, "that this pompous buffoon now has more cannon, more powder, and more lead than the combined armed forces of all Alta California. More than that, my Indians tell me the first American immigrants are in the Sierra. Many of them! It was not serious before but soon now Sutter will have a hundred and fifty men there—all expert riflemen."

Straightening, Vallejo clasped his hands in prayerful pleading. "In the name of God, Juan, you must send a messenger to Mexico City immediately. Demand professional soldiers and money to garrison them. The immigrants must be controlled or in six months there will be no sovereign state of Alta California in the Mexican Republic."

Worn, and tortured by doubt, Alvarado closed his eyes. Finally, he rose slowly and walked to Vallejo. "I will send you, Mariano. No man can make stronger representations. No man is more respected."

Incredulous, the sleek, elegantly dressed *ranchero* backed away. "But that is impossible, Juan! Impossible! I cannot be spared from the rancho. It would be dangerous, even fatal to us all for me to be away for three months. My men are all that stand in Sutter's path if he should decide to move. Who knows what that fanatic will accomplish in three months with Russian assistance? Look at what he has done already, with no assistance. No, Juan, the very best thing I can do is to write a letter to Santa Anna. I will say that I support you in everything you do. Completely!"

John caught the veiled smile as Alvarado turned away. "Very good, Mariano. I will think. It is not a mission that can be entrusted to just anyone." He glanced at Castro. "I will think who will be the right man and I will tell you."

Castro's face mirrored sudden scorn. "There is no proper man! Santa Anna is a fool. No man of reason can talk with him. Do you know why I was humiliated? Because the Central Government was trembling with fear that our courageous act in banishing foreigners from Alta California would anger the Americans still more." He flung an arm outward. "Forget Santa Anna! You will see. Not one soldier will be sent. Not one!" Castro wheeled and stomped over to the window.

Castro wished that Vallejo would stop his stupid raving. He wanted help not because he feared an invasion but because he was afraid Sutter would build a stronger personal empire than he. But what difference? Alvarado was deaf and blind to the realities now. Mexico's only hope would be to sell Alta California to England in settlement of unpaid bonds. Or even better to establish a Mexican monarchy under French protection. France, too, was a Catholic country. There were many in Mexico City who favored that course.

Castro pictured himself again in Alvarado's chair and a great frustration overcame him. How different history would have been if he himself had dared, as Pico once had, to declare himself governor general of the Department of the North. There would have been no waiting to act then. The foreigners would have been turned back or tried and executed as illegal immigrants.

A week passed and Alvarado had failed to make a choice. Filled with disgust, Vallejo returned to his rancho north of the Bay of San Francisco. Three days later a Chilean bark dropped anchor at Monterey. Aboard was Eugene Duflot de Mofras, the young attaché from the French Embassy in Mexico City. Acting under confidential orders from Paris, de Mofras was visiting the Alta California capital and other settlements along the coast, including the Russian installations, for the purpose of assessing their probable strength.

Monterey, knowing that such visits usually signaled a fiesta by official decree, began brushing up its best clothing. However, de Mofras departed for Yerba Buena two days later after no more than a coldly correct reception by Secretary of State Casarin and Governor Alvarado. Waves of speculation surged up and down the Calle Principal.

The truth reached John Lewis by way of Maria after she and her parents had endured another frightening tirade from Goyo. Bundled against the chill, the couple walked along the narrow tidal margin at sunset. Maria looked at John gravely. "I do not know what will happen, Juan Lewis. But it will not be good. Alvarado did not make a fiesta for the Frenchman because he believes that he came to Monterey to spy. The Frenchman asked to see the fort but Alvarado refused. Then he asked questions that were an impertinence. A young attaché does not innocently ask such questions of a governor. The Frenchman was very angry when he sailed yesterday. And do you know where he goes now? To visit Sutter. Do you know what he said to Alvarado? He said, 'It will be a pleasure to visit Señor Sutter who understands how to be courteous to the representative of a great friendly power.' Goyo is certain the Frenchman is here to spy. He has convinced

Alvarado." She tugged at Lewis' arm and they both retreated from the incoming tide. "It is not good, Juan Lewis. I am frightened."

He wanted to reassure her but there was nothing to say. As adroitly as possible he diverted the conversation to their own immediate plans. He knew that they too might well be swept aside by the mainstream of larger events. "No matter who dictates the operation of the government in Alta California, Maria, there can be no government without the trade that is controlled here by Cooper, Spear and Larkin, and by Abel Stearns in the south. Larkin and Stearns are partners in many ventures just as Cooper works with Spear and Hinckley and Leese too when it is necessary. Cooper needs me now just as Larkin needs his clerk, Talbot Green." He leaned down and kissed the end of her nose lightly.

Maria loved her big yanqui more than she had ever imagined possible. She reached up and put her arms around his neck and he half lifted her from her feet. A moment later she was kissing him with a hungriness that brought a muffled protest from his smothered lips. Before Maria realized what was happening he had carried her to the fine, dry sand beneath a low, grass-covered bank and had put her down gently.

There, in a timeless world, they lay together, talking, caressing, allowing themselves to imagine as much as they dared of that time of coming together when they would be coupled with no barriers of conscience and clothing between them and they could convey without words all that they felt, each for the other.

At about the same time, in Monterey's most imposing residence on the Calle Principal, Thomas Larkin was painstakingly phrasing a confidential message to the American President in Washington apprising him of the current French interest in Alta California.

Goyo Robles was confronting Juan Alvarado on the hillside high above the Mexican fort commanding Monterey Bay. They had ridden there at Goyo's request after he had pleaded so passionately for an opportunity to talk that Alvarado had not been able to deny him.

"The time for waiting has passed, Juan. There is no more time. You yourself have said there is nobody you can trust to send to Mexico City. Castro is finished there,

147

Mariano will not go. So, as ridiculous as it may be, Juan, you have no choice. You cannot go yourself. You must send me."

Alvarado rested a paternal arm across Goyo's shoulders. "Friend of my childhood, your passion makes a volcano of your mouth. In one moment you tell me you will die for me and in the next you call me a drunkard. But I understand you and even as I tell you that my heart is full of gratitude, I must also tell you that you do not have the qualifications. You do not know how to persuade. If a man does not listen and agree with you immediately, your temper flares." A thought made Alvarado laugh softly. "You would not be the first one to call Santa Anna a fool, but I fear you would be the first one to call him that to his face, if he did not agree with you." He shook his head sadly. "No. I cannot risk another Castro. And Castro holds his temper better than you do."

The outburst Alvarado anticipated did not come. Instead, Goyo lowered his arms slowly and turned away. When he spoke, after a long silence, his voice was strangely subdued. "So, Juan, the one who has the most need in his heart to help you has the least to offer here." He tapped his temple.

Alvarado moved close again. "Goyo, the man who goes to Mexico City must be one whose reason dominates his passion. He must have infinite patience. He must make indifferent men see the immense value of Alta California and convince them that a strong Department of the North is necessary to their own safety. Now, my dear friend, can you swear in truth that you are such a person?"

Another long silence passed, then Goyo shook his head. "I have not been such a person, Juan. I admit it. But I swear by the Holy Mother that I can learn to do it because it must be done. All we need is three hundred trained, well-armed men, a half dozen field pieces, a sloop of war, and enough powder and shot for our heavy cannon. For certain, Juan, the Republic of Mexico can spare that."

"Nothing is for certain in Mexico City, Goyo, except that the government will not act in time, and that it will do the wrong thing for Alta California. Remember the sixth of July, eighteen thirty-one, when Mexico secularized our missions and destroyed the only established order in the land? Do you think a government fearful enough of the church to destroy a state is concerned about a few deserters and trappers and an occasional foreign warship?"

"But, Juan, there are different men in the central government now. They know the American threat is real. Let me try, Juan. Please. I can do no worse than Castro. They will believe me and act or throw me in jail as they did Castro. You have nothing to lose, Juan. My father has many friends in Mexico City. He will help me, too." Goyo paused, waiting.

Alvarado knew that Goyo was correct when he said that his father could help him. In Mexico, Don José's reputation as an excellent military man still endured. His son would certainly meet men of influence. Perhaps, after all, it was not so important *who* told them and *how* they were told. If only he could rid himself of an abiding conviction that even now it was too late.

Alvarado straightened abruptly. "God help us all, Goyo, I have decided. If you wish, you may try." Alvarado regarded Goyo sternly. "But you must make me a sacred promise that you will do exactly as I ask. Can you make such a promise, Goyo?"

"Juan, the vow is made now, before God!"

On the eve of the new year, Goyo Robles and two companions quietly departed from Los Robledos with extra mounts and five pack animals carrying provisions for a long overland journey. Following Governor Alvarado's instructions, the party entered the San Joaquin Valley then headed south to the Old Spanish Trail. Alvarado hoped that this route would keep the south from knowing about the expedition. Once Goyo's mission became known to the Angelenos, they would race to Mexico City to protest help for the northerners. Pico would see in the move a direct threat to his own political ambitions.

On the morning of January twentieth, Goyo and his party crossed the Colorado, entered the state of Sonora, and began the long dangerous southward traverse of the *Gran Desierto*.

At noon on that same day, in the presidio chapel at Monterey, John Lewis and Maria Robles were joined in marriage. The event was celebrated by one of the gayest fiestas in the capital's recent memory. For three days old animosities were set aside and even the presence of a

149

transformed Isaac Graham, scrubbed, shorn of all but his mustache and side whiskers, did little to dampen the festivity.

For seventy-two hours no doors were closed and no lights were extinguished in the Robles adobe, and Doña Encarnacion Cooper, with the enthusiastic approval of her husband, who had been John Lewis' *padrino de boda* at the wedding, opened her home to the overflow.

When at last the couple was escorted by serenaders to La Casita, a tiny two-room adobe presented to them by Don José, they slept exhausted in each other's arms.

John awakened from a sleep made deeper by endless responses to endless toasts to find Maria watching him. His brief befuddlement and start of realization made her laugh and she snuggled close, urging the warm contours of her naked body against his. Excitement, almost unbearable, filled Maria. And unfamiliar fear too, as she lifted to let her husband's left arm slide beneath her and felt herself being pressed closer still—face to face as they had fallen asleep. Her lips, moist and giving, moved to his as she felt him growing against her. For Maria, time lost both measure and meaning. If, in his eagerness, this man of hers who meant to be gentle caused her pain, then it was a pittance to pay for the pleasure she would give him, for the pleasure that she herself would be given, pleasure sensed beyond the pain. The end of frustration, the freedom to cling naked to the massive, hard-muscled frame of her man carried Maria to the unknown, far beyond the limits of her most vivid fantasies. She crossed the new threshold joyously, passionately; and John, for whom love always had been a taking, found in giving a depth of joy beyond expression.

Through the early part of the new year and on into the summer very little happened to ruffle the placid surface of the capital. At La Casita, John and Maria went about the engrossing task of settling into domesticity. In the process they learned much of one another that confirmed still further their conviction that each had chosen well.

Maria had become pregnant early in the winter. As often as he was able between summer trading trips, John worked at the addition to La Casita being built to accommodate the child. In early autumn, the expanded house was the scene of a celebration as Doña Gertrudis and the older wives brought their hand-made presents of swaddling clothes, down blankets, crocheted caps and boots, and a feeding spoon fashioned from a Spanish silver coin.

As Maria neared her time, John's concern increased. At least once each day he would look at his wife, rest a hand gingerly on her protruding stomach and regard her gravely. Once, when he had said, "But there's so little of you and so much of—*it!*" she had looked at him reproachfully. "—*of him!*" She had laughed to conceal her own mounting apprehension.

During the last week of September, Monterey's outward calm was shattered by the arrival from Mazatlán of the bark *Annie Gleason*. The ship's master, a gaunt, weather-worn New Englander named Enoch Blakely, brought the astounding news that Santa Anna had dispatched Brigadier General Manuel Micheltorena and three hundred prison-recruited irregulars to Monterey. Governor Juan Alvarado and Commanding General José Castro were to be replaced and their respective duties combined in Micheltorena as Governor-General.

Dumbstruck, John Lewis joined the merchants in the master's cabin aboard the *Annie Gleason* to hear Blakely elaborate on the sources of his information. Larkin, Spence, and Cooper hardly needed the New England ship's master to spell out their mutual dilemma. Even if Pio Pico moved the customhouse to Los Angeles, they would be better off. A Mexican military governor would be under instructions to squeeze every last penny of duty and tax from the merchants to help Santa Anna finance the war with the United States that was considered inevitable. Under the circumstances Yankee trading ships could expect the most stringent application of the regulations.

All of this John understood in those first brief minutes. What panicked him was that the new restrictions in trade would make it all but impossible to establish his own business. Moreover, the new order could multiply the obstacles already making it difficult to earn a proper living. John found some relief in the knowledge that he could, if necessary, take over the active running of El Rancho de los Robledos. Properly supervised, the rancho could keep the two families. But even that future could be bleak since the hide trade was beginning to diminish as new South American cattle markets closer to New England cut into the California trade.

Don José had seen clearly when he pointed out that Alta California's weakness lay in its dependence on a single economy. "This is a rich land," he had said, pointing to the lush natural growth in the valleys, "but we do nothing to use those riches but slaughter our animals who feed on them. Our valleys are like our cattle. We strip them and waste the greater inner riches." John recalled that Sutter had deplored the waste of fine land too.

After Captain Blakely's account of the essential facts, he turned to other equally disturbing aspects of the change. The assembled men struggled to find a way to work out an accommodation with Micheltorena. Blakely was no happier than they. "I have no need to paint it darker than it is, gentlemen. You'll find out soon enough, if you care to send someone south, that these three hundred soldiers of Micheltorena's are nothing but thieves and cutthroats that he either can't or won't manage. He had only one officer worthy of the name, a Safardic I believe, a colonel named Raphael Telles. The captains and subalterns are as bad as the rabble they're supposed to command. They are robbers and they committed cold-blooded cuttings in San Diego. They molest women on the streets in broad daylight. They are in Los Angeles by now, and I'll wager that the story's the same there." He looked from one to the other, ominously, and his flat Down-East voice sharpened. "They'll fall on this town like a plague, gentlemen, and there'll be no safety unless you take strong measures to defend what is yours!"

Thomas Larkin spoke up. "Castro gives Micheltorena full credit for securing his release. Through Castro we might have an avenue to the general's ear." He turned to Captain Blakely. "If what you say is correct—and I have no doubt that it is, sir—I can see Pico speeding him on his way to us posthaste."

Cooper expelled a long breath. "Aye! We can count on Pico's generosity this time. What concerns me now is what will happen when Alvarado and Castro hear this news."

At La Casita, John deliberately withheld the news from Maria. When she fell asleep he pulled on his watchcoat and walked down the hill. He wanted to think. The change that he had sensed was upon them now.

The moon, approaching its new half, cast a luminous glow over the mesa. At the customhouse boat landing,

John found one of the official dories with oars still in place in the lock pegs. On an impulse he climbed in. Sculling with one oar, he worked the small boat clear of the crude stone jetty and headed westward. It was the first time he had taken a boat alone out on the bay. It was the first time he had wanted to and he was not aware that the act was an unconscious manifestation of the desire to repeat an old pattern of escape.

Earlier, when he had been pulled out to the *Annie Gleason* in its boat with the other merchants, he had imagined himself an important merchant about to be entertained aboard. He had seen himself, the director of the John Lewis Company, dressed in the latest conservative Boston fashion, examining critically the great array of goods displayed on deck for his approval. He had walked along slowly, fingering samples of linen and colored cotton cloth and rich wool carpeting at one dollar and fifty cents a yard on which he would gladly pay seventy-five cents a *vara* duty—if necessary. He had picked his way through stacks of iron try pots that would bring a five hundred percent profit and past heavy iron hoes and picks and spades, cases of chinaware, turpentine, kegs of nails, stacks of Morocco-tanned calfskins that had been shipped raw from this very port two years earlier, and cases of boots and shoes made from California hides that had brought two dollars each at Monterey. Each hide represented a dozen pairs of footwear that would bring ten dollars a pair. He examined cards of lace and cases of rifles, tortoise shell combs from the Caribbean, barrels of coffee, ground rhubarb, almond oil, quinine, and cannisters of salmon from Nova Scotia. It went on and on, the array of items that were so prized on the California coast. Then he had climbed the *Annie Gleason*'s Jacob's ladder and that part of the vision became reality. It was spread out before him, and somehow the dream had assumed substance even though he was nothing more than a clerk for one of the merchants who would use him now but surely would abuse him later if his personal ambitions were to run contrary to theirs.

A half mile off shore John shipped the oars and let the dory drift. Off to his left he heard the deck watch strike eight bells aboard the bark. A few cables distant, to the northwest, the tolling couplets set a dog to barking aboard a coastal schooner out of Yerba Buena. Facing the shore, he could make out the dark outline of the customhouse and the pitch torch burning atop the pole in front of Larkin's new home. The merchants would be there, engrossed in more speculation, and he was glad he had not been asked to join them. He felt now that his guess was as good as theirs. Once again he fought off the dark specter of disaster. Half aloud he commanded himself to "hold her steady as she goes." If you are alive and well and have food to eat, he told himself, you are ahead of the game. And if you have a wife and an in-law family that regards you with affection and respect, you are in the way to become rich. He pictured the cargo again: the merchants trading sharply, taking advantage of Captain Blakely's inexperience in the California trade. Blakely had forsaken the China trade in the face of impossible competition from faster clippers to try the California hide trade as a commission master.

Soon enough, Blakely would learn the value of holding out for his price. Monterey merchants were flinty bargainers. John found himself more grateful than ever to them for the lessons he was learning. Making a place for himself would not be easy but it would be far less difficult than the challenge that had confronted him at Graham's camp. As he rowed back, he wondered about Will Chard. He had seen little of him since his release. Maria had said he was spending most of his days near Branciforte, ostensibly to look into the possibility of starting a water-powered saw mill but Maria suspected and hoped that the presence of her cousin, Dolores, was the real reason. In any event, Chard's earlier plans that had included him were over the side now. Somehow the thought relieved him. He liked Will Chard. He would always be his friend. But he knew now that the relationship would rest there.

The day after the arrival of the *Annie Gleason,* Joaquin de la Torre, chief customs clerk, received the packet from Captain Blakely and took it to Alvarado at his place at Alisal. Within hours Monterey was aflame with rumors.

"Alvarado is preparing a statement of resignation!"

"Castro is going to remove the guns and powder and shot from the fort and hide it in some inland cache where he will recruit a revolutionary army!"

"José owes his freedom to Micheltorena. He will become

154

the commander of all government soldiers in Alta California!"

"Isaac Graham and Will Chard are raising an army and will join Sutter in declaring California an independent republic." And so it went, endlessly.

Finally, a week after the news had been received and confirmed, David Spence volunteered to ride to Los Angeles on behalf of the Monterey merchants and petition the new governor-general to leave the capital and customhouse at Monterey.

A month later Spence returned and the merchants met at Cooper's house to hear about the trip. Reporting that he had met Sutter's new emissary, a German named Flugge, Spence turned to John. "You've been replaced in your friend Sutter's affections, I fear. The young fellow, Bidwell, is now his personal secretary and Flugge seems to be his ambassador at large." Somewhat apologetically, he braced his hands on the back of the chair. "I was not as successful in my mission as Sutter apparently was. The best the general would promise is to discuss the matter. Otherwise, I found him agreeable and very presentable, a pleasant surprise, I must admit. His second in command, Colonel Raphael Telles, is a thoroughgoing gentleman and professional soldier. But the description of the troops was no exaggeration. They are little more than legally armed criminals. I am told they have not been paid for months. Micheltorena has provisioned them from his own purse."

Larkin, appearing more dour than usual, asked, "What was Sutter up to? Could you find out from this Flugge?"

"The German is a very stubborn man, not given to talking. But it's fair to say that Sutter sent him to Los Angeles to speak well of New Helvetia before some of his enemies —Vallejo for instance—could speak otherwise. The best I could get out of Flugge was that Micheltorena has written Sutter a very friendly note assuring him that he will cooperate in helping to keep foreign immigration under control."

A laugh burst from John and the older men regarded him uneasily. "I'm sorry, gentlemen, but that's like setting the cat to guard the mice. To Sutter, keeping foreign immigration under control means welcoming all who come to the country and putting them to work at the fort for good wages and better promises. If Micheltorena criticizes him for that, Sutter will haul out the grant agreement made with Alvarado stating he should encourage settlement by the heads of families. The governor-general will find a greased eel easier to take a hold on!"

The room was quiet until Cooper reached out to rest a hand on Lewis' knee and said, "Lad, none of us here likes the idea of keeping things covered too much. So, before you hear it spoken by others who may not wish you so well, I must confess that we've been wondering what part your brother-in-law, Goyo, may have had in bringing Micheltorena north."

Much to John's relief, Cooper had said it straight out. Despite the fact that his own ill-feeling for Goyo was well known, John had been suffering from a sort of guilt-by-association. Maria had told him about Goyo's mission to Mexico but had sworn him to secrecy. He had never believed anyway that Goyo would be able to convince any responsible official of the central government of anything. Alvarado must have been desperate when he sent him but the question had now been posed.

"I have only hearsay to go on. But if Goyo did go to Mexico City to ask for military help, he surely would not have suggested deposing his own blood brother. You gentlemen know him better than I. I would judge then that you'd give him even less chance to succeed."

Amused by John's artful dodging, Spence said, "We should try to discover upon whose recommendation he came."

John sought refuge in the obvious. "I'd wager a lot more on Castro than Goyo."

Spence nodded. "But it's unlikely Castro would petition Mexico City to replace him as commandant of the Department of the North.

"Any plea for help might be taken by Santa Anna as a sign of general incompetence requiring drastic changes."

"Aye! You've a point," Cooper said. "Why would Santa Anna send a distinguished general up here in command of a band of undisciplined ruffians?"

Spence snorted. "Either Mexico is far worse off for manpower than we believe, or its leaders lack common judgment and I do not rule out the possible existence of both conditions."

Larkin, listening attentively, made a mental note as the meeting broke up to transmit the reasoning to the United

156

States Department of State for whatever use it might be in making an evaluation of Mexico's condition.

On the way back to La Casita, John found himself wondering about the merchants. For some reason he was convinced that his relationships with them were changing. No single incident nor any overt action supported his conviction. But the change was real enough. John searched for a cause, perhaps he was assuming an unjustifiable burden of guilt because his marriage might have signified a possible shift in loyalty. Early in his association with Cooper, John detected an underlying anxiety. He had sensed it in Spence, too. Certainly Sutter troubled them. And Leidesdorff. In fact, anyone who could conceivably threaten their-sinecures troubled them.

Only Larkin really worked at building his business. John was sure that Larkin's energy was impelled by interests far beyond trade and merchandise. He had become the government's principal banker. The Department of the North borrowed interim financing at high interest to carry it between the peak revenue seasons. Earlier in the year, Cooper had observed dryly to John that if Larkin chose to call rather than refinance the loans, he could foreclose on the government and own Alta California.

John found the governor's inconsistent attitude toward outsiders ironic. If Alvarado was truly concerned about a foreign takeover, he should not have wasted time on Graham and Chard or become involved with Sutter. Instead, he should have examined the motives of the Yankee merchant to whom he was now so deeply beholden.

Alvarado's reaction, when it came, saddened his friends and delighted his detractors. On the twenty-fourth of September he posted a proclamation that Micheltorena had come to Alta California at his request to be relieved as governor because of poor health. The explanation produced derisive laughter from those who knew about Alvarado's penchant for *aguardiente* when the burdens of state grew too heavy. John deplored the transparency of the excuse. The portion of the proclamation congratulating the people upon the appointment of a successor, "so well spoken of for heroic military ability and blessed excellence of character," disgusted him. He was certain that Alvarado did not lack physical courage. He had risked execution for treason when he overthrew the Mexican gov-

157

ernor, Gutiérrez. He had been governor for six years and now it seemed he had suddenly become a weakling.

John voiced his doubts to Don José. "I foresee no future here now but confusion. There is no consistency in this government. There can be no salvation for us except in other than Mexican hands."

Don José laughed tolerantly and said, "Miguel de Cervantes wrote, 'It is a far cry from speech to deed.' Soon enough we shall find out Alvarado's reason. But do not become too quickly disillusioned. I have known Juan Alvarado since he was a child. He is deeply hurt now. But he loves this land and he will fight for it when the time is right."

Monterey's official business continued to be transacted with remarkable efficiency by Secretary of State Casarin. David Spence acted as both *alcalde* and justice of the peace, arbitrated civil controversies, and within a few weeks Alvarado's as well as Castro's absence was all but forgotten.

On the nineteenth of October, shortly after dawn, a lookout from the Mexican fort reported a sail on the horizon beyond Point Pinos. John and Cooper rode to the hilltop commanding the southwest to identify the ship. They saw not one but two sets of sails in addition to Captain Snook's coastal schooner, *Joven Guipuzcoana*. Cooper handed the glass to Lewis. "See what you make of them, John. One seems square-rigged and the other could be a sloop."

"I make out the one on the left to be a Cape Horner. The other is surely smaller but I can't put a name to her."

Cooper took the glass again. Without removing his eye from the objective, he reported, "Both vessels have altered course to the west. I can see the big one's—" he paused and looked at Lewis strangely "—and if you want to know something, lad, those stays'ls are a mite too square for a merchantman. She's flying no colors so I can't make her out, but she's a frigate, by God, as sure as she's a foot high. And that's a sloop of war with her. Those are not bogus gunports painted on her sides. Those are the real thing."

The words were scarcely out when they saw a puff of smoke erupt from one of the sloop's forward guns. A moment later, the muffled report reached them. Cooper focused on the smaller, faster vessel. "The sloop's put a shot across the *Joven*'s bow and Snook's coming about to lower sail." They watched in silence as the sloop of war approached the schooner, came about herself, and lowered a small boat.

"By God, lad, that's an armed boarding party!" The same fear struck both men simultaneously. A pair of privateers were on the California coast. While Cooper's eye remained trained on the smaller vessel, John had been watching the frigate, still holding a course that would bring her into the shelter of the Bay of Monterey. She had passed the sloop and its victim now and was standing inshore of her, no more than two miles off the headlands. Suddenly from the tip of her spanker gaff, the colors streamed. John saw them and called out. Cooper retrained the glass and let out a startled cry. "Mother of Jesus, lad, that's the Jack! She's British! Here!" Stepping aside, he held the glass in place until John grasped it. In an instant John confirmed the warship's identity. Quickly he swung the glass to the more distant vessels and saw that the *Joven Guipuzcoana* was preparing to make sail again. The sloop was staying outside of her, obviously intent upon convoying the captured craft back into port.

In less than a half hour all three ships were at the anchorage. Within the hour, John and Cooper watched with the stunned crowd that had gathered at the customhouse point. An outcry rose as the frigate struck the British colors and ran up the Stars and Stripes of the United States. Cooper, along with Spence and Hartnell, who had breathed private sighs of relief at the sight of the Union Jack, quickly made for the office of Jimeno Casarin. They learned there that Casarin had already dispatched a messenger to Juan Alvarado demanding his immediate return. Ashen, the frail Casarin looked from one to the other helplessly. "Governor-general Micheltorena's order commanded Alvarado to retain this office until he arrived in Monterey from Los Angeles. Under that order I have no legal right to act. He must come! And so must Castro—if he can be found. We have no defense!"

Unknown to Casarin, when the frigate first appeared in the harbor flying the British colors, Thomas Larkin had dispatched his own messenger to Alvarado demanding his return to organize defenses against a British takeover. Larkin's private information from the American charge d'affaires in Lima had included an assessment of relative foreign naval strength in the Pacific. The United States had five vessels in its Pacific squadron totaling one hundred sixteen guns. The French had eight vessels bearing two hundred forty-two guns. But it was the opinion of both the United States officials and the experienced naval men that the British warships would be more than a match for either of the other two powers. Larkin knew Monterey could be taken from the sea, but if strong resistance was organized inland he doubted that it could be held, particularly if the United States Pacific squadron could be summoned in time to bottle up the British.

Larkin was in his office writing a hurried dispatch describing the action, which he hoped somehow to get to the United States, when word came that the vessels had changed their colors. He shredded the papers and burned them. Then he hurried to the customhouse and overtook the other merchants returning from the government house.

Cooper, his face a study in anguish and confusion, laid a hand on Larkin's arm. "What do you make of this, Tom? Two of your warships using the tactics of a privateer to take the capital?"

The inference annoyed Larkin. "I have no idea what to make of it, unless the United States and Mexico have gone to war. Who is the commander? Does anyone know yet?"

Cooper glanced up at the old Mexican fort. "No. We have no intelligence and no means to resist if their intentions are not peaceful and there's nothing very Goddamned peaceful about a shot over your bow—under any colors!"

By the time they had reached the customhouse a small boat from the frigate was seen pulling for shore under a flag of truce. At the landing, Cooper ordered the bystanders back as an American officer in full dress blues and cockade accompanied by his guard stepped ashore. He looked about, questioning, until Jimeno Casarin advanced timorously and identified himself in Spanish. The American saluted and introduced himself as Captain James Armstrong of the U.S. Navy, acting under orders from the Commander of the Pacific Squadron. With military correctness, he demanded a conference with responsible officials. An awkward pause ensued as the puzzled Casarin turned to those around him. Thomas Larkin stepped for-

160

ward immediately, identified himself as a United States citizen, and offered his services as a translator.

"We are much obliged, Mr. Larkin. Will you please accompany us to a suitable place where I may transmit to the governor or his representatives the commodore's demand for a peaceful surrender. It is his wish to avoid the sacrifice of human life and the horrors of war." Larkin spoke briefly with Casarin, then led the official party up the slope to the government building. The apprehensive crowd strung along in its wake.

John Lewis, seated on the low stool beside the bed, tried to reassure Maria, whose labor seemed about to begin. Doña Gertrudis and the Indian midwives hovered silently in the background. Driven by duty, he had returned to La Casita when the ships were first identified. "There is no need to worry, *chiquita mia*. They are only British warships paying some sort of official visit. When your father returns he will tell us." Leaning close, he smoothed the moist hair from her temple and kissed her forehead.

"We have much more important welcoming to do here!" Maria managed a smile that did little to conceal her mounting fear. The first pains had begun just before dawn and she knew then that if these were the mild ones, as the women had said, then it would take all of her courage to face the ordeal ahead. She clung to her husband's hand and pressed the back of it against her neck.

"You are the president of this welcoming delegation. So you will be certain to be here when your son arrives?"

John smiled reassuringly and cursed the poor timing that prevented him from being with the others at the embarcadero. "I shall be here. After all, I am not Juan Alvarado." He regretted the joke instantly, but it was said. Maria closed her eyes and released his hand.

Doña Gertrudis tiptoed over to the side of the bed. "Let her rest now. I will tell you when to come in again."

John wandered into the parlor and stood in the doorway looking down over the pueblo. The sloop and the frigate were anchored a quarter of a mile out from the customhouse landing. Both ships were flying the United States colors from the gaffs but he could see them only as silhouettes against the lowering afternoon sun. Below him, a figure detached itself from a knot of people gathered before Thomas Larkin's house and he recognized his father-in-law

161

heading with unusual haste toward La Casita. Minutes later, Don José, puffing from the exertion, broke the news of the vessels' real purpose. "Larkin is going out to meet the American commodore now. Casarin has sent him as interpreter. Pedro Navaez and José Abrego will arrange the terms of surrender. Alvarado will arrive by six o'clock. He has sent word that he wishes all to be at his home at that hour."

John followed Don José inside. Still short of breath, the older man settled gratefully onto a chair. "So, my son-in-law, you were first an americano, then a mexicano, and now you will be an americano again." He pointed toward the bedroom. "And whether or not it is an honor remains to be seen, but you may also be distinguished for having fathered the first new citizen born here under the United States flag."

By half past five Maria's pains seemed to have eased again. When he looked in she was sleeping. Doña Gertrudis took his arm affectionately and led him out of the room. "I do not think anything will happen for some hours, Juan. She still keeps her water. Go with Don José to Alvarado's home. If you are needed I shall send one of the women."

John slipped an arm around his mother-in-law's slender shoulders. "Promise?"

She crossed herself hastily. "I promise. You shall know when to greet your child."

John and Don José arrived several minutes after the frigate *Cyane*'s gig put Larkin and Casarin's two emissaries ashore. Obviously both weary and worried, Larkin greeted them on the porch. "This is a very unfortunate business. I have told Commodore Jones that I do not think a state of war exists between the United States and Mexico. He disagrees. But when I asked for proof he could present none except the considered opinions of his officers and some Department of State underlings in Panama and Callao. I am certain a grave mistake is being made."

It was almost dark when Alvarado appeared on the San Jose road. He greeted Jimeno Casarin and his staff and the principal citizens of Monterey, about thirty in all. Quickly, Casarin recounted the events, then handed Alvarado the penned draft of the United States provisional surrender terms. Alvarado read the single sheet scornfully and pushed it aside. "Jimeno—what is the state of our defenses?"

Casarin turned questioningly to Captain Mariano Silva, who had been given temporary command of the garrison in the absence of both Castro and the subaltern, Chavez.

Silva glanced around unhappily. "Our defenses are of no consequence, as everybody knows. We have twenty-nine unpaid soldiers including twelve officers and four subalterns. There are twenty-five volunteer militiamen—if all can be found. We have no more than one hundred muskets that will fire. And eleven cannon—" He paused. "And sufficient powder to fire salutes only."

A heavy silence fell over the room. Alvarado straightened in his chair and reached for the surrender terms. Holding the paper aloft in a hand that trembled as much from fatigue as suppressed rage, he looked around, then let his eyes come to rest deliberately on Thomas Larkin. "None who know me would doubt that I would rather sign my own death warrant than this paper. I have lived through no darker hour. I shall not rest in peace until those who would destroy us are themselves destroyed." He lowered the paper. "We shall agree to these terms to prevent the futile shedding of blood. Our situation is temporarily without hope. We shall agree because we are men whose compassion matches our courage. This surrender will be signed with our hands but never with our hearts. In the sight of God, I swear in your presence, gentlemen, that one day soon our terms for the retrocession of Alta California shall be written in their blood!" He took the quill from its stand and offered it to Casarin. "You will write first, Jimeno."

The acting governor accepted the quill reluctantly and signed. Alvarado inscribed his signature below. Then he looked up at those before him. "I wished to have you here to counsel me and to pledge your loyalty in the battle to drive off the invaders. For the present that is not possible so now there is nothing for me to ask of you. Under the flag of the United States some of you will find yourselves favored." His eyes roved the unhappy faces until they came to John Lewis. "When that day arrives I ask you who know our hearts to give to those whose land this once was the same consideration you received from them when you first came among us."

John returned the governor's level gaze and wondered

163

how Isaac Graham and Will Chard and the others would have responded. Chard would have been amused, he thought. But Graham would have gloated openly and hailed the just providence that evened the score for him. John supposed that he might be doing that already, with the help of a demijohn. Graham's revenge would be as sweet as Alvarado's defeat was bitter.

John could guess what was going on in Larkin's mind, too. There would be relief that the warships were flying the Stars and Stripes and not the Cross of Saint George. The opposite would be true of Spence, Cooper, and Hartnell. As for his own destiny, much would depend upon how the officers left to administer the country viewed adventurers who had given up their birthright.

At eleven o'clock on the morning of October twentieth, Commander Stribling and one hundred fifty spit-and-polish U.S. Marines and sailors stepped ashore at the customhouse landing and prepared to take formal possession of Alta California. As the band struck up, the Mexican colors were lowered to make room for the Stars and Stripes of the United States. Two men were notably absent: thirty-three-year-old Governor Juan Bautista Alvarado, who, unable to face the public humiliation of surrender, had ridden inland to Alisal before daybreak, and John Lewis, twenty-two-year-old father of an infant daughter, who had fallen into a deep sleep after the terrifying experience of witnessing his first birth. When the guns at the fort, manned now by U.S. sailors, answered the salutes from the two warships lying at anchor in the bay, John scarcely stirred. Don José told him of the ceremony, of the pride and gallantry of the ragged Mexican soldiers as they trailed their arms to surrender them at the government house, and of the courtesy of the American officers and men who disarmed them and assured them that not a single person in Monterey would be deprived of his rights of liberty.

John awoke shortly after two o'clock in the afternoon and went in to see his wife. Her appearance shocked him and filled him with a strange guilt. He wandered that any woman ever survived the ordeal of giving birth to a new life. He had heard tales of Indian women who paused during their labor in the mission fields, dropped their child in a furrow, and resumed their work within the hour. He

164

could believe it of a broad-hipped, bandy-legged savage female but not of such fragile-appearing women as Maria and her mother or even his own sister, Alice. As he sat looking down at Maria's lips, swollen and bruised from teeth that had bitten through the thick, wet cotton labor braids, he resolved quietly not to expose her to such a trial again, not forever, but surely not for a long time. Holding his infant daughter, who would be christened Felicia Gertrudis Lewis, he looked up at Doña Gertrudis with frank disbelief as she reassured him that the wizened, beswaddled little creature in his arms was a faithful reflection of the best features of its mother and father.

At four o'clock, still heavy with fatigue, he walked down the hill with Don José to receive the congratulations of his friends in the upstairs sitting room of the Larkin house.

The merchant was cordial and generous with his good wishes and his brandy but he seemed distracted and hardly touched his drink. John was grateful to these men who had taken time in the midst of their own concerns to observe the custom. Ordinarily it would have been a long, relatively exuberant affair, but Larkin himself asked to be excused at five o'clock. The others soon departed, David Spence to write a long dispatch to the British attaché at Santa Barbara and another to Colonel Mariano Vallejo at Sonoma; John Cooper to continue reassessing his position as senior merchant; David Hartnell to ride to his inland college blessing his good fortune, for he knew that under American occupation, as under British, there would be a growing demand for the secular education of the young.

John returned to his new household, now so strangely rearranged. Alone, with only an oil lamp burning erratically above the adobe fireplace, he tried again to plan a future. Silently he thanked God that the war vessels lying at anchor were American. Immigration would come faster now, and with it a growing opportunity to establish himself in trade. Inadvertently, Captain Cooper had pointed the way when he had said, "There is only one great harbor on the entire California coast, lad, and that's the Bay of San Francisco. It is central and large enough for the greatest fleets. But the Mexicans will never make anything of it. Spain would have. But the Mexican government will not listen to the *californios* here who have vision and enterprise. Vallejo and Martinez and Peralta know what Yerba

165

Buena could be. So does Alvarado and Don José Robles. These men are done pleading with the central government. Their arguments are disputed in Mexico City by their own self-minded people in Los Angeles who mouth idiot dreams about a great harbor at San Pedro. San Diego, yes. But never San Pedro! There isn't a day the British don't kick themselves for having settled on the Columbia first. The future of this land lies in the great central harbor and in the rivers that empty into it from the lower valleys to the north and south. Those waterways are vital. Sutter understands that. In some ways that ridiculous little man is very clever. That's why he has chosen to fortify a position that commands them in every direction. But mark what I say, the key is the Bay of San Francisco."

John had recalled Cooper's reasoning often during his troubled ruminations over the past months. William Leidesdorff's name kept recurring in his thoughts; but each time he resolved to visit him some unexpected obstacle had intervened. Now, in the next room, he heard the hungry squalling of still another obstacle and smiled to himself as the muted cooing of the *mestizo* nurse reached his ears. Perhaps little Felicia should not be counted an obstacle at all. The infant and the war vessels in the harbor that portended such great changes might be the incentives he needed to rearrange his life.

Late in the afternoon of the twenty-first, Commodore Thomas Ap Catesby Jones made a tour of the Mexican fort and the customhouse, then he proceeded to the government building to meet with and reassure Monterey's leading citizens.

John Lewis and Don José fell in with Cooper and Spence and joined them. The Scotsman who would remain as *alcalde* indicated the streets and smiled wryly. "If you are looking for California officials, do not look in Monterey. This is a deserted town. Only Casarin remains at his post. A sorry spectacle, gentlemen. A sorry commentary."

Following a brief formal address to the small crowd gathered before the government building, the commodore asked to confer with Thomas Larkin. Embarrassed at being singled out, Larkin accompanied the officer inside the building. Captain Cooper remained grim and silent but David Spence laughed good-naturedly. "Before you think too harshly of Tom just remember that if the colors had

166

not been changed, 'twould have been one of us to receive the dubious honor."

Cooper and Spence returned to their offices and John strolled down to the landing to strike up a conversation with members of the shore party standing by the commodore's boat. The few Boston and New Orleans papers that reached the California coast were so prized that they never remained long in anyone's hands. He wanted to hear the details of President "Tippecanoe" Harrison's untimely death. After only thirty-one days in office, Vice President Tyler's banking orders had succeeded in making all but Webster in the old cabinet resign. People wondered if Tyler could keep the country on an even keel. John was also eager to know about the cargo of slaves who had mutinied aboard the *Creole* and had sailed the American vessel to the West Indian port of New Providence. The Navy had been asked to intervene but he had read that the British had given the slaves asylum. And he wanted late news about the trouble in Texas. He was badly disappointed that the seamen he talked to could tell him very little. They had been on the Pacific station for many months.

At the customhouse he talked with the commodore's executive secretary, a man named Reintrie, and with the squadron chaplain, the Reverend Bartow. John answered their questions about Monterey and its people and thought he detected a touch of wistfulness in their queries. In answer to his queries about the probable future of California under American rule, Reintrie was composed and confident. He told John, "The commodore is convinced, and his staff officers agree, that California may well become as important as the Louisiana Territory. If there was any risk in taking Monterey, the commodore feels it was justified to insure that England does not lay claim to it first."

Their conversation was interrupted by a messenger from Thomas Larkin asking John to come to the governor's office as quickly as possible. After the briefest introduction to Commodore Jones, John was dispatched to get the latest editions of American newspapers making the rounds of the merchants' offices and homes.

Within a half hour he was able to find seven editions of Boston and New Orleans papers. None was dated more recently than mid-July. Lewis delivered them to Larkin and was asked to wait. He watched as the merchant and the commodore pored over the editions side by side searching

167

for dispatches that would indicate the possibility that the United States and Mexico had gone to war. They found none. Commodore Jones, still unconvinced, pushed back his chair and looked up at Larkin. "I respect your opinion that a state of war might not yet exist, Mr. Larkin, but I cannot afford to risk letting Monterey fall into British or French hands in view of the fact that I have positive knowledge that Rear Admiral Thomas sailed with three British men-of-war under sealed orders from London. Moreover, the French Pacific Squadron left Valparaiso in June to cruise off the California coast. It is the sense of our attachés and of my officers, and I concur absolutely, that these fleets had been informed that war between the United States and Mexico was imminent and that they sailed under orders to take up positions enabling them to move against Monterey on a moment's notice. Buying our own continent back piecemeal from foreign invaders has become a fine proposition—for *them,* Mr. Larkin. It is my intention to forestall any more such unfavorable bargains."

Larkin peered at John. "Are you certain that John Cooper and David Spence have no later editions tucked away? It seems to me that somewhere I saw dispatches dated as late as the first of August that made no reference to a declaration of war."

John was unable to understand Larkin's curious behavior. He had been certain that the American merchant would have done all he could to support the United States' position. He replied with caution: "I did not see them myself, sir. But it is possible they've gone inland to Mr. Hartnell. Late news is very much in demand, as you know."

Commodore Jones stirred impatiently. "I think we've about come to the end of doubt-casting, gentlemen." He indicated a packet of papers. "I have here a letter from our man, John Parrott, in Panama, dated the twenty-second of June. It says without equivocation that war with Mexico can be expected momentarily. Parrott is very dependable. His view is supported by the June four issue of *El Cosmopolita* in which the Mexican Minister of State makes the most violent threats to Secretary Webster. These same dispatches were reprinted in the *New Orleans Advertister*. They also appeared in the Boston papers in July." He gathered the correspondence.

"I appreciate your concern, Mr. Larkin, and I understand that it is exercised in our behalf. But on evidence, I feel we have acted prudently." He rose and prepared to leave.

Suddenly John remembered that he had seen a bundle of newly arrived dispatches from Mexico City at the customhouse, including copies of *El Diario* with new customs regulations.

"Mr. Larkin—"

"Yes, John?"

"If the commodore would wait, I may be able to find some late dispatches from Mexico at the customhouse. I don't know that they'll shed much light on the matter."

Annoyed that there could still be any doubt as to the wisdom of his action, Commodore Jones gestured impatiently. "Fetch them, Lewis! And don't bother us with anything not pertinent to the issue."

"Yes, sir." John turned to go but Larkin stopped him. "It might be best if one of the commodore's men did the actual searching." It took a moment for John to realize that Larkin was attempting to protect him from being in the position of a Mexican citizen who had given aid and comfort to the enemy.

Pausing uncertainly, he looked from Larkin to Jones. The officer scowled. "All right. That is proper. My secretary and the chaplain are at the landing. Ask them to do the deed, if you're timid about it."

Accompanied by Reintrie and Bartow, Lewis entered the deserted customhouse. On the second floor, in the room occupied by the collector of customs, Reintrie found a bundle of official communications in a metal box. He called down to Bartow and Lewis. "Come up here and have a look. Some of these papers are dated as late as four August."

In the most recent of them, dated in mid-August, John found several references to United States ships bound for Alta California. The documents warned the collector of customs at Monterey that certain items in the cargoes were to be taxed at new rates. Lewis translated aloud as best he could. Reintrie, unmistakably upset, took the papers. "I'll carry them. I think the commodore will be interested in these."

Twenty minutes later Thomas Larkin, doing his best to resist a smug smile, finished translating the letters and dispatches to Commodore Jones. "I'm afraid, sir, that if a

169

state of war existed between us these communications from the Minister of Finance in Mexico City would not be reminding Señor Torre to be sure to collect duty at the new rate on American carpeting and iron try pots."

The commodore all but snatched the paper from the merchant's hand. "Damn it, Mr. Larkin, I do not need to be reminded of the obvious. Let me have the lot. I wish to take them aboard for study. We shall talk further."

At nine o'clock on the morning of October twenty-second, Commander Stribling stepped ashore and dispatched a runner to Thomas Larkin's residence. A half hour later Larkin dismounted at the customhouse and greeted the commander. Stribling came straight to the point. "I have here a letter from Commodore Jones to Governor Alvarado. He asks that you deliver it in person. I have been directed to inform you that it expresses the Commodore's deep regrets for any inconvenience he may have caused the government of Mexico, and the state of Alta California and its citizens. The commodore himself will sail for San Pedro this evening to deliver a similar letter to Governor-general Micheltorena. I have been instructed to arrange the ceremonial return of the government and of certain documents to Mexican officials and to strike our colors at four o'clock this afternoon. The commodore asks that you be good enough to so advise the responsible authorities here. Thank you, sir!"

Before Thomas Larkin could reply, Commander Stribling saluted, swung about abruptly, and ordered his men into the boat.

Late that afternoon, in the presence of a stunned citizenry, Secretary of State Jimeno Casarin, acting for Governor Alvarado, accepted the American commander's official apology. The United States flag was hauled down from the staffs at the customhouse and at the fort and Mexican colors returned to their rightful position. A salute was fired by the reinstated Mexican garrison and answered offshore by the sloop *United States*. A bugle sounded, commanders barked orders, drums ruffled, and two companies of disgusted U.S. sailors and marines marched back to their boats.

5 THE YEAR 1842 came to a close and Alta California moved into the new year—wondering. In the south, at Los Angeles, Abel Stearns, the merchant-banker who functioned for Pio Pico and his political *junta* in much the same manner that Thomas Larkin functioned for Juan Alvarado and therefore exerted much influence on local affairs, began moving carefully to hasten the departure of Governor-general Manuel Micheltorena. The outrageous behavior of the Mexican general's *cholos* had disaffected every decent citizen. As rumors of the excesses continued to filter northward, the Montereyans hoped against hope that the more agreeable climate in the south would tempt the general to tarry longer.

Micheltorena had received Commander Jones graciously, reciprocated with an official fiesta of his own, echoed protestations of friendship, and handed the American commander an exorbitant bill for damages. Commodore Jones had proceeded to the United States, where his action was creating a great controversy in Congress. He had been temporarily relieved of his command.

In Monterey the established merchants with the exception of Thomas Larkin went about their business as usual. Cooper had summed up their position in naval metaphor: "We're committed to a fixed course now, gentlemen. The Yankees have turned our battle line and the British and French are in a position to. No matter which way we steer we are in a fair way to take their broadsides. 'Tis only a matter of time—and whose shot you prefer to take."

Maria Lewis, fully recovered, had set about fixing up La Casita, renamed Casa Alta because of its sizeable addition and because it occupied the highest lot on the mesa. But John's activities in the capital had not expanded. He continued to receive thirty dollars a month from Captain

Cooper, most of which was taken out in trade at a favorable discount. As a consequence, the new family got by. In the spring of 1843 there was a flurry of excitement in Monterey. Abel Stearns had received a draft for three hundred and forty dollars from the United States Mint in return for a pouch of coarse gold nuggets he had forwarded for a *ranchero* named Francisco Lopez. Lopez had discovered the nuggets clinging to the roots of some wild onions in the San Gabriel Mountains. There were rumors that many citizens of Los Angeles were in the mountains prospecting along the streams and reports that traces of gold had been found in the Los Angeles river. At first, John was tempted to gamble a few months' time and try his luck but the sobering reality of his responsibilities soon put an end to that golden dream.

On the morning of April second, an American brig out of Charleston dropped anchor in Monterey Bay. The master lingered at the customhouse only long enough to ask the way to the home of Thomas Larkin. Two hours later all Monterey knew President Tyler had named Thomas O. Larkin as United States consul at Monterey. Well aware of the change this would work in his relationships with both Cooper and Spence, Larkin immediately went to Los Angeles to establish a commercial alliance with Abel Stearns, who he felt would be a dependable unofficial source of information useful to Washington.

In June, three fully laden merchant vessels, two from the United States by way of Honolulu and one from Peru, put into Monterey only long enough to secure permission to trade in Yerba Buena. To the consternation of both Cooper and Spence, the cargoes of the two American vessels were consigned in their entirety to William Leidesdorff. It was the first time that any major merchant vessel had failed to discharge the bulk of its cargo in Monterey.

On August thirteenth, Governor-general Micheltorena, having tired of Los Angeles, marched into Monterey. He was received with meticulously correct ceremony by Alvarado, who had assessed all those able to pay to raise the money for a proper official fiesta.

On the very night of the celebration, Micheltorena's *cholos*, officers and men, began confirming their reputations by molesting women at the cantinas and by stealing

172

livestock. On the third day Monterey was outraged by the raping and knifing of a young Indian mother by two irregulars crazed with *aguardiente*. John, afraid for the first time to allow Maria and little Felicia to walk unaccompanied, even in broad daylight, joined in a general protest to the governor.

General Micheltorena was contrite. He promised there would be no repetition of the outrages. Within the week, however, two young Monterey girls were dragged screaming between buildings and were saved by sheer chance from assault by a group of sailors on leave. This incident made John Lewis resolve to go to see William Leidesdorff in Yerba Buena.

A pre-dawn start brought him to San Jose, a hot, unprepossesing pueblo. He reached Yerba Buena late the next afternoon and went immediately to the home of Jean Vioget. The Swiss, who had just returned from New Helvetia, was delighted to see him.

They talked late and John learned that Sutter, little by little, was seeing his dream materialize. The walls and bastions of the fort were finished and cannon were in place. Several thousand acres were under cultivation in wheat and a German named Theodore Cordua had become the first of several settlers to take up land under the requirements of Sutter's grant. Harry Morrison had turned over his distillery to one of the new immigrants and had begun the establishment of a pharmacy and dispensary, a task more to his liking. But credit problems still dogged Sutter.

Vioget felt that Sutter would not be able to meet his own needs and still make a payment to the Russians. John learned that twenty-two-year-old John Bidwell had taken over more and more of the clerical detail, freeing Sutter to oversee his field workers. John wished young Bidwell better luck than his own in collecting wages. Despite numerous letters, Sutter still had not sent the promised draft for nearly two hundred dollars. John asked Vioget if he should extend his trip and ride to New Helvetia to collect. Vioget laughed and said that he had never been paid for surveys and maps and his efforts to collect had been met with protestations of undying friendship, gratitude, stupefying hospitality, and an avalanche of promises.

The conversation turned to Leidesdorff then. Vioget said intently, "Do you know this man? Have you met him?"

173

"No. I have not met him properly. I saw him when he sailed into Monterey for his customs clearance."

"Then, my friend, you may anticipate an extremely rare experience. Leidesdorff—his father was a Dane and his mother a West Indian Negro—is one of the most dynamic and cultured men I have met in many years. He speaks three languages fluently and soon will have mastered Spanish. His enterprise is changing Yerba Buena."

Suddenly more hopeful, John decided to confide in Vioget. "I have described conditions in Monterey under Micheltorena. I must remove my family. The real reason I came here was to try to find employment with Leidesdorff."

Vioget's alert blue eyes brightened. "And you will. I am certain. Our friend Jacob Leese is not very happy about it, but Leidesdorff has constructed the largest warehouse in the cove." He pointed to the southwest. "And he is building our first real hotel. A very imposing one. Also, he is talking of a school and he and William Richardson speak of sawmills in the forests north of the bay. He has many enterprises, *mon ami*. And unlike our friend Sutter, he has the means to see them through. I shall be happy to introduce you in the morning."

The meeting with the thirty-three-year-old Leidesdorff took place in his cottage not far from the center of the little pueblo. Jean Vioget managed the introductions and excused himself on the pretext of business with Captain Richardson. Everywhere in Leidesdorff's home John saw evidence of the man's excellent taste, and here and there he thought he detected a woman's touch. After a very few minutes, Lewis was convinced that those who had proclaimed William Leidesdorff's charm had not exaggerated.

Somewhat shorter than John, the young merchant was a strong, active man with very dark, compactly curled hair. His broad face was open and friendly and his restless, wide-set topaz eyes mirrored a quick, searching intellect. Only the most subtle trace of the man's mulatto origin was evident to John. The well-moulded features suggested a Polynesian strain if anything. Leidesdorff wore his hair somewhat shorter than most, the modest side whiskers being his principal concession to current fashion. Except for a black mustache, he was clean-shaven. His skin was a luminous light brown, no darker than Maria's or little Felicia's. In a voice that was deep and rich, he beckoned Lewis to a back door. "When you visit me again, soon I hope, Mr. Lewis, I shall be able to invite you into our garden. Its principal distinction now lies in the fact that it is the first one to be attempted in Yerba Buena."

Leidesdorff traced the course of the paths between the beds that soon would be lined with ornamental stones. "California is an extraordinary land. It suggests the possibility of becoming all things to all men." He laughed and the sound was infectious. "I must have a profusion of flowers around me, Mr. Lewis. As a child in the Caribbean, my first recollections were of blooms. And later, in New Orleans, my uncle's garden was the delight of my school years. Flowers and music. For me they are life's only indispensables!"

Inside again, Leidesdorff indicated a chair. "Please sit. We shall have some good coffee, French coffee with a bit of chicory, and you will tell me what I may be able to do for you."

John reviewed his own background and was surprised that Leidesdorff seemed familiar with much of it. Before he could come to the point of asking for a position, Leidesdorff interrupted him with a gracious smile. "Mr. Lewis, you and I are two of the richest young men in Alta California." He lifted a long, symmetrical finger. "Ah, but you are the richer because you are younger. Shall I say why? Because it has been given to us to help build this land. I should think that no pioneer since Moses himself has been given such a glorious opportunity. A man born to wealth may be sorely tempted to take his riches for granted. He may abuse them. Perhaps that is the fault we find in most *californios*. Not in all, I must hasten to add. Our friend General Vallejo knows what nature has conferred upon us here. And there are others." He paused thoughtfully. "But we who have come more recently see it all more clearly than most. Certainly that is true of Captain Cooper, David Spence, and your own father-in-law." Leidesdorff paused again, frowning slightly. "You have been working for Captain Cooper and I've been told that you perform services for him well beyond the call of the average clerk. Am I correct?"

"My training as a supercargo is of some use to him. Buying and droghing are occupations conducive to quick learning."

Leidesdorff laughed. "Indeed, the latter can place a dangerous burden on a man's brain and catch him his death of cold as well. But I interrupt. Go on, please."

"I know something of ships and stowing cargoes and the months I spent with Captain Sutter were an education."

"Yes. It is possible to learn much from Captain Sutter about grand enterprise, and equally to learn how not to undertake the financing of same." There was no suggestion of malice in the good-natured observation. "When you look ahead, Mr. Lewis, what do you perceive as your destination?"

The question was troublesome. Much had happened in short order since he had jumped ship in April a little more than four years earlier. There had been many plans and more dreams, but none had really promised viability.

"I think I have some talent for business, sir, but I'm afraid I've not yet thought of my particular niche."

Leidesdorff smiled sympathetically. "At your age, neither had I. But I was more fortunate than most. I was raised by an extraordinary man, an uncle. Quite early on in my rearing he was at great pains to make me see that the most favored place for a man whose bent is commerce lies between the producer and the consumer. My father was a planter. My uncle was a wholesale merchant. My father died in debt. My uncle died a man of wealth. The lesson he taught me was inscribed indelibly: If a man positions himself well between the producer and the consumer, he will seldom be unbalanced by the ill fortune of either and will most certainly be enriched by the good fortune of both. Indeed, if he manages his credit well, such a merchant may find both parties working for him!"

Leidesdorff set his cup aside and John sensed that it was time to make his proposal but the merchant anticipated him. "Mr. Lewis, I know you are well connected in Monterey. But I cannot help but wonder if someday soon we might not hope that you will see the possibilities in this most remarkable of all the world's harbors and decide to cast your fortune with us?"

At a loss for words, John sat desperately groping for some graceful way to answer in the affirmative. Smiling, Leidesdorff rose from his chair. "My hotel will be finished soon. I am expanding my warehouse and I plan to buy another schooner for coastal trade. I even dream of steam-powered craft on the bay so we shall not be at the mercy of wind and tide on our runs up the rivers. Those great interior valleys will be settled very soon now, Mr. Lewis. The trade will go to the firm that is ready. It strikes me that I could count myself extraordinarily fortunate if I were able to persuade one such as you to assist me in these ventures."

John Lewis scarcely saw his surroundings as he rode back to Monterey. For the entire journey his mind reeled with the bright prospects offered by Leidesdorff. He and his family would be given a house befitting the assistant manager of the William A. Leidesdorff Company. The salary would be seventy-five dollars a month with the promise of substantial increases as business warranted. Leidesdorff had also hinted that a successful association might also hold the promise of a partnership one day. Dazzled by new visions of a great merchandising empire, John rode into Monterey an hour after sundown.

Over dinner at the Robles adobe he recounted the details of his visit. Even the news that Goyo was rumored to be on the way home from Mexico did nothing to dampen his enthusiasm. Plans were made to move his family as early in the new year as possible. He explained that while Leidesdorff had set no definite time for the employment to start, it had been apparent that the earliest practical date would be the most agreeable.

It was decided that John would go north again after the first of the year to get established. No later than early March he would return for Maria and Felicia. These decisions having been settled, the family turned its attention to the coming holidays.

Monterey saw little of Juan Alvarado during the Christmas season. He had retired to Alisal swearing that he would never accept public office again. For a time an estrangement existed between the former governor and his military commander, José Castro. When Micheltorena came to Monterey he offered Castro the commission of lieutenant colonel and revoked the commission of colonel held by Mariano Vallejo. Alvarado expected Castro to refuse the commission. Instead he accepted with no apparent injury to his pride and pledged total loyalty to the new military governor.

More and more frequently word reached Monterey that parties of immigrants were on the overland trails from the United States. Early in January, Governor Micheltorena

177

became so concerned about the situation that he discussed issuing a decree requiring each able-bodied citizen in Monterey to join the militia and undergo training. Worried that he might be detained in the capital, John made plans to leave for Yerba Buena immediately.

In his arms on the night before his departure, Maria whispered contritely and sought to reassure him. "But, Juan Lewis, it was I who kept you here. You should have gone two weeks ago." She snuggled closer and rested her cheek against his bare chest. "But I was selfish, no?"

Lewis cradled her head in his palm and looked down at her. "You were selfish, *yes!* And so was I. You are a very difficult woman to leave, Maria."

"But you will go now and I shall come as soon as you tell me. If it is dangerous to return, you will write to me and I shall ask father to make the journey with us."

"You will not make the journey alone. If necessary I will offer to serve in the militia in Yerba Buena. But I myself shall ride down to bring you to your new home." He drew her warm, willing body closer and moved his big hand down along the smooth, curving furrow of her back and up over the gentle swell of her hip. The specter of militia service and the separation paled and were forgotten.

John returned to Yerba Buena in mid-January to advise Leisdesdorff that he would be able to move his family north by mid-February and to inspect the house the merchant had written about. The house, made of plastered adobe, stood above Leidesdorff's cottage and commanded a view of the cove, the islands to the east, and the hills of the *contra costa* on the far side of the bay. Furniture for the new home would present no problem. A shipload was en route from New England to furnish the planned hotel. The few interim things they might require would be available in the Leidesdorff warehouse.

John stayed on for ten days, familiarizing himself with the stock and with the lists of merchandise on order. When he returned to Monterey he was certain no obstacles to an early move remained.

Captain Cooper continued John's employment until the first of February. When the day came to tote up the bill for the things drawn against salary, John found that very little cash was left. Cooper frowned over the balance sheet and clucked with concern. "I was afraid of that, lad. Yes, I was. A worthy reputation's mighty hard to come by. I believe I've a good one and I wouldn't care to lose it." He reached for the handle of a small desk drawer. "So, rather than risk being called a tightwad for sending a young father away with an empty purse, I've thought to provide you with this." He removed a fist-sized deerskin pouch and handed it to Lewis. "There's a month's bonus in silver in there, John. It's small enough to be a practical amount to give and large enough to be a respectable amount to receive. Take it with the blessings of the missus and me. If the good Lord had given us another son we'd have wished for the likes of you."

Surprised and touched by the unexpected generosity, John managed what he knew were inadequate thanks. That night, at dinner with Don José and Doña Gertrudis, he presented the pouch to Maria. To his astonishment her eyes filled with tears. It would have been more like his fiery little wife, he thought, to straighten indignantly and demand why her husband's worth had not been properly acknowledged by twice the amount.

In bed, after evening prayers, Maria snuggled her head beneath his chin and urged him to describe again their new home and the sort of furniture that would come from the United States. John obliged and wondered vaguely at her quietness.

Several times in the ensuing days he found Maria slumped on a chair resting. When he asked her if she felt ill, she pretended annoyance and brushed the query aside. "It is a very difficult thing for a wife to pull up her roots and organize everything that must go in the *carretas*. Men do not understand how fatiguing it can be. Now get out, my dear husband. Go talk to my father, who is going to miss you very much, because when I am busy I don't miss you at all."

The following day, Doña Gertrudis took John aside and asked if the trip north could be postponed for a week. Her reasons were not convincing. "Many things remain to be done, Juan. I am making things the baby must have for the trip. They are not ready yet. Would a week make a great difference to Señor Leidesdorff?"

"Probably not. He is anxious for me to begin work but I am certain he would accept a reasonable explanation."

The next morning, under questioning, Doña Gertrudis

179

told John that she was certain now that her daughter was pregnant again. "She is happy and sick at the same time, Juan. She is happy to have another child for you. A son this time. She prays for it. But she knows that it makes things very difficult. She will not tell you herself, but she is afraid to have her baby at Yerba Buena."

Completely upset, John struggled to control his voice. "In the name of God, *why?* It is not against the law. Babies have been born there."

"Indian babies and *mestizo* babies. But it is not like Monterey where there are many women who understand these things and know what to do. Maria had a very difficult birth with Felicia."

"Second births are easier. Everybody's said that. All my life I've heard it."

Doña Gertrudis shrugged. "It depends. Perhaps yes. Perhaps no. Only the Holy Mother knows. If you really wish it, Maria will go with you and have her baby there and I shall come with the women to help. But you must send for me in time."

John was still in a turmoil when Don José found him seated on an outside bench. "So, Juan, we are about to be blessed again. This is very happy news." He sat down beside his son-in-law and peered at him curiously. "But it comes at an inopportune moment. Yes?"

"In truth, yes."

Resting a hand on John's shoulder, Don José nodded sympathetically. "You have very strong ammunition, my son!"

"And your daughter is a very easy target!"

Don José's eyes widened and he threw back his head and laughed. "My boy, my boy, there are few things in this life more comforting than a passionate woman who comes to her husband's arms willingly and often. Do not curse your luck for that. Thank God that you are not married to," an impatient gesture embraced the town, "to a dozen women I can think of, and that in your old age you will have a loving family to care for you. That is the ultimate wealth." He rose and moved away a little. "May I tell you what I think you should do?"

John got to his feet with an exasperated sigh. "It would have been better if you had told what not to do!"

"That is counsel no man listens to, my son. But I will give you some now that you should listen to. Do not disappoint Señor Leidesdorff. Go to Yerba Buena. Begin your work. Finish your house. He is reasonable and sympathetic. Also, I am told that he is a lonely man who has known hurt. He will understand. Maria says the baby will come in September. Surely you will be permitted to welcome it and be with your wife. And then, when your family is able to travel, you will all go to your new home." His arms extended appealingly. "What could be better? Now you know what your new house must accommodate. You can prepare in time. It is a blessing, really."

John returned to Yerba Buena in mid-February and began his duties with William Leidesdorff. Soon he was at work devising a departmentalized system of storage that would make order out of the confusion of general merchandise stacked at random in the large warehouse. Leidesdorff ran one six-hundred-ton schooner regularly between Yerba Buena and Honolulu. He had several small bay schooners making frequent trips to landings along the bay. John was obliged to make these local trips during March and April to contract for hides and tallow. He looked forward to them since they gave him an opportunity to visit old acquaintances.

Leidesdorff had extended Sutter's credit to the limit and now refused further requests for supplies. But John exchanged letters with Harry Morrison and kept abreast of the progress. Morrison reported that the white population in and around New Helvetia had grown to more than fifty but there were still only a half dozen settlers on Sutter's land and most of them had not complied with the terms set forth in the grants. He wrote that Sutter "closed an eye and accepted as the head of the family any man with fortitude enough to take an unwashed squaw into his bed."

In May, John managed a three-day visit with his family when he sailed south to attend some customs clearances. He was shocked at Maria's size. She seemed to have grown large much earlier than when she was carrying Felicia. She laughed at his concern. "That is because I am making a very large son for you, Juan Lewis. All of the signs are right. The Indian women say so."

Privately, Doña Gertrudis was not so sanguine. "She is very uncomfortable much of the time, Juan. Don José and

181

I wish you were here. When the time comes she will need you."

"I will be here. I have already spoken to Señor Leidesdorff. There will be no difficulty. But you must send word to me in time."

On the night before he sailed, John and Maria dined at the Robles adobe and John learned that Goyo had been seen in Los Angeles. Don José also reported that President Santa Anna had formed a new constitutional government that called for the immediate expulsion of all Americans who were not naturalized. Governor Micheltorena, charged with executing the order, was extremely unhappy because of the obvious exceptions that would have to be made. A second order had followed barring all but naturalized merchants from trading in Alta California. Don José told John that the new edict had been met with outright derision by Monterey's merchants. Larkin suggested to Micheltorena that since he was unable to comply, as an official of the United States Government, he would be happy to dissolve his business if the central government would immediately pay up, in full, all of the loans he had advanced to the Alvarado administration, Micheltorena had chosen to ignore Santa Anna's directives as impossible to enforce without destroying the economy of Alta California.

Don José outlined the situation. "Santa Anna merely hastens the inevitable, Juan. By insisting on impractical conditions here he earns the disrespect of his own officials and shows the entire world the weakness of the Mexican government. It is an open invitation to those who would take us over. The next time we are captured our liberty will not be returned." He glanced around to make certain he would not be overheard by the servants. "And we do not deserve to have it returned. Who can respect us? Micheltorena is intelligent, but he has no backbone. He openly refuses to discipline his *cholos* and he has even made his only responsible officer, Colonel Telles, ill with the fever by refusing to support him when he disciplines the men. 'It is wrong to flog men whose bellies are empty because they have not been paid. They steal to eat. And so would we if we were hungry.' That is what he said. And so we do not walk about Monterey unarmed. Soon now, serious trouble will come."

Far from reassured, John Lewis returned to Yerba

Buena. He had been gone less than three weeks when Goyo Robles reappeared in Monterey. He seemed unmoved by his family's emotional greeting and he evinced no interest in his infant niece. Except for visits to the customhouse and the government buildings he kept to himself.

His son's appearance and manner distressed Don José. Goyo seemed to have aged. His lean body was travel-worn and his narrow, sun-darkened face had assumed an even more hawkish aspect. Traces of gray had appeared at his temples and the bravado that once had been appealing had been displaced by a surly arrogance. Each day at mass, Doña Gertrudis implored the Holy Mother to save him.

Goyo spent several days at the family adobe before riding inland to see Alvarado at Alisal. Within the week he returned, grimmer than ever. He stopped at his home for food, then rode to the government house to see Micheltorena. Two days later he left Monterey for New Helvetia carrying a letter from Governor Micheltorena commissioning Sutter a captain in the Mexican militia. The commission was accompanied by an order to form and train a company of soldiers under a promise that a government draft would be forthcoming to provide arms, uniforms, and wages.

Shortly after Goyo's departure, the French whaler, *Ganges*, put into Monterey late in the afternoon for provisioning. The second mate and six sailors on temporary leave from the ship visited a waterfront cantina. Almost immediately they were confronted by a group of Micheltorena's drunken *cholos*. Within minutes, one of the French sailors had three fingers lopped off, one of the *cholos* was killed outright, and another was impaled on a harpoon. Then the battle spilled onto the street and down to the landing.

Colonel Telles, still recovering from a fortnight of high fever, managed to get his men under control. Word of the vicious brawl reached Micheltorena and he accused the whaling captain of breaking the law by letting his crew remain after the sundown curfew. The Frenchman, convinced by reports from his own crew that the *cholos* had provoked the fight, refused to accept the responsibility.

David Spence had been present when the master faced down the Mexican general, accused him of laxity in discipline, and threatened to sail immediately to intercept the

French fleet cruising off the California coast and report the outrage. Spence persuaded Micheltorena not to press the charge and finally the matter was settled without creating an international incident. The *Ganges* sailed on the early tide the next morning but Monterey had had its fill.

At Larkin's home, Spence expressed his disgust to his fellow merchants. "We have reached the end, gentlemen. Every reasonable person knows that the *cholos* started it. They've been spoiling for weeks now. Micheltorena's been warned a dozen times. This is enough to bring responsible men to the edge of treason."

The statement was greeted by a chorus of "Ayes." Spence reached into his pocket and removed a folded paper. "As though we have not had sufficient trouble, Micheltorena has called the new departmental assembly into session here. They have chosen a slate for the position of civil governor that will be submitted to Mexico City. Then Santa Anna will choose one of the men for the post.

"But the trouble is Micheltorena did not notify Pico or Carrillo or any of the men from the south."

William Hartnell whistled softly and gazed at the paper Spence was holding. "Who was chosen for this exalted position, David?"

"Micheltorena is his own first choice. Alvarado is second—"

Spence was interrupted by expressions of disbelief. Cooper swore quietly as he got to his feet. "Micheltorena cannot be fathomed. He's about to make the only reasonably competent governor we've had a poor second choice. Who were the others?"

"Telles, Osio, and Casarin in that order. The general's tactic is obvious. By making certain he is named civil governor, he would no longer be responsible for his rats. Alvarado, as ranking colonel of the government troops, would have the responsibility."

When the men had gone, each to nurture his own misgivings, United States Consul Thomas O. Larkin went to his desk to pen a memorandum noting that the time was fast approaching when the United States should redouble its efforts to bring about a peaceful annexation of Alta California and the adjacent territory. Reasonable Californians, both native and naturalized, would favor it, he was certain.

In Los Angeles former governor *pro tempore*, Pio Pico, lost no time in venting his outrage in an official communication. Micheltorena offered the lame excuse that the dispatch to the southern delegates had gone astray. Apologies were made and the assembly was convened in August. In Monterey and Yerba Buena the merchants waited on tenterhooks the outcome of the voting.

Pico refused to approve a slate unless Micheltorena would agree that both capital and customhouse be transferred to Los Angeles. Evidently his memory of the *cholos'* depredations had paled before the personal advantage he could derive from having the customs money under his control. Reports came that the assembly was hopelessly deadlocked, the five northern delegates against the five southern ones. Micheltorena, as governor, could cast the deciding vote. The meeting was adjourned and rescheduled at a later date. John Lewis, in Monterey to see Maria, who had grown enormously, was present when word came that the matter of the customhouse had been put to a ballot and Governor Micheltorena had broken the tie by casting his vote to leave the office in Monterey.

There was high jubilation in the capital and a great tirade in Los Angeles. Pico charged fraud and collusion in the north and threatened to journey to Mexico City to have the decision reversed.

At the Robles house Don José sat musing aloud in the garden with Maria and John. "Pico will not go to Mexico City. Instead he will ask Alvarado and Castro to support him in a revolt against Micheltorena. It is his way. First a revolt against Chico, then Gutiérrez and soon, Micheltorena. No Mexican governor ever will be acceptable. But once they are driven out, it will be Los Angeles against Monterey again, and always the same issue, that accursed customhouse. Pico never has recovered from the grandeur of being governor, even for only twenty-two days, twelve years ago. He will never rest until he holds that office again. And no man can trust him because his personal ambition is greater than his love for California. He is a pompous little man, often generous, more often selfish, and still more often, ridiculous."

John was about to add an observation of his own when Maria reacted to a sudden, sharp discomfort. He put an arm around her anxiously.

185

"It is all right, Juan. It is only that your son grows impatient now," Maria said.

John wished profoundly for an end to the confusion that kept him from concentrating all of his energies on his new position. Things were going well in Yerba Buena. The house was completed and some of the furniture had arrived. He was well pleased with that aspect. But the disruptive politics of Alta California were having their effect upon Leidesdorff and also upon Leese, Richardson and the others around the Bay. Few cared to plan too far ahead. Only Sutter remained supremely confident about the future and continued scheming, blithely indifferent to the burdensome past obligations upon which his dreams were based.

John excused himself to go down to visit Will Chard and Josiah Belden in their new store. Part of the modest credit for its establishment had been advanced by Leidesdorff upon John's personal assurances. He paused behind Maria and rested his hands protectively on her shoulders as he addressed Don José. "More than one *californio* now wishes this fiasco could be put to an end. Even Vallejo told Leidesdorff that he has given up hope of keeping out the settlers. He has changed completely. He would welcome an orderly government under the United States. Most of us at Yerba Buena agree."

Don José smiled and lapsed into Spanish. *"Mas vale tarde que nunca, Juan.* The change should happen now, but better it happens late than never."

Maria's eyes darted heavenward and she whispered a quick prayer. "Thank God Goyo does not hear you. He is a *nacionalista* now. He would destroy us all."

Doña Gertrudis glanced at her husband apprehensively, then lowered her head to stare in unhappy silence at her folded hands.

On September 19, 1844, Maria Robles Lewis gave birth to twin boys. They were christened Steven Joseph and William Eugene. The birth was difficult and no doctor was available. Skilled midwives and earnest prayer, plus liberal doses of laudanum, finally assured the twins' arrival. Badly torn and weak from loss of blood, Maria lingered in a comatose state for nearly a week.

John, fighting off panic induced by Maria's terrible outcries and her sobbing prayers, had wanted to ride to

186

Mount Diablo for Dr. John Marsh, but the men had intervened and convinced him that a doctor without proper qualifications would be far more dangerous than midwives who had successfully presided over a hundred difficult births. In the end, the weight of Doña Encarnacion Cooper's opinion dissuaded him. "She will live, Juan. We Spanish women have twenty children as easily as your woman have two. Now Maria has double the pain. But the Holy Mother is just. For all of her life, she will have double the pleasure also."

The midwives and Doña Gertrudis did not consider Maria to be out of danger until the beginning of October. Much relieved, John returned to Yerba Buena, where he was greeted by William Leidesdorff as a man of special distinction, the father of healthy twin sons.

Leidesdorff told him that forty-six more American men, women, and children had come into central California during his absence and that over a hundred more were on the trail south from Oregon territory. Few people had any doubt now that Washington planned to duplicate the Texas strategy when enough settlers arrived.

As autumn deepened, Micheltorena's two hundred *ratereos* grew bolder. Hardly a night passed without incident—another robbery, another woman accosted on the street, another drunken brawl with foreign sailors on shore leave, and endless internecine squabbles. The ultimate outrage came late in October when one of La Mofeta's prostitutes, a *mestizo* girl not yet sixteen, was found sexually ravaged on the beach. Before she died the next morning, she told Father Varga that she had been raped by seven of the *cholos*. One of them had paid and the others had been waiting in the darkness when she had accompanied him to the shore. Once again action was promised but not taken. Then La Mofeta herself attempted to stab the original offender and was disemboweled with her own knife in front of the cantina.

Goyo Robles, just returned from New Helvetia, learned that a delegation of *californios* who had stood with Alvarado had ridden inland to Alisal. These men were seeking Alvarado's leadership in a revolt. Goyo was torn between loyalty to a blood brother and a new and deep conviction that complete cooperation with Mexico was necessary to control the flood of Yankee immigrants. He returned to his father's house and prepared to ride to Alisal.

187

Goyo's brief presence at the family adobe occasioned nothing but hurt and anger. Once again he completely ignored little Felicia; but when he was forced by his mother's pleading to see the twin boys, he stood smirking beside the *doble cuna*, tracing his forefinger along the jagged scar that disfigured his temple. "Incredible! But I have seen that yanqui. He could not have made two sons at once without help!" Leaving his family mute with shock, he departed quickly.

He arrived at Alisal shortly before midnight and immediately confronted Alvarado. "Juan, I came to talk here because the whole town has ears. Do you know what is being said about you?"

"No. But nothing will surprise me."

"It may not. But it should worry you. Telles has been informed that you were asked to lead a revolt against Micheltorena. He has told that to the governor. They are saying that many *californios* and some foreigners have asked you to command an army to drive Micheltorena and his soldiers out of the country."

"Soldiers? Criminals!"

"What difference? I want to know whether or not they speak the truth."

Alvarado looked up unhappily. "I was asked if I would be governor again."

"And you answered?"

"Again I am administrator of customs. Also, I am colonel of the First Regiment. But most of all, I am tired. There is no persuasion in heaven or earth that could induce me to become governor again."

Sensing that Alvarado was being evasive, Goyo pressed his point. "Micheltorena respects you, Juan. It would be fatal to divide us now. What you must do is convince him of your loyalty."

Alvarado's eyebrows lifted. "And why should he doubt my loyalty with all of the honors he has conferred on me?"

"Because, Juan, he remembers that you led the revolt against his friend, Gutiérrez. Telles has told him that you would do it again."

Alvarado looked at Goyo incredulously. "Would I really? With no Isaac Graham? With no William Chard? I am a colonel without troops. From whom would I get help? Sutter? That one will march with the highest bidder—and

188

only the governor can give him what he wants. Sutter is desperate for the *soberante* grant, all the land between the two grants I gave to him. Micheltorena could buy Sutter's soul if he promised him that."

Goyo was surprised that Alvarado seemed to be aware of Sutter's attempts to bargain with the governor for the additional land. "But Juan, if you know that about Sutter, then you also know that the governor has not responded to his maneuvering."

Alvarado nodded. "For now, yes. But if he needs Sutter, he will."

Goyo eyed him suspiciously. "Why would he need Sutter unless someone threatened him?"

"Goyo, for years you have stood with me for independence. You fought beside me when we drove out Gutiérrez. But now, I feel that you stand with a Mexican governor. Could it be, as with José Castro, that you also are in Micheltorena's debt?"

Goyo fought to control his voice. "I told you that I had no influence in the choice of the commander or the soldiers to be sent. I told Santa Anna only what you told me to say, that Alta California cannot be defended against the foreigners without many more troops."

Alvarado turned away and Goyo knew he was responsible for the undoing of his closest friend. On the journey home from Mexico City he had confessed to himself that he had been flattered into saying what President Santa Anna wanted to hear. Goyo's answers to questions about Alvarado's relationships with Pio Pico and Mariano Vallejo made it imperative for Santa Anna to place Alta California under central government domination. Santa Anna knew he was facing a war with the United States and his advisers were urging him to move before the Americans could muster a strong force.

Alvarado turned back again but avoided Goyo's eyes. "There is only one reason why Santa Anna would choose Micheltorena to come north with the troops instead of sending a detachment under a Telles, for instance." Then he looked directly at his friend. "Can you imagine what the reason is?"

Goyo flung his hands out. "There could be a hundred reasons! Who knows the mind of Santa Anna?"

"There is only one reason, my dear friend, just one. A

189

civil governor is replaced in time of danger only when the central government feels that he is incompetent."

"That is a lie, Juan. You have been governor longer than any other in our history. Six years. And we have prospered. I told Santa Anna there would be no foreign problems here now if Bustamente had sent soldiers when they were requested four years ago. I told him that you are a brilliant field commander. I said the same of Castro. I said that you hold the loyalty of every man in Alta California." There was a leaden silence. Then Goyo heard Alvarado asking the question he had feared for so long.

"Tell me, on your oath as my blood brother, did Santa Anna speak of Pio Pico and Mariano Vallejo? Did he speak of our differences?"

He forced himself to nod. "Yes."

The weariness in Alvarado's face gave way to a smile that at once was sad and triumphant. "He asked you about them because Pico had written to him saying that I was too ill to discharge my duties and that I had been a party to collusion in the matter of keeping the customhouse and capital here. Mariano has written many times complaining of my lack of military initiative."

Goyo seemed pale beneath his deep tan and the scar line on his temple and cheek shone dead white in the firelight. "He knows Pico. He cannot take him seriously. Pico has bombarded him with letters accusing everybody of collusion ever since the assembly. Most of all, Pico has accused Micheltorena himself. And he knows Mariano."

"Ah, yes, Goyo, I have no doubt of that. But the truth remains that Santa Anna was forced to conclude that men who fight among themselves while foreigners invade their land are not fit to defend the Department of the North because they are wanting in common patriotism."

Goyo hoped that the heat of his anger would obscure the light of Alvarado's truth. Santa Anna had tricked him into confirming the central government's suspicions that there could never be unanimity of purpose in Alta California, had tricked him into admitting the truth. "No man would accuse you of that, Juan. Your patriotism is a shining inspiration. Even in defeat Gutiérrez agreed."

"But the fact remains, Goyo, my friend, that the poor opinion of me is the one that prevailed. I am certain that you were asked if you thought I was fit for the responsibility of commanding federal troops in an action against the foreigners. I want to hear now how you defended me."

"You have heard. I have just told you. In God's name what more can I say?"

Alvarado's incipient anger had gone. His face reflected only the great burden of inner weariness and his voice was soft. "It is not necessary to tell me anything more if you can say again, on your oath, that you did not know Micheltorena would be sent to replace me."

Goyo all but leaped across the space that separated them. "I did not know, Juan, I swear it. Santa Anna said only that he recognized the seriousness of the situation here. He said only that five hundred men would be sent. He said nothing about a commander, for certain, nothing about a military governor!"

Alvarado dropped heavily onto the chair again. "It is the will of God. What has been done has been done. And because of that—*much more remains to be done*, again." He leaned forward to look deeply into Robles' eyes. "Where would you stand now, Goyo, if I were to ask you to join me in another revolt? Where would your loyalty be? And what would your counsel be?"

"For God's sake, Juan, is it necessary to ask? I would stand with you, always. But I beg you to go to Micheltorena first. Offer to help raise money to pay the men. Remove his suspicion by your loyal action. There is no other way now to save Alta California. It is too late for independence. We can survive only by cooperation with Mexico. It is our last hope. If we do not act instantly we shall be overrun by the yanqui pigs or by the British."

"You echo Santa Anna's words. Yes?"

"In the name of God, no. *He echoes mine!*"

Stunned by his own inadvertent admission, Goyo stood rooted. The implication in his outburst was irrefutable. In the past Juan Alvarado had led him to undo himself with his own temper. Now he had done it again. Goyo knew that in three thoughtless words he had revealed all that he himself had said in Mexico City. "Micheltorena will be your friend, Juan. Your friend, if you will permit him."

The questioning in Alvarado's eyes turned to reproach and then to sadness. "I have no friends, Goyo. I have only one well-intentioned, politically inept blood brother."

John returned to Monterey in early November to attend

191

the wedding of William Chard and Dolores Robles. Maria, radiant again and sufficiently recovered to take part in the festivities, danced the waltzes, the quadrilles, the jotas, and the boleros with all of her former abandon. Amazed, John sat out most of them, claiming less grace than a hamstrung ox.

The next morning, Don José returned from the calle principal. He was unusually perturbed. "It is very bad. Very bad. Sutter's serving boy, Pablo, rode in this morning and was seen entering the governor's residence. A messenger came for Goyo. He left me and went to Micheltorena immediately. Then both of them rode out of the pueblo. Goyo would not say where he goes. There is talk everywhere that Micheltorena has ordered Colonel Telles to arrest Juan Alvarado."

Before Don José could answer John's stream of questions, Will Chard pulled up in front of the house and strode unceremoniously into the parlor. "I'm not sure this is anything I personally will shed tears over, but Micheltorena is going to arrest Alvarado for treason. Telles is gathering his men at the presidio now. They will leave at dark." He laughed ironically. "I seem to recall that they excel at these nocturnal surprise parties."

Doña Gertrudis was aghast. "But Juan Alvarado has nothing to do with politics now. He has sworn it. If Micheltorena is arresting him, then it is only to save face—to invent some excuse to make us forget his own irresponsibility. Juan must be warned and someone must tell the truth to the governor."

She looked from Chard to John and back to Chard again. Chard warded off the tacit request with upraised hands. "Don't look to me. I went to prison for eighteen months for doing nothing. How long do you think Micheltorena would let me rot at Tepic if he knew I really had assisted a revolutionist leader? Alvarado's record is against him. Anyone who even pleads for him will be suspect."

Don José nodded in agreement and glanced meaningfully at his son-in-law. "A deputation has appealed to Alvarado to lead them. There is no question. That in itself is sufficient. Alvarado will never receive a fair trial from General Micheltorena. The only way he could clear himself would be to name those who approached him. Juan would never do that. It would put their necks in nooses

192

immediately. I know Juan. He would rather risk his own neck. He must be told."

Maria laid a hand on John's sleeve. "Juan, please? We cannot know this and not warn him. He has promised to be the godfather of our sons. And he has been your good friend, too. I know the danger, but please, Juan, please?"

At noon that same day, on the pretext of departing for Yerba Buena somewhat earlier than planned, John said his farewells to the merchants and left the capital by way of the *Camino Real*. When he was well beyond the pueblo he turned inland and rode toward a difficult short-cut cattle trail through the canyon that would give him a six- or seven-hour advantage.

Shortly before midnight, John, worn almost to exhaustion, dismounted heavily in the courtyard of Juan Alvarado's adobe ranch house and handed the reins to a sleepy servant. Moments later, Alvarado listened dumbstruck to the story of his blood brother's treachery.

"Incredible!" he murmured over and over as John filled in the details. When John finished, Alvarado stepped forward and embraced him. "Juan Lewis, it is not possible to repay what I now owe you. Until this moment I have not permitted myself to believe what my heart knows to be the truth. Goyo has become a *fanatico*. No man is his brother who will not believe as he does. For him it is easier to hate than to reason. He has no loyalty now other than to his passion. He will reject those who love him and embrace those who despise him if he thinks they will serve his cause. He has not the wisdom to know that his cause will be the end of Alta California. Now we must fight to save Alta California from our own brothers."

The following morning, just before dawn, Colonel Telles and one hundred *cholos* surrounded El Alisal and found it deserted except for servants who purported to know nothing until one of the officers took an infant Indian child and threatened to throw it into the blazing adobe oven. The terrified mother broke down and revealed that Alvarado and an unidentified man had left for Castro's adobe in San Juan.

At San Juan, Telles discovered that Alvarado and Castro had left for San Jose. He stopped only long enough to strip the adobe of supplies, then set out in pursuit well aware that he was certain to be frustrated still again, for

193

Micheltorena had ordered him to pursue Alvarado no farther north than the Rancho de Laguna some five leagues south of the pueblo of San Jose. Micheltorena's excuse was that he didn't want Telles and his men too far from the capitol. Telles arrived in a soaking downpour and made camp near some abandoned adobe outbuildings. There, fuming at the inconceivable stupidity of the governor-general, he and his men watched Alvarado supervise the establishment of his camp less than a half league to the north.

Alvarado invoked the protection of his personal saint and named the place El Lugar Seguro de Santa Teresa. John helped to set up a headquarters hut and reflected with grim amusement that poor Santa Teresa would be taxed to the limit should Telles and his men decide to attack.

After a miserable night the rain stopped and Alvarado, expecting Telles to move, alerted his little force. The day passed with no evidence of preparation on the other side. John, Alvarado, and Castro speculated that Telles must have sent for reinforcements even though he would be opposed by fewer than half as many men. Reconciled to his involvement, John agreed to scour the countryside to the north for recruits. Within the week he was back at Santa Teresa at the head of forty-eight well armed *rancheros* and their workmen, all of whom had suffered enough of Mexican rule under Gutiérrez and preferred a return to Alvarado's orderly but more lenient home administration.

Two more days passed and still there was no indication that Telles was preparing to do more than make his camp livable. Each side watched the other slaughter and butcher appropriated range cattle. Each side could see the silhouettes around the large bivouac fires at night. And each side could hear the laughter and occasional singing as the men sought to relieve their discomfort and boredom.

During the week Alvarado had added a small company of expert French Canadian riflemen to his army. He promised them liberal pay and the necessary supplies to return to the Oregon country. On the eighth day of the uneasy truce, a rider arrived from San Jose bearing news that Telles himself had not been in the camp for a week. He had left his officers under orders not to attack until he had protested to General Micheltorena and received permission to advance. Immediately Alvarado and Castro began plans for a surprise attack. Alvarado was elated. "Telles is the only able officer in the force. If we strike them very early and surprise them sleeping, it will be finished within the hour. We will force Micheltorena and his *rateros* to leave the country."

Castro was reluctant to move too hastily. "Telles can call on one hundred reserves in the Monterey garrison. How can we be certain that he did not return for them? Man to man we are better than they. But if they outnumber us two to one and bring cannon it can go very hard unless our first attack is a complete success."

Castro's resistance puzzled Alvarado. It was unlike him to argue for inaction. "But of course we shall have a quick victory, José. I do not understand your reasoning. Their reinforcements are not here. It is foolish to wait. After midnight we will deploy riflemen on both flanks, the best shots we have. Then we will rout the *cholos* at first light. We can move the flanking companies down the ravines on both sides while we charge directly from the front. That is why I chose this place. It is perfect. We must move now."

John, who held no official rank but was accorded the courtesy due an officer, witnessed the argument but avoided taking sides. Hoping that the confrontation would be over soon, he tried not to dwell on the possible consequences and on those that undoubtedly had already accrued. Unless William Leidesdorff had received word of the revolt from Larkin, who would understand the nature of his involvement, John knew that his employer would be at a loss to explain his overlong absence. He hoped that Don José would have had the foresight to advise Leidesdorff also, but these aspects of his involvement paled before the other possible consequences. He forced himself to face the truth that in revolts men must die and leave behind them weeping women, defenseless children, and a legacy of unfulfilled dreams.

Alvarado prevailed and covert preparations were begun in the late afternoon. Bores were cleaned. Powder and shot and wadding were checked and rechecked and the snipers were chosen for the two flanking companies and surreptitiously moved into position to deploy.

An hour before sunset, the sentry who had been posted at the top of the hill spurred his mount down the slope, pulled up at Alvarado's headquarters shelter, and passed the word that Micheltorena and twenty riders were ap-

195

proaching the Rancho de Laguna camp from the southwest.

Upon hearing the news, Castro proposed that he himself ride out to meet Micheltorena under a flag of truce and offer the general an honorable settlement. "It is the only sensible action, Juan. We must preserve Mexican lives to defend our land against the Americans."

Alvarado, facing Castro across the makeshift table on which was spread his battle plan, stared at his old revolutionary commander in disbelief. "I do not know you, José." He snapped his fingers sharply. "Like that, you have become a stranger."

Castro's face mirrored the conflict that had prompted his irresolute behavior. "I know Micheltorena. He has no stomach for battle, Juan. We can win honorably without wasting a man. The issue is clear. You were asked to lead this revolt to get rid of the *cholos*. If he will send them to Mexico we can place ourselves under his command. The problem will be settled. You do not wish to be governor again. Micheltorena enjoys the task. It is an excellent solution. Once more you will be a brilliant hero and every *californio* with us will return to his land and his family. Allow me to arrange it, Juan. Then you talk with Micheltorena. He is an honorable man, an intelligent man. He must accept our terms."

Alvarado tried to speak but Castro cut him off. "Do not forget, Juan, that Micheltorena's heart is with us in the north. When the vote was tied in the assembly, it was he who braved Pico's anger to cast his vote for us. He kept the capital and customhouse in Monterey. We *all* owe him something for that. Most especially you. It was not necessary for him to commission you a colonel—nor to make you administrator of customs. These general decisions were made out of consideration for your devotion to this land."

John watched Alvarado's surprise turn to suspicion. "How do you know these things, José? Tell me! How do you know Micheltorena has such great respect for me?"

"Because I have spoken with him about you."

"Ah! Then you *have* visited him in Monterey?"

"No. I have not."

"Then he had visited you at San Juan."

"No. I have not seen him since Mexico City. But when I was a guest at his home we often discussed Alta Califor-

196

nia. If ·he has tried to make good laws and establish schools, it is because I have told him what is needed here."

Alvarado found himself staring incredulously at his one-time supreme military commander. It was impossible, he knew, that Castro could not have understood the consequences of such discussions. Suddenly it was clear to Alvarado that his deposal as governor had far deeper roots than he had imagined. Trusted friends had praised him but none had protected him. If Goyo had done the reaping, then surely Castro had done the sowing. As he searched Castro's pained face, Juan Bautista Alvarado felt completely alone. Realization gave way to leaden hopelessness. "It is well, José. Take the white flag and go to your friend, Micheltorena." He nodded toward the men seen through the open doorway. "These men are loyal to me; but now I find it futile to ask them to shed their blood."

On the morning of December 1, 1844, Juan Alvarado met with General Manuel Micheltorena and the general agreed to send his *cholos* back to Mexico within ninety days. Micheltorena would continue as civil governor of Alta California. In turn, Alvarado agreed to place himself and his revolutionary army under the general's command as soon as proof was offered that the *cholos* had departed.

Monterey exploded in rejoicing. In Los Angeles, the relief was no less apparent. Pio Pico ordered his brother Andres to begin immediate preparations to deport the small garrison of *cholos* still quartered there and a brig was chartered for the purpose.

At Santa Teresa, Alvarado and Castro prepared to move their troops to a winter camp near the mission at San Jose where Micheltorena promised they would be maintained at government expense until the ninety-day grace period had expired.

John, released from further duty, sent word to his family in Monterey that he would come for them as soon as he had made his explanations to William Leidesdorff in Yerba Buena.

Leidesdorff was generous and understanding, but he was far from sanguine about Alta California's future political stability. He spoke of it as he and John sat at supper in his cottage.

"At any rate, perhaps we shall know some peace.

Micheltorena does seem less concerned about immigration than he pretends to be."

John grinned. "And he has the questionable support of Captain Sutter."

Leidesdorff chuckled. "He wants immigrants and so do we merchants. He is a remarkable man with a remarkable lack of conscience, that may in the end serve our purpose."

Very shortly, the observations about the confused state of affairs in Monterey turned to business matters. Leidesdorff spoke of the need for an additional man after the first of the year, one whose duties would be to supervise his growing transportation operation. John, anxious to have Harry Morrison easily accessible to the family because of his wide knowledge of pharmaceuticals and his practical knowledge of simple medical procedures, recommended him. Leidesdorff knew of him and expressed interest and John agreed to sound him out. Satisfied that at last some stability seemed in the offing, he prepared to return to Monterey for the holidays.

During the previous summer, Sutter's long-time serving boy, Pablo, had visited Micheltorena. He was followed a few weeks later by Sutter's clerk, young John Bidwell. Both couriers had carried secret letters from Sutter, composed in grandiloquent French, pledging his absolute loyalty to the new governor. They emphasized the private militia Sutter was drilling and assured Micheltorena that these "perfectly trained" men were ready to march on any mission the governor might require.

In the government house, Micheltorena was reviewing those letters again. He was still smarting from the humiliating accusations he had endured from Colonel Raphael Telles. The colonel had called the general's conduct at Santa Teresa cowardly and disgraceful. He had charged the general with criminal stupidity in requiring that the pursuit be stopped at Rancho de Laguana.

News of the violent personal confrontation spread quickly through the capital. Micheltorena tried to save face by ordering Telles back to Mexico on the next ship. Monterey's citizens were outwardly polite but Micheltorena knew that he was a private laughingstock. Now he must demonstrate that he was not to be judged hastily, that he intended to enforce to the letter the central gov-

198

ernment's directives. The crux was Sutter and perhaps Isaac Graham also. Both men wanted land. Graham demanded a legal grant as part of his indemnity and Sutter wanted the vast *soberante* grant. Additionally, there was other land, huge stretches of fertile valley land in the interior that could be used to buy loyalty. Sutter's letters had the ring of sincerity. But against that Micheltorena knew that he must balance the evidence of the man's past performance. In the end, he realized that he had no choice but to trust Sutter and hope that promises compatible with the man's boundless personal ambition would constrain him to behave loyally.

Micheltorena dispatched Goyo Robles to New Helvetia for a second time with orders directing Sutter to form an army consisting of infantry, snipers, cavalry, and the brass field piece and march immediately to surprise Alvarado and Castro at San Jose. At the same time, he dispatched a letter to Castro. He reassured his "dear friend" that he was complying with the terms of the treaty and he was dedicated now to bringing a blessed peace to the hearths of every *californio*. Telles, he wrote, had been sent back to Mexico to arrange for the return of the soldiers. The letter ended with news that Sutter had reported a new immigrant train of eleven wagons and a hundred and forty men in the central valley and that still another party of sixty wagons was on the trail west. He urged Castro to send every soldier he could spare from San Jose into the interior to help Sutter turn back the foreign invaders.

On January 3, in response to a letter from John Lewis, Harry Morrison stepped ashore in Yerba Buena from the deck of William Davis' bay schooner. Morrison was overjoyed at the prospect of a position with Leidesdorff, but he also harbored some misgivings. When John heard them, he brought Morrison to Leidesdorff, who listened with mounting concern to the account of preparations underway at New Helvetia.

"Are you certain, Mr. Morrison, that Sutter is not raising this army to repel the immigrants as he is required by law?"

"Aye, sir. I'm certain. He's recruiting the settlers themselves to march wi' him, on the promise of a league of land."

The implications were unmistakable.

John whistled softly and looked across at Leidesdorff. "I know Sutter. He is the complete opportunist. But I cannot believe that Micheltorena would deliberately break this new treaty." He shook his head. "As a matter of fact, I even find it difficult to admit that Sutter would ever be an accomplice to such a vile plot."

Morrison gazed at them uncomfortably, then turned to John. "I hate to be the one to say it, laddie, but your own brother-in-law is there again. It's the second time in six weeks and he's making no bones about being there for Micheltorena. One of the reasons I was anxious to be coming here was to warn ye that I wudna be counting too much on Micheltorena's friendship. I mind well your real-tionship wi' Robles. And I'll wager he's found occasion to remind the governor ye were wi' Alvarado. I'm not one for alarrums, lad, but I'd be a wee bit concerned aboot your family if I were you."

John had been worried about Maria and the children. His anxiety increased to near panic as he listened to still more evidence of Sutter's conniving. The family had agreed that in case of trouble they would accompany Don José to the pueblo of Santa Cruz at the north end of Monterey Bay. It was a quiet settlement, off the main road to Yerba Buena and relatively free from the possibility of trouble. There, with Secundino Robles and her father, the family would be safe. John decided to leave for Los Méganos the following morning.

At the headquarters adobe in San Jose, Alvarado stared at Castro. "You may take Micheltorena's letters for the truth. I do not trust him as you do. It is not reasonable to send Telles to Mexico by himself to arrange for the deportation of the *cholos*. That could have been done in Monterey."

José Castro patted the penned sheets. "Of course it is reasonable, Juan. How could Micheltorena explain to Santa Anna the sudden arrival in Mexico of his second in command, and his soldiers, at a time when it is well known that yanquis are pouring over the sierra?" He gestured violently. "Of course it is reasonable."

Muttering, Alvarado rose from the chair. "Then I suppose the rumor that Sutter is preparing his forces to march on us is unresponsible."

"I do not believe it. We have only the word of Doctor

200

Marsh's man, who said that Sutter had sent Pablo there to ask Marsh for supplies and to join him. I do not believe Sutter would ask an enemy to join him, and Marsh is his enemy."

Alvardo snorted. "I believe Sutter would ask the devil to join him, if he could profit by it!"

Deliberately, Castro changed his tactics and assumed the air of a tolerant elder brother. "Juan, you are tired. Now, you see an enemy behind every tree. Do not be suspicious. Micheltorena is glad to be done with his *cholos* and have an honorable army at his disposal. We have pledged that. He will send proof soon that the *cholos* have sailed. He has commissioned Sutter on the understanding that Sutter's men will be available to us to stand against the yanquis. Sutter likes to drill and maneuver. But he will not march and never against you, his benefactor. All he has, he owes to you."

Alvarado looked askance. "Paying debts occupies Sutter least of all."

The fruitless discussion ended with Castro complacent and Alvarado unconvinced. Alvarado had not told Castro about Captain John Gantt, a former United States military officer who had ridden for New Helvetia. Gantt had gathered together a private militia of long riflemen. Alvarado was sure Gantt would join forces with Sutter if he was promised enough. Gantt, by nature a mercenary, had for years been friendly with Isaac Graham. Alvarado himself had considered recruiting Gantt but dismissed it as unrealistic. Alvarado knew that if he had irrefutable evidence that Isaac Graham also had committed himself and his men to Sutter, the last shred of doubt as to Sutter's motives would be gone. He had secretly dispatched his own spy to look around Branciforte. Ostensibly the man had been released to return to his home because of ill health and Castro himself had been maneuvered into signing the release.

Several miles along the way to Los Méganos, John Lewis met one of Dr. Marsh's Indian *vaqueros*. Terrified, the man was fleeing northward. John stopped him long enough to hear that Sutter had marched into the ranch that morning, had requisitioned all of the usable horses, mules, and oxen and had given Marsh the choice of joining his army

201

as a foot soldier or being imprisoned at the fort with the other Alvarado sympathizers.

John also learned that Sutter's personal courier, Pablo, and Goyo Robles were on their way to Monterey. If, as John supposed, Goyo's purpose was to coordinate a surprise move against Alvarado's forces at San Jose, then he had no choice but to try to reach Alvarado first. John rode straight south for San Jose.

Shortly before noon on the second day, five miles north of the eastern end of the Pájaro Valley, he saw a group of horsemen in the distance. Moving cautiously, he approached until he could identify the men as Alvarado's eight *exploradores* under Corporal Chavez. Relieved, John rode up to them at full gallop. He was about to hail the corporal when he saw the dangling body of Pablo, minus boots and pants, convulsing at the end of a *reata*. Before he could demand an explanation for the hanging of an apparently harmless young courier, Chavez thrust a paper up at him. "This stinking traitor was carrying this to Micheltorena. Your brother-in-law, Goyo, was riding with him. But his horse was faster."

John read the brief note signed by Sutter assuring Micheltorena that his army would be converging on Alvarado not later than the seventh of January.

"Why didn't you take him to Colonel Alvarado for trial?"

The corporal, who had made no effort to conceal his disdain for Yankees, smirked up at John. "Here, señor, I am the judge. He had the message sewn in the lining of his boot and he confessed that he knew it was there. He had a just trial and now he pays a just penalty."

John forced himself to look up. The boy's face had turned blue-black and red-flecked spittle was drooling from his blood-gorged tongue. "He is too young to hang."

Chavez shrugged. "He is too old to spank."

Turning away, Lewis pointed to the paper. "This boy's death gains nothing. Robles will be delivering that same message before sundown."

Chavez bowed with mock respect. "It is the truth, señor. He has an hour's lead because we attended to this business. And if you are dedicated to the cause of your sons' godfather, you will warn him of Sutter's approach with two hundred men," he smiled maliciously, "the best of whom are yanqui riflemen. Then, señor, you will continue to Monterey and attend to your wife's traitorous brother."

There was a murmur of agreement from the soldiers. John studied each in turn, then addressed himself clearly to the subaltern. "When I am ready to take advice, Chavez, it will not be from a fool nor will it be from a corporal." John reined his mount sharply and the livid Chavez stumbled and fell in an effort to avoid being struck by its plunging hindquarters.

At the mission headquarters in San Jose, Alvarado and Castro listened grimly to John's report of Sutter's activities. Alvarado wasted no time rebuking his duped second-in-command. Murderous rage smoldered on Castro's coarse-featured face and for the first time John felt that Alta California's military commander might actually fight without restraint.

Camp was struck, a messenger was dispatched to warn Pico at Los Angeles, and Alvarado's revolutionary army, numbering fewer than two hundred men, set off before dawn for the south. John Lewis was commissioned a captain and made responsible for the acquisition and transportation of supplies. He could also continue to help recruit men as they marched through the Salinas Valley for Gaviota Pass and Santa Barbara. It was Alvarado's intention to increase his strength to at least three hundred men en route. These, joined to Pico's force, which he estimated at one hundred men, would give him numerical superiority. The two five-pound cannon mounted on *carretas* would be supplemented, he hoped, by other pieces that could be adapted for the field in Los Angeles.

John rode to Santa Cruz to warn that community and to secure Secundino Robles' help in getting the family from Monterey. That accomplished, he cut back through the mountains and rejoined Alvarado and Castro at Santa Rita. Moving rapidly along the narrow Salinas River valley, the force paused at San Miguel, then turned southwestward for Gaviota Pass.

Amid the cheers of the populace, on the morning of January twenty-first, the insurgents made camp on the plaza in Los Angeles.

Pico embraced Alvarado as a brother and pledged the support of his troops and his people "to the last man."

On the twenty-ninth of January, John attended the extraordinary session of the departmental assembly and listened as Alvarado and Castro voiced the north's complaints against Micheltorena. It was obvious to John after listening to Alvarado's lurid predictions of the sack of Los Angeles that the people needed little convincing after their own experience with the general's *rateros*. He had gone to the meeting expecting an immediate declaration of revolution. Instead, after three days of deliberating, the Californians voted to appoint a committee of citizens to ride north in the hope of intercepting Micheltorena and Sutter and working out an accommodation short of war. Ignoring Alvarado's plea that any compromise would work to their disadvantage, the assembly stood by its decision.

On February seventh Micheltorena's combined regiments, totaling fewer than four hundred men because of desertions, reached Santa Barbara. Micheltorena, at Sutter's insistence, unequivocally turned down the commission's plea for peace. Defeated, they returned to Los Angeles. On February fifteenth, convinced at last that war was inevitable, Pico had himself declared temporary governor for the second time. He issued a call for volunteers to fight Micheltorena's despised *rateros* and elevated Alvarado to General of the Armies.

Meanwhile, Micheltorena, suffering from a nagging bowel ailment, had been so impressed with the strength and armament of Sutter's trained forces that the opportunistic Swiss had been able to secure approval of the *soberante* grant. In the absence of Telles, Micheltorena had also commissioned Sutter a full colonel.

Inflated with power, position, and wealth beyond his most extravagant fantasies, Sutter argued successfully against any further delay.

The proclamation announcing Pico's appointment as governor had scarcely been penned and posted in the plaza when a rider announced the arrival of Micheltorena's army in the San Fernando Valley. Governor Pico's brother, Andrés, an able soldier and commandant of the Los Angeles militia, departed immediately to place his men under Alvarado and Castro who were deploying troops in defensible positions on the hills overlooking the Plains of Cahuenga.

At daybreak on February twentieth the two armies found themselves face to face on opposite sides of the narrow southern end of the valley. On the previous night,

Sutter had chosen an exposed hilltop for his camp. The men had been bedded down less than an hour when the wind had changed and turned into an abrasive dust storm. All through the night, the men had struggled to keep their flimsy shelters intact. Morning, so long in coming, found them red-eyed and hungry.

Alvarado and Castro had deployed their men more sensibly. Sheltered in the lee of a spur of hills, they had spent a relatively comfortable night.

As the sun's upper limb shot its first blinding rays across the snow-dusted crests of the San Gabriel Mountains, John watched Sutter, in full regalia, mount his mule, Katy, and canter awkwardly to his forward artillery positions. Four cannon were visible, including the prized polished brass field piece. The other three pieces had been taken from the fort's battlements and mounted on ox-drawn *carretas*. Lewis looked down at their own positions and saw the gunners waiting with lighted linstocks beside the two cart-mounted ten-pounders. He was certain, despite his lack of experience, that the range was too great for either side.

Alvarado was deploying the skirmishers and the main body of infantry in much the same manner he had proposed at Rancho Laguna.

Castro and Andrés Pico were maneuvering their dragoons into position for a flanking sweep down across the flatter terrain, a move that would pen Sutter's cavalry between them and the sheer, crumbly banks of the Los Angeles River.

Pico and a small detachment of men, a dozen of whom were former American trappers and plainsmen, began to work to the left on the higher ground leading into Cahuenga Pass.

Sutter studied the maneuver anxiously, then jounced back on his mule to Micheltorena's command carriage. Moments later, John saw Captain John Gantt and his riflemen taking up positions to oppose a charge. To the north, a moving cloud of dust concealed a party of horsemen who appeared to be heading for Micheltorena's ranks. When the dust cleared, Lewis recognized Isaac Graham at the head of a column of men. He estimated their number to be about twenty. Alvarado also had recognized the trapper. He sent riders to call Castro and Pico in for a tactical conference. John joined them behind the artillery.

Alvarado inspected Micheltorena's position through his

glass, then handed it to Castro. Both men were remembering that during their revolt against Governor Gutiérrez, it had been the threat of action by Graham and Chard and their men that had decided the bloodless day. Together the Americans constituted a formidable force. Andrés Pico, short, tough, wily and without fear, watched them closely for a time.

"We have the advantage. We are higher. We must keep them on the defensive with cannon shot. I suggest that one gun be laid on them," he indicated the rise, "from a higher position."

Castro agreed and Alvarado gave the order. A yoke of oxen was brought up and the ten-pounder lumbered to the top of a low ridge. There it commanded the gullies and dry washes that would have provided cover for Micheltorena's men should they attempt their own flanking sweep.

John was directing the relocation of the ammunition cart that would serve the repositioned cannon when the first shot was fired by Sutter just after sunrise. The blast shocked the preoccupied men and panicked the animals. An instant later a warning shout went up and John, peering over the ridge, could see the ball from Sutter's Russian field piece. It struck short and began a series of erratic leaps over the hard-packed earth. Alvarado's loosely formed riflemen scrambled right and left as the ball continued up the slope, skidded past the main gun position, and smashed into the left front wheel of a commissary cart. The impact splintered the solid wooden wheel and the cart dropped down on its axle. The spent ball came to rest in some brush a few feet beyond.

John heard Alvarado order the gunners to prepare for action. Fifty yards to the left he saw the men swinging the newly positioned cannon back toward Micheltorena's battery. The first of their own ten-pounders shattered the silence and the wind blew the acrid smoke back into the faces of the men. John watched the ball arc, almost lazily, and strike the ground in a sandy explosion twenty yards in front of the enemy's position. Before he could follow its unpredictable course, Sutter fired one of his own ten-pounders. The heavy charge jumped the gun off line and the ball went short and wide. Deflected by the slope, it skittered harmlessly in a great curving path and came to rest in a nearby ravine, scattering a score of jackrabbits and raising a cloud of brush birds. Castro's high cannon fired then, slewing the makeshift cart carriage in a half circle. But the gun had been well laid. The ball fell a bit short, skipped high, and killed the mount beneath one of Sutter's dragoons. John saw men rush to the fallen cavalryman but he was unable to tell whether or not the man had been injured. Just then a shout went up from Castro, who was pointing toward the positions occupied by Gantt and Graham. It appeared that Graham and his men were trying to block any possible retreat to Los Angeles through Cahuenga Pass.

John ran to the higher gun and helped swing it toward the riders, who were partially concealed in a shallow defile. Castro ordered a dangerously heavy powder charge and a number of short lengths of iron chains rammed into the muzzle. The linstock touched the fuse hole. There was a deafening crash and the chain, clustered at first, began to separate into writhing, whirling metal serpents that could cut a man in half. They rained down among the riders, who broke ranks and sought cover behind the sage and the scrub oak. Convinced now that Alvarado's well-placed guns could reach them, Graham and Gantt gathered their force and circled to the rear out of range.

Rattled, Sutter began to spend his ammunition unwisely. When he elevated the guns, the balls fell short and imbedded harmlessly in the sandy soil. When he depressed them, the balls skipped and skidded past the men, who had ample time to avoid them. Sutter fired ten shots to every one ordered by Alvarado and Castro. The fruitless exchange lasted for an hour before Sutter's firing slackened. Lewis saw John Bidwell come running to Sutter. The pair talked earnestly for a time, then the young American and a party of men rode toward the nearly dry riverbed, pulling two *carretas* with lariats. Through the glass he could see that they were searching for cobblestones small enough to load into the muzzles. Alvarado also watched, then smiled and ordered another shot from his own lower ten-pounder. He laid the gun himself, aiming it for Sutter's prized Russian field piece. The ball carried over the gun position, forcing the men to throw themselves flat, and crashed into the brush behind. Above the reverberating echoes, they could hear Sutter shouting hysterically in bad Spanish and thick, broken English. He was ordering his men to drag the cannon back to safer positions. For the first time in months,

Alvarado laughed outright. "Splendid!" He turned to John. "We have nothing to fear from them now. I will have our riflemen give your men cover. While Sutter shoots stones at us, let us gather his iron shot and prepare to return it with our compliments."

John took a detail of men to retrieve a half dozen five-pound balls and three times that many ten-pounders. Several times musket balls from Sutter's snipers whooshed dangerously close. But answering fire from Alvarado's men on higher ground kept them pinned down and hurried their aim.

The rest of the morning saw only desultory firing. John watched as Sutter deployed and redeployed his men, seemingly uncertain as to how to mount an attack. Well to the rear, Micheltorena, suffering from the intestinal ailment that had become aggravated by anxiety, remained in the carriage that had brought him from Santa Barbara. He was apparently more than willing to let his new colonel improvise the field strategy.

Shortly before midday, with one horse killed and one cart wheel smashed, both sides suspended action and prepared to eat. From time to time John would see a small puff of smoke and hear the delayed report as a sniper tried for a target at long range. Alvarado's men returned the fire but neither side scored any hits.

The Battle of Cahuenga Plains wore on through the afternoon with only sporadic small-arms fire. Knowing Sutter, Lewis was able to tell from his antic dashing about that the man's anxiety was mounting. One minute he was examining the stones that had been gathered to supplement his ammunition. The next minute he was conferring with the officers commanding his various units. Then he was at the new gun positions again, inspecting them and gesticulating officiously.

Late in the afternoon, as the sun was beginning to lower, some of Graham's men, unarmed, appeared on foot and began advancing cautiously up the shallow draw toward the left side of Alvarado's position. John and the others watched curiously and could see no purpose in the maneuver. One of the men—John thought it might be Bill Dockey—moved out in front of the others with a hand raised in an obvious gesture of truce. They could hear him shouting but could not make out the words. Then they heard an answering shout from their own ranks and two

208

unarmed men, each with a hand raised, moved out of their cover cautiously. While Castro, Alvarado, and the officers and men watched in astonishment, the three men ran toward one another. They met on the open slope, embracing, slapping one another on the back and shoulders and whooping with the uninhibited delight of plainsmen reunited with old friends.

The first exchanges had scarcely taken place when Pico came charging along the contour of the slope and pulled his bloody-mouthed mount up, rearing. He leaped from the saddle with an arm outflung in the direction of the men who were being joined now by others from both sides. There were close to a hundred of them and in their midst was Dr. John Marsh.

"These gringos are deserting! They have seen friends. They will not fight them." Before Alvarado and John realized what was happening, Pico leaped to the gun trail and tried to pull the piece around so he could bring it to bear on the exposed Americans. John seized the linstock from the startled gunner and tossed it out of reach as Castro restrained the cursing Pico. "We are men of honor. We do not shoot unarmed men."

Pico continued to struggle, protesting in a torrent of unintelligible Spanish. John Marsh, holding both hands aloft, called for silence. Leaving Alvarado and Castro to restrain Pico, Lewis ran toward the gathering. Others were still joining it. At Marsh's request the men began squatting in a semicircle. An eloquent man, the doctor was in his element now. "Boys!" We are a hundred or more here and every one of us has two important things in common. We were born Americans and we have suffered the hardships of plain and mountain to get here. That binds us together as brothers."

An exuberant chorus of approval followed.

"We have been tricked into this war with promises that will not be kept. None of these posturing generals or colonels would dare fight a battle without us. They declare the wars and bribe us to fight. And if we do fight, do you know what the end would be? We'd kill each other off and leave them this land, which may be their true purpose!"A growl of agreement rose from the crowd.

Isaac Graham, dressed in buckskins agains, climbed laboriously up the short slope and stood beside Marsh. He was still armed. "I heard ye, Doc, and ye're very learned and sensible, like always. So I ask ye, boys, why don't we just go back and shoot them? That'll put a proper end to all of the trouble."

The men guffawed and Graham, eyes glittering with malicious humor, raised a hand and turned his snaggle-toothed grin on them. "Only thing is, I'm a selfish man as ye know, but I'd admire to reserve to myself the honor of exterminating Alvarado and Castro."

A chorus of sympathetic expressions went up from all who knew of the trapper's imprisonment. Graham looked around and his eyes fell on John Lewis squatting on the periphery of the group. He pretended great delight. "Wa'l now! Damn me, if I don't pleasure myself from seeing another old friend." He pointed to John and the men craned to look. "Boys, if ye're ever in need of a smart runaway sailor who's as handy with his hands and feet as he is with his tongue, then I'd be pleased to recommend him to ye." Graham hooked a thumb in his broad hunting belt and smiled. "Course, I'd be a mite more quick to do that if the lad were a little choosier about his friends lately—"

John sensed that silence might lead to difficulties. He decided to put an end to any baiting before it could begin. He stood up, "Captain Graham, I judge that I don't want a part of this war any more than most of you. It isn't going to settle anything that'll make for lasting peace of mind for any of us. If our side wins, the same things will happen."

Dr. Marsh, who had been sizing up the temper of the men, interrupted. "Young Lewis is right, boys. I came here with Sutter's gun in my back. How many of you came here on promises of free land?"

Half the men raised their hands. Marsh looked at the others. "Why are you here?"

When no immediate response came from the Americans on Alvarado's side, John spoke up for them. "Some of us with families prefer the bickering of Alvarado and Pico to a general from Mexico City whose troops rape and steal."

Shouts of confirmation came from a number of the men who had been recruited in and around Monterey. Marsh nodded, smiling smugly. "Before I'd take promises too seriously, boys, I'd remember that Micheltorena promised Alvarado he'd send his *cholos* out of the country. Then he tried to trick Alvarado into sending his men against you in the Sacramento so he and his *cholos* could stab Alvarado in the back."

Isaac Graham hooted derisively. "Ho! That makes him an evil one all right! But I'd say he's fair matched with Alvarado, who promised Will Chard and I his undying brotherly love just a few months before he sicked that mad dog Castro on us in the middle of the night and carted us through town in chains." He spat viciously. "I'm not trusting the best of them as far as a frothing wolf."

Marsh moved a half step forward. "Isaac may speak the truth. I hold this war to be none of our business. Thomas Larkin in Monterey would agree. He is the American consul now. He thinks that very soon we'll all be American citizens again without stirring a foot from here. So why fight for these men who are soon to be put aside? Land you buy with an eye or an arm or a leg is not *free* land. Have you thought that Sutter, who is really asking you to die for him, may not have the authority to grant you land?" Marsh asked. "Micheltorena keeps his promises no better than Alvarado. And Sutter is worse than either." He stood silently searching their faces, then turned away.

The men sat musing—Graham as silently as the others. He had seen enough of Micheltorena to judge him both incompetent and too ill to deal with troops. As for Sutter, nothing he had seen of the man and nothing he had heard about him changed his conviction that Sutter was no more than "a big blow of sour wind." Sutter had been many times a failure and there were numerous stories about the way he had cheated partners in the past and failed to pay his debts.

Graham, roused from his musing, rose and rested the musket in the crook of his arm. "You boys'll be making your own judgments. I reckon it won't surprise ye none to learn that I come a helling down here because I saw a change to weigh out accounts with Bautista. But I figure now that what I'm risking for what I'd likely be getting just ain't worth the aggravation. So I'm calling on my men to light out for Branciforte with me at first light if they be of a mind to."

Several rose to leave with him. John could see Sutter standing halfway down the slope, watching them through his glass. On the opposite slope, Alvarado and Castro were also watching. Hoping to avoid having to make an imme-

211

diate decision, John started to leave but Dr. Marsh challenged him. "Tell me, John, where do you stand?"

"I want no part of this either. But I promised Alvarado I'd help see Micheltorena and his *cholos* gone. I promised my family that too. I guess I'd best say to you that I stand on my word."

Marsh smiled thinly. "That is your right, John, but I think you'll find it lonesome territory to stand on just now."

John rejoined Alvarado and the other officers and watched all but a few Americans on each side retrieve their arms and mounts and retire from the field. Sutter brandished his sword and screamed threats at them. When they ignored him, he mounted Katy and jogged stiff-legged down the slope in pursuit. In the dim light, Andrés Pico slipped away and whispered to four of his dragoons. Moving quietly, they mounted and pretended to ride to the upper gun position. Alvarado glanced at them questioningly and continued to watch the main force of his American recruits in the act of deserting. Minutes later, Pico and his men dashed down the hill in a reckless gallop, caught Sutter by surprise from the rear, and made him their prisoner. It had happened so quickly that Sutter's men, unable to believe their eyes, stood by without lifting a gun. Graham and his men had gone. Gantt and his men had joined them.

Confronted by two simultaneous disasters, General Micheltorena left the comfort of his carriage and rode to the forward positions. For an hour, with no real hope for success, he did what he could to reorganize his stunned officers and men.

Alvarado established his headquarters in a nearby adobe. Sutter, stripped of his sword and pistols, was taken there, protesting tearfully that his captors were rough and were not observant of the rules of war pertaining to the capture of distinguished officers.

Shortly after dawn on February twenty-first, Alvarado ordered the batteries to begin firing Sutter's own ammunition at Micheltorena. The blast of the first round was still reverberating in the pass as Andrés Pico and José Castro led twin flanking attacks with their companies of dragoons.

Micheltorena's men, working frantically, succeeded in loading one ten-pounder with cobblestones. The blast knocked the cannon off its mount and the stones disintegrated in a spray of small gravel and showered down on the slope no more than a hundred feet to the front. A cry went up as one of Alvarado's ten-pounders, well aimed, sent a ball whistling between two of Micheltorena's guns to crash into the wagon carrying the water jars. A second ball from the higher gun landed short, peppered the front line troops with sand and small stone, then screamed down the far side of the hillock smashing brush and cactus until it came to rest within yards of Micheltorena's carriage and field tent.

Frantic officers relayed a cease-fire to the few disorganized skirmishers who were still trying to shoot down the hard riding dragoons in the process of outflanking them.

Desperate and in severe pain from his outraged intestines, Micheltorena untied his large white silk neckerchief, secured it to a ramrod and hastily waved it aloft. For several minutes Alvarado's troops continued to fire. Then suddenly the truce flag was seen and a victorious shout went up along the battle line. A few more shots echoed across the barren plain and the bloodless Battle of Cahuenga was over.

John Lewis spent ten days in and around Los Angeles before returning to Monterey in mid-March. The evening after the emotional reunion with his family he was asked to Thomas Larkin's home to describe the happenings at Cahuenga.

Over coffee and sweets the merchants and their wives and friends listened raptly as John recited the first eyewitness account of the battle. Taking pains not to exaggerate, since the bare facts alone were strange enough to tax the credulity of the most gullible, he told his tale matter-of-factly to the moment of surrender and indulged in personal speculation only when he came to the summing up: "So, under the circumstances, I would say that Alvarado treated Micheltorena with uncommon consideration. The general was escorted to Don Abel Stearns' adobe. When he tried to surrender his sword, neither Alvarado nor Castro would take it. But our friend Sutter was quite another story." In spite of himself, John smiled. "They took his sword on the field. And his mule. Then Pico put him in manacles, the same ones, by the way, that Sutter had

213

boasted he would lock on Alvarado." The irony delighted his listeners.

"Castro was really rough with him. He trussed him up like a pig and I do believe it is the only time anybody has heard Doctor Marsh laugh. In the cart, all the way to the Los Angeles jail, Sutter kept crying, 'I am innocent! I am the pitiful victim of another man's ambition. I am the pitiful victim of my own loyalty and patriotism. I had no choice but to follow the command of my governor. Any of you who love California would have done likewise.'

"I thought Andrés Pico would slit his throat, but Alvarado, and in the end Castro too, were very tolerant.

"They seemed to pity the man more than hate him for what he revealed himself to be, an unscrupulous fool whose vanity had endangered the lives of nearly a thousand men.

"In the end, Sutter asked me to plead with them to allow him to keep his sword so he could break it in honorable ꜱurrender."

There were more involuntary exclamations and Thomas Larkin lifted a hand in protest. "Surely they did not permit it?"

Struggling now to keep a straight face, John shook his head. "No. Castro came to see him, all solicitude, and explained that Governor Pico would not permit him to break the sword over his knee for fear he might injure himself."

An explosive roar of laughter went up.

"The man was truly a pathetic sight," John said, "and if the people of Los Angeles had been his judges, they would have hanged him within the hour. They saw Sutter as the cause of all their trouble. It's not a just charge, but strangely enough they do not blame Micheltorena. Both of the Picos behaved very honorably; some will say too honorably. They have pardoned Sutter and he is now on his way back to New Helvetia under orders to disband his army and never take up arms again."

Captain Cooper threw up his hands. "By God, we deserve what we get. And we deserve what we're about to get, too. Aye. And it won't take a mystic to divine it either." He rose so abruptly his brandy glass slopped over. "Pico will have himself voted constitutional governor again, the capital will be moved south and the customhouse with it, and until Divine Providence or the United

214

States saves us from our folly, we'll be the commercial mendicants of Alta California "

John Lewis disagreed. "I do not think we will lose the customhouse, Captain."

Cooper challenged him. "Pray tell us why not, lad?"

"Because Castro has promised Alvarado that he and his men will take the Monterey customhouse and hold it. He means to do that."

The news was greeted with both relief and apprehension. William Hartnell and David Spence foresaw the possibility that Andrés Pico would march north.

Cooper disagreed. "No, gentlemen. There's an easier way. Pico can open a customhouse at San Pedro or San Diego and drain off a good half of our comerce. We're in for some trying times—but I discount civil war."

Thomas Larkin rose to indicate the evening's approaching end. He seemed to be in an unusually good mood. "So, ladies and gentlemen, it would seem that in some respects we have concluded a very satisfactory war. No lives were lost save a horse; Sutter has been chastised and shrunk to size; Micheltorena and his rabble are to be deported immediately, and," he smiled cryptically, "Alvarado will no doubt continue as collector of customs and continue to interpret impractical law in a most practical manner."

John and the merchants suspected that the light, almost gay recapitulation would be couched in more sober tones when it was reported, as it certainly would be, to the Department of State in Washington.

A letter from William Leidesdorff to John said that everything was in order for his return to Yerba Buena with his family and thanked him for the account of events at Cahuenga. Leidesdorff concluded by writing, "Now, more than ever, most thoughtful Californians agree that the historic schism between the Picos and the Alvarados will never be permanently bridged and that the United States of America, by virtue of its geographic and political disposition, is the logical successor to the beleaguered central government of Mexico. Should American warships arrive carrying the prospect of peace and stability it is conceded without exception hereabouts that they would be welcomed with huzzahs."

Since Sutter's despicable behavior at Los Angeles, few

215

people regarded him seriously even though his fort, in itself, was an impressive accomplishment. He was thought of as a charlatan with a flare for grandiose schemes. But in early June this tolerant attitude suddenly changed. Still smarting from humiliation, Sutter perpetrated a brutal act that outraged the north.

Some immigrants had settled on an Indian tribe's traditional hunting ground. In reprisal, an Indian named Raphero led a party of braves against the settlers and killed one of the immigrants and stole some livestock.

Sutter took a squad of riflemen in search of Raphero and his Indians. When they caught up with him they killed most of the war party. Raphero was taken alive and returned to the fort. There, in a grisly ceremony presided over by Sutter, Raphero's head was hacked off with an ax and impaled upon a sharpened pole outside the gate as a reminder that the days of benevolent justice in New Helvetia were at an end.

Within a fortnight, Cooper's pessimistic predictions were beginning to came true. Pico had designated Los Angeles the capital and opened a competitive customhouse at San Diego. He sent messages to the New England traders inviting them to direct their principal cargoes to that port, where he promised they would find a plentiful supply of hides and tallow. He had drawn up a new set of ordinances prohibiting the import of all intoxicating beverages for five years and threatened to put the local distilleries out of business by holding them strictly accountable for their gallonage and placing a new, heavier tax on it. He exempted only the importation of fine liqueurs for which he had a personal preference. Monterey laughed when Isaac Graham, hurt by the new restriction on alcohol, publically rued his decision not to have stayed on long enough to whip Pico's forces.

However Pico's detractors grudgingly conceded the wisdom of demanding more teachers for the inclusion of both French and English as required studies. The teachers would be paid from the public treasury and schools established in all places were there were sufficient pupils to justify one. Parents would be required by law to send their children.

Pico also issued regulations prohibiting the destruction of the forests. He was aware of a survey that attributed the devastating series of droughts and floods to indiscriminate denuding of the mountains and hills. The Yankees, who were felling Monterey's great redwoods as fast as saws and axes could swing, derided the new law. The state government had no means to enforce it, so they continued to construct even more efficient sawmills.

John Lewis prepared at last to take his family to Yerba Buena. On the eve of their departure they had dinner with Don José and Doña Gertrudis. Will Chard and his wife Dolores joined them. One of Micheltorena's last official acts had been to give adjoining grants of land in the upper Sacramento Valley to Chard and Josiah Belden. As soon as the store and boarding house in Monterey began to show a profit, both men planned to build on their land and establish their families there. Chard, who had persistently urged John to apply for land, could not resist reminding him that his last chance was fast fading. "Micheltorena would not have approved your grant. But Pico will now, if you ask for it. Certainly he must have felt some gratitude when you refused to desert with the others. Remind him of that."

"Take Will's advice, Juan," Don José agreed. "Ask for the land. We are now attending the death vigil of our beloved Alta California. This beautiful land is dying and we are responsible. Mexico has tried every form of government from absolute monarchy to dictatorship. And now, Santa Anna is in prison too. It is too late. Mexico does not know what to do with us. We are remote. We are an enigma. The signs are unmistakable. Already an American flag has flown here. Do as Will says, Juan. Apply for a grant."

Don José's angular face saddened. "Goyo was wrong about many things, but he was correct when he predicted we would be strangers in our own land. I do not know where he is now, or even if he lives. But I hope that the men he tried to warn in Mexico City will remember and credit him for that, at least.

"But we were speaking of land, Juan. You must have a grant now. Another government will not be so willing to give away its wealth. Write to Pico now."

After they left the Robles adobe, John and Maria climbed to the scene of their first meeting alone. Maria linked an arm through her husband's. She loved this strong Yankee who was her man, a mate, a lover, the father of

217

her children. "What do you think about, Juan Lewis?" she asked.

John continued to reflect a moment longer, then smiled down at her. "I was thinking of my sister in the United States and how I wish she could know you and our children."

"Once you told me, Juan, that you would send for Alicia. We'll be in Yerba Buena soon. It is no longer necessary to wait. Why not write to her now and ask her to come?"

He considered briefly. "There are two good reasons."

Maria scoffed. "I can think of none!"

Lewis put his arm around her and brought her closer. "Dear one, you California women are accustomed to all this commotion, this changing of governments with every new moon. That is one reason. The other? You and I are accustomed to our adobe houses with no proper doors and few glass windows. But Alice is twenty-three years old now. She has been raised differently, in a great city. I do not know how she would take to our primitive ways."

Maria glared at him indignantly. "You speak as though we are Indians! We are *gente de razon*." Lewis chuckled indulgently, then mocked her.

"Indeed we are the best people. The very best. We have almost as many servants as we have fleas. And we live on the cleanest dirt floors. And we walk on Spanish tile to our elegant outhouses. Also we eat food only with clean fingers after we've set it afire with the hottest red pepper that sends its fire from one end of us to the other. Even so, my precious one, I would like to be able to do still better for my sister when I bring her here. I would like to show her that I have been able to provide for my own wife and children as much comfort as any man in this country. So, we will wait. There is time. Next year perhaps."

Maria did not argue, for she had come to understand the nature of her husband's quiet pride and she loved him all the more for it. They sat until the night chill began to seep through her shawl. Then they picked their way back down the slope to Don José's adobe.

While Maria was tending to the children, John stood in the doorway gazing down the length of the Calle Principal. Twice before he had thought he would be leaving Monterey. Twice before the internal troubles that plagued the capital and the state had diverted him. He did not doubt that still more disruptions would plague him and his family in the future and he tried again to guess what they might be—from what quarter they might come and when.

That night in bed, John thought about the future of Alta California. Larkin would figure prominently in any future change. He knew there would be increased sub rosa involvement on the part of the United States. Lieutenant Charles Wilkes' assessment of California's inland waterways was said to have prompted Senator Thomas Benton of Missouri to suggest the Frémont surveys. These took place in 1842 and 1843. Frémont had led a party of United States government surveyors westward to the Columbia River country and down the length of Alta California. His reports to Washington were said to be a factor in Commodore Jones' premature decision to take Monterey. Unquestionable, the United States was moving to prevent Britain from expanding its influence along the Pacific slope. The Republic of Mexico owed the British Crown twelve million pounds in bonds. Many felt the British bankers who held the obligations would be willing to accept repayment in California land.

This brought up the question of the loyalties of the merchants. Larkin, through his personal loans to the Alta California Government, had been trying to minimize the economic influence of Cooper and Spence that might conceivably prejudice Alvarado and Castro toward Britain. Cooper and Spence both had deep Anglo roots; Hartnell would be a realist about events beyond his control. As for Vallejo, his hero was George Washington and his hobby was the United States Constitution. Sutter had forfeited his right to any influence and Castro would accept France but, like Goyo, he would die before swearing allegiance to the United States.

John did not know the men in the south well. But they were predominantly Yankee. Whatever they might decide, the balance was tipping slowly toward Washington and its doctrine of Manifest Destiny.

Weary from a long day and what seemed to be an endless, perhaps hopeless, attempt to sort out the future probabilities, John tried to push the specter of the future from his mind. At daybreak they would say their last farewells and begin the slow journey that would take Maria and the three children to their new home in Yerba Buena.

6 San Francisco, Alta California
August 31, 1847
(Until this past January 30,
Yerba Buena!)

Mr. and Mrs. Edward James
Number 3 Washington Mews,
Manhattan Island,
New York State, U.S.A.

My dear sister and brother-in-law:

It is beyond my capability to express the joy I felt upon receiving the news of your wedding last June. From your letters, Alice, I suspected that ever since Ned delivered my first letter to you an eternity ago, you two had enjoyed each other's company. I am sure now that Ned would have resigned from Bryant-Sturgis to pursue you if his firm had not transferred him from Boston to the Battery. My congratulations, Ned, on becoming head clerk.

On my table here in Mr. Leidesdorff's new City Hotel stands my most prized possession, the Daguerreotype made at your wedding at the Gladwyns' home in Brooklyn. What a beautiful bride my old shipmate Ned has! Ten thousand blessings on you both. I can only wish for you the deep happiness Maria and I take in each other and in our offspring.

You say that the news journals there are filled with accounts of the varying fortunes of the Mexican War but that little of California finds its way into print. The same is true here since our two journals, *The Star* and *The Californian* (the latter first published as a weekly at Monterey), are scarcely a year old and consist of only four pages. Two of the *Californian*'s pages are printed in Spanish. Both journals are devoted to small kernels of late news that have been winnowed from the embellished adventures of overland immigrants plus some editorial opinion of dubious value.

In answer to your request for personal news I'll try to tell you all that has happened since I left Monterey and started with Mr. William A. Leidesdorff.

I could not have chosen a more unsettled time, for California itself has now embarked on a new course. As a result, I find myself being swept along in its wake.

Previously I told you about Captain Frémont and his company of men who set out to map the western portion of the continent. It now appears that the Captain (who rose to Lt. Colonel in a matter of months) was really mapping plans for the conquest of California. I believe his personal ambitions overburdened his judgment because last week we heard that he is being returned to Washington under arrest for refusing to obey the orders that would have deposed him as governor of California. He had aspired to that post with great passion.

Personally I would not regret his conviction for a legend has been growing, as a result of Frémont's own self-advertisements, that he and his crew of mountain men are superior beings. History will be the judge of that but if history judges Frémont's right bower, Kit Carson, other than a cold-blooded murderer then it will be in flagrant disregard of the facts.

Without elaborating, just let me say that during the infamous Bear Flag Revolt, Carson and two of his men shot down in cold blood the adolescent twin sons of Francisco de Haro, former Alcalde of Yerba Buena and one of the most respected men in California, on the pretext that they were carrying secret messages to Vallejo from Castro. The twins' uncle, Don José Berryessa, was also killed as he stepped from a small bay boat with his nephews. None of the three was armed. It was purely and simply an act of vengeance, the viler still because Don José was shot through the heart as he was pleading for the life of his nephews, offering proof of their innocence. If there is such a thing as Judgment Day, this man Carson may expect no more mercy than he has shown.

Maria, with Felicia and the twins, are still with her ailing father for whom I have the greatest affection. I have managed several visits with them to satisfy myself of their safety and comfort.

Late in January, shortly after Will Chard (now the fa-

221

ther of a fine boy), left to investigate the Santa Clara quicksilver mines, this fellow Frémont appeared in Monterey with eight of the roughest plainsmen imaginable, including the infamous Kit Carson. Frémont's declared purpose was to petition General Castro for permission to rest his men in Alta California before undertaking the return journey to St. Louis.

Castro, who is capable of passion which occasionally verges on valor, met him under the most correct auspices and granted him permission. But hardly had the meeting ended when word came that several of Frémont's ruffians had made vulgar overtures to two native girls who fled in terror.

Moreover, a *californio* contended that two of the mounts ridden by Frémont's men bore the brands of his ranch and had been stolen. Castro naturally tried to investigate but Frémont arrogantly dismissed him. Whereupon Castro summarily ordered Frémont and his men from the department.

The captain, who by this time had augmented his company to sixty-odd, marched his men from Monterey to Gabilan Peak, that overlooks the mission at San Juan Bautista. In a fury, he threw up a breastwork of logs, raised the American flag (the second time this has occurred in Alta California), and dared Castro to dislodge him.

Castro marshaled several hundred heavily armed soldiers and the chastened Frémont, acting something less than a hero, slipped away in the night and beat a hasty retreat to New Helvetia.

Sutter, ever the complete opportunist, greeted him warmly, repaired his outfit, and sent him on his way to Oregon from whence the party planned to return to the United States.

Then, unexpectedly, a mysterious stranger entered the scene. This was Archibald Gillespie, a Lieutenant in the United States Marine Corps, traveling incognito, who was carrying dispatches for Frémont from his father-in-law and from President Polk and Secretary of State Buchanan. Immediately after Gillespie intercepted Frémont's party, Frémont turned right around, marched to New Helvetia, and commandeered Sutter's entire establishment.

Sutter protested and was threatened with confinement in his own fort. Frémont has no respect for Sutter since the fiasco at Cahuenga although it appears to many that

Frémont himself was wanting in both strategy and valor at Gabilan.

Our impression is that Frémont is an ambitious and arrogant adventurer, fortunate to be the favored son-in-law of Senator Thomas Hart Benton, who has expedited his career. However, Frémont's impudence and mutinous insubordination have seriously aroused General Kearny and Senator Benton may have to salvage his son-in-law's career. Kearny is demanding his court martial.

In spite of Frémont's bickering over power and position, California is secured for the United States under the military governorship of Colonel Mason and is now, for practical purposes, one of its territories.

Frémont, who appears to have some instincts in common with Sutter, may, like that remarkable Swiss, not only survive but thrive. Both seem to have a rare gift of survival—of turning adversity to advantage.

I visited New Helvetia several weeks ago. Sutter seems to have totally forgotten about his inglorious defeat at Cahuenga and still insists upon the use of the title "Captain." He is playing the role of benevolent Lord of New Helvetia to the hilt, demanding near papal fealty from all who depend upon him. *"Qui se laudari gaudet verbis subdolis, sera dat poenas turpes paenitentia,"* was not spoken of Johann Augustus Sutter.

In spite of everything, Sutter is a most extraordinary fellow whose enterprise has wrought a miracle. There is a veritable city at the once primitive New Helvetia now, and another city is growing at nearby Sutterville. I counted nearly fifty structures and six mills, two distilleries, two heavy forges, and one blanket mill. There are several commercial shops and a large tannery. Also, Captain Sutter has an agreement with a New Jersey man named James Marshall. He is under contract to build a sawmill on the American River at Culuma, an Indian settlement about 40 miles away from New Helvetia.

Since most lumber is whipsawn by hand in the Bodega forest and brought by schooner to the Bay of San Francisco and then up the river, it is exceedingly hard to come by and very expensive. Sutter plans to raft the cut lumber from the mill most of the way to New Helvetia, then ship it by water down river to San Francisco. There is a growing demand for it here. Our interest lies in the possibility of acting as agent for the resale. Mr. Leidesdorff has

223

a spacious warehouse on the waterfront where a substantial quantity of prime material can be dry-stored. But to return to Sutter, taken with 12,000 head of cattle, 200 horses and mules, 15,000 sheep and 1000 hogs, and several hundreds of workers, he must be a wealthy man even though I know from our own account books that he owes substantial amounts on which little if anything has been paid in recent years.

He is also said to be badly defaulted on his debts to the Russians. However, a year or two of clement weather and political tranquility may well see his great dream fulfilled. Since in this land one man's victory is every man's hope, we can do naught but wish the captain well.

A tremendous transformation is taking place in San Francisco. The mere change of name seems somehow to have exorcised the spirit of lethargy from the entire community. We know, of course, that the prospect of orderly government is the true cause. There is more construction and bustle here than anywhere else in California. Some speculators are paying as much as $100 cash for lots in favored positions along the waterfront. Mr. Leidesdorff owns several well-situated plots including some on the lane that bears his name. His residence stands on one of these.

As I gaze down the hill to the Bay I can see the *U.S.S. Portsmouth* lying at anchor surrounded by a fleet of naval transports and several sloops of war. Indeed, Commodore Stockton seems to prefer this bay to all other harbors in the Pacific. He has said repeatedly that he will urge Washington to establish a major military installation here with a naval yard to rival the one at Boston. It is only rumor but it sends a fever of excitement burning through all those whose faith in Yerba Buena has been constant during the long, lean years.

I see also three deep-water merchantmen at anchor. Two are Boston ships and we hear on good authority that many more have San Francisco as their destination.

And now I hope that one day before many more months pass I shall be standing with Maria and the children watching the arrival of the vessel carrying you both here. How appropriate that you both, whose love so brightens my heart, should arrive through the Golden Gate, the entrance to the great bay quite aptly named by the controversial Frémont.

(some hours later)

Maria and Felicia are now at Santa Cruz with Dolores Chard, attending the latter's mother whose days seem to be nearing an end. The twins, William and Steven, remained at Monterey with their grandfather, Don José, who is doing his best to make *californios* of them and is succeeding. His *vaqueros* carry the 3-year-old boys in front of them in the saddle.

I must go now to visit a merchantman with Mr. Leidesdorff. More and more he depends upon me to bid in the cargoes that we need for resale here in the north.

With enduring affection, I am,
Your devoted brother and "in-law"
John

San Francisco, Alta California
February 14, 1848

Señora Maria Robles Lewis
El Rancho de los Robles
Monterey, Alta California.

My dearest wife and family:

Harry Morrison is returning to Monterey on business for our firm so I am writing again rather sooner than I had intended.

I regret to say that Mr. Leidesdorff has been suffering from an indisposition which the doctors diagnose as "an imbalance in the system" caused by a fall from one of his racing horses on the flats near the mission several weeks ago. Unfortunately, he struck his head sharply on a corral railing in an effort to save himself.

Both he and the doctors are predicting a full recovery soon. In the meantime a great deal of additional responsibility has devolved upon me. I welcome it both for the opportunity to learn by administering the more complicated aspects of international trade and for the additional salary. He is surely one of the most considerate and gracious of men and one must forciably remind one's self that he is of Negro extraction. My admiration for him and my feelings of fellowship are unparalleled. His strength of character commands the respect of all men and his unusually attractive personality have set many a feminine heart aflutter.

I happened to be at Rancho Petaluma with General Vallejo when a rider brought a startling letter to the General from Captain Sutter. Sutter claims to have discovered an extraordinarily rich gold mine near his fort. That should be the best of news, particularly to a long-time and unsatisfied creditor; but General Vallejo fell into a thoughtful silence. When he spoke he just kept shaking his head and muttering, *"¡No es bueno. No es bueno!"* He refused to discuss the news beyond saying that gold has been the downfall of nations of men and that in a fertile land such as ours the only true wealth is in the crops and in the herds. He was really much disconcerted.

Your last letter was most reassuring, my dearest wife. Your account of the children's progress is gratifying. I am pleased that Will's family is enjoying his presence again and that he will not start to drive his livestock north until the spring. The failure of his store and boarding house was predictable but it will yet prove to be a blessing both for him and for Josiah Belden.

Harry is impatient to be on his way so I shall seal this now, knowing you understand that it contains not just a modicum of news but the full measure of my affection.

Your devoted husband,
J.L.

San Francisco, Alta California
June 8, 1848

Mr. and Mrs. Edward James,
Number three Washington Mews
Manhattan Island, New York State,
U.S.A.

My dear Alice and Ned:

I suppose one should not reckon it news that in a topsy-turvy world a man is apt to find himself upset and unsettled. Scarcely a month ago disaster struck our city. The Mormon apostate, Sam Brannan, fired with avarice (and other raw spirits), rode through the streets displaying a bag of nuggets and shouting like a banshee that gold had been found on the American River. In a matter of hours he had emptied the town like some infernal Pied Piper. Every able-bodied man, including several preachers of the Gospel (who apologized for their desertion by saying they were going to gather golden riches to erect still greater cathedrals to the glory of God), has departed this town leaving only the timid, the skeptics, the infirm, and a handful of others such as myself who are bound to their posts by conscience.

And now I have a personal tragedy to report. My benefactor, William Leidesdorff, is dead. He died on May 18 after having lain in a coma for several days. The doctors have said the cause of death was "brain fever." He was 38 years of age, in the prime of his life, on the threshold of accomplishments that would have enriched this city beyond measure commercially and culturally.

Some measure of the respect in which he was held during the brief time he lived here may be judged from the fact that his mortal remains have been interred in the actual Mission itself, an honor accorded only to the highest churchmen and the most revered California Catholic families.

Mr. Leidesdorff spoke little of his own early life, but papers discovered by the executors indicate that he was actually the ward and heir of the man he referred to as his uncle. The executors are presently trying to establish the legal heirs, if any, since Mr. Leidesdorff, who could have indulged his fancy among the most beautiful women of the region, charmed all but apparently favored none. Mystery shrouded him in life, and the shroud is gathered still closer in death. It is rumored that he left his one true love in New Orleans when her father objected to the union with a mulatto. A delicately made lady's gold and jeweled crucifix was found among his personal effects and it was apparently intended as a gift, for it was still in the original vendor's silk-lined box. Enclosed with it was a card in Mr. Leidesdorff's fine script on which were inscribed the words, "To my beloved Hortense." There were also several extremely tender notes from the lady.

I sorely miss Mr. Leidesdorff and my plans once again are in a state of confusion. Sometimes I despair of ever finding my proper place in this land.

Several days after the uproar about the gold, Mr. Leidesdorff ordered me to get his schooner, *The Rainbow*, ready for a trading trip to Sutter's embarcadero. Sam Brannon reported that Sutter's store was stripped of everything of use by the men rushing to the mines. Judging by those who left here, most have undertaken a dangerous journey ill equipped. Indeed, most lack even elementary knowledge of mining.

If it's true that unlimited quantities of pure gold may

227

actually be garnered from streambeds with one's bare hands, then I may have the opportunity to establish my own trading firm. It would be possible for me to pay full market price here for supplies and still turn a handsome profit by transporting the material to or near the mines. One man is said to have been paid $100 a barrel to pack in two barrels of flour and one of salt meat.

Another received $4.00 a pound for flour, the same price for poor quality sugar and the cheapest chicory and coffee. Sardines sell for $5.00 the tin and a hundred-pound wheel of New York common yellow cheese sold for $500! Shovels, picks, and rakes and knives are in great demand as are pans of all descriptions, particularly frying pans and shallow winnowing pans with which the particles of gold apparently are separated from the lighter overburden by washing in some fashion.

I am singularly fortunate in that the executors have allowed me to pursue this trip on the same generous terms agreed to by Mr. Leidesdorff. Also, I am doubly blessed by having my old friend Harry Morrison, a most competent Scotsman, as my companion and assistant.

We leave for the Sacramento and American Rivers tomorrow morning. I shall try to keep in communication with you during these coming weeks but long silence is not to be construed as a mute harbinger of trouble since it may well be impossible to find a dependable person who is journeying back to civilization.

God grant us all safety and may He bless you both and grant me my most cherished wish, to greet you and embrace you here as,

Your affectionate brother and in-law
John.

San Francisco,
Alta California
February 10, 1849

Mr. Edward James,
Bryant-Sturgis & Company,
The Battery,
Manhattan Island,
New York.

Dear Ned:
We have just received word of President Polk's confir-

mation of the gold strike in his Congressional Message. The reports reaching us from the mining regions east of Sacramento are so extravagant that I would not have believed them if I hadn't witnessed the spectacle with my own eyes.

I have seen an entire land in turmoil. I have seen the gentlest of men turn monstrous with greed and the strongest of men weep with disappointment. Prudent men gamble away a fortune on a turn of the cards with the thought that another fortune might be plucked from the earth on the morrow. And on all sides men vie with one another to hand over their pokes of gold dust and nuggets for the simplest necessities of life—an ounce of pure gold worth $16.00 in exchange for a 25¢ washing pan; $20.00 for a simple physic; Laudanum at $1.00 the drop and pills for any ailment man is heir to (and some not yet discovered) at $1.00 each. Potatoes and apples, when available, bring $2.00 to $5.00 *each*. A poor shovel brings two ounces of gold. A pick the same. Any knife or usable piece of metal for seaming or sniping brings an ounce of gold or more. I saw an Indian exchange a half pound of gold for a half pound of raisins and if he had not done so an equally savage appearing "white" would have willingly done the same.

All civilized instincts seem to have been stifled except one. Curiously enough, there is almost no dishonesty among the miners, it being as easy to glean a fortune from the streambeds as to steal it. I have seen $3,000 in gold nuggets left unattended in a miner's lean-to with never a qualm for thieves. Precious tools left overnight in the claims are unmolested, as are the claims themselves.

At the end of last year it was estimated that ten thousand men were digging and had recovered $10,000,000 in gold. But the price exacted by nature for this bounty is dreadful to contemplate. Men suffocate of pneumonic lung congestion with no one to care or help. Others die raving of ague and burning fever, with malnutrition the cause, abetted by lack of shelter and hours upon hours soaked to the waist in icy water in their quest for a "strike."

The proven diggings now extend well north and eastward along both the Feather and American Rivers and along a number of lesser streams. The same is true to the south of Culuma where the first strike was made. Some miners have worked well up into the Sierra and found ex-

229

ceedingly rich pockets in the alluvial washings. One company of men mining have regularly taken up to $1,500 per day in the finest specimens of coarse gold. All this I have witnessed with my own eyes. And it is upon this evidence that I wish to make the following proposal:

I enclose a guarantee of credit for $5,000 signed by Messrs. Spear and Hinckley. I should like you to purchase and forward by the fastest vessel the appended list of tools, iron rod, and strap. If you cannot find copper washing pans resembling the slope-sided one indicated in my rather poor sketch, then I shall take any approximation of same in any durable metal whatever. Should some of the listed items be found in short supply there, make up an equal dollar value in the more plentiful items requested.

I should like you to act as my purchasing agent on a full and equal partnership basis. At present the gold seems inexhaustible. The mines will surely produce for another five years or more. I have talked with General Vallejo and also with Dr. John Marsh, who made $40,000 in trade and mining before ill-health forced him from the diggings. Both men agree that the wasteful methods presently employed will leave fortunes for those willing to mine more scientifically. But the number of farmers-turned-miners who have become discouraged and turned to the land hereabouts will increase as will those who made their stakes in order to purchase choice valley land. This will greatly speed up the process of settlement.

In this event, the firm of Lewis & James (it has a true ring to it) could well adapt its imports to farm implements and other vital necessities.

Poor Sutter is quite overwhelmed by events. He made a fortune as a trader and supplier but the first gold strike emptied his factories and fields of all help. In the end he himself was forced to try the mines. But his mining expeditions ended in defeat and his properties have now been signed over to his son, John Augustus Jr., to keep all that remains from being foreclosed by claimants. This occasions some embarrassment for me since Mr. Leidesdorff acted as collection agent for the Russian transactions on which, like all others, Sutter was in arrears. His trustees are unforgiving and somehow the captain—he seems to have aged greatly and has quite fallen apart from ill-fortune and brandy—connects me with his misfortune though heaven knows he found a far abler clerk and general factotum in the conscientious Bidwell than ever was to be found in the restless Lewis.

Sam Brannan has established a mercantile store and so has Thomas Larkin, who admits quite freely that his profits from supplying miners along the Feather River exceed 300%. (Knowing Mr. Larkin, I suspect him of undue conservatism in this connection also.) Bidwell has opened a store and somehow manages to keep a few men working at the fort's forges making improvised mining tools. Only the distillery continues unabated.

The warehouses here are stripped of merchandise since everyone who could garner a dollar or wheedle credit marched off to the mines with items to exchange for gold dust. Transient opportunists arrive each day and anyone who sets up his trade in a business-like manner will thrive.

I would not presume to tell you how to come to a decision on my proposal and I freely confess the underlying selfishness that motivates me for I see in this venture an opportunity to bring you and Alice, and my as yet unseen niece, to California, here to make a permanent home.

A man named Peter H. Burnett, lately a territorial judge in the Oregon country, has come to the new Sacramento City to supervise its sale and development for John Sutter Junior. Since the announcement on December 16th last, the lots have been selling like johnny cakes at $500 each.

Commercial situations along the river bank are going for well over $1,000. I have taken the precaution of securing a splendid lot directly opposite the embarcadero and adjacent to the new hotel for the firm of John Lewis and Company in the event Lewis and James is unable to use it. I have also placed options on a number of other sites well back from the water since my experience at Sutter's Fort has made me respectful of these rivers during the spring.

Again, poor Sutter. His plans for a city around the fort have come to naught and Sutterville lots are selling at half the Sacramento City price. Sacramento is becoming a hub of commercial activity in the central interior.

I shall try to curb my impatience and shall pray for a favorable reaction to my proposal. If you decide not to gamble in this uncertain land I will understand and harbor none but the most sympathetic attitudes along with my own keen disappointment.

Tomorrow I return to the mines with a schooner filled

231

with the most varied merchandise procurable. I shall pack merchandise into the lower hills where most of the mining is now concentrated due to snow above the four-thousand-foot level. It is bitterly cold, even in the lower reaches. I far prefer gold from merchandise than from marrow-freezing labor in the icy streams.

We send expressions of deepest affection to you; and how Maria wishes to embrace the distant un-met in-laws for whom she has quite as much affection as I.

Ever thine,
John Lewis

San Francisco, State of California
October 29, 1850

Mr. and Mrs. Edward James
13 Monroe Street
Brooklyn, New York
My dear Ned and Alice:

I hope I do not risk immodesty by reporting that the partnership of Lewis and James now holds in deposit another 1000 fine ounces of gold represented by 16,000 in certificates. These comprise the net proceeds of Harry Morrison's recent trip to the mines. In the appended list I have requisitioned supplies that will seem somewhat different, occasioned by a change in mining methods. The "easy pickin's" of the past seem to be exhausted. For months now fortune seekers have filled and emptied this city every twenty-four hours. No sooner are they off the ships that now clog Yerba Buena Cove than they find transport to Sacramento City. The same streams now support five times the number of prospectors, but the increase in profitable strikes is not proportionate. Many now make their way 200 miles north to the Trinity River country where new alluvial deposits are being found. But in the old diggings it is now necessary to search every nook and cranny that might conceal a nugget. These miners, now called "skimmers" and "snipers," still require the conventional picks, shovels, etc. But men are now engaged in a new form of burrowing called "coyoting" after the dens dug by that animal. It is still another technique introduced by the industrious Sonoran Miners from Mexico from whom we have learned most of our tricks.

It is extremely dangerous work but often results in high

yields. It is for this hazardous tunneling along the bedrock that the men need the smaller tools I have specified since most of these men live (and too often die) on their hands and knees in dark, wet passages seldom more than three feet in diameter. We anticipate no difficulty in disposing of all you can supply at from 1 to 2 ounces of gold per shovel or pick.

Harry Morrison made an arrangement with a blacksmith to make sniping irons from scrap metal scavanged from the abandoned ships that rot in this harbor. Even these primitive tools sell instantly for ½ oz. each. They cost us no more than $1.00 to make and transport, giving us a net profit of $7.00. Harry disposed of 500 in two days at Mormon Island. We now have three blacksmiths engaged full time manufacturing them for us. The "sniping irons" are used for seaming—scratching in the deep cracks in bedrock along streambeds where some of the finest nuggets are often trapped.

In July, I tried my hand at sniping on the Feather River. I located some likely looking seams and gouged deeply into the cracks with the spatula end of the iron. In a matter of minutes I had scraped out several handfuls of gritty clay which I then deposited in my washing pan. I kneaded the mass until the clots were broken up.

Even before the final washing I was able to pick out several beautiful nuggets the size of my fingernail. When I finished (and ineptitude doubtless cost me many small particles), I had over a half ounce of beautiful gold. I confess that the ease of obtaining it gave me a touch of "the fever." But a few minutes of sober reflection dispelled that.

That gold, by the way, will shortly reach you in the form of two rings, one for each, which your local goldsmith can adapt for correct size.

The spring floods that inundated Sacramento City earlier this year did little damage to our supplies stored there. Thanks to my experience with Captain Sutter, I took the precaution of making a heavy dry-storage loft in the back of the building a good ten feet above the floor. I stored all articles that might be damaged by moisture there.

I must say that while others were wading chest deep in the muddy swirl trying to rescue floating cases and barrels I felt inclined to smugness. The envious looks of some who earlier had scoffed at my precautions more than made up for the discomfort suffered from their derision.

Despite the annual inundation, optimistic merchants continue to repeat the old mistakes. Sacramento City may not have any secure future without a series of enormous levees or dikes to contain the rivers. Even old Sutter, now living at his new Hock Farm on the Feather River, even he, often called a "fool," had sense enough to heed the advice of the Indians when he chose the site for his fort.

Fire and flood! Sometimes it appears that Almighty God is displeased with our stewardship of this land. Since early May of this year five great fires have consumed over 500 houses and countless flimsier structures in San Francisco. This is despite valiant efforts of the new fire brigades and elaborate precautions that require each dwelling to have on hand at least five water buckets filled at all times. (We have profited by supplying these too.)

I do not think it an exaggeration to count the fire losses at close to $10,000,000 during the past six months alone. While carelessness (usually drunkenness) caused one or more, others may have been set by incendiarists bearing malice toward some gambling hall or saloon.

A great fire that destroyed over 100 structures was still smoldering when we turned out for the official Admission Celebration on October 29th. If these fires continue to destroy our flimsy structures, we may find that there is more true wealth in bricks of clay than in bricks of gold. A ship arrived here last month with 50,000 Belgian paving blocks as ballast. They were sold for $1.00 each. Such stones are hard to plumb and mortar but they won't burn.

You, of course, know all the details of California's admission to the Union, at long last, and about the seating of John Frémont and William Gwin as our first senators. We did not think statehood would come this soon because of Southern opposition to the anti-slavery provision in our new state constitution.

There is much dissension here still. Many from the slave states are sympathetic to that obnoxious system. Senator Gwin himself quietly favored slavery for California; but since political ambition overbalances his Southern conviction he "ran before the wind" most agreeably. Even so, the Southern sympathizers will leave their mark on California.

We have seen the passing of an era, and while my cheers were loud when the *S.S. Oregon,* all dressed in pennants, sailed into the bay on the 18th inst. carrying the

234

glad tidings, I could not help shedding an inward tear for General Vallejo and the 6 other *Californios* who sat with 41 Anglo-Americans at last year's constitutional convention. They were forced to expedite their own extinction. Vallejo's vast holdings have suffered the severest depredations and still he was the first to urge that California's constitution contain nothing inimical to the United States Constitution that he holds so sacred.

And Juan Alvarado. As I sit here in this thriving city in which so much has been accomplished in so short a time, those days ten years ago seem to belong to eons past. Alvarado the idealist was not proposed for the convention because (the kindlier said) he was intractable. The truth is he was ill from disappointment and from the spirits he took to assuage it. In recent years his uncle Vallejo told me he had persuaded Alvarado of the hopelessness of unifying an independent California and the inevitability of union with the United States. He would have added great insight to our deliberations. History must not be allowed to rob him of his place or give more to those who, through selfishness or short-sightedness, would not support him. The bromide "while Alvarado and Pico fought over the pail, the Yankees stole the cow" is a cruel jest.

Perhaps California is destined always to be split asunder, for even now, only 11 days a confirmed state, there is a group of separationists led by Abel Stearns who would petition Congress to divide us into Upper California with the capital to remain at San Jose, and Central California with the capital at Los Angeles.

The Angelenos deny it, but they are jealous of our gold, our harbor, our energetic population and of our leadership in social and political affairs. They are divided from us by both the Tehachapi and the Santa Barbara Mountains. Perhaps they should go their own way, for surely they provide us with little but distraction.

As you know, a company of Mormons sailed here from N.Y. with Sam Brannan, now a competitor merchant. Brannan, whose capacity for hard liquor is exceeded only by his rapacity, has been cut off from the tithes he had been collecting from the faithful. The money, long since diverted to his personal use, should have been forwarded to the Deseret Colony at the Great Salt Lake to sustain the Mormons there. The story goes that when Brigham Young said that it was God's money and demanded that it

235

be sent forthwith, Brannan replied, "I shall forward it in-
stanter upon receipt of a draft signed by God."

It is Brannan, by the way, who is now attempting to organize here a vigilance committee to insure law and order.
But the lay Mormon is a devout, skilled, hard-working,
and honest man and there are those who regret the departure of many of the original Brannan group to their settlement in Utah Territory. At the worst, one may say that
these Mormons are a mixed blessing.

As you know, Peter H. Burnett was selected as the first
civil Governor of California. Earlier, Burnett assumed the
responsibility for selling Sacramento City lots and in so
doing established a personal fortune solely upon his ability
to spellbind. He is a persuasive man, indeed. It was he
who sold me the embarcadero lot and others on J Street.
At $750 I overpaid by a third. But I did haggle him down
from $1,000.

I do not impugn him. His commission was only 25% of
the purchase price. (We customarily make far more than
that.) But most of the city lots sold by Burnett will be
worthless when the gold fever abates, for surely
Sacramento City must go the way of Sutterville. The miners are now reduced to scarcely an ounce of placer gold
for a day's digging and inevitably they will return to their
homes in the East or drift back to San Francisco. Many
are skilled in the crafts, arts, and sciences and have the
potential to enrich the California economy and culture
once the powder-flare of greed and adventure has subsided.

Though we build them up and burn them down as fast,
we are hopeful that theaters, hotels, schools, hospitals, a
library, and an art museum will soon outnumber the
saloons and bordellos that thus far have provided the principal diversions for our restless society.

What a commingling we have here. Every race, religion, vice and virtue. But still, through the strange al-
chemy, these disparate elements are slowly harmonizing.
Unfortunately, the unusual trust displayed at the mines has
given way to general crime and violence. This, paradox-
ically enough, may prove to be the making of our society.
I see in the conflict around me a principle, that extremism
in any form carries within it the seeds of its own destruction. Even an engaging rogue like Brannan calls upon the
"decent element" of San Francisco to establish an orderly

236

and safe society. Crimes of violence lately have rendered our streets a jungle. We need more virtuous women, more families, more teachers and preachers to demonstrate, by example, the blessings of the good Christian life.

One bit of news that I pray for Maria's sake and her father's is only erroneous assumption: you remember that my brother-in-law Goyo disappeared shortly before the so-called Battle of Cahuenga. Word has reached us that one of the bandit leaders harassing settlers from San Fernando to San Diego may be Goyo Robles. He (the man reputed to be Goyo) and his men are known as *Los Lobos,* the wolves. A vigilance committee riding after them lost one of its men to an ambush. An unseen assailant snared him around the neck with a *reata* and strangled him before he could utter a sound.

The only rancho victims of *Los Lobos* are *californios* who have lately sold to Yankees. Their cattle is taken, women and children are terrified, and loyal Mexicans and Indians are often murdered. There are a number of such outlaw bands operating, mostly in the south below the Tehachapis. Many raid and plunder indiscriminately, choosing as their victims both native and Yankee alike.

We have highwaymen here also but they seem to operate singly for the most part and attack messengers and the few stagecoaches that supply transport to mining areas.

This is the result of the great abuse that obtains against the Mexican miners. With the end of easy diggings, gangs of Yankees are falling upon peaceful and industrious professional Mexican miners and driving them off their rightful claims. Several hundreds have been illegally evicted in the past month. They have gone south of the Tehachapis and are joining outlaw bands to wreak vengeance on all Yankees wherever they find them. It is a tragic situation.

Knowing of the bitterness of my brother-in-law against all "foreigners," I fear there could be more fact than rumor in this news. We have taken the greatest possible precautions not to let it reach Don José in Monterey, for lately he has been in poor health.

He entertained such lofty aspirations for Goyo. But his son has remained a hotspur, a victim of irresponsible patriotism that he uses as an excuse to avoid the responsibility of managing his father's holdings. I can only pray that the apparent similarity between my brother-in-law and the leader of Los Lobos is merely cruel coincidence.

Maria and the children question me impatiently with each letter, so eager are they for word that you will at least visit us in California. Their concern helps to bind us all with love and affection in which I pledge I shall never be found wanting.

John

Book II

The Steven Lewis Story

7 CALIFORNIA WAS CHANGING. Its principle wealth was no longer founded on its mining claims, even though millions were occupied in the hard rock and hydraulic works and in new dredging operations that had superseded the ragged companies of men with picks and pans. The drifters and snipers of fifteen years earlier were now dead of privation or dying in the Comstock's fetid, silver-lined labyrinth. Some were freezing beside a remote stream in Colorado's new strikes.

In the beginning, California's wealth had been its land. Then had come furs, hides, and gold. In that era the merchants had prospered. Now the Central Pacific Railroad was lining its pockets with federal money. Huntington, Hopkins, Stanford, and Crocker had forced Theodore Judah, the scrupulously honest engineer whose concept the railroad had been, out of the corporation. Crocker, a ham-fisted, bull-necked slave driver, mercilessly pushed his huge army of Chinese coolies. It was true that Crocker and his coolies were chipping away at the granite escarpment of the Sierra but most men claimed this was merely a gesture. "The Big Four" were milking millions in taxpayer dollars by awarding the building contract to their own construction company. Both John Lewis and Billy Ralston doubted that Congress would continue to pay forty-eight thousand dollars a mountain mile for a road that progressed eastward less than a dozen miles a year after the first easygoing valley miles had been completed.

Steven Lewis was convinced that the transcontinental railroad was a necessity at any price. He was sure it would absorb or drive out the California Central, push northward toward Oregon and southward down the San Joaquin toward Los Angeles and San Diego. He remembered the persuasive argument by an old river captain who had seen it happen in the East.

"By the time you're as old as your father, Steve, there

241

won't be a freight wagon left, and there'll be damned few river boats. This ain't a frontier any more. This here's a civilized place with hotels and art museums and opera theaters and saloons and Turkish baths and gambling halls and fancy whorehouses with velvet settees. My God, they even got staterooms on the *Queen City.* You can sleep all night for five bucks and wake up in Sacramento. There's even a floating barber shop aboard. If you're looking for frontier in Californy, son, go to Los Angeles. They ain't come alive yet down there. But those cow counties are going to pretty quick now. There are thousands of Confederate folks that lost everything in the war heading west across Texas and New Mexico by now."

Steven had heard the same prediction the year before, the second year of the present drought. Most of the huge ranchos whose titles had finally been confirmed by the California Land Commission were delinquent with tax payments or mortgaged at usurious interest rates. Pasture land was being offered at ten cents an acre in the San Gabriel Valley. There were few takers. The best cattle were being sold for one dollar a head. Thousands of range cattle were sold for half that price and those not already dead were slaughtered on the parched ranges for their hides. Steven remembered Governor Low's recent wry observation: "The Almighty is punishing the *rancheros* for their indolence and for the willful squandering of His riches. The southern counties will never amount to a hill of beans until good small farmers come in who know what to do with land."

If the rumors were true, Steven thought, then before long the southern part of the state would come into its own. Most Southerners from the defeated Confederacy were, above all else, good small farmers. In southern California they could find modest parcels. Unlike his father and twin brother, Steven Lewis wanted land.

In his father's office, Steven broached the matter that had occupied most of his waking hours for nearly a year. John Lewis smiled across the polished expanse of walnut table. "Now, Steve, what's on your mind? Land again?"

"Yes."

"Where now?"

"In the south. Around Los Angeles."

John regarded his son unhappily. "Still Los Angeles? I thought I had convinced you that John Lewis & Sons has no opportunity there. It's too late. Phineas Banning's trying to do there what we're already doing here, and he has to build his own harbor to do it. I know that south country, son. Nothing will ever come of it until a railroad connects it to us, and that's at least fifty years away, in my estimation."

"A lot of people don't believe that, father. When Crocker blasts through the Donner Pass summit, it will go fast. Once Central Pacific rams through those mountains, freight and passengers will start moving overland at twenty miles an hour instead of ten knots at sea. We can steal land in the south now, and hold it for taxes."

John spread his hands. "To what purpose?"

"I have no definite purpose now, father. I want to go to Los Angeles and San Diego to look around. If the company does not want to invest in land there, I should like to borrow money from it, at the usual interest, and take up a league or two on my own account."

John was grim now. "Only eight thousand people in Los Angeles. Only five thousand in Santa Barbara. Half that number in San Diego with the second best harbor on the Pacific Coast. Arid mountains with little timber. Arid plains with fewer streams. No minerals of any account and some stinking black ooze called brea that the Mexicans smear on their roofs. A history of droughts. Right now, this minute, thousands of cattle are dying. Two *years* with no rain."

John bent forward, almost pleading. "Can't you see, Steve? Man must not only remake the land down there, he must control the climate as well. Nothing of any consequence grows without rainfall. The south is a desert. I've known it for twenty years. It has not changed, except in small patches where a few men have scratched out colonies."

There was a suggestion of tried patience in his tone as he continued. "No Steve. I'll give you *ten thousand* dollars to buy land in the Sacramento Valley, or anywhere else in the state, except south of the Tehachapis. You'll run afoul the same problems there that Will Chard and I had here twenty years ago. And what's more you'll have to bend God to your will to boot."

Short, stocky, elegantly groomed, delicately manicured Francis L. A. Pioche, the French-born financier and *bon vivant*, sat across from Steven in his new three-story Stockton Street mansion.

Pioche's obviously admiring appraisal made Steven ill at ease. He seemed to be devoting the same degree of intensly personal attention to him that he might lavish on a young lady. Steven had the feeling that Pioche was listening with two distinctly separate minds. One noted his request for a cash loan and the other was weighing the business merits of the request against the personal satisfaction Pioche might derive. Even the voice—women described it as "charmingly accented"—seemed to caress rather than question. *"Mais, mon ami, combien faut-il payer?* And what shall I ask of you as security for a personal loan?"

"I need a credit draft for five thousand dollars. For security you have my hand and my word and a deed for half of the land I purchase, with the understanding that I may buy it back from you at the market upon satisfaction of the entire loan."

Pioche's small shrewd eyes widened, then he threw back his head and a delighted laugh echoed through the room. *"Sacre bleu!* You will give me half of the land you buy with my money *if* I promise to trade the deed for the satisfied personal note." He leaned across the table and rested a hand on the younger man's knee.

"Steven, my dear young friend, if I am as shrewd as my competitors credit me with being, I should pay you fifty thousand dollars to desert your family, move in here with me, and become my business protégé. We would soon have M'sieur Ralston for our office clerk, would we not?"

Pioche gave Steven's knee an affectionate squeeze and leaned back to contemplate the fragile snifter of Sazerac poised in his left hand. "Let me ask you another question. Why did you not go to your father?"

"I did, sir. He will give me any amount I want to purchase land north of Monterey, but none for Los Angeles or San Diego."

"Then why did you not ask M'sier Ralston? Many times on the Champagne Route when we have discussed the new generation of promising young men he has had flattering things to say about the Lewis twins."

"Sir, my father is financed by Mr. Ralston now. I prefer to act independently of my father's business associates."

"Most particularly since you act in opposition to his wishes?"

"Yes, sir."

Pioche sniffed the Sazerac. "You wish to be your own man, then?"

"Yes, sir."

"And, forgive me, you find it doubly difficult being a twin."

"My brother and my father see with the same eyes. I don't say they are wrong. I don't say I am smarter."

Pioche nodded. "But you are different, Steven. Even I, who will never have sons, can see that. Somehow, at your tender age, you have concluded that great opportunities are to be found in solutions to great problems. *Très bien! Vous avez perception.* But what will you do with the land?"

Steven considered his answer. "It is possible, sir, that I will not make a purchase at all. I do not want just any land."

"So?"

"I want cheap land. And I want land that controls the headwaters of a river or a year-round stream with a good flow."

Incredulousness shone in the financier's eyes. "But my dear young man, you will sooner find virtue in a *demimonde* than abundant water in the south. What you are really asking me to do is finance an outing."

Steven Lewis answered by taking from his pocket a folded paper. "I have sufficient personal funds to finance my own journey and a stay of several months, sir." He unfolded the paper and revealed a crude hand-rendered map of the area from Santa Barbara to San Diego. The map showed outlines of the confirmed Mexican grants, the direction of the principal mountain ranges, and the approximate courses of their few known dependable streams. Pointing to the general area of San Diego and San Bernardino, Steven moved his finger along the base of the mountains whose slopes drained onto the coastal plain. "There are two dependable streams here."

The finger moved toward sprawling Los Angeles County. "And there are three more here. The ranchos with ranges in the foothills are best off. Even so, much of that rough land is for sale. That is what I propose to buy."

Recently Pioche and Bayerque, the name of Pioche's

245

trading firm, had speculated in several huge tracts of land for subdivision including hundreds of acres in San Francisco's Visitation Valley. Pioche had had a hand in building the San Jose Railroad, the line on which hordes of settlers traveled to buy home sites for ten times what Pioche and his partners had paid for the raw acreage. Still other Pioche money went to subsidize and eventually control cattlemen in Monterey County and Santa Barbara County. He was said to have as many fingers in as many pies as the dynamic banker Billy Ralston.

Steven Lewis watched him studying the map. The man was an enigma. He pretended as much interest in hunting and fishing as he did in San Francisco's blossoming arts. In contrast to Billy Ralston's preference for male companions in his favorite saloons, Francis Pioche preferred artists, writers, actresses, and singers. He was adored by women of the theater for his lavish gifts of jewelry. He asked nothing in return but their companionship, always in the company of others. No whisper of scandal ever touched him. His gifts were understood to be precisely what they were, expression of the donor's gratitude for the distinction, and occasionally the notoriety, the performers lent to San Francisco. If William Ralston called San Francisco "his city" Francis Pioche could make the same claim with equal justice. Their civic rivalry was amiable.

Steve's disquieting personal appraisal of Pioche was interrupted by a sudden question. "Tell me, Steven, where did you get the notion to control water?"

"From my grandfather, Don José Robles, and from my father's partner, Harry Morrison. When I was still in school Harry took me on a trip up the American River. I saw the ditches and flumes there. The men who brought the water from the high country and sold it by the inch made more money than the miners who used it for washing gold."

Pioche tapped the map. "And you propose to corner the market in liquid gold?"

"I propose to control a good share of it, sir."

"To what purpose?"

"For whatever purpose the people below will buy it."

"Have you thought out what some specific purposes might be?"

Steven Lewis met Pioche's questioning gaze for a time, then lowered his eyes. "Not exactly. They want water for

246

their cattle and their gardens, and everything else water's needed for."

Steve knew that he had replied lamely, because he had acted too hastily. His arguments in favor of the loan should have been thought out in more detail. He should have financed his own exploratory trip to Los Angeles and San Diego, seen the streams and the lower lands dependent upon their water. The map should have been drawn from personal observation. He was about to confess his dereliction when Pioche leaned back and laughed softly.

"Do you know something, my dear Steven? I am twenty-six years your senior but I am not so old that I cannot remember my own impetuosity, my own errors of haste." He lifted a hand. "Oh, I made them. I made them, *mon cher ami*. Some of them were *maqnifique*. I remember, while suffering from a multitude of youthful misjudgments that had compounded themselves into a *désastre*, the advice I received from Victor Hugo, who was my dear mother's close friend. He said, 'Great blunders are often made, like large ropes, of a multitude of fibers.' *Il y a du vrai*. But it is also true that strong ropes are made of many fibers well laid. Your concept is excellent. It has daring and majesty. What it lacks, my dear boy, is research and order. Not only must you see the opportunities for yourself, but also you must anticipate the problems. I do not refer to mechanical problems. Engineering brains and strong backs can be hired. I refer to the curses that are certain to accompany the blessings you propose to confer upon the southerners."

Pioche leaned forward earnestly. "Do you know anything about human nature, Steven?"

"Some, sir."

"Have you learned that the most feared and detested man is the émigré, the outsider, who would exploit the opportunity the natives have failed to see?"

Suddenly his father's warning whispered across the intervening days. "You'll run afoul the same problems there that Chard and I had here twenty years ago." Steven understood that while the situations were not precisely analagous the underlying principle was the same. The Anglo-Americans now outnumbered the native *californios*. They controlled the politics and economy of the state. Steven understood that he could expect no consideration from his own, most particularly if he seemed to threaten their

247

dwindling supply of lifeblood. In an instant he anticipated all that Pioche would say. But he listened attentively.

"The trick, *mon ami*, is to convince people, while you are gaining control over them, that you are their benefactor, that you have anticipated their need and you have come to help. Do that successfully and you become a hero. Fail and you become a corpse. There's only one principle. *Écoutez bien*, Steven. All men secretly desire something for nothing. But offer it to them and they grow suspicious. So? How do we take advantage of this human perversity? We do *not* offer them something for nothing. We offer them something of value, at a bargain. *C'est tout!*"

For a moment Pioche regarded his young visitor with undisguised admiration. "Of course, *mon jeune protégé*, there are other more obscure aspects of human nature that concern man's relationships, especially where the passions are involved. But that is another lesson. *N'est-ce pas?*"

Pioche removed an ornately carved gold watch from the pocket of his brocaded waistcoat and snapped open the diamond inset cover. "Forgive me, Steven, but my valet, Louis, will be reminding me that I have dinner guests this evening."

Pioche walked around the small table to guide Steven by the elbow. "I'll see you to the door and make you a proposal, yes?"

Steven's mouth twisted into a wry grin. "You'll propose that I have all of the facts before I waste your time again?"

Pioche linked his arm affectionately through Steven's. "You have not wasted my time, Steven, or yours. Invest your own funds in an exploratory trip, make a decision supported by facts. If I find it attractive you and I shall become partners, *equal* partners in the venture."

Steven turned to face his host. "But I am already a partner in Lewis & Sons . . ."

Pioche laughed. "My dear young friend, I am a partner in twenty ventures. That is how one covers his wagers."

He opened the door and extended his hand. "Thank you for choosing to come to me, Steven. I like you very much. The world deals kindly with smart attractive young men. You will do well. You shall see. *Au revoir, mon partenaire.* And if you do not sail for the south too soon I

248

should like you to join us at a little *soirée*. I think my friends might amuse you. And I know that you will prove devastatingly attractive to the ladies."

The invitation was extended as Steven made his way down the flight of marble stairs to Stockton Street. He acknowledged it awkwardly, from the walk. As he turned down the hill he sensed that Francis Pioche was still watching him through the half-opened door.

On the cabin deck of the thirty-three-hundred-ton side-wheeler *Persia,* Steven wondered if he should not have managed a more considerate family leave-taking, one that would have spared them all the unhappiness of the past few days.

Since, in business, Pioche was known to be a man of his word, Steven had not hesitated to tell his father and brother in detail the conditions of the agreement. They had listened in disapproving silence and his father had spoken first. "You are a goddamned fool, Steven. You will not be Pioche's partner. You will be his prat boy. He will flatter you and bend you around his fat little fingers and you will wear out a dozen pairs of boots running for him and pay for the privilege from your own purse."

His brother had been hardly less direct. "Pioche doesn't mix with our kind, Steve. He's too fancy. If you won't listen to father, the least you can do is talk to Billy Ralston. He's our kind. You don't have to ask him for money. Ask him for advice. Tell Pioche you've changed your mind."

The scene with his mother and sister at the big house in the country out near Mission Dolores had been the most difficult.

Doña Maria, withdrawn, graying but still lovely, in her forties, enlisted the assistance of twenty-two-year-old Felicia who, four years earlier, had married Lambert Lee, a San Francisco lawyer fifteen years her senior. Most of his practice had come to relate to the manifold affairs of John Lewis & Sons. It was understood that a partnership was in prospect.

Steve's brother-in-law, a tall, humorless man, had come in during the most turbulent moments with Doña Maria and Felicia. The scene had not been helped by Steven's nephew and niece, two-year-old Lewis Lee and six-month-old Emily. Sensing the emotional conflict, they add-

249

ed to the clamor. The hardest thing to bear had been his mother's quiet weeping. *"Por que? Por que, Steven?* Why do you leave us? You are of the north. The south is not good. *Es una tierra abandonado, un gran desierto.* Only death lives there. You will never return. I know. A woman knows these things about her own."

Felicia's pleading, done apart from the others, had been more subtle. "It is not that you shouldn't go, Steve. You are a man now. You have a right to make a man's choice. But wait until after the holidays. I know how you feel. Lambert and I have talked about your place in the company, about how Will seems more suited to work with Father."

Steven could not resist a retort. "I do not think your Lambert will shed many tears if I vacate my place in the company."

He had endured the torrent of angry denials but he knew it was true. There was still another truth to be faced. Steven had long sensed his twin brother's rivalry for parental affection. Will would make, indeed had made, a great show of protest too. But however convincing it had been to others, Steven knew that his absence would relieve rather than grieve his brother. He did not understand his own compulsive need to move out, to pit himself against the brotherhood of selfish giants who were shaping California. In five years, perhaps less, steel rails would bind the Atlantic to the Pacific. That meant the railroads would hold the entire continent in a gleaming skein.

After a half hour on deck a chill drove Steven below to the main saloon. For the most part the passengers were an affluent-appearing group.

In the dining saloon he was seated to make a fourth with a Reverend and Mrs. Lathrop and their daughter, a young woman in her mid-twenties whose pallid face was saved from the commonplace by arresting, long-lashed, slightly up-tilted peridot green eyes. During the discreet mutual questioning initiated by the preacher, Steven found those eyes disconcerting. He would not have called the young woman's gaze bold but certainly it was more than the polite pretense of attention usually accorded him by genteel, match-minded young ladies.

At twenty-five Arlene Lathrop Baker was a widow. Her

mother had volunteered the information. "It has been very difficult for her and I think she is very brave."

The daughter disagreed firmly. "I have not been brave, mother. Gerald's untimely death was part of his karma, just as it is part of mine. It does not require bravery to accept the workings of Divine Law. One needs only faith and understanding."

She turned to Steven. "I hope, Mr. Lewis, that I do not sound callous, most particularly since my late husband was taken from me by his own hand just a year ago. I do make the conventional concessions. I did go into mourning, of course. But I feel a bit of a hypocrite since only a person wanting in understanding would indulge in such self-pity."

The Reverend Lathrop laughed nervously. "I must say, Lewis, that my daughter and I do not see precisely eye-to-eye on this matter. I tend to stay with the literal word of the scriptures. I see no need to question their intent."

Arlene Lathrop Baker directed a benign smile at her father, then patted Steven's hand lightly. "The eternal schism between the generations, Mr. Lewis. Father only recently heard his call to the ministry and founded his own church against all of the remonstrances of his own father. But now that I've heard my call he uses all of the arguments that he himself renounced."

She laughed for the first time. The sound was gay and musical. Suddenly her face seemed quite beautiful. Then she was serious again and Mrs. Lathrop's expression of hurtful reproach faded, but the Reverend Lathrop continued to make soft lamenting sounds. "My daughter does not believe in sin, Mr. Lewis. She calls it the 'immutable law of cause and effect.' But I say a sin by any other name—"

"Oh, father!" Arlene raised long white fingers to her temples. "You will never understand. Of course I believe in what you call 'sin.' Almighty God knows I saw enough of it first hand with poor benighted Gerald. And he paid his penalty, didn't he? The law of cause and effect. He was punished on earth, not in Heaven. His hell was here."

Suddenly contrite, she turned to Steven and rested a cool hand on top of his. The sensation was not unpleasant. "You see, Mr. Lewis, I believe we learn our lessons here. Life is a classroom. If we do not learn here, then we are sent back a grade, reincarnated and made to face the same problems over and over until we solve them. Then and only then are we sent on to higher, more noble lessons."

Jabez Lathrop snorted. "Nonsense!"

Arlene was not to be stopped. "Only through the cultivation of true understanding, Mr. Lewis, can we make our bodies perfect instruments for the joyous expression of God's Will."

She punctuated the statement with a little pat and removed the hand that had rested on Steven's. The discussion unsettled Steven. He turned the conversation to more familiar matters. "Sir, will you start your church in Los Angeles?"

The preacher placed the palms of his hands together and bowed over them.

Arlene gave Steven a knowing smile. "Even though father denies its existence, father is using the Great Inner Eye—the symbolic term for Divine Perception. Yours, Mr. Lewis, by the way, is very marked. That means you are a very highly evolved soul in this incarnation."

Jabez Lathrop groaned and shot an appealing look at his wife. "Dammit, Arlene!"

Mrs. Lathrop gasped and lowered her head as he continued: "It's all right for you to spout that heresy in the privacy of our home. But not here, not in public!"

Mrs. Lathrop lifted a hand in a wan, helpless gesture. Arlene smiled triumphantly. "Father, the very fact that you can lose your self-possession in public and that I can keep mine, even when my deepest convictions are being attacked, is that not proof that my understanding is as great as yours?"

"Why don't you say, 'greater'?"

Arlene smiled compassionately. "Because I understand, dearest father, the misgivings felt by parents even though I was not fulfilled as a woman by poor Gerald."

It was obvious that the Reverend Lathrop could not cope with his daughter's forthrightness. He turned, almost desperately, to Steven. "You asked if I am going to found a church in Los Angeles? The answer is yes! I am going to found a church, somewhere in or around Los Angeles. I will call it the Tabernacle of Truth, a name I put together from the Scriptures by searching them with the *only* two eyes I'm certain the Almighty gave me." The Reverend's daughter smiled with long-suffering forbearance.

Warming to his subject, Lathrop's long, knob-knuckled fingers clutched at the satin-faced lapels of his frock coat. "Through no desire of my own, Mr. Lewis, I prospered during the late war. I resolved I would show my gratitude to the Almighty by devoting the rest of my days to His service."

Before he could launch into a full dissertation, Arlene interrupted. "Father—forgive me, please, but I do think the sea is getting a bit tippy." She patted her brow and looked about apprehensively. "I really feel the need for some fresh air, rather quickly, I fear!"

Steven's chair was back before she had finished. "It is cold on deck, Mrs. Baker. You'll need a wrap."

"I have one in my cabin, Mr. Lewis. If you would be good enough to see me there, then show me to the upper deck, I'd be most grateful."

Steven guided her into the main saloon and onto the weather side of the deck. Clutching his arm tightly, Arlene led him past several staterooms as she rummaged in a small bag for the key.

Steven let her in and stepped aside. "Do come in, Mr. Lewis. I shall need help getting this heavy old coat around me."

On deck again, Steven guided her to the ladder leading to the top of the deck house. So far as he could tell there had been no noticeable change in the sea. Even so, Arlene seemed very unsteady and he found himself supporting most of her weight whenever the vessel heeled. At the railing forward of the stack she realized that he was not wearing a topcoat. "Oh, Mr. Lewis, I'm so sorry. How utterly selfish of me. You'll take a chill."

"I'm quite all right, really, Mrs. Baker. You'll feel better in a moment and I'm used to the sea air."

Arlene turned, rested her hands against the front of his coat, and looked at him contritely. "I have a confession, Mr. Lewis. It was not the motion that made me ill. It was the thought of another hour of listening to father's pious pretensions. Oh, *I* could endure it all right. But I saw no earthly reason to ask an intelligent young gentleman like you to suffer through it."

She turned away. "Besides, it is so beautiful up here, so clean." She turned back to him. "I'll be all right alone for

253

a moment. Do get your coat and come back and talk with me. I'm hungry for intelligent conversation, Mr. Lewis."

He returned in several minutes and they found a seat on the frame of a closed skylight. Seated close to him, Arlene was silent but Steven could feel her studying him. The ship's bell sounded and she shivered and inched closer. Without warning, she laughed a bit wildly and turned to Steven. "It's amazing! Truly amazing." Puzzled, Steven looked around the deck.

"Nothing here, Mr. Lewis. I meant meeting someone such as you." She leaned closer. "Don't ask me to explain. As much as I know, there is more I do not know." She continued to search his face. In the starlight, framed by the soft knitted shawl that protected her smooth dark brown hair, Steven thought her face was serenely beautiful, almost madonna-like.

She smiled and seemed about to rest her cheek against his shoulder but changed her mind and looked away again. "It isn't necessary to talk, Mr. Lewis. Communing in silence is the highest form of communication."

She seemed to retreat into herself then and Steven watched her from the corner of his eye. Daughter of a well-to-do food processor, undoubtedly a war profiteer, offspring of a self-ordained minister of a nonexistent church, a free thinker, and a new widow to boot, all this troubled him far less than it intrigued him.

Neither he nor his twin brother were strangers to young ladies, proper ones and some not so proper. But this woman beside him did not conform to familiar patterns. He could not put a name to the curious quality in her that seemed at once both prim and promiscuous. In the midst of this speculation, she startled him again.

"I don't believe in senseless convention, Mr. Lewis. I can only hope that you find me direct but not bold." She turned to him. "For some reason I cannot explain I feel the need to be very truthful with you."

Uneasy, Steven started to protest but she cut him off. "I must say it. Father did not heed God's Call, so to speak, until he was tried and found innocent for want of sufficient evidence in the death of ten Union soldiers. They died of poisoning after eating food tinned on our factory.

"The prosecution proved contamination was present. But it could not prove that the contaminants had been present in the unopened tins. Father won the case because

254

there is always some spoilage in tinned foods. The jury agreed that the army cooks were negligent in not examining the contents before they emptied them into the common mess.

"But the boys' deaths shocked father. So he is trying to find his way to self-forgiveness by devoting all of his very great energy to God's work as he understands it. I see through him, Mr. Lewis, but I admire him. Most men would do nothing to make amends. And I share his burden of guilt."

Steven frowned. "I doubt that somehow. Why do you say it?"

"Because I often helped in the canning house. Several times I had the feeling that the tins were not being boiled long enough before the women filled them. And I often saw poor work in lidding and waxing. I did speak on these things. But not insistently enough."

"That's ridiculous. We are merchants in San Francisco. We handle thousands of cases of tinned food. I would say more pickled salted and dried stuff spoils. You are too hard on yourself."

His indignation seemed to touch the young woman. Gently, she captured his upper arm between her hands. "Dear Mr. Lewis. May I ask how old you are?"

The question took him by surprise. "Why I'm twenty-one, almost twenty-two. Why?"

"Oh. I would have said older. I'm sorry. I never should have started this dull personal conversation."

"It isn't dull. It's interesting. Do you mind if I ask you a question?"

She laughed softly and let her cheek brush against his shoulder. "Heavens, no. I'm flattered by your interest."

"It's about your late husband."

"Gerald?" She drew a deep breath and gazed upward. Her large eyes seemed luminous and her profile was lovely. "What do you want to know about poor Gerald?"

Steven struggled. "I can't guess why a man married to a wonderful woman like you would want to—to—"

She finished the question for him. "Take his own life?"

After a troubled silence, she released his arm, turned and folded her hands in her lap. "Are you certain, Mr. Lewis, that you will not find it too much if I speak with utmost candor?"

He nodded. "Quite sure, Mrs. Baker."

255

She moved an inch away and sat up very straight. "I married Gerald when I was nineteen. Our families had been friends for years in Springfield and in Pennsylvania before that. I guess we all just took it for granted that we would always be together."

"From the very moment the vows were taken, Mr. Lewis, something happened. We had always been comfortable with each other. Suddenly we were strangers, or at least Gerald was. We went to Lime Rock Springs for our honeymoon."

She paused and looked directly at Steven. "It was never consummated, Mr. Lewis. Whatever had happened between Gerald and me not only made us strangers, but almost enemies. We went back to Springfield and moved into the house father had built for us. I thought perhaps there, with our own things around us, it would be ..." she shrugged, ". . . better.

"I thought perhaps I had been too aggressive. I wasn't timid because our theosophy teacher in Chicago taught me that there should be nothing but joy and beauty in all natural relationships. So I took the initiative and let Gerald know that I was eager to receive him. But my honesty seemed to make him withdraw even more and he found excuses to stay at his father's hardware store where he worked later and later.

"We went on that way for a year. He avoided me as much as possible and I made excuses and pretended that I was bursting with pride in my industrious husband. But some people saw through it. In time even mother began to wonder why I did not become pregnant."

Arlene broke off and turned away. When she was in possession of herself she continued. "I sublimated my passionate nature in the making of a home. I worked long hours in the canning shed. I even got a citation from our mayor for outstanding patriotic war assistance. Gerald signed up with the Illinois Volunteers and went into training and that helped with appearances. He was gone a year but he didn't see any fighting. When he came back I welcomed him, Mr. Lewis, I welcomed him eagerly and without shame."

Again Arlene paused and turned away. "When Gerald returned from the war he was often in his cups. Sometimes it was so bad he could not work. Had he known combat it might have been taken as a reaction to the hor-

256

rors of the battlefield. But he had no such excuse, Mr. Lewis.

"Then one night when he had been drinking heavily I felt I had to be completely candid with him. I asked him if he did not find me attractive, as a woman, as a sex object. He—he smirked in a most revolting way." She stopped for a moment. When she continued her voice was subdued. "He told me that he found *all* women physically revolting, and that he always had. He told me that his needs were taken care of by others who felt the same way."

There was a long silence. Then Arlene rose abruptly. "On the first day of January he was found in a thicket out near the lake. The ball from his gun had passed upward and blown away the back of his head. It was set down as a hunting accident, Mr. Lewis. But it was not. It was suicide. Gerald was not a hunter. He took his own life. But he did it in a manner that left doubt. And in so doing he spared all but me the pain of absolute knowledge. A family can learn to live with the grief of an accident. But a suicide? No!"

Without quite realizing what he was doing, Steven reached out in an attempt to comfort her. She avoided him.

"I must go now, Mr. Lewis. I do not know how I shall ever thank you for your sympathetic understanding. I have needed," she closed her eyes tightly and her voice seemed near to breaking, "my God, how I have needed to say it all out loud to some compassionate soul. We were led to one another, Steven Lewis. Believe that."

Before he could respond she had kissed him full on the mouth and was hurrying toward the ladder. He overtook her and helped her down to the main deck. There, without a word, she brushed past him and ran to her stateroom.

Steven saw the Lathrops briefly when the launches from the new San Pedro pier carried them ashore. Since he was not delayed by a wait for luggage, he was able to take the first Banning stagecoach to Los Angeles. After a wild three-hour ride the shaken passengers alighted at the Bella Union Hotel advertised as "the finest hotel south of San Francisco." It wasn't, Steven observed as he let himself into a small third-floor room, much of a recommendation. Former governor Pio Pico's promised "great hotel," Pico

257

House, was still only a rumor. That afternoon he arranged for a saddle horse for the following morning.

When he returned to the hotel, the stagecoach bearing the Lathrops had just pulled up. Smiling, he walked over to them. "Allow an old-timer to welcome you to Los Angeles. I've been here for three hours."

Arlene laughed and gave him her hand. Mrs. Lathrop greeted Steven with a bleak smile. Jabez Lathrop looked about critically, studying the dung-littered, unpaved street, the motley ranks of pretentious two- and three-story brick buildings with arched windows overlooking low, flat-roofed, fly-infested adobe cantinas. "The Promised Land does not look full of promise, Lewis."

Steven was about to agree when the stage driver cut in. "You're dead right, Reverend. And some folks are thinking of suing the Almighty for *breach* of promise!"

Arlene Baker turned to Steven. "I'll forgive this land anything so long as I can enjoy this wonderful sun. And in November. It's a miraculous land, Mr. Lewis."

Steven nodded. "Perhaps. But I'm reserving my opinion until I've looked around a little."

"How long will you be here?"

"Long enough to ride out to the San Gabriel Valley. And perhaps down to San Diego."

Arlene pouted. "I envy you your adventures." She indicated the hotel. "Is this place at all bearable?"

"It's reasonably clean and a lot quieter than its namesake in San Francisco."

The stage driver spat and chortled. "You sure wouldn't have said that if you'd been here last July sixth, mister. No, sir! You would have been cut down by a dozen slugs when Bob Carlisle shot it out with the King brothers." He wagged his head. "Carlisle dead. Frank King dead. Sam King still hanging on with three slugs in him and likely to die from rope poisoning if the lead don't get him first. I don't think you folks up in Frisco get much noisier than we do, what with ordinary friendly shootings and lynching greasers and chinks. Last week them Lobos outlaws rode down from the hills and shot up two squatter shacks outside of El Monte, killed the men, stole the stock and the women, and set fire to the whole shebang."

Mrs. Lathrop listened with oval-eyed horror. Steven was puzzled by Arlene's smile until she explained. "Violence al-

258

ways results when man frustrates his true nature. Hatred is simply dammed up love."

Frowning, the perplexed driver examined Arlene and said, "Then there sure must be a lot of damned love in this here town, ma'am!" He looked expectantly from one upturned face to another. He saw a twinkle in Steven's eyes and slapped his thigh, succumbing to a paroxysm of self-appreciation. Arlene Baker shot a scornful look up at him and rested a hand on Steven's arm.

"I hope you will join us for dinner this evening, Mr. Lewis. It would seem strange not to have you with us now."

After the dinner, which Steven Lewis insisted upon hosting, the Lathrops retired to their rooms to finish unpacking. Steven went to the bar and struck up some easy conversations over a brandy. The information he gleaned was consistent: the entire southern region of the state was verging on disaster unless the rains came quickly. Even so, the rains would probably cause as much damage and suffering as the drought itself. They always had.

"It's the nature of this country," an older man with holdings in the Santa Ana River Valley volunteered. "There's not enough cover on these mountains to hold back a heavy run-off. It's been a plague on the land ever since earliest times. We've all seen it. A heavy rain comes, the run-off starts in the canyons, then all of a sudden a wall of mud and water flashes down onto the plains. Nothing can stand up to it. I've seen boulders as big as Phineas Banning's coaches come bouncing down a wash, smashing trees and houses and squashing cattle and horses like sow bugs. A river of liquid mud can pour out of a canyon and sweep everything away, completely burying houses without a minute's warning to the folks inside."

The man next to him nodded gravely. "What this country wants is dams to slow down the run-off and save the water for irrigation. Every time a flood comes it contaminates the drinking water."

The bartender's eyes turned limpid with exaggerated sorrow. "The pathetic part is, it drives a lot of otherwise temperate folks to come in here to disinfect theirselves and ward off the fevers. 'Course I admit we get a few such folks that prays for floods."

The customers laughed but there was no mistaking the anxiety underlying the levity. The drought was in its third year and if the rains were going to come they would come soon. They were, in fact, seven or eight weeks overdue.

After the second raw native brandy, Steven said his goodnights and climbed the stairs to his room. His window overlooked the Church of Nuestra Señora la Reina de Los Angeles de la Porciuncla. The original pueblo was founded in 1781 and it stood next to the Los Angeles River. The river had rampaged over its banks a dozen times.

In 1800, following a two-year drought, the floods had destroyed most of the thirty original adobes and washed away sections of the pueblo walls. The plaza had been moved then to its present location on high ground, but the river still took its toll in the outlying regions. No one dared to predict that it would remain in its present course for more than a season. Like a great, sandy serpent, it writhed over thousands of acres, brimful for weeks at a time, then diminished to a trickle, often so slight that people were forced to dig holes in its bed to find water.

Steven heard an echo of the stranger's words. "What this country needs, mister, is dams to slow down the run-off."

He lowered the shade and began emptying his pockets. As he was about to light the coal-oil lamp he heard a soft tapping on the door. For a moment he thought he'd been mistaken but it came again, more insistently. At the door he hesitated.

"Who is it?"

The whispered reply, so close it startled him, came through the ill-fitting jamb. "It's Arlene. Mrs. Baker. May I talk with you for a moment?"

Steven unlocked the door, intending to open it sufficiently to look out, but he found himself being moved backward as Arlene pushed into the room and closed the door behind her.

"Forgive me, Steven, I know my behavior must shock you, but I need your help desperately. I knew you'd be up and gone tomorrow before I had time to see you alone."

Flustered to be found undressing, he moved toward the lamp again.

"No. Don't bother with a light or your coat or tie, Steven. I'll just stay a moment. I realize what my presence

260

here could do to both our good names, but I simply had to risk it."

Steven turned back to her. "I'll put up the shade a bit."

The room lightened sufficiently to reveal that she was still dressed in the embroidered gray silk blouse and the mauve wool skirt she had worn at dinner. But the jacket was missing and her hair, usually pinned in a severe roll at the back of her head, now hung in loose soft curls that spilled from a band of ribbon secured low on her neck. The disconcerting eyes seemed wide and curiously luminous as she moved close and rested a hand lightly on his upper arm.

"I warned you, Steven, that you and I are old souls who have been through much together in other times. But as candid as I was on deck last night, there simply wasn't time to confide all that needed saying."

Steven flopped his hands awkwardly.

"I hope you don't mind if I call you Steven and you must please call me Arlene. Steven, I can't explain it, but once you must have been very dear to me for I still feel the light and warmth of that earlier relationship. Perhaps we were brother and sister, or husband and wife, or perhaps just sweethearts."

She looked around, obviously seeking a chair. Steven pulled a small cane-seated rosewood rocker from beside the washstand.

"You sit on that, Steven. I'll just perch here for a moment."

Before he could protest, she seated herself on the edge of the bed. In doing so the braided hem of her skirt hiked above the tops of her expensive shoes, revealing the curves of her surprisingly full-turned lower calves. She made no effort to conceal them and leaned toward him urgently.

"I told you father had decided to devote the rest of his life and his modest fortune to God's service. What I did not have time to tell you last night is this: I too am going to dedicate my life to the service of mankind. I told you what I believe in, the free expression of Divine Love with man and woman its perfect creative instruments. I am determined to found a colony for Free Thinkers, an earthly paradise where others who believe as I do may devote their lives to the passionate pursuit of the true Truth!"

Less uneasy and frankly curious, Steven controlled a smile. "I didn't know there was any other kind."

261

"Of course there is. Have you never heard of a half-truth?"

"Yes, ma'am. I've even heard myself saying a few."

Arlene Baker frowned. "Be serious, Steven. We have very little time."

"I am serious. But how can I help? What can I do?"

"While you're looking around for whatever you're after, Steven dear, you can seek a likely place for me to found my colony. I will call it Brindavanam."

Steven fidgeted on the small chair and it creaked alarmingly beneath its six-foot, one-hundred-eighty-pound burden. "But I don't know what you need."

"You'll be led to it, dear Steven, a beautiful secluded location, a place of beauty with trees and meadows and a stream of cold pure water."

"But you'll have to buy the land. You'll need buildings."

"And they will all come, Steven, everything I need."

Arlene Baker's long fingers explored the narrow waistband of her skirt. Steven watched amazed as she removed an envelope, opened it and slipped out five one-thousand-dollar bank notes.

"I brought these to you for a start. I have much more on deposit with Lazard Freres in San Francisco, enough to purchase Brindavanam, build our first temple, and our first fifty residences."

She rose from the bed and handed Steven the notes. "I want you to take them. I have perfect trust. I want no receipts or any evidence that I have given them to you."

Steven tried to return them. "But Arlene, this is a lot of money. I can't take it. Suppose I lose it or get held up or suppose I'm really just a crook. You don't know anything about me, really."

She put her hands behind her in a gesture of stubborn refusal. Steven stood up and reached around to force her to take the money. Laughing and protesting, she stepped back quickly. As she did the heavy wooden bedframe struck her behind the knees. Uttering an alarmed little cry, she began to topple backward. Caught off-balance himself, Steven tried, too late, to steady her. They fell together on the bed.

Arlene, lying with her sparsely fleshed, naked body flattened along Steven's upper belly and chest said in a whisper, "Are you shocked knowing I have no shame, Steven?"

He rolled his head away to look at her in the half light. "If you're not shocked, why should I be?"

"Because I let everything happen so fast, dearest one."

Steven smiled inwardly and thought that "*made* everything happen so fast" would have been more accurate. The time with her, whatever its length, had been a dark fantasy, a *tentación del diablo,* a fulfillment of all those hebetic, libidinous longings that had so often been vented in an agony of repression lest his twin brother, sleeping nearby, might hear and suspect. Now in this strange bed this voracious, man-hungry woman had twice drained him with her insatiable demands. Steven felt her hands move up the sides of his head. The thin fingers found the thick tangle of his dark brown hair and probed sensuously.

After a time, Arlene inched her body up along his side to kiss him. For several minutes he failed to respond to the new assault of caresses. Annoyed, she stopped. "Steven—"

"Yes?"

"Kiss, darling. Do things. Anything. Make me feel."

Before he could reply she was on his lips again, first probing in his mouth with her tongue, then once more moving her body urgently against his. She clung to him, pleading, coaxing him up and into her. For the third time, Steven could feel himself responding and then, too quickly, he came. Her narrow hips started to batter his and she began to plead again, "Stay—stay—stay—stay, Steven, oh—stay, darling."

They remained locked together until Steven heard her gasp, suck in a huge breath, then felt her body, cold with perspiration, collapse on his. After that he became aware of a new wetness, on her cheek. He touched the tears lightly and she began to quiver in his arms.

Holding the thin, vibrant body close, Steven struggled to sort out from the welter of feelings the dominent one that had possessed him for the past hours. There had been, still was, excitement, passion, tenderness, and a strange uneasiness too. Steven had never been in love. There had been, on several occasions, the not unwelcome pain of a brief infatuation early in his teens. But tonight something new had been awakened, some feeling he could not put a name to.

He had shared two wracking orgasms with Arlene Baker. After them had come an unaccountable tenderness.

She had satisfied him deeply and he wondered if what he was feeling might be the beginning of love, but he mistrusted his emotions. Certainly what he had felt was not what he had felt for the prostitutes who had taught him well. And it was not that added something he had felt for the fresh young whore who had claimed she had fallen in love with him and offered to renounce the profession if he would marry her. It was strange to remember her now and remember Harry Morrison's earthy observation: "Sometimes whores make the best wives, laddie. Aye. They say the first ladies of two of our states were madams who worked their way up. If you want a wife that knows how to meet a mon on his own terms ye couldna do better than a good clean professional lady."

Steven felt Arlene lift her head to look at him. When he opened his eyes her face was inches from his. "I could feel you thinking. Would a penny buy your thoughts?"

He smiled and attempted to bring her face closer, but she pulled away. "Please don't, darling! I've waited so long to find you and gotten so hungry in the process. You'll think I'm insatiable!"

She let her body roll from his then and curled up beside him. "Were you worried about," she rested a palm lightly on her lower stomach, "about complications?"

Steven lied. "I should. But I guess I wasn't."

The palm moved to his face and two fingers pressed against his lips. "You shouldn't. So don't. It is too late to explain now. I must dress and get back down to my room before it's light. But one day you'll understand."

She cocked her head quizzically. "By the way, when will I see you again?"

Steven pushed half upright. "It depends on when I get back. The night after tomorrow, maybe. I came here to find some land, remember?"

"And you promised to find me some, too."

He nodded and swung his legs off the bed. "What was the name of that place I'm going to be *led* to?"

Her eyes rebuked him. "Brindavanum." She repeated it slowly. "Brin-*dah*-van-um. It's ancient Tamil for Garden of Love. I first heard the name when I was studying Oriental theosophy in Chicago. I love the sound of it."

She saw Steven eyeing her narrowly. He grinned and said, "It sounds like Greek to me."

Arlene wrinkled her nose at him and without a trace of

264

false modesty stepped into the first of several petticoats thrown across the back of the little rocker. She finished dressing in silence. After a final deft poke at her hair she moved to the door, listened, then opened it a crack to peer out. Before Steven could say a word, she had slipped into the hall and closed the door behind her.

8 A FOUR-HOUR RIDE brought Steven Lewis to Anaheim, a communal German settlement. It had been founded by a group of emigrants from San Francisco, under the leadership of George Hansen, who had become discontented with the life there. It was the most successful new community in southern California.

Over a thousand acres were divided into a forty-acre townsite and twenty-acre plots. The cooperative Los Angeles Vineyard Company was the heart of the colony. The neat homes were entirely surrounded by a dense living willow thicket. It was as effective a barrier as John Sutter's adobe battlements.

Steven stopped at one of the four gates and asked permission to ride through. Hoping to find a bed for the night, he attempted to visit with several of the families preoccupied with pruning. Identifying himself as a San Franciscan warmed them but little. He inquired about the nearest stopping place and was told that travelers usually could find a bed at the Garcia y Perez Rancho de Los Nidos fifteen miles to the southeast. The vinculturist at the winery warned him that in any event it would be easier to find food for himself than his mount.

Saddle-weary and aching to the marrow, Steven pressed on along the base of the eroded, tinder-dry foothills of the Santa Ana Mountains. By dusk he was within a mile of a large hilltop, tile-roofed adobe surrounded by lacy pepper trees.

He crossed the yard-wide trickle of the sluggish Rio Claro. He had been told that the usually dependable river disappeared into the sand now several miles before its waters reached a confluence with the larger Rio Santa Ana.

All along the way, he passed heaps of bleached bones, unending evidence of the disaster that had overtaken the once great herds. The third year of the drought had extinguished the *rancheros'* hope that they could replenish their herds and find a new market in the north.

Steven knew well the futility of that hope. The northern California market no longer held much promise for the local ranchers and it simply had ceased to exist for those in the south.

Steven passed several mounds of smoldering carcasses from which hides had been stripped. The *vaqueros* had had to cremate the dead and dying animals in an effort to discourage the plague of flies, the scavenger birds, and the hordes of nocturnal carnivores.

Following a worn bridle path, he came to twin ranks of scrubby native junipers flanking the graveled driveway. At the top of the hill a leathery old *vaquero* stepped from the doorway of an outbuilding. Steven waved, reined up, and called to him. *"Buenas tardes. Dígame por favor. El patrón? Está en casa?"*

"Si, señor." The man pointed to the house and indicated that Steven should ride on.

Several minutes later, he dismounted. The house was impressive. Except for Mariano Vallejo's Casa Grande, it was the largest and most carefully made he had ever seen. It was one-story, planned in the classic U-form with the wings enclosing the sides of a patio. The thick walls were clean white and well-plastered. Hewn rafters protruded from beneath a red tile roof. The wooden window casements, two feet deep, shone with the gleam of hot tallowed oak and the door, placed in the center of a tree-shaded main facade, was much larger than most, at least six and a half feet high and four feet wide. Steven saw that it was beautifully constructed and divided into twelve carved panels, each representing some heraldic symbol that he assumed pertained to the Garcia y Perez families. The hinges and ring bolt were forged of hammered black iron.

While he was tying the halter lead to a ring post, a slender, fine-looking man in his late forties appeared in the doorway and addressed him in excellent English. "Good evening, my friend." He approached, smiling, and extended a hand. "You are a welcome visitor. I am Emilio Garcia y Perez." With his left hand he swept the horizon. "The owner of this *carniceria* that once was El Rancho de los Nidos. But whatever we have we are happy to share with you."

267

Steven found the proffered hand hard and good to the feel. "Thank you, sir. My name is Steven Lewis. I am from San Francisco."

Don Emilio's eyebrows lifted. "From San Francisco? Why would one leave that blessed land for this desert?" He laughed apologetically. "I do not intend *impertinencia*, my friend. It is only that in these times we are seldom visited except by worried money-lenders and neighbors who need to share their own troubles."

Steven smiled sympathetically. *"De nada, señor!* The whole south has my sympathy. But I assure you, I am not a worried money-lender."

As he spoke, Steven was aware of Arlene Baker's folded banknotes in his pocket. A money-lender, no. But perhaps in the end something worse, a man who would steal land that the unfortunate owner could not hope to recover.

"Pedro will care for your horse, Señor Lewis. Come in. Meet my family. Share our poor table. And we shall find you a bed. Should you bring us any news at all, we shall be in your debt."

Doña Paula Garcia y Perez was a slender, elegant woman not much younger than her husband. The only child in evidence was their fifteen-year-old daughter, Anita, a shy, extremely attractive girl who was built along her parent's finely drawn lines.

Steven wondered if there were other, older offspring, but the couple volunteered nothing beyond the name of the majordomo, Pedro Morales, whose wife, Gracia, headed a modest staff of household servants.

During the exchanges he managed a casual examination of the house. Its interior appeared to be as impressively built as the exterior. Inside the rafters had been squared and the ceiling carefully plastered between them. The effect was a finished interior, unusual in California ranch houses. A long, rectangular, tile-floored room was divided in two; the area to the right was a study filled with expensively bound volumes. A highly polished oak table dominated the center of the room.

The twin room in which they were standing was elaborately furnished with fine tables, sofas, and chairs that had come around the Horn from Spain. At the end of the room was a huge stone fireplace ornamented with inset Talavera tile. Logs were in place on heavy andirons and a small caldron hung from its hook at the end of a soot-blackened bracket.

Steven could see into the *comedor* with its long dining table, chairs, and side board. And on beyond in the *despensa* that adjoined the *cocina* he could hear the animated chatter of the women at work preparing food.

The evening meal was a simple dish known locally as *carne en camisa*. It consisted of inch-thick strips of lean barbecued beef wrapped in large corn tortillas and generously garnished with a fiery concoction of peppers and tomato that sent Steven repeatedly to his wine glass.

After the meal they retired to the far end of the room, where Pedro and an Indian helper had set a log fire. Purposely, Steven kept his chair well back in a futile attempt to combat a numbing drowsiness.

When he told them that he had Spanish blood and that his mother was of the Robles family, the last of Don Emilio's reserve melted. For more than an hour the ranchero described in detail the plight of the drought-plagued south. His realistic grasp of events and their inevitable consequences surprised Steven.

"We are finished, my friend. We *californios* had our time in the first half of this century. Fifty years in *paradiso*. That is what God gave us. But as it was in Eden, we abused His trust." The *patrón* of El Rancho de los Nidos extended his hands in eloquent resignation. "So now we dance at the Devil's fandango and feast on brimstone and bones."

His dark eyes challenged Steven. "Do you know how it will end? The great *ranchos* will be shattered like *ladrillos* in a *temblor de tierra*. The *yanquis* will pick up the pieces to make *ranchitos*. The traders who own the south, men with money, will bring water to the land and plant it with orchards, grain, and vineyards. They will take over. That will be the end and, *quien sabe*, perhaps a beginning too, but not for us."

Doña Paula and Anita excused themselves and the two men continued to talk over brandy. Steven had provided news of San Francisco and the beginnings of the railroad. His news from Los Angeles was meager. He could only confirm something Don Emilio knew: the migration of Southern farmers ruined by the Civil War had already begun to arrive in San Bernardino. They were coming in droves with their wagons and their meager belongings,

269

searching for cheap, fertile land. Nomads, living out of boxes and buckets, foraging for food, seeking work that did not exist. Both Steven and Don Emilio understood that out of the poor, polyglot mass would come whatever future southern California would have. Such a migration had been the strength of the north. But there was a vital difference and the *ranchero* stated it clearly:

"The miners who came to California were young and strong and single and free and filled with hope. These new émigrés in the south are not the same. They are defeated men. Even a drowning man has hope as long as he struggles. But hope without confidence makes nothing. So, we shall wait and see what these people make of our land."

Although he longed to collapse on a bed, Steven forced himself to stay alert. Don Emilio had provided the opening he needed to broach the matter of land acquisition. "The purpose of my visit is to buy some good land. For cash."

Don Emilio's face did not betray the sudden flicker of hope. "That should not be difficult now."

Steven smiled sympathetically. "Perhaps not. But I am looking for special land. Land with a guaranteed year-round supply of irrigation water."

Don Emilio's head rested against the chair back and he traced the contour of his lower lip thoughtfully with a forefinger. "That is more difficult."

"I am told it may even be impossible. But I felt I should ride out and see for myself."

"It is not necessarily impossible, Señor Lewis. We have land below the mouth of our canyon that receives water regularly, even in dry years. This year is unprecedented. But there is still some sub-surface water."

"Would you yourself sell such land?"

Don Emilio considered briefly. "Perhaps—"

"How many acres would be available?"

The question seemed to trouble the *ranchero*. "That would depend upon what a buyer planned to do with them."

Steven was evasive. "For the time being I have in mind an investment in fertile land that will support agricultural crops. I do not know just what crops yet. I do not want range land. I have little interest in cattle."

"And you have no market for agricultural crops, Señor

270

Lewis. You may have to wait until you are as old as I am to see such a market develop here in the south."

"I am in no hurry, Don Emilio. I wish to buy land at the lowest possible price and hold it. If enough of the right land was available, I would even be glad to buy it and let the original owner continue to use it, perhaps for just the taxes."

Don Emilio could no longer conceal his interest. His eyes narrowed thoughtfully. Steven decided to press a bit more. "Of course I would wish some assurance that the water would not be diverted, perhaps by a new owner of the headwaters."

"That will never happen as long as I live to control this rancho, Señor Lewis. The water and awareness of its value are my only real assests."

Suddenly less hopeful, Steven realized that he had been naïve to entertain even a small notion that any *ranchero* would part with the life source of his holdings, regardless of the pressures imposed by drought and economic setbacks. Those who were managing to survive were those who still had some water. But there might be some way. Billy Ralston had said often enough that there always was, for the man who wanted something badly.

"Perhaps tomorrow I could ride over the land you have in mind, Don Emilio."

"Indeed. If you wish we can ride to the source of the river and follow it down so you can reassure yourself."

"I would like that. Would there be as many as two thousand acres available, along the lower river?"

"It is possible. Yes."

In spite of himself, Steven yawned and Don Emilio smiled sympathetically. *"Ahora, dormimos, amigo!"* He rose and indicated the doorway. "I will show you to your room. Tomorrow we will leave early and we will have the day for questions."

That night Steven slept in a dreamless, timeless void. The barking of dogs awakened him shortly after sunrise. When he had washed and dressed, Don Emilio knocked on the door and escorted him to a breakfast of coffee, beans, and tortillas.

Young Anita, petulant with disappointment, watched her father and their guest mount and, followed by Pedro and several dogs, begin the ride into the wild *Cañada Laberinta* in whose upper reaches rose the Rio Claro.

271

For a mile or so the trail was well defined. They left it from time to time when Steven expressed curiosity about some of the side canyons that showed evidence of flash floods in the past. El Rancho de los Nidos suffered no less than the others from the uncontrolled run-off that continually changed the semi-barren canyons.

As they picked their way up the boulder-strewn Rio Claro the water became abundant. Two miles from the ranch the canyon widened abruptly and they came to the first of the *nidos*, the nest-like meadows that characterized the rancho. *Nido Mejor*, the largest, measured nearly forty acres. Despite the third year of drought, it was a beautiful spot, a miniature sheer-walled valley whose once lush grasses had been grazed to bare earth by wild cattle and deer.

The dogs, frustrated by a plethora of scents, went yelping off in every direction, then converged on a tangle of brush. Steven and Don Emilio rode over. There they found the half-eaten carcass of a doe. Pedro glanced at the carnage and grunted. "Pumas!"

One of the dogs began nosing the bluish blob of fly-covered entrails and Pedro sent it howling with a blow from a rawhide quirt. Don Emilio shook his head. "The big cats do not feed on dead cattle, but they are so bold now they come within a few yards of the main house to kill the dying." He turned to Steven with a sad smile and indicated the carcass. "Perhaps there is a lesson here that we *californios* should remember. No?"

As he relaxed in the shade, admiring the beauty around him, Steven recalled Arlene Baker's prediction that he would be led to Brindavanam. During the morning he had scarcely given her a conscious thought. But now, as her words echoed almost audibly, he was aware of a growing eagerness to be with her again.

Don Emilio had said there were many things to be seen within the eight square leagues of the rancho. Moreover, the Rio Claro's headwaters appeared to fit perfectly into a subdivision plan that had begun to assume real substance when he had visited the Anaheim colony the previous afternoon. There he had seen dramatic proof of the value of careful planning and the miracle of water at work in this arid land.

It was well past midday when they returned to the big

272

adobe by way of a series of lesser *cañadas* where Don Emilio had shown him other, smaller *nidos*, several of which were watered by dependable springs during normal years.

After the noon meal, Steven asked if he might ride over the lower ranges. Don Emilio, eager to prolong his visitor's stay, readily agreed after extracting a promise that his guest would not leave until the following morning.

A whispered family conference ended in a delighted squeal when Anita was permitted to join them. Instead of the circumspect side-saddle, and the full skirts usually worn by female riders, she appeared in full-cut brown woolen *pantalones* and beautifully handcrafted boots. Before Steven could assist, she was mounted astride a small, spirited gelding.

A wind, unseasonably warm, was blowing down from the mountains. Off to the south, beyond the low hills that concealed the ocean, Pedro called their attention to a formation of cirro-stratus clouds.

Don Emilio dismissed them. "How many times can a man's heart leap with hope?" But Pedro insisted the dry Santa Ana wind was the sign of a major weather change.

A half mile beyond the point where the Rio Claro disappeared beneath its bed, Steven dismounted and dug. Less than a foot down, moisture began to seep into the hole. The *ranchero* nodded. "It is there. The coyote dig for it. But there is not enough water for cattle. We have dug waterholes for two years now. But it is useless. The animals trample it full of sand again in five minutes. So we no longer try."

Steven frowned. "Couldn't you put a dam across here to hold back water when the stream runs?"

"Yes. It would be possible. But not for us. We do not have the tools or labor or the skill to build the stone dam required to hold against flash floods. And who is to say that the water might not make a new course? Look." He indicated the meandering streambed and the evidence of older watercourses that had scarred hundreds of acres of good land on either side. "These rivers are like temperamental women. With every storm they change their minds."

Steven laughed with Anita as Don Emilio turned his horse. "Come. Let us go back now. We have seen it all. One pile of bones looks much like another."

By five o'clock, darkness was closing in and Steven was relieved to be out of the saddle. Bootless, he stretched out on the laced rawhide bed and wished that he had brought another change of clothing.

Early the following morning, remounted and provisioned by Doña Paula and Gracia, he repeated the instructions given by Don Emilio. After sincerely meant expressions of gratitude, he reined his mount down the steep, tree-bordered lane.

El Monte, where he spent that night as a paid guest in a settler's home, disappointed him. The location was beautiful but it lacked the planning and order of the Anaheim colony. But even in the third year of the drought it was an oasis filled with promise. Several small orchards had been set out and there was evidence that the trees were being regularly irrigated from the main *zanja* that led down from the mountains. Here again he found confirmation of a pattern for the future.

Steven arrived at the plaza in Los Angeles shortly after noon the next day. He resisted an urge to ask the hotel desk clerk if the Lathrops were in and went directly to his room. There, filled with impatience, he stripped off his dusty clothing and managed an effective but unsatisfactory cat bath from the large porcelain basin. Clean and changed, he went to the bar and ordered food to be served at a corner table.

Arlene Baker's five one-thousand-dollar banknotes were still safely tucked in his pocket. No commitments had been made. But he was certain now that El Rancho de los Nidos offered the brightest promise despite its remoteness. Arlene had said that the right place would attract her followers regardless of its location. He wondered if the same principle, if that's what it was, would apply to the settlement that was beginning to assume detailed form in his mind.

While he was waiting, Steven tried to anticipate the questions Pioche would ask. Even if he could assure Pioche that they could effectively control the source of the water, transportation was a vital need before a country could really boom. The southland seemed years away from that. He would have to convince Pioche that immigrants could be diverted to the south by the lure of an easy living from some of the cheapest, most fertile land on the continent.

Steven tried unsuccessfully to keep his thoughts from straying to Arlene, to the part she might now be playing in his determination to produce answers that Pioche could accept. He both welcomed and resented her intrusion.

His introspection was interrupted by the flat, abrasive voice of Jabez Lathrop calling from the doorway. "There you are, Lewis. They told us at the stable that you'd come back sooner than you expected." As he crossed to the corner table, Lathrop glanced distastefully at the card players and the men drinking at the bar. "My daughter and I would like to talk to you. I wonder if you'd care to join us in a more genteel atmosphere in the dining room?"

Concealing his eagerness, Steven managed a token show of reluctance and agreed. The first course was being served when they joined the two women. Mrs. Lathrop favored Steven with her usual bleak smile but Arlene was flushed and effusive.

"Oh, Mr. Lewis. How fortunate for father that you were here. He wants so badly to ask your advice about some lots the real estate person showed us this morning."

The broker was a young man named Warden whose office consisted of one dingy room and a sidewalk bulletin board.

Nonetheless he had impressed Lathrop. "Mr. Warden seems like an honest Christian."

Steven smiled blandly. "If the real estate profession is beginning to attract *good* Christians, perhaps there's some hope for the country after all." The mild sarcasm was lost on Lathrop.

"When I get my ministry started, Lewis, I intend to see to it that all so-called Christians within the sound of my voice become good Christians. Mr. Warden is unaffiliated, it seems. He has already promised to be one of my deacons. He assures me he will concentrate his company's sales effort in the area to help me build our congregation."

Steven made a mental note to look up Warden. Obviously he was bringing to his profession some unusual imagination. Arlene leaned forward. "Mr. Warden was kind enough to interest himself in me, too, Mr. Lewis. He said that he lives with his sister and they would be happy to put themselves out to see that I am introduced into the proper young circles in Los Angeles."

275

Steven suffered a pang of anxiety mixed with jealousy. "I shouldn't think that would take much putting out."

Arlene glowed with appreciation.

Lathrop flopped his napkin, "Let's get on with this real-estate proposition. I'm imposing on Mr. Lewis as it is."

"It is no imposition, sir. Where is the property?"

"It's south, about a mile out of town where Main Street and Spring Street come together."

Steven shrugged. "I don't really know, but the other evening I heard one of the natives betting that Los Angeles would grow toward the south. He said in the beginning towns usually did."

"The price seems right and the places across the road are all about to get city water. Warden is offering me eight quarter-acre city lots at a hundred dollars an acre. He says I can hold it for a couple of years and sell off half of it for more than I paid for the whole shebang."

Steven nodded. "That's been happening in San Francisco too. But there's a lot more going on up there."

Lathrop laid his fork aside. "Warden says he'll guarantee title at the courthouse. As to taxes, I don't really care."

Steven spread his hands expansively. "Good, I envy you. Sounds like southern California has a useful new citizen, sir. My mother's people would say: *'Viva aqui en buen salud.'* Live here in good health."

For the first time during the discussion, Mrs. Lathrop spoke. "I didn't know you were Spanish, Mr. Lewis."

"I'm not. I'm half Mexican."

The woman attempted to conceal her stricken expression.

Arlene relieved the awkward silence. "Live here in good health. What a civilized benediction. Thank you, Steven."

Jabez Lathrop pushed back his chair. "Lewis, I have a feeling that if anything had been seriously wrong, as I've related it to you, the Lord would have moved you to speak out."

Steven stirred uncomfortably. "Sir, I'd feel a lot better if you took my silence for ignorance of the subject."

Arlene placed her napkin in its ring. "Nonsense, Steven. I want to hear about your land. Would you like to walk in the plaza and tell me about it?"

Several minutes later they were strolling in the warmth of the winter sun. Taking his arm, Arlene said, "I'm dying

276

to know whether you found anything approaching my Brindavanam."

Steven described his journey and the effects of the drought. Deliberately, he emphasized the negative aspects of the search, devoting only a brief comment to the Anaheim and El Monte colonies, avoiding entirely any mention of the vital part ample water was playing in their success. "There was only one place that looked promising, Arlene. But it's at least forty miles southeast of here, in the Santa Ana Valley."

"Forty miles? Is there nothing closer? Surely you can do better than that."

It was less rebuke than disappointment but it annoyed him. "Why not ask Warden? He's your official real-estate agent. He'll probably be able to do better. I'm just an amateur looking for land for myself."

Arlene Baker lowered her head. "I deserved that, Steven. I'm selfish and impatient. Please forgive me? And Mr. Warden is not my real-estate agent. He's just an eager young man who will promise anything to make his sale."

She studied him intently for a time. "Steven, I do believe you're jealous. Are you?"

His lips compressed stubbornly then twisted into a sheepish grin. "A touch, maybe."

Arlene looked around quickly and leaned close to brush her lips against her ear. "Darling Steven. Do you know what you've done to me, here?" She indicated her heart. "You have so unsettled me that I haven't slept a wink since you left. I've created the most wondrous fantasies about us." Arlene was tempted to tell him and watch his face as she described them. She had lain naked on her bed in the warm, dry night and with her hands had brought her nipples to hard little points and then aroused all of the senses in her body until she had been forced to cry out his name. Instead, she leaned away and regarded him gravely. "Do you know something else, dearest, darling? The only peace I've ever known was in your arms and I would rather die than never to be there again."

His annoyance faded as quickly as it had flared. He was seized by an overwhelming urge to be alone with her again, to shed the last of his reserve. But he remained frustrated and inarticulate. He could feel her, waiting, almost able to hear his unspoken words, but they were the wrong words and he could not bring himself to say them.

277

"Steven, if you can, please tell me one thing. Do you feel perhaps even a little of the same thing I feel?"

He nodded. "I do."

He heard her whisper something to herself. A moment later she drew in a deep breath and sat up straight. "One place looked promising. I said you would be led to the right place, Steven. If I suggest that it is the wrong place simply because it is not close by I shall be denying everything I profess. Tell me about it. Please."

Steven described El Rancho de los Nidos and spoke warmly of Don Emilio, Doña Paula, and their young daughter, Anita. He described *Nido Mejor*. But again he did not emphasize the water beyond saying it was adequate for limited use even after three years of drought. He said nothing about Don Emilio's reluctance to sell and omitted any mention of the Rio Claro. The ambivalent nature of his feelings distressed him. He wanted to be open with Arlene, to confide in her the tremendous possibilities he saw in the land. But when his own enthusiasm brought him to the verge of superlatives he restrained himself and did not know why.

"The *cañada* may be too wild and remote for you. The whole little valley might become a trap. And you may have trouble getting your followers to travel that far."

Arlene refused to acknowledge obstacles. "Oh no, darling. If it's the *right* place I shall know in an instant. They'll follow me. Why don't we go there together and look? We can hire a surrey and leave in the morning."

"It's not possible, Arlene. There are no proper places to stay between here and there. It's a two-day drive."

"Surely, if we say that I am your fiancée something decorous could be managed. I'm a farm girl, remember, darling? I can make do with very little, really."

"It won't work. We would have to drive it in one long day and Don Emilio would have to know we are coming. There may be more rooms but I saw only three bedchambers."

"I could share a room with the child. Gracious, Steven, I shared with three others at college in Chicago."

Steven remained adamant. "There's no guarantee that he will sell any land."

Arlene's eyes widened. "You didn't even discuss the possibility?"

"Not openly. Not the first time. The best way to run the

278

price up is to be an eager buyer. Besides, Don Emilio may not be in the same straits as the other *rancheros*. But the thing to find out is whether or not he has needed to borrow to pay taxes, and if so, how much and from whom."

"How can you find out?"

"I'm not sure I can, unless he'll confide in me. If he's been delinquent recently there may be a record at the courthouse. If so, chances are he borrowed. And if he did, the party that holds the note is taking his land from him a piece at a time."

"How could we get the land then?"

"Make an arrangement with Don Emilio to satisfy the note in exchange for acreage."

Arlene lapsed into a thoughtful silence. Suddenly she lifted a defensive hand. "Steven, I don't want to wave a red flag in front of you but perhaps Mr. Warden would know. He seems to know everything about land in the whole county."

Steven reacted instantly. "No! Don't do that. First of all. I'm not sure the rancho is in the county. And second, if you tell Warden that you're interested he'll run down there and try to make a deal to figure himself in. He would double the price, if he didn't crab the deal altogether."

Arlene laughed sympathetically. "Poor Mr. Warden, you don't trust him, do you?"

"I don't trust any professional real-estate boomer."

"I find that curious, darling."

"How so?"

"Well, isn't it a bit of the pot calling the kettle black?"

Stung, Steven moved to face her. "There's one whale of a difference, Mrs. Baker, between buying land and improving it, so others can use it productively, and running around peddling other people's land like—like tamales from a cart!"

Arlene was amused. "If I'm to be spanked, Mr. Lewis, I'd much prefer it be in private, over your knee." Her mocking laugh made him splutter as she rose and slipped a possessive arm through his. "Come, darling. Let's go to the courthouse and see what we can find." She glanced across the plaza. "If their silly siesta is over."

An apathetic clerk said the records were privileged information and he needed written authorization from the owner. A ten-dollar gold piece revised the rules long

279

enough for the records to reveal that a tax lien against El Rancho de los Nidos had existed for the previous year. It had been satisfied, with interest penalties, a few weeks earlier. There was no way of determining how Emilio Garcia y Perez had come by the lump sum of '$987'. Since there was no record of a recent land sale it seemed reasonable to assume that the money had been borrowed.

On the way back to the hotel, Steven insisted upon returning the bank notes, at least until he had seen Don Emilio once more. At the bottom of the lobby staircase they were still arguing.

"I don't want the responsibility. If there *is* a chance for making a purchase, I'll go to San Francisco immediately and discuss the proposition with my financial partner."

Arlene continued to shake her head in stubborn refusal. "No, Steven. Too many things can happen by then. It could take three weeks to get up north and back here again if there's no steamer connection. Please, darling, use my money for now. If you're going to be so everlastingly fussy, you can give me a note for it. Wouldn't that make it better?"

Steven realized it would indeed make it better. It would spare him the time and expense of two trips but would reduce the risk that the drought might break during his absence and relieve the pressure on the distressed land market.

"Why do you hesitate, Steven? Surely your overdeveloped male pride won't keep you from accepting a sound business proposal simply because a woman makes it."

After what he considered a decent show of reluctance, he nodded. "All right, as a business loan, and at the prevailing interest rate. But I won't need all of it. Half will do."

He glanced at the desk. "I'll get some paper and draw up a note for twenty-five hundred dollars, at three percent per month, compounded. That's the rate the *rancheros* have to pay."

Arlene gasped. "Why, that's usury. I don't want such interest."

"Want it or not, that's the deal. If I'm able to option the land the loan will be worth twice that in the time saved."

In the end she agreed. To avoid the problem of change,

he signed the note for three thousand dollars. The paper was callable in ninety days, a period Arlene still insisted was too short.

Steven dined with the Lathrops again that evening and Lathrop monopolized the conversation with his plans. "I'll have the biggest congregation in Los Angeles within a year. Know why? Because I'm not going to be dunning the congregation for cash. I'm going to ask for contributions of home-preserved food items to sell in the Lord's Pantry. Modest profits from good wholesome food will do the trick and it won't cost them anything because the church will buy their offerings at a fair price. And every penny of the profit from resale will go into blessed benefits they all can enjoy—a fine community house, a care home. Each member of my tabernacle will be God's working partner."

As they left the dining room, Arlene moved close to Steven. She whispered, "Darling, if I promise not to stay, may I stop by later?"

Before he could reply, she extended her hand. "Good night, Steven. And I do wish you every success tomorrow. Father and Mother will be as interested to hear about it as I."

Lathrop turned with a puzzled frown. "Hear about what?"

Arlene laughed apologetically. "I forgot to tell you. Steven thinks he may purchase some land too."

"Oh? Splendid. Splendid. Where?"

Arlene looked contrite. "Perhaps I shouldn't have said anything, Steven?"

Uneasy at such by-play, Steven dismissed the matter brusquely. "No, no. I'm just looking for cheap ranch land."

Lathrop snorted. "Good God, Lewis, if it's range land you want, for ten thousand dollars you can buy enough to make your own state."

Before he went to the hotel bar, Steven hired a mount for five in the morning. During a brandy he learned that a local Indian was predicting an end to the drought before the next full moon. At nine o'clock he went up to his room.

Impatient to hear the light tapping on his door, he occupied himself by listing the arguments intended to persuade Don Emilio to sell the high eastern portions of the ranch. The *ranchero* was an exceptional man, well-educated, and unlike many of his contemporaries he appeared to have staved off complete financial disaster even during the worst of the drought crisis.

When the clock passed ten and then ten-thirty, Steven could no longer concentrate. Anticipation had become anxious need and was now complicated by doubt and apprehension. Something must have happened to keep her in her room against her will. A few minutes after eleven he stepped quietly into the hall. At the stairwell he could hear the usual cacophony from the saloon. Ashamed of himself for behaving like a lovesick schoolboy, he moved cautiously down the stairs and along the second floor until he was within a few feet of the Lathrops' rooms. Clearly, from the nearer of the two, came the unmistakable sound of a strange man's voice followed by Arlene's sudden laughter. Shaken with anger and disappointment, he bolted down the stairs to the bar.

The bartender, mildly surprised to see him back again, moved two cataleptic regulars along to make room and served him a double whiskey. Several drinks later, grim and drunk, Steven climbed the stairs, let himself into his room, removed his boots and coat, and fell face-up on the bed.

Pistol shots, loud guffaws, and the panicked clatter of hoofs awoke him shortly after daybreak. At the window, he saw a knot of drunken men across the street in front of the Lafayette Hotel. All were clouting one another and laughing hysterically. In full flight, going south on Main Street, was a lone, hatless figure on horseback, obviously the victim of some malicious prank. He recognized the ringleader as the well-known bar bully, Jack Pollard.

Steven doused his head in cool water and blotted his face on the coarse towel. What he needed now was coffee. He'd be late to pick up his horse, but that mattered little at the moment. In the dining room, after drinking several cups of strong, hot black coffee, he forced down an unappetizing breakfast of limp, fat bacon and greasy, overdone eggs.

It was ten of six when he arrived at the stable. Since he intended to push through without stopping, he arranged for a second mount. Feed was in such short supply that both animals were little better than bone bags. At two dollars a day, the spare animal was cheap insurance.

282

Familiar now with the country southeast of the pueblo of Los Angeles, Steven headed directly for El Rancho de los Nidos. The spare mount proved to be the sturdier of the two and he arrived at the rancho corrals while it was still light. As he started up the long driveway, the dogs set up a welcoming clamor. Pedro saw him first and ran inside to alert his *patrón*. In a matter of moments both Don Emilio and Young Anita were greeting him, filled with delight at his early return.

Somehow, on short notice, Gracia and her *cocineras* managed a festive meal and for the first time since his futile wait for Arlene, Steven began to relax. It was strange, he thought as he listened to Anita's animated chatter and Doña Paula's gentle remonstrances, how much more at home he felt in the Garcia y Perez *casa grande* than in his own father's modern brick home in San Francisco. Here he felt the same ease that he had known as a child in his grandfather's adobe at Los Robledos.

Don Emilio carefully avoided any inquiry about his early return. Later, in bed, Steven forced himself to turn his mind from Arlene and review the points he would discuss with the *ranchero* in the morning. But the more he reviewed them the more he was afraid he would reveal himself as a naïve young fool.

Finally, the vague outlines of a proposition began to form. It would be based on Don Emilio's own assessment of the future of southern California and if the *ranchero* really believed his pessimistic predictions, it would be a plan, the only plan the man could agree to if he wished to survive.

Steven arose filled with excitement. He washed quickly, eager to breakfast with Don Emilio and get out on the land again. It troubled him not at all that the plan would be implemented with Arlene's cash. Since he had given her his personal note, his conscience was free. He felt refreshed and pleased with himself and he smiled as he recalled one of Harry Morrison's endless homilies: "Most of the good things happen to a mon when his head is oop and the bad things happen when his pants are doon."

Don Emilio suggested they ride eastward to explore several small *cañadas* that in normal times provided additional water from springs. He looked across the lowlands to the south and said, "When a cattleman is forced to sell

283

his range to keep his water, he sells himself out of business."

His pulse pounding, Steven waited. Then, almost casually, he broached the matter of a partnership in which the *ranchero* would pledge the land and its water sources, and he and Pioche would pledge the financing to divide several thousand acres of the rich grazing land into forty-acre *ranchitos* with irrigation guaranteed from reservoirs filled by the waters of the Rio Claro.

Don Emilio's long silence worried him. He wondered if he had moved too quickly, had suggested too radical a change even though the *ranchero* himself had referred to Anaheim with its engineered water supply as the shape of things to come.

Finally, Don Emilio turned to him. "Señor Lewis, if the proposal had been made by one who was not a *californio,* I would have taken only a minute longer to say yes. For you see, my young friend, there is no alternative."

As they rode down into the bottom land they talked of the best sites for the farms, the means for transporting water from *Cañada Laberinta* and of the financing. They would be equal partners, subject to Pioche's approval. Don Emilio would also provide the labor and facilities to board the workers. Steven and his backer would provide the sales organization. Two hundred forty-acre parcels would be platted and the farmers who purchased or leased the land would be chosen for their ability and the variety of the crops they were qualified to raise.

Don Emilio brought up the subject of crops. "When everyone raises the same crop and there is no big outside market, there can be no trade. In recent years that has been southern California's curse, a single crop, cattle, and a failing outside market."

At the house that afternoon, Steven put a rough agreement on paper and asked about liabilities against the ranch. Don Emilio confessed unhappily that he had to borrow to pay taxes. Then both men signed and Steven left immediately to buy up the note held by merchant John Temple in Los Angeles.

The Lathrops were not in the dining room when he came down late after changing and packing for an early morning departure for San Pedro Bay. An inquiry revealed that they had not been in earlier. At the desk he was told that they had been picked up in a surrey late that afternoon. The clerk did not know by whom. Restless and unwilling to stay alone in his room, he went to the bar.

Arlene found him there at ten o'clock and sent the clerk to summon him. She was pale, weary, and obviously distressed. "Darling, we can't talk here and I've been sick with worry about you. Go up to your room. I'll be there in less than an hour. I promise."

When he looked at her skeptically, tears threatened. "Please, Steven—" She turned away and hurried up to the second floor.

Curious, angry, and impatient, he returned to the bar for another half hour, then made his way up to his room. He had been there only a few minutes when he heard the light tapping. Arlene made no pretense of caution as she rushed to his arms, buried her face in his shirtfront, and began to sob. He held her for several minutes, cradling her cheek. Warm tears collecting between his fingers made him reach for his handkerchief. He attempted to blot them but she pushed away and moved to the foot of the bed.

"Darling, I couldn't come to you because Mr. Warden came to see father. They talked late and mother insisted on coming in to sleep with me. There was simply no chance to slip away and no possibility of getting word to you. I nearly died, darling. I wanted so to be with you. I knew we couldn't again, until much later." She reached out and rested her hands atop his. "Please forgive me, Steven darling. I don't know how much you suffered but I went through the torture of the damned." She lowered her head and Steven could see tears threatening again. "I'm so sorry, darling, really." A deep breath helped and she straightened. "Now then, tell me about the land for Brindavanam. Did you get it?"

Steven had been half dreading the question. "I worked out an agreement with Don Emilio and paid off his note late this afternoon. For all practical purposes we have what we want."

Arlene Baker brightened instantly. "Wonderful, darling. When can we go to see the little valley?"

"Not for a while. I have to catch the first stage for San Pedro in the morning. Banning's agent got me deck passage on a small Chilean coaster that steams at noon."

Arlene could not conceal her disappointment.

"How long will you stay in San Francisco?"

Steven evaded the question. "That depends on how long it takes me to convince my money people that I've done the right thing."

"But of course you have, darling. I don't know what you paid for the land but I'd trust your judgment in anything."

Steven looked at her hard. "When the deal is finally worked out I may remind you of that."

Concerned, she drew away and regarded him narrowly. "What do you mean, Steven dear?"

"I mean that Don Emilio refused to sell the land I really wanted. I had to work out an arrangement. Actually in the end it will make the financing a lot easier."

Perplexed now, Arlene freed her hands and folded them in her lap. "Then you paid off his note with my—with the money you insisted on borrowing—and now you hold the mortgage?"

Steven nodded. "That's what it amounts to, at the moment."

He reached into his pocket for the signed agreement. "This is Don Emilio's note for three thousand dollars. There are thirty-five thousand five hundred and twelve acres in El Rancho de los Nidos. If he should default on the note, we own six thousand acres of his best lower range. But that is not what we want." He glanced at the paper again. "What we want is equal partnership in the whole ranch. He puts in the land. My job is to get the money to develop dependable water and bring it down to subdivided forty-acre farm plots."

He knew that Arlene was following his reasoning. Her deliberate attitude of vague befuddlement annoyed him. "But Steven darling, I'm afraid I'm dense where business is concerned. Perhaps I don't really understand. The money hasn't bought anything, has it?"

"If you mean, has it bought you your valley, or has it bought me acreage, the answer is no. But it has bought us time and at no risk to your money."

She gestured impatiently. "I'm not worried about my money, darling. But that's not the point."

Before she could continue, Steven broke in. "The point is, Arlene, what you and I both want is secure now, rain or no rain. When I get back from San Francisco the first

286

thing I'm going to do is return the money to you. I'll feel easier. The risk should be all mine."

She sat in silence for a time, nervously flexing her long white fingers. Then she looked up. "I love you even more, Steven, because of your concern for me. But I am a little disappointed."

He pretended surprise. "Why?"

"Because I know that if it was just up to you there would be no question of my getting Brindavanam. But if you have a partner, I mean *another* partner, he might not feel the same way."

Steven raised his right hand. "Arlene, I promise you that if you want your settlement in *Nido Mejor* you shall have it there."

Her eyes, still tear-glazed, searched his face. Then she drew in a long breath, straightened, and stood up. Steven rose to take her by the shoulders and draw her close but she held him away with her palms against his chest. "Please, darling." Rising on tiptoe, she kissed him lightly on the lips. At the door she paused. "Mr. Warden is trying to find us a house, Steven. If we have moved before you return, they'll know at the desk where we are." She opened the door part way. "If you're going to be away longer than you expect, please write, darling."

The stay in San Francisco proved to be both triumph and trial for Steven. Francis Pioche was delighted to see him and evidently satisfied with his answers to questions about construction and developmental costs. However, Pioche was amazed at the way Steven had managed the interim financing. He confounded Steven by explaining that the so-called "bank notes" Arlene Baker had given him in return for his note were merely certificates and not generally negotiable. But before the enormity of his ignorance of finance overwhelmed Steven, Pioche said quickly, "It was not a *faux pas, mon jeune ami*. It was a very inventive and expeditious thing to do. The certificates were endorsed by Lazard Frères. They are better than any paper money. It will be some time before that sophisticated means of transaction is legally authorized and accepted out here. We Californians are accustomed to hard money. Too many of us remember the Southerners and their own recent paper currency."

Pioche promised that he would deposit ten thousand

287

dollars in gold for Steven's use. He looked up at Steven approvingly. "You inspire confidence, *mon ami*. Certainly you inspired Madame Baker. I only hope that her husband does not come to resent that."

Steven did not tell Pioche that Arlene Baker was a widow only three years older than he. It was not that he intentionally withheld the information, but neither could he explain why he could not bring himself to volunteer it. Instead, he dismissed the possibility of a jealous husband as "very unlikely."

Shortly before Christmas he received a letter from Arlene. Its impassioned, explicit, highly personal nature touched off an emotional conflagration whose effects were evident to everyone in the family. His mother, attempting to conceal her hurt with humor, chided him about his behavior, said that he seemed as nervous as an unwilling bridegroom.

Will merely laughed and said, "If I was responsible for ten thousand dollars of Pioche's gold, I'd be jumpy too."

John Lewis chimed in with unconvincing heartiness. "If you feel edgy because you're walking out on us, Steve, don't let that worry you. Will and I intend to demand exclusive rights to trade with your new southern empire. But try to comfort your mother. Women take the loss of sons harder. Daughters bring home husbands but sons seldom bring home wives."

Steven wrote to Arlene, but while he suffered the same physical longing to be with her, he could not bring himself to put on paper the same explicit statements about their infatuation. He had never read such a letter as hers. Each re-reading aroused him unbearably.

The holiday helped to divert him but shortly after Christmas, his impatience became insupportable. On January second he booked passage south on a rickety sidewheeler a week earlier than he had promised. His action precipitated a family scene that destroyed the last vestige of the old unity. Doña Maria clung to him, weeping. His sister Felicia and brother-in-law Lambert Lee made no attempt to conceal their deep disapproval. His father's transparent attempt at philosophical acceptance and his twin's token "God Speed" only worsened the situation.

Aboard the *Sea Bird*, a wheezing, under-powered, ill-maintained floating fire trap, he spent most of the time huddled in the lee of the deckhouse desperately impatient to be with Arlene.

Darkness had settled over Los Angeles when Steven arrived. He immediately inquired about the Lathrops. To his vast relief, they were still registered at the Bella Union and were, at the moment, in the dining room.

Upstairs, Steven finished a hasty wash and change. Then, filled with the turmoil of anticipation, he hurried down to meet them. When she saw him, Arlene, seated facing the entrance, gasped and turned dead white. An instant later she was on her feet, running toward him. For a moment she just stood looking up, her lips forming silent questions. Finally she managed a half whisper. "My God, Steven! We didn't expect you. I thought I had seen a ghost."

The color came flooding back to her cheeks. She caught his arm with both hands and led him to the table. As he allowed himself to be led, he realized that the fourth place was occupied.

Jabez Lathrop rose, beaming. "Steven. Steven, my lad. Our prayers are answered. You have returned safely. Welcome. Welcome." He motioned to a waiter as he spoke. "Come join us. We're hungry for news."

Steven acknowledged Mrs. Lathrop's restrained greeting and turned to the guest who was sharing the table. Arlene's hand went to her mouth. "I'm so sorry, Steven. I thought you two had met. Mr. Tom Warden, Mr. Steven Lewis."

Warden, a lean, clean-featured, blonde man in his early thirties, rose to greet Steven with the open smile of a salesman accustomed to meeting people. "I'm happy to know you. I've heard so much about you from the Lathrops."

Steven shook the proffered hand and nodded. "Happy to know you, Mr. Warden. I've heard a lot about you too."

Seated across from Arlene and Tom Warden, Steve was aware of a sense of unreality. The eager, beautiful young woman who had invested his restless nights with excitement seemed to bear little resemblance to the pale, almost too-thin, somewhat over-eager young woman smiling at him. Only the eyes were as he remembered them, dark-lashed, peridot green, and bright with fanatical intensity. The soft caressing voice of his fantasies bore little resemblance to the high-pitched one, strident with forced gaiety, that he was hearing now.

"We'll spoil you shamefully by telling you this, Steven, but you just can't imagine how we have tried to follow every step of your journey and how we've prayed that it would be successful. Do tell us how everything is in San Francisco."

Disappointed and upset to find Warden present, for he had sensed earlier that his interest in the Lathrops was not limited to just real estate, Steven spoke of San Francisco's phenomenal growth and the scandalous misappropriation of Central Pacific construction funds by the Big Four. He told them about the free-running operation of William Ralston's new Bank of California and the ominous rumblings in the Comstock mines in Nevada despite the wild speculation in shares and about Adolph Sutro's wild proposal to dig a tunnel through Mount Davidson to save the drowning Virginia City mines.

As he talked, Steven was aware that Warden was covertly watching the effect of his account on Arlene. He noticed that she was not unmindful of the effect her own enraptured attention was having on Warden.

To Steven it seemed that there were two selves at the table. One was speaking and the other was observing. The observant Steven was certain that Warden was a serious contender for Arlene's affection and that she enjoyed the competition between them and was deliberately exploiting it.

He wondered just how much progress Warden might have made during his absence, just how much he had been aided by Arlene's loneliness and candidly confessed needs. He wondered too about her eagerness to make love to him and he tried to find reassurance in accepting that devastating phenomenon, love at first sight. His own mother had often confessed to being its willing victim.

Steven said good night to the Lathrops with courtesy but with almost perfunctory haste. He did not know whether or not he wanted to give Arlene the chance to arrange a later tryst.

In the bar some minutes later, a Mexican bus boy brought Steven a note from Arlene—three hastily scribbled words: "Don't lock door."

He dozed off, naked under coarse sheeting. The lock

turning awakened him, but before he could rouse himself, Arlene was bending over him kissing his face hungrily, whispering his name over and over in a tight, strangely harsh voice. In the half light he saw her long, small-waisted skirt drop to the floor. She wore nothing beneath it. She straightened up and the blouse joined the skirt, discarded where she stood. An instant later she was in his bed.

Their first coming together was a violent, agonized clash. She clung to him, her lean arms and legs spidered around him, contracting to keep him in place, possessing him completely, a ravening, lubricious, lavender-scented sensualist.

When their breathing became more measured, her arms relaxed and slid along his body until her fingers found the abrasive stubble of growing beard. For a time they moved slowly, exploring the planes of his cheeks and the turn of his jaw and chin. Each finger seemed to possess an animus of its own. But once his mouth was on hers again her mood changed and she kissed him with restraint and tenderness, brushing her lips lightly around the periphery of his own.

The caresses continued for several minutes. Then her tongue forced its way between his teeth. In seconds her lean body began to quicken and moisten again. He felt himself beginning to burgeon inside of her and her body, sensing his mounting passion, began to move beneath him convulsively. There was no sound in the room save the faint creaking of the heavy bedframe and the whispery, rhythmic chafing of undulating flesh against the coarse cotton sheets.

Just before dawn, Steven started from a half slumber to find Arlene propped on an elbow, smiling down at him. "What's the matter, dearest?"

He blinked and looked at the rectangle of diffused light bordering the drawn shade. "Hadn't you better go?"

Her smile broadened. "No, darling, I rumpled my bed shamelessly last night and left a note for mother saying I had gone for an early walk."

He grunted sleepily and turned his head away from the window but she caught his face between her hands. "Dearest, we have an hour yet. Please, please, Steven?" Her right hand left his face and trailed down along the hard plaques of belly muscle. "Please, darling, make love to me again. Please—"

Moments later she was trying to urge him still deeper. As the time of her coming approached her moaning became beseeching and she gasped as she felt the first surge of him in her. Her body went rigid and turned clammy again with beads of perspiration. When his own orgasm had subsided she rolled from him and lay silent and still. Although he still held her, it seemed to him that Arlene had somehow moved away.

It was mid-February before Steven could make arrangements to take Arlene to El Rancho de los Nidos. He believed that he was deeply in love with her and to meet her endless need he willingly took her to his bed almost nightly. She laughed at his anxiety and made the most devious pretexts for their meetings seem rational and free of risk.

Some weeks earlier she had brought to his room a dozen neatly handwritten pages copied at the Chicago academy. They were excerpts from Vatsyayana's book of love precepts, the *Kama Sutra.*

He had been excited and troubled by her shameless pleasure in implementing the detailed techniques for heightening their physical responses. Some of them seemed ridiculous and made him self-conscious.

Skillfully, she had instructed him in what she called, "the bliss of self-awareness." Finally he came to accept her eroticism. Carried along by her uninhibited joy in love-making, he set aside most of his guilt and restraint. In the midst of their orgiastic couplings he was convinced that some force outside himself had brought about their meeting. Never had there been a woman, never would there be a woman whom he wanted as much as this one.

Brindavanam seemed to have become as much his commitment as hers as she pictured it over and over in their quiet times together. "Our mission is to found a pure society of teachers. In our library we shall have the entire *Kama Sutra* and Rousseau's *Confessions,* to show that the greatest of the philosophers were made to walk through a hell of self-doubt and guilt before experiencing perfect freedom."

He worried that Arlene was leaving him little time or energy to devote to Pioche's business interests. But that troubled him less than Pioche's own laxity in forwarding the promised funds. The bank in Los Angeles had not yet received the credit draft from Pioche. Steven could rationalize the delay just as he was rationalizing the enormous, unbridgeable chasm between his own Catholic-inculcated moral precepts and Arlene's equally passionate conviction that the human spirit could not thrive in a body constricted by superstition, guilt, and fear.

Steven and Arlene arrived by surrey at El Rancho de los Nidos after a properly arranged overnight stay at the new Ferguson Fruit Ranch. ("It was absolute hell, Steven darling, knowing you were on the other side of that thin wall. I could hear you tossing and turning!") Don Emilio and Doña Paula greeted them with unrestrained warmth.

During dinner, Doña Paula, who had been watching them with ill-concealed romantic delight, drew a pained expression from young Anita and a gentle rebuke from Don Emilio for saying, "How beautiful you two are together," she indicted her heart, "here!"

Arlene laughed and reached for Steven's hand. "Thank you. I know it shows with us, just as it does with you two."

They talked of the increasing number of Southern immigrants whose wagons rolled in a steady stream across the southwestern desert to San Bernardino. Arlene said that building lots were in demand in Los Angeles for the first time since the founding of the pueblo. She spoke also of the new home her parents were building and her own eagerness to see *El Nido Mejor*. The reference, in context, brought a fleeting look of puzzlement to Don Emilio's benign face and Steven was disturbed that he had not been more open with the older man who was now his partner. When the women retired he would attempt to explain more fully. But there was no chance. Don Emilio confessed to weariness and he was forced to wait for another opportunity.

When Arlene, riding Doña Paula's side-saddled mount, first saw *El Nido Mejor* she gasped and remained speechless until the party, which included Don Emilio, Anita, Steven, and Pedro, were well inside its western rim. Her voice was scarcely audible and the morning sun struck her widened, uptilted fanatic's eyes with pale green fire. "It *is* Brindavanum! It *is* Brindavanam!"

293

Steven noticed that the hand clutching the reins was trembling, and her face was white and drawn. He moved his horse next to hers. "Are you all right, Arlene?"

The cataleptic stare, the moist, half-parted lips, the rigidly braced body frightened him. He questioned her again but her only response was a mechanical nod.

"There's more to see, Arlene. Let's ride up to the springs."

Steven moved off, leading the way. A half hour later he approached Don Emilio and Anita, who were seated on a fallen log watching the lesser of the two cascades tumble into the stream from the lower pool. "Would Doña Paula be offended if Mrs. Baker and I stayed here for a while to talk?"

"*Por cierto, no, amigo.* She would be the first to understand. But what about food?"

Steven dismissed the matter. "We'll be back by mid-afternoon. Perhaps Gracia's women can give us a tortilla and some frijoles then."

The sun slipped behind the rocky shoulder of the canyon leading into the secluded oval valley and the air turned chilly. Steven rose and pulled Arlene to her feet. She let the momentum carry her into his arms and they stood laughing and kissing and resenting the barrier of clothing. Arlene pulled away and looked up at him.

"Darling, we must be together again. Somehow, we must. Every particle of me aches to let you know how grateful I am for this miracle, the miracle of you," her arm swept through an arc, "and the wonder of this." She moved closer again. "We must consecrate it with love, Steven," she whispered with her mouth on his. "Take me now, dearest Steven, like this. Let us consecrate Brindavanam a Garden of Love with our bodies and our souls!"

In the waning sunlight, a thousand feet above *Nido Mejor*, a quarter of a mile away, a lone figure stood at the mouth of the sandstone cavern. The cavern had been his hideout for the past five years, ever since *Los Lobos* had been driven from the mining country north of the Tehachapis. His right index finger was hooked through the handle of a tin cup and a rifle rested in the crook of his

294

left arm. For a time he squinted at the two distant figures on the canyon floor below. Puzzled, he called to one of his lieutenants and pointed.

The man watched curiously, then shook his head. "It is too far. I cannot see. There are two people and two horses. Do you want them?"

Goyo Robles considered for a moment and shrugged. "No, not now. There were others earlier. They may still be around. If they are from Los Nidos I have no quarrel with them. Don Emilio has resisted the gringos."

The man known and feared throughout the ranchos of California turned away. As he did, the last slanting rays of the sun struck the tin cup with a burst of light.

Steven started at the flash far above them on the mountainside. Arlene looked at him curiously. "What's the matter, dearest one?"

Steven continued to study the almost sheer face of the mountain rising above the springs. "I thought I saw something move up there." He slipped his hands beneath her arms to lift her from him. "Let's go now, Arlene. It will be dark in an hour and they'll be looking for us."

9 ON THE FIFTEENTH of February of the new year, 1866, Steven Lewis transferred the principal sum of three thousand dollars in gold, plus accrued interest, from the Pioche funds in the new Hellman Bank to an account opened for Mrs. Arlene Lathrop Baker. Arlene seemed both annoyed and distracted. Steven interpreted her behavior as disappointment that she herself no longer had a tangible tie to his undertakings at El Rancho de los Nidos.

Several days later he was shocked to the brink of panic to discover that she did indeed have a tangible tie to all of his undertakings. She was pregnant.

When he asked with undisguised anxiety if she had seen the "old Mexican woman" whose potions could often abort the fetus, she admitted that she had not. Moreover, she stated that she had no intention of attempting to rid herself of the child. Naked in his arms, where she had insisted she must be, she explained her reasoning. "Dearest darling, my refusal has nothing to do with our so-called high and mighty morals. I am not being inconsistent. Not at all.

"Our child was conceived at Brindavanam. I'm certain of it. Don't ask me how I know, darling, but when we were together there almost eight weeks ago and I felt you flooding in me, I knew then that what you call 'God' and I call 'Divine Force' had touched us and blessed us." She clung closer, urging her body against his. "Oh, darling, darling, no child will ever be more wanted and loved. This is the one I shall keep. *This* is the one."

Puzzled, Steven drew away. "The one you'll keep?"

She caressed his face and nodded. "Yes, darling. Once there was another that I did not want. I lost it."

"But I thought you told me your marriage was never consummated?"

"I did tell you that, dearest one. My marriage to Gerald was never consummated."

Steven gasped involuntarily. Arlene laughed gently and took his face between her hands. "Kiss me, Steven, and I'll see if I can't make that dreadful expression go away. Steven, my dear, you must have known that I was not a virgin, just as I knew that I was not the first woman you had known. I was fortunate, darling. I had my first experience with the professor of philosophy at the academy in Chicago. He was a much older man, a very sensitive, skillful married man and I did not realize that, well, I wasn't really aware of his intentions until it was too late." She turned her head away. "I got caught, the very first time.

"The poor man was so distraught I thought he was going to destroy himself. I did everything. I sat in hot epsom salts baths until I thought my skin would peel away. I drank a dozen horrible concoctions. Then one night, after three months, something seemed to let go inside and the problem was gone."

She glanced at Steven. His face was expressionless. "Oh, I don't mean it was gone as though some miracle had taken place. I was ill, dreadfully ill. Not until much later did the meaning of everything I'd learned at the academy come through to me. I understood that God intended the body to be a temple, that so long as the spirit abiding in it does not prostitute it, every act done in the name of love sanctifies it and makes it holy."

She edged closer and slipped an arm beneath his neck to force him to look at her. The intensity of her gaze, the wildness in the large, uptilted eyes fascinated him. She was whispering now, her lips moving against the corner of his mouth.

"That is why I want our child, my dearest darling Steven. I want it with every ounce of my being, just as I have always wanted you and always will."

Steven Lewis, 22, and Arlene Lathrop Baker, 25, were married in a small informal ceremony performed by the Reverend Jabez Lathrop, pastor of his newly founded, not yet completed Tabernacle of Truth.

The ceremony took place in the home of Tom Warden on South Fort Street in Los Angeles. Isobel, his sister, acted as maid of honor.

The couple had given no reason for the sudden decision.

Arlene dismissed her parents' startled inquiry by replying, "We just love each other, completely. We feel there's no point in waiting. Steven has so much to do. I am going to help him."

No formal wedding certificate was required. An affidavit was drawn up by Jabez Lathrop stating that on February 26th, 1866, he had, in the sight of God, joined the couple in holy wedlock. All present witnessed the document and the names and date were inscribed in the Lathrop family bible.

The bank account name was changed from Mrs. Arlene Lathrop Baker to Mrs. Steven Lewis, a fact that was noted by Isaias Hellman in some general correspondence addressed to Francis Pioche. The banker also complimented the financier on having such a bright and stable young associate in the south.

Two weeks later, Steven received a letter from Pioche demanding that he come to San Francisco at the earliest possible moment to report on progress.

The welcome at the Stockton Street mansion was cold and correct. Pioche was brusque as he led Steven into the sitting room and indicated a chair. "I must say, my young friend, that the news has distressed me." He lifted a cautioning finger when Steven seemed about to defend his action. "Not that your happiness is of no concern to me, for of course it is. But I am distressed that a partner upon whom I place so much reliance to handle my money sagaciously could make such a critical decision so quickly."

"It was not a sudden decision, sir."

"No? Three months? To choose a life partner? That seems a bit precipitous."

"Sir, if you have ever been married, then you know that the right woman can be a stabilizing influence on a man."

Pioche sighed and glanced heavenward. "My dear young man, I have not been married and I shall never comtemplate it. I much prefer the independence of my mode of living. My only obligation is to myself and my energies are always free to accomplish the things that you seem to admire so much. Your apparent interest in them was one of the reasons for my interest in you."

Steven knew he must hold his tongue and curb his temper.

As Pioche continued, a note of childish petulance crept into his irritable bill of complaints. "I am aware of the fact that President Washington and General Lafayette, and Benjamin Franklin too, were all happily married men. But history is replete with their peccadillos. Come with me on the Champagne Route, my young bridegroom, and I shall show you a dozen restaurants and saloons filled with happily married men and those *other* female companions with whom they are more often happy. And I presume that you would not be overly shocked to find your father among them from time to time."

Anger flushed Steven's face but Pioche dismissed the matter with a wave and continued. "*Ce n'fait rien.* What is done is done; what is not done concerns me much more."

He indicated some sheets of paper on the marble coffee table. Steven recognized them as his own written prospectus and the estimates of how the work would progress.

"Shall I be concerned now, my young friend, that you will be so preoccupied with your marital obligations that you will lack the time and perhaps the energy to implement these?" He tapped the papers with the back of his perfectly manicured nails.

Still smarting at the allusion to his father's occasional but well-known infidelities, Steven determined to stand firm. "I say that everything will be done as promised and barring acts of God—"

Pioche interrupted, smiling. "Such as children?"

Steven controlled himself and continued. "—barring acts of God, everything will be done on time despite the fact that funds were two weeks late arriving in Los Angeles."

Pioche was amused and Steven detected a trace of admiration in his shrewd little eyes. "*Bien. Bien.* Go ahead. I admire a man who makes his points."

"Don Emilio is hiring men. A good blacksmith shop is being built and I have ordered steel strap for scraper boxes, new harness, picks and shovels, crow bars and heavy tools." He pointed to the papers. "Everything in there is either on the job or on order. And I have a line on a good engineer. He is one of the men who planned the water system for El Monte."

Pioche spread his hands. "*C'est ça!* I am reassured. Now then, as my junior partner I am going to assign you a duty. Tomorrow night I am having a modest soirée in honor of Signor Guido Morelli of the Carlo Rosa Opera Company whom I hope to entice to San Francisco this coming year, There will be a cosmpolitan group of guests. I want you to meet some of them. I have in mind your education in cultural matters and even more practical considerations. If our sphere of influence in the south is to grow in all civilized respects, you should be making contacts in the arts." He paused with a questioning look. "One hopes that the new Madame Lewis has interest in this direction too?"

Steven realized that he didn't know the answer. "I think her chief interest lies in religion."

Pioche closed his eyes and touched his fingertips to his temple. "If this is true, my dear boy, I promise you that many a night you shall know cold feet!"

Steven managed a smile. "So far she seems to have a very warm heart." The attempt to relieve the tenseness with humor failed to move Pioche. "Remember, tomorrow night. Seven o'clock. Did you bring formal clothing?"

"No. But unless my brother has a formal engagement I'll manage to be presentable."

Pioche nodded slightly and rang for the butler. When the imperturbable German appeared, he rose and indicated the hall. "Louis, will you please show our young bridegroom out."

Steven could give only a day to visiting his family. The steamer *Goliath* sailed on the midday tide and he expected to board it without sleep after the Pioche soirée.

He found his mother suspended between delight and desolation. Doña Maria's tears of joy at seeing her son were mixed with those of anguish over news of his marriage. She seemed unable to comprehend the need for a ceremony that did not include the family. Nothing Steven could say seemed to assuage her hurt.

His brother-in-law, Lambert Lee, had been blunt. "The family considers this marriage a disaster!" His long arm had swept the city. "There are a dozen young ladies from splendid and affluent families here in San Francisco who have set their caps for you. And one of them would have been a brilliant match and useful to your career. But a precher's widowed daughter nearly four years older than yourself? Great God!"

Steven stunned Lambert by telling him to "go fuck himself."

Reluctant to attend though he had been, Steven was

caught up in the unconventional gaiety of the Frenchman's party. Many of the guests were members of the Parisian *Tableaux Vivants* troupe currently at the California Theater. In white make-up, some of the women entirely nude, the artistes simulated masterpieces of Greek and Roman sculpture. Ignoring the Eastern taboo against movements on stage, the members changed poses in full view of the audience. Steven had seen the performance several times. One of the girls had appealed to him. He met her now and was disappointed. Offstage she seemed much older and the seductiveness that projected across the footlights had vanished.

Several of the men in the troupe were professional acrobats and two of the younger ones appeared to be Pioche's favorites. He insisted that the guests admire their well-developed physiques as they obediently struck a number of poses.

At four o'clock the serving staff was still icing and uncorking champagne. Steven sought out Pioche and, pleading extreme weariness, thanked him and said his goodnights. The little Frenchman, mellowed by wine, feigned desolation.

"Alors. Je suis inconsolable, mon cher ami. Mais, c'est la vie. The best part of my soirée begins now. And later we all go to breakfast at Le Coq Rouge." One of the financier's sycophants laughed. Pioche frowned and raised a finger to his lips. Then he pointed to Steven. "It is demanding work, this new career as a husband. *Bon nuit, mon ami. Dormez bien et bon voyage."*

Arlene greeted Steven passionately when he arrived and promptly announced she had a surprise for him. The surprise was the house that Tom Warden had found. Warden's attempt at humor was less than convincing. "I've got to admit, Steven, that I hadn't expected to hunt out a house for Arlene and somebody else." He grinned. "But that's life in the Far West. I hope you like it."

They climbed down from the surrey and entered a small brick cubicle with a mansard roof. Warden's gesture encompassed the fenced-in dust-dry yard, two scraggly saplings struggling to survive, the woodshed and the outhouse.

"It is not Mr. Temple's Los Cerritos, but it does have an attic, if Steve doesn't mind stooping a bit."

Arlene slipped an arm around Steven. "I saw it the other day, darling. Tom took me to a half dozen houses. But I think this is the best one for us, for now."

Inside Steven found a parlor, a small dining room, a kitchen with a narrow attic staircase and one bedroom. It was hardly enough space for an expanding family. Arlene read his disappointment.

"Darling, for the time being this is all we need. The attic is quite adequate. It is warm and snug and I can fix it up very nicely as the extra bedroom."

Warden regarded Arlene with mock concern. "Remember, Mrs. Lewis. You've married a native son. These Californians have a tradition of large families."

Arlene smiled archly. "I know. Steven and I have agreed that we'll limit our family to twelve children."

Charlotte Lathrop Lewis was born in the Hill Street house shortly after midnight, October 11, 1866. The birth was not difficult even though the infant exceeded eight pounds.

Steven had arrived home from El Rancho de los Nidos in time for the birth. He had been working harder at Los Nidos than ever before in his life.

The days at home were trying. For practical reasons Arlene and the child were bedded in the single sleeping room and he had taken the makeshift attic. As much as he delighted in his first-born, Steven wanted to get back to the rancho that had become his second home. He felt responsible for overseeing the more than one hundred Mexicans and Indians who were laboring on the ditches and flumes to bring water to the new reservoir. Arlene, aware of his unrest, urged his early return. Apologetically, after ten days, he did.

From *La Cueva de Los Lobos*, the cave to which Goyo Robles and his followers retired after each successful plundering expedition, the *bandido* and his men watched the work proceeding far below.

"Don Emilio tries to outwit nature. That is good. But I wonder where he gets the money?"

The lieutenant, a pockmarked Sonoran called Bruto, patted the stock of his carbine. "From the gringos. There are two there now, both *jefes*. We Mexicans work. They

302

boss." He spat. "And we are the ones who taught them how to transport water at the mines."

Goyo smiled tolerantly. "Perhaps they are Don Emilio's engineers. There are no *americanos* here. We'll wait. We have time. If the *zanja* is for the gringos, the more work, the greater the loss."

As he continued to study the figures below, a plan began to evolve. "Bruto."

"Yes?"

"Cinco and Tito tell me they pay one dollar fifty, American, each day for work on the *zanja*. Perhaps one of them should work for a week and listen."

The lieutenant scowled. "I know a faster way to get their money."

Goyo laughed. "Those are Sonorans working. If we take their money on payday we take from our brothers."

A crude oath escaped from Bruto's lips. "If they work for gringos they are not *my* brothers."

Goyo nodded. "Let us be certain first. Tell Cinco to come here."

By the twentieth of December the main ditch had been pioneered to the mouth of *La Cañada Laberinta*. In several places it had become necessary to use black powder to remove boulders. Because of the scarcity of lumber, Steven was trying to hold the use of flume to a minimum.

After Christmas at home, Steven returned to his work on the aqueduct. Don Emilio and Doña Paula, enthusiastically supported by sixteen-year-old Anita, urged him to bring Arlene and the baby back with him.

"They will be much more comfortable here in this big house, Steven. They are as welcome as family."

Don Emilio agreed and offered a most convincing argument. "It will not be an imposition, as you suggest. It will be a great pleasure for all of us. But most important, my friend, Doña Arlene should be here when we begin your house. She must be here to help us bless the lintel. Pedro has already cut it from oak."

Despite Arlene's concern that they would have little or no privacy as guests, she found the arrangement agreeable. Her resilient body had returned to normal, and physical contact with Steven was resumed with undiminished passion. Although the idea repelled Arlene at first, she

303

came to enjoy the peace of mind that resulted from the moistened square of finely scraped lamb's intestine that Steven insisted they use to guard against a second pregnancy before he could provide a suitable home. Also, it seemed to intensify the sensation for her.

"But darling," she made a distasteful face, "we must find another name for it. I refuse to think of it as a 'gut bucket.' "

Privately, Steven agreed. At fourteen, he had first seen the sporting-house girls hanging their newly washed diaphanous membranes on roof lines and the explicit answer to his question both repelled and fascinated him.

The weather favored Steven's work. When word spread that the new *zanja* at El Rancho de los Nidos needed more workers at one dollar and fifty cents a day and found, men began drifting down from as far north as drought-parched Santa Barbara. By June two hundred Sonorans and Indians were at work building the stone sub-surface dam and the twenty-five-foot earthen embankment on top. It would be the first of three such dams that would, it was hoped, keep the river from rampaging over the lowlands during the flood season.

Several times during the summer, parties of horsemen stopped nearby to watch the work. Each time, after brief consultations, they rode off without attempting to identify themselves.

Quite by chance, on a visit to San Pedro Bay, Steven discovered that the anonymous visitors were ranchers from the Santa Ana River. He ran into Senator Phineas Banning, the dynamic pioneer developer. Banning was responsible for the huge Wilmington pier and there was a rumor that the road Banning's men were grading from his Wilmington slough to the south end of Main Street awaited only city and county approval of a two-hundred-and-twenty-five-thousand-dollar bond issue. It would then become the first railroad in southern California. Steven believed the rumor. Nobody had taken Banning seriously when he said that mankind would have to create what nature neglected. Now the Wilmington Harbor at San Pedro was a fact.

Banning told Steven that the ranchers at the lower end of the Santa Ana River, Steven's unidentified visitors, were afraid that Steven's project might cut into their

304

water supply. Steven went out of his way to reassure Banning that this was not so. Banning also said that while the water situation in Los Angeles was bad, farmers in the San Gabriel valley had hit artesian flows at sixty feet. Both the Danes and the British were looking for farmland for their settlers who were finding it easy to borrow in order to buy. This meant abundant new farm settlers and cash customers for Steven's parceled land.

There was no question about it. A feeling was in the wind. A good feeling.

By October most of the exterior work on their home in *Nido Mejor* had been completed. It was the first of the redwood and native stone structures designed for Arlene Lewis' cherished Brindavanam. Arlene became defensive when Steven cautioned her about expounding her unorthodox beliefs in the presence of the Garcia y Perez family. Don Emilio, at great trouble, had sent for the Catholic father at San Juan Capistrano to bless the lintel. In a heated scene, their first, Steven got her promise to endure "that barbarous ritual" on the reciprocal understanding that he would join her in their own private dedication. He wondered why she suddenly had become so intolerant. The religious ceremony was a small concession in view of the lavish hospitality they had enjoyed for months at the rancho.

Her fanaticism, gone briefly after the birth of Charlotte, had returned and increased in intensity. The compulsive drive was back. On the slightest pretext Arlene would turn the conversation back to Brindavanam and her plans to attract a congregation. Steven tried to forewarn her that Southern "hard shell" Baptists might resist, but she dismissed the possibility.

Arlene's financial independence added to Steven's uneasiness. Her funds had built their home; her funds had paid for the plans depicting an architectural potpourri of Egyptian and Oriental structures. She would also pay for the materials, many of them exotic even for a cosmopolitan community. In subtle ways she used her independence as a goad.

Arlene suggested that Steven appoint Tom Warden exclusive Los Angeles sales agent for the forty-acre tracts. Steven did so with some reluctance and found Warden

305

surprised and obviously pleased. The sales would begin from surveyors' plats immediately after the new year.

Arlene hugged him when he told her about Warden's acceptance. "Oh, darling, I'm so glad. I had no idea that Tom would be interested, really. But it's one less responsibility for you. That's my only concern. I didn't mean to butt into your business, dearest one, but I've been so worried that you are overdoing. And I couldn't bear to see you go on. Forgive?"

Moments later, in her arms, he forgave her.

Another irrigation project was started on the Rancho San Pasqual, ten miles northeast of Los Angeles, by the rancho's new owners, Benjamin Wilson and Dr. John S. Griffin. "Wilson Ditch," as it was being called, was an impressive feat. It represented new competition for *Los Nidos* and the defeat of Don Emilio's old friend, Manuel Garfias, who had lost the rancho. Don Emilio looked at the construction of the ditch and sighed. "God forbid, but what you see here is the story of our ranchos, *Esteban!* We will see it repeated many times. The land belonged to Manuel by right of inheritance but taxes and loans forced him to sell. He sold it all to Wilson for one thousand eight hundred dollars."

Steven whistled softly. "My God, that's thievery!"

Don Emilio shrugged. "Some would call it charity. A drowning man does not inquire into the motives of his rescuer."

Sad as it was for the *ranchero*, the trip to the project was useful, for it confirmed the soundness of their own project. While Steven conceded that the Wilson-Griffin development offered the attraction of proximity to Los Angeles, he felt that El Rancho los Nidos, by offering greater arable acreage and in normal seasons more dependable water, would be able to offset the competitor's advantage.

The main ditch at Los Nidos was completed in February of 1868. It was christened *Zanja Paula* in honor of Don Emilio's wife. The central reservoir, also completed, was named Lake Arlene. From it would radiate sub-canals designed to carry water to the various parcels of land. Arlene, who was pleased, did not resist the dedication ceremony.

A week after the dedication, during a three-day rain-

306

storm, Steven and the engineer, Monteith, opened the head gates in *El Nido Mejor* and diverted the flow into the long ditch. They followed the course of the controlled floods as fast as horses could travel the rough service trail. In less than an hour, first water from Rio Claro flowed into Lake Arlene.

Steven had deliberately chosen the rainstorm as the best time to divert water. The anxious downstream ranchers who had watched the work would see no appreciable change in the level of the flooded Santa Ana River below its confluence with the smaller Rio Claro.

In a week the lake was full and the excess water was released over the dam's cobblestone spillway. Another week found the surface of the big reservoir black with waterfowl—the tag end of the winter migration. Don Emilio, looking at the newly shot mallards and the brace of Canadian geese hanging from the side porch, was moved to silent thanks for the events that had brought this new bounty so close to home.

Steven had managed to postpone a meeting with Pioche on the plea that he wanted to wait until he was certain that the basic engineering was completed and sound and he had worked up a carefully audited set of books. The ledgers would show the banker that the work, thus far, had come in substantially below the original estimates and, due to the continuation of the drought, well ahead of schedule. On the basis of Tom Warden's leads, he could safely predict the lease of twenty or more parcels to the new bee farmers on a sharecrop arrangement.

Over a thousand double-supered beehives had been set out the previous October. Assuming a normal spring, the total harvest of fine comb honey would exceed one hundred thousand pounds. The rancho's share was twenty-five percent of the gross profits. It would be a fine report and he was eager to make it.

The trip would also be a chance to introduce Arlene and Charlotte to his family. Steven was sure Arlene's bright charm and little Charlotte's promise of great beauty would overcome any family resistance. To Steven's dismay, Arlene refused to leave until the house was finished and more work was done on the cottages and the temple.

"Dearest Steven, I have subordinated my own wishes to yours because of that ditch and the lake. It was not a sac-

307

rifice. It was an act of love. I ask you to repay me in kind by not insisting that I make that dreadful journey now. I promise that we shall make it in the fall. Nothing will be allowed to stand in the way. I do want to meet your family and show off my little lovely one. But I cannot go now, please don't insist."

Steven refused to allow her to stay alone in *Nido Mejor*. She agreed to go back to the main ranch house for the three weeks that he would be away. Disappointed and morose, he left for Los Angeles.

Warden and his sister asked him to dine and stay over with them. The following morning Tom drove him to the Bella Union to catch the stage. "Isobel sensed that you're worried about leaving Arlene and the baby alone."

"They won't be alone. They'll stay with Don Emilio until I get back."

"Would you feel easier if Isobel and I drove down and looked in on them?" When Steven did not answer immediately, Warden slipped a folded paper from his inside pocket. "Actually, I will have to go down there within a week or so anyway. Three deals look certain, but I want the parties to see the land first. I promised them their pick."

Steven thanked him for his thoughtfulness and wished him well with sales. Warden's optimism was unshakable. "Wait until spring, Steve. When the wildflowers are out and the ranges are green again we'll have to arm ourselves to keep from being stampeded by customers."

Aboard the northbound sidewheeler, *Golden City*, Steven concentrated on the report to Pioche. He wanted to tell him that his original investment would be returned within the following year—that the partnership would then begin operating at a profit. But he was leery of "supposings" as Harry Morrison called them. In this case, however, the odds seemed to favor him. Phineas Banning's new road would be a railroad all right. He was promising a spiking ceremony before the end of the coming summer, with free train rides to Wilmington and return and a huge barbecue, "all on the road."

Good transport had always been the magic, and the great magic was yet to come. Word from the Sierra said Charley Crocker's horde of coolies was successfully leveling every obstacle God and man could place in their path. Central Pacific was predicting a link-up with Union Pacific, probably in eastern Utah territory, within a year. Montague, the chief construction engineer, would have chosen an earlier date but the carnage from mishandled nitroglycerine was too much even for tough Charley Crocker. He ordered "the damn stuff buried" and a return to black powder and Oriental persistence. Relieved, Montague revised his date and the mangled bodies were replaced by other Chinese shipped into California, despite popular outcry, under a new treaty negotiated by Ambassador Anson Burlingame. The terms unmistakably favored the railroad.

Steven did worry about the effect of the transcontinental railroad on his father's business. Billy Ralston had suggested that Lewis & Sons hedge its bets by starting a chain of warehouses. His father and Will had built three. But there was little doubt in Steven's mind about the effect of the railroads on the southland.

At 806 Stockton Street, Steven was confronted by valet Louis Reiff. "Mr. Pioche is ill. He asks that you return at the end of the week."

When Steven did get to see the little Frenchman, it was only for a few minutes. Pioche was pale and distracted. He dismissed the proffered report: "It is commendable that you are meticulous, my dear boy. If you say things go well and we shall be even within a year, c'est ça. That is quite enough assurance for the present. You are doing your work well. Write me. Later, I shall ask you for a full report when I am less pressed."

Unsettled by Pioche's sudden turnabout, Steven went down to the Bank Exchange bar to meet his father and brother. He knew that Pioche was involved in a great many businesses, all successful. He had also finally arranged an appearance of the famous Carlo Rosa Opera Company in San Francisco in the fall. Whatever the Frenchman's problem, it did not seem to involve those aspects of his unusual life.

Steven was diverted from his preoccupation with Pioche by Will's announcement of his impending marriage to Mary O'Dwyer. The bride-to-be was the daughter of Sean O'Dwyer, who had brought a considerable fortune to San Francisco and had added to it through his successful foundry operations. The vows would be taken before Christmas in the Mission Dolores. Steven promised his twin

309

that he would bring Arlene and the baby to the wedding and that he would be Will's best man.

As eager as he was to return to the south, Steven took advantage of an enforced wait for the old sidewheeler, *Senator*, to visit family and friends and to get caught up on news.

Harry Morrison, who had been north assessing business possibilities for the freight lines, had attended the ceremony transferring Alaska from Russian ownership to the United States. The half million square miles had been officially proclaimed a Territory on the previous October eighteenth. As usual, the Scotsman's cogent observations amused Steven.

"I dinna ken the reason for calling it 'Seward's Folly.' The Secretary of State stole the world's largest curling court for seven million two hundred thousand dollars, which reckons doon to aboot tuppence an acre!"

Neighboring Canada had become a British dominion three months earlier and Steven, along with most Californians, felt easier with Alaska secured. He listened to a great deal of speculation about the refusal of Congress to concur in President Johnson's suspension of Secretary of War Stanton. There was a lot of wagering along the Champagne Route that Johnson would be impeached.

John Lewis joined with Harry Morrison in some anxious speculation about the machinations of the New York speculators, Jay Gould and Jim Fisk, and their domination of major railroad stock. But in the end, they agreed that the Big Four were more than a match for the most ruthless manipulators.

Steven was particularly interested in the impending transformation of the College of California at Berkeley into a state university to be known as the University of California. The move was long overdue in the north and a similar move was needed in the south. John Lewis quoted Governor Low on the subject: "Now, here you have scholarship, system, organization, reputation, everything but money; but we, the state, have none of these things, but we have money. What a pity they cannot be brought together." However, John had one misgiving. "I personally see nothing wrong with the plan, but a lot of good people, your mother included, see danger in allowing young men and women to study side by side in the same classes. It could make for distraction. For instance, when I was courting your mother I had the devil's own time keeping my mind on business and it was a matter of life and death that I should."

Steven found himself impatient to be back in the southland again doing what he could to help it move forward toward new progressive ideas in whose implementation San Francisco seemed always to stay ahead.

When the steam lighter came alongside the *Senator* in San Pedro Bay, Steven heard an urgent voice calling his name. He located the man and hailed him from the deck. Minutes later he was handed an envelope.

"The Reverend Lathrop sent this, sir. And I'm instructed to get you to his home by carriage. It's been terrible, waiting, sir. I've met the last three steamers."

Steven broke the wax seal and ripped open the message. Numb with apprehension, he read: "Steven: You will need all of the courage God can give to a mortal to bear up under the news you are about to receive. Hurry, you are desperately needed."

Tom Warden and Isobel had journeyed to El Rancho de los Nidos three days after Steven sailed for San Francisco. They were welcomed and settled for the night. Arlene Lewis, who had seemed depressed and impatient after Steven's departure, had been driving herself and the workmen at Brindavanam. With the Wardens' arrival her mood brightened and Doña Paula concluded that company and diversion were the best remedies for a lonely wife.

The following morning, full of enthusiasm for her own project, Arlene insisted that Tom and Isobel ride into *Nido Mejor* to see the new home. The three spent the night there, talking until late. Tom promised to help build her colony as he had built her father's congregation, but urged her to tell him more about it. "Farmers are sort of peculiar philosophers, Arlene. I'd better know what you're going to promise them. If you can give me a story that'll scare them real good you'll probably have more converts than you need. They won't admit it but they like that sort of excitement. The threat of hellfire and damnation is working wonders in your father's Tabernacle."

Arlene laughed. "I'm afraid you've got your work cut out for you, Tom dear. I promise them nothing but peace, understanding, and Divine Love."

311

The following day when the brother and sister were ready to leave for the return trip to Los Angeles, Tom admonished Arlene. "You know I'll do anything I can to help. But you still haven't told me enough to go on."

She laughed and winked at Isobel. "I'm not certain that a man who sells real estate can be trusted with the truth."

Two weeks later Tom Warden reappeared alone, ostensibly to check progress for Steven. When he returned from the subdivision with Don Emilio shortly after noon, Arlene called him aside. "Tom, were you serious about wanting to help me?"

"Never more. What can I do?"

"It's difficult to talk here. Will you ride up to Brindavanam with me? I want to see after things at the house. We can talk on the way."

Tom Warden's apparent hesitancy caused her to reassure him. "If you are worried about appearances, Tom dear, don't be. I have already told Doña Paula and Don Emilio that you promised Steven to counsel me while he was away."

They spent the afternoon walking to the springs and meandering along the headwaters of the Rio Claro while Arlene explained, a bit cautiously at first, the philosophy she would espouse at Brindavanam.

Warden did not appear surprised. On the contrary, he seemed to follow easily and agree completely with her reasoning. His rapt attention and discreet questions encouraged her. By the time they had returned to the new house Arlene was deep in the persuasive yet somehow defensive justification of her beliefs.

"All religions talk of love of God or Gods and love of fellow man. They all have their commandments. They are all the same, Tom." She regarded him closely. "And do you know why man has made no real progress—except in the art of annihilating his own kind? Because he has never learned the meaning of true love. Frustrated love becomes hate. We are ashamed of love, Tom. We are ashamed of our bodies. How monstrous. Our bodies are God's own temples of love."

Arlene's voice had become strident and her eyes glowed with a fanatical intensity that excited Tom. He noticed the strange whiteness around her mouth and the fine beading of perspiration on her forehead and at the corners of her

312

eyes. When he dismounted to assist her from the side-saddle her body was tense.

Inside the house she busied herself at pretending to see that things were in order. "I'm dreadfully nervous about leaving everything unattended like this, Tom. I don't trust our workmen as Steven does."

Warden chuckled. "How about using a little of that love on them?"

Arlene stiffened and turned to face him. "Oh, Tom Warden, I hope I've not misjudged you."

Embarrassed, he grinned. "I didn't mean to poke fun, Arlene. I know it didn't sound serious, but I meant it to be. You told me that like begets like. I took that to mean that love begets love." He managed a woebegone smile. "But I ought to know it doesn't always work, oughtn't I?"

Arlene moved closer. "Indeed you should not, dear Tom, if you're talking, as I presume you are, about my choice of Steven, even after you made your intentions clear."

Warden laughed with unconvincing good humor as he recalled his frustrated courtship. "I did everything but paint 'Tom loves Arlene' on the courthouse walls."

Arlene rested a hand on his forearm and looked up at him. Reproach turned to sympathy. "Poor Tom. Did you really think I was blind? Of course I could see. And how I loved you for caring. How lucky I was, two wonderful men in love with me."

The doleful expression on Warden's face turned to a wry smile and he shrugged. "So, the best man won. That's as it should be."

Arlene pouted. "Not the *best* man, but *another* wonderful man." She looked up at him intently for a moment then sighed. "Who can say what might have happened if it had been you on that boat? You and Steven have so many of the same fine qualities. Who can say? Who, really, can say? For I do love you both."

The words were half whispered, wistfully, and then unresisting, she allowed Tom Warden to take her in his arms.

Numb with shock and rage, Steven Lewis stood by the side of the bed in Jabez Lathrop's new home looking down at the sleeping Arlene. Dr. Rebecca Foley, a severe-visaged,

313

competent physician in her early fifties, touched his sleeve.

"She's been sleeping like this for three days now, Mr. Lewis. I have given her heavy doses of laudanum. I want her under sedation until some of the outraged tissues begin to heal." She indicated the door. "Let's go downstairs now. I want to talk with you."

Alone in the sitting room with the doctor, Steven listened. "The events are substantially as reported in unnecessary detail in the newspaper." She nodded distastefully at the issue of the *Los Angeles Evening Star* spread open on the side table.

Jabez had told him of the graphic coverage but he had not had time to read it.

"If you would prefer to hear this from a male physician I can arrange it, Mr. Lewis. Dr. Hoffman assisted me."

"No. You tell me. I want to know everything, right now."

Dr. Foley nodded. "All right. I'll tell you what I know for certain. The rest you'll get from Mr. Garcia at the ranch. When the Morales man rode here trying to get word to you, your father-in-law and I immediately followed him back to the ranch in a fast rig. There was nothing we could do for Mr. Warden. But Mrs. Lewis was still alive.

Dr. Foley paused. "Are you sure you want to hear all of the details, Mr. Lewis?"

Steven exploded. "Goddammit, yes! I want to hear it all. Every last word of it. Now tell me and quickly. Enough time has been lost already."

"Very well, Mr. Lewis. Your wife and Mr. Warden were found in the house. Both were naked. Mr. Warden had been emasculated in the crudest possible manner. He died from loss of blood and one can only hope that the brutality rendered him unconscious quickly. Mrs. Lewis was ravished and left for dead. There was no way to tell how many men used her. Speaking medically, I can only guess from the washings I was able to obtain that there must have been several. Perhaps as many as a half dozen. She was badly torn. I am sure she too was unconscious, very early on.

"Jabez has told you that Mr. Garcia and his men have been looking for the outlaws who did this for several days now. Deputy Sheriff Jim Bowdon took six men down there

314

the day before yesterday. Mr. Warden's body was brought here this morning and interred in the Protestant Cemetery on Fort Hill."

She nodded toward the staircase. "There is no way for me to tell for certain how long it will be before your wife recovers. She may never recover fully, Mr. Lewis. You must be prepared for that. She is blessed with a strong young body. Physically she may recover in time. Dr. Hoffman and I have done all of the surgery possible. But there may be other after-effects, mental damage. What she has suffered, Mr. Lewis, is the most traumatic experience a woman can endure. How she comes out of it will depend on factors we doctors simply cannot judge. I tell you this quite honestly. Be prepared for the worst and hope. Your daughter is safe. She is being cared for at the ranch and I would suggest, Mr. Lewis, that you leave her there for the time being. Your in-laws are too distraught to attend to her here."

Later, when the doctor went upstairs for a final look at Arlene, Steven read the lurid account in the newspaper. Its innuendo enraged him.

"It would be unfair to conclude that the victims were surprised in *flagrante delicto*, for it may well have been the intention of the leader of the murderous gang to leave precisely that impression for some evil motive of his own."

The account sickened Steven, all the more for the pernicious doubts that he couldn't exorcise by his rational attempts to counter them. Every aspect of his life with Arlene argued against deliberately contrived circumstantial evidence—her erratic behavior, her unashamed aggression, her eroticism and insatiable physical need, the open almost eager admission of her unconventional moral code, the abortion, and, hardest of all to admit, her obvious attraction to Tom Warden and Tom's scarcely concealed interest in her. True, as husband and wife they shared a common affection for little Charlotte. Even so, Steven recalled that he had often thought that Arlene regarded the baby daughter as a possession rather than a part of herself. And it had been at Arlene's insistence that Tom had become the sales representative for the development. Steven recalled his strange uneasiness when she said, "Oh darling,

315

I'm so glad. I had no idea that Tom would be interested, really. . . ." And there was the matter of Tom's eagerness to look in on Arlene and the baby, a friendly service perhaps, one he himself would likely have offered in similar circumstances, but again, there had been the uneasiness. Tom had left him no reasonable excuse to object, by reassuring him that business would be the prime reason for the trip.

Steven flung the paper aside and went upstairs again. For a time he stood looking down at his heavily drugged wife. Then, after advising Jabez Lathrop of his plans, he drove to the livery stable to arrange for a pair of fast saddle horses at dawn.

At El Rancho de los Nidos the shock of the rape and murder was evident in everyone. Don Emilio and Doña Paula had greeted him mutely while Pedro and Gracia and the other help stood back and watched with impassive faces and lowered heads.

Later, when he had settled in his room and washed, Don Emilio led him outside. "Before we eat, I want you to see this. Pedro found it on the nightstand in the sleeping room of your new house, when he found them."

Steven took the torn paper. He recognized it as half of a sheet of the notepaper that he had brought to Arlene from San Francisco on an earlier trip. The scrawl in Spanish was hasty and ink splattered. Half aloud, he translated: "Thank you, señora, for accommodating us as you accommodated your gringo lover."

Don Emilio watched as Steven's eyes remained fixed on the paper. "Only Pedro and I have seen this, Steven. Pedro does not read. It is the most despicable perfidy. I do not believe it. I was uncertain about showing it to you. I wanted to burn it immediately, and then I thought perhaps—"

Steven crumpled the paper and handed it back. "Burn it now, and forget you ever saw it."

While they were still at the table, Deputy Sheriff Bowdon rode up with six men. Some minutes later Los Angeles Marshal Bill Warren joined them with eight more men. Despite the fact that at least a score of lesser bandits continued to roam the state, raiding ranches and holding up coaches, the outrage in *Cañada Laberinta* was credited without a single dissent to *Los Lobos*.

Deputy Sheriff Bowdon reported. "We've been over the country for two days now. There's no way those men could have ridden into *Cañada Laberinta* without being in sight of this house. Pedro and his men have sworn on the Bible that no horsemen rode anywhere near the mouth of the canyon on the afternoon of the killing. They were working on the ditch all day. And there's no way that riders could get down into *Nido Mejor* from the east. Even deer keep away from those slopes. Whoever did this thing came in on foot. That don't square with the habits of most badmen around here. There was no fresh prints or signs around the house that could not of been made by Don Emilio's own men."

Don Emilio stiffened. Quickly, Bowdon raised a hand. "Don't take me wrong. I'm not trying to put a doubt on anyone. I know all your men. I'd as soon suspect the *padres* at Capistrano. There was no robbery as far as we could tell. Only the cutting and the—" He glanced at Steven and blinked in confusion. "—and—uh—the terrible thing with Mrs. Lewis. And there was the burning of the outbuildings."

Steven had learned earlier that three of the nearly completed residence cottages had been put to flames and an attempt to burn the main house had apparently been interrupted, for no flame had been touched to the leaves and sticks that had been heaped under the wooden front steps.

"The reason has to be revenge for something. And so far as we can tell there isn't a soul could have cause to hate Mr. Warden or the Lewises. There just ain't sense to it. If they were after livestock or money or food or clothes they would have struck this place here, not a house that wasn't even properly finished yet."

Steven, controlling hurt and cold rage aggravated by impatience, forced himself to listen. Don Emilio and Pedro had persuaded him not to ride into the canyon at night.

They were right. No purpose would be served. Much more sensible would be a coordinated search beginning before dawn. He prayed that the night would pass quickly.

The men spent an hour planning and checking their rifles and pistols before spreading their blanket rolls wherever they could. Shortly after four o'clock, Gracia turned them out. A half hour before daylight they were on their

317

mounts, breakfasted and provisioned with bundles of ropy, peppercoated dried beef.

Steven, Don Emilio, Pedro, and Marshal Bill Warren with four men rode along the *zanja* trail to *Cañada Laberinta*. Jim Bowdon with his own men and the remainder of Warren's men turned north along the base of the mountains. At Diablo Canyon they would divide into two smaller groups to comb the area for signs. At midday they would work their way south again and converge at *El Nido Mejor* to report their findings and coordinate further actions.

They let Steven go alone to look at the evidence of violence in his home. Upstairs in the bedroom that he and Arlene had shared, he stood gazing mutely at the big bed and its rumpled confusion of coarse sheeting and down-filled quilts. Just inside the door, where Tom Warden had been mutilated, the floor and the circular rag rug, a wedding gift from his mother-in-law, were covered with a yard-wide stain of dried blood. The room was permeated with the flat acridity of a charnel house. Steven walked to the bed and examined it briefly. Disgusted, he pulled the covers back in place. His throat was constricted and suppressed rage blurred his vision. Nothing seemed to be missing. Arlene's dresses were still in place on the peg hangers. Tom Warden's clothes were neatly folded and draped on the chair. The Congress gaiters he affected were side by side beneath it.

Steven forced himself to stay in the room until every detail of its condition was fixed in memory. Then he went downstairs. Things were also in order in the living room. But in the kitchen he noted that several utensils were missing as well as some containers of preserved food. The old Sharps rifle had been taken from its pegs over the hall door.

Outside again, he confronted Don Emilio. "When you and your men were here did anybody touch anything? Was anything moved or rearranged?" He indicated the house. "In there, or upstairs?"

Don Emilio, understanding the implications in the question, responded with an emphatic denial. "Nothing was moved, my friend. We brought Warden's body out of the house. Then we went back for your wife. Nothing else was touched. Absolutely nothing was rearranged."

When Steven continued to search his eyes appealingly,

318

Don Emilio lowered his head. "I am sorry. But it is the truth. Nothing was disturbed. Everything in the room is as we found it. *Verdaderamente. Lo siento mucho.*"

At the headgates there was evidence that several men had investigated its works. Pedro examined the footprints with two of his Indian *vaqueros*. After several minutes he called to Don Emilio and Steven. "These tracks were made by Mexican boots. But the men did not wear spurs. They came in on foot. We will look for more signs along the stream." He pointed to the steep slope between the two springs. "There are fresh slides there. Probably from animals. But we will look."

Shortly after noon, Jim Bowdon returned to *Nido Mejor* with all but two of his men. Seated on rocks beside the streams, they gnawed on the tough jerky and speculated on the absentees. "Merlhof and Parsons are good men. They're on to something. They'll show up."

Two hours later, just before the now uneasy Bowdon was about to go searching for them, the two men appeared. Parsons was excited.

"We found something all right, in that little canyon south of Diablo. There's a slide trail on the south slope so it's hard to tell one track from another. We could read the game and cattle droppings real clear. But what couldn't be seen was the horse turds, some no more than two days old. We didn't see them because all of them was kicked off the trail down into the brush, on purpose. Max and me went more than a mile up the slope. Droppings has been kicked off the trail for months now. Whoever's using it for damn sure don't want to be follered."

Bowdon looked around. "Anybody here been up that little canyon?"

Pedro nodded. *"Si, señor.* That is *Cañada Ciega.* In the first dry year we used to ride in looking for cattle. It only goes maybe two miles. It cuts down from a small mesa. There used to be feed there and a little spring."

Parsons nodded. "We didn't cross the mesa. The trail splits to a couple of other little canyons. But it was late. We didn't want to go up them alone. We figgered they's got to be six or eight men up there somewhere."

Steven tensed as he recalled Dr. Rebecca Foley's words: "There was no way to tell how many men used her, Mr. Lewis ... perhaps as many as a half dozen ..."

He rose and addressed both men. "Did you see any

319

place where they could have left horses if they did come in here on foot?"

Parsons shook his head. "Nope. We looked for that, too. The trail cuts southwest over to the river. If they were going to ride down out of the hills and fetch up, then walk here, they'd leave their animals close by. There are no signs that's what they done. We looked real good."

Jim Bowdon listened carefully, then turned to Pedro. "If anybody's holed up in those canyons they got to have water for themselves and their animals. Are there any other springs back up in there?"

The majordomo shook his head. "There is no other water. Only the little spring on the mesa above *Cañada Ciega*."

Pedro shrugged. *"No sé, señor.* There was not much in the summer. Maybe enough for men, if they dig."

The deputy tugged at his ear thoughtfully. "But maybe now with snow on top there's enough for men and animals?"

"It is possible señor."

Bowdon grunted and squinted at the sun. "We've got three hours of light. Let's take a look."

Pedro and his four men stayed with Don Emilio and Steven. They would rejoin at the rancho at dark. By tacit agreement Steven was left alone with his thoughts. For the first hour he wandered around assessing the damage to the cottages. Don Emilio watched him standing in the midst of the ruins and understood the conflict in him.

"He belongs to the yanquis and to us," he had said to Doña Paula. "There will always be a tempest in him. It is in his blood."

Anita disagreed. "He is more like us, father." Don Emilio had laughed. "To believe that, *hijita mia*, would be to make a grave mistake!" He had sensed from the beginning that while Steven Lewis might have inherited his mother's dark good looks he had also inherited his Yankee father's compulsive need to acquire, to build, and to innovate. No *californio* would have done or even cared to do what had already been accomplished by Steven at El Rancho de los Nidos and by the other yanquis at Anaheim, El Monte, and San Bernardino.

"Seedlings do not produce good wood or good fruit," he had observed. "They grow stronger and more fruitful

320

when they are crossed with other stock. It is the same with animals and with men."

After a second visit to the new house, Steven wandered back toward the springs. On his right he caught a glimpse of something that brought a quick stab of recognition. It was the fallen log against which he and Arlene had leaned on their first visit alone to *Nido Mejor*. It was the place, bathed in warm sunlight, where Arlene's emotion had overwhelmed them both. It was the place where Charlotte had been conceived. Steven turned away and the pressure of tears that would remain unshed blurred his vision.

A few yards along he stopped abruptly. For a moment he stood thinking. Then he made his way back to the log, leaned his carbine against it, and sat down. For a long time he rested his chin in his palms. Then he looked up at the steep, brushy slope that formed the vast north wall of the hidden valley. Something had disturbed him on that first afternoon. He remembered now. It had been a stabbing glint of sunlight on some bright object. The flash had endured for mere seconds but it had been bright enough to attract his attention.

Without taking his eyes from the slope, Steven pushed his back up against the smooth weathered log until he was seated on it. For several minutes he studied the mountain side, too steep in most places for a man to find secure footing. What could have caused a flash up there? A slab of gleaming basalt disturbed by an animal? But basalt is dark. The flash would have been muted. The one he had seen was intense enough to have come from a bright reflecting surface, a piece of mirror, a bottle or a piece of metal. It seemed preposterous to suppose that such items would be found several thousand feet up that precipitous slope. And still there was no ready explanation for the flash and he had not imagined it. He rose suddenly, picked up the carbine, and walked a hundred yards or so to the place where Don Emilio was standing with Pedro watching the action of the water on the large bend in the ditch.

"Do you have a telescope at your house?"

"I do. But it is very old."

"May I get it?"

Don Emilio was puzzled. "But of course. Pedro will get it for you. But why a telescope?"

"I want to look at something on that mountain." He

321

pointed well up the slope. "I remember seeing something up there when Arlene and I first came here alone. It's probably nothing, but I'd like to look anyway."

In a half hour Pedro returned with a tarnished brass ship's telescope bound in cracked leather.

Steven located a tree near the log that would give him approximately the same point of view and rested the instrument on a limb. Methodically he began scanning the slope from the top down, moving the lens across the narrowed field, then lowering it to sweep a new area directly below. For some minutes he continued the scrutiny. Don Emilio grew curious enough to question him.

"What do you see?"

"Nothing . . . ground squirrels mostly."

"What do you expect to see?"

Steven folded the instrument and shrugged. "I don't know. I recall seeing a strong flash of sunlight up there."

Don Emilio gazed up thoughtfully. "It could have come from water seeping down the face of a rock."

Steven nodded. "I thought of that. But Pedro says there is no water up there."

Don Emilio gazed at the patches of snow still visible in the rocky clefts at the summit. "There is melting snow up there every year, as there is now. Sometimes the water wil trickle down." Steven nodded. "That's possible. But if the sun was catching a wet surface it would shine on it a while. What I saw flashed quickly and disappeared." He pointed. "As I remember, it was up near the top of those big rocks, better than halfway up."

Both men studied the steep terrain. Don Emilio was skeptical. "It could have been many things, a big bird, a condor perhaps, turning into the sun just right." He shrugged. "Who knows? But if you wish, I can send the men up there."

Steven shook his head. "Let's hear what the others say first." The recollection was vivid now, even to the start it had caused. But there was no question that the flash had come from some highly reflective surface, not from the fleeting iridescence of plumage. Perhaps he and Arlene had been watched. The possibility produced a stab of guilt and embarrassment even now. And perhaps too he was attaching to it far more importance than it deserved.

At sundown, Steven and Don Emilio remounted. Followed by Pedro and his *vaqueros,* they wound westward along the ditch toward the *casa grande.*

As they emerged from the tortuous canyon they saw Marshal Bill Warren approaching across the mesa that led up to the base of the hill on which stood the Garcia y Perez home. He was riding flat out. Don Emilio, a superb horseman, spurred his mount and motioned the others to follow. The trio intercepted the marshal about a mile from the main trail. Warren shouted as he pulled up.

"We found them! By God, we found them! We stumbled onto their *caballada* in the south canyon just off the little mesa. Three men were fixing to take the animals to water. We got two of them and hazed their horses out of the canyon. I think we nicked the third one. I'm not sure. But there's more of them up there. Jim's got the men strung out but it'll be black dark tonight. We couldn't guarantee that someone who knows the lay couldn't slip through afoot. We need more men."

Don Emilio dismissed the possibility. "It will take until tomorrow afternoon to get more men here. By that time the matter will be settled. My men know these *cañons.* Let us talk with Señor Bowdon and see what it is possible to do."

A twenty-minute ride brought them to the mouth of *Cañada Ciega,* where the chief deputy was waiting for them. After Don Emilio explained the problem of getting settlers to leave their families alone at night in the scattered *ranchitos* Jim Bowdon agreed to make do. In all they were twenty-three men. Most carried Colt pistols in addition to an assortment of single-shot rifles and some repeating rifles. Steven was best armed. In San Francisco he had paid a premium for a new .44 caliber Henry Repeating Carbine.

While the men were planning, Pedro Morales moved beside Steven. "Señor Lewis, I think the *bandidos* came down the mountain into *Nido Mejor.* The prints go out of the cañada but none come in."

"You and your men saw slides. But you said animals made them."

"*Si, señor.* It is possible. But I looked again. There are little slides from those big rocks all the way down to the springs almost straight down. Animals move across a slope. There are many such trails."

323

"What about a rock falling down? That would make slides."

"It is possible. But a small rock would not come all the way down. A large one would make more marks. No, I think maybe two or three men came down."

Steven knew that if he accepted Pedro's theory the conclusion was inevitable: men barred from escape by the usual route would resort to any alternate route, no matter how difficult. He motioned to the majordomo. "Stay with me. Let's listen to Bowdon's plan first."

The deputy was an imaginative strategist. "It's going to be cold and black tonight. If they can see us I want them to think that we're going to wait until daylight to make our move. We'll make two camps—one here on the south side of the canyon mouth and the other on the north side. If they're cornered, they'll probably try to sneak out after midnight. They'll figure we're split, because of our fires, and they'll try to slip between us down this draw or they may try to angle up over the north rim and head for the San Gabriels.

"After dark I want four men with repeaters to split two and two and sneak up on both rims. Stay fifty feet or so above the trail. I want one man down here at each fire and keep moving around. Make it look like many men. If they're going to try to come out afoot this way, they'll either try the ridge or come straight down the draw. If you hear anything, challenge it. If it don't talk right, shoot it."

Bowdon began deploying his men and Steven moved beside him. "Pedro thinks some men came down the mountain into *Nido Mejor*. He may be right. Anyway, I want to go back up there with him tonight and wait. If they do try to come out that way and there's nobody there to stop them they could get to the ranch, steal horses and food and God knows what else, and be gone before we know it."

Bowdon was skeptical. So was Bill Warren. "If they're in those caves, they'd have a sheer drop of fifty feet or more to get to the slope. It'd be hard enough to make it down through the slides and brush in daylight let alone pitch dark."

Bowdon looked reflective. "But I wouldn't want to be the one to say a desperate man wouldn't try it."

Steven nodded. "And I don't like the idea of leaving the women alone at the ranch with no men around."

The deputy grunted. "All right. We'll make out. Let Don Emilio go back to the house with the women. You and Pedro go on up to the *Nido*. But leave the Indians with me."

Under protest, the *ranchero* agreed to return to the *casa grande*. Steven and Pedro rode with him. There they checked ammunition and ate a hasty supper. Then amid protestations and prayer, Steven and Pedro left for the hazardous night ride into *Cañada Laberinta*.

By eight o'clock the canyon was enveloped in velvet darkness. The sheer wall of the Santa Ana escarpment was a presence felt rather than seen. They tied their horses in the shelter lean-to and as an added precaution they hobbled them. Then they entered the big house.

In the kitchen Steven found some short tallow candles and a container of Allin matches. Working with a hooded light, he located two boxes of Sharps cartridges that the marauders had overlooked. These he gave to Pedro. After thinning the wax shipping lubricant on his own metal cartridges and making certain both weapons were clean, he extinguished the candles and took two ponchos from their pegs by the door.

"Let's go up to the springs. I'd rather be early and freeze than take a chance."

Thoroughly familiar with the trail, they made their way on foot to the upper pond and climbed over the rubble of boulders at the base of the slope. The mountain was referred to by the natives as *La Colema*, the beehive.

Steven had questioned Pedro about it. Now, as they huddled for wind shelter behind a large boulder, he visualized the three-thousand-foot slope rearing above them and pressed Pedro for more details. "Could men get to those caves by climbing up here?"

"No, señor." He pointed to his left. "It is necessary first to go over that ridge into *Cañada Ciega*. Above the *mesita* men can climb to the west face of La Colema and follow the shelf around to the caves." He pointed into the darkness above them. "Pumas used to live there. I have hunted them."

Steven reviewed what little he had seen of the mountain from the canyon to the north where the lawmen were now deployed. From experience he knew that it was easier to scale than descend a very steep incline. If the killers had managed somehow to come down several thousands of feet of rubbled, cactus-studded slope, then they must have devised some way to get down the sheer, fifty-foot cliff.

"If men came down this way, how did they get to the slope from the caves?"

"Who knows, señor? There must be some way. I am certain they did come down."

"How high would we have to climb the south slope to see across to the caves?"

Pedro considered briefly. "Perhaps five or six hundred *varas*, señor."

That meant a fifteen- to eighteen-hundred-foot climb. Under ordinary circumstances it would not be too difficult. But the plan he was formulating would call for the climb to be made in total darkness and in absolute quiet.

"If you were Los Lobos, Pedro, and you knew men were coming after you at first light, what would you do?"

The majordomo laughed softly. "I have been thinking of this, señor. Unless I had enough food and water and bullets for a week, I'd be trying to go away."

"What would you try first?"

"I would try to come out by *Cañada Ciega* and get by the men. Then I would steal horses from Don Emilio and take my chances."

"If you were El Lobo himself, would you be the first to try?"

"No, señor. He is the leader. He will be the last man."

"What happens if his men do not get out of the canyon tonight?"

"Probably they will try to come down this way, later when they can see."

Steven grunted. "That's what I think too."

He reached over and rested a hand on Pedro's forearm. "*Amigo.* I am going to climb until I can see the caves."

Pedro started. "No, señor. If they see you they will have a great advantage. They will be able to shoot across and down at you. Much better to wait here."

"They won't see me, Pedro. I'm going to climb now. By daylight I should be high enough."

The audacity of the decision left the Mexican momentarily speechless. "It is not possible. You will fall. You will break bones. You will be cut by a thousand cactus spines."

Steven peered around him, reaching out to touch nearby objects. He called each by name. "I can see well enough. My eyes are accustomed to the night now. I will

326

go one step at a time. You stay here. When it gets light you can cover me."

"Señor. It is not possible. When you are on the mountain they can shoot across at you and I will never see them from here. If you go, I go too."

Steven concealed his relief. "All right, let us go."

Goyo Robles, haggard and increasingly aware of his fifty years despite a spare, rawhide-tough physique, knelt on a sandstone ledge overlooking *Cañada Ciega*. Beside him was his second-in-command, the Sonoran, Bruto. Behind them were the remaining five members of the nine-man band.

Below them was the small mesa and the spring at the head of the blind canyon. Also below them were the bodies of their dead companions. The sloping plain was a mile or more beyond. In the distance two fires flickered, one on the north side of the dry *barranca*, the other on the south.

At first the loss of the horses and the two men had seemed to present an insurmountable obstacle. But the more Goyo thought about it the more hope he found. With horses, the temptation to try to run the cordon might prove irresistible. Goyo knew that would be fatal for most of his men.

Softly, he called them closer. "They expect us to come out on foot. The *barranca* will be guarded. If we follow it we will be slaughtered like pigs in a ditch."

He tapped Bruto's leg with the back of his hand. "Take Diego, Lucio, and Frasco. Go down along the top of the ridge almost to the mouth of the *cañada*. They'll be spread out, waiting. So go down the south face of the ridge and make your way over to *Los Nidos*. Wait down the road from the corral. Tito and Cinco will come down the *reatas* with me into *Nido Mejor*. We will take horses and food from Don Emilio and meet you."

Bruto grumbled unhappily. "You are going down the cliff in the dark. What if you don't make it?"

"Wait a half hour after first light, then take horses for yourselves. Ride to our place in *Cañada Santa Margarita*. If we do not meet you there in one week," he paused, "then, my friends, good luck and go with God!"

Bruto was reluctant. "Let us all go down the *reatas* as we did when we entertained the gringo woman."

A murmur of agreement came from the three who were chosen to go down the ridge. Goyo silenced them. "It will take too much time. You know that, besides, it is possible that Morales and his *indios* are waiting there. Already that one is suspicious. I have watched him studying the signs."

Still not convinced, Bruto pressed his point. "Everybody believes there is only one way out of *Cañada Ciega*. They will all wait for us." He pointed toward the mouth of the canyon. "We will have no chance."

Goyo turned to face his lieutenant. He spoke quietly and coldly. "Bruto, my friend, I will make a bargain with you. You and your men go down the *reatas*. Tito, Cinco, and I will go this way. But if you are not waiting with the horses after one half hour, then you and I will use these," he patted the haft of his knife, "to decide who leads *Los Lobos*. Yes?"

In the strained silence no sound was heard but the breathing of the men. Then abruptly, without a response, Bruto began to climb to the ridge a hundred yards above. Reluctantly, the others followed. When the four had disappeared Goyo called to the remaining men. "We go now."

Although he was burdened by pockets bulging with ammunition and handicapped by darkness, Steven climbed quietly and well. At times Pedro was forced to plead for a respite.

Except for the occasional rattle of light rubble that a listener might have taken for the sound of small night animals scurrying across the slope, the pair did nothing to arouse suspicion. After an hour Steven paused. "How high do you think we are?"

Pedro sucked in several labored breaths. "Not high enough, señor. Maybe two hundred *varas*."

Steven was about to speculate aloud on the time remaining until first light when the majordomo tapped his arm and signaled for silence. For a long moment neither man seemed to breathe.

"What was it?"

"I don't know. I thought I heard something." Pedro pointed upward and across to the invisible opposite slope. "It was a voice maybe." He listened again. "I am not certain."

They resumed the climb. Several times they barely avoided placing their free hands in stray clumps of low pincushion cactus. Once when Steven attempted to steady himself by grasping the branches of a scrub mesquite the shallow roots gave way. Only Pedro's instant reflex spared them both a painful slide.

For another half hour they struggled for each foot gained. Finally they rested, listening intently. Nothing could be heard but the muted rush of wind moving up the steep, flue-like ravine.

As they were about to resume the climb, Steven clutched Pedro's arm. "Look!" He pointed across the ravine somewhat above their position. "There's a fire. You can see its reflection on the rocks above."

For a long moment both men strained to see, but whatever Steven had glimpsed seemed to have disappeared. Then, just as suddenly, the dull red reflection flickered again. Taut and still, they watched until the flare subsided. Pedro moved closer.

"It is a little mesquite fire in one of the caves. Somebody put more wood on it. What we see is the light on the roof of the opening. Quick. Let us climb higher. A few *varas* more and we can see it."

They climbed for another ten minutes, pausing every few yards to peer across the narrowing ravine. In several more minutes the flames themselves were visible as were the silhouettes of three men who appeared to be busy at some urgent task.

Steven estimated his own position to be approximately three hundred line-of-fire yards from the cave. The men were well within the killing range of his Henry carbine. Even so, a moving man was a small target, especially in poor light. While Pedro Morales watched in mute disbelief, he raised the weapon, rested it on the slab and sighted along the barrel. After a moment he relaxed.

"I could pick off one of them."

"No, señor! If you kill only one, the other two will get us. We have no good cover here. They can stay on the mountain longer than we can."

Steven grunted. "Unless Bowdon's men follow them up, which they'll do at first light. They'll be in a cross-fire then."

Pedro remained adamant as Steven considered the options. *Los Lobos* must have some alternate route out of their hiding place by way of *Nido Mejor*. He turned to Pedro. "What do you think they're doing?"

"It is too far, señor. The fire does not make enough light. But I think they are tying something."

As Pedro spoke, two of the distant figures moved a bulky object to the edge of the outcropping and a third began securing something to it. Shortly afterward the object disappeared. The obvious explanation struck both watchers at once.

"They're lowering supplies. They're going to try to get out this way." Steven blurted the words half aloud and silenced himself instantly. Pedro confirmed his observation. "I think it is blankets and rifles. It may not be possible to carry anything down but their pistols."

Fingering the Henry carbine as he watched, Steven wished for the greater accuracy of the more cumbersome, long-barreled Henry rifle.

Pedro, sensing his impatience, again urged caution. "It is not possible for them to go in the dark, señor. They will go at first light and pick up their things at the bottom of the cliff. Then we can get all of them, before they have their own rifles. Let us wait and be ready."

Steven pressed his right cheekbone against the smooth, oiled stock and lined the sights on the distant silhouettes. "Why take a chance? They may get away. They'll still be dangerous with pistols. And they're sure to take our horses. Nobody will be there to stop them."

As he spoke, Steven could see the fire lowering again. The once clean silhouettes were becoming diffused. In a matter of minutes there would be no target. His thumb moved atop the hammer and forced it back. When it clicked into place, a sound of protest escaped from Pedro and he swung down the trigger guard on his own single-shot Sharps. Into the breech he slipped a linen-cased ball and powder charge. The block, sliding up into place again, sliced the tip off the powder bag exposing the charge to the primer.

The two rifles discharged almost simultaneously. The echo of the double blast drowned out the cries across the ravine as the man who had been standing erect in the center of the group threw up his arms and fell backward. The blast reverberated between the sheer walls and died in the upper reaches of the canyon.

No other sound was heard but the startled screeching of a night-bird and then a silence, so deep it was oppressive, settled over the *cañada*.

Steven saw Pedro reloading. His own repeating carbine,

with second cartridge chambered, was already sighted on the spot where a waning flicker still marked the mouth of the hideout. But no target remained.

Protected by their meager cover, they waited, expecting return fire. Instead, the glow of smoldering embers died suddenly, leaving only the deep, pre-dawn darkness. A large owl brushed the air nearby and startled them. Steven leaned close to whisper. "You were right, Pedro. They must have lowered their rifles down the cliff."

"Perhaps, señor. But I think they did not see where our shots came from. It is possible they are afraid to shoot because we could mark their position again. We wait now, until it is light."

They had scarcely settled down for the cold, cramped vigil when the flat splat of distant rifle fire reached them. The reports were coming from the northwest, across the rocky spine separating them from *Cañada Ciega*.

"They're trying to get out the other way. They've run into Bowdon's men."

Before Pedro could reply, they heard more shooting. The rapid fire of repeating rifles was unmistakable. "That sounds like fifty-six caliber fire from Spencers. I don't think we have to wait up here now. I think it's finished. Listen."

The fusillade had diminished to sporadic firing. Steven nudged Pedro. "Come on. Let's start down."

After a few steps they discovered that it had been much easier to ascend than descend the slope quietly. Heels, rammed into loose shale, gave way and started alarming cascades. From there on, neither man made a serious attempt to move in silence. Above him, Steven could hear Pedro's quiet cursing. Within a half hour, they had descended more than halfway. Below, they could hear the distant spill of water from the large pool at the head of *Nido Mejor*. Several times on the opposite canyon wall that reared to the foot of the cave shelf, they heard the running rattle of small stones. Pedro guessed that the noises came from animals moving down to water.

Well below the base of the opposite cliff, Steven paused. "We should have cut across higher up and picked up their blankets and guns."

Pedro glanced up at the gray seam of pre-dawn light that was beginning to outline the high, ragged ridges to the east.

"Nobody will take them, señor. Let us get the horses first. It will be light soon."

At the rim of the upper pool, both men flattened on their bellies to drink. Pedro rose first and, moving with care, he began to follow the path around the pool, moving toward the ruins of the three burned cottages and the animal shed beyond. Steven sheathed his knife and set out to catch up. He had gone only a few yards when Pedro called out, Señor Lewis! Señor! Our horses are gone. Someone has cut the halter leads."

For an instant, Steven had trouble understanding. Then, vaulting over the boulders, he began to retrace his steps up the slope. It was growing light now but he knew there was no need for caution. In long, driving strides he forced his way up the steep incline, losing a foot for every two he gained. When he was close enough to the base of the cliff to see clearly, he paused to look around.

Several more minutes of reckless climbing brought him to the base of the cliff and then he saw it, the slack rawhide line dangling from the lip of the ledge far above him. What had happened was obvious now. His shot had driven the bandits down the slender line. They had recovered their weapons, stolen the horses, and headed out of the canyon, probably as much as an hour before he and Pedro had been able to descend.

Also obvious was the manner in which *Los Lobos* had entered *Cañada Laberinta* the afternoon they had mutilated, raped, and burned.

Steven whirled and started back down the slope in reckless leaping strides. When Pedro met him he gasped a brief explanation and ordered him to follow. On foot, they reached the *casa grande* just as the sun broke over the ice-glazed crests of the Santa Ana Mountains.

Inside, they found Doña Paula unconscious on the tile floor. Anita, racked with sobs and bleeding from a head wound, was crouched over her mother as though to protect her.

Don Emilio, dazed and rigid, sat at the table clutching at a blood-soaked gash in the left shoulder of his short jacket. Before him were the scattered contents of his strongbox. When he saw Steven and Pedro he attempted to speak but no sound would come from his gray lips.

A moan sent Pedro diving for the kitchen door. There he found Gracia and her three Indian *cocineras* sprawled on the floor. Pedro's outcry brought Steven to the doorway just as he dropped to his knees appealing to his wife.

Another groan followed by a heavy thud sent Steven racing back to the big room. Don Emilio had fallen forward on the table. A film of blood had begun to stain the edges of the scattered papers. Steven pushed them aside, propped the *ranchero* upright, and ripped away the vest and shirt. What he saw sickened him. A blade had slashed from the top of the left shoulder across the base of the neck just above the collarbone. From deep in the wound a finger-thick stream of blood was pulsing from a severed artery.

Bill Warren and two of his men found Steven trying to staunch the flow with a balled-up remnant of Don Emilio's cotton shirt. The marshal pushed the cloth aside gently, studied the wound, and shook his head. "I doubt if a surgeon could fix it, Lewis. It's cut through under the bone. You can't get at it. Leave him now and let's get the women to another room. It'll be over in a minute or two."

They carried Anita and Doña Paula to a bedchamber and returned to the kitchen, where the Indian women were beginning to stir. Still on his knees, Pedro was blotting blood from his wife's left temple. When Steven knelt beside him, he nodded. "Thank God they live! For that thank God!"

Warren leaned over and placed a hand on Pedro's shoulder. "Pedro, my men will stay with the women for a few minutes. I want you and Lewis to come with me. We've got two of them tied up down at the corral. Jim Bowdon wants you to see if you know who they are."

As they rode down the knoll from the *casa grande*, Bill Warren explained the capture. "We heard your two shots so we started over. When we came up out of the riverbed we saw two men riding like hell southeast from the house. They had a big lead so we shot their horses. When they went down we couldn't figure out why they didn't shoot back. But we found out soon enough. One of them had three rifles wrapped in blankets. The other one had a bag of food and some gold money and jewelry. They both had knives but no pistols. They claim they don't speak English."

He looked across at them. "By the way, what were you shooting at?"

Steven recounted the events in the canyon and Bill War-

333

ren grunted. "I thought there was only six. We got four killed and two hogtied. If you hit him, you left one on the shelf. You better come with us for a look after Pedro sizes up these men."

At the corral they looked at the two trussed Sonorans who were squatting with their backs to the adobe wall. Beyond observing that both men were half-breed *yaquis* Pedro could do nothing to identify them. After another futile attempt at interrogation during which the Marshal's vicious kicks had no effect, Pedro returned to the house and Steven and the Marshal rode to *Cañada Ciega* to join Jim Bowdon and the others.

The bodies of the four who had attempted to escape down the south ridge were laid out by the nearest of the fires. Merlhof and Parsons wore pleased expressions as they contemplated their handiwork. Merlhof chuckled and kicked one of the lolling heads. "Now, lads, there's the perfect picture of three good greasers!"

He guffawed with the others and kicked the corpse again. "We gotta do this about three thousand more times afore the score is anywheres near even."

Bowdon snapped an order. "All right! Leave 'em be now! There's one more up the canyon. Lewis thinks he hit him. We're going to have us a look."

Bill Warren and several of his men accompanied them to the edge of the little mesa. From there Bowdon, Steven, and two of the marshal's deputies started up the trail. They climbed along the rocky spine until they reached a sandstone projection that led to an adjacent ravine. After several hundred feet, the treacherous path turned sharply eastward. Steven could see across to the rocks that hours earlier would have shielded them in the event of a rifle duel. He called to Jim Bowdon, who was in the lead. "The cave is just around the next shoulder. Maybe another fifty yards."

The ledge was no place for a nervous man. At times it was scarcely more than two feet wide and the drop, in most places sheer, was at least one hundred feet. The first wind cave they came to was not more than five or six feet deep with a portal a scant four feet high. A few yards farther along they found another cave. Moving quietly, Bowdon edged around the shoulder, then signaled them to follow. Certain that they were within a few yards of the

334

hideout, Steven tapped the deputy on the arm. "Not far now. Be careful."

They moved a step at a time until they were able to look directly across to the opposite slope. Steven studied the canyon wall and marveled that he and Pedro had managed to scale it at all, much less in the black of night. Jim Bowdon had eased back the hammer on his seven-shot Navy-model Spencer rifle. At close range its heavy ball could blow a man to pieces. Crouching, he moved to the projection of sandstone and peered around it. After a few seconds, he stood upright and beckoned to them. "Come on, boys. He's there all right and he ain't going to hurt anybody now."

They found the man sprawled in the mouth of the cave. He had been struck high on the right side of the chest. The slug had entered cleanly but the lead had mushroomed against the inner surface of the shoulder blade and ripped out of the body through a ragged six-inch wound. Slight but strongly built, the man appeared to be well into his middle years. The black hair had gone gray. The tanned skin was deeply lined from years of exposure but the features were clean and strong. Even the *serape* and the much-patched clothing could not conceal the unmistakable evidence that once this man had been of the *gente de razon*, a well-born *californio*.

He had died neither quickly nor easily. After his companions deserted him, he had managed to crawl into the cave. The blood-stained floor indicated that he had lain beside one of the two water *jarros*. Then later, perhaps in an attempt to reach the warmth of the still smoldering fire, he had tried to crawl back outside. Death had come within the present hour.

Bowdon made no effort to conceal his exultation. "By damn, Lewis, right through the lung at three hundred yards. But before we give you the sharpshooting medal I want to see you do that again with my own eyes."

Though he took little pleasure from it, Steven smiled as the others added their grudging admiration. He had never wanted to kill before Arlene's rape and Tom Warden's emasculation. These bandits were the ones who had done it, and still, kneeling beside the man who surely was their leader, he found no satisfaction in retribution. When Bowdon extended a foot to turn the face in order to get a better look at it, Steven fended off the boot.

335

Bowdon's smile was puzzled. "I wouldn't dirty my hands on these murdering greasers if I was you, Lewis. If you can keep remembering what they done at your house maybe you'll start cursing God for not letting you get here in time to uncock the son of a bitch while he was still alive."

Steven did not attempt to defend his action. It had been impulsive. If Bowdon did not understand it, neither did he. Somehow the dead bandit did not look evil and he did not look strange. In his mid-years his own grandfather, Don José, had been such a man, finely made, well-proportioned, differing very little in appearance. He would have found it less disturbing had the man resembled the two captives and their dead companions. They were villainous and looked the part.

A Colt Dragoon revolver had fallen from the bandit's belt. Bowdon shoved it with his boot. "It's yours if you want it, Lewis. You earned it right enough. I'm going to have a look around this place before we fix to carry this bastard out of here."

A brief inspection of the cave turned up a small buckskin bag containing some Mexican pesos and an oiled skin envelope containing several papers. Bowdon tossed the packet to Steven. "They're written in spic. You speak their lingo. Why don't you look at them?"

Of the four papers, three were personal notations and comments on conversations with various officials in Mexico City, Tepic, San Blas, and Mazatlán. All were dated in the mid-Forties. The fourth sheet, carefully folded in its own inner wrapper, was an official document dated Monterey, Alta California, 20 Septiembre, 1841. It was a single page. Steven started when he saw the governor's seal and the signature of Juan Bautista Alvarado. Quickly he translated to himself. The document commissioned Gregorio Robles as special envoy to Mexico City whose duty it was to convince General Santa Anna, President of the Republic of Mexico, that he should bolster the garrisons in the Department of the North.

Unmindful of the activity around him, Steven seated himself on a small boulder and studied the paper. In the prolix, formal prose, he could follow Goyo Robles' mounting frustration as one after another apathetic Mexican official temporized with him.

Frantically, Steven raked over the past. The possession

336

of the papers would not necessarily prove that the man he had killed was his uncle. There had never been any hard evidence that the man called *El Lobo* was actually Goyo Robles. For as long as he could remember, the family had believed that Goyo, while on a vital mission for his blood brother, had fallen ill or had been murdered. Alvarado himself had done little to dispel the assumption. Indeed his tacit agreement seemed to reinforce it.

Steven glanced at the men who were using *reatas* to lash the body to a carrying pole. He felt empty. There was no remorse, just a strange emptiness and a sense of deep futility. The body looked almost frail now, far older than its years, as the men bound the neck to keep the head from hanging. He had resumed his study of the papers when Bowdon approached. "What do you make of them?"

Steven shook his head. "Nothing. Just some personal letters."

Bowdon frowned and pointed to the separate envelope. "What's that fancy one?"

"Nothing important, some sort of commission, army or something. Written before I was born."

Bowdon glanced back at the body and grunted. "The son of a bitch was probably respectable once, huh? I saw a name on it, didn't I?"

Steven nodded. "Governor Alvarado signed it."

The chief deputy snorted. "*That* drunken greaser!"

Without intending to, Steven Lewis stood up abruptly. The move startled Bowdon. "What's eating on you?"

"*That* 'drunken greaser' is my godfather."

Bowdon's mouth slacked open. "*Your* godfather? Bullshit, Lewis, You ain't no spic."

For a long moment the two eyed one another. Then Steven brushed past him. At the fire he squatted, stirred the gray to uncover the glowing coals, and touched the edge of the papers to them. Bowdon watched but made no attempt to interfere as the yellowed papers cindered to brittle black flakes. "Why did you do that?"

Steven smiled deceptively. "No use keeping papers signed by a drunken greaser, is there?"

The chief deputy flushed, then addressed his men. "All right, boys. Let's get that bastard out of here. Maybe we can tease his two friends into telling us who he is."

The man supporting the rear end of the pole looked at Bowdon. "Jim?"

"Yeah?"

"In view of the shortage of trees and all, do you suppose we could arrange a little horizontal hanging for them other two?"

Bowdon glanced at Steven before he answered. "We just might save the state the nuisance. But killing seems to upset our friend here, so I figure we won't invite him."

As the trail narrowed, the ledge angled abruptly to the right. The eight-foot pole with its cumbersome burden created a dangerous problem. As the lead man began the turn, his companion lost his balance. Bowdon jumped to assist him but too late. The man released the rear end of the pole as he cried out. "I'm letting go, Pete! Dry gulch the bastard before we go with him!"

A fraction of a second apart, they heaved the pole over the edge and braced themselves against the cliff. Mutely, all four men watched the trussed body vault end over end down the gorge. Halfway, the corpse tore loose from its bindings and began a series of grotesque, loose-jointed cartwheels until, a broken thing, it came to rest in the jagged boulders far below.

Filled with an unexplainable outrage, Steven leveled his rifle at the men, then lowered it. If Bowdon saw the move he gave no evidence of it. Instead, he broke the taut silence with an unconvincing rebuke. "That was a wrong thing, boys. You should have clung on." He looked down at the partially visible body. "But I guess it ain't killing when a body's already dead."

An hour later, at the rancho's corrals, City Marshal Bill Warren, satisfied that the job was finished, ordered his own men to mount. "We can get back by sundown if we go now. Besides, Jim, this is your job—rightfully." He indicated the prisoners.

Bowdon looked at the two bound men propped against the adobe barn wall. "Did you get them to talk?"

Warren shook his head as he mounted. "We didn't try. None of my boys speak their lingo." He turned his horse. "Shall I noise it around that you're bringing them in?"

Bowdon straightened and dismounted slowly. "I dunno, Bill. We're all pretty tuckered out. If I decide to, suppose we let it be a surprise?" He considered a moment. "If we do come in, though, I won't show favorites. We'll stretch one on John Gollar's gate and the other at the lumber yard."

The men rode off laughing and the deputy turned to Steven. "All right, Lewis, you savvy their lingo. Suppose you try and get a little testimony about what happened to your wife," he paused intentionally, "and her friend."

Steven slipped Goyo's Navy Colt revolver into the saddlebag and walked over to confront the prisoners. Addressing them in Spanish, with the Henry repeater cradled in his hands for added emphasis, he asked a series of blunt questions. When it became obvious that the men would not answer, Bowdon dismounted and kicked them viciously. "You talk up, Goddamn you, or I'll snap your fucking heads off right here." He delivered two more kicks. "Now talk."

Steven motioned the deputy away and tried persuasion, holding out hope but not promising leniency, if they would identify the men who had committed the atrocities. But the two *yaquis* maintained their stolid silence. They would not so much as nod in answer. When Steven saw them look past him, he turned to find Bowdon uncoiling a lariat. Three of the men were doing the same. He pointed to them and questioned the *yaquis* again. *"Saben que van ahora?"*

The men's eyes remained expressionless. Bowdon dismounted. "They savvy all right! Let's pick a couple of good posts. This trial's over!"

In Los Angeles, lynchings and vigilante hangings were both tourist attractions and social events. Steven had witnessed several and if formal trials had preceded the execution he had not heard about them. Once, when the hanging of a minor outlaw was announced for a time two hours later than the newspaper was scheduled to go to press, the editor earned the admiration of the entire community by publishing the event in accurate and lurid detail forty minutes before the execution took place. The doomed man's terror and the audience's enjoyment was made all the more exquisite when he was forced to stand and listen to a reading of the gruesome account of his own hanging.

Whatever mercy Steven may have felt for the pair at his feet was dissipated by a new tide of revulsion as he looked at them and envisioned their filthy bodies on top of Arlene. Hang them and be done with the whole rotten

339

business. There were a dozen stout rafter poles protruding from the sheds that would serve very well.

To Steven's surprise, Bowdon ordered the men dragged to the corral fence. There they were laid out face up with their feet toward the rails. Quickly the men lashed their ankles securely to the bases of the posts. Before Steven realized what was intended, lariat loops had been tightened around their necks and the free ends tossed up to the two riders. Ten feet of slack was left coiled on the ground. Bowden raised his revolver to signal and said to Steven. "You best back off some too. This kind of neck snapping can get a bit squirty!"

Steven saw Bowdon's finger begin to tighten on the trigger. In an instant he foresaw the horror about to happen. He heard Bowdon's revolver crash above him and the riders whoop at their mounts. The Henry repeater jumped to his shoulder. Its first slug tore the back off the head of the *yaqui* farthest from him. The second ripped into the skull behind the ear of the man at his feet. A split second later both bodies seemed to leap into the air as the horses reached the end of their tethers. Steven heard the inhuman groan and the sickening rending that followed as the heads parted from the bodies. Suddenly ill, he turned away and staggered to the fence.

He did not know how long he'd been standing with eyes closed and his body bathed in chill sweat when he heard someone move up close beside him. Bowdon's voice echoed both pity and scorn. "You know something, Lewis? Fer a young fellow whose wife was ruined by these son-of-a-bitching animals, it seems to me you hold a lot of pity fer greasers." He smirked and wagged his head. "Fer the life of me, I can't figure you out."

Steven fought down a dry heave and turned to face him. When he spoke, his voice was little more than a whisper. "Maybe you'll savvy, Bowdon, when I tell you that I'm half greaser." His voice grew stronger. "And maybe you'll savvy that even to a half-breed like me, bastards who can do what you've just done are worse than animals."

He spat out the words and dizziness overcame him. Grasping the top rail, he cradled his glistening forehead in the crook of his arm. As he did he caught sight of the nearest headless body twitching. A fit of nausea seized him and the men laughed as dry heaves convulsed him.

10 EARLY IN JUNE, on a warm afternoon, Steven sat with Francis Pioche on the veranda of Pioche's mansion. He was there to tell Pioche about the tragedy at Los Nidos.

"In the opinion of the doctors, Mrs. Lewis has been afflicted mentally by her experience. She and the child are with her parents in Los Angeles. Recently she has turned devoutly religious, which seems to help. Her life now is wrapped up in her studies and speaking selfishly that is best for me too. There's much to do. Doña Paula insisted on my being the trustee of the ranch even though I have, we have, a proprietary interest in the estate. Legally, I am accountable to Judge William Vance, an old friend of Don Emilio's, and of course, to you."

Pioche smiled blandly. "Very interesting. Very interesting. All at once now you have much responsibility for a young man of—how many years now? Twenty-four?"

Steven nodded. "In September, sir."

"So, my young friend, I hope you will not think me insensitive if I also express concern over the state of our business affairs? I gather the late Mr. Warden had concluded some transactions. Correct?"

"Yes, sir. We have twenty-two leases for apiaries and leases, with water contracts, on fifteen forty-acre parcels."

If the Frenchman was impressed, he gave no sign of it. "Who will replace Mr. Warden?"

"For the time being I will, sir."

"Perhaps now, dear Steven, that you have learned the folly of haste in matters of the heart you are ready to listen to mature counsel."

Steven nodded.

"Good. Then I suggest you engage a competent salesman. Save your energy for administration. I shall add another ten thousand to your account. Do you have a prospective sales person?"

341

"Yes, sir. There is a new man named Widney. He is five or six years older than I. I had thought about interesting him." Pioche frowned, apparently searching his memory. "What is Widney's first name? Could it be Robert?"

"Yes sir. Robert M. Do you know him?"

"*Oui.* I know of him. He read the law here. Also in Sacramento. I recall that he made an excellent impression on men of judgment. But he does not strike me as a candidate for land salesman."

The extent of the little Frenchman's knowledge never failed to disconcert Steven.

"He has opened a law office in Los Angeles, sir. But I think his first interest is real estate. In the past several months he's ridden over most of southern California. I have not met him formally, but I shall. The Hellmans are kindly disposed toward him."

Pioche inclined his head and pursed his lips. "*C'est bien. C'est bien.* By all means talk to him."

The steamer schedule allowed Steven only two days with his family. Again, it was a difficult visit, made more so by the hurt in his mother's eyes, his brother's studious indifference, and the avoidance of any reference to Arlene. Hardest of all to endure were the congratulations tendered on all sides for what was considered his heroic part in the extermination of the *Los Lobos* gang. The San Francisco newspapers had made much of the story.

Billy Ralston had treated him like a hero and had insisted on standing drinks for all at the Bank Exchange Bar. It was an uncomfortable experience. While most of the men took his reticence for becoming modesty, he knew his father suspected that more lay behind his reluctance to disclose details not reported in the press. Steven's evasive reply to the obvious question had been the same to everyone. "There was no way to positively identify the man. Apparently he had been in Mazatlán and Mexico City some years ago. His men were all *yaqui* Indians, Rock Wolves, they call themselves in Mexico. Even the Mexican government has a bounty on their heads."

He knew that while his parents would always deny the possibility that Goyo and El Lobo were one and the same, nonetheless, their lingering doubt could never be dispelled.

In Los Angeles again, Steven, made contact with Rob-

342

ert Widney. But Widney's reaction to his proposal was somewhat disappointing.

"I'm very flattered, Mr. Lewis. I know the Los Nidos *ranchitos*. I have already surveyed your development and given it a high rating. There is no doubt that I could make much money with you. But I do not want to commit myself to any one area.

"There is such variety of land in southern California. When its richness is discovered, people in the East will want to use it in many different ways. I am now acting as a consultant as well as a broker and will publish a journal called the *Real Estate Advertiser* to circulate throughout the East. I will solicit inquiries for certain types of land and attempt to supply it. You may be sure that I'll sell many parcels for you, unless, of course, you do decide on an exclusive agent."

Steven was so impressed with Widney that when he engaged Henry Vance, a cousin of Judge Vance, to head sales, he did so on a non-exclusive basis.

Henry Vance, a man in his forties, had been in land sales in Missouri for some years. Borrowing a page from Robert Widney's plan, he prepared a prospectus and had copies printed and mailed to his acquaintances back home. The response was immediate. By the first of September, Steven was able to report to Pioche a substantial number of leads.

In October, he moved Arlene and little Charlotte to a larger home on Fort Street where he hoped to resume a normal relationship, if indeed, he thought, it ever had been normal. The physical damage did not preclude a union. But Arlene made it clear that the thought of being entered by a man, any man, under any circumstances had become utterly repugnant.

"God, in His Divine Wisdom, often demands terrible sacrifices so that we may grow strong in spirit. Earthly pleasures are given to be taken away. I have grown from this ordeal, Steven, far beyond what I thought I could."

He smiled bitterly at the bleak prospect his marriage held for him when his need finally drove him to visit the best new house in town.

Steven found himself spending more and more time at Los Nidos. As trustee for Doña Paula and actual head of the ranch, he had to supervise its operation.

Anita Garcia y Perez, now seventeen, was, as Doña

343

Paula observed somewhat sadly, "a beautiful young lady when she isn't masquerading as a *vaquero*." Pedro Morales was devoted to her. "She knows how this rancho works. She is a better majordomo than I." When Anita was with Pedro, Steven was forced to agree that she was as effective on horseback working with the *caballada* as she was at the *casa grande* playing a gracious hostess.

Since Don Emilio's death, Doña Paula had gone into semi-retirement, preferring to leave most of the day-to-day decisions to Anita and to Gracia, who had mothered the girl since infancy. There was no filling the void left by the tragedy. But a very real happiness prevailed at the big ranch when Steven was there. Neighbors had taken to coming by more often. The Yorbas, Lugos, Bandinis, Serranos, Sepulvedas, Del Valles, members of all the great southern families, were frequent visitors. Some came out of friendship and concern. Some came to see the new land usage and to learn. By late fall most of them had accepted Steven not only as a member of the family but as its actual head.

Steven came to think of himself as an integral part of the rancho and his ties to the north hardly seemed to exist as a part of his present.

He was forcibly reminded when Will's letter arrived setting the date of his marriage to Mary O'Dwyer. Steven had promised his twin that he would be the best man and that he would bring Arlene and the baby to meet the family.

After a consulting the doctor, who found Arlene well able to travel, Steven booked passage to San Francisco. The fact of the actual departure date made Arlene retreat into deliberate vagueness. Steven reminded her of her promises, but she denied any clear recollection, indeed, seemed to have forgotten that her in-laws even existed.

"I cannot possibly go. I am not up to it. Neither is Charlotte. It's very dangerous for her to travel by ocean this time of year. The doctor will agree. And more than that, I'm far behind with my studies. There is so much to learn before I am fit to teach others and that is to be my life, a life of service."

Arlene resorted to hysterics when Steven threatened to take the child without her. Alarmed, he called the doctor, who placed her under sedation. On the fifteenth of December he sailed for San Francisco, alone.

Even in the midst of the wedding confusion, Steven found it impossible to rid himself of the nagging problem of Arlene, a problem in which the Lathrops had been no help at all.

However much he once resented his brother-in-law calling into question the legality of a union performed by a self-ordained minister, he now found himself grateful when Lambert Lee broached the subject: "In the name of God, Steve, you've got to get out of this mess. Even when you stood up with Will yesterday, you were somewhere else. This situation is preying on you. We all see it. Call me presumptuous, but I've half a notion to return to Los Angeles with you and confer on your behalf with Judge Vance. There are legal remedies for your difficulty. Also moral and ethical ones. You've got to find them before you are destroyed."

It was strange, Steven thought, but he would have jumped at the offer had it been made by almost any other person. "I appreciate the family's interest. I suppose it's been discussed?"

"It has. But only with your father and brother. If your mother dwells on it, she does so in silence. She does not confide in your sister or in me. But I'll wager it occupies much of her introspection. Really you must do something."

After a moment, Steven said, "I will. I'll talk with the judge when I return home."

As soon as he returned to Los Angeles, Steven arranged a meeting with Judge Vance. As he sat in the judge's outer office, Steven wondered at the restlessness he felt when he was in Los Angeles and the deep contentment that came over him whenever he was at the ranch. He sensed a profound rightness in it. In the midst of his introspection, Judge Vance loomed in the doorway and motioned him inside. "Sorry to keep you waiting, Steven." Indicating a chair, he returned to his own behind a formidable, carved oak desk. "Now then, lad, what are we discussing this morning?"

After a full review during which he kept virtually nothing from the judge, including the reason he had moved into the marriage so hastily, Steven spread his hands. "That's it. I'm at some sort of crossroads. I don't know which way to turn. There's no use denying it."

345

Judge Vance leaned back and clasped his long fingers across a belly surprisingly flat for his years.

"It's a familiar problem, Steven. You are not lost, my boy. Your brother-in-law was correct when he said there are legal remedies. But you need some emotional remedies too. Before we undertake any steps, have you been completely candid about your feeling for this marriage?"

"So far as I know, yes."

"When Cervantes talked about 'absence, that common cure of love' he was not talking about the sort of love I have known with Mrs. Vance for thirty-five years. If I thought you and Mrs. Lewis shared something like that, I'd urge you to stay with her through all adversity. But from your commendable candor I suspect your relationship was principally grounded on other important, forgive me, earthier emotions."

Steven nodded. "I am not looking for any easy excuses."

"Nor am I putting any in your mouth, my boy. Go on."

"I'm here with my guard down, Judge Vance, because I'm only certain of one thing. Whatever happened was at least half my fault."

Judge Vance laughed softly. "In someone else I might take that as a sign of spurious virtue. In your case it is the reaction of a responsible person. But it's not necessarily true. You were pursued, were you not? You haven't said so in precisely those words, but—"

Steven interrupted. "Let's just say that I was a willing victim."

Judge Vance nodded. "Most men are. You were in the hands of a somewhat older, very attractive, highly experienced woman who wanted you and who knew how to get you. I doubt that any single young man in your position could have resisted her.

"Also I have no doubt that her extraordinary needs drove her to exert those same wiles on poor Warden. That is not a cruel observation, Steven. But you were not dealing with what your church and the law consider a moral young woman."

Steven started to rise. "Look. If we have to drag Tom Warden back from the grave to witness her guilt, to hell with it. We'll let things ride."

Judge Vance smiled compassionately. "Easy, lad. Easy. We'll do no such thing. But I want to make certain that

346

you carry no unjust burden of guilt if we decide to dissolve this marriage, which, by the way, was never legal in the eyes of the law or the church. You know that, don't you?"

Reassured, Steven settled back. "I've wondered about it."

"And now you hope I'm right. Is that it?"

"To be honest, yes."

Judge Vance plopped his big palms flat on the desk. "All right. Let's see what can be done. Obviously, we can't leave your daughter in the position of having been legally born out of wedlock. I will need testimony from the two attending physicians and one other. If they agree on the probable permanent nature of your wife's aberrations, we may be able to arrange a divorce with proper provisions for the child and your wife's expenses and so on. As to where this will leave you with your church in the event you wish to remarry," he shrugged, "I shall have to have a talk with your bishop."

When Steven grew concerned again, Judge Vance smiled reassuringly. "A discreet one, of course."

The settlement and dissolution of the marriage was arranged so quietly that several months passed before it became common knowledge in Los Angeles.

In San Francisco the news was greeted with vast relief by the family, including Doña Maria who had never considered them truly married anyway. Felicia wrote that Lambert had begun laying plans to introduce him to San Francisco ladies of acceptable social status. Both Felicia and Lambert were at pains to let him know that, as a divorced man, he would no longer have *carte blanche entré*.

Annoyed at the prospect of having to resist Lambert's maneuvering to reinstate him socially, he wrote to the family and to Francis Pioche delaying his promised visit several months because of difficulties in constructing the expanded irrigation system at Los Nidos.

By June first, 1869, Robert Widney's *Real Estate Advertiser* was able to report land sales of one hundred fifty thousand dollars with sales of two hundred thousand dollars a month predicted as soon as California began to feel the full impact of the completed transcontinental railroad. The joining of the Central Pacific and the Union Pacific Railroads had demolished the last barrier to the opening of the west. Few southern Californians understood this better than Judge William Vance and Steven Lewis.

The judge pushed a sheaf of land contracts across the desk to Steven and said, "When Banning gets his railroad finished we're going to see the beginning of the biggest migration since the Gold Rush. You have enough contracts now to more than carry your notes, Steven. Mind if I make a suggestion?"

"Certainly not."

"While there are still large parcels of distress acreage around, buy more land. And from here on out consider going back to your original plan of long-term leases on a minimum cash and maximum share crop basis with the Los Nidos Ranch Company furnishing water and seed and perhaps, in some cases, basic tools. Of course, your share of the percentages would increase with the services rendered. This tenant farmer idea is not new to Southerners. They may not like it. But the ones without capital or credit will settle for it. If they're good farmers, everyone will prosper. Be fair with them. If you are, they'll go along. But remember human nature too. Make them feel like partners, not tenants. Make their stiff-necked Southern pride work for you."

Steven outlined the plan in a long letter to Pioche, in which he did not credit Judge Vance with the concept. Pioche replied immediately and arranged for a new line of credit with Hellman's bank. By mid-July the company had purchased five thousand acres on the northwest boundary of San Diego County at two dollars an acre cash and had secured an option on another three thousand.

The engineer, Monteith, solved the water problem by using a new redwood stave and hoop pipe to transport it. Steven ordered enough for a half mile test run through John Lewis & Sons in San Francisco. The pipe was among the first freight transported from the harbor to Los Angeles on Banning's new Los Angeles-San Pedro Railroad.

Steven brought Doña Paula and Anita into the Bella Union as the senator's guests. In his private coach they made the gala inaugural trip with Banning from Los Angeles to the harbor and return and participated in the great barbecue and ball at the combination freight terminal and passenger station.

A late Indian summer, with dry, unusually gentle Santa Ana winds, expedited the work of expanding the irrigation

348

projects. Steven found himself spending more and more time at Los Nidos. He seldom went to Los Angeles except for his monthly visits to little Charlotte, a duty made painful by Arlene's growing jealousy and overprotectiveness. As much as he loved the child, Steven was relieved when Arlene told him that she and her father had purchased land in Santa Monica canyon where they would establish a philosophical study colony.

In early November, Steven brought Pedro and a crew into *Nido Mejor* to begin the careful demolition of his former home. Anita, in her unconventional masculine riding habit, watched the first cart load of furniture began its journey down the precarious ditch road. Steven studied the expression on her face, then rode over and dismounted beside her.

"*Chica mia,* I'd rather share happier days with you. Why don't you stay with Doña Paula and learn how to make a dress?"

Anita looked at him sternly. "I know how to make a dress. And if you had watched the gentlemen at Señor Banning's fandango you would have seen that I also know how to wear one."

"I watched them, Chica. I watched them. And men who look at you that way are not gentlemen."

Anita grinned wickedly, "In that case, Señor Lewis, it is not very flattering to me that you consider yourself to be such a grand *caballero.*"

She narrowly avoided an ungentlemanly swat. But she had made him laugh, a difficult thing to accomplish these past few months.

After Steven had shown the men how to pry the priceless redwood siding from the front of the house without damaging it, Anita moved beside him. "What will you do with all of this?"

He shrugged. "Store it for now. Then sell it, perhaps, to recover some of the money."

She winced as a hand-forged nail screeched loose from a stud. "Store it, Esteban. Someday you may wish to build another house."

The words were spoken innocently enough but something in her tone made Steven turn to look down at her. For a long moment he returned her deliberately ingenuous gaze and it annoyed him that his own eyes were the first to waver.

"Mind your own business, *Chica*, or I may use those boards to build a *calabozo* just for you."

Later, at the dinner table, Steven was aware again that the now eighteen-year-old girl he called *Chica*, because it annoyed her, was much more woman than she appeared to be in her practical riding clothes. He wondered whether his hearing had tricked him. Had her retort, half spoken as she had turned away, really been, "If you were my jailer, it's possible I would like it."

In late June the test section of redwood pipe had been completed on the new agricultural subdivision and filled with water to swell it tight. For Steven, now twenty-six, and called, *"El Patrón,"* the burden of past troubles began to give way to a sense of well-being and accomplishment.

One great sadness marred the summer. Word reached Steven that Harry Morrison had died suddenly in Sacramento. He prepared to leave immediately to attend the funeral in San Francisco.

Anita, in a scooped-neck Mexican blouse and embroidered skirt, stood in the doorway to Steven's room. She was carrying his freshly laundered clothing that she insisted on taking to him herself.

"How long will you be gone, Esteban?"

He motioned her to bring the laundry over to a heavy cowhide *maleta*. "I don't know, *chica*. It depends on the steamer schedule. If I can't get one immediately, I'll take the stage."

The girl grimaced. "That is so difficult and so dirty. And much too slow."

Steven took the shirts from her and dropped them on top of the other clothing. Anita let out an anguished little cry. "You'll spoil them, clumsy." She pushed him aside gently and rearranged them.

A half smile tugged at the corners of Steven's mouth and there was an expression in his eyes that she had not seen before. A flush spread upward from the expanse of smooth tanned skin above the flattering curve of the low neckline. She retreated to the safety of indignation. "Why do I trouble to pack them so nicely? When you come back you'll just throw them in again."

350

Steven grinned agreeably. "That's right, *chica.*"

She stood her ground with her fists braced against her hips. "You are like a child. You need somebody to look after you. Every man does. So if you're smart, you'll ask your sister to pack for you."

Steven reached out, captured her head gently between his hands and pulled her close. It was the first he had touched her deliberately. The jet hair, parted in the middle and drawn back in a glistening knot low on her neck, felt silky and warm. The young flesh of her cheeks was petal soft against his palms.

"Do you know something, *chica?*" His hands dropped to her half-bare shoulders. "I've a mind to . . ." He broke off and released her. "Never mind what I've a mind to."

After the funeral and a sad visit with his family, Steven sent a note to Pioche asking for a meeting. Pioche was morose, distracted, and uninterested in the affairs of Los Angeles. Steven had been told that Pioche was under increasing pressure by both the Central Pacific Railroad's Big Four and his own partners to sell out his interests in the Placerville–Sacramento Valley Railroad and the San Francisco–San Jose Railroad. Similar pressure was being put on Billy Ralston in an effort to gain control of his steam navigation interests on the bay and rivers. Clearly, Stanford, Crocker, Huntington, and Hopkins were determined to dominate all transportation in the state.

Steven persisted in attempting a progress report. Pioche cut him off. "*Mon vieux,* it should be clear to you by now that so long as each new operation is profitable I shall make funds available. *C'est tout.*"

Steven left the three-story Stockton Street mansion, troubled not so much by the Frenchman's brusqueness as by something strange that he detected beneath it. This time there had been no unsettling personal overtures and none of the annoying petulance.

On the top deck of the *Goliath* that was taking him back to San Pedro, Steven's thoughts turned back to the night four and one half years ago when he had met the Lathrops. Thoughts of Arlene and her uninhibited performances in bed still excited him but there were other equally vivid, more painful ones: the frustration, the hurt, and the futility of their union.

He banished the ghosts of those times by deliberately

351

thinking about Anita. He was startled to find that he had whispered *"Chica"* aloud. He could see her now, in the doorway of his bedchamber, her arms loaded with his fresh shirts. He smiled as he remembered how she had tried to mask her embarrassment with indignation. He could feel again her silky hair beneath his fingers and could see again the serious little face and the huge, luminous eyes, half eager, half defiant. The vision made him laugh silently and he spoke her name again with great tenderness.

"Chica, amorita. I miss you. I'll be damned if you haven't grown very dear to me."

Judge Vance attended to the complicated matters with the church. The bishop had been understanding and eager to return to the fold a stray as promising as Steven Lewis. One week before Christmas, after signing an agreement that their children would be raised in the faith, Anita Garcia y Perez and Steven Eugene Lewis were married in the family chapel at El Rancho de los Nidos.

The impetus of Robert Widney's real-estate publication, in addition to the Eastern promotion efforts by both the Union Pacific and Central Pacific railroads, was starting to give Henry Vance more business than he could handle. Vance was hampered though by Steven's refusal to sell the land in fee simple. "Hell's bell's! If we could let them buy outright, I could make five times as many deals."

Since he was limiting Vance's income, Steven asked Widney if Vance could help sell the new land Widney had bought in the San Fernando Valley. A few months later, Vance's anxiety to earn more money was explained by the arrival of his wife, a woman twenty years younger than he, and their five-year-old son Thomas.

For a while Vance seemed content. Then, using money borrowed from his cousin, Judge Vance, he asked Steven if he could buy twenty acres outright to start an experimental lemon grove. Partly to repay the Judge, Steven agreed to one last exception to his policy and soon afterward, Vance moved his family to a modest prefabricated house. It was one of the first, sent by ship from Oregon.

It seemed to those around Henry Vance that each boat brought packs of literature to him from the United States Department of Agrictulture. One communication particularly interested Steven; it described a new, seedless orange from Brazil that government horticulturists felt showed great promise. It was described as somewhat larger than the native Spanish or Mission orange, with a navel-like aperture at its apex.

Steven was excited by the possibilities in the new product but realized there would be a problem in the lease-partnership agreements. Since they did not own the land, farmers would be reluctant to invest in orchards that took years to bring to maturity. This meant that if orchards were to be set out in large numbers, they would have to be undertaken by the Los Nidos Ranch Company itself.

Steven began to study the situation with Pedro. Together they rode the highlands above the fog belt that were usually free of killing frosts. In less than two weeks they had staked out several likely eighty-acre parcels.

On September 30, 1871, Alicia Paula Maria Lewis was born at El Rancho de los Nidos. In a matter of hours the *casa grande* was filled with more joy than it had known since the arival of Anita herself.

Within days, Anita was up and around the house and grounds. In less than a month, to Steven's dismay and to the amusement of Gracia and her *cocineras*, Anita was astride a horse again, eager to be shown the new work.

In October, Steven persuaded his mother and father to come south for the holidays. John Lewis was eager to see his son's new family and wanted to catch up with the market in the south. He and Doña Maria arrived late on the afternoon of October twenty-third and went to Pico House, the former governor's elegant new eighty-room hotel. In addition to a fine restaurant and bar, it was the first hotel to offer inside plumbing with hot and cold water piped to the rooms.

Steven put an arm around his mother, saw his parents registered, and conducted them to their second-floor room. A short time later, as he freshened for dinner in his adjoining room, he heard the sound of horsemen and loud voices. Below, he saw a lighted, heavily guarded hack being driven westward toward Spring Street. In the lantern glow he could see that it was surrounded by pigtailed Chinese in their loose black trousers and jackets. He raised the heavy sash window to get a better view and saw his father peering from the adjoining window.

"What's going on, Steven?"

"I don't know. Let's go down and find out." Then Steven caught sight of Sheriff Frank Burns and his chief deputy coming out of the Lafayette Hotel with City Marshal Bill Warren. They were moving quickly toward the center of the disturbance.

In their haste to reach the staircase, Steven and his father nearly collided in the hall. On the street they were caught up in the surge of curious citizens hurrying toward the corner to follow the progress of the surrounded hack that was now moving along the south side of the Plaza.

In the crowd Steven saw Jesse Yarnell, one of the men who had just founded a newspaper called the *Los Angeles Evening Express*. Motioning to his father to follow, he forced his way across the street and caught up with him. "What's the trouble, Jesse?"

"They're bringing that Chinese slave girl back from Santa Barbara, the one Wong Lee's Chinamen charged with stealing jewelry."

A man behind Steven spoke up. "She belongs to Lum Fong's Chinaboys. They been claiming they'll take her back by force."

By the time the carriage reached the southwest corner of the plaza the local law-enforcement men had converged on it with rifles and drawn pistols and the Chinese had begun to disperse. Most of them were retreating toward the old Coronel adobe, the central structure in the labrynthine slums, known locally as Nigger Alley. Steven and his father returned to Pico House just as the carriage turned the corner on Spring Street.

They were about to order dinner when a man burst into the dining room shouting for Judge Widney. "If anybody knows where he is, get him! Those 'slants' just shot Jesus Bilderrain and Bob Thompson in Nigger Alley. Thompson's dead and the boys are fixing to lynch every chink in town. Burns, Warren, nobody can stop them."

By the time Steven and his father reached the street, hundreds of men were converging on the plaza. All were armed. Many were carrying ropes. In a matter of minutes every hitching post and rail in that area was jammed with mounts abandoned by their owners. Above the angry clamor some shots were heard. Sporadic at first, the firing soon turned into a barrage.

John Lewis grasped his son's arm. "This crowd's losing

354

its head, Steve. Let's keep back. Pretty quick they'll be shooting at each other. I've seen it happen before."

Steven dismissed the advice. Followed by his father, he had managed to move about halfway across the Plaza when a shout went up. Moments later several small groups of men, each dragging a terrified Chinese by his pigtail, began clubbing their way back through the crowd.

"Get out of the way, boys. We're going to start hanging the slant-eyed bastards." The cry was taken up by the onlookers, who began clearing a corridor. Soon, well over a thousand men and boys were screaming for a mass lynching. Beside Steven a man shouted, "Here he comes." Turning to look back, Steven saw Robert Widney's tall, powerful figure forcing a way through the crowd. In his wake were several members of the Law and Order Party. All were armed.

Steven pushed in behind Widney. Sheriff Burns and four deputies with guns drawn were retreating slowly as they shouted orders to disperse. The hysterical crowd ignored them.

Steven could hear Widney offering the lawmen the services of his civilian group. At the same time, Robert Widney's younger brother, William, forced his way through followed by a man named John Lazzarevich and a half dozen more armed members of the Law and Order Party. The younger Widney recognized Steven and shouted. "Steve? Are you armed?"

"No. My gun's at the hotel."

William Widney thrust a Colt Navy revolver into his hand. "Come on. Help us break this up."

Ignoring his father's warning, Steven disappeared into the crowd. The worst of the shooting had stopped but scattered fire continued to come from the north side of the low compound where a new mob was now intent upon a mass lynching. Behind him somebody yelled, "They're stringing up the first ones at Slaney's Shoe Store!"

There was a moment of uncertainty, then the more timid turned back and began running toward the intersection of Commercial and Los Angeles Street. Steven estimated that half the town of six thousand had gathered. Several times each minute small bands of shouting men dragged shrieking Chinese to the lynching sites. Somewhere off to the side, Steven could hear Sheriff Frank

355

Burns shouting, "It's too late to help those. Just keep them from taking any more."

Steven was beside Captain Cameron Thom and Sam Caswell. Behind them was the principal access to Nigger Alley. Captain Thom took up a position with his back to the opening. His new Winchester repeating carbine was leveled at the crowd. He motioned Steven in beside him. On the far side of the plaza an ugly cheer went up each time another writhing Chinese was hoisted from the awning braces at Slaney's store or from the gate post at John Goller's Wagon Shop. Every place that could accommodate a mob became a killing ground and every support that would hold a body became a gallows. In the flickering lantern light the contorted faces reflected mass hysteria bordering on madness. Hoarse voices were demanding the life of every last one of the one hundred seventy-five Chinese men, women, and children in Los Angeles. Steven guessed that a score were already dead or dying. The mob had no single ringleader. In each group at each possible access to the Chinese quarter smaller mobs formed around the loudest of the rabble rousers. The leader of the group confronting them now was the known troublemaker, Jack Pollard, whose chief delight was testing the courage of newcomers at Buffum's Saloon.

Pollard, mean drunk and flanked by his gang of sycophants, confronted Captain Thom. "We got no fuss with you, Cap. When chinks shoot down decent men and the law won't do nothing about it, then it's time we learned them." Pollard was seconded by a shout of agreement but the clamor subsided when a familiar voice boomed behind them. Judge Widney, followed by his brother, came pushing through and motioned Steven to follow. They held drawn revolvers. The judge shoved some bystanders aside and faced Pollard. Those nearest began edging away from the possible line of fire.

"Stand where you are. All of you!" Widney's voice was calm but clearly audible for fifty feet around. "Pollard, I've never used a pistol to take a man's life. But I promise you, the first one who tries to go into Nigger Alley to get another Chinaman is going to die. What's been done in this town tonight is the worst disgrace ever to happen here. And every one of you standing here is guilty of aiding and abetting it. You are under arrest, Pollard. Take his gun, Steve."

Steven moved to obey. Pollard's fingers locked around the grip and began to tremble.

The judge raised his pistol. "Don't try it, Pollard."

Steven, hoping his knotted gut wouldn't betray him, took the weapon and stuck it under his belt.

After that, Pollard's men were the first to turn away. Others followed on their heel. It was past midnight when the last of the ringleaders had been jailed and the corpses cut down. Twenty Chinese had been lynched at John Goller's Wagon Shop. Nineteen had strangled on gallows improvised in front of William Slaney's Shoe Store. Four more had died suspended from the stakes of a freight wagon on Fort Street. The next morning three more bodies, one a badly mutilated female Chinese, were found dumped in the *zanja* that still supplied most of the Los Angeles city water.

John Lewis, helpless and unable to stop his son, had returned to Pico House. Steven had come in for a few minutes shortly after three in the morning. His clothing was dirty and torn from working with the Widneys, Frank Burns and Bill Warren, and their men. Burns had sworn him in as a deputy on the spot and charged him with protecting the remaining Chinese. Now, over coffee and brandy, Steven and his father listened as eyewitness fragments were fitted together into a picture of unprecedented mob brutality.

Sheriff J. Frank Burns and his men had done what they could but everyone agreed that Judge Robert Widney and his Law and Order Party were the force that had turned the tide. The shoe merchant, William Slaney, was toasted as a hero. Even though the mob had used his iron awning support as gibbets, Slaney himself had saved the lives of his own Chinese employees and their families by locking them in his storeroom and defying the lynchers with a shotgun.

Some newspaper accounts of the Los Angeles Massacre were so incomplete they amounted to distortions. Others were deliberately slanted to discredit Los Angeles and leave the city in the position of sharing with other cities around the country equally deplorable records of mob violence.

Forty-six Chinese had died. Nearly one hundred Caucasian ringleaders had been taken to jail. Of these only a dozen were being held. The solid, law-abiding segment of

357

citizenry was confident that justice would be done and Los Angeles would redeem itself. Their optimism was based on the foregone conclusion that the men indicted would be tried before Judge Robert Widney.

John and Maria Lewis' visit to Los Nidos was marred only briefly by the mob violence in Los Angeles. As Steven had anticipated, Doña Paula and his mother were in immediate rapport. Anita was embraced as a daughter and the child as an "angel." He could not remember ever having seen his mother so happy and animated. As for Doña Paula, the visit had a miraculous effect. For the first time since the days with his grandfather, Steven experienced a keen sense of family. He was relieved that his parents had resigned themselves to his commitment to the southland.

Shortly after the beginning of the new year, Steven escorted his mother and father to Los Angeles and saw them safely aboard the steamer for San Francisco. Before returning to Los Nidos the following morning, he stopped by Judge Vance's home to pick up some books and pamphlets newly arrived from the East.

The judge handed Steven a book by a Charles Loring Brace.

"Widney's been accused of being too optimistic in his *Advertiser,* but compared to this man, who was here two years ago, for only one week mind you, Robert's an incurable pessimist."

Steven opened the book to the foreword and found himself laughing. "Why he uses the Holy Bible to prove that southern California is the New Palestine. According to this, we've got everything from mustard to Moses with the hot springs of Tiberius thrown in."

The Judge snorted. "Sodom and Gomorrah maybe. But the Holy Land, never." The judge thrust another book at him. "And try this one. The eminent Dr. Edwards, whoever he is, credits our 'salubrious ozone' with curative properties that even our Lord Jesus didn't have."

Disbelief spread across Steven's face as he read aloud: "—and also incipient phthisis, tuberculosis, liver disease, constipation, enlarged glands, scrofulus infections, scarlet fever, diphtheria, kidney malfunctions, jaundice, and most female disturbances." He shook his head. "It is a bit far-fetched, all right."

"Far-fetched? Southern California has a clear-cut case

358

of defamation of character chargeable to these boomers. Sooner or later we're going to have to do something to counteract this, Steven, or these rascals will turn us into the world's largest sanitarium and asylum."

The optimism of the land boomers was infectious, but it was predicated on the completion of a railroad down the San Joaquin Valley from San Francisco to Los Angeles, a difficult engineering feat, not due for at least five years. Immigrants found their way south by steamer and stagecoach but they were not arriving in the great numbers predicted.

On the ride back to the ranch, Steven reflected on the need to temper optimism with caution.

In early April, Steven saw Judge Vance again to show him a letter from Pioche. Pioche had written:

I watch with apprehension the increasing imbalance between your credits and debits. I fear you have undertaken more than you can manage. Or perhaps, having again assumed the responsibilities of a family, you find yourself without sufficient energy to devote to my interests. I will not be reassured until you present me with a practical plan for reconciling this disparity between receipts and disbursements. Meanwhile I urge you not to commit my credit to more acquisitions and have so advised Messrs Temple and Hellman of my wishes in this regard.

Judge Vance studied the letter, then laid it aside. "You've got to do what Widney suggests, sell them when they get off the trains before they get switched into the Sacramento and San Joaquin valleys. We're getting the dregs down here now. You've got to sell them on the idea of coming straight south to take up a going thing. Think about sending Henry up to San Fransisco to open a land office there."

The solution was obvious and Steven damned himself for not having thought of it. But he did not want to send Henry Vance to San Francisco immediately. "I have to talk with Pioche myself anyway. Let's not bother Henry now. I'll stay up there a few days longer and see what the prospects are."

He arrived in San Francisco on the last day of April, surprised his family, and sent a note requesting a visit with Pioche. For the second time in their association the Frenchman sent word that he was indisposed and would be unable to confer for several days.

Steven spent a day visiting land offices along Market Street. He found a firm advertising San Bernardino and Los Angeles land. When he asked about it, the clerk was vague. Steven discovered the vagueness was due to the man's pretense of not knowing the location of the holdings because he wanted to divert sales to land in the north near San Jose.

The time had been well spent. That evening, with his father, his twin, and his brother-in-law, Steven outlined a plan whereby John Lewis and Sons could diversify and open a land office exclusively representing the Los Nidos leases. Fearful that Pioche would summon him before the plan could be committed to paper, Steven sat up late with Lambert drawing up a prospectus.

He was in the office ahead of everyone else the next morning to review the plans. At a quarter of eight his father burst in. "Steve! Steven! Have you heard? Pioche is dead! He killed himself less than an hour ago."

Without waiting to hear more, Steven charged past his father, raced to the corner, and ran westward three blocks to Stockton Street. At the three-story mansion a crowd had already gathered around the steps. Two marshals were keeping the entrance clear. Steven recognized a surrey belonging to Pioche's personal physician whom he had met on several occasions.

Questions produced a variety of rumors ranging from a heart seizure to a lurid murder. Someone said the valet had killed him. Another said a jealous companion had stabbed him.

Shocked at losing his benefactor and filled with apprehension over the possible consequences, Steven was about to leave when Louis Reiff appeared at the doorway escorted by two men. Steven tried to force his way close enough to question the valet but was stopped by a policeman. Then J. B. Bayerque appeared. As Pioche's partner reached the bottom of the stairs, Steven shouldered through him.

"Mister Bayerque, I'm Steven Lewis. We've only met once briefly, but—"

Bayerque cut him off without stopping. "I know you, Lewis. Glad you're here. You may have to testify at the coroner's inquest. We'll be in touch with you."

Steven left the Stockton Street mansion as the black morgue van arrived and made his way back to the John Lewis & Sons offices. His father, flanked by Will and Lambert, listened with deep concern to the news that he might have to appear as a witness. Steven knew that his brother-in-law was struggling to resist saying, "I told you so." Instead, Lambert paced the office, thinking aloud. "There is no earthly reason why Steven should have to appear. We all know where he was when the man killed himself and there's little doubt that it was a suicide. That much seems clear.

"The point of the matter is this. The motives behind Pioche's suicide, which I suspect were purely personal, could be messy. Few people know of Steven's *business* association with Francis Pioche." Lambert Lee deliberately emphasized the word. "But if he's forced to take the stand at the inquest and reveal that the business connection also included some social obligations, his soirées and so forth, then," he threw up his long hands, "God alone knows what may come out. Never mind that Steven knows nothing of Pioche's private life. He might very well suffer guilt-by-association with those who did."

John Lewis rose suddenly and reached for his hat. "They're not going to call Steve. I don't give a goddamn who I have to buy off. He had no connection with Pioche's private affairs and he's not going to be made a party to a Roman circus."

At the door, Lambert laid a hand on his arm. "Where are you going to start, John?" John angled past him. "I'm going to start where the power is, with Billy Ralston."

A half hour later Steven edged through the crowd at the Bank Exchange Bar and joined his father at Ralston's table. Billy Ralston, profoundly unhappy, rose to greet him. "Sit down, Steven. And don't worry about being called to testify. That will be taken care of. There's little doubt that whatever drove Francis to take his own life was a personal matter, if he *did* take his own life."

Ralston ordered a coffee for Steven and continued. "I do not have full information about Pioche's dealings, of course. There may have been pressures on him that we don't know about. But as far as his financial situation in California is concerned, we can rule out money troubles. Pioche and Bayerque is one of the most solvent financial houses in the state and I do not except my own Bank of California."

Billy Ralston worried the heavy links of his gold watch chain. "I simply don't understand it. He was an impeccable business man. He was that above all, no matter what else he may have been called by lesser men. I personally have heard Francis express himself on the subject of suicide. He did so at Bleyer's funeral and at Wexler's, when they checked out of the game after the Silver Queen stock mess. He called suicide 'cowardly and unchristian.' He said a man should fight life's battles, that he has no right to kill himself."

Ralston exhaled heavily. "I'm a man who deals in facts. I can accept a proven one. But I still find it hard to believe that Francis Pioche would blow his own brains out." He looked across at Steven. "In any case, you can rest easy, Steve. I don't think Pioche has any family in this country. But nonetheless I intend to see that the newspapers don't make a sensational thing of this. I would like Francis remembered in this city for his contributions to it, particularly to the colleges of medicine and law and to the museum and the theater. Francis Pioche was the real successor to Leidesdorff, no matter how much I may blow about what I've done."

Steven smiled at the banker's oblique modesty, for he was certain that no single man had done more for San Francisco than William Ralston.

Ralston rose to return to the bank. "If this leaves you in a bind down south, Steve, let me know. Perhaps we can help tide you over there, too."

Later, at his father's office, Steven listened and was grateful for the family counsel. He had wanted to go his own way, to be a loner, but it was good now to have them on his side. He was particularly grateful to Lambert, who was reasoning with his unusual impersonal clarity.

"One thing you can expect now, Steven, is a power struggle to see who takes over Pioche's assets. The paper he holds on Los Nidos may be a very small factor. But Bayerque, or whoever comes out on top, will be just as greedy to acquire it. Do you have an idea how much your obligation is?"

362

Steven reached for a folded paper tucked away in an inside pocket. "About eighteen thousand dollars."

His father settled back, relieved. "Not as much as I thought it would be."

Steven smiled ruefully. "It's about half as much as it would have been if I'd seen Pioche before he died."

Lambert settled on the edge of a chair. "Work up an exact statement. It may be a blessing that most of us up here have little faith in southern California. Pioche's successors might be willing to make an accommodation on those debts." John Lewis agreed.

"And Billy will not be discommoded by eighteen thousand, if you need to refinance with him."

But what troubled Steven the most was not the immediate cash. What he needed was the continuing enthusiasm and willingness to go along that Pioche had displayed. Ralston could not be expected to share that. But perhaps Isaias Hellman would.

In Los Angeles again, Judge Vance listened while Steven recounted the Pioche tragedy and analyzed his own position: "... that's why I feel I'd be better off trying to work something out with the Hellmans. They know the land and they know me. What I do would be important to them. To Billy Ralston I'd be little more than a favor to my father."

"Hellman's your man, Steven. Your head's on straight. But there are times when I'm not so sure about Ralston's. He's keeping some very questionable company when he partners with that Kentucky scamp, Asbury Harpending."

Steven had forgotten about the flamboyant young soldier of fortune. "I remember the name all right, but that's all."

Judge Vance grinned and rubbed his nose. "Well, he's the lad who formed the private army and damn near got away with a plan to steal the whole west coast and hand it over to the Confederacy. He got put in Alcatraz for that."

Steven laughed. "They accused my father's good friend, Will Chard, of trying to steal California from Mexico. He was put in prison too, but now he's one of the most respected ranchers in the Sacramento Valley. And look at poor old Sutter. He plotted against everybody and wound up being honored by them all. I'm not sure a little insurrection counts much against a man in California."

363

"It didn't used to, son. But today it's a different game."

Steven's examination of his financial position proved that income from leases and the ranch's share of profits from partnerships fell short of covering overhead and note obligations. Enough new leases to cover the discrepancy could not be negotiated immediately. Steven would have to borrow or some land would have to be sold. He agreed with Judge Vance that a decision to spin off acreage should be kept as quiet as possible.

"Have a confidential talk with Robert Widney. He'll understand your situation. Set a bargain price on some good acreage and he'll probably see you out of trouble."

Widney was sympathetic but not helpful. "I would undertake the purchase myself, Steve, but I'm committed heavily in the San Fernando Valley. You need cash and I doubt that any of my prospects would be in a position to do you much good. But I'll be glad to speak with Isaias and see what he can do."

Hellman was also sympathetic but not interested in a small parcel of a few thousand acres. The only way Steven could get his hands on enough cash to satisfy the Pioche notes would be to mortgage the entire Los Nidos holding to the new Temple-Hellman bank. He was unwilling to do that even if Doña Paula agreed.

He could see nothing but trouble from the men who took over from the dead financier. If they demanded an immediate liquidation, it could mean the passing of Los Nidos from his control. Another down-turn in the southland's economy could have the same result if he was beholden to John Temple or Isaias Hellman or for that matter, even Billy Ralston, for all of his outward show of benevolence. After a pride-damaging reappraisal of his situation, he decided to seek help from his father.

Steven could imagine his father sitting at his big roller-top desk examining the projections he had set down and agreeing with them. But he knew that his father would not break the promise that Lewis & Sons would not lend him any money to buy land south of the Tehachapi mountains. So while he tried to give the request the appearance of a routine business proposal, it was actually a request for a personal loan. He could only hope that his father wasn't cash short because of recent purchases of Comstock shares as an expression of confidence in Billy Ralston.

364

Several anxious weeks of waiting passed before the steamer brought word from San Francisco. Two letters arrived. A brief but friendly note confirmed his father's cash short position. A longer letter from his brother-in-law was far more encouraging.

Lambert's business associate, Moses Berwin, had been considering the purchase of good agricultural land in the southern part of the state against the day when the population boom would materialize. On a long-term investment basis, he was offering six dollars an acre for the three thousand acres of El Rancho de Los Nidos land.

By mid-July, somewhat apprehensive because Moses Berwin had not been able to come south to see the land, Steven had Judge Vance attend to the final details of the sale and to the transfer of the eighteen thousand dollars to the late Francis Pioche's executors in return for a release from all obligations to the financier's estate plus a covenant not to sue.

As the year 1872 neared its end, Steven Lewis had good cause to thank God a thousand times that he was not beholden to the bankers. There was grave concern over the failure of the completion of the Central Pacific Railroad to San Francisco to produce a boom and mounting anger over the high-handed tactics of the Big Four. In spite of the fact that they held title to millions of acres of Federally deeded land, the railroad was blackmailing important farming centers along the San Joaquin Valley into donating land and pledging bond issues to pay for the construction of depots and rail yards. Communities that refused would be by-passed and left without means of getting their produce to the major markets.

The presentiment of trouble spread over the entire country. The Union Pacific's involvement in the Credit Mobilier scandal had repercussions that reached all the way to Congress. Speaker of the House James G. Blaine had ordered an inquiry.

As a result, the expulsion of Congressman Oakes Ames of Massachusetts and James Brooks of New York, who was also a government director of the railroad, was recommended. The report named several Congressmen and demanded the impeachment of the Vice President.

In the end the cynics were right. The two Congressmen

were merely reprimanded and the impeachment proceedings against Colfax were dropped.

To offset this, San Francisco morale received a big boost from the new Clay Street tram line. Suddenly the steepest hillside lots became the most desirable.

Steven managed to keep the development going along at Los Nidos and continued to send cautiously optimistic reports north to his family. He was concerned because land values had not risen as quickly as predicted and as grateful as he was, he could not rid himself of regret over the need to dispose of acreage. A dozen times Anita had heard him say, "I'll buy that land back one day or die trying!"

Twice during the first half of the year Steven and Anita thought they were to have an addition to their family. Gracia blamed the false pregnancies on too much horseback riding. *"Tu tienes el corazón del vaquero!"* she would say when she saw Anita going off to ride with Steven and Pedro.

Late in the summer, reports began reaching the Pacific Coast that a number of banks in the East were closing their doors. Steven's anxiety over the family's interests brought reassuring letters from Will and from Lambert. "The Bank of California is prospering as never before." They said, "Now it virtually dominates the Comstock mines and the stamp mills."

Then on the twentieth of September the transcontinental telegraph flashed the news that the New York Clearing House was closing its doors for ten days. Panic spread across the nation and gripped California. By the end of the year a dozen Los Nidos leases were in default. Cash from the partnership crops dwindled and dried up altogether. Steven had to slow down work on the subdivisions to conserve cash. For the time being, he would trim and retrench and produce working capital, if possible, through resale of the foreclosed parcels.

A score of times Steven recalled his father's gloomy assessment of the southern part of the state. He could not deny that when times were poor in other parts of the country they were worse in the southland. For the first time, he found the bright flame of faith in his dream beginning to flicker. He could only hope that San Francisco's banks, heavily involved in the Comstock's precious metals, would remain solvent.

11 PORTLY, HEAVILY BONED MERCHANT Moses Berwin, president of San Francisco's Congregation Emanuel, left the temple fifteen minutes later than he had intended. He set out on foot for the Dupont Street offices of his wholesale grocery business, the Golden West Processors. Half running, sometimes skipping with a curious grace, Berwin hurried along and was annoyed that in the seven years of his association with the punctual Lambert Lee this would be the first time that he was late.

In the outer office, before he could attempt an unaccustomed apology, Lambert greeted him with a placating hand upraised. "I was late too, Moses. I stopped to check another rumor that Stanford and Huntington and Hopkins are selling the Central Pacific from under Crocker."

Berwin, puffing a bit, showed his new corporate counsel into his office. Indicating a chair, he walked around his own cluttered desk and sat down. "When Charlie Crocker said he'd trade his interest in the Central Pacific for a clean shirt, believe me, it would have been safer to take the shirt instead of personal notes from his partners." Berwin laughed. "So why am I feeling sorry? Crocker's made four million dollars on construction profits alone. And I'll tell you something else, Lambert, he's the only man in the country who knows how to build a railroad through the mountains. When they start to build north from Redding, Sam Montague will need Charlie again. Anybody else will go broke trying to drive crews through the Siskiyous. You'll see."

His mood sobered. "But we have troubles of our own." He picked up a letter and handed it across the desk. "It's from your father-in-law's friend, Will Chard, at Tehama. Read it."

After Lambert had scanned the letter, Berwin's eyes flashed with indignation. "Would you believe it? In a free country? Would you believe the nerve of those *gonovim?*

367

Those thieves?" He retrieved the letter, searched out a paragraph and read aloud.

The only way the railroad can set realistic freight rates for its customers in the Sacramento Valley is to have as thorough an understanding of production costs for your various field crops as we have of their marketing costs. Consequently we are asking our division freight agent to call on each agricultural shipper so that he may go over your books first hand and give you the benefit of our experience.

He slammed the letter down. "Thieves! Damned, miserable thieves." Berwin punctuated each epithet with a hard fist. ". . . . the benefit of our experience! Do you know, Lambert, that they even asked to see *my* books? Would you believe that?"

Lambert smiled. "It doesn't surprise me. They asked to see our books too, so they can offer what they call a 'fair price' for our two freight boats."

"I know. Yesterday I saw Stanford at Bancroft's book store. Such a bad trader, that man. Huntington should keep him quiet." He paused. "I want to talk about that offer, Lambert. Already you and John have sold the long overland freight lines. What's left?"

"We still have the grain warehouses and the short freight lines from Red Bluff and Redding into Shasta, Whiskey Town, and Trinity mines."

Berwin's eyes narrowed. "Tell me. Do you think John would sell his boats?" Lambert considered briefly. "I don't know. When it looked like Congress would give the entire San Francisco waterfront franchise to the Central Pacific we thought we'd have to. But we're all right for now."

Berwin seemed relieved. "Lambert, there isn't a damned thing anybody can do to help Chard or Tyler or Thomas or Wilson or Kimball and the other ranchers up in the Sacramento Valley. So long as they have to ship by rail, they're going to pay with blood. Those monopolist bandits squeeze and leave men like Chard just enough to stay alive. Even me. They do it to me. From day to day I never know what I'm going to pay to get produce down here. The reason I want to talk is to find out if you think John will sign a two-year shipping contract with Golden West at the present rates? If he will, I'll guarantee that everything I buy in the valleys will be shipped here on your boats. What do you think?"

A long silence ensued. Lambert Lee knew that the man opposite him understood the alternatives. The mere rumor of such a contract, if noised around Montgomery Street, would place John Lewis & Sons in a more favorable bargaining position with the Big Four. If an improved offer should be refused, it was inevitable that the railroad would undercut the competitors' rate, thereby leaving John Lewis & Sons and Golden West Provisioners caught in an unfavorable contract. Eventually competition would force the Lewis rates down to impossible levels and Berwin would be obliged to seek a release and ship with the Central Pacific to maintain his own competitive position.

The proposal puzzled Lambert. He prided himself on his ability to see through obscure tactics and he was certain that Moses Berwin was thinking well beyond the immediate consequences. He always did. Whatever they were made Lambert wary. "Moses, let's dispense with obvious answers. Why do you want John Lewis and Sons to go out of business?"

Berwin didn't hesitate. "Because, Lambert, John Lewis is my friend. Things are stirring that could break his company. When men like Newhall, Ralston, and poor Francis Pioche sell their railroads and steamship lines to the Central Pacific," he gestured in annoyance, "or the Southern Pacific or whatever they call their railroad today, then you have no one to stand with you to fight them. John is fifty-four now. If he gets out at his price he can live very well for the rest of his days. The longer he waits the lower the price will be. He's got to think of himself. He shouldn't worry about the twins. The foundry business has a great future. If Will had a choice, he would go to work for his father-in-law. He's as much as told me so. And Steven is on his own."

Lambert waited expectantly as Moses Berwin frowned and seemed to gather his thoughts. "So now I'll tell you what I'm up to, Lambert. But first, if you wouldn't mind, I've got a personal question. How big a stockholder is John in the Bank of California? Would he get hurt if Billy got hurt?"

Lambert knew the question was not as ingenuous as it sounded. He hedged his reply. "If you mean would he get hurt like Ralston got hurt when Sharon touted him off

Consolidated Virginia, yes. He'd get hurt for well over a hundred thousand. That's a lot for John. Why?"

"We'll get to 'why' in a minute. How close is John to D. O. Mills?"

"Mills is president of the Bank of California. He and John are original stockholders."

Berwin verged on impatience. "This I know, Lambert. But are they close otherwise? Do they drink together or womanize?"

Lambert frowned. "If you mean do they talk off the record, no. You know John. He's not a social man. He doesn't inspire confidences. Look, Moses, what in hell are you driving at? Come out with it. Please."

Moses Berwin opened his arms. "All right. All right. I'll come out. I was hoping maybe you and John had heard the talk, then I wouldn't be spreading rumors. God forgive me if I'm wrong."

Alerted, Lambert leaned forward as though unconsciously bracing himself against an impact. "What rumor, Moses?"

"That Mills is forcing Billy to buy him out. He wants out of the Bank of California before it's too late."

Lambert settled back, remembering the long list of men who had lacked the courage and vision to play in Ralston's league. He smiled with relief. "Moses, how many times have we attended funerals for Billy only to find out that the corpus delicti has turned the predicted disastor into another fortune and he's gone for a swim in the bay?"

Berwin agreed. "I know. I know. I'm not exactly sitting shiva, Lambert. A professional mourner I'm not. But always in those times Billy had a mountain of Comstock ore to back him up. It's different now. They say he's tearing scrap paper into little bitty pieces again. That's a bad sign. He only does it when he's got deep trouble. Everybody on the street knows that. It's how Billy tips his hand." The merchant nodded, hoping that he was wrong. "Sure, he's got assets, the New Montgomery Street Real Estate Company," a thought struck Berwin and he threw up his hands in horror, "with that mad man Asbury Harpending for a partner. And he's got the woolen mills and the carriage company. Good businesses, maybe. But he doesn't have a mountain of gold and silver—and he needs one just to keep up that nice cozy little Belmont place of his, a nice little country house with rooms for a hundred guests yet,

370

and solid marble stables. And now, God help him, the Palace Hotel. Two years he's been building it and no roof yet. Do you know what it will cost? Give or take a few hundred thousand? It will cost five million."

Berwin closed his eyes and groaned. "Five hydraulic elevators. Eight hundred rooms. Eight hundred toilets. Four hundred and thirty-seven bathtubs. Thirty thousand special dishes."

Berwin paused for breath and reached for a paper at the side of his desk. "Personally, Lambert, I'm not complaining. This is Golden West's bid to supply the kitchens. And I'll get the business but believe me, it will be C.O.D."

Lambert frowned. "Mills doesn't make many mistakes. I admit that, Moses."

Berwin stifled a small belch. "Look, Lambert. A deal is only good when everybody makes out. It's only a matter of time, two years, three years, before the railroad puts the squeeze on John. They can bleed him and force him to sell at their price. Also they'll squeeze me because I can't figure freight costs for my goods. I need two years to get my packing plants built near the crops so I don't have to ship perishables so far. What I want John to do is make a contract with me, at today's top rates. If he does that, I'll have a guaranteed freight cost for the two years I need and he'll have another asset worth a quarter of a million dollars that he can use to boost the price of his business to the Central Pacific. They're still dickering, aren't they?"

Lambert nodded. "Stanford talked to John again the other day."

"Good. Good for everybody. Especially for you, Lambert. For you I've got some very exciting plans."

Moses Berwin was one of the few who anticipated the railroads' stranglehold on the economy. Lambert stated the case clearly. "It's still cheaper to ship fabricated and processed goods around the Horn from the east coast than it is to ship overland by rail. And no doubt, when we're ready to ship canned goods down here to the city, the railroad will invent a new higher rate for that, too."

Berwin's concept of distributing by growing and packing in close proximity to major markets was extremely profitable. Berwin had used John Lewis to good advantage to buy the time he needed to establish his packing plants and, as Berwin predicted, John prospered. In the fall of 1874,

371

independent at last, and tired by the unequal battle for survival with the Central Pacific railroad, John retired. He planned to take Doña Maria to Europe for the Grand Tour.

So far, Berwin's operation had not helped Steven Lewis. The sluggish southern economy made it impractical. Furthermore, the Southern Pacific was fighting the congealed seismic agony of the Tehachapi barrier. Engineer William Hood was driving four thousand of Charlie Crocker's China Boys through tunnels and over torturous summits. Steven was amused when Berwin growled, "California isn't one state. It's two *countries.*"

In the summer of 1875 Moses Berwin expanded by constructing the first flour mill near Chico and a new packing plant in the Santa Clara valley. He arranged the financing through individual investors; he did not trust the Bank of California.

Rumors of trouble around Ralston were rampant. Ralston's South Montgomery Street real-estate project was in jeopardy because Ralston had committed huge sums to extend the street to the waterfront without acquiring all the rights of way. The owners of two key parcels waited until nearly two million dollars had been sunk in excavations before they announced their flat refusal to sell.

Ralston had given Lucky Baldwin a personal note for three and a half million dollars for shares in the Ophir mine and he had been forced to borrow two million in gold from William Sharon, who had demanded the New Palace Hotel as security.

Berwin was bluntly suspicious. "I told you. Sharon never gives up. Ralston can't keep him from buying a Senate seat this time. Billy had better watch out. That beady-eyed little rat is liable to dump his stock."

Lambert Lee was incredulous. "For God's sake, he wouldn't dare. He'd squeeze Ralston, but he'd never risk wiping out the Nevada legislature. Most of them invested their personal fortunes in the Ophir on his say-so."

One week later, Nevada's U.S. Senator-elect, William Sharon, wiped them all out. Ralston, desperate for cash, began selling his remaining holdings in distress.

The exchange collapsed and once again panic pervaded San Francisco's financial district. Lambert and Moses watched in disbelief as one hundred and ten million dollars in paper value in the Ophir and other Comstock mines disappeared in the Eastern money markets and small investors went broke by the thousands.

Lambert Lee supposed that there wasn't a man in San Francisco who wouldn't pay for the chance to murder Sharon in cold blood; but he had concealed his stock manipulations so well that no proof of them existed.

Within a few short weeks, Golden West Processors was beset by nervous investors. Reassuring them as best he could, Moses called Lambert into a private emergency meeting.

"I told our investors that we'd be slowed down for a while, not to worry, that the state is built on more substantial assets than the Comstock."

Lambert smiled bleakly. "The state is, but not San Francisco. Damn near everybody here has a few shares of some mine."

"Sure, I got some too, but to me it's only a poker game. What I'm worried about is not the Comstock. I'm worried about Ralston's bank. If his investment ring starts cashing in its stock, Billy will have trouble like he's never seen." Berwin fingered his cravat. "Tell me, Lambert, do you have power of attorney for John's bank stock?"

The question surprised Lambert. "Yes. Why?"

"I'll tell you in a minute. Also, how much cash can you raise? Your own, I mean?"

"Fifty thousand, I'd guess." Immediately Lambert knew what Berwin was driving at. "You're going to ask if I would act to protect John by selling him out of the Bank of California?"

"Would you?"

"I would."

"If you did, could you recommend that John come in with us?"

The query troubled Lambert. He did not relish the responsibility it would place on him. "I don't know. Why?"

"Because I want to pay off our notes. I got maybe eighty thousand. You got fifty and maybe with something from John, we could do it. Our investors are asking too many questions. When I don't have good answers, I don't like questions." Berwin rested a palm on his breast. "Believe me, it hurts to use our own money. But I think we should buy them out. So maybe we'll expand a little slower. This morning I heard Ralston's sold his interest in the San Joaquin ranches, too. I don't like what I'm hearing, Lambert."

That afternoon, after telling Moses that he would sleep on the proposal, Lambert spent several hours on San Francisco's Champagne Route drinking moderately and listening.

Ralston had gained control in the city's Spring Valley Water Company. He had bought up the necessary shares with a three-and-a-half-million-dollar loan engineered through his wife's father, Colonel Fry. It was puzzling news because Ralston had always called his interest in the water company philanthropic; he was happy to subsidize it for San Francisco's benefit. But now it could be Ralston's most valuable asset, one for which he could demand any price. Still, such a tactic seemed uncharacteristic. Sharon might try it but not Billy Ralston.

Lambert also heard a rumor that William S. Chapman, California's largest rancher, had advanced funds that would indirectly bolster the Bank of California. Chapman, owner of a million fertile acres in the San Joaquin valley, did most of his financing through the bank.

That afternoon, with Will Lewis' consent, Lambert took John Lewis out of Bank of California stocks.

The next morning Lambert and Berwin stared at the morning newspapers. There it was, in print: William Chapman Ralston was accused of attempting to bribe the city council in order to insure the sale to San Francisco of the Spring Valley Water Company. The price Ralston was alleged to have asked was just under fifteen million dollars.

Repercussions in San Francisco and in the Eastern money markets were immediate. Lambert, with John Lewis' endorsed stock, did his best to be inconspicuous as he presented the certificates at the cashier's cage. He was relieved when Ralston, preoccupied at his big desk in the glass-enclosed corner cubicle, did not look up. Screened by two clerks from Golden West who had accompanied him to help transport the gold proceeds from the sale, Lambert did his business quickly.

For a week or more San Francisco financial circles were riddled with rumors. Some said Fisk and Gould had raided the Central Pacific, that Collis P. Huntington, the railroad's actual head, was mortally ill in New York. Another rumor said the Ophir had made a secret strike and

374

was keeping its miners underground, well-fed from the best restaurants and well entertained by the most accomplished prostitutes, to prevent word from leaking.

A few days later, a new rumor swept along California and Montgomery Streets. The silver kings were quietly unloading shares in some of their holdings. The veins had petered out. Was the Consolidated Virginia in trouble? What about the California? Nobody knew how many of the various Comstock shares Ralston still held, but most of the mines on Mount Davidson were said to be in deep trouble. Now, they were saying, even Adolph Sutro's tunnel could not save the Comstock. If Mackay and Fair were selling out, the mines were finished.

What that could mean to San Francisco was clear to Lambert and Moses. Banks that had invested heavily in the mines and mills and the ore-transporting railroad were finished, too. One hopeful report had it that the once rich California mine had struck a new vein. But San Francisco's attitude was negative now and Mackay and Fair would say nothing to dispel the gloom.

The rumors persisted through July and on into August. Billy Ralston continued to present his usual cheerful front to the city. At the Bank Exchange Bar he presided at his corner table exuding the same durable optimism. When questioned about Spring Valley, he replied, "There's lots of room for misunderstanding in a complicated deal, especially when politicians are involved." Questions about the bank were parried with a cryptic smile. "I think I'm known in this city. I have taken all of the precautions necessary to protect my depositors. I have an obligation to each one of them. After all, they are also my friends."

On the morning of August twenty-fifth, word came that the Bank of California was overextended. By noon every depositor in San Francisco and a number of correspondent banks on New York's Wall Street had been alerted. Nearly two million dollars had already been withdrawn. Stock certificates were being endorsed and presented for sale. By mid-afternoon the anxious customers had become a panic-ridden mob. In a frenzy, men and women charged the lobby and crushed against the huge bronze doors. Police were called as scores of carriages continued to arrive. Horses and vehicles were left unattended for blocks around as their owners raced on foot to withdraw their money.

At a quarter past four a great shout went up and the doors, each manned by a dozen club-wielding guards, began to swing shut. By four-thirty, the Bank of California had failed.

On Saturday morning Billy Ralston was seen entering the bank with William Sharon and other directors. The restaurants and bars were filled with men who had gathered early to speculate on the meaning. Ralston's deserted table at the Bank Exchange Bar somehow seemed an ominous portent and the rumors Lambert heard changed in tone. Gone was any hint of underlying optimism.

At three o'clock that afternoon, Ralston emerged from his massive neo-classic California Street Bank alone, entered his carriage, and drove northward toward the Golden Gate Straits. A half hour later, Sharon and the others pushed through the crowd with police help and departed in their carriages.

Shortly before five o'clock, word flashed through the city that Billy Ralston was dead. An hour after he had plunged into the Bay for his usual constitutional swim in the rough, icy currents, his body washed ashore a few yards east of the pier.

Lambert, Will, and Moses Berwin joined the incredulous crowd that rushed to Neptune Beach at the throat of the "narrows." They kept hearing the word "suicide" whispered, but they refused to believe it. Although the weather had been unseasonably hot for a week, the San Francisco Bay never warmed.

Berwin was indignant. "He was overheated. He had a seizure. A big man in his fifties should slow down, but Billy didn't know the meaning of the word. He'd never take his own life. Never."

On Monday the suicide theory gained some credibility. Sharon, already sole owner of the Palace Hotel, had also acquired ownership of Ralston's fabled Belmont country estate with its dining room that could seat three hundred guests. In addition he had forced Ralston to sign over every last personal possession including the great Pine Street mansion, his furniture and art collection, and his stable of one hundred thoroughbred race horses. In the name of depositors, Sharon, the conscienceless bachelor and manipulator, had left the grieving, much respected Mrs. Ralston a destitute widow.

376

In the office of Golden West Processors, Lambert Lee paced before his partner's desk. "That son-of-a-bitch. That miserable, ruthless, immoral, benighted son-of-a-bitch. There's got to be a special hell for him or I'll stop believing in an Almighty."

Moses Berwin studied the papers he held without really seeing them. When he spoke, it was softly and to himself. "Where is it written that in California the bad ones are always punished? Francis Pioche a suicide, and who knows the name of his tormenter? Now Billy Ralston dead and with him dies part of San Francisco, and Sharon winds up with everything Ralston built."

Hatred for Sharon reached an exploding point when San Francisco learned that Ralston had pleaded for a chance to pay off the bank's creditors with four and a half million dollars that had been promised to him, but Sharon forced the directors to refuse. Sharon fled San Francisco. There were reports that he had gone to Virginia City in Nevada where the power of his money remained undiminished.

Doña Maria and John Lewis, returned from their European Grand Tour, were profoundly shocked by Ralston's death. They too refused to accept the suicide theory. Grateful though he was to Lambert for saving what amounted to a fourth of his modest personal fortune, John Lewis declined to take an active part in Golden West. But he arranged confidentially to purchase for twenty thousand dollars in gold the three thousand Los Nidos acres Berwin had bought. In a secret holographic codicil to his will, he left the land to his son Steven. The cash gave Moses and Lambert the one hundred fifty thousand dollars needed to buy out the dissident stockholders and bring Golden West completely under their control. In the end, John Lewis bought one thousand shares of stock and accepted a seat on Golden West's board of directors.

The first evidence that the entire Comstock was in serious trouble reached San Francisco's exchanges in mid-January. Consolidated Virginia failed to pay a dividend. The silver kings had quietly disposed of their Consolidated and California shares at the top market price the moment they learned the rich veins were pinching out. To secure their own huge fortunes, they ruined the market.

No disaster that had ever befallen San Francisco could match the disruption caused by the failure of the Consolidated Virginia and the California. "Gentlemen" fought like savages on the floors of the exchange to divest themselves of shares at any price. The repercussions were felt up and down the entire state. Small businesses were the first hit. Then the companies that supplied the mines. Throughout all of California men were being laid off.

Even Moses Berwin felt pinched. Bitterly, he complained to Lambert and John Lewis. "Look at this city. In three weeks, millionaires are standing on breadlines. A professional whore can't find a patron because waitresses and chambermaids are servicing them for two bits to buy a dinner."

John tried to interject a note of optimism he did not feel. "Maybe it's not as bad as it looks. Will says there is good ore still coming from both mines. The speculators are the ones who got hurt."

There was no humor in Berwin's laugh. "My friend, that's all of San Francisco and half of Wall Street in New York."

For the average man, the future seemed to hold little. On the classic assumption that things must be better elsewhere, the philosophy that had brought many of them to California in the first place, a number of San Franciscans began to go south. The majority were defeated families with few assets beyond a determination to never again speculate in mining shares.

It was a time of disillusionment and many were forced to return to the East. They departed, filled with bitterness, and Eastern editors printed their experiences in full and lurid detail.

Steven Lewis' Los Nidos parcels began to lease well even though farmers in general were going broke. Guaranteed water and liberal share-crop terms attracted the most far-sighted among the newcomers and Steven was careful in his choice of landholders.

New orchards and groves were set out. Henry Vance did not get the new Brazil navel orange first. But he was among the first. His avocado trees were bearing but new growers were wary of the fruit from Central America that they called "the vegetable that grows on a tree."

As often as time permitted, Steven visited eleven-year-old Charlotte, but the child had been completely absorbed by Arlene. Jabez Lathrop and his wife had moved to Arlene's Santa Monica colony. She had abandoned the Oriental in favor of Greco-Roman. The faithful in evidence appeared to be in their late middle years. All affected togas and sandals.

Steven enjoyed his visits when he was alone with Charlotte. He tried bringing along six-year-old Alicia, hoping a friendship would develop, but he felt a return of Charlotte's earlier restraint. On the next visit, Arlene made it clear that she did not want "a confusing emotional problem" to develop between Charlotte and her half sister. "Divided loyalties are very damaging, Steven dear. I know your motives are the noblest but I must forbid any more such visits."

Steven welcomed the business responsibilities that diverted his thoughts from his personal trials. In San Francisco, things worsened. The Southern Pacific, pursuing its policy of squeezing the last cent the traffic could bear, had raised freight rates down the valley to the point where it was cheaper to order mine machinery shipped around the Horn. The bulk of the business had gone to Eastern agents.

Earlier, Moses Berwin had made contacts in Los Angeles through Steven to supply canned goods to the Cerro Gordo silver and lead mines in Inyo County's Owen's Valley. Memi Nadeau's freight lines provided the transport. That portion of the business allowed Golden West Processors to weather the Comstock debacle.

Things had changed in the city by the Golden Gate. It had built, burned, and rebuilt on an average of once every five years since it had cast off the humble name of Yerba Buena. Now, with a wild Irish teamster named Dennis Kearney shouting for "bread or blood" in the streets, with William Coleman back again at the head of a vigilante group armed with pick handles, there were those who said the city named for the gentle saint would have been more aptly named for Satan since San Francisco was "in a hell of a fix."

A Los Angeles news editor remarked with questionable civic loyalty that no change would be needed in the southland. "The City of the Angels is quite appropriately named in view of the fact that it's dead!"

379

The new opera *Carmen*, for which the late Francis Pioche had planned a gala for the history books, failed in an excellent production at the Baldwin Theater. Despite the international stars, Marie Rose and Anna Louise Carey, in the leading female roles, the presentation went twenty thousand dollars in the red. San Francisco had no stomach for grand opera. It needed plain bread.

In 1882, the Lewis twins lost their godfather, and California lost its first native-born Mexican Governor, Juan Bautista Alvarado. Several months later, the state lost another symbol of its pastoral days when their grandfather, Don José, died. For the Lewises and the Lambert Lees, it was a sad year relieved somewhat by the pleasure of growing children but by little else.

In 1883, optimism began creeping back. With typical ambivalence where Leland Stanford, Mark Hopkins, Charles Crocker and Collis Huntington were concerned, San Francisco celebrated the news that the Southern Pacific's two-way construction crews had met on the west bank of the Pecos River. Finally, the transcontinental rails of the SP's Sunset Route were joined, linking the South with the West.

The Berwins and the Lees went to Oakland to cheer the departure of the first train from California to New Orleans. Once again the Big Four were given conditional absolution in the expectation of another boom. Moses watched Stanford's Palace Car pull away from the terminal shed. "Every sabbath I do *bruchas* that the Santa Fe Railroad will get into Los Angeles first. What that *momser* Huntington needs is competition. That's the best steam for any man's engine."

No discernible good resulted from the completion of the Sunset Route in 1884; but as 1885 approached, a definite aura of hope was in the air. Even Santa Barbara, twice wooed and twice jilted by the railroad, felt a resurgence of hope as some of the immigrants made their way down the coast by steamer and stagecoach. As mid-year approached, there were other more significant indications that the corner had been turned. At a directors' meeting, Lambert Lee speculated aloud.

"We're bound to the rest of the country now by hundreds of miles of steel rail. In another couple of years we

380

should have California fruit and vegetables under our labels in every important city in the country. While we're damning them, we can thank the Big Four for that." Berwin and his son Samuel, who had just joined Golden West Processors, nodded in agreement.

Moses Berwin, speaking of Leland Stanford, wagged his head compassionately. "He's not the smartest of the Big Four, but Stanford has more heart. To found a great university in memory of Leland Junior is a wonderful thing. By this, he does for us all. He's cost me, this Stanford. But I like him." Moses looked across at twenty-four-year-old Samuel. "To have a son live and turn out, there's no greater blessing for a father. Believe me, for Stanford, I can weep."

Samuel Berwin smiled uneasily and wondered what his father and Lambert would say, or think privately, when they heard that he and Lambert's daughter, Emily had fallen in love and wished to marry. Samuel Berwin and Emily Lee had anguished singly and together over the ancient customs that divided them when they had fallen in love. Samuel had told Emily firmly, "I don't think I'd go as far as Benjamin Disraeli's father, Isaac, and turn Christian, Emily, but I do not intend to let my faith stand in the way even though our decision may disappoint my parents and yours."

Emily had tried to reassure him. "We both pray to the same God, darling Samuel. And we use at least half of the same book. I really think I shall be able to spare you for an hour on Saturday if you do the same for me on Sunday."

At first both families were deeply concerned when Emily and Samuel told them. In the end, convinced that Emily and Samuel knew their own hearts, they reconciled their prejudices by refusing to acknowledge them. Cheerfully, they made plans to celebrate the engagement at the Poodle Dog in early November. The parents admitted their decision had not been made free of all reservations but it was one that had been unselfishly made.

At the engagement party, Moses raised his glass and the other parents followed. "*Shalom*, dear children, *Shalom!* Together you had the chicken pox, the measles, and a thousand runny noses. Now, together may you have a lifetime filled with happiness and good health and with charity for all who have less!"

12 The invitation read:

SOCIETY OF CALIFORNIA PIONEERS
You are cordially invited to attend
a formal banquet and testimonial
honoring our beloved founder-member and native son,
STEVEN EUGENE LEWIS
on the occasion of his Eightieth Birthday,
Friday evening, September Nineteen,
One Thousand Nine Hundred and Twenty Four,
at Seven O'Clock.
Hotel Lankershim
Seventh and Broadway
Los Angeles, California

R.S.V.P.

Twelve hundred copies were mailed. Over six hundred responses arrived at the Founders' Parlor of the Society. Steven Lewis' widowed daughter, Alicia Lewis Vance, and her son Steven and daughter-in-law, Catherine Coulter Vance, were in charge of arrangements for the head table. Steven was still a remarkably rugged and active man despite his approaching eightieth birthday. He leaned forward and peered at the guest list and the names of the family members and others who had promised to attend the testimonial. After a brief inspection he grunted in mock displeasure and said to his daughter Alicia, *"Hijita mia,* if the object of this affair is to praise me, the job had best be left to them who don't know me." Alicia started to dispute him but he dismissed her. "All I ask is that the family be seated down front so they can't see the guests behind them smirking at the pack of lies they'll be hearing."

Alicia smiled indulgently and kissed him on the temple. "All right, father. I promise."

Steven reached up and cupped his hand around the cheek of the daughter who resembled so much his own beloved Anita, who had died of cervical carcinoma in 1892—"from riding like a vaquero instead of a lady!" old Gracia had insisted to the day of her own death three years later.

For the most part, the guests at the banquet were in their middle to late years. Here and there were tables of young people whose parents birthright entitled them to attend. From his seat to the right of the podium Steven could catch glimpses of some of the younger guests surreptitiously pouring the contents of silver hip flasks into the Prohibition punch. He did not disapprove; in fact he found himself amused and somewhat pleased that several of his livelier contemporaries were also spiking their drinks. A lot of things had happened since the turn of the century that had not pleased him. Foremost among them were the involvement in the European War and the passage of the Eighteenth Amendment while the country was preoccupied with something that was "none of its damned business." Once Los Angeles had been the toughest town in the nation. Now Steven reflected on what it had become—a city of blue noses from the middle west who had opened two churches for every saloon they closed. San Francisco voted ninety thousand strong *against* Prohibition. But what did this damned town do? It voted a hundred and forty-five thousand *in favor* of the Eighteenth Amendment and voted its vineyard industry, the biggest in the state, right the hell into bankruptcy.

While the guests were still occupied with their breast of chicken Mornay, Steven gazed thoughtfully over the room. Southern California had changed with the coming of the Santa Fe Railroad and the "Boom of Eighty-seven." It had changed again with the beach boom. They'll talk about all that tonight, he thought, but they won't say it like it really was.

The change had been very profitable for the Los Nidos Ranch Company and also for the state-wide Golden West Processors. With the help from Amadeo Gianinni's new Bank of Italy, the packing firm had thrived under the management of his niece's husband, Samuel Berwin, after Moses had retired in 1902.

Both Moses and Lambert had been gone now for more than twenty years. Their bones were resting in San Francisco. But at the long family table below him were their offspring, that once new generation which had taken over and the present newer generation that was in the process of taking over. Steven Lewis gazed at them now and smiled to himself. There it was, the history of California, spread before him, recorded in the blood and sinew and aspirations of as diverse a population as ever had gathered into one society—or was it in danger once again of succumbing to its repetitive pattern and dividing into two separate societies?

The nine persons seated along one side and at the ends of the long table directly in front of the dais were his family, Will's, and Felicia's.

Felicia had died three years earlier, but her daughter Emily and son-in-law Samuel and their son Lee and daughter-in-law Lillian Cohn Berwin, were there. He thought about Lambert now—humorless, a bit stuffy, but completely dependable, as he had proved again when he had straightened out the tangle left when no wills could be found after the deaths of his father and his twin brother. They had died on that April morning in 1906 when most of San Francisco had quaked, buckled, and burned and much of what had been left had been deliberately destroyed to halt the spread of the flames.

After the first quake shock his father and twin brother Will had rushed downtown to try to save the records in the modest family office they had rented after the company's sale. An exploding gas main had trapped them on the second floor of the building. Later the family had held a memorial service. There had been nothing left to inter. He recalled how he had stood disbelieving at the top of the hill on Valencia Street with his mother and Emily at his side and looked down along the contorted cable car line that ran for twenty buckled blocks through devastation no city in modern times had known. City Hall had become a twisted skeletal dome and the Nob Hill mansions were gutted. The after shocks and the distant detonations of dynamite signaled the final destruction of the tottering, fire-blackened walls that were destined to take the lives of a score of rescuers searching the rubble for victims. In

384

time the dead would number four hundred and fifty-two. But ten times that many had fled the cremated ruin that once had been the most boisterous, recklessly speculative, and ingenuously culture-conscious city in the world. Many of San Francisco's pioneers seated in the room to pay homage to Steven's years had been among the great scoffers and deriders. From the security of their small estates in outlying districts, they advertised the ambivalence of their loyalty by defining Los Angeles as "six suburbs in search of a city." But it was all *habladuria*, empty talk. If any of the palaver he was about to hear concerning his own accomplishments was deserved it would be that which acknowledged his devotion to southern California.

Southern California could be spelled with a capital "S" now, thanks to fellow pioneers and believers like the Chaffees and the Smiley twins and the Norths—and later Henry Huntington—expiating the sins of old C.P. who had died eight months into the new century and had never enjoyed a penny of his railroad millions that had not been spent in trying to add to them—and Van Nuys and Lankershim and Eli Clark and William May Garland and General Harrison Gray Otis—and Harry Chandler, William Randolph Hearst, and the Workmans. And E. L. Doheny who had drilled his damned stinking oil wells right in people's front lawns! And C. C. Julian who just might be the biggest of them all if he kept on going. He could put a name to every older face in the room and to some of the younger ones, too—Spanish, Mexican, and Anglo names—and many Italians—names that identified a living history more distinguished than his own, Steven thought.

On the left side of the room, he could see Guillermo Morales and his family. Guillermo was thirty-five, the youngest of four sons born to Pablo Morales and Gracia Lopez at the ranch. Pablo was the son of old Pedro and old Gracia, now long gone. The parents were at the table, self-conscious in their borrowed formal wear, unable to conceal their pride in Guillermo, who had earned his law degree, and immensely flattered because Don Esteban had said there would be no party without them.

There had been some resistance. Young Steven's wife, Catherine Coulter Vance, who would always remain a patrician from New England, had questioned the propriety of having servants present as guests. The twenty-two-year-old bride had acknowledged begrudging admiration for the Mexican American's achievements but seemed disconcerted to find dignity, intelligence, and innate grace in a person in whom "the Indian still shows." Those qualities, she had inferred, were reserved for "one's own kind." Steven smiled inwardly, remembering his grand-daughter-in-law's consternation when he had reminded her that she had married a *mestizo*.

The room quieted down and the president, Anthony La Barbera, was beginning to spout those ceremonial platitudes that always annoyed him if they did not lull him to sleep first. He would have to guard against dozing. The older he got, the easier it had become. He would try to listen now.

"... a man who, in the eight decades of his life so far—and I know we'll be here to celebrate his hundredth—has seen California progress from an unexplored wilderness in the north, and in the south a land of *ranchos* that provided their few hundred Mexican owners with a fine living, to a state with better than three and a half million souls who have dedicated their lives to making this the greatest state in the union ..."

Steven's gaze wandered to his grand-niece, twenty-four-year-old Maria Carla, now Mrs. Maylon McCarrell. Young Maylon had already become a promising San Francisco investment banker and Republican state committeeman. Steven liked him well enough but he resented the young man's cocksure attitude where the south was concerned. "You southern Californians will never get over your inferiority complex. With money you can imitate but never duplicate what God has given us up north." Annoyed all over again, Steven forced his attention back to the speaker.

"... Steven Lewis' story is the story of our glorious state. Born in Monterey five years before the Gold Rush, he moved to San Francisco. Very early, he saw his mistake and came to Los Angeles. And from the time Steven Lewis set foot south of the Tehachapis, he moved unerringly toward his goal ..."

Steven's eyes continued to wander along the family table. There should have been at least two and possibly four more chairs there. But some mistakes cannot be undone—

"... and his devotion to civic duty from the time he stood with Judge Widney and his brother against the lynch-

386

mob that decimated our great Chinese community right where this great hotel now stands . . ."

Inwardly, Steven Lewis snorted. And they hung the wrong goddamned men! Not one of those murdering bastards was punished. Oh, yes, one man was fined one hundred dollars by Judge Widney because the jury would not bring in a conviction against men who had done what each jury member himself had wanted to do.

Trivial human errors? How trivial is an error a man pays for privately for the rest of his days? He raised his head to look at the table again and wondered where Charlotte would be. In another month, she would be fifty-eight. After Arlene had died from an overdose of the laudanum she'd been taking for years, Charlotte had disappeared. He had tried for years to locate her, had spent thousands to no avail. One tracer had her in South America with a missionary group. A second indicated that she had died there. But somehow a part of him rejected the idea of her death. The thought of it brought back an old pang of guilt. He managed to rid himself of it as he usually did—like mother, like daughter. Arlene had done her work well on Charlotte all right and there was nothing anybody could do, no matter how much . . .

La Barbera's measured intonation separated again into meaningful thoughts. ". . . he has walked among us, sharing with us, blessing us with his compassion and wisdom . . ."

Steven Lewis jerked his head up and fixed the master of ceremonies with an incredulous stare. He wished he could say "Bullshit!" after each new extravagance. What about those sixty lease partners the judge and I dispossessed when all of the damned subdivision foolishness busted in our faces? In two years, from Eighty-seven to Eighty-nine, those "Escrow Indians" platted sixty new towns on eight thousand acres of land, some of it right in the middle of the Los Angeles River. I sold them a few thousand acres of that land. And that pious son-of-a-bitch, Jabez Lathrop, had made himself another fortune with Island View Harbor, boasting that he preached the Promised Land on Sunday and sold it on the other six days. Two thousand lots at five hundred dollars apiece and all he gave them was a church and a bathhouse. He had bought five thousand dollars' worth of lumber and bricks and had salted a bunch of lots so people would think homes were being built on them. Later, the boys who promoted Hollywood used that same trick. At least, by God, he could say with a clear conscience that every man who did business with him got a fair shake.

While La Barbèra droned on, Steven pondered the state of affairs in Sacramento. The population balance was moving southward but political power remained in the north. For forty years the Democratic Party had been in a decline because the stupid bastards kept fighting among themselves. It started back in "Seventy-three" about the time William Sharon began to double-cross Billy Ralston. That's when the People's Independent Party got hot enough under the collar to have a convention and publicly damn the Southern Pacific's corruption.

Both old-line parties started singing the same tune: "Get the railroad and its graft money out of politics."

La Barbera was winding up. ". . . it is my pleasure to introduce the President of State College, Dr. James Denton Boyce."

Steven applauded obediently and smiled at the distinguished educator. How long had it been? Twenty years?

". . . that was the year I first met our honored guest. If I have made some progress in understanding our complex society, a good part of the credit must go to my first practical instructor, the man who . . ."

Steven remembered. Who taught you that you don't help people by feeling sorry for them? Who taught you how to use a shovel on an irrigation ditch? A derisive snort escaped Steven. He pretended to clear his throat, muffling the sound with a napkin.

". . . I had a lot of thinking to do about the whole relationship between labor and management and what I learned from Steven Lewis in those days has colored my thinking down through the years."

Old misgivings made Steven shake his head and a newer worry sent his glance to where Guillermo Morales sat. He gazed at the young lawyer who was concentrating with the intentness of a well-disciplined mind, evaluating, agreeing, rebutting. What was going on in that smart brown head of Guillermo's just might stir up trouble someday.

Two weeks earlier, Guillermo had finally won his client's suit against the East County Land Company. A Mexican-American laborer had lost an arm in a bean

388

thresher. Negligence, they said. But Morales took the whole damned jury out into the field and showed them. The jury agreed that a safety screen should have been bolted around the conveyor chain pully and they set the price of an arm at ten thousand dollars. Los Nidos wouldn't order machines without them. More than that, the foremen had always been careful about the men they put on those platforms.

". . . a man's earnings are directly related to the quality of his training and in the case of agricultural workers, his basic education."

Steven Lewis tried to resettle himself on the chair and hoped to God that the remaining speakers didn't run on and on. This whole business about educating workers was fine in theory. But how often had he argued, "so you educate them to want to do something else. Then who in hell is going to work in the fields? Even stoop labor takes more than a strong back and that's why Los Nidos has always paid a few cents more and seen that they have a little better housing. Of course, in time, machines are going to take care of all that and farm laborers are going to agitate themselves out of jobs."

". . . that is why I have chosen this occasion to announce that next Saturday State College with confer upon Steven Lewis an honorary degree in economics."

Steven blinked in genuine surprise. Well, now. What do you know about that? It was difficult to swallow and the expression that had set on his face felt like a stupid grin. He was glad when Jim Boyce stopped the applause.

The next speaker was Harry Rhoades, a fellow rancher who with Steven had lived through the boom of "Eighty-seven" and the collapse that followed it.

Harry described the real beginning of growth in southern California shortly after the turn of the century. The "Pullman Crowd" arrived and bought up the orange groves along the sheltered citrus belt that runs from Santa Barbara to San Diego.

Quite properly, Rhoades credited J. W. North's pioneer experiments at Riverside in the Seventies for the citrus boom. North had proved with oranges what Anaheim had proved with grapes and what Los Nidos had proved with field crops: If you bring water down from the streams you can turn land that people call "nothing but desert" into profitable acreage.

"The point was—and Steve saw it—diversification. Steve had the acreage, the water, and the variety of terrain to prove the point. If you wanted insurance against crop failure you planted a variety of crops, you didn't put all your oranges in one basket. The ones that did got cleaned out in the big freeze eleven years ago and most of them wouldn't have had a dime if Steven hadn't offered to buy them out for cash."

Steven shot a baleful glance at his old friend. He wanted to remind Harry that he had bought them out for ten cents on the dollar because the little operators couldn't stay in the citrus business any more. It was a two-hundred-million-dollar industry now, no place for the little guys. If Harry had been really honest he would have reminded them that the others around the country accused him of stealing the ruined groves, even spread the rumor that he had sent his Los Nidos boys around with cash and loaded guns to force them to sell. It was a damned lie but it was the sort of thing you had to expect when you got big.

As the rancher continued to talk of the old days Steven watched him through heavy-lidded eyes. Rhoades had come from Idaho to work in sugar beets with the Spreckles interests. There were some things in southern California that Harry just never could adjust to. To himself Steven said, "You're a hell of a lot better rancher than a speaker, my friend, unless you're chewing the ass off your Mexicans. You never will learn how to handle a Mexican. You always figure they're talking behind your back in their own lingo. It ought to be a state law in California, Arizona, New Mexico, and Texas, by God, to have every kid learn both languages. We had to learn their language when we came here to steal California from them."

He glanced over at the table where Guillermo was seated. Maybe, by God, they'll steal it back some day.

Steven's musing was interrupted when Harry Rhoades turned to address him.

"So, Steve, nobody will argue much when I say that you and Los Nidos showed this part of the state how modern ranching works. And finally, I want to say this: You're not always lovable, Steve, but by God, we love you!"

There was another ovation and Steven flushed, trying not to look pleased. Then he entertained the futile hope that the next speaker would be brief and end his misery.

Anthony La Barbera brought the room under control and introduced State Senator George Kenny. Steven looked at the lawmaker with some affection. A good Democrat, a winner in a losing party, a realistic politician who knew what the trouble was.

"If there's anything scarcer in California these days than a genuine native son it's an honest-to-goodness Democrat. Our honored guest today has always voted for the man, not the party."

Steven nodded his approval openly and thought, "That's a very respectable way to get around the truth. But what I really voted for was a chance to let southern California catch up with San Francisco's forty-year head start."

". . . so I guess it shouldn't surprise us that politics in the state of California just don't seem to follow the old classic blueprints. My father told me that when President Lincoln was assassinated, San Francisco went into mourning and Los Angeles staged a celebration."

The statement brought back a troublesome memory for Steven. Not everybody celebrated. Only the Confederates. I didn't see any black man in that parade. And I didn't see any brown men, slant eyes, or otherwise. But I did see a hell of a lot of men who would have started the Civil War all over again if the police hadn't stepped in. And every one of them was an outsider, with his own set of notions. Steven wondered why people always seemed to bring their pet peeves to California. The place could only get built by pulling together. Lately they'd been doing a little better, but the friction with San Francisco was getting hotter with all of the "Keep the White Spot White" open-shop business and the growing harbor competition.

". . . We are a state of individualists. We are a state filled with immigrants and some of us brought our eccentricities with us. If you don't believe me, then listen to Sister Aimee on your crystal set—"

The reference to Aimee Semple McPherson turned Steven's thoughts momentarily to Arlene and an old sadness settled on him that the senator quickly dispelled.

"So that's all the more reason for paying homage to our few remaining, stabilizing old-timers who did the pioneering so that the rest of us can enjoy the most turbulent political climate and the most salubrious geographic climate in the United States. Now I want to extend to Steven Lewis the gratitude of the California State Senate in the form of this official resolution."

Senator Kenny held up a large plaque bearing the state seal at its top. The chairman whispered to Steven, who rose stiffly to receive it.

"Whereas the Senate of the State of California, being mindful of the valuable and unselfish services rendered to the state by Steven Lewis. And whereas . . ."

It felt good to stand. If it took a load of official horse manure from the boys in Sacramento to get a man to his feet, then it was the first useful thing that legislative body had done in some time.

Then, almost before he realized it, Steven Lewis found himself accepting the rarely given honor from the California State Senate and it was time to say some of the things that he'd been thinking about.

His words would be reported in the press, not all of which was friendly. But whether the press liked it or not he intended to set the record straight about what kind of men it took to build a country. There is no progress and opportunity without doers. California found that out from men like Ralston and Sutro and Huntington and Stanford and Crocker and Widney and Irvine and Newhall and Chandler and Hearst and Earl and Mulholland and Hellman and Gianinni, especially Gianinni, who was showing guts enough to buck Governor Richardson. He might have let the church get mixed up in his Bank of Italy as rumor had it. But he wasn't about to let the state stand in the way of expanding his branches to serve the little merchants and farmers in every small town in California.

All along the line you've got to fight the people you're trying to help, unless you promise to do it for nothing and give them all of the profits, like that damned lanky, starry-eyed dreamer, Upton Sinclair.

". . . so, if I were to say now, Steven Lewis, that you needed no introduction, it would be the truest thing I've ever said in public or in private. Ladies and gentlemen—Steven Lewis."

Steven Lewis waited through another ovation, then deferred briefly to those on either side of him.

"Mr. Chairman, honored guests, fellow members, and friends, I've had a lot of praise heaped on me tonight and you can see that I'm flowering under it. It also happens when we heap the same stuff on the beans down at the ranch." The deliberate scatalogical jibe brought raucous laughter from the men and subdued tittering from the ladies. He could see that Alicia was flushing as she exchanged horrified glances with Catherine.

"First, I want to say that no man deserves all this praise. But I bow to my peers in this decision because, after all, I'm only a doctor, not a judge." The laughter which had not quite subsided swelled again and was interspersed with a scattering of applause.

"And before I get down to speaking, I want to say that any man who would risk making me a doctor has got to be desperately sick."

Steven Lewis knew that he could keep them laughing and they would go away chortling over his bucolic humor and repeat it in clubs and in speakeasies. But it was time to try to put into words some of those things that he had been thinking about while seeming to listen. He would set the record straight and he would tell them that even in the free swinging year of 1924, they would have to fight for survival.

Book III

The Howard D. Goodwyn Story

13 FOR SEVENTY MILES in either direction there was nothing but the blinding, furnace-like heat—searing, bone-drying heat that had made the wasteland an unmarked graveyard for unnumbered millennia. And with the heat there was the wind. Day and night, the hot desert wind.

Engineers on the Santa Fe trains that roared and rattled along a roadbed made treacherous by the shifting sands swore that tumbleweed, racing the wind, had passed their locomotives going better than a mile a minute. No one who knew the land doubted it.

Howard Goodwyn, encumbered by shabby high-heeled cowboy dress boots, ran awkwardly across the sage-tufted flatland. He still had several hundred yards to travel when he heard the deep whistle of the westbound locomotive. A moment later the headlight stabbed the darkness as the train emerged from behind a range of low dunes.

He reached the right of way a scant thirty seconds before the ten-wheeler lumbered by. Crouching behind some scrub sage, he watched the passing string of dark cars. Despair turned to hope when he realized the train was slowing. Keeping low, his left hand gripping the black leather guitar case on which faded sequins spelled out "Howdy" and clutching a suitcase in his right hand, he came to within three feet of the moving train. His ankles turned dangerously in the heavy rock ballast as a rush of hot night air flapped his jacket. Then hope turned to despair again when he realized the box-car doors were sealed.

The train was traveling slowly enough for him to keep pace. Squinting anxiously into the darkness, he searched for the tell-tale darker rectangle that would mean an open box-car door. Finally he saw it—an automobile car with both doors wide open. If he could toss his stuff inside, he felt certain he could catch the rear edge of the opening and hoist himself in.

Ignoring the danger, he began to run, staying opposite the door. Just as he was about to swing his suitcase up, a silhouette appeared in the opening. For an instant he thought it was a trainman. A harsh voice called out: "If you're gonna hop it, Willie, toss me your bindle!"

Howdy swung the suitcase up. The guitar case struck the steel door guide, knocking him off balance. Recovering his footing in the treacherous ballast, he caught up again. The man was shouting something but his words were lost in the hiss of air brakes bleeding off. The couplings clanked taut down the line as the freight train began to gather speed again.

Panic knotted Howdy's guts. His clothes were in the car, everything he owned except the guitar. He had to get aboard. Running wildly now, he reached out, shouting. "Gimme a hand, pardner! Gimme a hand!"

"Heave your other stuff in first, pal. Hurry on!"

"I can't! It's my guitar! It'll smash! Just gimme a hand——" Racing along inches away from the rails, Howdy's right hand groped frantically for something to hold on to. His fingers found the cloth of the man's trouser leg and closed around it. Instantly there was a warning shout. "Let go, you goddamn fool! Let go."

Ignoring him, Howdy clung desperately and tried to pull himself up. "Christ—gimme a—"

The words were cut off and the cloth tore free from his grasp as a heavy work boot crashed into his left temple.

Howdy heard the man's curses but the voice trailed away, lost in the din of clanking trucks and couplings. His body seemed weightless and the night closed in around him black and silent.

The old man who ran the only filling station in the one-hundred-and-forty-five-mile stretch of Mojave desert that went from Needles, on the Arizona border, to Barstow, California, had been in bed for an hour when shouts awakened him. He got out of bed and with the aid of a flashlight found Howdy collapsed beside the water tank. He managed to get the injured stranger on his feet and into The Oasis, as he called the furnace-hot little wooden structure that also doubled as his home.

It took an hour and most of his meager first-aid supplies

to patch up the tall, thin, redheaded youth who managed between winces to explain that he was an out-of-work cowboy singer from Guymon, Oklahoma, trying to get to the West Coast.

When the deep abrasions and gravel scratches were finally bandaged, Howdy managed his sincerely felt thanks.

"I figured I could hop another freight like I been doing and get to Barstow tonight. Most often I can find a place that'll swap me some food and a place to flop for a few songs."

"Why a place like Barstow?" the old man asked. "Ain't nothin' much there—"

Howdy made a noncommittal gesture. "The way things is going I gotta take one town at a time. But I sure want to get to L.A. as fast as I can. They got a lot of radio stations there and I figure to catch on with Stuart Hamblin or the Sons of Pioneers or one of those outfits." He glanced down at his badly torn trousers and managed a grin. "You might say I'll be starting from scratch—"

The old man cackled. "Y'also might say that you haven't lost your sense of humor, son."

In the morning, Howdy stripped down to his boots. Clad in a grubby towel cinched around his lean middle, he washed his things in an iron tub. When they were hung on the line, drying almost visibly in the hundred-degree morning heat, he returned to the shack and studied his reflection in an unframed mirror. His left eye had purpled during the night and the worst of the abrasions had begun to scab over. A soundless laugh made his shoulders heave. "Howdy boy, you got nowhere left to go but up!"

During the morning, two trucks, both eastbound, had stopped to fill their radiators. The only westbound traffic had been a 1930 Buick bearing Arkansas license plates. It carried a gaunt man and wife, four worn children, and an unbelievable load of poor household articles. Howdy did not even consider asking for a lift. The old man said it all as the car pulled away after buying a dollar's worth of gasoline. "That car's so loaded that if one of the kid's ate a square meal the axle'd bust. They're all the same. More Arkies going west. Heading from bad to worse."

In mid-afternoon a gleaming passenger train blurred eastward but Howdy didn't even bother to look up. Then, during the hottest part of the day, around four-thirty, a middle-aged couple in a dust-covered touring car drove up.

The old man filled the car with gas and walked to the front of the car. The red fluid in the Moto-Meter on the radiator cap warned that the motor was dangerously close to boiling. The driver, a slightly paunchy man, opened the door and climbed out stiffly. "Better let her simmer down a little. Been pushing right along at forty for the last hour."

The old man glanced at the empty back seat. "You folks going as far as Barstow tonight?"

The driver nodded. "We live there. Been up to look over Hoover Dam. Decided to come home by way of Kingman."

The old man nodded, "I'll get some fresh water."

As he moved off toward the faucet Howdy came out of the small dug toilet a few yards behind the living quarters. The old man motioned to him urgently. "There's a couple out front going to Barstow. Whole empty back seat. Better go ask them."

Howdy moved instantly, but stopped a few steps away. A hand explored the bruised left cheek and temple. The old man urged him on. "Tell them the truth—you tripped and fell. They look like nice folks."

A few minutes later, Howdy and his guitar were in the back seat. As they pulled out onto the highway the woman, a comfortable, motherly type, turned to give Howdy a concerned inspection.

"Sure you didn't break anything?" she asked.

"The only thing I broke, ma'am, is a string of good luck I had finding rides. But you good people put that back together by taking me in. I thought for a while the Good Lord didn't want me to get to Barstow—or anywhere else, for that matter."

The husband called over his shoulder. "You got folks in Barstow?"

"Nope. Never been there."

The man nodded and probed again. "I expect you've got a job waiting for you."

There was no way to avoid an answer. "I have prospects."

The woman, still turned on the front seat so she could keep an eye on Howdy, frowned. "Way things are these days, I sure hope your prospects pan out." She looked at

400

the ornate guitar case and unaccountably began to giggle. "Frank and I saw a gangster picture in San Bernadino last month. Fellow had a violin case with a machine gun in it."

"I've got to tell you, Mrs.—" Howdy left the words suspended in a question.

"Persons. Melba Persons. My husband's Frank."

"I've got to tell you, Mrs. Persons, that if anybody looking like me asked for a ride, I'd be a touch skittish, too. But I sure appreciate your taking me in."

Melba clucked, "Well, my goodness, decent folks don't expect to be paid back for simple Christian charity. Truth is, it's nice to have company. At home we just sit and read and listen to music on the radio mostly on KNX and KOA and WLW."

Howdy rested a hand on the guitar case. "What's your favorite song, Mrs. Persons?"

"Mine? Oh, well, mostly old ones. But I like that new song the Crosby boy sings, 'How Deep Is the Ocean?' "

Howdy unsnapped the latches on the case. "Would you and Mr. Persons like a little music now?"

Melba looked at him, wide-eyed. "Well, my goodness." She watched, fascinated, while Howdy positioned the instrument and tuned it softly.

"Now then, I'm not up to the regular popular music. But I know a couple of new hillbilly songs that just came out. Ever hear 'I like Mountain Music'?"

Without waiting for an answer, Howdy went into a rhythmic vamp and started to sing. When he finished and looked up, the rapt expression on Melba Person's face made him chuckle inwardly. It never hurt to have an edge.

For an hour Howdy held the couple with the same spell he had cast on audiences in Texas and Oklahoma as the young featured soloist with Spike Spangler's Radio Rustlers.

The long summer twilight had fallen and in the distance Barstow's first lights were visible as he finished the last song. It was a new cowboy ballad by Billy Hill called "The Last Roundup."

Howdy smiled to himself as Mr. and Mrs. Persons awkwardly attempted to thank him. Some minutes later, in the heart of town, Frank Persons angled the car into a parking space in front of a grocery store. "Is there any particular place you were going, son? Soon as Melba picks up some food, we can drop you off."

Quickly Howdy decided what his approach would be. "No. No, thanks, sir. This'll be fine, right here."

Melba Persons was fussing with the door handle. Before she could move it, Howdy was out and helping. The attention embarrassed her. "My goodness, Mr. Goodwyn. I'm not used to all this fuss."

Howdy handed her down from the running board to the street and helped her over the curbing. "Well, that makes us even, Mrs. Persons, 'cause I'm not used to all the kindness you folks showed me."

Frank Persons got out and walked around to join them. "What are you going to do now? You said you don't know anybody in town. That right?"

"Yes, sir. I figured I had such good luck singing for my ride that I might look around for a likely café and see if I can't sing for my supper, too." He hoped he had gauged correctly the amount of casual bravado needed to elicit sympathy.

Melba reacted instantly. "Frank, this boy isn't going to look for a place to sing for food, for heaven's sake! You keep him right here while I go inside and get some things."

The Persons home was a neat one-bedroom cottage well away from the highway and the railroad. The screened front porch was furnished with weathered wicker chairs and a couch covered with faded green burlap. The interior consisted of a small living room with an adjoining alcove fitted out as an office, a small dining room, a good sized kitchen, and, down the hall, a double bedroom and bath. Howdy could easily have predicted the cretonne-covered overstuffed furniture, the plum-colored silk shade on the mahogony floor lamp, and the occasional tables with their assorted trinkets, dishes, and momentoes that included a piece of polished wood from the Petrified Forest and a winter scene captured in a crystal ball.

During the half hour it took Melba to throw together a salad of cold meat and hard-boiled eggs, Howdy sat on the porch with Frank. Choosing his words carefully, he answered most of the older man's queries about his past, but was grateful when Melba called them to the table.

After supper, when he offered to dry the dishes, she

402

shoo'd them both out. Pleasantly full for the first time in almost two weeks, Howdy would have been more than content to return to the cool porch. Instead, Frank diverted him to the tiny alcove.

"I want to show you that lots of good people got caught in this damned depression bind."

Three years earlier, Frank Persons had closed his downtown real-estate office, sold the furniture he couldn't get into the alcove to a San Bernadino dealer for ten cents on the dollar, and transferred his California real-estate-broker's license to the wall above the alcove's roller-top desk.

In the early Twenties, when Wall Street was heading toward its zenith, Frank had been one of the most active brokers in the country. There had been a lot of "new highway talk" and "railroad talk" and "new industry talk" and everybody around town agreed that Frank had been pretty damned smart to tie up pieces of land near the town for well under a hundred an acre, even though that was four or five times more than it sold for when the war ended.

In the end, at Melba's insistence, Frank had sold for cash the land that was to have become his ideal subdivision—"Garden Estates"—put the money in Postal Savings and thanked God that he'd had sense enough to listen to his wife. But Frank kept the dream alive by talking about it every chance he got and this unexpected young visitor offered the first chance in some months.

Laden with tubes of blueprints, Frank led Howdy back to the dining room table, where things could be spread out. For half an hour Howdy listened quietly, interrupting from time to time to ask questions that surprised Frank. Howdy's interest was no token act of politeness. Frank reluctantly rolled up the fading prints when Melba, carrying a pitcher of cold lemonade, joined them.

"All I ever wanted was to give young folks a subdivision with decent streets and good square lots with common sense restrictions on building." Frank sighed the sigh of a contented man who has been privileged to have a good listener.

Looking at her husband and Howdy with indulgence, Melba said, "You two talk a while longer and I'll make up the porch couch."

Howdy started to protest but Melba shushed him and

403

said, "Look young man, you don't have a bed and Frank and me are not going to turn you out to sleep under a bridge like some common hobo."

Melba left and after a brief reflective silence, Frank looked up. "I was thinking, Howdy. I know everybody in town. Maybe I can turn up a job that'll get you a few dollars for new clothes."

For once Howard Goodwyn had no calculated response. He mumbled his thanks and tried to cope with an unaccustomed pang of conscience. How could he keep his edge with people like these? Remembering Al Phelps advice, he said to himself silently. "Bet on human nature. It don't never change. That's your edge, kiddo!"

Bathed and freshly bandaged, Howdy stretched out in the first clean bed he had slept in for weeks. Ten hours later, voices—deliberately raised a bit—roused Howdy from his slumber. He pulled up the sheet as a screen and slipped into his B.V.D.'s. A clatter of dishes being scraped and stacked came from the back of the little cottage. He reached for the trousers that he had tossed on the chair the night before and found them neatly folded over its back.

Puzzled, he discovered that the ripped leg had been repaired with remarkable skill. Frank Persons appeared in the doorway as he was stuffing his shirt tails beneath the broad western belt.

"Thought I heard you. How'd you sleep?"

Howdy blinked in woozy wonderment. "Like a babe." He pointed to the trouser leg. "And while I was sleeping some ministering angel mended me."

"I know. The wife said if I was going to take you job hunting she'd better get you fixed up. Come on, chow's ready." In the kitchen, Melba was bending over the sink running water into a pot. Self-consciously, Howdy placed an arm around her shoulders.

"Thank you, ma'am, for putting me up—and patching me up." He turned to Frank, who was grinning at Melba's confusion.

Melba moved away, settling her hair, then shook her apron at him. "You two sit down. Things are getting cold."

After coffee they talked for an hour. Grown expansive, Howdy responded openly to most of their questions. He

404

told them about Al Phelps and how Phelps had promoted him as a teenage cowboy singing star with Spike Spangler's Radio Rangers.

Frank Persons' eyes gleamed as he listened to the story of Phelps' rise from an eighteen-dollar-a-week operator of a pill-stamping machine to the creator and distributor of the famous Phelps' Formula.

The patent medicine had become a household word by the simple expedient of promising listeners everything short of eternal life. The pills, mostly bismuth subcarbonate and bicarbonate of soda flavored with mint, cost Phelps fifty cents a thousand to manufacture, bottle, and sell. He sold bottles of one hundred for five dollars.

Howdy explained that Phelps had spread the news that he was going to build a big plant a few miles out of town to manufacture medicines. Phelps bought up four sections of land fronting on the highway for less than $25 an acre. Next, he had large amounts of building materials dumped near the highway in full sight of all passersby. After that, he put up signs indicating where the "town's" churches, market stores, and club houses would go. Even with the advertising, surveying, engineering, and the land cost that had been split with a partner, Phelps had less than fifty dollars an acre invested.

"And," Howdy continued, "he sold quarter-acre homesites like tortillas for nine hundred and ninety-five dollars with only one hundred dollars down."

The older man whistled. "Six hundred and forty acres to a section—makes at least ten thousand quarter-acre sites—"

Howdy nodded. His expression clearly spelled out appreciation of Phelps as an operator. "—makes better than ten million dollars."

Phelpsville had collapsed with Al himself. On the eve of his biggest promotion, Al had dropped dead of a coronary. After the funeral, a large sign appeared on the site of Phelpsville. The crude lettering read: "TO ALL US PHOOLS WHO BELIEVED IN PHELPS' PHOLLY."

Frank really tried to find work for Howdy, but there were no offers beyond a vague promise of some garden work "for two bits an hour." Apologetically, he reported to Howdy and then said, "Look, I've got to put a new

405

roof on the porch. If you can stick around to help me, I'll pay you six bits an hour plus room and board."

Howdy sensed that he would risk nothing by being honest. "I don't know a thing about shingling, Mr. Persons."

Frank laughed. "Name's Frank. Neither do I. We'll figure out how to do it together."

Late Friday afternoon, Melba Persons ended her anxious vigil at the bottom of a ladder as Frank and Howdy locked into place the last square of green asbestos roofing. She backed off and examined their work. "My goodness—what a difference!"

A pleased grin warmed Howdy's face as he stepped off the last rung. "Well, between singing at night and shingling by day I oughta be able to pile up a fortune."

Quickly, Melba said, "Don't ever give up your music, Howdy. I know two people right on this street that wouldn't be eating right now if they hadn't stayed with their music."

Frank Persons, still cautiously descending the ladder, said, "You talking about Angie and her mother?"

"I'm talking mostly about Angie Woodward. She may be playing piano in a saloon, but she's earning an honest living."

Howdy directed an amused glance at Melba Persons. "Playing in a saloon sounds like she's teetering on the brink."

Melba ignored the remark. "Angie's mother is the one that made the girl stay with her music. She's been teaching piano for twenty years or more—ever since Ralph went to war and never came back. He wasn't killed, either. He just plain deserted her and little Angie."

The men carried the ladder around to the side of the house and Melba followed, still talking. "That two-story place on the corner is theirs."

Howdy was surprised that it looked so well tended. "Except for yours, it's the nicest house on the street."

Later, when Howdy had pocketed the eight dollars earned for the work on the roof and they had settled on the porch, he asked, "Where's the place the girl's playing piano in?"

Melba pointed west. "Right outside of town—on the highway. Food's Italian and so are the owners. Rosa and Mike Rinelli. They call it the Roadside Inn."

Standing in the shade on the scraggly lawn at midmorning the following day, Howdy said his farewells to Frank and Melba. Tucked inside the guitar case were a pair of Frank's socks and a change of underwear they had insisted he take. As the couple watched him stride down the street, Melba shook her head. "If anybody'd told me I'd cotton to a total stranger I'd of said they were crazy."

Howdy headed westward toward San Bernadino. On the opposite side of the highway, less than a half mile distant, he saw the Roadside Inn. He decided that the best way to get a ride would be to invest in a beer and strike up a conversation.

Blinking in the darkness, he made his way to a small bar at the right of the entrance and rested his guitar case against a vacant stool. He asked the small, wiry, workworn woman who tended bar for a beer. He assumed she was Rosa Rinelli. As Rosa poured his beer, a young woman entered. The man at the bar nearest to Howdy swung around on the bar stool. "Hi there, Angie, how are you doing?"

Obviously, this was the girl Melba Persons had spoken about. Without appearing to, Howdy tried to get a look at her. As she exchanged easy banter Howdy was aware that her voice was low and pleasant. She seemed a bit taller than average—a good figure with wide, well-turned shoulders. Her hair was light brown, probably tending toward red in the sun. Unobtrusively, he tried for a glimpse at her legs but the guitar case leaning against the stool blocked his vision. He didn't catch the remark that made her laugh and step back but as she did, her leg struck the unseen guitar case. Too late he saw it start to topple. An instant later it slammed to the floor with a reverberating crash.

Stricken, the girl whirled around. "Oh my gosh! *Please* excuse me. I didn't see it."

"My fault, ma'am. I shouldn't have left it there."

"You mean, I should have looked where I was going." She glanced at the case. "It made an awful whang. Maybe you better look at it."

The other men had slipped off the stools and were moving toward the door. The younger of the two paused. "See you tonight, Angie. You got some new ones for us?"

Still upset, she replied over her shoulder. "Unless that piano tuner gets here we won't have any music."

Howdy had opened the case and quickly stuffed the

407

spare clothing out of sight beneath the instrument. The girl turned back and looked at the guitar, first with conern, then with admiration. "That's a beauty! Golly, I hope nothing broke—"

"It's okay. It's been banged around before—worse." He had positioned himself between the case and Angie when he picked up the guitar. He could see that while she wasn't a beauty she was certainly more than attractive. He liked her hair—the soft natural curl and the way she had it pulled back and caught at the nape of her neck. Apparently preoccupied with a bass string, his eyes focused briefly on her legs. From what he could see of them, they were clean-lined and symmetrical. But it was really her face that held him—an open face with regular features—a straight nose with a suggestion of an uptilt at the tip, and an agreeably full mouth. He assumed that in the light her wide-set eyes would be more green than gray.

After a moment he looked directly at her and smiled. "I could use a new bass string. But that didn't have anything to do with the fall." Effortlessly his hands drifted into position on the frets and he struck a soft chord. Listening critically, his fingers moved through a series of intricate rhythmic progressions.

Angie raised her eyebrows, "Mister, you certainly get around on that thing."

Howdy lowered the guitar and turned on a modest grin. "Let's say I did once, ma'am. But my fingers are sure out of practice."

She eyed him skeptically. "If yours are out of practice, then mine," she raised her hands for his inspection, "are in plaster casts."

They laughed together and Howdy put the instrument on top of the case. The girl sighed with exaggerated relief. "I'm sure happy nothing happened. By the way," she extended her hand, "I'm Angie Woodward. You must be the one Melba Persons was telling us about. You've been staying with them—helping Frank? Howdy Goodwyn. Right?"

Her hand had a good feel. "I was up on the roof with Frank, all right—but I don't know how much help I was."

"Don't be so modest. Melba told Mom you're one of the best workers they've ever had. And the best musician they've ever heard." A wistful note crept into her voice.

Howdy released her hand. "Well, from what I hear, you haven't got anything to worry about, Miss."

Rosa Rinelli returned from the dining room. "He's-a no come yet?"

Angie glanced outside unhappily. "Nope. He said he'd be here by lunchtime and work on the piano first thing this afternoon so I can practice. I want to run down some new things."

Rosa's eyes glowed with black malevolence. "He's-a no come last time too. If he's-a no come this time he's through. *Ecco! Finito!*"

"If he doesn't come this time I'll be through, too! Who's going to sit through an hour of me playing out-of-tune hits?"

Howdy craned to see the piano. On the way in he had caught a glimpse of the battered old upright in the far corner of the big dining room. "Is it just out of tune or is something busted?"

"Like everything and everybody around here, Mr. Goodwyn, it's dried up and coming apart." Outside, the sound of a car slowing made Howdy turn. The girl watched him curiously. "Melba said you're heading for Los Angeles."

"If I can find a ride." His fingers touched the bruised flesh beneath the gauze on his cheekbone.

Angie smiled sympathetically. "Melba told us about your accident. The only thing is, Mr. Goodwyn, if you're looking for an entertainment job I hope they told you things aren't very good in Hollywood either." Howdy nodded glumly and rested the guitar on the bar. "You've got my curiosity up. Would it hurt if I looked at the piano?"

"It's impossible to hurt it. Come on. I'll show you." She led him to the rear of the dining room where a bilious green lattice had been constructed. The whole graceless effect was punctuated here and there by great clusters of apoplectic artificial grapes. An old, ornately carved upright piano sat in the corner.

Angie executed a skillful arpeggio. Half a dozen notes were all but toneless. Howdy winced. "Mind if I look inside?"

He swung out the front panel and examined the strings, hammers, and pin block. "Could you fool around a little and let me see if I can find the worst ones?"

Angie began improvising. Working together, they isolated the worst of the offending strings and in half an hour managed passable repairs to more than a dozen of them. Angie let a tinny chord sustain. Satisfied, she turned to him. "Such versatility! First shingling, then piano tuning." She began the opening vamp of the "Peanut Vendor." Automatically, Howdy reached for his guitar and joined her in a series of improvisations. They traded the lead back and forth until finally, in complete accord, they went into a deliberately corny ending and burst into laughter.

Angie studied him appraisingly. "You *are* good."

Howdy bowed with mock dignity. "Likewise, ma'am." With the guitar braced on his knee, he began to sing. Angie turned back to the keyboard. Toward the end of the chorus Howdy saw Rosa Rinelli and her husband move quietly into the hall doorway. At the end of the song, Angie whirled around on the stool and began applauding. "Wonderful! Wonderful!" Rosa and her husband Mike joined in. Rosa was so pleased she lapsed into excited Italian superlatives. *"Bella, Bella. Que bella canzone. Que bella voce. Mi piace molto. Moltissimo."*

Mike Rinelli, grinning broadly, simply said, "Sing more."

By the time Howdy and Angie got through "Tiptoe Through the Tulips," Nick Lucas' huge success from the talkie, *Gold Diggers of Broadway,* the Rinellis were overcome with enthusiasm. Glowing from the excitement of working with her first professional, Angie got up and linked her arm through Rosa's. "Excuse us a moment, will you? I want to talk to the Rinellis privately. We'll be right back."

Within minutes they returned and Angie said, "Mr. Goodwyn, I hope you won't think I'm a butinsky—but I had an idea while you were singing. Tonight and tomorrow night are our big nights—Friday and Saturday always are. I know you're anxious to get into L.A., but I told Rosa and Mike that I thought our customers would really go for a little change—something like you and I working up a few more of the things we just did for a sort of special weekend show. I don't mean for nothing, of course. Anyway, the Rinellis are agreeable—if you are. How about it?"

Howdy had trouble concealing his relief. "Well—I sure am obliged to you." He wagged his head. "But you don't have to pay me. I know all about hard times. I'd be plenty glad just to sing for my supper."

"You no sing fo' you suppa!" The woman's work-corded

arm swept the room. "You sing for money. Understand? Atsa only way people work for Mike and Rosa Rinelli." She jabbed a forefinger at Howdy. "Atsa *only* way. You get food and you get five dollars—two-fifty each night. Okay? And you get a room back there." Her tightly pinned, gray-streaked head jerked in the direction of the kitchen. "Okay?"

Softly, Howdy expelled a held breath. "Mrs. Rinelli—nothing has been this okay for months."

More than a hundred persons jammed into the Roadside Inn for the special show. By the time the second duet ended the customers were keenly aware that they were witnessing an unusual performance. Angie was their favorite and they were loyal to her. But the tall, spare, somewhat shabby young man with the scuffed-up forehead and the shy, ingratiating manner possessed a quality that not only compelled their attention but seemed to inspire Angie as well. His two solos produced an unprecedented hammering on glasses.

Later, seated on the front steps of the Woodward house, Angie summed up the opening. "Howdy, we need a lot more material. I'll bring everything I've got here at the house and you try to remember some of the new ones you lost in the suitcase."

Howdy got up, moved down a couple of steps, and faced her. "I think I can remember three or four. I'll let you turn in now and I'll start picking my brains."

Angie would have preferred to sit in the warm night air longer, re-living their success. Reluctantly she said, "See you tomorrow," and rose to go inside.

Saturday night was a standing-room-only repetition of Friday's triumph. Howdy and Angie had worked hard for most of the day adding new numbers and polishing the old ones. Howdy was at a disadvantage when it came to reading music, but Angie had managed to improve the harmony in his simpler chords with a number of tactful suggestions. For Angie, Howdy found himself feeling something akin to affection. It troubled him in a way that he did not comprehend because he sensed that she threatened his edge.

After the show, Howdy returned a wave to the last of the patrons and was about to step outside to join Angie and the Persons when Rosa called him to the bar.

411

She came right to the point. "Hey, Howdy. You want to work one more week?" When he made no immediate response, Rosa added, "Bed, food, and twenty dollars, including Saturday."

It was a windfall and Howdy accepted as gracefully as he could. With a little more luck, he would be able to stretch his stake to a job in L.A.

It was nearly one in the morning when Howdy undressed and stretched out on the bed in the stuffy little room at the rear of the Inn. At peace for the first time in months, he thought about Angie and how friendly she was—how good she looked—how fresh and sweet she smelled. Suddenly he gave his head a violent shake. "Screw that stuff, Goodwyn! That's how it always starts. You got other fish to fry."

Angie had asked him if he wanted to go to church with her. Howdy thought about church—what it would be like—and begged off. He could still see his own mother and father anesthetizing themselves with a simplistic, do-nothing faith in God when the first of the real hard times hit the small ranchers in Oklahoma. The recollection brought a bitter half-laugh in the darkness.

He remembered the day his father's faith had failed and he had left the ranch to search for wage work—a search that soon ended in death from exposure and pneumonia. He recalled the old man's words just before he left: "God's turned his back on this land. The Dust Devils is stealing it. If I set here and wait for Him to see what's going on, the Devil himself will steal the lot of us, too."

Drowsy, Howdy forced his mind back to the present. "No church," he told himself. "I'll just thank God for every day I'm alive and well."

That weekend Howdy and Angie repeated their previous success before an over-capacity audience. Customers unable to find a table or a place at the bar jammed the doorway and lined the walls. The last customers drifted from the bar shortly after midnight. Rosa and Mike asked Howdy to stay on for a second week. Gratefully, Howdy accepted but he knew that would be his last.

Once again, Howdy and Angie sat on the porch at the Woodward home, talking about the performance, and speculating on new material.

412

Suddenly Angie laughed. "Just listen to us. You'd think Mike and Rose had signed us to a five-year contract."

There was an uneasy silence that Howdy broke by rising abruptly. "It's getting late. I better go now."

Controlling her voice, Angie nodded. "When are you going to leave? Town, I mean?"

"Tomorrow, I suppose—I've got almost forty bucks now. I better move on before I wear out my welcome."

Angie struggled against the nagging sense of loss. At times during the last few nights the feeling had come close to desperation. Before, she had never really measured the monotony of her life. The word "love" had not entered her mind—but Angie had come to understand Howdy's appeal. It did not incite romantic fantasies, but when she was close to him she was acutely aware of him and the feeling was akin to excitement.

Reaching up, Angie rested what she hoped was a comradely hand on his shoulder. "Well, Howdy, I'm lousy at weddings, funerals, and farewells. So," she removed her hand and offered it directly, "I want to thank you for everything and—and—good luck!"

Before Howdy could respond, she gave him a quick peck on the cheek and hurried inside.

After he left, Angie castigated herself for the fool she had been. Dismally, thinking of the success of their performances together, she felt she was stealing the ten dollars Rosa and Mike were paying her. The thought of going back, to face that relic of a piano, chilled her. Even more chilling was the realization that Howdy had ignored her as a female. Against all reason during the two weeks Howdy stayed, bright hope had displaced drab uncertainty—the same uncertainty she knew would prevail even if she did break down and marry one of the several steady and dull suitors who really wanted to take care of her.

14 Howdy Goodwyn stepped down from the electric car at the corner of Hollywood Boulevard and Vine Street. He had boarded the loaf-like street car at the Hill Street Station in downtown Los Angeles thirty minutes earlier.

For several minutes he stood looking around, a tall, lean young man dressed in worn show-business cowboy clothing, clutching a scuffed, sequin-embossed guitar case. He was inconspicuous among the regulars who haunted the notorious corner—an old-time fighter whose kneaded dough face and erratic speech and movements betrayed a lifetime of more punishment received than given, a dilapidated one-time silent-screen bathing beauty whose orange hair seemed about to burst into flame, a ferret-eyed taxi driver who augmented his uncertain income by pimping, a part-time, toga-wearing recluse known as the Hollywood Holy Man, and a benignly visaged eccentric with flowing white hair who claimed to be the incarnation of Edwin Booth.

The human menagerie made Howdy uneasy. He headed for a filling station on the Boulevard and asked the attendant where he could find a cheap boarding house. He was directed to one on Franklin Avenue, run by Georgette Follansby, where he took an attic room for six dollars a week. The price included breakfast. After he paid for his room Howdy returned to the filling station on the pretext of thanking the attendant and got directions to the radio stations he planned to start visiting the following morning. The man also told Howdy about the principal points of interest along the Boulevard.

It took him an hour to walk the circuit. Except for Grauman's Chinese Theater and the older Egyptian Theater, the Warner Brothers' Theater and the Pantages, which also boasted a stage show, he found the celebrated Boulevard shoddy. The famed Hollywood Hotel seemed

414

little different to him than a dozen turn-of-the-century resort hotels he had seen in the South and Southwest. Beyond the Broadway-Hollywood, the only stores that attracted him were The Columbia Outfitting Company and the "Walk-upstairs-and-save-ten-dollars" men's store, Foreman and Clark. By the time he climbed the hill to Franklin Avenue it was nearing eight o'clock.

Lingering on the broad front steps, he gazed out over the carnival sprawl that is Hollywood and Los Angeles after dark. The day had been hot, but sundown had brought a cool breeze from the ocean. It moved like a benediction over the coastal plain, bearing a trace of salt freshness and the perfume of numberless flowers and trees. A transformation had taken place in both the land and its beholder. He was responding to a magic well known to natives and frequent visitors—a phenomenon not to be ignored and never to be taken for granted. Commonplace by day, the squat stucco "California Cottages" and the newer, uninspired box-like brick apartments and commercial buildings had undergone a metamorphosis with the setting sun. Evening did for the city what candlelight does for a once-beautiful woman. The illusion cast its spell on the young man from the Southwest and he fell in love.

Breakfast was served at seven in the dark-paneled dining room across the hall from the sitting room. It was a gloomy room suffused with the odor of greasy cookery and dominated by an old oak dining room suite.

Of those at the table only one old man seemed to possess any animation or interest in his surroundings. Howdy watched him test the grease-curled margin of over-fried egg white. In a remarkably young, vibrant voice the senior boarder pronounced judgment on the temporarily absent landlady's cooking. "This egg, dear fellow sufferers, is not fried. It is ossifried!" He detached a portion, maneuvered it onto a wedge of limp toast, and regarded it philosophically.

Howdy glanced around at the two middle-aged women and the other two men, waiting for a reaction. There was none. At his father's funeral Howdy had seen several other corpses awaiting interment. These people reminded him of those dead ones—empty human husks. He smiled across at the old man. "The only time I eat worse eggs is when I cook 'em myself."

The old man lifted the paper napkin to stifle a small

415

belch, murmured an apology, and looked up at Howdy again. "What perverse fortune brings you here—if I may ask?"

"Prospects for some radio work. I play guitar—and sing a little."

"One of those, eh? Well, it's honorable work. I concede that—even if it does strike a blow at my livelihood."

"What do you do, sir—if I'm not prying?"

"Son," the word filled the dining room with a commanding resonance, "*I*—am a voice coach. Name's Neville Newton. When the Brothers Warner introduced their Vitaphone process and poor Jack Gilbert's career came to a squeaking end, I left Chicago and caught the express west to make a fortune salvaging the careers of silent stars." His expression turned forlorn. "However, during the past two years most of my energy has been devoted to an attempt to salvage my own career."

The two women looked at him reproachfully and excused themselves. Amused, Neville said, "They've heard this before a time or two. But it will be new to you."

"I'd be glad to hear anything about this town that'll help me get next to it."

"In that case, son, why don't we put my counsel on a professional basis?"

Howdy looked startled. "I could use the help but I can't pay for it. Anyways, not right now."

A resigned sigh deflated the old voice coach. "Well, like FDR said on the radio—the only thing we've got to fear is fear itself."

Howdy watched the old man closely, for what he was hearing induced in him pity not unmixed with fear. Abruptly, he rose from the table. " 'Long's you got hope you're not whipped." He indicated the street. "And if I want to stay alive I'd better get hitched up and start peddling me."

Neville nodded. "Good luck, my boy."

In three days Howdy visited five radio stations in Hollywood and Los Angles. At each, the answer was the same, "We're not putting on any new acts now, thanks." His attempts to talk with the leaders of the established groups were frustrated by receptionists, switchboard operators, and gate guards. It seemed to Howdy that the minute they saw him coming with his guitar, they automatically slammed the door.

For all his determination, Howdy's financial situation became acute. The condition was aggravated by the need to spend some of his meager cash to have his boots half-soled and to spend five dollars and change for a pair of frontier trousers at a Western outfitter's shop on Santa Monica Boulevard patronized by the cowboy actors and stunt men who appeared in Poverty Row's quickie Westerns.

The owner of the shop, a man called Harry, gave Howdy his first solid encouragement. "I know most of those producers. I'm carrying two of them on the books now for costumes. If you don't care how busted up you get for five bucks a day, I might be able to get you in. Can you ride?"

"I'm no bronc buster. But I know one end of a horse from the other."

"Know how to take a fall?"

Howdy smiled ruefully. "I've taken my share."

An hour later, wearing refurbished boots and carrying the new trousers in a bag under his arm, Howdy entered the shabby Sunset Boulevard office of Running Iron Productions. The producer, Bronc Bronston, who had been pressured by phone into seeing him, did little to hide his antagonism. "What do you want to do, kid—get paid to learn how to act?"

Howdy's smile lingered but his reply was flat. "What I want to do is eat. I don't know a solitary thing about acting, but I can ride a horse and do what you tell me. Is that good enough?"

The producer let Howdy endure a long, resentful appraisal. "Okay, kid. What's your name?"

"Goodwyn. Howdy Goodwyn."

The man made a note on a slip of paper and shoved it aside. "Okay, Goodwyn. You get three days at five bucks a day with grub on location."

Howdy tried not to show his relief. "Thank you, sir. I appreciate it."

"You better. When you pick up your dough just remember you took it from one of them pro's out there on the street who needs the job."

Howdy's eyes glazed and he had a hard time keeping his voice even. "Where do I show up? What time?"

417

Bronston jerked a stubby thumb at a cast notice tacked to the wall. "Six A.M.—Monday—out front. The bus don't wait."

As Howdy turned to leave, the man stopped him. "What work clothes you got?"

"What I got on—plus some pants."

"No Stetson? Or Levis?"

Howdy sensed the producer was about to find the out he had been looking for. "Hell, yes. I got them—if that's what you want."

Bronston smirked. "That's what I want."

Outside, Howdy walked through the knot of gaunt movie cowboys waiting by the door. He could feel their eyes following him—questioning. Edging through them with a friendly nod, he turned west and then south toward Santa Monica Boulevard and the Western outfitter's store. Harry greeted him expectantly. "How'd you make out with Bronc?"

Howdy laughed mirthlessly. "Like a skunk at a picnic. But I got the job—if you'd go me a little credit."

The clothier's pleased smile grew a bit strained. "What do you need?"

"A hat. Some Levis. That's all."

Harry shrugged. "So pick 'em out."

Howdy left the store in debt for eight of the fifteen dollars he had hoped to collect for the three days' work.

The Reo bus left Running Iron Productions at six in the morning. Howdy was among the last to board. Behind him during the long ride to the San Fernando Valley the men, all veterans of a hundred quickies, talked in the flat-voiced, nasal jargon of the movie cowboy. There was no laughter and what humor he could overhear in the rattletrap transport was grim.

When the bus pulled up to the false-front Western street the wranglers had horses saddled and ready to work. Immediately it was apparent to Howdy that each man had his favorite mount. The two stunt men had their own personal horses.

"Pick yourself out one, kid." Bronc Bronston indicated the corral with a jerk of his beefy head. Most of the other men were mounted and waiting.

Howdy walked over to the remaining half dozen animals. He knew Bronston and the others were watching.

Affecting the easy nonchalance of a man used to saddle animals, he moved smoothly from one to the other, sliding a gentling hand along each neck. He chose a sorrel mare who had the compact look of a good cutting mount. He led the animal into the clear, checked the bridle carefully as well as the cinches and the stirrups. Satisfied, he swung up easily and waited.

Some minutes later the director called for attention. "You men follow Charley." He indicated the assistant director. "You ride up that draw over there until you get to those big rocks, then keep your horses out of camera sight until Charley gives you the signal. We shoot all the long chase scenes this morning. I don't want more than one take or any busted animals. So don't bunch up—and watch the slide on this side. Most of you hands have been over the ground a hundred times."

He waved them away. "Okay, men—take your cues from Charley."

It took an hour to show the men the chase route and mark the paths the two stunt men would follow. At the top of the ridge the assistant marshaled the riders. "We'll run it once, real easy now, for Bronc and Lou. Let's go."

At the end of the line of horsemen, Howdy urged the sorrel into an easy lope down a long and sloping trail. The stunt men stopped at the point where they would take their falls. The others continued another hundred yards then pulled up on the brink of what seemed to be an almost vertical fifty-foot slope of loose rubble. The assistant director raised an arm, cavalry fashion. "Hold up here! No use going down now." He motioned to one of the men. "You and I'll ride down it for the camera, Joe. The rest of you guys go on back to the top."

After half an hour of tedious waiting, Charley and Joe rode back and joined the others behind the rocks at the top of the hill. Dismounted, Charley waited for the action signal. It came quickly, a puff of smoke followed by the sharp report of a forty-five. Charley wheeled and thrust an arm at the waiting men. "Okay—action!"

Howdy was not prepared for what followed. The riders—a dozen of them—urged their mounts into leaping starts and plunged headlong down the ridge. There was no time to think. All around him the men were yipping and rebel-yelling, oblivious of the dangerous mesquite, the manzanita brush, the loose rocks, and the choking cloud

419

of dust. Off to his right he saw the two stunt men separate and crash with their mounts in the prepared areas. He was dimly aware of scrambling bodies and thrashing animal legs and then they were behind him.

Seconds later, the men and animals immediately in front of him began to go over the edge. An animal on his left dug in stiff-legged and balked at the brink. The rider cursed and spurred it over. His own animal held up and tossed its head wildly. Howdy gouged his heels into its flanks but before he could regain control there was a shout behind him. Nearly blinded by dust, he half turned to look just as the animal behind him crashed into his own mount's hind quarters. He saw the rider leave the saddle, saw cartwheeling arms and legs disappear over the edge and then he, too, was on his way down in the heart of a sheer hell of sliding rock and ochre dust. Twice the mare's hind legs buckled so badly that his stirrups dug into the slope. The rubble flared out a bit and then he was on hard level ground again. Without thinking, he wheeled the animal and turned to look back. Through the ochre dust he could see the two stunt men running across the slope toward a riderless horse floundering in the brush.

In the distance he heard a voice bellowing. "Keep coming, you stupid son-of-a-bitch! Keep coming!"

Unaware that the order was intended for him, Howdy urged the mare into long leaps back up the slide. When it was no longer possible to find footing, he jumped off, dropped the reins, and scrambled toward the fallen man. The stunt men reached the fallen rider first. One of them turned to him. "What kind of jackass are you, pulling that mare up short at the top with Foy here right behind you?"

Howdy looked up from the unconscious rider. "I didn't pull up. Even if I wanted to, I couldn't. I couldn't see the drop. He was riding too close onto me."

The man spat. "Riding too close? Shit!" He pointed at the downed rider. "This man was a rodeo champ when you was still setting down to pee!" He spat again. "Riding too close. Shit!"

Howdy was still kneeling helplessly beside the man when Bronc Bronston came scrambling up. Without warning, he kicked Howdy on the thigh with the side of his boot. "Get off this goddam hill and wait for me, you dumb bastard!"

Howdy restrained an impulse to strike out at the man and scrambled back down to the mare.

It took ten minutes to get Foy off the slope. As far as they were able to tell, his most serious injury was an ugly gash on his forehead but he was still unconscious when Charley, drove off with him to the doctor in San Fernando.

As soon as the car was on its way, Bronc Bronston turned to Howdy. He jabbed a blunt forefinger into Howdy's shirtfront. "You get the hell out of here. You get in that goddam bus and sit there. Then you collect one day's pay and I don't ever want to see you around a movie company again. You lied to me, you bum. You said you knew how to handle horses."

The regret Howdy felt was submerged in a wave of cold anger. "Before you start calling me a liar, you better remember I told you I can ride a horse and follow orders and that's *all* I said."

Bronston thrust his face into Howdy's. "What you cost me today could break this picture. When a man asks me to pay good money to handle a horse, I expect him to know all there is to know about it."

"Look, mister, I was raised on a ranch outside of Guyon, Oklahoma. I been around animals all my life. If you expect a guy to know a new animal then you better give him time to learn it before you tell him to jump it off the edge of the Grand Canyon."

Somewhere in the crowd a man guffawed. Bronston, his face purple, struggled to keep from smashing a fist into Howdy's impassive face. He jerked a thumb toward the bus. "Get your ass over there and keep out of the way."

He turned on his heel and Howdy continued to stand. He wanted to make it clear that he had not been intimidated and would take his time getting to the bus.

By mid-afternoon some of the anger had leached out of Howdy. One of the cowboys had told him the report on the injured rider had been encouraging—some stitches and a possible concussion.

In the producer's office Howdy waited while Bronston drew a check for five dollars. Bronston shoved it to the front of the desk without blotting it. "You ought to give me five hundred. That's what you cost me today."

Howdy picked up the check, blew on it, and smiled pleasantly. "You claim to be an expert on lies, so you ought to know one when you tell it."

421

Bronston's eyes widened. Suddenly he laughed. "If you could ride as well as you talk, kid, you'd make one hell of a cowboy. Just get the hell out of this office."

On the corner of Hollywood and Vine an old man was handing out leaflets. Absent-mindedly, Howdy took one and was about to throw it in the gutter when his eye caught the phrase, "Free Refreshments." It was an announcement of a meeting of the Utopian Society, a secret fraternal organization being founded "to acquaint southern Californians with the New Economic Order that will guarantee every American a life of plenty."

The private residence on Fountain Avenue below Hollywood Boulevard, where the meeting was held, proved to be more impressive than Howdy had expected. More than thirty persons, most of them past middle age, listened to one of the three organizers, a persuasive man named Rousseau, expound on the society.

Rousseau exuded an infectious brand of confidence. "We have friends in high places—enough friends to unseat the 'Power Gods' downtown and their political machine in Sacramento. Their days are numbered. Today 'Plenty for All' means plenty for them and *nothing* for us. But we're going to change all that. We're going to end this depression. There is more than enough wealth in southern California to take care of every man, woman, child—*and old person*—in the entire state—when it's distributed fairly. We aim to see that done by exercising our power in a truly democratic way."

Rousseau went on to reveal a well-kept secret: the notorious Socialist, Upton Sinclair, was changing his political stripes and becoming a Democrat. His intention was to organize a new wing of the party that would support him for governor.

"I know Sinclair. He wants the same things we want. We foresee the possibilities of combining forces. Then, my friends, you will see an end to the Republican stranglehold on the politics of this state—an end to the domination of Harrison Gray Otis and his newspaper—and end to the Merchants and Manufacturers' lobby—an end to greed and an end to want!"

Rousseau spoke for thirty minutes and painted such a convincing picture of Utopian possibilities that every person present donated as much as he could spare. Even

Howdy put in two dimes while making mental note once again of a fundamental principle of salesmanship: *"Find out what they don't like and be against it. Then find out what they want and promise it to them."*

The free refreshments turned out to be watery, sparingly sweetened orange juice squeezed from fruit donated from the backyard tree of one of the charter members.

The Utopian hostess apologized: "It's not very thick, but it's a lot thicker than the milk of human kindness that flows in the veins of those Scrooges downtown."

Howdy managed to slip away unobtrusively after a half glass of the indifferent refreshment. Several minutes later he joined a crowd of people at the corner of Hollywood Boulevard and La Brea Avenue which had gathered to attend a radio show called *The Hollywood Barn Dance* featuring The Crockett Family, The Stafford Sisters, and Sheriff Loyal Underwood and the Arizona Wranglers. The largest letters on the banner across the front of The Hollywood Woman's Club Auditorium spelled out ADMISSION FREE.

The KNX radio show was one of the best country and western presentations Howdy had seen. After the show, he lingered with autograph seekers and managed a word with the Master of Ceremonies. Something about Howdy caused Sheriff Loyal Underwood to listen instead of smiling sympathetically and dismissing still another out-of-work musician. Without quite knowing why he did it, Underwood promised to arrange an appointment for Howdy with KNX's program director, Dave Layne.

Monday morning Howdy retraced the familiar route to the Western outfitting store. Harry was surprised to see him. "Hey! How'd it go, cowboy?"

Howdy slipped the check from his shirt pocket and handed it to the puzzled merchant. "I'll sign this over to you now. As soon as I get a job I'll pay you on the balance —with interest. I have an appointment with the program director at KNX."

Harry ignored the check. "What happened with Bronston?"

Howdy shrugged. "He's your buddy. You better let him tell it."

"You tell me. Bronc and I do business and we go to the same temple. But that don't make him a buddy. You can't

423

keep friends when you make pictures on his budgets. What happened?"

Simply, without rancor, Howdy recounted the events leading up to his dismissal. Harry listened, nodding silently. Then he took the check from the counter and stuffed it into Howdy's pocket.

"Bronston's going to eat tonight and so am I. I'll digest better if I know you're eating, too. Pay me a buck or two every so often and I'll do better with you than I do with most of them—including Bronston." He indicated the door. "So go get something to eat and Mozl-Tov with your audition." Howdy's puzzled expression made Harry chuckle. "Kiddo—'Mozl-Tov' is good luck the hard way. Now go eat."

Howdy's second break came when he persuaded Dave Layne to at least let him audition. Later, Layne like Underwood, was at loss to explain why. But Howdy knew he had his edge back when he saw the amiable, one-time Irish tenor waver.

"Okay, Goodwyn. Go ahead and sing. But I'm not making any promises."

When the first song was over Layne motioned through the glass of the control room for another. After the third song, a hymn, he returned to the studio.

"Remember. No promises. But I may be able to fix you up singing hymns on the Reverend Matt Simon's Wednesday prayer show." In spite of the program director's caution, Howdy's heart leaped. While waiting for Layne to arrange the studio he had seen the big, burly revival preacher in the front office counting an impressive stack of greenbacks sent in by the grateful radio congregation. To Howdy the sign augered well.

Howdy got his start and his first break when he persuaded David Layne, the program director for KNX, to let him audition. At the time, KNX had a money-making property in the Reverend Matthew Simon. Simon, an unctuously pious man in his forties, had assiduously studied the lesson implicit in Sister Aimee McPherson's spectacular success.

Howdy had his first glimpse of Simon sitting at a desk littered with mail. On his right side was a shoebox filled with currency, on his left was the lid, brimming with coins. The Reverend was busy slitting open envelopes and emptying their contents. His appearance repelled Howdy as surely as David Layne's engaging Irish face had attracted him.

David Layne literally coerced Simon into taking Howdy on his program as the hymn-singing guitar player. The Reverend put up with Howdy for exactly four weeks and then insisted on booting him off. As a man used to a recently achieved spotlight, who over the past two years on the air had acquired "a more suitable residence in which to meditate on the Word of God" along with a new La Salle touring car, he was not in a position to welcome anyone who might compete with him on his own program. Further, he was terrified of anything that might interfere with his tidy arrangement with KNX. The station made unproductive radio time available in return for a portion of the "Free Will" offerings.

After the Matthew Simon impasse Layne was hard-pressed to find anything for Howdy. But he had an innate faith in him and he finally succeeded in giving the young singer his own show when Howdy solved the problem of the Urinex commercial. Urinex wanted radio listeners with bladder trouble to send a specimen and twenty-five cents to Little Rock for a free urinalysis. The purpose was to hook the specimen-senders into buying the product. KNX's owner had shown an unexpected conservative streak by flatly refusing to permit such advertising on his station. Even David Layne admitted it took guts to turn down $500 a week during a depression. Howdy's solution, arrived at by poring over the ads in a cheap movie magazine, consisted of couching the ad in delicate terms, never quite coming to grips with specifics and asking the radio listeners to write in for "a personal symptom and diagnosis chart" at a cost of a dime. The chart, not the station, would suggest that the recipient send the specimen to Little Rock.

The news that Howdy would have his own radio program flabbergasted Angela Woodward. She was reading the Sunday papers and glancing through the entertainment section noting reports of the success of the Ballet Russe de Monte Carlo in the East and the proposed dramatization of Erskine Caldwell's controversial novel, *Tobacco Road*. There were several reviews of new radio programs. The *Romance of Helen Trent* was dismissed as "sentimental dime-novel trash" that would, mercifully, be short-lived. *The Woman in White* was given a better chance because

425

of its noble central figure. *The Lady Esther Serenade* was promising and Walter Winchell would reveal some earth-shaking news on Monday.

Angie turned to Carol Nye's column. Her eyes moved down to an item about Lanny Ross, Showboat's singing star, to a paragraph about a radio series starring "The old Maestro, Ben Bernie." Then a name jumped out of the page at her. An instant later she was running upstairs calling to her mother.

Ana Woodward stopped struggling with an awkwardly placed hook and eye. "For pity's sake, Angie, you'll have an attack!"

"Mom! Howdy's in Carol Nye's column. Listen!"

She gulped in a deep breath and began reading: "Dave Layne, progam director for Station KNX, tells us that new singing star Howdy Goodwyn, an import from Oklahoma and Texas, where he was featured with Spike Spangler's Radio Rustlers for three years on Station KALP, is going to be given his own show on Wednesday evenings. Will Thatcher and the KNX staff orchestra will support the personable young performer."

Angie gasped through the rest of the column. When she finished, she flung the paper aside and said dramatically, "*I* have discovered a star!"

Ana nodded indulgently. "I wouldn't brag about it yet. Wait a few months and we'll see. But I don't mean to dampen your enthusiasm just because I seem to have lost mine."

"How much proof do you need? Wait for what? You heard him on one of those dreadful religious shows. He was wonderful."

Ana smiled tolerantly and moved toward the door. "All right, dear, he was wonderful. Come on now, let's go. I want to run down the hymns before the service."

Angie was still standing outside the church when Melba Persons came bustling up, waving a folded clipping. "Did you see the news?"

Inside, they talked in excited whispers until Frank, who had arrived late, hushed them as the service began. Afterward they resumed their speculations. Angie swore that she would not work that night. Melba suggested they all listen at their house.

Frank Persons, who had been secretly sharing their pride and excitement, broke in, "Maybe we could find out if there's an audience and drive to L.A. to see the first show. But you better remember it'll make it a late night. Very late."

Excitedly they agreed to try it. There would be five of them; in addition to Frank and Melba, Angie and her mother, Rosa Rinelli was asked to join them.

Frank and the four women pulled up to the studio at ten of eight. It was twenty after eight before they found a parking place and began the walk to the auditorium. When they turned the corner they were confronted by a line of several hundred persons.

Melba stared at the scene open mouthed. "For heaven's sake! I wonder if we can get in."

Angie urged them on. "Let's get in line quickly and we'd better all stay together."

The front doors opened and by the time they were inside, the only unoccupied chairs were scattered singles on the sides and a sprinkling of two's and three's well toward the rear. Angie steered her mother and the Persons into the ones at the rear. Then, taking Rosa Rinelli by the arm, she hurried to a pair of singles one behind the other midway down the side. In a matter of minutes the auditorium was filled to capacity. Disappointed people lined both side walls. Others were still clogging the lobby. The house lights snapped off and a spotlight suspended from the ceiling projected a blazing circle on the drab monk's cloth curtain. The chatter subsided at once. A moment later an attractive young man slipped through and stepped out onto the apron. There was a scattering of applause, which he silenced with a good-natured gesture.

"Good evening, ladies and gentlemen. My name is Tom Smith. I'm your announcer and it's my pleasant duty to welcome you to the première of the Howdy Goodwyn Show."

A cheer went up all over the auditorium. The announcer consulted his wrist watch and held up his hand. "We have only three minutes until air time. Before I introduce our orchestra leader and our star, I just want to tell you that we hope you will applaud after each number. Let's try it now."

The engineer checked sound levels and signaled his okay. The musical director, Will Thatcher, appeared followed by his four boys. The audience applauded as Tom

427

Smith beamed his approval. "And now, ladies and gentlemen—the brightest new star in our Western skies——Howard D.——Howdy——Goodwyn!"

The din of cheering, applauding, stomping people filled the auditorium. It continued as Howdy deliberately delayed his entrance. Then, pretending he'd been unavoidably detained, he came running onstage with his guitar, glancing apologetically off into the wings.

For a full minute he stood acknowledging the ovation with the same underplayed engaging shyness he had affected in front of audiences in Oklahoma.

The Persons, Ana Woodward, Rosa Rinelli, and Angie sat open-mouthed, trying to reconcile the tall, well-groomed young man with the scarred, bedraggled hitchhiker they had first known two months earlier.

Rosa turned to Angie in the chair directly behind her. *"Mama mia!* Wotsa hoppen to Howdy?"

The girl leaned forward and grasped the older woman's shoulders. "Oh, he looks *wonderful,* Rosa! Look at him! He looks like a star already."

"He looks pretty good, all right. Beautiful."

Angie let her chin rest lightly on Rosa's shoulder. Her voice was scarcely audible. "He's not beautiful, honey, but he looks better than any man I've ever known."

Rosa Rinelli leaned away and twisted to look Angie in the eye. "Oh, ho, ho! *Amore,* eh?"

Howdy began to speak: "I only got about thirty seconds, but I want to thank you all for coming out tonight." He looked offstage, received a warning signal, and acknowledged it. "I have to go now. I sure hope you like what we do and I'll talk to you some more later."

From the wings an unseen cue was thrown to Will Thatcher and the theme song began. It was a bright, catchy original opening number entitled "Howdy" that had been composed by Thatcher several days earlier as a welcoming signature song. Accompanying himself on the guitar and backed by an infectious arrangement, Howdy smiled over the footlights and addressed the simple lyrics directly to the audience. When the theme ended the reaction was so enthusiastic that Howdy had to help Tom Smith quiet the house. The theme had performed its intended function perfectly. Dave Layne summed it up halfway through the show: "That kid could blow his nose onstage and get an ovation. I may send in a dime myself."

Compared to the big network musicals, or even to the *Saturday Night Barn Dance, The Howdy Goodwyn Show* was a simple affair. It relied on Dave Layne's pet formula—"Good shows are made by good performers, not gimmicks." Howdy proved to be a better performer than any of them had suspected.

Neil Carlson, the station's sales manager, listening to Howdy's Urinex pitches being delivered with such deceptive ease and confidential sincerity that they seemed to be the spontaneous after-thoughts of a dear friend concerned for the welfare of an intimate, shook his head in disbelief.

When the show was over, the front aisles were clogged with eager fans. Howdy, flushed with elation, scarcely saw individual faces as he squatted on the apron greeting acquaintances and strangers with both hands, scribbling autographs on proffered scraps and soaking up adulation. After ten minutes his legs began to cramp. Apologetically, he struggled upright and rubbed his knees. It was his first real chance to look out into the auditorium. The crowd had begun to thin as Howdy's eyes swept across it to the aisle on his right. Suddenly the set smile turned to amazement. An instant later, with a joyful whoop, he leaped into the startled crowd and began pushing his way toward the Persons, the Woodwards, and Rosa Rinelli. Halfway up the aisle Angie was in his arms. Too full of emotion to trust her voice, she clung to him. When Melba reached them, blinking and daubing her nose, Howdy freed an arm and reached out to embrace her too. Finally he pushed Angie gently from him and grasped the hands of each in turn. Over and over he kept expressing his wonderment.

"I don't believe it! I don't believe it! I just don't believe it . . ."

Angie, smiling broadly, said, "You'd better. Frank drove like Barney Oldfield for five hours to get us here on time."

Neil Carlson, looking apologetic, came up and tapped Howdy's shoulder. "Sorry to interrupt, but the boys from the agency want to talk to us over coffee at the Roosevelt." Sensing Howdy's keen disappointment, he added, "You come along when you can. We'll wait for you there." Before Howdy could introduce him he slipped into the crowd.

Frank Persons took out his pocket watch. "If we don't have a flat tire, it'll still take us until after two in the

429

morning to get home. So we'd better hit the road. And besides—Howdy shouldn't have anything but business on his
mind right now."

They left the auditorium and walked down La Brea
toward the car. Angie clung to Howdy's arm, visibly at
least more elated than he at the obvious success of the
first program. At the car, Melba, enthusiastically supported by Angie, urged Howdy to come down for a weekend.
The rest of them took up the coaxing. Rosa Rinelli's invitation was more of a command. "You coming down,
Howdy. That's all! We make you *pranzo speciale.*"

Howdy slipped his arms around Melba and Rosa. "As
soon as I know what'll happen next I'll mail you a card
and see when's a good time for you."

Melba patted his back maternally. "Any time's a good
time for us, young man. Your bed's waiting on the porch."
At Frank's insistence they got into the car. As it started to
pull away, Melba leaned out. "And you bring your mend-
ing too."

Howdy watched them depart and returned their waves
until the car disappeared around the corner. He did not
envy them the one-hundred-twenty-five-mile drive and he
was more grateful than he had been able to say that they
had come.

As he walked back to Hollywood Boulevard to join
Dave he wondered again at the curious loneliness he felt.

Howdy's success on the Urinex show was sufficient
to make other sponsors interested in signing him up. But
none of them wanted any part of him as long as he was
connected with a product whose commercial catchline
read, "Make your bladder gladder. Use Urinex." The
agencies bluntly told KNX "Get rid of Urinex and we'll
give you more quality business than you can handle."

As a result, KNX offered Howdy a five-year contract—with a catch. The station had the right to terminate
at the end of each twenty-six-week period but the contract
for five years was firm as far as Howdy was concerned.
Throughout the meeting with the KNX executives and David Layne, Howdy seemed to be understanding, even
agreeing. At the end of the session, he nodded to himself
and grinned amiably. "Everything's clear now. But I was
just wondering if I could have a little pondering time in

430

view of the fact that you good folks are taking the gamble and all you're asking me to take is the risk."

The next night David Layne surprised Howdy by asking him to stop by his apartment. The invitation carried the sound of urgency. Dave's apartment turned out to be a cottage in one of the ubiquitous garden courts scattered throughout Hollywood. Through Neville Newton Howdy had found a professional whore whom he used once or twice a week. David's place depressed Howdy because it reminded him of the cottage Renée occupied. There was the same shabbiness and cheap hominess about them both. Dave indicated a ratty easy chair and said, "How about a beer?" He disappeared into the kitchenette and returned in a moment carrying two bottles of Miller's High Life, and two jelly glasses. "Thought I'd better break out the family crystal. This might be a momentous meeting."

Howdy's wondering gaze made him grin.

"I may sound like Benedict Arnold when I tell you what I have in mind, Howdy—but that's a chance I'll have to take."

Howdy poured the beer without looking up. "Never did hear Arnold talk—so I wouldn't know. Shoot."

"Ever since you talked me into listening to you sing I expect you've noticed that I have done all I could, consistent with my job as program manager, to help you along."

Howdy confirmed the truth with a nod.

Dave paused to light a cigarette. "I told you I used to be a performer—a legit tenor. For almost six years I was a pretty big name on WLS in Chicago. I got signed to the same kind of contract our station is offering you. And I signed it—gladly—for security. Then along came a guitar twanger named Autry and the next thing I knew I was off the air. I couldn't find another show so I did what I could. I sang leads in *The Student Prince* in high school auditoriums and sang 'Danny Boy' and 'Mother Machree' at Women's Club luncheons. I got so damned hungry I had to lean on the piano. In the middle of all that my wife left me."

Dave smiled bitterly at the recollection. "Finally I got a job in the chorus of *The Desert Song* and wound up out here. Three years ago I started singing on the station for nothing and selling electric heaters door to door. Have you ever tried to sell an electric heater in Los Angeles in August—during a depression? My best month I made

431

sixty dollars but I guess God must have been listening because a local dairy bought my quarter-hour *Sunset Serenade Show* across the board and I got paid five bucks a program—twenty-five a week. I did every mother song, every sweetheart song, every dying child song and every non-denominational hymn in the books. I wrenched more hearts than Little Eva. Then one day, for another twenty-five, Neil Carlson asked me if I wanted to take over his programing duties. He'd been holding down both jobs.

"The rest is easy—I don't sing much anymore—I get seventy bucks a week—and if I'm lucky I may get up to a hundred bucks a week someday. And this, my friend, is not what Dave Layne calls a future."

Dave paused and gave Howdy a searching glance. "Look, do you think you should sign that contract?"

Howdy examined his guitar calluses. "I don't know."

Thoughtfully, Dave lit another cigarette. "First, the radio business is changing. The change is going to make a fortune for a few people who understand it and break those who don't. Those big advertisers may buy participation with Urinex out, but you've got Urinex now and they bought you awful damned cheap. Right now there are two things you ought to think about—upping the ante and holding the contract down to one year. Then everything possible should be done to build Howdy Goodwyn—releases, public fund raising appearances, phonograph records, later pictures, westerns . . ."

"To hell with that!" Howdy's explosive dismissal of the idea produced a surprised laugh.

"What's the matter?"

"Just put it this way, Dave—*I hate horses!*"

For the moment Dave could think of no rejoinder. Howdy filled the breach. "Also—I hate movies."

To Dave the statement had the shock impact of heresy. "Movies too? Why, for God's sake?"

"I tried 'em. I wound up on my ass, out of a job. I'm not that good a rider and I sure as hell can't act."

"What the hell do you think you're doing on stage when you turn on that bashful bumpkin charm? Don't tell me that's the real you."

"Shit, no. But it ain't acting either."

Dave gestured impatiently. "Quit crapping me, Howdy. If that isn't acting, neither is Will Rogers' cow-licking, turd-kicking bit. All you need is a little time, a little slick-

432

ing up ... you do for Billy Hill what John Charles Thomas does for Oley Speaks, what Lanny Ross does for Jerome Kern. I'm not a very bright man, Howdy, but this I know—it's a safe gamble that I can build you into a star. I'm willing to put my job on the line that I can build you into America's 'pop *corn* king.' " He fixed Howdy with a speculative look.

The silence that followed was long enough to make Dave uneasy but he waited. Finally, Howdy spoke. "I'm listening. Keep talking."

"All I need is enough percentage of the gross to live on. But our deal wouldn't start until you're making at least seven fifty a week. Then we can start with ten percent, plus reasonable expenses to pay for the buildup and you can approve every dime in advance."

Dave stopped long enough to let the point sink in. "First I want you to package your own radio show. Next we should look into syndicating a recorded radio show. Third, I want you to start making phonograph records of your best songs like Bing Crosby, Donald Novis, and Russ Columbo are doing. I've got a friend, a genius, Danny Mayer, who has a little recording studio. We set up Danny with a mail-order department and plug these records on your show. After a while the big distributors will be coming to us to make deals.

"I say 'we' because I expect to be an equal partner with you in all of the subsidiary enterprises. That means I'll work my butt off for ten percent of the radio show because that's what makes the rest possible."

Dave paused and looked at Howdy expectantly. "Well—that's about it. How does it strike you?"

Sprawled in the big chair, twisting a sandy-red forelock, Howdy was silent for almost a minute—a minute in which Dave's misgivings began to mount. He wondered if Howdy was hesitating because he felt that a man who would double deal with one employer might in time do the same with a partner.

Abruptly Howdy got up from his chair and stood looking down at Dave. He stepped close and extended his right hand. "I'd say it strikes me just fine."

15 SINCE THEIR VISIT to the premiere of Howdy's show in early October, Angie Woodward had been struggling against the run of emotional tides that left her alternately exhilarated and exhausted. Even though Rosa Rinelli, with a repressed romantic's perception, had diagnosed the symptoms correctly, Angie herself dared not acknowledge either infatuation or love as the cause. Scores of times she had shaken her head viciously to dispel the fantasies, not all of them pleasant, and had scolded herself half aloud in those wakeful pre-dawn hours.

Howdy seemed to have built an invisible wall between them. Still, Angie could reassure herself when she thought about her ability to help Howdy. "He's been using those chords and combinations I taught him. I didn't really begin to teach him all I can. More than that, when I ran down the aisle and threw myself into his arms like a simp, he hugged me like he was glad to see me. It wasn't just a friendly hug. Something happened because—anyway, he put as much into it as I did."

Some hours later she awakened and grimaced painfully. Downstairs one of her mother's beginning piano pupils was committing musical mayhem on "The Jolly Farmer." After a meal eaten mechanically, she bundled up against the gray December chill and went out onto the porch. Across the street Melba Persons was walking homeward burdened by two large food bags. Angie waved and called out. "What's the matter? Car broken down?"

Melba shook her head. "No. Frank's out looking at land with some fellow from San Bernardino. Say—come on over when you can. I want to talk to you about something."

Angie ran down the steps and crossed the unpaved street.

"Let me have one of those bags. I need the exercise."

Over coffee in the Persons kitchen Angie tried to con-

434

ceal her excitement as Melba revealed her plan to invite
Howdy down to spend Christmas.

Howdy seemed glad to receive Melba's telephone call
but reluctant to accept her invitation. When she persisted,
Howdy explained that his partner, Dave Layne, would be
alone, too. Melba promptly invited them both.

Angie was waiting with Frank and Melba at the railroad station in Barstow when Howdy and Dave stepped
down from the train Saturday evening. Melba's greeting to
her guests was motherly and effusive. Frank's welcome
was warm and a bit self-conscious. Angie's greeting was
restrained. She linked her arm through Howdy's, smiled
warmly at Dave, and pressed her cheek fleetingly against
Howdy's shoulder. "Everybody's so anxious to see you.
Rosa and Mike have invited us all to a special dinner
tonight."

Frank Persons reached for Howdy's valise. "But Rosa
won't let you sing, Howdy. You're her guest and she and
Mike would just feel awful if you thought they'd invited
you so you'd entertain."

The Rinellis' uninhibited greeting became a minor ovation. Howdy was certain he had never seen many of those
who pushed through the crowd at the bar to clutch his
hand. The Twenty-first Amendment had gone into effect
two and a half weeks earlier so the hard liquor Mike dispensed now stood in open array on the back bar. Mike
leaned across the bar and indicated Dave standing down
from Howdy with Frank Persons. "He's your boss, eh?"

Howdy nodded.

"Good. He's a nice guy."

After the most elaborate Italian dinner Howdy and
Dave had ever eaten, they followed Angie into the main
dining room to listen to the show.

The show was difficult for Angie. Since mixed drinks
had been legalized the crowds had become noisier and less
attentive. To counter their expansive good humor she had
added more rhythm numbers to her repertoire only to learn
that they produced the exact opposite effect. She had sent
for new sheet music. Among the selections was a piano solo
arrangement of Duke Ellington's biggest hit, "Sophisticated Lady." She had worked on it at home perfecting the
harmonies and rich modulations.

Angie played it beautifully, impressing both Howdy and

435

Dave with her complete competence. She was not a natural performer but she was an accomplished one.

Christmas dinner was to be held at the Woodward home. The house was all Melba and Frank had said it was. A well-kept old frame structure, it had been planned along the generous lines of the two-story family homes misleadingly labeled California Cottages. The front door led directly into a spacious, wainscoted, high-ceilinged living room. Directly opposite the front door stood a large red brick fireplace topped by a massive oak mantel.

An old, carefully preserved Knabe grand piano dominated the plain, turn-of-the-century furniture. The floor was covered with a worn but clean pseudo-Persian carpet. To the right of the entrance, through a square-pillared archway was the large dining room. A flight of banistered stairs led to the bedrooms and bath on the upper floor.

Howdy realized that most of the Persons' bright, homey little cottage down the street could have been placed in these two main rooms. But somehow the place depressed him. Then he remembered that he had felt much the same when he had first entered Georgette Follansby's old Franklin Avenue boarding house.

After welcoming Dave and Howdy, Angie and her mother excused themselves and went out to the kitchen. For the moment, Howdy and Dave were alone. Dave gave Howdy a knowing look. "Now I understand why you wanted to come down here. Nice! Very nice people. Especially your friend Angie."

Howdy avoided his eyes. "They're all nice. They sort of feel like family."

Dave fixed Howdy with a skeptical look. "Don't try and tell your Uncle David that she's your adopted kid sister."

Howdy was spared an awkward reply when Frank Persons appeared in the doorway.

They completed decorating the Christmas tree a few minutes before midnight. The garlands of cranberries and popcorn, Melba's Santa Claus cookies plus an assortment of old German tinseled globes resulted in an old-fashioned Christmas tree that filled everyone except Howdy with pleasant pangs of nostalgia. It was the first such tree he had ever known.

Dave Layne lit a cigarette and leaned against the mantel inspecting their work. "You know—for about three hundred years I've been telling myself that I didn't miss all this fuss." His pleasant Irish face grew pensive. "And you know something else? I'm a liar! I'd forgotten how nice it can feel to be domesticated."

His mood brightened again and he took Angie by the hand and led her toward the piano. "As long as I'm getting maudlin, why don't we do it up right with a quick chorus of 'Silent Night'?"

Acting as conductor, Dave led off with a downbeat and, somewhat timidly at first, they all began to sing. As Dave's clear lyric tenor began to dominate the other voices, Angie glanced up, her face mirroring surprise and admiration. When the singing ended Angie clasped her hands at the base of her throat. "Dave! You're wonderful! Tomorrow you've got to sing for us." The others took up the demand until he held up his hands in surrender. "I will. I will. In fact, I insist on it."

Sometime later, as they readied for bed on the Persons' porch, Dave looked up from undoing a shoe. He seemed unusually serious.

"Howdy—thanks for insisting I come down. This is the best medicine for me that you can ever imagine, pal. Your friends are great. And that Angie. She's for dreaming about, that one."

On Christmas Day, they were at the table for almost two hours before they pushed back their chairs and the men were sent to the living room to wait until the dishes were stacked. Good food and easy conversation spiced with Dave Layne's lively humor and interesting anecdotes of his days as a would-be musical comedy and operetta star produced an aura of well-being.

Ana Woodward, a bit less withdrawn, wondered at her daughter's radiance. She wondered, too, at the transformation that had taken place in the old house. Happiness was not a state a person could trust.

Shortly, everyone settled into a state of euphoria. Bored by the drowsy, sporadic talk, Angie moved inconspicuously to the piano with the package of new lead sheets Howdy had brought. Without striking the keys, her fingers sought out the chords as she went leisurely through the copies, putting aside those that appealed to her the most.

Dave Layne, slumped luxuriously in the Morris chair beside the fireplace, puffed on his cigarette and watched the play of light and expression on Angie's face as she responded to the beauty of silent harmonies. Behind her, the rickety wrought-iron bridge lamp created a pale halo that accented the highlights in her red-blonde hair. His eyes moved languidly to Howdy, who was listening with somnolent patience to still another of Frank Persons' land deals that "looked good." There comes a time in a man's life, Dave thought, when most dreams become absurdities—as unattainable as the mill donkey's dangling carrot. For at least a third of his thirty-five years, Dave Layne had followed the dangling carrot of his own unrealistic ambition until—blessed day—circumstances had cornered him and forced him to face the fact that his true talent lay in recognizing in others the ability he himself lacked. He was determined that time would not turn this new dream into an absurdity. The partnership with Howdy was the first move to use his own talent realistically. Nothing in Howdy's manner, or in Angie's, had suggested that the girl was to be a part of that dream. And still, as he watched her, he could not rid himself of the unsettling notion that somehow she would be. A compulsion brought him from the chair to the piano.

"Louder, lady! The customers in the back row can't hear."

Startled, Angie looked up, returned his broad grin, then compressed her lips and grimly began to play Mendelssohn's "Spring Song," deliberately interpolating into it a number of sour notes. Dave winced and plugged his ears. Imitating the New York Capitol Theater's famous Major Bowes he droned, "All right! All right!"

Angie made a defiant face and positioned her hands over the keys. "All right then, try this."

Her fingers began flying through her own arrangement of "Twelfth Street Rag." Ana Woodward's expression of shock turned to reproach. "Angela! Not on Christmas night, honey—please."

The sudden barrage of rhythm interrupted Frank Persons' aimless monologue and gave Howdy a chance to escape. In the hall he uncased his guitar and joined Angie and Dave at the piano.

For a half hour they went through old familiar numbers while Dave Layne listened and watched closely. His initial expression of pleased surprise faded a bit but did not vanish entirely as Angie and Howdy moved easily from one number to another. After Melba had coaxed Angie to play

438

her version of "Sophisticated Lady," Dave's expression changed to open admiration.

"Bravo! Pretty good for a backwoods kid from Barstow. In fact, pretty good for Hollywood and New York, too."

Angie grimaced. "Flattery will get you nowhere. But I love it. Anyway—Howdy and I have done our act. We want to hear you sing, Dave."

Angie pulled out a large paper-covered album and set it on the music rack. Dave moved in beside her and together they flipped the pages until they came to one of the favorite ballads from Sigmund Romberg's operetta, *The New Moon*. The show had opened in New York in 1928 and after five years it was still playing in road-company productions around the country.

Dave placed his hand over a page. "One of my fondest dreams was to sing this song in the show. I never got beyond the chorus—but—you will now confirm operetta's tragic loss as David Layne sings, 'Softly As in a Morning Sunrise.'"

The room filled with spontaneous applause as Dave ended the last note. None was more enthusiastic than Angie's. "You're wonderful, Dave! I can't imagine why. on earth you quit your career."

"I quit my career, young lady, because I have the voice of a romantic leading man and the appearance of an underweight bartender."

Later, when the dishes had been done and they were all gathered in the living room, Dave Layne spoke of something that had been on his mind most of the evening. "Angie, have you ever thought of training your voice?"

The question was greeted with an incredulous stare. "Heavens no!"

"I don't mean studying to develop a 'legit' sound. I mean coaching and learning phrasing so you can develop a popular sound, actually not a great deal more than you've been doing here tonight."

Angie looked astonished. "Me? Another Ruth Etting? It's never entered my mind."

"Well, it should. You don't have a big voice—but with the new amplifying equipment you don't need one. All you need is a pleasant sound and a good sense of rhythm. A coach can take care of developing style. Think about it. Like Howdy, I have a hunch you may be a natural."

439

Angie gave him a narrow, sidelong look. "I have a hunch Melba put too much brandy in the hard sauce."

The matter rested there until Howdy and Dave were getting ready for bed.

"Hey, Dave—were you serious about Angie singin'?"

"I sure as hell was. Why?"

Howdy settled himself beneath the covers. "Nothing. Just wondered."

"We've got to be ready to make our move by March first."

Dave Layne flipped to the month on the desk calendar resting on the arm of the easy chair at his apartment. "The brass wants national prestige sponsors and these new station sales representatives are promising spot business from people like Ford, Chevrolet, Nash, Franklin, and the soaps like Fels Naphtha and Gold Dust."

Bootless, Howdy was sprawled on the sofa. "How much time do we have?"

"A month—at the most. We've got to start right now figuring out what we want in the show."

Howdy stifled a yawn and reached for his boots. "If the station doesn't go for our idea—then what happens?"

"We're not even going to think about that. You've proved you're a talent that can draw. Chaffee isn't going to like our proposal, but he's going to have to buy it. You've got the hottest western family show on the air now."

He pushed the calendar aside and leaned back. "—and we're going to make it easy for him to keep you—for the time being. Now then, let's start building a basic family around you for the air. I don't mean a Crockett Family where one of the sisters is really that youngest Stafford kid, Jo. I'm talking about a group of friends who get to be as close as family—like you and the Persons—and Angie."

Dave broke off and frowned thoughtfully. "How would you like to have Angie on the show?"

For a long time Howdy stared at the program director without blinking. Then he broke into a slow grin. "Well, hell, yes. Why not?"

During the ensuing week the program director put together, complete with suggested music, three sample programs.

"There's nothing new about a family group in show

440

business, Howdy. The Foys, the Stones, the Barrymores are real families. The new radio shows are beginning to use the family idea too. Rudy Vallee, Fred Allen, Burns and Allen, Jack Benny."

He rattled a sheaf of papers. "Here's what I have in mind. A basic musical group of eight men called The Golden Westerners. Next a good singing chorus—The Campfire Choir—seven voices, four men and a girl trio. We'd call the four guys, 'The Menfolk,' and the girls, 'The Womenfolk.' They'd start the show off with a theme song probably built around the Golden West. Together they'd sing as a chorus to back you up in addition to their own ensemble.

"Then I'd take a guy like Steve Stone—the best harmonica player—and call him something like, 'The Maestro of the Mouth Organ.'

"The last regular on the list would be Angie. If you still think it's a good idea, I suggest we sound her out. If she wants to go along on a tryout basis we can bill her as, 'Our Lady Angela, Queen of the Keyboard.' Angela— Queen of the Angels—the thing works subjectively with Los Angeles—our Lady, the Queen of the Angels," he dismissed the idea with a wave, "etcetera. It may have some memory value."

Howdy had made no comment during the explanation. Dave watched him closely now as he studied the pages. There was a hard-headed, stubborn thoughtfulness behind the easy-going onstage personality. As Howdy's continued silence showed no sign of letting up, Dave became openly uneasy. To break the mounting tension he went to the kitchenette and returned with a bottle of cheap bourbon. He set a glass in front of Howdy and poured himself a stiff jolt. "Look, Howdy—if you and I are thinking in opposite directions all of a sudden, we can throw that out and start all over. But goddamn it, we better do it tonight. I want to break the news to Chaffee on Friday. That will give him the weekend to think it over."

Howdy nodded. "There's only two things I'd do to this show, Dave—"

"What are they?"

"I'd want me and the Campfire Choir to end each show with a hymn—and I wouldn't look for a new theme song. I'd keep the one Will did—'Howdy.' "

441

A broad, relieved smile spread over Dave Layne's face. "Is that *all* you'd change?"

Howdy looked up. "That's all. I think you've done a hell of a job. I think we're going to have a pissing winner with this show. But I wouldn't say anything to Angie 'til we know we got a deal. I'd sure hate to get her all riled up then have to call back with bad news. Okay?"

"Hell *yes!*"

Later, alone with a third drink, Dave spoke aloud to himself. "So far so good. But just protect yourself in the clinches, Davey. If this Okie's got what it takes to be a star, he's also got what it takes to be a son-of-a-bitch."

At four o'clock Friday afternoon Dave met with station owner Roger Chaffee. The meeting was not long and Chaffee was not happy. Dave had braced himself for recriminations and charges of ingratitude that did not come. Although the executive would not have admitted it, one of the reasons he had chosen Dave Layne to be program director was the very initiative that the man was demonstrating.

Chaffee kept them waiting until the following Wednesday, Howdy's show day. The delay had not been intentional. It was occasioned by the need for meetings with Neil Carlson and the treasurer of the company. He had told them, "I want the best deal I can drive them into—but I don't want to lose them. Also—I want a long-term contract—a tough one. I don't want to build up *The Howdy Goodwyn Show* just to have the networks outbid us at option time."

Dave Layne, with Howdy's approval, brought a lawyer into the negotiation. Everett Moses had been chief counsel for a group of independent stations in the Chicago area but illness had forced him to bring his wife to southern California.

"He's as smart as they come," Dave told Howdy. "He negotiated for the station when they signed me in Chicago—so I know. Also, he'll string along with us until we've got money coming in."

It took ten days for the contract to be finalized and signed. Moses secured all that Dave Layne had asked for—ownership of the package for Howdy and a guaranteed basic fee plus an override on sales above a specified base figure.

Howdy and Dave waited until they were certain Angie would be home before they phoned. The effect of the call was just short of cataclysmic in the Woodward home, the Persons' home, and at the Roadside Inn.

Angie arranged to commute by train each Tuesday morning to Los Angeles and return to Barstow on Thursday. The show would pay her transportation and her room at the Hollywood Roosevelt. She would continue her job at the Roadside Inn, at least until it was certain that the tryout had proved successful.

The opening of the new show was set for Wednesday, March nineteenth. The only person not happy over the prospect was the Reverend Matt Simon, who had been put on notice that any evidence of a substantial increase in his audience would automatically result in a larger percentage of his take for the station. He blamed Howdy for the turn of events that he refused to concede were inevitable.

The Howdy Goodwyn Show got off to a good start. Carroll Nye of the *Los Angeles Times* thought the show was "the equal in entertainment value of any variety hour on the air." Bernie Milligan of the *Los Angeles Examiner*, usually harder to please, was equally enthusiastic and singled Angie out for special mention. Zuma Palmer, whose column in *The Hollywood Citizen-News* reflected the women's point of view, found Howdy and his radio family "—an engaging, wholesome group, talented, well-mannered, and a credit to the medium." Several radio fan magazines asked for picture stories and Dave Layne engaged a part-time public relations man named Don Bellew.

Georgette Follansby was stricken when Howdy told her that he and Dave Layne had taken a two-bedroom furnished apartment. She offered them her two best rooms, but Howdy explained that the bachelor quarters in the new seven-story building on Las Palmas and Hollywood Boulevard would also serve as their production office.

From the beginning, the one-hundred-dollar-a-month apartment was more office than home. After a week of wading through their own confusion, by mutual consent they budgeted another twenty dollars a month for a part-time maid.

At the end of the first thirteen-week cycle in mid-June,

443

there was no doubt about the future of the show. The demand for seats was so great that tickets had to be printed and holders admitted on a first-come first-served basis.

Howdy's first recording on his own HG label was a double release featuring his theme song "Howdy" on the A-side and Billy Hill's "The Last Roundup" on the B-side. Copies of it were sent to every radio station of consequence in the eleven Western states together with a personal letter from the star saying they could expect other records to use for "program fillers." The letter made a point of thanking the station program directors for using them. The innovation caused a flurry in the trade and brought some angry press comment from the big record companies who saw the move as a threat to their retail market.

Dave, working with Everett Moses and Danny Meyer, set up a distribution company to service retail music stores with HG records. When the first sales returns became generally known in the trade, Dave Layne's prediction proved correct. Several national record distributors contacted them, offering distribution deals. The attorney looked at the offers and compared them with their own costs for doing the job. "Let's hold off a while. Their offers aren't that good yet. Let's get more records out and plug them on the show. If we can't handle the job as cheaply as they can, we'll do business."

Ev Moses brought to HG Enterprises a big-time knowledge of business practices and a shrewd understanding of the temper of the times. Howdy, applying lessons well learned from Al Phelps, understood how to relate his talent to the most cherished hopes and aspirations of the common man. Dave Layne's intimate knowledge of the performer's problems and a sure grasp of radio's mechanics and its new national dimension completed the know-how necessary for a firm foundation. Radio prospered and the new company prospered even though the country as a whole did not.

Upton Sinclair provided the nation's incredulous reactionaries with a shocking lesson. He fully demonstrated the power of radio as an educational medium by creating a wave of lower-middle-class white-collar discontent with the status quo. Eight hundred EPIC clubs, determined to End Poverty In California by implementing Sinclair's doctrine of "Production for Use," listened intently and heard

444

not only promises but something they had yet to hear—
and never would hear from the pleasure-loving Utopians
—a definite plan.

In one of the dirtiest campaigns in California's brief and
turbulent political history the reactionary establishment
defeated Upton Sinclair by slightly more than a quarter of
a million votes. No sooner had the socialist retired from
the arena than the lugubrious Dr. Francis Townsend appeared on the scene in Long Beach with his Old Age
Revolving Pension Plan. The doctor had received the
blessings of Governor Frank Merriam during his campaign
and no less a literary great than Kathleen Norris was en-
dorsing him. Everett Moses saw the meaning behind the
repetitive pattern of these something-for-nothing movements.

"Every penny we can divert to building the show and
your subsidiaries should be spent right now. There are
three hundred thousand unemployed people here in Los
Angeles. And God knows how many there are in the state.
I saw figures this morning showing that, all told, fourteen
million people are out of work in the country. Figure out
what it costs to get into a movie. Radio's free. That should
tell us where our audience is. If you entertain them now
and make them forget their troubles for a while, when
times get better they'll remember you. Keep giving them
more and tell them you're doing it. You're never going to
get the Park Avenue crowd in New York, Howdy, or the
Wilshire-Beverly Hills crowd here. There are a hundred of
your kind of people for every one of them. Right?"

Howdy agreed. But he kept to himself certain reservations that he would never quite be able to articulate.

Frank and Melba Persons and Ana Woodward came to
Hollywood as company guests for the Wednesday show
preceding Thanksgiving. The next day a special dinner was
arranged at the hotel. On an impulse Howdy invited Neville Newton. He had run into Georgette Follansby on the
street and, in making a routine inquiry, had found out the
old man was despondent.

After the old man left, Angie looked saddened. "I feel
so sorry for that sweet old man, Howdy. I can't tell you
why, but when he talks about going back to coaching
movie stars I get a lump in my throat."

Howdy's noncommittal "Yeah—" troubled her.

445

Christmas came on show day, making it impossible for Howdy and Dave to accept the Persons' invitation. The cast and engineers had a tree onstage and Howdy initiated a custom that would endure through all of his years as a performer. He gave each member of the show a personally inscribed gift. In return, members of the cast pooled their money and bought him a handsome new guitar case.

Angie spent New Year's eve with Howdy and Dave at the Moses' modest hillside home in the Silver Lake district midway between Hollywood and downtown Los Angeles. Throughout the three-and-a-half-hour vigil during which they listened to Ben Grauer describe the arrival of the New Year in New York's Times Square, Dave Layne drank constantly. He dominated the conversation, speaking with amusing and, to Angie, somewhat pathetic self-depreciation about his years as a singer on radio and in road-company musicals.

Shortly after twelve-thirty, following the ceremonial New Year toast and a thickly emotional prediction of good fellowship, they said their thanks to Ev and Miriam Moses and made their way from the front porch to the car.

On the way home, Dave slept with his head on Angie's shoulder. After they had put him to bed, Howdy drove Angie back to the Roosevelt Hotel and parked at the side entrance on Orange Drive. As he was about to open the door she rested a hand on his arm. "Howdy—"

"Yes."

"Would you think I was crazy if I asked you to do something?"

His puzzled expression made her laugh. "My goodness! All I had in mind was a little drive."

She peered through the windshield into the perfect night sky. "Ever since the show started I've always wanted to go up on top of the hills and look down at the city. Could we drive up there tonight—please?"

They drove out Hollywood Boulevard to Laurel Canyon, followed it to the summit, and turned onto Mulholland Drive, passing a dozen parked cars before Howdy found a place to pull off.

Spread below them like a random display of unset gems were the lights of Hollywood, West Los Angeles, Beverly Hills, Inglewood, and half a dozen other suburbs. On beyond were Culver City and Sawtelle and in the extreme distance, strung along the edge of the Pacific, they could see the lights of Santa Monica, Ocean Park, and Venice. For a time the spectacle held them speechless. Angie slipped her arm through Howdy's. A moment later her fingers laced through his and he felt her shiver.

"You cold?"

"No. That did it."

She indicated the panorama of lights with a nod. "It must be the most beautiful sight in the world."

Angie saw only the beauty below them. She did not relate the twinkling points to homes still ablaze with party lights, to blinking signs above sleazy clubs, cafés, and restaurants or to the moving headlights of homeward-bound revelers.

Howdy saw only a sprawling city over which he was rapidly getting an edge. The smallest of the suburban clusters scattered below him was larger than Guymon, Oklahoma. If things kept going the way they were, there wouldn't be a person in any one of those houses that wouldn't know the name, Howdy Goodwyn. The same would go for the thousands of other cities and towns within reach of the station's powerful transmitter.

He was not aware that Angie was speaking until she gave his sleeve an impatient tug. "I said, What are you thinking about? Or is it none of my business?"

Howdy looked at her and smiled. "Sure it's your business, Angie. Why wouldn't it be? If you hadn't spoke for me with the Rinellis—"

" 'Spoken. ' "

"What?"

"If I hadn't *spoken* for you . . ."

She laughed apologetically and pressed her cheek against his shoulder. "I'm sorry. But sometimes those words hit me just like a sour note. Go ahead. I didn't mean to interrupt."

Howdy chuckled. "You taught me an awful lot about music, Angie. But I don't know if you can teach me your kind of English. I figure if folks can understand me, that's good enough."

Angie slipped her arm free. "They understand—but they'll respect you even more if you get rid of some of the worst of those back home-isms. They're not very classy, you know."

Howdy studied her for a moment then turned back to

447

the panorama of lights. "I'm trying—like I told you I would."

Something in his tone made Angie uncomfortable. "Forgive me, Howdy. Here we are in a perfect place after a perfect night and I'm sounding like a prissy old school teacher."

When he did not respond she reached over and turned his chin toward her with a forefinger. "Forgive? Please—?"

For a long moment he looked down into the appealing face and then they were together, Angie so eagerly that her lips and caressing fingers brought a muffled protest. Tears that had welled up an instant before trailed down her cheek and across the back of his left hand. When he felt them he eased her away, questioning but she freed her head and pressed it hard against his chest. "Oh, darling— I'm sorry. I'm sorry. But I thought it would never happen."

Renée's hands pressed upward frantically against Howdy's lean, hard shoulders. "Hey! For God's sake, what's the matter with you?" His teeth found the soft fold of flesh at the base of her neck and she cried out again. "For Christ's sake—what are you trying to do, screw me or kill me? You'll split me open, you goddammed savage! Howdy! Howdy—"

Her plea died in a long, anguished moan as she felt him start and clamped her full legs around the small of his back to share with him a pain more to her liking.

Later, after she had done what she could to ease her own discomfort, she returned and slammed the wet towel onto his relaxed belly. "What the hell got into you tonight?"

"Nothing got into me. I got into you. Remember?"

Renée snorted. "Look. This may come as a big surprise, buster, but you're not the first man I've known. So don't crap your Aunt Renée. You weren't laying me. You were fucking some other dame—to get even for something. Right?"

Howdy shook his head. "Wrong."

Renée eyed him skeptically and started to sit on the bed. The edge buckled and she gave his hip a shove. "Move your cracker-ass!"

When he complied, she inspected him unhappily and using the towel still wadded on his belly, absently finished

448

the task he had left uncompleted. Viciously, a long polished nail flicked against the offending part. His pained outcry made her look apprehensively toward the living room. An instant later she yelped as he flicked her protruding left nipple. Before she could defend herself he had pulled her on top of him and locked her in a suffocating embrace. When her fingernails threatened to draw blood he released her and she rolled over beside him. For a time they lay panting, their faces inches apart. Renée broke the silence and her voice was strange. "I'm a professional whore—and you play with me like I was your girl-friend. How come?"

Howdy yawned. "Don't know, Reeny—I feel easy with you."

She pulled away and sat up. "Get out of here now. I've got to sit on an ice bag or take the night off."

She started to get up and Howdy caught the back of her wrapper.

"Hey. I'm hungry. Why don't we both go out?"

Renée jerked the garment free and turned on him. "You stupid yokel. If you start running around this town with me you'll be off the air and back on that freight train faster than Sam Hayes can say, 'That's thirty for tonight!' Now scram! Get out of here and leave me *alone* for a while, will you?"

Dave Layne was still at the desk working on a Lincoln's Birthday program when Howdy let himself in. He looked up and whistled softly. "Boy, you sure look relaxed! What did you do—get your gonads emptied?"

Smiling, Howdy collapsed in an easy chair and draped a leg over the arm. "Pumped dry—right down to the gravel."

Dave shook his head unhappily. "I told you to have the girl come here. Just let me know and I'll fade out for a hamburger, like you do for me. I don't think it's a good idea for you to be seen going into that place."

They sat in silence for a time, then Dave held up a memo slip. "I don't suppose this is the most appropriate time to tell you that Angie called about an hour ago? She wants you to phone her at the hotel."

Howdy looked at his wrist watch. "It's too late now."

She said any time you came in would be okay."

Clearly, the message annoyed Howdy. Dave studied him

449

briefly, pushed the papers aside, and got up. "Howdy, is this a good time to talk to you, about Angie?"

The singer looked up slowly and shrugged. "Sure. Good a time as any. Go ahead, only—"

"Only what?"

"Only don't start pacing like a caged cat. Just come right out with it."

Half defiantly, Dave began. "Look, Howdy. I'm your friend, not your keeper. We both understand that. So what I'm about to say has nothing to do with you or me personally—but only with the show and our business plans."

Howdy pulled himself erect. "You trying to tell me you want to can Angie?"

"Hell, no! Don't you read the mail? She's getting to be one of the best attractions on the show. What I'm trying to say is—everyone connected with the show—and some who aren't—knows the girl is nutty in love with you. And they know something else, too—"

Howdy's eyes challenged him. "What?"

"They know that you don't give a two-bit shit about her that way."

"That is a fucking lie! I never did think as much of any woman as Angie. And that's the by-God truth."

Dave waved his arms helplessly. "Then why don't you let *her* know—even just a little? Goddammit, Howdy—she stays in town—she comes over here and cooks for us—she tidies up the place—buys curtains—even buys flowers—for *us*—"

He broke off, frustrated by words that wouldn't come. "Why, Holy Mother! I'd get out of here and leave you two alone—but if I did, you'd probably walk out with me."

Howdy relaxed and the promise of a smile played at the corners of his mouth. "Dave, old buddy, I been wondering how long it would be before you and me got around to this, so lemme just tell you how I feel about things. Okay?"

Dave Layne braced himself against the desk. "Okay."

"I know how her hair feels—I know how she smells—I know how her mouth feels—and I know how her body feels—"

Pretending he had not, Howdy saw Dave stiffen. "—and I know that she loves me—she told me so—early New Year's morning—up on the mountain where she asked me

450

to take her. We stayed up there in the car until almost light—talking," he broke off and his eyes appealed for understanding, "and I could have done anything I wanted with her—right then. She didn't say so—but I just could . . ." He compressed his lips and shook his head. "But I didn't . . . I couldn't. I'm a son-of-a-buck if I know why . . . but I couldn't. Why, Goddamn—I slept with a woman back home that I respected. It didn't stop me. But with Angie it's different."

A thought remained suspended in silence and Dave Layne, his voice tight, prompted him. "Why?"

"I don't know!"

"Because you think of her as a sister?"

If Howdy had not been deafened by the clamor of his own conflict he would have detected the note of hope in Dave Layne's question. "Sister? Crap, no! Boy, I'll tell you something—when she come into my arms that night and kissed me with her mouth all loose I rassled with the Devil to keep from spreading her on the seat right then and there."

Dave was unable to conceal a pained grimace.

"But, like I said, I couldn't do it . . ."

"Because you respect her too much?"

Howdy deliberated, then looked up. "That's right. She's wonderful, Dave." He vented his frustration in an explosive sigh. "Oh, shit. I don't know why."

"I know why, Howdy."

"Sure you do. Sure you do."

"I do. I'm serious. You like the girl and you're annoyed with yourself because it gets in your way right now—and you've got too many other things on your mind that are more important. Right?"

Howdy jumped up from the chair with jaw out-thrust. "You're fucking well right! We're just getting a good start now—thanks to you." Howdy began pacing between the chair and the desk. "I know Angie loves me—but I don't know if I love her. I don't know what love feels like. So far I been able to get along without it. But I can't live without singing, and money, and what money gets me. I *know* that."

He turned to face Dave. "And I know something else too—I know I can't even think about love until we get everything we planned done. Angie can go back home if she

451

doesn't like it. We don't have to waste any more time discussing it because that's how it's going to be."

Dave stared at Howdy. "Jesus, Mary, and Joseph! And I was the idiot who suggested she join the show . . ."

Howdy crossed to the desk. "Dave, if you hadn't I would of."

A year slipped by. Howdy Goodwyn was too absorbed in achieving his own objectives to be aware of the national travail. He knew about President Franklin Roosevelt's "Brain Trust" and that it was radically altering the old order because Ev Moses was concerned about the possible effects on the radio business. The original Federal Radio Commission had been replaced with a seven-man Federal Communications Commission that promised far more stringent rulings about programing and commercials to protect the public interest.

When John Dillinger was shot to death outside a Chicago theater on July twenty-second the news reached the Pacific Coast while Howdy and his troupe were entertaining at a rodeo in Bakersfield. The announcement over the public address system brought a medley of cheers and boos from the stands.

In September, when the Morro Castle fire cost one hundred thirty-four lives off the New Jersey coast, Howdy joined the others in deploring the tragedy, and when three Army balloonists set a new altitude record of sixty thousand six hundred thirteen feet over Rapid City, South Dakota, Howdy was rebuked by Angie for saying, "That's almost as high as old Dave was last Saturday night." Only the continuing dust storms in Oklahoma, Texas, Arkansas, and in the middle west seemed to trouble him seriously. "Son of a bitch, I think God must have it in for us Sooners."

Then, when the full force of the Okie migration began to spill over the state line into California, Howdy reacted positively. "Good. Pretty quick we'll outnumber the tight-assed bastards that run this state. Dave—find out where my people are camping and let's give them a show to cheer them up."

Howdy's new La Salle sedan was loaded with cases of canned food, coffee, and candy bars. They entertained where they could and the stories they heard brought tears to Angie's eyes and a bitter set to Howdy's mouth. In an

452

unplanned comment, he vented his outrage on the air: "Seems to me that a country that can find money to send a flock of seaplanes flying from here to Hawaii and set up post offices at the south pole could find a few dollars to help dustbowl farmers."

Later, Ev Moses cautioned him. "God knows, Howdy, we understand how you feel. But don't get into that problem on the show. The sponsors are raising hell. And another thing—if you start championing causes you won't be able to turn any of them down."

Dave Layne concurred. "Keep out of controversy. Our job is to entertain. That's the best way we can help everybody."

When Howdy came to understand that his emotional involvement with the Okies, displaced by long years of soil abuse, could jeopardize everything he was trying to build, he contented himself with unpublicized appearances, with modest gifts of money and generous gifts of food. Angie, matching her own interest to his, worked beside him. Before Howdy realized what was happening, they became identified as a "couple."

Between Christmas and the New Year Howdy took the entire show to Barstow to entertain at the Roadside Inn. The decision was made when Frank Persons called to say that Mike Rinelli had suffered a heart attack.

After the party, Rosa clung to Howdy in a mute effort to convey the fullness of her gratitude.

Angie's Christmas present from Howdy and Dave was a new contract for one hundred dollars a week. Shortly afterward, Ana rented the old house in Barstow and moved into a Hollywood cottage with Angie. Aside from a few personal belongings, the only furniture taken to the new home was the grand piano. Before long Howdy and Dave found themselves eating several meals a week at the Woodwards' place. And quite often, without premeditation, they also found themselves involved in building shows with Angie helping at the piano.

Twice a week for almost a year Angie had been investing part of her earnings with a vocal coach. Dave had been right. Angie was a natural. Her first duets with Howdy and as a soloist on the show proved the point. Confirmation had come immediately in the fan mail. Very soon the duets with Howdy were established as a regular feature.

By early autumn six duets had been cut on the HG Record label. On-the-air promotion sold them out all over the West. Such new songs as "When I Grow Too Old to Dream," "Red Sails in the Sunset," and "East of the Sun and West of the Moon" were instant hits for them.

The Golden Westerners scored a regional triumph with their version of "Tumblin' Tumbleweeds" and Steve Stone's harmonica arrangement of the nostalgic ballad, "I'll See You in My Dreams" brought ovations and shouted demands for encores. The Howdy Goodwyn radio family was growing and prospering.

The longest new step ahead, the one that filled Howdy with the most misgivings, came early in 1938. Marty Wolfson, an engaging energetic fellow in his forties, owned a chain of independant movie theaters in the Northwest. He was an old friend of Harry Liebowitz, the clothier to whom Howdy remained loyal. Liebowitz arranged to have Marty and Howdy meet at his store but the meeting was to appear coincidental.

Wolfson's conversation was casual. He complained about current movie making. "The trouble is they're going to keep on making *Golddiggers, Broadway Melodies,* and *Big Broadcasts* until nineteen thirty-nine. All we're getting is Winnie Lightner, Eleanor Powell, Ruby Keeler, Dick Powell, Jack Benny, Fred Astaire, Ginger Rogers, Deanna Durbin, Shirley Temple, Bing Crosby, and the rest. They're great in cities but we need more product for our kind of audiences. They like westerns and they'd eat up western musicals. The outfit who starts making them is going to have a gold mine."

Howdy, apparently preoccupied with the fitting, nodded mechanically. "Maybe. But where can you show westerns with a lot of singing and dancing—Texas, Oklahoma, Arkansas, California, Washington? There isn't enough audience to make it worthwhile."

"That's the argument I've been getting from the majors. But they're asleep, like they are when they say their stars going on radio hurts box office." Wolfson pulled a folded paper from his coat.

"Take a look at these figures, Mr. Goodwyn. I'm in the theater business. I can't afford to kid myself about what my audiences want even if the big studios think they can."

He interpreted the figures as Howdy glanced at them

454

casually. "In forty of the forty-eight states most people outside the cities like country entertainment. Look at the *National Barn Dance*. And your show. The group I represent has put up hard cash to finance a couple of western musicals to prove the point. Most of the names we've approached don't have the imagination to see the potential. It's not classy enough for them. If Dick Powell put on a ten-gallon hat and Levis and made a western musical the sales would break records." He wagged his head mournfully. "But they won't let him! All they want him to do is stand in front of a bunch of wet bathing beauties and sing, 'By a Waterfall.'"

Wolfson reached for the paper. "I'm sorry, Mr. Goodwyn, I didn't mean to bother you with my troubles. Someday I'll be able to sell this idea."

When Marty Wolfson left after making a minor purchase, Howdy walked up behind Harry Liebowitz, who was on his knees measuring a trouser length. "Thanks, Harry, for plotting that accidental meeting."

Harry looked up, all innocence. "So what are you talking about?"

Howdy's reply was an unblinking stare and a disconcerting smile. Harry maintained the deception briefly then broke into a sheepish grin. "Okay, so I'm a fixer."

He rose and held a stubby finger under his star customer's nose. "But don't laugh, Howdy! I wasn't wrong about you and I'm not wrong about Marty. What's it gonna cost to talk to him?"

Howdy gave Harry's arm an affectionate squeeze. "... a nickel, if *I* call him. Ask him to call Dave in the morning."

It took Ev Moses and Dave Layne a week to work out a preliminary agreement with Marty Wolfson. Wolfson's group would provide fifty thousand dollars worth of financing in return for distribution rights. Howdy and Dave would form a corporation—the Rolling-O Corporation—that would sign a five-year agreement with Wolfson's company, Northwest, and Howdy would agree to appear personally in two out of every five musical productions. Choice of stories and stars to be vested in Rolling-O.

At the end of that week, Dave Layne came in with a contract for Howdy and the Golden Westerners to appear at the National Orange Fair in San Bernardino. In addition there was a bid for the cast to appear at the Roundup

455

in Pendleton, Oregon, during the coming summer with the provision that Howdy would also serve as Grand Marshal of the parade.

"Son of a bitch! This on top of the pictures." Howdy tossed the contract across the desk to Dave. "This means I haul my ass up on a horse. Does that have to be part of the deal?"

"Thirty-five hundred dollars and all expenses for five songs and a quarter-mile jog around the track. What do you say?"

Howdy jammed his hands into his pants pockets. "I say for thirty-five hundred bucks plus expenses I haul my ass up on a horse."

Early in March Howdy received a telephone call from Georgette Follansby. "I hate to bother you. I know how busy you are—but could I come and see you this afternoon? I'm in the neighborhood."

Several minutes later the boarding-house keeper met Howdy at the apartment. She was distraught. "I don't know what else to do. The old man's just plain down and out. I don't know what happened to the income he's supposed to have, but I know his daughter's husband has been out of work for a long time. She hasn't been able to send money for two months now. I've been carrying him. But I just can't any longer. I'm barely making ends meet myself."

For a time Howdy seemed to be studying the papers on his desk. "What about relief?"

Mrs. Follansby shook her head. "He's not eligible . . . some technicality. I called and asked. He's got so darned much pride he wouldn't do it for himself." She spread her arms. "How can I turn him out on the street? I even offered to borrow money and pay his bus fare back to his daughter but he wouldn't hear of it. I just don't know what to do, Mr. Goodwyn. I know it's none of your business but you're the only one . . ." The words trailed off.

Howdy rose from the chair. "Where is he now?"

"At the house. He just sits in his room."

"Okay. You go on back and tell him I want to see him. Ask him if he can come here just before noon tomorrow."

The change in Neville Newton shocked Howdy. The durable optimism that had made the old man seem younger than his years was gone. Howdy made an effort to greet

456

him optimistically. "What I want to know is—why haven't I seen you all year?"

The old man kept his eyes averted. "You're a big star now, son." For the first time since he had known him, there was bitterness in the old man's voice. "We're pretty stupid, most of us. If we weren't we'd understand that the possibility of success should be more frightening than the possibility of failure."

Howdy, envisioning Al Phelps, smiled ruefully. "That's the damn truth. Every time I think I need a larger Stetson I try to remember the time I got a hobo's boot in my face."

Exploring the still visible tracery of scars along his left temple, he asked, "What are you figuring on doing now?"

The old man shrugged. "For a while I kept telling myself that I'd find some voice or speech coaching at the studios. But they don't want an old has-been . . ." He broke off and lifted his head and looked Howdy directly in the eye. There was no self-pity in his voice. "I don't know what I'm going to do, son, I just don't know."

Howdy nodded. "When you heard I wanted to see you, did you think maybe I could help?"

Neville Newton replied without hesitation. "I could try to save face and say, 'No, it didn't occur to me.' But it did, Howdy. I'm ashamed that Mrs. Follansby had to speak to you. I thought about asking you to see me but I didn't have courage enough to come begging."

There was more than a trace of satisfaction in Howdy's smile. "You wouldn't call it begging if you come to offer your services, would you?"

"Of course not, son. But what on earth could I do for you?"

"The first thing you can do is get paid up with Mrs. Follansby. Then you come back here and start teaching me how to talk the English language better'n I do. Angie says I got everything but class." Howdy reached into his pocket.

The old man stared blankly at the proffered bills. His mouth slacked open and his lips began working in a futile attempt to form words. Howdy spared him the need. "If you think you can do anything to change a hopeless mess like me, then you got a job for thirty-five bucks a week for as long as it takes. How does that strike you?"

457

forced his eyes back up to Howdy again. They were oddly glazed. "Son—I think it would be a mistake to change you very much."

When Angie heard about the addition of Neville Newton to the staff from Dave, she waited until they were alone then hugged Howdy and rested her cheek against his chest. "You are a most surprising man. Just when I think you're made of ice you go out of your way to do something like this."

She looked up at him in the way that made Howdy want to take her face in his hands. "I love you for doing that for the old man."

She went up on her toes to kiss him. When their lips parted he held her away and grinned down at her. "I done it—I mean, 'I did it'—for you so you wouldn't have to go around holding your ears when I say more'n ten words. I'll get me some class if it kills me."

Angie looked into his eyes and spoke in a half whisper. "You've already got class, Mr. Goodwyn, but you're a faker. You're really as sentimental as a lace valentine."

Neville Newton tried hard to earn his money by disciplining Howdy to regular sessions. Inevitably the phone rang or Dave called him to discuss an urgent matter. In the end the old man was reduced to following Howdy around and recording in a notebook the worst of the singer's lapses. At odd moments he would try to point out the errors and correct them. Though Howdy's patience was often tried by the old man's persistence, he was never short with him.

Early in April, Rolling-O Productions got underway with its first western musical. It was completed in ten days and was moderately successful. Preview comment cards proved there was no question about the popularity of Howdy and Angie. People wanted "the songs Howdy and Angie sing on the radio show."

This led to the start of the HG Music Company, organized at Ev Moses' suggestion to avoid paying huge music royalties to other publishers. The company would hire its own song writers. Howdy and Angie would push their material on the radio show and use it later in the motion pictures. The increase in staff made it necessary to take new offices and the apartment became a deserted place used only for sleeping and occasional love-making with professional call girls. Howdy had taken Dave's advice and allowed Renée to come to the apartment. But he wasn't happy about it. He told David, "I don't like sleeping where I've been fucking."

David's reply had not amused him. "That could create a hell of a problem when you get married, buddy."

The second picture, openly exploited on the radio show, was a hit. Rolling-O Productions expanded. Dave urgently recommended hiring an executive producer and once again Liebowitz had a suggestion. "Look, Howdy, a sweetheart you're not looking for. Only a son-of-a-bitch can run a production company like that. And who's a bigger one than Bronc Bronston?"

It took Harry's logic a week to triumph over Howdy's prejudice. Bronston had been out of work for almost a year and pride had kept him from coming to the extra he had once fired. When they met, Bronston smiled crookedly. "Okay, Mr. Goodwyn, it's your turn now."

From the beginning Bronston ran an efficient operation. One of the pictures was based on Howdy's early career and filmed in Barstow. To save the Persons from hurt feelings, Howdy and Dave stayed with them during the shooting. On their last night, from his cot on the porch, Howdy watched a new street light cast a vivid moving tracery of cottonwood limbs on the wall. He was disturbed by a feeling that the clock had turned back to August of 1933 and nothing had changed. Dave interrupted his reverie. "We've got five pictures working for us. I predict this will be the biggest grosser of them all. We've got to find something to do with our money. I never thought I'd say this, but we're suffering from an embarrassment of riches."

Dave spelled out their sources of income ". . . and the radio show, bringing in over two thousand a week. . . . Howdy, you can afford about any damn thing you want right now."

At breakfast the next morning with the Persons, Howdy surprised Frank. "With your license, can you buy real estate anywhere in California?"

Something in Howdy's voice made Frank begin to hope again. "Sure, I'm licensed in the whole state. Why?"

"Just thinking."

Despite Melba's warning look, Frank began talking about the questionable land outside the city limits.

459

Howdy stopped him. "I don't want any land this far away. But if I do decide to buy land someday, maybe you could handle it."

After the picture was finished Howdy brought the subject up to Ev Moses. Ev was emphatic. "There has never been a better time to buy. There is good distress land close in. Use cash and you can probably get it for half its present market value."

In November, Howdy made his first purchase, a five-acre peach orchard in the San Fernando Valley. The land lay close to the main state thoroughfare to downtown Los Angeles. Frank Persons handled the transaction.

The year had been a good one for Rolling-O Productions. *The Guy From Guymon*, the movie based on Howdy's career, was a box-office success and a turning point for old Neville Newton. He was given a small part and although he had never been a professional actor, he had a flare for comedy. As a result of his performance, "Judge Habeas Q. Corpus" would become a regular in Howdy's and Angie's musical westerns. The old man's ecstasy in his late-in-coming success was pathetic to witness.

Don Bellew, the publicity man for Rolling-O Productions, was responsible for a three-page photo story in a national magazine featuring Howdy and Angie and their "radio and screen sweetheart relationship." As a result, a number of fan magazines and columns began conjecturing about the "*real* relationship." Don Bellew, quite honestly, denied any responsibility for the rumors and innuendo.

"I don't give a shit how you do it," Howdy exploded, "but I want those lies stopped. They're trying to dirty us up. If you can't get them stopped, then by God I'll find me a man that can."

Several weeks later Bellew was fired when in desperation he attempted to stop the rumors of an illicit relationship between Howdy and Angie by leaking to a nationally syndicated woman gossip columnist an exclusive story that the co-stars were secretly engaged.

In the meeting that followed the firing, Ev Moses summed up the situation to Howdy and Dave. "You can't deny the story. If you do you'll just confirm it. For now, play it cool. Keep smiling and don't say anything. As long as the people think you two are secretly engaged they'll read a lot more into your relationship on the air and on the screen than there really is."

After a long, unhappy silence he added, "Of course, if Angie were to actually fall for somebody else and get engaged or married, that would be another kettle of fish altogether."

Dave Layne's grim rejoinder said it all. "That, my friends, will be the day."

Renée raised up on an elbow and let her breasts, grown still heavier, spill over Howdy's naked chest. When he refused to look her in the eye, she reached down and gave him a painful jerk. "Cheer up, lover boy. Either marry her or fire her. What's so tough about that?"

In the end, it was Angie who brought up the problem. They had appeared together at a hospital benefit. On the way home they stopped at a drive-in for sandwiches and malts. Angie twisted on the seat to face him. "Let's talk about it, Howdy. Most of it's my fault, you know."

"There's nothing to talk about, Bellew's gone. He was a damned fool. Things'll die down."

Angie sighed impatiently and rested a hand on Howdy's shoulder. "I don't care about those lies, honey, but the more things we do together the more people are going to wonder about us. If we just go on like we have been, the gossips will say they were right. And if we don't go on the way we are—then—well—"

"Well, what?"

Her hands flopped down to her lap. "Nothing, really. In a situation like this the girl doesn't have any alternatives." Angie straightened and let her head fall back against the seat. She could feel Howdy eyeing her and she waited for the question she had made inevitable. It came quickly.

"Angie, I wish you'd talk straight. You're talking in circles and I'm not bright enough to follow you. Just tell me what's bothering you."

"Not here, Howdy. I didn't mean to bring it up here but I do want to talk to you. Could we go up to the mountain again, where we first talked?"

Twenty minutes later, the car pulled off Mulholland Drive and stopped on the crest of a hill worn barren by countless tires. Outwardly composed, Angie curled a leg beneath her and leaned against the door. "Howdy, I don't want to go over everything between us. I helped you a little bit and for one reason or another, perhaps gratitude, you helped me a lot. But something else has happened and

461

that is, I've become a problem . . . because I went and fell in love with you and had the shameless gall to tell you so."

She turned away to look out over the panorama of lights diffused beneath the thin blanket of low fog. When she turned back, Howdy saw the determined set of her jaw. "When this next contract cycle ends, Howdy, I'm leaving the show. You can make it sound like I'm just taking a vacation, that I'll be back. But I won't. I'm not being heroic. You don't need me. I've screwed up enough courage to say this tonight, so please just accept it. I'm more grateful to you than I can ever say. I always will be. You've never made any demands on me but I've made a lot on you. I told you how I felt," her arm swept across the horizon, "right here, where we are now . . . and I took a dirty lowdown sneaky female advantage of you. I'm sorry about that."

She paused and extended her hand. "Forgive me?" An instant later, Howdy's hand shot out and clamped around her wrist. He pulled her to him so roughly that her half reclining body slammed against his. His right arm locked her to him and she could feel the fingers of his left hand working deep into the thick tangle of her hair. And then his mouth smashed against hers so viciously that his teeth bruised her lips and brought a muffled cry. As eagerly, her arms moved around him as his hand cradled the back of her head.

"Little Angie—you're my luck, honey. You ain't goin' no place. I need you. Hear me?"

16 No ADVANCE ANNOUNCEMENT was made. Howdy and Angie were married in November in a quiet ceremony at the church in Barstow. Dave Layne stood up for the groom and Frank Persons gave the bride away. A small luncheon reception for family and old friends was held at the Roadside Inn. In mid-afternoon Dave drove the couple to the San Bernardino airport where a chartered four-place Beechcraft was waiting to take them to Palm Springs.

Dave had made the arrangements for everything—the wedding, the honeymoon and the Hollywood apartment. The latter was a large new place on Hollywood Boulevard at the east end of the county territory known familiarly as The Strip. It lacked nothing, including a Steinway grand piano.

In a private cottage at the Desert Inn, they were surrounded by evidence of Dave Layne's thoughtfulness. The flowers, the fruit, the refrigerator filled with champagne, guest cards to fashionable La Quinta and a letter of introduction to the management of The Dunes, the gambling resort and restaurant south of the town.

Howdy and Angie, inconspicuous in a town frequented by the great and near-great of a dozen arts and industries, dined quietly at the Inn then strolled along the picturesque Palm Canyon Drive.

It was after nine when they walked back along Indian Avenue. The town and all that was in it seemed miniaturized by the overwhelming, snow-mantled presence of eleven-thousand-foot Mount San Jacinto. Dead-black against the luminescent night sky, it sheered straight up from the neo-Spanish winter homes.

As they walked through the gardens of the Desert Inn, Angie tugged at Howdy's arm to stop him. "I wish we could climb up there sometime and just look down on things."

Howdy reached out and turned her to him. "No more mountains for a while, Angie—please?"

They laughed and went inside.

Howdy skinned out of the cashmere pull-over, tossed it on a chair, and went to the refrigerator. "Mrs. Goodwyn—would you like some bubbly?"

"What year is it, Mr. Goodwyn . . . darling?"

Howdy frowned. "What year is it? It's nineteen thirty-eight. November eleventh. We got married today. Remember?"

Angie giggled. "I mean the champagne, silly. What's the year printed on the label?"

Howdy examined the bottle. "It says, 'Piper' something or other and the date is nineteen twenty-eight. Why?"

Angie shrugged. "Because you're supposed to ask. Anyway, that's a perfect year—if we drink it right now."

They sat diagonally across from one another. Howdy, dressed in a sport shirt and gray flannel slacks, had one leg slung over the arm of the easy chair. Angie, in a new silk negligee that covered her slip and bra, sat curled on the near corner of the sofa. The bottle and two glasses stood on the end of the coffee table between them.

After the second glass, she patted the sofa beside her. "If you don't come over here and prop me up, I may just keel over from happiness."

She held up the glass to watch the bubbles rising from the hollow stem. For no apparent reason she giggled. "That's what I feel like—bubbly."

When he settled beside her, Angie made an elaborate ritual of cuddling. "Nineteen thirty-eight is a good year too, darling. You'll never know how good."

She put the glass down and a moment later, in his arms, her fingers had undone the buttons on his shirt. Inside, her hand explored the smooth skin of his rib cage and moved up to the hard pectoral swell and the small, protruding nipple. She disengaged her lips long enough to tease, "You've got a goose bump, darling," then twisted her waist to ease the way for the hand that was groping for the hook on her bra.

They undressed each other on the sofa, scarcely parting their lips. Angie found him first. Her fingers explored his length and girth, then closed around him and her thumb began smoothing the lubricious secretion in small circles. When she felt him responding, she quickened and widened

464

the movement, sensing and seeking out the pleasure points until he began to protest. Her free hand urged him to find her and when he did, she encouraged the fumbling fingers and guided him into her. For a moment she lay motionless—feeling. The wanting, the imagining so short of reality came flooding against the long-built barrier of sublimation. Howdy felt Angie's body, flushed and moist, begin to convulse beneath him. He called out and tried to control her—called out again. Seconds later, unable to hold off, he succumbed to a racking orgasm.

Rigid, Angie lay beneath him, her arms locked around his rib cage, her senses congealed in the penultimate agony of frustration. When she began to stir he moved as though to leave her.

"No darling! No! Stay with me. Stay with me. It's all right, darling. You were wonderful. Just let me feel you."

Howdy lifted his face from the soft wedge of her neck and shoulder. He wore a stricken expression. "I'm sorry. I shouldn't of let you horse around so long. I just couldn't hold back."

Grasping him in place, Angie tried desperately to reassure him. "Its all right darling. You were wonderful. Really. You were wonderful."

Ever since Dave had chartered the Beechcraft for the honeymoon flight, Howdy had been toying with the idea of learning to fly. Over mild objections from both Angie and Dave, he took a trial flight. Howdy found his medium in the air. After six and a half hours of dual instruction at the Grand Central Air Terminal in Glendale, he made his solo flight. Six weeks later he qualified for a private license.

For good luck he bought the plane he had soloed in but in midsummer he traded it for a new four-place Stinson. Every hour he could spare was spent in the air. On two occasions Howdy and Angie flew to Barstow and landed on a dirt road just outside of town.

Frank Persons was as enthusiastic as a boy when Howdy took him up and let him handle the plane from the right hand seat. "Damn! If they'd only got these things perfected while I was young enough I might have joined the Lafayette Escadrille." But Frank's principal preoccupation was the great living relief map spread out below them. "You can see tomorrow from up here, Howdy. You can see tomorrow."

He pointed southwest to the ribbon of Union Pacific tracks snaking over Cajon Pass toward San Bernardino and Colton. "It doesn't take much to figure out how real estate gets valuable."

Off to the west they could see the barren summit of Mt. San Antonio. "Old Baldy," the natives called it. It was the peak whose snow-shrouded crest and slopes provided dramatic contrast to the myriad rows of manicured orange groves at its base.

In the distance random mile-square patches of pasture and rich grain and bean fields—some green and some fallow—stretched away to the low, ocherous hills bordering the sea. Beyond shone the haze-screened silver sheet of the Pacific. Dimly, they could make out the mountainous profiles of Santa Catalina and San Clemente Islands.

Northeast of the mountain range, the oven brown expanse of the Mojave Desert disappeared into the distant shimmering scrim of late summer heat. At eleven thousand feet, as they prepared to cross the San Gabriel Mountains, Frank Persons' breathing became labored. Howdy's concerned look brought an apologetic smile. "I'm glad we're riding, not hiking. Air's a little thin up here." A moment later the craft's nose dipped below the horizon as they began a slow descent.

Reassured by the older man's grateful nod, Howdy studied the terrain. No map could depict southern California's growth as graphically as the panorama below them. Certainly no book could recite the decade-by-decade development of the fertile plain reaching from coastal ranges to the sea. It was all there—the groves turned to towns, the towns turned to cities, the cities spreading their suburbs outward along new arterials to merge with neighboring cities—an amorphous metropolis-in-the-making that Howdy could put no name to. But he understood, instinctively, the nature of the miracle and the opportunities inherent in it.

Risking some turbulence, he crossed back to the desert side of the mountains by way of a low saddle and made a big circle toward the southeast. Coming in over town, he landed on a hard dirt street less than two blocks from the Persons' cottage.

466

On the return flight, Howdy's preoccupation troubled Angie. When a half hour had passed with little more than a token grunt or the briefest laconism she made no attempt to conceal her annoyance. Surprised, Howdy looked away from the controls. "Nothing's wrong, I was just thinking about Frank and why that promotion of his never had a chance."

The inference seemed unfair to Angie. "He got caught in the depression, Howdy—that's all."

"Nope, that isn't all. Frank didn't bother to figure out the facts and work with them. He worked backward. He bent the facts to fit his idea."

"Oh? How do you know?"

They had recrossed the mountains. Ahead lay the neat cross-hatch pattern of the pioneer community of San Bernardino.

"It's all spelled out down there. Everything's growing along the highways leading toward L.A. People are going to fill up the good land between here and L.A. and from the beaches in toward L.A. before they start moving out from the edges. Barstow's way out beyond the main edge now. It's on the wrong side of the mountains, too. Most folks'll want to stay in close to the city like we did." He sensed Angie was beginning to bristle with hometown loyalty. "I'm not knocking Barstow. It's just that in the San Fernando Valley the towns are spreading together already. So far nothing's spreading around Barstow except sagebrush and it won't until a hell of a lot more people settle in southern California or unless some big industry comes to the area. If the railroad didn't do it for the town, then nothing will, not for years."

Howdy's logic was too clear to admit reasonable rebuttal. Her hometown was out on the edge, the ragged, hot, windy, sometimes icy edge, with none of the allure that brought people to southern California. Southern California really *was* the land on the ocean side of the mountains. But someday that would change.

They flew westward into the wind. Angie wondered now at Howdy's unusual interest in the southern California growth pattern. As well as she thought she knew him— "I'm not sure anybody'll ever really know him," she had confided to her mother—she was unprepared for this new preoccupation. In a strange way she feared it.

In September, when Hitler's armies invaded Poland, Everett Moses began to stockpile product in the recording division. He presented, and had accepted by Howdy and Dave, an accelerated overall schedule. A research division was incorporated and staffed and by the end of the year, Meyer Laboratories had applied for patents on a system of turntables and pickups with new amplifying circuits that was efficient in truly portable equipment. Shortly afterward came a portable field radio telephone.

When the first Nazi bombers dropped their explosives on Britain the U.S. Army Signal Corps contracted for an experimental order of one hundred field units. This was followed by an order for five hundred portable play-back units to be used for audio training. Meyer Laboratories became one of the most profitable of the several H.G. Enterprises.

In February Howdy and Angie moved into their new ranch-style home built on the five-acre peach farm in the San Fernando Valley. Fan magazines covered every aspect of the showplace, the house, the swimming pool, the stables and tack room, the corrals, the one-acre peach orchard saved for family use and the private airstrip and hangar that ran along the back line of the acreage and extended into additional acreage that Howdy had acquired the following year.

In April, Angie, three months' pregnant, suffered a miscarriage after a riding sequence in a picture—a sequence that she had insisted upon doing herself. After two weeks in the hospital Angie was permitted to complete the shooting. When Hedda Hopper revealed the reason for her absence, Dave Layne was forced to hire three temporary secretaries to answer the thousands of notes of condolence and to acknowledge the dozens of crocheted bed jackets, home knitted sweaters, afghans and shawls and the "lucky baby clothes" that were sent to insure a successful second attempt to establish a family. After what Angie insisted was a "decent time" the clothing was given anonymously to local charities.

John Steinbeck's novel, *The Grapes of Wrath*, had attracted attention to the plight of itinerant families: the crop followers. The state legislature passed restrictive immigration legislation aimed at stopping the mass migration of dust-bowl immigrants, the so-called "bum blockade," and the farmers and ranchers, to whom the unexpected cheap labor pool was a boon, often obstructed real help by minimizing the problem in Sacramento through their lobbying organizations.

"We're taking good care of them. It's the goddamn commies that are riling them up," was the explanation most frequently offered. Howdy himself succumbed to the propaganda after he and Angie attended a rally in a neighborhood school one evening. During the meeting an agitator berated capitalism for an hour and they were shocked to discover that a huge Soviet hammer and sickle flag hung opposite the American flag.

Howdy summed up his disenchantment: "If· they're dumb enough to swallow that crap then they're too dumb to bother with. To hell with them."

For Angie, the edict brought both relief and disappointment. She watched the man who was the dearest person in her life, so often now a stranger, and realized that despite the western clothing and the more than residual speech mannerisms, Howard Dennis Goodwyn, the genial "Howdy"—star of radio, screen, and rodeo—had crossed a border and in doing so had renounced any real identity with the "back home folks." She understood also that her husband would continue to give lip service to his background, but it would be done deliberately out of promotional need rather than emotional necessity. To her his repudiation was, in its way, an act of "honesty" but the answer did not satisfy her. More and more there were areas in their relationship in which she, too, was becoming an outsider. She assumed her husband's disenchantment stemmed from an absolute lack of sympathy for any person who would not fight for what he wanted. She knew that Howdy felt that his own circumstances seven years earlier had been no more promising than the least fortunate Okie of Arkie who got dusted off his land. Howdy had often talked of those places, incredibly poor spreads, on which a man and his family thought of security only in terms of an unlikely prayer answered or a patient creditor at the store, a little stringy meat on the hoof, and a patch of dry-land hay.

"If he ain't bedridden, any man can do better if he'll just get up and go. Better to die trying than die crying," his father had said. That was the son's philosophy, too.

Angie knew what Howdy wanted and she knew what she wanted and the price of keeping what she had was helping her man get where he wanted to go. If getting there meant closing her eyes and echoing his "To hell with them!" that would be part of it too ... if she could forget the scrawny, shivering, snotty-nosed kids in tar-paper lean-tos. A gentle, unsolicited observation from Dave Layne had surprised her on the set one day, shortly after she returned to work: "It's tough to be married to two men, Angie."

Three hundred and thirty-seven thousand British, French, and Belgian troops were a part of the miracle of Dunkirk that took place between May 28 and June 4. The Royal Air Force's pitifully outnumbered, haywired Spitfires took on the German Luftwaffe and were killed at the rate of four to one. In southern California Douglas and Lockheed became the cornerstones of the free world's airpower. The state's immigration laws tumbled and the itinerant fruit pickers and stoop-crop harvesters traded clippers and hoes for rivet guns and wrenches and a lot of consciences eased and a lot of money flowed.

Rolling-O Productions completed two musicals and three straight westerns during the year. Meyer Laboratories won state and Federal approval to go public and two hundred thousand shares of common stock were issued.

At the initial stockholders' meeting, coming after Roosevelt had been elected President for a unprecedented third term and the Lend Lease act was signed into law, Howdy wore a dark blue chalk-stripe flannel business suit. Angie, straightening his four-in-hand tie, found herself forcing a smile. "From now on I'm going to call you 'Ty.' That's short for tycoon, darling."

Neville Newton, doggedly persisting in his frustrated attempt to keep his end of the bargain between parts in pictures, shook his head and made no effort to conceal his misgivings. "Behold. We attend the rising of an industrial colossus. You fairly reek of class now, son—until you joust with the King's English. But I warn you, in the Jonathan Club it would be better to drop your pants than your G's."

The stockholders were impressed by President Daniel Meyer and Secretary Everett Moses, who projected an upcoming clutch of new government contracts. Moses surmised the book value of Meyer stock would, within two years, increase to ten times its present market value.

Howdy surprised the officers of his companies by not

staying with the script. He indicated the printed reports that each person had been given as they entered the convention room at the Ambassador Hotel. "You can look at it. Study it. It's a pretty good job of work for something under six years. The only thing is—except for Meyer Labs—it's all entertainment."

Behind him on the platform Ev Moses leaned unobtrusively toward Dave Layne. "We didn't talk about this. What's he up to?"

The partner shrugged. "I don't know. Whatever it is, it's too late now. He keeps a lot to himself these days."

Assuming his professional mantle of humbleness, Howdy grinned shyly at the gathering. " 'Course, I don't mean to downgrade entertainment, because that's how this all started. I just want to assay it for what it's really worth in times like I'm sure we're coming to."

He half turned and directed an apologetic little smile at his officers. "If the boys behind me are fidgeting a little, it's because we didn't rehearse this part of the show."

He watched them until the ripple of laughter subsided. "I've always been scared of big business because I thought it was like high religion—something only preachers could understand. But thanks to Dave and Ev and others, I found out that good business is just good horse-sense."

There was an instant murmur of approval throughout the big room. Dave and Ev smiled dutifully too, and tried to conceal their growing apprehension. "So, after looking at what Dan Meyer has been able to do with his amplifiers and communications gadgets and knowing that the government will spend billions on lend-lease war material, I figured it would make sense if everybody who's got a stake in our operations started to look for ways that we can catch some of our own money on the first bounce and keep it at home."

Howdy paused and pretended to ponder his next words. "What I'm trying to say—and you stockholders are probably way ahead of me—is that we should welcome suggestions about other things we can do—like certain kinds of defense work—to help the war effort along. We can't build shipyards and steel mills and things. But we can build electrical equipment and I figure there must be a lot of real smart guys around that have other good ideas but don't have the money to get 'em going yet."

Howdy concluded his impromptu remarks with a re-

471

quest that stockholders submit the names of any small companies interested in products geared to the war effort who might welcome an assist from H.G. Enterprises. Although they were taken by surprise, in the end Dave Layne and Everett Moses could find no fault with the request. The attorney summed up his reaction? "What he's doing is telling the stockholders that we are flexible and progressive, that our credit can be put to work for them—if they'll go to work for us. It's not the usual way to handle things—but then Howdy's not the usual man."

When the meeting ended, an attractive man in his early forties approached the platform and extended a hand. "My name is Vance. Steven Vance, Mr. Goodwyn. Our family votes twelve hundred shares of Meyer Laboratories. It's reassuring to hear management ask stockholders to do something more than sign proxies. I'd like to make a date with you gentlemen to tell you about a small company I've been watching that has a tremendous potential if this war business spreads. I think it could be taken in under the Meyer umbrella or as another subsidiary. It has a government contract but needs financing and management. About an hour of summary would give you enough background to decide whether or not it's worth exploring."

A luncheon meeting was set up for the following week. Steven Vance turned out to be a man of considerable means, the direct descendant of the California pioneer family that still owned the huge Los Nidos Company. He explained that he had come across the Lucifer Flare Company in Santa Ana. "Louis Siefert has been running a little fireworks factory there for years. He had a Coast Guard contract to make emergency flares and last year he went to Washington and got a Navy contract for an advanced type. But he's a scientist. No business man. I considered buying in, but aside from looking after our ranching operation and teaching the children something about it, I don't want any managerial responsibility. I find this sort of investing more fun, frankly."

After two months of meetings, H.G. Enterprises acquired controlling interest in the Lucifer Flare Company, incorporated it, and installed Steven Vance on the board of directors.

In July, Howdy and Angie spent a weekend with the Vances at El Rancho de Los Nidos some sixty miles southeast of Los Angeles. The house, most of which incorporated the original Garcia-Perez adobe built by the peons and Indians in the 1840's, was a spacious one-story U-shaped California-Spanish structure with heavy tile roofing. It was situated on top of a large knoll bordering the Santa Ana Mountains. The property, almost ninety thousand acres of it, reached from the serrated mountain slopes to the shoreward lowlands whose southern boundaries were obscured in the coastal mist.

Nineteen-year-old Steve, Junior, and thirteen-year-old Jennifer became the Goodwyns' principal guides. While the father watched with transparent pride, young Steve explained the complicated operation of the ranch.

Jennifer, an olive-skinned, arrow slim, violet-eyed tomboy, wore her glossy black hair in double braided "Indian pigtails" and rode like a Comanche. At the corral Howdy and Angie watched her throw a hackamore on a half-broken cow pony, leap on bareback and drive the animal to a lather. The girl was fearless, but not reckless.

Steven Vance regarded his daughter with wonder. "I think she's a throwback to her paternal great-grandmother. My own mother told me that Grandmother Anita Lewis was one of the best horsewomen in Southern California. She used to scandalize all the proper ladies by riding astride with the *vaqueros*," he indicated the backlands, "over these same hills."

They watched Jennifer pull her mount to a dust-exploding stop, dismount, slip off the rope, and whack the subdued animal on its glistening rump.

"Sometimes Catherine is convinced that they got the babies mixed up at the hospital. But one thing *is* for sure—she's a lot harder to break than any cayuse in this corral."

Catherine Coulter Vance, a somber beauty who seemed to resist smiling, though she was properly cordial to her guests, nodded. "Steven Junior runs truer to my New England breed."

Although her husband understood the oft-repeated observation for the veiled reproach it was, Howdy and Angie took it for another manifestation of the usual mother's-pride-in-handsome-son syndrome.

Later, alone in the guest room, Howdy and Angie discussed the younger Vances. "I thought you avoided having to demonstrate your horsemanship very gracefully, darling."

Howdy balanced on one leg and tugged at the bottom

473

of his narrow frontier pants. "I don't know anything about clay feet. But I'd rather have one of them than a busted back."

Angie laughed and returned to the mirror. "That's all right, she adores you anyway. I think you've made another conquest."

After a reflective silence she turned to him again. "But I'm not so sure either of us made a hit with Catherine Vance. She's very nice, but a little too," she shrugged, "I don't know—a little *too nice*, I guess."

Howdy finished removing his trousers. "She's all right. Cold-assed—but all right."

A surpressed explosion of mirth came from the dressing alcove. "Good God! I hope nobody ever describes *me* that way!"

Howdy took a shade too long to reply. "Not a chance."

They breakfasted in the patio on Sunday. After much urging, Howdy got into a pair of young Steve's trunks and joined the others in the pool. Both he and Angie felt naked white in contrast to the glowing, sun-bronzed bodies of the Vance family. Howdy was uncomfortably self-conscious about his color and about his awkwardness in the water. He could swim with a thrashing sort of stroke that Angie described as a "drowning dog paddle."

The senior Steven Vance was an excellent athlete, still quite lean despite his forty-three years. The boy was a beautifully coordinated natural athlete who performed matter-of-factly with no trace of exhibitionism. Jennifer was equally at home in the water and, again, she was the unabashed show-off on the diving board. Catherine Vance neither dressed for swimming nor made any explanation. In a smart sun dress she presided over after-breakfast coffee, asked all the correct questions, listened to Angie's replies with apparent interest, and turned the conversation to books and the theater. When Angie confessed that she had little time to read because of rehearsals, Catherine recommended the new Martin Dies book, *Trojan Horses in America*, which warned of the subversive forces at work in the land. She also spoke highly of Thomas Wolfe's newest outpouring. *You Can't Go Home Again*. Angie fibbed and said she had heard fine things about both books and intended to read them.

They toured the principal units of the ranch and at four o'clock returned to the main house for a patio barbecue prepared by Rosita and Pablo, the Vances' Mexican house couple.

"They are absolutely devoted to us," Catherine explained. "They're such jewels. Their grandparents worked for Steven's grandparents so they are really family, you see."

Later in the afternoon, rolling leisurely through the citrus groves that lined most of the highway to Norwalk on the first leg of their two-and-a-half-hour drive home, Angie and Howdy compared notes.

"I like Steve Senior and those kids, but frankly, Howdy, even if business depends on it, I can't warm up to Catherine. We sat there smiling—or rather I did—talking about all those silly la-dee-da things. I was bored stiff. She made me feel like a yokel when she talked about books and plays: *The Male Animal, Separate Rooms, Johnny Belinda, Charley's Aunt!* She must have done a cram course the last time they were in New York."

"Guess we'll just have to go get our brains gussied up too. Why didn't you put her on about pictures and music?"

"I did talk about music. She's up on that too. I didn't exactly feel uncouth, but I sure as hell felt uncultured."

Howdy was amused. "I'm a country bumpkin from Guymon, but it doesn't bother me. If you want to get a quick culture job, why don't you go on back to New York this winter? Maybe you can catch some culture for both of us."

Angie put her left arm along the backrest so her fingers could trace affectionate little figures on Howdy's shoulder. "I'd love that, but not without you."

At 7:55 A.M. Hawaiian time, on December 7th, one hundred Japanese carrier-borne aircraft and a school of vest pocket submarines changed the plans of one hundred and thirty-two million Americans. In order to directly involve the United States in Europe and divert defensive forces from the Pacific, Italy and Germany joined Japan by declaring war. In a matter of weeks the Allies were engaged on a half dozen fronts in two hemispheres. The United States became the world's largest military hardware manufactory and training camp. California became one of its most vital elements. President Roosevelt and General Hershey began sending their official greetings to

475

ten million young Americans, among them Steven Vance, Junior, and Howard D. Goodwyn.

When the classification card arrived placing Howdy in 1A, Everett Moses undertook a series of legal maneuvers to keep the young star-businessman out of service. But Howdy had other plans.

Company pilot, Chad Chadwick, a major in the Army Air Force Reserve, was called up for duty with the Second Air Force and assigned to Lowry Field near Denver. He had been on duty for less than two weeks when Howdy received a call from the commanding officer offering him a position as civilian instructor. On February 10th, 1942, Howard D. Goodwyn received orders to report to the base for indoctrination by midnight, March 15th.

During the scant five weeks a multitude of decisions had to be made. The practical business of running H.G. Enterprises would be left with Dave Layne and Everett Moses. Bronc Bronston was elected a vice president. Meyer Laboratories and Lucifer Flare remained under the active management of Layne and Moses with Danny Meyer and Louis Siefert running the companies. None of these men understood Howdy's decision. The most persistent protests came from Everett Moses and it was to him that Howdy addressed the blunt explanation that silenced them all:

"I know that strings can be pulled to get me reclassified 2A or 4B. I'm an entertainer, not really a business executive or an official. H.G. Enterprises is first and foremost an entertainment operation. I'm it's chief product and I'm twenty-nine years old. I don't intend to stand up there looking healthy in a cowboy suit singing back-home songs to a couple of hundred thousand guys in uniforms. If I was too old or had one leg or one eye that'd be different.

"When this war is over I don't want anybody to remember that I took the easy way out. When I come home I know I can hold my head up and there'd better not be any crap from anybody. That's how it's going to be and I'll thank you to just take it that way and keep on doing a great job.

"Just in case I don't come back, everything's been arranged by you, Ev. But I'm coming back. And I'll tell you why. Because any kid that gets into an airplane and can't do what I tell him the first time is going to get busted out of the Army Air Force on his ass and put in the infantry."

476

At the going-away party after the last radio show with Howdy as star, Angie tried to convey her feelings to Melba Persons. "I don't know how to say it."

She rested moist palms beneath her rib cage. "I feel so empty in here. But it isn't really emptiness because I'm filled with fears, Melba. Awful fears."

The older woman who had been trying to reassure her reached over and placed her fingertips over Angie's lips. "Honey, let me tell you something about men like Howdy. They live in another world that has another set of rules. They make them. So when he tells you he's coming back you better believe him because it's true. He *will* come back."

"I know he's coming back, Melba. I feel that. It's something else. I don't know what it is."

The written examination arrived. For Howdy it required little more effort than the completion of a form. At Lowry Field, as one of six civilian instructors who received officers' pay, allowances, and privileges, he was assigned to Bachelor Officers' Quarters. In his element and oblivious of his fame as a show-business personality, Howdy Goodwyn immersed himself in the demanding routine of training cadets. Seven days a week, fourteen hours a day, Howdy drove his four charges to the limits of their endurance. Consistently they were in the top ten of the class. Most pleased of all was Major Chad Chadwick. The major summed up his staff's appraisal of Howard Goodwyn. "That drawling Okie son-of-a-bitch can pound more flying savvy down their throats, into their ears, and up their trembling asses than any flight instructor I've ever seen . . . and for this we pay him less money a month than his corporations make for him every sixty seconds. That's what I call dedicated service."

As well as he knew his former boss, Major Chadwick could be forgiven for underestimating the true nature of Howdy's dedication.

At the end of the first two months of training, during a bull session in the BOQ, Howdy found out, inadvertently, that Gene Autry had made plans to establish a contract training school for military pilots at Gilpin Field near Phoenix, Arizona. As he lay awake, too tired to sleep, Howdy went over the advantages of such an arrangement. Half aloud, he voiced his conclusion: "God love him, that Gene's really got an edge."

477

Two nights later, Howdy was sitting in the bar at the Brown Palace Hotel in Denver with Dave Layne and Ev Moses: "I want you to look into every angle of this flying-school deal. Find out what Autry and his partner had to do, then get back to me in one hell of a rush."

The shy smile that misled or concealed so much from those who didn't know him tugged at the corners of his mouth. "I figure if we had our own school we could pay me more than three hundred and twenty-six bucks a month, and turn out more good combat pilots one hell of a lot faster than I'm turning them out here. If we can do the job better, it's unpatriotic not to do it, isn't it?"

A week later, Ev Moses telephoned. "I have the figures, Howdy. There's no question about getting a contract with either the Army or the Navy and maybe both. We have to acquire the right land, get priorities on material and equipment, and satisfy Stimson and the War Department."

"What about financing?"

"We can get Federal funds."

"How much of our own dough do we need to get going?"

"Whatever the payment on the land turns out to be."

"You and Dave do any thinking about locations?"

"Not yet. But that won't be a problem. Good land's going begging. Incidentally, we keep hearing rumors that the Navy is going to establish a major air base for the Marines on the Irvine Ranch back of Santa Ana. Some place called El Toro."

"Isn't that near Vance?"

"Not too far."

There was a thoughtful pause. "Ev."

"Yes?"

"Get hold of Steve. Ask him if he can fly over here with you this coming weekend. And bring Angie, too."

Steven Vance, Everett Moses, and Angie flew to Denver the following Saturday morning. Bone-tired and still in his flying clothes, Howdy met them that evening at the hotel. Angie was shocked by his appearance. "Honey, you've lost *pounds*! You look awful."

Howdy grinned. "You ought to see the kids I'm teaching."

Angie slipped an arm around his waist and rested her

head on his shoulder. "I don't care about them. I bet you're working twice as hard as they are."

Howdy tousled her hair. "Four times'd be more like it. Come on. I need some coffee."

For two hours Howdy drove himself through the details of his plan, describing what he felt would constitute an ideal training installation in terms of location, hangar facilities, classrooms, commissary, quarters, flight line equipment, and aircraft. ". . . But the key to the whole shebang is location, Steve. What I want to know is, have you got a square mile of land on your ranch that I can buy? I'll lease it if I have to, but I'd rather buy it."

"We have a lot of square miles, Howdy, but I can't answer that offhand. The best bottom land, the level stuff, is under tenant lease on crop percentages."

The spectacle of the tidy mosaic of fields and groves spread before them as they stood on the terrace of the Vances' Casa Grande was as vivid in his memory as it had been when he was there. Howdy understood the problem he was posing. "I don't mean the best bean field you got, Steve. What about some of that fallow stuff lying next to the hills? That's still lease-free, isn't it?"

Steven Vance nodded. "We have about six sections along the edge of the hills. But that's no place to build an airport."

Howdy chuckled mirthlessly. "By the time I get through with these kids, if they make it, any place they can get a ship into will look like an airport. All I need is two strips of about three thousand feet with enough room for a tower, hangars, and offices."

Later, when Vance had promised to see what he could do and Howdy and Angie were alone, she sagged on the edge of the bed beside him. "Does this mean you'll be coming back home instead of me coming here?"

Howdy yawned and began unlacing a shoe. "It sure as hell does."

He kicked off the other shoe and collapsed backward with his legs dangling. "There's no point in you coming. You'd be sitting alone all day and half the night in some dump here in Denver. You stay put. We'll have our own operation going in ninety days. You and I can start living like white folks again."

Angie undressed him and all but led him to the shower.

Later, in bed with him for the first time in eight weeks, she cuddled his head in the crook of her arm and watched him sink into the deep sleep of the exhausted. It was well past midnight before she slept.

Steven Vance agreed to lease, with option to buy, two contiguous half sections. The strip ran for ten thousand feet along the base of the low hills.

Early in August, when Howdy's first four cadets graduated from Lowry as numbers one, two, four, and five, respectively, most of the paper work had been processed and assurances had been received from Washington that the field as planned would be approved. On December first, when Howdy's second group graduated, again in the top ten, his own release was expedited by Chad Chadwick, now a Lieutenant Colonel.

Once again Howdy and Angie spent Christmas with their adopted family. The celebration on Christmas Day at the ranch house in the San Fernando Valley was the happiest gathering in several years. Only one thing marred the holiday—the worsening condition of Neville Newton. With grim humor the old man sought to dismiss his ailment. "Son, if you will forgive my indelicacy, whether it's sick or well man's prostate gland is a pain in the ass. It would seem that it's now exacting full payment for the alleged pleasure it has given me during the past sixty years."

The ground-breaking was celebrated on January fifteenth and the first group of cadets reported for primary training on June 20. Five weeks later Benito Mussolini resigned as Premier of Italy following the Albanian and North African debacles.

On June twenty-first a race riot broke out in Detroit. When the toll of 34 dead and 700 injured had been tallied, Howdy glowered at the newspaper story then tossed it aside. "I can tell you right now who's behind it. The commies. They'll never miss a chance to stir up trouble inside the country and they'll always wait until we're in some other big trouble. Just like they stirred up the fruit pickers right in the middle of the depression. It isn't easy to be a nigger. But I sure wish they'd wise up. As bad as they got it, it's better here."

Ev Moses smiled sadly. "We have a long way to go, friend. A long way. Everybody's motto seems to be:

480

*'If you aren't the same as me,
then you are my en-em-y!' "*

Howdy shook his head. "It isn't just the color. They think different too."

Ev Moses braced himself against the chair back. "My brother was the only Jew in medical school in San Francisco. Some of the guys gave him a bad time because they thought he was 'different' until the doctor in charge of dissection straightened them out."

Concealing his surprise, Howdy listened intently as Moses continued. "The surgeon got three hearts from the cooler and laid them on the table. Then he pointed to them and said, 'One of these came from a Catholic, one from a Protestant, and one from a Jew. I'll give any student a passing grade in the course who can tell me positively which one is which."

The attorney laughed softly. "That settled that."

Without knowing why, Howdy was suddenly self-conscious. He tried to conceal it with an elaborate casualness. "Hey now, I didn't know you were a Jew, Ev."

Ev smiled sadly and shrugged. "I didn't either until I was six."

During a July Fourth exercise Lt. Colonel Chadwick and a Navy cadet, simulating a carrier take-off from a marked area, encountered a dust devil in their AT-6. The ship stalled, sideslipped into the ground from fifty feet, and crashed. The left wing absorbed most of the impact. The cadet went on to combat duty but Chad was given a medical discharge after an operation on his smashed left hip. Two months later Chad was placed in charge of operations at Vance Field.

Howdy's growing preoccupation with the flying school made it impossible for Angie to keep her apprehension to herself. He was so deeply involved he was seldom home. Loneliness drove Angie into confiding in Dave Layne.

Dave's own status in H.G. Enterprises had changed. He was still involved in the recording business and the film productions but the top-level business decisions had fallen on Ev Moses. Dave had also begun to feel a nameless uneasiness. To avoid spending long, idle days alone at home, Angie continued to come to the office and three

481

nights a week she helped out at the Hollywood Canteen but even so she found herself welcoming the impromptu cocktails with Dave and Ev. As time went on she found more often than not that her evenings were spent with Dave alone. Since neither had home obligations, the cocktail hour was often prolonged to include dinner at the Vine Street Derby.

Angie refused a third martini. "No more. I'll start getting little blue veins in my nose."

Dave looked at the waitress. "One more, and a couple of menus, please."

He turned back to Angie. "Is Howdy going to stay at the Vances' again tonight?"

The answer was a glum nod. "Yes. That'll make four nights this week. He might as well have his mail sent there."

Dave impaled the martini olive and ate it. "In a way it's a good thing he can stay there. I'd hate to make that drive twice a day."

"I know. I know. But that doesn't help much when I'm alone in the house."

"Why don't you go down and stay with him more than you do?"

"Two reasons. I can't impose, and I don't feel easy with Catherine Vance." She worried the doily. "There's another reason too, Dave. I don't want Howdy to think that I'm always under foot. I compromised my pride enough when I threw myself at him."

Her eyes challenged him. "I *did*, you know."

Dave shrugged. "You went after something you wanted. There's no law against that."

The waitress returned with the drinks and the menus. Several minutes passed in silence as they studied a bill of fare they both knew by heart. Finally Dave folded the menu.

"Steaks are back on tonight. Want to splurge?"

Angie nodded and recited by rote: "Medium. Baked potato. Sour cream. Cobb salad. Same old routine."

Dave frowned. "I think you need one more drink."

"One more drink and I'll start crying on your shoulder."

"Unless I beat you to it and start crying on yours."

Angie smiled crookedly. "No, Dave, you wouldn't, I know. I can depend on you."

482

"Dear old dependable Dave." He emphasized each word disparagingly. Before Angie could protest he held up a hand. "I know. I said it. You didn't. Actually, I don't mind the 'dear' or the 'dependable'—it's the 'old' that gets me."

His eyes rolled upward. "Forty. Holy Mother!" He shuddered. "A terrible thing happens to a man about to be forty, Angie. It's called 'middle age' and it scares the hell out of me."

Angie studied him sympathetically. "Dave . . ."

"Yes?"

"Why do you go on like this?"

"You mean—a bachelor? Well, for one thing my first wife walked out on me. For another, you know perfectly well that if I had met you before you met Howdy, I'd have given him one hell of a run for his money."

Angie rested a hand lightly on top of his. "I know that, Dave, and I hope you know how important it's been to me. But tell me something—why did she walk out? Do you know?"

"Not really. I loved her. I did everything I could for her . . . I thought we did everything well together."

Angie finished the thought. "But she didn't love you, and love doesn't work when it's a one-sided affair."

Dave Layne smiled cryptically.

"You loved a girl who didn't love you—and—" Angie broke off briefly. "—and I love a guy who doesn't really love me—and—"

Very gently Dave shook her shoulder. "Wait, Angie. Don't jump to conclusions about Howdy. Remember when I said that it was tough being married to two men?"

Angie nodded.

"I thought you understood what I meant then."

"I did, then."

"Well, it means the same thing now. When you say Howdy doesn't really love you, you're wrong. I suspect he loves you as much as he can love any yoman. But you're not in competition with something as simple as another woman. You're in competition with—oh hell, 'dream' is a corny word—but that's what he's in love with—his dream of who and what Howard Goodwyn is going to be."

Angie smiled sadly. "I know. I've known that for years. But I always thought I fitted into that picture."

"You did and I did as long as the radio show was the

483

big dream. But he's dreaming bigger now, and that's going to call for some understanding, and some compromises, too."

Angie looked at him sharply as he continued.

"That means that you, as his wife, and I as his partner, will have to accept the fact that our roles are less important and may get even less so as time goes by."

Angie sat rigidly while the truth of Dave's words hammered her insides to pulp.

Howdy called early the next morning and asked Angie to come down for the weekend. She protested but Howdy told her the Vances expected her and that he had to be at the field both Saturday and Sunday.

Angie arrived at El Rancho de los Nidos at mid-afternoon on Saturday. It was blustery and a cold dry Santa Ana wind streaming from the northeast drove Angie and Catherine Vance indoors.

Catherine served tea before the fireplace in the study. The masklike beauty of her face remained undisturbed as she spoke. For the first time, Angie began to sense the undercurrent of loneliness in the woman. But Catherine Coulter Vance let no one see what lay behind her impassiveness. Beyond saying that she had been born in New Canaan, Connecticut, and had met Steven on a ski holiday at Mount Mansfield in Vermont, she revealed little. Perhaps, Angie thought, there was a clue in her statement that Steven had really planned to make a career in finance in the East. From Howdy, Angie had learned that the accidental death of Steven's father, Thomas Vance, in 1922, three years after his own marriage to Catherine Coulter, had necessitated the return to California to share the management of the Los Nidos Ranch Company with his grandfather, Steven Lewis.

By nature Steven Vance was neither a rancher nor a gentleman farmer. Even so, after his grandfather's death in 1930 at age eighty-six, he could not bring himself to dispose of the huge holdings. The dilemma deepened for Steven and Catherine when the depression made it impossible for them to return East even for the prolonged visits that Catherine thought would make her California stay endurable. So they stayed on at Los Nidos, telling each other that perhaps in a year or so they could return East, until so much time had passed and Steven had become so deeply involved that a second major uprooting was impossible.

As they talked about trivial, impersonal things Angie began to understand the older woman's disappointment and some of her resistance melted. She made no effort to encourage Catherine to reciprocal confidences. To do that she would first have to admit and then define her own dilemma. For the time being all she could do was admit that for those who compromise, time can become a relentless trap.

Always unsure on the slippery surfaces of small talk, Angie was relieved when Jennifer came home from the Glenborough School for Girls in Los Angeles. She greeted Angie with a friendly wave and gave her mother a perfunctory kiss. "Where are Dad and Howdy? Still down at Vance Field?"

Catherine felt no need to answer the obvious. Instead she countered with a question of her own. "How did your examinations go, Jennifer?"

Jennifer made a distasteful grimace. "If I told you, Mother, you'd wash my mouth out with soap. The only subject I got A in was dressage."

Catherine raised her eyebrows. "I do not want to hear any of those earthy protests from you when your father orders you to start tutoring on weekends once more."

Jennifer curtsied and smiled with vicious sweetness. "Yes, Mummie dear. You shan't hear a single four-letter word out of me. But you may hear one heck of a yowl from the frustrated tutor." She turned and marched out.

Howdy and Steven returned from the field at six. Jennifer launched herself at her father and gave him an affectionate kiss. She hugged Howdy quickly and pushed him away at arm's length. "If you have any excuses, sir, I refuse to hear them. Tomorrow, sick or well, you take me up in one of the new Stearmans and wring it out. Right?"

Howdy extricated himself and tousled the girl's neatly combed pageboy bob. The way he did it, and the girl's playful protest, brought a stab of loneliness to Angie. Howdy didn't answer and Jennifer persisted. "Howdy Goodwyn, you promised." She backed off, eyes snapping. "I've forgiven you for hating horses—almost. But I'll never forgive you for breaking your word. You made a solemn promise to Stevie and me. And by God, sir—you'll keep

485

it." Her mood changed abruptly and she stepped close to him, wheedling. "Won't you? Please?"

Howdy looked helplessly at the others. "I'll keep my promise if the mechanics keep theirs. If they don't, you'll get a nice straight and level ride in an Interstate. But, if you don't stop being sassy, you might get grounded for the duration."

He took her shoulders, turned her around, and whacked her backside. Howling with mock indignation, Jennifer whirled on him. "I'm pretty old to do that to. I'm beginning to shape up a bit, in case you hadn't noticed."

She assumed an exaggerated pose with one hip jutted out provocatively and her small, well-shaped bosom thrust forward beneath the loose cashmere sweater.

Catherine Vance gasped. "Jennifer!"

The girl held the pose defiantly then stalked contemptuously to the big oak coffee table where she scooped up a handful of small Spanish peanuts. Child-like, she crammed them into her mouth and brushed away the salt. Steven Vance laughed indulgently and moved toward the sideboard bar. *"Amigos! Beberemos! Vamos a beber!* What will you have?"

Young Steven arrived in the midst of cocktails. He was dressed in Levis and boots and an old corduroy field jacket. After booming greetings and placing a quick kiss on the cheek that softened Catherine's face for a fleeting moment, he poured himself a Coke and pointed to the residue of mud and straw on his heels. "Fine thing! Future Rickenbacker mucking around in manure."

Howdy chuckled. "I don't know if it's any kind of a recommendation, Stevie, but that's how I got my start."

Catherine Vance smiled bleakly as the others laughed outright.

Dinner was planned around a magnificent standing rib roast butchered from one of the ranch's prize steers. Steven Vance was apologetic. "I feel a little unpatriotic, knowing that most of us have to save about three weeks worth of stamps for this." He paused, smiling guiltily. "But I don't intend to let that spoil my enjoyment."

The dinner conversation was good. Steven Vance was proud of his California heritage and he understood it well. "We don't have any rickety salt-cellar houses with DAR plaques in front of them saying, 'George Washington Slept Here,' and we don't have a Grant's Tomb or a Lincoln

Memorial or a Jefferson Memorial. But we do have a Mount Whitney and a Mount Lassen and a Mount Dana and a Frémont Peak. I'd call them rather imposing monuments to some remarkable men, wouldn't you?"

Catherine Vance protested. "You know, dear, I think I rather resent the implication that your Western monuments are more imposing and therefore more important than ours. It is a sort of *nouveau snobisme*, really."

Steven laughed. "You can take the girl out of New England but you can't take New England out of the girl."

"That's not true. I consider myself as much a Californian as you. It is just that I refuse to become parochial about it. It's very important to remember and honor the traditions of our forebears no matter where they may have come from. After all, your own great grandfather, John Lewis, came from a very good family in Middletown, Connecticut. He was a direct descendent of Welsh royalty."

Steven Vance looked at his wife in astonishment. "He was a rabble-rousing merchant sailor who jumped ship, and if he hadn't got all hot and bothered about the sixteen-year-old Mexican girl who was my great grandmother, he probably would have ended up a jailbird like his friends, Isaac Graham and Will Chard."

For the first time since Angie and Howdy had known her, Catherine Vance displayed something close to animation. "Your great grandmother, Maria Robles, was not a *Mexican*. She was a *Californio*—and there's a vast difference, as you should know. Her people came to Monterey from Burgos in Castille—*by way of* Mexico City."

Her husband winked at Howdy. "I defer to you, darling. But I prefer my version. It makes them all seem more colorful."

Catherine Vance was not amused. "You are quite colorful enough for the lot of them."

Jennifer, who had been listening with a vixenish smile, giggled. "You two remind me of the girls at school." She mimicked them. " 'My family came over on the Mayflower.' 'Oh, really? So did mine.'—'Pipe down you two. My family *sold* it to them!' "

She looked from one parent to the other. "I think the only thing that's important is that they got here, and married fertile women."

Young Steven applauded his sister. "Bravo. If that's

487

what they're teaching you, Pop's getting his money's worth."

"The only thing I've learned in girl's school, brother dear, is how bitchy females can be."

For a moment Catherine Vance's mouth clamped into a bloodless line. When she spoke her repressed outrage was evident. "Jennifer. You are to be excused from the table ... immediately."

In spite of herself a small sound of protest escaped from Angie. The girl returned her mother's chill stare then turned appealing eyes to her father. Steven Vance seemed to waver. Then he nodded almost imperceptibly and Jennifer understood that the edict had been sustained. Slowly, defiantly, she left the table and went to her room. Young Steven broke the strained silence.

"You're pretty rough on her, Mom. After all, she's no kid any more. And she's heard you use a word or two."

Catherine Vance glared at her son. "Never in front of strangers."

Howdy and Angie exchanged startled looks, then Howdy scratched his head and turned on his shy country boy smile. "Stranger? Well dawgone, Mrs. Vance," the drawl was exaggerated deliberately, "we're sure hoping to work up to being acquaintances, at least, right after the bookkeepers get done with the half-year profit statement."

Suspended somewhere between mortification and amusement, Steven Vance busied himself with his coffee. As quickly as it could be gracefully managed, he terminated the strained evening.

Sunday dawned bright and clear and Jennifer with it. She paced the family room like a caged cheetah until the others appeared for the buffet breakfast, then hovered over Howdy so persistently that he felt obliged to chide her. "Miss, one of the first things you learn about flying is not to take off until you get into the airplane."

Jennifer was ready with her own ingenious logic. "It's not that I'm impatient, Howdy. It's just that the weather report says 'variable cloudiness' and this time of year a clear morning can turn into a typhoon or something by ten o'clock!"

All but Catherine Vance arrived at the field shortly after nine. For twenty minutes Howdy wrung out the trainer, putting it through every aerobatic maneuver he and the ship were capable of. Before each one he called

488

through the gosport to the girl in the front cockpit to alert her. On the ground Angie, Steven Jr., and his father watched as the craft sequenced from power-on and power-off stalls through spins, a single loop, three linked-up loops, chandelles, snap rolls, and finished with a half loop with a roll out at the top.

The altimeter read six thousand feet when Howdy finally called through the gosport. "Had enough?"

Jennifer twisted in the harness to look back at him. Her face glowed and Howdy could see her eyes literally dancing behind the big goggles. She shook her head and her lips formed an emphatic No.

Howdy laughed. "How would you like to fly it?"

He headed the ship out across the crazy quilt of cultivated fields in the direction of the Newport-Balboa area. Jennifer followed his instructions to the letter. She made predictable control errors causing the ship to dip and meander but Howdy was forced to concede that in an unusually short time Jennifer Vance would probably be as much at home in the cockpit as in the saddle.

On the ground, Howdy smiled at Jennifer and gave her neck a playful squeeze. Watching, Angie was annoyed again by the little twinges of jealousy she always felt when Jennifer monopolized Howdy. She would not have admitted it but increasingly his frequent affectionate little gestures toward Jennifer had produced painful stabs. She could count on the fingers of one hand the times she herself had been the object of such gentle demonstrations.

On D-Day plus two Ev Moses called a general meeting of the executives, announcing that the recent good war news made it necessary for the company to have a reappraisal. They would have to prepare to phase out some operations or fit them into a peace-time economy.

None of the operations was in immediate jeopardy except the flight school. Chad Chadwick put it bluntly: "Vance Field will be a dead duck as soon as the war is over. We can look into civilian light plane sales and into surplus and salvage. But whatever happens, I don't think we should exercise our option to buy the Vance Field land."

On the way out of the meeting Ev Moses stopped Howdy. "Has Angie told you about Neville?"

"Nope. What's the trouble?"

"They found a malignancy in his prostate. He's got to have an operation."

After a brief silence Howdy nodded. "It figured. The old guy's been drawin' his cards from a stacked deck all his life. It's that way with some people."

The attorney's eyes narrowed inquisitively.

Howdy tapped his chest. "The old man had something I needed so I used it. Now it looks like we've got something he needs so let's make sure he's got enough money to pay for things from here on out."

"From here on out?"

"That's what I said. If those sawbones are right, they just gave the old boy his last piece of bad news. I want to make it easy for him, if there is such a thing."

When the attorney appeared reluctant, Howdy brushed past him to enter the room. "You work it out, Ev, with my dough."

Chadwick's recommendation to drop the option on the land made Steven Vance uneasy. He had hoped that Howdy would exercise the option for a sizeable chunk of cash and relieve some of the financial pressure that had been mounting ever since the depression.

Nineteen forty had seen a little easing. Crop control and allotments had helped meet interest payments and taxes. But the war-time freeze had caused both the Vance operations and those of their tenant farmers to slow to a virtual standstill. Some tenants' taxes were in arrears, taxes that the main ranch was obligated to pay to avoid a lien. Steven Vance could see little to be hopeful about even after the lifting of war-time freezes.

He had given vent to his private worries to Catherine shortly before D-Day, "The damned trouble with our kind of ranching is very simple: when you have no crops to sell, you borrow on land to pay taxes. Then you sell other land to pay off the loans. In California the process is historic and self-defeating."

On July Fourth Howdy and Angie did their first show together since the war had begun. The appeal had come from Major Sam Garrett, thirty-two-year-old holder of the Victoria Cross, the Distinguished Flying Cross, and two Air Medals received for his exploits in the European Theater. The major had been fished from the Channel with a shattered left arm and sent back to the States for

490

more corrective surgery. Finally, with a minimal disability, he was assigned to the Santa Ana Air Base as regional public relations officer—"A hell of a letdown, you'd better believe," he said, openly admiring Angie, "not without its compensations!"

Dave Layne, filled with his old efficiency and enthusiasm, assembled many of the original radio gang and pulled together a ninety-minute show that was received with the deafening cheers of five thousand overworked, yearning-for-home recruits.

Acting on an impulse, Howdy asked how many of the airmen had come from out of state. More than three fourths of the crowd raised hands. When he asked how many would like to return to live in California after the war, the show of hands was virtually unanimous. Howdy filed the information in the back of his mind, unaware at the time of its eventual importance.

The huge success of the Santa Ana show, presented under the auspices of the USO, brought requests for performances at other installations. Major Garrett turned on his own brand of persuasiveness and Howdy finally agreed to do two more shows, one at March Field near Riverside and another near Sacramento.

Each time the response was overwhelming. But when Sam Garrett suggested a half dozen more appearances for the *Howdy Goodwyn Show,* Howdy balked.

"I got a flying school to run, and a flare outfit and an electronic outfit. I'll do some more if I can but you better not count on it for a while."

Garrett had anticipated the refusal. "What about letting Dave revamp the show around Angie? The boys will settle for a beautiful girl any time!"

Angie gasped and broke into surprised laughter. "Beautiful girl? Me? At thirty-three?"

On August first the *Howdy Goodwyn Caravan,* with Angie and Steve Stone as the stars, left in a chartered bus for a tour of a half dozen bases extending as far north as Chico. Major Garrett, traveling with the troupe, coordinated the transportation and living details with the public relations officers of the individual bases.

Shortly after Labor Day, Angie received a request to fly to Smoky Hill Army Air Field near Salina, Kansas, to entertain the men of the Second Air Force. After a futile attempt to induce Howdy to appear, she resigned herself to

491

the four-day tour that would also include the bases at Great Bend and Walker.

Angie and Sam Garrett took off early one morning in a war-weary B-17 Flying Fortress. They were squeezed down through the hatch behind the pilots' seats into the bombardier's "greenhouse." Hunched against the chill, they sat on their chutes and tried to play cards. When Angie's hands grew too cold to manage the cards they sat close together in their makeshift flying clothes and talked. The Major revealed a great deal about himself. There was an engaging wistfulness about him as he told Angie about a girl who had thrown him over just before he went overseas, first with the RAF then with our air force.

Angie studied him as he talked and was struck by a curious similarity to Howdy. Not feature by feature but they both possessed a certain reserved boyishness and while Howdy tended to be self-conscious around women, Sam was not. Sam was still young and had kept his dark good looks, the kind Angie was sure some women would call "sexy." In subtle ways, he appeared older than his years, the price for having endured some of the most vicious aerial combat in history.

When he finished the story of his abortive romance, Sam rumpled his hair and regarded Angie with an uncertain smile. "You're one of those warm, lovely females with big sympathetic green eyes and an understanding heart who doesn't have sense enough not to ask a man not to talk about himself. Now, what about you?"

Deliberately Angie tried to look mysterious. She was aware that she was responding to flattery. Casting her eyes down, she said, "Honestly, there's not much to tell. I grew up in a small town and married an entertainer. I'm, well, not very fascinating."

Sam gazed at her intently. "No woman can have as much wisdom and as much sadness in her eyes as you have without having felt things deeply and without having understood them. I know that, even though I'm an ageing boy who's done a lot of superficial feeling and not much thinking."

"That's a bid for sympathy if I ever heard one. I suppose you're going to tell me you haven't thought about the war you've been fighting either?"

Sam's expression was bleak. "Of course I have, Angie.

492

That's what scares me. I think this war may have taken the hope out of me just like it has for a lot of men in my old squadron. I can't explain it—but maybe what's happened is that a lot of us who have been living from day to day for so long are not going to be able to break that pattern. Maybe by now we're like the professional soldiers who say they hate war but die of boredom if they aren't fighting one. I don't know, Angie," he shrugged, "sometimes everything seems pointless."

"I think I understand. Women can feel that way too."

Abruptly, Sam changed the subject. "Ears pop yet? We're beginning to descend."

On the strip at Smoky Hill, the B-17 taxied to the administration building. It took an hour to get the men and women assigned to quarters. Angie was to be the guest of Colonel and Mrs. Price at their home just off the base. The others were quartered in the BOQ and in the WAC's barracks.

The show that night was held on an improvised stage set up in a huge hangar doorway facing a semicircle of portable bleachers. Although scheduled to run for just under two hours, the entertainment-hungry troops kept the performers on stage until they were literally out of material. It was close to midnight when Angie said good-by to Sam and the others and got in the car with Colonel and Mrs. Price.

When she finally climbed into bed in the little guest room she felt happy and pleasantly exhausted. For a time she stared up at the ceiling thinking about Howdy. Then, unaccountably, her thoughts drifted to Sam Garrett. As she thought back to their conversation in the plane she smiled sympathetically. "Poor baby," she said to herself. The thought made her laugh softly. "Poor *baby?* At thirty-two?"

After her third appearance with the troupe at the air base Angie called Howdy. When she could not reach him at home or at the Vance's ranch she called Dave Layne.

"He's up in Sacramento, Angie. He and Steve flew up to see Mr. Berwin this morning. He won't be back until Sunday afternoon. Can I pick you up somewhere?"

Angie hesitated. "No. No thanks, Dave, we're coming into Lockheed. Major Garrett's arranging for an Air Force car. He'll drop me off, then drive back down to Santa

Ana. We'll be in around noon tomorrow. You can take me to dinner tomorrow night if you haven't anything better to do."

From the corner of her eye Angie saw Sam's frantic signaling. "Hold it a moment, Dave."

Placing her hand over the mouthpiece she turned, questioning. Sam jabbed at his chest with a forefinger. "*I'm* taking you to dinner. That's an order."

Angie turned back to the instrument. "Dave, I'm sorry, but I forgot. I promised the Army Air Force they could do the honors, do you mind?"

After a brief silence Dave's voice grated in the earpiece. "Of course I mind. But that's not going to do much good, is it?"

Early the next morning, short of sleep and somewhat worse for the revelry at the farewell party given for them the night before, the troupe climbed aboard the Flying Fortress for the return flight to California.

Again occupying their exclusive place in the bombardier's greenhouse, Angie settled down for the six-hour flight. Sam tried to keep up a conversation but soon was dozing with his head on her shoulder.

Colonel Price, checking on his passengers as an excuse to stretch his legs, peered down through the greenhouse hatch at the couple. A minute later he was back in his seat at the controls wearing a cryptic smile. Lieutenant Colonel Sharpless glanced at him quizzically. "Why the Mona Lisa puss?"

The colonel grinned broadly and indicated the greenhouse with a nod. "Young 'Major Makeout' is pulling the sleepy boy bit again."

The two pilots exchanged knowing looks. Colonel Sharpless made a move toward his pocket. "I'll bet you twenty he doesn't make it this time."

The commanding officer shook his head. "Not with me you won't. I'm betting the same way."

The B-17 touched down in L.A. at 2:47 P.M. Both Angie and Sam were ravenous. When they got to the lunch counter in the terminal building, G.I.'s with duffle bags were three deep. Garrett was annoyed. "We'll have to try the one-arm joint across the street."

Angie glanced at the clock. It read twenty after three. "I've got a better idea. Let's go home and have Helga fix

494

us some sandwiches. It won't take any longer to drive out there than it will to find another place."

Twenty-five minutes later the staff car pulled into the long, secluded driveway. Sam began unloading baggage as Angie rang the bell. After several rings that produced no response she fumbled in her purse for the key. She called through the open door but the house was empty. Then she remembered. "Oh, for heaven's sake. I forgot. I gave Helga the day off."

In the hallway she pointed to the wall. "Leave the bags there, Sam. Helga will unpack for me later. Come on. Let's see what we can find in the kitchen."

From the well-stocked pantry Angie threw together a salad of avocado from their own trees while Sam, perched on a breakfast stool, watched her and made no attempt to conceal his admiration.

Moments later she placed the salads on the counter. "Let's eat out by the pool." She indicated a patio table and umbrella visible through the large sliding window-wall. "Think you can get them out without a disaster?"

"Just open the door for me, ma'am."

Angie followed, bearing a tray with place settings, rye crisps, and tall glasses of ice tea. It was hot in the patio. Occasionally the lightest breath of a breeze ruffled the peridot-green surface of the pool and rattled the tattered leaves of the travelers' palms. Sam slipped his cotton tie through its knot, opened the top two buttons of his khaki shirt, and heaved a huge sigh of relief. "Hope you don't mind?"

"Of course not. Why don't you roll up your sleeves too? In fact, you can take off your shirt if you want. We don't stand on ceremony around here."

Sam looked longingly at the pool. "Then I wouldn't be out of order if I invited myself for a swim after lunch? We've got to do something to work off this food, you know. We have a dinner date tonight. Remember?"

Her woeful expression alarmed him. "Don't tell me you forgot."

"No, I didn't forget—really. I just forgot how late it was."

"Lady, we have four hours of daylight left. An hour in the pool and I'll be ravenous again."

495

Angie was dubious. "An hour in the pool and I'll be three hours fixing my hair."

"In that case you'd better just go wading. I can't face three hours alone."

Angie clucked sympathetically and shook her head.

A white brick, tile-roofed dressing cabana and shower stood at the deep end of the pool. She pointed to it. "There are some shorts and towels on the right hand side. Help yourself and I'll join you after I clear off these things."

Ten minutes later she found him seated on the edge of the pool, smoking. His left shoulder was covered with a gaudy towel. He smiled at her apologetically. "You know, every once in a while I forget." He tapped the covered upper arm. "The surgeons patched this wing all right, but it's not the prettiest thing to look at." He saw the fleeting look of sympathy as she knelt beside him and removed the towel. The incision, less than a year old, still bore traces of proudness.

Angie slipped her hand beneath his arm and moved it toward her. As her free hand traced the course of the surgery she became aware of the ridges of scar tissue on the soft underside of the biceps. Lifting his arm, she peered at the second scar.

"That's where the slug came out. It ripped across the front of my flying suit, smashed the right corner of the instrument panel, and went clear through the fuselage on the other side."

She released his arm and shuddered. "My God! It missed your chest by inches."

"By millimeters. That's why I figure that no matter what happens to me from here on out, I've got it made."

She stood up and looked at him sternly. "Well, don't be silly about that scar, Sam. It has a very neat, business-like look to it—nothing to be self-conscious about. Now go and swim. I'll be with you in a few minutes."

"I'll wait for you. When you come out I'll demonstrate my world's championship one-arm Australian crawl."

"That's a blatant bid for sympathy. I watched you carry my bags in. Remember?"

"Okay, lady, but don't shoot me down for trying. I like a little sympathy."

When Angie reappeared and slipped off the thick white terry robe, Sam surfaced at the pool's edge. His wavy dark hair was plastered in water-set bangs and the thick matting that spread across his muscular chest veed into a gleaming curtain of dripping curls that divided his hard brown belly and disappeared beneath the drawstrings of Howdy's snug-fitting trunks.

Grinning, he subjected her to a deliberately impertinent appraisal. With the assurance of a woman who knows that an attractive male finds her desirable, Angie took more time than necessary to tuck her red gold hair underneath her bathing cap. But she kept her gaze cool and aloof.

After a long moment Sam let out two sibilant whistles. "Are you ever beautiful! I've always known you had a face for breaking hearts, but this," his hand swept the length of her, "this is more than mortal man can bear."

Angie tucked the last stray strand beneath the cap and regarded him archly. "You are an accomplished liar, Major."

"And you are Miss America. That girl Jean Bartel is an impostor."

"You *are* shameless." With that Angie entered the water with a flawless shallow dive. Turning seal-like beneath the surface, she emerged beside him.

"All right, Major, I'm ready to watch a demonstration of your world famous one-arm crawl."

"I'd rather watch you."

Angie splashed at him with the heel of her hand and an instant later found herself being pulled under by the ankles. They surfaced together spluttering and laughing and to Angie's amazement Sam pushed off and headed toward the deep end propelled by a powerful kick and a curious sort of circular one-armed sculling stroke that carried him along faster than she could swim using both arms. When he reached the end of the pool he executed an expert racing turn and swam back. Clutching the spill gutter, he drew in several deep breaths. "This is wonderful, Angie. But I'm out of shape. I used to be able to do fifty laps without puffing. Next to beautiful women swimming pools are the world's greatest inventions. I envy you and Howdy . . . a lot of things."

Angie let herself drift away a few feet and treaded water. "I'm afraid the pool is my pet plaything. Howdy doesn't have much use for it."

Angie swam several laps with Sam pacing her. On the last lap she veered to the ladder and climbed out. "Go on.

Swim as long as you like. I'm going inside and rest a bit. If you want to shave you'll find everything you need in the cabana. If you want to snooze in the sun, use the mats."

"If you're going to desert me I may do that. By the way, do you like Chinese food?"

"Love it."

"Okay. I know a place on Vineland near San Fernando Road."

"But that's way out past the terminal . . ."

"I don't care if it's east of the sun and west of the moon, I want to share it with you. I know the family. I'll phone and tell them we're coming. They'll really shoot the works."

It was dark by the time the staff car pulled up in front of Fat Lee's restaurant. When Sam introduced Angie, both the husband and the wife recognized her immediately. The obsequious flurry embarrassed her.

During the course of the dinner that was delicious but endless, she signed autographs on menus and odd slips of paper brought from the kitchen by the apologetic proprietor.

It was after ten when they left for Angie's home. The night was unusually still and warm and a hot, dry Santa Ana wind was promised by the forecasters. The valley-ites could feel it. Always, before it began to blow, something happened. The air seemed to have an ominous, positive electrical charge that made them restless. Some Californians called it "earthquake weather" but it was really fire weather. If the dry wind prevailed for more than a day or two the dehydrated foothills and mountains would be certain to suffer a series of disastrous blazes.

When the car pulled into the driveway, Angie rolled her head toward Sam with a satisfied smile. "It's been a wonderful day. Thank you for taking pity on a deserted wife.

"They've given medals to people for doing less than you've done for my morale these past few days."

He reached over to take her hand and she made no effort to pull it away. "You're something special, Angie Goodwyn." He looked at her intently for a moment. "I'm going to see you to the door now. I don't want to forget that all officers are supposed to be gentlemen."

He got out and ran around to help her. At the front

498

door Angie inserted the key in the lock and paused. "Sam, what time do you have to report back to Santa Ana?"

"Not until morning. Why?"

"Would you like another swim and a nightcap?"

Gently Sam took her by the shoulders and they walked through to the patio without turning on the lights. The night was so clear that the stars seemed three-dimensional. A host of cicadas pursued their raspish courtships and the frogs continued croaking with mechanical monotony. They listened in silence for a while, then Angie shivered. "I love it."

She started back toward the house. "Go change if you want. I'll get our nightcaps."

She returned with the drinks shortly and sat down beside him on the grass by the pool. She had changed into a two-piece swim suit. Conversation that had come so easily before now seemed forced.

Tacitly, Angie recognized their self-consciousness and said, "I'm sorry. It's my fault. I have a guilty conscience because I've been having so much fun."

"You? Guilty about having fun?"

"Oh, it's not fair to say it but even with all the things Howdy and I have, we don't seem to have—just fun. We enjoy a lot together but we don't . . ."

He clasped her face gently between his palms. "Angie, I've known a lot of women but you've got something they didn't have. I don't know what it is but it makes me feel like the high school boy who finally has a date with the girl he's been admiring all year and then when he's alone with her he gets all fouled up and tongue-tied."

Sam's eyes were serious. "If I said the things I want to say, Angie . . . Well, I can't. Not to you. Anyway, I'm suspect as a single officer. That means I'm automatically on the make."

Angie lowered her head. "And I'm a married woman who's alone tonight and I asked you in. What does that make me—automatically?"

The silence between them felt far longer than it was. Sam broke it. "Lonesome? Like I am . . . ?"

Angie looked down at the hands covering hers. After a moment she nodded almost imperceptibly. An instant later she was kissing him willingly. When she finally drew away he forced her head beneath his chin and pressed her cheek against the coarse cushion of hair that covered his chest.

The strange feel of it shocked Angie at first; but as he gently forced her head closer the sensation grew pleasant. She felt secure. Her left hand withdrew from beneath his right arm until the fingers found wiry hair. Like sensuous shuttles, they began to weave random patterns and her mouth, gone moist and slack, pressed against him.

Certain that she had been wanted ever since their first moments alone after the show in Kansas, Angie cast off the last of her reserve and allowed compulsion to drive back anxiety and guilt. Beneath the clinging shorts she could feel him growing and the realization started a burning between her legs. The sensation, pervasive and pleasant, was not unlike the slow beginning of the auto-erotic release to which she sometimes resorted after Howdy, consumed by his own spasmic pleasure-pain, had ground himself empty and withdrawn, convinced that his own coming was reward enough for both.

When it happened now, whatever it was, Angie knew that it would be more than she had known before. Her hand slipped between them, seeking the drawstring, and he twisted away to help. While she inched the cold, clammy cloth down over his hard buttocks, he unhooked her bra top and pulled it free. When the soft coils of his chest hair touched her erect nipples an involuntary little sound welled up. Angie's right arm tightened around his neck to pull her upper body closer still. Half recumbent, she began to undo the buttons at the side of her shorts. Quickly, Sam shucked off his own and kicked them free. He pushed Angie's hand aside gently and took over the task.

She lifted her bottom to free the shorts, slipped them down, and let him pull them over her ankles. Naked and free, their bodies came together in eager collision. For a time they clung to each other. Then Sam lowered them until they reclined on their sides on the mat, still embracing. Angie opened her eyes and lifted her lips to find his but he avoided them and his own lips brushed her cheekbone and her ear and moved down to the base of her neck. He felt her shoulders hunch against his mouth and his right hand caressed the smooth blade of her left shoulder. When his palm, surprisingly soft and gentle, reached the swell of her hip, Angie's left hand began to move down the muscled rib cage, out beyond the splay of dark hair that began at the base of his strong neck. She shifted to free him and cupped his testicles. Suddenly

Sam's head was up and his mouth was hard on hers. When his fingers found her and pressed slowly inward Angie began to move against them as she did during those tortured times alone when she was forced to fantasize a faceless lover. Sam's tenderness and finesse were bringing her close. She arched her back away urgently. "Sam, darling! Slower! Slower! Slower! please . . ."

Ignoring her plea, his lips moved down to her breast. Angie flung her head back and struggled to control the pulsing waves of electric sensation that seemed about to engulf her. Sam continued until she cried out again. Then he stopped suddenly and caught her face between his palms. "Angie . . ."

"What, darling?"

"Don't hold back, sweetheart. Let yourself feel. Feel clear into your guts. Don't be afraid. You've built up an awful debit. I'm going to pay it off in full now for that insensate idiot who doesn't know what a feeling, giving, out-of-this-world wonder you are."

Before Angie realized what he intended, Sam had swung his body around so that his lower belly was pressed against her face. She felt the stubble of his beard moving down along the soft swell of her middle and the abrasiveness excited her in still another way. She began to shiver. An instant later came the shock of realization. Her entire body went rigid. Her hips heaved in a violent spasm and she felt his arms locking hard around her to hold himself in place. She cried out and her head rolled frantically against his thighs. She felt him moistening against her cheek and still caught in the terminal tension of her own first uninhibited orgasm, Angie groped frantically, found him, and forced him to her mouth. She worked at him hungrily until he reached down to push her head away. Controlling himself, he persisted until the last of Angie's convulsions had subsided. He released her then, straightened, and took her in arms again. "That one was for you, darling Angie, so you can learn to feel all there is."

She buried her face against his chest. "Oh God, Sam, I didn't know there could be such a feeling. I didn't know . . ." Her voice trailed off and tears began to trickle down her face. Lovingly, he cradled her, caressing her cheek and kissing her hair, whispering reassuringly.

From the very first Sam had promised himself that sooner or later he would sleep with this woman. But there had been no intention to make it more than that, no intimation that it would not be just another lay. Angie was different.

In the warm, luminous Southern California night, Major Sam Garrett, decorated pilot, dedicated cocksman, held Angela Goodwyn in a way he had not held a woman since his first love. For a reason he did not fully understand, he wanted to make up to her for those numberless nights of frustration that he had been able to read out so clearly in all that she had said and had left unsaid. Without obstacles, with no sense of urgency, he would, in the next hours, bring them both to orgasm several times until no unfelt ecstasy remained. Then his own silent voice began to mock him. He heard it saying, "Bullshit, you greedy humpmaster!" He tried to squelch it and reaffirm as truth the strange conviction that after this neither of them would ever be quite so alone again.

Her fingers closed around him gently. "Oh, Sam, I'm in the world's most exquisite agony and I don't want it to end for hours. I want to do it all, everything. With you. I've never wanted to before, like this. You've let me out of my own jail. I used to dream of something vaguely like this and then I'd masturbate and feel sick with guilt and pray for forgiveness. But it's different now. I want to do it for you, Sam. Howdy uses me. He doesn't mean to but he just does. He doesn't know how a woman feels, that she's got to give back. It's not his fault. He is what he is and part of him is wonderful—his ambition, his determination, his courage, his loyalty to his friends. But he could never love me like you do because he could never stop thinking about himself long enough."

He searched her face for a moment, then braced himself, half reclining, and Angie moved down over him. He endured the torment of her body straining onto his until he was close to the limit of endurance, then, reaching down, he caught her face and forced her away until he could turn, inverted, to her again. Moments later the darkened patio was filled with the suppressed sound of his wordless outcrying.

Naked, they lay in a rigid embrace until a breath of breeze made Angie shiver. They sat up then, saying nothing, and slipped into the water. There, they played together like two satisfied, unashamed animals. When they were through swimming, Sam supported her with an elbow hooked onto the rung of the ladder, kissed her, and moved to bring her to him again. Angie let herself come to him willingly. As her arms went around his neck a light blazed in the upstairs room at the far corner of the patio.

Howdy crossed the Santa Monica Mountains at ten thousand feet and called the Burbank tower. Seven minutes later he touched down the Lodestar and let it run out to clear the following traffic.

It was three-thirty P.M. when he called Angie from the public phone outside the hangar. Helga answered. "She's not here now, Mr. Goodwyn. She called Mr. Layne early this morning and they went to lunch someplace to talk about the new camp shows, she said."

Howdy glanced at his watch. "When did she get back from Kansas?"

"Late Friday, I guess. She and Major Garrett were swimming when I got home just before midnight."

A brief silence followed. There was an edge of annoyance in Howdy's next question. "Did she say when she'll be back this afternoon?"

Helga nodded vigorously at the phone. "Yes, sir. She said she would be home by the time you got here, Mr. Goodwyn. She wants to cook out on the patio tonight."

"Okay, Helga, I'll be home in about an hour."

"Yes, sir. I tell her, Mr. Goodwyn."

"You do that."

Howdy checked the logs in the operations office, made his own entries in his personal log, and called a taxi. Dave and Angie had driven up to the front door moments before. The intensity of Angie's greeting puzzled and embarrassed him. He disengaged himself and put her away almost roughly. "You going to give me time to get showered and cleaned up before supper?"

Angie pouted indulgently. "Honey, I'll give you all the time you need. I've been trying to urge Dave to stay and have a steak with us."

Howdy turned a questioning look at his partner.

Dave smiled apologetically. "I'd love to but I can't. See you tomorrow, Angie."

A short while later Angie was sitting on the edge of the bed watching Howdy change. "I missed you, honey. I don't like doing out-of-state shows without you."

Howdy worked off the second Wellington flying boot.

503

"You don't have to do it any more. Trouble is, you let Garrett con you."

The accusation was innocent enough but it made her start inwardly. She laughed. "Is it con when he says those kids in camp need entertaining?"

Howdy turned to face her. "What time did you get back Friday?"

Angie frowned. "Late. Why?"

"Nothin'. I just happened to check the traffic log in operations and it showed Army 402 down at two forty-seven."

A quick stab of panic unsettled Angie. She knew she sounded defensive. "When I said 'late' I meant a lot later than we planned. You know we had a bad engine going over. Colonel Price wanted to be sure everything checked out."

Howdy gave her a long, penetrating look and resumed undressing.

Angie stood up. "You're acting funny. What's the matter?"

"What?"

"I said, 'What's the matter?' You're acting like I'm keeping something from you. I promised Sam he could take me to dinner because you weren't home. We didn't have any lunch so I fixed something here about three-thirty and he borrowed your trunks and had a swim while I took a rest. Then we went to a Chinese place on San Fernando Road. I signed a dozen autographs, we got home late, Sam jabbered about himself for a while, then we had a swim and he drove back to Santa Ana. What on earth is wrong with that?"

His amiable grin made her uncomfortable. "Nothing's wrong with *that*—but as of now no more shows. You've done your share of boosting GI morale. You tell your major that for me. Right?"

Guilt fueled a sudden indignant flare-up. "If that's how you want it, you tell him. And while you're at it, you might give him a good reason." She stalked to the door and turned back. "And you might give *me* one too!"

Howdy's grin faded until only a suggestion of a smile lingered. "What the hell do you mean by that crack?"

"Just what it sounds like. I've been out there in the sticks grinning like an idiot, pounding my fingers to stumps, shouting myself hoarse for the good old USA

504

while you've been doing," she tossed her hands. "God knows what! Then all of a sudden you cut me a new set of orders. I want to know why? Is that too damned much to ask—after ten years of saying, 'Yes, sir'?"

They studied each other coldly, then Howdy exploded in forced laughter. "God love you, honey, I expect I've been leaving you too much alone." He started toward her to take her in his arms and for the first time in the decade they had been together, Angie experienced a spasm of revulsion. "Howdy, I told Helga to start a fire—"

"Honey baby, you just tell her to hold off with her fire until we get this one put out." Slipping the silk tie free of the miniature gold guitar she had given him on their first anniversary, he reached out for her.

Angie backed away. "Howdy."

"Yes?"

"Couldn't we wait until after supper? I, well . . ." She continued to retreat and gestured helplessly. "It might be better. I mean without Helga sitting down there waiting."

Howdy unbuttoned the neckband of his expensive Western shirt and pulled her close. She was stiff and unyielding but he seemed not to notice. "Baby, what she's got'll keep. What I brought home ought to be used while its fresh."

Almost frantically, Angie pushed away from the embrace and resisted an impulse to cry out. Controlling herself, she managed a contrite smile. "Howdy, I'm afraid it's going to have to keep. I'm really in no condition."

Puzzled, his hands stopped manipulating the ornate silver buckle. "What the hell's eating on you?"

Angie grimaced. "Nothing's *eating* on me, as you call it. But please don't make me go into details. Later, Howdy, later."

Before he could protest, she fled from the room.

As the war drew to a close the housing shortage on both the east and west coasts became acute. In California huge industries such as Kaiser and Lockheed had increased over a thousand percent in the past five years. The government was anxious to persuade private industry to construct emergency housing. The President's special assistant on emergency housing had approached H.G. Enterprises. Howdy, along with Ev Moses, Danny Meyer, and Louis Siefert, agreed that as tempting as "cheap" govern-

505

ment loan money was, the risk was great. When the war was over, what would happen to the hordes of people out of jobs? Would they wander back East and if so, how many of them would go. Certainly, historically, Southern California could expect a large number to stay but how many workers would be willing to return to their old pursuits, in agriculture for instance, for far less money? And of those who stayed, how many could survive the readjustment period without defaulting on their new home mortgages?

During lunch at the Vine Street Derby, Ev Moses summed up his position: "There's no question about the need or the fact that it's nice to have cheap government money. But if we get into this on a long-term basis and the war ends in a year or so, we'd be caught with an unsettled market and thousands out of work."

Howdy turned to Danny Meyer and Louis Siefert. "What about you guys?"

Meyer responded without hesitation. "I see a great new market for electronic gadgets in this housing, but about the housing itself we better be sure the market is there and what kind. If we're going to build more civilian barracks like most of this emergency housing, then I say, no dice."

Abruptly, Howdy turned to Angie. "What about you, honey? Got any of that female intuition you always talk about?"

Angie was not amused. "I defer to all of this executive judgment. I don't have any feminine *intuition*. I do have a female *opinion*. I've been in a lot of that emergency housing. It's all around the bases we play. The women hate it. If that's what you have to build, then I wouldn't bet that many wives can be sold on buying."

Howdy listened with half-closed eyes, then sat up straight and pushed his steak sandwich aside. "Okay, I'll tell you what we're going to do. We pass this government deal and find out where folks really want to live if they get to stay. And we find out what kind of houses they want. Then if it adds up, we'll buy some land and try a few with our own dough."

He looked at each in turn. "We're going to get Steve Vance to work. He's a native. He knows what outsiders want in California. He told me once that, except for oil,

506

the land is getting to be worth more than anything you can raise on it. I want to know which land and why."

On October twenty-second the first contact was made in the battle for Leyte Gulf. In the three engagements that took place during the next six days and nights the effective power of the Japanese navy was destroyed. Experts felt optimistic about a victory in the Pacific and Howdy was glad that he had not succumbed to the temptations to get into Federally financed emergency housing.

At a Sunday barbecue in the patio of the Vances' Casa Grande, Howdy was fired with enthusiasm. The first surveys on probable migration patterns after the war had come in. They confirmed Howdy's hunch that G.I. home financing would, in time, create a huge market for better housing. Angie was annoyed by Catherine Vance's supercilious amusement as she watched Steven being caught up in Howdy's plans for the future development of H.G. Enterprises' land purchased from the Los Nidos Ranch Company

"Hot damn, Steve! If these projections are only half right we're going to build a whole new town right where we're training those kids to fly now. By golly, we might even call it Vanceville." Aiming a finger at a large artist's rendering propped on a chair, he continued to articulate his dream. "We'll give them two-bedroom, two-bath houses with a double garage on a quarter acre of flat ground. The whole place'll be fenced and seeded for lawn and we'll throw in an orange tree, a lemon tree, maybe even a peach or apricot tree and a bunch of rosebushes and things. And no two houses will be alike—on the outside, at least. There are all kinds of jimcracky tricks that can make a basic house look different."

Howdy walked to the end of the patio. "As soon as priorities are lifted on building materials we'll go into production with the first hundred houses. Ev will set up the subsidiary and I want you to head it, Steve, with Lloyd Wilson. He's the best city planner and architect around."

For an hour Angie watched and listened as Howdy carried Steven Vance along by sheer force of enthusiasm. Nobody she had ever known could conjure such an arresting vision with so few basic words. Even Catherine's thinly disguised skepticism seemed to have dissipated somewhat.

Seated on a cushion with legs crossed and chin propped

507

on clenched fists, Jennifer listened too. From time to time her violet eyes would widen with wonderment. The jet-haired girl's devoted concentration fascinated and troubled Angie. Try as she would, Angie could neither ignore nor dismiss Jennifer's girlish infatuation for Howdy. In another age in California, Angie could see the darkly beautiful Spanish-speaking girl eagerly awaiting the arrival of the New England traders from the far-off United States. Dimly she recalled her mother's occasional references to such girl-women in her own family. And they had been women—setting out to catch their men—often as early as fifteen—Jennifer's age—bearing their first child at sixteen. Perhaps they had not all been as beautiful as Jennifer Vance, Angie thought, but their blood, undiminished in passion and intensity, clearly coursed through this girl's veins.

Late that night, in the guest suite, Angie sat propped up in the oversized Spanish bed watching Howdy sprawled face down in deep slumber. With each new move away from show business, where she could participate, her anxiety had mounted. It was not for material reasons. She was, in her own right, a wealthy woman. It was not the disparity in their ages. She was three years older than Howdy but she had kept her figure and her good looks.

For a long time she studied the man who had always been more stranger than husband. She tried again to will more meaning into the present and the future by recalling the times in the past that had really been theirs together, refusing to acknowledge them as times of mutual need. She was deathly afraid of the terrible emptiness she had watched her mother become a victim of when her father left—the kind of emptiness that even a Sam Garrett could fill only momentarily. Cautiously she bent over Howdy until her lips brushed the unruly hair on his temple. "Slow down, honey—please," she breathed. "You're going too fast. I can't keep up—"

Alone in the family room well past midnight, Steven Vance studied the artist's rendering of the first unit of Howdy's proposed subdivision. After a minute or so he moved to a nearby window.

Midway across the dark expanse of fields were three scattered patches of light. Well within the heartland of the ranch, they marked the parcels that had been sold by his

508

grandfather, Steven Lewis, during hard times in the Eighteen sixties. They were the only areas within the original boundaries that still lay beyond ranch control and they had changed the nature of the ranch, for with the growth of population they had become the first urban incursions.

Steven Vance could visualize the shape of the future and the change he would preside over would be the most far-reaching of all. The proposed airport and commercial complex were irresistible magnets for urbanization. There was no doubt the plan would be successful—everything Howdy Goodwyn touched was. It would grow and consume hundreds of prime agricultural acres. In this vision of the future Steven Vance could see clearly the pattern of California's past.

Steven's great-grandfather, John Lewis, had had the same kind of pioneering instinct and shrewdness of vision that marked Howdy Goodwyn. Both had started with nothing, both were immigrants to California, and both would bend the land to the making of their personal empires.

*

Book IV

The Howard D. Goodwyn Story

(Continued)

17 AT THIRTY-EIGHT, after seventeen years of marriage to the man she had vowed to get from the time she was sixteen, Jennifer Vance Goodwyn was, by any measure, still an unusually attractive woman. Gleaming black hair framed large violet eyes. Her figure was usually described as sensational. Her weight hadn't varied by more than a pound or so in the fifteen years since she had added several pounds following the birth of their son, Howard Junior.

Seated alone in the bar in Sacramento's imposing new Eureka Tower Hotel, she was a familiar and welcome distraction both to the waiters and to the confidentially hunched huddles of lobbyists. The waiter, Joe, whom Jennifer addressed by his correct given name, José, glanced solicitously at the nearly empty brandy snifter. *"Desea Usted mas, Señora?"*

Her grateful smile was like a benediction. *"Nada mas ahora, gracias, José. Estoy esperando a mi esposo. Mas tarde, por favor."*

She had downed two and a half doubles already. Her Spanish, spoken infrequently now, was alcohol-thick and Yankee rough. It saddened her that she was losing it and she wondered morosely if it was going the way of certain other inherited advantages that she had taken for granted. In the midst of her moody introspection she saw Howdy's silhouette appear in the doorway. He was straining to penetrate the gloom.

"Over here, darling."

Howdy maneuvered onto the banquette beside her. Jennifer scarcely had time to give him a perfunctory little peck on the cheek when José appeared with a bottle and a set up.

"Buenas tardes, Señor Goodwyn." He indicated the tray. "We got your Crown Royal in this morning. Straight? With the water back?"

513

Howdy looked up, smiling. "Hi, Joe. Make it a double on the rocks and skip the water. You know us southern Californians. We'll save water for you northerners every chance we get."

Flattered to be a party to the joke, José laughed. He filled an Old Fashioned glass to the brim. When he turned questioning eyes to the snifter, Jennifer capped it. "Let's let Mr. Goodwyn catch up first."

Howdy waited until the man was beyond earshot, then gave Jennifer a searching look. "Catch up with you? God love you, Jenn, lately I'd have a better chance of catching up with Dean Martin." He put his glass to his lips and drained it.

"That wasn't a drink, darling. That was first aid. What happened at the hearing?"

Ignoring the question, Howdy reached for the check. "I'll tell you upstairs. I want to take a shower and stretch out for a while. Lew and Myra are going to pick us up here at seven."

A quarter of an hour later, wearing an oversized towel, Howdy came out of the bathroom. In spite of the air-conditioning the suite was hot. With less than three hours of daylight left, the nearby downtown buildings and streets were still shimmering in the mid-August heat. In the distance, trees and rooftops danced fluidly, distorted, apparently disconnected from the earth. The distant ones were all but obscured in the brown smog smear. Somewhere beyond was Sutter's Fort, reconstructed now as a state museum. Still farther to the east were the soaring peaks of the Sierra Nevada. The despised immigrants had crossed them. Many, like the Donner Party, who had waited until too late in the autumn, had died in them.

Howdy liked Sacramento. Much of the frontier spirit still lingered in its mixture of old and new. In a vague way Howdy wished that he could have shared the turbulent early times when Sacramento City's population exploded from several hundred adventurous trappers to five thousand permanent residents in less than three years. He envied Jennifer's great-great grandfather, John Lewis, who had founded his fortune there.

Jennifer wasted little time in the shower. Glowing from the sting of icy jets and the rough toweling, and still pleasantly high from the brandy, she stood beside the bed studying her husband's spare, long-muscled frame. He was

514

too white. He always had been, except for his forearms, neck, and face. She called it his farmer's sunburn and had long since given up trying to persuade him to swim. He was a Gemini, complex, mercurial, not one to be penned in. Her father had warned her of that. "A bird bath for Kiwis!" Howdy called the pool. It penned him in the same way a home penned him in. Lately, even a ninety-thousand acre ranch, Los Nidos itself, seemed to pen him in. The only time he seemed completely at ease was in one of their aircraft.

Cautiously, Jennifer lowered herself to the edge of the bed and sat with legs tucked under, watching him sleep. In repose Howdy's face seemed remarkably young. It betrayed almost nothing of the strain of responsibility that he seemed to absorb so well.

She let her eyes wander in critical approval from the wiry, short-cropped hair, shot with gray now, along the bridge of the clean-lined nose and slightly prominent cheekbones that betrayed a trace of Cherokee down to the well-formed lips, determined even in repose. Upon occasion they could widen into a wonderfully open smile or compress into a grim line that revealed a streak of latent cruelty.

She resisted an urge to wake him up with a kiss. Lately she had spent too little time with him. She had even had to ask to accompany him on this trip. The excuse had been a chance to visit with Lew and Myra Berwin, their closest friends.

Lew was the power behind the Association of California Grower-Packers. He had made Howdy listen to their pleas that the only way the state's five-billion-dollar agribusiness could survive was to put a man in the United States Senate who knew how to help. George Murphy was trying, but he needed help and the industry could settle on nobody until Lew suggested Howdy. At first the idea had seemed exciting. But now she was suffering misgivings.

At her dressing table she studied herself in the mirror and undertook, without enthusiasm, the minimal things her face required. Beyond her own reflection she saw Howdy beginning to stir. She'd let him have another ten minutes. The thought produced a smile. Another ten minutes? She'd let him have anything he wanted, just as she always had from their first time together after the messy divorce from Angie. To her mother's horror and her father's secret de-

515

light, she had taken the initiative then and the conquest had been easy.

It had been a good bargain, Jennifer reflected, this marriage that so few had held out hope for. Howdy had brought to it the tough-minded determination and vision that her father had lacked. And she had brought to it the elements that Howdy had needed to extend his own considerable success. She had brought to it also a savage passion and loyalty that awakened in him a latent ability to love. When Jennifer thought about the word love she was never certain that anyone could define it. Mutual need? Selfish need? "Whatever the hell it is," she had told herself a thousand times, "if it works, it's love."

Above all, Jennifer needed to see that the bargain endured. Family and friends had been far from unanimous in approving the romance. Catherine Coulter Vance's objection had not been based on anything more substantial than what she termed Howdy's lack of "Family Continuity." That had always been a favored euphemism for social background. Unlike Jennifer and her father, Catherine seemed to find more shelter beneath the spreading branches of the Vance-Lewis family tree than she found security in their huge ranch that straddled three county lines.

Catherine Vance's attitudes had imposed some painful burdens on the family. At the fashionable Glenborough School in Los Angeles, Jennifer had been in trouble from the outset because of her mother's insistence upon good form and her antipathy to the predictable nickname "Jenny."

Jennifer dubbed the whole business a "pain in the rump." She rebelled by standing rock-like against the current of the social swim, bringing her mother close to distraction. She had derived a perverse satisfaction from the knowledge that a girl less prominently placed would not have got away with it. Her satisfaction was short-lived, however, when she realized that her mother also derived a sort of inverse satisfaction for the same reason. The tacit competition had endured throughout the years.

In June of Nineteen fifty-two, with her own successful pursuit of Howard Goodwyn had come the crowning victory of her long rebellion. Jennifer made it secure by conferring on her successful but socially controversial husband all of the acceptance he really needed in California, a po-

516

tential net worth, including her own certain inheritance, of a half billion dollars.

From the moment of her engagement, Jennifer had worked to interest Howdy in the operation of Los Nidos Ranch Company's varied activities. Under her father it had managed to survive. Under Howard Goodwyn it had begun to reach a potential that nobody but her own great-grandfather had ever envisioned. Jennifer expected the process of indoctrinating Howdy to take years. Instead, the wartime death of her brother and the tragic death of her father and sixty others in a commercial airline crash placed the responsibility in her husband's hands overnight.

Quitely and surely, much to the dismay of some ranch officials whose long-tenured sinecures had been tolerated by her father with characteristic forebearance, Howdy cut away the dead wood. Within a year the huge operation had come completely under his control.

Jennifer had come to accept as inevitable the eventual erosion of their prime agricultural holdings in Southern California by the tidal bore of new space-age immigrants. It was California's history repeating itself for the fourth time. But this time the change would be as profound as the one wrought by the original Forty Niners. The optimists said the destruction of the present California would take fifty years. The pessimists gave the state twenty-five years. After that there would be no large agricultural holdings left. Jennifer knew the man on the bed would never accept the inexorable change that was coming unless he could control it.

She suspected and feared that Lew Berwin would finally convince Howdy that direct political action was the only way to control the state's future. The continuing urbanization of Southern California's best land had shifted the power to a new breed of opportunistic politicians, men untroubled by tradition. Jennifer knew it would take a knowledgeable and secure man to assure leadership of such a group. Howdy was both.

The question was where would all of this leave her? She recalled a frightening dream she had suffered through several nights earlier. Howdy was taking an unnamed oath of office and she was standing only slightly behind him, smiling bravely but her face had changed. She had aged and there was a haunted look in her eyes. It was the same look she had seen in the tired eyes of those other once

517

young, pretty political wives who stood in the glare of the television lights.

A restless thrashing on the bed made her turn, Howdy was sitting up, befuddled and blinking. "I musta corked off . . ."

Jennifer walked over to him. "That's what you musta did, honey. But I've been watching the time. You've got twenty minutes to get dressed."

He yawned and rubbed his head roughly between his hands. "Thanks."

Jennifer reached out and tweaked his ear. "Ever wonder what you'd do without me?"

He looked up at her, confused, then broke into a broad grin. "Nope. But I'd think of something."

In the private dining room at Sam Lee's Restaurant overlooking Sacramento's historic waterfront, Jennifer relished an array of Cantonese dishes along with the Berwins. Whatever Myra Berwin may have been, Jennifer agreed with Howdy that she liked her. The dinner was fun.

Myra had married Lew in 1952, the same year she and Howdy had been married. Both marriages had sent mild shock waves through their respective social strata in Los Angeles and in San Francisco. The marriage of long-divorced Lew Berwin to an obscure actress twenty years his junior had caused the most comment. The general tone of it was summed up by an older, unfulfilled actress who was reputed to have been a part-time occupant of the commodious Berwin bed. "That Mansfield dame's got sprocket holes in her back from so many continuous performances."

Privately, Jennifer conceded the possibility that Myra might have tripped an occasional producer and beat him to the floor, as one gossip columnist put it. But so far as she was concerned, Jennifer could find no fundamental difference in the tactics she herself had used to commit Howdy. Whatever the facts of Myra's former life, Jennifer had quickly accepted her as an ideal mate for Lew, an outgoing, "no crap" female who was doing her job. Watching Myra now, she acknowledged the uneasy kinship that exists between women who know what they want, know how to get it, and hope that they know how to keep it.

518

In the beginning Jennifer had been afraid of Lew. The balding, stocky, hard-handed, fifty-seven-year-old packing company president who drove head on at everything he undertook seemed dangerous to her in ways that she did not comprehend. He had expanded the packing company founded in San Francisco by his great-grandfather, Moses Berwin, in the post Gold Rush days until it had become a hydra-headed complex of related businesses under the Golden West trade mark.

Once, when Lew had been particularly insistent about Howdy's participation in a deal, she had expressed that fear. Howdy had made her dismiss it by saying, "Honey, you don't ever have to be afraid of people who need you. It's the ones *you* need that you've got to look out for." Howdy had also pointed out that Lew Berwin's companies had a vital need for the great variety of produce grown on the ranch. Los Nidos could do business with many packers but there were too few large, diversified ranches close in to the big markets to give the packers any real independence. Lew Berwin's extensive company-owned ranches were not sufficient and economic pressure to spin them off for suburban homesites was increasing. South of San Francisco, five hundred acres of Golden West's richest artichoke land had been over-run by a new suburb-city. The survival of all that both men had built had to be protected. That would mean they would have to fight. Dimly Jennifer feared that she and Myra could become casualties of that war.

At thirty-six, Myra was still a very desirable female. The startling resemblance to Rhonda Fleming that had been a disadvantage in Hollywood had become an asset in their segment of San Francisco's society. Her engaging informality disarmed the socially orthodox in the Berwin circle of friends and the conquest had become complete when her love for the children she could not have led to the discovery of a latent talent for portraiture. She had worked hard to develop it and over the years her pastels of friends' offspring had brought in an impressive number of commissions. Myra had accepted the money gladly and promptly endorsed the checks over to the Children's Hospital.

The dinner conversation managed to stay away from business until the subject of increased assessments and the Farm Workers Organizing Committee came up. Scowling,

519

Lew Berwin gestured impatiently. "That's all we need! Reassessment and Chavez. The goddamned fools. They'll put us out of business and themselves too. Their leaders don't tell them that. They don't tell them that you can't fill warehouses full of perishables and then negotiate while the inventory is being depleted, like they do with steel. All they say is, 'Screw the big big boys. Screw the growers.' And those soft-headed liberal politicians who brown nose them keep egging them on. What the hell do those Ivy League dudes care about agriculture? All they want is votes. And if they have to crawl in bed with those red-eyed liberals to get them, believe me they will—no matter who get's screwed!"

Myra gave Lew a pained look. He ignored it. "It's pretty clear now that we've got to have more help in Washington. If we don't get somebody back there who understands California's special ag problem, you and I are going to wind up in the real-estate business in another five years."

A long ash from his cigar cascaded down his shirtfront. Before Myra could stop him, he had brushed the mess away, leaving a long gray smudge. Moaning, she closed her eyes. "I hate those damned cigars!"

Lew shot her an annoyed glance and resumed his tirade. "And we're going to be put there by a bunch of brown-skinned idiots. On your ranches and on ours they average a hundred and twenty-five bucks a week each. Fifty weeks a year, with free permanent housing, free language schools and bonuses. That's pretty goddamned good for a bunch that can't write their own language and probably haven't bothered to learn to speak ours."

Despite the certain knowledge that Lew could never be either chauvinist or bigot, Jennifer found herself growing angry. In his frustration she knew that what he was really asking for was a willingness on the part of the present-day minorities to work as hard as his own immigrant forebears had, not only at their jobs but at the task of getting an education, pulling themselves up to positions of importance. She understood Lew's and Howdy's anger at the short-sightedness of the uneducated and uninformed workers who listened to the emotional appeals of Cesar Chavez. Chavez promised them power. But the workers would lose if they listened and in the end Chavez would be killing the patient to cure the ailment.

Much later Howdy and Jennifer lay naked watching their reflections in the mirror-wall beside the bed. Jennifer stirred first, rolled her head toward him, and kissed the end of his nose. "You're very good, darling, when you've saved up for a week."

Pulling away, Howdy peered down his nose at her. "What d'ya mean, a week?"

"I mean one week. I keep track. I used to be your CPA. Remember?"

He laughed suddenly at the half-forgotten private joke. In the early months of their marriage CPA had stood for "Cute Piece of Ass." Howdy studied her and thought that seventeen years and one kid later she still was. "You're not trying to tell me that I'm hanging you up these days?"

She nodded curtly and affirmatively.

For a long moment she returned his defiant gaze. "You're oversexed."

"I know. But we'll just have to try to live with it. Won't we?" When Howdy didn't respond, she frowned. Suddenly her mood changed from playful to serious. "It's not that I get hung up in the hay, darling. It's some other kind of damned hangup that's bugging me."

"Like what?"

"I don't know." She shook her head. "It's almost like I'm getting lost or falling by the wayside." She turned to him, appealing. "I can't put my finger on it."

Troubled now, Howdy pushed upright. He reached down and squeezed her neck gently. "You drink too much. You're beginning to imagine things."

Jennifer freed herself and gazed at the empty brandy snifter on the bedstand. "I won't deny that I've been slurping a little more than usual lately. But you've got it backwards, Howdy. I'm not feeling lost because I'm drinking. It's the other way around." Pushing herself up beside him, she appealed for understanding. "I don't know what it's all about, except that I don't feel it when we're buried in ranch problems and figuring things out together."

She was silent for a moment, reflecting. "In a way I'm to blame for this political thing. It sounded exciting when Lew first started working on you. It seemed like the best way to fight off those damned annexations and all the rest. Maybe it still is, but—" She broke off and groaned. "Oh damn it, Howdy. If you do run, you'll win. You always do. So I guess what I'm trying to say is, 'Where will that leave me?' Lost in some sort of a middle-aged female limbo?"

Jennifer's tanned body sagged and her eyes lowered to her hands. They had fallen, upturned, cupped between her thighs. Suddenly they seemed like beggar's hands, supplicating. She laughed half aloud and the harshness of the suppressed sound made Howdy turn to look at her closely. Jennifer slipped her hand beneath his upper arm and curled her fingers around his biceps.

"It's asking a lot of you, Howard Goodwyn, to expect you to know what it feels like to be alone. I haven't felt this way since boarding school when I used to sit alone in my room trying to get myself together after Mother and Father took off on another of their trips." She reached up to lace the fingers of both hands over the still hard muscle of his upper arm.

"I'm sorry, honey. How on earth can I expect a man like you to know how it feels to be—" she searched for the right word and settled again for "—lost."

With unusual gentleness Howdy reached for the tangle of dark curls and rumpled them. "I expect I don't know. But do·you want to know what I think about it?"

Her eyes narrowed suspiciously. "I'm not sure I do, if you're going to tell me that it's self-pity and I ought to be ashamed of myself."

"No. I'm not going to tell you that. But I am going to tell you again that you're imagining things." He grasped her neck gently and pretended to shake her head. "There's never been a time when you weren't part of everything I do. And there never will be. No matter what I do, here, in Sacramento, in Washington, and down at the ranch. You'll be with me, doing it with me, Jen. Now go to sleep. You've had too much of everything, but rest. And so have I."

Howdy slid down flat again. A moment or so later Jennifer worked her long, sun-browned legs beneath the cover and inched over beside him. Within minutes she slept.

As he watched her, the possibility that Jennifer would ever find herself in any sort of limbo struck Howdy as ludicrous. Still, her plea had betrayed genuine anxiety. It troubled him. There was a pattern to it, something familiar. It reminded him of Angie during those months following the end of their radio show and the beginning of the war years. She had complained of a lost feeling too, of being out of it, not needed. Howdy stirred restlessly at the thought and wished that he could turn his head off. Making love to Jennifer stimulated him with a sense of well-being. Other men said sex wound them down and caved them in. But it never had with him, except perhaps after a twenty-dollar session with Renée. But that was different. Renée's job was to take the pressure off. With Jennifer, sex added something. As he lay still, staring up into the lumimous darkness, Howdy wondered why it had not been that way with Angie. She had been loving and giving, but in her giving she had been strangely demanding, as though the gift of her body had carried an implied obligation to appreciate. An old twinge of guilt returned. He dismissed it with a silent curse. He had no reason to be concerned Angie was a wealthy woman. For almost ten years she had been Mrs. David Layne. The infrequent pangs of conscience he felt about Dave were quickly allayed by the knowledge that what he had paid for his partner's share in Goodwyn Enterprises, added to Angie's settlement, more than compensated them for the fact that both had run second best. He had been generous despite the fact that if he had used what Helga had seen from her window that night as evidence in court, he could have gotten off with a quarter of the offer Angie finally accepted.

After the first angry sting of wounded pride, Howdy convinced himself that Angie's need for another man was less physical than emotional. After the separation he had moved to an apartment in Hollywood's Sunset Towers and was surprised to find how little he had really needed Angie for anything. He welcomed the relief from a burden that had been unrecognized until removed.

Howdy looked down at the youthful nude body beside him and marveled again that Jennifer could have produced a child and kept herself in such lean-hipped, flat-bellied shape. That was something they had given each other, a son and heir. Howdy wondered at times if it had been a mistake not to have had another child, but one child seemed to have fulfilled Jennifer. Any number of times she had complained good-naturedly, especially as Howie entered his teens, that he was more than enough for one mother to handle.

At fifteen, Howard Junior was a rawhide lean, handsome boy with the easy grace of a dedicated surfer.

Howdy had once confided to Ev Moses, "The kid's well hung."

It troubled Howdy that he had so little time to spend with the boy. Politics would leave even less. He had convinced himself long since that in the main he was discharging his principal responsibilities as a father. He had discussed it with Ev Moses, but for once the attorney had not been so sanguine.

"A good-looking kid like Howie could become a mark for some scheming mother, somebody who knows how to figure the angles. It can be worth a lot to a prominent family to avoid a paternity suit." But the kid had kept his pants zipped and his nose clean. At least so far, Howie's dedication to surfing seemed to absorb most of his energy. Jennifer had said, "After those kids have been out there freezing their little ornaments off they don't have enough pizzazz left to get into trouble." Jennifer had forgotten just how much raw energy kids have. At Howie's age, Howdy could remember no time when sheer physical fatigue had not suddenly given way to a surge of new energy at the prospect of an imminent piece of tail.

Unable to will himself to sleep, Howdy let his mind drift to the ranch. His involvement in Los Nidos had begun immediately after Angie's Mexican divorce in the winter of Nineteen fifty-one. He had flown Jennifer up to Monterey. They had wandered along the same streets that her paternal great-great-grandfather, John Lewis, had known as a young man. Jennifer had just turned twenty-one that September. From the time she had been thirteen he had never been able to deny her anything and the trip was another promise kept.

He reserved separate suites, amusing Jennifer by his propriety. Later, when she knocked on the adjoining doorway and he opened it somewhat timidly, she laughed at the expression on his face. In a matter of minutes, he discovered in his arms the most intensely passionate female he had ever known. Howdy wondered again as his hand involuntarily reached out to touch her whether or not the decision to marry had really been his or hers.

The next step had been to exercise the option to buy the airport and to develop an ambitious light industrial complex there. Reluctantly, Steven Vance had accepted the move as inevitable. In the midst of the industrial development Steven Vance had died in a take-off crash at New

524

York's Idlewild Airport. Within weeks, as Jennifer's husband, and with Catherine's grateful consent, Howdy had taken over Los Nidos.

Quickly he had perceived one underlying principle: In a population explosion and in the sellers' market it produced, it was far better to own land than to have to bid for it. His marriage to Jennifer had placed ninety thousand acres of California's best diversified land under his control and it was located in the heart of the state's fastest-growing area.

When it was developed it would easily be worth a billion dollars. Los Nidos' position was very different from that of the smaller ranches. Almost without warning, residential developers had become the new sharecroppers with middle-income housing the most profitable crop. Howdy couldn't blame ranchers, beset with rising taxes, threats of annexation, and soaring labor costs, for taking their capital gains profits and heading for the nearest cruise vessels. Unexpectedly, the farmers had won economic survival but California's agriculture was the victim.

Subdivisions ate up more than one hundred thousand acres of the state's best agricultural land every twelve months. Much that remained of the state's arable land was polluted. And nobody seemed to give a damn. Mass builders who had learned from the Levitts cared about nothing but profits. They were being helped at every turn of the bulldozer by vote-hungry local politicians with an eye to new families in their districts. Howdy knew there was little chance that agribusiness executives could stand against the tide. California had become the only place on earth where bulldozers customized the land so builders could rubberstamp their houses.

Howdy had commissioned the architect and city planner Lloyd Wilson to update Los Nidos' original master plan. Wilson had recommended developing a revolutionary new plan whose guidelines would anticipate the urbanization of the entire southern California coastal basin. Lloyd Wilson was due at the ranch the following afternoon to demonstrate his preliminary concept.

Howdy damned the ambient thoughts that were keeping him awake. At three A.M. sleep finally came while he was still trying to reason a safe way into the future.

Lloyd Wilson and two associates arrived at Los Nidos

shortly before two P.M. Ev Moses, Jennifer, and Catherine Vance were all on hand to attend the conference.

Wilson wasted little time on preliminaries. He immediately set up aerial photographs with overlays and spread a rough plat of the general plan on a large table. Using a pencil, Wilson indicated the meandering boundaries of nearly twenty thousand acres of rich agricultural land that lay between the mountains on the northeast and the coastal hills on the south and west.

"This is the land that must remain in highly diversified agricultural production for at least twenty-five years before it can be rezoned."

The pencil moved to the foothills. "This land, currently producing citrus, avocado, oats, and barley, remains as is until it is moved to the company's new acquisitions in the San Joaquin Valley. When the demand warrants, it can be released to private developers working within our guidelines for single-family and multiple residences. Eventually there will be about twenty residential complexes, each accommodating about twenty thousand residents."

Wilson turned to indicate a large cardboard tube. "Now we're going to show you our artist's visualization of the entire project as it could look about the year Ninteen ninety."

While the others watched with growing wonderment, Wilson helped his men unroll a huge mural visualization and pin it to the top of a drapery valance. When the men stepped aside, Jennifer's hand went to her mouth. "My God, Lloyd. It's ... it's stupefying. It makes Reston and Columbia look like crossroads villages."

Wilson allowed himself to be cautiously encouraged. He would wait to believe that Jennifer's reaction signified approval. What he was visualizing for her was the end of El Rancho de los Nidos. She might call the rendering "stupefying" now, for it was a daring piece of exhibitionism, but he could hardly expect her to wholeheartedly accept this metamorphosis from Los Nidos to MICROPOLIS 2000.

Howdy ran the operation but when the chips were down it was not likely that he would do anything that was seriously contrary to her will. Wilson glanced at Howdy, whose face betrayed no visible emotion beyond concentration on the painting. Neither could he guess Ev Moses' reaction. He decided it would be wise to anticipate some inevitable questions.

"Our rough figures show income from real-estate leases and sales matching net earnings from agriculture in three years and leveling out at about four times the agricultural gross within ten years. After that we'll need computer projections on the cost of finance money and rising labor and material costs." He waved a hand. "And all the other factors. We feel it can be made economically sound because it has to be. The only alternative is to fight a losing battle against annexation and urban sprawl to tract-built super-slums."

Then Lloyd went on to interpret the remarkable rendering. In addition to the twenty complete communities with their own service centers and regional shopping centers, there were schools, a two-year junior college, a neighboring state university, and two private vocational schools. One was for general training and the other for training maintenance mechanics needed for the sophisticated new automatic farm equipment soon to replace much of the field labor.

The Casa Grande would still stand, with the adjoining hill and mountain country, undisturbed by the proposed development. The reservoir-lake would remain, surrounded by a naturalistic public park just as it had during Steven Lewis' time.

Reversing what he considered an idiotic modern concept, Wilson would build as many communities as possible on non-food-producing land. They would be built on otherwise useless hills. There would be no single dwellings, only apartments. There would be no streets for vehicles in the conventional sense. The steep, stone-paved streets that would provide access to the buildings were to be used mainly for walking. Only vehicles needed for heavy deliveries, emergencies, and services would be permitted. Residents would park private cars in disguised common garages engineered into the face of the hill and take moving side-walks along a tunneled subway to a point directly beneath the hill towns. There they would use express elevators that would let them off at various levels. Then, through common passageways, they would reach the lobbies of their respective buildings. Wilson's sketches created the aesthetically pleasing impression of an ancient tile-roofed Spanish balcony village constructed mainly of stone and great hand hewn beams.

Connecting all the MICROPOLIS 2000 communities

527

and serving the two major shopping centers would be a high speed monorail. Shoppers would leave their automobiles in underground parking lots beneath the central plazas of each community and walk under cover to the stations. The shopping centers themselves would be climatized and the monorail coaches would enter, discharge, and receive passengers in a fully protected environment.

Wilson planned a cultural center with an amphitheater for summer musical productions, a civic playhouse and auditorium, and, on a mesa at the extreme top of the coastal hills, a great circular convention center with exhibition galleries for the arts and crafts. A spectacular cantilevered restaurant was located on the top floor. The complex would be reached from parking areas at the base of the hills by escalators.

Wilson turned from the physical layout of MICROPOLIS 2000 and its main design innovations to the large forest that had been created by the artists. "One of the most exciting aspects of this concept to me is this reforestation project. If we can prove the practicability of the plan on Los Nidos land, I'm sure we can secure the cooperation of conservation groups and the Federal government. What we are proposing could radically improve the bionomic state of Southern California's coastal shelf. It could control the fire and flood cycle that has plagued this land. Wildlife would be preserved to proliferate again. In another generation, Southern Californians would not have to travel hundreds of miles to campsites only to find them jammed."

Grinning, Howdy pointed to the illustrated mountains. "If you start covering our national forests with trees, you're liable to confuse a lot of natives."

Jennifer shot him an oblique glance. "That's our normal state, darling."

Howdy ignored her. "Lloyd, if you and your boys have invented a new rain-making machine that really works, I'll buy the trees for Los Nidos and the whole damned national forest around us, too."

Wilson smiled and pointed to a network of light blue broken lines. "These are all sewage. About twelve acre feet of fluid per day per twenty-five thousand population, would be processed to potable purity, pumped to these storage reserviors, then repumped to these plastic line sprinkler systems deployed along these mountain slopes. In time they would cover five thousand acres of our highland.

In case of a brush fire, that particular sector could be turned on and fires brought under control. Our plan proposes that we plant a million trees that were indigenous to the area hundreds of years ago. We know the trees will root if they receive water. We provide irrigation with our portable treated sewage. The solid waste from the treatment plants will pay for itself as fertilizer, both for the reforestation project and for certain Los Nidos crops.

"In short, we are proposing to return this land to the pristine state that geologists and ecologists say existed up until several thousand years ago. If the surrounding cities could be induced to join in, perhaps with state and Federal grants, we could begin to transform Southern California's coastal shelf by the year 2000. The Israelis have done it with their tamarisk and eucalyptus forests and with cedar groves. They've planted over a hundred thousand acres. Why can't we?"

Lloyd Wilson looked intently at his listeners for a moment, then tossed the pencil onto the table. "That's it in sketchy form. There's a lot more detail to come. I'd like to leave this material here for you to go over. If this concept is too far out, we'll haul back a little. I'm not married to anything here except the basic premise that Los Nidos is not going to be able to survive much longer unless it beats the invaders to the bridge. Protect your best agricultural land at the heart of your township and phase it out according to your own time-table."

Jennifer was waiting for Howdy when he returned from the office at half-past six. As he entered the family room, she handed him a bourbon and water.

"I've got a glorious piece of news for you, darling."

Howdy gave her a perfunctory kiss and flopped into his favorite chair. "I could use some."

She perched on the chair arm and began massaging the back of his neck. "Well, see how this grabs you. Mother's coming here for the holiday, and—"

Howdy glanced up, frowning, and said, "That's news?"

"—and darling, she's bringing me a new step-daddykins who's only five years older than I and you're getting a new father-in-law who's only twelve years *younger* than you." She leaned away and looked down her nose at him.

Howdy had arrested the drink at his lips during Jennifer's revelation. He lowered it untouched. "Well, I'll be a son-of-a-bitch. So she went and did it?"

"That's right."

Howdy took a good swallow. "I never thought she would." He looked up, questioning. "Rob?"

"Yup."

"You met him in New York, didn't you?"

"I did."

"What's he like?"

Jennifer pursed her lips thoughtfully. "He's like, well, charming, attractive in an *Esquire Magazine* sort of way. Got gobs of his own sun-bleached hair, his own Ultra-Brite teeth, good New England background, plays excellent tennis and squash, golf, sails, flies light planes, knows all of the important maitre d's—" She pantomimed empty pockets. "—and hasn't got a pot to pee in."

Howdy rumpled his hair in annoyance and Jennifer smoothed it, clucking. "Everything'll be all right, darling, because mother's only sixty-six and has forty million dollars. That should be enough for them to get by on if they're careful." She kissed the end of his nose. "So now you know. And while I'm at it, I might as well tell you that Robert Westbrook Williamston III also expressed an interest in Los Nidos. He's absolutely champing at the bit to be given the short course in how to make millions at ranching between golf games and other urgent engagements."

While Howdy sat slumped, immersed in his own forebodings, Jennifer went to the bar, poured a brandy, and resettled on the chair arm. Howdy swung the ice in his drink. "I never took the bastard seriously."

"Neither did I, baby. But we better start now."

He slapped the left chair arm. "Yeah. But what stalls me is how your mother, who's a reasonably sensible woman, could fall for a lounge lizard."

"Mother is a reasonably sensible, very *lonely* woman, darling. And Rob's not a lounge lizard. He's got all of the right credentials. His grandfather had blanket mills in New England. His father was very well off but got wiped out in Twenty-nine. He went to all the proper schools and graduated from Yale. After college he dabbled in the brokerage business, just like Dad did. Then he tried gentleman farming in Virginia, like Dad did. I think he was also involved in banking and oh yes, he was a male model, for a gag, he says. Mostly he's made himself a very attractive and ac-

530

ceptable odd man, the kind hostesses dote on. That's how Mom met him. I was at the same dinner. In fact, he squired us both."

Deliberately, Jennifer needled a sore spot. "And, darling, I almost forgot, he's a divine dancer."

Howdy's muttered obscenity made Jennifer laugh. She patted his cheek. "Don't worry, sweetie. I promise not to dig his type until I'm a hundred and sixty-six. Meanwhile, back at the ranch, we've got ourselves a fox in the hen house, as your senator friend, Clair Engle, used to put it."

A clatter in the adjoining room interrupted their musing and young Howie appeared holding a soft drink bottle tipped up to his mouth. When he lowered it, some of the fluid spurted on the floor. He swabbed it with a bare foot and collapsed on the sofa. Jennifer regarded him critically and sat down beside him. "Cat got your tongue?"

Puzzled, Howie looked at her, then grinned. "Oh, hi, Mom."

"Hi." She indicated Howdy. "That man over there is your father. Say something nice to him, son."

Howie hoisted the bottle. "Hi, Dad." When Howdy grunted an acknowledgment without looking up, Howie turned from one to the other curiously. "Wow. You two are sure breaking it up. What's the matter?"

Jennifer answered, "We got a little surprise today. Your grandmother got married Saturday in Connecticut."

A broad grin spread over Howie's tanned face. "Hey. How about that? Grandma got stoked on some old guy and blew her mind. Wow. Crazy!"

Jennifer sniffed. "The 'crazy' part I agree with. But he isn't some old guy. Your new grandfather is only five years older than your *young* mother."

"Wowww—"

The exclamation made Jennifer grimace. "For God's sake, is 'wow' the only word you kids know?"

Howie ignored her. "I'll bet he's after her dough, huh?"

Jennier tried unsuccessfully to stifle a smile. "You may have put your finger on a good reason, son. You just may have."

"Are we going back East to meet him?"

"No. They're coming here for Christmas."

Howie looked pleased. "Hey! Neat!"

Jennifer patted his knee. "Isn't that nice? You'll have someone your own age to play with."

Howie finished off the last of the soft drink and wiped his mouth on his forearm. "Forty isn't very young."

She glowered and shook a finger under his nose. "You just watch it, young man, or I'll cut you off without an acre." Then she gave him a playful swat. "Now go put your hair up and polish your beads. We're about ready to eat."

When Howie rose with a pained expression, she caught a glimpse of his soles. "And while you're at it, sterilize those feet and put on your shoes. I know they kill you natives, but you've got to learn our Christian ways." The boy gave his mother a withering look that translated "Square" and ambled out.

Howdy stifled a yawn and slid his legs out straight. "If I didn't know how good that kid is, I'd worry about him with that long hair and those goddamned beads."

She nodded toward the door. "Don't worry about him. He's just being fifteen. Worry about his new grandfather." Serious again, she set the brandy snifter aside. "Honey, there's no way Rob can get his hands on any part of the ranch, is there?"

Howdy looked up, amazed. "God, no! No way. Your mother can't lay a hand on anything but the income from her trust stock. The way your father set it up she's got to leave it to you and your offspring, except the percentage that goes to the foundation."

"Thank heaven for that. The worst he can do is spend her income. It's only money and if I thought this marriage was really going to make Mom happy I could bear it. But it's such a damned ungraceful situation. Sure, we can say, 'to hell with money,' but I can't say 'to hell with pride.' "

For all of his preconceptions, Howdy was not prepared for Robert Williamston III. As Chad Chadwick eased the Los Nidos chopper down onto the pad, Howdy could see Rob in the right-hand seat, apparently questioning the pilot about the operation of the ship. By the time Howdy and Jennifer were out of the car, the bride of three weeks was hurrying toward them with outstretched arms. The groom, a remarkably handsome man, was exacting a last-minute promise from Chad. "I'm going to hold you to that. I really want to try it. I got a little Benson up and down in one piece once, in Florida. But this is another bird." Howdy could see that Chad was pleased.

The feared introduction was made so effortless by Rob Williamston's easy graciousnes that Howdy was nonplused. Helpless to resist, he found himself being carried along by the force of the younger man's personality. After greeting Jennifer warmly, Rob turned back to Howdy.

"Catherine's eagerness to see you all is a good part of my enjoyment. But I'm going to be honest. I admit to a selfish desire to meet a living legend. Take my word for it, Howdy, you have no public relations officer like your mother-in-law." The man's handshake was good. And so was the face although it seemed older than Jen had led him to believe. But he looked right at you, and there was no discernible embarrassment at being married to a woman old enough to be his mother.

Howdy found himself thinking, "How in hell do you dig out of a snow job like this?" Uncomfortable, he excused himself and walked over to Chad, who was dragging luggage from the chopper.

The pilot greeted him with a perfunctory "Hi" and jerked his head in the direction of Santa Ana. "There's more at the airport. I couldn't handle it all this trip."

Howdy helped lower a wardrobe case: "What do you think?"

A warning look silenced him.

"Forgive me, fellows, I didn't mean to leave you with the baggage." Rob clamped a small case under each arm and grasped a larger one in each hand and headed toward the car. When he was beyond earshot, Howdy repeated the query. "Well, what's the verdict?"

"He's a bright son of a gun." The pilot studied the retreating figure. "I can't figure him out. But he looks to me like one of those characters who can steal a dame right out from under you."

Before a week had passed, Rob Williamston had made himself a welcome presence at Casa Grande. His ability to divert and amuse seemed endless. By the end of the third day he had won over young Howie. "Know what he did, Dad? He rode Bandido bareback and jumped him over the sawhorse and over the logs on Laberinta trail. He rides better than I do. He's like—wow!"

The boy's quick enthusiasm for the Easterner annoyed Howdy. Apparently in a matter of hours Rob had established a rapport that he had never quite managed himself.

Howdy felt certain that Howie respected him and admired his ability on the guitar and as a pilot. But his day-to-day relationship with young Howie was filled with too many long periods of detachment.

Howdy was the victim of a curious ambivalence; he welcomed and at the same time resented the presence of Robert Westbrook Williamston III.

It also annoyed Howdy that a time or two Jennifer had come close to defending Rob. But if Rob was attracted to Jen he concealed it well. A new brightness permeated the house and Jennifer was pleased to see Howdy responding. She wondered how much of his uncharacteristic animation stemmed from an unconscious need to compete, to keep his everlasting edge.

Late one night, after a stubborn session in the bedroom, Jennifer conviced Howdy that a cocktail party should be given during the Thanksgiving week. "Look, honey, we can't keep him in an isolation ward. All our friends know he's here. If we don't make some gesture they'll be completely convinced that the gossip is true, that we're embarrassed to present him. If we take the initiative he'll bedazzle them and they won't be so rough on Mother. The worst they can say is 'if she had to have a plaything at least she bought a delightful one.' "

Jennifer's prediction proved accurate. Some of the younger men, whose attractive, aggressive wives failed to conceal their admiration for Rob, left the party privately hating his guts. The more secure ones were unanimous in their acceptance. During the ensuing weeks the invitations to other parties reached such a peak that Jennifer was forced to plead prior family commitments.

They did accept a New Year's Eve party at the club. Rob had danced every other dance with Catherine until she herself had begged off. During his half dozen dances with Jennifer, Howdy had watched with transparent indifference.

Later, Howdy, half-joking, had accused her of encouraging Rob's attention. The subject would not have come up if both of them had not drunk full measures of champagne and brandy. But it had come up. And in a perverse way it pleased Jennifer. She would have liked Howdy to make love to her but he was either too far gone or too obstinate to respond to her tacit invitation.

Several days before the Williamstons' departure, Jennifer insisted that Howdy go over the M-Two plan with her mother. Another meeting would take place the following week.

After dinner, they drove down to the administration building, where the exhibits were still on display. For two hours Howdy went over the plan in some detail. Catherine said little but it was evident that she was impressed. Rob, who had followed the concept avidly, found it difficult to contain his enthusiasm. When Howdy finished, he bounced to his feet.

"Marvelous, Jennifer, Howdy, marvelous! The most exciting plan I've seen."

Howdy resisted the urge to ask how many such plans he'd been exposed to and contented himself with a noncommittal, "Good."

At Casa Grande, Rob paused at the front door to look back out over the broad lowlands. The fields were dark, but around their periphery the lights of the coastal communities and the newer, closer-lying suburban areas sparkled in the clear winter air. Rob indicated the dark panorama and said, "It must take real will-power to hold all those acres in agriculture when you could quadruple your worth by urbanizing the entire ranch."

Jennifer held her breath but Howdy's reply was deceptively mild. "Not if you're a country boy like me and want a little fresh air and elbow room as much as you want dollars."

An hour later, in bed, Catherine looked up from her book as Rob came out of the bathroom. He perched on the edge of the bed and kissed her temple lightly. "I can't help but think how fortunate you are, and Jennifer too, to have Howdy in the family. Apparently he's done wonders with the ranch. Certainly he's added vast amounts to the value of the stock."

"Indeed he has. I think that's why Steven was very happy to give him an interest when he married Jennifer and turn over most of the practical management to him."

Rob put his hand on hers. "It was a wise choice. But after looking at the potential in the MICROPOLIS plan, I wonder what Steven would have said about holding out the best subdivision acreage for bean and asparagus fields."

"I don't know, Rob. Generally he approved of Howdy's

535

decisions. Why do you ask? Do you question his judgment in this?"

Rob smiled deprecatingly and squeezed Catherine's hand. "I wouldn't say I disapprove. Good heavens, Catherine, I'm hardly qualified to criticize. Moreover, even if I were it would not really be my place to, would it?"

"I should think it would be your duty now, if you truly think you see a serious flaw in the plan."

Frowning, Rob rose from the bed. "I'm sure there's some explanation for it, Catherine. But I can't help wondering about two things. First, why doesn't Howdy consider planning the entire ranch into communites when urbanization is so much more profitable than farming? And second, why would he borrow millions and pay interest to build the shopping centers and recreation areas when the company could go public and raise money, interest-free, from the sale of stock?" He squinted quizzically. "I've had some experience in finance, just as Steven had. There'd be no trouble underwriting an issue. Heavens, love, with my, our connections in the East—" He broke off and shook his head. "More than that, there'd be no danger of your losing control of Los Nidos. You've said that between the three of you and the Vance Foundation, you control the company absolutely."

Rob turned to her with a disarming smile. "In any case, my love, I repeat, it's none of my business, really. And you'd be quite within your right to remind me that I'm being presumptuous." He slipped into the bed beside her. "Forgive me. Put it out of your mind. It was thoughtless of me to raise such a question when you're trying to woo Morpheus." He peered at the dust jacket of the book on the bed. "On the other hand, if you're reading *that* for entertainment, my lovely, I'm sure I can come up with at least one better suggestion."

Catherine allowed herself a small smile and inclined her head toward him playfully. "I don't think you're presumptuous, my sweet. But sometimes I think you're insatiable. Now go read until you're sleepy."

Grateful that she had not responded to his contrived overture, Rob laughed softly, kissed her again, and settled down with a copy of *Motor Age*.

In the master bedroom in the opposite wing of Casa

536

Grande, Howdy sprawled naked on top of the seven-foot-square bed and watched Jennifer undressing.

"I know we had to include your mother's little playmate. But I can tell you one thing—I felt a hell of a lot easier before he knew so much about our plans."

"Relax, darling. That wasn't real interest he expressed. It was just lip service to make Mother happy. Actually, I think it may have been helpful."

Howdy snorted. "About as helpful as a skunk at a picnic." He raised his head to watch her before she spoke again. "You know something? I think the little boy in old Robbie appeals to your mother instinct, now, don't it?"

Jennifer stiffened and glared at him icily. "No. It doesn't, you bastard." A moment later she went into the bathroom nursing a feeling of righteous injury. But it was not unmixed with guilt.

The MICROPOLIS 2000 meetings lasted a week. Thirty-three million dollars in financing would be required to initiate the first phase that would include three complete communities. A San Francisco firm was employed to analyze the charters of each adjacent city. Ev Moses, knowing the startled sister cities would fight to keep their boundaries from being frozen by the new community that would dominate them, intended to incorporate in the M-Two Charter everything that was good in neighboring charters. It was one way to soften their opposition.

Both Ev Moses and Lloyd Wilson agreed that the beach cities in particular would be foolish to oppose MICROPOLIS 2000. They would be the ones to profit the most because they had a monopoly on shore-side recreation. There was no way for a developer to duplicate beaches for inland population.

The agruicultural land that would remain was the crucial bone of contention. Once those preserves were incorporated into the new city that would be beyond county control. The hostile communities could be expected to protest violently the moment they found out that might happen. The strategy the Los Nidos Ranch Company was using was precarious but it was the only way a sizeable portion of Southern California's best agricultural land could be kept out of the hands of the developers. It wouldn't work though, unless strong political power was applied at both state and national levels. Jennifer knew

537

this and she knew that such strong pressure could be brought by only one man.

In mid-February they received a long-distance call from Catherine and Rob from their leased estate in Pound Ridge, New York. They had purchased a brownstone on East 71st Street in Manhattan. Rob's bachelor 'digs,' as he called them, had been in the building. They had also purchased the identical building next door. Both would be converted into a single three-story town house. Work would begin in early spring. Meanwhile, as soon as the plans were completed and the contract let, they would go to Palm Beach and hope to be in California again before the decorator began his work.

After one hour of conversation, Jennifer replaced the phone and retrieved the brandy snifter. Her face was a study of conflicting emotions. "Who says money can't buy happiness?"

Howdy looked up from the newspaper. "People who don't have money. Old Rob, for instance. He's parlayed a winning smile into a multimillion dollar purse, and on a slow track too."

Jennifer grimaced. "God Almighty! Isn't there any compassion in you at all? Is everyone wrong who doesn't play by your rules?"

Howdy smiled amiably. "Nope. But if they don't play by my rules, they don't play on my team. And I can't buy the notion that any able-bodied kept man has a right to play games with an old lady's misery and money. One day your mom will wake up and find out that the screwing she's getting isn't worth the screwing she's getting."

Jennifer studied her husband with icy detachment. "You know, darling, there are times when you can be a vulgar, dispicable, cob-wiping yokel."

Grinning broadly, Howdy rose. "That's not the latest news, is it?" Then his tone changed abruptly. "Look, I know what Catherine's income amounts to, after taxes. And I know pretty well what a double town house in Manhattan costs to buy and remodel. Little Robbie-boy is going to put your mom in hock. I like Catherine. And I feel sorry for her. I felt that way the day I first met her when you were a snot-nosed pigtailed kid showing off on a horse. I can see what's going to happen, and the mess is going to land on us. It sure as hell isn't hard to figure a he-whore like Williamston."

He crossed the room and looked at her hard. "So, I'll bet you a Rolls-Royce pickup truck that she'll be yelling for help before next Christmas. And six months after that, she'll be needing more income because Prince Charming will be pissing it down the drain faster than the treasury can print it." Howdy leveled a finger inches from her nose. "Mind my words. Everything I'm telling you is going to happen. It's not that I don't want her to be happy. I sure as hell do. But for God's sake, how much is a now-and-then piece of tail and a stream of charming horse shit worth?"

Jennifer eyed him coldly and deflected the offending finger. "I don't know about Mother, darling. But you ought to know what they're worth to me. And you're not even charming."

18 IN MARCH there were rumors in the press that the ranch was going to ask the county to place several thousand more acres of prime development land in the agricultural preserve. It was enough to make militant members of surrounding school boards bring great pressure on county supervisors to rescind the law that protected the preserves.

Howdy knew that some fears could be allayed by publicizing the fact that thirty million dollars would be spent in the county on labor and materials. But that was a drop in the bucket compared with the astronomical figures the aerospace industry tossed around. The county, in fact the entire state, was too dependent on Federal contracts for moon shots, satellites, and supersonic transports. California history might well be repeating a disastrous pattern. In the pastoral days, too much dependence on a single activity had cost Mexico all of Alta California.

Now, out of a total manufacturing labor force of one hundred and thirty thousand persons, fifty-seven percent of the workers were employed in aerospace or related industries. With more than fifty percent of the county's work force in one industry, Howdy had asked Ev Moses, "Who in hell can predict what will happen in a recession?" He knew it worried Ev, too, that savings and loan outfits were multiplying like rabbits.

Dr. Carl Werner of OMNISPACE had said, only half facetiously, "Don't worry too much about instability in the aerospace program, Mr. Goodwyn. We are the ultimate show business, the science that has made honest men out of Flash Gordon and Buck Rogers. We do not add to the gross national product in obvious ways; but we make our contribution in the one thing man cannot live without, high adventure and dreams of a great tomorrow. We will never be taken for granted."

Ev Moses, who had been present at the meeting, had scarcely waited for the outer door to close behind the brilliant scientist. "I'd say the good doctor knows more about elliptical orbits and propellants than human nature. There are only three things man does not take for granted—health, sex, and money."

Howdy recalled how he had started at the words. They were an almost exact paraphrase of Al Phelps' credo of thirty years earlier. The words had been worth remembering. In his own father's time it was taken for granted that the foundation of the nation's wealth was agriculture, the horse-sensical, grass roots society that set the political balance. And now, in his own time, he reflected unhappily, college boys only lately farm boys were banding together as a new generation of suburban householders to vote some of the nation's richest and most productive agricultural land out of existence to make room for what Lloyd Wilson called, "the Great Rubber-Stampede to Suburbia."

Perhaps rape was inevitable, Howdy conceded, but he'd be damned if he'd relax and enjoy it. He'd hold out at Los Nidos as long as possible, then turn pro and charge the bastards an arm and a leg for the privilege of overrunning the ranch's best land.

Alone in the Bell helicopter, Howdy cruised aimlessly over the ranch's vast acreage for the better part of an hour. Avoiding the subdivisions and schools, he dropped down to five hundred feet over the fields where men, recognizing the familiar red and white craft, signaled casual greetings.

Climbing, he inspected the firebreaks that the forestry crews had hacked out some years before. Lloyd Wilson hated those firebreaks. "They help," he had admitted, "but they're liabilities when we get flooding rains. They give up enough mud to start a small avalanche. When that's combined with the debris from the burned areas, you've got an incipient disaster. Fifty percent of the water we import to urban areas is returned as raw sewage and better than half of that can be processed. Billions of gallons a year. We should be pumping it up to irrigate new forests."

Lloyd had his chance to prove his theory now, with MICROPOLIS 2000. If it worked, even the Federal government might see that it was cheaper to prevent the fires than to fight them and lose thousands of acres of water-

541

shed each summer. Last summer, Southern California had been lucky. Only twenty-three thousand acres had been lost to fire in three counties. But this was a dry winter. The worst could still be in the offing.

On the way back to the hangar, Howdy dropped down again to inspect new acreage being set out in Valencia oranges. He spotted a parked bright green Jeepster. That would be his new superintendent, Carlos Morales.

Thinking about the progress Carlos had made gave Howdy a deep sense of satisfaction. The Mexican American was a third generation Los Nidos employee whose first language had been Spanish. He had gone to school and then on to the University of California at Davis. Later he had taken post-graduate work in viniculture and citrus growing at Redlands. There was no more promising young employee on the ranch. Howdy recalled that it had been Morales' suggestion to teach the ranch's Mexican-American workers English and to pay them a bonus for proficiency in the second language.

Quite logically, Morales had followed up with a suggestion that non-Mexican foremen and superintendents be taught Spanish and urged to learn the language with the same bonus incentive. As a result, communications had improved dramatically and with them had come a corresponding rise in productivity.

Howdy wondered if Cesar Chavez would do half as much for those workers who bought the crap being dished out by his United Farm Workers Organizing Committee. Howdy had been grimly amused by a quietly derisive remark Morales had made: "Those initials read backward spell COWFU. And that's what Chavez is peddling, all right!"

The thought of Chavez and the havoc his organizers could raise by making Los Nidos an undeserving example made Howdy curse aloud. "The blackmailing sons-of-bitches!" As he headed toward the hangar, he silently reaffirmed Lew Berwin's logic: The only solution would be a strong bi-partisan agricultural team in the senate. New laws were needed. Agriculture had no protection against labor.

It was bright twilight when Howdy returned to the house a few minutes past seven. Jennifer met him at the door with a drink. "How's the Sky Spy?"

He took the glass and gave her a peck on the nose. "I

542

wasn't snooping. I was trying to get far enough from the battle to see what I'm fighting."

Jennifer let him settle down before she spoke again. "I've got ginger-peachy news for you, darling."

"What's that?"

"Mother called today. When Rob finishes handling the remodeling of the town houses he wants to come out and help us with M-Two."

Howdy gave her a searching look. "He can help me by staying right the hell where he is, three thousand miles from here."

"He's serious."

"So am I."

"Mother said he wants to help arrange financing in New York. She says he has good contacts there."

"So do we." After a moment he looked up, questioning. "What did you tell her?"

"I said I'd talk to you and I said to tell Rob that you're grateful for the offer."

"I'm not. I want him to keep his effing nose out of M-Two. And I sure as hell don't want him yakking about it around Wall Street, or wherever his alleged connections are."

"Actually, Howdy, Mother says he's had as much finance background as Dad had. I didn't know that."

Thoroughly annoyed, Howdy arose abruptly. "I don't give a Goddamn if he was personal adviser to J. P. Morgan and Andrew Mellon. I don't want that fucking widow lover meddling in our business."

On the verge of anger, Jennifer got up to face him. "Really, Howdy. You don't have to like the man, but we can be civil to him for Mother's sake. That's all I'm thinking about, really. It's very important to her self-respect to feel that Rob is making some sort of a contribution."

"He is making a contribution by keeping Catherine happy, horizontally, vertically, any old way. That's all he's cut out for."

"Howdy! What went wrong today?"

Howdy swung the ice in the drink and took a swallow. "Nothing went wrong today, until now. In some areas I'll put up with him for your mother's sake. But you both better know that I'll go back to strumming a guitar and selling Peruny before I'll let him get mixed up in management or finance and I'm saying that knowing full well that you and your mother can outvote me and vote me out."

Suddenly very upset, Jennifer set her brandy glass down hard. "Shit, Howdy. Don't play showdown with me. I'm your loving wife, remember? I'm with you. And don't make the poor man into a monster. At least pretend to consider his offer, for Mom's sake. It'll make things a hell of a lot easier all around. She said Rob's been studying M-Two. He thinks it's the most exciting thing he's seen since—"

"Since your mother's checkbook?"

"Goddammit, Howdy, stop it. He thinks it's the most exciting new living concept he's ever heard of and he's been working on a financing plan that may save the company several millions in interest. He wants to talk to you about it."

For the first time in a long while, Jennifer saw the strange veil-like change in Howdy's eyes. Ignoring her words completely, he rose from the chair, set the unfinished bourbon on the side table, and looked off into the dining room where Rosita was setting the table for two. "Why only two? Where's the boy?"

Jennifer took her time replying. "He and Doug have something doing at school tonight. I told him he could have dinner at Doug's. Doug's father will drive him home about nine-thirty."

Dinner was eaten in uncharacteristic silence and Jennifer retired shortly after Howie came in. The boy sprawled with his father in the family room for a while watching a network special on the redwood forest controversy in northern California. When the commercials came on he excused himself and went off to his room. Neither father nor son was aware that the only exchange between them had been, "Hi, Dad."

"Hello, son!"

And, 'G'night, Dad."

"Good night, son."

It was enough for Howdy that his son was there with him for a while during the evening. Young Howie seemed to expect no more.

Early in May, Jennifer and Howdy received a letter from Catherine saying that she and Rob were flying out within the week. There had been a change in plans that they wished to discuss. Beyond that there was no hint. Speculation that had ranged from hopeful to the impossible ended at dinner on the night of their arrival, Catherine turned to Jennifer.

"I'm sure you will think that I've suddenly lost my mind, dear. But after Rob and I got into the bother of doing over those two brownstones it suddenly seemed like an unwise thing to do, particularly since Rob has really fallen in love with Southern California."

Jennifer shot an anxious glance at Howdy, whose impassive expression revealed nothing. "But Mother, I thought you had already bought the two buildings."

"When I wrote to you, dear, Rob planned to open the escrow the following week. There had been a very inexpensive option. But when we got the first estimates we felt it was the better part of wisdom to sacrifice the five thousand dollars rather than commit for a hundred times that. Actually, what we plan to do is keep the apartment at the hotel and look into building a little place out here. Both Rob and I love the solitude of Laberinta."

Jennifer looked at her mother in open disbelief. "But ever since I can remember, you have always said you could not abide canyon living. You said they made you feel penned in, that they were dark and dangerous."

Smiling tolerantly, Catherine placed her napkin beside her plate. "Dear, that was before I found out how dangerous the city has become. Rob and I are agreed on that."

Still unable to accept the change, Jennifer persisted. "Even so, mother, if you do want to live on the ranch again, even part of the year," she indicated the room, "this house is really yours. We should be the ones to find another place."

The thought seemed to horrify Catherine. "Indeed not. When you and Howard married I gave this place to you. There were no strings or reservations, Jennifer. If Rob and I decided this is a wise thing to do," she said as she turned to Howdy whose expression still revealed nothing, "after we've had everybody's counsel, then we shall do a modest place in the canyon. When Howie drove us up there last trip I was amazed at how beautiful it is. I'd forgotten, really. And we found an absolutely ideal place just below the old springs. Lovely old trees. Close to the stream." She turned to Howdy again. "Is there any reason why we should not do a little place up there, Howard?"

Howdy pushed back his chair and pondered briefly. "Nope. You can build there. If you're sure you want to. But I wouldn't put up anything too elaborate. And I sure as hell wouldn't pick a site too close to the creek."

Rob stirred uneasily. "Is there danger up there? From floods?"

Howdy nodded. "And fires. But you'd be all right with a little common-sense precaution. Since the burn-offs the stream gets pretty wild in the spring. Four cabins got washed away last year. They weren't ours. They were on the land Steve leased years ago. There are quite a few permanent people up in there now. If you and Catherine are really looking for solitude, you should know that the state stocks the stream and we allow fishermen in there from spring through the fall."

Sensing Rob's growing anxiety, Jennifer sought to reassure him. "Actually there are not all that many transients up where you two would be. And the regulars in the canyon love it and do a great deal to protect its natural beauty. They have their own private fire brigade and one of the men is a deputy fire warden for the county. I think there'd be little danger if you resist the temptation to build right on the stream like so many others do." Jennifer hoped she had sounded convincing and she silently upbraided Howdy for the subtle negativity that he had managed more by tone and manner than by words.

Later, in their bedroom, she chided him. "You were very clever, darling. You made Laberinta canyon sound like the setting for the apocalypse."

Howdy addressed the image of Jennifer reflected in the dressing table. "There are a lot of things I am. But there are a lot of things I am not and one of them is a hypocrite. I told you I don't want that son-of-a-bitch within three thousand miles of this ranch. And you'd better believe that I'm going to do everything I can to see that he's entertained elsewhere."

Jennifer turned on the flounced stool to face him. "Don't be a bastard. Mother has no family in the East anymore. Just a couple of distant cousins in Massachusetts. In spite of all of her talk I think she really feels this is home now. I mean California, the ranch." For a moment she studied him narrowly. "You know something? I'm beginning to think you're afraid of poor Rob."

Howdy came out of bed in one violent move. "You're

546

goddamned right I'm afraid of 'poor' Rob. I'm afraid of any poor blue blood with so little talent and pride that he'd rather play lapdog to a rich old lady than try for an honest living." He moved closer, shaking a finger in her face. "He may not be as bad as some. And that makes him even more dangerous to us. He may just have enough pride to feel that he's got to get involved in things here to justify the dough Catherine's lavishing on him. And all that can do is make my job twice as hard and cause a split in this family because sooner or later, if he starts fucking around with M-Two or anything else here, I'll kick his ass right off the place."

As angry as she was, the faint whiteness around Howdy's compressed lips warned Jennifer not to pursue the subject further for the time being. Cocking her head to one side, she eyed him critically. "Well, if that day ever comes, darling, I hope for Rob's sake that his 'ass' is better protected than yours is at the moment."

Suddenly aware that he was without pajamas, Howdy stalked back to the bed and pulled up the covers. Out of the corner of his eye, as he turned away from the light, he caught a glimpse of Jennifer back at the mirror again. He detected a touch of smugness in her amused smile. The picture of himself standing there shaking his finger under her nose in a towering bare-assed rage brought a reluctant smile to his face too. He'd handle Rob all right. And he'd do it without laying anything he valued on the line.

The site Catherine and Rob chose in Laberinta Canyon was in a clump of fine old sycamores well back from the stream. By early summer Lloyd Wilson Associates had completed preliminary plans for Casa Escondida, Hidden House, and the contracts for excavating and foundations had been let. In late September Catherine and Rob held a house warming. The cottage was large, rustic, made of stone and redwood and perfectly suited to its surroundings. Several days later young Howie remembered a number or ornate handcarved doors and panels that had been stored for years in one of the old adobe buildings near the ranch offices. When Rob and Catherine saw them they immediately arranged to have them cleaned up to replace the ones Wilson's architects had called for.

"Someone must have spent a fortune on these." Rob had been caressing the deeply stained oak *dibujos* the designs, primitive and unmistakably Mexican, that had been beautifully carved into the richly grained wood.

Catherine studied them, concentrating. "You know, I vaguely remember Steven's father referring to these once. But for the life of me I can't recall in what connection."

The architect's assistants, with some reluctance, had altered dimensions and two of the carved doors, used as a pair, were adapted for the front entrance. The entire job, not counting the token land lease, had cost seventy-eight thousand dollars.

Summer passed to fall. The days were occupied with furnishing, decorating, and landscaping. Several times Howie managed to coax Rob down to the beach for surfing lessons. But for the most part, he seemed intent on devoting as much time as possible to Catherine and Casa Escondida.

Howdy watched this devotion with distaste. "What a hell of a way for a guy to live. I'd rather haul garbage than be an old lady's kept man."

Jennifer winced but said nothing. One of the qualities that made Howard Goodwyn attractive was his independence, his insistence upon pulling more than his share of the load. He was rough, she admitted, often ruthless, frequently crude, and, lately, absorbed in ranch projects to the exclusion of all else, even politics. But for all of his neglect of family he was the most responsible man she had ever known.

Immediately after Labor Day, Catherine and Rob left for the East. After a look at the fall theater season, they would return to California for the holidays. Then they would go to Puerto Vallarta or Acapulco for a month or so. There had been some talk about possibly getting a place down there.

Lew and Myra Berwin flew down from San Francisco to spend New Year's Eve at Casa Grande. They drank, watched the New Year arrive in New York, danced to the Hi-Fi, toasted their own new year with champagne, finished off with stingers, and retired by one-thirty.

Jennifer had just gotten undressed when the silence was shattered by the flatulent flutter of a souped-up exhaust. A sports car skidded to a stop in the driveway, a door slammed, and a moment later she heard the car roar away down the hill. Howie let himself in the back door and went to the kitchen. Jennifer hesitated, then pulled on a robe and waited for him in the hallway. When he appeared, after several minutes, she could see that he was walking unsteadily. Her greeting startled him. A bit bleary, he stopped in his tracks and grinned at her stupidly.

"Hi, Mom. You still up?"

Jennifer folded her arms and inspected him deliberately.

"No. I'm sound asleep. And I'll give you three guesses. Where are you?"

Howie feigned injury. "What do you mean, 'where am I'?" He pointed to himself. "I'm home, before two o'clock, just like you and Dad said." He stood his ground defiantly as she came closer and sniffed.

"You smell funny. Have you had anything to drink?"

"Yeah, some milk." Jennifer reached out and unsnapped the formal tie that was dangling by one side of its clip.

"I can see that. It's dribbling down your chin. I mean did you kids fool around with any liquor?"

Howie pulled away and made an attempt at righteous indignation. "Wow! You're putting me on, Mom. I didn't drink a thing except punch at the club. They won't give us anything. You know that."

Jennifer frowned and sniffed again. "You smell funny just the same, kind of sicky sweet." Howie's face contorted with long-suffering innocence. "Well, cripes, Mom. We danced and ate a lot of stuff. I could smell like a lot of things."

"You do, baby." She turned him around and gave him an affectionate swat. "And you look plenty beat. Go shower and we'll talk in the morning." As he reached the bedroom door she called to him again. Howie stopped with an "Oh God, what now?" expression.

"Yeah, Mom?"

"Happy New Year, sweetie."

Lifting an arm wearily, he nodded. "Yeah, Happy New Year, Mom." In the doorway he paused again. "Mom."

"Yes?"

"Do grandma and Rob have anybody at the house?"

"No. Why?"

Howie shrugged. "I thought maybe Rob'd do something with me tomorrow. Like ride, maybe."

"Try him, darling. I'm sure he will. Good night, now."

In the master bedroom Jennifer climbed in beside

549

Howdy. "Your son came in looking a little goofy and smelling funny."

Howdy squinted at her sleepily. "When he stinks he's my son, huh?"

"I'm serious, Howdy. He's been drinking or something."

"Him and fifty million others." Grinning lasciviously, he moved his hand beneath the blanket until it came to rest on her bare stomach. Smoothing her skin lightly, he moved his nose inches from hers. "It's the 'or something' he's too young for." When Jennifer continued to regard him with a baleful stare his grin broadened. "Did anyone ever tell you you're beautiful?"

She shook her head. "Not when he's sober, no. And did anyone ever tell you that you get horny when you drink and you're a hell of a lot better when you don't?"

Howie appeared while they were having brunch in the patio beside the pool. He acknowledged the greetings perfunctorily and flopped down on a lounge chair. Jennifer turned to him, questioning.

"Didn't you see Rob, honey?"

"Yeah."

"Well?"

"He didn't want to do anything."

"Really? He should be feeling okay this morning. What's the trouble?"

Howie shrugged. "Don't know. He acted funny."

"You mean he wasn't his usual ebullient self?"

"He wasn't happy, if that's what you mean. He was down. Real down."

"Did you see your grandmother?"

"Yeah, for a second. She was down too. She didn't look very good."

"Oh?" Jennifer turned back to the chafing dish as Myra came up holding out Lew's plate for seconds.

"Trouble in paradise?"

A sharp look made Myra glance around furtively. "Sorry." She whispered the apology and continued in a barely audible voice. "I hope it had nothing to do with the way he was dancing with me last night. Your mother was watching him like a hawk."

Jennifer turned, perplexed. "For heaven's sake, what does that mean? I was there and reasonably alert, or so I thought."

550

"After the second dance your pretty blonde stepdaddy was 'ready Eddie.' "

Jennifer closed her eyes and groaned. "Oh, dear Jesus. Not *that* sick bit!"

Myra took the plate. "Pray for him, sweetie, not me. I didn't hang him up."

"I'm not inferring that you did. It's just beginning to happen, that's all. These March-November things never work out."

Myra arched a brow over a long green, knowing eye. "Especially not when March has a hard-on for somebody else. It's none of my business, but you and Howdy were batty to bring them out here."

After brunch Howdy turned to his son. "Your Uncle Lew and I are going to take the chopper and nose around. Care to come along?"

For a moment Howie thought it might be something to do. "No, Dad. I don't think so—I'll just goof off around here."

Howdy was visibly disappointed. "How about driving us down to the hangar?"

The boy brightened. "Hey! Cool! Call me on the short wave when you're coming back? I'll come get you, too."

Ten minutes later, at the hangar, Howdy repeated the usual caution. "Remember what I told you about driving?"

"Sure, stay on the ranch roads. Don't drive on the county roads until I get my learner's and be careful I don't flip it."

Howdy and Lew watched Howie wheel the jeep expertly around. Lew laughed sympathetically. "That's a horrible age to be. Remember?"

"I remember all right. I had to work my ass off. Back in Guymon in the late Twenties I didn't have time to wonder what I wanted to do. These kids have it too easy."

"Howdy, can I tell you something? It's the same with every generation. I remember Grandfather Sam giving me hell when I was Howie's age because I wanted to go out to Stinson Beach swimming on Sunday. He said Sunday was the start of the week and I should be learning the business."

Howdy smiled dryly. "The week never had a beginning or an end when I was a kid."

"For you, my friend, they still don't."

A half hour later Myra arrested a half-finished vodka and tonic inches from her lips and watched the helicopter rise from behind the trees and swing away to the west. "There go our men. Bless their throbbing middle-aged hearts."

Jennifer retrieved the brandy snifter. She and Myra had decided that what they needed was "a hair of the dog." "Howdy enjoys himself when Lew comes down."

Myra nodded. "Ditto for my guy. If you hadn't invited us, Lew would have dreamed up some excuse anyway. Those two have more fun than anybody with their grown-up games." She took a long swallow of her drink. "You know something, Jen? I think if I'd had my druthers I'd have asked God to fit me out with whiskers and a set of balls. How about you?"

Jennifer smiled pensively. "No. I like it this way."

"Sure you do. Why not? You're a part of the action around here. If I envy any female in the world, honey, I envy you."

"Why me?"

"Because you're on the inside." A note of sodden persistence had crept into Myra Berwin's voice. She regarded the empty glass petulantly and rose to refill it.

Jennifer took it from her. "Stay where you are, I'll fix it."

When Jennifer returned with the drink, Myra reached up eagerly. "Thanks, angel. You know something? I'm getting smashed again. Goody." She took a sip and approved it. "Now then, where the hell were we? Oh, yes. I remember. You're 'in' and I," she broke off and studied the glass, "and I'm nowhere."

Jennifer was in no mood for female confidences but Myra was determined. "It surprises you when I say I'm nowhere, doesn't it, Jen? I mean . . ." Myra gestured spastically and her cigarette flew from her fingers onto the glazed patio tile. She regarded it with unfocused eyes and shrugged. "To hell with it. Even if I am scared. Shitless."

"You? Scared? What on earth for?"

"Because Lew's reached the age where hard drivers like him begin to drop over."

Jennifer's first impulse was to reassure her, to say, "That's ridiculous. He's in wonderful shape." But she couldn't manage it.

Somehow it seemed to be tempting fate. A similar

vagrant worry had troubled her at times, usually after the death of one of Howdy's contemporaries. A distressing number of them were gone now. The most recent was Bronc Bronston, whom she had met while they were negotiating the sale of Rolling-O Productions to a television network. "But Myra, it could happen to any of us, an accident, anything." She pointed toward the lower fields. "Lately I worry about Howdy in the chopper."

Jennifer could see the glaze gathering in Myra's eyes, a moist alcoholic luminance not unlike the look of the imminent tears that had blurred her own vision in boarding-school days when her mother and father would come to say good-bye before they left for Europe There were times when Jennifer still could feel the tight lump of loneliness that had constricted her throat

The freshening breeze had begun to ruffle the surface of the pool. Myra shivered.

"How about going inside?" Jennifer asked.

Myra passed through the patio door a bit unsteadily. In the family room she perched precariously on a bar stool while Jennifer freshened their drinks. "Excuse me, doll. This isn't Mrs. Lewis Berwin's best day. As you can see. I am slightly fucked up on this first day of the new year of our Lord One Thousand Nine hundred and, screw it. I've had too many of them." She smirked at the glass. "Years. Not drinks."

Concerned, Jennifer suggested coffee. "No thanks, angel. I dont want the novacaine to wear off."

Usually Jennifer enjoyed her gabfests with Myra. Now she wished to God Howdy and Lew would come back.

Cautiously, Myra guided herself to Howdy's big chair at the side of the fireplace and collapsed into it. Several minutes passed in silence and then her mood seemed to change. "Baby, forgive me for crapping up the afternoon. Maybe I'd better go see a head-shrinker. I've made it all the way from Castoria to Serutan. I don't want to blow it now. I don't know what the hell's with me." She slumped cornerwise against the arm and the back of the chair. "Let's put it this way, if I don't have Lew, I don't have anything. His doctor told him to slow down and lose weight two years ago. He goes on a diet for two weeks. Then comes this bright idea that he can save California if he can promote Howdy. So now he's right back in the grind again, worse than ever. As sure as I'm here, I know

553

that he's not going to make it to the Social Security office. I just know it, honey." She lowered her head. "Up until a few years ago, we had a ball. Do you know when I started playing with colored crayons? When Lew began to speed up in the office and to slow down in the sack. We haven't had a real old-fashioned tooth-and-nail orgasm for two years. I used to love it when Lew would slide his 'Jolly Pink Giant' into me. I used to grab the hair on his shoulders, yell, 'Hallelujah,' and kick his galloping ass black and blue. For me, he was the greatest. And baby, I've known a few. Now, if we make it twice a month I have to work him like a Polak sausage stuffer. And half the time he leaves me hung higher than a hooker's douche bag." She broke off and raised both palms to her cheeks.

"Oh, Christ! There I go again. I'm sorry, Jen. I truly am. I had no intention of getting into this. I don't even know how it happened. It's just that I'm so—so—" She took her hands from her cheeks and held them out, appealing. "—so goddamned *lost*."

"Last night when Super-Jock tried to dry-fuck me on the dance floor, I almost got excited, for a second. Then I remembered what he is, a mother-grabbing he-whore. Is that what's going to happen to me when I'm a loaded widow? Next year? Five, ten years from now? All I know is, women who are over the hill have a lousy deal. Criminal."

Without warning, Myra leaped from the chair, drew her arm back and shattered her glass against the fireplace. Horrified at her own action, she stared open-mouthed at the debris, shot Jennifer a wild look, and fled from the room.

After dinner alone, Jennifer and Howdy retired to the family room. There Jennifer told Howdy about Myra's outburst. Slouched in his chair, Howdy listened with half-closed eyes. "Would you like to know what I think?"

Jennifer eyed him skeptically. "From the look on your face I'm not sure."

"I think she's setting up a real good alibi for a little screwing on the side."

Jennifer stared at him. "For God's sake, Howdy!"

"Don't brush it off too fast. Look at her track record. Do you think Myra trapped Lew with her beautiful brain?"

"Of course not. He wanted a stunning, sexy woman who wouldn't get under foot unless he asked her to. In return for that he's given her everything she ever wanted."

Howdy inclined his head in conditional agreement. "Could be. Except now that she's getting a little older maybe she feels she's not getting as many of Doctor Berwin's famous beef injections as she needs to convince herself that she's still a young swinger."

Jennifer concealed her surprise and her eyes narrowed to sapphire slits. "You're vicious, Howdy. Myra would rather cut off her head than cheat on Lew. She knows what she's got. And she truly loves him. She's told me so a dozen times. She'd do anything to make him happy."

Howdy's knowing grin was exasperating. "That's right. She'd do anything to make him happy and everything to convince herself she's not beginning to fall apart. Maybe I don't know much about women, but I know a hell of a lot about people. And the safest bet in the world is that under certain conditions they're going to behave mostly alike."

An outraged wail escaped Jennifer. "That is the vilest mess of mixed-up male reasoning I've ever heard. If you believe that crap then one of these days I expect you'll be having me trailed to make sure I'm not orgasming under some young stallion to prove I'm still good."

Howdy grinned broadly and deliberately lapsed into his old Southwestern drawl. "Not a chance. I figure to keep you squirming happily until I'm going on ninety. Then I'll just roll over with a smile on my face and check out. And since my last thoughts'll be about what a great lay you've been, I reckon the undertaker'll have a tough time getting the lid on the coffin."

"Oh God! How corny can you get?" But in spite of herself, Jennifer laughed and crossed to his chair, then slipped into his lap. For a time she wondered what he was thinking. After he had been a couple of hours with Lew, Jennifer felt pretty sure she'd be right on the first guess.

"Lew's hard to say 'no' to, isn't he?"

"Not really, if you want to."

"Oh? And here I've been feeling guilty about starting this idea." When Howdy did not respond she looked up at him curiously. "*I did* start it, you know."

"If it makes you feel guilty to believe that, go ahead."

She pushed away to look at him. "You always need that

damned 'edge' of yours, don't you? What are you so insecure about?"

"Nothing, if I don't make any stupid moves."

"You won't. You can't lose."

He eyed her sternly. "Don't *ever* say that! It's just plain inviting the stuff to hit the fan. Don't even think it."

"The power of positive thinking?"

"The weakness of being cocksure."

Jennifer nodded. "Where you're concerned, darling. I confess that's one of my weaknesses. I'm absolutely cocksure that if you really want to run for a Senate seat you'll make it. But even if you didn't win the primary, you couldn't lose."

"Why?"

"Because you'd have piled up enough support to make a deal for agriculture with the front runner."

Mildly surprised, Howdy studied Jennifer. "You've got things pretty well figured out, haven't you?"

"All but the most important thing."

"What's that?"

"Whether there'll be room for little old Jenny on the bandwagon."

During the long, deliberate silence Howdy watched Jennifer's uncertainty turn to hurt and then to defiance. His sudden laugh startled her and he reached out to draw her close again.

"Jen, I'm not going anywhere without you. You're my luck. Remember?"

Catherine and Rob returned from their Mexican vacation during the first week of February. The trip had been a huge success for Rob. Deeply tanned, he was filled with enthusiasm for the things his wife's money and his own vitality made possible. Catherine announced that they had put down an option on a house in the Mexican resort village. Howdy was barely able to conceal his disgust. Later he exploded to Jennifer.

"A twelve-hundred-a-month hotel apartment in New York, a seventy-thousand-plus house in the canyon, and now a sixty-five-thousand dollar shack in Acapulco. Christ on the cross! What next?"

"For heaven's sake, Howdy. What difference does it make? It's her money. She can afford it. And besides, you yourself said that you want him as far away from the ranch as possible."

Howdy turned on her and the flames crackling behind him in the huge family-room fireplace seemed to Jennifer to intensify his rage. "I do. God knows I do. But did you get a good look at your mother's face? Did you?"

"I hardly expect her to be a radiant bride at her age."

"Bride, hell. That's not what I'm talking about. I mean that look in her eyes that says she's just along for the ride, to pay the bills."

Jennifer gestured impatiently. "For God's sake, Howdy, who ever thought this was a love match? Least of all Mother. She's made a bargain she thinks she can live with, money for companionship. As long as she can live with it it's none of our business.

Howdy shook his head impatiently. "Look, Jen. It's not the dollars I'm thinking about. It's Catherine. The day is coming when she's going to be hurt a lot worse by the bargain she's made than by being lonely. Every time she writes a check she writes off some of her pride and self-respect. When that's all gone she'll still have money. Sure. But we're going to have a bitter old lady on our hands."

Suddenly grateful for a glimpse of an unexpected capacity for compassion in Howdy, Jennifer's eyes misted. "You're an incredible man. Just about the time I think I've got you all figured out at last, you take off the mask and show me still another man." Reaching up, she smoothed the graying hair at his left temple. "Gemini. I should have suspected that."

"You believe in that astrology junk if you want to. I'll just go along with human nature." He squeezed her shoulders. "Years ago, when you were a kid, I didn't understand Catherine. I thought she had all the warmth of a lemon popsicle. But after Steve was killed and I got to know her, I liked her a lot better. That's why I don't want to see her get kicked in the guts. For your sake, too."

Jennifer regarded him with wonder. "You're not twins. You're triplets!"

Howdy knew that when news of MICROPOLIS 2000 reached the labor union he could expect a strong reaction. He was not surprised when he learned that Luis Morales, chairman of the Regional Agricultural Workers Organizing Committee, appeared at the ranch. His cousin, Carlos

557

Morales, alerted Howdy. "Luis came by our house last night, Mr. Goodwyn. He was not at all subtle in his attempt to find out how much acreage would be taken out of agriculture production."

"What did you tell him?"

Carlos shrugged. "The truth, that I know nothing."

"Was there any union talk?" Howdy asked.

"Not in front of me. He knows how I feel about that for Los Nidos workers."

Howdy had never doubted that when the time came for a confrontation at Los Nidos it would be Luis Morales and his group who would make the first move, with Chavez backing. "Carlos, so you don't have to guess any more, no Class One land will be taken out of production for the M-Two Plan. Before your cousin gets our workers riled up and nervous, why don't you find a way for him to accidentally run into me?"

A knowing smile warmed Carlos' face. "I understand, Mr. Goodwyn."

Howdy left then for a leisurely walk through the packing house. Following the Valencias, Howdy watched the oranges drop through the rollers onto conveyors. There the ranch's force of Mexican-American women separated, graded, and boxed the fruit. They were a happy and contented lot. Most of them had husbands and children. Some had grown grandchildren who finished their schooling and were working on the ranch. Many of them were living rent-free in modern ranch cottages and receiving a good portion of their fresh food. They were good people making a good living under good conditions: the best in the country. But because human nature is so predictable, Howdy knew that many of them would listen to a Cesar Chavez and a Luis Morales and that a certain number would agree to go along with union demands. What they might get would actually amount to less than they were getting now, but there was no effective way to convince them of that.

Frustration and growing anger at what he conceived to be the injustice and plain stupidity of indiscriminate union demands filled Howdy as he drove through several miles of the ranch's citrus groves to the Casa Grande and lunch with Jennifer and Ev Moses.

Shortly before two o'clock the phone rang. Howdy took

the call in the family room. It was Carlos Morales, calling on the ranch's short-wave radiophone system.

"Mr. Goodwyn."

"Yes, Carlos?"

"I thought you'd like to know that the first two mobile home units have been trucked in for the new employee housing project. We'll be setting them on their pads this afternoon."

Howdy was mildly surprised that Carlos would trouble him at hime about a fairly routine matter until he heard Carlos say, "I thought you'd be interested to know that my cousin is down here. I don't think he really believes this fancy housing is intended for field foremen and their families."

Howdy hesitated briefly. "I'll drop by. If Luis is still there maybe we can convince him. Thanks, Carlos." As he hung up Howdy smiled to himself and reaffirmed the wisdom of his choice for ranch superintendent. Behind the man's easy-going manner was a trained mind that missed nothing.

On the pretext of inspecting some experimental asparagus fields, Howdy circled around the housing to approach the project.

As he pulled into the area, he saw Carlos and his cousin standing by the excavation for one of the oversized septic tanks that would serve groups of four of the thousand-square-foot mobile homes. Carlos immediately excused himself and came over.

"I'm glad you happened by, Mr. Goodwyn." He indicated the first of the new units. "My wife got a look at the kitchens and bathrooms in these homes. She wants to know if we can trade the superintendent's house for one of them."

Silently Howdy blessed Carlos for scoring a telling point within earshot of his cousin.

"If we have any more trouble with the plumbing at Casa Grande, Carlos, we may put one of these up on the hill, too."

Turning to look over the area where twenty-five mobile homes would be installed in the first unit of the project, Howdy pretended to discover Luis Morales. Looking pleasantly surprised, he strode over and offered his hand. "Hello, Luis. I heard you were down visiting your family. Glad to see you." His arm swept the construction area,

559

which included a community playground, a half-finished swimming pool, and two service buildings to house social rooms and laundromats. "Pretty fancy new housing, eh?"

Luis Morales smiled guardedly. "It will be a showplace all right."

The subtle inference that the ranch was constructing the model project as much for political advantage as for needed new housing was not lost on Howdy, but he betrayed none of the quick annoyance he felt. Instead, he smiled broadly. "I found out a long time ago in show business that the setting isn't the whole show. These houses are better, more livable, and about thirty percent cheaper than conventional construction of the same square footage." He pointed to a stack of components. "When they get the porches and lanais and carports hung on them, they'll equal any good tract house."

Luis Morales nodded. "At least they don't have the barracks look of a trailer park."

Howdy chuckled amiably. "They're for employees, not just ordinary people."

From Carlos, Luis Morales had already learned that there would be complete integration of Anglos and Mexican-Americans in the housing project. He had to conceal his overt approval of the plan, but privately he was delighted with it. He was seeing the sort of luxury quarters his own parents and grandparents dared not even dream about in their day. Howard Goodwyn was speaking the truth when he claimed that the housing represented an innovation, even for foremen. Nevertheless, he found it easy to convince himself that the ranch might have settled for inexpensive renovation of the present fifty-year-old cottages if the unions hadn't put increasing pressure on agribusiness. The pressure would be continued and increased.

Luis Morales was familiar with every detail of labor conditions on the ranch. Los Nidos would be the union's toughest organizing challenge. The old abuses were negligible. Living conditions, pay scales, and benefits were the best in the industry. This meant that excuses would have to be found, or situations created, that would justify organizing these workers. Los Nidos was the prime target in California's five-billion-dollar agribusiness. If it could be organized, all of the others would be forced to fall in line.

Luis' satisfaction at the progress he was seeing was lessened somewhat by the certain knowledge that when the confrontation came a lot of good workers on this ranch, including members of his own family, would be hurt. But that was part of the price. Time and time again he had told himself it had always been so in labor wars because it was the only way.

During the next week rumors about MICROPOLIS 2000 were so rampant that Howdy and Ev Moses were forced to hold a press conference. Lloyd Wilson and his associates gave the press the same exposure to the plans that he had given the Los Nidos owners. Afterwards, for a week, there was apparent press approval, even enthusiasm. Then the *La Playa Gazette* and the *Los Campos Journal*, the two widely read weeklies, published full-page editorials expressing grave concern. The changed point of view seemed inexplicable until Howdy learned that something called The Five Cities Committee had been formed to study the effect of MICROPOLIS 2000 on adjacent communities and take whatever legal action was indicated to avoid being "gobbled up by a new metropolitan monster."

Immediately, the local dailies and the regional editions of the Los Angeles papers began probing members of the Five Cities Committee. It didn't take long for the big papers and wire services to start interviewing local businessmen and city officials. Flattered and puffed with importance, some of them began issuing belligerent statements.

Howdy called a strategy meeting. He held up an advance copy of a statement by Mayor Howell for Ev, Jennifer, Lloyd Wilson, and others to see.

"I want you to listen to this. It'll give you a rough idea of what we're going to be up against."

GRAVE CONCERN EXPRESSED
OVER LOS NIDOS PROJECT

"Mayor Henry Howell of La Playa, recently elected chairman of the new Five Cities Committee, last night expressed fear over MICROPOLIS 2000, the huge city complex planned on more than fifty thousand of the Los Nidos Ranch Company's ninety thousand acres.

"Mayor Howell, said, 'As long as the friendly green giant in our backyards was content with his farming, we neighboring communities had little to

561

fear from him and much to thank him for. But now that Howard Goodwyn seems determined to become the Howard Hughes of city building, we must reassess our positions.

"We do not quarrel with Mr. Goodwyn's contention that much good can come to our resorts from the influx of a half million people over the next two and a half decades. But we are disturbed by the very real possibility that many people attracted by the MICROPOLIS concept will come to look but will wish to settle in our long-established beach communities instead. The question arises—where will we put them within the confines of our present city limits? The answer is clear. We can't take many—unless we expand. The proposed boundaries of MICROPOLIS 2000 block inland annexation. That leaves only the coast, which is still in agriculture and grazing. Since it is potentially the most valuable commercial land they have they're not likely to let it go. It seems then that we may soon run the risk of becoming the fenced-in beach playground for a new city the size of Cincinnati or Denver or Atlanta."

Howdy flapped the clipping impatiently and tossed it aside. "Howell's just talking for the record. He's always been friendly. But he's going to start up some of the others."

Ev Moses smiled thinly. "I don't think we're going to have to play the card yet, but something tells me that open ocean-front acreage is going to come in very handy soon."

Jennifer, who had picked up the clipping, finished scanning it. "Oh, God, I suppose this means another battle with the Board of Supervisors?"

Howdy nodded. "And the County Planning Commission." He glanced over at Peter Daniels. "Two of the five supervisors are friendly but that fellow Evans was on the fence until the last minute with the preserve decision. Do we know where he stands now?"

The public-relations expert weighed his answer. "No. But we helped his campaign, quietly."

Howdy draped a leg over the chair arm. "Yeah, so quietly he might not've known who the money really came from."

562

Dousing his cigarette, Daniels looked around the table innocently. "If necessary we can find a way to whisper in his ear. But if we're going to have a fight the best thing to do is to start getting public opinion on our side."

It amused Howdy that the public relations and political patterns always seemed to be so similar. Daniels was suggesting the same strategy that Lew Berwin wanted to employ to fight agriculture's battle politically. "Okay. What I want one week from today is a preliminary plan of action on three fronts. I want an approach from Hamlett and Daniels on the best way to secure county planning approval, the best way to keep the Five Cities Committee from pushing the panic button, and the best way, on a long-term basis, to sell the taxpayers on the idea that M-Two will be good for their health and their pocketbooks."

Hamlett and Daniels struggled to conceal their chagrin. The problem Howard Goodwyn had just handed them was all but impossible to solve in so short a time. Hamlett, the elder of the partners, mentally canceled a long-planned weekend trip and resigned himself to a succession of eighteen-hour days.

The general discussion of alternatives continued for over an hour. No one present felt any real optimism. Jennifer and especially Ev Moses were aware of the flaws in the argument that new homes would provide more tax income for both county and state than the revenue received from agribusiness. Former governor Pat Brown had proved it simply was not true. But nobody had listened when he showed that taxes from farm income actually underwrote the cost of providing community services to lower-income housing. It cost an average of up to fifteen thousand dollars a family. Only homes at thirty-five thousand dollars and above paid their way. If urban sprawl overran the few remaining ranches in the county, the local government would go bankrupt trying to provide basic living amenities. The alternative would be to hike property taxes still more. The preserves had been a last ditch stand against the politicians who wanted to hold the line for urban home-owners by taxing the ranches out of existence. Supervisors who could get away with that could also be assured of a lot of sympathetic new voters in their districts. They won both ways, temporarily.

At the Casa Grande following the meeting, Howdy

563

paced the family room clutching a drink. Worried, Jennifer, perched uneasily on a bar stool, watching him. Seldom had she seen him so preoccupied.

"Howdy."

"What?"

"You always say 'fight the ones we can win and negotiate the ones we may lose.' It seems to me that's what we have to do."

"Yes, ma'am. But first we've got to figure out which ones we can win."

Disgustedly, Howdy switched on TV. NBC's meteorologist, Gordon Wier, appeared on the screen. Jennifer referred to him as "Weeping Wier, the Voice with a Tear." Wier announced that a new storm front was approaching from the Gulf of Alaska. Behind that was still another front that promised to supplement the rainfall that had begun shortly after the new year and had scarcely let up since. Howdy turned off the set and Jennifer grimaced. "If you don't like that forecast, darling, why don't you try Bill Keane on CBS? Maybe he's got better weather."

Howdy rebuked her with a glance. "Any weather's better than this. Dust clouds or rain clouds and nothing in between."

Jennifer did not discount the graveness of the problem. "Rita Morales came by today to show me that new Montel rug hooking needle she's been raving about. She said that Carlos is worried about the run-off from the national forest burn area above Laberinta."

"He should be. There's twenty-five thousand acres of ashes and not a damned stick of new growth."

"I know. Are the creeks up?"

"Some."

"What about Claro?"

"It's okay. But all the springs are running again for the first time in a couple of years. That means the moisture on top has gone in real fast, like a sponge."

"Were you up there before the meeting today?" Jennifer asked.

"Yeah."

"Did you stop at Mother's?"

"For a minute."

"Was everything all right?"

A humorless grin twisted Howdy's mouth. "Rob's pacing

around like a caged cat. He's a sun boy. He sure doesn't like this."

"Who does? Mother's got cabin fever, too. I called her to see what night they'd like to come down for dinner and she sounded as though she couldn't wait to take off."

"That's a hell of an idea. Why don't you talk them into trying Hawaii, too? We're going to start dozer crews clearing the flood channel. About the time they crank up under her window at seven in the morning, we'll all catch hell."

"When are they going to start?"

"Monday."

"That soon? We'd better have them down for dinner tomorrow night and sell them on an island trip. She hasn't been there and there's a lot more for her to do than at a ski lodge."

"You had enough of their glooms too?"

A bit stiff-necked, Jennifer paused midway to the bar. "You know damned well I'm not thinking of us. That marriage is only going to work as long as they keep up the happy playtime bit." She continued to the bar and leaned against it wearily. "I wish to God I'd talked Mom out of building the house. They should have taken a condominium down on the bay. If either of them ever really stops to think, the whole deal is going to fall apart." Howdy's mirthless chuckle deepened her frown. "What was that about?"

"That? That was about the truest thing you've ever said. Incidentally, where's Howie? Isn't he going to have dinner with us?"

"No. I told him he could stay out. They're having some kind of a special get-together tonight at the teacher's house."

"Oh? Where does this teacher live?"

The question reminded Jennifer that Howie had not said where. All he had volunteered was, "It's just a thing, Mom, at Mr. Smith's house."

She thought Alan Smith taught social science and knew that the kids thought he was groovy and that he lived at the beach. Beyond that she was hazy. The realization brought a twinge of guilt.

Before she could answer Howdy's first question, he interjected another. "I suppose Doug Connolly's with him?"

"Yes." Jennifer paused. She knew Howdy's chief concern was young Connolly's driving. "They're at the beach someplace. He promised to be home by midnight. Doug's got his license now and he'll drive Howie home. I said he could stay out on Fridays and Saturdays if his homework was done."

Howdy seemed satisfied. He set his glass aside as Rosita appeared to announce dinner.

Young Doug Connolly racked his MGB-GT into the yellow loading zone in front of the Beady Eye, the most garish of La Playa's several hippie shops. Howie swung the right-hand door open cautiously to clear the high curb. Huddled in the doorway were a half dozen soggy young people in the weird, improvised costumes that, with their long hair, beards and trappings, had become the *sine qua non* of their sub-society. Doug, who knew them all, scrunched down to greet God's Mother, a lummoxy, barefooted, immense-bosomed young woman of uncertain age. Her gross body was swathed in a triple sari of purple cotton. A nondescript undergarment was girdled by a tarnished golden drapery cord from whose ends dangled clusters of small brass elephant bells. Chicklet was there too, her stringy silver blonde hair hanging in tatters like the shredded remnants of a cheap bridal veil. She was sixteen, an Army brat whose tissue-thin blouse and plaster-snug Levis left no contour of her child-woman's body unrevealed. Her dead-white complexion had a bluish cast from the cold and from two bad trips on Syndicate acid. Doug had told Howie that the "bad shit" had nearly killed her the last time when God's Mother administered thorazine to bring her down. She had been strung out for two weeks. The dealer who had copped three grams in Tijuana and sold the caps all along the coast had managed a ten-grand burn before the good people wised up. At the Beady Eye the word was around that some of the faithful had let a contract on him. But the narcos nailed him in San Diego and probably saved his skin.

Howie, not as fluent in hippie jargon as Doug, misunderstood. He thought "contract" referred to the Mafia. His apprehension had brought a derisive laugh.

"No, man! They wouldn't do anything merciful like shoot him. They'd just stash him in a pad and make him drop a dozen mikes of his own poison. That's enough to zonk him for six months. He'd come down a slobbering freak."

Doug had been smoking pot on weekends for over a year but had never dropped acid. "I would if I could be sure of the dealer. But Capso says there's not very much righteous stuff around anymore. I don't want to blow my mind so I don't drop. But I dig a good grass high. Like, wow! It's a groovy trip, especially with a chick. If I ever do get some good acid, like Big-O, I'm going to drop it with a chick who'll ball. I've listened to the guys rapping. They say it's freaky. You blow nothing but screamers."

Howie had been in the Beady Eye only once before and it had been fun, cool. Everybody had been friendly. Like Doug said, you could feel love. Even the Hairy Gross Dinger, in the greasy Tibetan yak jacket that made him look like a cross between Big Foot and a petroglyph from a French cave, had been real loving. His thing was scaring the dung out of tourists with an overwhelming brotherly love bit. "I'm a living lesson, Gold Kid," he had said to Howie. I'm proof incarnate that horrible can be beautiful."

The hirsute monster had given Howie the name "Gold Kid." It had stuck. Howie was not certain that it had been intended as a compliment, but it made him feel he belonged. "Pot's all right, maybe better than liquor or beer," he had told Doug after the New Year's party. "But it doesn't stoke me like shooting a tube or body surfing at the wedge. Wow! That's something else! You can ride a big one clear out of yourself. You're unhitched!"

Doug had disagreed. "Man, that's a bad scene. Escape is not good. You got to get *into* yourself. Not out—that's what grass does and hash and acid. Speed too, if you geeze it right. You climb inside yourself and walk around in your own brains, like in *The Fantastic Voyage*."

Howie had scoffed at the idea. "How do you know all this? You talk like a real head, or whatever you call those freaky characters."

"Who said I knew it from doing it? Alan told us. We rapped all night Friday. Man, he's a brain. He'll shock you maybe. But he's turned on like a big light. That's why I want you to come."

"Be any chicks there?"

"Sure. But different ones. Nobody from school."

So, he was going to make that scene now. No passing

567

roaches around on back roads. Howie eased into the sports car with the roll of psychedelic posters Swazy had given him. Swazy ran the Beady Eye. His wife Newfy helped. Newfy had freaked out on magic mushrooms. They took her to the community hospital and when she was let loose the good ones at the Beady Eye passed a sack and paid the bill. The young doctor never sent his. Two months later he delivered the baby boy. They passed the bag again and made him take the money. Swazy had cracked up when the doctor was surprised that hippies had money. He didn't know that one gram of LSD would split into four thousand one-mike caps that retailed at from one to three bucks a dose. There was money all right. Doug's father was an officer in the bank where the Beady Eye kept its commercial account. It was always good for a solid five figures. Howie remembered that he himself had seen one hippie cash a seven-hundred-dollar postal money order from back home. The guy had said it was his monthly allowance. Lots of kids got money from home. Howie wondered aloud why kids from good homes wanted to do the Haight-Ashbury thing, flop in crash pads, and live worse than animals. Doug looked at him amazed.

"When you're turned on, man, you don't care about material comforts and stuff. Rapping and tripping and getting punked and making out—that's real. All the material stuff is for hang-ups. That's the matter with your folks and mine. Their values are unreal, man."

Alan Smith's canyon house was in La Playa set back from the highway by a hundred yards or so. It was an old bungalow with an air of shabby respectability. A cement-lined storm ditch ran in front of the house, necessitating access by a small bridge.

It was the first time Howie had been to the "Peda-gogues' Pad." Alan Smith, an academic type who affected owlish glasses, long hair, and wedged "chops" met them at the door with a great show of pleasure. Doug had been there several times but his allusions to the strange rites conducted in the pad struck Howie as being exaggerated, if not pure fiction. As soon as they were in the house, Doug made for a nearby staircase, leaving him alone with Alan.

"Take a look around, Howie. Then we'll go upstairs and meet the people. How do you like the vibes so far?"

Howie squinted to accommodate his eyes to the dim light. "Groovy! Real neat."

"It's a real super power house, a natural accumulator. Do you know about the orgone force, Howie?"

Howie looked at him blankly. "No."

Alan put an arm around his shoulders and laughed sympathetically. "Don't let that bother you, friend. If you dig what we do here, you'll learn. You'll learn fast. Doug says you're sharp, right on about most things that have to do with nature."

As he talked, Alan steered Howie inside. The transition was a shock. Howie could see into a forty-foot room with a fireplace that had been converted into a shrine. Above the mantel was a poster depicting the Hindu Trimutri of the Three-in-one God with Brahma, the Creator, as the central figure. Soft Indian music was coming from concealed speakers and the air was suffused with sandalwood incense.

Alan moved toward the stairs. "Come on up. I want you to meet some turned-on friends." He stopped at the top in front of a curtained doorway. "I don't think you'll know these friends, Howie. But they know you. Accept that. The fact that we're all here means that we have all known each other in other times, in other worlds."

The back of Howie's neck prickled and a strange excitement filled him. Alan reached to pull the drape aside. "Any light from the outside disturbs our meditation. It's part of our self-awareness discipline. So we've draped all of the windows with black sateen. We wish we could keep out all the noise too but the highway's too close."

The room, almost totally dark to unaccustomed eyes, was not much smaller than the one downstairs but Howie sensed the difference in feeling. The walls were completely covered with psychedelic posters, similar to the ones Doug had brought. Some were animated by special black lights with slowly revolving shutters. A low crescent-shaped table was in the center of the room, whose floor was entirely covered with thick black cotton shag. Howie could make out about a dozen persons seated around it on cushions. He saw Doug settling a cushion at the far end of the semicircle. No sound could be heard except the muted stereo, the gentle drubbing of rain, and the occasional frying of speeding tires on the wet canyon pavement.

As Alan Smith led Howie to the open end of the low table, the tiny lights dimmed out, leaving the room in total darkness. He felt the teacher move behind him.

"When the light comes on say, 'My name is Howie. I love you.' "

An instant later a small red spotlight illuminated his face and he found himself repeating the words. The spot snapped off and a flashlight beam shone on the face of a pretty Negro girl. She smiled, a wonderful, open smile.

"My name is Lorna. I love you, Howie."

The beam glided to the next face, a young man with a beard. "I'm Alex. I love you." Again and again the beam pinpointed the faces as each boy and girl said the words unself-consciously and with unaffected sincerity.

Finally, the spot snapped off and the tiny lights came up a bit. Alan was beside him, slipping a necklace of small beads over his head. He pointed to a vacant cushion between two girls and said, "Sit there, Howie, and know that I love you, too."

As he picked his way around the table, Howie wanted to ask how they could say they loved him when, aside from Doug Connolly and Alan, who knew him as a student, they didn't know a thing about him except his first name. He sat down and the girl on his right, Mala, took his hand in both of hers. They felt warm and strong.

"Slip off your shoes, Howie." She reached down to help. Cross-legged on the cushion, his Levi slims bound his knees. A girl named Jill on his other side, who sounded younger than Mala, leaned close and made a sympathetic little sound. "Next time wear something loose. You have to let your body be free. Do you surf?"

"Sure."

"Good. Then you know a lot about what we're learning already. You know about one kind of free." She reached up and pressed a cool palm to his cheek. "You have good vibes, Howie."

Mala heard and inclined her head toward him. "I could feel it when he came in." She let her temple rest against his. She smelled good. It was funny, he thought. She was warm and Jill was cool. But they both felt good, only in different ways. Around him the others were talking quietly while Alan arranged things on a taboret in the open end of the crescent-shaped table.

A match flared and Howie looked around. Alan Smith was applying the light to a squat brazier. Almost instantly,

small varicolored flames leaped to life. Finger cymbals tin-
kled and Howie felt the girls reach for his hands. He
clasped them and glanced around the semicircle. All were
holding hands. Then the beginning of a strange chant, a
mantra. Mala enunciated the unfamiliar words clearly in
his ear, encouraging him to join in.

"*Hare Krishna, Hare Krishna, Krishna Krishna, Hare
Hare, Hare Rama, Hare Rama, Rama Rama, Hare Hare
. . .*" The words were half sung softly to a simple monot-
onous minor strain that quickly became hypnotic. Mala
felt Howie relaxing and abandoning himself to the rhythm.
She fondled his hand and moved closer.

". . . Hare Krishna Hare Krishna Krishna Krishna Hare
Hare, Hare Rama, Hare Rama, Rama Rama, Hare
Hare . . ."

After a quarter of an hour Howie was no longer aware
of the beginning or ending of the *mantra*'s cycle. His sense
of place had ceased to exist and in its place came an expanding sense of peace. He was no longer aware of Mala's
hand or Jill's. They were part of a larger sensation. With
eyes closed he seemed to be suspended in a velvety void.
He was not even aware that the *mantra*'s rhythm had
changed subtly when, by subjective assent, the group had
begun to complete two full cycles in a single breath. Then
the little cymbals were ringing again. The sound seemed to
come from a great distance and the chant slowed down
and began to fade.

He felt Mala's hand disengage and suddenly he realized
that he was no longer holding Jill's. He could hear the rain
again and, in the distance, tires were still frying on the
wet pavement. Somewhere, down the canyon toward the
beach, one siren was wailing and a second was sounding a
steady "woop, woop, woop."

The flames were still flickering in the brazier and the air
was heavy sweet with incense. The music had started
again, louder, twangy, Oriental. Alan lifted his head and
nodded. Zurin took an inlaid box from the table and
passed it to Rick. It went along the line. When it got to
Mala, Howie saw it was filled with joints of marijuana. She
placed her lips against his ear in a half-caress that excited
him and whispered, "Take one, love. It's Aggie." "Aggie,"
he knew, was Acaupulco Gold, the most righteous grass
around. He took one and passed the box to Jill. A mo-

ment later a small aromatic candle was passed along. Aware that Mala was watching, he lit up casually.

Close beside him, Mala savored a long drag, exhaled with her eyes closed then lifted herself on her arms to free the cushion and reclined, facing Howie. He was surprised that she was such a large girl. And older, too.

He smothered a glowing cinder on his leg and resumed smoking. After several drags he began to feel it but it was different now. The first time it hadn't sent him any higher than a couple of beers would. The joint he'd smoked in Doug's car on New Year's Eve had done more. He'd gotten pretty zonked but not like now. This time he was going way out. He felt Mala snuggling her bare thighs against him. The feel of her made a freaky thing happen inside him. It went clear down to his crotch. Her voice jazzed him too.

"Alan told me how old you are, Howie. I still don't believe him. You're at least one thousand and sixteen, an old soul." She reached up and took his chin between long, warm fingers and turned his face toward hers. "I'm only one thousand and nineteen. At our age that's no difference at all." The way she looked at you was crazy, he thought. Real groovy.

"Even-before I knew you I asked Alan to put us together."

The thing inside of him got worse, or better. When he squirmed uncomfortably, Mala smiled knowingly and pasted her body right up against him. She could turn you on by doing nothing. What a chick. She was not like some he had necked with until he had creamed his shorts from dry grinding. This chick was like the ones when he was dreaming. You felt them all over you until a come racked you numb and you didn't want it to quit. Mala was making him feel that way just being close and smiling and doing little things. And she was real, real, real.

Howie's body felt light. He tried to push himself upright, then dropped back giggling. Mala laughed softly and leaned over him. When she kissed his eyes he could feel the swell of her breasts on his neck and chin.

He had trouble with the joint. It was getting short. Mala had two roach clips fastened to the low neck of her shift. She took one and nipped it on, close to his lips, and asked for the shell ash tray. He reached out but stopped halfway. Jill was half on top of Doug rubbing his bare

572

belly down under his belt. He straightened to look past Mala. He wanted to see the others. It was too dark but he knew everybody was flaked out.

He exhaled the last drag. Mala held out the shell while he worked the scorched paper free of the roach. "What are you going to be?"

He shrugged. "I don't know. I'm only a sophomore in High now." Then he felt like going goofy again. He was very funny, very funny. "I'm going to be a surfing major. I'm going to open my own college of surfing." The image of himself in a gown and mortarboard hanging ten at Doho and San Clemente gave him the dry hysterics. Then Mala—her face was beautiful so close—put him down flat with his head on the cushion and the mood passed. He started to ache inside and her face above his made him want to cry and tell her he loved her but it was all mixed up, like he wanted *her* to hold *him,* tight, against her big soft jugs that were hanging half out most of the time. Anyway, they made his mouth feel funny. He wanted to suck and kiss her all over, everywhere. Everything was all crazy, wanting to be mothered, wanting to get lost inside of her and come, so the ache would stop. But he couldn't do anything and she was bending over him, holding his face between big, warm hands and she sounded like his Mom . . .

"Are you all right, Howie?" Only it wasn't his mother's voice. It was Mala's, soft like her skin and smooth like the inside of her leg against the back of his hand. Thinking about that made it worse and he tried to hold off but it started and he couldn't do anything but let it go and when it did he didn't want it to stop and he didn't care because Mala was whispering and doing something and there were other sounds. Voices but no words; everything was changing, even the music. It had a beat now, like knocking, and there were more voices coming from somewhere down below. Mala had stopped rubbing him and she was sitting up. Alan was up too, moving toward the door. Then the strobe lights went out and there were scurrying sounds and nobody was close to him anymore. People were moving in the dark, quiet and quick, and the rain was louder, and there were heavy feet on the stairs, coming up, and men's voices, coming nearer and calling.

Howdy played the pick-up's spotlight over the basin

spillway at the mouth of Ciega Canyon. The powerful beam couldn't penetrate the sheeting downpour by more than fifty feet. It was enough to tell Howdy that the men had to get the portable clam dredge there immediately. Mud and debris from the saturated, burned-off slopes was beginning to clog the overflow channels. In the basin itself the water, three feet above any previous mark, was threatening to spill over an earthen barrier that was never intended to be a conventional dam. Temporary barriers farther up the canyon had been breached the previous day. Boulder-strewn slurry had slithered down the slopes and demolished the man-made obstacles as though they had been built of sand.

Carlos Morales stepped out for a closer look. In a moment he was back. "It's not going to hold much longer, Mr. Goodwyn. It's beginning to cut through this end already. If you'll take me back, I'll get the boys to run the dredge over here right away. If we can pick up enough to take the pressure off maybe we can work the rig out on the fill far enough to clear the south spillway."

Howdy played the spotlight on the flood again. "Okay, Carlos, try it. But don't take any chances. I'd rather dynamite a channel on this end and lose some fields than risk lives."

Twice on the way back to the ranch headquarters Howdy switched on the spotlight to examine water damage to new orange groves. The run-off had been too heavy for the culverts. In several places the water had boiled out of the drainage ditches paralleling the roads. One such overflow had cut a three-foot-deep channel at least fifty feet wide diagonally through a new grove. Carlos cursed quietly.

"That one's cost us a hundred trees already!"

The zero-graded, terraced strawberry fields, undamaged as yet, were rain-pocked lakes. Below them an experimental field of asparagus had been half destroyed and the water was still spreading. Howdy tried to alert equipment foreman Ruiz on the short-wave radio in the truck but the frequencies were jammed with cross talk. When they arrived at headquarters they found Ruiz in the service yard refueling the equipment.

"We could use six of these buckets tonight, Mr. Goodwyn. We can't get the blades close enough to the trouble anywhere." Howdy nodded and pointed toward the smaller of the two canyons.

"Get the mouth of Ciega clear first if you can. I'll see if we can rent anything from the construction companies."

The private line to Casa Grande was ringing insistently as Howdy entered his office. It was Jennifer. "God! I thought I'd never reach you. Hal Halstead wants you to call him at the La Playa police station as soon as you can. He called an hour ago, at least."

"What the hell for?"

"I'm not sure, Howdy." Her voice was close to breaking. "It's something about Howie. Some trouble. Hal won't tell me what it is. He wants to talk to you. Please call him right now, darling. I'll hold on this line."

Howdy punched in an outside line before Jennifer had quite finished. "This is an emergency, operator. Howard Goodwyn calling. I want Lieutenant Halstead at the La Playa station." After a brief delay the police operator came on. "The Lieutenant's on another line, Mr. Goodwyn."

"Well, tell him I'm waiting. He's been trying to get me."

"Yes, sir."

Howdy heard the line click on hold. He had just seated himself on the corner of the desk when the office lights blacked out. There had been annoying, but not serious power failures during the early part of the storm. Transformers had blown and the sub-stations had switched to alternate circuits. For some reason now it was taking longer, he thought. He peered through the second-story window but no lights were visible anywhere. The line clicked then and a heavy voice, flat with fatigue, grated in the earpiece.

"Halstead speaking."

"Hal. This is Howdy Goodwyn, Jen says you've been calling me."

"Yeah, I sure have. I've got some rough news for you. We've got your boy in custody."

"For Christ's sake what for?" The words were close to a shout.

"He was picked up in the canyon, sort of by accident. Two of my boys went in to help the county road crew warn the people to be ready to evacuate. They stumbled into a pot party, two teachers, some guys and girls from State College, and your kid and the Connolly kid."

"What the hell were they doing there?" Even as he asked it, Howdy knew it was a damn fool question.

"They were lying around on mats smoking marijuana and doing a lot of other stuff. We got Smith and a guy named Zurin charged with possession and contributing. We're holding the others in protective custody."

Howdy could think of nothing to say. Hung between shock and anger, he blurted the classic defense. "Shit, Hal. Howie doesn't know his ass from his elbow about dope and things. He's not even sixteen."

Although Los Nidos Ranch was out of the La Playa jurisdiction, Lieutenant Halstead and his men were often grateful recipients of unsolicited favors from the ranch. They liked Jennifer and Howdy and the boy had never been in any trouble more serious than surfing on a restricted city beach an hour too early, a misdemeanor that in the department's opinion should never have been made part of the local law. The lieutenant couldn't bring himself to say that the kid had been so stoned they had to pick him off the deck, that his shirt had been up around his ears, that his pants had been open and that he'd shot a juicy teenage wad as big as a Frisbee.

"He got into bad company, Howdy. We've been watching this Smith character for a couple of months. We're pretty certain he's got some connection with this New Left trouble on the State campus. And we know damn well now that we've got him cold on possession of narcotics. He had two kilos stashed in his place, plus about twenty made-up joints. Jack Connolly's over here now with his kid. I wish you'd get over here, too."

Jennifer was on the phone with her mother when Howdy pulled the pickup in close to the front of the blacked out Casa Grande and ducked inside. A flashlight lit the phone table. Relieved, she motioned to him to wait as she tried to hasten an end to the conversation. "Mother, Howdy's just come in. I've got to go out with him for a while. Tell him about it while I get into my things." She broke off then and held out the instrument.

Howdy grabbed it impatiently. "Yes?"

"Oh, thank heavens. I'm so glad you're there, Howdy. Rob's worried about our being up here in this storm. We have no lights. We're using candles."

576

"Don't worry about it. They'll get the power on soon. You'll be okay where you are."

"I've told him. But he says a couple of cottages below us are in danger. The Barker place is undercut. He says one whole corner of the house is hanging over the stream."

Howdy was not surprised. He had little sympathy for the canyonites who courted disaster by building right on the banks of the Rio Claro. Most of them had been in Laberinta long enough to know that a prolonged storm could swell the stream enough to endanger them. But even the old-timers who had lost their places years before went right on taking chances. Catherine and Rob would be safe because they were back fifty yards at least and twenty feet above the river road on the inside of a bend. The water could erode the opposite bank but it would never get to them.

Howdy glanced toward the door where Jennifer stood pulling on her rain gear. "Catherine, what transportation do you have up there?"

"Just the little open Jeep. The Mercedes is at the dealer's getting serviced and Rob's car is still in the shop being fixed."

"Why don't you two drive down here? Maybe he'll feel safer."

Jennifer was holding the door, beckoning impatiently.

"I've got to go now, Catherine. You have a key. Come on down. We'll be back in about an hour."

"Where on earth do you two have to go on a night like this?"

Howdy tried to keep the annoyance from his voice. "The boy is stuck down at the beach. Jen and I are going to get him. I don't want him driving through the canyon tonight."

"I don't blame you. Go along. We'll be all right. I'll tell Rob you said we're perfectly safe here."

In the car, Howdy repeated his conversation with Halstead. Jennifer just sat there beside him saying nothing. He suspected that she too was indulging in self-recrimination. Suddenly it occurred to him that all those "where did we go wrong?" jokes on TV weren't so damned funny any more. Something had gone wrong. But for Jesus Christ's sake, what? he asked himself. He'd never gotten into mess-es when he was a kid and most of the time, nobody was looking after him either. And for sure, nobody'd handed him things on a platter—motor scooters, hundred-and-fifty-dollar surf-boards, a quarter horse, two of them, or anything else. What he'd gotten he'd slugged for, and that included his guitar. Sure, he'd gotten his nuts popped a dozen times by the time he was sixteen. By now, Howie had probably managed a few pieces of tail, too. Who in hell cared as long as he didn't get a dose or some babe in trouble? But dope? Christ. Kids smoking that stuff. What the hell's the matter with them? But Howie, Holy Jesus, being held for—what the hell did Hal say? Just being in a place where marijuana was found? If that's the only charge it probably wouldn't be too bad. Hell. Most likely the kid didn't know what he was getting into. The Connolly kid probably did. He was older. A smart ass.

Howdy speeded up the wipers and leaned forward. The rain had been coming down for hours. The third storm in ten days had put so much water in the side ditches that it was hard to see the edge of the pavement. Half aloud, Howdy wished to God they'd get around to widening the canyon road.

His thoughts returned to the Connolly kid. At eighteen, he couldn't cop out by saying he didn't know what was going on. He'd hate like hell to be in Doug's father's place. Jack Connolly was a wheel in the Rotary and Chamber of Commerce besides being a bank official.

Jennifer remained huddled in the corner of the front seat. It was funny, Howdy thought, but she looked so small. He had never seen her look so defenseless.

"Don't get all hacked, honey, until we find out what's really playing. It's probably not as bad as it sounds."

Curling her legs up on the seat she angled toward him. "It's probably worse. Remember New Year's Eve when Howie came in acting strange and smelling funny? I told you about it."

"Yeah."

"I know now what he was doing. He smelled of marijuana."

"You don't know that for sure."

"I do know it for sure. He was stoned, glassy-eyed. I thought he was just pooped from all that spastic dancing they do. But I know now. He was high. And that little son-of-a-bitch, Doug, is the one who put him up to it."

578

"How do you know it wasn't that Smith guy?"

"Maybe so, but I've never trusted that Connolly boy. Howie's big for his age. He's always wanting to be taken for eighteen. He could pass for it easy. He'd probably go along with anything the older boys were doing, just to be part of the gang."

"Sure. Up to a point. But I still don't think he'd walk into a thing like this with his eyes open."

Jennifer snorted. "We're about to find out, aren't we?" She uncurled her legs and turned front again. "Oh Lordy, this is going to look just peachy dandy in the papers." Another thought struck her and she moaned. "And what about Myra and Lew?" She imitated Myra's effusively sincere manner. "You two have been such *wonderful* parents. Howie's one of the dearest, nicest kids we know.' "

"For God's sake, stop it, Jen! The boy didn't kill anybody. He hasn't knocked up some girl. He hasn't stole anything. All that's happened is that he got picked up, accidentally, in a place where marijuana was being used. That doesn't prove anything so far as he's concerned. They can't give him life for that."

"Oh, God, Howdy! Quit trying to make it sound like a parking ticket. Don't you realize that he's being held in a jail? Behind bars? He's been arrested. That means he's got a record."

"It mean no such damned thing! Howie's got no record until he's been convicted. And they're not going to convict a kid of mine on a dope charge if I have to turn the whole fucking state upside down. Bet on it."

When Howie was brought into Lieutenant Halstead's office, Jennifer broke into silent tears. The boy's appearance shocked Howdy, too, but quickly the feeling turned to outrage. Obviously, the kid was still high. His eyes were glazed, unfocused, and his face, usually tanned, was pallid and twisted in a sickly grin. When Jennifer saw the semen stain, she closed her eyes and turned away. Howdy stared at it with disgust.

"Hi, Mom. Hi, Dad." The greeting was mumbled with eyes downcast.

Lieutenant Halstead dismissed the officer who had brought him in and led Howie to a chair. Steadying him, he looked from one parent to the other.

"I could tell you that I'm sorry we caught him up there.

579

But I'm not. He tells me he's never been with this bunch before. But how do we know that he wouldn't have been hooked if he'd gotten away with it a few times? Of his own accord, I don't think he would." He indicated another part of the building. "But he was in pretty fast company tonight. Except for young Connolly, most of them are confirmed 'heads.' "

Howdy rested a hand on his son's shoulder. "Let me take him home now, Hal, and I promise you, as God is my witness, you'll never see him here again, for any reason."

"I'm sorry, Howdy. He's been booked. He's got to stay here. But we'll try to set a hearing for the first thing Monday morning."

"Monday?" The question was a cry as Jennifer came to her feet. "Hal, this is Friday. Do you mean he has to stay in jail until Monday?"

"There's nothing I can do about it now, Jennifer. It's rotten luck that a kid like Howie had to get picked up and booked on a narcotics offense the first time. But it's rotten that any of them have to be brought here, even on traffic violations." He glanced at Howie and shook his head. "Most of these kids come from good homes. They're not bums. That's why we try to make it as light as we can." He indicated the boy. "We had no choice but to allege 'under the influence.' Look at his eyes." Halstead indicated the boy's dilated pupils.

"I can't brush that off. He was caught in a place where narcotics were being used. More than that, we're certain Smith's on an acid trip. He must have dropped earlier. He's worse now than when we booked him."

Jennifer moved in front of the officer, her eyes moist and appealing.

"There's nothing I can do, Jen. My boys didn't know who he was. But it wouldn't have made any difference. We'd have brought in the chief's kid if we'd caught him."

Howdy released his son and put an arm around Jennifer. "Hal, look. Who do I have to call to get the boy released into my custody? What's it take? Reagan? Who'll hear this thing on Monday?"

"It'll be heard in Juvenile Hall, Howdy. I don't know who's sitting Monday. But it doesn't matter. There's not a damn thing we can do anyway. The boy is booked, just like the others. They're all in the same boat. Jack Connolly's been trying, too."

Jennifer slipped free of Howdy's arm and moved to Howie, who had dropped heavily onto a straightbacked chair. His head fell against her waist. "I'm sorry, Mom, Dad." The words were thick, barely audible. Unblinking, his eyes remained fixed on the floor. Jennifer cradled his face against her coat and her own tears coursed, unashamed.

Howdy turned away, his face whitened with suppressed rage. There was no edge now and he knew it. This was one of the few times he couldn't handle a situation. It irked him all the more that his inability was apparent to both his wife and son. He needed to get out now, away someplace where he could think. There was always an angle. But who in hell could figure one with Halstead standing looking like an acid indigestion ad and Jen clutching Howie, fighting to control her sobs and the kid staring at the floor with his mouth hanging open like some kind of a drooling idiot.

A few minutes before 11:00 P.M. the weather instruments on top of the 6000-foot Diablo Peak recorded an unusual increase in temperature. Two feet of new snow had fallen during the past forty-eight hours. Most of the ski resorts had been isolated for more than a week. Road crews were fighting drifts thirty feet deep. Earlier, helicopters had been able to evacuate several ill persons and drop emergency supplies to stranded motorists. But for three full days now they had been grounded.

The snow turned from feathery flakes into heavy, wet blobs. By 11:15 P.M. it was mixed with rain and the marginal packs had begun to decay at the edges. By 11:30 P.M. slanting rain was slicing into the slopes turning the surface into an emulsion of mud, ash, shale, and small twigs. Soon the fluid mass began separating into sluggish streams. As they moved down the slopes above Laberinta Canyon they began to gather small boulders and burned chaparral. The heavier material gave way and added its weight to the force sluicing down the ridges.

The light rattle of rubble deepened as the larger debris was carried down. By the time the mass draining into the upper end of the canyon had reached the cliffs beneath the old caves, it had become a roiling torrent thundering downward toward the rising headwaters of the Rio Claro.

Rob Williamston turned from the front windows of Casa Escondida. "For God's sake, Catherine, I'll never understand you Californians. In the midst of the most disaster-prone area on earth you refuse to take even the most elementary precautions."

The toe of his rubber-soled shoe caught the edge of the throw rug by the front door. He kicked at it viciously. "Even at Fire Island we had sense enough to keep a Coleman lantern handy during the hurricane season. But here?" He pointed scornfully at the candles flickering on the mantel. "Here we have to ransack old storage boxes to find three stinking penny candles."

Catherine, curbing a growing annoyance at the harangue that had been increasing with the storm's intensity, turned back to the fireplace and poked at the last of the logs. Rob crossed the room and stood behind her. "You wouldn't get out of here earlier when the rain let up. So now we're both going to be drenched in the wretched rattletrap Jeep. That's if I'm lucky enough to get it started. I want to get you out of here, Catherine. That miserable creek is beginning to look like the Johnstown Flood."

Still outwardly calm, she turned to him. "I've told you, Rob, the lights will come on any minute now. We often lose them during storms, but not for long. And you do not have to worry about the creek. We are well out of its way, even if it does go rampaging a bit."

"Well out of its way, indeed." He strode to the door and opened it. "Listen. It sounds like Niagara. It's changed just in the last hour." He pointed toward the black upper reaches of the canyon. "And I don't like the sound of things up there. The water is moving boulders the size of Volkswagens." He closed the door and glanced at his watch. The radio was dead without power and the electric clock had stopped nearly a half hour earlier at three minutes to midnight.

"I want us out of here now, Catherine, while we can still get out. If you're going to be stubborn about it then by God you can stay here alone."

Patience and self-possession were two of Catherine's acknowledged virtues. She prided herself on them. One did not give in to adversity and fear. But another's poor deportment might be vexing enough to try one's patience.

582

"I am not being stubborn, my dear. I am simply telling you that we shall remain safe so long as we stay here. The site for this house was chosen with great forethought. I have no doubt that some of those cottages along the stream will be in danger if this keeps up. But we shall be perfectly safe here. I resent the inference that we Californians left our common sense in the East when we migrated." She turned away from the petulant face that anxiety had robbed of the quality she had mistaken for strength. "Although I must say that you're beginning to make me wonder whether or not I brought mine along this time."

Rob's reaction was instantaneous. "What does *that* mean, precisely?"

"It means, dear Robbie, that I'm having some second thoughts about you in this environment. Your behavior these past several hours has not exactly been reassuring." She started to move away but he reached out and turned her back roughly.

Repressed indignation strictured his throat. "If it's reassurance you want, I wonder if you know just how fortunate you are to have a husband who is willing to commit himself to solitary confinement in this backwoods prison? And as to my behavior, it is the result of your own. I have no doubt about my judgment. I am not worried about this house. But I am worried about that buckled asphalt goat trail out there that you call a road. I suspect Jennifer was worried too when she told us to come down to Casa Grande. Now get your things and let's get out of this dead end before we die in it."

Catherine regarded her husband with exasperating detachment. "All right, Robbie. We shall prepare for another soaking. But I think you should know that this melodramatic behavior of yours forces me to some odious comparisons."

Rob's retort died on his lips as the headlights of a car raked the room and a horn began blowing insistently. He got to the door as an older man in bright yellow rain wear was preparing to knock.

"Mr. Williamston?"

"Yes?"

"We need help down the road. Two houses have washed away. There was a grandmother and a small child in the Schmidt place. Could you get somebody from the ranch to

583

help us? We need trucks with work lights. We've got nothing but flashlights and some kerosene lamps. The phone lines are down too. We can't raise anybody." At the door Catherine recognized Ed Simmons, one of the old-time canyonites. "Come in. Let me try our phone."

She hurried to the end of the long breakfast bar and picked up the instrument. It was dead. Standing in the doorway to keep from dripping on the carpet. Simmons pointed to the Jeep in the carport. "You got a Citizens Band radio there. It's a slim chance here in the canyon, but maybe we can raise somebody outside."

Bareheaded and belted in an expensive trenchcoat, Rob stood with Simmons and alternated at the microphone. There was no response. All they could hear was unintelligible "hash" from emergency crews communicating on the same frequencies several miles beyond the effective range of their transmitter.

Catherine joined them. She had changed into slacks, black hooded rainwear, and boots. Somewhere she had found a large umbrella. The Jeep's engine, the wind gusting up the canyon, the rain pelting the carport roof, and the roar of the Rio Claro made it difficult to be heard. She leaned close to Simmons and shouted.

"Will this be any help?" She was pointing to the small accessory light clamped to the Jeep's windshield. The man flipped the switch and elevated the reflector. The beam penetrated a scant twenty yards through the sheeting downpour.

"It's a lot better than nothing, Mrs. Williamston."

"Good. Let's go down and see if we can help. You go on down in your car and show us the way. We'll be right behind you." After hasty thanks the man hunched and ran to his old sedan. Rob stared at her.

"For God's sake, Catherine, you're not serious? It's one thing to run for Jennifer's place." He pointed toward the river. "But to go out on a fool's errand? Those people are dead by now if they were washed away. Risking more lives doesn't make any sense."

She started to reply but Simmons was shouting from his car. "Don't try to cross the bridge. I just made it. The far side is undercutting. Pull up alongside me. I'll shine my lights where I think they need them."

Catherine turned to wave but Rob grasped her arm. "I

584

tell you, we are not going to get involved in any rescue operations. Those people don't mean a thing to us."

The last vestige of patience gone, Catherine wrenched her arm free. "Shut up and drive this car down there. Those people may not mean anything to you, but as the undisputed queen of over-age damned fools they mean something to me." She slid over to the passenger side. "Now drive it or I will."

The force of the storm stunned them as they pulled clear of the carport. Windshield wipers were useless. His face a mask of repressed rage, Rob leaned out the left side to follow the tail lights of Simmons' old sedan, frustrated and furious at the water streaming into his eyes.

For a half mile the two vehicles crept along the tortuous canyon road. On the right, dangerously close, they could make out the churning torrent. Here and there, huge sycamore trees had crashed into the stream, forming dam-like obstacles that caught and held small debris. Behind them the waters had risen to within inches of the right of way. Rob was badly shaken by the sight. "This is a fool's mission. Nobody could live in that." Catherine didn't bother to reply.

Ahead, Ed Simmons stopped the old sedan, flashing a warning with the brakes. Then he activated the left-turn blinker and angled to the opposite side of the road. Following cautiously, Rob probed the right edge with the spotlight. The narrow beam shone on angry back-pressure waves whose crests rose above the level of the road. Catherine could feel the rising panic in Rob as he dropped the Jeep into gear and gunned past the place, almost overrunning Simmons' car. It was a foolish thing to do in the short-coupled vehicle. They had been lucky. But she kept silent. Safely past the danger spot, he allowed the other car to pull ahead again. Twigs snapped off by the force of the wind whipped down on them. Another large tree crashed nearby on the opposite bank. Simmons' brake lights flickered again. On the far side of the stream, a flashlight winked on, made frantic circular signals directing their attention to the wooden bridge that provided sole access to the cottages strung along a quarter mile reach of river. It had collapsed at the far end.

An involuntary cry escaped Catherine. "They're stranded completely! We couldn't help them if we wanted to."

Rob cursed the blinding downpour. "Good! Then let's

get out of here while we're still able." He started to put the Jeep in gear again, but Catherine forced his hand away. Her strength surprised him.

"We stay right here. They need our lights. Pull up along side of Mr. Simmons and shine ours across with his."

"Catherine, so help me God . . ."

She whirled on him. "So help *me*, Robert Williamston, if you try to leave here without doing what you can to help, you will have forfeited the last shred of respect I have for you."

Before he could recover she was out of the Jeep. Standing in the headlight beams she directed him to turn right and pull up to the edge of the torrent. For an instant he considered leaving her with Simmons and making a run down the canyon for help but he couldn't bring himself to do it. Using extreme caution, he inched the vehicle into place. Simmons had tied an old cloth around his neck to keep the water from draining inside his collar. With one hand holding the cloth in place, he signaled with the other and shouted. "Try the spotlight, Mr. Williamston."

Rob switched it on but it was hardly more effective than the head-lamps, but car lights on the opposite side flared and a man stepped into their beam and waved frantically, pointing downstream. Simmons played the spot along the crumbling opposite bank. It rested on the grotesquely tilted ruins of a summer cabin. The sound of shattering lumber reached them. Rob swung the light upstream again in time to catch a section of screen porch breaking away. As they watched, it cartwheeled into the torrent and disappeared. Simmons turned to them in open-mouthed amazement. "I've never seen it like this, in twenty-five years."

Catherine called in an anxious tone. "Where are your people, Mr. Simmons?" He pointed directly across the torrent.

"In a woodshed on high ground. They'll be all right." His arm swung downstream toward the wreckage of the other cottage. "But if old lady Schmidt and the child didn't get out I don't think there's any hope for them."

Rob, indicating the torrent, pushed Catherine aside, saying, "There's no hope for any of us unless we get out of here and get some help. It's ridiculous to stand here shining lights. You can get in or stay here but I'm going to try to get down to the ranch."

Simmons peered across the bank. The figure in the headlights was still making futile attempts to communicate. "I think he's right, Mrs. Williamston. I don't know's anybody can do anything tonight. But it would be comforting if somebody outside knew the fix we're in."

Once again, above the din came the sound of heavy wood splintering and a tangle of gnarled sycamore limbs laced solid with debris swept through the corridor of light from the two vehicles. It pushed over the bank and Rob shouted and gunned the jeep in reverse. Catherine and Simmons cried out as the vehicle shot backward across the pavement, slammed tail first into the muddy bank, and stalled. Catherine got to him first.

"What an idiotic thing to do! What an absolutely idiotic thing. You might have killed both of us."

Grim-faced, Rob ignored her and restarted the engine. Catherine climbed in beside him, struggling to regain her composure. "All right! I'm satisfied there's nothing we can do here. We'll go to the ranch and see if we can get help there."

Drenched to the skin, Rob glowered at her, then bowed elaborately. "Very good, Madam, at your service."

About a mile above the Casa Grande, the canyon widened. When they rounded the last looming shoulder of high canyon wall they could hear the roar of the water again, very close to the road. Puzzled, Rob slowed to a crawl and repositioned the spotlight. Its beam, more effective now, swept across the torrent. The main channel, usually thirty yards or more west of the road, had cut in much closer. As they inched along in low gear, they could see great slices of undercut banks collapsing.

They were out of the canyon proper now, but the road stayed with the stream. They knew this area well. At the point where the pavement angled toward the center of the delta-like expanse of gravel and boulders, Rob stopped. His humiliation forgotten for the moment, he turned to Catherine, "I don't like this. The channel looks like it's moved quite close to the road up ahead." With his left hand he swept the beam slowly back and forth.

Catherine leaned out the other side. "It seems all right for the next few yards." Rob looked uncertain, then eased out the clutch; but after a minute, he stopped again.

"If you want to drive, Catherine, I'll walk ahead. This is

587

treacherous along here. I can't tell how close the water has cut to the road."

She shook her head. "No, let me get out. I'm more familiar with this road than you." He made no effort to dissuade her as she climbed out and hurried forward into the headlight beams.

Twice he came to a stop when Catherine signaled to him. The rain had lessened to a light drizzle and to the west he could see the diffused glow of a large subdivision. Behind him he could make out the mouth of Laberinta Canyon, a gaping, dead black maw. A chill shook him and for the first time he was aware of the cold.

Catherine was cautioning him to keep well to the left of center. He swung the spotlight more to the right and its beam revealed a fifty-foot-wide torrent. Startled, he tried to attract Catherine's attention by shining the light on her. She turned and for an instant looked full into it. He saw her squint and raise a hand to block out the glare. Then she waved and started to walk again. He directed the beam to the right briefly, then returned it to the road. Catherine had vanished. Puzzled, he jammed the brakes and swept the spotlight back and forth. He was certain that she could not have moved beyond its range. He probed the road again but there was only darkness. Panicked, Rob bore down on the horn. When Catherine did not appear he leaped out of the Jeep and ran forward, shouting. A few yards beyond he let out a cry and stopped inches short of the caved-in pavement. A large segment of undermined asphalt had broken off and dropped ten feet into the flood. Running wildly around the break, he shouted Catherine's name over and over. Continuing his search, he sucked in great sobbing breaths, screaming her name. Half-crazed, he raced back to the Jeep and gunned it recklessly around the cave-in. Sweeping the crumbling banks of the swollen stream he drove back and forth, searching, listening, until he had covered every foot of the quarter-mile reach where the waters had cut closest to the road. Then, abandoning the vehicle, he searched the edge of the torrent on foot.

The emergency crew found him an hour after daybreak slumped on the crumbled bank unconscious and chilled blue. Out of gas, the Jeep stood with its battery dead, within inches of the washed-out segment. They did not look for Catherine then because there was no one to tell them until later, when they notified Howdy, that Robert Williamston III had not been alone.

19 IN THE OLD HOLLYWOOD OFFICES of H.G. Enterprises, expanded and redone several times during the past two decades, Howard Goodwyn sat slumped in his comfortable old executive chair. It had been the first symbol of his new affluence as a radio star and it remained unchanged, a haven for concentrating on particularly sticky situations. Ev Moses sat nearby leafing through a stapled sheaf of legal papers. The intercom buzzed discreetly and Howdy lifted the key. "Yes?"

"It's Ward Roberts of the *Probe Show* again, Mr. Goodwyn."

Howdy was plainly annoyed. "Tell him I'm out, that I'll call him back."

"Yes, sir."

Ev Moses flipped over a page and glanced up. "Still after you to do an interview?" When Howdy grunted affirmatively, Moses smiled. "I suppose they want to probe your political future. Isn't that what Roberts said?"

"The hell they do." Howdy aimed a finger at the papers. "The son-of-a-bitch wants to probe into that, into the mess with Howie." His lips twisted. "And if they ever try to, they're going to get a public kick right in the ever-loving *cojones!*" Deliberatly, he had lapsed into his Western Movie Star drawl. Moses tapped the sheets.

"Considering the possible consequences of a narcotics charge, this isn't too bad, Howdy, if the boy keeps his nose clean."

"He will, Ev. Bet on that. I must have been blind. I didn't know what a great kid I had until I saw him in jail. In jail, for God's sake. And I'm the one that probably put him there, because I fluffed the kid, always too goddamned busy." He looked up and his eyes mirrored the first deep pain he had known in some time. "Now ain't that one hell of a speech for a father to make?"

Howdy had given Ev Moses the court papers decreeing

Howie a ward of the court. Ev agreed to look them over even though he knew it was pointless. The legal process was automatic and, as Howdy had discovered, not subject to influence or pressure. Ev Moses knew it had been a disquieting revelation to Howdy to discover that personal power couldn't always be relied upon. He knew, too, what it meant to his partner to lose an edge. Howdy had lost an important one with his own boy. As a friend and counselor, Ev knew there was little he could do but try to comfort Howdy.

He handed back the sheaf. "Personally I think the court showed excellent judgment. Howie could have drawn thirty days in Juvenile Hall. That could be rough. But as a ward, he's on formal probation until he's eighteen. At that time, if he doesn't violate, the probation will be terminated. If he stays clean until he's twenty-one, they'll seal his record."

Howdy straightened. "That means no criminal record, right?"

"Exactly. His record is sealed. Legally, it doesn't exist any more. It can't be brought up and used against him, ever."

Relieved, Howdy relaxed and redraped his leg over the chair arm. The typical mannerism made Ev smile inwardly. It was a good sign, that leg dangling. So was the grin.

"Ev, old pal, even if I am fifty-six years old, you'll see Reagan selling soap again before you see that record being used against my kid."

"Never say never. By the way, how did the boy take the downfall of—you should excuse the expression—his fairhaired hero?"

"Williamston?" Howdy smiled a bit sadly. "It's a funny thing, Ev. The kid seems to want to take part of the blame for his grandmother's death. He said if he hadn't gotten into trouble he'd have taken one of the ranch trucks and a portable generator into Laberinta as soon as the power went out. I guess we'll never know what really happened. Anyway, Howie's turned cold as a well-digger's butt where that prick is concerned. He won't even talk about him."

The attorney snorted. "I'm still not so sure a criminal charge wouldn't have been sustained."

Howdy's face darkened. "I know. I know. Something

591

could have been made out of Ed Simmons' testimony. It was no secret that Catherine and Rob were squabbling toward the end. But as far as I'm concerned we're best out of it this way. Lew Berwin agrees too, because of the possible political thing. A trial could have been messy. A two-hundred-thousand-buck settlement and the house in Acapulco in lieu of any claims on her estate was a hell of a bargain. So he'll piss it down the drain playing in the sun then look for another old broad. Let him. All I care is that we've heard the last of the sonuvabitch."

Ev nodded and rose to go. At the door he paused with his hand on the knob. "What are you going to do about this political talk?"

Howdy shrugged. "I don't know yet. Why?"

"If Lew talks you into running it might be smart to remember that the opposition has some strong ammunition now, with the circumstances surrounding Catherine's death and with the narcotics thing. It could get pretty rough."

Howdy regarded his friend with a humorless smile. "That's the wrong way to scare me, Ev. Rough I don't mind, as long as I've got the odds on my side."

Ev nodded obliquely. "You're not the first public figure whose kid has been busted on a narcotics charge. It hasn't seemed to hurt them—politically, at least." He shrugged again. "Who knows? Everybody's got a skeleton in the closet, I guess. As far as odds are concerned, gambling's not my long suit. Talk to Lew about that."

Howdy grinned. "You gambled on me."

Ev nodded vigorously. "A gamble you weren't, Mr. Goodwyn. You were a sure thing."

After Ev left, Howdy locked the probation papers in his personal file and placed a call to Lew Berwin in San Francisco. Lew brought up the subject on Howdy's mind first. "Believe it or not, Howdy, I was working for you. I had some of the valley growers in. I wanted to find out what they think of the picture so far."

"What picture?"

"The political picture. Your chances. The whole *schmiere*."

"That's what I called you about, Lew. Ever since the thing with Catherine, and then the rotten business with the kid, I'm not so damned pure any more. Better think about that before you talk it up too much."

Lew's laugh boomed through the earpiece. "We have." To Howdy his tone sounded a bit too reassuring. If a man had no alternatives he could bring a lot of rationalization to bear on his choice. Howdy suspected Lew Berwin was doing just that.

"Lew."

"Yeah? What?"

"I'm serious. That business cost me a lot of edge."

"I'm serious, too. By the time you file, it will be forgotten. It won't mean a thing." His manner changed. "Look, Howdy, you're the man the boys want. We've talked it over, upside down, ass-end to, every which way. Nothing's going to change that."

"Maybe you're right, Lew. But Ev says they'll dig all the way back to Angie."

"Let them, for Christ's sake! It wasn't you they caught fucking the major. You were the righteous, injured husband. The cuckold. Ev's off base on that. I think he's special pleading. He doesn't want you to leave the ranch and dump all of that responsibility on him."

Howdy's eyes grew troubled. "That's part of my concern too, to tell you the truth."

There was a charged silence at the other end of the line.

"Howdy. Let me tell you the truth. If you, or somebody who's as effective as you, doesn't get in and fight this thing from the inside a hell of a lot of ranches are going to be in bad trouble. I'm asking you to think about Los Nidos, sure. But we're asking you to think about the whole picture. You don't have to make up your mind now. You've got months yet. Just think it over some more."

Howdy was seriously concerned about his chances and the effect recent events might have on them. He understood well the fickleness of California's electorate. The "Capricious Million" could not be counted on by either party. Nobody could say that they might not take it into their heads to elect S. I. Hayakawa, John Wayne or Sam Yorty as United States Senator. The only predictable thing about California's politics was its utter unpredictability.

Nobody who really understood what westward migration was doing to California's agribusiness believed that the status quo could be maintained. Because Howdy knew this, he made an impulsive decision that could buy him some time.

593

"Lew, I'll tell you what. For a starter I'll agree to do *Probe*. But I'm going to take the questions cold. No set-ups. No bullshit. If I can handle them on that basis, and Ward Roberts is tough, then I'll make up my mind. How's that?"

Lew could not conceal his disappointment, but he understood. You let Howard Goodwyn do things his way. If he committed on his own terms you had a tiger by the tail. "Okay, Howdy. When you know the date send me a memo."

Howdy left the office early. Howie had made the La Playa swimming team and he had promised the boy that he would go to the first meet. He had also promised the boy that they would go down to the Hotel Cabo San Lucas in Baja California for some deep-sea fishing. Somehow he had to find ways to spend more time with Howie. The whole damned youth business was baffling. That morning Jen had read him a piece in the *Times* about a psychiatric association in New York that had been studying youth problems. The survey showed that most hippies came from well-educated, well-to-do families. The report said: "Many of them lack confidence in the structure of society." Most of the hippies described their parents as "non-authoritarian" and their main beef seemed to be that the establishment was hypocritical. He remembered bellowing at Jen, "Well, for Christ's sake, what's new about that? Don't they read history?"

On the TV screen the animated letters PROBE zoomed away from the viewer, formed an arrow and impaled themselves in the chest of the live silhouetted figure of Howard D. Goodwyn. Off camera, the announcer introduced the new show that "Probes to the heart of our most controversial issues."

The title dissolved to a close-up of Ward Roberts, an alert, attractive newsman in his late thirties, who began the introduction. Howdy moved to his chair and saw Camera 2 light up. A medium shot of them both appeared on the portable monitor. "Mr. Goodwyn, do you object if I report to our viewers the conversation we had when I contacted you about this appearance?"

Howdy smiled easily. "No. Go ahead."

Roberts looked earnestly into the lens. "Ladies and gentlemen, it is customary for *Probe's* editorial staff to get together with our guests for a brief discussion of the subjects to be covered on the show.

"But Mr. Goodwyn insisted, as the condition of his appearance, that he be allowed to come on the air cold, without any knowledge of the direction or the the depths *Probe* would explore. We agreed to that gladly." He smiled across the desk at his guest. "So, Mr. Goodwyn, no holds barred, question one." Howdy saw Camera 3 zoom in for a tight close-up of his face. "It's been rumored around that you are considering a return to public life, this time in politics. Is that true?"

"It's true that it's been rumored around, all right."

Roberts was only mildly amused. "Let's put it this way, Mr. Goodwyn. Does the office of Governor of the State of California interest you?"

"Yes. Very much."

"In what respect, Mr. Goodwyn?"

Howdy frowned reflectively. "Well, three years ago I was one of two million seven hundred and forty-nine thousand voters who thought somebody else could do the job better."

Roberts ignored the stifled off-camera chortling and managed a grin.

In his penthouse condominium overlooking San Francisco Bay, Lew Berwin smiled conspiratorily at the half dozen leading growers from the Sacramento and San Joaquin valleys who were his guests. He turned back to the television set as the newsman prepared to pursue his point.

"Mr. Goodwyn, it is not *Probe*'s intention to turn this into a *de facto* impeachment proceeding but don't you think it's a trifle naïve not to take everything any candidate says with a grain of salt?"

"Generally, yes." Howdy looked into the lens and smiled engagingly. "Trouble is, so many of us got so used to believing our own soap commercials."

Again there were subdued snickers from the floor crew. In the control room overlooking the stage, the director turned to his technical director. "This guy's cool and slippery."

"There's twenty-five minutes yet, pal. Roberts'll skewer him," was the reply.

Roberts pressed ahead. "Mr. Goodwyn, you are operating head of one of California's largest and most success-

595

ful ranches. Recently you announced a radical change in some of your plans, the incorporation of a new city of a half million people to be called Micropolis Two thousand.

"Mr. Goodwyn, the fact that the plans show that most of your ranch's prime agricultural land is protected in the heart of the over-all city suggests that you are building a sort of urban fortress around it. Have you done this because you feel a huge ranching operation such as yours is doomed to extinction and you are trying to save it?"

Smiling tolerantly, Howdy nodded. "That's partly right, Mr. Roberts. But I resent the inference that we are about to be put out of business. On the contrary." His manner turned serious. "Have you ever heard the expression, 'the highest and best use of land'?"

The question caught Roberts by surprise. "I have. Yes."

"Do you understand what that means in terms of land management?"

Roberts smiled smugly. "Suppose you tell our viewers, Mr. Goodwyn. I'm sure your explanation will be more expert than mine."

"The value of certain land changes as the economy changes, Mr. Roberts. In our case land that was once most valuable as hilly grazing land is now more valuable as residential subdivision land. As the people kept coming into the county we kept phasing out our cattle operation, moving to other states where land as cheap as this once was makes it more economical for us. Incidentally, that also helps keep beef prices at reasonable levels. So you can see that we are not building a fortress as you call it. On the contrary, we're actually making an economically sound move while opening up some beautiful land to a higher use. In order to make certain the people who will live there will live better than they can live elsewhere, we're spending millions to make certain we don't repeat the old mistakes that we see around us that have created urban sprawl and slums."

"On the face of it, Mr. Goodwyn, that sounds like a very benevolent attitude for a huge corporation. Los Nidos is to be congratulated."

"We're not a benevolent society. What we're doing is just good business."

Roberts smiled without humor. "In my business, Mr. Goodwyn, it's good business to keep probing until I get at the real truth of the matter." When he saw his guest's eyes narrow he broke into a broad smile. "I'm not suggesting that you are being evasive. But I'm betting that we're not going to get it all from you this afternoon. After all, this is not a court of law, Mr. Goodwyn, and it *is* your business. So let's get back to your political aspirations."

A resigned expression settled over Howdy's face. "I haven't admitted that I have any."

The director smiled at his technical director. "Okay, pal, who's doing the skewering now?" The response was a guttural obscenity.

On camera, Ward Roberts laughed self-consciously. "On the other hand, Mr. Goodwyn, you haven't denied that you have any either, and your presence on *Probe* suggests some sort of interest on your part, for the rumor is that you may enter politics in order to fight more effectively for agriculture or for agribusiness as it's called in this day when most small ranchers are being forced to quit." He emphasized the words to make clear the inference that big ranching was to blame and his guest was one of the principal defendants.

"Agriculture's my special interest politically speaking and every other way, Mr. Roberts. But let's get something straight first. What's driving small farmers and ranchers out of business in California is not competition from the large operators. It's rising taxes and rising labor costs and people pressure, urbanization of our best land. We big ranchers are just as worried as the small ones. We have close to a five-billion-dollar business to protect. That's not much compared to what the space boys will need this year to leave their footprints on the moon. But our billions put food on the table, here and around the world. Four counties in California grow more crops than forty-four other states put together. One of our counties, Kern, is the most productive farm county in the United States and acre for acre the most productive farm county in the world. Isn't that worth protecting? Doesn't that suggest that agriculture had better be my special interest and yours and every other Californian's?"

"Somehow, Mr. Goodwyn, I get the impression that *I'm* being interviewed, by an expert." Both men laughed. "But I'm going to keep trying, Mr. Goodwyn. Let's talk about labor. Cesar Chavez and his United Farm Workers Organizing Committee have forced some big agribusiness to sign—Di Giogio, Christian Brothers, Pirelli-Minetti, Alma-

597

den Vineyards, and others. Is this the beginning of a labor domino effect? And if so, what will it do to California Agriculture?"

Howdy pondered briefly. "I don't have a crystal ball, but unless we get some special protective legislation like the rest of industry has, or unless the unions make provisions for the small grower, it could mean increased costs, higher wholesale prices, and higher retail prices." He looked directly into the camera. "Anything Chavez does to us he'll do to you and every one of your viewers out there."

"What is the answer, Mr. Goodwyn?"

"Fair legislation to protect us, or mechanization, more machines, less labor, more efficient growing methods. The prices won't come down, of course. But perhaps we can hold them."

Roberts smiled shrewdly. "You wouldn't be overstating the case just a bit, for rather obvious reasons?"

"If we don't get some relief I'll invite you to ask that question again, in a couple of years."

Roberts laughed good-naturedly. "So, it boils down to more machines and fewer bodies?"

"It could. But some crops, like strawberries for instance, can't be machine-picked."

"What will happen with them?"

"We'll have to grow them in areas where climate and hand labor makes it economically possible. Probably in Mexico."

Roberts appeared genuinely interested. "Mr. Goodwyn, there are a great many things that you and I could probe if time permitted. For instance, pollution and its effect on crops, the hippies phenomenon, the whole tragic narcotics problem, but for now I'd like to move along to something all prospective candidates for any high office should give top priority to. Our minority groups. For instance, what do you think should be done about the demands of our black Americans and our Mexican Americans?"

In San Francisco, Lew Berwin threw up his hands. "It's anybody's ball game now. The bastard reminds everybody of the narcotics thing with the kid and then slips him this. Take it easy, Howdy."

For Jennifer, at Los Nidos, anxiety turned to dismay. She glanced uneasily at young Howie, who was watching with no apparent interest from his father's easy chair.

598

Resorting to his old Will Rogers "shy bit," Howdy lowered his head. His smile was deceptively uncertain. "Mr. Roberts, you're too smart to expect an off-the-cuff answer to such tough questions. An honest man wouldn't pretend to know the answers." He looked directly into the lens. "I sure don't. But let me say this, some of my best friends are Mexican Americans." Roberts' startled expression turned to relief when Howdy added, "My wife's grandmother, for instance.

"At Los Nidos, Mr. Roberts, about ninety-five percent of our help is Mexican American. I've gotten to know them well. Some of them are fourth generation. I even manage to speak a little Spanish, though I've got to admit they understand me better when I speak English.

"I think every franchised American citizen should be given an honest-to-God chance at the best education we can give him. And he should get an equal chance at any job's he's qualified for and feels he wants to try."

Howdy paused for effect. "But if he gets that chance and doesn't take full advantage of it and still keeps crying about a lousy deal, then the taxpayers should say, 'Shut up, brother, and step down.' "

"What about those who for one reason or another simply can't bootstrap themselves up, even if they try?"

Howdy delayed his answer until the close-up was on him again. "We have to take care of them just as we care for any of our less fortunate people."

"Mr. Goodwyn, you have a reputation of putting your money where your mouth is. Would a man win or lose if he bet that Los Nidos did not always practice what its head man preaches?"

"The best way to answer that, Mr. Roberts, is to send you a half dozen of my Mexican American superintendents and foremen to be interviewed. Would you like that?"

Roberts matched Howdy's smile. "They'd be more than welcome. But wouldn't I find out that they have been coached and that they don't really have the same chance to work into the upper echelon of management that their non-minority opposite numbers have?"

Howdy answered with a flat, "You would not. As a matter of fact, you'd find out that my top ranch superintendent, a young man who graduated from the University

599

of California's Agricultural College at Davis, is going to be made a vice president soon."

Roberts was genuinely surprised. "What's the man's name, if I may ask?"

"Carlos Morales."

"But Mr. Goodwyn, can you say in all honesty that promoting an exceptional minority member is not a form of tokenism?"

Howdy managed to conceal his anger. "I just wish I had twenty more like him."

Roberts eyed his guest a bit skeptically, then consulted his notes. "How many Mexican Americans do we have in California now?"

"About a million and a half."

"What's going to happen when this mechanization you predict begins to phase out their jobs?"

"We'll find other productive work for them. We already have a retraining program at Los Nidos—and not, by the way, at taxpayers' expense. At ours."

"What will they do?"

Howdy's concern was more than convincing. "It's hard to say, exactly, at this point in time. We are teaching some of them to operate the new machines. Others are learning maintenance. As the nature of our operation changes, there'll be plenty of chances for those who'll trouble to get an education. We're helping with that, too, by staggering work hours for those who want to attend schools."

"Do Mexican Americans have an equal opportunity at an education? Equal, for instance, to our black Americans?"

Howdy considered briefly. "No. I don't think so."

Roberts' eyebrows lifted. "Oh? Why?"

The furrows on Howdy's brow deepened and he explored his chin briefly. "Well—for one thing, the black American's first language is English. That's generally not true of the Mexican American kid. So the black kid and the brown kid don't start even on that score. The black kid has a definite edge." On the floor the black cameraman in the crew grinned and whispered over the intercom. "Man! I didn't know how lucky we is!" The director shushed the crew good-naturedly and called for a close-up on the newsman.

Unperturbed, Roberts smiled. "You've got ready answers, Mr. Goodwyn. Now let's talk some more about just how real this alleged threat to California agriculture is.

600

Are you saying that the time may come when there would no longer be any significant agriculture in the state?"

It was precisely the question Howdy had been waiting for. "California has roughy one hundred million acres. One half of the state is in National Parks and other public lands; one third is covered with forests; one fifth is desert and mountains. This year more than ten percent of our prime land is polluted. Subdividers are taking one hundred thousand acres a year out of production. Mr. Roberts, if the present rate of immigration holds, there will be no worthwhile agricultural land left in California by the year two thousand and thirty. In two thousand and fifty, all those people who haven't moved back to Iowa, Illinois, Nebraska, and Arkansas for their rich farm land will have to start digging for the most precious thing of all, good soil that'll grow cheap food."

Howdy leaned forward. "And do you know where they'll find it? Under the asphalt streets and deserted tract slums that were the biggest subdivisions of the Nineteen Sixties and Seventies."

Roberts eyed his guest a bit skeptically, then let an ingenuous smile spread over his face. "One can't help observing that you'd make a very persuasive man in politics, state or national . . . but then it's never *Probe's* purpose to add fuel to rumors. So let me use the remaining time to cover a couple of highly sensitive areas.

"Mr. Goodwyn, what do you think is behind the growing use by young people of marijuana, LSD, amphetamines, and worse? Is it a rebellion similar in some respects to the one your generation staged when it drank illegal liquor in Prohibition days? Or is it something more serious?"

"Roberts," Howdy's voice was flat, "I had hoped you'd have the decency to avoid this subject under the circumstances. But if you have the poor taste to ask, I have the guts to answer. You know the answer to that question as well as I do. Did you ask it so your audience could watch a father who loves his son hurt a little in public, or do you really want a meaningful answer?"

Roberts was visibly flustered. "I know that we've been accused of deliberately staging Roman Holidays here, Mr. Goodwyn, but I assure you that was *not* my purpose in posing the question. What *Probe* is after is new insight

601

into problems. Our editorial staff felt you were uniquely qualified to shed some light on a very grave one."

Howdy allowed his manner to soften slightly. "Anybody who's thought twice about it knows that the violation of the Eighteenth Amendment was no kids' revolt. It was an adult rebellion against a stupid and unrealistic law. Today these kids are rebelling against lack of discipline. We haven't taught them who they are or given them a place of their own and made them stay in it until they earn their way out. So some of them take up causes—anti-war, conservation, brotherhood—to give them identity. The ones who take dope are retreating, copping out."

Eager to keep the exchange heated, Roberts broke in. "Agreed. But what about the adults who feel that the narcotics law is unrealistic in so far as marijuana is concerned? What about the ones who say it's no worse than liquor and should be legalized in compliance with the wishes of the majority? Do you agree?"

"I do not."

"Do you drink, Mr. Goodwyn? And do you smoke?"

"I take an occasional drink. I haven't smoked since I tried a five-cent cigar forty years ago."

"But if it was proved that marijuana is actually less harmful than tobacco and alcohol, would you condone its legal use?"

"I would not."

"Why?"

"Because there's no such thing as the unvarnished truth where billions of dollars in potential profit are concerned. The American Medical Association says it's proved that tobacco is the prime cause of lung cancer. The tobacco companies are spending millions to prove that it's not, that the real cause is air pollution. So the automobile companies and the power companies and oil companies spend more millions to prove that they're not to blame. Everybody's trying to blame everybody else in order to save their profits. I've been told, and I find it easy to believe, that an international syndicate is moving to control the principal marijuana-growing areas in the world, that they have an underground public relations campaign selling the idea that pot is easier on you than aspirin.

"And I've heard that the big tobacco companies are quietly making plans to get into the marijuana cigarette business the minute it's legalized and that they are quietly

602

helping to sell the notion of legalization. With that much money power behind an idea do you really think we'll ever hear the truth?"

Plainly apprehensive, Roberts sought refuge in professional righteousness. "Mr. Goodwyn, those are very serious allegations—like some that have been aimed at your interests."

Howdy laughed. "Those aren't allegations, Roberts. Those are rumors like the one you used to open this show."

Roberts managed a fleeting smile. "All right, Mr. Goodwyn, but I want to make it clear that, rumor or not, your point of view does not necessarily represent the point of view of this station or this commentator. Unfortunately, our time is up now. I want to thank you for appearing on *Probe* today, cold, the way you insisted, and submitting to some questions that were as difficult for us to ask as they must have been for you to answer. Thank you."

The director called for a close-up on Roberts and began pantomiming a flowing violin passage. The technical director began imitating him. A moment later a young assistant director followed suit. The sound man looked across at his companions. "What the fuck are you nuts doing?"

The director gave him a pitying look. "Are you deaf? We're playing the Establishment Waltz, you middle-aged peasant."

In San Francisco, Lew Berwin switched off the set. Myra, who had watched with him, stood up. "He was wonderful, Lew. Do you think he'll really buy the idea?"

"Who knows? But he's only got two choices, get into politics and try for protective legislation or resign himself to subdividing Los Nidos piecemeal. That's one more alternative than the rest of us have. That's why I've got two million in pledges if he'll enter the primary."

Myra thought about Jennifer. She also had two alternatives: lose her husband piecemeal to politics or lose the ranch the same way. "Does Howdy know about the pledges?"

Lew shook his head. "Nope. But he will next week."

At Casa Grande, Jennifer switched off the set and turned to Howie. "I thought your father was wonderful, didn't you?" The boy let his leg slip from the broad chair

603

arm in unconscious imitation of Howdy and got up heavily.

"Sure, Mom. Dad was neat."

In Los Angeles, at the public-relations offices of Hamlett and Daniels, Earl Hamlett turned to the staff assembled in the conference room.

"What do you think?"

A young political copywriter grinned smugly. "Wow! If that beautiful corn ball will just stick to the script we can get him elected God."

Jen's Jenny II, the new seven-passenger Aero Commander "Shrike," waited at the County Airport's private passenger ramp with its twin 290 horsepower, fuel-injected Lycoming engines idling while Howdy said his farewells to Jennifer and Howie.

In the pilot's seat Chad Chadwick was quietly explaining cockpit procedures to Howie—explaining what Howdy had promised his son he'd explain when the boy flew "co-pilot" with his father.

Howdy rested his hand on his son's shoulder and leaned in behind the front seat to speak with Jennifer. "I'm sorry as hell, honey. But I'll be down on Mexicana Friday noon. Ev and I have no choice. We've got to look into this new ag preserve hassle."

Jen nodded. "Do you believe Lew now, when he says the only way to fight is inside the political machine?"

"Not necessarily, hon. At least not nationally. Washington's not giving us trouble. It's the county and state level that's shooting us down. There's where the fight's going to be."

Jennifer tried to conceal her relief. At least they would not be uprooted and moved to Washington. "It would be a break if we could fight them on our own ground, wouldn't it?"

Howdy grinned at her reassuringly. "We'll fight on our ground. If we win—and by God we will—we'll still be 'millionaire ranchers.' He straightened and winked. "And if we lose, we'll just be *billionaire* subdividers!"

Chad Chadwick lifted an earphone and turned to address Howdy. "Which ain't exactly a fate worse than debt."

Jennifer grimaced and plugged her ears. "On that note, dear husband, we're off to Baja."

604

Howdy stepped back. "—noon, Friday—for sure!" Turning to Howie, he checked the boy's seatbelt and shoulder harness unnecessarily. "You and Chad charter a boat and get in a couple of days fishing. When I get down there we'll go out again and you two can lie about the big ones that got away. Okay?"

Howie ducked his head without turning away from the instrument panel. "Sure, Dad. See ya."

Over La Playa, the altimeter read fifty-five hundred feet. Chad was following the coast and still climbing, Jennifer could look down on the entire ranch. She had seen it many times from the air and always it had been beautiful.

The barren, corrugated slopes of the San Gabriel Mountains in the background, the velvety green foothills, in summer beige velvet, the corduroy rectangles of furrowed fields, all were contrasting but compatible elements in a panorama that never ceased to thrill her.

Scattered through the original boundaries she could see the troublesome "windows," small privately held farms that great-grandfather Steven Lewis had sold in fee simple during times of tight money. Some of them were still in citrus groves. The most troublesome ones were those closest to the encroaching urban sprawl on the north and west, well within the boundaries of MICROPOLIS 2000. Because they would be bidding against private subdividers. they would pay dearly to buy them back.

For years she had watched Howdy stave off the intruders seeking a foothold within the ranch itself. He had given those old-timers private loans, company loans, and often equipment and advice. But it was only a stop-gap measure. With MICROPOLIS 2000 announced publically, the day of reckoning was on them. Jennifer did not doubt at all that outside subdividers were behind attempts to nullify the ranch's agriculture preserves. She could still see Ward Roberts' incredulous expression when Howdy had said that if people kept coming into California at the present rate for the next fifty years there would be no more major agriculture left in the state.

Jennifer looked down at the freeway and the network of expressways crisscrossing the ranch. Each led to a creeping cross hatch of subdivision streets with their rubber-stamp tract houses, each rectangular backyard set with its own shimmering turquoise gem of a heated pool. The spectacle was breath-taking and ludicrous. Thousands of pools, including their own, were within minutes of the finest swimming beaches in the world. Enough was never enough for California!

In Steven Lewis' time they had dynamited their way three hundred miles northeastward across the mountains to the Owens Valley to drain off water for the first southern California developments. There had been a minor civil war over that.

In her father's time they had blasted and tunneled two hundred miles southeastward to the Colorado River to help slake southern California's insatiable thirst. There had been a major political war over that. But southern California had won again.

Now they were about to bring still more water from the north. For five hundred miles it would flow through the Feather River Project canals. Two and a half billion dollars would only be the beginning because another million people had come to live the good life in southern California since it had been planned. Like the freeways, it would be obsolete before it could be completed. And they were talking quietly about another huge project, the Dos Rios Dam on the middle fork of the Eel River in Mendicino County. Conservationists had pressured Reagan to stop it and, thank God, he had. That one would have turned Round Valley, the second largest level agricultural valley along the northern coast, into a reservoir. Covolo, a lovely little community that they had often visited, would have been inundated by three hundred feet of water that would flow to southern California to fill still more swimming pools and water still more lawns and private golf courses and irrigate fewer and fewer acres of the world's finest farmland.

The vision brought a contemptuous laugh from Jennifer. It was idiotic, immoral, the perverted act of a sick society. She glanced at young Howie, all concentration in the co-pilot's right-hand seat. Of his own volition, to please his father, he had been shorn of his long surfer's locks. Howie, their son, would be living in this world fifty years from now. And the odds were that he would be enormously wealthy. The subdividers were offering more per acre for land around Los Nidos than Howie's great-great-grandfather, Steven Lewis, had paid for the entire original

606

rancho. Howard D. Goodwyn Junior and his family would be able to buy and sell Onassis.

If, at some time in the future, Howie did sell the remaining prime agricultural land, he would have traded for inflated dollars—Ev called them "phony mazuma—a place of beauty and a way of life that could never again be created in California. Jennifer closed her eyes and moaned inwardly. "Why does it always have to be the same?" she asked, not quite aloud. "Why don't people learn? We've made the mistake enough times here, going all the way back to my Mexican ancestors."

Rousing herself forcibly, she reached for the newspaper Howdy had tossed on the seat beside her. Idly, she glanced at the headlines and found in them little to cheer her up. She turned to the entertainment section and scanned Maggi Joyce's column. Hollywood society was still raving about the Jules Stein party. Lynn Carlin was starring opposite Jim Brown, whose handsome black face had reportedly been the undoing of Raquel Welch. Toward the bottom of the column a familiar name caught her eye and the single sentence made her gasp:

"The loneliest lady in Acapulco this season no longer is gilt-edged, double divorcee, Angela Goodwyn Layne—poor angel!"

Something inside Jennifer's middle crumbled. She read the line again and again and the loneliness and rejection she had grown up with flooded back, making her vaguely sick at her stomach as a curious transmigration took place. It was her name she was reading and it was she sitting alone under the beach *ramada*. Sick and frightened, she pushed the paper aside violently enough to make Chad direct a quizzical glance over his shoulder. But the act had broken the fantasy pattern. And it was a pattern. The goddamned paper had not printed an item. It had reproduced a pattern, a too-familiar one whose beginning was lost in the miscellania of all of her yesterdays. Patterns were made for repeating.

Then, unaccountably, a voice that she scarcely remembered, that of her *abuelita*, her "little grandmother," Alicia, whispered and the voice was clearer than all the others:

"Donde una puerta se cierra, otra se abre—"
Whenever a door closes, another door opens.

607

20 HOWDY WATCHED UNTIL THE TWIN ENGINE Shrike was just a speck in the sky over Los Nidos. Then he crossed to the parking lot at the County Airport and headed for the southbound Interstate 5 freeway and the ranch.

From his study he called headquarters and asked the operator to contact Carlos Morales. He was told the young superintendent was in one of the fields supervising the installation of a new drip irrigation system. He left word to have Morales call him on the shortwave phone. Thirty minutes later the young Mexican-American was at the door of the Casa Grande.

Howdy ushered him into the study and indicated a chair.

"How about a cold beer?"

"Thanks, Mr. Goodwyn. I'm so dry I'm spitting cotton."

Howdy opened two bottles of Carta Blanca. After each man had drunk directly from the bottle, Howdy perched on the arm of his easy chair and pointed to the outside with his thumb.

"Carlos, Immigration is starting to make surprise field sweeps again. They haven't hit us yet, but they will when the new picking crews come on. Tell me—how do you see the illegal Mexican labor problem now—so far as Los Nidos is concerned?"

The superintendent's smooth brown brow furrowed thoughtfully.

"Not too bad, Mr. Goodwyn. It will get worse before it gets better though. We've had a few illegals in our picking crews—mostly in strawberries and some in asparagus. So far none in the citrus groves—or in avocados. I got rid of them. A lot of our green card workers are probably carrying fakes, but I can't prove it."

Howdy nodded. The counterfeit green cards and social security cards were almost impossible to detect. The ranch's position was that they were hired in good faith.

"What are we paying on straight time now, Carlos?"

"Two fifty—"

"And piece workers average about the same?" Howdy knew the

figures but he asked anyway. It was a part of his endless practice of double checking.

"The old timers average better than that, but on a per unit basis they are more cost-effective." Howdy knew that too, and once in a while it troubled his conscience a bit when he saw women in the fields bent over the rows driving themselves unmercifully.

"What about Caesar Chavez? Has he been checking us on hours—and illegals?"

"He's got some hired 'eyes' in our crews. I'm sure they keep him filled in."

Howdy took a long swig and blotted his mouth on the back of his hand.

"At what point do you think he'll begin pushing again?"

"I don't think he will—right now."

"Why not?"

"A lot of illegals are managing to stay because there's no practical way for the immigration people to get them out. The more who get away with it, the more potential members and muscle for his union. And that goes for the Teamsters too."

Carlos was confirming his own reasoning, Howdy thought, as he lowered himself into the chair.

"I smell a lot of trouble ahead," he said.

The young Mexican-American's grunt was noncommittal. After a brief lapse Howdy leaned forward.

"You know your Mexican history, Carlos. How up on the Mexican economy are you now? How do you see it?"

The superintendent nodded toward a stack of newspapers and magazines on a coffee table.

"I'm as up as they are. I see it pretty much like they do. It's not good. There's talk about devaluating the peso—fifty to the dollar some say."

Howdy shook his head sadly. "Mexican workers can't live on that. They'll start coming across the border like rats deserting a sinking ship."

Carlos Morales was tempted to say, "—or even like human beings—?" Instead, he said, "When California was made a state in 1850, the population of Mexico was seven and a half million.

"Sixty years later—in 1910, the year of the revolution—the population had only doubled. But in the thirty years from 1940 to 1970 the population has gone from nineteen million to over forty-eight million. And the way the mortality rate is declining and the birth rate is climbing there'll be a hundred million Mexicans by the turn of the century. Counting women, the work force is projected to be around twenty percent."

A mirthless laugh bounced Howdy's shoulders.

"A hundred million? But only, my friend, until fifty million of them pay coyotes to smuggle them across the border." He shook his head again. "Holy Jesus! They'll outnumber us!"

Carlos Morales managed a wry smile. "Yes sir—just like we outnumbered the *Californios* when we Yankees came pouring over the mountains from the east in 1849 looking for gold."

Howdy's smile was thin and his eyes narrowed as he looked across at his superintendent.

"We—?"

Carlos' smile was innocent. "After three generations in California, ten years of marriage to a redheaded Irish girl, and two kids, I think of my family as bona fide Yankee."

Howdy looked down and toyed with his beer bottle. He knew that young Morales had as much right to regard himself as a Yankee as any other third generation immigrant. The difference with the minority groups was in distance. Their original home boundaries were thousands of miles away, not just down the San Diego Freeway an hour's distance by car.

After a moment he looked up. Returning his superintendent's smile, he said,

"Carlos—give me your educated Yankee guess as to what we're going to be up against in the next ten years or so."

The Mexican-American brushed a strand of coarse, straight black hair from his forehead and gazed past Howdy to the big window that overlooked several hundred acres of the lower fields that were being readied for planting. The workers were all Mexican. At least half of them were Mexican nationals carrying work permits.

"There are no good figures on the ebb and flow of immigrants across our state borders from here to Texas," he said. "Some immigration people say it could be as high as a quarter of a million a year—estimating the illegals." He paused to recall some figures. "California's population is around twenty million now. It's coming up about eleven percent a year according to the last census. But there's no way to tell what it really is because of illegal immigration."

"The census figures say about four percent," Howdy interjected. Carlos seemed surprised. "I thought it would be higher."

"My guess is, it's twice that," Howdy said.

610

"Could be," Carlos conceded. "But we've got to study those figures. A lot of illegals stay here for less than a year. They earn enough money to buy some land of their own in Mexico and they go home. For the time being, the more cropland we lose to subdividers here, and the more acreage we have to lease south of the border, the fewer immigrants will come north."

He straightened and set his bottle aside.

"Unless I change my name to Smith or Jones, I'll be accused of special pleading, but I can see a time when we'll be damned glad to have a steady supply of Mexican pickers who are willing to do the field labor that machines can't do—and do it for what we can afford to pay."

Howdy gestured impatiently. "But for how long? Chavez is losing workers to the Teamsters because they promise to squeeze more out of the growers. Now Chavez is calling them and raising. Cheap field labor will be a thing of the past before we know it." Obviously growing more impatient at the prospect, he continued.

"When we get more mechanical pickers we'll have to use them down there too, to cut costs. So will the Mexican growers. The escalation is built in. When Mexican workers are displaced here—and in Mexico—they'll come back north looking for new kinds of work. Because our pay and living is better, and because they'll do work our own spoiled rotten domestic workers won't do, they'll stay in the States and be integrated in our population."

A thought passed through his mind—when rape is inevitable relax and enjoy it. The cynical aphorism twisted his mouth into a grim smile. But the word was not rape. It was change!

"It will only take two generations of little brown Mexican kids educated in our schools to do the trick!"

Carlos Morales laughed softly and regarded his boss. Howdy Goodwyn was not really a chauvinist—not at heart. There was no way he could be. At a time when most growers were exploiting Mexican and Mexican-American workers, he was helping them: Mexican blood flowed in Howdy Junior's veins. His Mexican roots ran back to the first Robles who came to Alta California with explorer Gaspar de Portolá in 1769.

"Those schools are important," he said softly. "That's how I made it—and how my father made it. My grandfather was a wetback."

Howdy nodded. "So—?" It was more a musing sound than a question.

"So—," Carlos Morales continued, "we'll have a bumper crop

611

of brown native sons and daughters whose surnames will sound as strange to you Anglos as Commodore Thomas ap-Catesby Jones' name sounded to my great, great grandfather in 1842 when that impatient navy officer embarrassed President Polk by sailing his warship into Monterey Bay and capturing California four years too early."

Howdy leaned back, smiling, and laced his fingers across his lean middle. "So," he said, "given a little more time, history is going to repeat itself. Is that it?"

Carlos Morales matched his smile and noted the resignation in his boss' tone. Both were reassuring as he replied,

"*Si, señor Goodwyn. El circulo estará completo—de neuvo.* The circle will be complete—again."